Primate Neuroethology

Primate Neuroethology

Edited by
Michael L. Platt
Asif A. Ghazanfar

2010

OXFORD
UNIVERSITY PRESS

Oxford University Press, Inc., publishes works that further
Oxford University's objective of excellence
in research, scholarship, and education.

Oxford New York
Auckland Cape Town Dar es Salaam Hong Kong Karachi
Kuala Lumpur Madrid Melbourne Mexico City Nairobi
New Delhi Shanghai Taipei Toronto

With offices in
Argentina Austria Brazil Chile Czech Republic France Greece
Guatemala Hungary Italy Japan Poland Portugal Singapore
South Korea Switzerland Thailand Turkey Ukraine Vietnam

Published by Oxford University Press, Inc.
198 Madison Avenue, New York, New York 10016

www. oup. com

Library of Congress Cataloging-in-Publication Data
Primate neuroethology / edited by Michael L. Platt and Asif A. Ghazanfar.
p. cm.
Includes bibliographical references and index.
ISBN 978-0-19-532659-8
1. Primates—Nervous system. 2. Primates—Psychology.
3. Primates—Behavior. I. Platt, Michael L. II. Ghazanfar, Asif A.
QL737. P9P67265 2010
599. 8—dc22

2009024115

9 8 7 6 5 4 3 2 1

Printed in China on acid-free paper

ABOUT THE EDITORS

Michael L. Platt is an Associate Professor of Neurobiology and Evolutionary Anthropology at Duke University, and Director of the Center for Cognitive Neuroscience. His research focuses on the neuroethology and neuroeconomics of human and nonhuman primate behavior and cognition. Michael received his B.A from Yale and his Ph.D. from the University of Pennsylvania, both in biological anthropology, and was a post-doctoral fellow in neuroscience at New York University.

Asif A. Ghazanfar is an Assistant Professor in the Neuroscience Institute and Departments of Psychology and Ecology & Evolutionary Biology at Princeton University. His research focuses on the neurobiology and evolution of primate vocal communication and how both aspects are influenced by body morphology and socioecological context. Asif received his B.Sci. in Philosophy from the University of Idaho and his Ph.D. in Neurobiology from Duke University. He was a postdoctoral fellow at Harvard University and a research scientist at the Max Planck Institute for Biological Cybernetics in Tuebingen, Germany, before moving to Princeton.

CONTRIBUTORS

Louise Barrett
Department of Psychology
University of Lethbridge
Lethbridge, Alberta, Canada

Elizabeth M. Brannon
Center for Cognitive Neuroscience
Department of Psychology
and Neuroscience
Duke University
Durham, NC

Charles Cadieu
Redwood Center for Theoretical
Neuroscience
Helen Wills Neuroscience Institute
University of California
Berkeley, CA

Matt Cartmill
Department of Evolutionary Anthropology
Duke University Medical Center
Durham, NC

Dorothy L. Cheney
Department of Biology
University of Pennsylvania
Philadelphia, PA

Yale E. Cohen
Department of Otorhinolaryngology: Head and
Neck Surgery
University of Pennsylvania School of
Medicine
Philadelphia, PA

Asif A. Ghazanfar
Neuroscience Institute
Departments of Psychology and Ecology &
Evolutionary Biology
Princeton University
Princeton, NJ

Herbert Gintis
Central European University
Budapest, Hungary

Katalin M. Gothard
Department of Physiology
University of Arizona
Tucson, AZ

Michael S. A. Graziano
Neuroscience Institute
Department of Psychology
Princeton University
Princeton, NJ

Jennifer M. Groh
Center for Cognitive Neuroscience
Department of Psychology and Neuroscience &
Neurobiology
Duke University
Durham, NC

Brian Hare
Center for Cognitive
Neuroscience
Department of Evolutionary
Anthropology
Duke University
Durham, NC

Nicholas G. Hatsopoulos
Department of Organismal Biology & Anatomy
Committee on Computational Neuroscience
University of Chicago
Chicago, IL

Benjamin Y. Hayden
Department of Neurobiology
Duke University Medical School
Durham, NC

Kari L. Hoffman
Department of Psychology
York University
Toronto, Ontario, Canada

William D. Hopkins
Department of Psychology
Agnes Scott College
Decatur, GA
Division of Psychobiology
Yerkes National Primate Research Center
Atlanta, Georgia

Atsushi Iriki
Laboratory for Symbolic Cognitive Development
RIKEN Brain Science Institute
Saitama, Japan

Karline Janmaat
Institute for Biodiversity and Ecosystem Dynamics
University of Amsterdam
Amsterdam, the Netherlands

Sarah M. Jones
Center for Cognitive Neuroscience
Department of Psychology and Neuroscience
Duke University
Durham, NC

Kerry E. Jordan
Department of Psychology
Utah State University
Logan, UT

Jon H. Kaas
Department of Psychology
Vanderbilt University
Nashville, TN

Daeyeol Lee
Department of Neurobiology
Yale University School of Medicine
New Haven, CT

Margaret S. Livingstone
Department of Neurobiology
Harvard Medical School
Boston, MA

Dario Maestripieri
Department of Comparative Human Development
University of Chicago
Chicago, IL

Julian A. Mattiello
Department of Surgery, Section of Neurosurgery
University of Chicago
Chicago, IL

Cory T. Miller
Department of Psychology
University of California, San Diego
La Jolla, CA

Tirin Moore
Department of Neurobiology
Stanford University School of Medicine
Stanford, CA

Yuji Naya
Center for Neural Science
New York University
New York, NY

Andreas Nieder
Department of Animal Physiology
University of Tübingen
Tübingen, Germany

Behrad Noudoost
Department of Neurobiology
Stanford University School of Medicine
Stanford, CA

Dinesh K. Pai
Department of Computer Science
University of British Columbia
Vancouver, British Columbia, Canada

Michael L. Platt
Center for Cognitive Neuroscience
Department of Neurobiology
Duke University Medical Center
Durham, NC

Todd M. Preuss
Division of Neuroscience
Yerkes National Primate Research Center
Emory University
Atlanta, GA

Drew Rendall
Department of Psychology
University of Lethbridge
Lethbridge, Alberta, Canada

Lizabeth M. Romanski
Department of Neurobiology & Anatomy
University of Rochester
Rochester, NY

Alexandra G. Rosati
Center for Cognitive Neuroscience
Department of Evolutionary Anthropology
Duke University
Durham, NC

Osamu Sakura
Interfaculty Initiative in Information Studies
The University of Tokyo
Tokyo, Japan

Maryam Saleh
Department of Organismal Biology & Anatomy
Committee on Computational Neuroscience
University of Chicago
Chicago, IL

Laurie R. Santos
Department of Psychology
Yale University
New Haven, CT

Robert J. Schafer
Department of Neurobiology
Stanford University School of Medicine
Stanford, CA

Daniel Schmitt
Department of Evolutionary Anthropology
Duke University
Durham, NC

Wolfram Schultz
Department of Physiology, Development and Neuroscience
University of Cambridge
Cambridge, UK

Robert M. Seyfarth
Department of Psychology
University of Pennsylvania
Philadelphia, PA

Stephen V. Shepherd
Neuroscience Institute
Princeton University
Princeton, NJ

Jeffrey R. Stevens
Center for Adaptive Behavior and Cognition
Max Planck Institute for Human Development
Berlin, Germany

Wendy A. Suzuki
Center for Neural Science
New York University
New York, NY

Doris Y. Tsao
Division of Biology
California Institute of Technology
Pasadena, CA

Jonathan D. Wallis
Department of Psychology
University of California at Berkeley
Berkeley, CA

Yumiko Yamazaki
Laboratory for Symbolic Cognitive Development
RIKEN Brain Science Institute
Saitama, Japan

Klaus Zuberbühler
School of Psychology
University of St. Andrews
St. Andrews, Scotland

CONTENTS

Primate Neuroethology

CHAPTER 1

Introduction

Michael L. Platt and Asif A. Ghazanfar

MOTIVATION FOR THE BOOK

Why do people find monkeys and apes so compelling to watch? One clear answer is that they seem so similar to us, and thus perhaps provide a window into our own minds and how they have evolved over millennia. As Charles Darwin wrote in his *Notebook M,* "He who understands baboon would do more toward metaphysics than Locke." Such similarities notwithstanding, Darwin recognized that behavior and cognition, and the neural architecture that support them, evolved to solve specific social and ecological problems (Darwin, 1872).

Darwin, and later the pioneering ethologists Konrad Lorenz, Niko Tinbergen, and Karl Von Frisch (who shared the Nobel Prize in 1973), argued that behavior must be understood in terms of its proximate causes, evolutionary origins, developmental sequence, and physiological and anatomical mechanisms (Hinde, 1982). Defining and operationalizing species-typical behaviors for neurobiological study and conveying neurobiological results to ethologists and psychologists are therefore fundamental to an evolutionary understanding of brain and behavior. Neurobiological, psychophysical, and ethological perspectives must be integrated. Unfortunately, behavioral scientists and neurobiologists rarely interact, and most practitioners remain experts in their own fields but maintain little knowledge of the others.

The "neuroethological" approach envisioned by Darwin, pioneered by the European ethologists, and finally refined by modern neurobiologists and biologists like Walter Heiligenberg, Fernando Nottebohm, Nobuo Suga, and others has provided rich insights into the minds of a number of different nonhuman animals. Research into the natural behavior of bats, for example, led scientists to discover that these animals use the acoustic and timing differences between the sound of an emitted vocalization and its subsequent echo to identify and localize a target prey (Simmons, 1989). With this behavioral foundation, neuroscientists used the temporal and spectral attributes of echolocation signals to reveal the specialized functional organization of the bat's auditory cortex (Suga, 1990). A similar story holds for the electric fish (Heiligenberg, 1991). In the early 1960s, ethologists discovered that certain species of fish emit electrical discharges for locating salient objects and can adaptively shift the frequency of these discharges so that they do not interfere with the discharges of other fish. Once this "jamming avoidance" behavior was characterized in more detail, neuroscientists were able to anatomically and physiologically map out the sensorimotor neural circuitry underlying it. Bats and electric fish continue to be popular model systems primarily because of this strong link between natural behaviors and brain function.

In stark contrast, neuroscientists who investigate the function and structure of *primate* brains often focus on more general cognitive processes and neglect their species-typical behaviors. Psychological and neurobiological studies of primates typically require them to discriminate simple stimuli whose salience or behavioral significance is arbitrarily assigned. Observational studies of primates in the wild, however, demonstrate that primates (like bats, electric fish, and numerous other species) are not "generalized" processors of information. Instead, specific stimuli, such as the facial gestures or vocalizations of others in their groups, are intrinsically salient, attract attention, and evoke species-typical responses. Many of these behaviors are shared by both human and nonhuman primates, although they may vary according to social structure, habitat, mating systems, and developmental processes. Ignoring the species-typical behavior of primates leads to the potentially erroneous idea that all primate brains are essentially different-sized versions of the same basic plan. A more promising, and biologically realistic, way to examine the neural bases of primate behavior would be to move beyond measures of brain size or neocortex size and investigate the anatomy and physiology of particular brain structures as they relate to species-typical behaviors. That is, we must develop a neuroethology of primate behavior and cognition.

The goal of this book is to do just that. Our aim is to bridge the epistemological gap between ethologists and neurobiologists who study primates by collecting, for the first time in a single book, both basic and cutting-edge information on primate behavior and cognition, neurobiology, and the emerging discipline of neuroethology. In this volume, leading scientists in several fields review work ranging from primate foraging behavior to the neurophysiology of motor control, from vocal communication to the functions of the auditory cortex. Written by some of the foremost experts in these fields, we hope this book will serve as an important resource for the professional and the student alike. The resulting synthesis of cognitive, ethological, and neurobiological approaches to primate behavior yields a richer understanding of our primate cousins that also sheds light on the evolution of human behavior and cognition.

ORGANIZATION

This book brings together the latest information on primate behavior, cognition, and neurobiology in chapters written by the foremost experts in the field. The book is roughly organized into three sections. The first section reviews our current understanding of key issues in primate taxonomy, behavior, and cognition. The second section reviews recent advances in our knowledge of the neural mechanisms underlying perception, motor control, and cognition. The final portion of the book covers work that explicitly attempts to bring together species-typical behavior and neurobiology—work that represents a new wave of neuroethological research on primate behavior and cognition. Our hope is that this synthesis will set the stage for an interdisciplinary dialogue between investigators on either side of the behavior–biology divide.

The first section begins with a discussion of current understanding of primate phylogeny by Cartmill. His thesis is that a cladistic approach based on genetics, supplemented by morphological and behavioral data, offers unique promise for organizing relationships among living primates as well as their pattern of descent from a common ancestor. Understanding evolutionary relationships within the Order Primates is a key starting point for the comparative study of primate behavior and neurobiology. This chapter is followed by an in-depth review of primate locomotion by Schmitt, who argues that primates (including humans) show patterns of locomotion and locomotor control that are different from all other mammals. Schmitt argues that changes in limb function associated with the adaptive diversification of locomotor patterns in the primate clade probably required the evolution of profound specializations in the neural control of locomotion. Most of these putative specializations remain unknown or unexplored. This realization suggests that comparative studies of the neuroethology of locomotion in primates may offer unique insights into motor control, and such insights may have implications for fields as diverse as robotics and the clinical treatment of paralysis with brain–machine interface devices.

Following this discussion of primate locomotion, Janmaat and Zuberbühler review recent studies of primate foraging behavior in the wild. The authors suggest that the information-processing problems primates encounter in foraging, particularly searching for ripe edible fruits, may have provided the impetus for the evolution of enhanced cognitive skills such as cognitive mapping and forecasting the ripeness of fruits based on the recent history of weather. Next, Cheney and Seyfarth review our understanding of vocal communication in primates. They contend that primate communication calls convey information about both the caller's affective state and objects and events in the world. Crucially, this mixed referential signaling mechanism appears to be fundamentally social in nature and thus crucial for the representation of goals, intentions, and knowledge.

Synthesizing the prior chapters on foraging and vocal communication, Stevens reviews decision-making behavior in primates. Evidence from both human and nonhuman primates demonstrates that decision makers often fail to behave as predicted by economic principles of rational utility maximization. Based on this evidence, Stevens contends that understanding decision making and its underlying mechanisms will be most successfully advanced by an evolutionarily informed framework termed "bounded and ecological rationality," which emphasizes the match between decision mechanisms and the natural environment. Central to understanding the mechanisms underlying decision making is defining the role of intentionality. Rosati, Santos, and Hare review evidence that monkeys and great apes understand the psychological states of others. They conclude that some apes, and perhaps some monkeys as well, understand behavior in terms of goals, intentions, and even knowledge.

Next, Brannon, Jordan, and Jones provide compelling evidence for a homologous cognitive and neural system supporting numerical approximation in lemurs, monkeys, apes, and humans. This approximate number system appears to form the backbone upon which symbolically mediated numerical computation and mathematical operations are built in humans. Finally, Gintis reviews contemporary models of human behavior in various fields including economics, biology, anthropology, sociology, and neuroscience. He argues that, although these models are often incompatible, they can be rendered more coherent by incorporating core principles that include an evolutionary perspective. Together, the chapters in the first section of the book clearly endorse the notion that understanding the neurobiology of primate behavior and cognition will profit from an evolutionary and ethological approach.

The second section of the book reviews our knowledge of the brain circuits that underlie behavior and cognition in human and nonhuman primates. First, Kaas outlines the major organizational features of the sensory and motor systems in primates. Comparison of these systems with respect to other mammals suggests their likely organization in ancestral primates and reveals an adaptive diversity in extant primates that is unique among mammals. Hayden builds on this comparative analysis of sensory and motor systems with a detailed neuroanatomical and neurophysiological account of visual processing in the primate brain. Hayden contends that consideration of the natural requirements for detecting and identifying behaviorally meaningful stimuli such as insects, fruits, and the facial identities and expressions of other individuals likely played an important role in the evolution of visual processing and, by extension, the evolution of cognition in primates. Moore and Noudoost focus on one salient aspect of visual behavior with a review of the neural mechanisms mediating selective visual attention. They describe a body of evidence that strongly implicates specific neural circuits in controlling visual orienting behavior.

Following this discussion of vision and attention, Miller and Cohen describe our understanding of how the primate brain parses vocalizations as auditory objects. The authors argue that in the primate auditory system, evolution selected for those neural mechanisms that bind the acoustic features of vocalizations into behaviorally meaningful units that can be acted upon, just like objects in the visual domain. Hatsopoulous, Saleh, and Mattiello build on the preceding reviews of sensory mechanisms with a review of the neural mechanisms underlying the production of movement. Based on current evidence, they contend that motor cortex does not encode any simple physical variable like velocity or direction but rather encodes elementary action fragments that can be assembled into simple behaviors. This hypothesis is consistent with a neuroethological approach, which predicts that neural mechanisms will be organized to produce adaptive behavior rather than follow arbitrary physical principles. Groh and Pai take this discussion one step further by looking at the neural transformations that mediate sensory-motor integration. They suggest that the brain transforms head-centered auditory information into a rate-coded format anchored to a hybrid reference frame that is suitable for guiding movements of the eyes, but may also permit extrapolation of sound location for guiding other types of movements.

The foregoing review of sensory and motor systems is followed by reviews of the neural mechanisms underlying cognitive processes including emotion, reward, memory, social behavior, numerosity, and executive control. First, Gothard and Hoffman examine the neural circuits that process emotion in the primate brain. They argue that two nested cortical and subcortical circuits mediate emotional evaluation of behaviorally meaningful stimuli. Although these systems are shared with other mammals, they appear to be further specialized for social behavior in primates. Schultz builds on this discussion by reviewing the neural mechanisms underlying reward learning in primates. He argues that structures involved in learning and reward, particularly midbrain dopamine neurons, appear to calculate the difference between expected and received rewards. This prediction error signal appears to be fundamental to learning and decision making in primates and other mammals. Following these discussions of emotion and reward, Naya and Suzuki review the role of the medial temporal lobe in associative memory. They argue that parallel, but distinct, mechanisms mediate the formation of long-term associations between stimuli versus associations between stimuli and action. Synthesizing the discussions of emotion, reward, and memory, Maestripieri reviews the neural mechanisms mediating species-typical social behavior. He concludes that neuromodulatory influences on specific neural circuits, including amygdala, orbitofrontal cortex, and hippocampus, underlie specific patterns of affiliation, dominance behavior, and social tolerance in different species and among individuals within a species.

The last two chapters of the second section of the book cover aspects of primate cognition that are often assumed to be uniquely human. First, Nieder reviews the evidence that neurons in the parietal cortex and prefrontal cortex selectively encode the numerical values of objects or events in the environment. The tuning properties of these neurons directly parallel the psychophysical properties of numerical judgments. Moreover, many of these neurons also appear to encode spatial extent—thus suggesting a single cortical system dedicated to representing approximate quantity derived from multiple features of particular stimuli or events. Finally, Wallis examines the role of prefrontal cortex in controlling complex, flexible behavior. He concludes that homologous mechanisms mediate the abstraction of rules, strategies, and task sets in human and nonhuman primates, and argues that this system likely evolved in concert with increasing behavioral complexity in primates.

The final section of the book sketches an outline of the neuroethology of primate behavior and cognition. This portion of the book is the most speculative, but builds upon the firm foundations of behavioral description and basic neurobiology described in the preceding sections of the book. The chapter by Preuss advocates an evolutionary approach to understanding comparative brain anatomy in primates. He argues that deep understanding of the relationships between brain and behavior requires determining how evolution modifies specific systems of neurons and their interconnections, and not just relating brain size to gross measures of cognition or behavior. Such neuroethological studies will require active management of captive and wild populations of primates needed for detailed comparison.

Following this charge, Graziano reviews evidence that motor cortex in primates is not organized according to topographic maps related to the body surface, but is organized according to species-typical motor behavior. He finds that microstimulation with behaviorally relevant time courses evokes basic movements such as bringing food to the mouth, climbing, or defensive responses. He concludes that primate motor cortex serves as an interface functionally specialized for producing species-typical actions. Tsao, Cadieu, and Livingstone then take a neuroethological approach to object and face recognition. They argue that the specialization of the primate brain for identifying and assigning meaning to faces—a ubiquitous and salient social stimuli—may provide a roadmap for understanding how the primate brain identifies and extracts information about other types of objects. Synthesizing what we know about face and voice processing in the primate brain, Romanski and Ghazanfar argue that understanding the evolution of human communication requires the recognition that communication is fundamentally multimodal in nature. They review behavioral, anatomical, and neurophysiological data to support this contention. Shepherd and Platt review the neural mechanisms underlying social attention in primates. They suggest that the neural systems mediating visual orienting behavior are intrinsically sensitive to social cues in the environment, thereby promoting the adaptive acquisition of behaviorally relevant social information.

Lee follows these sensory-level discussions with a review of the neuroethology of decision making in primates. He argues that primates evolved more complex decision-making circuits to deal with the increasing complexities associated with social interactions. He reports evidence that primates can treat interactions with a computerized opponent strategically, a behavior that requires the representation of current goals as well as prior rewards and prior actions. Neurons in the prefrontal cortex, in particular, seem to encode these variables. Barrett and Rendall present an alternative view to the notion that complex social behaviors in primates require complex brain processes. They argue that social behavior in primates may be mediated by relatively simple rules that use the structure of the social environment as a scaffold. This is in opposition to the notion that social knowledge must be explicitly represented by specialized neural circuits. The complexity of the social environment is, in essence, an emergent property of these simple rules of social engagement.

The final two chapters of the book examine the ability of primates to use tools. Hopkins reviews behavioral and neurobiological data on tool use in primates. He finds that great apes, in particular chimpanzees, excel at tool use—especially generalizing principles to new tool-using tasks and contexts. He finds limited evidence that monkeys, even highly manual species such as capuchins, do so as readily. Hopkins argues that this behavior is strongly associated with neuroanatomical changes that include the expansion of the cerebellum and interhemispheric connectivity. Finally, Iriki, Yamazaki, and Sakura review neurophysiological studies of how primates learn to extend their actions with tools. They find that learning to use tools modifies not only the response properties of neurons involved in motor planning and sensory-motor transformation, but also their anatomical connections. Moreover, they contend that tool learning prepares and adapts the primate brain to learn more complex combinatorial tool use techniques. The authors speculate that tool use learning in primates may provide the scaffolding upon which other more complex aspects of cognition are built.

THE WAY FORWARD

The primate brain is not a generalized information-processing device, or simply a differently scaled version of a prototypical mammalian brain. Our hope is that the juxtaposition of these various ideas from the ethological, cognitive, and neurobiological literature will lead to the recognition that we cannot understand the evolution of primates (and humans, in particular) without understanding the sophisticated relationships between species-typical behaviors and neural processes.

Recognition of this relationship leads to several ideas about the way forward. We offer a few of these ideas here. First and foremost (and as illustrated in our cover illustration), the Order Primates is a very diverse taxon, comprising numerous different species with different habitats and social systems. Unfortunately, very few primate species are used as subjects in behavioral and neurobiological studies. That is, the "comparative approach" has not been taken to heart, and without it, we cannot accurately construct the evolution of any particular trait or identify what is unique to our species (Preuss, 2000).

Another omission in most discussions of the evolution of primate brains and behaviors is the role of development. To understand the evolutionary origins of a phenotype, we must understand the relationship between ontogenetic and phylogenetic processes (Gottlieb, 1992; Gould, 1977). This relationship can inform questions about homology and help determine whether putative homologies reflect the operation of the same or different mechanisms (Schneirla, 1949). Are the developmental processes leading to the emergence of particular behaviors similar or different across primate species? The answer will likely determine to what extent homologies at the neural level make sense. For example, the rate of neural development in Old World monkeys and humans differs considerably—all sensorimotor tracts are heavily myelinated by 2 to 3 months after birth in rhesus monkeys, but not until 8 to 12 months after birth in human infants. These differences are paralleled at the behavioral level in the emergence of species-specific motor, socioemotional, and cognitive abilities (Antinucci, 1989; Konner, 1991).

Between brains and the environment, there is a body. How the body shapes brain processes and vice versa during development and experience with the environment is almost completely ignored (see Schmitt, this volume, for an exception). For example, whereas other New World monkeys are not very dexterous and possess a poorly developed area 5, Cebus monkeys are the only New World primate known to use a precision grip, and thus have an extended repertoire of manual behaviors. Unlike other New World monkeys, but much like the macaque monkey, Cebus monkeys possess a proprioceptive cortical area 2 and a well-developed area 5, which is associated with motor planning and the generation of internal body coordinates necessary for visually guided reaching, grasping, and manipulation (Padberg et al., 2007). These types of data suggest that parallel evolution of brain areas and behaviors can be driven (or at least paralleled) by changes in body morphology (Rose, 1996).

Finally, a real synthesis of the emerging ideas from ethology and neurobiology will require better experimental paradigms for the latter. As it currently stands, most primate neurophysiological studies are carried out under conditions of restraint while the subjects view static presentations of stimuli, trial after trial. This, of course, is nothing like the real world. Future studies will get around this artificiality (at least partially) in two ways. First, the use of interactive paradigms between two (albeit restrained) primates holds great promise for understanding the neural bases of social interactions, including dominance interactions (Fujii et al., 2008). Related to this, a second method of simulating dyadic interactions is through the use of synthetic agents, either computer animations or robots. These afford the experimenter the ability to control one side of the social interaction (and thus explore experimentally different questions in a tightly controlled manner). Finally, telemetric technology is getting increasingly more refined, allowing the recording of multiple channels of neural signals remotely with lightweight, and long-lasting, battery packs (Obeid et al., 2004). This allows the monitoring of free-ranging primates in a limitless variety of scenarios. The added realism of these emerging techniques offers great promise for full realization of an integrated, evolutionarily motivated neuroethology of primate behavior and cognition.

ACKNOWLEDGMENTS

Many people were instrumental in the translation of a germ of an idea into a full-fledged volume on primate neuroethology. This project began as a symposium at the Society for Neuroscience (SFN) meeting in Washington, DC, in 2005. Our editor, Catherine Carlin, eagerly solicited this book from us at that meeting (and later sealed the deal with fine wine, food, and conversation at the following SFN meeting in Atlanta). Of course, the members of our labs contributed in many ways to the ideas guiding this project, and also read and commented on various chapters. For this, Michael extends his gratitude to Robert O. Deaner, Jamie Roitman, Karli Watson, Jeffrey Klein, Heather Dean, Alison McCoy, Michael Bendiksby, Sarah Heilbronner, Rebecca Ebitz, John Pearson, Arwen Long, and Amit Khera. Michael also thanks his colleagues for their strong support and encouragement, including Dale Purves, Larry Katz, Rich Mooney, David Fitzpatrick, Scott Huettel, and Ron Mangun. Asif expresses his gratitude to his lab, including Joost Maier, Chandramouli Chandrasekaran, Hjalmar Turesson, and Stephen Shepherd, as well as to his mentors (in the temporal order of their impact), Matthew Grober, Miguel Nicolelis, Marc Hauser, and Nikos Logothetis.

REFERENCES

Antinucci, F. (1989). Systematic comparison of early sensorimotor development. In: F. Antinucci (Ed.), *Cognitive structure and development in nonhuman primates* (pp. 67–85). Hillsday, NJ: Lawrence Erlbaum Associates.

Darwin, C. (1872). *The expression of the emotions in animals and man.* London: John Murray.

Fujii, N., Hihara, S., & Iriki, A. (2008). Social cognition in premotor and parietal cortex. *Social Neuroscience, 3,* 250–260.

Gottlieb, G. (1992). *Individual development & evolution: The genesis of novel behavior.* New York: Oxford University Press.

Gould, S. J. (1977). *Ontogeny and phylogeny.* Cambridge, MA: Belknap-Harvard.

Heiligenberg, W. (1991). *Neural nets in electric fish.* Cambridge, MA: MIT Press.

Hinde, R. A. (1982). *Ethology: Its nature and relations with other sciences.* New Zealand: Harper Collins.

Konner, M. (1991). Universals of behavioral development in relation to brain myelination. In: K. R. Gibson & A. C. Petersen (Eds.), *Brain maturation and cognitive development: Comparative and cross-cultural perspectives* (pp. 181–223). New York: Aldine de Gruyter.

Obeid, I., Nicolelis, M. A. L., & Wolf, P. D. (2004). A low power multichannel analog front end for portable neural signal recordings. *Journal of Neuroscience Methods, 133,* 27–32.

Padberg, J., Franca, J. G., Cooke, D. F., Soares, J. G. M., Rosa, M. G. P., Fiorani, M., et al. (2007). Parallel evolution of cortical areas involved in skilled hand use. *Journal of Neuroscience, 27,* 10106–10115.

Preuss, T. M. (2000). Taking the measure of diversity: Comparative alternatives to the model-animal paradigm in cortical neuroscience. *Brain Behavior and Evolution, 55,* 287–299.

Rose, M. D. (1996). Functional morphological similarities in the locomotor skeleton of miocene catarrhines and platyrrhine monkeys. *Folia Primatologica, 66,* 7–14.

Schneirla, T. C. (1949). Levels in the psychological capacities of animals. In: R. W. Sellars, V. J. McGill, & M. Farber (Eds.), *Philosophy for the future* (pp. 243–286). New York: Macmillan.

Simmons, J. A. (1989). A view of the world through the bat's ear: The formation of acoustic images in echolocation. *Cognition, 33,* 155–199.

Suga, N. (1990). Biosonar and neural computation in bats. *Scientific American, 262,* 60–68.

CHAPTER 2

Primate Classification and Diversity

Matt Cartmill

A TAXONOMY OF CLASSIFICATIONS

There are basically two ways of grouping things: by their properties or by their connections. *Descriptive* groupings are defined by shared properties of their members. An example is the class of diamonds, which includes all and only those objects composed mainly of carbon atoms arranged in a cubical crystal lattice. *Historical* groupings, by contrast, are defined by causal linkages among their members—for example, the class of all the ancestors of George Washington, from the Precambrian down to Washington's parents. These are different kinds of sorting criteria. Diamonds are not connected with each other in any way, and Washington's ancestors have no special properties in common apart from those shared with other organisms.

Biologists have tried classifying organisms by their observable properties, by their genealogical connections, and by a mixture of both. Each approach has both merits and defects. The chief difficulty with a purely descriptive system is that each descriptor defines a different set of organisms. These sets overlap, and it is not clear why some descriptors should have priority over others in making sequential cuts. For example, in the first (1735) edition of *Systema Naturae*, Linnaeus classified the whales as fish because they had glabrous skin and lacked feet; but in the canonical 10th edition of 1758, he reclassified them as mammals because they had milk glands and bore live young. This correction seems warranted to us, but there was no way of justifying it in the purely descriptive framework imposed by Linnaeus's own creationist assumptions. We have no reason for thinking that milk glands take precedence over feet in the mind of God.

The underlying justification for Linnaean classification became apparent when historical and causal linkages were introduced into the system after the emergence of Darwinism in the mid-1800s. Linnaeus's nested sets were real entities because they corresponded to successive branching points on the phylogenetic tree. Milk glands were relevant properties for defining the 1758 class Mammalia because they were inheritances from a common ancestor shared by all mammals and by no other organisms. Glabrous skin and fins—ancient vertebrate traits lost in the ancestors of the first mammals, and subsequently re-evolved in the Cetacea as special adaptations to life in the sea—did not reflect the geometry of evolutionary relationships, and therefore were not relevant properties for defining a taxon (the 1735 class Pisces).

For a century after Darwin, biological classification was dominated by mixed systems that combined phylogenetic and essentialist criteria. In such systems, sometimes referred to as *evolutionary systematics*, taxa were defined by the acquisition of key evolutionary novelties. Within a taxon, subtaxa were distinguished from each other by clusters of shared adaptive novelties that evolved later than the key traits defining the larger taxon. One of these subtaxa usually constituted a basal "wastebasket taxon" comprising early, little-differentiated members of the larger taxon, together with later forms that had remained persistently primitive. Such wastebasket taxa included Amphibia within Tetrapoda, Reptilia within Amniota, Prosimii within Primates, and so on.

In these mixed systems, taxonomic practice was constrained by phylogenetic or historical criteria as well. Taxa were usually required to be "monophyletic," which meant that they were supposed to contain only descendants of the first species having the taxon's defining properties. The lower boundary of each taxon was defined by the point of acquisition of those properties on the lineage leading to its last common ancestor (LCA). If the key properties were acquired independently in two separate lineages, the descriptive grouping that they defined was "polyphyletic." Such taxa were generally forbidden, though some systematists accepted low levels of polyphyly (Simpson, 1945, 1961).

During the 1950s and '60s, the primacy of this mixed approach was challenged by two rivals, one of which ultimately swallowed the other and replaced evolutionary systematics in a classical Kuhnian paradigm shift (Cartmill, 1999). The defeated challenger was a strictly descriptive system known as "numerical taxonomy" (Sokal & Sneath, 1963), which excluded phylogeny as a classificatory criterion on the not unreasonable grounds that it is not an observable property. Numerical taxonomists began by enumerating the species included in the classification and then analyzing their properties into characters with varying states. The resulting character-state matrix was processed to generate taxonomic groupings that maximized the overall sum of shared intrataxon resemblances.

The victorious challenger and currently regnant style in systematics, known as *cladistics* or *phylogenetic systematics* (Hennig, 1966), groups organisms solely on the basis of phylogenetic relationships. The determination of those relationships begins with a character-state analysis of the species in the group being classified, as in numerical taxonomic practice. All character states that are primitive within the group (*symplesiomorphies*) are discarded as irrelevant to the determination of phylogeny. The remainder are sorted out into *synapomorphies* (traits of uniquely shared derivation) and *homoplasies* (parallelisms, convergences, and coincidences that constitute noise in the phylogenetic signal), using criteria of maximum parsimony (minimizing the number of assumed changes from primitive to derived states) or maximum probability. This process generates hierarchically arranged groupings defined by nested synapomorphies. Each group is required to be *holophyletic*, encompassing all and only the descendants of its LCA. Such groupings are called *clades*. Wastebasket taxa, which contain only the LCA's descendants but not all of them, are termed *paraphyletic* and are not admitted to the system. All taxa are defined exclusively by synapomorphies, and every grouping must be distinguished from its nearest relative—its "sister group"—by at least one synapomorphy represented in the sister by a more primitive state of the same character. The nonintersecting sets generated by this sort of analysis can be read out either as hierarchical Linnaean classifications or as atemporal branching tree diagrams known as *cladograms*. The classification is isomorphic with the cladogram.

SOME PROBLEMS WITH CLADISTICS

The theoretical rigor and fascinating technical intricacies of the cladistic approach have helped to give it a position of unchallenged dominance in modern systematic practice. Apart from its esthetic appeal, cladistics has contributed to biology by bringing the logic of phylogenetic reconstruction into sharper focus. But a strictly phylogenetic systematics encounters problems—in theory, in systematic practice, and in practical utilization.

The deepest theoretical problem is that some organisms are in fact wholly primitive relative to others, and thus lack defining apomorphies (nonprimitive traits). Therefore, they cannot be classified. For example, all character states in an ancestral species are by definition primitive relative to other states of the same characters in its descendants. This problem is particularly irksome for paleontologists, who occasionally discover extinct organisms that left recognizable descendants. Cladistic systematists tend to deal with this difficulty by seeking unique apomorphies in apparently primitive organisms that exclude them from an ancestral status, or by sneaking wastebasket taxa back into the system in various ways—erecting unranked taxa called "plesions," referring to primitive extinct forms informally as "stem groups," or using old wastebasket-taxon names like "reptile" inside quotation marks to show that they are being naughty.

Another deep problem involves the identification of morphological characters and character states. The words we use to describe morphology are not objectively determined by nature. Choosing such words involves what can only be called an act of poetic imagination; and different terminologies with similar information content can yield different cladograms when they are fed through the machineries of cladistic analysis (Cartmill, 1982, 1994a). In primate systematics, a simple example of this sort of problem is provided by the postorbital septum, a bony partition separating the orbital contents from the chewing muscles in monkeys and apes (Anthropoidea). The small carnivorous Asian primates called tarsiers have a septum with a gap in it. Other primates lack the septum (Fig. 2.1). Not having a septum is primitive. If we score the septum as "present" versus "absent," it counts as a synapomorphy linking anthropoids to tarsiers (Cartmill, 1980; Cartmill & Kay, 1978). But if we score it as "complete" versus "incomplete" (Beard & MacPhee, 1994), the septum counts as a unique synapomorphy of the anthropoids that distinguishes them from tarsiers, thereby skewing the analysis in the opposite direction. Neither dichotomization seems obviously preferable. Analyzing the character into multiple states arrayed in a linear or branching transformation series involves arbitrary choices of other sorts.

The weighting of morphological characters also raises serious questions for cladistic practice. Most cladistic methodologies apply equal weighting to all characters, so that the difference between (say) "tail long" and "tail short" counts as much in judging phyletic affinities as that between "six cervical vertebrae" and "seven cervical vertebrae." But variation in some characters is known to be far more significant than in others. Among mammals, the number of caudal vertebrae is highly variable, but the number of cervical vertebrae is virtually invariant and appears to be controlled by regulatory genes acting at a much deeper level early in the process of segmentation. Interspecies differences in the number of neck vertebrae should therefore count much more heavily in assessing group affinities than differences in the number of tail

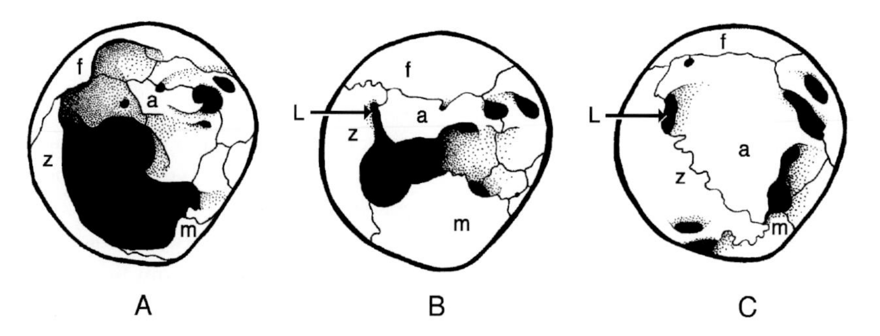

Figure 2.1 The inside of the right orbit in three primates: semidiagrammatic front views. In the primitive condition (*Galago*, A), the frontal, zygomatic, and maxillary bones (f, z, m) form a postorbital bar. In the anthropoid arrangement (*Saimiri*, C), outgrowths from the alisphenoid (a) and zygomatic combine with flanges of the maxillary and frontal bones to form a complete bony postorbital septum separating the orbital contents from the temporal fossa. The condition in *Tarsius* (B) is intermediate. L, lateral orbital fissure. From Cartmill, M., & Smith, F. H. (2009). *The human lineage.* New York: Wiley-Blackwell. Used with permission.

vertebrae. But how much more? We can feel certain that the arbitrary imposition of equal weighting here is a mistake that introduces error into the reconstruction, but alternative weightings seem equally arbitrary. When the unknown errors due to arbitrary weighting are multiplied across the hundreds of characters commonly tallied up in morphological character-state matrices, it is hard to have much confidence in the details of the resulting cladogram.

WHALES AND HIPPOS AND COWS, OH MY

Many of these problems can be obviated by giving up on morphology and going directly to the genome. Although identifying homologous parts of different genomes is not a simple matter, the conceptual basis of homology is clearer for DNA data than it is for morphological data (Cartmill, 1994b), and the boundaries of characters (nucleotide positions in a homologous sequence) and character states (the four nucleotides) are unambiguous. Over the course of the past two decades, DNA data have increasingly supplanted phenotypic data in reconstructing the phylogenetic relationships of living organisms. For the most part, phylogenies inferred from the genome have corroborated those inferred from the phenotype; but some intractable disputes of long standing have been decisively resolved, and there have been surprises.

The use of SINEs ("Short Interspersed Nuclear Elements") as lineage markers has enhanced the consistency and reliability of molecular phylogenetic analyses (Cook & Tristem, 1997; Shedlock et al., 2004). A SINE can be thought of as a transfer RNA molecule that has become parasitic by introducing a retrotranscribed DNA copy of itself into the nuclear DNA. The introduced sequence is capable of making new RNA copies, which then reproduce themselves at new target loci. SINE insertions are particularly reliable markers of lineage relationships for three reasons: (1) the primitive state at the parasitized locus is always "SINE absent"; (2) the template DNA copy is not excised during replication, so that it remains indefinitely at the parasitized locus as a permanent marker of a unique character-state change; and (3) parallel mutations to the derived character state and reversions to the primitive character state are vastly less likely than they are in the case of single-nucleotide mutations.

Molecular analyses of mammalian phylogeny have revealed some highly corroborated clades that had gone undetected by morphologists (Fig. 2.2). Primitive eutherian (placental) mammals first appear in the Lower Cretaceous of Asia around 125 million years ago (Mya), and were present in both Asia and North America by 110 Mya (Ji et al., 2002). Molecular-clock analyses suggest that the divergence of the extant eutherian orders dates back to about this time (Eizirik et al., 2001; Kumar & Hedges, 1998; Murphy et al., 2001). The initial split seems to have been between the South American edentates (Xenarthra) and all other eutherians, followed by the divergence of an Africa-based supraordinal clade (Afrotheria) comprising the elephants, sea cows, hyraxes, tenrecs, and some other originally African groups. The remaining eutherian orders form a clade (Boreotheria) with two major subdivisions: (1) primates and related groups plus the rodents and rabbits (Euarchontoglires), and (2) everybody else (Laurasiatheria, including carnivores, ungulates, whales, shrews, hedgehogs, and bats).

Parts of some of these clusters had been glimpsed by morphologists and paleontologists—for example, the primate-treeshrew-colugo group (Archonta: Gregory, 1910) and the elephant-seacow-hyrax group (Paenungulata: Simpson, 1945)—but even in these cases the affiliations of the tenrecs and the rodent-rabbit group (Glires) were unexpected. Perhaps the biggest surprise was the deeply imbedded position of the whales and dolphins (Cetacea) within the even-toed hoofed mammals or Artiodactyla, where they fall out as the sister group of hippopotamuses and as successively more distant relatives of ruminants-plus-pigs and camels. The molecular identification of the cetaceans as artiodactyls (Arnason et al., 2000; O'Leary & Gatesy, 2007) has since been corroborated by the discovery of early fossil whales that retained hindlimb bones with distinctively artiodactyl

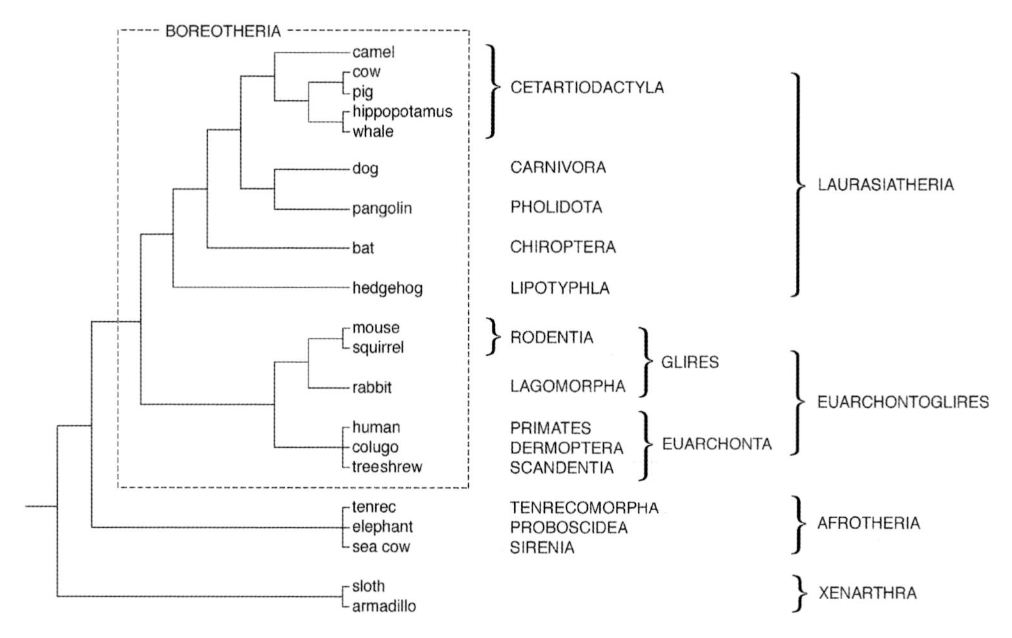

Figure 2.2 Phylogenetic relationships of major clades of extant eutherian mammals, as inferred from retroposon insertions and other molecular data. Based on Kriegs, J., Churakov, G., Kiefmann, M., Jordan, U., Brosius, J., & Schmitz, J. (2006). Retroposed elements as archives for the evolutionary history of placental mammals. *PLoS Biology, 4*(4), e91. (Cetartiodactyl branching details added from O'Leary, M. A., & Gatesy, J. [2007]. Impact of increased character sampling on the phylogeny of Cetartiodactyla (Mammalia): Combined analysis including fossils. *Cladistics, 23,* 1–46.)

morphology (Gingerich et al., 2001; Thewissen et al., 2001).

These revelations bring into focus some questions about the utility of classifications that are based solely on genealogical connections. Phylogenetic relationships are usually correlated with the distribution of phenotypic properties, but not always. Whales, cows, pigs, and camels are all artiodactyls in a cladistic sense, but there is virtually nothing that they have in common apart from some shared DNA sequences. A killer whale and a sheep are as different in anatomy, ecology, and way of life as it is possible for two placental mammals to be. Classifying them together as "cetartiodactyls" encodes no useful information about their biology. It is not entirely clear why this is supposed to be a desirable outcome of biological systematics. In a mixed or evolutionary system of classification, this sort of situation would be handled by drawing an ordinal boundary across the artiodactyl lineage leading to the ancestral whales and treating Artiodactyla and Cetacea as cognate orders, with Artiodactyla being retained as a paraphyletic but adaptively coherent grouping ancestral to the Cetacea. But in a cladistic classification, we are obliged to erect a sequence of nested taxa of continually diminishing biological import to express the successive furcations of the cladogram: Superorder hippos-and-whales, Hyperorder hippos-whales-and-cows, Grandorder hippos-whales-cows-and-camels, and so on. The resulting groupings are not very useful for talking about anything except genealogy.

The foregoing exposition on systematics is offered by way of an apology for what follows—namely, an annotated partial classification of the order Primates (Table 2.1) that is neither entirely descriptive nor entirely genealogical. The classification includes extinct primate groups, but I have given them short shrift, for two reasons. First, their phylogenetic relationships are often unclear and inferable only from morphological data, with all the inherent defects

Table 2.1 A Partial Classification of the Order Primates

SUBORDER STREPSIRRHINI

Infraorder †Adapiformes
  Superfamily †Adapoidea
    Family †Adapidae (*†Adapis, †Adapoides, †Cryptadapis, †Leptadapis, †Palaeolemur*, etc.)
    Family †Notharctidae (*†Notharctus, †Cantius, †Cercamonius, †Periconodon*, etc.)
    Family †Sivaladapidae (*†Sivaladapis, †Guangxilemur, †Indraloris, †Sinoadapis*, etc.)

Infraorder Lemuriformes
  Superfamily Lemuroidea
    Family Daubentoniidae (*Daubentonia*)
    Family Indriidae (*Indri, Avahi, Propithecus*, etc.)
    Family Lepilemuridae (*Lepilemur*)
    Family Lemuridae (*Lemur, Eulemur, Hapalemur, Varecia, †Pachylemur*)
    Family Cheirogaleidae (*Cheirogaleus, Microcebus, Mirza, Allocebus*)
    Family †Megaladapidae (*†Megaladapis*)
    Family †Paleopropithecidae (*†Paleopropithecus, †Mesopropithecus, †Babakotia*, etc.)
    Family †Archaeolemuridae (*†Archaeolemur, †Hadropithecus*)
  Superfamily Lorisoidea
    Family Lorisidae (*Loris, Nycticebus, Perodicticus, Arctocebus, †Karanısıa*, etc.)
    Family Galagidae (*Galago, Galagoides, Euoticus, Otolemur, †Saharagalago*, etc.)

SUBORDER HAPLORHINI

Infraorder Tarsiiformes
  Superfamily †Omomyoidea
    Family †Omomyidae (*†Omomys, †Tetonius, †Necrolemur*, etc.)
  Superfamily Tarsioidea
    Family Tarsiidae (*Tarsius*)

Infraorder Anthropoidea
  SECTION PLATYRRHINI
  Superfamily Ceboidea
    Family Callitrichidae (*Callithrix, Cebuella, Callimico, Saguinus, Leontopithecus*)
    Family Cebidae (*Cebus, Saimiri*)
    Family Aotidae (*Aotus, †Tremacebus*)
    Family Atelidae (*Ateles, Lagothrix, Brachyteles, Alouatta, †Protopithecus*)
    Family Pitheciidae (*Pithecia, Cacajao, Chiropotes, Callicebus, †Homunculus*, etc.)
  SECTION CATARRHINI
  Superfamily †Propliopithecoidea (*†Propliopithecus, †Aegyptopithecus, †Moeripithecus*)
  Superfamily †Pliopithecoidea (*†Pliopithecus, †Epipliopithecus, †Crouzelia*, etc.)
  Superfamily †Proconsuloidea
    Family †Proconsulidae (*†Proconsul, †Afropithecus, †Kenyapithecus*, etc.)
    Family †Sugrivapithecidae (*†Sivapithecus, †Ankarapithecus, †Gigantopithecus*, etc.)
    Family †Dendropithecidae (*†Dendropithecus, †Micropithecus, †Simiolus*, etc.)
  Superfamily Hominoidea
    Family †Dryopithecidae (*†Dryopithecus, †Pierolapithecus, †Oreopithecus*, etc.)
    Family Hylobatidae (*Hylobates*)
    Family Pongidae (*Pongo, †Khoratpithecus?*)
    Family Hominidae (*Homo, †Australopithecus, Pan, Gorilla*)
  Superfamily Cercopithecoidea
    Family †Victoriapithecidae (*†Victoriapithecus, †Prohylobates*)

(*continued*)

Table 2.1 (Continued)

Family Cercopithecidae (*Cercopithecus*, *Chlorocebus*, *Macaca*, *Papio*, etc.)
Family Colobidae (*Colobus*, *Presbytis*, *Trachypithecus*, *Nasalis*, etc.)

Anthropoidea *incertae sedis*:
Families †Proteopithecidae, †Oligopithecidae, †Parapithecidae

HAPLORHINI *INCERTAE SEDIS*:
Family †Eosimiidae

PRIMATES *INCERTAE SEDIS*:
Families †Plesiopithecidae, †Amphipithecidae.

Extinct groups (including paraphyletic stem groups) are indicated by daggers.

of such data magnified by the scrappy nature of the fossil record. Second, they are of secondary interest in the context of the present volume. Recent reviews of the systematics, diversity, and evolution of extinct primates can be found in Hartwig (2002) and Cartmill and Smith (2009).

The following classification is generally conservative and utilizes taxon names that are commonly employed and should be widely understood. It follows phylogenetic relationships where it seems convenient and biologically useful to do so, but employs paraphyletic groupings where phylogeny is unclear or where deeply imbedded groups seem sufficiently distinct adaptively to warrant distinction by elevating their rank (e.g., Callitrichidae). However, I have followed cladistic principles and current fashion in sinking gorillas and chimpanzees into the human family (Hominidae) rather than adopting the now quaint-seeming practice of using the orangutan family (Pongidae) as a taxonomic wastebasket for the great apes. What seems to me to be the most probable pattern of phyletic relationships among the extant primates, as judged from molecular evidence, is presented in Figures 2.3 and 2.4.

ORDER PRIMATES

The primates are a moderately diverse order of mainly arboreal eutherians known as fossils from the earliest Eocene onward in both the New and Old Worlds. Almost all extant primates inhabit tropical and subtropical forests and woodlands, though humans and some Old World monkeys range into drier and colder habitats. Distinctive morphological synapomorphies of the crown group (living primates and extinct descendants of their LCA) include large, forward-facing eyes set in complete bony rings; elaboration of the visual apparatus and visual parts of the brain; a tympanic bulla formed by an extension of the bone (petrosal) surrounding the membranous labyrinth; and grasping hind feet with a divergent hallux bearing a flat nail. The claws of the other digits are also modified into flattened nails in most primates, which is probably another crown-group synapomorphy. Grasping specializations of the hand have evolved secondarily in several primate lineages. Molecular data show that primates are particularly close relatives of treeshrews and "flying lemurs" or colugos (Fig. 2.2). Most paleontologists regard the Plesiadapiformes, an extinct (Cretaceous-late Eocene) group of primarily arboreal, vaguely rodent-like mammals, as stem-group primates (that is, the extinct sister of the crown group). However, plesiadapiforms lack the primate cranial synapomorphies listed previously, and share apomorphies of their own that exclude them from the direct ancestry of the crown-group primates (Bloch et al., 2007). In what follows, the term "primate" will be restricted to members of the crown group.

For most practical purposes, primates can be divided into a wastebasket "prosimian" group (Prosimii) of so-called "lower" primates comprising early and persistently primitive forms,

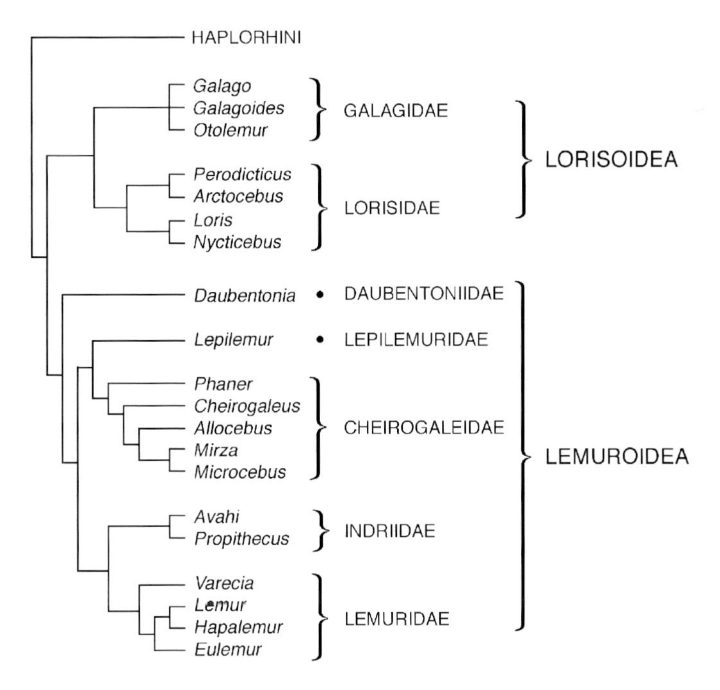

Figure 2.3 Phylogenetic relationships of major clades of extant strepsirrhine primates (lorises and lemurs), as inferred from molecular data. After Roos, C., Schmitz, J., & Zischler, H. (2004). Primate jumping genes elucidate strepsirrhine phylogeny. *Proceedings of the National Academy of Sciences, 101,* 10650–10654, with compatible addenda from Horvath, J., Weisrock, D., Embry, S., Fiorentino, I., Balhoff, J., Kappeler, P., et al. (2008). Development and application of a phylogenomic toolkit: Resolving the evolutionary history of Madagascar's lemurs. *Genome Research, 18,* 489–499.

and a holophyletic clade of "higher" primates (Anthropoidea) comprising monkeys, apes, and humans. One prosimian group, the tarsiers (Tarsiidae), is linked to the anthropoids by various morphological synapomorphies (Fig. 2.1: Cartmill & Kay, 1978; Kay et al., 1997, 2004) and by molecular apomorphies including patterns of SINE insertion (Schmitz & Zischler, 2004). A subordinal cut between Haplorhini (tarsiers-plus-anthropoids) and Strepsirrhini (nontarsier prosimians), rather than a cut between Prosimii and Anthropoidea, is adopted here on cladistic grounds.

SUBORDER STREPSIRRHINI

Extant strepsirrhine primates can be described as looking rather like monkeys with the heads of dogs (Fig. 2.5). Most of the traits in which they differ systematically from other primates are primitive (plesiomorphic) states of various characters of the central nervous system and the visual and nasal apparatus. Strepsirrhine symplesiomorphies include the comma-shaped nostrils and wet dog-like rhinarium that give the taxon its name. However, the living strepsirrhines are bound together as a clade by a few apparent synapomorphies, the most obvious being the modification of the lower incisors and canines into a "toothcomb" (Fig. 2.6) used for grooming the fur (and for specialized feeding activities in some species). Strepsirrhines also differ from other primates in having an epitheliochorial rather than a hemochorial placenta. This is counted as a synapomorphy by some and a symplesiomorphy by others (Wildman et al., 2006).

† Infraorder Adapiformes

This extinct group of lemur-like primates includes most of the larger and more herbivorous primates known as fossils from the Eocene. Adapiforms lack the toothcomb, and it is not certain that they are

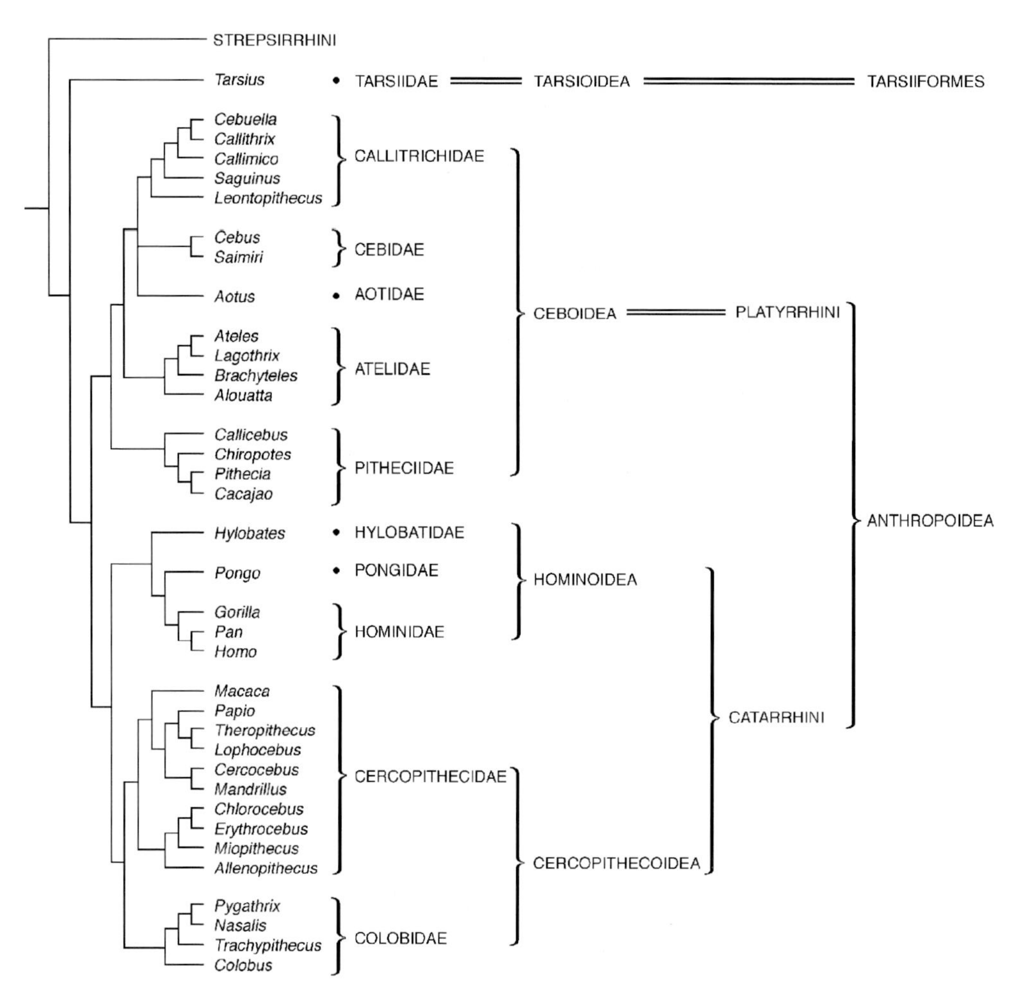

Figure 2.4 Phylogenetic relationships of major clades of extant haplorhine primates (tarsiers and anthropoids), as inferred from molecular data. After Xing, J., Wang, H., Han, K., Ray, D., Huang, C., Chemnick, L., et al. (2005). A mobile element based phylogeny of Old World monkeys. *Molecular Phylogenics and Evolution, 37,* 872–880, with compatible addenda from Goodman, M., Porter, C., Czelusniak, J., Pages, S., Schneider, H., Shoshani, J., et al. (1998). Toward a phylogenetic classification of primates based on DNA evidence complemented by fossil evidence. *Molecular Phylogenics and Evolution, 9,* 585–598, and Tosi, A., Disotell, T., Morales, J., & Melnick, D. (2003). Cercopithecine Y-chromosome data provide a test of competing morphological evolutionary hypotheses. *Molecular Phylogenics and Evolution, 27,* 510–521.

more closely related to lemurs than they are to monkeys (Gebo, 2002; Rasmussen, 1986). If they are not, then Strepsirrhini as defined here is a wastebasket taxon.

Adapiforms constitute a single superfamily, Adapoidea, divisible into three families. The **Adapidae** comprise several genera of medium-sized, primarily folivorous primates, including some heavily built, slow-moving forms that may have resembled sloths in their ecology. **Notharctidae** were more active, lemur-like, running and leaping animals. The **Sivaladapidae** were an Asian radiation of adapiforms that culminated in some largish (ca. 7 kg), deep-jawed, monkey-convergent prosimians known from the Miocene of Southeast Asia.

Figure 2.5 The face of a strepsirrhine primate (*Varecia* juvenile). From Cartmill, M., & Smith, F. H. (2009). *The human lineage*. New York: Wiley-Blackwell. Used with permission.

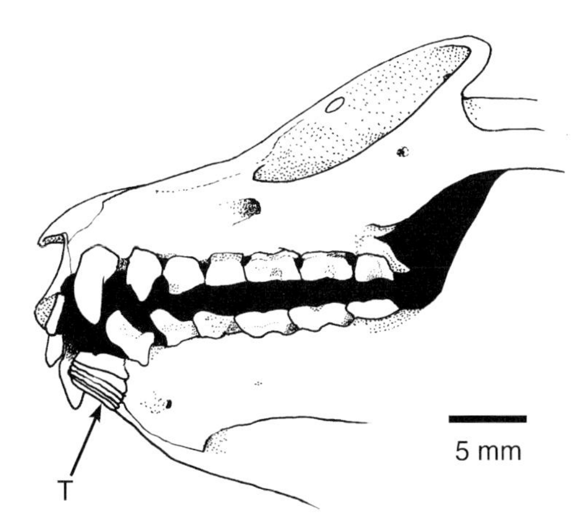

Figure 2.6 Dentition of the cheirogaleid lemuroid *Cheirogaleus medius*, showing the toothcomb (T). From Cartmill, M., & Smith, F. H. (2009). *The human lineage*. New York: Wiley-Blackwell, after James, W. (1960). *The jaws and teeth of primates*. London: Pitman Medical Publishing. Used with permission.

Infraorder Lemuriformes

This grouping comprises the toothcomb primates, divided into the lemurs of Madagascar (Lemuroidea) and the lorises and galagos or bushbabies (Lorisoidea) of Africa and Asia. The two groups differ systematically in a number of anatomical details, particularly of the cheek teeth and the middle ear. Lorisoids are known from late Eocene deposits in the Fayum Depression of Egypt (Seiffert et al., 2003) and from the Miocene of Africa and southern Asia. There is essentially no fossil record of lemuroid evolution; the oldest extinct lemuroids are Holocene species that overlapped in time with the early human colonists of Madagascar.

Superfamily Lemuroidea

The Malagasy primates constitute a holophyletic clade of considerable antiquity. Recent molecular-clock estimates date the first divergence within the lemuroids (between *Daubentonia* and the others) to around the time of the K-T boundary (Horvath et al., 2008). The available data point to a late Cretaceous or early Tertiary colonization of Madagascar by a single ancestral lemuroid species, presumably derived from the same undocumented proto-lemuriform radiation in Africa that gave rise to the ancestral lorisoids (Yoder et al., 1996). Like other small, isolated land masses, Madagascar has a taxonomically impoverished fauna whose endemic groups have radiated to fill ecological spaces preempted by other taxa on the major continents. Most of the extant lemuroids occupy niches broadly similar to those filled by primates in other tropical forests, but *Daubentonia* and some of the large extinct lemurs developed unusual specializations.

Family Daubentoniidae The aye-aye *Daubentonia madagascariensis* is the sole primate representative of a foraging guild that has been labeled "mammalian woodpeckers" (Beck, 2009) or "woodpecker avatars" (Cartmill, 1974). The extant members of this guild—the aye-aye and the Australasian marsupials of the genus *Dactylopsila*—inhabit areas devoid of birds that feed on wood-boring insect larvae. They have developed similar convergent adaptations for exploiting this resource, including enlarged incisors for cutting into infested wood and an elongated, clawed finger used for probing exposed tunnels and snagging the grubs. The aye-aye's complex foraging habits (Erickson, 1994) are probably related to the large size of its brain, which is as big relative to body size as those of some monkeys (Stephan, 1972; Stephan & Andy, 1969). The aberrant morphological specializations of the aye-aye (including inguinal nipples, claws on all digits except the hallux, and permanently growing incisor teeth like those of rodents) and the antiquity of its estimated divergence from the other Malagasy primates lead some systematists to assign it to a separate infraorder of its own, Chiromyiformes.

Family Indriidae The indriid genera *Indri*, *Propithecus*, and *Avahi* comprise several species of lemurs specialized for a diet of leaves and a "vertical clinging and leaping" pattern of positional behavior. They have long, powerful hindlimbs capable of propelling them from tree to tree in 10-meter leaps, and their long hands and feet have robust, widely divergent first digits adapted to grasping large vertical supports. The indriids include the largest extant strepsirrhine, *Propithecus diadema* (>7 kg), as well as one of the smallest folivorous primates, *Avahi laniger* (≈ 1 kg). *Avahi* is nocturnal in its activity, but *Propithecus* and *Indri* are mainly diurnal animals, and some species have strikingly patterned and colored pelage, presumably correlated with color vision.

Family Lepilemuridae Seven or more species of *Lepilemur* are distinguished by taxonomists, mainly on the basis of pelage variants. All are small (500 to 1,000 g), long-tailed, nocturnal, arboreal folivores with a moderate degree of specialization for leaping locomotion. Despite its adaptive convergence with *Avahi*, *Lepilemur* shows no special morphological similarity to the indriids, and recent molecular data (Horvath et al., 2008) position it as the phyletic sister of the mouse- and dwarf-lemur family Cheirogaleidae (Fig. 2.3).

Family Lemuridae Of the four genera of extant lemurids, three (*Lemur*, *Eulemur*, and *Varecia*) are rather similar-looking medium-sized animals with long foxy muzzles, generally fruit-centered dietary preferences, and daily activity cycles ranging from semi-nocturnal (*Eulemur mongoz*) to strictly diurnal (*Lemur*). *Lemur catta*, the sole species in its genus, has a more terrestrial activity pattern and larger group sizes (up to 20+ animals) than other living lemuroids. Its phyletic sister *Hapalemur* is another small folivore (700 to 2,400 g), with a suite of anatomical, behavioral, and biochemical specializations for feeding on bamboo. Although the Lemuridae as a whole lack obvious morphological synapomorphies, SINE insertions (Roos et al., 2004) identify them as a holophyletic clade and the sister group of the indriids (Fig. 2.3).

Family Cheirogaleidae This family comprises five genera of small nocturnal primates ranging in size from ≈300 g (*Mirza coquereli*) to 30 g (*Microcebus myoxinus*, the smallest living primate). All feed on high-energy foods—fruit, nectar, gums, and animal prey—and some species of *Microcebus* and *Cheirogaleus* conserve energy during lean seasons of the year by estivating or by metabolizing fat stored in the tail. Many experts have pointed to *Microcebus* as a plausible living model for the last common ancestor of the crown-group primates. Cheirogaleids share certain potentially apomorphous anatomical traits with galagos, including elongation of the tarsal bones and peculiar specializations of the carotid arterial circulation. Some systematists have accordingly suggested that they may be more closely related to the lorisoids than to the other Malagasy lemurs (Cartmill, 1975; Schwartz, 1986; Szalay & Katz, 1973). However, molecular data consign the cheirogaleids to a deeply imbedded position within the lemuroid clade (Fig. 2.3: Goodman et al., 1998; Horvath et al., 2008; Roos et al., 2004; Yoder et al., 1996). Any features they share with galagos must therefore be convergences, or symplesiomorphies retained from the lemuriform LCA.

Three families of large-bodied lemuroids became extinct around the time of the initial human colonization of Madagascar some 2,000 years ago. Paleopropithecidae (*Paleopropithecus, Mesopropithecus, Babakotia*) were suspensory arboreal leaf eaters convergent in their ecology and locomotor anatomy with the tree sloths of the New World. The largest of them, *Archaeoindris*, is known mainly from a single skull, which was as big as that of a male gorilla. This gigantic "sloth lemur" may have been a terrestrial form resembling a ground sloth. Another huge extinct lemur, *Megaladapis* (family Megaladapidae), appears to have been a slow-moving (but not suspensory), great-ape-sized (40 to 80 kg) folivore. It is sometimes likened to a giant koala. The Archaeolemuridae (*Archaeolemur, Hadropithecus*) include the most terrestrial offshoots of the lemuroid radiation, convergent in various respects with certain ground-feeding Old World monkeys.

Superfamily Lorisoidea

Although far less diverse than the primates of Madagascar, the continental lemuriforms are nevertheless divided into two sharply differentiated subgroups. Both molecular and fossil data point to a late Eocene divergence of the two. The family Lorisidae (lorises) comprises four genera of African (*Perodicticus, Arctocebus*) and Asian (*Loris, Nycticebus*) prosimians that range in size from 1,200 g (*N. coucang*) to around 200 g (*A. aureus*). All four lorisid genera share nocturnal habits, diets featuring animal prey, and a suite of striking apomorphies of locomotor anatomy and behavior—vestigial tails, vise-like hands and feet, and cautious, often weirdly slow patterns of locomotion in which at least one hand or foot remains in contact with the support at all times, even during running (Schmitt et al., 2006). The members of the other lorisoid family, the Galagidae (*Galago, Otolemur, Galagoides, Euoticus*), are also nocturnal animals with mixed diets, but their locomotor specializations are just the opposite: galagos are adapted to leaping and have correspondingly long tails, long and powerful hindlimbs, and elongated tarsal bones that add leverage and speed at takeoff. Ranging in size from *Otolemur* (≈1,100 g) to the diminutive *Galagoides* (≈60 g), galagos are limited in distribution to Africa.

SUBORDER HAPLORHINI

Although tarsiers and anthropoids are united by a substantial list of shared derived features of morphology, many systematists continue to resist their assimilation into a common suborder Haplorhini (Shoshani et al., 1996.) There are three main reasons for this resistance: (1) some of the candidate haplorhine synapomorphies (e.g., the hemochorial placenta) may in fact be symplesiomorphies, (2) most of the candidate synapomorphies are either unknown or lacking in the fossil prosimians often regarded as possible tarsier relatives, and (3) some molecular data suggest that tarsiers have closer ties to lemurs and lorises than to anthropoids (Eizirik et al., 2004). However, the bulk of the molecular data, including SINE insertions, confirm the monophyly of the extant Haplorhini (Schmitz & Zischler, 2004).

Infraorder Tarsiiformes

The only animals universally admitted to this infraorder are the five to eight species in the extant genus *Tarsius* (tarsiers), which inhabit several islands of the Malay Archipelago including Sumatra, Borneo, Sulawesi, and the Philippines. The affinities of tarsiers are obscured by their grotesque specializations, which collectively justify their German name of "goblin lemurs" (*Koboldmaki*). Visual predators par excellence, these small animals (≈100 g) have enormous eyeballs—each as large as the brain, in some species. Their retinas lack cones but have foveas (Wolin & Massopust, 1970), suggesting descent from a diurnal ancestor with smaller eyes. Tarsiers are the only exclusively faunivorous primates, and their antemolar dentition is uniquely specialized for seizing and holding prey, with stabbing, dagger-like incisors supplementing the canines and a battery of pointed premolars (Fig. 2.7). The tarsier postcranium is highly specialized for vertical clinging and leaping, with long hindlimbs; elongated tarsal bones (which give tarsiers their name); highly stabilized, hinge-like hip and ankle joints; and long stiff tails that serve as props in clinging to vertical supports. Tarsiers' bulging eyes, elongated legs and feet, and long fingers and toes tipped with expanded pads give them a vaguely froggy appearance despite their long tails and big, galago-like ears.

Living tarsiers are here placed in a **Superfamily Tarsioidea** of their own, distinguished from a **Superfamily †Omomyoidea** containing an array of extinct Paleogene prosimians often regarded as tarsiiforms. Omomyoids comprise most of the nonadapoid primates known from the Eocene. Most of them are small animals known exclusively from teeth and jaws, which evince a spectrum of dietary adaptations ranging from insectivory to frugivory. A few of the larger omomyoids may

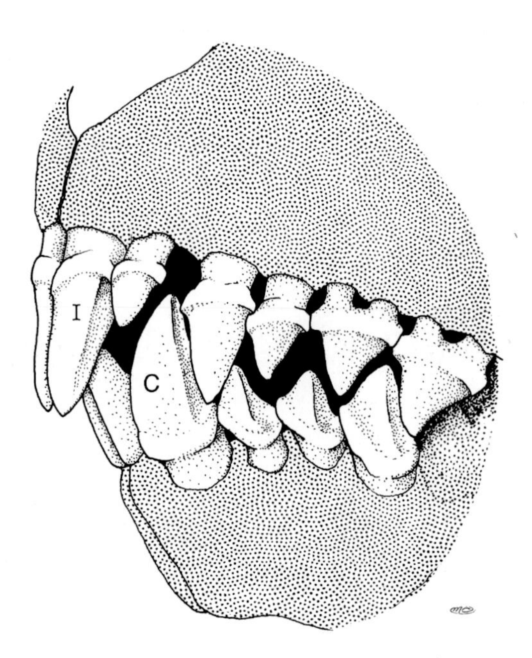

Figure 2.7 Antemolar dentition of *Tarsius*. I, upper central incisor (I^1); C, lower canine. From Cartmill, M., & Smith, F. H. (2009). *The human lineage.* New York: Wiley-Blackwell. Used with permission.

have been folivores. Known skulls and postcranial remains variously resemble those of cheirogaleids and galagos, with varying degrees of limb specializations for leaping and generally large eye sockets that suggest nocturnal habits. Several omomyoid genera exhibit apomorphies peculiar to tarsiers among living primates, but these are scattered in distribution and are not hierarchically nested in a way that would imply ordinated degrees of phyletic relationship to *Tarsius*.

The first omomyoids known, from the early Eocene, are very similar in molar morphology to the earliest adapoids (Ni et al., 2004; Rose & Bown, 1991), intimating that these fossils may be sampling the initial stages in the divergence of haplorhines from strepsirrhines. However, this inference conflicts with molecular-clock estimates that date the haplo-strepsirrhine split to some 80 Mya, well back in the Mesozoic (Horvath et al., 2008; Martin, 1993). Some tiny but strikingly tarsier-like molars and a fragmentary maxilla from the middle Eocene of China have been attributed to the genus *Tarsius* (Beard et al., 1994; Rossie et al., 2006). If *Tarsius* itself was already present in the Eocene, then most of the isolated tarsier-like features found among omomyoids must be convergences or parallelisms. At present, the haplorhine affinities of these extinct "tarsiiforms" remain uncertain (Beard, 1988).

Infraorder Anthropoidea

Living anthropoids are distinguished from tarsiers and other prosimians by a host of synapomorphies, including a complete postorbital septum, fusion of the two halves of the lower jaw, vertically implanted spatulate incisors, a foveate retina with a cone-rich area centralis, enlargement of the brain and elaboration of the visual centers, and numerous details of the cheek teeth, ear region, and postcranium. Molecular phylogenetic studies uniformly confirm the holophyly of the anthropoid clade. The earliest fossils universally acknowledged as anthropoids come from Eocene deposits of the Egyptian Fayum (Seiffert et al., 2005; Simons & Rasmussen, 1994). The known specimens show or imply that these stem-group anthropoids (Families †Oligopithecidae, †Proteopithecidae, and †Parapithecidae) had distinctively anthropoid teeth, ear regions, and orbits, but lacked some of the synapomorphies of the crown group (e.g., big brains and fused mandibular symphyses). Other candidate Eocene anthropoids include two groups from eastern Asia, the †Amphipithecidae and †Eosimiidae, which are dismissed by some authorities as anthropoid-convergent adapoids and omomyoids, respectively (Beard, 2002; Beard et al., 2005; Ciochon & Gunnell, 2004; Ciochon et al., 2001; Gunnell & Miller, 2001; Rasmussen, 2002).

Section Platyrrhini = Superfamily Ceboidea

Living anthropoids are divided into two groups: the monkeys of the New World tropics (Platyrrhini) and the monkeys, apes, and humans of the Old World (Catarrhini). The platyrrhines are almost wholly plesiomorphous relative to the crown-group catarrhines, lacking such distinctive catarrhine synapomorphies as a tubular extension of the bony ring (ectotympanic) around the eardrum and the loss of the anterior premolar. Nevertheless, there are a few candidate synapomorphies of the New World monkeys, mainly in details of the dentition, and they consistently sort out as a clade in analyses based on molecular data.

The origins of the New World monkeys are mysterious. They first appear in Oligocene deposits of South America, at a time when South America was separated from other continental land masses by oceanic gaps of hundreds of kilometers. Some systematists who think that anthropoids originated in Asia postulate an entrance into South America via Beringea and North America, but the general consensus is that they probably descend from some African basal anthropoid that somehow managed to get across the South Atlantic. Floating rafts of coastal vegetation torn loose by a tropical storm are a popular fantasy vehicle for this sea crossing (Chiarelli & Ciochon, 1980). Origin by vicariance (continental rifting yielding a cladistic split in the attached fauna) is unlikely, because estimated molecular dates for the platyrrhine-catarrhine divergence (32–36 Mya) postdate the Cretaceous zoogeographic isolation of South America from the other continents by at least 30 million years (Bocxlaer et al., 2006; Glazko & Nei, 2003).

Morphological and molecular data concur in segregating the platyrrhines into five clusters (Fig. 2.4). The marmosets and tamarins (Family Callitrichidae) are the smallest living anthropoids, ranging in size from 500 g (*Callimico*) down to ≈110 g (*Cebuella*). Their small size correlates with a dietary preference for insects, plant gums, and other high-energy foods. All their digits except the first toe are furnished with sharp claws, which facilitate clinging to thick tree trunks in feeding on exudates. The callitrichids' small size is also correlated with uniquely high reproductive rates for anthropoids—up to four offspring annually per female. This high birth rate is made possible by consistent twinning and by a cooperative breeding system in which males and juveniles carry and tend infants, handing them over at intervals to the group's reproductive female to nurse. *Callimico*, which lacks the twinning pattern and the reduced dental formula seen in the rest of the callitrichids, looks like it ought to represent a primitive outgroup of the others—but the molecular data belie this interpretation (Fig. 2.4).

At the opposite end of the platyrrhine size spectrum, the Atelidae are a family of medium-sized (≈5 to 10 kg) arboreal plant eaters, including the New World's only primarily folivorous monkey, *Alouatta*. The platyrrhine stock has not given rise to a radiation of arboreal leaf eaters like those that evolved among the lemurs of Madagascar and the Old World anthropoids, perhaps because that niche was preempted in South America by the tree sloths. All the atelids have prehensile tails, and all but *Alouatta* spend a significant amount of time hanging and swinging underneath branches, suspended by their tails and hands. As adaptations to this sort of arm-swinging locomotion, they have evolved some convergently ape-like apomorphies of the limbs and trunk (Erikson, 1963). Some extinct Pleistocene atelids (*Caipora*, *Protopithecus*) approached the great apes in body size, with weights of up to 25 kg.

Of special interest to neurobiologists, the night or owl monkey *Aotus* is the only nocturnal anthropoid. Its eyes are enlarged for purposes of light-gathering under scotopic conditions, and its retinas are afoveate and devoid of cones. This small monkey (≈1 kg) is sometimes classed in the family Cebidae, along with *Cebus* (capuchin monkeys) and *Saimiri* (squirrel monkeys), but is here assigned to its own family (Aotidae) because of its markedly divergent specializations. *Aotus* has an average-sized brain for a New World monkey, whereas *Cebus* and *Saimiri* are the most highly encephalized of the living platyrrhines. This may be a synapomorphy linking the two. However, brain growth follows different ontogenetic trajectories in the two genera, occurring mainly before birth in *Saimiri* and mainly after birth in *Cebus* (Hartwig, 1999). The large brain of *Cebus* correlates with some "advanced" behavioral apomorphies, including tool use in both captivity and the wild (Fragaszy et al., 2004).

The remaining platyrrhine genera (*Pithecia*, *Chiropotes*, *Cacajao*, *Callicebus*) constitute a fifth family of New World anthropoids, the Pithecidae, distinguished by several synapomorphies including specializations of the anterior teeth for feeding on hard, unripe fruit. Some systematists regard *Aotus* as part of the pithecid clade, but this assignment is not supported by the molecular data (Fig. 2.4).

Section Catarrhini

The earliest fossil anthropoids from the Old World are no more clearly related to modern catarrhines than they are to platyrrhines. The oldest taxon widely accepted as a catarrhine stem group is the early Oligocene family †Propliopithecidae from the Fayum. Propliopithecids have the reduced dental formula characteristic of later Old World anthropoids (2.1.2.3/2.1.2.3) and share some other catarrhine dental synapomorphies (e.g., loss of the paraconid cusp on the lower molars). Otherwise, they appear to have been persistently primitive anthropoids, with relatively small brains, ring-shaped ectoympanics, and generalized arboreal-quadruped postcranial anatomy.

Modern catarrhines fall into two holophyletic superfamilies: apes-plus-humans (Hominoidea) and Old World monkeys (Cercopithecoidea), distinguished from each other by conspicuous synapomorphies peculiar to each group. The oldest known cercopithecoids occur as fossils in the early Miocene, whereas the first fossil apes known to exhibit the distinctive shared peculiarities of the living hominoids are *Dryopithecus* and related genera (family †Dryopithecidae) from the late Miocene of Europe. The remaining Miocene catarrhines are sometimes described as "dental apes" because they had ape-like dentitions but monkey-like limb and trunk skeletons, adapted to quadrupedal locomotion rather than arm swinging. They are assigned here to two extinct superfamilies: the paraphyletic **†Proconsuloidea** (which have a complete tubular ectotympanic) and the probably holophyletic **†Pliopithecoidea** (which do not, and presumably branched off at an earlier point from a stem catarrhine resembling the propliopithecids). Both cercopithecoids and hominoids probably originated from proconsuloid ancestors.

From the Miocene onward, catarrhines have exhibited two evolutionary tendencies not evident in other primates: (1) a tendency to evolve mating systems involving high levels of male–male competition and marked sexual dimorphism and (2) a tendency to occupy terrestrial niches. The two tendencies are probably adaptively correlated with each other. Substantial radiations of terrestrially adapted forms have appeared in all three of the post-Oligocene catarrhine superfamilies—in the large-bodied, thick-enameled proconsuloids here classified as the family †Sugrivapithecidae, in the cercopithecids among the Cercopithecoidea, and in the hominids among the Hominoidea.

Superfamily Hominoidea

The living apes and humans are distinguished by a suite of postcranial synapomorphies thought to have originated as adaptations to an arm-swinging form of suspensory locomotion. Hominoids have long, limber forelimbs with modifications of the joints that maximize flexibility and permit the arm to be held over the head with the elbow fully extended. The thorax is transversely broadened and the clavicle elongated (Fig. 2.8), redirecting the glenoid socket of the shoulder joint to face more laterally and thereby facilitating swinging and hanging from supports above and behind the head. Because flexion and extension of the back no longer contribute to locomotion as they do in a galloping quadruped, the lumbar part of the hominoid vertebral column is reduced in length and its epaxial extensor muscles are reduced in volume and cross-sectional area, resulting in characteristic changes in the morphology of the lumbar vertebrae.

Extant hominoids are here divided into three families. The lesser apes or gibbons (Hylobatidae) are medium-sized (5 to 12 kg), strictly arboreal inhabitants of tropical forests in Southeast Asia and Indonesia. Hylobatids are swift, ricochetal arm-swingers that often hurl themselves in acrobatic, arm-propelled leaps across gaps of several meters between trees. They seldom descend to the ground, and walk bipedally with their hands in the air on the rare occasions when they do so.

The remaining apes are considerably larger animals and correspondingly more cautious in their arboreal locomotion. Orangutans (Pongidae) are the largest strictly arboreal primates, with body weights ranging from ≈36 kg in females up to around 80 kg in big males. Their evolutionary history is disputed. The Miocene sugrivapithecid *Sivapithecus* had a strikingly orangutan-like facial skeleton, and most authorities regard it as an early representative of the Pongidae. However, known postcranials of *Sivapithecus* seem to have been generally monkey-like, lacking key hominoid apomorphies of the hand and shoulder (Pilbeam, 2002). Either the orangutan-like cranial features of *Sivapithecus* are convergences or else most of the hominoid postcranial apomorphies evolved separately three times—in the ancestors of gibbons, of orangutans, and of the African apes and humans. The jury is still out on this issue.

The final hominoid family, the Hominidae, comprises humans (*Homo*) and their Plio-Pleistocene relatives (†*Australopithecus*),

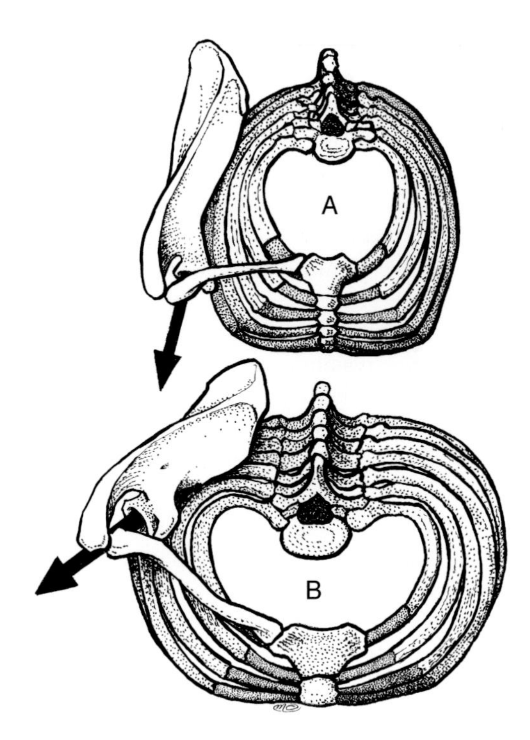

Figure 2.8 Cranial views of the thorax and right shoulder girdle of a cercopithecoid (*Papio*, **A**) and hominoid (*Homo*, **B**), showing the elongated clavicle, transversely broad thorax, and reoriented shoulder socket (arrows) characteristic of the crown-group Hominoidea. From Cartmill, M., & Smith, F. H. (2009). *The human lineage.* New York: Wiley-Blackwell, after Schultz, A. (1969). *The life of primates.* London: Weidenfeld and Nicholson. Used with permission.

chimpanzees (*Pan*), and gorillas (*Gorilla*). Chimpanzees and gorillas exhibit a so-called "knuckle-walking" pattern of quadrupedal locomotion, in which the weight of the upper body is borne on the backs of the middle phalanges, and they share some functionally related weight-bearing specializations of the hand and wrist not seen in humans. Despite these apparent synapomorphies, molecular data strongly indicate that chimpanzees are more closely related to humans than to gorillas, and all three genera are now conventionally lumped into the human family on cladistic grounds.

Superfamily Cercopithecoidea

Molecular-clock studies suggest a date of 23–25 Mya for the divergence of Old World monkeys from the "dental ape" ancestors of the Hominoidea. Cercopithecoids first appear in the late early Miocene of North and East Africa and spread into Asia around the end of the Miocene. The first cercopithecoids (†Victoriapithecidae) are already distinguished from hominoids by several characteristic cercopithecoid synapomorphies, including a "bilophodont" molar pattern (in which the four main cusps of the molar teeth are joined together by transverse crests) and various postcranial traits that appear to be functionally related to an increase in the amount of time spent on the ground (Benefit & McCrossin, 2002; McCrossin et al., 1998).

The extant cercopithecoids are divisible into two differently adapted families. The Colobidae or leaf monkeys are primarily folivores, with correspondingly specialized teeth and sacculated, somewhat ruminant-like stomachs. Most extant colobids are largely or exclusively arboreal. The Cercopithecidae are more diversely adapted, omnivorous animals that manifest a wider range of life habits and body sizes, from the small arboreal form *Miopithecus* (≈1,200 g) up to the large-bodied, mainly terrestrial quadrupeds *Papio* and *Mandrillus* (over 30 kg in big males). Colobids find their greatest diversity in Asia, with one genus (*Colobus*) native to Africa; conversely, cercopithecids constitute a largely African radiation that has given rise to a single cosmopolitan genus (*Macaca*) extending beyond Africa eastward across southern Asia into Indonesia. Perhaps the most successful and speciose of living primate families, the cercopithecids comprise a bewildering variety of variously adapted arboreal and terrestrial forms, including some 20 recognized species in each of the genera *Macaca* and *Cercopithecus*.

REFERENCES

Arnason, U., Gullberg, G., Gretarsdottir, S., Ursing, B., & Janke, A. (2000). The mitochondrial genome of the sperm whale and a new molecular reference for estimating eutherian divergence dates. *Journal of Molecular Evolution, 50*, 569–578.

Beard, K. C. (1988). The phylogenetic significance of strepsirrhinism in Paleogene primates. *International Journal of Primatology, 9*, 83–96.

Beard, K. C. (2002). Basal anthropoids. In: W. Hartwig (Ed.), *The primate fossil record* (pp. 133–149). Cambridge: University of Cambridge Press.

Beard, K. C., Jaeger, J-J, Chaimanee, Y., Rossie, J., Soe, A., Tun, S., et al. (2005). Taxonomic status of purported primate frontal bones from the Eocene Pondaung Formation of Myanmar. *Journal of Human Evolution, 49*, 468–481.

Beard, K. C., & MacPhee, R. D. E. (1994). Cranial anatomy of *Shoshonius* and the antiquity of the Anthropoidea. In: J. F. Fleagle & R. F. Kay (Eds.), *Anthropoid origins* (pp. 55–97). New York: Plenum.

Beard, K. C., Qi, T., Dawson, M., Wang, B., & Li, C. (1994). A diverse new primate fauna from middle Eocene fissure-fillings in southeastern China. *Nature, 368*, 604–609.

Beck, R. M. D. (2009). Was the Oligo-Miocene Australian metatherian *Yalkaparidon* a 'mammalian woodpecker'? *Biological Journal of the Linnaean Society of London, 97*, 1–17.

Benefit, B., & McCrossin, M. (2002). The Victoriapithecidae, Cercopithecoidea. In: W. Hartwig (Ed.), *The primate fossil record* (pp. 241–253). Cambridge: University of Cambridge Press.

Bloch, J. I., Silcox, M. T., Boyer, D. M., & Sargis, E. J. (2007). New Paleocene skeletons and the relationship of plesiadapiforms to crown-group primates. *Proceedings of the National Academy of Sciences, 104*, 1159–1164.

Bocxlaer, I., Roelants, K., Biju, S., Nagaraju, J., & Bossuyt, F. (2006). Late Cretaceous vicariance in Gondwanan amphibians. *PLoS ONE, 1*, e74.

Cartmill, M. (1974). *Daubentonia, Dactylopsila*, woodpeckers, and klinorhynchy. In: A. Martin, G. Doyle, & A. Walker (Eds.), *Prosimian biology* (pp. 655–670). London: Duckworth.

Cartmill, M. (1975). Strepsirhine basicranial structures and the affinities of the Cheirogaleidae. In: W. Luckett & F. S. Szalay (Eds.), *Phylogeny of the primates* (pp. 313–354). New York: Plenum.

Cartmill, M. (1980) Morphology, function and evolution of the anthropoid postorbital septum. In: A. B. Chiarelli & R. L. Ciochon (Eds.), *Evolutionary biology of the New World monkeys and continental drift* (pp. 243–274). New York: Plenum.

Cartmill, M. (1982). Assessing tarsier affinities: Is anatomical description phylogenetically neutral? *Geobios, mémoire special, 6*, 279–287.

Cartmill, M. (1994a), Anatomy, antinomies, and the problem of anthropoid origins. In: J. F. Fleagle & R. F. Kay (Eds.), *Anthropoid origins* (pp. 549–566). New York: Plenum.

Cartmill, M. (1994b). A critique of homology as a morphological concept. *American Journal of Physical Anthropology, 94*, 115–123.

Cartmill, M. (1999). Revolution, evolution, and Kuhn: A response to Chamberlain and Hartwig. *Evolution and Anthropology, 8*, 45–47.

Cartmill, M., & Kay, R. F. (1978). Craniodental morphology, tarsier affinities, and primate suborders. In: D. J. Chivers & K. A. Joysey (Eds.), *Recent advances in primatology* (vol 3, pp. 205–214). London: Academic Press.

Cartmill, M., & Smith, F. H. (2009). *The human lineage*. New York: Wiley-Blackwell.

Chiarelli, A., & Ciochon, R. (Eds). (1980). *Evolutionary biology of the New World monkeys and continental drift*. New York: Plenum.

Ciochon, R., Gingerich, P., Gunnell, G., & Simons, E. L. (2001). Primate postcrania from the late middle Eocene of Myanmar. *Proceedings of the National Academy of Sciences, 99*, 7672–7677.

Ciochon, R., & Gunnell, G. (2004). Eocene large-bodied primates from Myanmar and Thailand: morphological considerations and phylogenetic affinities. In: C. Ross & R. F. Kay (Eds.), *Anthropoid origins: New visions* (pp. 249–282). New York: Kluwer/Academic.

Cook, J. M., & Tristem, M. (1997). 'SINEs of the times' — transposable elements as clade markers for their hosts. *Trends in Ecology and Evolution, 12*, 295–297.

Eizirik, E., Murphy, W. J., & O'Brien, S. J. (2001). Molecular dating and biogeography of the early placental mammal radiation. *Journal of Heredity, 92*, 212–219.

Eizirik, E., Murphy, W. J., Springer, M. S., & O'Brien, S.J. (2004) Molecular phylogeny and dating of early primate divergences. In: C. Ross & R. F. Kay (Eds.), *Anthropoid origins: New visions* (pp. 45–64). New York: Kluwer/Academic.

Erickson, C. J. (1994). Tap-scanning and extractive foraging in aye-ayes, *Daubentonia madagascariensis. Folia Primatologica, 62*, 125–135.

Erikson, G. (1963). Brachiation in the New World monkeys. *Symposia of the Zoological Society of London, 10*, 135–164.

Fragaszy, D., Izar, P., Visalberghi, E., Ottoni, E., & Gomes de Oliviera, M. (2004). Wild capuchin monkeys (*Cebus libidinosus*) use anvils and stone pounding tools. *American Journal of Primatology, 64*, 359–366.

Gebo, D. (2002). Adapiformes: Phylogeny and adaptation. In: W. Hartwig (Ed.), *The primate fossil record* (pp. 21–43). Cambridge: University of Cambridge Press.

Gingerich, P., ul Haq, M., Zalmout, I., Khan, I., & Malkani, M. (2001). Origin of whales from early artiodactyls: hands and feet of Eocene Protocetidae from Pakistan. *Science, 293*, 2239–2242.

Glazko, G. V., & Nei, M. (2003). Estimation of divergence times for major lineages of primate species. *Molecular Biology and Evolution, 20*, 424–434.

Goodman, M., Porter, C., Czelusniak, J., Pages, S., Schneider, H., Shoshani, J., et al. (1998). Toward a phylogenetic classification of primates based on DNA evidence complemented by fossil evidence. *Molecular Phylogenics and Evolution, 9*, 585–598.

Gregory, W. K. (1910). The orders of mammals. *Bulletin of the American Museum of Natural History, 27*, 1–524.

Gunnell, G., & Miller, W. (2001). Origin of Anthropoidea: Dental evidence and recognition of early anthropoids in the fossil record, with comments on the Asian anthropoid radiation. *American Journal of Physical Anthropology, 114*, 177–191.

Hartwig, W. (1999). Perinatal life history traits in New World monkeys. *American Journal of Primatology, 40*, 99–130.

Hartwig, W. (Ed.). (2002). *The primate fossil record.* Cambridge: University of Cambridge Press.

Hennig, W. (1966). *Phylogenetic systematics.* Urbana: University of Illinois Press.

Horvath, J., Weisrock, D., Embry, S., Fiorentino, I., Balhoff, J., Kappeler, P., et al. (2008). Development and application of a phylogenomic toolkit: Resolving the evolutionary history of Madagascar's lemurs. *Genome Research, 18*, 489–499.

James, W. (1960). *The jaws and teeth of primates.* London: Pitman Medical Publishing.

Ji, Q., Luo, Z.-X., Yuan, C.-X., Wible, J. R., Zhang, J.-P., & Georgi, J. A. (2002). The earliest eutherian mammal. *Nature 416*, 816–822.

Kay, R., Ross, C., & Williams, B. (1997). Anthropoid origins. *Science, 275*, 797–804.

Kay, R., Willams, B., Ross, C., Takai, M., & Shigehara, N. (2004). Anthropoid origins: A phylogenetic analysis. In: C. Ross & R. F. Kay (Eds.), *Anthropoid origins: New visions* (pp. 91–135). New York: Kluwer/Academic.

Kriegs, J., Churakov, G., Kiefmann, M., Jordan, U., Brosius, J., & Schmitz, J. (2006). Retroposed elements as archives for the evolutionary history of placental mammals. *PLoS Biology, 4*(4), e91.

Kumar, S., & Hedges, S. B. (1998). A molecular timescale for vertebrate evolution. *Nature, 392*, 917–920.

Linne, K. (1735). *Systema naturae sive regna tria naturae systematice proposita per classes, ordines, genera, et species.* Leiden: T. Haak. 1964 facsimile reprint. Nieuwkoop: de Graaf.

Martin, R. (1993). Primate origins: Plugging the gaps. *Nature, 363*, 223–234.

McCrossin, M., Benefit, B., Gitau, S., Palmer, A., & Blue, K. (1998). Fossil evidence for the origins of terrestriality among Old World higher primates. In: E. Strasser, J. Fleagle, A. Rosenberger, & H. McHenry (Eds.), *Primate locomotion: Recent advances* (pp. 353–396). New York: Plenum.

Murphy, W., Eizirik, E., O'Brien, S., Madsen, O., Scally, M., Douady, C., et al. (2001). Resolution of the early placental mammal radiation using Bayesian phylogenetics. *Science, 294,* 2348–2351.

Ni, X., Wang, Y., Hu, Y., & Li, C. (2004). A euprimate skull from the early Eocene of China. *Nature, 427,* 65–68.

O'Leary, M. A., & Gatesy, J. (2007). Impact of increased character sampling on the phylogeny of Cetartiodactyla (Mammalia): Combined analysis including fossils. *Cladistics, 23,* 1–46.

Pilbeam, D. (2002). Perspectives on the Miocene Hominoidea. In: W. Hartwig (Ed.), *The primate fossil record* (pp. 303–310). Cambridge: University of Cambridge Press.

Rasmussen, D. (1986). Anthropoid origins: A possible solution to the Adapidae-Omomyidae paradox. *Journal of Human Evolution, 15,* 1–12.

Rasmussen, D. (2002). Early catarrhines of the African Eocene and Oligocene. In: W. Hartwig (Ed.), *The primate fossil record* (pp. 203–220). Cambridge: University of Cambridge Press.

Roos, C., Schmitz, J., & Zischler, H. (2004). Primate jumping genes elucidate strepsirrhine phylogeny. *Proceedings of the National Academy of Sciences, 101,* 10650–10654.

Rose, K., & Bown, T. (1991). Additional fossil evidence on the differentiation of the earliest euprimates. *Proceedings of the National Academy of Sciences, 88,* 98–101.

Rossie, J., Ni, X., & Beard, K. C. (2006). Cranial remains of an Eocene tarsier. *Proceedings of the National Academy of Sciences, 103,* 4381–4285.

Schmitt, D., Cartmill, M., Griffin, T., Hanna, J. B., & Lemelin, P. (2006). Adaptive value of ambling gaits in primates and other mammals. *Journal of Experimental Biology, 209,* 2042–2049.

Schmitz, J., & Zischler, H. (2004). Molecular cladistic markers and the infraordinal phylogenetic relationships of primates. In: C. Ross & R. F. Kay (Eds.), *Anthropoid origins: New visions* (pp. 65–77). New York: Kluwer/Academic.

Schultz, A. (1969). *The life of primates.* London: Weidenfeld and Nicholson.

Seiffert, E., Simons, E., & Attia, Y. (2003). Fossil evidence for an ancient divergence of lorises and galagos. *Nature, 422,* 421–424.

Seiffert, E., Simons, E., Clyde, W., Rossie, J., Attia, Y., Bown, T., et al. (2005). Basal anthropoids from Egypt and the antiquity of Africa's higher primate radiation. *Science, 310,* 300–304.

Shedlock, A. M., Takahashi, K., & Okada, N. (2004). SINEs of speciation: Tracking lineages with retroposons. *Trends in Ecology and Evolution, 19,* 545–553.

Shoshani, J., Groves, C., Simons, E., & Gunnell, G. (1996). Primate phylogeny: Morphological vs. molecular results. *Molecular Phylogenics and Evolution, 5,* 102–154.

Simons, E., & Rasmussen, D. (1994). A whole new world of ancestors: Eocene anthropoideans from Africa. *Evolution and Anthropology, 3,* 128–139.

Simpson, G. G. (1945). The principles of classification and a classification of mammals. *Bulletin of the American Museum of Natural History, 85,* 1–350.

Simpson, G. G. (1961). *Principles of animal taxonomy.* New York: Columbia University Press.

Sokal, R. R., & Sneath, P. H. A. (1963). *Principles of numerical taxonomy.* San Francisco: W. H. Freeman & Co.

Stephan, H. (1972). Evolution of primate brains: A comparative anatomical investigation. In: R. H. Tuttle (Ed.), *The functional and evolutionary biology of primates* (pp. 155–174). Chicago: Aldine-Atherton.

Stephan, H., & Andy, O. (1969). Quantitative comparative neuroanatomy of primates: An attempt at a phylogenetic interpretation. *Annals of the New York Academy of Science, 167,* 370–387.

Szalay, F. S., & Katz, C. (1973). Phylogeny of lemurs, galagos, and lorises. *Folia Primatology, 19,* 88–103.

Thewissen, J., Williams, M., Roe, L., & Hussain, S. (2001). Skeletons of terrestrial cetaceans and the relationship of whales to artiodactyls. *Nature, 413,* 277–281.

Tosi, A., Disotell, T., Morales, J., & Melnick, D. (2003). Cercopithecine Y-chromosome data provide a test of competing morphological evolutionary hypotheses. *Molecular Phylogenics and Evolution, 27,* 510–521.

Wildman, D., Chen, C., Erez, O., Grossman, L., Goodman, M., & Romero, R. (2006). Evolution of the mammalian placenta revealed by phylogenetic analysis. *Proceedings of the National Academy of Sciences, 103,* 3203–3208.

Wolin, L., & Massopust, L. (1970). Morphology of the primate retina. In: C. Noback and W. Montagna (Eds.), *The primate brain* (pp. 1–28). New York: Appleton-Century-Crofts.

Xing, J., Wang, H., Han, K., Ray, D., Huang, C., Chemnick, L., et al. (2005). A mobile element based phylogeny of Old World monkeys. *Molecular Phylogenics and Evolution, 37,* 872–880.

Yoder, A., Cartmill, M., Ruvolo, M., Smith, K., & Vilgalys, R. (1996). Ancient single origin for Malagasy primates. *Proceedings of the National Academy of Sciences, 93,* 5122–5126.

CHAPTER 3

Primate Locomotor Evolution: Biomechanical Studies of Primate Locomotion and Their Implications for Understanding Primate Neuroethology

Daniel Schmitt

Primate locomotor diversity is extraordinary when compared to most other vertebrate groups with at least three forms of locomotion not seen among any other mammals alive today. One of those unique forms of locomotion is our own upright striding bipedal gait. The 65 million–year story of how our lineage departed from a tiny mammalian ancestor and evolved the locomotor variation we see today involves a series of profound shifts in the way primates use their limbs. These dramatic changes in limb function reflect major adaptive shifts and locomotor innovations during primate and human evolution. These changes in limb function may also reflect profound differences in the neural control of locomotion between primates and almost all other animals.

In the past 50 years primate models have played an important role in studies of brain and spinal cord injury and pathology (i.e., Courtine et al., 2005a,b, and Xiang et al., 2007, for some of the most recent work, and Capitanio and Emborg, 2008, for a review of the history of this research). Beginning with the pioneering work of Harvard neurobiologist Derek Denny-Brown, studies of locomotor patterns in healthy primates and those with spinal and brain lesions have led to significant insight into the neural control of human locomotion (see Gilman, 1982, and Vilensky et al., 1994a, 1996 for a review of this work). Primates are critical models for human clinical research not only because they are our closest relatives, but also because primates (including humans) show patterns of locomotion and locomotor control that are different from all other mammals. For example, the important (albeit controversial) theory that primates as a group lack central pattern generators and rely more on supraspinal control of motion compared to other mammals has stimulated research that has led to a deeper understanding and treatment of spinal injury (see Duysens & van de Crommert, 2007; Grillner and Wallen 1985; Vilensky & Larson, 1989; Vilensky & O'Connor, 1997, 1998).

In addition to the profound clinical and basic science insights, biomechanical studies that compare primates to other mammals have allowed us to better understand and reconstruct the evolution of locomotor behavior in our Order and the underlying adaptive foundations of patterns of primate gait. Table 3.1 provides a summary of biomechanical studies of primate locomotion that may be a resource for those neuroethologists interested in the underlying differences both within primates and across orders. This chapter describes the extreme locomotor diversity found within primates and describes

Table 3.1 A Representative* List of Experimental Studies of Primate Locomotion

Source	Taxa	Data	Movement(s)
Cartmill et al., 2002, 2007a,b	All	T	TQ, AQ
Hildebrand, 1967	All	T	TQ
Larson, 1998; Larson et al. 1999, 2001	All	K	TQ, AQ
Lemelin & Schmitt, 1998	All	K	TQ, AQ
Reynolds, 1985	All	T, FP	TQ
Reynolds, 1987	All	T, K	TQ, TB
Vilensky, 1987, 1989; Vilensky & Gehlsen, 1984; Vilensky & Larson, 1989	All	T, K, EMG	TQ
Aerts et al., 2000	Hom	T	TQ, TB
Chang et al., 1997, 2000; Bertram & Chang, 2001	Hom	FP	AS
D'Août et al., 2002	Hom	T, K	TQ, TB
Elftman, 1944; Elftman & Manter, 1935	Hom	K, T	TB
Jenkins, 1972	Hom	K	TB
Kimura, 1990, 1991, 1996	Hom	T, En	TQ
Larson & Stern, 1986, 1987	Hom	EMG	TQ, AQ, R
Larson et al., 1991	Hom	EMG	AS, TQ, R
Larson, 1988, 1989	Hom	EMG	AS
Okada & Kondo, 1982; Okada, 1985	Hom	EMG	TB
Prost, 1967, 1980	Hom	K, T	TQ, TB, VC
Shapiro et al., 1997	Hom	EMG, T	TQ
Stern & Larson, 2001	Hom	EMG	TQ, AS
Stern & Susman, 1981	Hom	EMG	TQ, TB, VC
Susman, 1983	Hom	K	TQ, TB
Swartz et al., 1989	Hom	BS	AS
Tardieu et al., 1993	Hom	K	TB
Tuttle & Basmajian, 1974a,b,c, 1977, 1978a,b; Tuttle et al., 1983, 1992	Hom	EMG	TQ, TB, AS
Vereecke et al., 2003, 2006	Hom	FP	TQ
Wunderlich & Jungers, 2009; Wunderlich & Ford, 2000	Hom	Pr	TQ, AQ
Yamazaki & Ishida, 1984	Hom	K, T	TB, VC
Jenkins et al., 1978	NWM	K, C,	AS
Prost & Sussman, 1969	NWM	K, T	IQ
Schmitt, 2003a	NWM	FP, K, T	AQ, TQ
Turnquist et al., 1999	NWM	K	AS
Vilensky & Patrick, 1985	NWM	T, K	TQ
Vilensky et al., 1994	NWM	T, K	IQ
Fleagle et al., 1981	NWM, Hom	EMG, BS	VC, TQ, TB
Ishida et al., 1985	NWM, Hom	EMG	TQ, TB
Jungers & Stern, 1980, 1981, 1984	NWM, Hom	EMG	AS
Stern et al., 1977, 1980	NWM, Hom	EMG	AQ, VC
Taylor & Rowntree, 1973	NWM, Hom	En	
Hirasaki et al., 1993, 1995, 2000	NWM, OWM	T, K, FP, EMG	VC
Prost, 1965, 1969	NWM, OWM	T	TQ
Kimura et al., 1979; Kimura, 1985, 1992	NWM, OWM, Hom	FP	TQ
Kimura et al., 1983	NWM, OWM, Hom	T	TQ, TB
Schmitt & Larson, 1995	NWM, OWM, Hom	K	TQ, AQ
Vangor & Wells, 1983	NWM, OWM, Hom	EMG	TQ, TB, VC

(continued)

Table 3.1 (Continued)

Source	Taxa	Data	Movement(s)
Wunderlich & Schmitt, 2000	NWM, OWM, Hom	K	TQ, AQ
Demes et al., 1994	OWM	BS	TQ
Larson & Stern, 1989; 1992	OWM	EMG	TQ
Meldrum, 1991	OWM	K, T	AQ, TQ
Polk, 2002	OWM	T, FP, K	TQ
Rollinson & Martin, 1981	OWM	T	AQ, TQ
Schmitt et al., 1994	OWM	EMG	TQ
Shapiro & Raichlen, 2005	OWM	K	TQ
Wells & Wood, 1975	OWM	K	TQ, L
Schmitt, 1994, 1998, 1999, 2003b	OWM	K, FP	TQ, AQ
Vilensky, 1980, 1983, 1988; Vilensky & Gankiewicz, 1986, 1990, Vilensky et al., 1986, 1990, 1991	OWM	K, T	TQ
Whitehead & Larson, 1994	OWM	K, C, EMG	TQ
Alexander & Maloiy, 1984	OWM, Hom	T	TQ
Shapiro & Jungers, 1988, 1994	OWM, Hom	EMG, T	TQ, TB, VC
Anapol & Jungers, 1987	Pro	EMG, T	TQ, L
Carlson et al., 2005	Pro	FP	TQ, AQ
Demes et al., 1990	Pro	T	AQ
Demes et al., 1998, 2001	Pro	FP	L
Franz et al., 2005	Pro	K, FP	TQ, AQ
Gunther, 1991	Pro	FP, EMG	L
Ishida et al., 1990	Pro	T, FP	AQ
Jouffroy, 1983; Jouffroy & Gasc, 1974; Jouffroy et al., 1974	Pro	K, C	AQ
Jouffroy & Petter, 1990	Pro	T, K,	AQ
Jouffroy & Stern, 1990	Pro	EMG	AQ
Jungers & Anapol, 1985	Pro	T, EMG	TQ
Schmidt & Fischer, 2000	Pro	K, C	AQ
Schmitt & Lemelin, 2002	Pro	FP	TQ, AQ
Shapiro et al., 2001	Pro	K	AQ
Stevens, 2001; Stevens et al. 2001	Pro	K, T	AQ, IAQ

Taxa: All, representative species from all of the taxa; Hom, hominoid; NWM, New World monkey; OWM, Old World monkey; Pro, prosimian.
Data: BS, bone strain; C, cineradiography; EMG, electromyography; En, energetics; FP, force plate; K, kinematics; Pr, pressure; T, temporal characters.
Movements: AQ, arboreal quadrupedalism; AS, arm-swinging; IAQ, inclined quadrupedalism (pole); IQ, inclined quadrupedalism (flat substrate); L, leaping; R, reaching; TB, terrestrial bipedalism; TQ, terrestrial quadrupedalism.

*This is not an exhaustive list of all studies on primate locomotion. I have included those studies that focus specifically on primate locomotor mechanics primarily in a laboratory setting. I apologize to anyone who was excluded. A review of many experimental studies can be found in Fleagle (1979), Jouffroy (1989), Churchill and Schmitt (2003), and Lemelin and Schmitt (2007). The above table does not include studies by anthropologists that focus solely on human bipedalism like those of Li and colleagues (1996), Schmitt and colleagues (1996, 1999), or Crompton and colleagues (1998).

laboratory and field studies that have illuminated significant differences between primates and other animals. These two aspects of primate locomotion—diversity and difference—are brought together to interpret the adaptive significance of fundamental changes in the role of the forelimb associated with the origin of primates, the evolution of suspensory behavior, and the evolution of upright, striding bipedalism.

PRIMATE LOCOMOTOR DIVERSITY

Classifying the locomotion of any animal is difficult. Primates are especially difficult. Stern and Oxnard (1973) argued that a system of classification must be an aid to thinking and communication. But with primates this is an almost impossible goal. There are many primates that can effectively walk, run, leap, climb, and swing through the trees. What name can we give to that repertoire? What category would such an animal fit into?

The Order Primates contains between 100 and 400 (see Chapter 2 for a review) recognized species (Table 3.2). Figure 3.1 shows a small sample of this diversity. Our Order includes many primates that walk and run in a manner that is superficially similar to dogs and cats. But there are also many that leap large distances between horizontal or vertical supports. There are some primates that swing like pendulums through the trees and others that walk on their fingers, their fists, or even their knuckles. Finally, a select group of primates walks on two legs exclusively. This wide is made more complex by the reality that many primate species can use several of those different modes equally adroitly. That flexibility was the key to the successful radiation of our Order.

At its core, the Order Primates is an arboreal radiation. Although several species in Africa and Asia spend most of their time on the ground (Table 3.1), the vast majority of primates, including all those in South America, restrict their movements to the trees, and all primates move into the tress to sleep and escape predators. The physical characteristics that we use to define primates—grasping, prehensile hands with nails instead of claws and forward-facing eyes (Figure 3.2)—have long been associated

Table 3.2 List of Primates and Their Most Commonly Used Substrates and Locomotor Modes

Common Name	Species Name	Substrate Used	Locomotor Mode
SUBORDER STREPSIRRHINI			
INFRAORDER LEMURIFORMES			
Superfamily Lemuroidea			
Family Cheirogaleidae			
Fat-tailed dwarf lemur	*Cheirogaleus medius*	Arboreal	Quadruped/leaper
Greater dwarf lemur	*Cheirogaleus major*	Arboreal	Quadruped/leaper
Lesser mouse lemur	*Microcebus murinus*	Arboreal	Quadruped/leaper
Coquerel's mouse lemur	*Mirza coquereli*	Arboreal	Quadruped/leaper
Hairy-eared dwarf lemur	*Allocebus trichotis*	Arboreal	Quadruped/leaper
Fork-crowned lemur	*Phaner furcifer*	Arboreal	Quadruped/clinger
Family Lemuridae			
Ring-tailed lemur	*Lemur catta*	Terrestrial/ arboreal	Quadruped/leaper
Black lemur	*Eulemur macaco*	Arboreal	Quadruped/leaper
Brown lemur	*Eulemur fulvus*	Arboreal	Quadruped/leaper
Mongoose lemur	*Eulemur mongoz*	Arboreal	Quadruped/leaper
Crowned lemur	*Eulemur coronatus*	Arboreal	Quadruped/leaper
Red-bellied lemur	*Eulemur rubriventer*	Arboreal	Quadruped/leaper
Gray gentle lemur	*Hapalemur griseus*	Arboreal	Quadruped/leaper
Golden gentle lemur	*Hapalemur aureus*	Arboreal	Quadruped/leaper
Broad-nosed gentle lemur	*Hapalemur simus*	Arboreal	Quadruped/leaper
Ruffed lemur	*Varecia variegata*	Arboreal	Quadruped

(*continued*)

Table 3.2 (Continued)

Common Name	Species Name	Substrate Used	Locomotor Mode
Family Lepilemruidae			
Weasel sportive lemur	*Lepilemur mustelinus*	Arboreal	Vertical clinger & leaper
Family Indriidae			
Indri	*Indri indri*	Arboreal	Vertical clinger & leaper
Eastern woolly lemur	*Avahi laniger*	Arboreal	Vertical clinger & leaper
Western woolly lemur	*Avahi occidentalis*	Arboreal	Vertical clinger & leaper
Diademed sifaka	*Propithecus diadema*	Arboreal	Vertical clinger & leaper
Verreaux's sifaka	*Propithecus verreauxi*	Arboreal	Vertical clinger & leaper
Family Daubentoniidae			
Aye-aye	*Daubentonia madagascariensis*	Arboreal	Quadruped
Superfamily Lorisoidea			
Family Loridae			
Golden potto	*Arctocebus aureus*	Arboreal	Quadruped
potto	*Perodicticus potto*	Arboreal	Quadruped
Slender loris	*Loris tardigradus*	Arboreal	Quadruped
Slow loris	*Nycticebus coucang*	Arboreal	Quadruped
Lesser slow loris	*Nycticebus pygmaeus*	Arboreal	Quadruped
Family Galagidae			
Greater galago	*Otolemur crassicaudatus*	Arboreal	Quadruped/leaper
Garnett's galago	*Otolemur garnettii*	Arboreal	Quadruped/leaper
Southern needle-clawed galago	*Euoticus elegantulus*	Arboreal	Quadruped/leaper
Lesser galago	*Galago senegalensis*	Arboreal	Quadruped/leaper
Southern lesser galago	*Galago moholi*	Arboreal	Quadruped/leaper
Allen's galago	*Galago alleni*	Arboreal	Quadruped/leaper
Demidoff's galago	*Galago demidoff*	Arboreal	Quadruped/leaper
SUBORDER HAPLORHINI			
INFRAORDER TARSLLFORMES			
Family Tarsiidae			
Philippine tarsier	*Tarsius syrichta*	Arboreal	Vertical clinger & leaper
Horsfield's tarsier	*Tarsius bancanus*	Arboreal	Vertical clinger & leaper
Spectral tarsier	*Tarsius spectrum*	Arboreal	Vertical clinger & leaper
INFRAORDER PLATYRHINI			
Superfamily Ceboidea			
Family Cebidae			
Dusky titi	*Callicebus moloch*	Arboreal	Quadruped
Northern owl monkey	*Aotus trivirgatus*	Arboreal	Quadruped
White-faced saki	*Pithecia pithecia*	Arboreal	Quadruped
Monk saki	*Pithecia monachus*	Arboreal	Quadruped
White-nosed saki	*Chiropotes albinasus*	Arboreal	Quadruped
Black-headed uakari	*Cacajao melanocephalus*	Arboreal	Quadruped
Red uakari	*Cacajao calvus*	Arboreal	Quadruped
White-throated capuchin	*Cebus capucinus*	Arboreal	Quadruped
White-fronted capuchin	*Cebus albifrons*	Arboreal	Quadruped

(continued)

Table 3.2 (Continued)

Common Name	Species Name	Substrate Used	Locomotor Mode
Weeper capuchin	*Cebus olivaceus*	Arboreal	Quadruped
Black-capped capuchin	*Cebus apella*	Arboreal	Quadruped
Common squirrel monkey	*Saimiri sciureus*	Arboreal	Quadruped/leaper
Family Atelidae Mantled howler monkey	*Alouatta palliata*	Arboreal	Quadruped
Red howler monkey	*Alouatta seniculus*	Arboreal	Quadruped
Brown howler monkey	*Alouatta fusca*	Arboreal	Quadruped
Black howler monkey	*Alouatta caraya*	Arboreal	Quadruped
Black spider monkey	*Ateles paniscus*	Arboreal	Arm-swinger/quadruped
Long-haired spider monkey	*Ateles belzebuth*	Arboreal	Arm-swinger/quadruped
Brown-headed spider monkey	*Ateles fusciceps*	Arboreal	Arm-swinger/quadruped
Black-handed spider monkey	*Ateles geoffroyi*	Arboreal	Arm-swinger/quadruped
Woolly spider monkey	*Brachyteles arachnoides*	Arboreal	Quadruped/arm-swinger
Humboldt's woolly monkey	*Lagothrix lagotricha*	Arboreal	Quadruped/arm-swinger
Yellow-tailed woolly monkey	*Lagothrix flavicauda*	Arboreal	Quadruped/arm-swinger
Family Callitrichidae			
Goeldi's marmoset	*Callimico goeldii*	Arboreal	Vertical clinger & leaper
Common marmoset	*Callithrix jacchus*	Arboreal	Vertical clinger & leaper
Geoffroy's tufted-ear marmoset	*Callithrix geoffroyi*	Arboreal	Vertical clinger & leaper
Buffy-headed Marmoset	*Callithrix flaviceps*	Arboreal	Vertical clinger & leaper
Silvery marmoset	*Callithrix argentata*	Arboreal	Vertical clinger & leaper
Pygmy marmoset	*Cebuella pygmaea*	Arboreal	Vertical clinger & leaper
Golden lion tamarin	*Leontopithecus rosalia*	Arboreal	Vertical clinger & leaper
Golden-headed lion tamarin	*Leontopithecus chrysomelas*	Arboreal	Vertical clinger & leaper
Midas tamarin	*Saguinus midas*	Arboreal	Vertical clinger & leaper
Black-mantled tamarin	*Saguinus nigricollis*	Arboreal	Vertical clinger & leaper
Saddle-back tamarin	*Saguinus fuscicollis*	Arboreal	Vertical clinger & leaper
Black-chested moustached tamarin	*Saguinus mystax*	Arboreal	Vertical clinger & leaper
Red-bellied tamarin	*Saguinus labiatus*	Arboreal	Vertical clinger & leaper
Emperor tamarin	*Saguinus imperator*	Arboreal	Vertical clinger & leaper
Pied tamarin	*Saguinus bicolor*	Arboreal	Vertical clinger & leaper
Cotton-top tamarin	*Saguinus oedipus*	Arboreal	Vertical clinger & leaper
INFRAORDER CATARRHINI			
Superfamily Cercopithecoidea			
Family Cercopithecidae			
Subfamily Cercopithecinae			
Allen's swamp monkey	*Allenopithecus nigroviridis*	Terrestrial	Quadruped
Talapoin monkey	*Miopithecus talapoin*	Arboreal	Quadruped
Patas monkey	*Erythrocebus patas*	Terrestrial	Quadruped

(*continued*)

Table 3.2 (Continued)

Common Name	Species Name	Substrate Used	Locomotor Mode
Vervet monkey	*Cercopithecus aethiops*	Terrestrial/ arboreal	Quadruped
Diana monkey	*Cercopithecus diana*	Arboreal	Quadruped
Greater white-nosed monkey	*Cercopithecus nictitans*	Arboreal	Quadruped
Blue monkey	*Cercopithecus mitis*	Arboreal	Quadruped
Mona monkey	*Cercopithecus mona*	Arboreal	Quadruped
Campbell's monkey	*Cercopithecus campbelli*	Arboreal	Quadruped
Crowned guenon	*Cercopithecus pogonias*	Arboreal	Quadruped
Wolf's monkey	*Cercopithecus wolfi*	Arboreal	Quadruped
Lesser white-nosed monkey	*Cercopithecus petaurista*	Arboreal	Quadruped
Redtail monkey	*Cercopithecus ascanius*	Arboreal	Quadruped
L'hoest's monkey	*Cercopithecus lhoesti*	Terrestrial/ arboreal	Quadruped
Preuss's monkey	*Cercopithecus preussi*	Arboreal	Quadruped
DeBrazza's monkey	*Cercopithecus neglectus*	Terrestrial/ arboreal	Quadruped
Barbary macaque	*Macaca sylvanus*	Terrestrial	Quadruped
Lion-tailed macaque	*Macaca silenus*	Terrestrial/ arboreal	Quadruped
Pigtailed macaque	*Macaca nemestrina*	Terrestrial	Quadruped
Tonkean macaque	*Macaca tonkeana*	Terrestrial	Quadruped
Celebes macaque	*Macaca nigra*	Terrestrial	Quadruped
Crab-eating macaque	*Macaca fascicularis*	Arboreal	Quadruped
Stumptailed macaque	*Macaca arctoides*	Terrestrial/ arboreal	Quadruped
Rhesus macaque	*Macaca mulatta*	Terrestrial/ arboreal	Quadruped
Japanese macaque	*Macaca fuscata*	Terrestrial/ arboreal	Quadruped
Bonnet macaque	*Macaca radiata*	Terrestrial/ arboreal	Quadruped
Gray-cheeked mangabey	*Lophocebus albigena*	Arboreal	Quadruped
Black mangabey	*Lophocebus aterrimus*	Arboreal	Quadruped
Hamadryas baboon	*Papio hamadryas*	Terrestrial	Quadruped
Guinea baboon	*Papio papio*	Terrestrial	Quadruped
Olive baboon	*Papio anubis*	Terrestrial	Quadruped
Yellow baboon	*Papio cynocephalus*	Terrestrial	Quadruped
Chacma baboon	*Papio ursinus*	Terrestrial	Quadruped
Gelada baboon	*Theropithecus gelada*	Terrestrial	Quadruped
Sooty mangabey	*Cercocebus atys*	Terrestrial/ Arboreal	Quadruped
White-collared mangabey	*Cercocebus torquatus*	Terrestrial/ Arboreal	Quadruped
Agile mangabey	*Cercocebus agilis*	Terrestrial/ Arboreal	Quadruped
Tana river mangabey	*Cercocebus galeritus*	Arboreal	Quadruped

(continued)

Table 3.2 (Continued)

Common Name	Species Name	Substrate Used	Locomotor Mode
Mandrill	*Mandrillus sphinx*	Arboreal/ Terrestrial	Quadruped
Drill	*Mandrillus leucophaeus*	Terrestrial/ arboreal	Quadruped
Subfamily Colobinae			
Black colobus monkey	*Colobus satanas*	Arboreal	Quadruped
Angolan black-and-white colobus	*Colobus angolensis*	Arboreal	Quadruped
Western black-and-white colobus	*Colobus polykomos*	Arboreal	Quadruped
Abyssinian black-and-white colobus	*Colobus guereza*	Arboreal	Quadruped
Western red colobus monkey	*Procolobus badius*	Arboreal	Quadruped
Olive colobus monkey	*Procolobus verus*	Arboreal	Quadruped
Hanuman langur	*Semnopithecus entellus*	Terrestrial/ arboreal	Quadruped
Purple-faced langur	*Trachypithecus vetulus*	Arboreal	Quadruped
Silvered leaf monkey	*Trachypithecus cristatus*	Arboreal	Quadruped
Tenasserim langur	*Trachypithecus barbei*	Arboreal	Quadruped
Dusky leaf monkey	*Trachypithecus obscurus*	Arboreal	Quadruped
Delacour's langur	*Trachypithecus delacouri*	Arboreal	Quadruped
Mitered leaf monkey	*Presbytis melalophos*	Arboreal	Quadruped
Douc langur	*Pygathrix nemaeus*	Arboreal	Quadruped
Snub-nosed langur	*Pygathrix roxellana*	Arboreal	Quadruped
Proboscis monkey	*Nasalis larvatus*	Arboreal	Quadruped
Superfamily Hominoidea			
Family Hylobatidae			
White-handed gibbon	*Hylobates lar*	Arboreal	Brachiator
Siamang	*Hylobates syndactylus*	Arboreal	Bbrachiator
Family Hominidae			
Orangutan	*Pongo pygmaeus*	Arboreal	Quadrumanous/ arm-swinger
Gorilla	*Gorilla gorilla*	Terrestrial	Knuckle-walker
Pygmy chimpanzee (bonobo)	*Pan paniscus*	Arboreal/ terrestrial	Knuckle-walker/ arm-swinger
Common chimpanzee	*Pan troglodytes*	Terrestrial/ arboreal	Kknuckle-walker/ arm-swinger
Human	*Homo sapiens*	Terrestrial	Biped

This list of primates is derived from a review of animals listed by Fleagle (1999), Szalay and Delson (1979), and Rowe (1996). This is not meant as a definitive or inarguable phylogenetic arrangement of primates (see Cartmill, this volume for a critical review of primate phylogeny). This list is serve as a basis for describing the variation in locomotor behavior in an organized list. I have chosen to list a subset of the nearly 400 species that one might be included in a comprehensive list. For example, the many identified species of night monkey are not listed because their locomotion does not differ. The locomotor classification is based on Rose (1973), Oxnard et al. (1990), and Fleagle (1999).

Figure 3.1 This painting ("Darwin and Friends," Stephen Nash, 1985) illustrates the remarkable locomotor diversity in the order. A terrestrial quadrupedal lemur and monkey are shown on the ground to the left, the knuckle-walking apes are on the ground to the right, arboreal quadrupeds are on the middle and high branches, an arm-swinging monkey and ape are hanging in the center, and the vertical clingers and leapers are shown on the primary trunk. Finally, our own bipedalism is represented on the right. Used with permission.

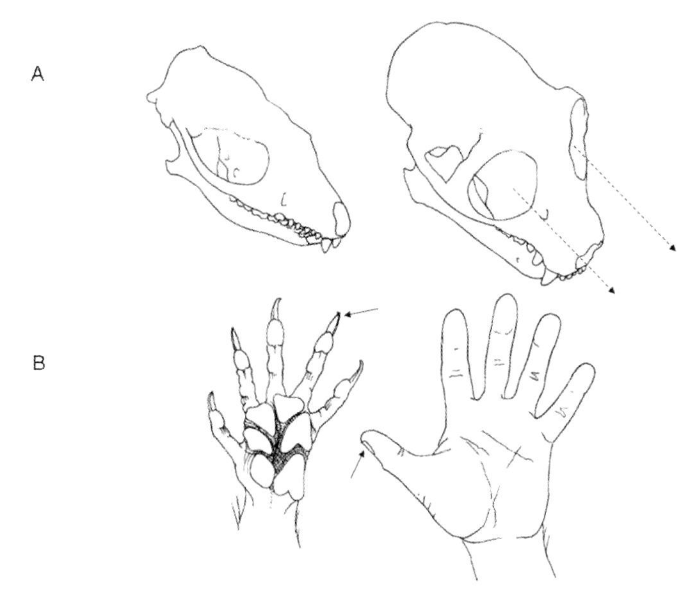

Figure 3.2 This figure highlights two pieces of a larger suite of features that define primates. (**A**) Forward-facing eyes in the primate (right) allow detailed stereoscopic vision and depth perception. (**B**) Grasping hands with nails in primates (right) allow effective gripping of branches as well as single-handed, prehensile capture and manipulation of objects. Modified from Cartmill, M. (1992). New views on primate origins. *Evolution and Anthropology, 1,* 105–111. Used with permission.

with arboreal habits. Smith (1912) and Jones (1916) were among the first to relate some of the unique anatomical and behavioral characteristics of primates to living in the trees. They suggested that grasping hands and stereoscopic vision met the critical need to leap great distances and land safely. Their views were further promoted in writings of LeGros Clark (1959) but later challenged and refined by Cartmill (1970, 1972, 1974a,b, 1992), who suggested that the forward-facing eyes and grasping extremities of primates should be interpreted not as adaptations for leaping, but rather as adaptations to cautious, nocturnal foraging for insect prey on thin, flexible branches. Recent morphological and laboratory-based data support the latter argument (Cartmill et al., 2002; Hamrick, 1998; Lemelin, 1999; Schmitt, 1999; Schmitt & Lemelin, 2002). Others have suggested that primates evolved their unusual physical characteristics in association with foraging for fruits rather than insects (Sussman, 1991). Exactly which aspects of arboreality influenced the development of primate anatomy remains a subject of continued debate (Cartmill, 1992; Cartmill et al. 2002, 2007a,b; Lemelin & Schmitt, 2007; Raichlen & Shapiro, 2006; Schmitt & Lemelin, 2002; Sussman, 1991; Szalay & Dagosto, 1988; Szalay et al., 1987). But the fact remains that the challenges of moving and foraging in a thin, flexible, terminal branch milieu appear to be at the core of the anatomical, behavioral, and neurological qualities that define primates. As Jenkins (1974: 112) noted concisely, "The adaptive innovation of ancestral primates was therefore not the invasion of the arboreal habitat, but their successful restriction to it." So it is reansonable to begin this exercise of classifying primate locomotion by saying that the underlying primate bauplan reflects a history of arboreal quadrupedal locomotion on relatively thin supports.

Another broad statement that can be made is that most primates rely primarily on a quadrupedal mode of locomotion (Fig. 3.3). Of the 149 living species of primates listed in Table 3.2, at least 107 them rely heavily on quadrupedal locomotion (Rose, 1973). However, as has been recognized by many authors, the term "quadruped" masks a remarkable diversity(Rose 1973,Hunt et al.1996). Quadrupedal primates can be broken down, as Fleagle and Mittermier (1980) did, into "quadrupedal runners" and "quadrupedal leapers." This difference is subtle but has important implications for anatomy and

Figure 3.3 Quadrupedalism is the most common mode of locomotion among primates (Rose, 1973; Table3.2). Quadrupedalism is found in strepsirrhines, New and Old World monkeys, and apes as is illustrated here moving counter clockwise from left by this ring-tailed lemur, yellow-tailed woolly monkey, diana monkey, and gorilla. Quadrupedalism is found in terrestrial and arboreal primates and in primates as small as 50 grams and as large as 500 kg. Drawings by Stephen Nash; courtesy of Conservation International.

ecology. This was made clear in a seminal study in which Fleagle (1977a,b) was able to show that two closely related species of Asian colobines that we readily classify as arboreal quadrupeds—the dusky leaf monkey and the banded leaf monkey—use different amounts of leaping between horizontal supports and show significant differences in the detailed anatomy of muscle markings and joint shape in their lower limb (Fig. 3.4).

The primates that we label as quadrupeds show greater diversity than simple variation in

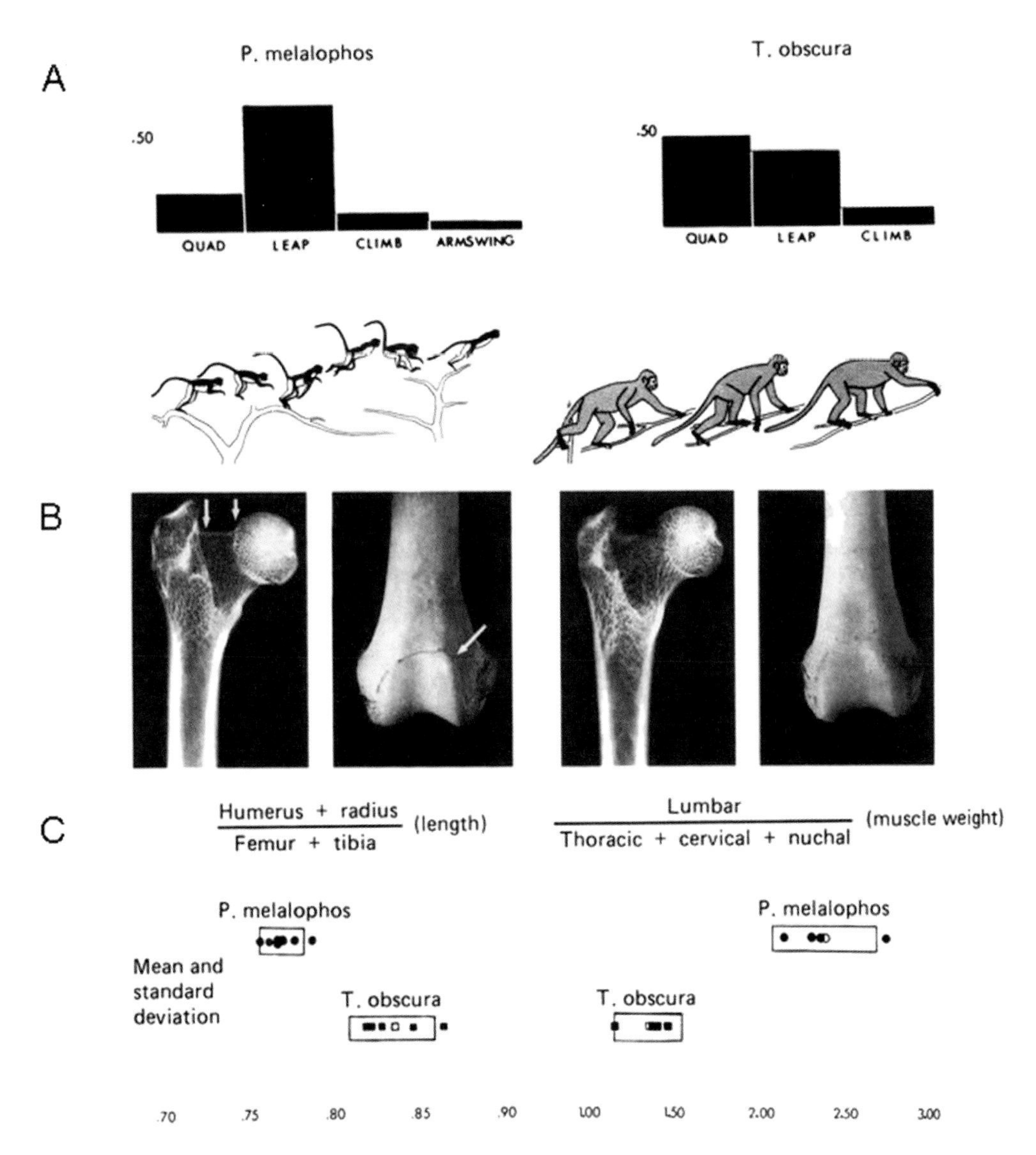

Figure 3.4 The banded leaf monkey (*Presbytis melalophos*) (left) and dusky leaf monkey (*Trachypithecus obscura*) (right) are sympatric and closely related but they exhibit dramatic differences in the percentage of time spent leaping relative to the amount of time spent walking quadrupedally **(A)**. These differences in the functional role of the hindlimbs are reflected in joint surface shape and area **(B)** as well as limb length and back muscle weight **(C)**. Modified from Fleagle, J. G. (1999). *Primate adaptation and evolution.* New York: Academic. Used with permission.

the amount of leaping. For example, Byron and Covert (2004) have recently reported that, despite having upper and lower limbs of nearly equal length, the rare red-shanked douc langur (Fig. 3.5a) can swing by its arms as elegantly as dedicated arm-swingers like the spider monkey of South America (Fig. 3.5a) and the gibbon of Southeast Asia (Fig. 3.5b). Even the more prosaic quadrupedal monkeys of the South American and African rainforest show distinct differences in which parts of the canopy they occupy (Fleagle & Mittermeir, 1980; Gebo & Chapman, 1995; McGraw, 1996, 1998a,b,c) and anatomical differences associated with those choices. Oxnard and colleagues (1990) and Dagosto (1995) have shown great variation in the nature of quadrupedalism in Madagascar lemurs. To try to make sense out of all this, Hunt and colleagues (1996) identified a wide variety of locomotor modes among quadrupedal monkeys. Though they have standardized the descriptions of these categories, their lengthy list of types of movement can be difficult to penetrate. That fact alone reveals how complex primate locomotion is.

There are of course some primates that are dedicated to specialized locomotor modes. I will simplify this down to four groups: (1) vertical clinging and leaping (VCL), (2) arm-swinging/brachiation, (3) knuckle and fist walking, and (4) bipedalism. The first two categories represent opposites on the spectrum of forelimb-dominated versus hindlimb-dominated locomotion (Fig. 3.6). The second two categories represent solutions to the problem of using forelimbs highly modified for the demands of arboreal movement during terrestrial locomotion (Fig. 3.7).

Vertical clinging and leaping behavior is found among at least five living strepsirrhine genera and in three living haplorhine genera (the tarsier, marmosets, and tamarins). In this remarkable locomotor mode, animals cling to a vertical support and launch themselves with powerful extension of the lower limb (Fig. 3.6a). In the case of small leapers like galagos, the leap is driven primarily by extension at the ankle, whereas in larger leapers the extension is primarily at the hip Demes and Gunther, 1995; (Jouffroy & Lessertisseur, 1979). Small

Figure 3.5 The douc langur (**A**) arm-swinging through trees (from Byron, C. D. and Covert H. H. (2004). Unexpected locomotor behaviour: brachiation by an Old World monkey (Pygathrix nemaeus) from Vietnam. Journal of Zoology, 263, pp. 101–106) in a manner similar to that of gibbons or spider monkeys (**B**) (from Turnquist, J. E., Schmitt, D., Rose, M.D., & Cant, J. G. [1999]. Pendular motion in the brachiation of captive *Lagothrix* and *Ateles*. *American Journal of Primatology, 48,* 263–281). Used with permission.

Figure 3.6 Primate locomotion exhibits extremes in a continuum of "hindlimb-dominated" to "forelimb-dominated" modes of locomotion. Sifaka (**A**) exhibit powerful leaps using long muscular legs. Gibbons (**B**) exhibit acrobatic brachiation with their relative long forelimbs. Drawings by Stephen Nash; courtesy of Conservation International.

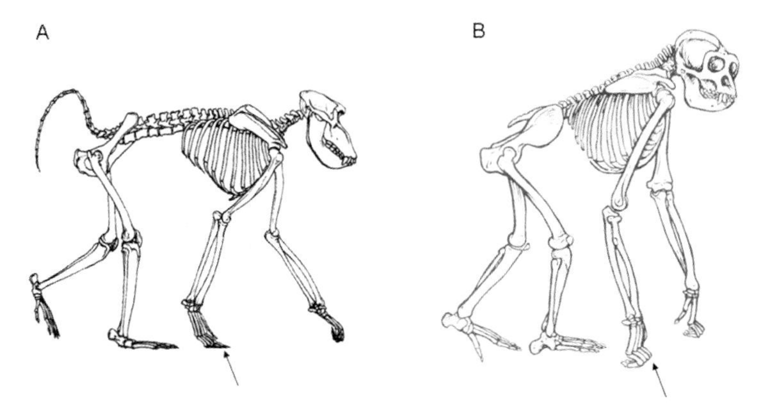

Figure 3.7 Differences in digit form and hand posture during terrestrial locomotion in the baboon (**A**) and the chimpanzee (**B**). The baboon has short, thick fingers and adopts a digitigrade posture. The chimpanzee has long, curved fingers and adopts knuckle-walking hand postures when walking quadrupedally (Images modified from Fleagle, J. G. (1999). *Primate adapation and evolution.* New York: Academic). Used with permission.

leapers, in contrast, have extended tarsal bones, whereas large leapers have exceptionally long femora (Demes and Gunther, 1995; Jouffroy & Lessertisseur, 1979). During the leap the animals rotate and land on the next vertical support, gripping it with their feet (Dunbar, 1988; Napier & Walker, 1967).

In contrast, arm-swinging locomotion is driven entirely by hand-over-hand motion of the forelimb. Though the distinction between arm-swinging and brachiation is imprecise, it is generally argued that brachiation involves whole body rotation (Cant, 1987a,b; Larson, 1988). Arm-swinging is seen among South American primates like the spider monkey (Fig. 3.5b) and wooly spider monkey, the douc langur, and orangutans. Brachiation is seen in gibbons and chimpanzees. Ricochetal brachiation, in which there is a whole body aerial phase, is found in gibbons (Fig. 3.6b) and sometimes chimpanzees. In all cases, the limbs are used in tension and are generally long and gracile. When such highly specialized animals walk on branches or the ground, they bear very little weight on their forelimbs compared to their hindlimbs (Kimura et al., 1979; Reynolds, 1985; Schmitt, 1994) or even none at all in the case of the bipedal gibbon.

Digit morphology is one of the aspects of primate anatomy most sensitive to locomotor demands. The feet and hands are in direct contact with the substrate and thus strongly reflect the complex interaction between locomotor behavior and the environment. The relationship between forces produced by the animal and substrate reaction forces is strongly influenced by the position and design of the hand. There is a strong need to balance bone length, which increases leverage and distance, against bone strength, which increases safety but also mass and cost. Animals that swing by their arms develop long and curved phalanges (Richmond et al., 2002) that are relatively gracile. Terrestrial quadrupeds like the baboon shorten and thicken their finger bones and bear weight directly at the base of the fingers and heads of metacarpal bones (Fig. 3.7a). That option is unavailable to arm-swinging primates. When they come to the ground or large supports, animals like the orangutan ball up their fists and bear weight on the outer edge of the metacarpals and phalanges. In contrast, chimpanzees choose to walk on the dorsal surface of the middle phalanx (See Richmond et al. 2002 for a review) (Fig. 3.7b).

Finally, we have our own characteristic form of locomotion: upright, striding bipedalism. Rather than bear weight on gracile limbs, our earliest ancestors removed their forelimbs completely from a weight-bearing role during locomotion and walked around on two legs. The exact nature of bipedalism in our early ancestors is a matter of serious debate (see Latimer et al., 1987; Schmitt, 2003a,b; Stern, 2000; Susman et al., 1984; and Ward, 2002 for a balanced set of arguments), but the fact that they did so and the adaptive pathway that primates followed to get there is really the key to understanding the underlying neural mechanisms of locomotion in primates.

DIFFERENCES BETWEEN PRIMATE AND NONPRIMATE LOCOMOTOR BEHAVIOR

The fact that we are unique among primates in using a striding, bipedal form of locomotion begs a number of critical questions, most of which are debated in detail elsewhere (see Schmitt, 2003c; Stern 2000; and Ward, 2002 for a review), not directly relevant to neuroethology, and well beyond the scope of this chapter. These questions include the following: Which apes are our immediate ancestors? In what environment did bipedalism first evolve? Why did humans become bipedal?

Embedded in these questions are additional questions that this chapter will address: What are the fundamental differences between the locomotion of primates compared to other animals? Did these differences, which probably accrued early in the evolution of the primate lineage, facilitate the evolution of this startling array of specialized locomotor behaviors, including our own, that we see today?

If we start with the model that the basic primate bauplan is that of an arboreal quadruped, we can ask what aspects of locomotion have changed over time to yield vertical clingers and leapers, brachiators, and bipeds. From this perspective, it becomes clear that primate locomotor evolution is characterized by dramatic changes in the functional role of the forelimbs. Rather than have a near-equal division of labor between forelimbs and hindlimbs as in almost all other legged vertebrates, there has been a change such that we might describe primates as "hindlimb dominated," relying heavily on the hindlimbs to power locomotion (Kimura et al., 1979; but see Demes et al., 1994 for a revision of the concept of "hindlimb-drive"). The forelimbs of primates, in contrast, may be described as "free" to provide stability and guidance ("steering") as well as grasping and manipulation. This changed functional relationship between the forelimb and hindlimb is highlighted in the many ways in which the walking gaits of primates differ from those of other mammals. In every case described later this division of labor between forelimbs and hindlimbs appears to relate to the biomechanical challenges of arboreal locomotion and reflects a forelimb used less in compressive weight support and more in complex movement of guidance and manipulation.

Figure 3.8 illustrates the ways in which most primates differ from most other mammalian quadrupeds (see Demes et al. 1994; Larson, 1998; Lemelin Schmitt, 1999, 2003a,b,c; Schmitt and Lemelin, 2002; Vilensky, 1989; and Vilensky and Larson, 1989 for additional review). The quadrupedal walking gaits of most primates can be distinguished from most other mammals by the presence of:

1. Diagonal sequence footfall patterns
2. High degrees of limb protraction
3. Relatively low peak forelimb forces
4. Relatively low forelimb spring stiffness
5. The lack of a running trot and the use of an ambling gait instead
6. Reduced oscillations of the center of mass.

All of these aspects of primate gait have, at one time or another, been argued to be part of a basal adaptation to locomotion on thin flexible branches (see Cartmill et al., 2002; Larson, 1998; Larson et al., 1999; 2001; Lemelin et al., 2003; Rollinson & Martin, 1981; Schmitt 1999, 2003a,b; Schmitt & Lemelin, 2002). These unusual gait choices appear to provide smooth movement, flexibility, and security for an animal with no claws walking on relatively thin arboreal substrates. These ecological/behavioral factors become critical targets for neural control because of their presumed adaptive value. By examining these one at a time, a picture develops of the functional role of each unusual aspect of primate locomotion.

Footfall Pattern

Diagonal sequence (DS) walks are symmetrical gaits in which the contact of each hindfoot is followed by that of the contralateral forefoot (Fig. 3.8a). During a DS walk the right hindfoot (RH) contact is followed by that that of the left forefoot (LF) so that a DS gait can be summarized as follows: RH, LF, LH, RF in series. Most primates consistently use DS walking gaits, although they can and do use lateral sequence (LS = RH, RF, LH, LF) gaits on occasion. Most other mammals exclusively use LS

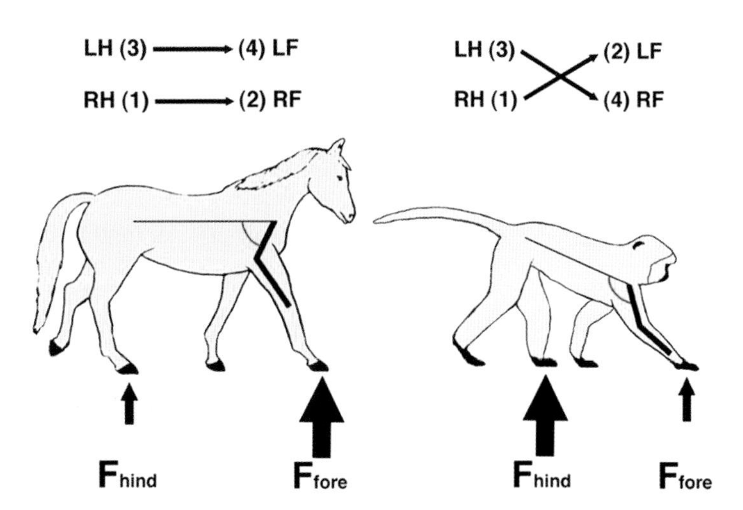

Figure 3.8 Summary of the commonly accepted differences that are believed to distinguish the walking gaits of most primates from those of most nonprimate mammals. Nonprimates generally use (**A**) lateral sequence walking gaits, (**B**) have a humerus that at ground contact is retracted relative to a horizontal axis passing through the shoulder, and (**C**) have greater peak vertical forces on their forelimbs than they do on their hindlimbs. Primates show the opposite pattern. From Schmitt, D., & Lemelin, P. (2002). The origins of primate locomotion: Gait mechanics of the woolly opossum. *American Journal of Physical Anthropology, 118,* 231–238. Used with permission.

gaits (Cartmill et al., 2002, 2007a,b; De La Croix, 1936; Dunbar & Badam, 2000; Hildebrand, 1967; Lemelin et al., 2003, 2008; Muybridge, 1887; Schmitt & Lemelin, 2002; Vilensky & Larson, 1989; White, 1990). This pattern has been recognized since Muybridge (1887) first filmed locomotion of the baboon and described what Magna De La Croix (1936) would later call the "pithecoid gait." Since then many arguments have been advanced to explain the preference for DS gaits among primates (Gray, 1959; Meldrum, 1991; Rollinson & Martin, 1981; Vilensky & Larson, 1989). The recent finding by Schmitt and Lemelin (2002) and Lemelin and colleagues (2003) that the woolly opossum (*Caluromys philander*), a dedicated fine-branch arborealist, uses DS gaits almost exclusively supports a link between fine-branch arboreality and footfall sequence. The closely related terrestrial marsupial, the short-tailed opossum (*Monodelphis domestica*), exclusively uses LS gaits (Lemelin et al., 2003). Based on these data on primates and marsupials, Cartmill and colleagues (2002, 2004, 2007a,b) articulated a model that suggested that DS walks were adopted by primates to ensure that a grasping hindfoot is placed on a tested support when the contralateral forefoot touches down on an untested support. Although there has been a small amount of debate on this subject (Shapiro & Raichlen, 2005, 2007), the recent findings of DS gaits in a highly arboreal carnivore (the kinkajou) and the increased prevalence of DS gaits on arboreal supports in capuchin monkeys (Wallace & Demes, 2008) further supports the association of this footfall pattern with locomotion and foraging on thin branches. Thus, the current explanation places a "testing" role for the forelimb and a "safety" role for the hindlimb.

Vilensky and Larson (1989) have argued previously that the presence of DS gaits reflects a fundamental change in the neural control of locomotion. They argued that other animals are restricted by central pattern generators to adopt only LS gaits. Primates, in contrast, are given credit for greater supraspinal control, as evidenced by the inability of decerebrate primates to initiate stepping patterns (Vilensky & Larson, 1989; Vilensky & Oconnor, 1997, 1998). It is argued that this greater control allows primates to select both DS and LS gaits. In this model primates choose to use DS gaits possibly for one of the advantages described previously. Other animals do not have that choice. This argument is intuitively appealing and corresponds with important well-documented differences in primate locomotor control that have implications for treatment of spinal cord injuries. But the presence of DS gaits in arboreal opossums (Lemelin et al., 2003; Pridmore, 1994), kinkajous (Lemelin et al., 2008; McClearn, 1992), and even, about half the time, the Virginia opossum (White, 1990) calls this model into question.

(2) Limb position at touchdown:

A second character believed to be typical of primate gaits—an arm (humerus) that is protracted much further than 90 degrees relative to a horizontal axis at touchdown of the forelimb (Fig. 3.8b)—has also been related to arboreal quadrupedalism (Larson, 1998; Larson et al., 1999, 2001; Lemelin and Schmitt, 2007; Schmitt, 1998, 1999, 2003a,d). It has been argued that early primates, having first evolved flattened nails, required long limbs with large excursions to reach above their head or around a trunk during climbing and to use long smooth, strides that would not oscillate thin branches (Demes et al., 1990; Larson, 1998; Larson et al., 1999, 2001; Schmitt, 1995, 1998, 1999). Lemelin and Schmitt (2007) recently compared arm protraction between *Caluromys* and *Monodelphis*. They found that *Monodelphis* protracts its arm slightly beyond 90 degrees relative to its body axis at forelimb touchdown during quadrupedal walking. Lemelin and Schmitt (2007) observed significantly greater arm protraction at forelimb touchdown in *Caluromys*, exceeding that of *Monodelphis* by nearly 20 degrees. From these data, it appears that opossums, like other marsupials in general, are characterized by higher degrees of arm protraction at forelimb touchdown than most nonprimate mammals (Larson et al., 1999; Lemelin & Schmitt, 2007). The fact that woolly opossums have much greater arm protraction at forelimb touchdown underscores its close link with arboreality, particularly locomotion on thin branches. Once again, this difference reflects a changing role of the forelimb as a primate reaches around to grasp. This has been argued to also reflect increased supraspinal control of locomotion in primates in general (Vilensky & Larson, 1989), but now we would have to argue the same for the woolly opossum.

Weight Distribution

A third feature (Fig. 3.8c) thought to distinguish primates from other mammals is the presence of higher vertical peak (Vpk) substrate reaction forces on the hindlimbs than on the forelimbs (Demes et al., 1994; Kimura, 1985, 1992; Kimura et al., 1979; Reynolds, 1985; Schmitt & Lemelin, 2002). There is variation in this feature, and not all primates show statistically significant differences between forelimb and hindlimb peak forces (Ishida et al., 1990; Polk, 2000, 2001; Schmitt & Hanna, 2004; Schmitt & Lemelin, 2002). However, most primates experience lower forelimb peak forces. Moreover, this disparity between forelimb and hindlimb peak forces is exaggerated on arboreal supports (Schmitt & Hanna, 2004). This pattern is profound and separates most primates from most other animals (Fig. 3.9).

The exceptions from this pattern of force distribution are also informative. Schmitt and Lemelin (2002) have shown that the woolly opossum has relatively low peak vertical forces on the forelimbs compared to the hindlimbs (Fig. 3.9) and concluded this was adaptively advantageous for locomotion and foraging on thin, flexible branches. Conversely, Schmitt (2003b) reported recently that the common marmoset (*Callithrix jacchus*), a clawed primate that spends much of its time clinging on large tree trunks, predominantly uses LS gaits, more retracted arm positions, and higher peak vertical forces on the forelimbs relative to the hindlimbs (Fig. 3.9).

These kinetic data support the widely held assumption that the difference in weight distribution between primates and nonprimate mammals represents a basal adaptation to arboreal locomotion (Larney & Larson, 2004; Larson, 1998; Schmitt, 1998, 1999, 2003a,b,c; Schmitt & Lemelin, 2002). Furthermore, variation in the ratio of forelimb to hindlimb Vpk within primates supports this intuitive assumption. In general, primates that spend most of their locomotor time in trees have the lowest FL/HL Vpk ratios, whereas the most terrestrial primates have a nearly equal distribution of forelimb and hindlimb vertical peak forces (Demes et al., 1994; Kimura, 1985, 1992; Kimura et al., 1979; Reynolds, 1985; Schmitt & Lemelin, 2002). Kimura (1985, 1992) recognized this pattern and argued strongly that the degree to which forelimb forces are reduced is directly related to substrate preference. It is easy to see the benefits that this pattern would provide. As early primates moved and foraged in the trees, they would need to reach out and grab food placed off of the path of locomotion and they would have to reach out in order to effect often abrupt changes of direction along branches. In those cases a highly mobile forelimb that was not responsible for bearing 60% of the animal's body weight would provide a distinct advantage.

Limb Stiffness

It is tempting to argue that this shift in weight-bearing role reflects an anatomical redistribution of weight, but there is no evidence to support such a claim. Studies of center of mass position show no differences across taxa (Vilensky & Larson, 1989). Moreover, short-tailed macaques show the same pattern and apes, which have no tails and huge forequarters, are among the most extreme in showing relatively high hindlimb forces. The difference between forelimb and hindlimb peak vertical forces must be an active shift of load posteriorly and must reflect different patterns of limb use.

So the question remains as to what the actual mechanism is. Reynolds (1985) argued that activity in the hindlimb retractor muscles while the hindlimb was highly protracted (reaching forward) actively shifted weight posteriorly. Schmitt (1998, 1999) and Schmitt and Hanna (2004)

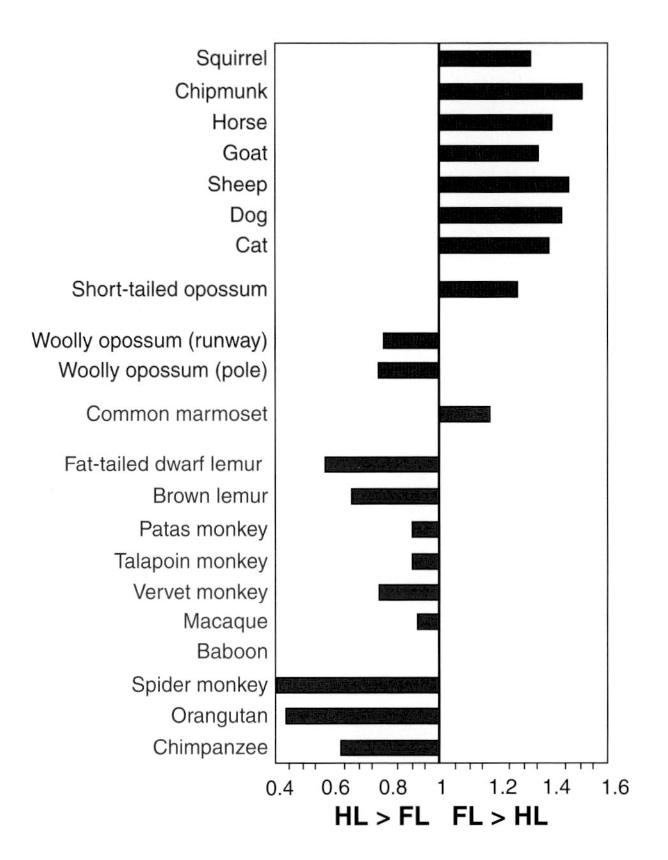

Figure 3.9 Bar graph of the ration of peak vertical forces on the forelimb divided by peak vertical forces on the hindlimb in representative primates and nonprimate mammals. A ratio of 1.0 represents equal weight distribution. Most primates (red) fall to the side that indicates increased peak vertical forces on the hindlimb, whereas most nonprimate mammals (yellow) fall to the opposite side. The exceptions to this pattern, the highly arboreal woolly opossum and the claw-bearing common marmoset, are seen as indicating the importance of arboreality in driving this functional differentiation. The figure is modified from Demes, B., Larson, S. G., Stern, J.T., Jr., Jungers, W. L., Biknevicius, A. R., et al. (1994). The kinetics of primate quadrupedalism: Hindlimb drive reconsidered. *Journal of Human Evolution, 26*, 353–374, to include data on marsupials (Schmitt, D., & Lemelin, P. [2002]. The origins of primate locomotion: gait mechanics of the woolly opossum. *American Journal of Physical Anthropology, 118*, 231–238; Lemelin, P., & Schmitt, D. (2007). The orgins of grasping and locomotor adaptations in primates: Comparative and experimental approaches using an opossum model. In: M. Dagosto & M. Ravosa (Eds.), *Primate origins* (pp. 329–380). New York: Kluwer.) and the common marmoset (Schmitt, D. [2003b]. The relationship between forelimb anatomy and mediolateral forces in primate quadrupeds: Implications for interpretation of locomotor behavior in extinct primates. *Journal of Human Evolution 44, 49–60*).

agreed that "Reynolds' mechanism" could also play a substantial role in shifting weight posteriorly but in addition, changes in limb stiffness could also play a role in reducing forelimb peak loads. This notion that arboreal mammals such as primates adjust their limb stiffness in order to moderate loads on arboreal supports has become an increasingly frequent area of research and debate in anthropology and has important and broad implications for neural control of locomotion (Franz et al., 2005; Larney & Larson, 2004; Li et al., 1996; Schmitt, 1995, 1998, 1999, 2003a,b; Schmitt & Hanna, 2004; Schmitt & Lemelin, 2002; Schmitt et al., 2007, 2008; Wallace & Demes, 2008). This is not a new concept for biology in general. Twenty years before physical anthropologists incorporated this idea, limb stiffness was recognized as a critical variable for understanding the relationship between limb design and locomotor behavior in a wide variety of animals (Ahn et al., 2004; Alexander, 1992; Cavagna et al., 1977; Farley et al., 1993; Full & Tu, 1990; Griffin et al., 2004; McMahon, 1985; McMahon et al., 1987).

Under experimental conditions, Schmitt (1998, 1999) showed that Old World monkeys reduce forelimb stiffness by increasing forelimb "compliance"—a term borrowed from general animal biomechanics (Alexander, 1992; McMahon, 1985; McMahon et al., 1987)—when walking on relatively thin poles compared to a runway. In these studies, forelimb compliance was measured in terms of joint yield. By increasing joint yield, it was argued that the forelimb operates relatively more like a spring, which increases contact time and reduces peak vertical forces on the limbs (Blickhan, 1989; Li et al., 1996; McMahon et al., 1987; Schmitt, 1995, 1998, 1999). Schmitt and Hanna (2004) found a similar but less pronounced pattern for the hindlimb of primates.

Recently, Larney and Larson (2004) provided an extensive dataset on forelimb and hindlimb joint yield for a wide variety of mammals. They confirmed the high values of forelimb compliance in most primates, but they also reported that marsupials have forelimb compliance as great, if not greater, than that of most primates. Schmitt and colleagues (n.d.) recently confirmed that woolly opossums, like primates, have high values of elbow yield, while the short-tailed opossum does not.

Lack of a Running Trot and the Use of an Amble

As most quadrupedal mammals increase speed, they shift from a walking gait with no aerial phase (Fig. 3.10a) to a running gait with a whole body aerial phase (Fig. 3.10b) (Cartmill et al., 2002; Gambaryan, 1974; Hildebrand, 1985; Howell, 1944; Muybridge, 1957). At their fastest

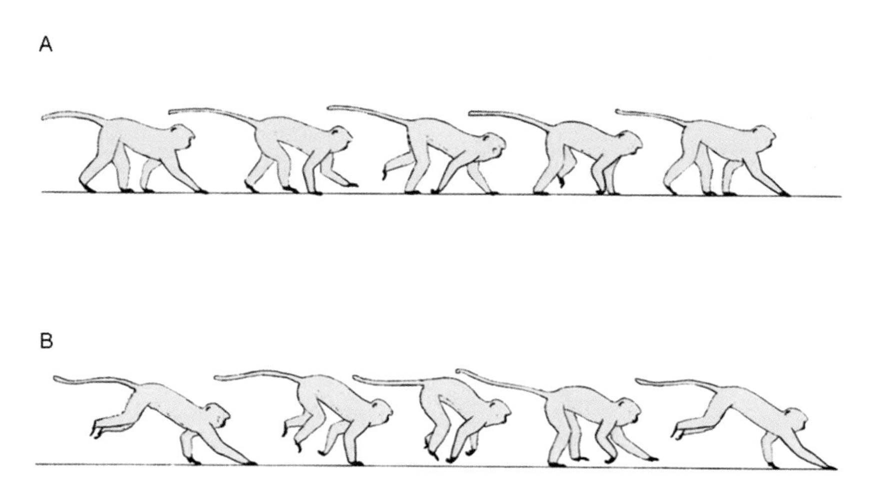

Figure 3.10 A vervet monkey using two common gaits. At slow speeds most quadrupedal primates adopt a walking gait (**A**) in which a forelimb contact is followed by a hindlimb contact and there is always at least two limbs on the substrate at one time with no whole body aerial phase. At faster speeds most quadrupedal primates adopt a gallop in which two hindlimb contacts are followed by two forelimb contacts and there is a whole body aerial phase. Many primates use a canter, which is a slow gallop with no aerial phase. Modified from Schmitt, D., Larson, S. G., & Stern, J. T., Jr. (1994) Serratus ventralis function in vervet monkeys: Are primate quadrupeds unique? *Journal of Zoology, 232,* 215–230. Used with permission.

speeds, quadrupedal mammals generally use a running gait such as a gallop (Gambaryan, 1974; Hildebrand, 1985; Howell, 1944; Muybridge, 1957). But at speeds between that of a walk and a gallop, quadrupedal mammals often use symmetrical running gaits that have an aerial phase and in which the feet strike down in diagonal pairs (trot) or unilateral pairs (pace) (Cartmill et al., 2002; Gambaryan, 1974; Hildebrand, 1985; Howell, 1944; Muybridge, 1887). These gaits are faster than walking gaits but still provide relatively longer periods of support by both a forelimb and a hindlimb than does galloping (Cartmill et al., 2002).

Unlike other mammals, primates almost never adopt a running trot or pace (Demes et al., 1994, 1994; Hildebrand, 1967; Preuschoft and Gunther, 1994; Rollinson & Martin, 1981; Schmitt, 1995; Schmitt et al., 2006; Vilensky, 1989). Instead, Schmitt and colleagues (2006) showed that primates adopt a highly unusual running gait called an "amble" (Fig. 3.11). These gaits are referred to as "grounded running gaits" (Rubenson et al., 2004) because they do not involve a whole body aerial phase. Ambles are exhibited by almost all primates (Schmitt et al., 2006) as well as certain breeds of horses (Barrey, 2003; Biknevicus et al., 2004; Muybridge, 1887) and elephants (Gambaryan, 1974; Hutchinson et al., 2003). Schmitt and colleagues (2006) argued that ambling ensures continuous contact of the body with the substrate while dramatically reducing vertical oscillations of the center of mass. This may explain why ambling appears to be preferable to trotting for extremely large terrestrial mammals such as elephants and for arboreal mammals like primates that move on unstable branches.

Reduced Oscillations of the Center of Mass

Both the changes in spring stiffness (#4) and the use of ambles rather than trots (#5) appear to have implications for the vertical oscillations of primates as they walk. The movements of the animal's body, reflected in the movements of the animal's center of mass (COM), have implications for loading, stability, and energy exchange. The latter represents an important target of selection that helps explain the postural and gait choices made by primates.

Schmitt (2003) argued that forelimb compliance moderates vertical oscillations of the body and peak vertical forces on the limbs. He and his colleagues (Schmitt et al., 2006) argued that moderating vertical oscillations of the center of mass are a critical control target during primate

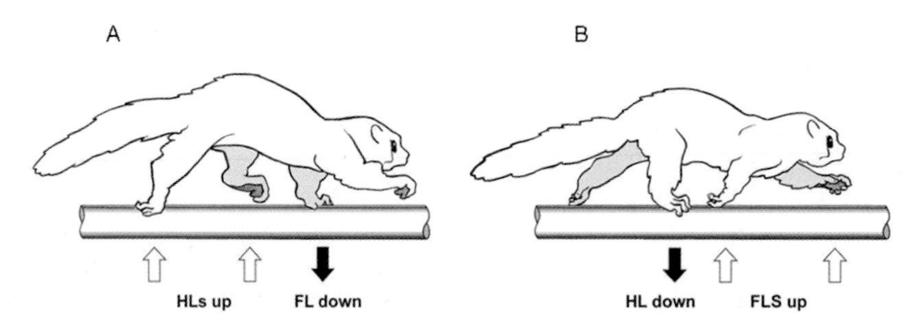

Figure 3.11 A primate adopting an amble, a gait in which a forelimb contact is followed by a hindlimb contact and there is always at least two limbs on the substrate. Unlike a walk, however, there are alternating aerial phases for the hindlimb **(A)** and forelimb **(B)** pairs. The amble is a rare gait adopted only by some breeds of horses, elephants, kinkajous, and primates. This gait reduces vertical oscillations of the center of mass and increases stability compared to a running trot. Schmitt, D., Cartmill, M., Hanna, J., Griffin, T., and Lemelin, P. (2006). The adaptive value of ambling in primates. *Journal of Experimental Biology 209,* 2042–2049. Used with permission.

locomotion at both normal and fast walking gaits and at slow running gaits. However, the movements of the COM discussed in these studies were assumed and had not been quantified. They were simply inferred from the limb stiffness itself, which in turn was inferred from joint yield.

The movements of the COM have important implications for understanding the targets of locomotor control. Any animal must balance needs of stability, speed, and energetic efficiency. The control of COM influences all three factors to varying degrees. The ways in which the COM oscillates in nonprimate bipeds and quadrupeds has been a critical variable both in defining gaits as speed increases and in understanding the adaptive tradeoffs between stability and efficiency of any gait (see Ahn et al., 2004; Biewener, 2003; Cavagna et al., 1977; Griffin et al., 2004). It appears that most animals follow the same the basic underlying governing principles for walking and running. Nonhuman primates, however, may represent an exception to this broadly conserved pattern (Schmitt, 2003).

Direct measures of COM movements in primates are very limited. Wells and Wood (1975) described the movements of the COM during leaping in vervet monkeys. Using videorecords, Vilensky (1979) provided data on the COM in macaques at a wide range of speeds on a treadmill. His data suggested that changes in limb kinematics minimized the movements of the COM at different speeds. Few studies, with the exception of Cavagna and colleagues (1977) and Kimura (1990, 1991, 1996), have used force platform data to infer the behavior of the center of mass for either whole animal or individual limb girdles.

In contrast to primate studies, the analysis of COM movements is a common method of analysis in biomechanical studies of other animals including cockroaches, crabs, frogs, lizards, ostriches, penguins, sheep, horses, dogs, and humans (Ahn et al., 2004; Alexander, 2003; Biewener, 2003; Bishop et al., n.d.; Blickhan & Full, 1992; Cavagna et al., 1976, 1977; Farley & Ko, 1997; Farley et al., 1993; Full, 1991; Full & Tu, 1990; Griffin, 2000, 2002; Manter, 1938). The data derived from these studies allow researchers to explore underlying mechanics of various mammalian gaits and allow for comparison across a wide variety of taxa. They have revealed that the walking and running gaits of most animals, regardless of phylogeny and morphological design, operate with the same basic mechanical principles and that gaits may be defined by those principles.

To understand the implications of this finding, it is worth reviewing gait definitions. Classifications of gait and analyses of gait choice have traditionally been based on visual rather than biomechanical criteria. This is useful and appropriate but also omits important information about the mechanical costs and benefits of specific gaits and the underlying principles that guide gait selection in animals.

Under visual schemes of gait classification, quadrupedal walking, trotting, and running gaits have been defined in the following way. A walk (Fig. 3.10a) is a symmetrical gait in which hindlimb footfalls alternate with forelimb footfalls and there are at least two feet on the ground at all times. Thus, in a walk the duty factor (contact time of any limb divided by total stride time) is at least 0.5. Gaits with a duty factor of less than 0.5 have an aerial phase.

Some gaits, like an amble (Fig. 3.11), are symmetrical gaits in which there is no whole body aerial phase but there is an aerial phase in either the forelimbs or the hindlimbs (Barrey, 2001; Gambarayan, 1974; Howell, 1944; Hutchinson et al., 2003; Schmitt et al., 2006). Thus, the body is always supported by at least one limb.

A running trot is a symmetrical gait in which diagonal limb pairs (right hindlimb and left forelimb) swing and contact the ground simultaneously and there is a whole body aerial phase. Finally, galloping is an asymmetrical gait with a whole body aerial phase

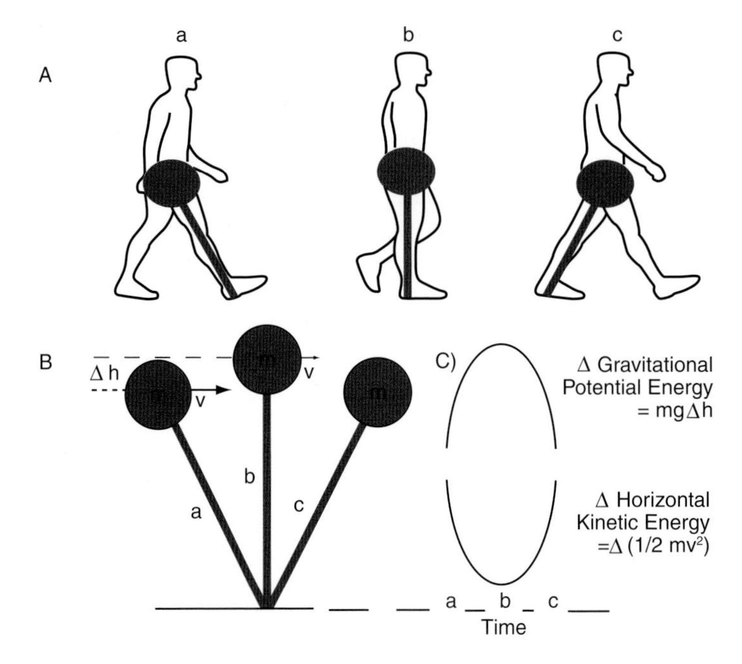

Figure 3.12 The inverted pendulum model of walking. The limb is modeled as a massless strut and with the center of mass (red circle) concentrated at one end. In both **(A)** and **(B)** the center of mass (COM) is at its low point at the beginning of the step (a). At the midpoint of the step (b) the center of mass is at its highest point with a large amount of potential energy. The change in height (Δh) is indicated in (B). In the second half of the step (c), as the body moves forward the center of mass drops and potential energy (PE) is converted into kinetic energy (KE). If fluctuations in PE and KE are out of phase and of the same amplitude (C), then there is a highly effective conversion of PE into KE. This energy exchange (recovery) reduces the work that needs to be done by muscles to accelerate and decelerate the center of mass. A high value of exchange of PE and KE (between 50% and 75%) is found in the walking gaits of a diverse array of quadrupeds and bipeds and is seen as a potentially important target of selection.

(Fig. 3.10b). During galloping, forelimb and hindlimb contacts do not alternate but rather contacts of the hindlimbs are followed by contacts of the forelimb.

These definitions work very well for visual distinction, but in reality gaits defined in this fashion are part of a continuum (Cartmill et al., 2002). Using the definitions given previously, gaits with a duty factor (contact time of a limb divided by total stride time) of 0.51 compared to 0.53 are both walks. However, if the duty factors are 0.51 and 0.49, one animal is running while the other is walking. It has been argued that gaits should not be defined in this way, but rather defined by discrete differences. It is possible, however, to define walking (duty factor >0.5) and running gaits (duty factor <0.5) in mechanical terms. Walking and running gaits show discrete differences in the behavior of the COM and in the exchange of potential and kinetic energy. This is most easily explained using a biped model but defines the gaits of quadrupeds as well (Fig. 3.12).

When modern humans walk, we vault over relatively stiff lower limbs in such a way that our center of mass is at its lowest point at heel-strike and rises to its highest point at midstance (Fig. 3.12a,b) (Cavagna et al., 1976; Lee & Farley, 1998). This type of inverted pendulum gait is common to almost all known quadrupeds and bipeds (Ahn et al., 2004; Alexander, 1977, 2003; Biewener, 2003; Blickhan & Full, 1992; Cavagna et al., 1976, 1977; Farley & Ko, 1997; Farley et al., 1993; Full, 1991; Full & Tu, 1990; Gatesy & Biewener, 1991; Griffin & Kram, 2000; Heglund et al., 1982). In contrast, when humans and other animals run, their limbs operate more like they were part of a spring-mass system in which the limb is compressed during support phase, and the center of mass is at its lowest point during the middle of stance.

Center of mass movements can be directly calculated from force curves of all three components of the ground reaction forces exerted by an animal. Forces reflect accelerations of the body's center of mass. Those accelerations can be easily converted to velocities and displacements from which both kinetic energy and potential energy can be calculated. The mathematics for calculating these movements are detailed in textbooks (Biewener, 2003) and recent publications (Ahn et al., 2004).

The relationship between the potential energy (PE) and kinetic energy (KE) of the COM provides information about the relative efficiency of different gaits. If the fluctuations of the PE and KE are mirror images of each other (have the same amplitude, shape, and opposite direction) (Fig. 3.12C), then PE stored at midstance can be converted to KE during the second half of stance. The recovered PE converted into KE can be used to drive the COM forward. If the leg is held stiff, as most animals do, then the COM will also be driven upward and regain PE than can be converted into KE. This mechanism of energy recovery reduces the amount of muscular work required to accelerate and decelerate the COM and thus may reduce the costs of locomotion. This amount of exchange is referred to as the percentage of recovery. This is a minimum estimate of the energy used by the system. In walking, PE and KE are equal in amplitude and shape but are largely out of phase. As a result, as much as 70% of the energy needed to move forward can be conserved through an exchange of PE and KE (Cavagna et al., 1977). During running, in contrast, KE and PE are in phase and cannot exchange energy. During running much of this energy is converted to elastic energy in spring elements of the lower limb.

This exchange mechanism works when fluctuations in PE and KE are equal in magnitude. Excessively large amounts of PE due to high vertical oscillations would be wasted. Similarly, excessively low amounts of PE due to a smooth, nonoscillating gait would be insufficient for generation of enough KE. These constraints may explain the relatively stiff-legged oscillating gaits that are found in many legged mammals and even explain the waddle of penguins (Griffin et al., 2000).

Primates, however, are different. In the case of ambling and compliant walking, previous studies have suggested that primate gaits are characterized by low oscillations of the COM. Thus, it has been assumed that primates have given up this energy-saving mechanism in exchange for security while running on branches. This notion received support from recent findings (Bishop et al., 2008) that cats show low values of recovery when executing stealthy, crouched gaits. It is expected that primates would show the same pattern. However, recent preliminary analysis of lemurs has shown that the story is more complex (Schmitt et al., 2008). Although it is true that lemurs show lower values of recovery than dogs (highest average value = 75%; mean recovery value = 60; Griffin et al. (2004). and other quadrupeds (Cavagna et al., 1977), the values are not always as low as would have been expected. The ring-tailed lemur, a highly terrestrial strepsirrhine, shows values of recovery near that of dogs with an average recovery of 49% and maximum values of 70%. Schmitt and colleagues (2008) argued that by combining long limbs and deep compliance, these lemurs achieved relatively acceptable levels of energy exchange. Had they not been compliant, then their long limbs would have caused a large amount of change in height of the COM (Δh in Fig. 3.12b) and PE would have been wasted. In contrast, the brown lemur, a much more arboreal primate compared to the ring-tailed lemur, showed excessively low values of energy recovery with an average value of 36% and no value greater than 50% recovery. These latter data suggest that much of the serious differences between primate locomotion and that of other animals are specifically related to requirements of locomotion in an arboreal environment.

PATHWAY OF PRIMATE LOCOMOTOR EVOLUTION

It is overly simplistic to try to draw a straight line from the origin of primates to the origin of bipedalism by connecting the dots with each new functional change in the forelimb. The reality is that there have been many detours and dead ends along the way. Nonetheless, when these data on locomotion are taken together, an intriguing picture emerges that may explain why our mode of locomotion evolved in our lineage only. The model I am sketching here is one in which adaptations to life in an arboreal environment, specifically on thin, flexible branches, drove fundamental changes in the functional role of the forelimb and may have involved significant changes in neural control. These changes allowed the exploration of new, previously unavailable locomotor niches. Although we now have more details and more data, this scenario is not new. It was first articulated by Jones (1916) and tested and elaborated on by later researchers (Clark, 1959; Jenkins, 1972; Larson, 1998; Lemelin, 1999; Lemelin and Schmitt, 2007; Schmitt, 1998, 1999, 2003a,c,d; Schmitt & Lemelin, 2002; Stern, 1976). In this scenario the earliest primates in the Paleocene 60 to 65 million years ago would have moved and foraged frequently on thin, flexible, terminal branches. Their locomotion would have been characterized by changes in the relationship between the forelimb and the hindlimb in order to promote smooth and secure locomotion. These early primates, unlike their nonprimate ancestors, would have relied on diagonal sequence footfall patterns in order to ensure the placement of the forelimb on a tested support. These animals adopted highly protracted limbs and deep elbow flexion to effect long strides and smooth out their gait. This was also associated with a reduction in forelimb compressive weight bearing that facilitated mobility of the forelimb. This posterior weight shift freed the limbs to be used in tension (Stern, 1976; Stern & Oxnard, 1973) and facilitated the evolution of suspensory locomotion.

As presented, it would follow that our prebipedal ancestors simply descended to the ground and, like gibbons, walked bipedally because their limbs were too gracile to support weight. This is a logical but untestable conclusion. But one question we can address is whether the bipedal locomotion of our ancestors was compliant like all nonhuman primates or relatively stiff like our own. Since nonhuman primates typically utilize compliant gaits when they walk either quadrupedally or bipedally, it seems plausible, then, that early bipedal hominins would have retained a compliant walking style typical of other nonhuman primates. Postcranial anatomy of early hominins suggests that some of them walked with a deeply yielding knee and hip (Stern & Susman, 1983). But beyond being simply a primitive retention, compliant walking in prehominins may have had several advantages. Among quadrupedal nonhuman primates, low peak forces and reduced stride frequencies make their locomotion relatively smooth, which helps them avoid shaking flexible branches, thus enhancing their stability and helping them escape the notice of predators (Demes et al., 1990; Schmitt, 1998, 1999). These features may have also allowed early human ancestors to maintain mobile, loosely stabilized forelimb joints. Kinematic, force plate, and accelerometer studies on human compliant bipedalism show that humans who adopted a complaint gait achieved longer stride lengths, faster maximum walking speeds, lower peak vertical forces, and improved impact shock attenuation between shank and sacrum compared to normal walking (Schmitt et al., 1996, 1999, 2003). These data are consistent with findings of several other studies (Li et al., 1996; Yaguramaki, 1995) and support claims by Stern and Susman (1983) that compliant bipedalism may have been an effective gait for a small biped with relatively small and weakly stabilized joints, which had not yet completely forsaken arboreal locomotion.

In 1998, Susan Larson suggested, based on electromyographic data (Larson, 1989; Larson & Stern, 1987, 1989, 1992; Schmitt et al., 1994), that primates were fundamentally different from all other mammals. She argued that while motor patterns were conserved across broad orders of vertebrates, primates had deviated from this pattern and had adopted different patterns of muscle recruitment (Larson, 1998). This was a profound conclusion that suggested that basal primates had not just slightly modified their locomotor behavior, but had instead adopted a derived locomotor pattern in association with arboreal movement. This conclusion was supported by data on footfall patterns (Cartmill et al., 2002), forelimb kinematics (Larson et al., 1999, 2001), and ground reaction forces (Demes et al., 1994; Kimura et al., 1979; Reynolds, 1985; Schmitt 2003a; Schmitt & Lemelin, 2002). Now it appears that even the fundamental underlying mechanisms that characterize walking gaits of most animals are not conserved by primates. This lack of conservation of basic locomotor behavior and control is not a trivial mathematical distinction, nor is it simply a difference of degree. Instead, the way in which primates different in their locomotor behavior from most other mammals represents an abrupt behavioral shift that suggests major ecological, anatomical, and neurological changes in our earliest ancestors that set the stage for own evolution and present condition.

REFERENCES

Aerts, P., Van Damme, R., Van Elsacker, L., & Duchene V. (2000). Spatiotemporal gait characteristics of the hind-limb cycles during voluntary bipedal and quadrupedal walking in bonobos (*Pan paniscus*). *American Journal of Physical Anthropology, 111*, 503–517.

Ahn, A. N., Furrow, E., & Biewener, A. A. (2004). Walking and running in the red-legged running frog, *Kasina maculata*. *Journal of Experimental Biology, 207*, 399–410.

Alexander, R. McN. (1977). Mechanics and scaling of terrestrial locomotion. In: T. Pedley (Ed.), *Scale effects in animal locomotion* (pp. 93–110). London: Academic Press.

Alexander, R. McN. (1992). A model of bipedal locomotion on compliant legs. *Philosophical Translations of the Royal Society of London Series B, 338*, 189–198.

Alexander, R. M., & Maloiy, G. M. (1984). Stride lengths and stride frequencies of primates. *Journal of Zoology of London, 202*, 577–582.

Anapol, F. C., & Jungers, W. L. (1987). Telemetered electromyography of the fast and slow extensors of the leg of the brown lemur (*Lemur fulvus*). *Journal of Experimental Biology, 130*, 341–358.

Barrey, E. (2001). Inter-limb coordination. In: W. Back & H. M. Clayton (Eds.), *Equine locomotion* (pp. 77–94). London: W.B. Saunders.

Bertram, J. E. A., & Chang, Y. (2001) Mechanical energy oscillations of two brachiation gaits: Measurement and simulation. *American Journal of Physical Anthropology, 115*, 319–326.

Blickhan, R. (1989). The spring mass model for running and hopping. *Journal of Biomechanics, 22*, 1217–1227.

Byron & Covert. (2003).

Carlson, K. J., Demes, B. and Franz, T. M. (2005). Mediolateral forces associated with quadrupedal gaits of lemurids. *J. Zool. 266*, 261–273.

Cartmill, M. (1970). *The orbits of arboreal mammals: A reassessment of the arboreal theory of primate evolution*. Ph.D. Thesis, Chicago, University of Chicago.

Cartmill, M. (1972). Arboreal adaptations and the origin of the order primates. In: R. H. Tuttle (Ed.), *The functional and evolutionary biology of primates* (pp. 3–35). Chicago: Aldine-Atheton.

Cartmill, M. (1974a). Pads and claws in arboreal locomotion. In: F. A. Jenkins Jr. (Ed.), *Primate locomotion* (pp. 45–83). New York: Academic Press.

Cartmill, M. (1974b). Rethinking primate origins. *Science, 184*, 436–443.

Cartmill, M. (1992). New views on primate origins. *Evolution and Anthropology, 1*, 105–111.

Cartmill, M., Lemelin, P., & Schmitt, D. (2002). Support polygons and symmetrical gaits in mammals. *Zoological Journal of the Linnean Society, 136*, 401–420.

Cartmill, M., Lemelin, P., & Schmitt, D. (2007a). Primate gaits and primate origins. In: M. Dagosto & M. Ravosa (Eds.), *Primate origins* (pp. 403–436). New York: Kluwer.

Cartmill, M., Lemelin, P., & Schmitt, D. (2007b). Understanding the adaptive value of diagonal-sequence gaits in primates: A comment on Shapiro and Raichlen (2005). *American Journal of Physical Anthropology, 133*, 822–825.

Cavagna, G. A., Heglund, N. C., & Taylor, C. R. (1977). Mechanical work in terrestrial locomotion: Two basic mechanisms for minimizing energy expenditure. *American Journal of Physiology, 233*, R243–261.

Cavagna, G. A., Thys, H., & Zamboni, A. (1976). Sources of external work in level walking and running. *Journal of Physiology of London, 262*, 639–657.

Chang Y. H., Bertram J. E., & Ruina A. (1997). A dynamic force and moment analysis system for brachiation. *Journal of Experimental Biology, 200*, 3013–3020.

Chang, Y., Bertram, J. E. A., & Lee, D. V. (2000). External forces and torques generated by the brachiating white-handed gibbon (Hylobates lar). *American Journal of Physical Anthropology,* 113, 201–216.

Churchill S. E. and Schmitt D. (2003) Biomechanics in paleoanthropology: engineering and experimental approaches to the investigation of behavioral evolution in the genus Homo. In C Harcourt and R Crompton (eds.): New Perspectives in Primate Evolution and Behavior. London: Linnaean Society, pp. 59–90.

Crompton, R. H., Li, Y., Wang, W., Gunther, M. M., & Savage, R. (1998). The mechanical effectiveness of erect and bent-hip, bent-knee bipedal walking in Australopithecus afarensis. *Journal of Human Evolution,* 35, 55–74.

D'Août, K. D., Aerts, P., De Clercq, D., De Meester, K., & Van Elsacker, L. (2002). Segment and joint angles of hind limb during bipedal and quadrupedal walking of the bonobo (Pan paniscus). *American Journal of Physical Anthropology,* 119, 37–51

Demes, B., Jungers, W. L., & Nieschalk, U. (1990). Size- and speed-related aspects of quadrupedal walking in slender and slow lorises. In: F. K. Jouffroy, M. H. Stack, & C. Niemitz (Eds.), *Gravity, posture and locomotion in primates* (pp. 175–198). Florence: Il Sedicesimo.

Demes, B., Larson, S. G., Stern, J. T. Jr., Jungers, W. L., Biknevicius, A. R., & Schmitt, D. (1994). The kinetics of primate quadrupedalism: hindlimb drive reconsidered. *Journal of Human. Evolution, 26*, 353–374.

Demes, A. B., Qin, Y., Stern, J. T. Jr., Larson, S. G., & Rubin, C. T. (2001). Patterns of strain in the macaque tibia during functional activity. *American Journal of Physical Anthropology, 116*, 257–265.

Demes, A. B., Stern, J. T. Jr., Hausman, M. R., Larson, S. G., McLeod, K. J., & Rubin, C. T. (1998). Patterns of strain in the macaque ulna during functional activity. *American Journal of Physical Anthropology, 106*, 87–100.

Elftman, H. (1944). The bipedal walking of the chimpanzee. *Journal of Mammals, 25*, 67–71.

Elftman, H., & Manter, J. (1935). Chimpanzee and human feet in bipedal walking. *American Journal of Physical Anthropology, 20*, 69–79.

Farley, C. T., Glasheen, J., & McMahon, T. A. (1993). Running springs: Speed and animal size. *Journal of Experimental Biology, 185*, 71–86.

Fleagle, J. G. (1977a). Locomotor behavior and skeletal anatomy of sympatric Malaysian leaf-monkeys (*Presbytis obscura* and *Presbytis melalophos*). *Yearbook of Physical Anthropology, 20*, 440–453.

Fleagle, J. G. (1977b). Locomotor behavior and muscular anatomy of sympatric Malaysian leaf-monkeys (*Presbytis obscura* and *Presbytis melalophos*). *American Journal of Physical Anthropology,* 46, 297–308.

Fleagle, J. G. (1979). Primate positional behavior and anatomy: Naturalistic and experimental approaches. In: M. E. Morbeck, H. Preuschoft, & N. Gomberg (Eds.), *Environment, behavior, and morphology: Dynamic interactions in primates* (pp. 313–326). New York: Wenner-Gren Foundation.

Fleagle, J. G. (1999). *Primate adaptation and evolution.* New York: Academic.

Fleagle, J. G., Stern, J. T., Jr., Jungers, W. L., Susman, R. L., Vangor, A. K., & Wells, J. P. (1981). Climbing: A biomechanical link with brachiation and with bipedalism. *Symposium of the Zoological Society of London, 48,* 359–375.

Franz, T. M., Demes, B., & Carlson, K. J. (2005). Gait mechanics of lemurid primates on terrestrial and arboreal substrates. *Journal of Human Evolution, 48,* 199–217.

Full, R. J., & Tu, M. S. (1990). Mechanics of six-legged runners. *Journal of Experimental Biology, 148,* 129–146.

Gambaryan, P. P. (1974). *How mammals run.* New York: Wiley.

Gatesy, S. M., & Biewener, A. A. (1991). Bipedal locomotion: Effects of speed, size, and limb posture in birds and humans. *Journal of Zoology of London, 224,* 127–147.

Griffin, T. M., & Kram, R. (2000) Penguin waddling is not wasteful. *Nature, 408,* 929.

Griffin, T. M., Main, R. P., & Farley, C. T. (2004). Biomechanics of quadrupedal walking: How do four-legged animals achieve inverted pendulum-like movements? *Journal of Experimental Biology, 207,* 3545–3558.

Gunther, M. M. (1991). The jump as a fast mode of locomotion in arboreal and terrestrial biotopes. *Zeitschrift für Morphologie und Anthropologie, 78,* 341–372.

Heglund, N. C., Cavagna, G. A., & Taylor, C. R. (1982). Energetics and mechanics of terrestrial locomotion. III. Energy changes of the center of mass as a function of speed and body size in birds and mammals. *Journal of Experimental Biology, 97,* 41–56.

Hildebrand, M. (1967). Symmetrical gaits of primates. *American Journal of Physical Anthropology, 26,* 119–130.

Hildebrand, M. (1985). Walking and running. In: M. Hildebrand, D. M. Bramble, K. F. Liem, & D. B. Wake (Eds.), *Functional vertebrate morphology* (pp. 38–57). Cambridge: Harvard University Press.

Hirasaki, E., Kumakura, H., & Matano, S. (1993). Kinesiological characteristics of vertical climbing in *Ateles geoffroyi* and *Macaca fuscata. Folia Primatologica, 61,* 148–156.

Hirasaki, E., Kumakura, H., & Matano, S. (1995). Electromyography of 15 limb muscles in Japanese macaques (Macaca fuscata) during vertical climbing. *Folia Primatologica, 64,* 218–224.

Hirasaki, E., Kumakura, H., & Matano, S. (2000). Biomechanical analysis of vertical climbing in the spider monkey and the Japanese macaque. *American Journal of Physical Anthropology, 113,* 455–472.

Howell, A. B. (1944). *Speed in animals.* Chicago: University of Chicago Press.

Hutchinson, J. R., Famini, D., Lair, R. and Kram, R. (2003). Are fastmoving elephants really running? *Nature 422,* 493–494.

Ishida, H., Jouffroy, F. K., & Nakano, Y. (1990). Comparative dynamics of pronograde and upside down horizontal quadrupedalism in the slow loris (*Nycticebus coucang*). In: F. K. Jouffroy, M. H. Stack, & C. Niemitz (Eds.), *Gravity, posture and locomotion in primates* (pp. 209–220). Florence: Il Sedicesimo.

Ishida, H., Kumakura, H., & Kondo, S. (1985). Primate bipedalism and quadrupedalism: Comparative electromyography. In: S. Kondo, H. Ishida, & T. Kimura (Eds.), *Primate morphophysiology, Locomotor analyses and human bipedalism* (pp. 59–80). Tokyo: University of Tokyo Press.

Jenkins, F. A., Jr. (1972). Chimpanzee bipedalism: Cineradiographic analysis and implications for the evolution of gait. *Science, 178,* 877–879.

Jenkins, F. A., Jr. (1974). Tree shrew locomotion and the origins of primate arborealism. In: F. A. Jenkins, Jr. (Ed.), *Primate locomotion* (pp. 85–115). New York: Academic Press.

Jenkins, F. A., Jr., Dombrowski, P. J., & Gordon, E. P. (1978). Analysis of the shoulder in brachiating spider monkeys. *American Journal of Physical Anthropology, 48,* 65–76.

Jones, F. W. (1916). *Arboreal man.* London: Edward Arnold.

Jouffroy, F. K. (1983). Etude cineradiographique des deplacements du membre anterieur du Potto de Bosman (Perodicticus potto, P.L.S. Muller, 1766) au cours de la marche quadrupede sur une branche horizontale. *Annales des Sciences Naturelles, Zoologie, Paris, 5,* 75–87.

Jouffroy, F. K. (1989). Quantitative and experimental approaches to primate locomotion: A review of recent advances. In: P. Seth & S. Seth (Eds.), *Perspective in primate biology.* (pp. 47–108). New Delhi: Today and Tomorrow's Printers and Publishers.

Jouffroy, F. K., & Gasc, J. P. (1974). A cine-radiographic analysis of leaping in an African prosimian (*Galago alleni*). In: F. A. Jenkins

(Ed.), *Primate locomotion* (pp. 117–142). New York: Academic Press.

Jouffroy, F. K., Gasc, J. P., Decombas, M., & Oblin, S. (1974). Biomechanics of vertical leaping from the ground in *Galago alleni*: a cineradiographic analysis. In: R. D. Martin, A. G. Doyle, & A. C. Walker (Eds.), *Prosimian biology*. Liverpool: Duckworth and Co. Ltd.

Jouffroy, F. K., & Petter, A. (1990). Gravity-related kinematic changes in lorisine horizontal locomotion in relation to position of the body. In: F. K. Jouffroy, M. H. Stack, & C. Niemitz (Eds.), *Gravity, posture and locomotion in primates* (pp. 199–207). Florence: Il Sedicesimo.

Jouffroy, F. K., & Stern, J. T., Jr. (1990). Telemetered EMG study of the antigravity versus propulsive actions of knee and elbow muscles in the Slow loris (*Nycticebus coucang*). In: F. K. Jouffroy, M. H. Stack, & C. Niemitz (Eds.), *Gravity, posture and locomotion in primates* (pp. 221–236). Florence: Il Sedicesimo.

Jungers, W., & Anapol, F. (1985). Interlimb coordination and gait in the Brown Lemur (Lemur fulvus) and the Talapoin Monkey (Miopithecus talapoin). *American Journal of Physical Anthropology, 67,* 89–97.

Jungers, W., & Stern, J. (1980). Telemetered electromyography of forelimb muscle chains in gibbons. *Science, 208,* 617–619.

Jungers, W., & Stern, J. (1981). Preliminary electromyography of brachiation in gibbons and spider monkeys. *International Journal of Primatology, 2,* 19–30.

Jungers, W. L., & Stern, J. T. (1984). Kinesiological aspects of brachiation in lar gibbons. In: H. Preuschoft, D. J. Chivers, W. Y. Brockelman, & N. Creel (Eds.), *The lesser apes: Evolutionary and behavioural biology* (pp. 119–134). Edinburgh: Edinburgh University Press.

Kimura, T. (1985). Bipedal and quadrupedal walking of primates, comparative dynamics. In: S. Kondo, H. Ishida, & T. Kimura (Eds.), *Primate morphophysiology, Locomotor analyses and human bipedalism* (pp. 81–104). Tokyo: University of Tokyo Press.

Kimura, T. (1990). Voluntary bipedal walking of infant chimpanzees. In: F. K. Jouffroy, M. H. Stack, & C. Niemitz (Eds.), *Gravity, posture and locomotion in primates* (pp. 237–251). Florence: Il Sedicesimo.

Kimura, T. (1991). Body center of gravity and energy expenditure during bipedal locomotion in humans, chimpanzees and macaques. *Primate Reports, 31,* 19–20.

Kimura, T. (1992). Hindlimb dominance during primate high-speed locomotion. *Primates, 33,* 465–474.

Kimura, T. (1996). Center of gravity of the body during the ontogeny of chimpanzee bipedal walking. *Folia Primatologica, 66,* 126–136.

Kimura, T., Okada, M., & Ishida, H. (1979). Kinesiological characteristics of primate walking: Its significance in human walking. In: M. Morbeck, H. Preuschoft, & N. Gomberg (Eds.), *Environment, behavior, and morphology: Dynamic interactions in primates* (pp. 297–312). New York: Gustav Fischer.

Kimura, T, Okada, M., Yamazaki, N., & Ishida, H. (1983). Speed of the bipedal gaits of man and nonhuman primates. *Annals of Scientific National Zoology of Paris, 5,* 145–158.

Larney, E., & Larson, S. G. (2004). Limb compliance during walking: Comparisons of elbow and knee yield across quadrupedal primates and in other mammals. *American Journal of Physical Anthropology, 125,* 42–50.

Larson, S. G. (1988). Subscapularis function in gibbons and chimpanzees: Implications for interpretation of humeral head torsion in hominoids. *American Journal of Physical Anthropology, 76,* 449–462.

Larson, S. G. (1989). Role of supraspinatus in quadrupedal locomotion of vervets (Cercopithecus aethiops): Implications for interpretation of humeral morphology. *American Journal of Physical Anthropology, 79,* 369–377.

Larson, S. G. (1998). Unique aspects of quadrupedal locomotion in nonhuman primates. In: E. Strasser, J. G. Fleagle, A. L. Rosenberger, & H. M. McHenry (Eds.), *Primate locomotion: Recent advances* (pp. 157–173). New York: Plenum.

Larson, S. G., & Stern, J. T., Jr. (1986). EMG of scapulohumeral muscles in the chimpanzee during reaching and "arboreal" locomotion. *American Journal of Anatomy, 176,* 171–190.

Larson, S. G., & Stern, J. T., Jr. (1987). EMG of chimpanzee shoulder muscles during knuckle-walking: problems of terrestrial locomotion in a suspensory adapted primates. *Journal of Zoology, London, 212,* 629–655.

Larson, S. G., & Stern, J. T. (1989). The role of propulsive muscles of the shoulder during quadrupedalism in vervet monkeys (*Cercopithecus aethiops*): Implications for neural control of locomotion in primates. *Journal of Motor Behavior, 21*, 457–472.

Larson, S. G., & Stern, J. T., Jr. (1992). Further evidence for the role of supraspinatus in quadrupedal monkeys. *American Journal of Physical Anthropology, 87*, 359–363.

chimpanzee: Scapular rotators revisited. *American Journal of Physical Anthropology, 85*, 71–84.

Larson, S. G., Schmitt, D., Lemelin, P., & Hamrick, M. W. (1999). The uniqueness of primate forelimb posture during quadrupedal locomotion. *American Journal of Physical Anthropology, 112*, 87–101.

Larson, S. G., Schmitt, D., Lemelin, P., & Hamrick, M. W. (2001). Limb excursion during quadrupedal walking: How do primates compare to other mammals. *Journal of Zoology of London, 255*, 353–365.

Latimer, B., Ohman, J. C., & Lovejoy, C. O. (1987). Talocrural joint in African hominoids: Implications for Australopithecus afarensis. *American Journal of Physical Anthropology, 74*, 155–175.

Lee, C. R., & Farley, C. T. (1998). Determinants of the center of mass trajectory in human walking and running. *Journal of Experimental Biology, 201*, 2935–2944.

Lemelin, P. (1999). Morphological correlates of substrate use in didelphid marsupials: implications for primate origins. *Journal of Zoology, London, 247*, 165–175.

Lemelin, P., & Schmitt, D. (1998). Relation between hand morphology and quadrupedalism in primates. *American Journal of Physical Anthropology, 105*, 185–197.

Lemelin, P., & Schmitt, D. (2007). The origins of grasping and locomotor adaptations in primates: Comparative and experimental approaches using an opossum model. In: M. Dagosto & M. Ravosa (Eds.), *Primate origins* (pp. 329–380). New York: Kluwer.

Lemelin, P., Schmitt, D., & Cartmill, M. (2003). Footfall patterns and interlimb coordination in opossums: Evidence for the evolution of diagonal-sequence walking gaits in primates. *Journal of Zoology, 260*, 423–429.

Lemelin, P., Schmitt, D., MacKenzie, A., George, G., & Cartmill, M. (2008). The effects of substrate type and size on the locomotion of kinkajous. *Integrated and Comprehensive Biology, 47*, Suppl. 1, e-71.

Li, Y., Crompton, R. H., Alexander, R. McN., Gunther, M. M., & Wang, W. J. (1996). Characteristics of ground reaction forces in normal and chimpanzee-like bipedal walking by humans. *Folia Primatologica, 66*, 137–159.

McClearn, D. (1992). Locomotion, posture, and feeding behavior of kinkajous, coatis, and raccoons. *Journal of Mammals, 73*, 245–261.

McMahon, T. A. (1985). The role of compliance in mammalian running. *Journal of Experimental Biology, 115*, 263–282.

McMahon, T. A., Valiant, G., & Frederick, E. C. (1987). Groucho running. *Journal of Applied Physiology, 62*, 2326–2337.

Meldrum, D. J. (1991). Kinematics of the Cercopithecine foot on arboreal and terrestrial substrates with implications for the interpretation of hominid terrestrial adaptations. *American Journal of Physical Anthropology, 84*, 273–290.

Muybridge, E. (1887). *Animals in Motion* [1957 reprint]. New York: Dover.

Okada, M. (1985). Primate bipedal walking: Comparative Kinematics. In: S. Kondo, H. Ishida, & T. Kimura (Eds.), *Primate morphophysiology, Locomotor analyses and human bipedalism* (pp. 47–58). Tokyo: University of Tokyo Press.

Okada, M., & Kondo, S. (1982). Gait and EMGs during bipedal walk of a gibbon (*Hylobates agilis*) on flat surface. *Journal of the Anthropological Society of Nippon, 90*, 325–330.

Polk, J. D. (2001). *The influence of body size and body proportions on primate quadrupedal locomotion*. PhD. Dissertation, State University of New York at Stony Brook.

Polk, J. D. (2002). Adaptive and phylogenetic influences on musculoskeletal design in Cercopithecine primates. *Journal of Experimental Biology, 205*, 3399–3412.

Pridmore, P. A. (1994). Locomotion in *Dromiciops australis* (Marsupialia: Microbiotheriidae). *Australian Journal of Zoology, 42*, 679–699.

Prost, J. H. (1965). The methodology of gait analysis and the gaits of monkeys. *American Journal of Physical Anthropology, 23,* 215–240.

Prost, J. H. (1967). Bipedalism of man and Gibbon compared using estimates of joint motion. *American Journal of Physical Anthropology, 26,* 135–148.

Prost, J. H. (1969). Gaits of monkeys and horses: A methodological critique. *American Journal of Physical Anthropology, 332,* 121–128.

Prost, J. (1980). Origin of bipedalism. *American Journal of Physical Anthropology, 52,* 175–189.

Prost, J. H., & Sussman, R. W. (1969). Monkey locomotion on inclined surfaces. *American Journal of Physical Anthropology, 31,* 53–58.

Raichlen, D., & Shapiro, L. (2007). The evolution of mammalian biomechanics: Adaptations or spandrels. *Journal of Morphology, 268,* 1122.

Reynolds, T. R. (1985). Stresses on the limbs of quadrupedal primates. *American Journal of Physical Anthropology, 67,* 351–362.

Reynolds, T. R. (1987). Stride length and its determinants in humans, early hominids, primates, and mammals. *American Journal of Physical Anthropology, 72,* 101–115.

Richmond, B. G., Begun, D. R., & Strait, D. S. (2002). Origin of human bipedalism: The knuckle-walking hypothesis revisited. *Yearbook of Physical Anthropology,* Suppl. 33, 70–105

Rollinson, J., & Martin, R. D. (1981). Comparative aspects of primate locomotion with special reference to arboreal cercopithecines. *Symposium of the Zoological Society of London, 48,* 377–427.

Rose, M. D. (1973). Quadrupedalism in primates. Primates 14: 337–358.

Rowe, N. (1996) The *pictorial guide to living primates.* New York: Pogonias Press.

Schmidt, M., & Fischer, M. S. (2000). Cineradiographic study of forelimb movements during quadrupedal walking in the brown lemur (Eulemur fulvus, primates: Lemuridae). *American Journal of Physical Anthropology, 111,* 245–262

Schmitt, D. (1994). Forelimb mechanics as a function of substrate type during quadrupedalism in two anthropoid primates. *Journal of Human Evolution, 26,* 441–458.

Schmitt, D. (1998). Forelimb mechanics during arboreal and terrestrial quadrupedalism in Old World monkeys. In: E. Strasser, J. G. Fleagle, A. L. Rosenberger, & H. M. McHenry (Eds.), *Primate locomotion: Recent advances* (pp. 175–200). New York: Plenum Press.

Schmitt, D. (1999). Compliant walking in primates. *Journal of Zoology, 247,* 149–160.

Schmitt, D. (2003a). Evolutionary implications of the unusual walking mechanics of the common marmoset (*Callithrix jacchus*). *American Journal of Physical Anthropology.* 122, 28–37

Schmitt, D. (2003b). The relationship between forelimb anatomy and mediolateral forces in primate quadrupeds: Implications for interpretation of locomotor behavior in extinct primates. *Journal of Human Evolution.* 44, 49–60

Schmitt, D., & Larson, S. G. (1995). Heel contact as a function of substrate type and speed in primates. *American Journal of Physical Anthropology, 96,* 39–50.

Schmitt, D., & Lemelin, P. (2002). The origins of primate locomotion: gait mechanics of the woolly opossum. *American Journal of Physical Anthropology, 118,* 231–238.

Schmitt, D., Lemelin, P., & Trueblood, A. (1999). Shock wave transmission through the human body during normal and compliant walking. *American Journal of Physical Anthropology,* Suppl. 28, 243–244.

Schmitt, D., Stern, J. T., Jr., & Larson, S. G. (1996). Compliant gait in humans: Implications for substrate reaction forces during australopithecine bipedalism. *American Journal of Physical Anthropology,* Suppl. 22, 209.

Shapiro, L. J., Anapol, F. C., & Jungers, W. L. (1997). Interlimb coordination, gait, and neural control of quadrupedalism in chimpanzees. *American Journal of Physical Anthropology, 102,* 177–186.

Shapiro, L. J., Demes, A. B., & Cooper, J. (2001). Lateral bending of the lumbar spine during quadrupedalism in strepsirhines. *Journal of Human Evolution, 40,* 231–259.

Shapiro, L. J., & Jungers, W. L. (1988). Back muscle function during bipedal walking in chimpanzee and gibbon: Implications for the evolution of human locomotion. *American Journal of Physical Anthropology, 77,* 201–212.

Shapiro, L. J., & Jungers, W. L. (1994). Electromyography of back muscles during quadrupedal and bipedal walking in primates.

American Journal of Physical Anthropology, 93, 491–504.

Smith, G. E. (1912). The evolution of man. *Smithsonian Institution Annual Reports,* 1912, 553–572.

Stern, J. T., Jr. (1976). Before bipedality. *Yearbook of Physical Anthropology, 20,* 59–68.

Stern, J. T., Jr. (2000). Climbing to the top: a personal memoir of Australopithecus afarensis. *Evolutionary Anthropology, 9,* 113–133.

Stern, J. T., Jr., & Larson, S. G. (2001). Telemetered electromyography of the supinators and pronators of the forearm in gibbons and chimpanzees: Implications for the fundamental positional adaptation of hominoids. *American Journal of Physical Anthropology, 115,* 253–268.

Stern, J. T., Jr., & Susman, R. L. (1981). Electromyography of the gluteal muscles in Hylobates, Pongo and Pan: Implications for the evolution of hominid bipedality. *American Journal of Physical Anthropology, 55,* 153–166.

Stern, J. T., Jr., Wells, J. P., Vangor, A. K., & Fleagle, J. G. (1977). Electromyography of some muscles of the upper limb in *Ateles* and *Lagothrix. Yearbook of Physical Anthropology, 20,* 498–507.

Stern, J. T., Jr., Wells, J. P., Vangor, A. K., & Fleagle, J. G. (1980). An electromyographic study of the pectoralis major in atelines and Hylobates with special reference to the evolution pars clavicularis. *American Journal of Physical Anthropology, 52,* 13–25.

Stevens, N. J. (2001). Effects of substrate orientation on quadrupedal walking in *Loris tardigradus. Journal of Morphology, 248,* 288.

Stevens, N. J., Demes, A. B., & Larson, S. G. (2001). Effects of branch compliance on quadrupedal walking in *Loris tardigradus. American Journal of Physical Anthropology, 113,* Suppl. 32, 142.

Susman, R. L. (1983). Evolution of the human foot: Evidence from Plio-Pleistocene hominids. *Foot and Ankle, 3,* 365–376.

Susman, R. J., Stern, J. T., Jr., & Jungers, W. L. (1984). Arboreality and bipedality in the Hadar hominids. *Folia Primatologica, 43,* 113–156.

Sussman, R. W. (1991). Primate origins and the evolution of angiosperms. *American Journal of Primatology, 23,* 209–223.

Swartz, S. M., Bertram, J. E., & Biewener, A. A. (1989). Telemetered in vivo strain analysis of locomotor mechanics of brachiating gibbons. *Nature, 342,* 270–272.

Szalay, E. S., and E. Delson, E. (1979). Evolutionary history of the primates. Academic Press, New York.

Szalay, F. S., & Dagosto, M. (1988). Evolution of hallucial grasping in the primates. *Journal of Human Evolution, 17,* 1–33.

Szalay, F. S., Rosenberger, A. L., & Dagosto, M. (1987). Diagnosis and differentiation of the Order Primates. *Yearbook of Physical Anthropology, 30,* 75–105.

Tardieu, C., Aurengo, A., & Tardieu, B. (1993). New method of three-dimensional analysis of bipedal locomotion for the study of displacements of the body and body-parts centers of mass in man and non-human primates: Evolutionary framework. *American Journal of Physical Anthropology, 90,* 455–476.

Taylor, C. R., & Rowntree, V. (1973). Running on two or four legs: Which consumes more energy. *Science, 179,* 597–601.

Turnquist, J. E., Schmitt, D., Rose, M. D., & Cant, J. G. (1999). Pendular motion in the brachiation of captive *Lagothrix* and *Ateles. American Journal of Primatology, 48,* 263–281.

Tuttle, R. H., & Basmajian, J. V. (1974a). Electromyography of Pan gorilla: an experimental approach to the problem of hominization. *Symposium of the 5th Congress of the International Primate Society,* 303–312.

Tuttle, R., & Basmajian, J. (1974b). Electromyography of the brachial muscles in Pan gorilla and hominoid evolution. *American Journal of Physical Anthropology, 41,* 71–90.

Tuttle, R., & Basmajian, J. (1974c). Electromyography of the forelimb musculature in Gorillas and problems related to knuckle walking. In: F. Jenkins (Ed.), *Primate locomotion* (pp. 293–383). New York: Academic.

Tuttle, R. H., & Basmajian, J. V. (1977). Electromyography of pongid shoulder muscles and hominoid evolution. I. Retractors of the humerus and "rotators" of the scapula. *Yearbook of Physical Anthropology, 20,* 491–497.

Tuttle, R. H., & Basmajian, J. V. (1978a). Electromyography of pongid shoulder muscles. III. Quadrupedal positional behavior. *American Journal of Physical Anthropology, 49,* 57–70.

Tuttle, R. H., & Basmajian, J. V. (1978b). Electromyography of pongid shoulder muscles II. deltoid, rhomboid and "rotator cuff". *American Journal of Physical Anthropology, 49,* 47–56.

Tuttle, R. H., Hollowed, J. R., & Basmajian, J. V. (1992). Electromyography of pronators and supinators in Great Apes. *American Journal of Physical Anthropology, 87,* 215–226.

Tuttle, R. H., Velte, M. J., & Basmajian, J. V. (1983). Electromyography of brachial muscles in *Pan troglodytes* and *Pongo pygmaeus. American Journal of Physical Anthropology, 61,* 75–83.

Vangor, A. K., & Wells, J. P. (1983). Muscle recruitment and the evolution of bipedality: Evidence from telemetered electromyography of spider, woolly and patas monkeys. *Annales des Sciences naturelles, zoologie, Paris, 5,* 125–135.

Vereecke E, D' Août K, De Clercq D, Van Elsacker L, Aerts P (2003) Dynamic plantar pressure distribution during terrestrial locomotion of bonobos (Pan paniscus). Am. J. Phys. Anthropol. 119, 37–51.

Vereecke, E. E., DAût, K. and Aerts, P. (2006). The dynamics of hylobatid bipedalism: evidence for an energy-saving mechanism? J. Exp. Biol. 209, 2829-2838.

Vilensky, J. A. (1980). Trot-gallop transition in a macaque. *American Journal of Physical Anthropology, 53,* 347–348.

Vilensky, J. A. (1983). Gait characteristics of two macaques with emphasis on the relationship with speed. *American Journal of Physical Anthropology, 61,* 255–265.

Vilensky, J. A. (1987). Locomotor behavior and control in humans and non-human primates: Comparisons with cats and dogs. *Neuroscience and Biobehavioral Review, 11,* 263–274.

Vilensky, J. A. (1988). Effects of size on Vervet (Cercopithecus aethiops) gait parameters: A Cross sectional Approach. *American Journal of Physical Anthropology, 76,* 463–480.

Vilensky, J. A. (1989). Primate quadrupedalism: How and why does it differ from that of typical quadrupeds? *Brain Behavioral Evolution, 34,* 357–364.

Vilensky, J. A., & Gankiewicz, E. (1986). Effects of size on Vervet (*Cercopithecus aethiops*) gait parameters: A preliminary analysis. *Folia Primatologica, 46,* 104–117.

Vilensky, J. A., & Gankiewicz, E. (1990). Effects of speed on forelimb joint angular displacement patterns in vervet monkeys (*Cercopithecus aethiops*). *American Journal of Physical Anthropology, 83,* 203–210.

Vilensky, J. A., Gankiewicz, E., & Townsend, D. (1986). Effects of size on Vervet (*Cercopithecus aethiops*) gait parameters: A cross sectional approach. *American Journal of Physical Anthropology, 76,* 463–480.

Vilensky, J. A., & Gehlsen, G. (1984). Temporal gait parameters in humans and quadrupeds: how do they change with speed? *Journal of Human Movement Study, 10,* 175–188.

Vilensky, J. A., & Larson, S. G. (1989). Primate locomotion: Utilization and control of symmetrical gaits. *Annual Review of Anthropology, 18,* 17–35.

Vilensky, J. A., Libii, J. N., & Moore, A. M. (1991). Trot-gallop transition in quadrupeds. *Physiology and Behavior, 50,* 835–842.

Vilensky, J. A., & Patrick, M. (1985). Gait characteristics of two squirrel monkeys with emphasis on the relationship with speed and neural control. *American Journal of Physical Anthropology, 68,* 429–444.

Vilensky, J. A., Moore, A. M., & Libii, J. N. (1994). Squirrel monkey locomotion on an inclined treadmill: Implications for the evolution of gaits. *Journal of Human Evolution, 26,* 375–386.

Vilensky, J. A., Moore-Kuhns, M., & Moore, A. M. (1990). Angular displacement patterns of leading and trailing limb joints during galloping in monkeys. *American Journal of Primatology, 22,* 227–239.

Ward, C. V. (2002). Interpreting posture and locomotion of *Australopithecus afarensis*: Where do we stand? *Yearbook of Physical Anthropology, 45,* 185–215.

Wells, J. P., & Wood, G. A. (1975). The application of biomechanical motion analysis to aspects of green monkey (*Cercopithecus a. sabaeus*) locomotion. *American Journal of Physical Anthropology, 43,* 217–226.

White, T. (1990). Gait selection in the brush-tail possum (*Trichosurus vulpecula*), the northern quoll (*Dasyurus hallucatus*), and the Virginia opossum (*Didelphis virginiana*). *Journal of Mammals, 71,* 79–84.

Whitehead, P. F., & Larson, S. G. (1994). Shoulder motion during quadrupedal walking in *Cercopithecus aethiops*: Integration of cineradiographic and electromyographic data. *Journal of Human Evolution, 26,* 525–544.

Wunderlich, R. E., & Ford, K. R. (2000). Plantar pressure distribution during bipedal and

quadrupedal walking in the chimpanzee (Pan troglodytes). EMED Scientific Meeting, August, 2000, Munich, Germany.
Wunderlich, RE and Jungers, WJ (2009). Manual digital pressures during knuckle-walking in chimpanzees (Pan troglodytes) American Journal of Physical Anthropology 139, 394–403.
Wunderlich, R. W., & Schmitt, D. (2000). Hindlimb adaptations associated with heel-strike plantigrady in hominoids. *American Journal of Physical Anthropology*, Suppl. 30, 328.
Yaguramaki, N. (1995). The relationship between posture and external force in walking. *Anthropological Science, 103*, 117–140.
Yamazaki, N., & Ishida, H. (1984). A biomechanical study of vertical climbing and bipedal walking in gibbons. *Journal of Human Evolution, 13*, 563–571.

CHAPTER 4

Foraging Cognition in Nonhuman Primates

Klaus Zuberbühler and Karline Janmaat

FORESTS AS PRIMATE HABITATS: COEVOLUTION OF PRIMATES AND FRUIT

In terms of total biomass, primates are very successful vertebrates in most undisturbed tropical forests (Chapman et al., 1999a; Fleagle & Reed, 1996). Many primate species are forest dwellers, and the forest habitat is likely to have had a major impact on primate evolution. This is especially true for the great apes, whose changes in diversity have followed climate-related retractions and expansions of wooded habitats since the late Miocene (Potts, 2004). Most primates, including typical leaf-eaters, consume considerable amounts of fruits as part of their daily diets (e.g., Korstjens, 2001). Fleshy fruits and the arthropods that associate with them are highly nutritious, which provide arboreal animals, such as primates, birds, and bats, with a stationary and relatively reliable source of energy (Janmaat et al., 2006a). Primates and fruiting trees have shared a long evolutionary history, and the arrival of angiosperm fruits and flowers may have been of particular importance in primate evolution (Soligo & Martin, 2006; Sussman, 1991, 2004). About 85 million years ago, a trend toward increased fruit size can be found (Eriksson et al., 2000), roughly coinciding with the radiation of early ancestors of today's primates, about 82 million years ago (Tavaré et al., 2002).

Compared to other groups of animals, primates possess a number of adaptations that make them particularly suited for arboreal foraging on fruit. Many primate species have opposable thumbs and toes, allowing them to grasp and reach fruit at the terminal tree branches, which are inaccessible to many other animals. Hindlimb dominance and grasping ability enable many primates to leap between trees in an energetically efficient way, in contrast other arboreal mammals such as most tree squirrels (Gebo, 2004; Sussman, 1991; Taylor et al., 1972). Other adaptations concern forward-facing cycs and stereotypic vision, which facilitates hand–eye coordination and foraging at high speed (Cartmill, 1972; Gebo, 2004). Similarly, diurnal activity, high visual acuity, and color vision enable spotting of fruit and their nutritional value from large distances (Barton, 2000; Polyak, 1957; Riba-Hernández et al., 2005; Sumner & Mollon, 2000). Diurnal foraging is also beneficial because ripening rates of fruits tend to be highest in the early afternoon following high midday incident radiation and ambient temperature (Diaz-Perez et al., 2002; Graham et al., 2003; Houle, 2004; Spayd et al., 2002).

EVOLUTIONARY THEORIES OF PRIMATE COGNITION

Primates, and especially humans, have relatively larger brains than other groups of mammals (Harvey & Krebs, 1990; Jerison, 1973). It has also been noted that a variety of brain size variables in primates correlate positively with measures of social complexity, such as group size, deceptive behaviour, or strength of social bonds (Barton, 1996, 1999; Byrne & Corp, 2004;

Dunbar, 1998; Dunbar & Shultz, 2007). This has been taken to suggest that large groups, and social complexity that emerges from them, have acted as a primary selection force favouring the evolution of increased brain size. This is because high social intelligence is likely to provide individuals with a competitive and reproductive advantage over their less socially skilled conspecifics.

As appealing as it is, the social intelligence hypothesis has a number of problems. Large promiscuous multimale/multifemale groups, the presumed breeding grounds for high social intelligence, are the exception in primate societies (Smuts et al., 1986) and it is often not specified how group size relates to social complexity. Moreover, although food competition is likely to increase with group size, larger groups also benefit from increased search swath and accumulated knowledge of individuals to locate food sources, avoid predators, and deal with neighboring groups (Garber & Boinski, 2000; Janson & Di Bitetti, 1997). Individuals are especially likely to benefit from older and more knowledgeable group members during periods of food scarcity when long-term experience is more crucial (Byrne, 1995; Chauvin & Thierry, 2005; van Roosmalen, 1988). The social intelligence hypothesis also struggles to explain how exactly primates were able to grow expensive large brains in the first place. Why did primates benefit more than other social animals from increased encephalization? The relationship between neocortex and group size is certainly real, but the causal arrow could also point the other way: Primates have evolved large brains for nonsocial reasons, which enables them to live in larger groups, form more complex social systems, and maintain more complex social relations than other smaller-brained species (Müller & Soligo, 2005).

A main contender of social intelligence is the "ecological intelligence" hypothesis developed by Milton (1981). Large brains, according to this idea, are the evolutionary products of extensive mental mapping requirements faced by frugivorous species, a hypothesis that emerged from empirical work comparing highly encephalized and frugivorous spider monkeys (*Ateles geoffroyi*) with less encephalized folivorous howler monkeys (*Alouatta palliata*). It is interesting that in diurnal frugivorous primates, relative brain enlargements are primarily found within the visual system, while in nocturnal species enlargements are in the olfactory structures (Barton et al., 1995), suggesting that the brain has directly responded to the demands of foraging. In addition, increases in the degree of orbital convergence (associated with stereotypic vision) correlate with expansion of visual brain structures and, as a consequence, with overall size of the brain (Barton, 2004).

Compared to other body tissues, brains are metabolically expensive organs, requiring a continuous and reliable flow of nutrients (Armstrong, 1983; Mink et al., 1981). According to recent analyses, relative brain size is positively correlated with basal metabolic rate, indicating that larger brains may be a reflection of being able to sustain higher basal energy costs (Isler & van Schaik, 2006a). Any increase in relative brain size, therefore, may only be possible in populations that have managed to either improve their access to nutrition or decrease other existing energy demands. Energy can be saved, for example, by reducing an organism's locomotor costs (Isler & van Schaik, 2006b) or reducing the metabolic requirements of other expensive tissues, such as the digestive system (Aiello & Wheeler, 1995). Higher-quality foods, such as fruit and animal matter, are easier to digest than other material, allowing the organism to reduce the size of its digestive tract. This hypothesis is supported by the findings that frugivorous primates usually have relatively larger brains and smaller digestive systems than folivorous primates (Barton, 2000; Clutton-Brock & Harvey, 1980; Hladik, 1967).

The various special adaptations for harvesting the fruits discussed in the previous section enabled primates to monopolize one of the most nutritious food sources in these forests. This may have allowed primates, especially haplorhines (see Chapter 1) that live in areas with relatively high fruit production, to afford larger brains than other groups of animals (Cunningham & Janson, 2007; Fish & Lockwood, 2003). What benefits they gain from this relatively costly trait and what selection pressures have favored its evolution is subject of an ongoing debate.

In sum, a more complete understanding for why primates have relatively bigger brains than other groups of animals requires evidence at the ultimate and proximate level (Tinbergen, 1963). The current literature favors social explanations, mainly because of what is available in terms of empirical studies, but we have outlined a number of reasons for caution. By contrast, we discuss recent empirical progress on understanding the impact of foraging problems on cognition. The studies we review all have been conducted with the intent to investigate the cognitive capacities employed by nonhuman primates in relation to finding food in their natural habitats, and we contend that some of these findings are of direct relevance to the ecological intelligence hypothesis.

Figure 4.1 A gray-cheeked mangabey (*Lophocebus albigena johnstonii*) feeding on purple flowers of *Milettia dura*. Picture by Rebecca Chancellor.

HOW DO FOREST PRIMATES KNOW WHERE TO FIND FRUIT?

A large-bodied monkey group's home range can contain as many as 100,000 trees (e.g., *Lophocebus albigena johnstonii*; Waser, 1974), yet only a small fraction of these trees will carry ripe fruit at any given time. Estimates for some forests vary anywhere from 50 to 4,000 trees per average home range (Janmaat et al., submitted). Are primates able to find these trees, and how efficient are they at doing so? A number of studies found that wild primates were more efficient in finding food than predicted by random search models, suggesting that individuals use some mental heuristics to locate food (e.g., Cunningham, 2003; Garber & Hannon, 1993; Janson, 2000; Milton, 2000; Valero & Byrne, 2007). In our own studies on gray-cheeked and sooty mangabeys (*Lophocebus albigena johnstonii*; *Cercocebus atys atys*), we found that monkeys were more likely to approach and search for fruit under or in trees that had produced fruits than empty trees of the same species (Janmaat et al., 2006b; Figs. 4.1 and 4.2).

We also found that target trees with fruit were approached significantly faster and with sharper angles than trees without fruit (Janmaat, 2006; Janmaat et al., 2006b; Figs. 4.3 and 4.4). When we measured the number of trees that were encountered while following individual monkeys, we found that they encountered or approached significantly more fruit-bearing trees than during control transects (i.e., when the observer walked a path parallel to the monkey's own route) (Janmaat, 2006; Janmaat et al., submitted).

It has been argued that the most efficient way to optimize foraging success is to mentally represent the location of all fruit trees in a home range as well as their fruiting state and overall temporal patterning, that is, to maintain a cognitive map (Milton, 1981, 2000). According to most definitions, cognitive maps are mental representations of the real world, as if viewed from above, a Euclidian representation of landmarks with vector distance and angular relationships between them (O'Keefe & Nadel, 1978; Tolman, 1948).

Figure 4.2 Terrestrial sooty mangabeys (*Cercocebus atys atys*) foraging for insects in dead wood. Picture by Karline Janmaat.

However, current evidence suggests that it is not very likely that nonhuman primates, or even humans, represent their home ranges in such a way (Byrne, 1979; Janson & Byrne, 2007). One crucial empirical test for the cognitive map hypothesis concerns the responses of individuals to obstacles on a foraging route (Bennett, 1996), that is, whether they are capable of finding an efficient detour. So far, wild primates have failed this test. In one study, the behavior of wild Chacma baboons (*Papio ursinus*) was studied when encountering neighboring groups on their habitual foraging routes. Individuals did not take detours around such groups to get to their foraging goal, but either waited for them to pass by or simply abandoned their goal completely (Noser & Byrne, 2007a). Some support for something like a cognitive map comes from a captive study with a young bonobo (*Pan paniscus*) that was tested with an artificial lexigram system, but it is unclear to what degree primates use this capacity in the wild (Menzel et al., 2002).

SPATIO-TEMPORAL MENTAL REPRESENTATIONS IN THE NATURAL HABITAT

Apart from the general difficulties of providing empirical evidence for a cognitive map, no one seriously doubts that primates are able to mentally represent space in some way, although in many cases it is not clear what exactly these representations consist of, especially in the

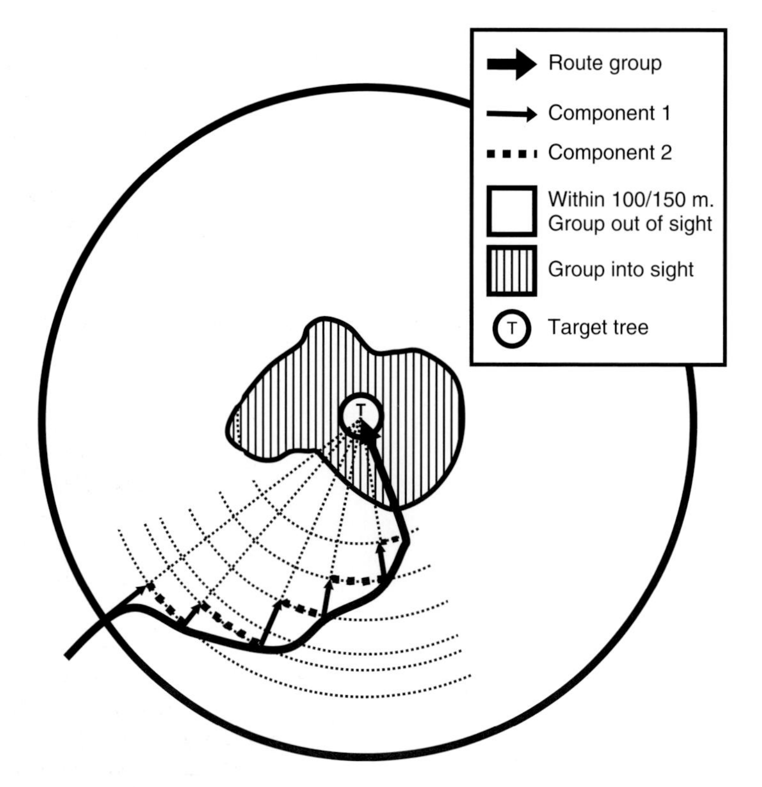

Figure 4.3 Illustration of the method used to measure speed toward a target tree. The observer follows the group while staying within a 5-m distance of the individual that is closest to the target tree. Following took place in two components of direction, either along (a) component 1 (arrow) that is directed toward the tree trunk or (b) component 2 (thick dotted line), which is directed along the imaginary circle around the tree trunk. Speed was determined by counting steps per minute when walking in the direction of the tree (component 1) only. The observer was updated on the direction of the tree trunk by the calling or clicking sounds produced by a second observer, who was waiting under the tree trunk. The outer circle has a radius of 100 or 150 meters dependent on the species. The shaded area represents the area in which the group comes into sight of the second observer waiting under the tree.

natural habitat. A major challenge in field studies is to determine if primates reach a resource by goal-directed travel, an indicator of mental representations of space, or by chance. It is important to consider that the shortest route is not always the most efficient one, and that animals could combine different goals in one single route, and that they could monitor food without exploiting it (Sigg & Stolba, 1981). Some researchers have generated geometric or step models combined with sophisticated statistics to determine the likelihood of whether spatial representations are involved in travel decisions (Bates, 2005; Cunningham & Janson, 2007; Garber & Hannon, 1993; Janson, 1998; Milton, 2000; Noser & Byrne, 2007a; Valero & Byrne, 2007).

In one experiment with Argentinean capuchin monkeys (*Cebus apella nigritus*), three feeding sites were arranged in a triangle and provisioned once per day. Once a monkey group had chosen a site, its next choice was between the two remaining sites, a close one with less food and a far away one with more food. The surprising finding was that capuchins generally chose the closer feeding site, even when the more distant site offered up to 12 times as much food (Janson, 2007). Should we conclude that the monkeys did not possess a mental

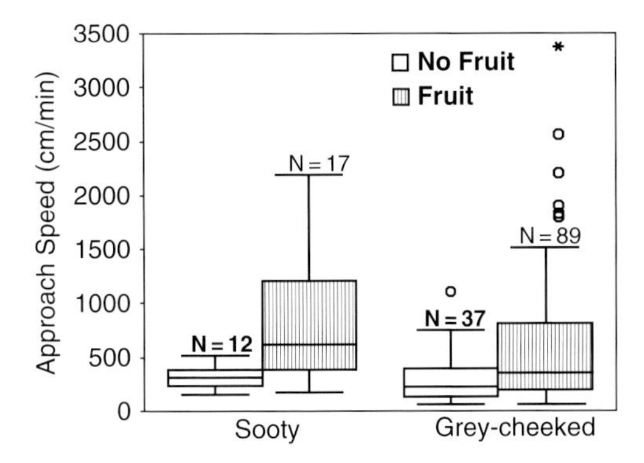

Figure 4.4 Speed of approach to trees with and without fruits. For both mangabey groups, bars represent the median speeds, while the top and bottom of the boxes represent the percentiles. The highest and lowest whiskers represent the highest and smallest values, which are not outliers. Circles and stars represent outliers and extreme values.

representation of the locations and value of the provided food? Alternatively, did they simply weigh up travel distance and likelihood of arriving at the food in time in the highly competitive situation of a rainforest? Further experiments will be required to determine what exactly influenced the monkeys' foraging decisions.

Similarly, a recent study on Chacma baboons (*Papio ursinus*) showed that the sleeping cliff, a presumably important goal, was not always approached fast and in a straight line, because the group was regularly foraging for seeds close to the sleeping site (Noser & Byrne, 2007b). Spider and woolly monkeys (*Ateles belzebuth*; *Lagothrix poeppigii*) travel through their home ranges along repeatedly used paths, which has been taken as evidence that spatial mental representations are in the form of route-based or network maps (De Fiore & Suarez, 2007). Nonrandom foraging patterns have also been reported from tamarins (*Saguinus mystax, S. fuscicollis*) in the Amazon of northeastern Peru (Garber, 1989). For great apes, the empirical evidence for spatial cognition is surprisingly weak. One study on tool-transporting behavior in wild chimpanzees (*Pan troglodytis*) concluded that subjects remembered distances between different nut-cracking sites and different stone hammers, as if using Euclidian space, but this interpretation is controversial (Boesch & Boesch, 1984).

The Role of Secondary Cues

One problem with field studies is that it is often difficult to make reliable assumptions about how far an individual can detect, using both visual and olfactory sensory information, a target resource. Moreover, travel decisions may be influenced by other secondary cues, such as food calls of other species. The availability of visual cues is particularly difficult to assess in a rainforest where fruit trees are sometimes visible over considerable distances, even from the ground. Humans are capable of spotting fruits in emergent trees from a distance of 150 m if the view is unobstructed, suggesting that other primates may possess comparable abilities (Golla et al., 2004; Janmaat, unpublished data).

Only a small number of field studies have been able to convincingly reject the use of such sensory cues to find resources (Garber & Paciulli, 1997; Janson, 1998; Janson & Di Bitetti, 1997; Sigg & Stolba, 1981). For example, departure latency in Chacma baboons was significantly shorter before traveling to scarce mountain figs compared with traveling to other more abundant fruit sources. Because the fig trees were approximately 700 m from the sleeping site, visible to human observers only from short distances, it was unlikely that the monkeys were guided by any secondary cues (Noser & Byrne, 2007b). In another study, the ranging behavior of sooty mangabeys was studied in relation to Anthonota trees with empty crowns (Janmaat et al., 2006a; Fig. 4.5). Monkeys approaching within 150 m of empty trees were more likely to approach if the tree was surrounded by fruits that had fallen to the forest floor than if the tree had not produced any fruit. The authors were able to rule out the possibility that the monkeys had seen any of the inconspicuous fallen fruits in the leafy substrate, indicating that the monkeys used spatial knowledge acquired during previous feeding experiences to relocate trees with fruit (Janmaat et al., 2006b).

In a similar way, gray-cheeked mangabeys that came within 100 m of an empty *Ficus sansibarica* tree were less likely to enter if the tree had recently been depleted than if the tree had not produced any fruits so far (Fig. 4.6). Since both tree types had empty crowns, with no differences in overall appearance, the visiting pattern was best explained by memories of previous visits (Janmaat et al., 2006b).

In sum, it seems safe to assume that navigation of primates in their natural habitats involves some kind of mental representations of space, but it is often unclear how enduring and rich these memories really are. Memories of spatial locations could be relatively short lived (a few days), and there is no good evidence for a geometric representation of space (Byrne, 2000; Janson, 2000).

EVIDENCE FOR FRUIT LOCALIZATION STRATEGIES

Another characteristic of forest fruits concerns their ephemeral nature. Temporal patterns of emergence can be complex, and fruits are often present for short periods only (Chapman et al., 1999b; 2004; Janmaat et al., submitted; Milton, 1981, 1988). Many fruit tree species rely on animals for seed dispersal and have evolved features that make their fruits appeal to a large number of species, leading to high levels of inter- and intraspecies competition (e.g., Hauser & Wrangham, 1990; Houle et al., 2006; McGraw & Zuberbühler, 2007; Sterck, 1995). Early arrival is therefore advantageous, and natural selection is likely to favor any cognitive strategy that makes this behavior possible. In the following final section, we discuss a number of behavioral strategies, and their potential underlying cognitive processes, that enable free-ranging primates to deal with these temporal constraints.

Monitoring Individual Trees

Gray-cheeked mangabeys have been observed to bypass about a third of all available fruit-bearing fig species (Janmaat et al., 2006b). The monkeys were more likely to revisit trees in which they had good feeding experiences before, compared to trees in which they were less successful. Similar

Figure 4.5 A sooty mangabey eating *Anthonota fragans* fruit (left). Picture by Karline Janmaat. Ripe *Anthonota fragans* fruits are harvested by the monkeys after they have fallen into the leaf litter underneath the tree (right). Picture by Ralph Bergmüller. Used with permission.

Figure 4.6 A gray-cheeked mangabey inspecting the ripening state of *Ficus sansibarica* fruit (left), *F. sansibarica* fruit (right). Pictures by Karline Janmaat. Used with permission.

patterns have been reported from a study on wild tamarins (Garber, 1989). For the mangabeys, intriguingly, this was also the case for fig trees with unripe fruits. Unripe figs are attractive to these monkeys because some of them contain weevil larvae or edible seeds. Monkeys have to inspect each fig individually, an interesting fact for the purpose of cognitive studies, because it effectively rules out the possibility that the monkeys responded to long-distance visual or olfactory cues (Janmaat et al., 2006a). In addition, the authors also found that the average speed with which the group approached such trees was significantly correlated with their average prior feeding experience in that tree (Fig. 4.7). Similar results have been reported from free-ranging sakis (*Pithecia pithecia*) in Venezuela. These primates bypassed a majority of fruit-bearing trees without feeding because they preferentially revisited specific trees, which they already knew as highly productive from prior visits (Cunningham & Janson, 2007).

In sum, there is good evidence that primates are able to distinguish between individual trees, which they assess in terms of quality, and that they use such memories in their daily foraging decisions.

Monitoring Meteorological Cues

Work on captive primates has shown that they can readily learn to anticipate delayed food rewards (Dufour et al., 2007; Ramseyer et al., 2005) and that they can trade off reward

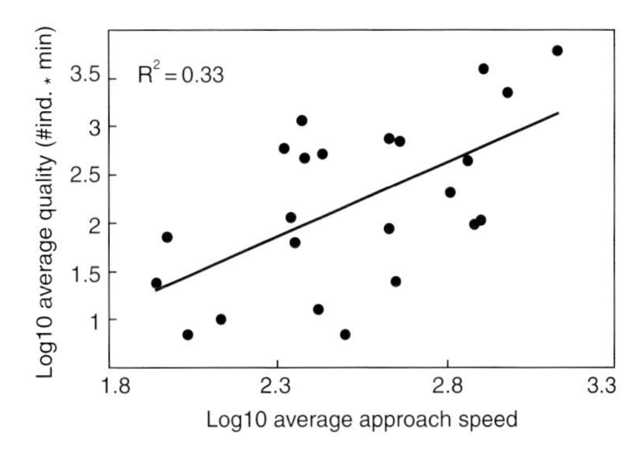

Figure 4.7 Approach speed to trees of different quality. The sum of quality values per tree is plotted in relation to the average of the total speeds with which the gray-cheeked group traveled toward that tree. Each dot represents the values of a target tree that carried unripe fruits. Values are based on an average number of three visits ($N_{min} = 1$, $N_{max} = 11$).

amount versus time delay (Stevens et al., 2005). Similar suggestions for a rudimentary ability to anticipate future events have also been made for wild primates, but only a few good empirical studies are available (Janmaat et al., 2006a; Janson, 2007; Noser & Byrne, 2007b; Sigg & Stolba, 1981; Wrangham, 1977).

Temperature and solar radiation influence ripening rates of fruits as well as the maturation of insect larvae inside them (e.g., Adams et al., 2001; Diaz-Pérez et al., 2002; Houle, 2004; Mazzei et al., 1999; Morrison & Noble, 1990), making the emergence of edible fruits somewhat predictable. A recent study on free-ranging mangabeys investigated whether these primates were able to take previous weather conditions into account when deciding to revisit particular fruit trees (Janmaat et al., 2006b). For this purpose, a study group was followed from dawn to dusk for three continuous long observation periods totalling 210 days, yielding an almost complete record of all revisit decisions toward 80 preselected fruit trees (Fig. 4.8).

The results were consistent with the idea that these monkeys made foraging decisions based on episodic-like memories of whether or not a tree previously carried fruit, combined with a more generalized understanding of the relationship between temperature and solar radiation and the maturation rate of fruit and insect larvae (Fig. 4.9). How exactly the monkeys managed to register the relatively subtle differences in average temperature values was not addressed, a topic for further research.

Monitoring Competitor Behavior

The presence of other fruit-eating individuals may also serve as a reliable indicator of the presence of edible fruits, especially for tree species that do not have predictable patterns of fruit emergence and that do not offer conspicuous secondary cues of edibility. In free-ranging tamarins (*Saguinus imperator* and *S. fuscicollis*), high-ranking individuals tended to monitor the activities of other group members, rather than to initiate their own food searches, providing evidence that these primates were able to associate social cues with the presence of foods (Bicca-Marques & Garber, 2005). Similarly, Tonkean macaques, *Macaca tonkeana*, kept in a large outdoor enclosure used food odor cues, acquired by smelling the mouths of other group members, to guide their own search for food (Chauvin & Thierry, 2005).

Primates also use auditory cues, such as feeding calls of group members, to find fruit

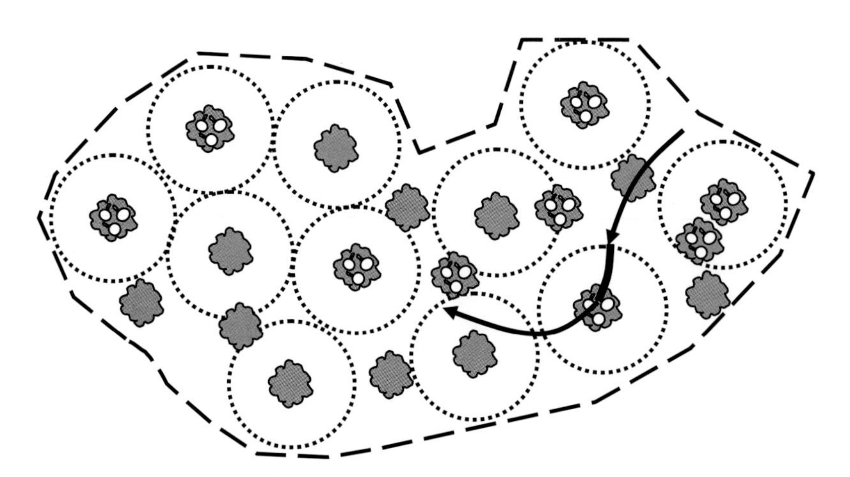

Figure 4.8 Measuring revisiting Behavior. The diagram illustrates an example of part of the study group's daily route (arrows) among target trees, each surrounded by an imaginary 100-m radius circle (dotted line). Once the group entered the circle, one observer rushed to the tree to determine the fruiting state and whether the group came into sight and entered the tree. In this example, the group visited one tree with fruit and bypassed one without fruit.

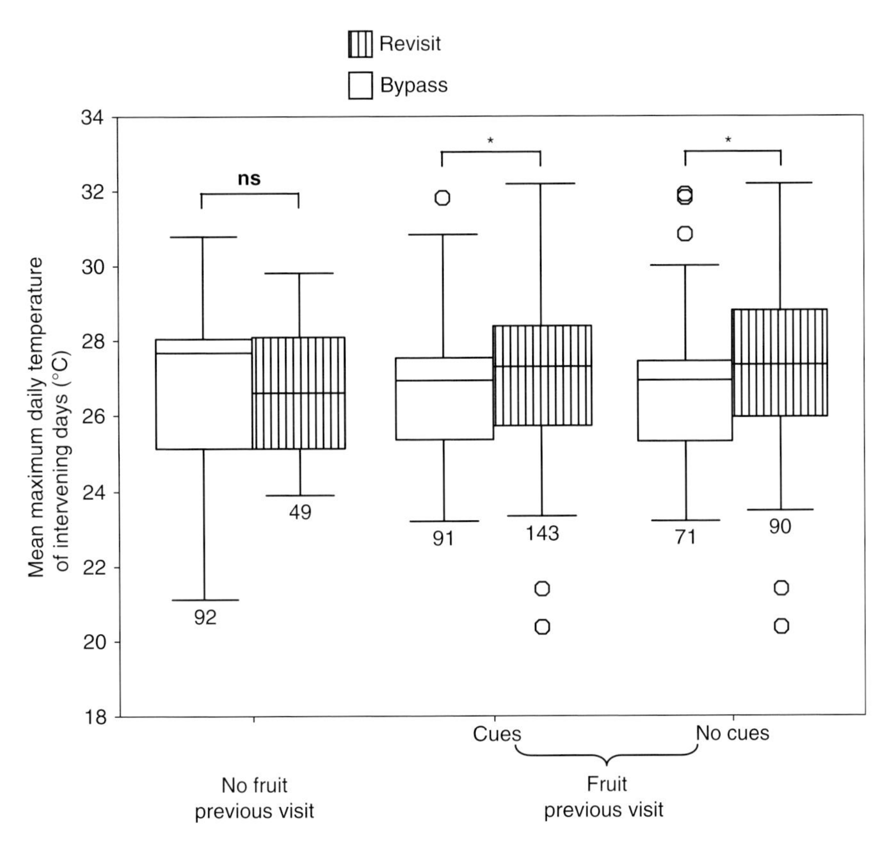

Figure 4.9 The influence of temperature on revisiting behavior. Average daily maximum temperature determined for the intervening period between the time the group entered the 100-m-radius circle and the time the group last visited the same tree. Shaded boxes represent average temperature values for revisits; white boxes represent bypasses. Different clusters refer to trees that (1) did not carry fruit at the previous visit, (2) carried fruit at the previous visit, and (3) carried fruit at the previous visit but no longer offered any sensory cues. Bars represent the median values of the average temperatures; top and bottom of the boxes represent the 75 and 25 percentiles. Whiskers represent highest and lowest values; circles represent outliers. Results showed that average daily maximum temperature was significantly higher for days preceding revisits than bypasses. These effects were found only for trees that carried fruit at the previous visit but not for trees that had carried none, providing empirical evidence that these primates were capable of taking into account past weather conditions when searching for food.

(e.g., tamarins, *Saguinus labiatus*: Caine et al., 1995; macaques, *Macaca sinica*: Dittus, 1984). Red-tailed monkeys (*Cercopithecus ascanius*), blue monkeys (*C. mitis*), and gray-cheeked mangabeys have been suggested to recognize the food-arrival calls of sympatric frugivores (Hauser & Wrangham, 1990).

In Kibale National Park, Uganda, fig trees that carried fruit contained a significantly larger number of noisy frugivorous animals, such as chimpanzees or hornbills, than fig trees that carried none, suggesting that primates could use the sound of sympatric foragers as an indicator for fruit availability (Janmaat, 2006). We thus analyzed the behavior of our mangabey study group on 10 different occasions when they discovered newly emerged or newly ripened fruits (Janmaat, 2006). In 2 out of 10 encounters, the tree was already occupied by a chimpanzee or hornbills feeding inside the tree, and interestingly in these cases the speed of approach was much higher than in the other eight cases (Fig. 4.10).

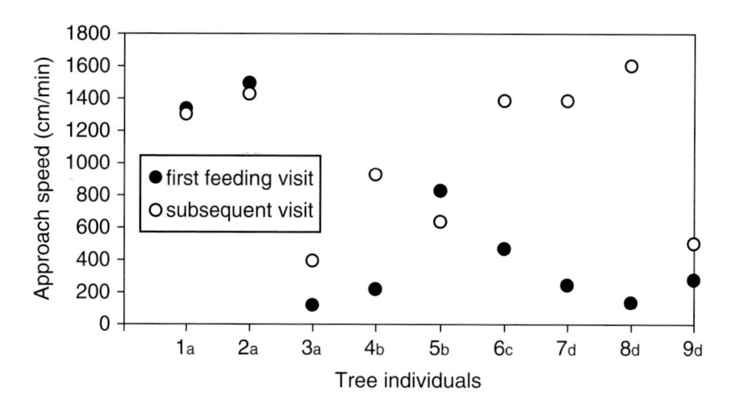

Figure 4.10 Speed of approach at discoveries of new edible figs. Closed circles represent the speed with which a mangabey group approached a fig tree in which the group was thought to discover newly emerged edible fruits. Open circles represent the speed with which the group approached the same tree at the subsequent visit. Trees 1 and 2 were occupied by other frugivores before the group's arrival. a, b, c, and d represent the type of discovery visit: type (a) a feeding visit that succeeds a visit in which the group entered but did not eat; (b) a feeding visit in which the tree had grown new fruits during the observation period (50, 60, or 100 days); (c) a feeding visit in which the tree had grown new fruits during the observation period and that succeeds a visit in which the group entered but did not eat; and (d) the first time that feeding was observed in a tree after a period of at least 40 days in which the group did not come within 100 m of the same tree.

Olupot and colleagues (1998) found that mangabeys were more likely to travel in the direction of areas from which hornbills (*Bycanistes subcylindricus*) were calling earlier in the day compared to other areas. Of course, it is possible that the monkeys already knew that the targeted area contained fruit from previous visits, regardless of the hornbills' behavior. Apart from these and other anecdotes (e.g., Kinnaird & O'Brien, 2000), little systematic experimental research on the use of auditory cues in fruit finding has been conducted.

To address the issue, we conducted a series of playback experiments in which we played different animals' sounds from fig trees that either carried no fruit or only unripe, inedible fruits. We used calls produced by hornbills or chimpanzees, which were recorded while individuals were feeding inside fig trees (Fig. 4.11). As a control, we used the territorial calls of a local bird species, the yellow-rumped tinkerbird (*Pogoniulus bilineatus*). KJ carried out all experiments with the help of field assistants. For each trial the speaker was positioned at an elevation of at least 12 m within a fig tree.

Our observations suggested that the presence of chimpanzees did not stop the mangabeys from approaching fruit trees, despite the fact that chimpanzees are notorious monkey predators. During continuous observation periods totaling 210 days, we observed seven times that the study group was feeding in a fig tree when chimpanzees arrived. Three times the mangabeys left the tree after being chased by male chimpanzees, which were in groups of more than four. Four out of seven times, however, the monkeys continued feeding together with the chimpanzees, but these were usually single individuals or small groups. Twice, one of the mangabey males even chased a female chimpanzee out of the tree. On a further six occasions, the mangabeys encountered chimpanzees that were already feeding inside a fig tree. In all cases, the study group eventually entered the tree, five times within 100 minutes after waiting at the same spot, and one time only after 6 hours, after some additional traveling.

To investigate systematically whether these monkeys took the presence of other frugivorous species into account when trying to locate food

Figure 4.11 Chimpanzees often produce pant-hoot and rough grunts before feeding on *Ficus capensis* fruits. Picture by Karline Janmaat. Used with permission.

trees, we conducted two playback experiments. The first consisted of a small number of calls (hornbill and chimpanzee), played from inside fig trees, with variable fruiting states, at a distance of 50 m to 200 m from five different mangabey groups. The second type consisted of a large number of hornbill or tinkerbird calls, played from inside fig trees, regardless of whether we knew of the presence of any monkeys nearby.

After playing back hornbill vocalizations, the experimental tree was reliably approached by other hornbills, suggesting that the playback stimuli were effective. However, we never managed to attract any mangabeys in response to playback of chimpanyee or hornbill calls compared to tinkerbird control calls. When comparing latency and duration of looking towards the speaker, we did not find any differences between the hornbill calls and tinkerbird control calls. We also failed to detect any differences in these measures when comparing playbacks of chimpanzee feeding grunts, played from within a fig tree, with chimpanzee pant hoot vocalizations, played from the forest floor. Finally, the monkeys did not respond differently to chimpanzee feeding grunts played from empty trees or trees with potentially ripe fruits (Janmaat 2006). An alternative explanation is that the mangabeys had previously visited the experimental trees and already knew that the tree did not carry any edible fruits.

Monitoring Synchronicity

Most rainforest trees produce fruits synchronously with fruit production peaking some time of the year (Chapman et al., 1999b; van Schaik et al., 1993). In these species, finding fruit in one tree can be a reliable indicator for the presence of fruit in other trees of the same species, potentially allowing primates to make predictions about where to find fruit without having to remember the fruiting states of individual trees. Japanese macaques (*Macaca fuscata*), artificially provisioned with fruits of the *Akebia trifoliate* vines prior to fruiting season, were more likely to inspect other *Akebia trifoliate* vines than if they were provided with other food items (Menzel, 1991). Intriguingly, the monkeys manipulated both *Akebia trifoliata* and *Akebia quinata*vines, although the leaves and fruits of this vine species look very different. Both *Akebia* species fruit simultaneously, suggesting that the monkeys were not simply searching for the original source of the presented fruit, but used the discovery of a fruit as an indicator for the presence of fruit in vines of the same or other simultaneous fruiting vine species.

Compared to temperate zones of the Japanese woodlands, seasonality is much less pronounced in African rainforests (Walter, 1984; Worman & Chapman, 2005). In Kibale forest, Uganda, a majority of tree species fruit synchronously (64%), but the percentage of trees that carry fruit during fruiting peaks differs substantially between species, and within species between years and areas (Chapman et al., 1999b; Janmaat et al., submitted). For example, *Strombosia scheffleri* produced fruits only four times within 12 years (Chapman et al., 2004), with variable peaks from 5% to 50%. In May 1996, 60% of the *Uvariopsis congensis* population at the Kanyawara research site carried fruit, while none did at three other research sites, all within a 12-km distance (Chapman et al., 2004).

In a recent study, Janmaat and colleagues (submitted) investigated the foraging behavior of gray-cheeked mangabeys in relation to different levels of synchronicity in rainforest fruit species. Results showed that active searching was only triggered if the monkeys encountered high frequencies of trees with ripe fruits in the same area. Thresholds for switching to an "inspect-all" strategy appear to vary between different tree species, perhaps influenced by the nutrition value and productivity of the trees. Such a strategy is likely to be adaptive for the monkeys, because it allows them to flexibly respond to frequent and irregular fluctuations in fruit production and differences in nutritional value between species (Chapman et al., 1999b, 2003, 2004; Janson et al., 1986; Worman & Chapman, 2005). The results were also consistent with the findings that primates generally forage on a relatively small number of commonly distributed species per time period (Eckardt & Zuberbühler, 2004; Janson et al., 1986) and "trap-line" trees of species that have a high density and high fruit production (Janson et al., 1986; Milton, 2000; Terborgh & Stern, 1987). At a proximate level, it is possible that monkeys develop a "search image," originally proposed to explain the behavior of predators (Tinbergen, 1960). Identifying trees by their visual features is not a trivial task, as trees of different fruit species can resemble each other to a high degree (Janmaat, 2006; Fig. 4.12). During the entire study period the mangabeys fed on 28 different fruit tree species, suggesting that the monkeys must be able to retain a long-term memory of the specific visual characteristics of a large number of fruit tree species, which could require substantial processing power (Barton, 2000; Fagot & Cook, 2006).

CONCLUSIONS

Several decades ago, Eisenberg (1973) and Napier (1970) described what sets primate societies apart from those of other long-lived animals. The suggestion was that primates were equipped with brains able to store and retrieve a great deal of independently acquired information about the environment and able to apply considerable degrees of behavioral plasticity in responding to specific situations. In searching for food, primates could be more skilled than other groups of animals in their abilities to combine and integrate different types of information. For example, in order to use synchronicity in fruit emergence, monkeys need to keep track of the local density of several fruit species, recognize the visual characteristics of fruits and trees that show local abundance, and/ or remember the location of the trees or patches of trees that have started fruiting (Janmaat et al., submitted). There is currently no strong evidence that primates use olfactory cues or the sounds made by other animal species to locate fruit trees, but it is likely that such cues are integrated with spatial knowledge of likely food sources (Janmaat, 2006). Before revisiting particular trees, primates appear to combine weather conditions with

Figure 4.12 Illustration of the similarity in appearance of *U. congensis* and *T. nobilis* trees. Leaf shape, color, and configuration of *U. congensis* (top left) and *T. nobilis* (top right) and a close-up of the (yellow marked) trunks of both tree species (bottom). Despite the similarity, mangabeys did not enter more *T. nobilis* trees in *U. congensis* season than out of season, or in areas with higher ripe *U. congensis* fruit densities, suggesting that the monkeys use a memory of the visual characteristics of the trees or the locations of the tree patches of each species when searching for fruit (Janmaat 2006). Used with permission.

memories of previous fruiting states of trees, and they may even have some rudimentary understanding that high temperature and radiation accelerates fruit ripening (Janmaat et al., 2006b). One emerging point from the studies reviewed is that successful foraging depends on various cognitive skills, much beyond simply remembering the spatial location of a number of food trees throughout a home range. The degree to which nonprimate species possess comparable abilities is an important question, but unfortunately the answer is largely unknown.

How are these foraging abilities relevant to the more general question of why primates have relatively larger brains than other animals? Fieldwork in different parts of the world has shown that primates have been exceptionally successfully in monopolizing the arboreal space of most tropical forests, much more so than other groups of animals. Moreover, primates possess a number of morphological adaptations that make them especially well suited for arboreal foraging. The brain is an expensive organ, and primates' reliable access to the highest-quality nutrition available in this habitat may have enabled them to afford unusually large brains.

Fruit-eating bats and birds also feed on arboreal fruit, but these competitors are perhaps more constrained by their specialized locomotor apparatus, which also prevents them from manipulating and harvesting difficult-to-open fruits (Isler & van Schaik 2007b; Ross, 1996). Parrots, often noted for their exceptional cognitive abilities, are somewhat of an exception as their tongue, feet, and toes are highly mobile, allowing them to manipulate and discard lower-quality portions of food items (Milton 2001). Isler and van Schaik (2007b) report a negative relationship between brain size and the relative mass of pectoral muscle in birds, which are crucial for taking in air. It is interesting that some forest birds appear to minimize these costs by climbing up trees with their hooked claws (Hoatzin bird; *Opisthocomus hoazin*). Primates are less constrained in these ways, which may have allowed them to evolve larger bodies and brains while accessing the most nutritious foods, including young leaves, fleshy fruits, and the arthropods associated with them (Kay, 1984; Martin, 1990). As a result, primates have been able to evolve more complex behavioral strategies and mental capacities when dealing with both environmental and social problems. Various studies reviewed in this chapter indicate that primates engage in a number of complex cognitive foraging strategies, which gives them an advantage in competing over food with other species, reinforcing their chosen strategy to invest in brain size. Although not very popular at the moment, the ecological intelligence hypothesis appears to be more parsimonious than its rivals in the evolutionary scenario it presupposes: Overall, primates have been more successful in exploiting sustained high-quality nutrition from their habitats compared to their competitors, which has allowed them to evolve an unusually large brain. The sophistication seen in primates' social behavior as well as other aspects of their cognitive sophistication may be a by-product of their highly encephalized neural system, afforded by their special adaptations to the ecological conditions of the forest habitat.

ACKNOWLEDGMENTS

KJ's fieldwork was funded by the following organizations: the Wenner-Gren Foundation, the Leakey Foundation, the University of St Andrews' School of Psychology, Primate Conservation International Inc., Schure-Bijerinck-Popping fonds of the KNAW, Stichting Kronendak, Dobberke Stichting voor Vergelijkende Psychology, Lucie Burger Stichting, and Fonds Doctor Catharine van Tussenbroek. The Tai Monkey Project obtained core funding by the ESF's "Origins of Man, Language and Languages" program, as well as the EC's FP6 "What it means to be human." In Ivory Coast, we thank the Ministère de la Recherche Scientifique and the Ministère de l'Agriculture et des Ressources Animales of Côte d'Ivoire, the Centre Suisse de Recherches Scientifiques (CSRS), PACPNT, Eaux et Forêt, and the Tai Monkey Project (R. Noë in particular) for logistic support and permission to conduct research in the Tai National Park. In Uganda, we thank the Office of the President, Uganda National Council for Science and Technology, Uganda Wildlife Authority, Makerere University Biological Field Station, and Kibale Fish and Monkey Project (C. A. Chapman in particular) for logistic support and permission to conduct research in the Kibale National Park. For invaluable assistance in the field we are indebted to R. Meijer, D. J. van der Post, L. B. Prevot, D. C. M. Wright, R. Samuel, J. Rusoke, and P. Irumba. We thank Gabriel Ramos-Fernandez, Malcolm MacLeod, and Dick Byrne for their comments and suggestions.

REFERENCES

Adams, S. R., Cockshull, K. E., & Cave, C. R. J. (2001). Effect of temperature on growth and development of tomato fruits. *Annals of Botany, 88,* 869–877.

Aiello, L. C., & Wheeler, P. (1995). The expensive tissue hypothesis. *Current Anthropology, 36,* 184–193.

Armstrong, E. (1983). Relative brain size and metabolism in mammals. *Science, 220,* 1302–1304.

Barton, R. A., Purvis, A., & Harvey, P. H. (1995). Evolutionary radiation of visual and olfactory brain systems in primates, bats and insectivores. *Philosophical Transactions of the Royal Society B, 348,* 381–392.

Barton, R. A. (1996). Neo-cortex size and behavioral ecology in primates. *Proceedings of the Royal Society of London, Series B,* 263, 173–177.

Barton, R. A. (1999). The evolutionary ecology of the primate brain. In: P. C. Lee (Ed.), *Comparative primate socioecology* (pp. 167–194). Cambridge: Cambridge University Press.

Barton, R. A. (2000). Primate brain evolution: Cognitive demands of foraging or of social life? In: S. Boinski & P. A. Garber (Eds.), *On the move. How and why animals travel in groups* (pp. 204–238). Chicago, London: University of Chicago Press.

Barton, R. A. (2004). Binocularity and brain evolution in primates. *Proceedings of the National Academy of Sciences of the United States of America, 101,* 10113–10115.

Bates, L. (2005). Cognitive aspects of travel and food location by chimpanzees (Pan troglodytes schweinfurthii) of the Budongo Forest Reserve, Uganda. Ph.D. thesis. University of St. Andrews, St. Andrews.

Bennett, A. T. C. (1996). Do animals have cognitive maps? *Journal of Experimental Biology, 199,* 219–224.

Bicca-Marques, J. C., & Garber, P. A. (2004). Use of spatial, visual, and olfactory information during foraging in wild nocturnal and diurnal anthropoids: A field experiment comparing Aotus, Callicebus, and Saguinus. *American Journal of Primatology, 62,* 171–187.

Boesch, C., & Boesch, H. (1984). Mental map in wild chimpanzees: An analysis of hammer transports for nut cracking. *Primates, 25,* 160–170.

Byrne, R. W. (2000). How monkeys find their way. Leadership, coordination, and cognitive maps of African baboons. In: S. Boinski & P. A. Garber (Eds.), *On the move: How and why animals travel in groups* (pp. 491–518). Chicago: University of Chicago Press.

Byrne, R. W., & Corp, N. (2004). Neocortex size predicts deception in primates. *Proceedings of the Royal Society B, 271,* 1693–1699.

Byrne, R. W. (1979). Memory for urban geography. *Quarterly Journal of Experimental Psychology, 31,* 147–154.

Byrne, R. W. (1995). *The thinking ape. Evolutionary origins of intelligence.* Oxford: Oxford University Press.

Caine, N. G., Addington, R. L., & Windfelder, T. L. (1995). Factors affecting the rates of food calls given by red-bellied tamarins. *Animal Behaviour, 50,* 53–60.

Cartmill, M. (1972). Arboreal adaptations and the origin of the order Primates. In: Tuttle R. H. (Ed.), *The functional and evolutionary biology of primates* (pp. 97–122). Chicago: Aldine-Atherton.

Chapman, C. A., Chapman, L. J., Struhsaker, T. T., Zanne, A. E., Clark, C. J., & Poulsen, J. R. (2004). A long-term evaluation of fruit phenology: Importance of climate change. *Journal of Tropical Ecology, 21,* 1–14.

Chapman, C. A., Chapman, L. J., Rode, K. D., Hauck, E. M., & McDowell, L. R. (2003). Variation in the nutritional value of primate foods: Among trees, time periods, and areas. *International Journal of Primatology, 24,* 317–333.

Chapman, C. A., Gautier-Hion, A., Oates, J. F., & Onderdonk, D. A. (1999a). African primate communities: Determinant of structure and threats to survival. In: J. G. Fleagle, C. H. Janson, & K. E. Reed (Eds.), *Primate communities* (pp. 1–37). Cambridge: Cambridge University Press.

Chapman, C. A., Wrangham, R. W., Chapman, L. J., Kennard, D. K., & Zanne, A. E. (1999b). Fruit and flower phenology at two sites in Kibale National Park, Uganda. *Journal of Tropical Ecology, 15,* 189–211.

Chauvin, C., & Thierry, B. (2005). Tonkean macaques orient their food search from olfactory cues conveyed by group mates. *Ethology, 111,* 301–310.

Clutton-Brock, T. H., & Harvey, P. H. (1980). Primate brains and ecology. *Journal of Zoology London, 190,* 309–323.

Cunningham, E., & Janson, C. (2007). A socioecological perspective on primate cognition, past and present. *Animal Cognition, 10,* 273–281.

Cunningham, E. P. (2003). The use of memory in Pithecia Pithecia's foraging strategy. Ph.D. thesis. The City University of New York, New York.

Di Fiore, A., & Suarez, S. A. (2007). Route-based travel and shared routes in sympatric spider and woolly monkeys: Cognitive and evolutionary implications. *Animal Cognition, 10,* 317–329.

Diaz-Perez, J. C., Bautista, S., Villanueva, R., & Lopez-Gomez, R. (2002). Modelling the ripening of sapote mamey (Pouteria sapota (Jacq.) H. E. Moore and Stearn) fruit at various temperatures. *Postharvest Biology and Technology, 28,* 199–102.

Dittus, W. P. J. (1984). Toque macaque food calls: semantic communication concerning food distribution in the environment. *Animal Behaviour, 32,* 470–477.

Dufour, V., Pele, M., Sterck, E. H. M., & Thierry, B. (2007). Chimpanzee anticipation of food return: coping with waiting time in an exchange task. *Journal of Comprehensive Psychology, 121*(2), 145–155.

Dunbar, R. I. M. (1998). The social brain hypothesis. *Evolutionary Anthropology, 6,* 178–190.

Dunbar, R. I. M., & Shultz, S. (2007). Evolution of the social brain. *Science, 317,* 1344–1347.

Eckardt, W., & Zuberbühler, K. (2004). Cooperation and competition in two forest monkeys. *Behavioral Ecology, 15,* 400–411.

Eisenberg, J. F. (1973). Mammalian social systems: Are primate social systems unique? In: E. W. Menzel, Jr. (Ed.), *Precultural primate behavior* (pp. 232–249). Basel, Switzerland: S. Karger.

Eriksson, O., Friis, E. M., & Löfgren, P. (2000). Seed size, fruit size and dispersal systems in angiosperms form the early Cretaceous to the late Tertiary. *American Natute, 156,* 47–58.

Fagot, J., & Cook, R. G. (2006). Evidence for large long-term memory capacities in baboons and pigeons and its implications for learning and the evolution of cognition. *Proceedings of the National Academy of Sciences, 103,* 17564–17567.

Fish, J. L., & Lockwood, C. A. (2003). Dietary constraints on encephalization in primates. *American Journal of Physical Anthropology, 120,* 171–181.

Fleagle, J. G., & Reed, K. E. (1996). Comparing primate communities: A multivariate approach. *Journal of Human Evolution, 30,* 489–510.

Garber, P. A., & Hannon, B. (1993). Modeling monkeys: A comparison of computer-generated and naturally occurring foraging patterns in two species of neotropical primates. *International Journal of Primatology, 14,* 827–852.

Garber, P. A., & Paciulli, L. M. (1997). Experimental field study of spatial memory and learning in wild capuchin monkeys (Cebus capucinus). *Folia Primatologica, 68*(3–5), 236–253.

Garber, P. A. (1989). Role of spatial memory in primate foraging patterns Saguinus mystax and Saguinus fuscicollis. *American Journal of Primatology, 19,* 203–216.

Garber, P. A., & Boinski, S. (2000). *On the move. How and why animals travel in groups.* Chicago, London: University of Chicago Press.

Gebo, D. L. (2004). A shrew-size origin for primates. *Yearbook of Physical Anthropology, 47,* 40–62.

Golla, H., Ignashchenkova, A., Haarmeier, T., & Thrier P. (2004). Improvement of visual acuity by spatial cueing: a comparative study in human and non-human primates. *Vision Research, 44,* 1589–1600.

Graham, E. A., Mulkey, S. S., Kitajima, K., Philips N. G., & Wright, S. J. (2003). Cloud cover limits net CO_2 uptake and growth of a rainforest tree during tropical rainy seasons. *Proceedings of the National Academy of Sciences, 100,* 572–576.

Harvey, P. H., & Krebs, J. R. (1990). Comparing brains. *Science, 249,* 140–146.

Hauser, M. D., & Wrangham, R. W. (1990). Recognition of predator and competitor calls in nonhuman primates and birds: A preliminary report. *Ethology, 86,* 116–130.

Hladik, C. M. (1967). Surface relative du tractus digestif de quelques primates: Morphologies villosites intestinales et correlations avec le regime alimentaire. *Mammalia, 31,* 120–147.

Houle, A., Vickery, W. L., & Chapman, C. A. (2006). Mechanisms of coexistence among two species of frugivorous primates. *Journal of Animal Ecology, 75,* 1034–1044.

Houle, A. (2004). Mécanismes de coexistence chez les primates frugivores du Parc National de Kibale en Ouganda. Ph.D. thesis. University of Quebec, Quebec.

Isler, K., & van Schaik, C. P. (2006a). Metabolic costs of brain size evolution. *Biology Letters, 2,* 557–560.

Isler, K., & van Schaik, C. P. (2006b). Costs of encephalization: The energy trade-off hypothesis tested on birds. *Journal of Human Evolution, 51*(3), 228–243.

Janmaat, K. R. L. (2006). Fruits of enlightenment. Fruit localization strategies in wild mangabey monkeys. Ph.D. thesis, University of St. Andrews, St. Andrews.

Janmaat, K. R. L., Byrne, R. W., & Zuberbühler, K. (2006a). Evidence for spatial memory of fruiting states of rain forest fruit in wild ranging mangabeys. *Animal Behavior, 71,* 797–807.

Janmaat, K. R. L., Byrne, R. W., & Zuberbühler, K. (2006b). Primates take weather into account when searching for fruit. *Current Biology, 16,* 1232–1237.

Janmaat, K. R. L., Byrne, R. W., & Zuberbühler, K. (Submitted). The use of fruiting synchronicity by foraging Mangabeys (Lophocebus albigena johnstonii). *Animal Cognition.*

Janson, C., & Byrne, R. W. (2007). Resource cognition in wild primates – opening up the black box. *Animal Cognition, 10,* 357–367.

Janson, C. (2007). Experimental evidence for route integration and strategic planning in wild capuchin monkeys. *Animal Cognition, 10,* 341–356.

Janson, C. H., & Di Bitetti, M. S. (1997). Experimental analysis of food detection in capuchin monkeys: Effects of distance, travel speed, and resource size. *Behavioral Ecology and Sociobiology, 41,* 17–24.

Janson, C. H. (1998). Experimental evidence for spatial memory in foraging wild capuchin monkeys Cebus apella. *Animal Behaviour, 55,* 1229–1243.

Janson, C. H. (2000). Spatial movement strategies: Theory, evidence and challenges. In: S. Boinski & P. A. Garber (Eds.), *On the move. How and why animals travel in groups* (pp. 165–204). Chicago, London: University of Chicago Press.

Janson, C. H., Stiles, E. W., & White, D. W. (1986). Selection on plant fruiting traits by brown capuchin monkeys: a multivariate approach. In: A. Estrada & T. Fleming (Eds.), *Frugivores and seed dispersal* (pp. 83–92). The Hague: Junk Publishers.

Jerison, H. (1973). *Evolution of the brain and intelligence.* New York: Academy Press.

Kay, R. F. (1984). On the use of anatomical features to infer foraging behavior in extinct primates. In: P. A. Rodman & J. G. H. Cant (Eds.), *Adaptations for foraging in nonhuman primates* (pp. 21–53). New York: Columbia University Press.

Kinnaird, M. F., & O'Brien, T. G. (2000). The Tana River Crested mangabey and the sulawesi crested black macaque. Comparative movement patterns of two semi terrestrial cercopithecine primates. In: S. Boinski & P. A. Garber (Eds.), *On the move. How and why animals travel in groups* (pp. 327–350). Chicago, London: University of Chicago Press.

Korstjens, A. H. (2001). The mob, the secret sorority, and the phantoms: An analysis of the socio-ecological strategies of the three colobines of Taï. In: *Behavioural Biology Group*: Utrecht.

Martin, R. D. (1990). *Primate origins and evolution: A phylogenetic reconstruction.* London: Chapman and Hall.

Mazzei, K. C., Newman, R. M., Loos, A., & Ragsdale, D. W. (1999). Developmental rates of the native milfoil weevil, euhrychiopsis lecontei and damage to Eurasian water milfoil at constant temperature. *Biological Control, 16,* 139–143.

McGraw, S. W., Zuberbühler, K. & Noë, R. (2007). Monkeys of the Taï Forest: An African Monkey Community. In: *Cambridge Studies in Biological and Evolutionary Anthropology (No. 51).* Cambridge: Cambridge Univ. Press.

Menzel, C. R. (1991). Cognitive aspects of foraging in Japanese monkeys. *Animal Behaviour, 41,* 397–402.

Menzel, C. R., Savage-Rumbaugh, E. S., & Menzel, E. W., Jr. (2002). Bonobo (Pan paniscus) spatial memory and communication in a 20-hectare forest. *Journal of International Journal of Primatology, 23,* 601–619.

Milton, K. (1981). Distribution pattern of tropical plant foods as an evolutionary stimulus to primate mental development. *American Anthropologist, 83,* 534–548.

Milton, K. (1988). Foraging behaviour and the evolution of primate intelligence. In: R.W. Byrne & A. Whiten (Eds.), *Machiavellian intelligence: Social expertise and the evolution of intellect in monkeys, apes and humans* (pp. 285–305). Oxford: Clarendon Press.

Milton, K. (2000). Quo Vadis? Tactics of food search and group movement in primates and other animals. In: S. Boinski & P. A. Garber (Eds.), *On the move. How and why animals travel in groups* (pp. 375–417). Chicago, London: University of Chicago Press.

Mink, J. W., Blumenschine, R. J., & Adams, D. B. (1981). Ratio of central nervous system to body metabolism in vertebrates – its constancy and functional basis. *American Journal of Physiology, 241,* 203–212.

Morrison, J. F., & Noble, A. C. (1990). The effects of leaf and cluster shading on the composition of cabernet sauvignon grapes and on fruit and wine sensory properties. *American Journal of Enology and Viticulture, 41*(3), 193–199.

Müller, A. E., & Soligo, C. (2005). Primate sociality in evolutionary context. *American Journal of Physical Anthropology, 128,* 399–441.

Napier, J. (1970). *The roots of mankind.* Washington, D. C.: Smithsonian Press.

Noser, R., & Byrne, R. W. (2007a). Investigating the mental maps of chacma baboons (Papio ursinus), using intergroup encounters. *Animal Cognition, 10,* 331–340.

Noser, R., & Byrne, R. W. (2007b). Travel routes and planning of visits to out-of-sight resources in wild chacma baboons (Papio ursinus). *Animal Behaviour, 73,* 257–266.

O'Keefe, J., & Nadel, L. (1978). *The hippocampus as a cognitive map.* Oxford: Oxford University Press.

Olupot W., Waser, P. N., & Chapman, C. A. (1998). Fruit finding by mangabeys (Lophocebus albigena): Are monitoring of fig trees and use of sympatric frugivore calls possible strategies? *International Journal of Primatology, 19*(2), 339–353.

Polyak, S. L. (1957). *The vertebrate visual system.* Chicago: Chicago University Press.

Potts, R. (2004). Paleo environmental basis of cognitive evolution in great apes. *American Journal of Primatology, 62,* 209–228.

Ramseyer, A., Pelé, M., Dufour, V., & Chauvin, C. (2005). Accepting loss: The temporal limits of reciprocity in brown capuchin monkeys. *Proceedings of the Royal Society B, 273,* 179–184.

Riba-Hernández, P., Stoner, K. E., & Lucas, P. W. (2005). Sugar concentration of fruits and their detection via color in the Central American spider monkey (Ateles geoffroyi). *American Journal of Primatology, 67,* 411–123.

Ross, C. (1996). Adaptive explanation for the origins of the Anthropoidea (primates). *American Journal of Primatology, 40,* 205–230.

Sigg, J., & Stolba, A. (1981). Home range and daily march in a hamadryas baboon troop. *Folia Primatologica, 36,* 40–75.

Smuts, B. B., Cheney, D. L., Seyfarth, R. M., Wrangham, R. W., & Struhsaker, T. T. (1986). *Primate societies.* Chicago: University of Chicago Press.

Soligo, C., & Martin, R. D. (2006). Adaptive origins of primates revisited. *Journal of Human Evolution, 50,* 414–430.

Spayd, S. E., Tarara, J. M., Mee, D. L., & Ferguson, J. C. (2002). Separation of sunlight and temperature effects on the composition of vitis vinifera cv. merlot berries. *American Journal of Enology and Viticulture, 53*(3), 171–182.

Sterck, E. H. M. (1995). Females, food and fights. A socioecological comparison of the sympatric Thomas langur and long-tailed macaque. Ph.D. thesis. Utrecht University, Utrecht.

Stevens, J., Rosati, A., Ross, K., & Hauser, M. (2005). Will travel for food: Spatial discounting in New World monkeys. *Current Biology, 15*(20), 1855–1860.

Sumner, P., & Mollon, J. D. (2000). Chromaticity as a signal of ripeness in fruits taken by primates. *Journal of Experimental Biology, 203,* 1987–2000.

Sussman, R. W. (1991). Primate origins and the evolution of angiosperms. *American Journal of Primatology, 23,* 209–223.

Sussman, R. W. (2004). Flowering plants and the origins of primates: a new theory. In: M. Bekoff (Ed.), *Encyclopedia of animal behavior* (pp. 967–969). Portsmouth: Greenwood Press.

Tavaré, S., Marshall, C. R., Will, O., Soligo, C., & Martin, R. D. (2002). Using the fossil record to estimate the age of the last common ancestor of extant primates. *Nature, 416,* 726–729.

Taylor, C. R., Caldwell, S. L., & Rowntree, V. J. (1972). Running up and down hills: Some consequences of size. *Science, 178,* 1096–1097.

Terborgh, J., & Stern, M. (1987). The surreptitious life of the saddle-backed tamarin. *American Scientist, 75,* 260–269.

Tinbergen, N. (1960). The natural control of insects in pine woods: Vol. I. Factors influencing the intensity of predation by songbirds. *Archives Neelandaises de Zoologie, 13,* 265–343.

Tinbergen, N. (1963). On the aims of methods of ethology. *Zeitschrift fur Tierpsychologie, 20,* 410–433.

Tolman, E. C. (1948). Cognitive maps in rats and men. *Psychological Review, 55,* 189–208.

Valero, A., & Byrne R. W. (2007). Spider monkey ranging patterns in Mexican subtropical forest: Do travel routes reflect planning? *Animal Cognition, 10,* 305–315.

Van Roosmalen, M. G. M. (1988). Diet, feeding behaviour and social organization of the Guianan black spider monkey (Ateles paniscus paniscus). 12th Congress of the International Primatological Society, July 24–29, Brasilia, Brazil.

Van Schaik, C. P., Terborgh, J. W., & Wright, S. J. (1993). The phenology of tropical forests: Adaptive significance and consequences for primary consumers. *Annual Review of the Ecological System, 24,* 353–377.

Walter, H. (1984) Zonobiome of the equatorial humid diurnal climate with evergreen tropical rain forest. In: H. Walter (Ed.), *Vegetation of the earth and ecological systems of the geo biosphere* (pp. 46–71). Berlin: Springer-Verlag.

Waser, P. M. (1974). Intergroup interacton in a forest monkey: The mangabey Cercocebus albigena. Ph.D. thesis. The Rockefeller University, New York.

Worman, C., & Chapman, C. A. (2005). Seasonal variation in the quality of a tropical ripe fruit and the response of three frugivores. *Journal of Tropical Ecology, 21,* 689–697.

Wrangham, R. W. (1977). Feeding behaviour of chimpanzees in Gombe National Park, Tanzania. In: T. H. Clutton-Brock (Ed.), *Primate ecology: Studies of feeding and ranging behaviour in lemurs, monkeys and apes* (pp. 503–556). London: Academic Press.

CHAPTER 5

Primate Vocal Communication

Robert M. Seyfarth and Dorothy L. Cheney

In 1871, Charles Darwin drew attention to a dichotomy in the vocal communication of animals that had perplexed philosophers and naturalists for at least 2,000 years. In marked contrast to human language, he wrote, animal vocalizations appeared to be involuntary expressions of emotion and movement: "When the sensorium is strongly excited, the muscles of the body are generally thrown into violent action; and as a consequence, loud sounds are uttered,... although the sounds may be of no use" (1871/1981, p. 83). However, two pages later Darwin wrote: "That which distinguishes man from the lower animals is not the understanding of articulate sounds, for, as every one knows, dogs understand many words and sentences.... Nor is it the mere capacity of connecting definite sounds with definite ideas; for it is certain that some parrots, which have been taught to speak, connect unerringly words with things, and persons with events" (1871/1981, p. 85).

The vocal communication of monkeys and apes appears to be no different from that of other animals. Production is highly constrained. Nonhuman primates have a relatively small repertoire of calls, each of which is closely linked to particular social circumstances and shows little modification during development (see Hammerschmidt & Fischer, 2008, for a review). Perception, by contrast, is more flexible, open ended, and modifiable as a result of experience. The difference between production and perception is puzzling because producers are also perceivers: Why should an individual who can deduce an almost limitless number of meanings from the calls of others be able to produce only a limited number of calls of his or her own? The difference between production and perception is also puzzling because it constitutes a crucial distinction between human and nonhuman primates. Why should monkeys and apes—so similar to humans in so many other respects—be so different when it comes to vocal production?

The contrast between vocal production and perception constitutes the starting point for this chapter. We wish to make three points. First, it is important to be clear exactly what we mean when we say that primate vocal production is "sharply constrained." Many scientists have taken this to mean that primate call production is fixed, uncontrollable, and involuntary. This conclusion is too extreme. In fact, both field and laboratory studies paint a more complex picture. Monkeys and apes can call or remain silent, modify the timing and duration of calling, and make subtle acoustic modifications to the calls they give in specific social contexts. However, although they can modify call production in many ways, they rarely create entirely new calls or call combinations, or sever the link between a particular call type and the circumstances in which it is normally given.

Second, the dichotomy between production and perception has important theoretical implications, because it draws our attention to the very different mechanisms that underlie the behavior of speakers and listeners, even when these individuals are involved in the same communicative event. Nonhuman primates present us with a communicative system in which a small repertoire of relatively fixed, inflexible calls, each linked to a particular social context, nonetheless gives rise to an open-ended, highly modifiable, and cognitively rich set of meanings.

Third, the contrast between production and perception in primates and many other mammals cries out for an evolutionary explanation. What selective pressures caused our human ancestors—and they alone among the primates—to develop flexible vocal production? Unconstrained by any definitive data that might help resolve the issue, we offer some speculations.

VOCAL PERCEPTION

As Darwin noted, primates—like other mammals—produce a small repertoire of acoustically fixed, species-specific calls that are closely tied to particular contexts and show little modification during development. By contrast, when it comes to perception and comprehension, primates and other animals display an almost open-ended ability to learn new sound-meaning pairs. They also appear to ascribe intentions and motives to signalers when assessing whether or not to respond to a given individual's calls. Consider baboons, for example.

Baboons are Old World monkeys that shared a common ancestor with humans roughly 30 million years ago (Steiper et al., 2004). They live throughout the savannah woodlands of Africa in groups of 50 to 150 individuals. Although most males emigrate to other groups as young adults, females remain in their natal groups throughout their lives, maintaining close social bonds with their matrilineal kin (Silk et al., 1999, 2006a,b). Females can be ranked in a stable, linear dominance hierarchy that determines priority of access to resources, and daughters acquire ranks similar to those of their mothers. Baboon social structure can therefore be described as a hierarchy of matrilines, in which all members of one matriline (e.g., matriline B) outrank or are outranked by all members of another (e.g., matrilines C and A, respectively). Ranks within matrilines are as stable as those between matrilines (e.g., A1 > A2 > A3 > B1 > B2 > C1, etc.) (Cheney & Seyfarth, 2007).

Listeners Extract Rich "Narratives" from Simple Call Sequences

Baboon vocalizations, like those of many other primates, are individually distinctive (e.g., Owren et al., 1997; Rendall, 2003), and listeners recognize the voices of others (reviewed in Cheney & Seyfarth, 2007). Baboons' vocal repertoire contains a number of acoustically graded signals, each of which is relatively context specific. The alarm "wahoos" given by adult males to predators, for example, are acoustically similar to the wahoos that males give during aggressive contests (Fischer et al., 2002). Nonetheless, listeners respond to the two call types as if they convey qualitatively different information (Kitchen et al., 2003).

Grunts, the most common call given by baboons, are given in a variety of social interactions and also differ acoustically according to context. *Move* grunts are given in bouts of one or two calls while the group is on a move or when one or more individuals attempt to initiate a group move, and they often elicit answering grunts from nearby listeners. Slightly acoustically different *infant* grunts are given in a variety of affiliative contexts and function to facilitate social interactions (Cheney et al., 1995; Rendall et al., 1999). If a high-ranking female grunts as she approaches a lower-ranking female, the lower-ranking female is less likely to move away than if the approaching female remains silent. Grunts also function to reconcile opponents after a dispute, increasing the likelihood that former opponents will tolerate each other's close proximity and reducing the probability of renewed aggression (Cheney & Seyfarth, 1997; Cheney et al., 1995).

Because calls are individually distinctive and each call type is predictably linked to a particular social context, baboon listeners can potentially acquire quite specific information from the calls that they hear. This applies not only to calls of a single type, like predator alarm calls (Fischer et al., 2000, 2001a,b), but also to the sequences of different call types that arise when two or more individuals are interacting with each other.

Throughout the day, baboons hear other group members giving vocalizations to each other. Some interactions involve aggressive competition, for example, when a higher-ranking animal gives a series of threat-grunts to a lower-ranking animal and the latter screams. Threat-grunts are aggressive vocalizations given by higher-ranking to lower-ranking individuals, whereas screams are submissive signals, given primarily by lower- to higher-ranking individuals. A threat-grunt-scream sequence, therefore, provides information not only about the identities of the opponents involved but also about who is threatening whom. Baboons are very sensitive to both types of information. In playback experiments, listeners respond with apparent surprise to sequences of calls that appear to violate the existing dominance hierarchy. Whereas they show little response upon hearing the sequence "B2 threat-grunts and C3 screams," they respond strongly—by looking toward the source of the call—when they hear "C3 threat-grunts and B2 screams." Between-family rank reversals (C3 threat-grunts and B2 screams) elicit a stronger violation of expectation response than do within-family rank reversals (C3 threat-grunts and C1 screams) (Bergman et al., 2003).

A baboon who ignores the sequence "B2 threat-grunts and C3 screams" but responds strongly when she hears "C3 threat-grunts and B2 screams" reveals, by his or her responses, that he or she recognizes the identities of both participants, their relative ranks, and their family membership. He or she also acts as if he or she assumes that the threat-grunt and scream have occurred together not by chance, but because one vocalization caused the other to occur. Without this assumption of causality there would be no violation of expectation when B2's scream and C3's threat-grunt occurred together.

Baboons' ability to deduce a social narrative from a sequence of sounds reveals a rich cognitive system in which listeners extract a large number of complex, nuanced messages from a relatively small, finite number of signals. A baboon who understands that "B2 threat-grunts and C3 screams" is different from "C3 threat-grunts and B2 screams" can make the same judgment for all possible pairs of group members as well as any new individuals who may join (Cheney & Seyfarth, 2007, Chapters 10 and 11).

Underlying the baboons' sophisticated social cognition is an almost open-ended ability to learn new sound-meaning pairs. This open-ended learning is found in many nonhuman primates, as well as other animals. Baboons and other primates learn to recognize the voices of new individuals as they are born or join the group from elsewhere, just as they learn to distinguish their own species' alarm calls (Fischer et al., 2000; Seyfarth & Cheney, 1986; Zuberbühler, 2000) and the different alarm calls of sympatric birds and mammals (Hauser, 1988; Hauser & Wrangham, 1990; Seyfarth & Cheney, 1990; Zuberbühler, 2001). Primates in laboratories readily learn to recognize the voices of their different caretakers and to associate different sounds, like the rattling of keys or the beep of a card swipe, with impending events that may be good (feeding) or bad (the visit of a veterinarian). In cross-fostering experiments, infant rhesus (*Macaca mulatta*) and Japanese macaques (*M. fuscata*) raised among the members of another species learned to recognize their foster mothers' calls—and the foster mothers learned to recognize theirs—even in contexts in which the two species used acoustically different vocalizations (Seyfarth & Cheney, 1997). Taken together, these results suggest that nonhuman primates are both innately predisposed to ascribe meaning to different sounds and always ready to learn new information from novel auditory stimuli.

These generalizations apply with equal force to other mammals. Consider Rico, for example, a border collie who learned the names of more than 200 different toys (Kaminski et al., 2004). Rico was able to learn and remember the names of new toys by process of exclusion, or "fast mapping," and—like small children—used gaze and attention to guide word learning. But of course Rico never learned to *say* any of the words he learned. In this respect, his vocal perception and production were similar to those of language-trained apes (Savage-Rumbaugh, 1986; Terrace, 1979), sea lions (Schusterman et al., 2002), and dolphins (Herman et al., 1993).

Listeners Ascribe Intentions to Signalers

In addition to making judgments based on social causation, baboons appear to recognize other individuals' intentions and motives. Baboon groups are noisy, tumultuous societies, and baboons would not be able to feed, rest, or engage in social interactions if they responded to every call as if it were directed at them. In fact, baboons appear to use a variety of behavioral cues, including gaze direction, learned contingencies, and the memory of recent interactions with specific individuals when making inferences about the target of a vocalization. For example, when a female hears a recent opponent's threat-grunts soon after fighting with her, she responds as if she assumes that the threat-grunt is directed at her, and she avoids the signaler. However, when she hears the same female's threat-grunts soon after grooming with her, she ignores the calls and acts as if the calls are directed at someone else (Engh et al., 2006). Conversely, when a female hears her opponent's friendly *infant* grunt soon after a fight, she acts as if she assumes that the call is directed at her and is intended as a reconciliatory gesture. She approaches her recent opponent and tolerates her opponent's approaches at a rate that is even higher than baseline rates (Cheney & Seyfarth, 1997). In contrast, hearing the grunt of an uninvolved dominant female unrelated to her opponent has no effect on the female's behavior. In this latter case, she acts as if she is not the intended target of the call and treats the call as irrelevant.

In some cases, inferences about the intended target of a call seem to involve rather complex and indirect causal reasoning about, among other things, the kinship bonds that exist among others. Playback experiments, for example, have shown that baboons will accept the "reconciliatory" grunt by a close relative of a recent opponent as a proxy for direct reconciliation by the opponent herself (Wittig et al., 2007). To do so, the listeners must be able to recognize that a grunt from a particular female is causally related to a previous fight even though she has not interacted recently with the signaler, but with the signaler's relative.

There are intriguing parallels between these results and recent neurophysiological research. In primates, faces and vocalizations are the primary means of transmitting social signals, and monkeys recognize the correspondence between facial and vocal expressions (Ghazanfar & Logothetis, 2003). When rhesus macaques hear another monkey's calls, they exhibit neural activity not only in areas associated with auditory processing but also in higher-order visual areas (Gil da Costa et al., 2004). Ghazanfar and colleagues explored the neural basis of sensory integration using the coos and grunts of rhesus macaques and found that cells in the auditory cortex were more responsive to bimodal (visual and auditory) presentation of these calls than to unimodal presentation (Ghazanfar et al., 2005; see Romanski & Ghazanfar, this volume). Intriguingly, the effect of cross-modal presentation was greater with grunts than with coos. The authors speculate that this may have occurred because grunts are usually directed toward a specific individual, whereas coos are more often broadcast to the group at large. The greater cross-modal integration in the processing of grunts may arise because, in contrast to a coo, listeners who hear a grunt must immediately determine whether or not the call is directed at them.

PRIMATE VOCAL PRODUCTION

Monkeys and apes have a relatively small repertoire of context-specific calls that show relatively little modification in their acoustic properties during development (Janik & Slater, 1997; McComb & Semple, 2005; Seyfarth & Cheney, 1997). Cross-fostering experiments with macaques suggest that the link between particular call types and the contexts in which they are given is difficult to break. For example, normally raised rhesus and Japanese macaques differ in their use of calls in several social contexts: Rhesus macaques use a mixture of coos and grunts, whereas Japanese macaques use coos almost exclusively. In a 2-year experiment involving four individuals who were raised in groups of the other species, infant rhesus and Japanese macaques adhered to their species-typical pattern of calling even though, in every other respect, they were fully integrated into their adopted social groups (Owren et al., 1993).

There is also little evidence that nonhuman primates adapt calls to different contexts or create new calls to deal with novel situations. And although they routinely hear different call combinations—combinations, like those described previously, whose meaning is more than the sum of their constituent elements—these combinations are created when two baboons are vocalizing to each other. With a few possible exceptions (see later), signalers never combine different vocalizations to create new messages. Thus, primate vocal repertoires are far from open ended. Production is very different from perception.

PRIMATES ARE TYPICAL OF MOST MAMMALS

In their highly constrained vocal production combined with flexible perception and cognition, nonhuman primates are typical of most mammals. Indeed, the ability to modify the acoustic features of calls depending on experience seems comparatively rare in the animal kingdom. As of 1997, when Janik and Slater published their review of the topic, vocal learning had been documented in only three orders of birds, cetaceans, harbor seals, and humans. True, we know much more about vocal communication in monkeys than in nonprimate mammals or even the great apes, and we may yet be surprised by novel evidence of vocal imitation (e.g., Poole et al., 2005) or creative call combinations (Arnold & Zuberbühler, 2006; Crockford & Boesch, 2003; Zuberbühler, 2002). For the moment, however, there is no reason to believe that mammals in general—including apes—differ from the baboons and other primates described previously. The question for primatologists then becomes: What selective forces gave rise to learned, flexible vocal production in our hominid ancestors? Below we offer a speculative answer to this question, but first we consider more closely the "fixed" nature of primate vocal production and explore the theoretical implications of communication between relatively constrained vocal producers and flexible, open-ended receivers.

VOCAL PRODUCTION, THOUGH CONSTRAINED, IS NOT FIXED AND INVOLUNTARY

Compared to the large, learned repertoires of many songbirds and the imitative skills of cetaceans and pinnipeds, the vocal repertoires of nonhuman primates are small (McComb & Semple, 2005[1]). Nonhuman primates also use their vocalizations in highly predictable social circumstances. These two observations, together with early neurophysiological studies showing that seemingly normal calls could be elicited by electrical stimulation of subcortical areas in the brain (e.g., Jurgens & Ploog, 1970; Ploog, 1981), have led many anthropologists (Washburn, 1982; Gardenfors, 2003), ethologists (Goodall, 1986), linguists (Bickerton, 1990), psychologists (Terrace, 1983), and neuroscientists (Arbib, 2005) to conclude that primate vocalizations are reflexive, involuntary signals—or, in Bickerton's words, "quite automatic and impossible to suppress" (1990:142). This characterization is misleading.

In both the field and the laboratory, nonhuman primates appear to be able to control whether they produce a vocalization or remain silent. Baboons, as already noted, may follow an aggressive interaction with a reconciliatory grunt or they may not, and like other primates they vocalize more to some individuals than to others (e.g., Smith et al., 1982). Even in highly emotional circumstances like encounters with predators, some individuals call at high rates, others call less often, and still others remain silent (Cheney & Seyfarth, 1990).

The "decision" to call or remain silent can have significant behavioral consequences. In experiments conducted on wild capuchin monkeys (*Cebus capucinus*) in Costa Rica,

Gros-Louis (2004) found that individuals who discovered food were more likely to give "food" calls when other group members were nearby than when they were alone. Furthermore, they were more likely to call if a higher-ranking, as opposed to a lower-ranking, bystander was nearby. Individuals who called when approached by a high-ranking animal were less likely to receive aggression than those who remained silent. Gros-Louis (2004) concluded that capuchin food calls function to announce both ownership and the signaler's willingness to defend his or her possession. As a result, unless they were strongly motivated to take the food, listeners refrained from harassing the signaler.

In more controlled laboratory settings, the timing, duration, and rate of calling by monkeys can be brought under operant control (Pierce, 1985). In a recent series of experiments, Egnor et al. (2007) exposed cotton-top tamarins (*Saguinus oedipus*) to intermittent bursts of white noise and found that subjects quickly learned to restrict their calling to the silent intervals. Clearly, then, primates can control whether they vocalize or not depending on variations in the ecological, social, and acoustic environments.

Within a given context, nonhuman primates can also make subtle modifications in the acoustic structure of their calls (reviewed by Hammerschmidt & Fischer, 2008). Wild chimpanzees (*Pan troglodytes*) in Uganda, for example, give long, elaborate pant-hoots either alone or in "choruses" with others. When two individuals have called together several times, the acoustic features of their pant-hoots begin to converge (Mitani & Brandt, 1994; Mitani & Gros-Louis, 1998). Apparently, they modify the acoustic structure of their calls depending on auditory experience. Crockford et al. (2004) tested this hypothesis on four communities of chimpanzees in the Tai Forest, Ivory Coast. They found that males in three contiguous communities had developed distinctive, community-specific pant-hoots, whereas males in a fourth community 70 km away showed only minor acoustic differences from males in the other three communities. By comparing the genetic relatedness between pairs of males with the acoustic similarity of their calls, Crockford and colleagues could rule out an explanation of call convergence based on shared genes. Instead, they propose that "chimpanzees may actively modify pant hoots to be different from their neighbors" (2004, p. 221). Such differences have functional consequences: Playback experiments conducted by Herbinger (2003) on individuals in the same West African community found that chimpanzees recognize the pant-hoots of other individuals, associate them with particular areas, and distinguish the calls of neighbors from strangers.

Like rhesus macaques (Gouzoules et al., 1984), wild chimpanzees who are receiving aggression produce acoustically different screams depending on the severity of the attack (Slocombe & Zuberbühler, 2005). Intriguingly, Slocombe and Zuberbühler (2007) also found that chimpanzee victims produced screams that appeared to exaggerate the severity of the attack, but they did so only when there was at least one individual nearby whose dominance rank was equal to or higher than that of the aggressor. These results suggest that chimpanzees both have a limited ability to modify the acoustic structure of their vocalizations and can recognize the dominance relations that exist among others.

Laboratory experiments confirm that primates can make subtle modifications to the acoustic features of their calls depending on experience. Elowson and Snowdon (1994) documented acoustic differences between the calls of pygmy marmosets (*Cebuella pygmaea*) housed in Washington, DC, and Madison, Wisconsin. When a group of marmosets was moved from Washington to Madison, the calls of the transplanted individuals changed to become more like their hosts'. In another experiment, Egnor et al. (2007) exposed cotton-top tamarins to bursts of white noise just as they produced their "contact loud call." The tamarins responded by producing calls that were shorter, with fewer pulses. Calls given immediately before or after white noise were also louder and had longer interpulse intervals. Egnor and Hauser (2004) review several other cases in which nonhuman primates make subtle modifications in the acoustic structure of their calls.

In sum, it is misleading and overly simplistic to describe primate vocal production as "fixed" and "involuntary." A more accurate conclusion is that the basic structure of nonhuman primate vocal signals appears to be innately determined, whereas the fine spectrotemporal features can be modified based on auditory experience and social context (Egnor & Hauser, 2004; Hammerschmidt & Fischer, 2008). The distinction between relatively innate and more modifiable components of phonation is important, because it has significant implications for future research on the neurobiology of primate communication (see Egnor & Hauser, 2004; Hammerschmidt & Fischer, 2008). For example, what brain areas are responsible for the innate and the modifiable components of primate call production, and how are these two aspects integrated at the neural level? Second, what contextual factors are most important in modifying primate vocal production: Age? Caller identity? The history of interaction between participants? And what neural pathways are responsible for this modulation? One recent study found differences between the neural mechanisms involved in spontaneous vocalizations and those involved in the production of calls that were elicited by calls from another individual (Gemba et al., 1999). Given the flexibility of human phonation, those interested in the evolution of language will be curious to know which social situations and areas of the brain are responsible for the limited flexibility that occurs in the phonation of monkeys' and apes' calls.

"AFFECTIVE" AND "SYMBOLIC" SIGNALS: A FALSE DICHOTOMY

Nonhuman primates present us with a communicative system in which a small repertoire of relatively fixed and inflexible calls, each linked to a particular social context, nonetheless gives rise to an open-ended, highly modifiable, and cognitively rich set of meanings. Listeners extract rich, semantic, and even propositional information from signalers who did not, in the human sense, intend to provide it (Cheney & Seyfarth, 1998).

The sharp distinction between signaler and recipient helps to clarify a theoretical issue that has deviled studies of primates'—and other animals'—vocalizations since Darwin first discussed them in *The Expression of the Emotions in Animal and Man*. Following Darwin, modern ethologists have typically assumed that vocal communication in animals differs from human language largely because the former is an "affective" system based on emotion, whereas the latter is a "referential" system based on the relation between words and the objects or events they represent (see, for example, Hauser, 1996; Marler et al., 1992; Owings & Morton, 1998; Owren & Rendall, 1997; Seyfarth et al., 1980). But this dichotomy is logically false.

A call's potential to serve as a referential signal depends on how tightly linked the call is to a particular social or ecological context. The mechanisms that underlie this specificity are irrelevant. A tone that informs a rat about the imminence of a shock, an alarm call that informs a vervet monkey about the presence of a leopard, or a sequence of threat-grunts and screams that informs a baboon that B3 and D2 are involved in a fight all have the potential to provide a listener with precise information because of their predictable association with a narrow range of events. The widely different mechanisms that lead to this association have no effect on the signal's potential to inform (Seyfarth & Cheney, 2003). Put slightly differently, there is no obligatory relation between "referential" and "affective" signaling. Knowing that a call is referential (i.e., has the potential to convey highly specific information) tells us nothing about whether its underlying cause is affective or not. Conversely, knowing that a call's production is due entirely to the caller's affect tells us nothing about the call's potential to serve as a referential signal.

It is therefore wrong, on theoretical grounds, to treat animal signals as either referential or affective, because the two properties of a communicative event are distinct and independent dimensions. Highly referential signals could, in principle, be caused entirely by a signaler's emotions, or their production could be relatively independent of measures of arousal. Highly affective signals could be elicited by very specific stimuli and thus function as referential calls, or they could be elicited by so many different stimuli that they provide listeners with only general information. In principle, any combination of results is possible. The affective and referential properties of signals are also logically distinct, at least in animal communication, because the former depends on mechanisms of call production in the signaler, whereas the latter depends on the listener's ability to extract information from events in its environment. Signalers and recipients, though linked in a communicative event, are nonetheless separate and distinct, because the mechanisms that cause a signaler to vocalize do not in any way constrain a listener's ability to extract information from the call.

Baboon grunts offer a good example. Rendall (2003) used behavioral data to code a social interaction involving *move* or *infant* grunts as having high or low arousal. He then examined calls given in these two circumstances and found that in each context certain acoustic features or modes of delivery were correlated with apparent arousal. Bouts of grunting given when arousal was apparently high had more calls, a higher rate of calling, and calls with a higher fundamental frequency than bouts given when arousal was apparently low. Further analysis revealed significant variation between contexts in the same three acoustic features that varied within context. By all three measures (call number, call rate, and fundamental frequency), infant grunts were correlated with higher arousal than were move grunts. Infant grunts also exhibited greater pitch modulation and more vocal "jitter," a measure of vocal instability (Rendall, 2003). In human speech, variations in pitch, tempo, vocal modulation, and jitter are known to provide listeners with cues about the speaker's affect or arousal (e.g., Bachorowski & Owren, 1995; Scherer, 1989).

It is, of course, difficult to obtain independent measures of a caller's arousal in the field. However, similarities between human and nonhuman primates in the mechanisms of phonation (Fitch & Hauser, 1995; Fitch et al., 2002; Schön Ybarra, 1995) support Rendall's (2003) conclusion that different levels of arousal play an important role in causing baboons to give acoustically different grunts in the infant and move contexts. This conclusion, however, tells us nothing about the grunts' potential to act as referential signals that inform nearby listeners about social or ecological events taking place at the time. As already noted, *move* grunts are given in a restricted set of circumstances, when the group is about to initiate, or has already initiated, a move. As a result, they have the potential to convey quite specific information to listeners. When one baboon hears another give a *move* grunt, he or she learns with some accuracy what is happening at that moment. By comparison, *infant* grunts are not as tightly linked to a particular type of social interaction. They may be given as the caller approaches a mother with infant, in answer to another animal's grunt, or as a reconciliatory signal following aggression. As a result, their meaning is less precise. When one baboon hears another's *infant* grunt, he or she learns only that the caller is involved in some sort of friendly interaction, but the precise nature of the interaction is unknown.

In sum, far from being a communicative system that is either affective or symbolic, vocal communication in nonhuman primates (and many other animals) contains elements of both. In their production, monkeys and apes use a small repertoire of relatively stereotyped calls, each closely linked to a particular context. This predictable association between call and context creates, for listeners, a world in which there are statistical regularities—regularities that allow them to ascribe meaning to vocalizations and to organize their knowledge into a rich conceptual structure (Cheney & Seyfarth, 2007, Chapters 10 and 11).

THE EVOLUTION OF FLEXIBLE VOCAL PRODUCTION

At some point in our evolutionary history—probably after the divergence of the evolutionary lines leading to chimpanzees and bonobos on the one hand and humans on the other (Enard et al., 2002)—our ancestors developed much greater control over the physiology of vocal production. As a result, vocal output became both more flexible and considerably more dependent on auditory experience and imitation (Fitch, 2007; Lieberman, 1991). What selective pressures might have given rise to these physiological changes?

Vocal communication in nonhuman primates lacks three features that are abundantly present in human language: the ability to generate new words, lexical syntax, and a theory of mind. By the latter we mean the ability of both speakers and listeners to make attributions about each other's beliefs, knowledge, and other mental states (Grice, 1957). These are the simplest, most basic features that distinguish human and nonhuman primate vocal production, and it is with these traits that speculations about the evolution of language must start. At the earliest stages of language evolution we need not worry about the more complex properties of language that probably came later—properties like case, tense, subject-verb agreement, open- and closed-class items, recursion, long-distance dependency, subordinate clauses, and so on.

How might the ability to generate new words, lexical syntax, and a theory of mind have evolved: simultaneously, in response to the same selective pressures, or more serially, in some particular order? We propose that the evolution of a theory of mind preceded language, creating the selective pressures that gave rise to the ability to generate new words and lexical syntax, and to the flexibility in vocal production that these two traits would have required (Cheney & Seyfarth, 2005, 2007). We make this argument on both empirical and theoretical grounds.

Empirically, there is no evidence in nonhuman primates for anything close to the large vocal repertoire we find even in very young children. Similarly, nonhuman primates provide few examples of lexical syntax. Recent work by Zuberbühler and colleagues on the alarm calls of forest monkeys provides intriguing evidence that the presence of one call type can "modify" the meaning of another (Arnold & Zuberbühler, 2006; Zuberbühler, 2002), and a study by Crockford and Boesch (2003) suggests that a call combination in chimpanzees may carry new meaning that goes beyond the meaning of the individual calls themselves, but these rare exceptions meet few of the definitions of human syntax. By contrast, there is growing evidence that both Old World monkeys (Cheney & Seyfarth, 2007; Engh et al., 2006; Flombaum & Santos, 2005) and apes (Buttelmann et al., 2007; Hare et al., 2001; Tomasello et al., 2005) may possess rudimentary abilities to attribute motives or knowledge to others, and engage in simple forms of shared attention and social referencing.

More theoretically, we suggest that the evolution of a theory of mind acted as a prime mover in the evolution of language because, while it is easy to imagine a scenario in which a rudimentary theory of mind came first and provided the impetus for the evolution of large vocabularies and syntax, any alternative sequence of events seems less likely.

Consider, for example, the course of word learning in children. Beginning as early as 9 to 12 months, children exhibit a nascent understanding of other individuals' motives, beliefs, and desires, and this skill forms the basis of a shared attention system that is integral to early word learning (Bloom & Markson, 1998; Tomasello, 2003). One-year-old children seem to understand that words can be mapped onto objects and actions. Crucially, this understanding is accompanied by a kind of "social referencing" in which the child uses other people's direction of gaze, gestures, and emotions to assign labels to objects (Baldwin, 1991; reviewed in Fisher & Gleitman, 2002; Pinker, 1994). Gaze and attention also facilitate word learning in dogs and other animals. Children, however, rapidly surpass the simpler forms of shared attention and word learning demonstrated by animals. Long before they begin to speak in sentences, young children develop implicit notions of objects and events, actors, actions, and those that are acted upon. As Fisher and Gleitman (2002:462) argue, these "conceptual primitives" provide children with a kind "conceptual vocabulary onto which the basic linguistic elements (words and structures) are mapped." Moreover, in contrast to monkeys, apes, and other animals, 1-year-old children attempt to share what they know with others (Tomasello & Carpenter, 2007). While animals are concerned with their own goals and knowledge, young children are motivated to make their thoughts and knowledge publically available.

The acquisition of a theory of mind thus creates a cognitive environment that drives the acquisition of new words and new grammatical skills. Indeed, the data on children's acquisition of language suggest that they could not increase their vocabularies or learn grammar as rapidly as they do if they did not have some prior notion of other individuals' mental states (Fisher & Gleitman, 2002; Pinker, 1994; Tomasello, 2003).

By contrast, it is much more difficult to imagine how our ancestors could have learned new words or grammatical rules if they were unable to attribute mental states to others. The lack of syntax in nonhuman primate vocalizations cannot be traced to an inability to recognize argument structure—to understand that an event can be described as a sequence in which an agent performs some action on an object. Baboons, for example, clearly distinguish between a sequence of calls indicating that Sylvia is threatening Hannah, as opposed to Hannah is threatening Sylvia. Nor does the lack of syntax arise because of an inability to mentally represent descriptive verbs, modifiers, or prepositions. In captivity, a variety of animals, including dolphins (Herman et al., 1993), sea lions (Schusterman & Krieger, 1986), and African gray parrots (Pepperberg, 1992), can be taught to understand and in some cases even to produce verbs, modifiers, and prepositions. Even in their natural behavior, nonhuman primates and other animals certainly seem capable of *thinking* in simple sentences, but the ability to think in sentences does not motivate them to *speak* in sentences. Their knowledge remains largely private.

This may occur in large part because primates and other animals cannot distinguish between what they know and others know and cannot recognize, for example, that an ignorant individual might need to have an event explained to them. As a result, although they may mentally tag events as argument structures, they fail to map these tags into a communicative system in any stable or predictable way. Because they cannot attribute mental states like ignorance to others, and are unaware of the causal relation between behavior and beliefs, monkeys and apes do not actively seek to explain or elaborate upon their thoughts. As a result, they are largely incapable of inventing new words or of recognizing when thoughts should be made explicit.

We suggest, then, that long before our ancestors spoke in sentences, they had a language of thought in which they represented the world—and the meaning of call sequences—in terms of actors, actions, and those who are acted upon. The linguistic revolution occurred when our ancestors began to express this tacit knowledge and to use their cognitive skills in speaking as well as listening. The prime mover behind this revolution was a theory of mind that had evolved to the point where its possessors did not just recognize other individuals' goals, intentions, and even knowledge—as monkeys and apes already do—but were also motivated to share their own goals, intentions, and knowledge with others. Whatever the selective pressures that prompted this change, it led to a mind that was motivated to make public thoughts and knowledge that had previously remained private. The evolution of a theory of mind spurred the evolution of words and grammar. It also provided the selective pressure for the evolution of the physiology adaptations that enabled vocal modifiability.

NOTE

1. One should, however, treat estimates of the size of a species' vocal repertoire with caution. Often the best predictors of repertoire size are the length, creativity, and ingenuity with which a species has been studied.

REFERENCES

Arbib, M. (2005). From monkey-like action-recognition to human language: An evolutionary framework for neurolinguistics. *Behavioral and Brain Science, 28,* 105–167.

Arnold, K., & Zuberbühler, K. (2006). Language evolution: Compositional semantics in primate calls. *Nature, 441, 303.*

Bachorowski, J. A., & Owren, M. J. (1995). Vocal expression of emotion: Acoustic properties of speech are associated with emotional intensity and context. *Psychological Science, 6,* 219–224.

Baldwin, D. (1991). Infants' contribution to the achievement of joint reference. *Child Development, 92,* 875–890.

Bickerton, D. (1990). *Language and species.* Chicago: University of Chicago Press.

Bloom, P., & Markson, L. (1998). Capacities underlying word learning. *Trends in Cognitive Science, 2,* 67–73.

Buttelmann, D., Carpenter, M., Call, J., & Tomasello, M. (2007). Enculturated apes imitate rationally. *Developmental Science, 10,* 31–38.

Cheney, D. L., & Seyfarth, R. M. (1990). *How monkeys see the world.* Chicago: University of Chicago Press.

Cheney, D.L., & Seyfarth, R. M. (1997). Reconciliatory grunts by dominant female baboons influence victims' behaviour. *Animal Behaviour 54,* 409–418.

Cheney, D. L., & Seyfarth, R. M. (1998). Why monkeys don't have language. In: G. Petersen (Ed.), *The Tanner lectures on human values* (vol 19, pp. 175–219). Salt Lake City: University of Utah Press.

Cheney, D. L., & Seyfarth, R. M. (2005). Constraints and preadaptations in the earliest stages of language evolution. *Linguistic Review, 22,* 135–159.

Cheney, D. L., & Seyfarth, R. M. (2007). *Baboon metaphysics: The evolution of a social mind.* Chicago: University of Chicago Press.

Cheney, D. L., Seyfarth, R. M., & Silk, J. B. (1995). The responses of female baboons to anomalous social interactions: Evidence for causal reasoning? *Journal of Comparative Psychology, 109,* 134–141.

Crockford, C., & Boesch, C. (2003). Context-specific calls in wild chimpanzees (*Pan troglodytes verus*): Analysis of barks. *Animal Behaviour, 66,* 115–125.

Crockford, C., Herbinger, L., Vigilant, L., & Boesch, C. (2004). Wild chimpanzees have group-specific calls: A case for vocal learning? *Ethology, 110,* 221–243.

Darwin, C. (1871/1981) *The descent of man, and selection in relation to sex.* Princeton: Princeton University Press.

Egnor, S. E. R., & Hauser, M. D. (2004). A paradox in the evolution of primate vocal learning. *Trends in Neuroscience, 27,* 649–654.

Egnor, S. E. R., Wickelgren, J., & Hauser, M. D. (2007). Tracking silence: Adjusting vocal production to avoid acoustic interference. *Journal of Comparative Physiology Series A, 193,* 477–483.

Elowson, M., & Snowdon, C. T. (1994). Pygmy marmosets, *Cebuella pygmaea,* modify vocal structure in response to changed social environment. *Animal Behaviour, 47,* 1267–1277.

Enard, W., Przeworski, M., Fisher, S. E., Lai, C. S. L., Wiebe, V., Kitano, T., et al. (2002). Molecular evolution of FOXP2: A gene involved in speech and language. *Nature, 418,* 869–872.

Engh, A. E., Hoffmeier, R. R., Cheney, D. L., & Seyfarth, R. M. (2006). Who, me? Can baboons infer the target of vocalisations? *Animal Behaviour, 71,* 381–387.

Fischer, J., Cheney, D. L., & Seyfarth, R. M. (2000). Development of infant baboon responses to female graded variants of barks. *Proceedings of the Royal Society of London Series B, 267,* 2317–2321.

Fisher, C., & Gleitman, L. R. (2002). Language acquisition. In: H. F. Pashler & C. R. Gallistel (Eds.), *Handbook of experimental Psychology, vol 3: Learning and motivation* (pp. 445–496). New York: Stevens Wiley.

Fischer, J., Hammerschmidt, K., Cheney, D. L., & Seyfarth, R. M. (2002). Acoustic features of male baboon loud calls: Influences of context, age, and individuality. *Journal of the Acoustical Society of America, 111,* 1465–1474.

Fischer, J., Hammerschmidt, K., Seyfarth, R. M., & Cheney, D. L. (2001a). Acoustic features of female chacma baboon barks. *Ethology, 107,* 33–54.

Fischer, J., Metz, M., Cheney, D. L., & Seyfarth, R. M. (2001b). Baboon responses to graded bark variants. *Animal Behaviour, 61,* 925–931.

Fitch, W. T. (2007). The evolution of language: A comparative perspective. In: G. Gaskell (Ed.), *Oxford handbook of psycholinguistics.* Oxford: Oxford University Press.

Fitch, W. T., & Hauser, M. D. (1995). Vocal production in nonhuman primates: Acoustics, physiology, and functional constraints on "honest" advertisement. *American Journal of Primatology, 37,* 191–220.

Fitch, W. T., Neubauer, J., & Herzel, H. (2002). Calls out of chaos: The adaptive significance of nonlinear phenomena in mammalian vocal production. *Animal Behaviour, 63,* 407–418.

Flombaum, J. L., & Santos, L. R. (2005). Rhesus monkeys attribute perceptions to others. *Current Biology, 15,* 447–452.

Gardenfors, P. (2003). *How homo became sapiens: On the evolution of thinking.* Oxford: Oxford University Press.

Gemba, H., Kyuhou, S., Matsuzaki, R., & Amino, Y. (1999). Cortical field potentials associated with audio-initiated vocalization in monkeys. *Neuroscience Letters, 272,* 49–52.

Ghazanfar, A. A., & Logothetis, N. K. (2003). Facial expressions linked to monkey calls. *Nature, 423,* 937–938.

Ghazanfar, A. A., Maier, J. X., Hoffman, K. L., & Logothetis, N. K. (2005). Multisensory integration of dynamic faces and voices in rhesus monkey auditory cortex. *Journal of Neuroscience, 25,* 5004–5012.

Gil da Costa, R., Braun, A., Lopes, M., Hauser, M. D., Carson, R. E., Herscovitch, P., et al. (2004). Toward an evolutionary perspective on conceptual representation: Species-specific calls activate visual and affective processing systems in the macaque. *Proceedings of the National Academy of Sciences, 101,* 17516–17521.

Goodall, J. (1986). *The chimpanzees of Gombe: Patterns of behavior.* Cambridge, MA: Harvard University Press.

Gouzoules, S., Gouzoules, H., & Marler, P. (1984). Rhesus monkey (*Macaca mulatta*) screams: Representational signaling in the recruitment of agonistic aid. *Animal Behaviour, 32,* 182–193.

Grice, H. P. (1957). Meaning. *Philosophical Review, 66,* 377–388.

Gros-Louis, J. (2004). The function of food-associated calls in white-faced capuchin monkeys, *Cebus capucinus,* from the perspective of the signaler. *Animal Behaviour, 67,* 431–440.

Hammerschmidt, K., & Fischer, J. (2008). Constraints in primate vocal production. In: U. Griebel & K. Oller (Eds.), *The evolution of communicative creativity: From fixed signals to contextual flexibility* (pp. 93–119). Cambridge, MA: MIT Press.

Hare, B., Call, J., & Tomasello, M. (2001). Do chimpanzees know what conspecifics know? *Animal Behaviour, 61,* 139–151.

Hauser, M. D. (1988). How infant vervet monkeys learn to recognize starling alarm calls: The role of experience. *Behaviour, 105,* 187–201.

Hauser, M. D. (1996). *The evolution of communication.* Cambridge, MA: MIT Press.

Hauser, M. D., & Wrangham, R. W. (1990). Recognition of predator and competitor calls in nonhuman primates and birds: A preliminary report. *Ethology, 86,* 116–130.

Herbinger, I. (2003). Inter-group aggression in wild West African chimpanzees (*Pan troglodytes verus*): Mechanisms and function. Ph.D. dissertation, University of Leipzig.

Herman, L. M., Pack, A. A., & Morrel-Samuels, P. (1993). Representational and conceptual skills of dolphins. In: H. L. Roitblat, L. M. Herman, & P. E. Nachtigall (Eds.), *Comparative cognition and neuroscience* (pp. 403–442). Hillsdale, NJ: Lawrence Erlbaum Associates.

Janik, V. W., & Slater, P. J. B. (1997). Vocal learning in mammals. *Advances in the Study of Behavior, 26,* 59–99.

Jurgens, U., & Ploog, D. (1970). Cerebral representation of vocalization in the squirrel monkey. *Experimental Brain Research, 10,* 532–554.

Kaminski, J., Call, J., & Fischer, J. (2004). Word learning in a domestic dog: Evidence for "fast mapping." *Science, 304,* 1682–1683.

Kitchen, D. M., Cheney, D. L., & Seyfarth, R. M. (2003). Female baboons' responses to male loud calls. *Ethology, 109,* 401–412.

Lieberman, P. (1991). *Uniquely human.* Cambridge, MA: Harvard University Press.

Marler, P., Evans, C. S., & Hauser, M. D. (1992). Animal signals: Motivational, referential, or both? In: H. Papousek, U. Jurgens, & M. Papousek (Eds.), *Nonverbal vocal communication: Comparative and developmental approaches* (pp. 66–86). Cambridge: Cambridge University Press.

McComb, K., & Semple, S. (2005). Coevolution of vocal communication and sociality in primates. *Biology Letters, 1,* 381–385.

Mitani, J. C., & Brandt, K. L. (1994). Social factors influence the acoustic variability in the long-distance calls of male chimpanzees. *Ethology, 96,* 233–252.

Mitani, J., & Gros-Louis, J. (1998). Chorusing and call convergence in chimpanzees: Tests of three hypotheses. *Behaviour, 135,* 1041–1064.

Owings, D. H., & Morton, E. S. (1998). *Animal vocal communication: A new approach.* Cambridge: Cambridge University Press.

Owren, M. J., Dieter, J. A., Seyfarth, R. M., & Cheney, D. L. (1993). Vocalizations of rhesus and Japanese macaques cross-fostered between species show evidence of only limited modification. *Developmental Psychobiology, 26,* 389–406.

Owren, M. J., & Rendall, D. (1997). An affect-conditioning model of nonhuman primate vocal signaling. In: M. D. Beecher, D. H. Owings, & N. S. Thompson (Eds.), *Perspectives in ethology* (vol 12, pp. 299–346). New York: Plenum Press.

Owren, M. J., Seyfarth, R. M., & Cheney, D. L. (1997). The acoustic features of vowel-like grunt calls in chacma baboons (*Papio cynocephalus ursinus*): Implications for production processes and functions. *Journal of the Acoustical Society of America, 101,* 2951–2963.

Pepperberg, I. M. (1992). Proficient performance of a conjunctive, recursive task by an African gray parrot (*Psittacus erithacus*). *Journal of Comparative Psychology, 106,* 295–305.

Pierce, J. (1985). A review of attempts to condition operantly alloprimate vocalizations. *Primates, 26,* 202–213.

Pinker, S. (1994). *The language instinct.* New York: William Morrow and Sons.

Ploog, D. (1981). Neurobiology of primate audio-visual behavior. *Brain Research Review, 3,* 35–61.

Poole, J., Tyack, P., Stoeger-Horwarth, A., & Watwood, S. (2005). Elephants are capable of vocal learning. *Nature, 434,* 455–456.

Rendall, D. (2003). Acoustic correlates of caller identity and affect intensity in the vowel-like grunt vocalizations of baboons. *Journal of the Acoustical Society of America, 113,* 3390–3402.

Rendall, D., Seyfarth, R. M., Cheney, D. L., & Owren, M. J. (1999). The meaning and function of grunt variants in baboons. *Animal Behaviour, 57,* 583–592.

Savage-Rumbaugh, E. S. (1986). *Ape language: From conditioned response to symbol.* New York: Columbia University Press.

Scherer, K. R. (1989). Vocal correlates of emotion. In: H. Wagner & A. Manstead (Eds.), *Handbook of psychophysiology: Emotion and social behavior* (pp. 167–195). New York: John Wiley and Sons.

Schön Ybarra, M. (1995). A comparative approach to the nonhuman primate vocal tract: Implications for sound production. In: E. Zimmerman, J. D. Newman, & U. Jurgens (Eds.), *Current topics in primate vocal communication* (pp. 185–198). New York: Plenum Press.

Schusterman, R .J., & Krieger, K. (1986). Artificial language comprehension and size transposition by a California sea lion (*Zalophus californianus*). *Journal of Comparative Psychology, 100,* 348–355.

Schusterman, R. J., Reichmuth Kastak, C., & Kastak, D. (2002). The cognitive sea lions: Meaning and memory in the lab and in nature. In: M. Bekoff, C. Allen, & G. Burghardt (Eds.), *The cognitive animal: Empirical and theoretical perspectives on animal cognition* (pp. 217–228). Cambridge, MA: MIT Press.

Seyfarth, R. M., & Cheney, D. L. (1986). Vocal development in vervet monkeys. *Animal Behaviour, 34,* 1640–1658.

Seyfarth, R. M., & Cheney, D. L. (1990). The assessment by vervet monkeys of their own and another species' alarm calls. *Animal Behaviour, 40,* 754–764.

Seyfarth, R. M., & Cheney, D. L. (1997). Some general features of vocal development in non-human primates. In: M. Husberger & C. T. Snowdon (Eds.), *Social influences on vocal development* (pp. 249–273). Cambridge: Cambridge University Press.

Seyfarth, R. M., & Cheney, D. L. (2003). Signalers and receivers in animal communication. *Annual Review of Psychology, 54,* 145–173.

Seyfarth, R. M., Cheney, D. L., & Marler, P. (1980). Vervet monkey alarm calls: Semantic communication in a free-ranging primate. *Animal Behaviour, 28,* 1070–1094.

Silk, J. B., Altmann, J., & Alberts, S. C. (2006a). Social relationships among adult female baboons (*Papio cynocephalus*). I. Variation in the strength of social bonds. *Behavioral Ecology and Sociobiology, 61,* 183–195.

Silk, J. B., Altmann, J., & Alberts, S. C. (2006b). Social relationships among adult female baboons (*Papio cynocephalus*). II: Variation in the quality and stability of social bonds. *Behavioral Ecology and Sociobiology, 61,* 197–204.

Silk, J. B., Seyfarth, R. M., & Cheney, D. L. (1999). The structure of social relationships among female baboons in the Moremi Reserve, Botswana. *Behaviour, 136,* 679–703.

Slocombe, K., & Zuberbühler, K. (2005). Agonistic screams in wild chimpanzees (*Pan troglodytes schweinfurthii*) vary as a function of social role. *Journal of Comparative Psychology, 119,* 67–77.

Slocombe, K., & Zuberbühler, K. (2007). Chimpanzees modify recruitment screams as a function of audience composition. *Proceedings of the National Academy of Sciences, 104,* 228–233.

Smith, H. J., Newman, J. D., & Symmes, D. (1982). Vocal concomitants of affiliative behavior in squirrel monkeys. In: C. T. Snowdon, C. H. Brown, & M. Petersen (Eds.), *Primate communication* (pp. 30–49). Cambridge: Cambridge University Press.

Steiper, M. E., Young, N. M., & Sukarna, T. Y. (2004). Genomic data support the hominoid slowdown and an Early Oligocene estimate for the hominoid cercopithecoid divergence. *Proceedings of the National Academy of Sciences, 101,* 17021–17026.

Terrace, H. S. (1979). *Nim.* New York: Knopf.

Terrace, H. S. (1983). Nonhuman intentional systems. *Behavioral and Brain Science, 6,* 378–379.

Tomasello, M. (2003). *Constructing a language: A usage-based theory of language acquisition.* Cambridge, MA: Harvard University Press.

Tomasello, M., & Carpenter, M. (2007). Shared intentionality. *Developmental Science, 10,* 121–125.

Tomasello, M., Carpenter, M., Call, J., Behne, T., & Moll, H. (2005). Understanding and sharing intentions: The origins of cultural cognition. *Behavioral Brain Science, 28,* 675–691.

Washburn, S. L. (1982). Language and the fossil record. *Anthropology UCLA, 7,* 231–238.

Wittig, R. M., Crockford, C., Seyfarth, R. M., & Cheney, D. L. (2007). Vocal alliances in chacma baboons, Papio hamadryas ursinus. *Behavioral Ecology and Sociobiology, 61,* 899–909.

Zuberbühler, K. (2000). Referential labeling in Diana monkeys. *Animal Behaviour, 59,* 917–927.

Zuberbühler, K. (2001). Predator-specific alarm calls in Campbell's guenons. *Behavioral Ecology and Sociobiology, 50,* 414–422.

Zuberbühler, K. (2002). A syntactic rule in forest monkey communication. *Animal Behaviour, 63,* 293–299.

CHAPTER 6

Rational Decision Making in Primates: The Bounded and the Ecological

Jeffrey R. Stevens

A young female rhesus macaque steals furtive glances at the male off to her right. He just arrived to the territory and therefore immediately piques her interest. The alpha male, however, sits a few meters off, basking in the sun. Being in estrus, the young female faces a choice: solicit a mating from the alpha male or follow the unfamiliar male into the brush to sneak a mating with him. Mating with the alpha male almost guarantees "good genes" for her offspring. But something pushes her toward the unfamiliar male. Mating with him reduces the probability of inbreeding and adds a bit of genetic diversity to her offspring. Additionally, mating with the new male could act as an investment in the future: The current alpha male is getting old, and befriending a prospective alpha male could yield future benefits. In addition, spreading the possibility of paternity may secure protection for the offspring. Yet, this mating also involves risks. Males often vocalize while mating, which attracts the attention of other males. If the unfamiliar male vocalizes, the alpha male may attack the female. Should she take the safe option or risk punishment for possible future payoffs?

Primates constantly face decisions that influence their survival and reproduction. Continue foraging in this tree or move on to another? Expose oneself to a hidden predator by straying from the group or enjoy the safety of having other potential victims nearby? Defend one's territory from invaders or abandon it and seek a new home? In all of these cases, primates must trade off the costs and benefits associated with uncertain and delayed decision outcomes. The outcomes of these choices influence survival and reproduction, and natural selection should favor those individuals whose choices lead to the propagation of their genes.

The vast majority of economic analyses of decision making define good or "rational" decisions as those consistent with a set of mathematical principles. Yet, this ignores the evolutionary pressures on decision making for the sake of mathematical elegance (Kacelnik, 2006; Stevens, 2008). Meanwhile, the standard psychological view of decision making seeks to empirically undermine the economic theory but cannot offer an alternative explanatory theory. Here, I emphasize an evolutionarily informed framework for studying decision making: the bounded and ecological rationality approaches (Gigerenzer et al., 1999). Though these approaches have traditionally focused on human decision making, they are just as relevant for other species, including other primates. To illustrate the relevance of bounded and ecological rationality to the study of primate decision making, I begin by introducing various visions of rationality found in the economic and psychological study of decision making. I then explore how primate studies inform three aspects of decision making: utility, uncertainty, and time. Together, these aspects will guide our understanding of the evolutionary origins of primate and human rationality.

VISIONS OF RATIONALITY

Rationality means different things to different people. Kacelnik (2006) proposed that rationality refers to decisions that are either (1) *consistent with expected utility maximization* for economists and psychologists, (2) *consistent within the self* for philosophers, and (3) *consistent with fitness enhancement* for biologists. Of particular relevance here are the economic, psychological, and biological views. A review of these visions of rationality will frame the question, "Are primates rational decision makers?"

Rational Choice

Which would you prefer: receiving two bananas with certainty or receiving either one or three bananas with equal probability? Depending on whether you like bananas and your level of hunger, this may be tricky to answer. These questions of decisions under risk mirror fundamental choices that we and other animals frequently face. Very little in life is certain, so all organisms choose between options without knowing the exact consequences.

Economists have approached the question of uncertainty by developing *expected utility theory* (reviewed in Wu et al., 2004). In expected utility theory, three features characterize all options: magnitude, utility, and probability. Magnitude (x) refers to the amount of the benefit (or cost) associated with the option. Utility (u) is the mapping of magnitude onto some measure of satisfaction or "goodness."[1] Finally, probability (p) is the chance that the outcome occurs. Thus, if one faces risky options, the expected utility is $EU = p \cdot u(x)$, where utility is some function of magnitude. Von Neumann and Morgenstern (1947) formalized expected utility theory to show that following specific mathematical principles maximizes expected utility. Therefore, expected utility acts as a normative standard for what decision makers should maximize when making risky choices.

Expected utility maximization assumes consistent choice, which requires a number of principles to hold (reviewed in Luce & Raiffa, 1957; Rieskamp et al., 2006b). First, choices must be *transitive*, meaning that a fixed order of preference exists. If an individual prefers bananas over apples and apples over oranges, then he or she must prefer bananas over oranges to maintain transitivity. Second, when transitive, choices are also *independent from irrelevant alternatives*, meaning the relative preference between options should not be affected by the presence or absence of other options. If bananas are preferred to apples, the addition of watermelons to the choice set should not affect the banana/apple preference. Finally, choices must be *invariant*, meaning that option A is preferred to B regardless of presentation format. If endowed with a banana and an apple and asked to give up one, preferences should be the same as if asked to choose freely between a banana and apple. Given these and a few other principles, one can show that preferences follow expected utility calculations.

Irrational Choice?

Not long after von Neumann and Morgenstern (1947) published the principles of expected utility theory, cracks began appearing in the mathematically elegant framework when data showed violations of expected utility predictions (Allais, 1953; Ellsberg, 1961). Soon afterward, Kahneman and Tversky started a cottage industry of demonstrating violations of the theoretical predictions (Kahneman & Tversky, 2000; Kahneman et al., 1982; Tversky & Kahneman, 1981). In both experimental settings and in real-world decisions, laypeople and experts made "irrational" choices.[2] Subjects showed intransitivity, irrelevant alternatives changed preferences, and the framing of decision questions greatly influenced preferences. Economists minimized the problem by calling these findings "anomalies," whereas psychologists emphasized their robustness and labeled them "biases." To psychologists, expected utility theory was deeply flawed because it rested purely on mathematically derived principles and not what we know about human behavior.

Kahneman and Tversky (1979) injected a bit of psychological realism into decision theory when they proposed *prospect theory*. Instead of using a utility function characterized over total wealth, prospect theory uses a value function $v(x)$ relative to a reference point. So with 100 bananas in the bank, a decision maker would view a choice between gaining one and losing three bananas as a choice between 101 and 97 bananas from the expected utility perspective but a choice between gaining one and losing three from the prospect theory account. In addition to altering the reference point of the value function, prospect theory integrated data about probability perception into the equation by adding a decision weight function to the probability. Thus, instead of having a linear relationship between the objective probability and the perceived probability, prospect theory assumes that people tend to overestimate low probabilities and underestimate high probabilities. The probability, then, is weighted by the function $\pi(p)$. Thus, prospect theory predicts that preferences depend on which option has the highest value $V = \pi(p) \cdot v(x)$. As a descriptive theory, it nicely fits people's preferences (Kahneman & Tversky, 2000; Wu et al., 2004). However, it does not explain why we have reference points or nonlinear value functions—it takes these as givens and describes how they influence decisions.

Bounded and Ecological Rationality

Expected utility theory is mathematically elegant but fails to account for many of the data. Prospect theory fairs much better descriptively but lacks explanatory power. Parameters in the models are fit to the data, with no a priori predictions about parameter values. Prospect theory therefore offers a slight modification of expected utility theory by patching a few of the holes that data have poked into the theory. But both theories face a more fundamental problem. Namely, neither of these theories adequately addresses two crucial components of decision making: the structure of the mind and the structure of the decision-making environment.

Early in the study of decision making, Simon (1955, 1956) highlighted not only the study of the mind but also the fit between the mind and the environment. He criticized the unrealistic assumption that decision makers have infinite time to decide, full knowledge of the problem, and unlimited computational resources to find an optimal solution to a decision problem. This vision of unbounded rationality contrasts sharply with what we know about human cognition and decision making, so Simon proposed the study of *bounded rationality*—the exploration of decision making given realistic assumptions about cognitive abilities. Real-world decision makers lack knowledge and cannot use optimization processes to make decisions. Thus, much previous research has ignored cognitive processes at work in decision making (but see Payne et al., 1993). The bounded rationality approach calls for realistic models of the decision process based on what we know about cognition rather than on a set of mathematical principles. Knowing the underlying process can help us better understand the decisions. Yet, Simon emphasized that studying only the mind gives you but half of the picture.

To fully understand decision making, we must embed the mind in the environment. Gigerenzer and colleagues (1999) have termed this *ecological rationality*—the match between a decision mechanism and the environment. The unbounded rationality approach assumes that expected utility works in all decision-making situations—it applies universally. Ecological rationality, however, appeals to the evolutionary idea that adaptations match the environment in which they evolved. Therefore, decision mechanisms should not be universal and domain general but specifically tailored to the environment in which they operate (Barkow et al., 1992). In fact, ecological rationality suggests that we do not possess a single, complex decision-making mechanism used in all contexts. Rather, we have an "adaptive toolbox" of mechanisms (including simple heuristics or rules of thumb) that, when used in the appropriate environment, perform quite well (Gigerenzer et al., 1999). This perspective, then, offers an explanation for the anomalies and biases seen by the experimental economists and psychologists. Rather than being evidence for flawed thinking and irrationality, we are simply putting these decision mechanisms in an unfamiliar and artificial environment—the experimental laboratory. In general, however, our decision mechanisms serve us quite well by exploiting critical aspects of the environment.

Both bounded and ecological rationality offer appealing alternatives to the standard unbounded approach because they rest on realistic evolutionary principles instead of mathematical formalizations. Comparative analyses provide a unique method for testing questions of bounded and ecological rationality because we have great variability across species in their ecological environments. Primates offer an ideal group of species for these investigations because their phylogenetic proximity allows us to test interesting hypotheses about the evolution of human decision making. With this framework in hand, we can now review the bounded and ecological rationality of primate decisions. Note that here I focus on what has traditionally been termed "individual decision making." Though it is likely impossible to completely extract an individual from the social environment (Stevens & King, in press), for the purposes of this review I will put aside the exciting and complicated world of social decision making in primates and refer the reader to Maestripieri (Chapter 19) or Rosati, Santos, and Hare (Chapter 7). Instead, I will focus on three important components of individual decision making[3]: utility, uncertainty, and time.

UTILITY AND PREFERENCE

Utility refers to the mapping of the magnitude of a benefit or cost onto some measure of "goodness" or goal achievement (Baron, 2000), and a utility function describes this mapping. Utility itself is difficult to assess, so typically we measure choices to infer preferences. For expected utility theory to work, preferences must follow the principles mentioned previously: transitivity, independence of irrelevant alternatives, and invariance. It is well established that humans violate these principles, deviating from the normative standard (Kahneman & Tversky, 2000). The question remains, "Is this the appropriate standard?" Examining choices in other species can address this question, because if other species also show the biases, then we likely are using the wrong standard. Do other primates violate these standards, suggesting deep evolutionary roots, or do only humans show these violations?

Transitive preferences[4] are not well studied in primates. Though other species seem to show intransitivity (Shafir, 1994; Waite, 2001b), few data exist for primates, so we will not consider transitivity here.[5]

The principle of *independence of irrelevant alternatives* implies that previously available options should not influence the current preferences (Simonson & Tversky, 1992). Waite (2001a), however, found that the background context does matter for gray jays (*Perisoreus canadensis*). When required to pay a lower relative cost for food in previous choices, the birds reduced their preference for that option later when required to pay a higher relative cost. Tinklepaugh (1928) found a similar result in rhesus macaques (*Macaca mulatta*) and long-tailed macaques (*Macaca fascicularis*). These monkeys observed an experimenter place a banana under one of two cups. The monkeys then immediately chose the correct cup and received the banana. However, in some trials, the experimenter substituted a piece of lettuce under the cup, unbeknownst to the monkeys. When the monkeys lifted the cups and found the lettuce, they rejected the less-preferred food item. Though outside of this context the monkeys readily consumed lettuce, they refused to eat it when expecting a banana. Thus, preferences are not fixed but depend on previous options.

Tinklepaugh's data suggest that the monkeys do not have absolute preferences. If they did, they would always consume the lettuce because it is always better than nothing. However, the preference for consuming lettuce is *relative* to the previous availability of the highly preferred banana. The preference is relative to an expectation of other options. There are good, adaptive reasons for avoiding absolute preferences. For instance, it is well known in foraging theory that optimal choices depend on the background environment. If the environment is rich, animals should "skim the cream" and choose to invest little time in extracting food from patches; if, however, the environment is poor, they should more thoroughly exploit the patches (Stephens & Krebs, 1986). Houston (1997) argued that since the current choice is incorporated into the estimate of the background environment for the next choice point, preferences can change for the same set of options depending on the previous background environment. Thus, relative preferences can result from an ecologically rational mechanism of adaptive decision making.

The classical economic approach to rationality also assumes that decision makers have strictly ordered preferences that are *invariant* to extraneous characteristics of the choice situation. If an agent prefers A to B, it should always prefer A, regardless of whether one is buying or selling A, or whether other choices are made before the A/B choice, etc. A common violation of invariance found in behavioral economics is the endowment effect. This phenomenon occurs when decision makers have a higher preference for an object when they own it (Kahneman et al., 1990). In humans, this is typically demonstrated by showing that subjects require a higher price to sell an object they possess than to buy the same object. Brosnan and colleagues (2007) tested similar effects in chimpanzees (*Pan troglodytes*) by offering them a choice between two different food items and recording their preferences. The experimenters then endowed the chimpanzees with one of the items and measured their willingness to trade for the other item. The choice preferences and trading preferences did not match, suggesting an endowment effect. Similar results have been found for brown capuchin monkeys (*Cebus apella*—Lakshminaryanan et al., 2008). Though anomalous to economists, an evolutionary perspective provides an explanation for the endowment effect. The question of why owners fight harder than intruders to maintain a resource has a long history in behavioral ecology (Hammerstein, 1981; Krebs & Davies, 1993; Maynard Smith & Parker, 1976). Ownership has its privileges, including additional knowledge about a resource. Even without direct benefits, as an arbitrary rule, respecting ownership can avoid costly conflicts. Thus, owning a resource can increase its value.

Utility Building Blocks

Simon's (1955, 1956) vision of bounded rationality emphasized realistic assumptions about the cognitive abilities of decision makers. Applying this perspective suggests that we need to consider both the cognitive building blocks or *evolved capacities* needed for exhibiting preferences and the limits placed on these capacities.

Magnitude Perception

A critical component of exhibiting preferences is the ability to perceive differences in magnitude between options. Preferring two bananas to one requires discriminating between the amounts two and one. There are many mechanisms used to quantify objects in the world, and many studies have explored these mechanisms in primates (reviewed in Brannon, 2005a,b; Hauser & Spelke, 2004). Brannon et al. (Chapter 8) provide a useful overview of primate quantification in this volume, but it is worthwhile to briefly describe two of these mechanisms. The first mechanism discriminates between quantities only approximately via the analog magnitude system. Importantly, the discriminations follow Weber's law: Variance in the representation increases with magnitude (Gallistel, 1990). This results in the ratio between quantities rather than the absolute magnitude driving the discrimination. The approximate number system yields a limit to the precision with which individuals can discriminate magnitudes, with larger magnitudes being more difficult than smaller ones (Fig. 6.1). In addition to the approximate system, both humans and other primates seem to have a precise system that tracks individual objects. With this system, individuals can enumerate discrete quantities but only up to a maximum of three or four objects (Hauser et al., 2000). Therefore, the precision that primates can exhibit in their preferences depends on the magnitude of the amount: They can precisely choose between small amounts and approximately choose between larger amounts.

The standard rationality approach would assume that if an organism has the more precise system, it should use it when making quantity

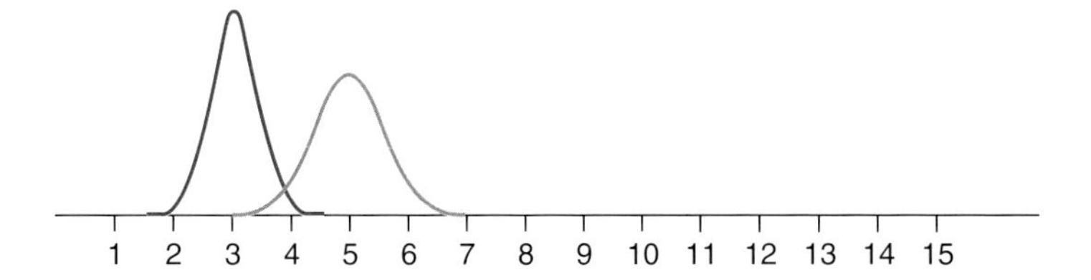

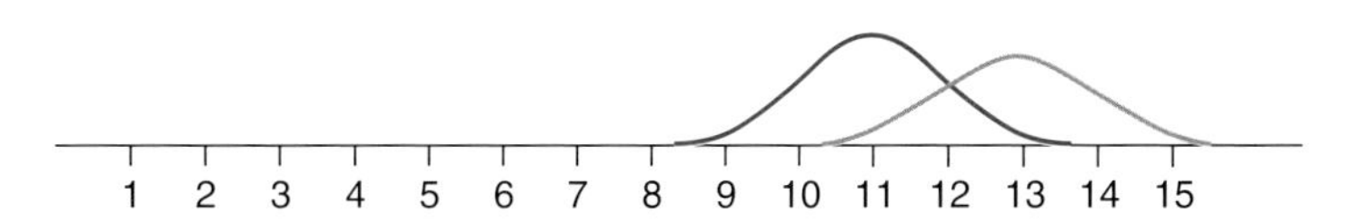

Figure 6.1 Weber's law states that the variance around estimates of quantity increases with magnitude. Smaller quantities therefore are easier to discriminate than larger quantities. For instance, there may be little variance for estimations of three and five objects, so discriminating between them is easy. However, the variance dramatically increases for 11 and 13 objects, and estimates greatly overlap for these durations, making them difficult to distinguish.

judgments. However, Stevens and colleagues (2007) showed that cotton-top tamarins (*Saguinus oedipus*) used the approximate system in a foraging task, even though this species can use the precise system (Hauser et al., 2003; Uller et al., 2001). The tamarins used the simpler, approximate system as a default mechanism unless the task demanded the more precise system. Thus, different aspects of the decision environment trigger different mechanisms of discrimination.

Valuation

Options differ not only in quantity but also in quality. In many cases, decision makers face choices between qualitatively different reward types, both within and between reward domains. To choose between different types of food or even between different types of reward (food, water, sex, social contact), animals must have a mechanism to evaluate the utility of these reward types; that is, they must have a valuation mechanism that converts different reward types into a common currency. Padoa-Schioppa and colleagues (2006) explored how capuchin monkeys traded off different amounts of various food types to generate a valuation function (e.g., one piece of apple may be worth three pieces of carrot). Deaner and colleagues (2005) pitted juice rewards against social information in rhesus macaques. Male monkeys chose between receiving juice and viewing images of either higher-ranking males' faces, lower-ranking males' faces, or female perinea (sexual areas). Interestingly, the valuation functions showed that the monkeys would forego juice to view high-ranking males and female perinea but had to be "paid" in juice to view low-ranking males. Comparing these kinds of qualitatively different rewards is a critical capacity for decision making, although we do not have good cognitive models for how these tradeoffs occur.

Inhibitory Control

Organisms must not only discriminate the magnitudes of benefits to establish a preference but also must favor the larger (positive) outcome. When motivated, this is not a problem for primates (except for very large rewards; see Silberberg et al., 1998). In fact, primates have a very difficult time going against this preference. In a task in which chimpanzees had to point to the smaller of two rewards to receive the larger reward, they failed miserably (Boysen & Berntson, 1995). In addition to the chimpanzees, bonobos (*Pan paniscus*), gorillas (*Gorilla gorilla*), orangutans (*Pongo pygmaeus*), rhesus and Japanese macaques (*Macaca fuscata*), squirrel monkeys (*Saimiri sciureus*), tamarins, and lemurs (*Eulemur fulvus* and *E. macaco*) all fail, at least initially, on this task (Anderson et al., 2000; Genty et al., 2004; Kralik et al., 2002; Murray et al., 2005; Silberberg & Fujita, 1996; Vlamings et al., 2006). Clearly, an unboundedly rational agent would adapt quickly to the contingencies of this task, but the preference for a large reward is so powerful that primates cannot inhibit their propensity to choose this. Of course, evolutionarily, it makes sense to employ the simple heuristic "choose the larger." When in an organism's ecology would they opt for a small reward when a larger is present? This must occur only rarely.

UNCERTAINTY AND RISK

In a letter to Jean-Baptiste Leroy, Benjamin Franklin stated that "in this world nothing can be said to be certain, except death and taxes" (13 November 1789). That leaves a lot of uncertainty in the world. As agents navigating in this world, we must deal with this uncertainty in an adaptive manner. Knight (1921) posited a useful distinction between uncertainty (not knowing the distribution of possible payoffs) and risk (knowing the distribution of payoffs but not knowing which payoff will be realized). Though we and other animals frequently face uncertainty, this is difficult to study in the laboratory, so we will focus on risk.

Risky Gambles

Much of the work undermining classical expected utility theory involved asking human subjects about their preferences in risky gambles. Rather than the banana examples suggested previously, subjects chose between risky reward amounts. Would you prefer a 50% chance of receiving $100 (and a 50% chance of receiving nothing) or a 100% chance of receiving $50? Though these two options have equal expected values, most people have a strong preference for the sure thing—they avoid risk. Risk-averse preferences arise with nonlinear utility functions, specifically when utility increases at a slower rate than the magnitude of the benefit (Fig. 6.2). In general, this seems to be the case for human risk preferences for intermediate to large gains (Tversky & Kahneman, 1992).

Interestingly, nonhuman animals seem to show a similar pattern. Animals typically are tested by repeatedly experiencing choices similar to the banana examples mentioned previously.

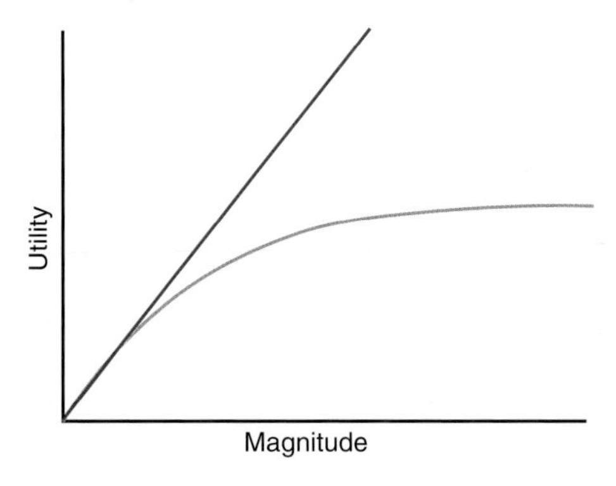

Figure 6.2 If utility is a linear function of magnitude (red line), then decision makers should be neutral to risk; they should be indifferent between a guaranteed banana and a 50/50 chance of no bananas or two bananas. If utility increases at a decelerating rate with magnitude (blue line), increments of utility are less valuable at larger magnitudes than they are at smaller magnitudes. Satiation offers an important biological example of diminishing utility because limited gut capacity constrains the utility of excess amounts of food. The additional utility of receiving three over two bananas is high, but the additional utility of receiving 103 over 102 is not as great, primarily because no one can consume 102 bananas. Diminishing utility implies risk aversion because the additional gain of the risky option is valued less than the loss.

In a review of risk sensitivity in over 25 species, Kacelnik and Bateson (1996) found that most species were either risk averse or risk neutral. Work on primates, however, has provided a more mixed result. Early tests of risk sensitivity in rhesus monkeys showed risk aversion (Behar, 1961). Yet, more recent studies have shown a preference for risky rewards in these macaques (Hayden & Platt, 2007; McCoy & Platt, 2005). Meanwhile, cotton-top tamarins and bonobos (*Pan paniscus*) seem to avoid risk, while common marmosets (*Callithrix jacchus*) ignore risk and chimpanzees prefer risk (Heilbronner et al., 2008, unpublished data). Why does such variation exist?

Variation in Risk Preferences

One of the first hypotheses proposed to account for differences in risk sensitivity was the "energy budget rule" (Caraco et al., 1980; Stephens, 1981). This rule suggests that hungry individuals should prefer risks because the safe option will not allow them to survive. Though this seems to work in some situations, there is no evidence for this rule in primates. Hayden and Platt (2007) tested an alternative idea proposed by Rachlin and colleagues (1986). Instead of preferring risky options because of hunger, animals may prefer risky options when they require low costs—specifically, when another choice will arise soon. Repeatedly choosing the risky option guarantees receiving the large payoff at some point. With short time delays between choices, waiting a few more seconds for this jackpot is not that costly. Rhesus macaques seem to follow this rule. The macaques preferred the risky option more when facing shorter delays between choices than with larger delays (Hayden & Platt, 2007).

The ecological rationality approach may also account for some of the patterns of risk preferences seen in primates. In particular, when species experience risk in their natural ecology, they may have decision rules that bias them toward risky options. With this hypothesis in mind, Heilbronner and colleagues (2008) predicted a species difference in risk preferences between chimpanzees and bonobos. Although their diets overlap quite a bit in their natural habitat, bonobos feed primarily on terrestrial herbaceous vegetation, an abundant and reliable food source, and chimpanzees rely more on fruit, a more temporally and spatially variable food source (Wrangham & Peterson, 1996). Moreover, chimpanzees face risks when they hunt monkeys and other small mammals; bonobos rarely hunt. Interestingly, wild chimpanzees engage in this risky activity more often when fruit is abundant rather than scarce, a direct contrast to the energy budget hypothesis for risk-seeking behavior (Gilby & Wrangham, 2007). Given the generally higher level of risky choice in chimpanzees, Heilbronner and colleagues predicted that this would select for risk-taking decision mechanisms. As predicted, chimpanzees preferred the risky choice in a laboratory experiment, whereas bonobos preferred the safe option. Therefore, to exploit risky options in their natural environment, natural selection has likely endowed chimpanzees with ecologically rational decision mechanisms, yielding preferences for risky outcomes even in captive laboratory situations.

The Framing of Risk

Though risk aversion and risk-seeking preferences do not pose a great challenge to expected utility theory, framing effects do challenge the theory. Prospect theory highlights two types of framing effects: reference dependence and loss aversion (Tversky & Kahneman, 1991). Reference dependence refers to viewing choices as gains or losses relative to a reference point rather than as absolute increases or decreases in utility. Thus, a set of outcomes could result in the exact same levels of wealth but be framed as a gain or loss. In the classic Asian disease problem, a medical treatment has a particular effectiveness in combating a disease (Tversky & Kahneman, 1981). However, when the outcome of a treatment is framed as number of people saved (a gain), subjects prefer the risky option more than when framed as number of people that die (a loss). Thus, framing the exact same outcome as either saving or losing lives greatly influences risk preferences. Yet, this reference dependence is not symmetric. People will try to avoid losses more than they will try to obtain gains—a phenomenon termed loss aversion (Tversky & Kahneman, 1981). We have already shown that, in risky gambles, people typically avoid risk over gains. When experiencing a loss (a sure loss of $50 or a 50% chance of losing $100), however, people prefer risks to avoid the guaranteed loss.

Hundreds of studies have documented the effects of reference dependence and loss aversion in humans (Kahneman & Tversky, 2000). If this is truly a bias, then we might expect to find it only in humans. But if framing effects offer an adaptive, ecologically rational advantage, other animal species may exhibit them. Though first demonstrated in European starlings (Marsh & Kacelnik, 2002), Chen and colleagues (2006) explored reference dependence and loss aversion in capuchin monkeys. To test reference dependence, the monkeys chose between two risky options. In one option, subjects saw one food reward and either received one or two rewards with equal probability. In the other option, they saw two rewards and received either one or two rewards with equal probability. Though identical in outcome, the reference point (number of initial rewards) varied, resulting in a perceived gain or loss. In this condition, subjects strongly preferred the gain option, showing clear reference dependence. Another experiment tested loss aversion. Here, one option consisted of seeing and then receiving a single reward, and the other option consisted of seeing two rewards but always receiving one. Again, the monkeys faced identical outcomes—a guaranteed one reward—but receiving that one reward could have been neutral or perceived as a loss. Again, the monkeys avoided the loss option, revealing the precursors to loss aversion in nonhuman primates. Thus, we share framing effects with other primates, suggesting deep evolutionary roots for this phenomenon.

Uncertainty Building Blocks

Which evolved capacities does an organism need to cope with uncertainty and risk? When given a choice between risky gambles, a decision maker must compare the probabilities of each outcome. In the human risk literature, subjects typically choose based on written descriptions of probabilities (e.g., a 50% chance) and/or on visual displays (e.g., a pie chart with half of the circle colored in). These techniques allow fairly accurate discriminations between probabilities. In the animal risk literature (and in some human experiments; see Hertwig et al., 2004), the subjects repeatedly experience the outcomes to gauge the level of risk. Therefore, to choose between gambles, animals must discriminate probabilities based on experienced outcomes. Though few studies have explicitly tested this, Weber's law may describe probability discrimination. Both Herrnstein and Loveland (1975) and Bailey and Mazur (1990) showed that pigeons' choices for the less risky option increased as the ratio between the small to large probability decreased (probabilities became less similar). In addition, Bailey and Mazur and Krebs et al. (1978) showed that pigeons and starlings (respectively) took more time to stabilize their preferences when the ratio between probabilities increased, further suggesting difficulty in discriminating similar probabilities. Thus, like numerical magnitude, probability discrimination likely follows Weber's law: Individuals can discriminate a 10% from a 20% chance better than an 80% from a 90% chance. This has important implications for how animals deal with risk. When facing unlikely events, animals may discriminate probabilities well and therefore respond appropriately to risk. For more likely events, however, animals may ignore the probabilities and simply focus on the payoffs to determine choice.

TIME

All decisions have a temporal component, from choosing to search for predators instead of searching for prey to delaying reproduction until the next breeding season. Delayed payoffs often have both benefits and costs (Stevens & Stephens, 2009). They may be beneficial when investing time in obtaining resources allows for the extraction of more resources. For instance, the more time chimpanzees spend fishing for termites, the more termites they will extract. Yet, delayed rewards often come with a cost.

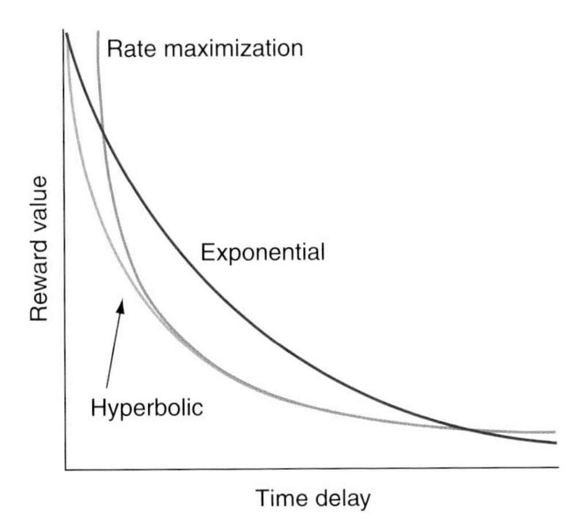

Figure 6.3 Models of intertemporal choice differ in their predictions about how the value of a reward decreases with the time delay to receiving the reward. Exponential discounting (red line) predicts a constant rate of decrease over time. Hyperbolic discounting (green line) predicts a decreasing rate of decrease over time, such that decision makers exhibit high discount rates at short-term delays but lower discount rates at longer delays. Rate maximization (blue line) predicts similar patterns as hyperbolic, albeit with strange behavior at very small time delays. The advantages of rate maximization models are that they include repeated choices and have biologically relevant parameters.

Temporal Discounting

When facing options with smaller, sooner payoffs and larger, later payoffs, animals must make an *intertemporal choice*; that is, they must trade off the magnitude of rewards with the delay to receiving them (reviewed in Read, 2004). In some cases, ignoring the temporal component and choosing based on magnitude is best, but in other cases, a long delay may prove too costly. How should animals deal with this tradeoff? They may discount or devalue delayed rewards because the future is uncertain. The risk of not collecting a reward grows with delay because some event may interrupt its collection. For instance, a predator may interrupt an extended courtship or a bank may collapse before an investment matures. Economists have modified the expected utility models to create a *discounted utility model* of delayed benefits (Samuelson, 1937). This model replaces p from the expected utility model with a discounting function that includes a constant rate of interruption λ per unit time. Thus, for a reward amount A delayed for t time units, $DU = e^{-\lambda t} \cdot u(A)$. Again, utility is difficult to assess, so most versions of this model drop utility and just discount the absolute reward amount: $V = A \cdot e^{-\lambda t}$. Because the value of a reward decays exponentially with time, this is called the *exponential model* of discounting (Fig. 6.3).

Though intuitively appealing, the data do not support the exponential model. Humans, pigeons, and rats violate predictions of this model in self-control experiments (Fig. 6.4) by choosing between a smaller, immediate reward and a larger, delayed reward (Ainslie & Herrnstein, 1981; Frederick et al., 2002; Mazur, 1987; Richards et al., 1997). In fact, when choosing between immediately receiving two pieces of food and waiting for six, rats and pigeons only wait a few seconds for three times as much food. Animals would have to face extraordinarily (and unrealistically) high interruption rates for discounting by interruptions to account for this level of impulsivity.

Psychologists proposed an alternative model that captures the data much better: the *hyperbolic discounting model* (Ainslie, 1975; Mazur, 1987). In the hyperbolic model,[6] $V = A / (1 + kt)$, where k represents a fitted parameter that describes the steepness of discounting. Rather than predict a constant rate of discounting over time, this model predicts a decelerating rate over

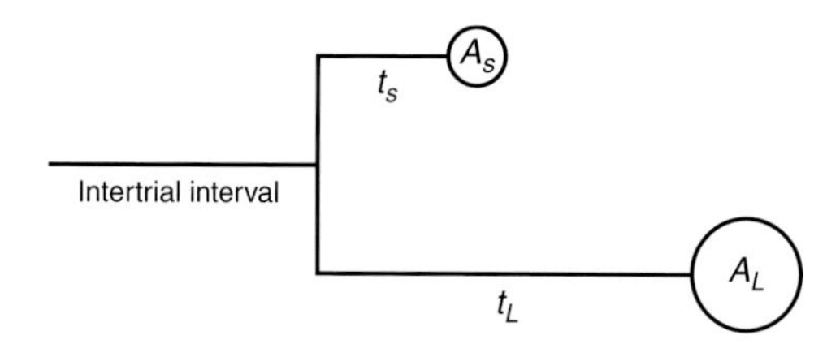

Figure 6.4 In self-control experiments, animals face a choice between a small reward (A_S) after a short delay (t_S) and a large reward (A_L) after a long delay (t_L). The animals wait for an intertrial interval, choose one option, wait the specified delay, consume their food, and then begin another intertrial interval.

time—decision makers steeply discount at short time delays, and the rate of discounting declines at longer delays (Fig. 6.3). This model nicely matches the data in humans, rats, and pigeons (Bickel & Johnson, 2003; Green et al., 2004). A number of studies have tested intertemporal choice in primate species, including cotton-top tamarins, common marmosets, brown capuchins, long-tailed and rhesus macaques, chimpanzees, and bonobos (Dufour et al., 2007; Ramseyer et al., 2006; Rosati et al., 2007; Stevens et al., 2005; Szalda-Petree et al., 2004; Tobin et al., 1996). Unfortunately, we have too few quantitative data on primates to distinguish between these two models. Nevertheless, in cotton-top tamarins and common marmosets, the rate of discounting slows with time, contradicting predictions of the exponential model (Hallinan et al., unpublished manuscript).

Though the hyperbolic model has achieved great empirical success, it suffers from a critical disadvantage: Like prospect theory, it is a purely descriptive model. It cannot make a priori predictions about intertemporal choices and thus lacks explanatory power (Stevens & Stephens, 2009). In addition, both exponential and hyperbolic models of discounting rest on the assumption of one-shot choice. In self-control experiments, however, animals face repeated choices between the same options, and the repeated nature of these experiments has important implications for models of intertemporal choice (Kacelnik, 2003; Stevens & Stephens, 2009).

Rate

The evolutionary approach to optimal foraging theory takes as its major assumption the fact that animals maximize their intake rate, that is, the amount of food gained per unit time (Stephens & Krebs, 1986). In the classic patch-choice scenario, an animal travels among the many patches of food in his or her environment and extracts resources from each patch. The question is: At what point should the animal stop extracting resources from the current patch and move on to search for a new patch? This scenario sounds quite similar to the self-control situation tested in the laboratory. Animals must choose between staying in a patch for a short time to extract a small gain and staying for a longer time to extract a larger gain. Optimal foraging theory makes predictions about how long to stay in a patch: stay until the intake rate drops below the background intake rate in the environment (Stephens & Krebs, 1986).

The rate-based approach to intertemporal choice has two key advantages over the standard hyperbolic approach. First, it is by definition a model of repeated choice. Each decision faced by animals is embedded in a series of decisions. Thus, the rate-based approach models choices in situations very similar to the self-control experiments. Second, rate models provide an explanation of the pattern of temporal preferences seen in animals rather than just a description. Animals that maximize their intake rates should survive and reproduce better than their counterparts that do not—rate models are built on a foundation of evolutionary fitness.

We have sound theoretical reasons to favor rate models, but how do they fare empirically? Actually, quite well. The short-term rate model makes similar predictions as the hyperbolic model (Fig. 6.3). Bateson and Kacelnik (1996) first demonstrated the effectiveness of the short-term rate model in describing choices by European starlings in a self-control experiment.[7]

Stevens and colleagues (2005) then tested a variant of this model: $A / (t + h)$, where h represents the time required to handle the food, an important component of the foraging timeline (Rosati et al., 2006). Cotton-top tamarins chose the option that maximized this intake rate. Like rats and pigeons, the tamarins waited only 6 to 10 seconds for three times as much food. However, the results do not appear so "impulsive" because we have an explanation for waiting such short delays that does not require unrealistic rates of interruption (Stevens & Stephens, 2009).

Data on bonobos also agree with rate maximization predictions, but this time long-term rather than short-term rate predictions (Rosati et al., 2007). The long-term rate $A / (\tau + t + h)$ includes τ, the time between trials (or between patches in the natural scenario). The long-term rate includes all of the relevant time intervals, and therefore this is the standard model used in optimal foraging theory. Why would some species ignore the intertrial interval while other species include it? An ecological rationality approach may help answer this question.

Ecological Rationality of Intertemporal Choice

Tamarins and bonobos match predictions of rate-based models of intertemporal choice. Yet, bonobos wait over a minute for three times as much food, whereas tamarins wait less than 10 seconds. Why might the tamarins may use short-term rates when making these types of choices? An answer lies at the heart of ecological rationality. Namely, the natural decision environment strongly shapes the decision mechanisms. Though the self-control experiments appear similar to natural patch-foraging scenarios, a key difference arises: Animals rarely face simultaneous choices in nature but often face sequential choices in patch situations. Rather than facing two binary options, animals regularly choose to stay or leave a patch. Stephens and Anderson (2001) argued that this represents the natural foraging decision that animals typically confront, and therefore this should be the situation for which decision mechanisms evolved. Interestingly, a decision rule that assesses short-term intake rates in a patch situation will automatically result in long-term rate maximization because they are logically equivalent (Stephens & Anderson, 2001).

If short- and long-term rules are logically equivalent, this then may explain why bonobos differ from the tamarins and marmosets. Both represent "adaptive peaks," so it does not matter on which peak a particular species rests. Each rule, however, has advantages and disadvantages. The short-term rule has the advantage of simplicity and possibly higher accuracy. It is simpler because it ignores information, namely, the intertrial interval. This may also increase accuracy because, by ignoring this time interval, animals can estimate much smaller intervals, which results in more accurate estimates (Stephens et al., 2004). Despite the benefits of the short-term rule, using the long-term rule provides advantages as well, primarily because it applies broadly and should result in the "correct" outcome in most cases. Unfortunately, this is not true for the short-term rule. Though the short-term rule works well in patch situations, this does not directly carry over to the self-control situation (Stephens & Anderson, 2001). In fact, as shown by the tamarins, the short-term rule results in "impulsive" choice in self-control situations, so the animals do not achieve the optimal long-term rate of gain. Thus, tamarins may use the short-term rule because it works well in a natural foraging task—patch exploitation. When placed in an artificial situation, the rule "misfires." Bonobos, in contrast, may possess more acute estimation abilities and therefore use the more powerful long-term rule.

Specialized diets also raise interesting questions about ecologically rational intertemporal choice. Common marmosets offer an intriguing case because they specialize on a rather unique food source: tree exudates. Marmosets have specialized teeth, as well as digestive morphology and physiology, adapted to gouging holes in tree bark and eating the sap and gum that exude from the holes (Rylands & de Faria, 1993; Stevenson & Rylands, 1988). Because this foraging strategy requires waiting for the sap to exude, Stevens and colleagues (2005) proposed an ecologically rational response to this type of foraging strategy: A decision mechanism that is biased toward waiting for longer delays. Compared to the more insectivorous tamarins, marmosets should show stronger preferences for delayed rewards. As predicted, the marmosets waited longer than tamarins in a self-control situation, suggesting that the natural foraging ecology shapes the decision mechanisms (Stevens et al., 2005).

Intertemporal Building Blocks

Time poses unique cognitive challenges for organisms. Intertemporal choice combines establishing preferences over rewards with tracking these payoffs over time. Now, we explore what kind of cognitive building blocks might be required to make intertemporal choices.

Time Perception

To cope with delayed rewards, an organism must perceive the delay. Unfortunately, we know little about primate time perception. We do, however, know a lot about time perception in rats, which likely applies to primates. Like magnitude judgments, time interval judgments seem to follow Weber's law. In fact, Gibbon (1977) showed that variance scaled with magnitude in time perception before applying it to quantification. Given that both primate numerical judgments (see Chapter 8) and human time perception (Allan, 1998) follow Weber's law, likely nonhuman primate time perception does as well. This has important implications for the study of intertemporal choice. First, logarithmic time perception (resulting from Weber's law) may result in the hyperbolic pattern of intertemporal choice typically seen in both humans and other animals (Takahashi, 2005; Takahashi et al., 2008). This finding could rescue the exponential model of discounting by overlaying a subjective time perception function. Second, as we have already seen, long time delays make discriminating the delays difficult. If time delays are viewed as equivalent, decision makers should opt for the larger (and therefore later) reward, resulting in more patient choice.

Delayed Gratification

In addition to perceiving time, delayed rewards require a motivational ability to wait—decision makers must delay gratification. Mischel and colleagues pioneered the study of delayed gratification by measuring how long children would wait for delayed rewards (Mischel & Ebbesen, 1970; Mischel et al., 1989). They recorded the ability to wait at different ages and showed that delayed gratification at a young age strongly predicted intelligence, academic success, standardized test scores, and drug use much later in life.

Beran and colleagues have tested primates in a similar paradigm by offering chimpanzees and rhesus macaques a stream of food rewards (Beran et al., 1999; Evans & Beran, 2007b). The experimenter placed rewards in front of the subjects one by one at a particular rate, say one every 10 seconds. If the subject began consuming the food, however, the experimenter stopped the flow of food. Chimpanzees performed quite well on this task and some waited for over 10 minutes for the stream of food to be completed (Beran & Evans, 2006). This corroborates findings in the self-control task in which chimpanzees wait longer than any other species tested so far, including bonobos (Dufour et al., 2007; Rosati et al., 2007). Interestingly, this ability can be influenced by attentional factors. Evans and Beran (2007a) found that offering chimpanzees a distraction during the waiting period could significantly increase their abilities to delay gratification. Thus, delayed gratification could provide a key building block required for patient choice, but it too is mediated by other factors such as attention.

CONCLUSION

Bounded rationality and ecological rationality are both firmly grounded in an evolutionary perspective on decision making. Natural selection places limitations on cognitive capacities and tailors cognitive mechanisms to the environment in which they are used (Barkow et al., 1992; Stevens, 2008). Darwin (1871, 1872) emphasized a continuity between the "mental powers" of humans and other animals, and the analysis provided here confirms a similar continuity for decision making. Nonhuman primates show many of the so-called biases or anomalies demonstrated by human decision makers. The human decision-making literature has highlighted these results as failures to achieve a normative outcome. The evolutionary approach suggests that the bias lies not in the behavior but in the normative criteria used. Organisms did not evolve to follow a mathematically tractable set of principles—rather, natural selection favored decision strategies that resulted in greater survival and reproduction. In some cases, the evolutionary and normative perspectives may overlap. However, the normative perspective often fails to properly account for the role of the environment in decision making, whereas this is a critical part of the evolutionary view. Natural selection shapes decision strategies to match the environment.

This lesson about the limitations of the normative approach is an important one because many fields, especially neuroscience, use tools borrowed from economics to greatly expand the study of decision making. Though this could prove a fruitful enterprise, the data on decision making in other animals caution against relying exclusively on the normative models of decision making. As a biological science, neuroscience should heed the warning of Dobzhansky (1964): "nothing makes sense in biology except in light of evolution" (p. 449). Of course, the emphasis on evolutionarily plausible models is nothing new to the field of neuroethology, and this tradition should continue in the study of decision making. Rather than looking for expected or discounted utility in the brain, perhaps we should test hypotheses about the bounded and ecological rationality of decision making. The bounded rationality approach makes clear predictions about what cognitive building blocks might be recruited, and neuroscience can help test this. In addition, neuroscience can allow us to explore what aspects of the environment are relevant for triggering specific decision rules. The neuroethological approach offers an ideal interface between the evolutionary and mechanistic approaches to decision making.

The study of primate decision making has much to offer the larger field of decision making, and prospects are bright for future contributions. To advance the field, we must begin exploring primate decisions in the wild or at least in more naturalistic situations. Currently, most studies of primate decisions (and animal decisions more broadly) occur in the laboratory with rather artificial scenarios. Though these studies provide valuable information on primate decision making, they may also lead us astray. If natural selection tailors decision mechanisms to the environment in which they are adapted, then the artificial nature of the laboratory might not trigger the appropriate mechanism. In this case, behavior seen in both humans and other animals may simply be spurious results elicited by an unnatural environment (Houston et al., 2007b; Stevens & Stephens, 2009). Thus, exploring natural behaviors in natural environments underlies our understanding of primate decisions.

Another important advance that primate researchers may offer is the development and testing of process models of decision making. David Marr (1982) introduced the idea of three levels of information processing. The computational level emphasizes the goal of the system, the algorithmic level emphasizes the processes used, and the implementational level emphasizes the neural circuitry required to process information. Most models of decision making in animals act at the computational level—that is, level of evolutionary function—in the field of behavioral ecology or the implementational level in the field of neurobiology. The algorithmic level has been greatly neglected as a relevant level of analysis. Yet, adding an analysis of the relevant cognitive processes and algorithms could constrain and improve evolutionary models of decision making. Only an integrative approach across these levels of analysis—from the evolutionary to the psychological to the neurobiological—will yield satisfying answers to questions of the nature of decision making.

NOTES

1. Utility does not necessarily increase linearly with magnitude. The difference between consuming one and three bananas is not the same as the difference between consuming 11 and 13 bananas. A difference of two bananas is much more relevant when there are fewer to begin with.
2. Kahneman and Tversky do not use the term "irrational." They describe deviations from the normative theory as "biases," "fallacies," or "cognitive illusions." Nevertheless, the emphasis on putative errors highlights the irrational nature of human decision making (Rieskamp et al., 2006a).
3. Most work on decision making in animals, including primates, involves food as the reward, though there are important exceptions (Deaner et al., 2005). Therefore, in this review, I will also focus on food as the reward domain. Nevertheless, many of the principles mentioned here apply to other reward domains, and the study of animal decision making needs more in-depth exploration of other reward types.
4. Though we have few data on transitive *preference*, transitive *inference* is well studied in primates. Transitive preference refers to an ordered preference over choices. Transitive inference refers to the ability to infer a transitive relation between objects (Vasconcelos, 2008). For instance, if individual A is dominant to B and B is dominant to C, can an individual infer that A is dominant to C without actually seeing A and C interact? Primates are quite good at these tasks, both with arbitrary objects in artificial laboratory tasks (reviewed in Tomasello & Call, 1997) and with social agents in the more naturalistic social inferences (Cheney & Seyfarth, 1990).
5. Despite the sparse data on animal intransitivity, Houston and colleagues (Houston, 1997; Houston et al., 2007a) have developed a series of models demonstrating that intransitivity might be a perfectly adaptive strategy when a decision maker is either updating his or her assessment of the environment or betting on a changing environment. Schuck-Paim and Kacelnik (2004) also assert that intransitive choices (particularly when food is the reward) can be perfectly biologically rational (Kacelnik, 2006) when the internal state changes (i.e., decision makers become satiated).
6. Though $V = A/(1 + kt)$ is called the hyperbolic model, it represents only a single instance from a class of hyperbolic models. In fact, most nonexponential models (including the rate models) are hyperbolic. Nevertheless, the term "hyperbolic discounting" typically refers to Mazur's (1987) version.
7. The key difference between the short-term and long-term rate models is that the short-term model ignores the travel time or intertrial interval. Thus, animals using the short-term rate only focus on the times between choice presentation and reward consumption.

REFERENCES

Ainslie, G. (1975). Specious reward: A behavioral theory of impulsiveness and impulse control. *Psychological Bulletin, 82,* 463–496.

Ainslie, G., & Herrnstein, R. J. (1981). Preference reversal and delayed reinforcement. *Animal and Learning Behavior, 9,* 476–482.

Allais, P. M. (1953). The behavior of rational man in risk situations: A critique of the axioms and postulates of the American School. *Econometrica, 21,* 503–546.

Allan, L. G. (1998). The influence of the scalar timing model on human timing research. *Behavioural Processes, 44,* 101–117.

Anderson, J. R., Awazu, S., & Kazuo, F. (2000). Can squirrel monkeys (*Saimiri sciureus*) learn self-control? A study using food array selection tests and reverse-reward contingency. *Journal of Experimental Psychology: Animal Behavior Processes, 26,* 87–97.

Bailey, J. T., & Mazur, J. E. (1990). Choice behavior in transition: Development of preference for the higher probability of reinforcement. *Journal of the Experimental Analysis of Behavior, 53,* 409–422.

Barkow, J. H., Cosmides, L., & Tooby, J. (1992). *The adapted mind: Evolutionary psychology and the generation of culture.* New York: Oxford University Press.

Baron, J. (2000). *Thinking and deciding.* Cambridge, UK: Cambridge University Press.

Bateson, M., & Kacelnik, A. (1996) Rate currencies and the foraging starling: The fallacy of the averages revisited. *Behavioral Ecology, 7,* 341–352.

Behar, I. (1961). Learned avoidance of nonreward. *Psychological Reports, 9,* 43–52.

Beran, M. J., & Evans, T. A. (2006). Maintenance of delay of gratification by four chimpanzees (*Pan troglodytes*): The effects of delayed reward visibility, experimenter presence, and extended delay intervals. *Behavioural Processes, 73,* 315–324.

Beran, M. J., Savage-Rumbaugh, E., Pate, J. L., & Rumbaugh, D. M. (1999). Delay of gratification in chimpanzees (*Pan troglodytes*). *Developmental Psychobiology, 34,* 119–127.

Bickel, W. K., & Johnson, M. W. (2003). Delay discounting: A fundamental behavioral process of drug dependence. In: G. Loewenstein, D. Read, & R. F. Baumeister (Eds.), *Time and decision: Economic and psychological perspectives on intertemporal choice* (pp. 419–440). New York: Russell Sage Foundation.

Boysen, S. T., & Berntson, G. G. (1995). Responses to quantity: Perceptual versus cognitive mechanisms in chimpanzees (*Pan troglodytes*). *Journal of Experimental Psychology: Animal Behavior Processes, 21,* 82–86.

Brannon, E. M. (2005a). Quantitative thinking: From monkey to human and human infant to adult. In: S. Dehaene, J-R. Duhamel, M. D. Hauser, & G. Rizzolatti (Eds.), *From monkey brain to human brain* (pp. 97–116). Cambridge, MA: MIT Press.

Brannon, E. M. (2005b). The numerical ability of animals. In: J. D. Campbell (Ed.), *Handbook of mathematical cognition* (pp. 85–108). New York: Psychology Press.

Brosnan, S. F., Jones, O. D., Lambeth, S. P., Mareno, M. C., Richardson, A. S., & Schapiro, S. J. (2007). Endowment effects in chimpanzees. *Current Biology, 17,* 1704–1707.

Caraco, T., Martindale, S., & Whittam, T. S. (1980). An empirical demonstration of risk-sensitive foraging preferences. *Animal Behaviour, 28,* 820–830.

Chen, M. K., Lakshminarayanan, V., & Santos, L. R. (2006). How basic are behavioral biases? Evidence from capuchin monkey trading behavior. *Journal of Political Economy, 114,* 517–537.

Cheney, D. L., & Seyfarth, R. M. (1990). *How monkeys see the world: Inside the mind of another species.* Chicago: University of Chicago Press.

Darwin, C. (1871). *The descent of man and selection in relation to sex.* London: Murray.

Darwin, C. (1872). *The expression of the emotions in man and animals.* London: Murray.

Deaner, R. O., Khera, A. V., & Platt, M. L. (2005). Monkeys pay per view: Adaptive valuation of social images by rhesus macaques. *Current Biology, 15,* 543–548.

Dobzhansky, T. (1964). Biology, molecular and organismic. *American Zoologist, 4,* 443–452.

Dufour, V., Pele, M., Sterck, E. H. M., & Thierry, B. (2007). Chimpanzee (*Pan troglodytes*) anticipation of food return: Coping with waiting time in an exchange task. *Journal of Comparative Psychology, 121,* 145–155.

Ellsberg, D. (1961). Risk, ambiguity, and the Savage axioms. *Quarterly Journal of Economics, 75,* 643–669.

Evans, T. A., & Beran, M. J. (2007a). Chimpanzees use self-distraction to cope with impulsivity. *Biology Letters, 3,* 599–602.

Evans, T. A., & Beran, M. J. (2007b). Delay of gratification and delay maintenance by rhesus macaques (*Macaca mulatta*). *Journal of General Psychology, 134,* 199–216.

Frederick, S., Loewenstein, G., & O'Donoghue, T. (2002). Time discounting and time preference: A critical review. *Journal of Economic Literature, 40,* 351–401.

Gallistel, C. R. (1990). *The organization of learning.* Cambridge, MA: MIT Press.

Genty, E., Palmier, C., & Roeder, J-J. (2004). Learning to suppress responses to the larger of two rewards in two species of lemurs, *Eulemur fulvus* and *E. macaco*. *Animal Behaviour, 67,* 925–932.

Gibbon, J. (1977). Scalar expectancy theory and Weber's law in animal timing. *Psychological Review, 84,* 279–325.

Gigerenzer, G., Todd, P. M., & the ABC Research Group. (1999). *Simple heuristics that make us smart.* Oxford: Oxford University Press.

Gilby, I. C., & Wrangham, R. W. (2007). Risk-prone hunting by chimpanzees (*Pan troglodytes schweinfurthii*) increases during periods of high diet quality. *Behavioral Ecology and Sociobiology, 61,* 1771–1779.

Green, L., Myerson, J., Holt, D. D., Slevin, J. R., & Estle, S. J. (2004). Discounting of delayed food rewards in pigeons and rats: Is there a magnitude effect? *Journal of the Experimental Analysis of Behavior, 81,* 39–50.

Hallinan, E. V., Stevens, J. R., & Hauser, M. D. (Unpublished manuscript). Rate maximization in common marmosets (*Callithrix jacchus*) and cotton-top tamarins (*Saguinus oedipus*).

Hammerstein, P. (1981). The role of asymmetries in animal contests. *Animal Behaviour, 29,* 193–205.

Hauser, M. D., Carey, S., & Hauser, L. B. (2000). Spontaneous number representation in semi-free-ranging rhesus monkeys. *Proceedings of the Royal Society of London Series B, 267,* 829–833.

Hauser, M. D., & Spelke, E. S. (2004). Evolutionary and developmental foundations of human knowledge: A case study of mathematics. In: M. Gazzaniga (Ed.), *The cognitive neurosciences III* (pp. 853–864). Cambridge, MA: MIT Press.

Hauser, M. D., Tsao, F., Garcia, P., & Spelke, E. S. (2003). Evolutionary foundations of number: Spontaneous representation of numerical magnitudes by cotton-top tamarins. *Proceedings of the Royal Society of London Series B, 270,* 1441–1446.

Hayden, B. Y., & Platt, M. L. (2007). Temporal discounting predicts risk sensitivity in rhesus macaques. *Current Biology, 17,* 49–53.

Heilbronner, S. R., Rosati, A. G., Stevens, J. R., Hare, B., & Hauser, M. D. (2008). A fruit in the hand or two in the bush? Divergent risk preferences in chimpanzees and bonobos. *Biology Letters, 4,* 246–249.

Herrnstein, R. J., & Loveland, D. H. (1975). Maximizing and matching on concurrent ratio schedules. *Journal of the Experimental Analysis of Behavior, 24,* 107–116.

Hertwig, R., Barron, G., Weber, E. U., & Erev, I. (2004). Decisions from experience and the effect of rare events in risky choice. *Psychological Science, 15,* 534–539.

Houston, A. I. (1997). Natural selection and context-dependent values. *Proceedings of the Royal Society of London Series B, 264,* 1539–1541.

Houston, A. I., McNamara, J. M., & Steer, M. D. (2007a). Violations of transitivity under fitness maximization. *Biology Letters, 3,* 365–367.

Houston, A. I., McNamara, J. M., & Steer, M. D. (2007b). Do we expect natural selection to produce rational behaviour? *Philosophical Transactions of the Royal Society Series B, 362,* 1531–1543.

Kacelnik, A. (2003). The evolution of patience. In: G. Loewenstein, D. Read, & R. F. Baumeister (Eds.), *Time and decision: Economic and psychological perspectives on intertemporal choice* (pp. 115–138). New York: Russell Sage Foundation.

Kacelnik, A. (2006). Meanings of rationality. In: S. Hurley & M. Nudds (Eds.), *Rational animals?* (pp. 87–106). Oxford: Oxford University Press.

Kacelnik, A., & Bateson, M. (1996). Risky theories: The effects of variance on foraging decisions. *American Zoologist, 36,* 402–434.

Kahneman, D., Knetsch, J. L., & Thaler, R. H. (1990). Experimental tests of the endowment effect and the Coase theorem. *Journal of Political Economy, 98,* 1325–1348.

Kahneman, D., Slovic, P., & Tversky, A. (1982). *Judgment under uncertainty: Heuristics and biases.* New York: Cambridge University Press.

Kahneman, D., & Tversky, A. (1979). Prospect theory: An analysis of decision under risk. *Econometrica, 47,* 263–292.

Kahneman, D., & Tversky, A. (2000). *Choices, values, and frames.* New York: Russell Sage Foundation; Cambridge University Press.

Knight, F. H. (1921). *Risk, uncertainty, and profit.* Boston: Houghton Mifflin.

Kralik, J. D., Hauser, M. D., & Zimlicki, R. (2002). The relationship between problem solving and inhibitory control: Cotton-top tamarin (*Saguinus oedipus*) performance on a reversed contingency task. *Journal of Compartive Psychology, 116,* 39–50.

Krebs, J. R., & Davies, N. B. (1993). *An introduction to behavioural ecology.* Oxford: Blackwell Science.

Krebs, J. R., Kacelnik, A., & Taylor, P. (1978). Test of optimal sampling by foraging great tits. *Nature, 275,* 27–31.

Lakshminaryanan, V., Chen, M. K., & Santos, L. R. (2008). Endowment effect in capuchin monkeys. *Philosophical Transactions of the Royal Society of London, Series B, 363,* 3837–3844.

Luce, R. D., & Raiffa, H. (1957). *Games and decisions: Introduction and critical survey.* New York: Wiley.

Marr, D. (1982). *Vision: A computational investigation into the human representation and processing of visual information.* New York: W. H. Freeman and Co.

Marsh, B., & Kacelnik, A. (2002). Framing effects and risky decisions in starlings. *Proceedings of the National Academy of Sciences, 99,* 3352–3355.

Maynard Smith, J., & Parker, G. A. (1976). The logic of asymmetric contests. *Animal Behaviour, 24,* 159–175.

Mazur, J. E. (1987). An adjusting procedure for studying delayed reinforcement. In: M. L. Commons, J. E. Mazur, J. A. Nevin, & H. Rachlin (Eds.), *Quantitative analyses of behavior: The effect of delay and of intervening events on reinforcement value* (pp. 55–73). Hillsdale, NJ: Lawrence Erlbaum Associates.

McCoy, A. N., & Platt, M. L. (2005). Risk-sensitive neurons in macaque posterior cingulate cortex. *Nature Neuroscience, 8,* 1220–1227.

Mischel, W., & Ebbesen, E. B. (1970). Attention in delay of gratification. *Journal of Personality and Social Psychology, 16,* 329–337.

Mischel, W., Shoda, Y., & Rodriguez, M. L. (1989). Delay of gratification in children. *Science, 244,* 933–938.

Murray, E. A., Kralik, J. D., & Wise, S. P. (2005). Learning to inhibit prepotent responses: Successful performance by rhesus macaques, *Macaca mulatta,* on the reversed-contingency task. *Animal Behaviour, 69,* 991–998.

Padoa-Schioppa, C., Jandolo, L., & Visalberghi, E. (2006). Multi-stage mental process for economic choice in capuchins. *Cognition, 99,* B1–B13.

Payne, J. W., Bettman, J. R., & Johnson, E. J. (1993). *The adaptive decision maker.* Cambridge: Cambridge University Press.

Rachlin, H., Logue, A. W., Gibbon, J., & Frankel, M. (1986). Cognition and behavior in studies of choice. *Psychological Review, 93,* 33–45.

Ramseyer, A., Pele, M., Dufour, V., Chauvin, C., & Thierry, B. (2006). Accepting loss: The temporal limits of reciprocity in brown capuchin monkeys. *Proceedings of the Royal Society of London Series B, 273,* 179–184.

Read, D. (2004). Intertemporal choice. In: D. Koehler & N. Harvey (Eds.), *Blackwell handbook of judgment and decision making* (pp. 424–443). Oxford: Blackwell.

Richards, J. B., Mitchell, S. H., de Wit, H., & Seiden, L. S. (1997). Determination of discount functions in rats with an adjusting-amount procedure. *Journal of the Experimental Analysis of Behavior, 67,* 353–366.

Rieskamp, J., Hertwig, R., & Todd, P. M. (2006a). Bounded rationality: Two interpretations from psychology. In: M. Altman (Ed.), *Handbook of contemporary behavioral economics* (pp. 218–236). Armonk, NY: M.E. Sharpe.

Rieskamp, J., Busemeyer, J. R., & Mellers, B. A. (2006b). Extending the bounds of rationality: Evidence and theories of preferential choice. *Journal of Economic Literature, 44,* 631–661.

Rosati, A. G., Stevens, J. R., Hare, B., & Hauser, M. D. (2007). The evolutionary origins of human patience: Temporal preferences in chimpanzees, bonobos, and adult humans. *Current Biology, 17,* 1663–1668.

Rosati, A. G., Stevens, J. R., & Hauser, M. D. (2006). The effect of handling time on temporal discounting in two New World primates. *Animal Behaviour, 71,* 1379–1387.

Rylands, A. B., & de Faria, D. S. (1993). Habitats, feeding ecology and range size in the genus *Callithrix.* In: A. B. Rylands (Ed.), *Marmosets and tamarins: Systematics, behaviour, and ecology* (pp. 262–272). Oxford: Oxford University Press.

Samuelson, P. A. (1937). A note on measurement of utility. *Review of Economic Studies, 4,* 155–161.

Schuck-Paim, C., Pompilio, L., & Kacelnik, A. (2004). State-dependent decisions cause apparent violations of rationality in animal choice. *PLoS Biology, 2,* e402.

Shafir, S. (1994). Intransitivity of preferences in honey-bees: Support for comparative-evaluation of foraging options. *Animal Behaviour, 48,* 55–67.

Silberberg, A., & Fujita, K. (1996). Pointing at smaller food amounts in an analogue of Boysen and Berntson's (1995) procedure. *Journal of the Experimental Analysis of Behavior, 66,* 143–147.

Silberberg, A., Widholm, J. J., Bresler, D., Fujita, K., & Anderson, J. R. (1998). Natural choice in nonhuman primates. *Journal of Experimental Psychology: Animal Behavior Processes, 24,* 215–228.

Simon, H. A. (1955). A behavioral model of rational choice. *Quarterly Journal of Economics, 69,* 99–118.

Simon, H. A. (1956). Rational choice and the structure of the environment. *Psychological Review, 63,* 129–138.

Simonson, I., & Tversky, A. (1992) Choice in context: Tradeoff contrast and extremeness aversion. *Journal of Marketing Research, 29,* 281–295.

Stephens, D. W. (1981). The logic of risk-sensitive foraging preferences. *Animal Behaviour, 29,* 628–629.

Stephens, D. W., & Anderson, D. (2001). The adaptive value of preference for immediacy: When shortsighted rules have farsighted consequences. *Behavioral Ecology, 12,* 330–339.

Stephens, D. W., Kerr, B., & Fernandez-Juricic, E. (2004). Impulsiveness without discounting: The ecological rationality hypothesis. *Proceedings of the Royal Society of London Series B, 271,* 2459–2465.

Stephens, D. W., & Krebs, J. R. (1986). *Foraging theory.* Princeton: Princeton University Press.

Stevens, J. R. (2008). The evolutionary biology of decision making. In: C. Engel & W. Singer

(Eds.), *Better than conscious? Decision making, the human mind, and implications for institutions* (pp. 285–304). Cambridge, MA: MIT Press.

Stevens, J. R., Hallinan, E. V., & Hauser, M. D. (2005). The ecology and evolution of patience in two New World monkeys. *Biological Letters, 1,* 223–226.

Stevens, J.R., & King, A.J. (in press). The lives of others: social rationality in animals. In: R. Hertwig, U. Hoffrage, & the ABC Research Group (Eds.), *Social rationality.* Oxford: Oxford University Press.

Stevens, J. R., & Stephens, D. W. (2009) The adaptive nature of impulsivity. In: G. J. Madden & W. K. Bickel (Eds.), *Impulsivity: Theory, science, and neuroscience of discounting* (pp. 361–387). Washington, DC: American Psychological Association.

Stevens, J. R., Wood, J. N., & Hauser, M. D. (2007). When quantity trumps number: Discrimination experiments in cotton-top tamarins (*Saguinus oedipus*) and common marmosets (*Callithrix jacchus*). *Animal Cognition, 10,* 429–437.

Stevenson, M. F., & Rylands, A. B. (1988). The marmosets, genus *Callithrix.* In: R. A. Mittermeier, A. B. Rylands, A. F. Coimbra-Filho, & G. A. B. Fonseca (Eds.), *Ecology and behavior of neotropical primates* (pp. 131–222). Washington, DC: World Wildlife Fund.

Szalda-Petree, A. D., Craft, B. B., Martin, L. M., & Deditius-Island, H. K. (2004). Self-control in rhesus macaques (*Macaca mulatta*): Controlling for differential stimulus exposure. *Perceptual and Motor Skills, 98,* 141–146.

Takahashi, T. (2005). Loss of self-control in intertemporal choice may be attributable to logarithmic time-perception. *Medical Hypotheses, 65,* 691–693.

Takahashi, T., Oono, H., & Radford, M. H. B. (2008). Psychophysics of time perception and intertemporal choice models. *Physica A: Statistical Mechanics and Its Applications, 387,* 2066–2074.

Tinklepaugh, O. L. (1928). An experimental study of representative factors in monkeys. *Journal of Comparative Psychology, 8,* 197–236.

Tobin, H., Logue, A. W., Chelonis, J. J., & Ackerman, K. T. (1996). Self-control in the monkey *Macaca fascicularis. Animal Learning and Behavior, 24,* 168–174.

Tomasello, M., & Call, J. (1997). *Primate cognition.* Oxford: Oxford University Press.

Tversky, A., & Kahneman, D. (1981). The framing of decisions and the psychology of choice. *Science, 211,* 453–458.

Tversky, A., & Kahneman, D. (1991). Loss aversion in riskless choice: A reference-dependent model. *Quarterly Journal of Economics, 106,* 1039–1061.

Tversky, A., & Kahneman, D. (1992). Advances in prospect theory: Cumulative representation of uncertainty. *Journal of Risk and Uncertainty, 5,* 297–323.

Uller, C., Hauser, M. D., & Carey, S. (2001). Spontaneous representation of number in cotton-top tamarins (*Saguinus oedipus*). *Journal of Comparative Psychology, 115,* 248–257.

Vasconcelos, M. (2008). Transitive inference in non-human animals: An empirical and theoretical analysis. *Behavioural Processes, 78,* 313–334.

Vlamings, P., Uher, J., & Call, J. (2006). How the great apes (*Pan troglodytes, Pongo pygmaeus, Pan paniscus,* and *Gorilla gorilla*) perform on the reversed contingency task: The effects of food quantity and food visibility. *Journal of Experimental Psychology: Animal Behavior Processes, 32,* 60–70.

von Neumann, J., & Morgenstern, O. (1947). *Theory of games and economic behavior* (2nd ed.). Princeton: Princeton University Press.

Waite, T. A. (2001a). Background context and decision making in hoarding gray jays. *Behavioral Ecology, 12,* 318–324.

Waite, T. A. (2001b). Intransitive preferences in hoarding gray jays (*Perisoreus canadensis*). *Behavioral Ecology and Sociobiology 50,* 116–121.

Wrangham, R. W., & Peterson, D. (1996). *Demonic males: Apes and the origins of human violence.* Cambridge, MA: Harvard University Press.

Wu, G., Zhang, J., & Gonzalez, R. (2004). Decision under risk. In: D. Koehler & N. Harvey (Eds.), *Blackwell handbook of judgment and decision making* (pp. 399–423). Oxford: Blackwell.

CHAPTER 7

Primate Social Cognition: Thirty Years After Premack and Woodruff

Alexandra G. Rosati, Laurie R. Santos and Brian Hare

In 1871, Darwin wrote, "The greatest difficulty which presents itself, when we are driven to the above conclusion on the origin of man (evolution through natural selection), is the high standard of intellectual power and moral disposition which he has attained." Since Darwin declared the mind as the province of biology as well as psychology, the human intellect has been a major challenge for evolutionary biologists, with some researchers emphasizing the continuity between humans and other animals and others emphasizing seemingly unique aspects of our psychological makeup. Increasing observations of nonhuman primate (hereafter, primate) behavior in both the wild and in captivity in the mid-twentieth century led to a number of proposals addressing the question of why primates seem to be so "smart." These proposals, and the comparative research they have sparked, have far-reaching implications for how we place human cognition in a broader evolutionary context—both in terms of how or to what degree humans are different from our closest relatives as well as whether broad taxonomic-level evolutionary changes in the primate lineage were necessary precursors to human evolution.

The most well-received proposal for the origin of primate intelligence argues that the social lives of primates is sufficiently complex—or predictably unpredictable—to have acted as a driving force in primate cognitive evolution. Alison Jolly (1966) set forth one of the earliest such proposals, musing on the "social use of intelligence" following her observations of wild lemurs and sifakas in Madagascar. A decade later, Nicholas Humphrey (1976) drew many of the same conclusions from watching captive rhesus monkey colonies, noting that it was navigation of the social world, rather than the physical world, that seemed to require the most complex skills. This basic thesis—that the sophisticated cognitive abilities of primates have evolved for a social function—has since taken several forms. For example, some researchers have emphasized the political maneuvering (de Waal, 1982) or "Machiavellian intelligence" (Byrne, 1988) that primates must use to succeed in their societies, while others have focused on the evolutionary arms race between intelligence and increasing social complexity (Dunbar, 2003) (for a different perspective on these issues, see Chapter 28).

The social world has therefore long been thought to be a major force shaping primate cognition—but, paradoxically, very little was known about the cognitive abilities primates actually use when interacting with other social agents. Most early proposals of the social intelligence hypothesis stemmed from observations of complex social behaviors across the primate taxon, but the psychological mechanisms underlying these behaviors were not well understood. For example, although human social behavior is supported by a rich belief-desire psychology through which we can represent and reason about others' subjective psychological states, it was unknown if primates possessed any similar representational capacities. In fact, when Premack and Woodruff (Premack, 1978) first asked their big question, "Does the chimpanzee have a theory of mind?" they argued that their single test subject had shown the ability to assess the intentions of another. However, after two decades of research following this pioneering paper, several major syntheses of primate cognition weighed the evidence and concluded that although primates can use observable phenomena to make predictions about the future behaviors of others, there was no convincing evidence that any nonhuman primates represent the underlying, unobservable psychological states of others' minds (Cheney & Seyfarth, 1990; Heyes, 1998; Tomasello & Call, 1997). Research over the past 10 years, however, has drawn this initial sweeping conclusion into question, revealing that at least some primates have some capability to assess the psychological states of others—while simultaneously showing striking differences between the social-cognitive capacities of humans and other primates (Call & Tomasello, 2008; Tomasello et al., 2003).

Here we address two aspects of primate social cognition—understanding of intentional, goal-directed action and understanding perceptions, knowledge, and beliefs—focusing on the newest comparative research since the last major reviews were written on the topic over a decade ago. We first review evidence suggesting that diverse species of primates understand the actions of others in terms of goals and intentions, and furthermore can reason about some, but probably not all, kinds of psychological states. We then examine the hypothesis that primates show their most complex social skills in competitive contexts, and suggest that inquiry into other aspects of primate social life, such as during cooperative interactions, may prove to be the next important step for experimental inquiries into primate social-cognitive skills. Finally, we examine primate social cognition in a broader evolutionary context that may allow us to better understand both primate and human cognitive skills.

REASONING ABOUT PSYCHOLOGICAL STATES

While studies of primate social cognition have until recently made it difficult to characterize the social skills of primates with confidence, studies of human infants and toddlers have mapped out with ever-increasing resolution the fundamental changes that occur in the way young children come to think about others. This research has pointed to the importance of social-cognitive skills for the development of normal functioning adult behavior. For example, without the normal development of social-cognitive skills, children cannot participate in all forms of cultural endeavors—including language (Tomasello, 1999). Starting in the first year of life, children begin to treat other people as *intentional agents* and come to organize other people's actions in terms of goals and desires (Behne et al., 2005; Carpenter et al., 1998; Gergely et al., 1995, 2002; Meltzoff, 1995; Repacholi & Gopnik, 1997; Woodward, 1998; Woodward et al., 2001). Secondly, children also come to realize that other agents will behave according to their perceptions and knowledge (Brooks & Meltzoff, 2002; Flavell, 1992; Moll & Tomasello, 2004; Phillips, 2002). By the time they are around 4 years of age, children begin to expect that another person will also act in accord with their beliefs, even when such beliefs conflict with the current state of the world (Wellman, 1990; Wimmer & Perner, 1983) (see also (Onishi & Baillargeon, 2005; Southgate et al., 2007; and Surian et al., 2007 for possible evidence at an even earlier age).

To what extent do primates share these human developmental achievements? Do they come to reason about others' behavior in terms of internal, unobservable psychological states? Many of the abilities that are of interest to developmental psychologists have been the topic of extensive research in nonhuman primates, and often the same paradigms used with children have been directly translated into primate studies (Tables 7.1 through 7.4). Here we first review evidence addressing what various primate species understand about *intentional action*, and then examine what primates understand about *perceptions, knowledge, and beliefs.*

Goal-directed Behavior and Intentional Action

Evidence that at least some primates treat the actions of others in terms of their underlying goals and intentions comes from several different sources (Table 7.1). Some of the earliest evidence that primates understand the goals of others emerged through studies of social learning.[1] Such research has revealed that apes may represent the actions of another individual specifically in terms of that person's goal. That is, when confronted with an individual engaging in a novel action, apes rarely engage in exact copying of that action, but rather are more likely to engage in behavior toward the same goal that the actor was pursuing, a process referred to as *goal emulation* (Tomasello, 1990; Tomasello et al., 1987). Chimpanzees (*Pan troglodytes*) also seem to react differentially depending on whether a human demonstrator's actions are relevant to his or her goal. For example, when confronted with a human demonstrator performing various actions to obtain food from a causally confusing opaque puzzle box, chimpanzees faithfully imitate the actor's complete sequence of actions. In contrast, when the box is transparent and thus the causal nature of the box and the actor's goal are clear, chimpanzees engage in goal emulation, excluding actions that were irrelevant to the goal (Horner & Whiten, 2005).[2] In addition to imitating only goal-relevant actions, other evidence suggests that apes are more likely to exactly copy a human demonstrator's behavior when that demonstrator successfully completes his or her goal than when he or she fails (Call et al., 2005; Myowa-Yamakoshi & Matsuzawa, 2000). Taken together, this work suggests that apes seem to naturally parse the behavior of others in terms of goals, and will only copy the superficial behavior when the link between the actions and goal at hand is not readily apparent, and no other

Table 7.1 Studies of Goal and Intention Understanding Across Nonhuman Primate Species

		Inferring Goals	Distinguishing Intentions	Ontogeny
Hominoids	Chimpanzees	Buttleman et al., 2007; Call et al., 2005; Horner & Whiten, 2005; Myowa-Yamakoshi & Matsuzawa, 2000; Tomasello & Carpenter, 2005; Tomasello et al., 1987; Uller, 2004; Uller & Nichols, 2000; Warneken & Tomasello, 2006; Warneken et al., 2007*;	Call & Tomasello, 1998; Call et al., 2004; Povinelli et al., 1998; Tomasello & Carpenter, 2005	Tomasello & Carpenter, 2005; Uller, 2004
	Other great apes		Orangutans: Call & Tomasello, 1998	
	Lesser apes			
Old World monkeys	Macaques	Rhesus		
	Baboons			
	Other			
New World monkeys	Capuchins	Brown: Kuroshima et al., 2008	Brown: Lyons & Santos, submitted; Phillips et al., 2009	
	Callitrichids	Cotton-top tamarins		
	Other			
Strepsirrhines	Lemurs			

* Indicates that the study involved both human and conspecific social partners.

appropriate means is available. To our knowledge, there is only limited work testing more distantly related species on similar goal emulation tasks, with mixed results (e.g., Kuroshima et al., 2008).

A second line of evidence suggesting that primates have some understanding of others' intentional action comes from studies in which similar or identical actions are performed, but the intention underlying these actions vary. Orangutans (*Pongo pygmaeus*) and chimpanzees can tell whether an action is intentional versus accidental (Call & Tomasello, 1998). Importantly, this capacity is likely not limited to apes—capuchins (*Cebus apella*) show similar abilities (Lyons & Santos, submitted). Moreover, Chimpanzees and capuchins seem to differentiate between different *types* of underlying intentions. When chimpanzees are confronted with a human who fails to give them food, they are more likely to produce begging and other relevant behaviors (and are less likely to leave the room) when the human is *unable* to give them the food (e.g., because he or she dropped it) than when the human is *unwilling* to give the food (e.g., because he or she is teasing). That is, the chimpanzees did not react only to the superficial result of the human's behavior—not getting any food—but also to the *reason* the human failed to give the food (Call et al., 2004). Capuchin monkeys also seem to discriminate between actors that are either unwilling or unable, remaining for a longer period in the testing area when a human is unable to give them food (because a second human keeps stealing it) than when a human teases them with food. Furthermore, capuchins make these distinctions specifically when the relevant actor is an agent (i.e., a human hand), but not when an inanimate object (e.g., a stick) enacts the same behavior (Phillips et al., 2009).

Understanding Intentional Communicative Cues

Although these findings support the idea that at least chimpanzees and capuchins perceive others' behaviors in terms of goals and intentions, such studies have been conducted with very few species; this limitation makes it difficult to assess whether these abilities represent convergent cognitive evolution between apes and capuchins, a distantly related New World monkey, or are rather a set of abilities that are widely shared across primates. A more widely used assessment of intention understanding in primates is a method referred to as the '"object-choice" paradigm, in which animals are presented with intentional communicative cues (Table 7.2). The goal of such studies is to examine whether primates can successfully use communicative gestures to locate hidden objects, typically desirable food items. In a typical version of this type of task, a human experimenter might point at one of several cups that contains a piece of food, and then allow the subject to choose between the cups (the subject only knows something is hidden but does not know where).

Many studies utilizing this sort of object-choice paradigm suggest that while apes *are* able to spontaneously use such gestures to find food, their performance is fragile and often only successful at the group level (see reviews of this object-choice work in Hare & Tomasello, 2005; Call & Tomasello, 2008). However, other evidence suggests that the fragility of the apes' performance may be less because apes cannot use gestures to find food and more due to the difficulty in understanding the cooperative-communicative intentions underlying these gestures. For example, chimpanzees are more successful when a human competitor reaches for a food cup that they also want than when a human simply points to a cup in a cooperative fashion (Hare & Tomasello, 2004). Similarly, apes are more successful using prohibitive hand gestures ("Don't touch that one!") to find food then they are using a standard cooperative pointing cue (Herrmann & Tomasello, 2006). This is surprising given that both the reaching and prohibitive gestures have nearly identical surface features to the pointing gesture. One interpretation of this pattern of performance is that primates are more successful at using these reaching and prohibitive types of cues because apes more often compete with others over food rather than cooperatively share information about its location with others. These competitive cues may therefore be more ecologically valid, and

Table 7.2 Studies of Social-Cue Use Across Nonhuman Primate Species

		Gaze	Points and Other Gestures	Touch, Body Position, and Physical Markers	Ontogeny
Hominoids	Chimpanzees	Barth et al., 2005; Braeuer et al., 2006; Call et al., 1998, 2000; Herrmann et al., 2007; Itakura et al., 1999; Okamoto-Barth et al., 2008; Povinelli et al., 1999*	Barth et al., 2005; Braeuer et al., 2006; Hare & Tomasello, 2004,* 2006; Herrmann et al., 2007; Itakura et al., 1999*; Okamoto-Barth et al., 2008; Povinelli et al., 1990, 1997, 1999; Wood et al., 2007	Barth et al., 2005; Call et al., 2000; Herrmann et al., 2006, 2007; Itakura et al., 1999*; Okamoto-Barth et al., 2008	Herrmann et al., 2007; Okamoto et al., 2002; Okamoto-Barth et al., 2008; Tomasello & Carpenter, 2005; Tomonage et al., 2004
	Other great apes	Bonobos: Braeuer et al., 2006 Orangutans: Herrmann et al., 2007	Bonobos: Braeuer et al., 2006; Herrmann & Tomasello, 2006; Orangutans: Herrmann et al., 2007	Bonobos: Herrmann et al., 2006 Gorillas: Herrmann et al., 2006 Orangutans: Herrmann et al., 2006, 2007	Orangutans: Herrmann et al., 2007
	Lesser apes	White-handed gibbons: Inoue et al., 2004	White-handed gibbons: Inoue et al., 2004		
Old World monkeys	Macaques	Rhesus: Anderson et al., 1996; Hauser et al., 2007	Rhesus: Anderson et al., 1996; Hauser et al., 2007; Wood et al., 2007		
	Baboons	Olive: Vick & Anderson, 2003; Vick et al., 2001			
	Other				
New World monkeys	Capuchins	Brown: Anderson et al., 1995; Vick & Anderson, 2000	Brown: Anderson et al., 1995; Vick & Anderson, 2000	Brown: Vick & Anderson, 2000	
	Callitrichids	Cotton-top tamarins: Neiworth et al., 2002 Common marmosets: Burkhart & Heschl, 2006	Cotton-top tamarins: Neiworth et al., 2002; Wood et al., 2007 Common marmosets: Burkhart & Heschl, 2006	Cotton-top tamarins: Neiworth et al., 2002 Common marmosets: Burkhart & Heschl, 2006	
	Other				
Strepsirrhines	Lemurs				

* Indicates that the study involved both human and conspecific social partners.

potentially more motivating for some apes (as reviewed in Hare, 2001; Lyons & Santos, 2006; Santos et al., 2007a). In the context of intention understanding, interpreting others' behaviors in terms of competitive goals ("I want the food too!") in these social cuing paradigms may be more transparent than interpreting their behavior in terms of cooperative goals ("I want to tell you where the food is for your benefit.").

In contrast to the studies with apes, however, many monkey species fail to use communicative cues in similar kinds of studies, at least in the absence of extensive training. For example, capuchins can learn to use a pointing cue to find food, but only following several dozens or even hundreds of trials (Anderson et al., 1995; Vick & Anderson, 2000). Rhesus macaques (*Macaca mulatta*; Anderson et al., 1996) and cotton-top tamarins (*Saguinus oedipus*; Neiworth et al., 2002) also perform poorly on these tasks. However, more recent evidence complicates this picture. For example, common marmosets (*Callithix jacchus*) are more successful using a pointing cue on a modified version of the task (Burkhart & Heschl, 2006), while rhesus monkeys tested with a more species-specific looking gesture are better able to determine the location of hidden food (Hauser et al., 2007). Wood and colleagues (2007) further argue that cotton-top tamarins, rhesus, and chimpanzees are sensitive to hand gestures when such hand gestures are indicative of an intentional component of a goal-directed action plan. As it is unclear why this result is discrepant with results from past studies involving chimpanzees, further investigations would profit from parsing out why primates may demonstrate understanding of intentions in some contexts but not others. Taken together, then, studies of communicative gesture use suggest that primates' performance may be fragile and context dependent, but primates do seem to readily use information regarding another individual's intentions and goals in more competitive paradigms.

Gaze Following and the Roots of Mind Reading

Early studies exploring what primates know about others' visual attention suggested that primates lack even a very gross understanding of the nature of visual perceptions. For example, inspired by Premack (1988), Povinelli and Eddy (1996c) taught young chimpanzees to use a visual begging gesture to obtain food from a human experimenter. The researchers then presented the chimpanzees with a situation in which they could choose one of two experimenters from whom to beg. The trick was that the two experimenters differed in their perceptual access to the chimpanzees: One experimenter could see the chimpanzees, whereas the other could not for a variety of reasons. Although the chimpanzees spontaneously chose the human with visual access to their gestures in the condition involving the most contrast between the two humans (e.g., preferring to beg from a human facing them than with her back turned), they failed to discriminate between the two humans in a variety of other, more subtle situations (such as a person with her face turned away versus one oriented toward the subject or one with a blindfold over her eyes versus another with a blindfold over her mouth). Early experiments such as these seemed to provide strong evidence that primates do not understand what others can and cannot see.

However, converging evidence from many different paradigms and species now appears to refute this early view of primates' understanding of others' perspective: Many primates are at least behaviorally responsive to the direction of others' gaze and attention, and there is a subset of these species that appears to have a flexible understanding of what others perceive. At the most basic level, diverse species of primates spontaneously follow the gaze of human experimenters or conspecifics. Gaze-following behaviors allow individuals to apprehend important objects and events that others have detected in the environment, including food sources, predators, and conspecifics. Thus, gaze following allows individuals to exploit the information that others have acquired about the world. Species including chimpanzees (Povinelli & Eddy, 1996a; Tomasello et al., 1998) and the other great apes (Braeuer et al., 2005; Okamoto-Barth et al., 2007); Old World monkeys such as various macaques (rhesus: *Macaca mulatta*; stumptail: *M. arctoides*; pigtail: *M. memstrina*;

Emery et al., 1997; Tomasello et al., 1998), mangabeys (*Cercocebus atys torquats;* Tomasello et al., 1998) and olive baboons: mangalys (Cercocebus atys torquats; Tomasello et al., 1998), New World monkeys including capuchins (Vick & Anderson, 2000), cotton-top tamarins (Neiworth et al., 2002), and common marmosets (*Callithrix jacchus;* Burkhart & Heschl, 2006); and even some lemur species (ring tailed: *Lemur catta*; brown lemurs: Eulemur fulvus; black lemurs: Eulemur macaco; Shepherd & Platt, 2008; Ruiz et al., 2009) all follow gaze, at least in some contexts. Although there is variation in the degree to which various species can successfully follow eye position alone (e.g., apes: Tomasello et al., 2007; olive baboons (e.g., apes: Tomasello et al., 2007) or rather can only follow shifts in the position of the face, head, or even entire body (e.g. capuchins: Vick & Anderson, 2000; cotton-top tamarins: Neiworth et al., 2002; ring-tailed lemurs: Shepherd & Platt, 2008), this variation may be due to variation in the amount of information that the eye carries due to differences in morphology across different taxa (Kobayashi & Kohshima, 1997, 2001).

Although gaze-following behaviors are widely shared across the primate order, the psychological basis of these co-orienting behaviors seems to vary widely. For example, the nature of gaze following in chimpanzees and other great apes suggests that individuals of these species follow gaze because they understand something about the nature of "seeing." Apes not only direct their own gaze in the direction of others but also follow gaze around barriers and past distracting objects that are not the target of another's gaze, sometimes by physically reorienting their own bodies (Povinelli & Eddy, 1996a; Tomasello et al., 1999). They may also "check back" with the actor in an attempt to verify the direction of the other's gaze or quickly stop following the gaze cues when they cannot locate the target of the other's gaze (Braeuer et al., 2005; Call et al., 1998; Tomasello et al., 2001). These flexible shifts in behavior across contexts suggest that apes follow the gaze of others because they expect there to be something interesting to see. Interestingly, those species most closely related to humans—chimpanzees and bonobos—appear to be especially sophisticated in these contexts even compared to other great apes (Okamoto-Barth et al., 2007).

The evidence for such behavior in more distantly related primate species is less complete, mostly because few studies have been conducted (Table 7.3). Macaques, like apes, habituate to repeated gaze cues when they repeatedly cannot locate the target of another's gaze (Goossens et al., 2008; Tomasello et al., 2001). However, studies of New World monkeys and lemurs suggest that the co-orienting behaviors in some of these species are more reflexive. For example, cotton-top tamarins will co-orient with conspecifics at high rates during natural interactions (although the cause of this co-orienting is unclear), but fail to follow the explicit gaze cues provided in controlled experimental settings (Neiworth et al., 2002). Similarly, some lemur species co-orient with conspecifics during their natural behaviors (Shepherd & Platt, 2008), but seem less able to follow gaze in experimental contexts (Anderson & Mitchell, 1999; but see Ruiz et al., 2009 for an experimental study using Conspecific Photographs). Thus, although behavioral co-orienting may be common to all primates, not all primates necessarily follow gaze because they understand that others *see* things.

Using Information About Gaze and Attention

Further evidence supporting the potential distinction between apes and other species comes from social cuing (or object-choice) studies. This paradigm is similar to those involving pointing gestures, although here the experimenter's cue involves looking at the correct option (Table 7.2). Overall, evidence suggests that apes are generally successful at spontaneously using gaze cues to find the food, although, like with gesture cues, the effects are often small (e.g. Call et al., 1998; Itakura et al., 1999). Notably, apes' performance may change dramatically depending on the specific paradigm utilized. For example, chimpanzees are much more successful using gaze cues when the experimenter looks *into* an object whose contents he or she alone can see (such as a tube) than when the

Table 7.3 Studies of Gaze Following Across Nonhuman Primate Species

		Follow Head Orientation	Follow Eye Orientation Alone	Follow Gaze Around Barriers	Check Back with Actor; Habituate in Absence of Target	Ontogeny
Hominoids	Chimpanzees	Braeuer et al., 2005; Call et al., 1998; Herrmann et al., 2007; Itakura, 1991; Okamoto-Barth et al., 2007; Povinelli & Eddy, 1996a; Tomasello et al., 1998*, 2001, 2007	Herrmann et al., 2007; Povinelli & Eddy, 1996a; Tomasello et al., 2007	Barth et al., 2005; Braeuer et al., 2005; Okamoto-Barth et al., 2007; Povinelli & Eddy, 1996a; Tomasello et al., 1999	Braeuer et al., 2005; Call et al., 1998; Okamoto-Barth et al., 2007; Tomasello et al., 2001	Barth et al., 2005; Braeuer et al., 2005; Herrmann et al., 2007; Tomasello & Carpenter, 2005; Tomasello et al., 2001; Tomonage et al., 2004
	Other great apes	Bonobos: Braeuer et al., 2005; Okamoto-Barth et al., 2007; Tomasello et al., 2007 Gorillas: Braeuer et al., 2005; Okamoto-Barth et al., 2007; Tomasello et al., 2007 Orangutans: Braeuer et al., 2005; Herrmann et al., 2007; Okamoto-Barth et al., 2007; Tomasello et al., 2007	Bonobos: Tomasello et al., 2007 Gorillas: Tomasello et al., 2007 Orangutans: Herrmann et al., 2007; Tomasello et al., 2007	Bonobos: Braeuer et al., 2005; Okamoto-Barth et al., 2007 Gorillas: Braeuer et al., 2005; Okamoto-Barth et al., 2007 Orangutans: Braeuer et al., 2005; Okamoto-Barth et al., 2007	Bonobos: Braeuer et al., 2005; Okamoto-Barth et al., 2007 Gorillas: Braeuer et al., 2005; Okamoto-Barth et al., 2007 Orangutans: Braeuer et al., 2005; Okamoto-Barth et al., 2007	Bonobos: Braeuer et al., 2005 Gorillas: Braeuer et al., 2005 Orangutans: Braeuer et al., 2005; Herrmann et al., 2007
	Lesser apes	Pileated gibbons: Horton & Caldwell, 2006†				

Old World monkeys	Macaques	Rhesus: Emery et al., 1997*; Itakura, 1996; Tomasello et al., 1998,* 2001 Stumptail: Anderson & Mitchell, 1999; Itakura, 1996; Tomasello et al., 1998* Pigtail: Itakura, 1996; Tomasello et al., 1998* Long-tail: Goossens et al., 2008 Tonkean: Itakura, 1996		Rhesus: Tomasello et al., 2001 Long-tailed: Goossens et al., 2008	Rhesus: Tomasello et al., 2001
	Baboons	Olive: Vick et al., 2001	Olive: Vick et al., 2001		
	Other	Sooty mangabey: Tomasello et al., 1998*			
New World monkeys	Capuchins	Brown: Itakura, 1996; Vick & Anderson, 2000 White-faced: Itakura, 1996	Brown: Vick & Anderson, 2000		
	Callitrichids	Cotton-top tamarins: Neiworth et al., 2002† Common marmosets: Burkhart & Heschl, 2006	Cotton-top tamarins: Neiworth et al., 2002		
	Other		Squirrel monkey: Itakura, 1996		
Strepsirrhines	Lemurs		Ring-tailed: Shepherd & Platt, 2008* Black: Anderson & Mitchell, 1999; Itakura, 1996; Ruiz et al., 2009* Brown: Itakura, 1996; Ruiz et al., 2009*		

* Indicates that the study involved conspecific social partners; unless noted, the study involved human experimenters as actors.
† Indicates that the study involved both human and conspecific social partners.

experimenter just looks *at* the external surface of a cup, an act that is divorced from actually seeing something (Call et al., 1998). In contrast to the results with apes, studies with monkeys suggest that whereas monkeys will often follow gaze, they tend not to use gaze as a social cue when searching for food in experiments. In many experiments, both Old and New World monkeys require extensive training with cues or fail to use gaze cues at all (e.g., olive baboons: Vick et al., 2001; rhesus macaques: Anderson et al., 1996; capuchins: Anderson et al., 1995; cotton-top tamarins: Neiworth et al., 2002; but see Hauser et al., 2007, for successful use of gaze cues in rhesus monkeys). This trend of failures suggests that, in contrast to apes, some monkey species may follow gaze without actually understanding anything about the nature of attention and visual perceptions. However, as mentioned previously, there is some evidence that modification of the standard two-option object-choice paradigm may improve the performance of some species (e.g., common marmosets: Burkhart & Heschl, 2006), so future research is warranted with a wider range of species and paradigms before any strong conclusions can be made about a clade-level distinctions between apes and monkeys groups.

More converging evidence that apes have some understanding of the nature of visual perception comes from studies examining their gesture use in response to others who vary in attentional state. One such study (Povinelli & Eddy, 1996c, described previously) suggested that chimpanzees understand very little about the nature of seeing in an experimental setting. However, other research by these researchers suggests that chimpanzees may be sensitive to head movements and eye contact in similar contexts (Povinelli & Eddy, 1996b), although it is not clear what factors drive this sensitivity. Nonetheless, more recent research has suggested that apes may have performed poorly in these early gesture-use studies because they favor head and body orientation over eye position as cues to what others are seeing (Tomasello et al., 2007). One possibility is that the low degree of contrast between the iris and sclera makes it difficult to discriminate eye direction in almost all primates but humans; humans also appear to be unique in our ability to move our eyes independent of our general head direction (Kobayashi & Kohshima, 1997, 2001). For example, chimpanzees, bonobos, and orangutans spontaneously adjust their gesture frequency to the attentional state of the observer (i.e., they produce more gestures when an experimenter can see them), but they treat body and face orientation, rather than eye position, as the most relevant factors (Kaminski et al., 2004). Chimpanzees do, however, attend to whether an experimenter's eyes are open when this is the only cue available (Hostetter et al., 2007). Furthermore, chimpanzees will adjust the location of their gesture depending on the focus of their partner's attention (Povinelli et al., 2003), and all four species of great ape will move to face an experimenter so that they can execute their gestures in that person's line of sight, rather than perform the gesture behind his or her back (Liebal et al., 2004b). Similar results have come from naturalistic observations of the gestures that apes use when interacting with each other; apes modulate their gesture use to the attentional state of their conspecific partner (Liebal et al., 2004a; Pika et al., 2003, 2005) and may use loud noises to attract attention before making visual gestures (Call & Tomasello, 2007; Poss et al., 2006).

Although monkey species do not produce gestures with the flexibility that apes do (Call & Tomasello, 2007), evidence that other primate species understand something about the nature of visual perception comes from studies looking at how the attentional state of others influences the predictions that monkeys make about the behavior of others after they look at an object. For example, when cotton-top tamarins saw a human actor look at one of two objects, they expected the actor to reach for and grab that object rather than another, previously unattended object, demonstrating longer looking at the unexpected outcome (Santos & Hauser, 1999). Diana monkeys seem to have similar expectations about the directed gaze of conspecifics (Scerif et al., 2004), but two other New World monkey species (tufted capuchins and squirrel monkeys) fail to demonstrate an understanding of the link between attention and behavior at least when tested using an expectancy violation looking method (Anderson et al.,

Table 7.4 Studies of Understanding Perceptions and Knowledge Across Nonhuman Primate Species

		Attention and Predictions About Seeing	Visual Perspective	Auditory Perspective	Deception	False Beliefs	Ontogeny
Hominoids	Chimpanzees	Herrmann et al., 2007; Hostetter et al., 2007; Kaminski et al., 2004; Liebal et al., 2004a,* 2004b; Povinelli & Eddy, 1996b, 1996c; Povinelli et al., 1997, 2002, 2003; Reaux et al., 1999; Theall & Povinelli, 1999; Tomasello & Carpenter, 2005; Tomasello et al., 1994*	Braeuer et al., 2007,* ; Hare et al., 2000,* 2001,* 2006; Hirata & Matsuzawa, 2001*; Melis et al., 2006a; Povinelli & Eddy, 1996c; Povinelli et al., 1990	Braeuer et al., 2007*; Melis et al., 2006a	Hare et al., 2006; Hirata & Matsuzawa, 2001*; Melis et al., 2006a	Call & Tomasello, 1999; Hare et al., 2001*; Kaminski et al., 2008; Krachun et al., 2009	Herrmann et al., 2007; Povinelli et al., 2002; Reaux et al., 1999; Tomasello & Carpenter, 2005; Tomasello et al., 1994*
	Other great apes	Bonobos: Kaminski et al., 2004; Liebal et al., 2004b; Pika et al., 2005* Gorillas: Kaminski et al., 2004; Liebal et al., 2004b; Pika et al., 2003*; Poss et al., 2006 Orangutans: Herrmann et al., 2007; Kaminski et al., 2004; Liebal et al., 2004b; Poss et al., 2006				Orangutans: Call & Tomasello, 1999	Orangutans: Herrmann et al., 2007
	Lesser apes	Siamangs: Liebal et al., 2003* Pileated gibbons: Horton & Caldwell, 2006					

(continued)

Table 7.4 (Continued)

		Attention and Predictions About Seeing	Visual Perspective	Auditory Perspective	Deception	False Beliefs	Ontogeny
Old World monkeys	Macaques		Rhesus: Flombaum & Santos, 2005; Povinelli et al., 1991 Long-tailed: Kummer et al., 1996	Rhesus: Santos et al., 2006		Rhesus: Santos et al., 2007b	
	Baboons		Olive: Vick & Anderson, 2003				
	Other	Diana monkeys: Scerif et al., 2004*					
New World monkeys	Capuchins	Brown: Anderson et al., 2004; Kuroshima et al., 2002; Kuroshima et al; 2003	Brown: Fujita et al., 2002*; Hare et al., 2003*;		Brown: Fujita et al., 2002 *		
	Callitrichids	Cotton-top tamarins: Santos & Hauser, 1999 Common marmosets: Burkhart & Heschl, 2007*	Common marmosets: Burkhart & Heschl, 2007*				
	Other	Squirrel monkeys: Anderson et al., 2004					
Strepsirrhines	Lemurs					Black: Genty & Roeder, 2006	

* Indicates that the study involved conspecific social partners; unless noted, the study involved human experimenters as actors.

2004). Thus, whereas there is robust evidence that apes understand something about the nature of attention, the results are more variable across monkey species, suggesting again the importance of the kind of behavioral task and context employed.

From Perspective Taking to Understanding of Knowledge and Beliefs

Together, the evidence from gesture use and looking-time work suggests that apes and possibly some monkeys may be sensitive to the visual perception of others. Perhaps the most conclusive evidence that some primates have an understanding of visual attention, however, comes from studies of perspective taking. As previously mentioned, several primate species tend to perform poorly in early studies testing their understanding of visual attention and perspective taking (e.g., Povinelli & Eddy, 1996c; Reaux et al., 1999). However, these studies typically used a cooperative-communicative paradigm in which a human experimenter shared food with the chimpanzees, a situation that may be highly unnatural or unmotivating for primates, as previously noted. Faced with this problem, researchers have more recently tried to develop more ecologically valid tests of perspective taking, ones that are designed around a context that may be more natural (and motivating) for primates: food competition. The basic setup of the original studies by Hare and colleagues (2000, 2001) using this logic involved two chimpanzees competing with each other for access to food. However, the two chimpanzees had differing knowledge about the food that was available. For example, in one series of studies, the more subordinate of the two chimpanzees could see two pieces of food, and the dominant individual could only see one (the second piece was blocked from her view). Researchers then measured which piece of food the subordinate targeted when she was released with a slight head start over the dominant. Using this technique, a series of experiments demonstrated that subordinate chimpanzees were more likely to choose the food that dominant individuals could not see. In addition, when the roles were reversed and now a dominant could see both pieces of food and was released before a subordinate who could only see one, the dominant targeted the visible (at-risk) piece of food before taking the second piece hidden from the subordinate's view. A number of controls ruled out the possibility that such strategies were due to behavioral monitoring of the dominant individual (e.g., the subjects were forced to make a decision before they ever saw their competitor make a move; Hare et al., 2000). A second set of studies indicated that chimpanzees demonstrated these preferences because they understood something about the link between seeing and knowing: When subordinates had to decide whether or not to approach a piece of food hidden from a dominant's view, they made more attempts to obtain the food when the dominant had not been present when the food was hidden than when she had been present during the baiting (Hare et al., 2001).

Following these initial studies, several experiments using competitive paradigms have demonstrated perspective-taking skills in both apes and monkeys. For example, Flombaum and Santos (2005) developed a paradigm in which rhesus macaques could choose to steal food from one of two experimenters, and then varied the degree to which those experimenters could see the food. In many ways this setup therefore parallels the preferential begging paradigm developed by Povinelli and colleagues, except that the decision was placed in a competitive context. Rhesus monkeys showed sensitivity to a wide variety of variations in visual access, even when the manipulations involved very subtle differences in eye position. Studies with chimpanzees have similarly shown that they prefer to retrieve a piece of food that a competitive human cannot see over one he or she can, even engaging in attempts to disguise their interest in the food as they approach it (Hare et al., 2006; Melis et al., 2006a). Moreover, some evidence suggests that the perspective taking that chimpanzees and rhesus monkeys engage in extends to the auditory modality. For example, when rhesus macaques are confronted with a human competitor sitting in front of two boxes containing food where one box has functional bells attached to it while the other box has nonfunctional bells, they preferentially steal food from the box that is silent, and do so only when the competitor cannot already see their actions (Santos et al., 2006). This suggests that rhesus monkeys recognize how their behavior will alter the psychological state of the human: If the human cannot see them, then the noise will alert him or her to their presence. If the human can already see them, then noise will have no impact on the human's knowledge about their behavior. Chimpanzees also prefer a silent approach over a noisy one when competing with a human over food (Melis et al., 2006a; but see Braeuer et al., 2008). Despite these successes, other monkey species have demonstrated poor performance in similar visual perspective-taking tasks, providing further converging evidence that the psychological mechanisms supporting social interactions vary across primates. For example, both capuchins (Hare et al., 2003) and common marmosets (Burkhart & Heschl, 2007) have been tested in versions of the conspecific competition paradigm used with chimpanzees, but appear to depend heavily on the behavior of the competitor, rather than reasoning about what the competitor sees or knows, when making food choices.

The current evidence suggests that at least chimpanzees and rhesus macaques know something about what others can and cannot perceive, and use this information to guide their own behavioral decisions. However, an open question concerns the issue of what primates are actually representing when faced with these kinds of social problems. Although recent studies provide strong evidence that rhesus monkeys and chimpanzees understand something about others' perception and knowledge, thus far there is no evidence that primates go beyond a distinction between knowledge and ignorance to actually represent the false beliefs of others. For example, chimpanzees perform at chance when confronted with two humans trying to direct them to food, one of whom had seen the food being hidden and the other who had originally seen the food being hidden, but had a false belief about its location due to a subsequent switching out of her view (Call and Tomasello, 1999). Competitive versions of false-belief tasks further confirm that chimpanzees use information about true but not false beliefs to find food hidden (Kaminski et al., 2008; Krachun et al., 2009; see also the informed-misinformed condition in Hare et al., 2001). Similarly, when tested in a looking-time violation-of-expectation false-belief test (see Onishi & Baillargeon, 2005), rhesus monkeys make correct predictions about where a human actor will search when they have a true belief about a food item's location, but make no predictions about the actor's behavior when the actor has a false belief (Santos et al., 2007b). These findings using various false-belief tasks suggest that while primates can represent whether others are knowledgeable or ignorant, they may not represent beliefs of others in cases where those beliefs conflict with the true state of the world.

Conclusions: Understanding Psychological States

Overall, research from the past decade has greatly illuminated the cognitive skills underlying the complex social behaviors of primates. First, both apes and at least some species of monkey seem to parse the actions of others in terms of underlying goals and intentions. Similarly, apes and some monkeys seem to understand something about the perspective of others. Most research addresses whether primates understand visual perspective, but other studies suggest that this capacity may also encompass perception in other modalities. These social-cognitive abilities may not be shared by all primates: Although some behaviors, such as gaze following, seem to be widely shared, some species engage in superficially similar behaviors but do not seem to understand the nature of seeing in the same way that chimpanzees and rhesus macaques do. However, the current research supports the conclusion that at least some primates understand others' behavior in terms of psychological states such as goals and knowledge, rather than merely in terms of observable behavioral features (but see Povinelli & Vonk, 2003, 2004, for alternative interpretations of these results).

However, there are, at present, still many limitations to our understanding of primate social cognition. As mentioned previously, many of the paradigms used to examine social-cognitive skills in primates have been adapted from the human developmental literature. As such, the "interesting" topics in primate social cognition tend to grow out of developmental studies of theory of mind. As there is some indication that primates look more skillful in studies involving ecologically valid paradigms, such as competition for food, directly adapting developmental paradigms for primates may not be the only productive way to study primate social cognition. Indeed, what these kinds of paradigms emphasize is that social-cognitive skills are functional, guiding effective behavior and allowing organisms to choose the most advantageous course of action. For example, a study varying the "intensity" of competition in the sort of conspecific-competition paradigm described earlier illustrates how perspective taking is an ability that chimpanzees use *strategically*. When chimpanzees only have time to retrieve one piece of food, perspective taking increases their payoff—they will therefore target the piece that their competitor cannot see. However, if the physical properties of the task are altered such that chimpanzees can potentially retrieve all the food regardless of what their competitor can see, they will simply use a "fast" strategy and race to take both pieces while choosing indiscriminately (Braeuer et al., 2007; see also Karin-D'Arcy & Povinelli, 2002). This finding emphasizes the importance of examining primate social-cognitive skills in a functional framework. Researchers will therefore profit from critically considering the kinds of skills that might allow primates to be more effective social decision makers in their natural environments, and when it actually benefits them to use the skills they possess.

FROM COMPETITION TO COOPERATION

Competition is just one example of an ecologically relevant domain—primates certainly do not spend all their time competing with others for food! Rather, primate social life is a complex patchwork of both competition and cooperation—but these two opposing forces may come into play in different contexts and differentially impact different kinds of social interactions. To take one example, wild-living male chimpanzees engage in several complex cooperative behaviors (Muller & Mitani, 2005), including meat sharing (Mitani & Watts, 2001), group hunting (Boesch & Boesch, 1989), coalitionary mate guarding (Watts, 1998), and territorial boundary patrols (Watts & Mitani, 2001). Other primates also have complex patterns of cooperation and alliance formation (e.g., de Waal, 1996; Kappeler & van Schaik, 2006). As such, primates may possess sophisticated social-cognitive skills to deal with both competitive *and* cooperative interactions, but the kinds of skills they use may be very different in these disparate contexts. Indeed, the cooperative-communicative paradigms (such as object-choice) used so often in primate research may fail to demonstrate robust social-cognitive abilities in various species not because these tasks cooperative per se, but because they utilize specific forms of cooperation (sharing information or sharing food) that may not be a part of species-typical social interactions. In fact, studies of human cooperation suggest many ways to approach the problem in nonhuman primates that might lead to a better understanding of breadth of possible social-cognitive skills beyond the competitive contexts studied thus far. A variety of social-cognitive skills play important roles in shaping human cooperation, including knowledge about the intentions of others, the social relationship between cooperative partners, and reputation management (see reviews in Gintis et al., 2005).

Do similar social-cognitive mechanisms underlie the cooperative behaviors of nonhuman primates? Increasing evidence suggests that they do, at least in some species and contexts. For example, apes appear to have some knowledge of the quality of the relationships they share with social partners as well as being able to remember how those partners behaved in past cooperative interactions. Chimpanzees will spontaneously cooperate to acquire food in an instrumental task requiring joint action with conspecifics that they share a tolerant relationship with, but will not cooperate with intolerant partners (Melis et al., 2006b; see Hare et al.,

2007, for a comparison of chimpanzees and bonobos in a similar task), and will also preferentially choose to cooperate with more skillful partners over less skillful partners (Melis et al., 2006c). Correlations of natural behaviors further suggest that chimpanzees prefer to cooperate with those who have cooperated with them in the past (Mitani, 2006). Although there are several very different types of mechanisms that could underlie such behaviors (see de Waal & Lutrell, 1988), experimental evidence supports the hypothesis that chimpanzees show calculated reciprocity in grooming (Koyama et al., 2006) and collaborative (Melis et al., 2008) contexts. Together, these results suggest that apes remember something about the behavior of others and use this information when making social decisions—that is, they are guided by something like direct reputation when deciding who to interact with (See also Subiaul et al., 2008). But primate social-cognitive skills are not limited to direct interactions with others: Some primates also seem to represent the ongoing relations of other members of their groups. For example, experiments with wild baboons suggest that this species understands not only their relations with others but also the third-party relationships between other members of their groups (see Cheney & Seyfarth, 2007), an ability that may function as a precursor to indirect reputation formation.

There is also evidence that primates use social-cognitive skills such as intention reading in cooperative contexts. For example, chimpanzees' use their understanding of both humans' and conspecifics' goals to help them when they fail to reach those goals (Warneken & Tomasello, 2006; Warneken et al., 2007). Chimpanzees also use information about whether a conspecific was the cause of their losing access to food when deciding whether to punish that individual (Jensen et al., 2007), which may involve some form of intention reading. If this is so, chimpanzees then possess an ability thought to be an important mechanism for sustaining cooperation across repeated interactions. Notably, however, there is little evidence from any of this work that chimpanzees or other species understand the potential of using overt forms of communication to enhance success in cooperative endeavors (See Melis et al., 2009 for Chimpanzee's lack of communication in a negotiation game). The lack of communication may suggest that there is a lack of motivation by non-human primates to assess the cooperative-communicative intent of others.

Altogether, these results suggest that many primates do engage in "cognitive" cooperation, using their social-cognitive skills to engage in more efficient and more successful forms of cooperation. Critically, the payoffs of many cooperative interactions depend not only on whether two individuals act together but also on their level of skill when performing the act. That is, if one partner cannot successfully perform his or her role, both members of the pair will fail to get anything. The, use of social-cognitive skills can increase the rewards associated with cooperation, so it can pay for individuals to sustain relationships with potential partners, selectively cooperate with good partners, and be adept at the mutualistic activity itself (e.g., coordinate with the partner and be sensitive to the other's intentions). Thus, future studies of the social-cognitive abilities that primates use in cooperative interactions will likely reveal that these abilities are different than those needed in cooperative interactions, but not any less complex.

THE EVOLUTION OF PRIMATE SOCIAL COGNITION

Despite the major inroads that research examining primate social cognition have made in the last decade, there are still some major limitations to current research. First of all, although we began by asserting that we would review the cognitive skills that primates use during social interactions, it is notable that the vast majority of studies we have reviewed involve primates interacting with humans (Tables 7.1 through 7.4; studies that involve conspecifics are marked studies with human partners are unmarked). Consequently, while we know a lot about the cognitive skills primates *can* utilize, we are less sure about when and how primates actually use these skills when interacting with conspecifics in natural contexts. Similarly, few studies have examined the ontogenetic development of these skills (Tables 7.1 through 7.4; the last column in each category references studies with a developmental component). Developmental studies have provided critical insights into human social cognition, so they could potentially do the same for nonhuman primate social cognition. For example, divergent developmental trajectories may be evidence of different underlying psychological mechanisms across species, even when adult behaviors appear similar (e.g., see Tomasello et al., 2001 for a developmental comparison of rhesus and chimpanzees). But perhaps the most salient limitation of current research into primate social cognition is the one easiest to remedy: whereas almost every major category of social-cognitive research has several studies examining that ability in chimpanzees, the existing data across other taxa are more patchy—with only one or two relevant studies—and often nonexistent (see Tables 7.1 through 7.4; the first row lists studies with chimpanzees). This missing evidence becomes all the more striking for tasks that do not involve gaze-following paradigms. The consequence of this imbalance is that most of what we know about "primate social cognition" is really "chimpanzee social cognition." This paucity of data on the social-cognitive skills of the vast majority of the Primate Order makes it difficult to draw any broad conclusions about either the social-cognitive skills of nonhuman primates or the evolutionary pressures shaping these skills. Consequently, many empirical tests of these models involve very rough quantifications of intelligence via morphological correlates such as brain size (Dunbar, 1992), making it difficult to assess the very evolutionary hypotheses that originally spurred interest in primate social cognition. However, several new approaches to the study of primate social cognition—including comparisons between closely related species and studies of convergence with other taxa—have begun to tackle this problem.

The Comparative Method: Identifying the Forces Shaping Social Cognition

The comparative method—examining the traits of different populations or species that have been shaped by differing ecological or social forces in order to better understand how natural selection proceeded—is one of the most important techniques in evolutionary biology (Mayr, 1982). The comparative method allows us to reconstruct a phenomenon (evolution via natural selection) that often cannot be directly observed, and therefore address not just *what* the differences are between different groups of organisms, but also *why* those differences arose. Consequently, it may be the most powerful technique we have to answer functional questions about social-cognitive abilities across primate taxa.

One such approach is to test closely related species on a battery of tests that can be used to identify whole suites of shared and derived traits across different domains of cognition (e.g., Herrmann et al., 2007). However, several more specific hypotheses about the role of social and ecological factors in the evolution of particular cognitive abilities can also be addressed with comparative data. For example, one prediction of the competition hypothesis described earlier is that there will be critical differences between the social-cognitive skills of more despotic, aggressive species compared to more egalitarian, tolerant species. Specifically, as despotic species face more intense competition for food, as well as a steeper dominance hierarchy limiting their access to that food (de Waal & Lutrell, 1989), they may more readily show sophisticated social-cognitive skills when competing with others. Conversely, more egalitarian species might show greater skills in cooperative contexts (e.g., Hare et al., 2007; Petit et al., 1992). Notably, the two species that have been successfully studied using competitive paradigms—rhesus macaques and chimpanzees—are both more despotic than closely related egalitarian sister species. Thus, comparing the social-cognitive skills of chimpanzees and rhesus to bonobos and Tonkean macaques, respectively, in a food competition paradigm would be helpful and could provide a direct test of this hypothesis.

This kind of framework raises additional issues about evolutionary interpretations of the comparative data that we do have. For example, recent studies have indicated that many of the social-cognitive abilities identified in chimpanzees, such as perspective taking and intention reading, are also present in more distantly related monkeys such as rhesus macaques and capuchins. One interpretation of these data are that such mechanisms are quite evolutionarily ancient, extending back to approximately 40 Mya (Steiper & Young, 2006) when the primate lineage leading to New World monkeys such as capuchins split from the lineage leading to Old World monkeys and apes. However, another possibility is that these similar behaviors actually represent instances of social-cognitive convergence, or parallel evolution, in different lineages. Capuchins—who engage in both sophisticated tool use (Visalberghi, 1990; see Chapter 29), and hunting behaviors (Rose, 1997)—are often considered behaviorally convergent with chimpanzees (Fragaszy et al., 2004). Similarly, if food competition is a critical selective force driving the evolution of perspective taking, then rhesus—with their highly despotic social system (de Waal & Lutrell, 1989)—might also represent a case of convergence. However, such instances of possible convergence are certainly not a problem for studies of social cognition—in fact, they provide a critical method for testing how and why these abilities evolve. Indeed, some of the strongest tests of the evolutionary forces driving social-cognitive evolution comes from outside the primates.

Using Convergence in Other Taxa as a Model for Primate Evolution

Studies of social-cognitive evolution in primates face two major problems: Often the critical taxa are extinct (e.g., we cannot compare humans to other hominid species to identify uniquely human cognitive traits) or most primates share the feature in question (e.g., most anthropoids are highly social to some degree, so it is difficult to use monkeys to address coarse-grained evolutionary questions about how the presence or absence of sociality impacts social cognition). Luckily, evolution has provided an alternative route—studies of convergence in other taxa can often remedy these kinds of difficulties that arise when looking within primates. Such studies also provide a critical check to primate-centric views of social-cognitive evolution, as some "general" principles of social-cognitive evolution do not seem to hold up very well in other taxa (e.g., Dunbar & Schultz, 2007).

For example, primates seem to be relatively unskilled at interpreting communicative behavior—making it difficult to assess how such abilities arose in humans. Consequently, some researchers have begun to use dogs and wolves as helpful models for understanding the evolution of communicative gestures such as pointing and gaze cues. Whereas wolves are not very successful at using pointing or gaze cues in the absence of extensive experience with humans, dogs from a very young age appear to be highly tuned to human communication, following such cues spontaneously (Hare et al., 2002; Riedel et al., 2007; Viranyi et al., 2008). These differences suggest that the changes that occurred during domestication may be important for some kinds of social cognition, and many psychological mechanisms have been proposed for the behavioral changes that resulted from this selection, including increased attention to faces of humans (Miklosi et al., 2003) and reduced fear responses (Hare & Tomasello, 2005). Studies of other domesticated species, such as an experimental population of domesticated foxes (Hare et al., 2005), domestic goats (Kaminski et al., 2005, 2006b), and cats (Miklosi, 2003) further support the possibility that domestication can influence some forms of social-cognitive abilities. These findings suggest that interpersonal tolerance may be a critical prerequisite for some kinds of human-like social-cognitive skills, particularly those involving cooperation (Melis et al., 2006b).

Studies of convergence can also illuminate the evolution of social traits that likely emerged in basal primate groups, such as in catarrhines, and thus are widely shared across large taxonomic spaces. For example, wild spotted hyenas (*Crocutta crocutta*) live in large social groups with Old World primate–like linear dominance hierarchies and engage in cooperative hunting behaviors (Holekamp et al., 2007). This suggests that these social mammals may possess sophisticated social-cognitive skills to deal with their social landscape much like those observed in some monkey species (e.g., Drea & Carter, 2009). As spotted hyenas have two closely related relatives with significant variation in their social structure—striped hyenas (*Hyaena hyaena*) appear to be solitary, and brown hyenas (*Parahyaena brunnea*) live in smaller, less gregarious social groups (Watts & Holekamp, 2007)—comparative studies of these species with an eye to variation in their natural ecologies could illuminate why such complex abilities emerge.

Arguably, the most sophisticated social-cognitive skills are actually found outside mammals—in corvids, a taxa that includes jays, ravens, and crows. Studies of these birds have revealed startling parallels with the abilities of primates (Emery & Clayton, 2004). Specifically, corvids appear to use many primate-like social-cognitive skills (such as perspective taking) to protect their food stores when they engage in caching behaviors. For example, ravens and jays employ protective strategies when they cache (Emery & Clayton, 2001), and seem to use information about the visual perspective of others when doing so (Bugnyar & Kotrschal, 2002; Dally et al., 2004; Heinrich & Pepper, 1998). Furthermore, they not only respond to the behavior of competitors but also seem to differentiate between some kinds of knowledge states, much like chimpanzees and rhesus macaques (Bugnyar & Heinrich, 2005; Dally et al., 2006). Ravens even appear to predict how humans will behave in a caching context based on their past interactions with the humans in a noncaching context, suggesting they represent the "reputation" of social partners (Bugnyar et al., 2007). Some corvids can even make social inferences from watching third-party interactions (Paz-y-Miño et al., 2004), suggesting that some of their social-cognitive skills are also employed outside of caching contexts. Taken together, comparative work examining social cognition in other taxa makes it clear that a complete understanding of the evolutionary pressures that led to the development of primate social cognition will require a more thorough understanding of the mechanisms in similarly sophisticated social cognition in distantly related taxa as well.

Human Evolution and Social Cognition

A final limitation of present work on the nature of primate social cognition involves what is possibly the toughest question of all— the question that Darwin (1871) defined as "the greatest difficulty" facing anyone interested in the evolution of human social cognition. Namely, what aspects of primate social cognition are truly unique to our own species? In recent years, primate researchers have gained some new traction on this question. Recent findings using more ecologically relevant tasks have led to a growing consensus that humans and at least some other primates share the capacity to represent the intentions, perceptions, and knowledge of others. Thus, several new or more specific hypotheses have arisen that attempt to pinpoint the major social-cognitive differences between humans and other primates. For example, there is currently little evidence that primates share the capacity to reason about others' belief states; indeed, there is some evidence that primates *fail* to reason about others' belief states even when tested using a variety of different methodologies (Call & Tomasello, 1999; Kaminski et al., 2006a; Krachun et al., 2007; Santos et al., 2007b)—which suggests that representing others' beliefs might be a capacity limited to our own species (e.g., Povinelli & Giambrone, 2001). Other proposals have focused on other aspects of intentionality, such as the ability to represent (and the motivation to share) joint goals and shared intentions (Tomasello et al., 2005). This proposal highlights that many human-unique behaviors, such as participation in cultural endeavors, are fundamentally collaborative in nature. Although the available work to date suggests that apes perform very differently than human children on collaborative tasks with shared goals (e.g., Tomasello et al., 2005), more work is needed to directly test both this hypothesis and the belief representation hypothesis. Indeed, such work will allow us to not only gain insight into socio-cognitive capacities that might be unique to humans but also discover why these purportedly unique capacities evolved in the first place.

Conclusions

The past decade has produced significant advances in our understanding of primate social cognition. The development of novel experimental methodologies has led to increasing evidence that some primates can assess the psychological states of others in some contexts. Thus, while human social-cognitive abilities may still be outstanding, they nonetheless appear to have deep evolutionary roots. However, researchers still have a multitude of fascinating questions to attack in the future, as research has suggested that even very superficially similar social behaviors (such as gaze following) can be supported by very different underlying psychologies. The question has therefore shifted from not just *if* the sophisticated social behaviors of primates are the consequence of sophisticated cognitive skills, but *why* they might be so. With increasing comparative data, researchers can begin to address the ultimate causes that shape social cognition in both human and nonhuman primates. Armed with a new appreciation of the importance of ecologically relevant tasks that can be used across species, the stage is now set for primate cognition researchers to answer Darwin's question.

ACKNOWLEDGMENTS

We thank Felix Warneken for comments on an earlier draft of the manuscript.

NOTE

1. Note that a complete review of the vast literature on social learning in primates is outside the scope of this chapter (see Tomasello & Call, 1997, for a review of this extensive work).
2. Interestingly, chimpanzees goal emulation differs from the performance of children in this task, who faithfully imitate all of the actions of a human actor even when some of those actions are clearly irrelevant to obtaining the goal (e.g., Gergely, et al., 2002; Horner & Whiten, 2005; Meltzoff, 1995; Nagell et al., 1993; see Lyons & Keil, 2007, for a discussion of this species difference).

REFERENCES

Anderson, J. R., Kuroshima, H., Kuwahata, H., & Fujita, K. (2004). Do squirrel monkeys (Saimiri sciureus) and capuchin monkeys (Cebus apella) predict that looking leads to touching? *Animal Cognition, 7,* 185–192.

Anderson, J. R., & Mitchell, R. W. (1999). Macaques but not lemurs co-orient visually with humans. *Folia Primatologica, 70,* 17–22.

Anderson, J. R., Montant, M., & Schmitt, D. (1996). Rhesus monkeys fail to use gaze direction as an experimenter-given cue in an object-choice task. *Behavioral Processes, 37,* 47–55.

Anderson, J. R., Sallabery, P., & Barbier, H. (1995). Use of experimenter-given cues during object-choice tasks by capuchin monkeys. *Animal Behaviour, 49,* 201.

Barth, J., Reaux, J. E., D.J. Povinelli (2005). Chimpanzees' (Pan troglodytes) use of gaze cues in object-choice tasks: different methods yield different results. *Animal Cognition, 8,* 84–92.

Behne, T., Carpenter, M., Call, J., & Tomasello, M. (2005). Unwilling versus unable: Infants' understanding of intentional action. *Developmental Psychology, 41,* 492–499.

Boesch, C., & Boesch, H. (1989). Hunting behavior of wild chimpanzees in the Tai National Park. *American Journal of Physical Anthropology, 78,* 547–573.

Braeuer, J., Call, J., & Tomasello, M. (2005). All great ape species follow gaze to distant locations and around barriers. *Journal of Comparative Psychology, 119,* 145–154.

Braeuer, J., Call, J., & Tomasello, M. (2007). Chimpanzees really know what others can see in a competitive situation. *Animal Cognition, 10,* 439–448.

Braeuer, J., Call, J., & Tomasello, M. (2008). Chimpanzees do not take into account what others can hear in a competitive situation. *Animal Cognition, 11,* 175–178.

Braeuer, J., Kaminski, J., Riedel, J., Call, J., & Tomasello, M. (2006). Making inferences about the location of hidden food: Social dog, causal ape. *Journal of Comparative Psychology, 120,* 38–47.

Brooks, R., & Meltzoff, A. N. (2002). The importance of the eyes: How infants interpret adult looking behavior. *Developmental Psychology, 38,* 958–966.

Bugnyar, T., & Heinrich, B. (2005). Ravens, Corvus corax, differentiate between knowledgeable and ignorant competitors. *Proceedings of the Royal Society B, 272,* 1641–1646.

Bugnyar, T., & Kotrschal, K. (2002). Observation learning and the raitding of food caches in ravens, Corvus corax: Is it 'tactical' deception. *Animal Behaviour, 64,* 185–195.

Bugnyar, T., Schwab, C., Schloegl, C., Kotrschal, K., & Heinrich, B. (2007). Ravens judge competitors through experience with play caching. *Current Biology, 17,* 1–5.

Burkhart, J., & Heschl, A. (2006). Geometrical gaze following in common marmosets (Callithrix jacchus). *Journal of Comparative Psychology, 120,* 120–130.

Burkhart, J., & Heschl, A. (2007). Understanding visual access in common marmosets, Callithix jacchus: Perspective taking or behavior reading? *Animal Behaviour, 73,* 457–469.

Buttleman, D., Carpenter, J., & Tomasello, M. (2007). Enculturerated chimpanzees imitate rationally. *Developmental Science, 10,* F31–F38.

Byrne, R. W. W. (1988). *Machiavellian intelligence: Social expertise and the evolution of intellect in monkeys, apes, and humans.* Oxford: Oxford University Press.

Call, J., Agnetta, B., & Tomasello, M. (2000). Cues that chimpanzees do and do not use to find hidden objects. *Animal Cognition, 3,* 23–34.

Call, J., Carpenter, M., & Tomasello, M. (2005). Copy results and copying actions in the process of social learning: chimpanzee (Pan troglodytes) and human children (Homo sapiens). *Animal Cognition, 8,* 151–163.

Call, J., Hare, B., Carpenter, M., & Tomasello, M. (2004). 'Unwilling' versus 'unable': Chimpanzees' understanding of human intentional action. *Developmental Science, 7,* 488–498.

Call, J., Hare, B., & Tomasello, M. (1998). Chimpanzees gaze following in an object-choice task. *Animal Cognition, 1,* 89–99.

Call, J., & Tomasello, M. (1998). Distinguishing intentional actions in orangutans (Pongo pygmaeus), chimpanzees (Pan troglodytes), and human children (Homo sapiens). *Journal of Comparative Psychology, 112,* 192–206.

Call, J., & Tomasello, M. (1999). A nonverbal false belief task: The performance of children and great apes. *Child Development, 70,* 381–395.

Call, J., & Tomasello, M. (2007). *The gestural communication of apes and monkeys.* London: Erlbaum.

Call, J., & Tomasello, M. (2008). Does the chimpanzee have a theory of mind? 30 years later. *Trends in Cognitive Sciences, 12,* 187–192.

Carpenter, M., Akhtar, N., & Tomasello, M. (1998). Fourteen- through 18-month old infants differentially imitate intention and accidental actions. *Infant Behavior and Development, 21, 315–330.*

Cheney, D. L., & Seyfarth, R. M. (1990). *How monkeys see the world.* Chicago: Chicago University Press.

Cheney, D. L., & Seyfarth, R. M. (2007). *Baboon metaphysics: The evolution of a social mind.* Chicago: University of Chicago Press.

Dally, J. M., Emery, N. J., & Clayton, N. S. (2004). Cache protection strategies by western scrub-jays (Aphelocoma californica): Hiding food in the shade. *Proceedings of the Royal Society B, 271,* S387–S390.

Dally, J. M., Emery, N. J., & Clayton, N. S. (2006). Food-caching Western scrub jays keep track of who was watching when. *Science, 312,* 1662–1665.

Darwin, C. (1871). The Descent of Man. Appleton and company: New York.

de Waal, F. B. M. (1982). *Chimpanzee politics: Power and sex among apes.* Baltimore: John Hopkins University Press.

de Waal, F. B. M. (1996). *Good natured: The origins of right and wrong in humans and other animals.* Cambridge, MA: Harvard University Press.

de Waal, F. B. M., & Lutrell, L. M. (1988). Mechanisms of social reciprocity in three primate species: Symmetrical relationship characteristics or cognition? *Ethology and Sociobiololgy, 9,* 101–118.

de Waal, F. B. M., & Lutrell, L. M. (1989). Toward a comparative socioecology of the genus Macaca: Different dominance styles in rhesus and stumptail monkeys. *American Journal of Primatology, 19,* 83–109.

Drea, C. M., & Carter, A. N. (2009). Cooperative problem solving in a social carnivore. *Animal Behaviour.* doi:10.1016/j.anbehav.2009.06.030

Dunbar, R. I. M. (1992). Neocortex size as a constraint on group size in primates. *Journal of Human Evolution, 20,* 469–493.

Dunbar, R. I. M. (2003). The social brain: Mind, language, and society in evolutionary perspective. *Annual Review of Anthropology, 32,* 163–181.

Dunbar, R. I. M., & Schultz, S. (2007). Evolution in the social brain. *Science, 317,* 1344.

Emery, N. J., & Clayton, N. S. (2001). Effects of experience and social context on prospective caching strategies by scrub jays. *Nature, 414,* 443–446.

Emery, N. J., & Clayton, N. S. (2004). The mentality of crows: Convergent evolution of intelligence in corvids and apes. *Science, 306,* 1903–1907.

Emery, N. J., Lorincz, E. N., Perrett, D. I., Oram, M. W., & Baker, C. I. (1997). Gaze following and joint attention in rhesus monkeys (Macaca mulatta). *Journal of Comparative Psychology, 111,* 286–293.

Flavell, J. H. (1992). Perspectives on perspective-taking. In: H. Beilin & P. B. Pufall (Eds.), *Piaget's theory: Prospects and possibilities* (pp. 107–139). Hillsdale, NJ: Erlbaum.

Flombaum, J. I., & Santos, S. (2005). Rhesus monkeys attribute perceptions to others. *Current Biology, 15,* 447–452.

Fragaszy, D. M., Visalberghi, E., & Fedigan, L. M. (2004). *The complete capuchin: The Biology of the genus Cebus.* Cambridge: Cambridge University Press.

Fujita, K., Kuroshima, H., & Masuda, T. (2002). Do tufted capuchin monkeys (Cabus apella) spontaneously deceive opponents? A preliminary analysis of an experimental food-competition contest between monkeys. *Animal Cognition, 5,* 19–25.

Genty, E., & Roeder, J. J. (2006). Can lemurs learn to deceive? A study in the black lemur. *Journal of Experimental Psychology: Animal Behavior Processes, 32,* 196–200.

Gergely, G., Bekkering, H., & Kiraly, I. (2002). Rational imitation in preverbal infants. *Nature, 415,* 755.

Gergely, G., Nadasdy, Z., Csibra, G., & Biro, S. (1995). Taking the intentional stance at 12 months of age. *Cognition, 56,* 165–193.

Gintis, H., Bowles, S., Boyd, R., & Fehr, E. (2005). *Moral sentiments and material interests: On the foundations of cooperation in economic life.* Cambridge: MIT Press.

Goossens, B. M. A., Dekleva, M., Reader, S. M., Sterck, E. H. M., & Bolhuis, J. J. (2008). Gaze following in monkeys is modulated by observed facial expression. *Animal Behaviour, 75, 1673–1681.*

Hare, B. (2001). Can competitive paradigms increase the validity of experiments on primate social cognition? *Animal Cognition, 4,* 269–280.

Hare, B., Addessi, E., Call, J., Tomasello, M., & Visalberghi, E. (2003). Do capuchin monkeys, Cebus apella, know what conspecifics do and do not see? *Animal Behaviour, 65,* 131–142.

Hare, B., Brown, M., Williamson, C., & Tomasello, M. (2002). The domestication of social cognition in dogs. *Science, 298,* 1634–1636.

Hare, B., Call, J., Agnetta, B., & Tomasello, M. (2000). Chimpanzees know what conspecifics do and do not see. *Animal Behaviour, 59,* 771–785.

Hare, B., Call, J., & Tomasello, M. (2001). Do chimpanzees know what conspecifics know? *Animal Behaviour, 61,* 139–151.

Hare, B., Call, J., & Tomasello, M. (2006). Chimpanzees deceive a human competitor by hiding. *Cognition, 101,* 495–514.

Hare, B., Melis, A. P., Woods, V., Hastings, S., & Wrangham, R. (2007). Tolerance allows bonobos to outperform chimpanzees on a cooperative task. *Current Biology, 17,* 619–623.

Hare, B., Plyusnina, I., Ignacio, N., Schepina, O., Stepika, A., Wrangham, R., et al. (2005). Social cognitive evolution in captive foxes in a correlated by-product of experimental domestication *Current Biology, 15,* 226–230.

Hare, B., & Tomasello, M. (2004). Chimpanzees are more skillful in competitive than cooperative cognitive tasks. *Animal Behaviour, 68,* 571–581.

Hare, B., & Tomasello, M. (2005). Human-like social skills in dogs? *Trends in Cognitive Sciences, 9,* 439–444.

Hauser, M. D., Glynn, D., & Wood, J. (2007). Rhesus monkeys correctly read the goal relevant gesture of a human agent. *Proceedings of the Royal Society B, 274,* 1913–1918.

Heinrich, B., & Pepper, J. W. (1998). Influence of competitors on caching behaviour in the common raven, Corvus corax. *Animal Behaviour, 56,* 1083–1090.

Herrmann, E., Call, J., Hernadez-Lloreda, M. V., Hare, B., & Tomasello, M. (2007). Humans have evolved specialized skills of social cognition: The cultural intelligence hypothesis. *Science, 317,* 1360–1366.

Herrmann, E., Melis, A. P., & Tomasello, M. (2006). Apes' use of iconic cues in the object-choice task. *Animal Cognition, 9, 118–130*

Herrmann, E., & Tomasello, M. (2006). Apes' and children's understanding of cooperative and competitive motives in a communicative situation. *Developmental Science, 9,* 518–529.

Heyes, C. M. (1998). Theory of mind in nonhuman primates. *Behavioral and Brain Sciences, 21,* 101–148.

Kuroshima, H., Fujita, K., Adachi, I., Iwata, K., & Fuyuka, A. (2003). A capuchin monkey (Cebus apella) recognizes when people do and do not know the location of food. *Animal Cognition, 6.*, 283–291.

Hirata, S., & Matsuzawa, T. (2001). Tactics to obtain a hidden food item in chimpanzee pairs (Pan troglodytes). *Animal Cognition, 4,* 285–295.

Holekamp, K. E., Sakai, S. T., & Lundrigan, B. L. (2007). Social intelligence in the spotted hyena (Crocuta crocuta). *Philosophical Transactions of the Royal Society B, 362,* 523–538.

Horner, V., & Whiten, A. (2005). Causal knowledge and imitation/emulation switching in chimpanzees (Pan troglodytes) and children (Homo saiens). *Animal Cognition, 8,* 164–181.

Horton, K. E., & Caldwell, C. A. (2006). Visual co-orientation and expectations about attentional orientation in pileated gibbons (Hylobytes pileatus). *Behavioral Processes, 72,* 65–73.

Hostetter, A. B., Rseel, J. L., Freeman, H., & Hopkins, W. D. (2007). Now you see me, now you don't: Evidence that chimpanzees understand the role of the eyes in attention. *Animal Cognition, 10,* 55–62.

Humphrey, N. K. (1976). The social function of intellect. In: P. P. G. Bateson & R. A. Hinde (Eds.), *Growing points in ethology* (pp. 303–317). Cambridge: Cambridge University Press.

Inoue, Y., Inoue, E., & Itakura, I. (2004). Use of experimenter-given directional cues by a young white-handed gibbon (Hylobytes lar). *Japanese Psychological Research, 46,* 262–267.

Itakura, S. (1991). An exploratory study of gaze-monitoring in nonhuman primates. *Japanese Psychological Research, 38,* 174–180.

Itakura, S. (1996). An exploratory study of gaze-monitoring in nonhuman primates. *Japanese Psychological Research, 38,* 174–180.

Itakura, S., Agnetta, B., Hare, B., & Tomasello, M. (1999). Chimpanzees use of human and conspecific social cues to locate hidden food. *Developmental Science, 2,* 448–456.

Jensen, K., Call, J., & Tomasello, M. (2007). Chimpanzees are vengeful but not spiteful. *Proceedings of the National Academy of Sciences, 104,* 13046.

Jolly, A. (1966). Lemur social behavior and primate intelligence. *Science, 153,* 501–506.

Kaminski, J., Call, J., & Tomasello, M. (2004). Body orientation and face orientation: Two factors controlling ape' begging behavior from humans. *Animal Cognition, 7,* 216–223.

Kaminski, J., Call, J., & Tomasello, M. (2008). Chimpanzees know what others know, but not what they believe. *Cognition, 109,* 224–234.

Kaminski, J., Call, J., & Tomasello, M. (2006b). Goats' behaviour in a competitive food paradigm: Evidence for perspective taking? *Behaviour, 143,* 1341–1356.

Kaminski, J., Riedel, J., Call, J., & Tomasello, M. (2005). Domestic goats, Capra hircus, follow gaze direction and use social cues in an object choice task. *Animal Behavior, 69,* 11–18.

Kappeler, P., & van Schaik, C. P. (2006). *Cooperation in primates and humans: Mechanisms and evolution.* Berlin: Springer-Verlag.

Karin-D'Arcy, R., & Povinelli, R. J. (2002). Do chimpanzees know what each other see? A closer look. *International Journal of Comparative Psychology, 15,* 21.

Kobayashi, H., & Kohshima, S. (1997). Unique morphology of the human eye. *Nature, 387.*, 767–768.

Kobayashi, H., & Kohshima, S. (2001). Unique morphology of the human eye and its adaptive meaning: Comparative studies on external morphology of the primate eye. *Journal of Human Evolution, 40,* 419–435.

Koyama, N. F., Caws, C., & Aureli, F. (2006). Interchange of grooming and agonistic support in chimpanzees. *International Journal of Primatology, 27,* 1293–1309.

Krachun, C., Carpenter, M., Call, J., & Tomasello, M. (2009). A competitive nonverbal false belief task for children and apes. Developmental Science, 12: 521–535.

Kummer, H., Anzenberger, G., & Hemelrijk, C. K. (1996). Hiding and perspective taking in long-tailed macaques (Maccaca fasicularis). *Journal of Comparative Psychology, 110,* 97–102.

Kuroshima, H., Fujita, H., Fuyuki, A., & Masuda, T. (2002). Understanding the relationship between seeing and knowing by tufted capuchin monkeys (Cebus apella). *Animal Cognition, 5,* 41–48.

Kuroshima, H., Kuwahata, H., & Fujita, K. (2008). Learning from others' mistakes in capuchin monkeys (*Cebus apella*). *Animal Cognition, 11,* 599–609.

Liebal, K., Call, J., & Tomasello, M. (2004a). Use of gesture sequences in chimpanzees. *American Journal of Primatology, 64,* 377–396.

Liebal, K., Pika, S., Call, J., & Tomasello, M. (2004b) To move or not to move: How apes adjust to the attentional state of others. *Interaction Studies, 5.*, 199–219.

Liebal, K., Pika, S., & Tomasello, M. (2003). Social communication in siamangs (Symphalangus syndactylus): Use of gestures and facial expressions. *Primates, 45,* 41–57.

Lyons, D. E., & Santos, L. R. (2006). Ecology, domain specifics, and the origins of theory of mind: Is competition the catalyst? *Philosophy Compass, 1,* 481–492.

Lyons, D., & Santos, L. R. (Submitted). Capuchin monkeys (Cebus apella) discriminate between intentional and unintentional human actions. *Developmental Psychology.*

Lyons, D. E., Young, A. G., & Keil, F. C. (2007). The hidden structure of overimitation. *Proceeding of the National Academy of Sciences, 104,* 19751–19756.

Mayr, E. (1982). *The growth of biological thought.* Cambridge, MA: Harvard University Press.

Melis, A., Call, J., & Tomasello, M. (2006a). Chimpanzees conceal visual and auditory information from others. *Journal of Comparative Psychology, 120,* 154–162.

Melis, A. P., Hare, B., & Tomasello, M. (2006b). Engineering cooperation in chimpanzees: Tolerance constraints on cooperation. *Animal Behaviour, 72,* 275–286.

Melis, A. P., Hare, B., & Tomasello, M. (2006c). Chimpanzees recruit the best collaborator. *Science, 311,* 1297–1300.

Melis, A. P., Hare, B., & Tomasello, M. (2008). Do chimpanzees reciprocate received favors? *Animal Behaviour, 76,* 951–962.

Melis, A. P., Hare, B., & Tomasello, M. (2009). Chimpanzees coordinate in a negotiation game. *Evolution and Human Behavior.* doi: 10.1016/j.evolhumbehav.2009.05.003

Meltzoff, A. N. (1995). Understanding the intentions of others: Re-enactment of intended acts by 18-month-old children. *Developmental Psychology, 31,* 838.

Miklosi, A., Pongracz, P., Lakatos, G., Topal, J., Csanyi, V. (2005). A comparative study of the use of visual communicative signals in dog–human and cat–human interactions. *Journal of Comparative Psychology. 2, 179–186.*

Miklosi, A., Kubinyi, E., Topal, J., Gacsi, M., Viranyi, Z., & Csanyi, V. (2003). A simple reason for a big difference: Wolves do not look back at humans, but dogs do. *Current Biology, 13,* 763–766.

Mitani, J. C. (2006). Reciprocal exchange in chimpanzees and other primates. In: P. M. Kappeler & C. P. Schaik (Eds.), *Cooperation in primates and humans. Mechanisms and evolution.* Berlin: Springer-Verlag.

Mitani, J. C., & Watts, D. P. (2001). Why do chimpanzees hunt and share meat? *Animal Behaviour, 61,* 915–924.

Moll, H., & Tomasello, M. (2004). 12- and 18-month-olf infants follow gaze to spaces behind barriers. *Developmental Science, 7,* F1–F9.

Muller, M. N., & Mitani, J. C. (2005). Conflict and cooperation in wild chimpanzees. *Advances in the Study of Behavior, 35,* 275–331.

Myowa-Yamakoshi, M., & Matsuzawa, T. (2000). Imitation of intentional manipulatory actions in chimpanzees (Pan troglodytes). *Journal of Comparative Psychology, 114,* 381–391.

Nagell, K., Olguin, R., Tomasello, M. (1993). Processes of social learning in the tool use of chimpanyees (Pan troglodytes) and human children (Homo Sapiens). *Journal of Comparative Psychology, 107,* 174–186.

Neiworth, J. J., Burman, M. A., Basile, B. M., & Lickteig, M. T. (2002). Use of experimenter-given cues in visual co-orienting and in an object-choice task by a New World monkey species, cotton-top tamarins (Sanguinus oedipus). *Journal of Comparative Psychology, 116,* 3–11.

Okamoto-Barth, S., Call, J., & Tomasello, M. (2007). Great apes' understanding of other individuals' line of sight. *Psychological Science, 18,* 462–468.

Okamoto, S., Tomonaga, M., Ishii, K., Kawai, N., Tanaka, M., & Matsuzawa, T. (2002). An infant chimpanzee (Pan troglodytes) follows human gaze. *Animal Cognition, 5,* 107–114.

Okamoto-Barth, S., Tomonaga, M., Tanaka, M., & Matsuzawa, T. (2008). Development of using experimenter-given cues in infant chimpanzees: Longitudinal changes in behavior and cognitive development. *Developmental Science, 11,* 98–108.

Onishi, K. H., & Baillargeon, R. (2005). Do 15-month-old infants understand false beliefs? *Science, 308,* 255.

Paz-y-Miño, G., Bond, A. B., Kamil, A. C., & Balda, R. P. (2004). Pinyon jays use transitive inference to predict social dominance. *Nature, 430,* 778–781.

Petit, O., Desportes, C., & Thierry, B. (1992). Differential probability of coproduction in two species of macaque (Macaca tonkeana, M. mulatta). *Ethology, 90,* 107–120.

Phillips, A. T. (2002). Infant's ability to connect gaze and emotional expression to intentional action. *Cognition, 85,* 53–78.

Phillips, W., Barnes, J. L., Mahajan, N., Yamaguchi, M., & Santos, L. R. (2009). 'Unwilling versus unable': Capuchins' (Cebus apella) understanding of human intentional action. *Developmental Science.* doi:10.1111/j.1467–7687. 2009. 00840.

Pika, S., Liebal, K., & Tomasello, M. (2003). Gestural communication in young gorilla (Gorilla gorilla): Gestural repetoire, learning, and use. *American Journal of Primatology, 60, 95–111.*

Pika, S., Liebal, K., & Tomasello, M. (2005). Gestural communication in subadult bonobos (Pan paniscus): Repertoire and use. *American Journal of Primatology, 65,* 39–61.

Poss, S. R., Kuhar, C., Stoinski, T. S., & Hopkins, W. D. (2006). Differential use of attentional and visual communicative signaling by orangutans (Pongo pygmaeus) and Gorillas (Gorilla gorilla) in response to the attentional status of a human. *American Journal of Primatology, 68,* 978–992.

Povinelli, D. J., Bierschwale, D. T., & Cech, C. (1999). Comprehension of seeing as a referential act in young children, but not juvenile chimpanzees. *British Journal of Developmental Psychology, 17,* 37–60.

Povinelli, D. J., Dunphy-Llii, S., Reaux, J. E., & Mazza, M. P. (2002). Psychological diversity in chimpanzees and humans: New longitudinal assessments of chimpanzees' understanding of attention. *Brain, Behavior, & Evolution, 59,* 33–53.

Povinelli, D. J., & Eddy, T. J. (1996a). Chimpanzees: Joint visual attention. *Psychological Science, 7,* 129–135.

Povinelli, D. J., & Eddy, T. J. (1996b). Factors influencing young chimpanzees' (Pan troglodytes) recognition of attention. *Journal of Comparative Psychlogy, 110,* 336–345.

Povinelli, D. J., & Eddy, T. J. (1996c). What young chimpanzees know about seeing. *Monographs for the Society for Research in Child Development, 61,* 1–152.

Povinelli, D. J., & Giambrone. S. (2001). Reasoning abut beliefs: A human specialization? *Child Development, 72,* 691–695.

Povinelli, D. J., Nelson, K. E., & Boysen, S. T. (1990). Inferences about guessing and knowing by chimpanzees. *Journal of Comparative Psychology, 104,* 203–210.

Povinelli, D. J., Parks, K. A., & Novak, M. A. (1991). Do rhesus monkeys (Macaca mulatta) attribute knowledge and ignorance to others? *Journal of Comparative Psychology, 105,* 318–325.

Povinelli, D. J., Perilloux, H. K., Reaux, J. E., & Bieschwale, D. T. (1998). Young and juvenile chimpanzees' (Pan troglodytes) reactions to intentional versus accidental and inadvertent actions. *Behavioral Processes, 42,* 205–218.

Povinelli, D. J., Reaux, J. E., Bierschwale, D. T., Allain, A. D., & Simon, B. B. (1997). Exploitation of pointing as a referential gesture in young children, but not adolescent chimpanzees. *Cognitive Development, 12,* 327–365.

Povinelli, D. J., Theall, L. A., Reaux, J. E., & Dunphy-Leli, S. (2003). Chimpanzees spontaneously alter the location of their gestures to match the attentional orientation of others. *Animal Behaviour, 66,* 71–79.

Povinelli, D. J., & Vonk, J. (2003). Chimpanzee minds: suspiciously human? *Trends in Cognitive Sciences, 7,* 157.

Povinelli, D. J., & Vonk, J. (2004). We don't need a microscope to explore the chimpanzee's mind. *Mind & Language, 19,* 1–28.

Premack, D. (1988). "Does the chimpanzee have a theory of mind?" revisited. In: R. B. A. Whiten (Ed.), *Machiavellian intelligence* (pp. 160–179). New York: Oxford University Press.

Premack, D. W. (1978). Does the chimpanzee have a theory of mind? *Behavioral and Brain Sciences, 4,* 515–526.

Reaux, J., Theall, L. A., & Povinelli, D. J. (1999). A longitudinal investigation of chimpanzee's understanding of visual perception. *Child Development, 70,* 275–290.

Repacholi BM, Gopnik A (1997) Early reasoning about desires: Eidence from 14- and 18-month-olds. Developmental Psychology 33:12–21.

Riedel, J., Schumann, K., Kaminski, J., Call, J., & Tomasello, M. (2007). The early ontogeny of human-dog communication. *Animal Behaviour, 75,* 1003–1014.

Rose, L. (1997). Vertebrate predation and food-sharing in Cebus and Pan. *International Journal of Primatology, 18,* 727.

Ruiz, A., Gomez, J. C., Roeder, J. J., * Byrne, R. W. (2009) Gaze following and gaze priming in lemurs. *Animal Cognition, 12,* 427–434.

Santos, L. R., Flombaum, J. I., & Phillips, W. (2007a). The evolution of human mindreading: How non-human primates can inform social cognitive neuroscience. In: S. Platek (Ed.), *Evolutionary cognitive nueroscience.* Cambridge, MA: MIT Press.

Santos, L. R., & Hauser, M. D. (1999). How monkeys see the eyes: cotton-top tamarins' reactions to changes in visual attention and action. *Animal Cognition, 2,* 131–139.

Santos, L. R., Marticorena, D., & Goddu, A. (2007b). Do monkeys reason about the false beliefs of others? In: Symposium presented at the 14th Biennial Meeting of the Society for Research in Child Development, Boston, MA.

Santos, L. R., Nissen, A. G., & Ferrugia, J. A. (2006). Rhesus monkeys, Macaca mulatta, know what others can and cannot hear. *Animal Behavior, 71,* 1175–1181.

Scerif, G., Gomez, J. C., & Byrne, R. W. (2004). What do Diana monkeys know about the focus of attention of a conspecific? *Animal Behavior, 68,* 1239–1247.

Shepherd, S. V., & Platt, M. L. (2008). Spontaneous social orienting and gaze following in ringtailed lemurs (Lemur catta). *Animal Cognition, 11,* 13–20.

Southgate, V., Senju, A., & Csibra, G. (2007). Action anticipation through attribution of false belief by 2-year-olds. *Psychological Science, 18,* 587.

Steiper, M. E., & Young, N. M. (2006). Primate molecular divergence dates. *Molecular Phylogenetics and Evolution, 41,* 384–394.

Subiaul, F., Vonk, J., Okamoto-Barth, S. & Barth, J. (2008). Do chimpanzees learn reputation by observation? Evidence from direct and indirect experience with generous and selfish strangers. *Animal Cognition, 11,* 611–623.

Surian, L., Caldi, S., & Sperber, D. (2007). Attribution of beliefs by 13-month-old infants. *Psychological Science, 18,* 580–586.

Theall, L. A., & Povinelli, D. J. (1999). Do chimpanzees tailor their gestural signals to fit the attentional state of others? *Animal Cognition, 2,* 207–214.

Tomasello, M. (1990). Cultural transmission in the tool use and communicatory signaling of chimpanzees? In: S. T. Parker & K. R. Gibson (Eds.), *"Language" and intelligence in monkeys and apes.* Cambridge: Cambridge University Press.

Tomasello, M. (1999). *The cultural origins of human cognition.* Cambridge, MA: Harvard University Press.

Tomasello, M., & Call, J. (1997). *Primate cognition.* New York: Oxford University Press.

Tomasello, M., Call, C., & Hare, B. (1998). Five primate species follow the visual gaze of conspecifics. *Animal Behaviour, 55,* 1063–1069.

Tomasello, M., Call, J., & Hare, B. (2003). Chimpanzees understand psychological states–the question is which ones and to what extent. *Trends in Cognitive Sciences, 7, 153–156.*

Tomasello, M., Call, J., Nagell, K., Olguin, R., & Carpenter, M. (994). The learning and use of gestural signals by young chimpanzees: A trans-generational study. *Primates, 35,* 137–154.

Tomasello, M., & Carpenter, M. (2005). The emergence of social cognition in three young chimpanzees. *Monographs of the Society for Research in Child Development. 70, 1–154.*

Tomasello, M., Carpenter, M., Call, J., Behne, T., & Moll, H. (2005). Understanding and sharing intentions: The origins of cultural cognition. *Behavioral and Brain Sciences, 28,* 675–735.

Tomasello, M., Davis-Dasilva, M., Camak, L., & Bard, K. (1987). Obervational learning of tool-ise by young chimpanzees. *Human Evolution, 2,* 175–183.

Tomasello, M., Hare, B., & Agnetta, B. (1999). Chimpanzees, Pan troglodytes, follow gaze direction geometrically. *Animal Behaviour, 58,* 769–777.

Tomasello, M., Hare, B., & Fogleman, T. (2001). The ontogeny of gaze following in chimpanzee, Pan troglodytes, and rhesus macaques, Macaca mulatta. *Animal Behaviour, 61,* 335–343.

Tomasello, M., Hare, B., Lehmann, H., & Call, J. (2007). Reliance on head versus eyes in the gaze following of great apes and human infants: The cooperative eye hypothesis. *Journal of Human Evolution, 52,* 314–320.

Tomonage, M., Tanaka, M., Matsuzawa, T., Myowa-Yamakoshi, M., & Kosugi, D. (2004). Development of social cognition in infant chimpanzees (Pan troglodytes): Face recognition, smiling, gaze, and the lack of triadic interactions. *Japanese Psychological Research, 46,* 227–235.

Uller, C. (2004). Disposition to recognize goals in infant chimpanzees. *Animal Cognition, 7,* 154–161.

Uller, C., & Nichols, S. (2000). Goal attribution in chimpanzees. *Cognition, 76,* B27–B34.

Vick, S. J., & Anderson, J. R. (2000). Learning and limits of use of eye gaze by capuchin monkeys (Cebus apella) in an object-choice task. *Journal of Comparative Psychology, 114,* 200–207.

Vick, S. J., & Anderson, J. R. (2003). Use of human visual attention cues by olive baboons (Papio anubis) in a competitive task. *Journal of Comparative Psychology, 117,* 209–216.

Vick, S. J., Bovet, D., & Anderson, J. R. (2001). Gaze discrimination learning in olive baboons (Papio anubis). *Animal Cognition, 4,* 1–10.

Viranyi, Z., Gacsi, M., Kubinyi, E., Topal, J., Belenyi, B., Ujfalussy, D., et al. (2008). Comprehension of human pointing gestures in oung human-reared wolves (Canis lupus) and dogs (Canis familiaris). *Animal Cognition, 11,* 373–387.

Visalberghi, E. (1990). The manufacture and use of tools by capuchin monkeys (Cebus apella). *Journal of Comparative Psychology, 101,* 159.

Warneken, F., Hare, B., Melis, A. P., Haunus, D., & Tomasello, M. (2007). Spontaneous altruism by chimpanzees and young children. *PLoS Biology, 5,* 1414–1420.

Warneken, F., & Tomasello, M. (2006). Altruistic helping in human infants and young chimpanzees. *Science, 311,* 1301–1303.

Watts, D. P. (1998). Coalitionary mate guarding by male chimpanzees in Ngogo, Kibale National Park, Uganda. *Behavioral Ecology and Sociobiology, 44,* 43–55.

Watts, D. P., & Mitani, J. C. (2001). Boundary patrols an intergroup encounters in wild chimpanzees. *Behaviour, 138,* 299–327.

Watts, H. E., & Holekamp, K. E. (2007). Hyena societies. *Current Biology, 17,* R557–R660.

Wellman, H. M. (1990). *The child's theory of mind.* Cambridge, MA: MIT Press.

Wimmer, H., & Perner, J. (1983). Beliefs about beliefs: Representation and constraining function of wrong beliefs in young children's understanding of deception. *Cognition, 13,* 103– 128.

Wood, J. N., Glynn, D. D., Phillips, B. C., & Hauser, M. D. (2007). The perception of rational, goal-directed action in non-human primates. *Science, 317,* 1402.

Woodward, A. (1998). Infants selectively encode the goal object of an actors reach. *Cognition, 69,* 1–34.

Woodward, A., Sommerville, J. A., & Guajardo, J. J. (2001). How infants make sense of intentional action. In: B. F. Malle, L. J. Moses, & D. A. Balin (Eds.), *Intentions and intentionality.* Cambridge: MIT Press.

CHAPTER 8

Behavioral Signatures of Numerical Cognition

Elizabeth M. Brannon, Kerry E. Jordan and Sarah M. Jones

INTRODUCTION

As adult humans, we rely on numbers for almost everything we do. We use numbers to measure time, weight, height, money, speed, and scores. We use numbers to organize the world around us by labeling things such as the classes we teach or the IP addresses of our computers. We also use numbers as ordinals to rank our place in a line or a race. Our mathematical capacities are honed in school and become highly symbolic and abstract over development. No animal will ever achieve the human capacity to precisely count the number of cars in a parking lot or use algebra to calculate the speed of a jet that travels 5,400 km in 9 hours against a jetstream of 190 km/hour. But what are the evolutionary foundations that support the uniquely human ability for mathematical thinking? There has been over a century of research on the numerical abilities of nonhuman animals. The ability to represent number approximately has been documented in animals as diverse as lizards, rats, pigeons, ferrets, raccoons, dolphins, lemurs, monkeys, and apes. Here we focus on findings that establish the nature of the approximate number system (ANS) in nonhuman primates, both because of the focus of this book and because of the extensive amount of work on this topic with monkeys and apes and more recently prosimians. Our review reveals that nonhuman primates and humans possess a shared ANS that functions to represent number as an abstract variable permitting computations about the quantitative aspects of the world.

IS IT NUMBER OR A SIMPLE TRICK?

The study of numerical abilities in animals got off to a fractious start with false claims that a horse named "Clever Hans" could add, subtract, multiply, and even divide. Nevertheless, over the last century there have been a multitude of carefully controlled studies that demonstrate that many species of nonhuman animals can ignore perceptual variables such as surface area, density, perimeter, and intensity and attend to the purely numerical attribute of visual, auditory, and, in rare demonstrations, even tactile stimulus arrays.

For example, with laboratory training rhesus monkeys are able to match arrays based on numerosity and ignore continuous dimensions of an array such as element size, cumulative surface area, and density (Jordan & Brannon, 2006; Nieder & Miller, 2004; Nieder et al., 2002). Performance has been shown to be equivalent when continuous variables are congruent or incongruent with numerosity (Jordan & Brannon, 2006). Studies of numerical ordering in monkeys have used similar stringent stimulus controls and have also demonstrated that monkeys can make purely numerical judgments (e.g., Brannon & Terrace, 1998, 2000; Brannon et al., 2006; Cantlon & Brannon, 2005, 2006a, 2006b; Judge et al., 2005; Smith et al., 2003; see Fig. 8.1 for example stimuli and a screenshot of a monkey solving the task). For example, Beran (2008) showed that rhesus monkeys could enumerate moving arrays of dots and attend to the numerosity of a subset of the dots within an array.

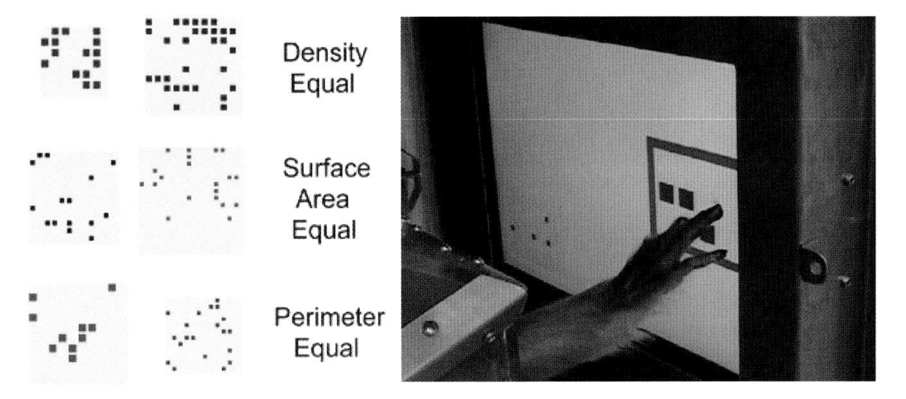

Figure 8.1 Example stimuli used in a numerical ordering task. On some trials number covaried with cumulative surface area, density, or perimeter, while on other trials these variables were equated for the larger and smaller numerosity. After Cantlon, J. F., & Brannon, E. M. (2006a). Shared system for ordering small and large numbers in monkeys and humans. *Psychological Science, 17*, 402–407. Used with permission.

Arrays contained red and black dots that moved independently and monkeys were rewarded for choosing the array with the larger number of black dots in each display (Fig. 8.2). Capuchin monkeys and rhesus monkeys were tested in this paradigm; however, only the rhesus monkeys were above chance on trials in which the correct answer involved choosing an array with a larger number of black dots but a smaller number of total dots.

Furthermore, a great deal of work also suggests that rats and pigeons, among many other nonprimate species, have numerical competence (e.g., Emmerton et al., 1997; Meck & Church, 1983; Olthof & Roberts, 2000; Roberts, 2005). Given these carefully controlled studies, there can be little doubt that nonhuman animals have the capacity to discriminate stimuli based solely on number and ignore nonnumerical dimensions that typically covary with number in nature.

The Behavioral Signature of Animal Numerical Competence

Symbolic representations of number (e.g., words and Arabic numerals) allow us to make precise discriminations, such as distinguishing between 33 and 34 children on a school bus. In contrast, a ubiquitous feature of the ANS is that discrimination depends on the ratio between numerical values. With count words, it is apparent that

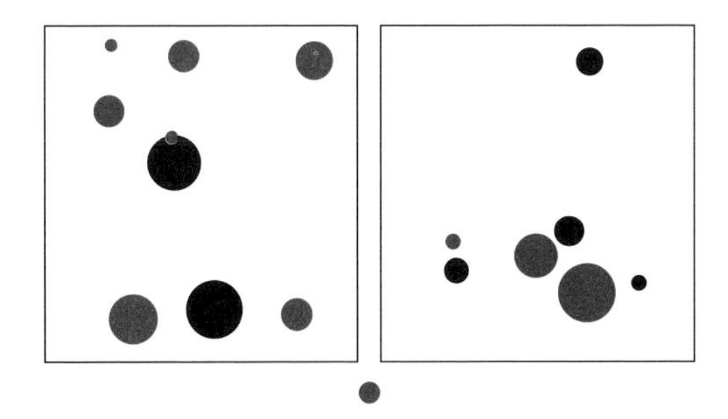

Figure 8.2 Stimuli used to test the ability of capuchin and rhesus monkeys to enumerate a subset of an array. Monkeys were rewarded for moving the cursor (the red dot below the display) to choose the array with the larger number of black dots. Beran, M. J. (2008). Monkeys (Macaca mulatta and Cebus apella) track, enumerate, and compare multiple sets of moving items. *Journal of Experimental Psychology: Animal Behavior Processes, 34*, 63–74. Used with permission.

1,053 and 1,054 differ by the same single unit as 3 and 4, whereas with the ANS such comparisons are markedly different given the discrepancy in the ratios for each pair. One explanation for ratio dependence is that the underlying representation for numerosity is an analog format (e.g., Dehaene, 1992; Feigenson et al., 2004; Gallistel & Gelman, 1992). In this format, numerosity is represented as a mental magnitude that is proportional to the quantity it represents; consequently, discrimination obeys Weber's Law. Weber's Law states that $\Delta I/I = C$, where ΔI is the increase in stimulus intensity to a stimulus of intensity I that is required to produce a detectable change in intensity and C is a constant. Therefore, if a graduate student detects a change when her advisor adds a 2-pound book to her 10-pound backpack full of books, she would need a 4-pound book to detect a change in a 20-pound backpack.

Ratio dependence can be seen when animals compare arrays of elements based on numerosity (e.g., Beran, 2008; Brannon & Terrace, 1998, 2000; Brannon et al., 2006; Cantlon & Brannon, 2005, 2006a, 2006b; Judge et al., 2005; Smith et al., 2003; Tomonaga, 2008), when they are required to match sets based on numerosity (e.g., Jordan & Brannon, 2006), and when they make comparisons of food quantities (e.g., Addessi et al., 2008; Beran, 2001; Beran & Beran, 2004; Beran et al., 2008; Call, 2000; Evans et al., 2008; Stevens et al., 2007). For example, Cantlon and Brannon (2006a) demonstrated that when rhesus monkeys were presented with pairs of arrays of 2 to 30 elements and required to choose the numerically smaller array, their accuracy and reaction time to respond was modulated by the ratio between the compared values (Fig. 8.3). Furthermore, when rats, pigeons, or monkeys are tested on numerical bisection tasks in which they are first trained to categorize small and large anchor values and then required to classify intermediate values as closer to the small or large anchor, their performance is best explained by models that assume they use a ratio comparison rule (Church & Deluty, 1977; Emmerton & Renner, 2006; Fetterman, 1993; Jordan & Brannon, 2006; Meck & Church, 1983; Meck et al., 1985; Platt & Davis, 1983; Roberts, 2005; Stubbs, 1976).

Ratio dependence has been found in all the systematic tests of numerical competence throughout the animal kingdom. This is especially noteworthy given that a shortcoming of research in this domain is that direct species comparisons have rarely been made. Testing multiple species with the same stimuli and tasks would allow a test of whether differences in socioecology have selected for differences in numerical sensitivity between genera or species. Thus, a next step for comparative research on numerical competence will be to standardize tasks and stimuli and test multiple species in an effort to understand the selective pressures that account for variability in numerical abilities.

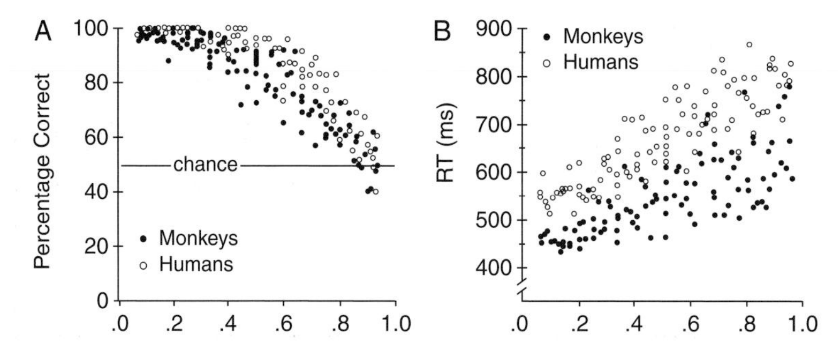

Figure 8.3 Accuracy (**a**) and reaction time (**b**) for monkeys and humans as a function of the ratio between values in a numerical ordering task. For both monkeys and humans, accuracy decreased and reaction time increased as the ratio approached 1. After Cantlon, J. F., & Brannon, E. M. (2006a). Shared system for ordering small and large numbers in monkeys and humans. *Psychological Science, 17*, 402–407. Used with permission.

THE ANS SHARED BY ANIMALS AND HUMANS ALIKE

Although adult humans from most parts of the world are highly adept at verbal counting and using symbolic number systems to perform complex mathematical operations, the ANS coexists with language-mediated number representations in adult humans. When adults are tested in nonverbal tasks that are designed to emulate animal tasks, their performance often looks virtually indistinguishable from their furry counterparts (e.g., Barth et al., 2006; Beran, 2008; Boisvert et al., 2003; Cantlon & Brannon, 2006a, 2007b; Cordes et al., 2001; Moyer & Landauer, 1967; Whalen et al., 1999). For instance, Whalen and colleagues (1999) and Cordes and colleagues (2001) adapted a classic paradigm from the animal literature for use with humans. In the original experiment, rats were required to press a lever a target number of times to gain reward (Platt & Johnson, 1971). The mean number of responses the rats made increased with the required number; importantly, the variability (standard deviation) in their response distributions was proportional to the required number of responses. In the human analog, subjects were either required to press at rates so fast they were prevented from verbally counting or were engaged in a verbal distracter task. Like rats, the mean number of key presses made by human subjects was proportional to the target number, and representations showed scalar variability in that the standard deviation of the response distributions increased with target magnitude (Cordes et al., 2001; Whalen et al., 1999).

In another example, Cantlon and Brannon (2006a) directly compared adult humans' and rhesus monkeys' abilities to order numerical arrays. As shown in Figure 8.3, the two species showed strikingly similar patterns in accuracy and reaction time. In a similar numerical ordering task, Beran (2006) uncovered another intriguing parallel between monkey and human numerical cognition. Monkeys, like humans, were sensitive to the regular-random numerosity illusion (RRNI). That is, both monkeys and humans systematically overestimated the number of items in sets of stimuli that were regularly (rather than randomly) arranged (see Fig. 8.4 for example stimuli).

Another striking similarity between the numerical representations of adult humans and rhesus monkeys is the effect of semantic

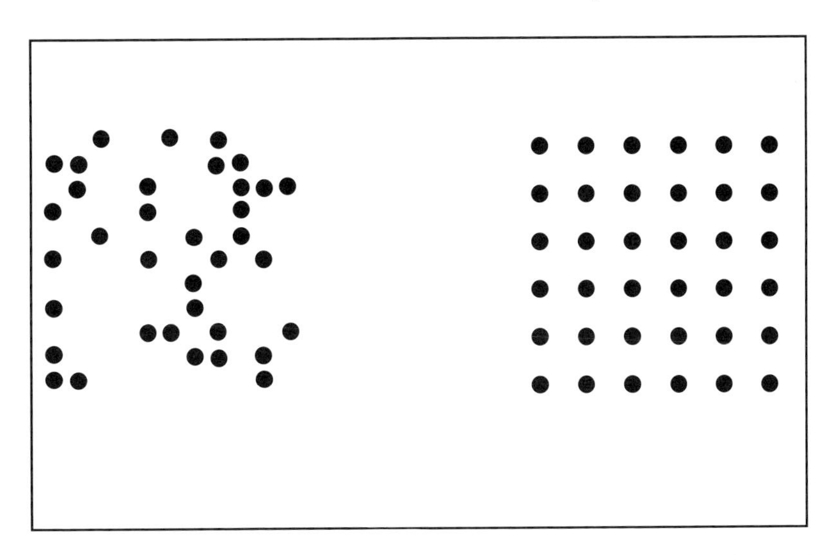

Figure 8.4 Dot displays used with monkeys and humans. Both species overestimated the number of items in sets of stimuli that were regularly (rather than randomly) arranged. After Beran, M. J. (2006). Quantity perception by adult humans (*Homo sapiens*), chimpanzees (*Pan troglodytes*), and rhesus macaques (*Macaca mulatta*) as a function of stimulus organization. *International Journal of Comparative Psychology, 19,* 386–397. Used with permission.

congruity on both species' numerical judgments (Cantlon & Brannon, 2005). The semantic congruity effect has been found whenever adults compare the relative magnitude of values along an ordinal continuum including numerical comparisons (Banks et al., 1976; Moyer & Bayer, 1976). For example, people are faster at judging which of two small animals (i.e., a rat and a rabbit) is smaller than they are at judging which of these two small animals is larger. In contrast, we are faster at judging which of two large animals (i.e., a zebra and an elephant) is larger than we are at judging which of two large animals is smaller (Banks et al., 1983; Shaki & Algom, 2002). Intuitively, this effect seems driven by language (Holyoak, 1978). However, Cantlon

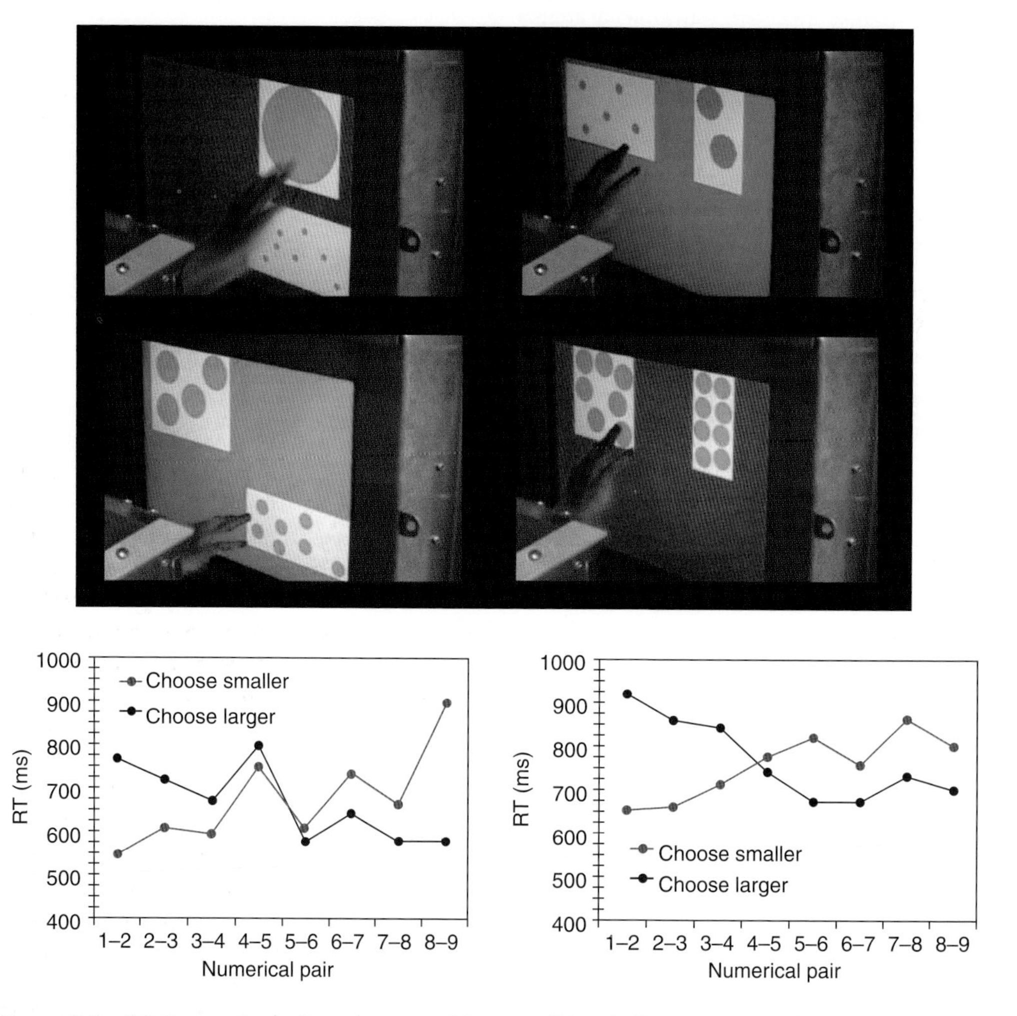

Figure 8.5 (**A**) Screen shots of monkeys tested in a conditional discrimination, whereby a blue screen indicated that they should choose the larger numerosity first and a red screen indicated that they should choose the smaller numerosity first. (**B**) Reaction time as a function of numerical pair for monkeys making an ordinal number judgment illustrates the semantic congruity effect. Monkeys were faster at choosing the smaller of two small numbers than they were at choosing the larger of two small numbers (red line). Conversely, when required to choose the larger of two large numbers, they were faster than when they were required to choose the smaller of two large numbers (blue line). After Cantlon, J. F., & Brannon, E. M. (2005). Semantic congruity affects numerical judgments similarly in monkeys and humans. *Proceedings of the National Academy of Sciences, 102*, 16507–16511. Used with permission.

and Brannon (2005) found a semantic congruity effect for rhesus monkey numerical comparisons. Monkeys were faster at choosing the smaller of two small numerosities (i.e., 2 vs. 3) compared to the larger of two small numerosities (Fig. 8.5A, B). Conversely, when required to choose the larger of two large numerosities, they were faster than when they were required to choose the smaller of two large numerosities. Thus, monkeys share with adult humans a susceptibility to the semantic congruity effect in numerical comparisons, suggesting that the effect may better reflect internal representations of ordinal continua rather than linguistic encoding.

In summary, when humans are tested in nonverbal tasks that circumvent verbal counting, their performance looks similar to nonhuman animals. This suggests that even educated adults with knowledge of higher-order symbolic mathematics simultaneously possess an ANS that is shared with other animals. Additional evidence for this conclusion comes from recent studies with two Amazonian groups, both of which lack verbal counting systems. Despite an inability to label a large collection with a precise numerical value, both societies possess nonverbal numerical representations with sensitivity that is roughly equivalent to Europeans with verbal counting systems (Frank et al., 2008; Gordon, 2004; Pica et al., 2004). Such data indicate that the ANS is likely universal among humans and not culturally specific.

ARITHMETIC OPERATIONS

As suggested by Gelman and Gallistel (Gallistel, 1990; Gallistel et al., 2005), nonverbal numerical representations are likely crucial throughout the animal kingdom insofar as they support calculations for behaviorally important decisions. Bolstering this idea, a handful of studies have demonstrated that nonhuman animals can perform mathematical operations on their numerical representations. In one compelling study, a chimpanzee named Sheba was first trained to map symbols to numerosities and subsequently tested on her ability to add symbols or sets of objects (Boysen & Berntson, 1989). The chimpanzee was led around a room to various hiding places where she saw Arabic numerals or sets of oranges. She was then required to choose the Arabic numeral that corresponded to the sum of the values she has seen. Amazingly, the chimpanzee performed well on the 14 test trials.

Taking a different approach, Hauser and colleagues adapted a paradigm, first developed by Wynn (1992) to study human infant numerical cognition, to test whether untrained monkeys spontaneously look longer at physical outcomes that violate their arithmetic expectations (Hauser et al., 1996). Monkeys observed as eggplants (or lemons) were placed on a stage. A screen was then raised to occlude the eggplants, and additional eggplants were then placed behind the screen. The screen was raised to reveal either the expected outcome or an arithmetically impossible outcome. In multiple studies using this paradigm, untrained monkeys have been found to look longer at mathematically impossible compared to possible outcomes (e.g., Flombaum et al., 2005; Hauser et al., 1996). Beran and colleagues have also found that nonhuman primates reliably choose the larger of two food quantities, even when this requires tracking one-by-one additions to multiple caches over time (Beran, 2004; Beran & Beran, 2004).

In the most parametric study of addition in nonhuman animals to date, Cantlon and Brannon (2007b) directly compared monkeys and humans in an explicit addition task. Subjects viewed two addend arrays presented in succession on a touch-sensitive screen and were required to choose the array (from two arrays) that was equal to the sum of the two addend arrays (Fig. 8.6A). The monkeys were trained on a few addition problems (1 + 1 = 2, 4, or 8; 2 + 2 = 2, 4, or 8; 4 + 4 = 2, 4, or 8) and then tested on novel problems. Monkeys performed at a level significantly greater than chance on each of these three problems within 500 trials. In the critical test, monkeys were presented with all possible addends of the novel sums 3, 7, 11, and 17. Monkeys were rewarded regardless of which of the two choice stimuli they selected as the sum to prevent learning during the experiment. Performance on the novel problems was

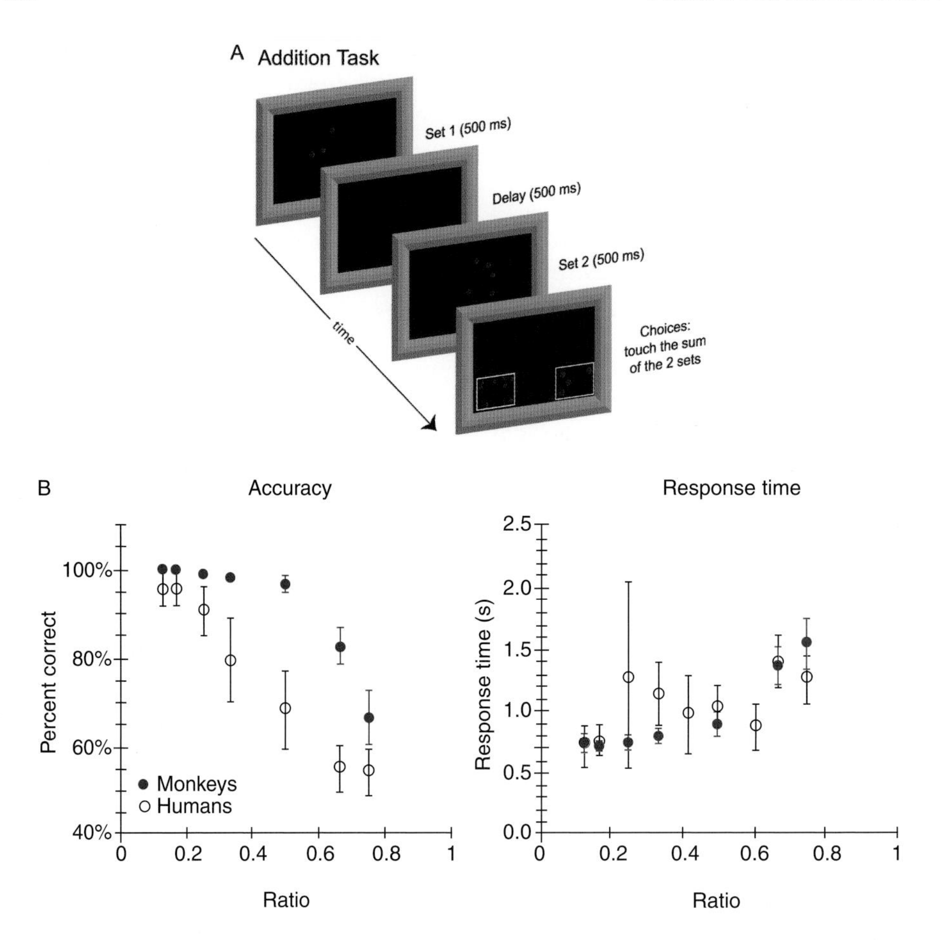

Figure 8.6 **(A)** Stimuli and task design used to test nonverbal addition in monkeys and college students. Monkeys and humans were presented with one set of dots (set 1), followed by a brief delay, after which a second set of dots was presented (set 2). Then, two choices (the sum and the distractor) were presented, and monkeys were rewarded for touching the choice that represented the numerical sum of the two sets. **(B)** Monkeys and humans exhibited ratio-dependent accuracy and response time when solving addition problems. Accuracy (left panel) and response times (right panel). Error bars reflect the standard error among subjects. After Cantlon, J. F., & Brannon, E. M. (2007b). Basic math in monkeys and college students. *PLoS Biology, 5*, e328. Used with permission.

significantly greater than that predicted by chance, and performance was modulated by numerical ratio. Importantly, as shown in Figure 8.6B, humans and monkeys tested on the same nonverbal addition task showed very similar performance (see also Barth et al., 2006). Thus, it appears that the ANS supports arithmetic, and an important function of the ANS system may be to make approximate calculations about the world around us.

IS REPRESENTING NUMBER A LAST-RESORT STRATEGY?

In this section we query the functional utility of the ANS. One view has been that animals only represent number as a last resort after extensive laboratory training (Breukelaar & Dalrymple-Alford, 1998; Davis, 1993; Davis & Memmot, 1982; Davis & Perusse, 1988; Seron & Pesenti, 2001). Under this view, animals may possess a latent capacity to represent number, but the capacity will only emerge with extensive training and only when they are prevented from using an alternative strategy. While there are some experimental results consistent with the idea that animals may under some circumstances avoid using number, by and large there is more support for an alternative view that many nonhuman animals spontaneously use numerical representations in the wild.

For example, American coots have been found to keep track of the number of their own eggs in clutches that contain both their own eggs and those of conspecific parasites (Lyon, 2003). Females appear to terminate the production of egg follicles based on the number of their own eggs, which are only a subset of the total eggs in the nest. Territorial defense is another domain in which number might be particularly useful. An animal might benefit from being able to compare the numerosity of her own group to a competing group. Wilson and colleagues (2001) designed a study to test the possibility that wild chimpanzees might base decisions about whether to defend a territory against an invading animal on the number of individuals available to help in the defense. They played the pant-hoot of a single male chimpanzee (who was not a member of the group being tested) to variable numbers of adult males in the test group. They found that groups of three or more males consistently chorused loudly in response to the playback and approached the speaker. In contrast, groups containing fewer than three males usually failed to make such responses. Chimpanzees may thus calculate whether they have "strength in numbers" before deciding whether to defend their territory.

Another important reason a nonhuman animal might need to represent number is to maximize food intake. Not surprisingly, multiple studies have found that animals spontaneously choose the larger of two food arrays (e.g., Beran, 2001; Beran et al., 2005; Boysen & Berntson, 1995; Call, 2000). However, when number and quantity are in conflict in such food choice situations, a recent study suggests that maximizing amount of food wins out over number (Stevens et al., 2007). Of course this makes adaptive sense; all organisms should be motivated to maximize food intake, not number of food items.

While most laboratory studies of numerical competence in animals employ extensive training and focus on the question of what animals are capable of rather than what they do spontaneously, a few laboratory studies have directly addressed the question of spontaneous numerical cognition. Hauser and colleagues (2002) tested whether untrained cotton-top tamarins could recognize the numerical equivalence between auditory stimuli presented in different formats. Using a habituation-discrimination procedure, they presented cotton-top tamarins with a mixture of either two- or three-syllable speech sequences varying in overall duration, intersyllable duration, and pitch. After the tamarins habituated to this repeated presentation of speech syllables of either numerosity, they presented the tamarins with novel two- or three-tone sequences in a test phase. They found that tamarins oriented to the speaker longer when the number of tones in the test phase differed from the number of speech syllables in the habituation phase. Because Hauser and colleagues controlled for various nonnumerical parameters such as duration and tempo that are often confounded with number in the real world, their study showed that without any reinforcement or training, tamarin monkeys could discriminate numerical values in the auditory modality—even when this required comparing sounds that were quite different (also see Hauser et al., 2003).

Prosimian primates have also been tested in spontaneous numerical cognition tasks. Santos et al. (2005) modified a task used with human infants and rhesus macaques (Wynn, 1992, and Hauser et al., 1996, respectively) and tested lemurs with a looking time experiment contrasting a numerically possible event with a numerically impossible event. The lemurs saw two lemons hidden one at a time behind a screen. The screen was lifted to reveal one, two, or three lemons. Lemurs looked longer at the unexpected compared to the expected arithmetic outcomes. The results showed that lemurs, like rhesus macaques and human infants, are able to spontaneously quantify a small set of objects and form precise expectations about a simple 1 + 1 addition problem.

In another task adapted from research with human infants (Feigenson & Carey, 2003), Lewis et al. (2005) tested the numerical capacities of mongoose lemurs. Lemurs observed as a number of grapes (up to eight) were dropped into a container; they were then allowed to retrieve all or a subset of the grapes. Results showed that lemurs continued to search when the bucket should have still contained grapes and that they required a 1:2 ratio to discriminate the number of grapes they obtained from the number that should have been remaining in the bucket.

Cantlon and Brannon (2007a) designed a study to directly assess the relative salience of number, color, shape, and cumulative surface area for monkeys. They trained monkeys in a match-to-sample task in which the sample and correct match were the same in both numerosity *and* in an additional dimension (color, shape, or cumulative surface area of the elements). The incorrect choice differed from the sample and match both in numerosity and on the alternative dimension. After the monkeys reached a criterial level of performance, they were tested on nondifferentially reinforced probe trials in which one choice matched the sample in numerosity and the other choice matched the sample on an alternative dimension (color, shape, or cumulative surface area of the elements). They found that, for all conditions (color, shape, or cumulative surface area of the elements), the probability that monkeys made a numerosity match increased with the numerical distance between the numerosity match and the numerosity of the alternative stimulus. Furthermore, in the case of the cumulative surface area condition, monkeys were more likely to match stimuli based on numerosity at all distances. The bias for number over area was found both for the monkeys that had prior numerosity training *and* a single monkey who had never been trained on a numerical discrimination task. Thus, even for a monkey, number can be more salient than other set summary statistics, such as cumulative surface area.

CROSS-MODAL REPRESENTATIONS OF NUMBER IN ANIMALS

Four cherries and four submarines have little in common except for their numerosity. Yet clearly, adult humans have no trouble recognizing such numerical equivalence. Number can be extracted from sets of simultaneously visible individuals or from sequentially presented sets such as a sequence of tones. Furthermore, numerical representations are not specific to the sensory modalities in which they are established. Three telephone rings, three taps on your shoulder, and three political candidates are equally good examples of threeness. The ability to represent number without regard to stimulus modality is integral to any notion of a truly abstract concept of number. Here, we explore whether the ANS is sufficiently abstract to represent number independent of stimulus modality. If so, is this a fundamental part of animals' numerical perception and cognition, or does it depend on extensive training?

Barth and colleagues (2003) have shown that in human adults the ANS operates independently of stimulus modality. Specifically, they found that adults showed virtually no performance cost of comparing numerosities across versus within visual and auditory stimulus sets. This leads to the prediction that nonhuman animals should also be able to detect the numerical correspondence between sets of entities presented in different sensory modalities.

A few studies have addressed this question. Church and Meck (1984) found that rats trained to discriminate two from four sounds or light flashes later responded to compound cues of two lights and two sounds as if four events had occurred, suggesting that rats can transfer numerical representations across modalities. Yet, when Davis and Albert (1987) trained rats to discriminate three sequentially presented sounds from two or four sounds and then exposed rats to sequences of two, three, and four lights, they found no evidence that the rats transferred their auditory numerical discrimination to the visual modality. The Davis and Albert (1987) results raise the possibility that the rats in the Church and Meck study (1984) made a dichotomous discrimination that was purely intensity based (i.e., they equated the less intense sound with the less intense light).

Field playback studies have yielded complementary evidence suggesting that animals predict the number of intruders they expect to see based on the number of vocalizing intruders they hear. In these studies, the probability that an animal from a focal group approached a speaker emitting vocalizations from foreign conspecifics depended on the relationship between the number of vocalizing foreign animals and the number of animals present in the focal group (Kitchen, 2004; McComb et al., 1994). For example, McComb and colleagues (1994) found that lions were more likely to approach a speaker emitting the roar of a single unfamiliar lion than a chorus of three unfamiliar lions, suggesting that lions decide whether to defend their territory based on the perceived number of intruders. However, such studies did not control for all possible nonnumerical auditory cues that covary with number (e.g., some aspects of duration), leaving open the question of whether the calculations made by the animals were in fact based on number.

A recent experiment capitalized on the social expertise of nonhuman primates by framing a numerical problem within a social context: Specifically, researchers tested whether rhesus monkeys spontaneously matched the number of dynamic conspecific faces they saw with the number of vocalizations they simultaneously heard by employing the sort of preferential looking paradigm that had been used extensively with human infants (Jordan et al., 2005; see Fig. 8.7A). The study employed strict stimulus controls for temporal attributes that often covary with number and used a between-subject design. Monkeys looked significantly longer at the matching display than the nonmatching display, and the effect held for individuals who heard two sounds or three sounds (Fig. 8.7A, B). Thus, monkeys recognize the numerical correspondence between two or three dynamic faces and two or three concurrent voices. However, these experiments left open the possibility that this ability was some context-specific capacity peculiar to social judgments. It was possible that these experiments succeeded because they tapped into a socioecologically relevant scenario; in their everyday lives, for example, territorial animals might be helped by being able to assess the number of individuals they will likely encounter by how many individuals they hear. In addition, the Jordan et al. (2005) study was restricted to the small values 2 and 3.

To ask whether monkeys could numerically match arbitrary stimuli across sensory modalities and to see whether the ability extended beyond the small values 2 and 3, Jordan and colleagues (2008) used a task that required an active choice to numerically match auditorially presented tones and visually presented squares. They trained two monkeys to choose a visual array of two, three, six, or eight squares that numerically matched a sample sequence of shapes or sounds. Monkeys numerically matched across (audio-visual) and within (visual-visual) modalities with equal accuracy and transferred to novel numerical values (see Fig. 8.8A for an example trial). Monkeys were also able to sum over shapes and tones and choose the visual array that numerically matched the total number of visual or auditory events in a sequence. As shown in Figure 8.8B, accuracy and reaction time depended on the ratio between the correct numerical match and incorrect choice. Thus, monkeys and humans appear to share an abstract numerical code that can be divorced from the modality in which stimuli are first experienced.

HOW ARE ANALOG MAGNITUDE REPRESENTATIONS OF NUMEROSITY FORMED?

A host of models have been proposed to explain how the ANS forms analog magnitude representations. Gelman and Gallistel (1978; Gallistel & Gelman, 1992) suggested that animals and young children use a nonverbal process that obeys three critical counting principles. The one-to-one principle states that there is a one-to-one correspondence between the number of labels applied and the number of to-be-counted elements in a set (labels can be nonverbal). The stable order

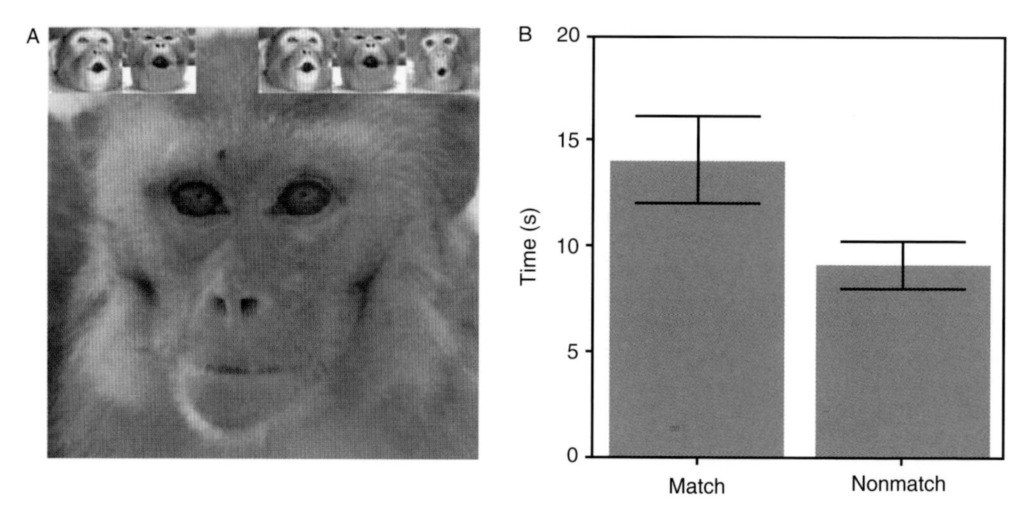

Figure 8.7 (**A**) Static images from video displays used to test whether rhesus monkeys spontaneously matched the number of dynamic conspecific faces they saw with the number of vocalizations they simultaneously heard in a preferential looking paradigm. (**B**) Monkeys looked longer at the display that numerically matched the number of voices they heard. After Judge, P. G., Evans, T. A., & Vyas, D. K. (2005). Ordinal representation of numeric quantities by brown capuchin monkeys (*Cebus apella*). *Journal of Experimental Psychology: Animal Behavior Processes, 31*, 79–94. Used with permission.

principle states that the labels are applied in a stable order across counting episodes (one cannot count 1-2-3 today and 3-1-2 tomorrow). The cardinal principle states that the last label applied serves to represent the cardinal value of the set.

The mode-control model, developed to describe how rats enumerate serial arrays (sequences of sounds or light flashes), follows these counting principles (Meck & Church, 1983). Under this model, a pacemaker emits a regular stream of pulses at a constant frequency. When a stimulus begins, pulses are gated into an accumulator, which integrates the number of pulses over time. Depending on the nature of the stimulus, a mode switch that closes after stimulus onset gates the pulses to the accumulator in one of three different modes: in the "run" mode the stimulus starts an accumulation process that continues until the end of the signal; in the "stop" mode the process occurs whenever the stimulus occurs; and in the "event" mode each onset of the stimulus produces a relatively fixed duration of the process regardless of stimulus duration. This mechanism can thus be used as a counter (the event mode) or a timer (the run and stop modes). In all modes, the value stored in the accumulator increases with duration or number. When the stimulus stops, the mode switch opens, stopping the gating of pulses to the accumulator. The final magnitude is placed into memory, while the accumulator is reset to zero and the organism is considered ready to time or count another stimulus. The representation of the magnitude in memory is noisy and obeys Weber's Law. Importantly, such a mode-control mechanism can represent and integrate sequentially processed numerical stimuli presented in any sensory modality, potentially accounting for highly abstract numerical representations.

The mode-control model works well for explaining how serially presented items are enumerated. But what happens when animals encounter visual arrays? It is possible to convert a simultaneous array into a sequential array by serially allocating attention to each element in the array. However, the behavioral and neurobiological evidence to date suggests that it is unlikely rhesus macaques take this approach.

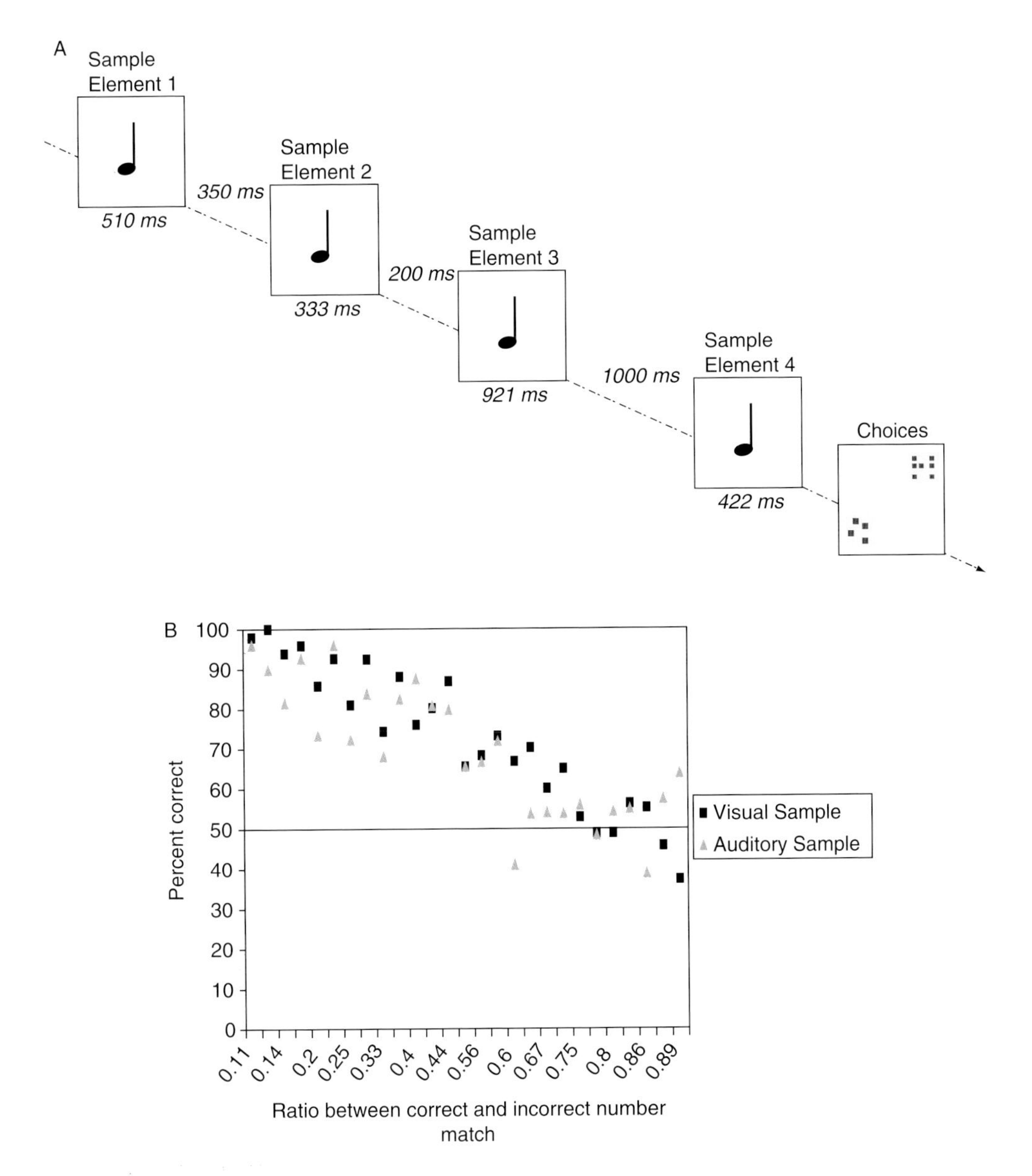

Figure 8.8 (**A**) Example sample stimuli from an auditory-visual trial. Monkeys heard four successive tones and were then rewarded for choosing an array of four visual elements. (**B**) Average accuracy as a function of the ratio between the numerical value of the two choice stimuli. Black squares indicate visual samples and gray triangles indicate auditory samples. Accuracy in both modalities was modulated by ratio. Dotted line indicates chance performance (50%). After Jordan, K. E., MacLean, E., & Brannon, E. M. (2008). Monkeys match and tally quantities across senses. *Cognition, 108,* 617–625. Used with permission.

Nieder and colleagues found that neurons in the prefrontal cortex (PFC) of the macaque brain that were selective for one, two, three, four, or five elements showed similar response latencies of approximately 120 ms (Nieder et al., 2002). Furthermore, if animals were serially allocating attention to elements in an array, we might expect to find that the number of visual saccades increases with the number of elements in an array. However, Nieder and Miller (2004) found that the number of visual saccades made by monkeys did not increase with the number of display items in a numerical same-different task as set size increased from one to seven. Similarly, analysis of reaction time for monkeys engaged in the numerical delayed match-to-sample task described earlier in this chapter provided no evidence that monkeys serially enumerate visual arrays (Jordan & Brannon, 2006). Since the monkeys controlled the duration of the sample array and hence the processing time for the sample numerosity, a serial process should be reflected by increased reaction times to larger numerical samples. Reaction time, however, did not increase with sample numerosity. Barth and colleagues (2003) also found that it took adults no longer to compare large sets than to compare small sets with comparable ratios, suggesting that adults may also use a parallel enumeration mechanism for nonverbally estimating and comparing large sets.

Dehaene and Changeux's (1993) neural network model achieves numerical representations through such a parallel process. This model is composed of four primary levels: the topographically organized input retina, an organized map of element positions from this retina, summation units that sum the input from the organized position map, and an array of numerosity detectors. These numerosity detectors respond selectively to certain numerosities (i.e., certain ranges of activity from the summation clusters). The numerosity detection system ultimately represents the numerosity of a set of simultaneously presented elements regardless of the size and position of the elements (see also Verguts & Fias, 2004).

CONCLUSION

Our review revealed strong behavioral parallels between human and animal numerical cognition such as ratio dependence, semantic congruity, cross-modal matching, and nonverbal arithmetic. Despite early skepticism about animal numerical abilities, it is now irrefutable that a wide variety of animal species are sensitive to the numerical attributes of the world around them. Elsewhere in this volume, Nieder describes the mounting evidence that homologous brain structures support numerical representations in humans and macaque monkeys (see Chapter 20). A strong possibility is that the ANS observed in nonhuman primates and humans represents an evolutionarily ancient system embedded in the same neural hardware. Future work should systematically study variation in numerical cognition between species throughout the animal kingdom to explore whether there are quantitative or qualitative differences that can be tied to socioecology or to phylogeny.

REFERENCES

Addessi, E., Crescimbene, L., & Elisabetta V. (2008). Food and token quantity discrimination in capuchin monkeys (Cebus apella). *Animal Cognition, 11*, 275–282.

Banks, W. P., Fujii, M., & Kayra-Stuart, F. (1976). Semantic congruity effects in comparative judgment effects in comparative judgments of magnitudes of digits. *Journal of Experimental Psychology: Human Perception and Performance, 2*, 435–447.

Banks, W. P., White, H., Sturgill, W., & Mermelstein, R. (1983). Semantic congruity and expectancy in symbolic judgments. *Journal of Experimental Psychology: Human Perception and Performance, 9*, 580–582.

Barth, H., Kanwisher, N., & Spelke, E. (2003). The construction of large number representations in adults. *Cognition, 86*, 201–221.

Barth, H., La Mont, K., Lipton, J., Dehaene, S., Kanwisher, N., & Spelke, E. (2006). Non-symbolic arithmetic in adults and young children. *Cognition, 98*, 199–222.

Beran, M. J. (2001). Summation and numerousness judgments of sequentially presented sets of items by chimpanzees (*Pan troglodytes*). *Journal of Comparative Psychology, 115*, 181–191.

Beran, M. J. (2004). Chimpanzees (*Pan troglodytes*) respond to nonvisible sets after one-by-one addition and removal of items. *Journal of Comparative Psychology, 118*, 25–36.

Beran, M. J. (2006). Quantity perception by adult humans (*Homo sapiens*), chimpanzees (*Pan troglodytes*), and rhesus macaques (*Macaca mulatta*) as a function of stimulus organization. *International Journal of Comparative Psychology, 19*, 386–397.

Beran, M. J. (2008). Monkeys (Macaca mulatta and Cebus apella) track, enumerate, and compare multiple sets of moving items. *Journal of Experimental Psychology: Animal Behavior Processes, 34*, 63–74.

Beran, M. J., & Beran, M. M. (2004). Chimpanzees remember the results of one-by-one addition of food items to sets over extended time periods. *Psychological Science, 15*, 94–99.

Beran, M. J., Beran, M. M., Harris, E. H., & Washburn, D. A. (2005). Ordinal judgments and summation of nonvisible sets of food items by two chimpanzees and a rhesus macaque. *Journal of Experimental Psychology: Animal Behavior Processes, 31*, 351–362.

Beran, M. J., Evans, T. A., Leighty, K. A., Harris, E. H., & Rice, D. (2008). Summation and quantity judgments of sequentially presented sets by capuchin monkeys (Cebus apella). *American Journal of Primatology, 70*, 191–194.

Boisvert, M. J., Abroms, B. D., & Roberts, W. A. (2003). Human nonverbal counting estimated by response production and verbal report. *Psychonomic Bulletin and Review, 10*, 683–690.

Boysen, S. T., & Berntson, G. G. (1989). Numerical competence in a chimpanzee (*Pan troglodytes*). *Journal of Comparative Psychology, 103*, 23–31.

Boysen, S. T., & Berntson, G. G. (1995). Responses to quantity: Perceptual versus cognitive mechanisms in chimpanzees (*Pan troglodytes*). *Journal of Experimental Psychology: Animal Behavior Processes, 21*, 82–86.

Brannon, E. M., Cantlon, J. F., & Terrace, H. S. (2006). The role of reference points in ordinal numerical comparisons by Rhesus macaques (*Macaca mulatta*). *Journal of Experimental Psychology: Animal Behavior Processes,* 32, 120–134.

Brannon, E. M., & Terrace, H. S. (1998). Ordering of the numerosities 1 to 9 by monkeys. *Science, 282*, 746–749.

Brannon, E. M., & Terrace, H. S. (2000). Representation of the numerosities 1–9 by rhesus macaques (*Macaca mulatta*). *Journal of Experimental Psychology: Animal Behavior Processes, 26*, 31–49.

Breukelaar, J. W. C., & Dalrymple-Alford, J. C. (1998). Timing ability and numerical competence in rats. *Journal of Experimental Psychology: Animal Behavior Processes, 24*, 84–97.

Call, J. (2000). Estimating and operating on discrete quantities in orangutans (*Pongo pygmaeus*). *Journal of Comparative Psychology, 114*, 136–147.

Cantlon, J. F., & Brannon, E. M. (2005). Semantic congruity affects numerical judgments similarly in monkeys and humans. *Proceedings of the National Academy of Sciences, 102*, 16507–16511.

Cantlon, J. F., & Brannon, E. M. (2006a). Shared system for ordering small and large numbers in monkeys and humans. *Psychological Science, 17*, 402–407.

Cantlon, J. F., & Brannon, E. M. (2006b). The effect of heterogeneity on numerical ordering in rhesus monkeys. *Infancy, 9*, 173–189.

Cantlon, J. F., & Brannon, E. M. (2007a). How much does number matter to a monkey (*Macaca mulatta*)? *Journal of Experimental Psychology: Animal Behavior Processes, 33*, 32–41.

Cantlon, J. F., & Brannon, E. M. (2007b). Basic math in monkeys and college students. *PLoS Biology, 5*, e328.

Church, R. M., & Deluty, M. Z. (1977). Bisection of temporal intervals. *Journal of Experimental Psychology: Animal Behavior Processes, 3*, 216–228.

Church, R., & Meck, W. (1984). The numerical attribute of stimuli. In H. L. Roitblat, T. G. Bever, & H.S. Terrace (Eds.), *Animal cognition* (pp. 445–464). Hillsdale, NJ: Erlbaum.

Cordes, S., Gelman, R., & Gallistel, C. R. (2001). Variability signatures distinguish verbal from nonverbal counting for both large and small numbers. *Psychological Bulletin and Review, 8*, 698–707.

Davis, H. (1993). Numerical competence in animals: Life beyond Clever Hans. In S. T. Boysen & E. J. Capaldi (Eds.), *The development of numerical competence: Animal and human models* (pp. 109–125). Hillsdale, NJ: Erlbaum.

Davis, H., & Albert, M. (1987). Failure to transfer or train a numerical discrimination using sequential visual stimuli in rats. *Bulletin of the Psychonomic Society, 25*, 472–474.

Davis, H., & Memmott, J. (1982). Counting behavior in animals: A critical evaluation. *Psychological Bulletin, 92*, 547–571.

Davis, H., & Perusse, R. (1988). Numerical competence in animals: Definitional issues, current evidence, and a new research agenda. *Behavioral and Brain Sciences, 11*, 561–651.

Dehaene, S. (1992). Varieties of numerical abilities. *Cognition, 44*, 1–42.

Dehaene, S., & Changeux, J. P. (1993). Development of elementary numerical abilities: A neuronal model. *Journal of Cognitive Neuroscience, 5*, 390–407.

Emmerton, J., Lohmann, A., & Niemann, J. (1997). Pigeons' serial ordering of numerosity with visual arrays. *Animal Learning and Behavior, 25*, 234–244.

Emmerton, J., & Renner, J. C. (2006). Scalar effects in the visual discrimination of numerosity by pigeons. *Learning and Behavior, 34*, 176–192.

Evans, T. A., Beran, M. J., Harris, E. H., & Rice, D. F. (2008). Quantity judgments of sequentially presented food items by capuchin monkeys (Cebus apella). *Animal Cognition, 12*, 97–105.

Feigenson, L., & Carey, S. (2003). Tracking individual via object-files: evidence from infants' manual search. *Developmental Science, 6*, 568–584.

Feigenson, L., Dehaene, S., & Spelke, E. (2004). Core systems of number. *Trends in Cognitive Sciences, 8*, 307–314.

Fetterman, J. G. (1993). Numerosity discrimination: Both time and number matter. *Journal of Experimental Psychology: Animal Behavior Processes, 19*, 149–164.

Flombaum, J. I., Junge, J. A., & Hauser, M. D. (2005). Rhesus monkeys (*Macaca mulatta*) spontaneously compute addition operations over large numbers. *Cognition, 97*, 315–325.

Frank, M., Everett, D., Fedorenko, E., & Gibson, E. (2008). Number as a cognitive technology: Evidence from Piraha language and cognition. *Cognition, 108*, 819–824.

Gallistel, C. R. (1990). *The organization of learning.* Cambridge, MA: Bradford/MIT Press.

Gallistel, C. R., & Gelman, R. (1992). Preverbal and verbal counting and computation. *Cognition, 44*, 43–74.

Gallistel, C. R., Gelman, R., & Cordes, S. (2005). The cultural and evolutionary history of the real numbers. In S. Levinson & P. Jaisson (Eds.), *Culture and evolution* (pp. 247–276). Oxford: Oxford University Press.

Gelman, R., & Gallistel, C. R. (1978). *The child's understanding of number*. Cambridge, MA: Harvard University Press.

Gordon, P. (2004). Numerical cognition without words: Evidence from Amazonia. *Science, 306*, 496–499.

Hauser, M., Dehaene, S., Dehaene-Lambertz, G., & Patalano, A. (2002). Spontaneous number discrimination of multi-format auditory stimuli in cotton-top tamarins. *Cognition, 86*, B23–B32.

Hauser, M. D., MacNeilage, P., & Ware, M. (1996). Numerical representations in primates. *Proceedings of the National Academy of Sciences, 93*, 1514–1517.

Hauser, M. D., Tsao, F. T., Garcia, P., & Spelke, E. S. (2003). Evolutionary foundations of number: Spontaneous representation of numerical magnitudes by cotton-top tamarins. *Proceedings of the Royal Society, London, B, 270*, 1441–1446.

Holyoak, K. (1978). Comparative judgments with numerical reference points. *Cognitive Psychology, 10*, 203–243.

Jordan, K. E., & Brannon, E. M. (2006). Weber's Law influences the numerical representations in rhesus macaques (Macaca mulatta). *Animal Cognition, 9*, 159–172.

Jordan, K. E., Brannon, E. M., Logothetis, N. K., & Ghazanfar, A. A. (2005). Monkeys match the number of voices they hear to the number of faces they see. *Current Biology 15*, 1–5.

Jordan, K. E., MacLean, E., & Brannon, E. M. (2008). Monkeys match and tally quantities across senses. *Cognition, 108*, 617–625.

Judge, P. G., Evans, T. A., & Vyas, D. K. (2005). Ordinal representation of numeric quantities by brown capuchin monkeys (*Cebus apella*). *Journal of Experimental Psychology: Animal Behavior Processes, 31*, 79–94.

Kitchen, D. M. (2004). Alpha male black howler monkey responses to loud calls: Effect of numeric odds, male companion behaviour and reproductive investment. *Animal Behaviour, 67*, 125–139.

Lewis, K. P., Jaffe, S., & Brannon, E. M. (2005). Analog number representations in mongoose lemurs (*Eulemur mongoz*): Evidence from a search task. *Animal Cognition, 8*, 247–252.

Lyon, B. E. (2003). Egg recognition and counting reduce costs of avian conspecific brood parasitism. *Nature, 422*, 495–499.

McComb, K., Packer, C., & Pusey, A. (1994). Roaring and numerical assessment in contests between groups of female lions, *Panthera leo. Animal Behaviour, 47*, 379–387.

Meck, W., & Church, R. (1983). A mode control model of counting and timing processes. *Journal of Experimental Psychology: Animal Behavior Processes, 9*, 320–334.

Meck, W. H., Church, R. M., & Gibbon, J. (1985). Temporal integration in duration and number discrimination. *Journal of Experimental Psychology: Animal Behavior Processes, 11*, 591–597.

Moyer, R. S., & Bayer, R. H. (1976). Mental comparison and the symbolic distance effect. *Cognitive Psychology, 8*, 228–246.

Moyer, R., & Landauer, T. (1967). Time required for judgments of numerical inequality. *Nature, 215*, 1519–1520.

Nieder, A., Freedman, D. J., & Miller, E. K. (2002). Representation of the quantity of visual items in the primate prefrontal cortex. *Science, 297*, 1708–1711.

Nieder, A., & Miller, E. K. (2004). Analog numerical representations in rhesus monkeys: Evidence for parallel processing. *Journal of Cognitive Neuroscience, 16*, 889–901.

Otlhof, A., & Roberts, W. A. (2000). Summation of symbols by pigeons (*Columba livia*): The importance of number and mass of reward items. *Journal of Comparative Psychology, 114*, 158–166.

Pica, P., Lemer, C., Izard, V., & Dehaene, S. (2004). Exact and approximate arithmetic in an Amazonian indigene group. *Science, 306*, 499–503.

Platt, J. R., & Davis, E. R. (1983). Bisection of temporal intervals by pigeons. *Journal of Experimental Psychology: Animal Behavioral Processes, 9*, 160–170.

Platt, J. R., & Johnson, D. M. (1971). Localization of position within a homogeneous behavior chain: Effects of error contingencies. *Learning and Motivation, 2*, 386–414.

Roberts, W. A. (2005). How do pigeons represent numbers? Studies of number scale bisection. *Behavioral Processes, 69*, 33–43.

Santos, L. R., Sulkowski, G. M., Spaepen, G. M., & Hauser, M. D. (2002). Object individuation using property/kind information in rhesus macaques (*Macaca mulatta*). *Cognition, 83*, 241–264.

Seron, X., & Pesenti, M. (2001). The number sense theory needs more empirical evidence. *Mind & Language, 16*, 76–89.

Shaki, S., & Algom, D. (2002). The locus and nature of semantic congruity in symbolic comparison: Evidence from the Stroop effect. *Memory and Cognition, 30*, 3–17.

Smith, B. R., Piel, A. K., & Candland, D. K. (2003). Numerity of a socially housed hamadryas baboon (*Papio hamadryas*) and a socially housed squirrel monkey (*Saimiri sciureus*). *Journal of Comparative Psychology, 117*, 217–225.

Stevens, J., Wood, J., & Hauser, M. (2007). Quantity trumps number: Discrimination experiments in cotton-top tamarins (*Saguinus oedipus*) and common marmosets (*Callithrix jacchus*). *Animal Cognition, 10*, 429–437.

Stubbs, D. A. (1976). Scaling of stimulus duration by pigeons. *Journal of the Experimental Analysis of Behavior, 26*, 15–25.

Tomonaga, M. (2008). Relative numerosity discrimination by chimpanzees (Pan troglodytes): evidence for approximate numerical representations. *Animal Cognition, 11*, 43–57.

Verguts, T., & Fias, W. (2004). Representation of number in animals and humans: A neural model. *Journal of Cognitive Neuroscience, 16*, 1493–1504.

Whalen, J., Gallistel, C. R., & Gelman, R. (1999). Nonverbal counting in humans: The psychophysics of number representation. *Psychological Science, 10*, 130–137.

Wilson, M., Hauser, M., & Wrangham, R. (2001). Does participation in intergroup conflict depend on numerical assessment, range location, or rank for wild chimpanzees? *Animal Behaviour, 61*, 1203–1216.

Wynn, K. (1992). Addition and subtraction by human infants. *Nature, 358*, 749–750.

CHAPTER 9

The Foundations of Transdisciplinary Behavioral Science

Herbert Gintis[1]

INTRODUCTION

The behavioral sciences include economics, biology, anthropology, sociology, psychology, and political science, as well as their subdisciplines, including neuroscience, archaeology and paleontology, and, to a lesser extent, such related disciplines as history, legal studies, and philosophy. These disciplines have many distinct concerns, but each includes a model of individual human behavior. These models are not only different, which is to be expected given their distinct explanatory goals, but also *incompatible*. This situation is well known, but does not appear discomforting to behavioral scientists, as there has been virtually no effort to repair this condition.

The behavioral sciences all include models of individual human behavior. Therefore, these models should be compatible, and indeed, there should be a common underlying model, enriched in different ways to meet the particular needs of each discipline. Realizing this goal at present cannot be easily attained, since the various behavioral disciplines currently have *incompatible* models. Yet, recent theoretical and empirical developments have created the conditions for rendering coherent the areas of overlap of the various behavioral disciplines, as outlined in this paper. The analytical tools deployed in this task incorporate core principles from several behavioral disciplines.

Evolutionary Perspective

Evolutionary biology underlies all behavioral disciplines because *Homo sapiens* is an evolved species whose characteristics are the product of its particular evolutionary history. For humans, evolutionary dynamics are captured by *gene-culture coevolution*. The centrality of culture and complex social organization to the evolutionary success of *Homo sapiens* implies that individual fitness in humans will depend on the structure of cultural life. Since obviously culture is influenced by human genetic propensities, it follows that human cognitive, affective, and moral capacities are the product of a unique dynamic of gene-culture interaction. This coevolutionary process has endowed us with preferences that go beyond the self-regarding concerns emphasized in traditional economic and biological theory and embrace such non–self-regarding values as a taste for cooperation, fairness, and retribution; the capacity to empathize; and the ability to value such constitutive behaviors as honesty, hard work, toleration of diversity, and loyalty to one's reference group.[2]

Evolutionary Game Theory

The analysis of living systems includes one concept that does not occur in the nonliving world, and is not analytically represented in the natural sciences. This is the notion of a *strategic interaction*, in which the behavior of agents is derived by assuming that each is choosing a *fitness-relevant response* to the actions of other agents. The study of systems in which agents choose fitness-relevant responses and in which such responses evolve dynamically is called *evolutionary game theory*. Game theory provides a transdisciplinary conceptual basis for analyzing choice in the presence of strategic interaction. However, the classical game theoretic assumption that agents are self-regarding must be abandoned except in specific situations (e.g., anonymous market interactions), and many characteristics that classical game theorists have considered deductions from the principles of rational behavior, including the use of backward induction, are in fact not implied by rationality. Evolutionary game theory, whose equilibrium concept is that of a stable stationary point of a dynamic system, must thus replace classical game theory, which erroneously favors subgame perfection and sequentiality as equilibrium concepts.

The Beliefs, Preferences, and Constraints (BPC) Model

General evolutionary principles suggest that individual decision making can be modeled as optimizing a preference function subject to informational and material constraints. Natural selection leads the content of preferences to reflect biological fitness. The principle of expected utility extends this optimization to stochastic outcomes. The resulting model is called the *rational actor model* in economics, but I will generally refer to this as the *beliefs, preferences, and constraints* (BPC) model to avoid the often misleading connotations attached to the term "rational."

Society as Complex Adaptive System

The behavioral sciences advance not only by the developing analytical and quantitative models but also by the accumulating historical, descriptive, and ethnographic evidence that pays close attention to the detailed complexities of life in the sweeping array of wondrous forms that nature reveals to us. This situation is in sharp contrast with the natural sciences, which have found little use for narrative alongside analytical modeling. By contrast, historical contingency is a primary focus of analysis and causal explanation for many researchers working on sociological, anthropological, ecological, and even biological topics.

The reason for this contrast between the natural and the behavioral sciences is that *living systems are generally complex, dynamic adaptive systems* with emergent properties that cannot be fully captured in analytical models that attend only to the local interactions of the system. The hypothetico-deductive methods of game theory, the BPC model, and even gene-culture coevolutionary theory must therefore be complemented by the work of behavioral scientists who adhere to more empiricist and particularist traditions. For instance, cognitive anthropology interfaces with gene-culture coevolution and the BPC model by enhancing their capacity to model culture at a level of sophistication that fills in the black box of the physical instantiation of culture in coevolutionary theory.

A *complex system* consists of a large population of similar entities (in our case, human individuals) who interact through regularized channels (e.g., networks, markets, social institutions) with significant stochastic elements, without a system of centralized organization and control (i.e., if there a state, it controls only a fraction of all social interactions, and itself is a complex system). A complex system is *adaptive* if it evolves through some evolutionary (genetic, cultural, agent-based silicon, or other) process of hereditary reproduction, mutation, and selection (Holland, 1975). To characterize a system as complex adaptive does not explain its operation, and does not solve any problems. However, it suggests that certain modeling tools are likely to be effective that have little use in a noncomplex system. In particular, the traditional mathematical methods of physics and chemistry must be supplemented by other modeling tools, such as agent-based simulation and network theory.

Such novel research tools are needed because a complex adaptive system generally has *emergent properties* that cannot be analytically derived from its component parts. The stunning success of modern physics and chemistry lies in their ability to avoid or strictly limit emergence. Indeed, the experimental method in natural science is to create highly simplified laboratory conditions, under which modeling becomes analytically tractable. Physics is no more effective than economics or biology in analyzing complex real-world phenomena in situ. The various branches of engineering (electrical, chemical, mechanical) are effective because they recreate in everyday life artificially controlled, noncomplex, nonadaptive, environments in which the discoveries of physics and chemistry can be directly applied. This option is generally not open to most behavioral scientists, who rarely have the opportunity of "engineering" social institutions and cultures.

In addition to these conceptual tools, the behavioral sciences of course share common access to the natural sciences, statistical and mathematical techniques, computer modeling, and a common scientific method.

EVOLUTIONARY PERSPECTIVE

A *replicator* is a physical system capable of drawing energy and chemical building blocks from its environment to make copies of itself. Chemical crystals, such as salt, have this property, but biological replicators have the additional ability to assume myriad physical forms based on the highly variable sequencing of its chemical building blocks (Schrödinger called life an "aperiodic crystal" in 1943, before the structure of DNA was discovered). Biology studies the dynamics of such complex replicators using the evolutionary concepts of replication, variation, mutation, and selection (Lewontin, 1974).

Biology plays a role in the behavioral sciences much like that of physics in the natural sciences. Just as physics studies the elementary processes that underlie all natural systems, biology studies the general characteristics of survivors of the process of natural selection. In particular, genetic replicators, the epigenetic environments to which they give rise, and the effect of these environments on gene frequencies account for the characteristics of species, including the development of individual traits and the nature of intraspecific interaction. This does not mean, of course, that behavioral science in any sense *reduces* to biological laws. Just as one cannot deduce the character of natural systems (e.g., the principles of inorganic and organic chemistry, the structure and history of the universe, robotics, plate tectonics) from the basic laws of physics, similarly, one cannot deduce the structure and dynamics of complex life forms from basic biological principles. But, just as physical principles inform model creation in the natural sciences, so must biological principles inform all the behavioral sciences.

THE FOUNDATIONS OF THE BPC MODEL

For every constellation of sensory inputs, each decision taken by an organism generates a probability distribution over fitness outcomes, the expected value of which is the *fitness* associated with that decision. Since fitness is a scalar variable, for each constellation of sensory inputs, each possible action the organism might take has a specific fitness value, and organisms whose decision mechanisms are optimized for this environment will choose the available action that maximizes this value.[3] It follows that, given the state of its sensory inputs, if an organism with an optimized brain chooses action A over action B when both are available, and chooses action B over action C when both are available, then it will also choose action A over action C when both are available. This is called *choice consistency*.

The so-called *rational actor model* was developed in the twentieth century by John von Neumann, Leonard Savage, and many others. The model appears prima facie to apply only when actors possess extremely strong information-processing capacities. However, the model in fact depends only on choice consistency and the assumption that agents can trade off among outcomes in the sense that for any finite set of outcomes $A_1, \ldots, A_n$, if A_1 is the least preferred and A_n the most preferred outcome, then for any A_i, $1 \leq i \leq n$ there is a probability p_i, $0 \leq p_i \leq 1$ such that the agent is indifferent between A_i and a lottery that pays A_1 with probability p_i and pays A_n with probability $1 - p_i$ (Kreps, 1990). Clearly, these assumptions are often extremely plausible. When applicable, the rational actor model's choice consistency assumption strongly enhances explanatory power, even in areas that have traditionally abjured the model (Coleman, 1990; Hechter & Kanazawa, 1997; Kollock, 1996).

In short, when preferences are consistent, they can be represented by a numerical function, often called a *utility function*, which the individual maximizes subject to his or her beliefs (including Bayesian probabilities) and constraints. Four caveats are in order. First, this analysis does not suggest that people consciously maximize something called "utility," or anything else. Second, the model does *not* assume that individual choices, even if they are self-referring (e.g., personal consumption), are always welfare enhancing. Third, preferences must be stable across time to be theoretically useful, but preferences are ineluctably a function of such parameters as hunger, fear, and recent social experience, while beliefs can change dramatically in response to immediate sensory experience. Finally, the BPC model does not presume that beliefs are correct or that they are updated correctly in the face of new evidence, although Bayesian assumptions concerning updating can be made part of consistency in elegant and compelling ways (Jaynes, 2003).

The rational actor model is the cornerstone of contemporary economic theory, and in the past few decades has become the cornerstone of the biological modeling of animal behavior (Alcock, 1993; Real, 1991; Real & Caraco, 1986). Economic and biological theory thus have a natural affinity: The choice consistency on which the rational actor model of economic theory depends is rendered plausible by biological evolutionary theory, and the optimization techniques pioneered by economic theorists are routinely applied and extended by biologists in modeling the behavior of a vast array of organisms.

In addition to the explanatory success of theories based on the rational actor model, supporting evidence from contemporary neuroscience suggests that expected utility maximization is not simply an "as if" story. In fact, the brain's neural circuitry actually makes choices by internally representing the payoffs of various alternatives as neural firing rates and choosing a maximal such rate (Glimcher, 2003; Glimcher et al., 2005). Neuroscientists increasingly find that an aggregate decision making process in the brain synthesizes all available information into a single, unitary value (Glimcher, 2003; Parker & Newsome, 1998; Schall & Thompson, 1999). Indeed, when animals are tested in a repeated trial setting with variable reward, dopamine neurons appear to encode the difference between the reward that an animal expected to receive and the reward that an animal actually received on a particular trial (Schultz et al., 1997; Sutton & Barto, 2000), an evaluation mechanism that enhances the environmental sensitivity of the animal's decision-making system. This error-prediction mechanism has the drawback of only seeking local optima (Sugrue et al., 2005). Montague and Berns (2002) address this problem, showing that the orbitofrontal cortex and striatum contains a mechanism for more global predictions that include risk assessment and discounting of future rewards. Their data suggest a decision-making model that is analogous to the famous Black-Scholes options pricing equation (Black & Scholes, 1973).

The BPC model is the most powerful analytical tool of the behavioral sciences. For most of its existence this model has been justified in terms of "revealed preferences" rather than by the identification of neural processes that generate constrained optimal outcomes. The neuroscience evidence, for the first time, suggests a firmer foundation for the rational actor model.

GENE-CULTURE COEVOLUTION

The genome encodes information that is used to construct a new organism, to instruct the new organism how to transform sensory inputs into decision outputs (i.e., to endow the new organism with a specific preference structure), and to transmit this coded information virtually intact to the new organism. Since learning about one's environment is costly and error prone, efficient information transmission will ensure that the genome encodes all aspects of the organism's environment that are constant, or that change only very slowly through time and space. By contrast, environmental conditions that vary across generations and/or in the course of the organism's life history can be dealt with by providing the organism with the capacity to *learn*, and hence phenotypically adapt to specific environmental conditions.

There is an intermediate case that is not efficiently handled by either genetic encoding or learning. When environmental conditions are positively but imperfectly correlated across generations, each generation acquires valuable information through learning that it cannot transmit genetically to the succeeding generation, because such information is not encoded in the germ line. In the context of such environments, there is a fitness benefit to the transmission of *epigenetic* information concerning the current state of the environment. Such epigenetic information is quite common (Jablonka & Lamb, 1995) but achieves its highest and most flexible form in *cultural transmission* in humans and to a considerably lesser extent in other primates (Bonner, 1983; Richerson & Boyd, 1998). Cultural transmission takes the form of vertical (parents to children) horizontal (peer to peer), and oblique (elder to younger), as in Cavalli and Feldman (1981); prestige (higher influencing lower status), as in Henrich and Gil-White (2001); popularity related, as in Newman and colleagues (2006); and even random population-dynamic transmission, as in Shennan (1997) and Skibo and Bentley (2003).

The parallel between cultural and biological evolution goes back to Huxley (1955), Popper (1979), and James (1880).[4] The idea of treating culture as a form of epigenetic transmission was pioneered by Richard Dawkins, who coined the term "meme" in *The Selfish Gene* (Dawkins, 1976) to represent an integral unit of information that could be transmitted phenotypically. There quickly followed several major contributions to a biological approach to culture, all based on the notion that culture, like genes, could evolve through replication (intergenerational transmission), mutation, and selection (Boyd & Richerson, 1985; Cavalli-Sforza & Feldman, 1982; Lumsden & Wilson, 1981).

Cultural elements reproduce themselves from brain to brain and across time, mutate, and are subject to selection according to their effects on the fitness of their carriers (Boyd & Richerson, 1985; Cavalli-Sforza & Feldman, 1982; Parsons, 1964). Moreover, there are strong interactions between genetic and epigenetic elements in human evolution, ranging from basic physiology (e.g., the transformation of the organs of speech with the evolution of language) to sophisticated social emotions, including empathy, shame, guilt, and revenge seeking (Zajonc, 1980, 1984).

Because of their common informational and evolutionary character, there are strong parallels between genetic and cultural modeling (Mesoudi et al., 2006). Like biological transmission, culture is transmitted from parents to offspring, and like cultural transmission, which is transmitted horizontally to unrelated individuals, so in microbes and many plant species, genes are regularly transferred across lineage boundaries (Abbott et al., 2003; Jablonka & Lamb, 1995; Rivera & Lake, 2004). Moreover, anthropologists reconstruct the history of social groups by analyzing homologous and analogous cultural traits, much as biologists reconstruct the evolution of species by the analysis of shared characters and homologous DNA (Mace & Pagel, 1994). Indeed, the same computer programs developed by biological systematists are used by cultural anthropologists (Holden, 2002; Holden & Mace, 2003). In addition, archeologists who study cultural evolution have a similar modus operandi as paleobiologists who study genetic evolution (Mesoudi et al., 2006). Both attempt to reconstruct lineages of artifacts and their carriers. Like paleobiology, archaeology assumes that when analogy can be ruled out, similarity implies causal connection by inheritance (O'Brien & Lyman, 2000). Like biogeography's study of the spatial distribution of organisms (Brown & Lomolino, 1998), behavioral ecology studies the interaction of ecological, historical, and geographical factors that determine distribution of cultural forms across space and time (Smith & Winterhalder, 1992).

Perhaps the most common critique of the analogy between genetic and cultural evolution is that the gene is a well-defined, discrete, independently reproducing and mutating entity, whereas the boundaries of the unit of culture are ill-defined and overlapping. In fact, however, this view of the gene is simply outdated. Overlapping, nested, and movable genes discovered in the past 35 years have some of the fluidity of cultural units, whereas quite often the boundaries of a cultural unit (a belief, icon, word, technique, stylistic convention) are quite delimited and specific. Similarly, alternative splicing, nuclear and messenger RNA editing, cellular protein modification, and genomic imprinting, which are quite common, undermine the standard view of the insular gene producing a single protein and support the notion of genes having variable boundaries and having strongly context-dependent effects.

Dawkins added a second fundamental mechanism of epigenetic information transmission in *The Extended Phenotype* (Dawkins, 1982), noting that organisms can directly transmit environmental artifacts to the next generation, in the form of such constructs as beaver dams, bee hives, and even social structures (e.g., mating and hunting practices). The phenomenon of a species creating an important aspect of its environment and stably transmitting this environment across generations, known as *niche construction*, is a widespread form of epigenetic transmission (Odling-Smee et al., 2003). Moreover, niche construction gives rise to what might be called a *gene-environment coevolutionary process*, since a genetically induced environmental regularity becomes the basis for genetic selection, and genetic mutations that give rise to mutant niches will survive if they are fitness enhancing for their constructors. The analysis of the reciprocal action of genes and culture is known as *gene-culture coevolution* (Bowles & Gintis, 2005; Durham, 1991; Feldman & Zhivotovsky, 1992; Lumsden & Wilson, 1981).

Neuroscientific studies exhibit clearly the genetic basis for moral behavior. Brain regions involved in moral judgments and behavior include the prefrontal cortex, the orbitofrontal cortex, and the superior temporal sulcus (Moll et al., 2005). These brain structures are virtually unique to or most highly developed in humans and are doubtless evolutionary adaptations (Schulkin, 2000). The evolution of the human prefrontal cortex is closely tied to the emergence of human morality (Allman et al., 2002). Patients with focal damage to one or more of these areas exhibit a variety of antisocial behaviors, including the absence of embarrassment, pride and regret (Beer et al., 2003; Camille, 2004), and sociopathic behavior (Miller et al., 1997). There is a likely genetic predisposition underlying sociopathy, and sociopaths comprise 3% to 4% of the male population, but they account for between 33% and 80% of the population of chronic criminal offenders in the United States (Mednick et al., 1977).

It is clear from this body of empirical information that culture is directly encoded into the human brain, which of course is the central claim of gene-culture coevolutionary theory.

GAME THEORY: THE UNIVERSAL LEXICON OF LIFE

In the BPC model, choices give rise to probability distributions over outcomes, the expected values of which are the payoffs to the choice from which they arose. Game theory extends this analysis to cases where there are multiple decision makers. In the language of game theory, *players* (or *agents*) are endowed with a set of available *strategies* and have certain *information* concerning the rules of the game, the nature of the other players and their available strategies, and the structure of payoffs. Finally, for each combination of strategy choices by the players, the game specifies a distribution of *individual payoffs* to the players. Game theory attempts to predict the behavior of the players by assuming that each maximizes its preference function subject to its information, beliefs, and constraints (Kreps, 1990).

Game theory is a logical extension of evolutionary theory. To see this, suppose there is only one replicator, deriving its nutrients and energy from nonliving sources (the sun, the earth's core, amino acids produced by electrical discharge, and the like). The replicator population will then grow at a geometric rate, until it presses upon its environmental inputs. At that point, mutants that exploit the environment more efficiently will out-compete their less efficient conspecifics, and with input scarcity, mutants will emerge that "steal" from conspecifics who have amassed valuable resources. With the rapid growth of such predators, mutant prey will devise means of avoiding predation, and predators will counter with their own novel predatory capacities. In this manner, strategic interaction is borne from elemental evolutionary forces. It is only a conceptually short step from this point to cooperation and competition among cells in a multicellular body, among conspecifics who cooperate in social production, between males and females in a sexual species, between parents and offspring, and among groups competing for territorial control.

Historically, game theory did not emerge from biological considerations, but rather from the strategic concerns of combatants in World War II (Poundstone, 1992; Vonneumann & Morgenstern, 1944). This led to the widespread caricature of game theory as applicable only to static confrontations of rational self-regarding agents possessed of formidable reasoning and information-processing capacity. Developments within game theory in recent years, however, render this caricature inaccurate.

First, game theory has become the basic framework for modeling animal behavior (Alcock, 1993; Krebs & Davies, 1997; Maynard Smith, 1982), and thus has shed its static and hyperrationalistic character, in the form of evolutionary game theory (Gintis, 2000). Evolutionary and behavioral game theory do not require the formidable information-processing capacities of classical game theory, so disciplines that recognize that cognition is scarce and costly can make use of game-theoretic models (Gintis, 2000; Gigerenzer & Selten, 2001; Young, 1998). Thus, agents may consider only a restricted subset of strategies (Simon, 1972; Winter, 1971), and they may use rule-of-thumb heuristics rather than maximization techniques (Gigerenzer & Selten, 2001). Game theory is thus a generalized schema that permits the precise framing of meaningful empirical assertions but imposes no particular structure on the predicted behavior.

Second, evolutionary game theory has become key to understanding the most fundamental principles of evolutionary biology. Throughout much of the twentieth century, classical population biology did not employ a game-theoretic framework (Haldane, 1932; Fisher, 1930; Wright, 1931). However, Moran (1964) showed that Fisher's Fundamental Theorem, which states that as long as there is positive genetic variance in a population, fitness increases over time, is false when more than one genetic locus is involved. Eshel and Feldman (1984) identified the problem with the population genetic model in its abstraction from mutation. But how do we attach a fitness value to a mutant? Eshel and Feldman (1984) suggested that payoffs be modeled game theoretically on the phenotypic level and a mutant gene be associated with a strategy in the resulting game. With this assumption, they showed that under some restrictive conditions, Fisher's Fundamental Theorem could be restored. Their results were generalized by Liberman (1988), Hammerstein and Selten (1994), Hammerstein (1996), Eshel and colleagues (1998), and others.

Third, the most natural setting for biological and social dynamics is game theoretic. Replicators (genetic and/or cultural) endow copies of themselves with a repertoire of strategic responses to environmental conditions, including information concerning the conditions under which each is to be deployed in response to character and density of competing replicators. Genetic replicators have been well understood since the rediscovery of Mendel's laws in the early twentieth century. Cultural transmission also apparently occurs at the neuronal level in the brain, in part through the action of *mirror neurons* (Meltzoff & Decety, 2003; Rizzolatti et al., 2002; Williams et al., 2001). Mutations include replacement of strategies by modified strategies, and the "survival of the fittest" dynamic (formally called a *replicator dynamic*) ensures that replicators with more successful strategies replace those with less successful ones (Taylor & Jonker, 1978).

Fourth, behavioral game theorists now widely recognize that in many social interactions, agents are not self-regarding, but rather often care about the payoffs to and intentions of other players, and will sacrifice to uphold personal standards of honesty and decency (Fehr & Gächter, 2002; Gintis et al., 2005; Gneezy, 2005; Wood, 2003). Moreover, human actors care about power, self-esteem, and behaving morally (Bowles & Gintis, 2005; Gintis, 2003; Wood,

2003). Because the rational actor model treats action as instrumental toward achieving rewards, it is often inferred that action itself cannot have reward value. This is an unwarranted inference. For instance, the rational actor model can be used to explain collective action (Olson, 1965), since agents may place positive value on the process of acquisition (for instance, "fighting for one's rights"), and can value punishing those who refuse to join in the collective action (Moore, 1978; Wood, 2003). Indeed, contemporary experimental work indicates that one can apply standard choice theory, including the derivation of demand curves, plotting concave indifference curves, and finding price elasticities, for such preferences as charitable giving and punitive retribution (Andreoni & Miller, 2002).

As a result of its maturation of game theory over the past quarter century, game theory is well positioned to serve as a bridge across the behavioral sciences, providing both a lexicon for communicating across fields with divergent and incompatible conceptual systems and a theoretical tool for formulating a model of human choice that can serve all the behavioral disciplines.

EXPERIMENTAL GAME THEORY AND NON–SELF-REGARDING PREFERENCES

Contemporary biological theory maintains that cooperation can be sustained based on *inclusive fitness*, or cooperation among close genealogical kin (Hamilton, 1963) by individual self-interest in the form of *reciprocal altruism* (Trivers, 1971). Reciprocal altruism occurs when an agent helps another agent, at a fitness cost to itself, with the expectation that the beneficiary will return the favor in a future period. The explanatory power of inclusive fitness theory and reciprocal altruism convinced a generation of biologists that what appears to be altruism—personal sacrifice on behalf of others—is really just long-run genetic self-interest. Combined with a vigorous critique of group selection (Dawkins, 1976; Maynard Smith, 1976; Williams, 1966), a generation of biologists became convinced that true altruism—one organism sacrificing fitness on behalf of the fitness of an unrelated other—was virtually unknown, even in the case of *Homo sapiens*.

The selfish nature of human nature was touted as a central implication of rigorous biological modeling. In *The Selfish Gene*, for instance, Richard Dawkins (1976) asserts, "We are survival machines—robot vehicles blindly programmed to preserve the selfish molecules known as genes. . . . Let us try to teach generosity and altruism, because we are born selfish." (p. 1) Similarly, in *The Biology of Moral Systems*, R. D. Alexander (1987, p. 3) asserts, "Ethics, morality, human conduct, and the human psyche are to be understood only if societies are seen as collections of individuals seeking their own self-interest." More poetically, Michael Ghiselin (1974) writes, "No hint of genuine charity ameliorates our vision of society, once sentimentalism has been laid aside. What passes for cooperation turns out to be a mixture of opportunism and exploitation. . .. Scratch an altruist, and watch a hypocrite bleed."

In economics, the notion that enlightened self-interest allows agents to cooperate in large groups goes back to Bernard Mandeville's "private vices, public virtues" (Mandeville, 1705) and Adam Smith's "invisible hand" (Smith, 1759). Full analytical development of this idea awaited the twentieth-century development of general equilibrium theory (Arrow & Debreu, 1954; Arrow & Hahn, 1971) and the theory of repeated games (Axelrod & Hamilton, 1981; Fudenberg & Maskin, 1986). So powerful in economic theory is the notion that cooperation among self-regarding agents is possible that it is hard to find even a single critic of the notion in the literature, even among those that are otherwise quite harsh in their evaluation of neoclassical economics.

By contrast, sociological, anthropological, and social psychological theory generally explain that human cooperation is predicated upon affiliative behaviors among group members, each of whom is prepared to sacrifice a modicum of personal well-being to advance the collective goals of the group. The vicious attack on "sociobiology" (Segerstrale, 2001) and the widespread rejection of the bare-bones *Homo economicus* in the "soft" social sciences (DiMaggio, 1994; Etzioni, 1985; Hirsch et al., 1990) is due in part to this clash of basic explanatory principles.

Behavioral game theory assumes the BPC model of choice and subjects individuals to strategic settings, such that their behavior reveals their underlying preferences. This controlled setting allows us to adjudicate between these contrasting models. One behavioral regularity that has been found thereby is *strong reciprocity*, which is a predisposition to cooperate with others and to punish those who violate the norms of cooperation, at personal cost, even when it is implausible to expect that these costs will be repaid. Strong reciprocity is other-regarding, as a strong reciprocator's behavior reflects a preference to cooperate with other cooperators and to punish noncooperators, even when these actions are personally costly.

The result of the laboratory and field research on strong reciprocity is that humans indeed often behave in ways that have traditionally been affirmed in sociological theory and denied in biology and economics (Andreoni, 1995; Fehr & Gächter, 2000, 2002; Fehr et al., 1997, 1998; Gächter & Fehr, 1999; Henrich et al., 2005; Ostrom et al., 1992). Moreover, it is probable that this other-regarding behavior is a prerequisite for cooperation in large groups of nonkin, since the theoretical models of cooperation in large groups of self-regarding nonkin in biology and economics do not apply to some important and frequently observed forms of human cooperation (Boyd & Richerson, 1992; Gintis, 2005).

Character Virtues in the Laboratory

Another form of prosocial behavior conflicting with the maximization of personal material gain is that of maintaining such *character virtues* as honesty and promise keeping, even when there is no chance of being penalized for unvirtuous behavior. Our first example of non–self-regarding behavior will be of this form.

Gneezy (2005) studied 450 undergraduate participants paired off to play several games of the following form. There are two players who never see each other (anonymity) and they interact exactly once (one shot). Player 1, whom we will call the Advisor, is shown the contents of two envelopes, labeled *A* and *B*. Each envelope has two compartments, the first containing money to be given to the Advisor, and the other to be given to player 2. We will call player 2 the Chooser, because this player gets to choose which of the two envelopes will be distributed to the two players. The catch, however, is that the Chooser is not permitted to see the contents of the envelopes. Rather, the Advisor, who did see the contents, was required to advise the Chooser which envelope to pick.

The games all begin with the experimenter showing both players the two envelopes, and asserting that one of the envelopes is better for the Advisor and the other is better for the Chooser. The Advisor is then permitted to inspect the contents of the two envelopes, and say to the Chooser either "*A* will earn you more money than *B*," or "*B* will earn you more money than *A*." The Chooser then picks either *A* or *B*, and the game is over.

Suppose both players are self-regarding, each caring only about how much money he earns from the transaction. Suppose also that both players believe their partner is self-regarding. The Chooser will then reason that the Advisor will say whatever induces him, the Chooser, to choose the envelope that gives him, the Chooser, the lesser amount of money. Therefore, nothing the Advisor says should be believed, and the Chooser should just make a random pick between the two envelopes. The Advisor can anticipate the Chooser's reasoning, and will pick randomly which envelope to advise the Chooser to choose. Economists call the Advisor's message "cheap talk," because it costs nothing to give, but is worth nothing to either party.

By contrast, suppose the Chooser believes that the Advisor places a positive value on transmitting honest messages, and so will be predisposed to follow whatever advice he is given, and suppose the Advisor does value honesty, and believes that the Chooser believes that he values honesty, and hence will follow the Advisor's suggestion. Then, the Advisor will weigh the financial gain from lying against the cost of lying, and unless the gain is sufficiently large, he will tell the truth, the Chooser will believe him, and the Chooser will get his preferred payoff.

Gneezy (2005) implemented this experiment as a series of three games with the aforementioned structure (his detailed protocols were slightly different). The first game, which we will write $A = (6,5)$, $B = (5,6)$, pays the Advisor 6 and the Chooser 5 if the Chooser picks A, and the reverse if the Chooser picks B. The second game, $A = (6,5)$, $B = (5,15)$, pays the Advisor 6 and the Chooser 5 if the Chooser picks A, but pays the Advisor 5 and the Chooser 15 if the Chooser picks B. The third game, $A = (15,5)$, $B = (5,15)$, pays the Advisor 15 and the Chooser 5 if the Chooser picks A, but pays the Advisor 15 and the Chooser 5 if the Chooser picks B.

Before having the subjects play any of the games, he attempted to determine whether Advisors believed that their advice would be followed, because, if they did not believe this, then it would be a mistake to interpret their giving advice favorable to Choosers to the Advisor's honesty. Gneezy elicited truthful beliefs from Advisors by promising to pay an additional sum of money at the end of the session to each Advisor who correctly predicted whether his advice would be followed. He found that 82% of Advisors expected their advice to be followed. In fact, the Advisors were remarkably accurate, since the actual percent was 78%.

The most honesty was elicited in game 2, where $A = (5,15)$ and $B = (6,5)$, so lying was very costly to the Chooser and the gain to lying for the Advisor was small. In this game, a full 83% of Advisors were honest. In game 1, where $A = (5,6)$ and $B = (6,5)$, so the cost of lying to the Chooser was small and equal to the gain to the Advisor, 64% of the Advisors were honest. In other words, subjects were loathe to lie, but considerably more so when it was costly to their partner. In game three, where $A = (5,15)$ and $B =(15,5)$, so the gain from lying was large for the Advisor and equal to the loss to the Chooser, only 48% of the Advisors were honest. This shows that many subjects are willing to sacrifice material gain to avoid lying in a one-shot, anonymous interaction, their willingness to lie increasing with an increased cost of truth telling to themselves, and decreasing with an increase in their partner's cost of being deceived.

Similar results were found by Boles and colleagues (2000) and Charness and Dufwenberg (2004). Gunnthorsdottir and colleagues (2002) and Burks and colleagues (2003) have shown that a social-psychological measure of "Machiavellianism" predicts which subjects are likely to be trustworthy and trusting, although their results are not completely compatible.

The Public Goods Game

The *public goods game* has been analyzed in a series of papers by the social psychologist Toshio Yamagishi (1986, 1988a,b), by the political scientist Elinor Ostrom and her coworkers (Ostrom et al., 1992), and by economists Ernst Fehr and his coworkers (Fehr & Gächter, 2000, 2002; Gächter & Fehr, 1999). These researchers uniformly found that *groups exhibit a much higher rate of cooperation than can be expected assuming the standard economic model of the self-regarding actor*, and this is especially the case when subjects are given the option of incurring a cost to themselves in order to punish free riders.

A typical public goods game consists of a number of rounds, say 10. The subjects are told the total number of rounds, as well as all other aspects of the game. The subjects are paid their winnings in real money at the end of the session. In each round, each subject is grouped with several other subjects—say three others—under conditions of strict anonymity. Each subject is then given a certain number of "points," say 20, redeemable at the end of the experimental session for real money. Each subject then places some fraction of his or her points in a "common account" and the remainder in the subject's "private account." The experimenter then tells the subjects how many points were contributed to the common account, and adds to the private account of each subject some fraction, say 40%, of the total amount in the common account. So if a subject contributes his or her whole 20 points to the common account, each of the four group members will receive eight points at the end of the round. In effect, by putting the whole endowment into the common account, a player loses 12 points but the other three group members gain in total 24 (= *8* × *3*) points. The players keep whatever is in their private account at the end of the round.

A self-regarding player will contribute nothing to the common account. However, only a fraction of subjects in fact conform to the self-interest model. Subjects begin by contributing on average about half of their endowment to the public account. The level of contributions decays over the course of the 10 rounds, until in the final rounds most players are behaving in a self-regarding manner (Dawes & Thaler, 1988; Ledyard, 1995). In a meta-study of 12 public goods experiments, Fehr and Schmidt (1999) found that in the early rounds, average and median contribution levels ranged from 40% to 60% of the endowment, but in the final period 73% of all individuals (N = 1,042) contributed nothing, and many of the remaining players contributed close to zero. These results are not compatible with the selfish actor model, which predicts zero contribution on all rounds, though they might be predicted by a reciprocal altruism model, since the chance to reciprocate declines as the end of the experiment approaches. However, this is not in fact the explanation of moderate but deteriorating levels of cooperation in the public goods game.

The explanation of the decay of cooperation offered by subjects when debriefed after the experiment is that cooperative subjects became angry at others who contributed less than themselves, and retaliated against free-riding low contributors in the only way available to them—by lowering their own contributions (Andreoni, 1995). This view is confirmed by the fact that when subjects play the repeated public goods game sequentially several times, each time they begin by cooperating at a high level, and their cooperation declines as the end of the game approaches.

Experimental evidence supports this interpretation. When subjects are allowed to punish noncontributors, they do so at a cost to themselves (Orbell, Dawes, & Van de Kragt, 1986; Sato, 1987; Yamagishi, 1988a,b, 1992). For instance, in the Ostrom et al. (1992) study subjects interacted for 25 periods in a public goods game, and by paying a "fee," subjects could impose costs on other subjects by "fining" them. Since fining costs the individual who uses it, but the benefits of increased compliance accrue to the group as a whole, the only Nash equilibrium in this game that does not depend on incredible threats is for no player to pay the fee, so no player is ever punished for defecting, and all players defect by contributing nothing to the common pool. However, the authors found a significant level of punishing behavior.

These studies allowed individuals to engage in strategic behavior, since costly punishment of defectors could increase cooperation in future periods, yielding a positive net return for the punisher. Fehr and Gächter (2000) set up an experimental situation in which *the possibility of strategic punishment was removed.* They used 6- and 10-round public goods games with group sizes of four, and with costly punishment allowed at the end of each round, employing three different methods of assigning members to groups. There were sufficient subjects to run between 10 and 18 groups simultaneously. Under the *Partner* treatment, the four subjects remained in the same group for all 10 periods. Under the *Stranger* treatment, the subjects were randomly reassigned after each round. Finally, under the *Perfect Stranger* treatment, the subjects were randomly reassigned and assured that they would never meet the same subject more than once. Subjects earned an average of about $35 for an experimental session.

Fehr and Gächter (2000) performed their experiment for 10 rounds with punishment and 10 rounds without.[5] Their results are illustrated in Figure 9.1. We see that when costly punishment is permitted, cooperation does not deteriorate, and in the Partner game, despite strict anonymity, cooperation increases almost to full cooperation, even on the final round. When punishment is not permitted, however, the same subjects experience the deterioration of cooperation found in previous public goods games. The contrast in cooperation rates between the Partner and the two Stranger

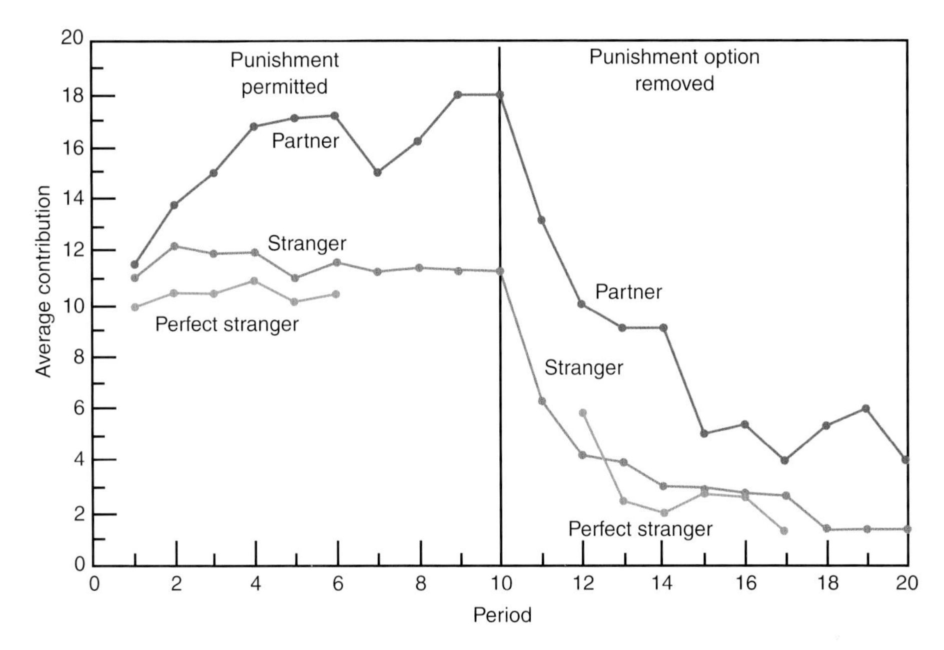

Figure 9.1 Average contributions over time in the Partner, Stranger, and Perfect Stranger treatments when the punishment condition is played first. Adapted from Fehr, E., & Gächter, S. (2000). Cooperation and punishment. *American Economic Review, 90,* 980–994. Used with permission.

treatments is worth noting, because the strength of punishment is roughly the same across all treatments. This suggests that the credibility of the punishment threat is greater in the Partner treatment because in this treatment the punished subjects are certain that, once they have been punished in previous rounds, the punishing subjects are in their group. This result follows from the fact that a majority of subjects showed themselves to be strong reciprocators, both contributing a large amount and enthusiastically punishing noncontributors. The prosociality impact of strong reciprocity on cooperation is thus more strongly manifested the more coherent and permanent the group in question.

CONCLUSION

I have shown that the core theoretical constructs of the various behavioral disciplines currently include mutually contradictory principles, but that progress over the past couple of decades has generated the instruments necessary to resolve the interdisciplinary contradictions. I have outlined several of the key ideas needed to specify a unified analytical framework for the behavioral sciences.

NOTES

1. The arguments in this paper are developed in greater depth in Gintis (2007).
2. I use the term "self-regarding" rather than "self-interested" (and similarly "non–self-regarding" or "other-regarding" rather than "non–self-interested" or "unselfish") for a situation in which the payoffs to other agents are valued by an agent.
3. This argument was presented verbally by Darwin (1872) and is implicit in the standard notion of "survival of the fittest," but formal proof is recent (Grafen, 1999, 2000, 2002). The case with frequency-dependent (nonadditive genetic) fitness has yet to be formally demonstrated, but the informal arguments in this case are no less strong.
4. For a more extensive analysis of the parallels between cultural and genetic evolution, see

Mesoudi et al. (2006). I have borrowed heavily from this paper in this section.
5. For additional experimental results and analysis, see Bowles and Gintis (2002) and Fehr and Gächter (2002).

REFERENCES

Abbott, R. J., James, J. K., Milne, R. I., & Gillies, A. C. M. (2003). Plant introductions, hybridization and gene flow. *Philosophical Transactions of the Royal Society of London B, 358,* 1123–1132.

Alcock, J. (1993). *Animal behavior: An evolutionary approach.* Sunderland, MA: Sinauer.

Alexander, R. D. (1987). *The biology of moral systems.* New York: Aldine.

Allman, J., Hakeem, A., & Watson, K. (2002). Two phylogenetic specializations in the human brain. *Neuroscientist, 8,* 335–346.

Andreoni, J. (1995). Cooperation in public goods experiments: Kindness or confusion. *American Economic Review, 85,* 891–904.

Andreoni, J., & Miller, J. H. (2002). Giving according to garp: An experimental test of the consistency of preferences for altruism. *Econometrica, 70,* 737–753.

Arrow, K. J., & Debreu, G. (1954). Existence of an equilibrium for a competitive economy. *Econometrica, 22,* 265–290.

Arrow, K. J., & Hahn, F. (1971). *General competitive analysis.* San Francisco: Holden-Day.

Axelrod, R., & Hamilton, W. D. (1981). The evolution of cooperation. *Science, 211,* 1390–1396.

Beer, J. S., Heerey, E. A., Keltner, D., Skabini, D., & Knight, R. T. (2003). The regulatory function of self-conscious emotion: Insights from patients with orbitofrontal damage. *Journal of Personality and Social Psychology, 65,* 594–604.

Black, F., & Scholes, M. (1973). The pricing of options and corporate liabilities. *Journal of Political Economy, 81,* 637–654.

Boles, T. L., Croson, R. T. A., & Murnighan, J. K. (2000). Deception and retribution in repeated ultimatum bargaining. *Organizational Behavior and Human Decision Processes, 83,* 235–259.

Bonner, J. T. (1984). *The evolution of culture in animals.* Princeton, NJ: Princeton University Press.

Bowles, S., & Gintis, H. (2002). Homo reciprocans. *Nature, 415,* 125–128.

Bowles, S., & Gintis, H. (2005). Prosocial emotions. In: L. E. Blume & S. N. Durlauf (Eds.), *The economy as an evolving complex system III.* Santa Fe, NM: Santa Fe Institute.

Boyd, R., & Richerson, P. J. (1985). *Culture and the evolutionary process.* Chicago: University of Chicago Press.

Boyd, R., & Richerson, P. J. (1992). Punishment allows the evolution of cooperation (or anything else) in sizeable groups. *Ethology and Sociobiology, 113,* 171–195.

Brown, J. H., & Lomolino, M. V. (1998). *Biogeography.* Sunderland, MA: Sinauer.

Burks, S. V., Carpenter, J. P., & Verhoogen, E. (2003). Playing both roles in the trust game. *Journal of Economic Behavior and Organization, 51,* 195–216.

Camille, N. (2004). The involvement of the orbitofrontal cortex in the experience of regret. *Science, 304,* 1167–1170.

Cavalli-Sforza, L. L., & Feldman, M. W. (1981). *Cultural transmission and evolution.* Princeton, NJ: Princeton University Press.

Cavalli-Sforza, L. L., & Feldman, M. W. (1982). Theory and observation in cultural transmission. *Science, 218,* 19–27.

Coleman, J. S. (1990). *Foundations of social theory.* Cambridge, MA: Belknap.

Darwin, C. (1872). *The origin of species by means of natural selection* (6th ed.). London: John Murray.

Dawes, R. M., & Thaler, R. (1988). Cooperation. *Journal of Economic Perspectives, 2,* 187–197.

Dawkins, R. (1976). *The selfish gene.* Oxford: Oxford University Press.

Dawkins, R. (1982). *The extended phenotype: The gene as the unit of selection.* Oxford: Freeman.

DiMaggio, P. (1994). Culture and economy. In: N. Smelser & R. Swedberg (Eds.), *The handbook of economic sociology* (pp. 27–57). Princeton, NJ: Princeton University Press.

Durham, W. H. (1991). *Coevolution: Genes, culture, and human diversity.* Stanford, CA: Stanford University Press.

Eshel, I., & Feldman, M. W. (1984). Initial increase of new mutants and some continuity properties of ESS in two locus systems. *American Naturalist, 124,* 631–640.

Eshel, I., Feldman, M. W., & Bergman, A. (1998). Long-term evolution, short-term evolution, and population genetic theory. *Journal of Theoretical Biology, 191,* 391–396.

Etzioni, A. (1985). Opening the preferences: A socio-economic research agenda. *Journal of Behavioral Economics, 14,* 183–205.

Fehr, E., & Gächter, S. (2000). Cooperation and punishment. *American Economic Review, 90,* 980–994.

Fehr, E., & Gächter, S. (2002). Altruistic punishment in humans. *Nature, 415,* 137–140.

Fehr, E., Gächter, S., & Kirchsteiger, G. (1997). Reciprocity as a contract enforcement device: Experimental evidence. *Econometrica, 65,* 833–860.

Fehr, E., Kirchsteiger, G., & Riedl, A. (1998). Gift exchange and reciprocity in competitive experimental markets. *European Economic Review, 42,* 1–34.

Fehr, E., & Schmidt, K. M. (1999). A theory of fairness, competition, and cooperation. *Quarterly Journal of Economics, 114,* 817–868.

Feldman, M. W., & Zhivotovsky, L. A. (1992). Gene-culture coevolution: Toward a general theory of vertical transmission. *Proceedings of the National Academy of Sciences, 89,* 11935–11938.

Fisher, R. A. (1930). *The genetical theory of natural selection.* Oxford: Clarendon Press.

Fudenberg, D., & Maskin, E. (1986). The folk theorem in repeated games with discounting or with incomplete information. *Econometrica, 54,* 533–554.

Gächter, S., & Fehr, E. (1999). Collective action as a social exchange. *Journal of Economic Behavior and Organization, 39,* 341–369.

Ghiselin, M. T. (1974). *The economy of nature and the evolution of sex.* Berkeley, CA: University of California Press.

Gigerenzer, G., & Selten, R. (2001). *Bounded rationality.* Cambridge, MA: MIT Press.

Gintis, H. (2000). *Game theory evolving.* Princeton, NJ: Princeton University Press.

Gintis, H. (2003). Solving the puzzle of human prosociality. *Rationality and Society, 15,* 155–187.

Gintis, H. (2005). Behavioral game theory and contemporary economic theory. *Analyze Kritik, 27,* 48–72.

Gintis, H. (2007). A framework for the unification of the behavioral sciences. *Behavioral and Brain Sciences, 30,* 1–61.

Gintis, H., Bowles, S., Boyd, R., & Fehr, E. (2005). *Moral sentiments and material interests: On the foundations of cooperation in economic life.* Cambridge, MA: MIT Press.

Glimcher, P. W. (2003). Decisions, uncertainty, and the brain: The science of neuroeconomics. Cambridge, MA: MIT Press.

Glimcher, P. W., Dorris, M. C., & Bayer, H. M. (2005). *Physiological utility theory and the neuroeconomics of choice.* New York: Center for Neural Science, New York University.

Gneezy, U. (2005). Deception: The role of consequences. *American Economic Review, 95,* 384–394.

Grafen, A. (1999). Formal darwinism, the individual-as-maximizing-agent analogy, and bet-hedging. *Proceedings of the Royal Society B, 266,* 799–803.

Grafen, A. (2000). Developments of price's equation and natural selection under uncertainty. *Proceedings of the Royal Society B, 267,* 1223–1227.

Grafen, A. (2002). A first formal link between the price equation and an optimization program. *Journal of Theoretical Biology, 217,* 75–91.

Gunnthorsdottir, A., McCabe, K., & Smith, V. (2002). Using the machiavellianism instrument to predict trustworthiness in a bargaining game. *Journal of Economic Psychology, 23,* 49–66.

Haldane, J. B. S. (1932). *The Causes of Evolution.* London: Longmans, Green Co.

Hamilton, W. D. (1963). The evolution of altruistic behavior. *American Naturalist, 96,* 354–356.

Hammerstein, P. (1996). Darwinian adaptation, population genetics and the streetcar theory of evolution. *Journal of Mathematical Biology, 34,* 511–532.

Hammerstein, P., & Selten, R. (1994). Game theory and evolutionary biology. In: R. J. Aumann & S. Hart (Eds.), *Handbook of game theory with economic applications* (pp. 929–993). Amsterdam: Elsevier.

Hechter, M., & Kanazawam S. (1997). Sociological rational choice. *Annual Review of Sociology, 23,* 199–214.

Henrich, J., Boyd, R., Bowles, S., Camerer, C., Fehr, E., & Gintis, H. (2005). Economic man' in cross-cultural perspective: Behavioral experiments in 15 small-scale societies. *Behavioral and Brain Sciences, 28,* 795–815.

Henrich, J., & Gil-White, F. (2001). The evolution of prestige: Freely conferred status as a mechanism for enhancing the benefits of cultural transmission. *Evolution and Human Behavior, 22,* 165–196.

Hirsch, P., Michaels, S., & Friedman, R. (1990). Clean models vs. dirty hands: Why economics is different from sociology. In: S. Zukin & DiMaggio (Eds.), *Structures of capital: The social organization of the economy* (pp. 39–56). New York: Cambridge University Press.

Holden, C. J. (2002). Bantu language trees reflect the spread of farming across sub-saharan africa: A maximum-parsimony analysis. *Proceedings of the Royal Society of London Series B, 269,* 793–799.

Holden, C. J., & Mace, R. (2003). Spread of cattle led to the loss of matrilineal descent in Africa: A coevolutionary analysis. *Proceedings of the Royal Society of London Series B, 270,* 2425–2433.

Holland, J. H. (1975). *Adaptation in natural and artificial systems.* Ann Arbor, MI: University of Michigan Press.

Huxley, J. S. (1955). Evolution, cultural and biological. *Yearbook of Anthropology,* 2–25.

Jablonka, E., & Lamb, M. J. (1995). *Epigenetic inheritance and evolution: The Lamarckian case.* Oxford: Oxford University Press.

James, W. (1880). Great men, great thoughts, and the environment. *Atlantic Monthly, 46,* 441–459).

Jaynes, E. T. (2003). *Probability theory: The logic of science.* Cambridge: Cambridge University Press.

Kollock, P. (1997). Transforming social dilemmas: Group identity and cooperation. In: P. Danielson (Ed.), *Modeling rational and moral agents.* Oxford: Oxford University Press.

Krebs, J. R., & Davies, N. B. (1997). *Behavioral ecology: An evolutionary approach* (4th ed.). Oxford: Blackwell Science.

Kreps, D. M. (1990). *A course in microeconomic theory.* Princeton, NJ: Princeton University Press.

Ledyard, J. O. (1995). Public goods: A survey of experimental research. In: J. H. Kagel & A. E. Roth (Eds.), *The handbook of experimental economics* (pp. 111–194). Princeton, NJ: Princeton University Press.

Lewontin, R. C. (1974). *The genetic basis of evolutionary change.* New York: Columbia University Press.

Liberman, U. (1988). External stability and ess criteria for initial increase of a new mutant allele. *Journal of Mathematical Biology, 26,* 477–485.

Lumsden, C. J., & Wilson, E. O. (1981). *Genes, mind, and culture: The coevolutionary process.* Cambridge, MA: Harvard University Press.

Mace, R., & Pagel, M. (1994). The comparative method in anthropology. *Current Anthropology, 35,* 549–564.

Mandeville, B. (1705). *The fable of the bees: Private vices, publick benefits.* Oxford: Clarendon.

Maynard Smith, J. (1976). Group selection. *Quarterly Review of Biology, 51,* 277–283.

Maynard Smith, J. (1982). *Evolution and the theory of games.* Cambridge: Cambridge University Press.

Mednick, S. A., Kirkegaard-Sorenson, L., Hutchings, B., Knop, J., Rosenberg, R., & Schulsinger, F. (1977). An example of bio-social interaction research: The interplay of socio-environmental and individual factors in the etiology of criminal behavior. In: S. A. Mednick & K. O. Christiansen (Eds.), *Biosocial bases of criminal behavior* (pp. 9–24). New York: Gardner Press.

Meltzhoff, A. N., & Decety, J. (2003). What imitation tells us about social cognition: A rapprochement between developmental psychology and cognitive neuroscience. *Philosophical Transactions of the Royal Society of London B, 358,* 491–500.

Mesoudi, A., Whiten, A., & Laland, K. N. (2006). Towards a unified science of cultural evolution. *Behavioral and Brain Sciences, 29,* 329–383.

Miller, B. L., Darby, A., Benson, D. F., Cummings, J. L., & Miller, M. H. (1997). Aggressive, socially disruptive and antisocial behaviour associated with fronto-temporal dementia. *British Journal of Psychiatry, 170,* 150–154.

Moll, J., Zahn, R., di Oliveira-Souza, R., Krueger, F., & Grafman, J. (2005). The neural basis of human moral cognition. *Nature Neuroscience, 6,* 799–809.

Montague, P. R., & Berns, G. S. (2002). Neural economics and the biological substrates of valuation. *Neuron, 36,* 265–284.

Moore, B., Jr. (1978). *Injustice: The social bases of obedience and revolt.* White Plains, NY: M.E. Sharpe.

Moran, P. A. P. (1964). On the nonexistence of adaptive topographies. *Annals of Human Genetics, 27,* 338–343.

Newman, M., Barabasi, A. L., & Watts, D. J. (2006). *The structure and dynamics of networks.* Princeton, NJ: Princeton University Press.

O'Brien, M. J., & Lyman, R. L. (2000). *Applying evolutionary archaeology.* New York: Kluwer Academic.

Odling-Smee, F. J., Laland, K. N., & Feldman, M. W. (2003). *Niche construction: The neglected process in evolution.* Princeton, NJ: Princeton University Press.

Olson, M. (1965). *The logic of collective action: Public goods and the theory of groups.* Cambridge, MA: Harvard University Press.

Orbell, J. M., Dawes, R. M., & Van de Kragt, J. C. (1986). Organizing groups for collective action. *American Political Science Review, 80,* 1171–1185.

Ostrom, E., Walker, J., & Gardner, R. (1992). Covenants with and without a sword: Self-governance is possible. *American Political Science Review, 86,* 404–417.

Parker, A. J., & Newsome, W. T. (1998). Sense and the single neuron: Probing the physiology of perception. *Annual Review of Neuroscience, 21,* 227–277.

Parsons, T (1964). Evolutionary universals in society. *American Sociological Review, 29,* 339–357.

Popper, K. (1979). *Objective knowledge: An evolutionary approach.* Oxford: Clarendon Press.

Poundstone, W. (1992). *Prisoner's dilemma.* New York: Doubleday.

Real, L. A. (1991). Animal choice behavior and the evolution of cognitive architecture. *Science, 253,* 980–986.

Real, L., & Caraco, T. (1986). Risk and foraging in stochastic environments. *Annual Review of Ecology and Systematics, 17,* 371–390.

Richerson, P. J., & Boyd, R. (1998). The evolution of ultrasociality. In: I. Eibl-Eibesfeldt & F. Salter (Eds.), *Indoctrinability, idology and warfare* (pp. 71–96). New York: Berghahn Books.

Rivera, M. C., & Lake, J. A. (2004). The ring of life provides evidence for a genome fusion origin of eukaryotes. *Nature, 431,* 152–155.

Rizzolatti, G., Fadiga, L., Fogassi, L., & Gallese, V. (2002). From mirror neurons to imitation: Facts and speculations. In: A. N. Meltzhoff & W. Prinz (Eds.), *The imitative mind: Development, evolution and brain bases* (pp. 247–266). Cambridge: Cambridge University Press.

Sato, K. (1987). Distribution and the cost of maintaining common property resources. *Journal of Experimental Social Psychology, 23,* 19–31.

Schall, J. D., & Thompson, K. G. (1999). Neural selection and control of visually guided eye movements. *Annual Review of Neuroscience, 22,* 241–259.

Schrödinger, E. (1944). *What is Life?: The Physical Aspect of the Living Cell.* Cambridge: Cambridge University Press.

Schulkin, J. (2000). *Roots of social sensitivity and neural function.* Cambridge, MA: MIT Press.

Schultz, W., Dayan, P., & Montague, P. R. (1997). A neural substrate of prediction and reward. *Science, 275,* 1593–1599.

Segerstrale, U. (2001). *Defenders of the truth: The sociobiology debate.* Oxford: Oxford University Press.

Shennan, S. (1997). *Quantifying archaeology.* Edinburgh: Edinburgh University Press.

Simon, H. (1972). Theories of bounded rationality In: C. B. McGuire & R. Radner (Eds.), *Decision and organization* (pp. 161–176). New York: American Elsevier.

Skibo, J. M., & Bentley, R. A. (2003). *Complex systems and archaeology.* Salt Lake City: University of Utah Press.

Smith, A. (1759). *The theory of moral sentiments.* New York: Prometheus.

Smith, E. A., & Winterhalder, B. (1992) *Evolutionary ecology and human behavior.* New York: Aldine de Gruyter.

Sugrue, L. P., Corrado, G. S., & Newsome, W. T. (2005). Choosing the greater of two goods: Neural currencies for valuation and decision making. *Nature Reviews Neuroscience, 6,* 363–375.

Sutton, R., & Barto, A. G. (2000). *Reinforcement learning.* Cambridge, MA: MIT Press.

Taylor, P., & Jonker, L. (1978). Evolutionarily stable strategies and game dynamics. *Mathematical Biosciences, 40,* 145–156.

Trivers, R. L. (1971). The evolution of reciprocal altruism. *Quarterly Review of Biology, 46,* 35–57).

Von Neumann, J., & Morgenstern, O. (1944). *Theory of games and economic behavior.* Princeton, NJ: Princeton University Press.

Williams, G. C. (1966). *Adaptation and natural selection: A critique of some current evolutionary thought.* Princeton, NJ: Princeton University Press.

Williams, J. H. G., Whiten, A., Suddendorf, T., & Perrett, D. I. (2001). Imitation, mirror neurons and autism. *Neuroscience and Biobeheavioral Reviews, 25,* 287–295.

Winter, S. G. (1971). Satisficing, selection and the innovating remnant. *Quarterly Journal of Economics, 85,* 237–261.

Wood, E. J. (2003). *Insurgent collective action and civil war in El Salvador.* Cambridge: Cambridge University Press.

Wright, S. (1931). Evolution in mendelian populations. *Genetics, 6,* 111–178.

Yamagishi, T. (1986). The provision of a sanctioning system as a public good. *Journal of Personality and Social Psychology, 51,* 110–116.

Yamagishi, T. (1988a). The provision of a sanctioning system in the United States and Japan. *Social Psychology Quarterly, 51,* 265–271.

Yamagishi, T. (1988b). Seriousness of social dilemmas and the provision of a sanctioning system. *Social Psychology Quarterly, 51,* 32–42.

Yamagishi, T. (1992). Group size and the provision of a sanctioning system in a social dilemma. In: W. Liebrand, D. M. Messick, & H. Wilke (Eds.), *Social dilemmas: Theoretical issues and research findings* (pp. 267–287). Oxford: Pergamon Press.

Young, H. P. (1998). *Individual strategy and social structure: An evolutionary theory of institutions.* Princeton, NJ: Princeton University Press.

Zajonc, R. B. (1980). Feeling and thinking: Preferences need no inferences. *American Psychologist, 35,* 151–175.

Zajonc, R. B. (1984). On the primacy of affect. *American Psychologist, 39,* 117–123.

CHAPTER 10

Sensory and Motor Systems in Primates

Jon H. Kaas

INTRODUCTION

The task of describing the sensory and motor systems of primates, and how they interrelate to mediate complex behaviors, is clearly daunting. There are a number of systems, all of which are variable in organization and function across primate taxa, and they are incompletely understood, some more than others. In addition, this review needs to be reasonably short, although much can be said, as a recent six-volume series on sensory systems demonstrates (Basbaum et al., 2008). Thus, this review is necessarily selective, with a focus on the evolution of primate sensory and motor systems, especially in regard to the evolution of a parietal-frontal cortical system that uses information produced by early stages of sensory processing to guide ongoing behavior. We start with an overview of the components of sensory and motor systems that have been retained from non-primate ancestors, and proceed to an attempt to reconstruct the organization of these systems from early primates to branches of the anthropoid radiation, including humans. Aspects of the present discussion can be found in previous reviews (Kaas, 2007ab; Kaas & Preuss, 2008; see also Preuss, this volume).

OUR LEGACY: SENSORY AND MOTOR SYSTEMS OF EARLY MAMMALS

The sensory and motor systems of early mammals are of interest because early primates inherited these systems around 60 to 80 Mya and modified them. Early mammals had small brains with proportionately little neocortex. We can infer a lot about how sensory and motor systems were organized in these mammals by seeing how features or traits of these systems are presently distributed across present-day mammals (Kaas, 2007b). It has been especially informative to study the organizations of these systems in those present-day mammals that have small brains and proportionately little cortex (Fig. 10.1). A survey of such brains indicates that early mammals likely had, as one might expect, rather simple sensory and motor systems, with few cortical areas devoted to each system. Remarkably, there was probably no motor cortex that was distinct from somatosensory cortex in the first mammals. Separate motor areas appear to have emerged with the evolution of placental mammals. The major features of sensory and motor systems of early mammals are outlined below.

The Visual System

Early mammals had most of the subcortical components of the visual system that are now found in primates. The retina had several classes of ganglion cells, and they projected to the superchiasmatic nucleus of the hypothalamus, the nuclei of the accessory optic system, the ventral lateral geniculate nucleus, the dorsal lateral geniculate nucleus, the pretectum, and the superior colliculus (Berson, 2008). The dorsal lateral geniculate nucleus provided the major activating input to visual cortex (Kaas et al., 1972). Retinal ganglion cell classes resembling the P and M classes of primates (see later)

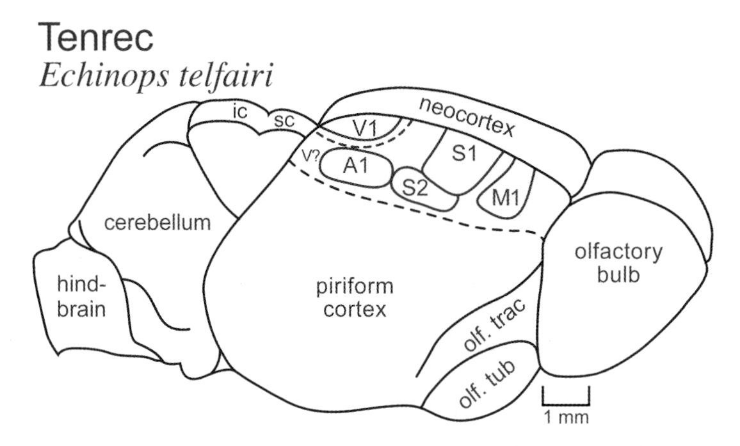

Figure 10.1 A dorsolateral view of the brain of a Madagascar tenrec (Echinops telfairi), a small Afrotherian mammal. Unlike most extant mammals, the neocortex is only a small cap on the brain. The primary visual (V1), somatosensory (S1), and auditory (A1) and motor (M1) areas have been identified, as well as a secondary somatosensory area (S2). Cortex caudal to A1 may be visual (V?) or multisensory. Notice how much of the forebrain is devoted to olfaction (olf trac = olfactory tract, olf tub = olfactory tubercle). Because of the limited extent of neocortex, the inferior colliculus (IC) and superior colliculus (SC) of the midbrain are exposed. Based on Krubitzer, L.A., Künzle, H., and Kaas, J. (1997). The organization of sensory cortex in a Madagascan insectivore, the tenrec (*Echinaps telfairi*). *Journal of Comparative Neurology, 379,* 399–414.

innervated about half of the neurons in the dorsal lateral geniculate, and retinal inputs were segregated to form cryptic contralaterally and ipsilaterally innervated layers. A more caudal portion of the geniculate received inputs from thin, more slowly conducting axons resembling those of the primate K-cell class (see later). Most of the geniculate outputs were to primary visual cortex (V1), which projected broadly to other visual and nonvisual areas of cortex, including somatosensory and auditory fields (Wang & Burkhalter, 2007). Major visual targets were the second visual area, V2, along the lateral border of V1, and a small visual temporal region caudolateral to V2. Cortex on the medial border of V1, a limbic or retrosplenial field, now identified in primates as prostriata, also received V1 inputs (Lyon, 2007). These areas also distributed visual information to other regions of cortex, including somatosensory, auditory, motor, frontal, limbic, and multisensory fields. Finally, the visual thalamus included a small pulvinar, usually identified in nonprimates as the lateral posterior nucleus or complex. Different parts of this small pulvinar received inputs from the superior colliculus and visual cortex, and this information was relayed to visual cortex. Another part of this poorly differentiated pulvinar received inputs from visual cortex and relayed back to visual cortex. Thus, the visual pulvinar possibly had three divisions, and visual cortex had three or four. A sector of the reticular nucleus of the ventral thalamus had inputs from visual cortex and the visual thalamus, projecting via inhibitory neurons into the visual thalamus (Crabtree & Killackey, 1989).

The Somatosensory System

Early mammals retained from reptilian ancestors most of the afferent systems that brought sensory information into the central nervous system (Kaas, 2007b). Obviously, new sensory opportunities arose with the evolution of body hair. Thus, early mammals had receptor afferents sensitive to touch, hair movement, vibration, muscle and joint movement, temperature, and painful stimuli. Longer sensory hairs (vibrissae) evolved to detect objects at short distances from the skin via receptors around the base of each hair. In early mammals, afferents from the face, facial vibrissae, and mouth were especially important, as they are in most mammals today.

Sensory afferents from the skin and deeper tissue entered the spinal cord or brainstem, where they terminated in specific layers of the dorsal horn of the spinal cord or in sensory brainstem nuclei (Fig. 10.2). The large fiber afferents related to tactile or muscle spindle receptors also sent branches that coursed in the brainstem or dorsal columns of the spinal cord to terminate in a complex of nuclei, the trigeminal–dorsal column complex, at the junction of the cervical spinal cord and brainstem. The traditional gracile, cuneate, and trigeminal nuclei for tactile inputs formed a medial to lateral sequence of subnuclei of a single functional unit that systematically represents the body from the hindlimb to the forelimb to the face and mouth. Neurons in this dorsal column–trigeminal nucleus projected to the ventroposterior nucleus of the contralateral thalamus. Neurons representing at least part of the mouth, the tongue, and teeth projected to the ipsilateral ventroposterior nucleus as well (Bombardieri et al., 1975). As in present-day mammals, the ventroposterior nucleus (VP) represented the body from hindlimb (and tail) to the mouth in a lateromedial sequence. In early mammals, and most mammals today, many or most of the tactile inputs were from the face and mouth, and a large subnucleus of VP, distinguished as the ventroposterior medial "nucleus" (VPM), represented the face and mouth. The remaining, more lateral part of VP, representing the rest of the body, is typically called the ventroposterior lateral "nucleus" (VPL). Functionally, VPM and VPL are subdivisions of VP.

In all mammals, VP projects to primary somatosensory cortex (S1). In primates, this cortex is distinguished as area 3b of anterior parietal cortex, and it is appropriate to use the anatomical term 3b for S1 in other mammals, although various combinations of architectonic terms for S1 are in current use. Comparative evidence indicates that S1 of early mammals represented the contralateral body surface from tail to tongue in a mediolateral sequence in parietal cortex, with the limbs facing forward (rostrally or anteriorly). As in mammals today, part of the ipsilateral mouth, the tongue and teeth, were also represented via the ipsilateral inputs to VPM.

Primary somatosensory cortex of early mammals activated three or four adjoining fields. Cortex along the rostral and caudal margins of S1 were activated via topographically ordered projections in mediolateral sequences so that they formed narrow somatosensory representations on the margins of S1. Here these areas are called the rostral (SR) and caudal (CR) somatosensory areas (Fig. 10.1). In primates, these bordering bands of cortex constitute area 3a, with a major involvement in proprioception, and area 1, a secondary tactile area. One or two areas lateral to S1 also received topographically organized projections from S1. One of these areas, the second somatosensory area (S2), has been described in all adequately explored mammals, and was certainly present in early mammals. Another adjoining area, the parietal ventral area (PV), has been more recently identified in mammals, and the comparative evidence indicates that it exists in a broad range of mammals, and most probably was present in early mammals. S2 and PV represent the contralateral body as two small mirror images of each other. Both S2 and PV received activating inputs from the ventroposterior nucleus, as well as from S1. These somatosensory areas distributed to adjoining insular and cingulate and frontal areas of cortex. A small fringe of more posterior cortex was all that could be considered posterior parietal cortex, and this region received somatosensory and other sensory inputs, and had connections with frontal cortex. Connections also reached perirhinal cortex and the hippocampus (Fig. 10.2).

In addition to cortical representations of tactile receptors, muscle spindle and joint receptor information reached the somatosensory thalamus from separate subnuclei in the dorsal column–trigeminal complex. In primates, the relayed proprioceptive information terminates in a separate nucleus of the contralateral thalamus, the ventroposterior superior nucleus (VPS), but this nucleus has not been commonly distinguished in nonprimates. Yet, the available evidence suggests that the proprioceptive inputs

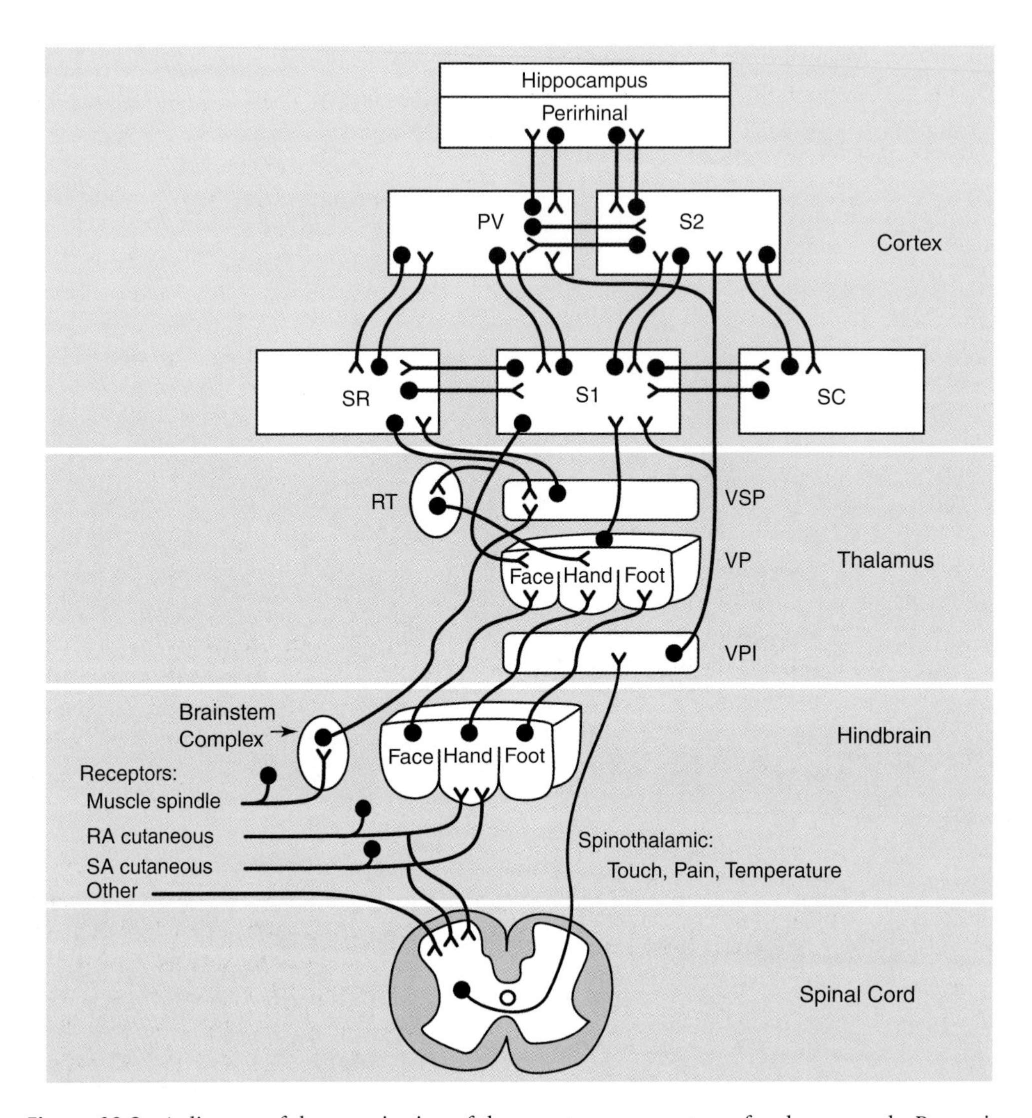

Figure 10.2 A diagram of the organization of the somatosensory system of early mammals. Processing starts with receptors and afferents of the skin, muscles, and joints. Rapidly (RA) and slowly (SA) adapting receptors in the skin mediate aspects of touch. Muscle spindle receptors help mediate proprioception. Other receptors include those that mediate pain, temperature, and aspects of touch and movement. Afferents terminate in the trigeminal–dorsal column complex where the face, hand, and foot are represented in a lateromedial sequence, and muscle spindle inputs terminate in separate nuclei. Afferents also activate spinal cord neurons that project to the contralateral thalamus. The somatosensory thalamus includes the ventroposterior nucleus for SA and RA inputs, and usually unnamed adjoining groups of cells corresponding to the ventroposterior superior nucleus (VPS) of primates for muscle spindle inputs and the ventroposterior inferior nucleus (VPI) of primates for spinothalamic inputs. The somatosensory segment of the reticular nucleus (RT) interacts with cortex and these nuclei. Somatosensory cortex includes the primary area, S1, with adjoining rostral somatosensory (SR) and caudal somatosensory (SC) areas, the second somatosensory area (S2), and the parietal ventral area (PV).

to the thalamus of early mammals was likely to have been at least partly segregated in the dorsorostral part of VP, becoming more distinctly separated later with the evolution of primates, carnivores, and perhaps other lines of descent. Here, a VPS is shown as present in early mammals (Fig. 10.2), although it was probably poorly differentiated from VP.

Another source of information to the somatosensory thalamus was from second-order neurons that form the spinothalamic tract for the body and the functionally equivalent part of the trigeminothalamic tract for the face and mouth. Much of the information relayed in this pathway is nociceptive and thermoreceptive, but tactile and proprioceptive information is included as well (Dostrovsky & Craig, 2008). These thalamic inputs, representing the contralateral half of the body, terminated in and around the ventroposterior nucleus. In primates, they largely terminate in the ventroposterior inferior nucleus (VPI) and cell-poor septa that extend from this nucleus well into the ventroposterior nucleus to isolate subnuclei. However, VPI has not been distinguished in most mammals, although a VPI is well developed in raccoons, and it can be identified in cats and a few other mammals. Here a VPI nucleus is shown for early mammals (Fig. 10.2), although it may not have been histologically distinct from VP. Probably, most mammals have a spinothalamic pathway that terminates largely on small neurons on the ventral margin of VP and scattered within VP. These small cells in VPI and VP appear to project to the superficial layers of S1, S2, PV, and the rostral and caudal somatosensory belt areas (e.g., Penny et al., 1982). Their major role may be to modulate neurons in these areas and signal stimulus intensity. Other neurons in this spinothalamic pathway terminated in poorly defined groups of cells ventral and caudal to VP, sometimes called the posterior group, where information about temperature and pain is relayed to cortex caudal to S2 and PV in insular cortex. In addition, somatosensory visceral information in spinothalamic pathways was relayed via a basal ventromedial thalamic nucleus to part of insular cortex, and from the parafasciculus nucleus to the striatum (Craig, 2002; Dostrovsky & Craig, 2008).

Taste

The gustatory sense is mediated by a specialized part of the somatosensory system. As in present-day mammals, primary taste afferents from the taste buds on the tongue terminated in the nucleus of the solitary tract, a nucleus that receives various viscerosensory inputs over its long extent through most of the medulla. One of the functions of the nucleus is to project to neurons in the adjacent reticular formation, where neurons in turn project to motor nuclei involved in reflexes concerned with the acceptance (licking) or rejection (tongue protrusion) of palatable or aversive tastes. The nucleus of the solitary tract also projected rostrally to the parabrachial nucleus that in turn relayed to the lateral hypothalamus and amygdala. The parabrachial nucleus also provided taste, tactile, and viscerosensory information to the gustatory nucleus of the somatosensory thalamus, the parvocellular ventroposterior medial nucleus (VPMpc). This nucleus projected to cortex just ventral to the mouth and face representations of S1, possibly including both the granular cortex belonging to the S1 tongue representation and the dysgranular cortex corresponding to a representation of taste and tactile inputs just ventral to the S1 tongue representation and dorsal to the rhinal fissure (Kosar et al., 1986; Norgren & Wolf, 1975). Thus, there may have been at least two targets of VPMpc in early mammals, one in the tongue representation in primary somatosensory cortex (see Remple et al., 2003) and one in adjacent insular cortex. However, the representation of the tongue in S1 is usually not considered to be part of gustatory cortex. Gustatory cortex of insular cortex is thought to project directly or indirectly to a higher-order multisensory processing zone in orbitofrontal cortex (Pritchard & Norgren, 2004).

Pain and Temperature

A number of ascending systems that carry nociceptive and temperature information have been described in mammals (see for review Dostrovsky & Craig, 2008; Lima, 2008). For our purposes here, the relevant pathways are those that reach the dorsal thalamus and are relayed to cortex, as the cortex is thought to be necessary for consciously appreciating the nature of painful stimuli and temperature change. Modifications and elaborations of nociceptive and temperature systems at the thalamic and cortical levels in primates allowed these systems to become involved in more behaviors (Craig, 2007). Limited comparative evidence suggests that in all mammals, pain and temperature information is relayed from peripheral afferents by trigemino- and spinothalamic projections to and around the ventroposterior nucleus of the thalamus, the poorly defined posterior complex or nucleus, the intralaminar nuclei, and other targets, such as hypothalamus, amygdala, and septal nuclei. Much of the information is relayed from thalamic targets to insular cortex just ventral to somatosensory cortex (S1 and S2), somatosensory cortex, and anterior cingulate cortex. Craig (2007) proposes that the primordial role of insular cortex was to participate in the sensing of noxious and temperature stimuli and to modulate and control brainstem homeostatic integration sites, including those associated with the autonomic nervous system. Somatosensory cortex probably played a role in sensing these stimuli as well, but possibly in terms of the intensities of noxious stimuli, rather than the painful aspects. Inputs to cingulate cortex were probably indirect, but important in limbic motor functions.

The Auditory and Vestibular Systems

The auditory systems of mammals share a number of components that likely were retained from an early mammal ancestor (Carr & Edds-Walton, 2008; Kaas & Hackett, 2008). All depend on a peripheral auditory system that includes an external ear that is generally mobile and a short canal ending at the tympanic membrane. The presence in the middle ear of a chain of three small bones that transmit vibrations from the tympanic membrane to the sound window of the cochlea of the inner ear is a characteristic of mammals. The cochlea is a complicated organ that allows the hair cells to be stimulated and activate the afferents of the auditory nerve that terminate on the neurons of the cochlear nuclei. These neurons provide information about sound intensity and frequency. Disparities in the information relayed by the afferents from each cochlea allow central circuits to extract information about sound location. Auditory processing starts in the three divisions or "nuclei" of the cochlear nuclear complex, which relay to nuclei of higher levels in the brainstem of both sides. These include the nuclei of the superior olivary complex, the nuclei of the lateral lemniscus, and subdivisions of the inferior colliculus of the midbrain. The divisions of the inferior colliculus project to the medial geniculate complex, where neurons relay to auditory cortex. The ventral nucleus of the medial geniculate complex (MGv) projects to the auditory core of auditory cortex, the primary area or areas, while other divisions, the dorsal nucleus (MGd) and medial or magnocellular nucleus (MGm), project more broadly to secondary auditory and multisensory areas of cortex.

Most investigated mammals have an auditory core of two to three primary or primary-like areas, one of which (but not always the same one) has been identified as primary auditory cortex, A1 (Fig. 10.1). The comparative evidence suggests that early mammals had a core of at least an anterior auditory field, AAF, and an A1, surrounded by a narrow belt of several secondary auditory fields. Each secondary field in turn involved, via connections, other regions of cortex in auditory and multisensory processing. Core areas were tonotopically organized, and AAF and A1 were distinguished by having mirror-image or reversed patterns of tonotopic organizations. In present-day mammals, the tonotopic organizations of secondary auditory cortex are less precise, when they are present, and often difficult to reveal.

The sense of balance is mediated by the vestibular system (see for review Graf, 2007), which is not well understood at thalamic and cortical levels of processing, partly because the system can be difficult to study and partly because we are largely unaware of its sensory functions in postural control, reflexes, and the perception of self-movement. The organ for balance is part of the inner ear, and it consists of semicircular canals and otoliths of the labyrinth. Hair cells in the labyrinth are stimulated by the movement of fluid in the canals and in the otoliths. Afferents in the vestibular nerve innervate vestibular nuclei, which provide inputs to a number of control systems, such as those controlling eye movement reflexes. Vestibular information also is relayed to the thalamus, and then to cortex, but this network is poorly understood in most mammals. The thalamic and cortical levels of processing, which are largely multisensory, have been more investigated in primates.

Motor Systems

Judging from somewhat limited comparative evidence (see for review Beck et al., 1996; Nudo & Frost, 2007), early mammals did not have a region of motor cortex that was distinct from somatosensory cortex. While more studies are needed, motor cortex apparently did not emerge as a separate area or areas until eutherian mammals evolved some 140 million years ago (Kaas, 2007b). Judging from results from members of the other two major branches of mammalian evolution, the monotremes and the marsupials, primary somatosensory cortex of early mammals received both somatosensory information from the ventroposterior nucleus of the thalamus and motor-related information from the cerebellum via projections from the ventrolateral thalamic nucleus, which is considered part of the motor thalamus. This early somatosensory cortex influenced motor behavior via projections to the basal ganglia and other subcortical targets, but the corticospinal pathway was rather poorly developed (Nudo & Frost, 2007; Nudo & Masterton, 1988). Comparative studies of placental mammals, in contrast, indicate that most or all have at least one distinct motor area, primary motor cortex (M1), and likely a second motor area (M2), which may be homologous to either premotor cortex or the supplementary motor area of primates. In placental mammals, M1 (Fig. 10.1) and S1 are distinguished by inputs from the ventral lateral nucleus of the motor thalamus and the ventroposterior nucleus of the somatosensory thalamus, respectively. In addition, M1 has a more developed corticospinal projection system, and thereby has a more direct impact on motor behavior. Most of the evidence for a second motor area in placental mammals comes from rats, where there is evidence for a second forelimb representation rostral to M1 that has cortical and subcortical connections that differ from those of M1 and are more similar to those of premotor cortex and the supplementary motor area of monkeys (Rouiller et al., 1993). Tree shrews, which are one of the closest living relatives of primates, have a strip of cortex along the rostral border of M1 that can be either considered a second parallel motor representation, M2, or possibly part of M1 (Remple et al., 2006). Microstimulation of M2 required higher currents than M1 for evoked movements, and M2 had few corticospinal neurons, which were densely distributed across M1, area 3a, and S1. Connection patterns suggest that tree shrews also have a motor area on the medial wall of the cerebral hemisphere, possibly a cingulate motor area. Thus, tree shrews appear to resemble primates more closely than rats in motor cortex organization. Studies of other mammals would be useful in determining how motor cortex varies and is similar across mammalian taxa.

Olfactory System

Structures for processing olfactory information were proportionately large and important for early mammals. Olfactory receptor cells connect directly to the olfactory bulb, which in early mammals had a surface area as large as all of neocortex, as in tenrecs today (Fig. 10.1). The olfactory bulb, in turn, projected to a huge expanse of olfactory (piriform) cortex, which had connections to frontal cortex via the dorsal thalamus (Wilson, 2008). In proportion to the rest of the brain, the olfactory bulb and cortex are small in primates, especially in anthropoid primates, where vision has become so important for the recognition of conspecifics, other animals, and food, and smell has become less important, especially in humans. Mammals also have an accessory olfactory system with inputs from the vomeronasal organ that is involved in pheromone detection and other chemosensory functions. In chimpanzees and humans, the vomeronasal organ and the rest of the accessory olfactory system are degenerate (Bhatnagar & Smith, 2007) and nonfunctional. Thus, males no longer sniff the urine of females to determine the state of their reproductive cycle.

SENSORY AND MOTOR SYSTEMS IN PRIMATES

The sensory and motor systems of early mammals have been modified and expanded in many ways in subsequent lines of mammalian evolution. The altered auditory cortex of echolocating bats serves as one well-known example. Yet, the greatest changes are likely to be found in the cortical components of sensory and motor systems in those mammals with large brains and greatly expanded cortex, and humans are exceptional in this regard. However, all primate brains are exceptional in that they have many more neurons than rodent brains of the same size (Herculano-Houzel et al., 2007). Primates, in addition, have a cortex that is subdivided into an unusually large number of cortical areas, and the number of areas appears to be greater in those anthropoid primates with larger brains. In addition, many of their cortical areas are involved in sensory processing and motor behavior.

Here, we consider sensory and motor systems, especially at the cortical level, in members of the major branches of the primate radiation, the prosimian primates, New and Old World monkeys, and apes and humans. The interesting tarsiers, with a highly differentiated and expanded visual system (Collins et al., 2005a), are not considered here, as little is known about their other sensory systems or their motor system. Tarsiers have evolved to be highly specialized as visual predators of insects and small vertebrates.

The Visual Systems of Primates

All primates are characterized by a well-developed visual system, including forward-facing eyes, a laminated lateral geniculate nucleus and a number of pulvinar nuclei in the visual thalamus, and a visual cortex with a large primary area and a number of additional visual areas (Kaas, 2003; Kaas & Collins, 2004; Kremers, 2005). Early primates were nocturnal, and many of the prosimian primates remain nocturnal, but the early anthropoid primates were diurnal, and all have remained diurnal except owl monkeys, which reverted to nocturnal life. Thus, all anthropoid primates, except owl monkeys, are specialized for diurnal vision. This includes an emphasis on cone- rather than rod-mediated vision, a fovea in the retina for detailed vision, and a great investment in the parvocellular pathway from the retina to the lateral geniculate nucleus, as this pathway is devoted to the detailed, color vision that is used in object recognition. In macaque monkeys, and probably most other anthropoid primates, 80% of the retinal ganglion cells are devoted to the parvocellular (P) pathway (Weller & Kaas, 1989). The retina of primates also has two other types of outputs that are named after the layers they project to in the lateral geniculate nucleus (LGN); the magnocellular or M-cell pathway and the koniocellular or K-cell pathway, each accounting for about 10% of the retina's output. The M-cell subsystem is specialized for the detection of changes in contrast, such as that caused by a blinking light or a moving object. The koniocellular system is not well understood, but it contains information from the blue (S) cones and thereby is involved in color vision (Casagrande & Xu, 2004).

The LGN of all primates is divided into two pairs of parvocellular layers, one with inputs from the ipsilateral eye and one with inputs from the contralateral eye (Kaas et al., 1978). In anthropoid primates other than owl monkeys, these layers are thick and they are subdivided, producing four or more sublayers in the part of the nucleus devoted to central vision. The thinner magnocellular layers are also paired, one for each eye. The LGN of nocturnal prosimian primates has two well-developed koniocellular layers, one for each eye, but the small cells of this system are not so distinct in anthropoid primates, and they are generally considered to be between layers (interlaminar cells) rather than forming layers. However, in nocturnal owl monkeys, a thick koniocellular region is found. Thus, the koniocellular pathway seems to be more important in nocturnal primates. Unlike some mammals, such as cats and other carnivores, the LGN of primates projects almost exclusively (but not completely so) to a large primary visual area, V1.

The retinal K and M cells (but not the P cells in anthropoid primates) also project to the superior colliculus, which in turn projects to the LGN and parts of the visual pulvinar (Kaas & Huerta, 1988). The superior colliculus of all primates is specialized for frontal, binocular vision, and unlike other mammals, each colliculus represents only the contralateral visual hemifield, rather than the whole visual field of the contralateral eye (Fig. 10.3). This remarkable modification of the visual system appears to be related to an emphasis on frontal vision, but the implication of this change for the behavior in early primates is not completely clear. Another major modification is the huge projection to the superior colliculus from the many visual areas of cortex (Collins et al., 2005b).

The pulvinar of primates contains inferior, lateral, and medial divisions (Kaas & Huerta, 1988). The medial division has mixed functions, judging from its cortical connections, but the lateral division of two nuclei and the inferior division of four nuclei are all visual (Kaas & Lyon, 2007). The superior colliculus provides K-cell and M-cell information to the pulvinar and K-cell information to the LGN. Most of the other activating inputs to pulvinar nuclei are from visual cortex. Thus, the pulvinar complex

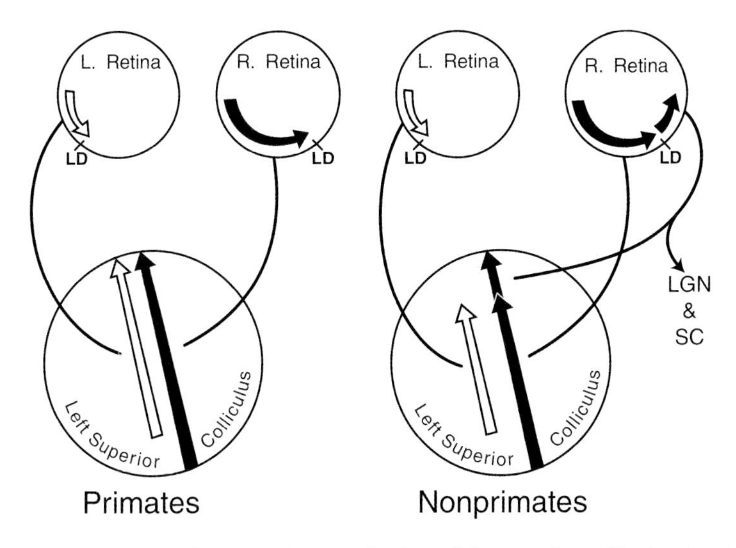

Figure 10.3 The unique type of retinotopic organization of the superior colliculus that is shared by all primates but differs from all studied nonprimates. The circles at the top represent the two eyes, while the large circle below represents a surface view of the left superior colliculus (SC). Arrows on the back of the eye indicate the temporal part of the retina that projects to the ipsilateral SC in mammals (white arrows). In primates, this projection is largely overlapped by a projection from the larger nasal retina from the contralateral eye (black arrow). The line of decussation (LD) marks the dividing point for these two projections. The LD runs vertically through the fovea. In nonprimates both the temporal segment of the retina of the contralateral eye (longer black arrow) and the temporal retina (short black arrow) project to the contralateral superior colliculus, creating a representation of the complete retina. This extended representation may aid in the detection of movement and predators from all parts of the visual field, while the primate SC is more devoted to frontal vision.

relays superior colliculus information to visual cortex, and distributes visual cortex information back to visual cortex. Some of the possible functions of such cortical-thalamic-cortical loops are discussed by Sherman and Guillery (2005). The larger nuclei of the pulvinar complex project to early visual areas in the cortical processing hierarchy, those with mixed visual functions or an emphasis on object vision, while a cluster of inferior pulvinar nuclei, including those with dense superior colliculus inputs, project to a cluster of temporal lobe visual areas concerned with visual motion and using vision for guiding motor behavior (Kaas & Lyon, 2007).

All primates have a greatly expanded cortical visual system. This elaboration is greater in anthropoid primates than in prosimians. However, many of the cortical visual areas that characterize anthropoid primates are present in prosimian galagos (Fig. 10.4), indicating that these visual areas emerged early in primate or preprimate evolution. Tree shrews, which are the closest living relative of primates that have been studied, also devote much of their cortex to visual processing (see Remple et al., 2006), suggesting that some of the specializations of visual cortex in primates predate primates and are shared by tree shrews. All primates have a large primary visual area, V1, that has an orderly internal organization so that cells are grouped by the particular orientation of visual stimuli, such as lines, that best activate them. V1 is divided into a number of pinwheels of such clusters of cells, which together represent a given region of visual space and all orientations. Tree shrews and carnivores also have such pinwheels of orientation selective cells, but rodents do not (Van Hooser et al., 2005), suggesting that tree shrews and primates shared an ancestor with orientation pinwheels, while carnivores evolved them independently. The main outputs of V1 are to the second visual area, V2, which is divided into a series of repeats of three types of band-like clusters of cells, two of which are devoted to different aspects of object vision and the third part of a processing stream for visual guidance of motor control (Casagrande & Kaas, 1994; Roe, 2003). This specialization of V2 is apparently unique to primates. Most of the other outputs of V1 are to a third visual area, V3, and the middle temporal visual area, MT.

Two of the classes of cell bands (modules) in V2 project to the dorsolateral visual area (DL), also known as V4. DL/V4 is a critically important visual area in the so-called ventral stream of visual processing areas devoted to object vision (Ungerleider & Mishkin, 1982). DL/V4 projects in turn to inferior temporal cortex (IT), which consists of several visual areas. Parts of IT cortex feed into memory-related areas that connect with the hippocampus, and to the prefrontal cortex of the frontal lobe for "working memory" and other functions. Overall, a huge amount of cortex is devoted to this object vision cortical stream, even in prosimians (Fig. 10.4), but much more in anthropoid primates (Fig. 10.5).

The third type of module in V2 projects to MT, as some classes of neurons in V1 do directly (Casagrande & Kaas, 1994). MT contains orderly arrangements of clusters of cells sensitive to stimulus orientation and direction of movement (see Xu et al., 2004). MT is one of a number of visual areas that appear to be unique to primates and collectively have the major role of further analyzing visual inputs to extract information about object and global motion, which is then sent to subdivisions of posterior parietal cortex (see Figs. 10.4 and 10.5). Thus, area MT is interconnected with bordering areas MST, MTc, and FST. FST has dorsal and ventral divisions with the dorsal division, FSTd, having interconnections with MT and the ventral division, FSTv, having connections with MTc (Kaas & Morel, 1993). These areas are also associated with band-like portions of V3, the caudal division of DL (DLc), and the dorsomedial area, DM. Areas MT, MST, and FSTd all project to portions of posterior parietal cortex. Together, these areas, including those in posterior parietal cortex, constitute most of the dorsal stream of visual processing that is concerned with visually guiding motor behavior. It is in posterior parietal cortex where transformations occur from analyzing sensory information to informing sensorimotor programs that induce ethologically relevant behaviors.

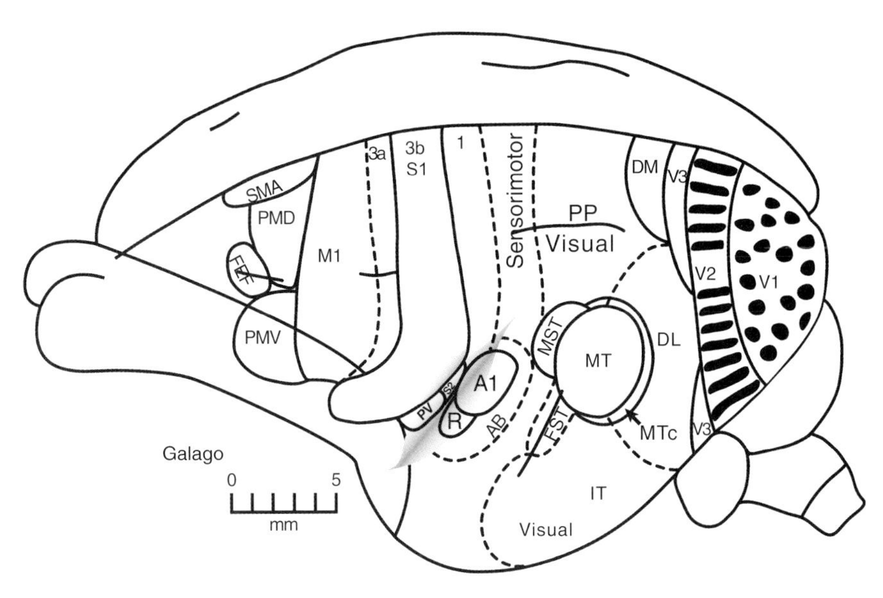

Figure 10.4 A dorsolateral view of the brain of a prosimian galago with major sensory and motor areas indicated. Visual areas include primary (V1), secondary (V2), and third (V3) visual areas, as well as the dorsolateral (DL), middle temporal (MT), fundal superior temporal (FST), MT crescent (MTc), medial superior temporal (MST), and dorsomedial (DM) visual areas. Much of posterior parietal (PP) and inferior temporal (IT) cortex contains additional visual areas. Auditory cortex includes primary auditory cortex (A1), the rostral auditory area (R), areas of the auditory belt (AB), and areas of the parabelt (not shown). Somatosensory cortex includes primary somatosensory cortex (S1 or area 3b), adjoining somatosensory areas 1 and 3a, the second area (S2), and the parietal ventral area (PV). Motor areas include the primary area (M1), dorsal (PMD) and ventral (PMV) premotor areas, a frontal eye field (FEF), supplementary motor cortex (SMA), and rostral and caudal cingulate motor areas (CMAr and CMAc) that are not shown. Dots in V1 and bars in V2 signify types of modular organization (see text).

Posterior Parietal Cortex and the Sensorimotor Transformation

Most mammals have very little cortex that can be called posterior parietal cortex (e.g., Fig. 10.1), but this region of the brain has greatly expanded in all primates (Figs. 10.4 and 10.5). The expansion may have preceded the emergence of primates, as tree shrews, close relatives of primates, have a strip of posterior parietal cortex that is somewhat enlarged, and has basic features of primate posterior parietal cortex (Remple et al., 2007). Thus, posterior parietal cortex in tree shrews receives both visual inputs from higher-order visual areas and somatosensory inputs from higher-order somatosensory areas, while projecting to motor cortex.

Prosimian primates have proportionately less posterior parietal cortex than anthropoid primates, but recent studies in prosimian galagos have revealed a lot about the organization of posterior parietal cortex (Fig. 10.6). Most notably, posterior parietal cortex is divided into rostral (PPr) and caudal (PPc) zones, with the rostral zone dominated by somatosensory inputs from higher-order somatosensory areas (S2, PV, area 1) and the caudal zone dominated by visual inputs from higher-order visual areas (V2, V3, DM, MT, MTc, MST, DLr). The caudal zone, PPc, projects to the rostral zone, PPr, so that both visual and somatosensory inputs reach PPr. Rostral PP is organized into a mediolateral series of functionally distinct sensorimotor zones that were revealed by electrical stimulation with microelectrodes. As described by Graziano and coworkers for motor cortex and even part of posterior parietal cortex in macaque monkeys

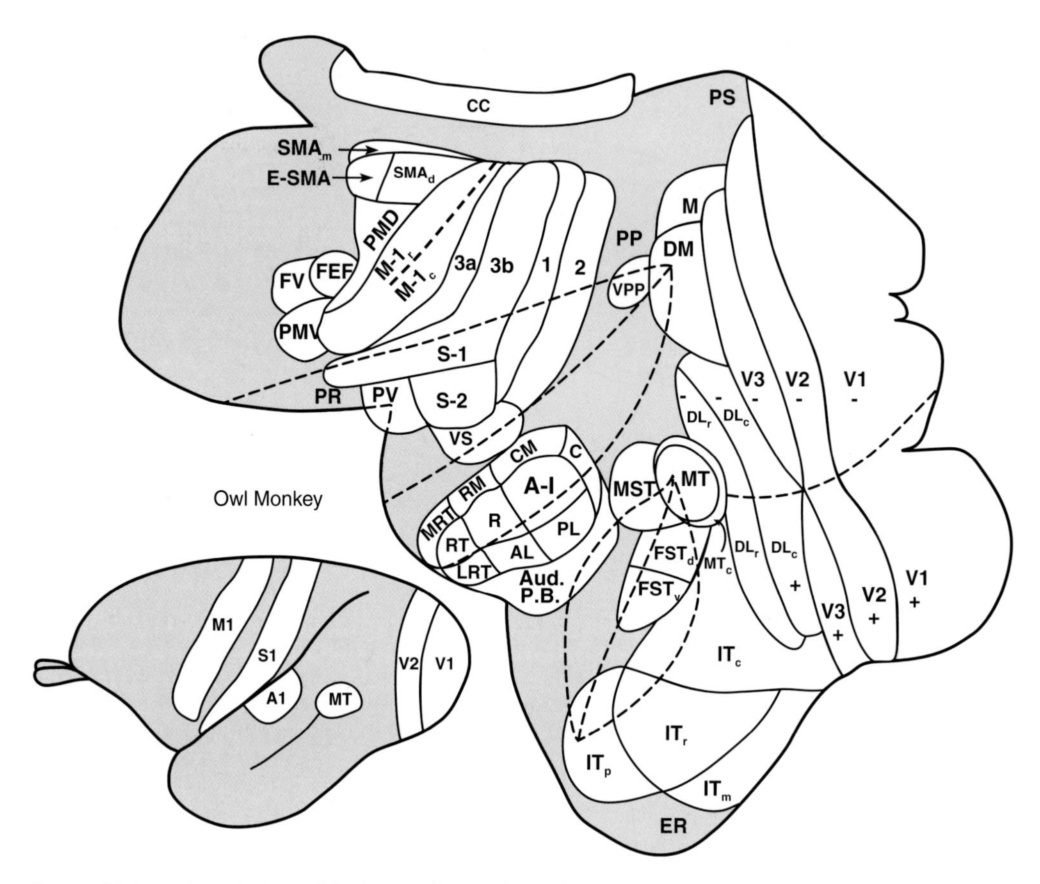

Figure 10.5 A lateral view of the brain of an owl monkey showing subdivisions of sensory and motor cortex (below) and a view of the cortex after it has been removed and flattened. See Figure 10.4 and text for abbreviations. Plus and minus notations in visual areas indicate representations of upper (+) and lower (–) visual hemifields. Rostral (DL_r) and caudal (DL_c) divisions of DL (V4) are shown, as well as caudal (IT_c), rostral (IT_r), medial (IT_m), and polar (IT_p) divisions of inferior temporal cortex. Dashed lines indicate opened fissures. M, medial visual area; VPP, ventral posterior parietal area; RI, retroinsular cortex; CC, corpus callosum.

(see Graziano, 2006; Graziano, this volume), half-second trains of electrical pulses, which are longer than those typically used for cortical stimulation, can evoke complex movements. In galagos, such half-second trains of brief electrical pulses evoked hindlimb and forelimb movements when medial sites in PPr were stimulated (hindlimb and forelimb in Fig. 10.6C). These movements, evoked in anesthetized galagos, resembled climbing movements. More laterally in PPr, defensive forelimb movements, reaching movements, and hand-to-mouth movements could be evoked from separate regions of cortex. Finally, in the most lateral portion of posterior parietal cortex, eye movements, defensive face movements, and aggressive face movements were evoked from separate cortical territories (Stepniewska et al., 2005). The existence of these different subregions for different complex, ethologically relevant movements suggests that a number of cortical networks exist for specific types of functionally important movements. As similar movements can be evoked from motor and premotor cortex, it appears that posterior parietal modules contain circuits that use visual and somatosensory information to activate and modulate outputs. These modules dictate specific categories of useful movements, which are then mediated via projections to motor and premotor cortex, where the movement patterns are likely adjusted and refined. Aspects of the final movement patterns would also depend on motor nuclei circuits in the brainstem and spinal cord. In support of this proposal, our ongoing experiments in galagos and New World monkeys indicate that any block of neural activity in primary motor cortex, M1, abolishes the motor behavior evoked from PPr.

While posterior parietal cortex is unlikely to be organized in the same way in all primates, our experiments have produced results very similar to those from galagos in New World owl and squirrel monkeys. The arrangements of areas and functional subregions may be somewhat different in posterior parietal cortex of Old World macaque monkeys, where current proposals include a number of areas, defined by responses to sensory inputs, connection patterns, and cortical architecture, that do not closely reflect their organization described here for posterior parietal cortex of galagos and New World monkeys. While there is not complete agreement on areas and names for areas, macaque areas include ventral (VIP), medial (MIP), lateral (LIP), and anterior (AIP) areas of the intraparietal sulcus, as well as a number of other areas (see Lewis & Van Essen, 2000). Yet, when half-second trains of electrical pulses were applied to VIP, defensive movements were evoked (Cooke et al., 2003), and eye movements have been elicited by electrically stimulating LIP (Kurylo & Skavenski, 1991; Thier & Andersen, 1998). Other parts of posterior parietal cortex in macaques appear to be involved in reaching and grasping (Calton et al., 2002; Snyder et al., 2000). Thus, the organization of posterior parietal cortex in Old World macaque monkeys may not be so different than in New World monkeys and prosimians. Some

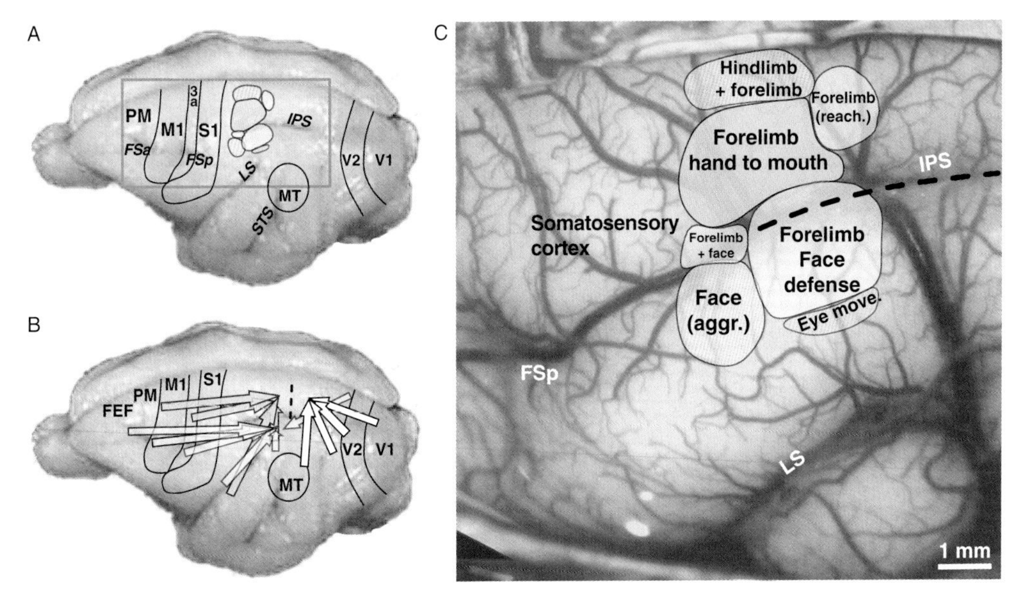

Figure 10.6 The organization of posterior parietal cortex in prosimian galagos. See Figure 10.4 and text for abbreviations. (**A**) A dorsolateral view of the brain with visual, somatosensory, motor, and posterior parietal areas outlined. (**B**) Arrows indicate visual areas projecting to the caudal half of posterior parietal cortex and somatosensory and motor areas projecting to the rostral half. (**C**) An enlarged view of posterior parietal cortex showing regions where electrical stimulation evoked different types of ethologically relevant complex movements. From Stepniewska, I., Fang, P. C., & Kaas, J. H. (2005). Microstimulation reveals specialized subregions for different complex movements in posterior parietal cortex of prosimian galagos. *Proceedings of the National Academy of Sciences, 102,* 4878–4883. Used with permission.

investigators, using functional magnetic resonance imaging, have results that suggest similarities between macaques and the more extensive posterior parietal cortex of humans (e.g., Swisher et al., 2007).

The Somatosensory System of Primates

While vision was obviously very important to early primates, most early primates were small and adapted to a nocturnal lifestyle that included feeding in the fine branches of bushes and trees on insects, small vertebrates, fruits, and leaves (Ross & Martin, 2007). This lifestyle required unusual sensorimotor abilities as these primates needed to hold on to moving branches while reaching for food. According to Whishaw (2003), visual guidance of hand movements is one of the most distinguishing features of primates. One of the reasons for reaching for food, rather than grasping it with their mouth, was to protect the large, forward-facing eyes. As an adaptation for greater hand use, primates have large concentrations of low-threshold mechanoreceptors in the glabrous skin of the hand, especially of the Meissner corpuscles, subserving the rapidly adapting type 1 afferents with small receptive fields and sensitivity to stimulus change (see for review Kaas, 2004). An enlarged representation of the glabrous skin of the hand is found in somatosensory nuclei and cortical areas of primates, especially Old World monkeys, apes, and humans.

In the thalamus, the ventroposterior complex of primates is well differentiated into a ventroposterior inferior nucleus (VPI) with spinothalamic inputs, a ventroposterior nucleus (VP) with inputs from cutaneous mechanoreceptors, and a ventroposterior superior nucleus (VPS) with inputs from muscle spindle receptors (Kaas, 2007). An anterior pulvinar (PA) can be identified, and it has connections with areas of somatosensory cortex. The nonprimary homolog of PA is not obvious, but PA possibly corresponds to the posterior nucleus of rodents.

Anterior parietal cortex organization varies across primates (Qi et al., 2008). In prosimian primates, three areas can be distinguished: a primary area, S1, which is clearly homologous with area 3b of anthropoid primates, and narrow strips of somatosensory cortex bordering S1 (3b) rostrally and caudally (Fig. 10.4). Area 3b gets inputs from VP in prosimians and all other primates. The more rostral somatosensory strip (SR) gets input from VPS, and is involved in proprioception in all primates and in at least some other mammals. This strip is clearly area 3a of anthropoid primates. The identity of the caudal somatosensory strip (SC) is less clear. It is in the position of area 1 of anthropoid primates, but unlike area 1, SC does not respond well to light touch on the skin. While area 1 gets dense projections from VP, the projections are sparse in SC. As a further difference, there is no evidence for an area 2 just caudal to SC, as an area 2 with inputs from VPS is caudal to area 1 in anthropoid primates. Possibly an area like SC differentiated into area 1 of anthropoid primates, or perhaps SC differentiated into both area 1 and area 2. Here we tentatively identify SC as area 1. Areas 3a, 3b, 1, and 2 of anthropoid primates (Fig. 10.5), including humans, contain parallel representations of the contralateral half of the body, from hindlimb to tongue in a mediolateral sequence. Together these fields interconnect with somatosensory areas of lateral (insular) parietal cortex, posterior parietal cortex, and motor cortex.

Lateral somatosensory of the upper bank of the lateral fissure and the insula contain additional somatosensory areas, including S2 and PV of other mammals, the ventral somatosensory area (VS), the parietal rostral area (PR), and likely others. The organization across primate taxa is not well understood, but differences are likely given the large extent of insular cortex in some anthropoid primates. Pathways through lateral parietal cortex are thought to be important in the recognition of objects by touch (Murray & Mishkin, 1984) and form the functional equivalent of the ventral visual stream of processing. Posterior parietal cortex, as discussed previously, also forms an important part of the somatosensory system, constituting much of the dorsal steam of somatosensory processing for guiding reaching and other actions.

Gustatory Cortex

The organization of the taste or gustatory system is not well understood in any mammals. Thus, modifications in primates are not clear. The standard view for primates is that the thalamic taste nucleus, VPMpc, projects to both the tongue representation of "S1" and to a large primary gustatory region in the cortex of the lateral sulcus (Fig. 10.7A). The gustatory area, G, in turn projects (apparently not directly) to orbitofrontal cortex, where hedonic or pleasurable aspects of taste are processed with other types of relevant information (see for review Kaas et al., 2006). More current evidence indicates that VPMpc projects to the tongue representation of area 3b, and possibly area 1, and that corticocortical connections implicate tongue representations in areas 3a and 1 in processing taste (Iyengar et al., 2007). While an area G may exist, another possibility is that the tongue portions of several areas of the cortex of the lateral sulcus are involved in taste (Fig. 10.7B).

Pain and Temperature

According to Craig (2003, 2007), primates differ from other mammals in having two specific regions of the thalamus that have differentiated as sites for the termination of nociceptive information from the spinal cord and brainstem, the posterior part of the ventral medial nucleus (VMpo) and the ventral caudal part of the medial dorsal nucleus (MDvc). VMpo provides projections in turn to a representation of painful stimuli in the dorsal portion of insular cortex, where other representations of body sensations, including temperature, may exist. Another projection is to part of area 3a for uncertain functions. MDvc projects to anterior cingulate cortex to motivate behavioral responses.

Auditory Cortex in Primates

Possibly due to limited study, the subcortical auditory system of primates is thought to be highly similar to those in other mammalian taxa. This cannot be quite true, as the greatly expanded cortical auditory system of primates would be reflected by changes in the thalamus, as cortical areas have thalamic interconnections and subcortical connections, such as those to the inferior colliculus. With this note of caution, present understandings of the organization of auditory cortex in primates are outlined.

All anthropoid primates appear to have a strip of auditory cortex that has the characteristics of a primary sensory field (Kaas & Hackett, 2008). This strip of cortex, generally called the auditory core, consists of three auditory areas, distinguished by their differing patterns of tonotopic organization (Fig. 10.5). The core has the well-differentiated layer 4 and other architectonic characteristics of primary sensory areas, as well as activating inputs from a thalamic relay nucleus, the ventral nucleus of the medial geniculate complex, MGv. Neurons in the core respond well to pure tones, and neurons in different locations across the three primary areas respond best, or at the lowest sound intensity, to tones of different frequencies. The auditory core in monkeys is in cortex of the lower bank of the lateral sulcus, where it forms an elongated caudorostral strip. The so-called primary area, A1, is the caudal-most area. The most caudal neurons in A1 respond to tones of the highest frequency, and neurons at progressively more rostral locations in A1 respond best to tones of progressively lower frequencies. Neurons in bands running perpendicular to this caudorostral frequency gradient respond best to tones of roughly the same frequency. These bands or rows of neurons constitute the lines of isorepresentation for tones in A1. The pattern of tonotopic representation in A1 reverses for the rostral auditory area, R, and again for the rostrotemporal auditory area, RT. The core has been histologically identified in a number of primates, including macaques, chimpanzees, and humans (Hackett et al., 2001). In prosimian galagos, only A1 and R of the core have been identified (Fig. 10.4). The three core areas seem very much alike, but a presumption is that they have at least somewhat different functional roles. One hypothesis is that they contribute differently to dorsal and ventral streams of auditory processing, with the dorsal stream more concerned with locating sounds in space and the ventral stream involved in deducing the meanings of the sounds (see Rauschecker & Tian, 2000).

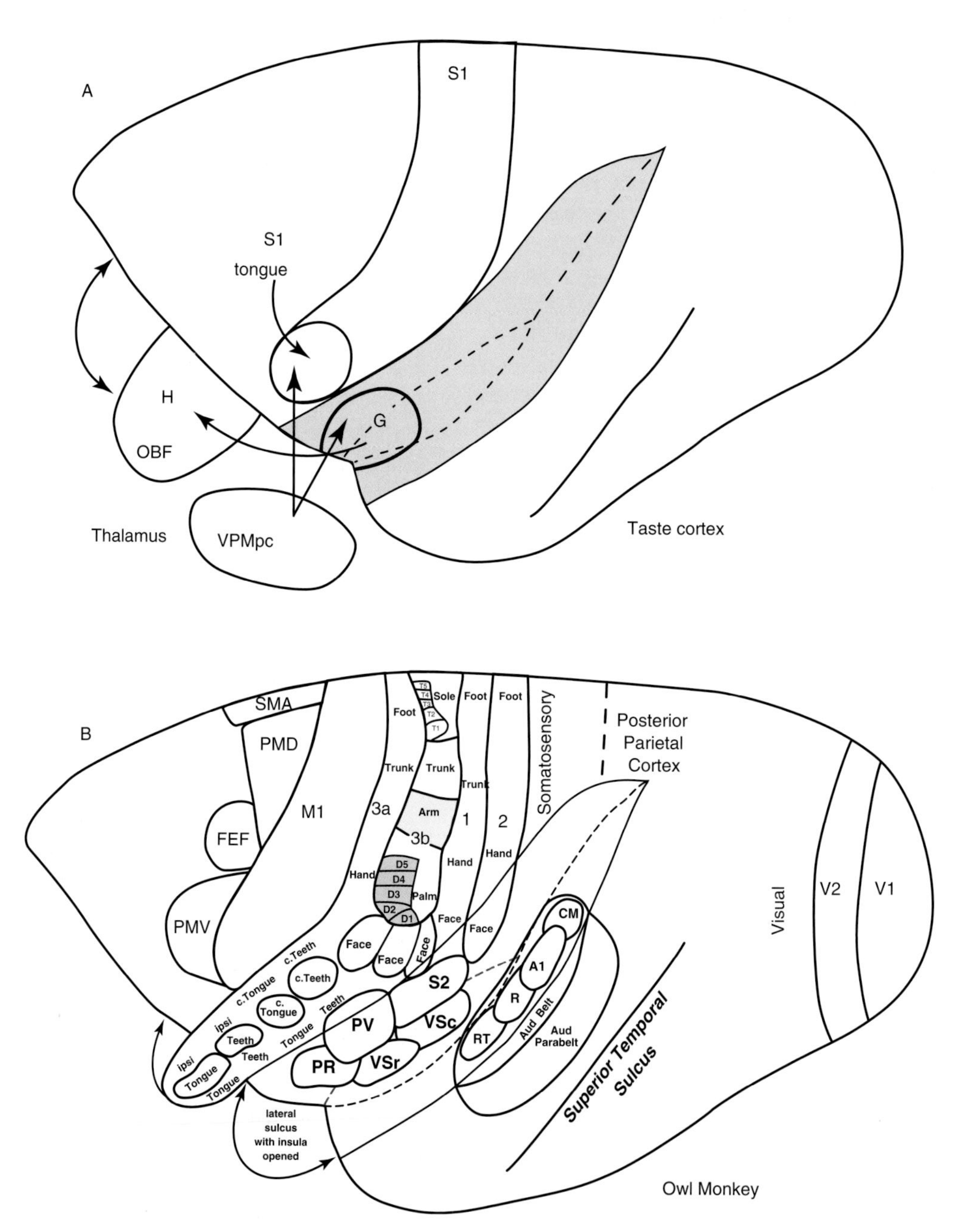

Figure 10.7 Proposed locations of cortical taste areas shown on a lateral view of the brain of a squirrel monkey. **(A)** Early investigators did not consider the architectonic fields 3a, 3b, 1, and 2 of anterior parietal cortex to be separate fields, and grouped them together in S1, where the lateral part represents the tongue and gets input from the parvocellular ventral posterior medial nucleus (VPMpc). However, the main projections of VPMpc were thought to be to the gustatory area, G, at the rostral end of the central sulcus. Area G was thought to project to a hedonistic (H) taste region in orbitofrontal cortex (OBF). **(B)** A contemporary view of somatosensory cortex organization in monkeys. The region of G appears to contain parts of several somatosensory areas (also see Fig. 10.5), suggesting that tongue representations in several fields may contribute to the processing of taste information. Note that the tongue is represented twice in area 3b. c, contralateral. Modified from Kaas, J. H., Qi, H-X., & Iyengar, S. (2006). Cortical network for representing the teeth and tongue in primates. *Anatomical Records Part A, 288A,* 182–190. Used with permission.

The auditory core is surrounded by a "belt" of adjacent auditory areas that at least partly depend on inputs from the core for activation. The cortical connections of the core, at least in monkeys, are almost completely with the belt. The number of areas in the belt is somewhat uncertain; the proposed number is eight, with two areas on the caudal end of A1 and pairs of inner and outer areas on each side of A1, R, and RT. Presently, there is evidence that at least three of the auditory belt areas are tonotopically organized, although this is difficult to determine as neurons in the belt areas generally respond much better to complex sounds than tones (Rauschecker et al., 1995). One of the belt areas, the caudomedial area (CM), has many neurons that respond to light touch (Fu et al., 2003), while another area, the middle lateral belt (ML), is influenced by vision (Ghazanfar et al., 2005), demonstrating a surprising substrate for bisensory integration at a very early level of auditory processing (see Romanski & Ghazanfar, this volume). Auditory belt areas connect broadly to core areas, other belt areas, the adjacent parabelt region, and even more distant cortical regions, such as prefrontal cortex where neurons responsive to visual and auditory stimuli are found (Romanski & Goldman-Rakic, 2002). Thalamic inputs are from dorsal (MGd) and medial (MGm) divisions of the medial geniculate complex, and other thalamic nuclei, suggesting multisensory or broader auditory functions. The more rostral belt areas appear to be more involved in a ventral stream for sound identification, with the more caudal belt areas more concerned with sound localization.

The auditory parabelt constitutes a third level of auditory cortical processing in primates. The parabelt in monkeys occupies the part of the superior temporal gyrus that adjoins the lateral belt (Fig. 10.5). The parabelt region gets dense inputs from the auditory belt areas, but practically no input from the core. Connections from belt areas are most dense with nearer portions of the parabelt, suggesting that the parabelt has functional divisions. Hackett and colleagues (1998) divided the parabelt into rostral (RPB) and caudal (CPB) regions with most dense connections with rostral or caudal belt areas, respectively.

The parabelt projections are to regions of cortex that we define as the fourth level of cortical auditory processing. Areas of the fourth level are diverse and distributed across the temporal, parietal, and frontal lobes. One of these regions is cortex of the upper bank of the superior temporal cortex in monkeys. This appears to be a region where neurons respond to auditory and visual stimuli (see Cusick, 1997). Neurons in this and other bisensory fields may function to localize a sound to a visual object (the so-called ventriloquist effect) (see Romanski & Ghazanfar, this volume). Other projections are to rostroventral parts of the superior temporal cortex. This cortex relays auditory information to orbitofrontal cortex that is involved in evaluating the rewarding value of stimuli (Rolls, 2004). Projections of the parabelt to cortex of the temporal-parietal junction and adjoining parietal cortex are to multisensory regions (visual somatosensory, auditory) where neurons project to the frontal eye fields (Huerta et al., 1987), perhaps to help direct the eyes toward sounds of interest. Parietal lobe multisensory areas also project to premotor areas of the frontal lobe, where they help guide motor behavior. Finally, parabelt projections to prefrontal granular cortex (Romanski & Goldman-Rakic, 2002) may be involved in working memory for auditory signals. Auditory and visual parts of prefrontal cortex project to premotor areas of frontal cortex, thus providing another source of sensory guidance of motor behavior.

Humans, of course, differ from other primates in that the left cerebral hemisphere is usually specialized for language. However, an asymmetry between the extent and shape of the lateral sulcus between the left and right hemispheres has been described, not only in humans, but also, from skull endocasts, in the brains of our extinct ancestors, suggesting that left hemisphere auditory specialization preceded the emergence of language (Galaburda et al., 1978). In humans, a region of cortex near the auditory core is enlarged in the left hemisphere, and this region, as part of Wernicke's area, appears to be important in language. In the frontal lobe, a premotor region is enlarged in the left hemisphere, constituting the main part of Broca's speech area (Foundas et al., 1996). These may be only some of the ways in which the human auditory system differs from that of monkeys. For further discussion on the issue of cerebral asymmetries, see Hopkins in this volume.

Vestibular Cortex

Multisensory integration in the vestibular system occurs as early as in the vestibular nuclei of the brainstem, and cortical areas with neurons sensitive to vestibular system activation are multisensory areas (Guldin & Grüsser, 1998). Probably the main cortical vestibular region in monkeys, the closest to a primary vestibular area, occupies the medial part of retroinsular cortex (Fig. 10.5). This region where half of the neurons respond to vestibular stimuli is called the parietal insular vestibular cortex (PIVC). A more posterior region of the cortex of the lateral fissure, the visual posterior sylvian area (VPS), is involved in visuomotor reflexes, and has neurons that are vestibularly activated or modulated. A third vestibular zone, the part of somatosensory area 3a where neck muscles are represented, also has neurons activated by vestibular stimulation. Parts of area 7 and area 2 have also been implicated in this extended, multisensory vestibular cortical system, and these vestibular cortical areas have a broader influence via connections with other cortical fields.

Cortical Motor Areas in Primates

Motor cortex has expanded and increased in number of areas in primates (Kaas, 2007b). Even prosimian galagos have an enlarged primary area (M1), a dorsal (PMD) and ventral (PMV) premotor area, a supplementary motor area (SMA), a presupplementary motor area (pre-SMA), a frontal eye field (FEF), and at least rostral (CMAr) and caudal (CMAc) cingulate motor areas (Figs. 10.4 and 10.6). These same areas are found in anthropoid primates (Fig. 10.5), where there is evidence for an increase in the number of premotor fields.

Primary motor cortex (M1) contains a mosaic of small regions, each devoted to a specific movement. These small regions are disbursed within a larger, gross somatotopic pattern that progresses from hindlimb, trunk, forelimb, and face to tongue in a mediolateral sequence across frontal cortex (Donoghue et al., 1992; Gould et al., 1986; Huang et al., 1988; Preuss et al., 1996). The gross pattern of somatotopic organization is similar across individuals of the same species, but the mosaic pattern varies in detail. M1 receives inputs from other frontal motor areas, including PMD, PMV, SMA, and cingulate areas; some of the somatosensory areas including areas 3a, 1, 2, S2, and PV; and posterior parietal cortex (see Fang et al., 2005; Stepniewska et al., 1993). Other important inputs are from the posterior ventrolateral nucleus (VLp) of the thalamus, which receives projections from the cerebellar nuclei (e.g., Stepniewska et al., 2003). M1 provides the majority of projections to brainstem and spinal cord motor circuits, and the projection to the upper spinal cord circuits that control the digits is enlarged in Old World monkeys and in the highly dexterous New World Cebus monkeys compared to most New World monkeys and prosimian primates (Nudo & Frost, 2007). Due to this more direct pathway, movements can be evoked at lower current levels from M1 than from other frontal motor areas, although stimulation thresholds from PMV can be nearly as low. Brief trains of electrical pulses in M1 have long been known to evoke simple movements, such as the extension of a digit. More recently, Graziano (this volume; Graziano et al., 2002) and coworkers have shown that larger, half-second trains of electrical pulses can evoke more complex, behaviorally relevant movements, such as hand-to-mouth movements, from M1 of macaque monkeys. In our unpublished studies led by Iwona Stepniewska, complex movements were evoked from M1 of galagos and squirrel monkeys. Complex movements have also been evoked from premotor cortex, as well as from posterior parietal cortex (see previous section), but inactivation of M1 abolishes or greatly modifies these movement patterns, suggesting that M1 is the final cortical target in the circuits for such behaviors. However, the circuits addressed by the subcortical projections of M1 must be important in organizing aspects of the movement patterns.

All primates have dorsal and ventral divisions of the classical premotor cortex. Mainly forelimb and mouth and face movements are evoked from PMV, while both forelimb and hindlimb movements are evoked from PMD. Thresholds for evoking movements in PMD are higher than those for M1, while those for PMV can be higher or similar (e.g., Preuss et al., 1996). Although both of these areas project to the spinal cord and brainstem, much of their influence on motor behavior may depend on their projections to M1. Both PMV and PMD receive inputs from frontal cortex, posterior parietal cortex, and higher-order somatosensory fields. PMD appears to be divided into two fields in New and Old World monkeys, and two or three divisions of PMV have been proposed for Old World monkeys (e.g., Matelli et al., 1985). Thus, the number of premotor areas appears to be greater in macaque monkeys, and this is likely the case for apes and humans, where even more premotor fields may exist. Broca's area in the left cerebral hemisphere of humans may be an elaboration of one or more of the PMV fields (see Preuss et al., 1996).

The supplementary motor area, SMA, has been described in prosimian galagos (Fig. 10.4), New World monkeys (Fig. 10.5), Old World monkeys, and humans (see Tanji, 1994; Wu et al., 2000, for reviews). SMA is located just dorsal to the hindlimb representation in M1, and it represents the hindlimb, forelimb, and face in a caudorostral sequence (e.g., Gould et al., 1986). SMA has been implicated in generating movement sequences and in bimanual coordination of movements (Tanji, 1994). SMA has dense projections to M1, while having some projections to the spinal cord. Inputs include those from posterior parietal cortex, pre-SMA, and cingulate motor cortex (Luppino et al., 1993). In addition, SMA receives inputs from both the basal ganglia and the cerebellum relayed through the motor thalamus (Akkal et al., 2007). The more rostrally located pre-SMA differs from SMA, being densely connected with prefrontal cortex, being involved in nonmotor tasks, and having a lack of projections to the spinal cord. Connections with SMA are not dense and direct projections to M1 are sparse or absent (Luppino et al., 1993; see Akkal et al., 2007, for review).

The cingulate motor areas are in frontal cortex of the medial wall of the cerebral hemisphere (Fig. 10.4). In macaques, three cingulate motor areas have been proposed, on the dorsal (CMAr and CMAd) and ventral (CMAv) banks of the cingulate sulcus (Picard & Strick, 1996, for review). These areas differ somewhat in connections, but collectively they include those with prefrontal cortex, M1, the spinal cord, and parietal cortex. Microstimulation of CMAd and CMAv evoke movements in patterns that suggest that these fields have somatotopic organizations. CMAv is less responsive to stimulation. CMAr is thought to not control movements directly, but signal errors, reinforcement, and conflict, and thereby be involved in supervisory control (Schall et al., 2002).

Finally, all primates appear to have a frontal eye field (FEF) and perhaps a supplementary eye field (SEF). Microstimulation of these areas produces eye movements, and both fields project to the deeper layers of the superior colliculus (Huerta & Kaas, 1990; Huerta et al., 1986). The FEF has cortical connections with SEF and the SMA, prefrontal cortex, visual areas of the temporal lobe, and posterior parietal visuomotor areas (Huerta et al., 1987). Neurons in FEF respond to visual stimuli or control eye movements in a decision-making process (Schall et al., 2002). The connections of the SEF are more extensive with cortical areas and subcortical structures related to prefrontal and skeletomotor functions (Huerta & Kaas, 1990).

CONCLUSIONS

This review outlines major features of the organizations of sensory and motor systems in primates in comparison with the likely organizations of these systems in the nonprimate ancestors of primates. The major premise of the review is that sensory and motor systems have changed in many ways with or before the emergence of early primates, and more variation subsequently occurred in the different lines of primate evolution. In general, prosimian galagos seem to have changed the least, but the sensory and motor systems of all primates are distinctly different from those of other mammals.

Alterations of sensory systems have been revealed most extensively in neocortex, which of course has expanded greatly in all primates, but especially humans. Alterations have been of two main types: those within areas common to all or most mammals, and the addition of areas that seem to be unique to primates, and even to specific lines of primate evolution. This review gives a global overview of the evolution of sensory and motor areas in primates, but much has been left out, most notably the implications of the huge increases in brain and especially neocortex size in some primates, especially apes and humans. A larger brain could mean larger sensory and motor areas, and at least the more easily identified sensory areas, such as V1, S1, and the auditory core are larger in larger brains. Yet, cortex has expanded more in the larger ape and human brains than the primary sensory areas, indicating that the larger brains also likely have more cortical areas. In addition, the functions of cortical areas relate to their sizes (Kaas, 2000), in part because cortical neurons do not vary much in size so that large areas have more neurons but less widespread intrinsic connections. Thus, large sensory areas are not well suited for global integration of sensory inputs from across the receptor sheet. As primates with larger brains and larger cortical areas evolved, the internal organizations and functions of cortical areas changed as areas increased in size so that they were less involved in global integration. Other, smaller sensory areas evolved to take over roles in global integration. Thus, the other major change that seems to have occurred in the evolution of primate taxa with larger brains is an increase in the number of areas, including the number in sensory and motor systems. This potentially increases the steps in serial processing, and it is the repetition of the local processing within columns of cells from area to area that allows complex outcomes from computationally simple steps. Adding cortical areas also increases the potential for functionally distinct parallel pathways to emerge, thereby adding functions and abilities.

In brief, primate brains differ from other brains by maintaining a similar level of neuronal density as brains increase in size. Primate brains also differ from nonprimate brains in the ways their sensory and motor systems are organized. Finally, the organizations of these systems vary within and across the major branches of primate evolution, probably much more than in any other mammalian order.

REFERENCES

Akkal, D., Dum, R. P., & Strick, P. L. (2007). Supplementary motor area and presupplementary motor area: Targets of basal ganglia and cerebellar output. *Journal of Neuroscience, 27,* 10659–10673.

Basbaum, A. I., Kaneko, A., Shepherd, G. M., & Westheimer, G. (Eds.). (2008). *The Senses: A Comprehensive Reference.* London: Elsevier.

Beck, P. D., Pospichal, M. W., & Kaas, J. H. (1996). Topography, architecture, and connections of somatosensory cortex in opossums: Evidence for five somatosensory areas. *Journal of Comparative Neurology, 366,* 109–133.

Berson, D. M. (2008). Retinal ganglion cell types and their central projections. In: R. H. Maslan & T. Albright (Eds.), *The Senses: A Comprehensive Reference* (vol 1, pp. 491–519). London: Elsevier.

Bhatnagar, K. P., & Smith, T. D. (2007). The vomeronasal organ and its evolutionary loss in catarrhine primates. In: J. H. Kaas & T. M. Preuss (Eds.), *Evolution of Nervous Systems* (vol 4, pp. 142–152). London: Elsevier.

Bombardieri, R. A., Johnson, J. I., & Campos, G. B. (1975). Species differences in mechanosensory projections from the mouth to the ventrobasal thalamus. *Journal of Comparative Neurology, 163,* 41–64.

Calton, J. L., Dickinson, A. R., & Snyder, L. H. (2002). Non-spatial, motor-specific activation in posterior parietal cortex. *Nature Neuroscience, 5,* 580–588.

Carr, C. E., & Edds-Walton, P. L. (2008). Vertebrate auditory pathways. In: P. Dallos & D. Oertel (Eds.), *The Senses: A Comprehensive Reference* (vol 3). London: Elsevier.

Casagrande, V. A., & Kaas, J. H. (1994). The afferent, intrinsic, and efferent connections of primary visual cortex in primates. In: A. Peters & K. Rockland K (Eds.), *Cerebral Cortex* (vol 10, pp. 201–259). New York: Plenum Press.

Casagrande, V. A., & Xu, X. (2004). Parallel visual pathways: A comparative perspective. In: L. M. Chalupa & J. S. Werner (Eds.). *The Visual Neurosciences* (pp. 494–506). Cambridge, MA: MIT Press.

Collins, C. E., Hendrickson, A., & Kaas, J. H. (2005a). Overview of the visual system of tarsius. *Anatomical Record, 287A,* 1013–1025.

Collins, C. E., Lyon, D. C., & Kaas, J. H. (2005b). Distribution across cortical areas of neurons projecting to the superior colliculus in New World monkeys. *Anatomical Record,* 285A, 619–627.

Cooke, D. F., Taylor, C. R., Moore, T., & Graziano, M. S. A. (2003). Complex movements evoked by microstimulation of the ventral intraparietal area. *Proceedings of the National Academy of Sciences, 100,* 6163–6168.

Crabtree, J. W., & Killackey, H. P. (1989). The topographic organization and axis of projection within the visual sector of the rabbit's thalamic reticular nucleus. *European Journal of Neuroscience, 1,* 94–109.

Craig, A. D. (2002). How do you feel? Interoception: The sense of the physiological condition of the body. *Nature Reviews Neuroscience, 3,* 655–666.

Craig, A. D. (2003). A new view of pain as a homeostatic emotion. *Trends in Neuroscience, 26,* 303–307.

Craig, A. D. (2007). Evolution of pain pathways. In: J. H. Kaas & L. A. Krubitzer (Eds.), *Evolution of Nervous Systems* (vol 3, pp. 227–235). London: Elsevier.

Cusick, C. G. (1997). The superior temporal polysensory region in monkeys. In: K. S. Rockland (Ed.), *Cerebral cortex: Extrastriate cortex in primates* (pp. 4356–4463). New York: Plenum Press.

Donoghue, J. P., Leibovic, S., & Sanes, J. N. (1992). Organization of the forelimb area in squirrel monkey motor cortex: representation of digit, wrist, and elbow muscles. *Experimental Brain Research, 89,* 1–19.

Dostrovsky, J. O., & Craig, A. D. (2008). The thalamus and nociceptive processing. In: M. C. Bushnell & A. I. Bashaum (Eds.), *The Senses: A Comprehensive Reference* (vol 5, pp. 635–668). London: Elsevier.

Fang, P-C., Stepniewska, I., & Kaas, J. H. (2005). Ipsilateral cortical connections of motor, premotor, frontal eye, and posterior parietal fields in a prosimian primate, *Otolemur garnetti. Journal of Comparative Neurology, 490,* 305–333.

Foundas, A. L., Leonard, C. M., Gilmore, R. L., Fennell, E. B., & Heilman, K. M. (1996). Pars triangularis asymmetry and language dominance. *Proceedings of the National Academy of Sciences, 93,* 719–722.

Fu, K. M., Johnston, T. A., Shah, A. S., Arnold, L. L., Smiley, J. F., Hackett, T. A., et al. (2003). Auditory cortical neurons respond to somatosensory stimulation. *Journal of Neuroscience, 23,* 7510–7515.

Galaburda, A. M., LeMay, M., Kemper, T. L., & Geschwind, N. (1978). Right-left asymmetries in the brain. *Science, 199,* 852–856.

Ghazanfar, A. A., Maier, J. X., Hoffman, K. L., & Logothetis, N. K. (2005). Multisensory integration of dynamic faces and voices in rhesus monkey auditory cortex. *Journal of Neuroscience, 25,* 5004–5012.

Gould, H. J., Cusick, C. G., Pons, T. P., & Kaas, J. H. (1986). The relationship of corpus callosum connections to electrical stimulation maps of motor, supplementary motor, and the frontal eye fields in owl monkeys. *Journal of Comparative Neurology, 247,* 297–325.

Graf, W. M. (2007). Vestibular system. In: J. H. Kaas & L. A. Krubitzer (Eds.), *Evolution of Nervous Systems* (vol 3, pp. 341–371). London: Elsevier.

Graziano, M. S. A. (2006). The organization of behavioral repertoire in motor cortex. *Annual Review of Neuroscience, 29,* 105–134.

Graziano, M. S. A., Taylor, C. S. R., & Moore, T. (2002). Complex movements evoked by microstimulation of precentral cortex. *Neuron, 34,* 841–851.

Guldin, W. O., & Grüsser, O. J. (1998). Is there a vestibular cortex? *Trends in Neuroscience, 21,* 254–259.

Hackett, T. A., Preuss, T. M., & Kaas, J. H. (2001). Architectonic identification of the core region in auditory cortex of macaques, chimpanzees, and humans. *Journal of Comparative Neurology, 441,* 197–222.

Hackett, T. A., Stepniewska, I., & Kaas, J. H. (1998). Subdivisions of auditory cortex and ipsilateral cortical connections of the parabelt auditory cortex in macaque monkeys. *Journal of Comparative Neurology, 394,* 475–495.

Herculano-Houzel, S., Collins, C. E., Wong, P., & Kaas, J. H. (2007). Cellular scaling rules for primate brains. *Proceedings of the National Academy of Sciences, 104,* 3562–3567.

Huang, C. S., Sirisko, M. A., Hiraba, H., & Murray, G. M. (1988). Organization of the primate face motor cortex as served by intracortical microstimulation and electrophysiological identification of afferent inputs and corticobulbar or projections. *Journal of Neurophysiology, 59,* 796–818.

Huerta, M. F., & Kaas, J. H. (1990). Supplementary eye field as defined by intracortical microstimulation: Connections in macaques. *Journal of Comparative Neurology, 293,* 299–330.

Huerta, M. F., Krubitzer, L. A., & Kaas, J. H. (1986). Frontal eye field as defined by intracortical microstimulation in squirrel monkeys, owl monkeys, and macaque monkeys: I. Subcortical connections. *Journal of Comparative Neurology, 253,* 415–439.

Huerta, M. F., Krubitzer, L. A., & Kaas, J. H. (1987). Frontal eye field as defined by intracortical microstimulation in squirrel monkeys, owl monkeys, and macaque monkeys. II. Cortical connections. *Journal of Comparative Neurology, 265,* 332–361.

Iyengar, S., Qi, H-X., Jain, N., & Kaas, J. H. (2007). Cortical and thalamic connections of the representations of the teeth and tongue in somatosensory cortex of New World monkeys. *Journal of Comparative Neurology, 501,* 95–120.

Kaas, J. H. (2000). Why is brain size so important: Design problems and solutions as neocortex gets bigger or smaller. *Brain and Mind, 1,* 7–23.

Kaas, J. H. (2003). The evolution of the visual system in primates. In: L. M. Chalupa & J. S. Werner (Eds.), *The Visual Neurosciences* (pp. 1563–1572). Cambridge, MA: MIT Press.

Kaas, J. H. (2004). Somatosensory system. In: G. Paxinos & J. K. Mai (Eds.), *The Human Nervous System* (2nd ed., pp. 1059–1092). London: Elsevier.

Kaas, J. H. (2007a). Reconstructing the organization of the forebrain of the first mammals. In: J. H. Kaas & L. A. Krubitzer (Eds.), *Evolution of Nervous Systems* (vol 3, pp. 27–48). London: Elsevier.

Kaas, J. H. (2007b). The evolution of sensory and motor systems in primates. In: J. H. Kaas & T. M. Preuss (Eds.), *Evolution of Nervous Systems* (vol 4, pp. 35–57). London: Elsevier.

Kaas, J. H., & Collins, C. E. (2004). *The Primate Visual System.* Boca Raton: CRC Press.

Kaas, J. H., Guillery, R. W., & Allman, J. M. (1972). Some principles of organization in the dorsal lateral geniculate nucleus. *Brain Behavior and Evolution, 6,* 253–299.

Kaas, J. H., & Hackett, T. A. (2008). The functional neuroanatomy of the auditory cortex. In: P. Dallos & D. Oertel (Eds)., *The Senses: A Comprehensive Reference* (vol 3, pp. 765–780). London: Elsevier.

Kaas, J. H., & Huerta, M. F. (1988). The subcortical visual system of primates. In: H. P. Steklis (Ed.), *Comparative Primate Biology* (vol 4, pp. 327–391). New York: Alan R. Liss.

Kaas, J. H., Huerta, M. F., Weber, J. T., & Harting, J. K. (1978). Patterns of retinal terminations and laminar organization of the lateral geniculate nucleus of primates. *Journal of Comparative Neurology, 182,* 517–554.

Kaas, J. H., & Lyon, D. C. (2007). Pulvinar contributions to the dorsal and ventral streams of visual processing in primates. *Brain Research Reviews, 55,* 285–296.

Kaas, J. H., & Morel, A. (1993). Connections of visual areas of the upper temporal lobe of owl monkeys: the MT crescent and dorsal and ventral subdivisions of FST. *Journal of Neuroscience, 13,* 534–546.

Kaas, J. H., & Preuss, T. M. (2008). Human brain evolution. In: L. R. Squire (Ed.), *Fundamental Neuroscience* (3rd ed., pp. 1027–1035). San Diego: Elsevier.

Kaas, J.H. The somatosensory thalamus and associated pathways. In: E. Gardner & J. H. Kaas (Eds.), *The Senses: A Comprehensive Reference,* (vol 6, 117–141) London: Elsevier.

Kaas, J. H., Qi, H-X., & Iyengar, S. (2006). Cortical network for representing the teeth and tongue in primates. *Anatomical Record, 288A,* 182–190.

Kosar, E., Grill, H. J., & Norgren, R. (1986). Gustatory cortex in the rat. I Physiological properties and cytoarchitecture. *Brain Research, 379,* 329–341.

Kremers, J. (2005). The Structure, Function and Evolution of the Primate Visual System. Chichester, UK: John Wiley and Sons.

Krubitzer, L.A., Künzle, H., and Kaas, J. (1997). The organization of sensory cortex in a Madagascan insectivore, the tenrec (*Echinaps telfairi*). *Journal of Comparative Neurology, 379,* 399–414.

Kurylo, D. D., & Skavenski, A. A. (1991). Eye movements elicited by electrical stimulation of area PG in the monkey. *Journal of Neurophysiology, 65,* 1243–1253.

Lewis, J. W., & Van Essen, D. C. (2000). Mapping of architectonic subdivisions in the macaque monkey, with emphasis on parieto-occipital cortex. *Journal of Comparative Neurology, 428,* 79–111.

Lima, D. (2008). Ascending pathways: Anatomy and physiology. In: A. I. Basbaum & M. C. Bushnell (Eds.), *The Senses: A Comprehensive Reference* (vol 5, pp. 477–526). London: Elsevier.

Luppino, G., Matelli, M., Camarda, R., & Rizzolatti, G. (1993). Corticocortical connections of area F3 (SMA-proper) and area F6 (pre-SMA) in the macaque monkey. *Journal of Comparative Neurology, 338,* 114–140.

Lyon, D. C. (2007). The evolution of visual cortex and visual systems. In: J. H. Kaas & L. A. Krubitzer (Eds.), *Evolution of Nervous Systems* (vol 3, pp. 267–306). London: Elsevier.

Matelli, M., Luppino, G., & Rizzolatti, G. (1985). Patterns of cytochrome oxidase activity in the frontal agranular cortex of macaque monkey. *Behavioral Brain Research, 18,* 125–137.

Murray, E. A., & Mishkin, M. (1984). Relative contributions of S-II and area 5 to tactile discriminations in monkeys. *Behavioral Brain Research, 11,* 67–83.

Norgren, R., & Wolf, G. (1975). Projections of thalamic gustatory and lingual areas in the rat. *Brain Research, 92,* 123–129.

Nudo, R. J., & Frost, S. B. (2007). The evolution of motor cortex and motor systems. In: J. H. Kaas & L. A. Krubitzer (Eds.), *Evolution of Nervous Systems* (vol 3, pp. 373–395). London: Elsevier.

Nudo, R. J., & Masterton, R. B. (1988). Descending pathways to the spinal cord: A comparative study of 22 mammals. *Journal of Comparative Neurology, 277,* 53–79.

Penny, G. R., Itoh, K., & Diamond, I. T. (1982). Cells of different sizes in the ventral nuclei project to different layers of the somatic cortex in the cat. *Brain Research, 242,* 55–65.

Picard, N. & Strick, P.L. (1996). Motor areas of the medial wall: a review of their location and functional activation. *Cerebral Cortex, 6,* 342–353.

Preuss, T. M., Stepniewska, I., & Kaas, J. H. (1996). Movement representation in the dorsal and ventral premotor areas of owl monkeys: A microstimulation study. *Journal of Comparative Neurology, 371,* 649–676.

Pritchard, T. C., & Norgren, R. (2004). Gustatory system. In: G. Paxinos & J. K. Mai (Eds.), *The Human Nervous System* (pp. 1171–1198). Amsterdam: Elsevier Press.

Qi, H-X., Preuss, T. M., & Kaas, J. H. (2008). Somatosensory areas of the cerebral cortex: Architectonic characteristics and modular organization. In:E. Gardner & J. H. Kaas (Eds.) *The Senses: A Comprehensive Reference* (vol 6, pp. 143–169). London: Elsevier.

Rauschecker, J. P., & Tian, B. (2000). Mechanisms and streams for processing of "what" and "where" in auditory cortex. *Proceedings of the National Academy of Sciences, 97,* 11800–11806.

Rauschecker, J. P., Tian, B., & Hauser, M. (1995). Processing of complex sounds in the macaque nonprimary auditory cortex. *Science, 268,* 111–114.

Remple, M. S., Henry, E. C., & Catania, K. C. (2003). Organization of somatosensory cortex in the laboratory rat (Rattus norvegicus): Evidence for two lateral areas joined at the representation of the teeth. *Journal of Comparative Neurology, 467,* 105–118.

Remple, M. S., Reed, J. L., Stepniewska, I., & Kaas, J. H. (2006). The organization of frontoparietal cortex in the tree shrew (Tupaia belangeri): I. Architecture, microelectrode maps and corticospinal connections. *Journal of Comparative Neurology, 497,* 1–154.

Remple, M. S., Reed, J. L., Stepniewska, I., Lyon, D. C., & Kaas, J. H. (2007). The organization of frontoparietal cortex in the tree shrew (Tupaia belangeri): II Connectional evidence for a frontal-posterior parietal network. *Journal of Comparative Neurology, 501,* 121–149.

Roe, A. W. (2003). Modular complexity of area V2 in the macaque monkey. In: C. Collins & J. H. Kaas (Eds.), *The Primate Visual System* (pp. 109–138). New York: CRC Press.

Rolls, E. T. (2004). Convergence of sensory systems in the orbitofrontal cortex in primates and brain design for emotion. *Anatomical Record, 281A,* 1212–1225.

Romanski, L. M., & Goldman-Rakic, P. S. (2002). An auditory domain in primate prefrontal cortex. *Nature Neuroscience, 5,* 15–16.

Ross, C. F., & Martin, R. D. (2007). The role of vision in the origin and evolution of primates. In: J. H. Kaas & T. M. Preuss (Eds.), *Evolution of Nervous Systems* (vol 4, pp. 59–78). London: Elsevier.

Rouiller, E. M., Moret, V., & Liang, F. (1993). Comparison of the connectional properties of the two forelimb areas of the rat sensorimotor cortex: Support for the presence of a premotor or supplementary motor cortical area. *Somatosensory & Motor Research, 10,* 269–289.

Schall, J. D., Stuphorn, V., & Brown, J. W. (2002). Monitoring and control of action by the frontal lobes. *Neuron, 36,* 309–322.

Sherman, S. M., & Guillery, R. W. (2005). *Exploring the Thalamus and its Role in Cortical Function.* Cambridge, MA: MIT Press.

Snyder, L. H., Batista, A. P., & Andersen, R. A. (2000). Intention-related activity in the posterior parietal cortex: a review. *Vision Research, 40,* 1433–1441.

Stepniewska, I., Fang, P. C., & Kaas, J. H. (2005). Microstimulation reveals specialized subregions for different complex movements in posterior parietal cortex of prosimian galagos. *Proceedings of the National Academy of Sciences, 102,* 4878–4883.

Stepniewska, I., Preuss, T. M., & Kaas, J. H. (1993). Architectonics, somatotopic organization, and ipsilateral cortical connections of the primary motor area (M1) of owl monkeys. *Journal of Comparative Neurology, 330,* 238–271.

Stepniewska, I., Sakai, S. T., Qi, H-X., & Kaas, J. H. (2003). Somatosensory input to the ventrolateral thalamic region (VL) in the macaque monkey: A potential substrate for Parkinsonian tremor. *Journal of Comparative Neurology, 455,* 378–395.

Swisher, J. D., Halko, M. A., Merabet, L. B., McMains, S. A., & Somers, D. C. (2007). Visual topography of human intraparietal sulcus. *Journal of Neuroscience, 27,* 5326–5337.

Tanji, J. (1994). The supplementary motor area in the cerebral cortex. *Neuroscience Research, 19,* 251–268.

Thier, P., & Andersen, R. A. (1998). Electrical microstimulation distinguishes distinct saccade-related areas in the posterior parietal cortex. *Journal of Neurophysiology, 80,* 1713–1735.

Ungerleider, L. G., & Mishkin, M. (1982). Two cortical visual systems. In: D. G. Ingle, M. A. Goodale, & R. J. Q. Mansfield (Eds.), *Analysis of Visual Behavior* (pp. 549–586). Cambridge, MA: MIT Press.

Van Hooser, S., Heimel, J., Chung, S., Nelson, S., & Toth, L. (2005). Orientation selectivity without orientation maps in visual cortex of a highly visual mammal. *Journal of Neuroscience, 25,* 19–28.

Wang, Q., & Burkhalter, A. (2007). Area map of mouse visual cortex. *Journal of Comparative Neurology, 502,* 339–357.

Weller, R. E., & Kaas, J. H. (1989). Parameters affecting the loss of ganglion cells of the retina following ablations of striate cortex in primates. *Visual Neuroscience, 3,* 327–349.

Whishaw, I. Q. (2003). Did a change in sensory control of skilled movements stimulate the evolution of the primate frontal cortex? *Behavioral Brain Research, 146,* 31–41.

Wilson, D. A. (2008). Olfactory cortex. In: S. Firestein & G. K. Beauchamp (Eds.), *The Senses: A Comprehensive Reference* (vol 4, pp. 687–706). London: Elsevier.

Wu, CW-H., Bichot, N. P., & Kaas, J. H. (2000). Converging evidence from microstimulation, architecture, and connections for multiple motor areas in the frontal and cingulate cortex of prosimian primates. *Journal of Comparative Neurology, 423,* 140–177.

Xu, X., Collins, C. E., Kaskan, P. M., Khaytin, I., Kaas, J. H., & Casagrande, V. A. (2004). Optical imaging of visually evoked responses in prosimian primates reveals conserved features of the middle temporal visual area. *Proceedings of the National Academy of Sciences, 101,* 2566–2571.

CHAPTER 11

Vision: A Neuroethological Perspective

Benjamin Y. Hayden

INTRODUCTION

Primates are fundamentally visual creatures (Le Gros Clark, 1959). It is estimated that up to half of the surface area of the macaque cerebral cortex, and one third of the human cerebral cortex, is specialized for visual processing (Drury et al., 1996; Felleman & Van Essen, 1991). Consistent with this elaboration of neural hardware, primates have greater visual acuity than almost all other mammals and birds (Kirk & Kay, 2004). Primates can classify images faster and more accurately than the best computers. Given the close parallels in primate brain evolution and specialization of the primate visual system, it has been argued that increasing demands for visual processing were the dominant force driving the evolution of the primate brain (Barton, 1998). Although the visual system is often seen as a marvel of elegant engineering (Purves et al., 2008; Wandell, 1995), like any other evolved trait it reflects the outcome of a long series of compromises and competing demands. This chapter focuses on how the visual system works, with an eye on how evolutionary demands shaped visual system function.

EVOLUTIONARY INFLUENCES ON OUR VISUAL SYSTEM

The remarkable visual acuity found among primates depends on several anatomical adaptations. Primates have unusually large eyes and pupils (Kirk & Kay, 2004) that point in the same direction (orbital convergence). Orbital convergence effectively doubles the number of photoreceptors devoted to a given location. In contrast to many other mammals, most primates have no tapetum lucidum. This reflective sheet behind the retina gives photoreceptors a second chance to catch any photons that pass into the eyes, but reduces acuity by scattering them. The tapetum is the reason feline eyes glow but human eyes do not. The absence of the tapetum is likely an adaptation to demands for greater acuity coupled with a diurnal lifestyle (Kirk & Kay, 2004). The high visual acuity found in primates is also facilitated by the presence of a retinal fovea, a centrally located specialization of the retina made up entirely of cone photoreceptors and possessing especially high visual acuity. As a consequence, most primates move their eyes several times a second to bring the fovea into register with an area of interest in the visible scene. The speed and efficiency with which we can shift gaze contributes to our acuity.

Orbital Convergence and Stereoscopic Vision

The eyes of primates exhibit an extreme degree of orbital convergence, causing the visual fields of the two eyes to overlap substantially (Barton, 2004; Le Gros Clark, 1959). Convergent vision improves discrimination at the cost of reduced visual field size (Barton, 2004). The reduction in the size of visual field may reflect reduced demands for avoiding predators (relative to ungulates, for example) and increased demands for locating animal prey. Convergent vision has other costs as well: Information from the two eyes must be combined, thus increasing computational demands on the cortical visual system, possibly demanding increased brain size (Barton, 2004).

In addition to enhanced visual acuity, orbital convergence permits more accurate calculations of object distance. This ability, known as stereoscopic (three-dimensional) vision, is beneficial for arboreal primates, especially during locomotion and hunting in the terminal canopy (Cartmill, 1970, 1974; Jones, 1916; Le Gros Clark, 1970; Smith, 1912).

Finally, orbital convergence facilitates night vision by effectively doubling the number of photoreceptors looking at a given segment of the world, suggesting that orbital convergence may be an adaptation to scotopic (low-light) visual conditions confronted by ancestral nocturnal primates (Ross et al., 2005). It is argued that, in general, visual acuity can be improved by either restricting pupil diameter or aligning the eyes. Reducing pupil diameter is especially detrimental to nocturnal animals because it reduces the amount of light entering the eyes. Therefore, the fact that primates have convergent orbits but large pupils suggests that the ancestral primate was nocturnal, and that orbital convergence represents an adaptation to nocturnality (Allmann, 1977; Pettigrew, 1978).

Trichromatic Vision

The human visual system, like that of most other Old World haplorhine primates, contains three types of color-detecting photoreceptor cells, known as cones, that selectively detect red, green, and blue wavelengths of light (approximately 430 nm, 530 nm, and 560 nm). Across the retina, cones compete for space with rods, which are color-insensitive, but which provide greater sensitivity to light and facilitate motion detection. Cones are especially dense at the fovea, the portion of the retina with the greatest visual acuity. It is the contributions of these three types of photoreceptor cells, and their interactions, that provide us with our rich color vision.

Apes, including humans, and Old World monkeys are predominantly trichromats (Surridge et al., 2003). New World monkeys exhibit great heterogeneity in color vision, both between and within species. Many are dichromats, most of which typically cannot discriminate red from green (just like 8% of human males) (Jacobs, 1993, 1995). Most other mammals have dichromatic color vision (Surridge et al., 2003). Despite the obvious benefits of color vision, the simple physical presence of extra receptors reduces the density of rods, thus reducing sensitivity to light and motion. Physical space on the surface of the retina is limited, so any increase in the number of cones demands a reduction in the number of rods. Therefore, our superior color vision comes at the expense of a reduction in visual acuity (Osorio & Vorobyev, 1996, 2008).

Trichromatic vision permits enhanced discrimination of colors and thus more efficient search for fruit and leaves, which could benefit both folivorous and frugivorous primates (Smith et al., 2003; Sumner & Mollon, 2000). The relative preponderance of folivory (over frugivory) among trichromatic Old World monkeys and New World howler monkeys suggests that the demand for leaves, not fruit, has been the dominant factor driving color vision (Surridge et al., 2003). The relative importance of color in sexual signals in Old World monkeys (e.g., mandrills) is likely to be a by-product, not a cause of trichromacy (Dixson, 2000). Alternatively, the close match between the color coding used by our cones and the colors of primate faces may reflect the outcome of demands for rapid assessment of the emotional state of conspecifics (Changizi et al., 2006). Finally, at night, when colors are less vibrant and the reduced light increases the demand for acuity, the benefits of color vision are reduced, and the costs increased. Thus, nocturnal owl monkeys (the New World Aotus monkey) lack color vision.

Why are cones, the source of rich color vision, so dense at the fovea? A hint comes from a comparison of the retinas of primates with those of carnivores such as wolves (Mech & Boitani, 2003), large cats, and ferrets

(Calderone & Jacobs, 2003). These animals have similar retinal structure to primates, with foveas and photoreceptor density gradients. However, cats and dogs have a much greater density of rods across the entire retina (about three to four times) and have rods in the fovea, whereas primates have only cones in the fovea. Moreover, for these predators, the density of cones in the fovea is between 10% and 20% that of primates (Mech & Boitani, 2003). Given that rods have better temporal resolution than cones, carnivores have a corresponding greater temporal acuity than primates. Thus, it appears that the fact that the primary foods of anthropoid primates, namely fruits and leaves, are immobile allows the luxury of superior color vision.

ORGANIZATION OF THE VISUAL SYSTEM

The human and rhesus macaque visual systems are remarkably similar. It is because of these similarities that the rhesus monkey is one of the most popular organism for understanding human vision. Indeed, much more is known about the visual system of the rhesus monkey than about the visual system of the human. Even less is known about the visual systems of other primates. Consequently, the present discussion will be biased toward information about visual processing in the rhesus monkey.

All visual information enters through the eyes and is transduced into electrical signals by photoreceptors in a six-layer sheet of neurons called the retina. Visual information from across the retina converges in the optic nerve, and travels from there to the rest of the brain. About 90% of optic nerve neurons project via the lateral geniculate nucleus (LGN) of the thalamus to the visual cortex. (Most of the remaining neurons project to the superior colliculus, a midbrain structure involved in simple visual processing and orienting movements of the eyes and head, and to the suprachiasmatic nucleus of the hypothalamus, which regulates circadian rhythms, and the pretectum, which regulates pupil size; Purves et al., 2008).

Within the cerebral cortex, visual information is processed in a variety of specialized visual cortical areas (Felleman & Van Essen, 1991; Wandell, 1995). Visual areas are defined by a combination of anatomical features and functional features. Functionally, they are discrete cortical regions containing retinotopic maps of all visual space. Single neurons in these areas respond to visual patterns in restricted regions of the visual field; the portion of the visual field in which visual patterns can activate a neuron is known as the neuron's receptive field. Adjacent neurons in each area tend to have adjacent receptive fields, and these collectively form a full representation of the visual field. Neurons in higher-level visual areas do not necessarily have receptive fields, and so are demarcated through architectonic differences, neural or hemodynamic responses to different types of stimuli, or even studies of homologous visual areas in other animals (Felleman & Van Essen, 1991; Purves et al., 2008; Wandell, 1995).

The first stage of cortical visual processing occurs in the primary visual cortex, V1. V1 projects directly to visual areas V2, V3, V4, MT, and other areas, and these then project to the rest of the brain. The rhesus monkey brain contains around 32 visual areas (Felleman & Van Essen, 1991), although few (if any) of these solely respond to visual stimulation. The functional neuroanatomy of the human visual system is not nearly as well understood, although many structures appear to be homologous, especially in the early visual areas (Van Essen, 2004).

Two Visual Processing Streams

The most well-known organizing principle for the visual system is called the "two-streams model" (Milner & Goodale, 1995; Ungerleider & Mishkin, 1982). According to this model, visual information travels along two divergent, hierarchically organized, processing pathways. These pathways are known as the dorsal ("where" or "how") stream and the ventral ("what") stream. The dorsal stream is concerned with identifying the location of visual stimuli and their direction of movement; the ventral stream identifying object form, shape, and identity. The dorsal stream travels along the parietal lobe, from V1 to the thick stripes of V2, to MT (possibly through V3), to the lateral intraparietal cortex (LIP), and then to the frontal eye fields (FEF). The ventral stream travels along the temporal lobe, from V1 to the thin stripes and interstripes of V2 to V4, to the posterior inferotemporal cortex (PIT), and then to anterior inferotemporal cortex (AIT), and, by some definitions, to the perirhinal cortex (Merigan & Maunsell, 1993; Murray & Bussey, 1999; Murray et al., 2007).

Although the two-streams model is still the dominant framework for thinking about the organization of the visual system, its status remains controversial (Felleman & Van Essen, 1991; Hegde & Felleman, 2007; Hilgetag et al., 1996; Lennie, 1998; Merigan & Maunsell, 1993; Schiller, 1993; Ungerleider & Haxby, 1994). Lesion studies often affect adjacent structures and fibers passing through lesioned areas, complicating interpretations of their results (Murray & Mishkin, 1998). Moreover, many studies that have directly compared dorsal and ventral stream lesions have reported effects inconsistent with the idea of distinct processing pathways (Merigan & Maunsell, 1993). More generally, a complete map of the interconnections between the visual areas does not reveal two clear pathways (Felleman & Van Essen, 1991). Of the approximately 30 visual areas, there are at least 300 (out of 900 possible) connections, supporting the idea that the visual system is, at best, a highly tangled hierarchy. Moreover, visual areas tend to get smaller as the hierarchies progress, and the strength of connections between them gets weaker, challenging the idea that these streams serve as major conduits of all visual information (Lennie, 1998).

In addition, the latencies of visual responses in sequential visual areas do not provide much evidence for a hierarchical sequence of processing, especially in the dorsal stream (Schmolesky et al., 1998; Vanni et al., 2004). Moreover, neurons in dorsal stream areas have been shown to be selective to form and shape (Janssen et al., 2008; Lehky & Sereno, 2007; Peng et al., 2008; Sereno & Maunsell, 1998), while neurons in the ventral stream are sensitive to visual motion (Schiller, 1993; Tolias et al., 2005). Finally, the extent to which areas within streams form hierarchies remains murky. For example, it is generally thought that there is a clear gradient of complexity in the form of information represented by neurons in the ventral stream areas.

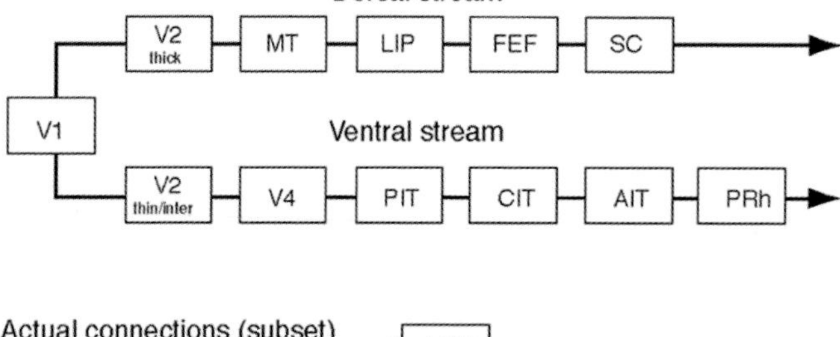

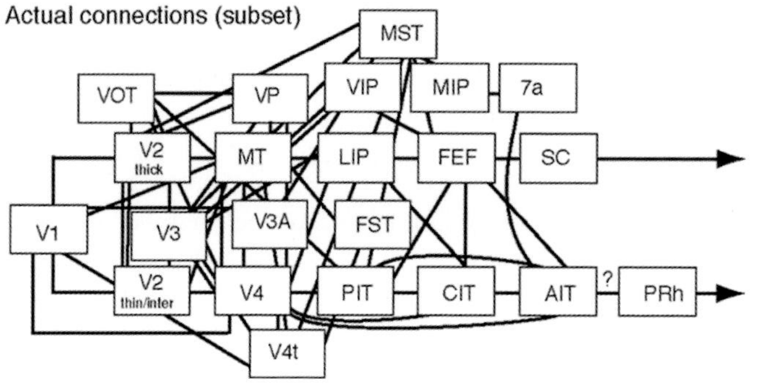

Figure 11.1 Organization of the visual system. The visual system is often thought of as an ordered set of areas arranged in a simple pair of functionally distinct hierarchies. Although this idea has some support, anatomical considerations hint only weakly at the idea of two streams.

However, most studies do not directly compare responses of single neurons in different areas using the same experimental paradigm. When they do so, it becomes difficult to qualitatively distinguish neuronal selectivity in different areas in the hierarchy (e.g., Chafee & Goldman-Rakic, 1998; Hegde & Van Essen, 2007).

Collectively, these findings suggest that the two-streams hypothesis is at best oversimplified, and at worst misleading (Fig. 11.1). Although there is no single dominant alternative hypothesis, future directions may emphasize the importance of recurrent processing (Hegde & Felleman, 2007), predictive encoding (Rao & Ballard, 1999), and distributed decision making (Lennie, 1998). Despite the lack of a clear explanatory framework, a great deal of information is known about the response properties of individual neurons in each of the visual areas.

Processing in Early Visual Areas

The visual system begins at the photoreceptors, with something akin to a pixel representation of the visual field, and ends with a categorical representation of different types of images (i.e., dog vs. cat). How does this transformation happen? We lack the detailed knowledge of the connectional and functional properties of visual cortex neurons needed to make strong theories about how this process happens. Moreover, we cannot get computers to perform most higher-level visual processes, such as image classification and segmentation, so we do not have a good computational model to show how it *could* be done. This section will discuss some of the large amount of information we do know about the problem.

In general, it is assumed that the goal of the visual system is to represent relevant features of the visual world with the greatest efficiency (Atick, 1992; Attneave, 1954; Barlow, 1961; Laughlin et al., 1998). In other words, the structure of the visual system reflects the outcome of evolutionary processes that emphasize the need for high-quality extraction of visual information with minimal wasted energy. The costs associated with vision probably involve two factors: the metabolic demands of spiking (which take up 20% of energy) and the metabolic and physical demands of a larger brain (Aiello & Wheller, 1995; Laughlin et al., 1998). From the perspective of information theory (Shannon & Weaver, 1963), transmission efficiency can be increased by reducing redundancy in visual representations (Barlow, 1961; Field, 1987; Rieke et al., 1997). Suppose, for example, that a large red square appears in the visual field. The visual system can represent each point in the square, or it can represent the edges only. Because the information inside the square is highly redundant, there is an opportunity for an efficient system to reduce metabolic costs by not representing it. Many aspects of visual representation can be thought of as serving this purpose.

Note that efficiency of neural coding is quite different from efficiency in image encoding as this term is used in computer science. Traditionally, a computer encoding algorithm, such as Huffman encoding, must not lose any information when it compresses data (i.e., it is lossless), whereas neural codes lose unimportant information (i.e., they are lossy; Simoncelli & Olshausen, 2001). A computer algorithm must be invertible, while the neural code does not have to be (Simoncelli & Olshausen, 2001). A computer algorithm is generally assumed to be processed in a relatively noiseless environment, while cortical processing is very noisy (Rieke et al., 1997). Finally, the efficiency of a neural code is determined, in part, by the tradeoff between the costs of creating more neurons and producing more spikes, an empirically defined set of parameters whose analogs are typically ignored in computer codes (Olshausen & Field, 1997).

The Retina and Lateral Geniculate Nucleus

The retina is a flexible sheet of neurons, about 2 cm in diameter, that is attached to the back of the orbits of the eyes. At its center is the fovea, a highly sensitive patch of photoreceptors with especially high color sensitivity. Photoreceptor cells in the retina consist of rods and cones. The retina is generally thought to serve as a simple photoreceptor array. However, many of its operations, including the decrease in responses when a stimulus appears close to a neuron's receptive field (center-surround inhibition) and the transformation of signals from three color inputs to opponent colors, reduce redundancy (Buschbaum & Gottschalk, 1983). Indeed, many of the computational processes of the retina can be described in terms of redundancy reduction, given the constraints of the visual world (Atick & Redlich, 1991, 1992). Underscoring the critical importance of redundancy reduction in the retina, consider that there are about 100 million photoreceptors and only 1 million optic nerve fibers (Thorpe et al., 2001).

The next step in visual processing is the LGN of the thalamus. Commonly thought of as nothing more than a relay linking the primary visual cortex to the retina, the LGN instead actually appears to perform some sophisticated computational processing. Like the retina, the LGN recodes visual information into a less redundant, more efficient form (Dan et al., 1996). The LGN therefore takes advantage of redundancies in the temporal (Dong & Atick, 1995) and spatial structure of natural images and eliminates these. Such redundancy reduction is known as image whitening. In fact, similar principles are used in standard computer image compression algorithms, such as jpeg and mpeg.

Area V1

V1, the first stop for all visual information entering the cortex, occupies 13% of the surface area of the macaque cerebral cortex, making it the largest single cortical brain region (Van Essen, 2004). Small lesions to V1 lead to scotomas (i.e., a small blind spot), while larger lesions can cause near-complete to total blindness (Sprague et al., 1977). Interestingly, the number of neurons in V1 is about 100 times larger than the number in the LGN (Wandell, 1995). This fact suggests that V1 "unpacks" visual information that has been compressed, making it more accessible for subsequent processing. This unpacking process makes the visual code more sparse—which is a more metabolically efficient code (Olshausen & Field, 1997; Zhao, 2004).

Most V1 neurons respond only to visual stimuli located within a small region of the visual field, the receptive field. They typically respond most strongly to lines oriented at a particular angle, known as the neuron's peak orientation tuning. According to legend, Hubel and Wiesel (who won the Nobel prize in Physiology and Medicine in 1978) were having no luck using small dots painted onto microscope slides to excite V1 neurons when one of them placed a cracked slide into the slide projector and as it fell into place, they heard a loud burst of neuronal activity. Indeed, action potentials in V1 neurons are elicited quite effectively by oriented lines placed within their receptive fields. Their responses are somewhat suppressed by similar stimuli just outside of their receptive field, a phenomenon known as surround suppression (Hubel & Wiesel, 1968).

Response properties such as orientation tuning motivated the idea that visual neurons (and sensory neurons more generally) are "feature detectors." Indeed, it is generally believed that V1 neurons respond most strongly to the appearance of a bar in the receptive field of a specified orientation and color (Hubel & Wiesel, 1959, 1962; Neisser, 1967). It is more accurate, however, to say that they represent not image features, but the local Fourier energy within a restricted orientation and spatial frequency domain (Albrecht et al., 1980, 1982; De Valois et al., 1982). A V1 neuron can thus be described as a band-pass filter (Campbell & Robson, 1968; Campbell et al., 1969; Enroth-Cugell & Robson, 1966), and V1 as a whole performs something akin to a localized Fourier energy analysis (i.e., a wavelet decomposition) of the retinal image (Olshausen & Field, 1996, 1997). The population of V1 neurons therefore contains a complete representation of the local spectrotemporal energy patterns of the visual field.

Although it is the best-studied visual area, V1 is far from completely understood. Indeed, one ambitious paper argues, with admirable precision, that we understand about 15% of what there is to know about V1 (Olshausen & Field, 2005). For example, typical studies sample neurons with a bias toward neurons with large somas and predicted responses (Olshausen & Field, 2005). Moreover, the results of our single-unit studies and functional magnetic resonance imaging (fMRI) studies, our two best methods, do not always agree (Maier et al., 2008). Finally, some recent studies indicate that many V1 neurons are tuned for non-Cartesian images (spirals, concentric circles, etc.) (Mahon & De Valois, 2001) and hermite functions (Victor et al., 2006), challenging the simple spectrotemporal energy models of V1 function (Mahon & De Valois, 2001).

Area V2

V2 is the largest visual area after V1, and is nearly as large as V1 itself. It occupies about 10% of the surface of the cerebral cortex (Van Essen, 2004). Receptive fields of V2 neurons are larger than those of V1 neurons at a given eccentricity, and eccentricities are greater, on average, than those in V1. The average latency of spiking responses to visual information is slightly longer (Schmolesky et al., 1998). Beyond this, however, response properties of V2 neurons are remarkably similar to those of V1.

Early studies reported that V2 neurons are primarily distinguished from V1 neurons by the fact that they respond to illusory contours, such as the imaginary line formed by two vertices of a Kanisza Triangle (Peterhans & von der Heydt, 1991; von der Heydt et al., 1984). Such coding could reflect a possible elaboration of the more veridical encoding patterns observed in V1. However, subsequent work has shown that illusory contour coding is also found in V1 (Grosof et al., 1993; Mahon & De Valois, 2001; Ramsden et al., 2001; Sheth et al., 1996), demonstrating that this property does not emerge in V2. It may instead emerge through local circuit activity within V1.

It is also thought that shape tuning in V2, which includes small contours, angles, and non-Cartesian stimuli, is more complex than that observed in V1 (Hegde & Van Essen, 2000). However, direct comparisons are rarely if ever made, and when they are, the differences between V2 and V1 are weak and unsystematic (Hegde & Van Essen, 2007). Other functional properties observed in V2 include relative (as opposed to absolute) retinal disparity (Thomas et al., 2002); stereoscopic edges, which can lead to depth information (von der Heydt et al., 2000); three-dimensional surface configurations (Bakin et al., 2000); and second-order edges (see Figure 1, Marcus & Van Essen, 2002). The majority of these properties have not been properly tested in V1, however, so it remains unclear whether these properties emerge within V2.

It is quite difficult to show that a given response property emerges in a particular visual area, such as V2. It must first be shown that the response property does not appear in V1 or even LGN. Second, it must be shown that that the property is not initially generated in a higher visual area, and then transmitted, via feedback, back to V2. Such feedback connections are quite fast, so it is nearly impossible to use timing information to identify the source of a brain signal. These possibilities are only compounded by the diffuse and tangled nature of the interconnections within the visual system—V2 receives direct projections from at least nine different visual areas, which could potentially support the formation of any representation pattern (Felleman & Van Essen, 1991).

Object Identification and Face Processing

Most primates are highly social, and most behave as if they identify others based on their faces and bodies (Cheney & Seyfarth, 1992). Given the importance of faces to social behavior, it has been hypothesized that specialized regions of the primate brain mediate face processing. Single neurons in the monkey brain respond selectively when a human or monkey face is presented (Desimone, 1991; Hasselmo et al., 1989; Perrett et al., 1982; Tsao et al., 2006). Neurons in the superior temporal sulcus (STS) analyze movable parts of the face, such as the eyes or mouth (Hasselmo et al., 1989; Perrett et al., 1984, 1985, 1992), while inferior temporal regions appear to represent stable face properties, namely identity (Hasselmo et al., 1989; Perrett et al., 1984, 1985). To a first approximation these same regions are present in the human (Haxby et al., 2000b; Rolls, 2007; see Chapter 24).

The neural processes that support the ability to recognize and respond to facial identity remain a contentious subject of debate (Bukach et al., 2006; Dekowska et al., 2008; McKone et al., 2007; Yovel & Kanwisher, 2004). The majority of research on this question has been performed using fMRI. It is clear that a specific region of the visual cortex in humans responds selectively to faces (Kanwisher et al., 1997). This region, known as the fusiform face area (FFA), in the temporal lobe, is hemodynamically activated when human subjects look at faces. A homologous area in the monkey brain is also activated by (monkey) faces, and contains large patches of single neurons that are highly face specific (Tsao et al., 2006; see Chapter 24). Interestingly, a separate region responds to bodies (Downing et al., 2001). As in the monkey, it appears that distinct face-selective cortical regions represent stable features such as identity and variable features such as facial expression (Gobbini & Haxby, 2007; Haxby et al., 2000a, 2002); mobile features are represented in the superior temporal sulcus, while the constant features are represented in the fusiform gyrus (FFA).

Is the FFA specialized for recognizing faces, or is it activated when we respond to any stimulus for which we are experts? The issue is complex because we are all experts at recognizing faces; those of us who are not are likely to have pathological conditions, such as autism (Dalton et al., 2005; Schultz et al., 2000). When bird experts are asked to identify birds, their FFAs are selectively activated (Gauthier et al., 2000). The same is true for car experts and experts trained to identify novel complex shapes with no inherent ecological validity ("greebles") (Gauthier et al., 1999). However, upon closer examination, it appears that trial-to-trial variations in performance on such tasks are well predicted by activity in the FFA only for identifying faces, whereas other types of judgments, such as car identification by car experts, are mediated by other brain areas in the ventral occipitotemporal cortex (Grill-Spector et al., 2004).

Grandmother Cells and Jennifer Aniston

What is the endpoint of visual coding? One extreme idea is that there are neurons at the temporal pole that represent fully elaborated concepts, irrespective of any particular stimulus configuration. This idea, known as the "grandmother cell" hypothesis, was popularized by Lettvin, who developed it as a straw man to demonstrate the necessity of combinatorial coding schemes (Barlow, 1995; see also Gross, 2002; Konorski, 1967). Indeed, the idea of the grandmother cell is often used as the basis for reductio ad absurdum arguments for dense population coding, in which the majority cells in a population contribute to representation by representing different aspects of images (Barlow, 1972; Kandel et al., 2000; Purves et al., 2008).

In fact, the grandmother cell is one endpoint along a continuum from dense to sparse coding strategies. Despite the intuitive appeal of dense coding strategies, much neural evidence supports the idea that representations are quite sparse, and may even approximate the ideal of grandmother cell encoding (Logothetis & Sheinberg, 1996; Perrett et al., 1992; Tanaka, 1996; but see Quiroga et al., 2008). Indeed, a recent study of single neurons in the human medial temporal lobe indicates that the idea of the grandmother cell may not be so far-fetched (Kreiman et al., 2000; Quiroga et al., 2005). Scientists placed electrodes into the medial temporal lobes of humans undergoing surgical treatment for epilepsy. They found that single isolated neurons responded to particular individuals, whether presented in photograph, caricature, or merely a written name. These neurons seem to represent the image of the person at the highest level of abstraction (i.e., the "concept" of the person). For example, one neuron in the left posterior hippocampus responded to seven photographs of Jennifer Aniston, but not to any of 80 other photographs, including one of Jennifer Aniston and Brad Pitt together. Another neuron responded to many images of the Sydney Opera House, and to a photo of a Baha'i Temple that the subject erroneously believed was another view of the opera house.

These results suggest that visual coding is sparse, and they are consistent with the idea that sparse codes are more efficient and thus adaptive than dense codes (Olshausen & Field, 1996, 1997, 2004; Zhao, 2004). It remains unclear whether these neurons are tuned for other stimuli that were not probed or whether they were indeed unique to individual faces and places (Quiroga et al., 2008). It also remains unclear how the coding properties of these neurons change over time, as more is learned or forgotten about an individual.

NATURAL VISION

Natural Images

Standard methods of characterizing response properties of visual areas involve presenting simple synthetic stimuli in their response fields. Such stimuli typically include spots, oriented bars, drifting gratings, Gabor patches (Fig. 11.2), white noise, and coherently moving dots. It is generally assumed that neuronal responses to these simple stimuli will predict those to other, more complex, naturalistic stimuli. However, recent work calls these assumptions into question, and instead, indicates the importance of characterizing neural responses to natural stimuli for understanding visual processing (David et al., 2004). Indeed, natural stimuli are those that the visual system has evolved to process, and are the ones that were seen most of the time in the environment of evolutionary adaptiveness (Geisler, 2008; Simoncelli & Olshausen, 2001).

The visual system is highly adapted to the statistics of the visual world we inhabit. Thus, a greater understanding of statistical properties of the visual world can shed light on the evolution of vision (Field, 1987; Geisler, 2008; Gibson, 1966). Although natural images are highly diverse, they have several consistent properties that make them suitable for study: although there is virtually no limit to the number of things we can see, the things we are likely to see come from a very small subset of all possible configurations. For one thing, colors and intensity of adjacent pixels are highly correlated (Simoncelli & Olshausen, 2001). It is estimated that the average amount of information carried by a single pixel in a color photograph image is 1.4 bits (Kersten, 1987), whereas the amount of information stored by a computer in an explicit representation of the same image (i.e., a bitmap) is 8 bits in a black and white image and 24 bits in a color image. The extra bits come from redundant information, and they provide a great opportunity for redundancy reduction in visual encoding. Such reductions appear to be achieved, in part, by surround suppression and related mechanisms (Barlow, 1961; Grigorescu et al., 2003; Olshausen & Field, 1996).

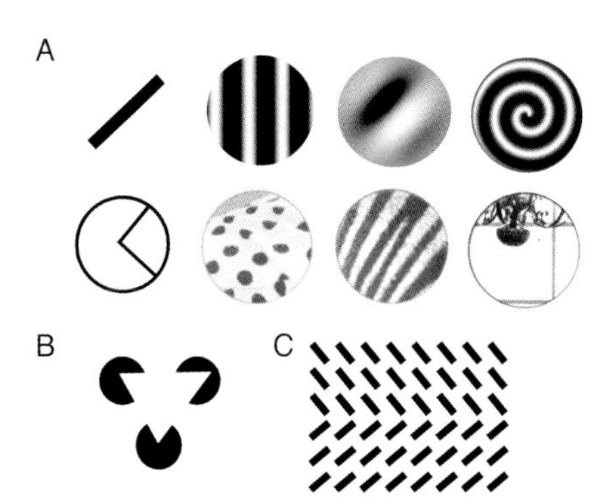

Figure 11.2 Visual stimuli. (**A**) Examples of visual stimuli used to probe neuronal response functions. Examples include, from left to right, oriented bars, oriented sinusoids, Gabor patches, non-Cartesian gratings, angled contours, and three examples of natural images. (**B**) Example of Kanisza triangle. Many people perceive that the edges of the triangle continue into the white space between the circles. Such contours are known as illusory contours. (**C**) Example of second-order edge. The transition between one pattern and the next forms a type of contour.

Interestingly, there is little correlation in luminance or contrast of sequentially fixated locations in a natural scene (Frazor & Geisler, 2006). This suggests that one function of saccades (rapid, reorienting movements of the eyes; see later and Chapters 15 and 26) is to increase the efficiency by which information is harvested. This is done by allowing quick shifts to more informative locations, and to help the visual system focus on the most informative points in a scene (Fig. 11.3) (Geisler et al., 2007; Mante et al., 2005). Further support for the idea that saccade targets are chosen to provide maximal information comes from the finding that saccade endpoints have more information than randomly chosen pixels (Reinagel & Zador, 1999).

Another important statistical property of natural images is that they have a characteristic power spectrum that is highly biased toward low frequencies (Field, 1987; Ruderman & Bialek, 1994). The power spectrum of natural images falls off as a function of $1/f^2$ (Ruderman & Bialek, 1994; Tolhurst et al., 1992). This finding applies to contrast as well (Ruderman & Bialek, 1994). Although individual images are highly variable, sizeable populations of images have consistent statistical properties. This regularity can be exploited by the visual system (Ruderman & Bialek, 1994). This particular power spectrum may reflect the size invariance of natural images—objects in the visual world can appear at many distances, sizes, and angles. Consequently, natural images have a fractal character, and thus possess a Fourier spectrum with an inverse power law distribution (Fig. 11.4) (Simoncelli & Olshausen, 2001). (Of course, such distributions characterize white noise as well.) Alternatively, natural images are replete with edges, which have a Fourier spectrum of $1/f^2$ (Simoncelli & Olshausen, 2001).

Indeed, the unique power spectrum of natural images is quite distinct from that of random images (white noise). It is likely that this idiosyncratic power spectrum is exploited by the visual system, and is therefore reflected in the tuning properties of neurons in the early visual system. For example, the aggregate tuning properties of V1 neurons appear to represent the independent components of natural images

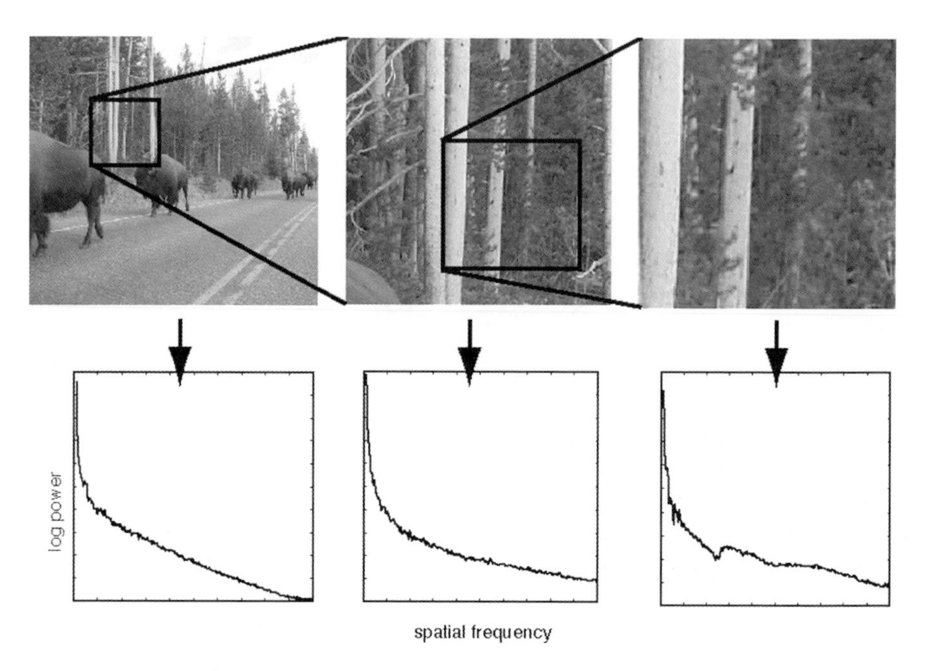

Figure 11.3 Example of natural image and naturalistic saccades. Dark lines represent paths between progressive saccades.

Figure 11.4 Example of natural image and associated power spectra. Three levels of magnification are shown. Power spectra are similar at each magnification level.

(Bell & Sejnowski, 1997; Olshausen & Field, 1996; van Hateren & Ruderman, 1998). The use of Gabor filters (spatially localized sine wave gratings; see Fig. 11.2), which are locally limited in both the spatial and spectral domains, may represent an optimal compromise between the spatial and spectral domains (Daugman, 1985). Indeed, theoretical studies support this hypothesis, and demonstrate that a Gabor wavelet-style decomposition may represent the most efficient way to represent the visual world (Field, 1987).

Saccades

Just as the visual system has not evolved to process simple synthetic images of the type generally used to characterize neuronal receptive fields, it also has not evolved to perform simple laboratory tasks, most of which require stable gaze for unnaturally long periods. Natural vision is distinguished from vision as it is studied in the lab in several ways. The most characteristic feature of natural vision is the occurrence of saccades, the natural scanning movements of the eyes that occur about three to four times a second. Saccades align the fovea with points of interest in the scene, and are directed by both internal and external factors (Yarbus, 1967).

Attempts to study the role of saccades on visual processing have recently begun. At the simplest level, visual images can be presented with the same or similar temporal dynamics as saccades. Neuronal responses in visual areas have characteristic temporal dynamics that are obscured by rapid presentation of visual images (David et al., 2004). These results suggest that the rate of saccades is determined to be as slow as possible (to reduce motor costs) while still yielding maximal information. Despite the added noise associated with uncertainty in eye position and uncontrollable nonclassical receptive field stimulation when experimental animals are free to move their eyes, visual tuning and patterns of attentional modulation can be measured under free-viewing conditions (David et al., 2008; Mazer & Gallant, 2003). These studies suggest that neuronal response properties are preserved during free-viewing and point to improvements in our understanding of visual processing that can be achieved by incorporating natural eye movements into future studies of visual processing.

Attention

Visual attention can be defined as the selection of part of the visual world, whether in mind (covert attention) or by gaze direction (overt attention) for enhanced scrutiny. Attention can select a location (spatial attention); a feature, such as all the green items in a scene (feature-based attention); or an object (object-based attention). Attention may either be directed by external events, such as a flash of light (exogenously cued attention), or inwardly, such as by one's long-term goals (endogenously directed attention; Egeth & Yantis, 1997). Behaviorally, attention enhances responsiveness to visual stimuli, including improving detection and discrimination thresholds (Desimone & Duncan, 1995; Egeth & Yantis, 1997; Maunsell & Treue, 2006; Pashler, 1999; Reynolds & Chelazzi, 2004).

Importantly, attention and vision are not discrete processes but instead are intertwined. In other words, attention does not affect complete representations formed at the end of visual processing, but influences the way visual information is processed at all levels of cortical processing (Mehta et al., 2000; Reynolds & Chelazzi, 2004; Treue, 2001), and may even alter responses in the LGN (O'Connor et al., 2002). Spatial attention enhances responses of neurons whose receptive fields match the location of the attended stimulus (McAdams & Maunsell, 1999; Moran & Desimone, 1985). Analogously, feature-based attention enhances responses of neurons whose tuning curves include the attended features (Hayden & Gallant, 2005; McAdams & Maunsell, 2000; Treue & Martinez Trujillo, 1999).

Nonetheless, the precise pattern of enhancement associated with attentional modulation is disputed. In some situations, attention can increase the gain of visual neurons, essentially turning up their volume (David et al., 2008; McAdams & Maunsell, 1999; Treue & Martinez Trujillo, 1999). In other situations, attention appears to enhance neuronal contrast response functions (Reynolds et al., 2000). More powerfully, attention can alter a neuron's tuning function, changing the stimulus that most strongly excites the neuron (David et al., 2008). For example, if one were searching for Waldo, neurons that did not normally respond most strongly to horizontal red and white lines would change their tuning to become Waldo detectors. It appears that the effects of attention are not limited to a single form, but instead depend on task demands and the type of visual processing performed within an area. These studies confirm the validity of models of neuronal responses for natural images and natural eye movements; nonetheless, it remains unclear how robust these findings will be in even more natural tasks.

There are several important remaining questions in vision and attention. First, the exact patterns of attentional modulation remain uncharacterized. Second, how does the microcircuitry of the brain support attentional processes? For example, it is possible that different neuronal subtypes mediate different aspects of attention. Third, there is a large discontinuity between blood oxygen level–dependent (BOLD)/fMRI results and single-unit physiology results. For example, spatial attention has large effects on the BOLD signal in V1, but negligible effects on single units (Yoshor et al., 2007). What is the reason for this discrepancy? These are just a sampling of the major issues in visual attention.

Working Memory

Working memory may be defined as the mental maintenance of task-relevant information for use in guiding subsequent behavior (see Chapter 18). Sometimes called the mental sketchpad, it is distinguished from long-term memory by its duration—typically a few seconds versus up to 80 years. It is thought that working memory is mediated by persistent changes in responses of individual neurons or neuronal populations in specialized brain areas (Goldman-Rakic, 1995). In contrast, long-term memory is thought to be mediated by molecular changes in specific neurons (Purves et al., 2008).

Incoming visual stimuli typically activate single neurons in visual cortex briefly, from 50 to 500 ms. This phasic response is thought to be identical whether the information is remembered or not. However, when information is maintained across a delay in working memory, responses of single neurons in the prefrontal cortex are typically elevated throughout the delay. Prefrontal cortex neurons are selectively tuned for specific remembered locations (Funahashi et al., 1989) and nonspatial features (Miller et al., 1996; Rainer et al., 1998; Scalaidhe et al., 1999). These persistent elevations are thought to be *the* memory trace (Courtney et al., 1998; Desimone, 1996; Funahashi et al., 1989; Fuster, 1973; Fuster & Alexander, 1971; Goldman-Rakic, 1995; Machens et al., 2005; Miller et al., 1996). Consequently, it is commonly believed that the visual cortex represents incoming visual information but does not participate in cognitive processes such as working memory and decision making (Constantinidis & Steinmetz, 1996; Desimone, 1996; Goldman-Rakic, 1995; Miller et al., 1996; Moody et al., 1998).

Thus, it is traditionally thought that vision and working memory are discrete cognitive processes, mediated by discrete brain regions. However, some evidence points to the idea that the same neurons that subserve vision are reactivated to maintain information in working memory (Pasternak & Greenlee, 2005; Postle et al., 2003). Neurons in many visual areas are activated during memory epochs of working memory tasks (Ferrera et al., 1994; Gnadt & Andersen, 1988; Haenny et al., 1988; Maunsell et al., 1991; Mikami & Kubota, 1980; Miller & Desimone, 1994; Pasternak & Greenlee, 2005; Super et al., 2001; Zhou & Fuster, 1996). Although these changes may reflect cognitive processes other than working memory (Bisley et al., 2004; Romo et al., 2002), there is very little distinction between neuronal response patterns that support delay modulation in visual cortical areas and prefrontal cortex. Moreover, direct comparisons of the functions of temporal cortex and prefrontal cortex suggest that prefrontal cortex mediates executive aspects of task performance, while visual cortex stores sensory information (Petrides, 2000; Lebedev et al., 2004).

Decision Making

Ultimately, the goal of vision is to guide behavior (Schall, 2001). However, not much is known about how visual processing is used to guide behavioral decisions. Nonetheless, the information we have about the visual system provides an excellent opportunity to study the neural mechanisms of decision making. The most well-understood system for visual decision making is the motion detection system, the middle temporal (MT)/lateral intraparietal (LIP) circuit (Shadlen & Newsome, 2001). MT responses are highly specific for motional direction; an energy model shows that responses are an approximately instantaneously linear function of motion energy (Simoncelli & Heeger, 1998). Motion energy can be manipulated along with the proportion of coherently moving dots in a random-dot stimulus. In a motion discrimination task with low coherence stimuli, it often takes up to 2 seconds, or more, to make an accurate decision. In such situations, there are single MT neurons that classify motion as well as, or even sometimes better than, the monkey whose behavior is under study (Britten et al., 1992, 1993; Shadlen et al., 1996). From this neurometric/psychometric match, it has been inferred that MT is the area that makes decisions about moving visual stimuli.

When visual motion is reported with an orienting saccade, responses of LIP neurons gradually rise as evidence is accumulated. More specifically, LIP neurons have localized contraversive response fields. As evidence is accumulated that a response must be made into the receptive field, neuronal activity rises, and as evidence is accumulated that a response must be made away from the receptive field, firing rates fall (Roitman & Shadlen, 2002; Shadlen & Newsome, 1996, 2001). These properties of LIP do not depend on the stimulus—the same effects are observed for abstract stimuli that provide explicit probability information (Yang & Shadlen, 2007) and for evidence that is accumulated over several trials about the weighting of a target based on learning rules. Moreover, these responses do seem to be specific to the output modality—when the task dictates a response reporting the judgment with an arm movement, neuronal activity in the parietal reach region (PRR), an arm movement analog of the LIP, accumulates in the same way (Musallam et al., 2004; Scherberger & Andersen, 2007).

Collectively, these results suggest that LIP serves to represent the accumulated weight of evidence that a saccade must be made toward a particular location. However, LIP function is more subtle. In a situation in which the amount of reward associated with a particular response is manipulated but the evidence is not manipulated (because there is only one option), firing rates of LIP neurons depend on saccade value (Platt & Glimcher, 1999). In fact, when the reward is reduced in a stochastic manner, thus providing a large reward only half the time, neuronal activity covaries with the economic *expected value* of the target (Platt & Glimcher, 1999). Furthermore, this pattern is obtained independent of the modality of the reward; LIP neurons are sensitive to both fluid and social rewards, such as the opportunity to view faces of attractive conspecifics (Klein et al., 2008).

Although the MT/LIP circuit is the most well-understood circuit for visual decision making, it is not the only one. When a saccade target is selected in a visual detection task, responses of neurons in the FEF gradually evolve toward a threshold. Variability in the times for single neurons to reach this threshold is correlated with variability in saccade onset times in the task, suggesting that this area plays a critical role in the mechanisms of visual decision making (Schall & Thompson, 1999). The responses of FEF neurons do not predict a saccade with certainty though—when a briefly presented stimulus is not perceived because another stimulus appears immediately afterward (backward masking), responses of FEF neurons are somewhat enhanced, but do not reach the threshold. This finding suggests that FEF does not mediate consciousness, but instead encodes a preconscious *decision variable*—a hypothetical construct that represents the likelihood that a saccade will be made. The decision variable gradually rises over time, and when it hits a threshold, the decision is triggered.

Visual Awareness

Consider two photographs of the beach. They are nearly identical, except that in one, a palm tree in the background has been digitally erased. If the two photographs are flipped back and forth, the tree will appear and disappear. However, if a white frame is inserted between the two images for a tenth of a second, you will see no change (Simons & Levin, 1997; Simons & Rensink, 2005). Even if you do detect the change, it does not pop out the way it does when the images flip immediately from one to the other. Our inability to detect surprisingly large changes when a brief blank divides two presentations of a similar scene is known as change blindness (Simons & Levin, 1997; Simons & Rensink, 2005). In fact, the blank period is not needed. Even small, task-irrelevant, distracting exogenous cues far from the location of the change can lead to change blindness (O'Regan et al., 1999). Moreover, change blindness is not limited to laboratory situations. Carefully constructed real-world situations reveal clear limits to our moment-to-moment awareness of our visual world (Simons & Levin, 1998).

The fact that even the slightest interruption is enough to erase any record of the world suggests that our internal representation of the world is highly fragile. Although very little is known about the mechanisms underlying change blindness (but see Beck et al., 2001, for a discussion of the neural correlates of change blindness), the phenomenon suggests that our representation of the visual world is not nearly as rich as we believe. Indeed, one natural interpretation of the phenomenon of change blindness is that it reveals how little information is available to conscious awareness at a time. Even though we *feel* like we perceive and are aware of much of the multitude of stimuli in our visual world, instead, our awareness appears to be ruthlessly efficient, only keeping track of the most important or salient features of a scene.

CONCLUSION

We are highly visual creatures. Vision dominates our mental life so much that, in English, we express comprehension by saying "I see." Primate behavior is controlled to a large extent by what can be seen. Primate vision is determined, in large part, by the accumulated forces of natural selection. These forces have determined the way our eyes are situated in our skulls, the construction of our retina, and structures of our brain. It is clear that a richer understanding of the statistical properties of our visual world can provide insights into how the visual system is constructed. By a similar logic, a richer understanding of the forces that have shaped evolution can provide information about how we see the world. Much remains to be determined. Evolutionary pressures on our visual systems must be more fully explored. The neural processes allowing rapid and accurate detection and discrimination of behaviorally relevant stimuli must be identified. Such information will provide a more complete picture of how behavior affects, and is affected by, our visual systems.

REFERENCES

Aiello, L. C., & Wheller, P. (1995). The expensive tissue hypothesis: The brain and the digestive system in human and primate evolution. *Current Anthropology, 36,* 199–221.

Albrecht, D. G., De Valois, R. L., & Thorell, L. G. (1980). Visual cortical neurons: Are bars or gratings the optimal stimuli? *Science, 207,* 88–90.

Albrecht, D. G., De Valois, R. L., & Thorell, L. G. (1982). Receptive fields and the optimal stimulus. *Science, 216,* 205.

Allmann, J. (1977). Evolution of the visual system in early primates. In: J. M. Sprague & A. N. Epstein (Eds.), *Progress in psychobiology and physiological psychology.* New York: Academic.

Atick, J. J. (1992). Could information theory provide an ecological theory of sensory processing? *Network: Computation in Neural Systems, 3,* 213–251.

Atick, J. J., & Redlich, A. N. (1991). Predicting ganglion and simple cell receptive field organizations. *International Journal of Neural Systems, 1,* 305–315.

Atick, J. J., & Redlich, A. N. (1992). What does the retina know about natural scenes? *Neural Computation, 4,* 196–210.

Attneave, F. (1954). Some informational aspects of visual perception. *Psychological Review, 61,* 183–193.

Bakin, J. S., Nakayama, K., & Gilbert, C. D. (2000). Visual responses in monkey areas V1 and V2 to three-dimensional surface configurations. *Journal of Neuroscience, 20,* 8188–8198.

Barlow, H. B. (1961). Possible principles underlying the transformation of sensory messages. In: W. A. Rosenblith (Ed.), *Sensory communication.* Cambridge, MA: MIT Press.

Barlow, H. B. (1972). Single units and sensation: a neuron doctrine for perceptual psychology? *Perception, 1,* 371–394.

Barlow, H. B. (1995). The neuron in perception. In: M. S. Gazzaniga (Ed.), *The cognitive neurosciences* (pp. 415–434). Cambridge, MA: MIT Press.

Barton, R. A. (1998). Visual specialization and brain evolution in primates. *Proceedings of Biological Sciences, 265,* 1933–1937.

Barton, R. A. (2004). From the cover: Binocularity and brain evolution in primates. *Proceedings of the National Academy of Sciences, 101,* 10113–10115.

Beck, D. M., Rees, G., Frith, C. D., & Lavie, N. (2001). Neural correlates of change detection and change blindness. *Nature Neuroscience, 4,* 645–650.

Bell, A. J., & Sejnowski, T. J. (1997). The "independent components" of natural scenes are edge filters. *Vision Research, 37,* 3327–3338.

Bisley, J. W., Zaksas, D., Droll, J. A., & Pasternak, T. (2004). Activity of neurons in cortical area MT during a memory for motion task. *Journal of Neurophysiology, 91,* 286–300.

Britten, K. H., Shadlen, M. N., Newsome, W. T., & Movshon, J. A. (1992). The analysis of visual motion: A comparison of neuronal and psychophysical performance. *Journal of Neuroscience, 12,* 4745–4765.

Britten, K. H., Shadlen, M. N., Newsome, W. T., & Movshon, J. A. (1993). Responses of neurons in macaque MT to stochastic motion signals. *Vision Neuroscience, 10,* 1157–1169.

Bukach, C. M., Gauthier, I., & Tarr, M. J. (2006). Beyond faces and modularity: The power of an expertise framework. *Trends in Cognitive Science, 10,* 159–166.

Buschbaum, G., & Gottschalk, A. (1983). Trichromacy, opponent colours coding and optimum colour information transmission in the retina. *Proceedings of the Royal Society of London B, 220,* 89–113.

Calderone, J. B., & Jacobs, G. H. (2003). Spectral properties and retinal distribution of ferret cones. *Vision Neuroscience, 20,* 11–17.

Campbell, F. W., Cooper, G. F., Robson, J. G., & Sachs, M. B. (1969). The spatial selectivity of visual cells of the cat and the squirrel monkey. *Journal of Physiology, 204,* 120–132.

Campbell, F. W., & Robson, J. G. (1968). Application of Fourier analysis to the visibility of gratings. *Journal of Physiology, 197,* 551–566.

Cartmill, M. (1970). *The orbits of arboreal mammals: A reassessment of the arboreal theory of primate evolution.* Chicago: University of Chicago.

Cartmill, M. (1974). Rethinking primate origins. *Science, 184,* 436–443.

Chafee, M. V., & Goldman-Rakic, P. S. (1998). Matching patterns of activity in primate prefrontal area 8a and parietal area 7ip neurons during a spatial working memory task. *Journal of Neurophysiology, 79,* 2919–2940.

Changizi, M. A., Zhang, Q., & Shimojo, S. (2006). Bare skin, blood and the evolution of primate color vision. *Biology Letters, 2,* 217–221.

Cheney, D. L., & Seyfarth, R. M. (1992). *How monkeys see the world.* Chicago: University of Chicago Press.

Constantinidis, C., & Steinmetz, M. A. (1996). Neuronal activity in posterior parietal area 7a during the delay periods of a spatial memory task. *Journal of Neurophysiology, 76,* 1352–1355.

Courtney, S. M., Petit, L., Maisog, J. M., Ungerleider, L. G., & Haxby, J. V. (1998). An area specialized for spatial working memory in human frontal cortex. *Science, 279,* 1347–1351.

Dalton, K. M., Nacewicz, B. M., Johnstone, T., Schaefer, H. S., Gernsbacher, M. A., Goldsmith, H. H., et al. (2005). Gaze fixation and the neural circuitry of face processing in autism. *Nature Neuroscience, 8,* 519–526.

Dan, Y., Atick, J. J., & Reid, R. C. (1996). Efficient coding of natural scenes in the lateral geniculate nucleus: Experimental test of a computational theory. *Journal of Neuroscience, 16,* 3351–3362.

Daugman, J. G. (1985). Uncertainty relation for resolution in space, spatial frequency, and orientation optimized by two-dimensional visual cortical filters. *Journal of the Optical Society of America A, 2,* 1160–1169.

David, S. V., Hayden, B. Y., Mazer, J. A., & Gallant, J. L. (2008). Attention to stimulus features shifts spectral tuning of V4 neurons during natural vision. *Neuron, 59,* 509–521.

David, S. V., Vinje, W. E., & Gallant, J. L. (2004). Natural stimulus statistics alter the receptive field structure of v1 neurons. *Journal of Neuroscience, 24,* 6991–7006.

De Valois, R. L., Albrecht, D. G., & Thorell, L. G. (1982). Spatial frequency selectivity of cells in macaque visual cortex. *Vision Research, 22,* 545–559.

Dekowska, M., Kuniecki, M., & Jaskowski, P. (2008). Facing facts: Neuronal mechanisms of face perception. *Acta Neurobiology Experiement (Wars), 68,* 229–252.

Desimone, R. (1991). Face-selective cells in the temporal cortex of monkeys. *Journal of Cognitive Neuroscience, 3,* 1–8.

Desimone, R. (1996). Neural mechanisms for visual memory and their role in attention. *Proceedings of the National Academy of Sciences, 93,* 13494–13499.

Desimone, R., & Duncan, J. (1995). Neural mechanisms of selective visual attention. *Annual Review of Neuroscience, 18,* 193–222.

Dixson, A. F. (2000). *Primate sexuality.* London: Oxford University Press.

Dong, D. W., & Atick, J. J. (1995). Statistics of natural time-varying images. *Network: Computation in Neural Systems, 6,* 345.

Downing, P. E., Jiang, Y., Shuman, M., & Kanwisher, N. (2001). A cortical area selective for visual processing of the human body. *Science, 293,* 2470–2473.

Drury, H. A., Van Essen, D. C., Anderson, C. H., Lee, C. W., Coogan, T. A., & Lewis, J. W. (1996). Computerized mappings of the cerebral cortex: A multiresolution flattening method and a surface-based coordinate system. *Journal of Cognitive Neuroscience, 8,* 1–28.

Egeth, H. E., & Yantis, S. (1997) Visual attention: Control, representation, and time course. *Annual Review of Psychology, 48,* 269–297.

Enroth-Cugell, C., & Robson, J. G. (1966). The contrast sensitivity of retinal ganglion cells of the cat. *Journal of Physiology, 187,* 517–552.

Felleman, D. J., & Van Essen, D. C. (1991). Distributed hierarchical processing in the primate cerebral cortex. *Cerebral Cortex, 1,* 1–47.

Ferrera, V. P., Rudolph, K. K., & Maunsell, J. H. (1994). Responses of neurons in the parietal and temporal visual pathways during a motion task. *Journal of Neuroscience, 14,* 6171–6186.

Field, D. J. (1987). Relations between the statistics of natural images and the response properties of cortical cells. *Journal of the Optical Society of America A, 4,* 2379–2394.

Frazor, R. A., & Geisler, W. S. (2006). Local luminance and contrast in natural images. *Vision Research, 46,* 1585–1598.

Funahashi, S., Bruce, C. J., & Goldman-Rakic, P. S. (1989). Mnemonic coding of visual space in the monkey's dorsolateral prefrontal cortex. *Journal of Neurophysiology, 61,* 331–349.

Fuster, J. M. (1973). Unit activity in prefrontal cortex during delayed-response performance: neuronal correlates of transient memory. *Journal of Neurophysiology, 36,* 61–78.

Fuster, J. M., & Alexander, G. E. (1971). Neuron activity related to short-term memory. *Science, 173,* 652–654.

Gauthier, I., Skudlarski, P., Gore, J. C., & Anderson, A. W. (2000). Expertise for cars and birds recruits brain areas involved in face recognition. *Nature Neuroscience, 3,* 191–197.

Gauthier, I., Tarr, M. J., Anderson, A. W., Skudlarski, P., & Gore, J. C. (1999). Activation of the middle fusiform 'face area' increases with expertise in recognizing novel objects. *Nature Neuroscience, 2,* 568–573.

Geisler, W. S. (2008). Visual perception and the statistical properties of natural scenes. *Annual Review of Psychology, 59,* 167–192.

Geisler, W. S., Albrecht, D. G., & Crane, A. M. (2007). Responses of neurons in primary visual cortex to transient changes in local contrast and luminance. *Journal of Neuroscience, 27,* 5063–5067.

Gibson, J. J. (1966). *The perception of the visual world.* Boston: Houghton-Mifflin.

Gnadt, J. W., & Andersen, R. A. (1988). Memory related motor planning activity in posterior parietal cortex of macaque. *Experimental Brain Research, 70,* 216–220.

Gobbini, M. I., & Haxby, J. V. (2007). Neural systems for recognition of familiar faces. *Neuropsychologia, 45,* 32–41.

Gold, J. I., & Shadlen, M. N. (2001). Neural computations that underlie decisions about sensory stimuli. *Trends in Cognitive Science, 5,* 10–16.

Gold, J. I., & Shadlen, M. N. (2002). Banburismus and the brain: Decoding the relationship between sensory stimuli, decisions, and reward. *Neuron, 36,* 299–308.

Goldman-Rakic, P. S. (1995). Cellular basis of working memory. *Neuron, 14,* 477–485.

Grigorescu, C., Petkov, N., & Westenberg, M. A. (2003). Contour detection based on nonclassical receptive field inhibition. *IEEE Transactions of Image Processes, 12,* 729–739.

Grill-Spector, K., Knouf, N., & Kanwisher, N. (2004). The fusiform face area subserves face perception, not generic within-category identification. *Nature Neuroscience, 7,* 555–562.

Grosof, D. H., Shapley, R. M., & Hawken, M. J. (1993). Macaque V1 neurons can signal 'illusory' contours. *Nature, 365,* 550–552.

Gross, C. G. (2002). Genealogy of the "grandmother cell". *Neuroscientist, 8,* 512–518.

Haenny, P. E., Maunsell, J. H., & Schiller, P. H. (1988). State dependent activity in monkey visual cortex. II. Retinal and extraretinal factors in V4 *Experimental Brain Research, 69,* 245–259.

Hasselmo, M. E., Rolls, E. T., & Baylis, G. C. (1989). The role of expression and identity in the face-selective responses of neurons in the temporal visual cortex of the monkey. *Behavioral Brain Research, 32,* 203–218.

Haxby, J. V., Hoffman, E. A., & Gobbini, M. I. (2000a). The distributed human neural system for face perception. *Trends in Cognitive Science, 4,* 223–233.

Haxby, J. V., Hoffman, E. A., & Gobbini, M. I. (2002). Human neural systems for face recognition and social communication. *Biological Psychiatry, 51,* 59–67.

Haxby, J. V., Ishai, I. I., Chao, L. L., Ungerleider, L. G., & Martin, I. I. (2000b). Object-form topology in the ventral temporal lobe: Response to I. Gauthier (2000). *Trends in Cognitive Science, 4,* 3–4.

Hayden, B. Y., & Gallant, J. L. (2005). Time course of attention reveals different mechanisms for spatial and feature-based attention in area v4. *Neuron, 47,* 637–643.

Hegde, J., & Felleman, D. J. (2007). Reappraising the functional implications of the primate visual anatomical hierarchy. *Neuroscientist, 13,* 416–421.

Hegde, J., & Van Essen, D. C. (2000). Selectivity for complex shapes in primate visual area V2. *Journal of Neuroscience, 20,* RC61.

Hegde, J., & Van Essen, D. C. (2007). A comparative study of shape representation in macaque visual areas v2 and v4. *Cerebral Cortex, 17,* 1100–1116.

Hilgetag, C. C., O'Neill, M. A., & Young, M. P. (1996). Indeterminate organization of the visual system. *Science, 271,* 776–777.

Horwitz, G. D., Batista, A. P., & Newsome, W. T. (2004). Representation of an abstract perceptual decision in macaque superior colliculus. *Journal of Neurophysiology, 91,* 2281–2296.

Hubel, D. H., & Wiesel, T. N. (1959). Receptive fields of single neurones in the cat's striate cortex. *Journal of Physiology, 148,* 574–591.

Hubel, D. H., & Wiesel, T. N. (1962). Receptive fields, binocular interaction and functional architecture in the cat's visual cortex. *Journal of Physiology, 160,* 106–154.

Hubel, D. H., & Wiesel, T. N. (1968). Receptive fields and functional architecture of monkey striate cortex. *Journal of Physiology, 195,* 215–243.

Jacobs, G. H. (1993). The distribution and nature of colour vision among the mammals. *Biological Review of the Cambridge Philosophical Society, 68,* 413–471.

Jacobs, G. H. (1995). Variations in primate colour vision: Mechanisms and utility. *Evolutionary Anthropology, 3,* 196–205.

Janssen, P., Srivastava, S., Ombelet, S., & Orban, G. A. (2008). Coding of shape and position in macaque lateral intraparietal area. *Journal of Neuroscience, 28,* 6679–6690.

Jones, F. W. (1916). *Arboreal man.* London: Edward Arnold.

Kandel, E., Schwartz, J., & Jessell, T. (2000). *Principles of neural science* (4th ed.). New York: McGraw-Hill.

Kanwisher, N., McDermott, J., & Chun, M. M. (1997). The fusiform face area: A module in human extrastriate cortex specialized for face perception. *Journal of Neuroscience, 17,* 4302–4311.

Kersten, D. (1987). Predictability and redundancy of natural images. *Journal of the Optical Society of America A, 4,* 2395–2400.

Kirk, E. C., & Kay, R. F. (2004). The evolution of high visual acuity in the Anthropoidea. In: C. F. Ross & R. F. Kay (Eds.), *Anthropoid origins: New visions.* New York: Springer.

Klein, J. T., Deaner, R. O., & Platt, M. L. (2008). Neural correlates of social target value in macaque parietal cortex. *Current Biology, 18,* 419–424.

Konorski, J. (1967). *Integrative activity of the brain: An interdisciplinary approach.* Chicago: University of Chicago Press.

Kreiman, G., Koch, C., & Fried, I. (2000). Category-specific visual responses of single neurons in the human medial temporal lobe. *Nature Neuroscience, 3,* 946–953.

Laughlin, S. B., de Ruyter van Steveninck, R. R., & Anderson, J. C. (1998). The metabolic cost of neural information. *Nature Neuroscience, 1,* 36–41.

Le Gros Clark, W. E. (1959). *The antecedents of man.* New York: Harper.

Le Gros Clark, W. E. (1970). *History of the primates.* London: British Museum of Natural History.

Lebedev, M. A., Messinger, A., Kralik, J. D., & Wise, S. P. (2004). Representation of attended versus remembered locations in prefrontal cortex. *PLoS Biology, 2,* e365.

Lehky, S. R., & Sereno, A. B. (2007). Comparison of shape encoding in primate dorsal and ventral visual pathways. *Journal of Neurophysiology, 97,* 307–319.

Lennie, P. (1998). Single units and visual cortical organization. *Perception, 27,* 889–935.

Logothetis, N. K., & Sheinberg, D. L. (1996). Visual object recognition. *Annual Review of Neuroscience, 19,* 577–621.

Machens, C. K., Romo, R., & Brody, C. D. (2005). Flexible control of mutual inhibition: A neural model of two-interval discrimination. *Science, 307,* 1121–1124.

Mahon, L. E., & De Valois, R. L. (2001). Cartesian and non-Cartesian responses in LGN, V1, and V2 cells. *Vision Neuroscience, 18,* 973–981.

Maier, A., Wilke, M., Aura, C., Zhu, C., Ye, F. Q., & Leopold, D. A. (2008). Divergence of fMRI and neural signals in V1 during perceptual suppression in the awake monkey. *Nature Neuroscience, 11,* 1193–1200.

Mante, V., Frazor, R. A., Bonin, V., Geisler, W. S., & Carandini, M. (2005). Independence of luminance and contrast in natural scenes and in the early visual system. *Nature Neuroscience, 8,* 1690–1697.

Marcus, D. S., & Van Essen, D. C. (2002). Scene segmentation and attention in primate cortical areas V1 and V2. *Journal of Neurophysiology, 88,* 2648–2658.

Maunsell, J. H., Sclar, G., Nealey, T. A., & DePriest, D. D. (1991). Extraretinal representations in area V4 in the macaque monkey. *Vision Neuroscience, 7,* 561–573.

Maunsell, J. H., & Treue, S. (2006). Feature-based attention in visual cortex. *Trends in Neuroscience, 29,* 317–322.

Mazer, J. A., & Gallant, J. L. (2003). Goal-related activity in V4 during free viewing visual search. Evidence for a ventral stream visual salience map. *Neuron, 40,* 1241–1250.

McAdams, C. J., & Maunsell, J. H. (1999). Effects of attention on orientation-tuning functions of single neurons in macaque cortical area V4. *Journal of Neuroscience, 19,* 431–441.

McAdams, C. J., & Maunsell, J. H. (2000). Attention to both space and feature modulates neuronal responses in macaque area V4. *Journal of Neurophysiology, 83,* 1751–1755.

McKone, E., Kanwisher, N., & Duchaine, B. C. (2007). Can generic expertise explain special processing for faces? *Trends in Cognitive Science, 11,* 8–15.

Mech, L. D., & Boitani, L. (2003). *Wolves: Behavior, ecology, and conversation.* Chicago: University of Chicago Press.

Mehta, A. D., Ulbert, I., & Schroeder, C. E. (2000). Intermodal selective attention in monkeys. I: Distribution and timing of effects across visual areas. *Cerebral Cortex, 10,* 343–358.

Merigan, W. H., & Maunsell, J. H. (1993). How parallel are the primate visual pathways? *Annual Review of Neuroscience, 16,* 369–402.

Mikami, A., & Kubota, K. (1980). Inferotemporal neuron activities and color discrimination with delay. *Brain Research, 182,* 65–78.

Miller, E. K., & Desimone, R. (1994). Parallel neuronal mechanisms for short-term memory. *Science, 263,* 520–522.

Miller, E. K., Erickson, C. A., & Desimone, R. (1996). Neural mechanisms of visual working memory in prefrontal cortex of the macaque. *Journal of Neuroscience, 16,* 5154–5167.

Milner, A. D., & Goodale, M. A. (1995). *The visual brain in action.* New York: Oxford.

Moody, S. L., Wise, S. P., di Pellegrino, G., & Zipser, D. (1998). A model that accounts for activity in primate frontal cortex during a delayed matching-to-sample task. *Journal of Neuroscience, 18,* 399–410.

Moran, J., & Desimone, R. (1985). Selective attention gates visual processing in the extrastriate cortex. *Science, 229,* 782–784.

Murray, E. A., & Bussey, T. J. (1999). Perceptual-mnemonic functions of the perirhinal cortex. *Trends in Cognitive Science, 3,* 142–151.

Murray, E. A., Bussey, T. J., & Saksida, L. M. (2007). Visual perception and memory: A new view of medial temporal lobe function in primates and rodents. *Annual Review of Neuroscience, 30,* 99–122.

Murray, E. A., & Mishkin, M. (1998). Object recognition and location memory in monkeys with excitotoxic lesions of the amygdala and hippocampus. *Journal of Neuroscience, 18,* 6568–6582.

Musallam, S., Corneil, B. D., Greger, B., Scherberger, H., & Andersen, R. A. (2004). Cognitive control signals for neural prosthetics. *Science, 305,* 258–262.

Neisser, U. (1967). *Cognitive psychology.* New York: Appleton-Century-Crofts.

O'Connor, D. H., Fukui, M. M., Pinsk, M. A., & Kastner, S. (2002). Attention modulates responses in the human lateral geniculate nucleus. *Nature Neuroscience, 5,* 1203–1209.

O'Regan, J. K., Rensink, R. A., & Clark, J. J. (1999). Change-blindness as a result of 'mudsplashes'. *Nature, 398,* 34.

Olshausen, B. A., & Field, D. J. (1996). Emergence of simple-cell receptive field properties by learning a sparse code for natural images. *Nature, 381,* 607–609.

Olshausen, B. A., & Field, D. J. (1997). Sparse coding with an overcomplete basis set: A strategy employed by V1? *Vision Research, 37,* 3311–3325.

Olshausen, B. A., & Field, D. J. (2004). Sparse coding of sensory inputs. *Current Opinions in Neurobiology, 14,* 481–487.

Olshausen, B. A., & Field, D. J. (2005). How close are we to understanding v1? *Neural Computation, 17,* 1665–1699.

Osorio, D., & Vorobyev, M. (1996). Colour vision as an adaptation to frugivory in primates. *Proceedings of Biological Sciences, 263,* 593–599.

Osorio, D., & Vorobyev, M. (2008). A review of the evolution of animal colour vision and visual communication signals. *Vision Research, 48,* 2042–2051.

Pashler, H. E. (1999). *The psychology of attention.* Cambridge, MA: MIT Press.

Pasternak, T., & Greenlee, M. W. (2005). Working memory in primate sensory systems. *Nature Reviews Neuroscience, 6,* 97–107.

Peng, X., Sereno, M. E., Silva, A. K., Lehky, S. R., & Sereno, A. B. (2008). Shape selectivity in primate frontal eye field. *Journal of Neurophysiology, 100,* 796–814.

Perrett, D. I., Hietanen, J. K., Oram, M. W., & Benson, P. J. (1992). Organization and functions of cells responsive to faces in the temporal cortex. *Philosophical Transactions of the Royal Society of London B Biological Sciences, 335,* 23–30.

Perrett, D. I., Rolls, E. T., & Caan, W. (1982). Visual neurones responsive to faces in the monkey temporal cortex. *Experimental Brain Research, 47,* 329–342.

Perrett, D. I., Smith, P. A., Mistlin, A. J., Chitty, A. J., Head, A. S., Potter, D. D., et al. (1985). Visual analysis of body movements by neurones in the temporal cortex of the macaque monkey: A preliminary report. *Behavioral Brain Research, 16,* 153–170.

Perrett, D. I., Smith, P. A., Potter, D. D., Mistlin, A. J., Head, A. S., Milner, A. D., et al. (1984). Neurones responsive to faces in the temporal cortex: Studies of functional organization, sensitivity to identity and relation to perception. *Humam Neurobiology, 3,* 197–208.

Peterhans, E., & von der Heydt, R. (1991). Subjective contours—bridging the gap between psychophysics and physiology. *Trends in Neuroscience, 14,* 112–119.

Petrides, M. (2000). Dissociable roles of mid-dorsolateral prefrontal and anterior inferotemporal cortex in visual working memory. *Journal of Neuroscience, 20,* 7496–7503.

Pettigrew, J. D. (1978). Comparison of the retinotopic organization of the visual world in nocturnal and diurnal raptions, with a note on the evolution of frontal vision. In: S. J. Cool & E. L. Smith (Eds.), *Frontiers of visual science.* New York: Springer-Verlag.

Platt, M. L., & Glimcher, P. W. (1999). Neural correlates of decision variables in parietal cortex. *Nature, 400,* 233–238.

Postle, B. R., Druzgal, T. J., & D'Esposito, M. (2003). Seeking the neural substrates of visual working memory storage. *Cortex, 39,* 927–946.

Purves, D., Augustine, G. J., Fitzpatrick, D., Hall, W. C., LaMantia, A-S., McNamara, J. O., et al. (2008). *Neuroscience* (4th ed.). Sunderland, MA: Sinauer.

Quiroga, R. Q., Kreiman, G., Koch, C., & Fried, I. (2008). Sparse but not 'grandmother-cell' coding in the medial temporal lobe. *Trends in Cognitive Science, 12,* 87–91.

Quiroga, R. Q., Reddy, L., Kreiman, G., Koch, C., & Fried, I. (2005). Invariant visual representation by single neurons in the human brain. *Nature, 435,* 1102–1107.

Rainer, G., Asaad, W. F., & Miller, E. K. (1998). Memory fields of neurons in the primate prefrontal cortex. *Proceedings of the National Academy of Sciences, 95,* 15008–15013.

Ramsden, B. M., Hung, C. P., & Roe, A. W. (2001). Real and illusory contour processing in area V1 of the primate: A cortical balancing act. *Cerebral Cortex, 11,* 648–665.

Rao, R. P., & Ballard, D. H. (1999). Predictive coding in the visual cortex: A functional interpretation of some extra-classical receptive-field effects. *Nature Neuroscience, 2,* 79–87.

Reinagel, P., & Zador, A. M. (1999). Natural scene statistics at the centre of gaze. *Network, 10,* 341–350.

Reynolds, J. H., & Chelazzi, L. (2004). Attentional modulation of visual processing. *Annual Review of Neuroscience, 27,* 611–647.

Reynolds, J. H., Pasternak, T., & Desimone, R. (2000). Attention increases sensitivity of V4 neurons. *Neuron, 26,* 703–714.

Rieke, F., Warland, D., de Ruyter van Steveninck, R., & Bialek, W. (1997). *Spikes: Exploring the neural code.* Cambridge, MA: MIT Press.

Roitman, J. D., & Shadlen, M. N. (2002). Response of neurons in the lateral intraparietal area during a combined visual discrimination reaction time task. *Journal of Neuroscience, 22,* 9475–9489.

Rolls, E. T. (2007). The representation of information about faces in the temporal and frontal lobes. *Neuropsychologia, 45,* 124–143.

Romo, R., Hernandez, A., Zainos, A., Lemus, L., & Brody, C. D. (2002). Neuronal correlates of decision-making in secondary somatosensory cortex. *Nature Neuroscience, 5,* 1217–1225.

Ross, C. F., Hall, M. I., & Heesy, C. P. (2005). Were basal primates nocturnal? Evidence from eye and orbit shape. In: M. J. Ravosa & M. Dagosto (Eds.), *Primate origins: Adaptations and evolution.* New York, NY: Springer.

Ruderman, D. L., & Bialek, W. (1994). Statistics of natural images: Scaling in the woods. *Physical Review Letters, 73,* 814–817.

Scalaidhe, S. P., Wilson, F. A., & Goldman-Rakic, P. S. (1999). Face-selective neurons during passive viewing and working memory performance of rhesus monkeys: Evidence for intrinsic specialization of neuronal coding. *Cerebral Cortex, 9,* 459–475.

Schall, J. D. (2001). Neural basis of deciding, choosing and acting. *Nature Review Neuroscience, 2,* 33–42.

Schall, J. D., & Thompson, K. G. (1999). Neural selection and control of visually guided eye movements. *Annual Review of Neuroscience, 22,* 241–259.

Scherberger, H., & Andersen, R. A. (2007). Target selection signals for arm reaching in the posterior parietal cortex. *Journal of Neuroscience, 27,* 2001–2012.

Schiller, P. H. (1993). The effects of V4 and middle temporal (MT) area lesions on visual performance in the rhesus monkey. *Vision Neuroscience, 10,* 717–746.

Schmolesky, M. T., Wang, Y., Hanes, D. P., Thompson, K. G., Leutgeb, S., Schall, J. D., et al. (1998). Signal timing across the macaque visual system. *Journal of Neurophysiology, 79,* 3272–3278.

Schultz, R. T., Gauthier, I., Klin, A., Fulbright, R. K., Anderson, A. W., Volkmar, F., et al. (2000). Abnormal ventral temporal cortical activity during face discrimination among individuals with autism and Asperger syndrome. *Archives of General Psychiatry, 57,* 331–340.

Sereno, A. B., & Maunsell, J. H. (1998). Shape selectivity in primate lateral intraparietal cortex. *Nature, 395,* 500–503.

Shadlen, M. N., & Newsome, W. T. (1996). Motion perception: Seeing and deciding. *Proceedings of the National Academy of Sciences, 93,* 628–633.

Shadlen, M. N., & Newsome, W. T. (2001). Neural basis of a perceptual decision in the parietal cortex (area LIP) of the rhesus monkey. *Journal of Neurophysiology, 86,* 1916–1936.

Shadlen, M. N., Britten, K. H., Newsome, W. T., & Movshon, J. A. (1996). A computational analysis of the relationship between neuronal and behavioral responses to visual motion. *Journal of Neuroscience, 16,* 1486–1510.

Shannon, C. E., & Weaver, W. (1963). *The mathematical theory of communication.* Urbana, Chicago: University of Illinois Press.

Sheth, B. R., Sharma, J., Rao, S. C., & Sur, M. (1996). Orientation maps of subjective contours in visual cortex. *Science, 274,* 2110–2115.

Simoncelli, E. P., & Heeger, D. J. (1998). A model of neuronal responses in visual area MT. *Vision Research, 38,* 743–761.

Simoncelli, E. P., & Olshausen, B. A. (2001). Natural image statistics and neural representation. *Annual Review of Neuroscience, 24,* 1193–1216.

Simons, D. J., & Levin, D. T. (1997). Change blindness. *Trends in Cognitive Science, 1,* 261–267.

Simons, D. J., & Levin, D. T. (1998). Failures to detect changes to people during a real-world interaction. *Psychonomic Bulletin and Review, 5,* 644–649.

Simons, D. J., & Rensink, R. A. (2005). Change blindness: Past, present, and future. *Trends in Cognitive Science, 9,* 16–20.

Smith, A. C., Buchanan-Smith, H. M., Surridge, A. K., Osorio, D., & Mundy, N. I. (2003). The effect of colour vision status on the detection and selection of fruits by tamarins (Saguinus spp.). *Journal of Experimental Biology, 206,* 3159–3165.

Smith, G. E. (1912). The evolution of man. In: *Annual report.* Washington, DC: Smithsonian Institution.

Sprague, J. M., Levy, J., DiBerardino, A., & Berlucchi, G. (1977). Visual cortical areas mediating form discrimination in the cat. *Journal of Comprehensive Neurology, 172,* 441–488.

Sumner, P., & Mollon, J. D. (2000). Chromaticity as a signal of ripeness in fruits taken by primates. *Journal of Experimental Biology, 203,* 1987–2000.

Super, H., Spekreijse, H., & Lamme, V. A. (2001). A neural correlate of working memory in the monkey primary visual cortex. *Science, 293,* 120–124.

Surridge, A. K., Osorio, D., & Mundy, N. I. (2003). Evolution and selection of trichromatic vision in primates. *Trends in Ecology and Evolution, 18,* 198–205.

Tanaka, K. (1996). Inferotemporal cortex and object vision. *Annual Review of Neuroscience, 19,* 109–139.

Thomas, O. M., Cumming, B. G., & Parker, A. J. (2002). A specialization for relative disparity in V2. *Nature Neuroscience, 5,* 472–478.

Thorpe, S., Delorme, A., & Van Rullen, R. (2001). Spike-based strategies for rapid processing. *Neural Networks, 14,* 715–725.

Tolhurst, D. J., Tadmor, Y., & Chao, T. (1992). Amplitude spectra of natural images. *Ophthalmic Physiology Optometry, 12,* 229–232.

Tolias, A. S., Keliris, G. A., Smirnakis, S. M., & Logothetis, N. K. (2005). Neurons in macaque area V4 acquire directional tuning after adaptation to motion stimuli. *Nature Neuroscience, 8,* 591–593.

Treue, S. (2001). Neural correlates of attention in primate visual cortex. *Trends in Neuroscience, 24,* 295–300.

Treue, S., Martinez Trujillo, J. C. (1999). Feature-based attention influences motion processing gain in macaque visual cortex. *Nature, 399,* 575–579.

Tsao, D. Y., Freiwald, W. A., Tootell, R. B., & Livingstone, M. S. (2006). A cortical region consisting entirely of face-selective cells. *Science, 311,* 670–674.

Ungerleider, L. G., & Mishkin, M. (1982). Two cortical visual systems. In: D. J. Ingle, R. J. W. Mansfield, & M. S. Goodale (Eds.), *The analysis of visual behavior.* Cambridge, MA: MIT Press.

Ungerleider, L. G., & Haxby, J. V. (1994). 'What' and 'where' in the human brain. *Current Opinions in Neurobiology, 4,* 157–165.

Van Essen, D. C. (2004). Organization of visual areas in macaque and human cerebral cortex. In: L. M. Chalupa & J. S. Werner (Eds.), *The visual neurosciences.* Cambridge, MA: MIT Press.

van Hateren, J. H., & Ruderman, D. L. (1998). Independent component analysis of natural image sequences yields spatio-temporal filters similar to simple cells in primary visual cortex. *Proceedings of Biological Sciences, 265,* 2315–2320.

Vanni, S., Warnking, J., Dojat, M., Delon-Martin, C., Bullier, J., & Segebarth, C. (2004). Sequence of pattern onset responses in the human visual areas: an fMRI constrained VEP source analysis. *Neuroimage, 21,* 801–817.

Victor, J. D., Mechler, F., Repucci, M. A., Purpura, K. P., & Sharpee, T. (2006). Responses of V1 neurons to two-dimensional hermite functions. *Journal of Neurophysiology, 95,* 379–400.

von der Heydt, R., Peterhans, E., & Baumgartner, G. (1984). Illusory contours and cortical neuron responses. *Science, 224,* 1260–1262.

von der Heydt, R., Zhou, H., & Friedman, H. S. (2000). Representation of stereoscopic edges in monkey visual cortex. *Vision Research, 40,* 1955–1967.

Wandell, B. A. (1995). *Foundations of vision.* Sunderland, MA: Sinauer.

Yang, T., & Shadlen, M. N. (2007). Probabilistic reasoning by neurons. *Nature, 447,* 1075–1080

Yarbus, A. L. (1967). *Eye movements and vision.* New York: Plenum.

Yoshor, D., Ghose, G. M., Bosking, W. H., Sun, P., & Maunsell, J. H. (2007). Spatial attention does not strongly modulate neuronal responses in early human visual cortex. *Journal of Neuroscience, 27,* 13205–13209.

Yovel, G., & Kanwisher, N. (2004). Face perception: Domain specific, not process specific. *Neuron, 44,* 889–898.

Zhao, L. (2004). Is sparse and distributed the coding goal of simple cells? *Biological Cybernetics, 91,* 408–416.

Zhou, Y. D., & Fuster, J. M. (1996). Mnemonic neuronal activity in somatosensory cortex. *Proceedings of the National Academy of Sciences, 93,* 10533–10537.

CHAPTER 12

Circuits of Visual Attention

Tirin Moore, Robert J. Schafer and Behrad Noudoost

Primate vision is severely constrained by the fact that fine details in a visual scene can only be resolved by the fovea, where acuity is greatest. This tiny portion of each retina, which amounts to less than half of 1 degree of visual angle, must be moved around and positioned on behaviorally relevant stimuli in order to facilitate visual perception. Saccadic eye movements (saccades) reposition the direction of gaze (and the fovea) some three to five times per second and provide the means by which detailed visual information is accumulated during visual scanning. The ability to move the eyes accurately and precisely among targets of interest is crucial to adaptive behavior. Moreover, understanding how visual and saccadic mechanisms interact to choose each successive target of saccades not only is central to understanding the physiology of primate vision but, as it turns out, is also important for understanding the neural basis of visual attention. Attention is typically focused on targets of interest by directly foveating them overtly with saccades. That is, saccades are the typical way in which information is selectively processed and other information is ignored. But attention can be directed to objects covertly as well (Sperling & Melchner, 1978). For example, when direct eye contact with another person of interest is avoided, one's attention can still be focused on him or her (see Chapter 26). As we shall see, the neural mechanisms involved in mobilizing the fovea and the focus of attention together are also involved in moving attention by itself. To begin, we discuss the factors that determine which targets we attend to, either overtly or covertly.

DECIDING WHERE TO LOOK

Numerous factors influence the decision of where to shift the direction of gaze, including factors that are both external and internal to the viewer. Consider a driver, maneuvering a car through traffic. Certain properties of the visual world might draw her attention involuntarily: a flashy billboard, or a sputtering street lamp. On the other hand, the driver can direct her attention voluntarily, such as to scan street signs to find a particular destination. Finally, the interaction between properties of the visual world and the driver's own intentions adds another layer of complexity to the attention problem: While she might direct her gaze toward restaurants along the side of the road if she is hungry, she won't if she has just eaten, despite being presented with the same visual scene.

One approach toward understanding how neural circuits direct attention has been to construct models that explain and potentially predict the type of scanning and search behavior exhibited by primates. In creating one of the first models of visual attention, Treisman and Gelade (1980) used human psychophysical data to demonstrate that very low-level features of a visual scene, such as intensity, color, and orientation, predict how easily a viewer finds a target in an array of simple visual stimuli. When a target differs from all nontargets in one or more feature dimensions (such as a red object in an array of green, or a horizontal bar surrounded by vertical ones), the target seems to "pop out" at the viewer, and the time required to find the target stays constant even if the number of distracters is greatly increased. This observation led Treisman and colleagues to hypothesize that the visual system conducts parallel "preattentive processing" of the simple features of visual scenes, and then directs attention serially to each stimulus to bind the simple features together. This model, known as the *feature-integration theory of attention*, became the foundation of numerous subsequent models of visual attention.

To extend the feature-integration theory and explain how it might be implemented in a biologically plausible model, Koch and Ullman (1985) incorporated the concept of a "saliency map" that encodes the conspicuity ("salience") of every part of the visual scene. First, consistent with the concept of preattentive processing, several topographic "feature maps" of visual space were organized in parallel, one for each of a number of low-level visual features (color, orientation, direction of movement, disparity, etc.). Within each of these maps, a region was marked as conspicuous if it differed along its feature axis from its immediate neighborhood: For example, in a color map, a bright red object surrounded by green would be especially conspicuous, and in a motion map, leftward motion surrounded by rightward motion would be as well. Finally, the conspicuous regions of each feature map were combined into a single feature-independent saliency map of the image, and a winner-take-all competition determined the most salient region. According to this model, the viewer's gaze would then be directed to this part of the scene.

Guidance of attention based on features of the image, or "bottom-up" guidance, captures one source of attentional signals, and is successful at mimicking how certain aspects of a visual scene attract a viewer's gaze. However, as in the case of the driver in her car, attention can also be guided voluntarily, from the "top down." Several models have maintained the concept of bottom-up processing of simple features, but have added a role for "top-down" attention to influence the saliency map or the feature maps themselves (Wolfe, 1994). Noting that a subject's knowledge of task demands can modulate how different visual features influence perception and behavior, Francolini and Egeth (1979) suggested that the many low-level feature maps might be combined with weights that can change according to the knowledge and intentions of the viewer. The *FeatureGate model* of visual selection (Cave, 1999) used a related concept, in which locations with features dissimilar to the known target were subject to top-down inhibition. Similarly, Navalpakkam and Itti (2005) used remembered representations of visual targets to bias the map weights according to the known features of a target, or to maximize differences in saliency between targets and nontargets (Navalpakkam & Itti, 2007).

NEURAL CIRCUITS FOR SHIFTING OVERT ATTENTION

These models of bottom-up and top-down influences on attention have found considerable success in describing the visual behavior of primates, but the question remains whether these models map onto circuits in the brain. To the extent that the models provide a conceptual description of how the visual and oculomotor systems work, one would like to be able to identify brain regions with the properties of the low-level feature maps, and perhaps with the feature-independent saliency map. (Note, however, that some other models do not require separate feature maps [e.g., Li, 2002] or an explicitly represented saliency map [e.g., Desimone & Duncan, 1995].) Furthermore, one would like to see pathways through which top-down influences, such as prior expectations or motivation, can affect the representations in the feature maps, saliency map, or both. Neurophysiological studies have only begun to provide candidate neural substrates for many of these maps and connections.

The primate visual system comprises several distinct cortical areas, each with its own retinotopic map of visual space (Felleman & Van Essen, 1991; Goodale & Milner, 1992; Ungerleider & Mishkin, 1982; Van Essen et al., 1992). A single visual cortical neuron only represents information about stimuli that fall within a limited part of the visual field, known as the neuron's receptive field (RF). Furthermore, neurons in each cortical area are responsive to certain features of a stimulus: For example, neurons in monkey primary visual cortex (V1) are selective to the orientations of edges and in some cases to directions of motion (Hubel & Wiesel, 1965, 1968); many neurons in area V4 are color selective (Schein & Desimone, 1990); and most neurons in the middle temporal visual area (MT) are selective for the direction and speed of visual motion (Dubner & Zeki, 1971; Zeki, 1974). Despite the fact that most of these areas are not as specialized as the low-level feature maps in the models of attention (e.g., most color-selective V4 neurons are not selective for color alone, but often also for spatial frequency and orientation of stimuli), they nonetheless represent features of the visual scene in a retinotopic manner, as described in the models. Moreover, local inhibition between neurons with neighboring receptive fields in the same visual cortical area enhances the activity associated with local contrast within a feature's dimension, as in several of the models.

To qualify as a saliency map, an area must satisfy several requirements in terms of connectivity and function (Fecteau & Munoz, 2007). First, since salience can be driven by low-level features, a saliency map should have access to the numerous visual cortical areas that represent various features of visual stimuli. However, neurons participating in a saliency map representation should not themselves be feature selective: Instead, they should represent the featureless conspicuity of a region of visual space. Next, a saliency map should have access to top-down signals and receive inputs from areas that represent motivation, expectation, and reward. In terms of function, if the saliency map is localized to one or a handful of areas, then lesions to these regions should result in deficits in shifting the focus of attention. Finally, in a brain area representing a saliency map, the activity of neurons should be related to the likelihood that the represented part of visual space will be the target of attention. In the monkey brain, an interconnected group of areas fits these criteria: the network of oculomotor structures, responsible for planning and executing saccadic eye movements. In particular, the frontal eye field (FEF) and lateral intraparietal area (LIP) in cortex and the superior colliculus (SC) in the midbrain each appear to meet the requirements of saliency maps. The FEF, LIP, and SC are interconnected, and all receive extensive projections from areas throughout much of visual cortex. Despite this connectivity, neurons in saccade-related areas are not themselves selective for low-level visual features. All three regions receive projections from prefrontal cortex, which is known to be involved in representing top-down, task- or goal-related information. The activity of neurons in these saccade-related structures also distinguishes the targets of a visual search from nontargets and represents the targets of impending shifts of attention (Kusunoki et al., 2000; McPeek & Keller, 2002; Robinson et al., 1995; Schall, 1995; Steinmetz & Constantinidis, 1995). Finally, the oculomotor structures are connected to brain regions that encode reward expectation and motivation: The caudate nucleus of the basal ganglia, which is known to encode reward expectation and motivation, projects through the thalamus to the FEF and LIP, and the SC receives projections from the caudate through the substantia nigra pars reticulata. Neurons in all three areas are strongly modulated by reward expectation (Ding & Hikosaka, 2006; Ikeda & Hikosaka, 2003; Platt & Glimcher, 1999; Sugrue et al., 2004).

NEURAL CORRELATES OF OVERT AND COVERT ATTENTION

Beginning in the 1980s and continuing throughout the 1990s, a large body of neurophysiological work established that the activity of neurons within many visual cortical areas is modulated by selective attention. That is, the magnitude of responses to visual stimulation (i.e., number of action potentials elicited) depends on whether an animal pays attention to that stimulus. For example, since a monkey can be trained to repeatedly fixate and hold its gaze on a spot within a stable display, the experimenter can study the responses of single neurons within the animal's visual cortex by presenting visual stimuli on the display as the animal fixates. The experimenter can probe the neuron's response to stimuli presented within the RF. Furthermore, the experimenter can train the monkey to covertly monitor the RF stimulus (e.g., by requiring it to detect changes in that stimulus), and then compare the neuron's responses to conditions in which the monkey monitors a stimulus at some other location (Fig. 12.1, left). In both cases, the monkey remains fixated and thus the RF stimulus stays constant. (The monkey doesn't do all of this work for free of course, but is instead motivated by the drop of juice it receives at the end of each trial.) Using this basic experimental design, numerous investigators have found that the visually evoked responses of visual cortical neurons are generally more robust when attention is directed to their RFs. The fact that the strength of visual representations is dynamic with respect to the locus of attention has altered the view of visual cortical neurons as passive filters to one in which they play an active role in constructing visual representations according to behavioral demands. But most important, as it is known that human visual perception is heightened at the locus of attention (Cameron et al., 2002), the observation of attention-driven modulation in monkey visual cortex provides a striking neural correlate of a fundamental cognitive function. This correlate allows us to link the activity of single neurons within the monkey brain to a basic human mental faculty.

Before it had been established that covert attention effects were widespread within visual cortex, it had been observed that visually evoked neuronal responses in a number of brain structures are enhanced when an animal targets an RF stimulus with a saccade. Wurtz and Goldberg (1972) were the first to note this *presaccadic* enhancement among neurons within the superficial layers of the SC, again a structure known to play a role in the programming of visually guided saccades. Subsequently, Mountcastle and colleagues (1975) observed a similar effect in posterior parietal cortex. Later studies demonstrated that the presaccadic visual enhancement is also observed among neurons in visual cortex, namely in area V4 (Fischer & Boch, 1981), and in inferior temporal cortex (Chelazzi et al., 1993). Although there is a

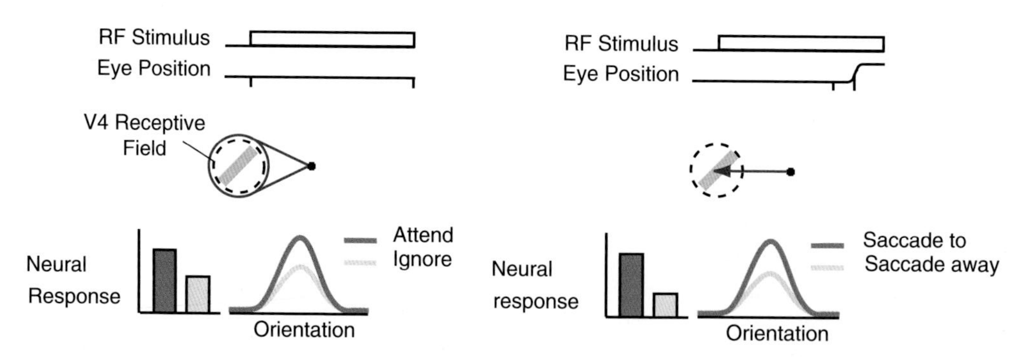

Figure 12.1 Similarity of the effects of covert and overt attention on the visual responses of neurons in visual cortex. Top left: In studies of covert attention, monkeys are trained to attend (red searchlight icon) to a peripheral stimulus (yellow bar) positioned within a neuron's receptive field (dotted circle) while remaining fixated. The event plot at the top indicates the period over which the activity of the cell is compared in the attended and unattended conditions (tick marks on stable eye position). Top right: In studies of overt attention a monkey is trained to make saccades to stable visual stimuli after a delay and the effects on the presaccadic responses are compared between the different conditions. The event plot at the top indicates the period during which the activity of the cell is compared (tick marks prior to time of saccade) in the overtly attended (saccade to receptive field) and unattended (saccade away) conditions. The bottom plots (left and right) summarize the basic results obtained: Visual responses are enhanced when either type of attention is directed to the receptive field stimulus, and the greatest enhancement corresponds to the stimulus that best drives the neuron.

striking parallel apparent between the presaccadic enhancement of visual responses throughout the visuo-oculomotor axis and the attention-related modulation in visual cortex (Fig. 12.1, right), a connection between the two phenomena was ambiguous for many years. This ambiguity was due primarily to the fact that the influence of saccade preparation (overt attention) on visual activity in some structures appeared to be contingent upon a saccade actually being made. In other words, the covert and overt attention effects seemed to occur in different neural structures. For example, in earlier experiments, there was little or no evidence that directing covert attention to visual stimuli presented to SC or FEF RFs had any effect on neuronal activity, whereas activity was found to be robustly modulated by saccades (e.g. Goldberg & Bushnell, 1981). However, more recent studies have indeed found modulation of SC and FEF neuronal activity during covert attention (Ignashchenkova et al., 2004; Thompson et al., 2005). Thus, there is now good evidence that visual responses, whether exhibited by neurons more closely related to saccades or more closely related to coding the parameters of visual stimuli, are modulated both by covert and overt attention.

SACCADE PREPARATION AND ATTENTION

A number of psychophysical studies have revealed evidence linking covert attention and the preparation of saccades. Given that the visual and saccadic systems of the primate brain are heavily interconnected (Fig. 12.2), such links should not be surprising. Generally, this evidence suggests one or both of two possibilities: that visual spatial attention influences saccade preparation or that saccade preparation influences the deployment of visual spatial attention. Perhaps the most classic evidence of the influence of directed attention on saccades is that provided by Sheliga and colleagues (1994). In this study, the authors examined how covert attention perturbs saccade trajectories. Subjects were instructed to initiate saccades to a location in one half of the visual field according to cues presented in the other half. The cues could be presented in one of several locations in the cued half of the visual field. The authors found that saccade trajectories were systematically deviated according to the location of the covertly attended (i.e., cued) location. Thus, the deployment of covert attention can perturb saccade programming. This observation and similar findings from other investigators are frequently cited as evidence for what is commonly referred to as the *premotor theory of attention*, a modern incarnation of earlier *motor theories of cognition* (see Moore et al., 2003, for review).

To address the role of saccade preparation in the deployment of attention, Hoffman and Subramaniam (1995) instructed subjects to make saccades to a location while also detecting visual targets presented prior to the eye movement. These authors found that target detection was typically best at the location of planned saccades and that subjects were unable to completely dissociate the attended and saccade locations. These results suggest that the preparation of saccades to a location deploys attention to that location. Deubel and Schneider (1996) observed similar results and concluded that a single mechanism must drive the selection of objects for perceptual processing and the information needed to drive the appropriate saccadic response.

Although these psychophysical studies provide a sound rationale for the hypothesis that oculomotor and attentional circuits have a common neural substrate, they cannot rule out the alternative view that these are two distinct and independent systems. The results from recording and lesion studies suggest that the direction of spatial attention is caused by neural activity within saccade-related structures (Moore et al., 2003). If this is the case, then it should be possible to direct attention by perturbing neural activity within these same structures. Recent studies have used electrical microstimulation to test the hypothesis that saccade-related circuits play a causal role in driving attention to particular locations in the visual field (Fig. 12.3). While it had long been known that microstimulation of the FEF and the SC elicits saccades, Moore and Fallah (2001) were

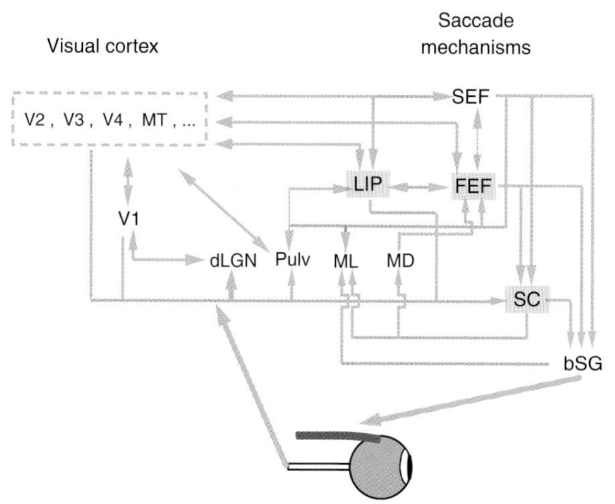

Figure 12.2 A schematic of the major pathways carrying visual information in from the retina and then back to the eye muscles in the form of saccade commands. Several oculomotor structures involved in planning and triggering visually guided eye movements, and their connections (gray arrows), are shown. Areas such as the frontal eye field (FEF), the lateral intraparietal area (LIP) and the superior colliculus (SC) (yellow boxes) are each involved in transforming visual information into saccade commands. Each of these areas is connected to each other, and to visual cortex (dotted box with sample visual areas and V1), and the SC, supplementary eye field (SEF), and FEF also have projections directly to the brainstem saccade generator (bSG). Recent neurophysiological studies have uncovered a role of the FEF, lateral intraparietal area (LIP), and SC in the allocation of visual spatial attention. IML, internal medullary lamina of the thalamus; dLGN, dorsal lateral geniculate nucleus; MD, medial dorsal nucleus of the thalamus; Pulv, pulvinar nucleus.

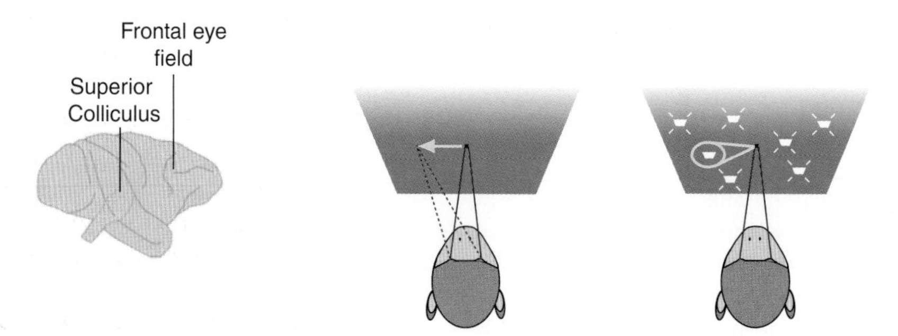

Figure 12.3 Electrical microstimulation of sites within the frontal eye field (FEF) and the superior colliculus (SC) can elicit both shifts in gaze and in visual spatial attention. Left cartoon indicates the location of the FEF in the anterior bank of the arcuate sulcus in prefrontal cortex, and the location of the SC in the midbrain. Microstimulation of either structure has long been known to elicit saccades (yellow arrow, middle diagram) that are identical to voluntary ones. Recent studies demonstrate that microstimulation using submovement stimulation currents increases covert attention (spotlight icon, right diagram) at the location to which suprathreshold currents shift gaze. The right diagram depicts the monkey performing a covert attention task in which it must ignore distracters (flashing squares) and respond to a dimming of a particular target (as in Moore, T., & Fallah, M. (2001). Control of eye movements and spatial attention. *Proceedings of the National Academy Sciences, 98*, 1273–1276.)

the first to examine the effect of microstimulation on visual perception. They found that when they stimulated sites within the FEF using subthreshold currents (i.e., currents below that needed to evoke a saccade), they could nonetheless enhance monkeys' abilities to detect luminance changes in a visual target. Importantly, the improvements observed with microstimulation were dependent upon the target of attention being positioned at the location to which suprathreshold microstimulation would shift the monkey's gaze. Furthermore, the enhancing effect of microstimulation depended on the temporal synchrony of the microstimulation and the luminance change. Specifically, the more temporal overlap, the bigger the improvement. Thus, it appeared that by increasing the likeli hood that the monkey would foveate a location in visual space, the experiments also enhanced the animal's ability to process visual events there. Subsequently, two other studies reported similar enhancements in visual spatial attention following subthreshold microstimulation of the SC (Cavanaugh & Wurtz, 2004; Muller et al., 2005). In both cases, the performance-enhancing effects of microstimulation were spatially dependent as in the FEF studies.

In all of the aforementioned studies, in which electrical microstimulation of a saccade-related structure produces spatially and temporally specific improvements in visual perception, it is important to consider how direct the effects are. Although one assumes that the results indicate that microstimulation directly strengthens the deployment of attention at the target, and therefore reveals an underlying role of the excited neurons (both at the site of stimulation and among neurons connected with neurons at the site), it is important to entertain any other possible explanations for the results. To date, the alternative explanation most entertained is that microstimulation of the FEF and/or the SC generates a visual percept (or other "experience") at the corresponding point in space, for example, a "phosphene" (Cavanaugh & Wurtz, 2004; Moore & Fallah, 2004; Muller et al., 2005; Murphey & Maunsell, 2008), and it is that percept that in turn strengthens spatial attention at the corresponding point in space. Because the experimenter is ultimately unable to measure the impact of microstimulation on monkeys' perceptual experiences, there is probably no way of completely eliminating this possibility. However, there are a few key observations that appear to make that explanation unlikely. First, as previously mentioned, there is a strict dependence of the enhancement effect on the synchrony of FEF microstimulation and the target event (Moore & Fallah, 2004). Specifically, when microstimulation is initiated more than 150 ms prior to the target event, there is no reliable improvement in performance, while the biggest improvement is seen when the microstimulation train temporally overlaps with the target event. This temporal dependence places severe limitations on the speed at which a perceptual experience could have indirectly shifted attention. Cueing with visual stimuli typically requires more than 50 ms between the cue and the target event to improve detection performance (Posner & Cohen, 1984). Moreover, when the cue and the target event occur within less than 50 ms of each other, performance can actually be worse (Yeshurun & Carrasco, 1999), presumably because of *masking* (Breitmeyer & Ganz, 1976). A second set of observations that seems to rule against the hypothesis that attentional improvements were indirectly due to microstimulation-driven perceptual experiences is that the use of artificial "phosphenes" have thus far failed to reproduce the effects of microstimulation (Cavanaugh & Wurtz, 2004; Muller et al., 2005). That is, simply replacing SC microstimulation with an equally long presentation of a Gaussian "blob" at the target location is not enough to produce the same improvements observed with microstimulation.

A third study was specifically aimed at measuring the perceptual impact of FEF microstimulation in monkeys. Murphey and Maunsell (2008) trained monkeys to detect electrical microstimulation of sites within the FEF using currents similar to those used in the attention studies and across the range used to evoke saccades. As in an earlier study conducted by the same authors (Murphey & Maunsell, 2007), it was observed that monkeys could detect microstimulation currents significantly *below* those needed to evoke saccades. Monkeys could reliably report currents that ranged between 60% and 70% of the saccade-evoking current. This indicates that although subthreshold currents do not result in an observable change in behavior, it does not mean that the animal does not experience them. *What* the animal experiences, however, is still very much an open question. Experiments conducted in humans with surgically implanted electrodes above and near the human homolog of the FEF seem to show that microstimulation at saccade-evoking sites does not produce any visual experiences (Blanke et al., 2000), only a sense that gaze is not under the subjects' control (O. Blancke, personal communication). However, by determining the current monkeys can detect, and by demonstrating that that current can be predicted by the current required to evoke a saccade (threshold current), the authors seemed to have validated the use of the "50% of threshold" rule adopted in studies of effects on attention (Cavanaugh & Wurtz, 2004; Cavanaugh et al., 2006; Moore & Fallah, 2001, 2004). This rule was originally adopted because it was found that currents that exceeded ≈75% of the movement threshold current tended to impede attention performance rather than improve it (see Fig. 2A, Moore & Fallah, 2004). The combined results of the two FEF studies suggest that whatever the perceptual "experience" induced by microstimulation, it is typically absent with currents that are 50% of the movement threshold, and when it does occur it impedes performance. Lastly, a recent study by Elsley and colleagues (2007) suggests that rather than visual "phosphenes," subthreshold "experiences" might instead be neck muscle proprioception. These authors report robust, short-latency neck muscle twitches with FEF microstimulation. Importantly, they found that these neck muscle twitches frequently occurred even in the absence of evoked saccades. This raises the possibility that even in the absence of measurable gaze shifts when delivering "subthreshold" currents to the FEF, proprioceptive sensations via neck muscle twitches may nonetheless be delivered.

The effect of electrical microstimulation of the FEF and the SC on visual spatial attention performance in monkeys suggests that during stimulation, visual representations of particular stimuli are enhanced above those of others. As mentioned earlier, a wealth of evidence has established that correlates of visual attention can be observed throughout much of macaque visual cortex; visual responses to RF stimuli are typically enhanced when a monkey is trained to covertly attend to them. Thus, one might expect that microstimulation of the FEF or the SC will likewise result in the enhancement of the responses of visual cortical neurons to RF stimuli. Indeed, this was the hypothesis posed by Moore and Armstrong (2003), who paired FEF microstimulation with single-neuron recording in extrastriate area V4. Specifically, they studied the impact of subthreshold microstimulation of FEF sites on the visual responses of V4 neurons with RFs that were either overlapping or nonoverlapping with the spatial location represented at the FEF site. They found that following very brief trains of FEF stimulation (20 to 50 ms), visual responses were enhanced at overlapping locations in the visual field, whereas they were suppressed at nonoverlapping locations. Furthermore, this attention-like modulation was greatest when there was another "distracter" stimulus presented outside of the V4 neuron's RF. A subsequent study also found that FEF microstimulation can alter the structure of V4 RFs and can bias V4 responses in favor of one of two RF stimuli (Armstrong et al., 2006). Importantly, these changes in V4 responses are very similar to what has been observed in the same area during voluntary attention (Armstrong & Moore, 2007) (Fig. 12.4). Together, these results suggest that microstimulation of saccade-related structures drives covert attention and its correlates in visual cortex, and that indeed, as speculated by Latto and Cowey (1971), visual spatial attention seems to be the result of feedback from the oculomotor system onto visual cortical areas, and this feedback influence may exist in order to modulate processing according to eye movement plans.

In summary, recent neurophysiological work has directly addressed and found much evidence of a role of saccade mechanisms in directing covert visual spatial attention. This evidence

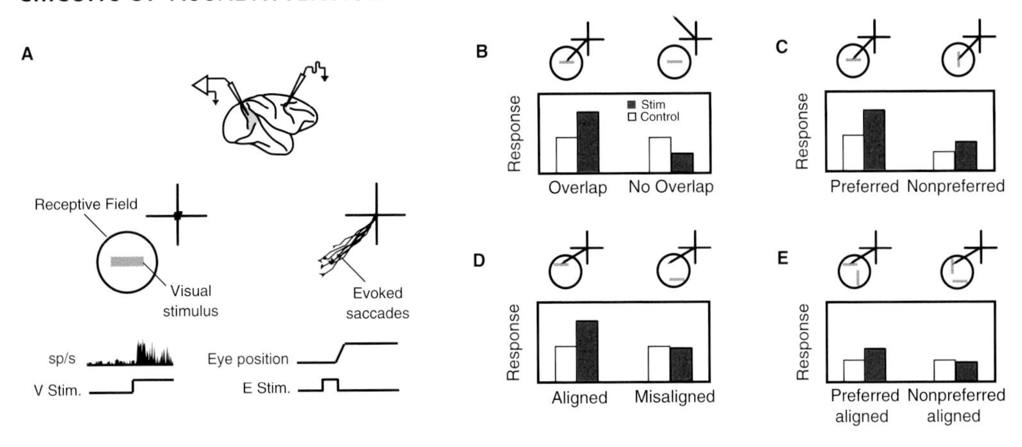

Figure 12.4 Modulation of visually driven responses of area V4 neurons with microstimulation of the frontal eye field (FEF). Sites within the FEF were electrically stimulated while recording from neurons in area V4. Cartoon shows a side view of the macaque brain. **(A)** Top: The locations of the FEF in the anterior bank of the arcuate sulcus and area V4 in the prelunate gyrus and below the inferior occipital sulcus are shown (shaded). Bottom left: The response of a V4 neuron when a visual stimulus (V Stim) appears in its receptive field (RF) while the monkey maintains central fixation (dots at origin). Bottom right: Eight saccades evoked to a location overlapping the V4 RF following microstimulation of an FEF site (E stim). **(B—E)** Subthreshold microstimulation of the FEF increases the visual activity ("Response") of V4 neurons with receptive fields (circles) that overlap with the saccade represented at the stimulated FEF site (arrow from crosshairs), but decreases visual activity when the receptive fields and the saccades do not overlap (B). The gray bar in each receptive field represents the visual stimulus used to drive the V4 response; in this example, the neuron prefers horizontal bars to vertical ones. The magnitude of enhancement is greatest for stimuli that evoke the largest response (C). The enhancement of V4 visual responses also depends on the spatial alignment of the receptive field stimulus and the FEF saccade vector *within* the RF (D). FEF microstimulation also biases the responses of neurons to two RF stimuli in favor of the one aligned with the saccade vector (E).

appears to support the conclusion that the neural mechanisms of covert and overt spatial attention are identical with respect to the influence of the saccade plan on visual representations. Although it should be assumed that a different set of mechanisms is in effect when attention is directed overtly (i.e., when planned movements are triggered and shifts in gaze are actually carried out) than when it is directed covertly, recent results suggest that those differences are less important than the similarities.

Simultaneous Control of Attention and Saccades

The link between attention and saccades, and the apparent dual role of saccade-related structures in controlling both of these phenomena, raises the question of how the two processes interact during visually guided behavior. Specifically, how does the brain select the target of visual attention, enhancing the perception of certain features within the visual scene, and simultaneously use information about these same features to specify an appropriate saccade plan? Depending on the answer to this question, one could imagine that perturbing neural activity in an oculomotor region might have one of several different effects: First, if visual selection and saccade specification are independent modules, it is possible that an experimental perturbation of neural activity could alter a saccade plan without noticeable effects on perception. Alternatively, the perturbation of oculomotor activity could affect the perception of a visual target, presumably by modulating feedback signals to visual cortex. According to this possibility, the altered perception could be independent of a concurrently developing saccade plan, and thus the perturbation might uncouple the link between perception and saccades. Finally, if perception and saccade selection are reciprocally linked, the perturbation might cause both a difference in the perception of a target and an altered saccade plan that is consistent with the new percept, rather than with the perturbation itself.

Schafer and Moore (2007) tested these possible outcomes by using a paradigm in which electrical stimulation of an FEF site pitted the potential perceptual effects of stimulation against the saccadic effects. When monkeys made voluntary saccades to a sinusoidal grating drifting within a stationary aperture, the endpoints of their saccades were biased in the direction of the grating motion, consistent with an "apparent position" illusion. This illusion, and the motion-induced bias of the saccades away from the center of the visual target, allowed the authors to separate the veridical position of the target in visual space from the perception of the target, and thus the endpoints of the saccades (Fig. 12.5).

Low-frequency electrical stimulation was then delivered via a microelectrode to an FEF site corresponding to the veridical position of the center of the grating while the monkey planned and executed its targeting saccade. The authors suggested at least two possible consequences of microstimulation: First, the effect of stimulation on saccade planning might

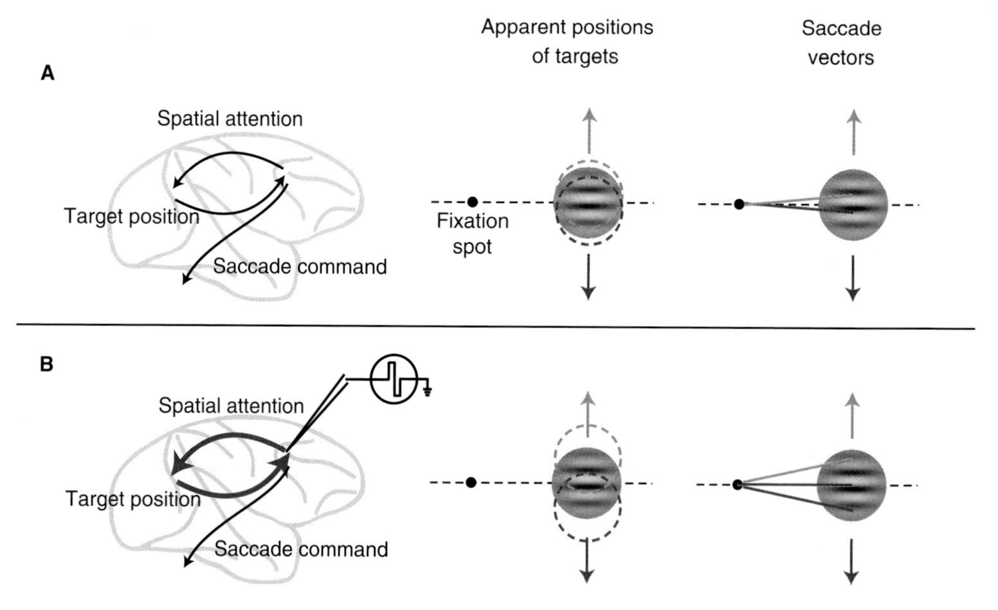

Figure 12.5 Interaction of attentional and saccadic signals during visually guided behavior. **(A)** Left: Projections from the frontal eye field (FEF) carry spatial information to visual cortex and saccade signals that ultimately redirect gaze, and visual target information is sent from visual cortex to the FEF. Middle: When a sinusoidal grating presented in the periphery drifts upward (green arrow) or downward (blue arrow), its apparent position is shifted in the direction of motion (green and blue dotted circles). Right: Voluntary saccades to the drifting gratings are biased in the direction of motion (green and blue lines), consistent with the apparent position illusion. **(B)** Left: Microstimulation of spatially aligned representations in the FEF during preparation of saccades seems to enhance the interaction between the FEF and visual cortex (red lines). Middle: During stimulation the apparent position illusion is increased, and the gratings appear to be located farther from the veridical position. Right: Voluntary saccades to the drifting gratings are influenced more strongly by grating motion, and are biased farther from the stimulation site's representative saccade vector (red line).

dominate any effects on attention and perception, thereby influencing the monkey's saccade plan alone. This potential outcome would cause the saccade to land closer to the center of the part of space represented by the FEF stimulation site and therefore toward the center of the grating, thus eliminating the motion-induced saccade bias. Alternatively, if the attentional role of the FEF interacts with and informs the saccadic role, microstimulation could lead to an enhancement of the apparent position illusion and a resulting increase in the motion-induced saccade bias. The authors observed the latter: When voluntary saccades were paired with low-frequency stimulation, the effect of the motion-induced illusion on saccade trajectories was enhanced, not decreased (Fig. 12.5). This outcome indicates that the attentional effects of this FEF perturbation govern the simultaneously planned saccades. More generally, the results suggest that the feedback connections from the FEF to visual cortex are integral both for appropriately perceiving the visual world and for preparing precise, target-guided saccades.

GAZE CONTROL AND MULTISENSORY ATTENTION

In describing the evidence to date on the mechanisms within the primate brain that guide the selection of targets of interest, we have focused solely on the visual modality. While this may be appropriate given the dominance of the visual modality in primate species, it is nonetheless important to consider whether the control of visual attention is a special case. How might the findings of a role of saccade mechanisms in visual attention generalize to other sensory modalities? Could it be that biasing spatially directed movements facilitates other types of sensory processing at the intended location of action? Winkowski and Knudsen (2006) provide evidence that the central nervous system uses a common gaze-related strategy for dynamically regulating sensory gain across modalities, even in nonprimate species. Neurons within the optic tectum of the barn owl comprise a map of auditory space (Cohen and Knudsen, 1999). The owl's gaze direction is controlled in part by neurons within the archistriatal gaze field (AGF), in the forebrain, which and is considered homologous with the FEF in primates. Microstimulation of sites within the AGF evokes short-latency head movements of a given direction (Knudsen et al., 1995). Winkowski and Knudsen (2006) examined the influence of subthreshold microstimulation of AGF on auditory responses of tectal neurons and found that responses to stimuli in the space represented by the stimulated AGF site were enhanced. They also found that AGF microstimulation sharpens the tuning of auditory receptive fields in the optic tectum and tends to shift spatial auditory tuning toward the auditory receptive field encoded by the AGF stimulation site (Winkowski & Knudsen, 2007). Since neurons in the optic tectum are multimodal and exhibit visual responses, the authors were also able to demonstrate that AGF microstimulation enhances visual signals as well. Thus, evidence that gaze control mechanisms provide spatial attention-like biases in sensory processing extends to the auditory domain, and to nonprimate species.

The possibility that mechanisms controlling the deployment of visual spatial attention also influence other sensory modalities is not entirely surprising. Psychophysical and electrophysiological studies have suggested that multisensory integration occurs at very early stages of sensory processing, and perhaps before attentional selection is accomplished. Driver (1996) exploited the well-known ventriloquism illusion in which the speaker's lips appear to "capture" a sound and translocate it when there is a spatial offset between the real auditory and apparent visual source of the sound. He showed that this cross-modal illusion can enhance selective spatial attention to speech sounds. It is based on the observation that under some circumstances, ventriloquism can apparently influence the ease of attentional selection. Participants had to repeat one of two concurrent auditory messages, while they were both played in mono from the same loudspeaker without any auditory cue that reliably distinguishes the relevant message from the concurrent distracting message. The distinction was by lip movements presented visibly, for the relevant message only. The performance was improved for recognizing the relevant message when this concurrent visual information was presented at a slightly different location from the mono sound source. Driver attributed this outcome to ventriloquism arising for just the relevant sounds toward the concurrent lip movements that matched them temporally in the spatially offset condition. It was argued that such visually determined illusory spatial separation could only aid auditory selection if it arose before that selection was complete, thus implying that the cross-modal integration underlying ventriloquism might be preattentive in this particular sense. The idea was supported later by studies showing that the ventriloquism effect did not depend on the extent to which the visual events were attended (Bertelson et al., 2000; Spence & Driver, 2000; Vroomen et al., 2001). Using event-related potentials in humans, Giard and Peronnet (1999) provided evidence that multisensory integration can take place very early in the processing of auditory and visual information.

SUMMARY REMARKS

For primates, the gathering of sensory information is an active process. This is particularly true for the dominant sense, vision, which requires the organism to sample discrete points of the visual environment in a sequential fashion. This sampling is accomplished primarily by moving the focus of gaze from item to item with head and eye movements, most notably saccades. Neurophysiological studies have recently revealed that the visual and saccadic systems of the primate brain are not only heavily interconnected but also that the influence of the two processes on one another is bidirectional. Most notably, it appears that the faculty of attention, a fundamental cognitive function, emerges from the reciprocal interaction of visual processing and saccade preparation. Above its significance for understanding the neural basis of cognition, this revelation has not only exposed the limitations of a strictly sensory or motor approach to neural systems but also reaffirmed the need for a more ethological approach to the question of how particular types of brains solve the particular classes of problems faced by an organism. One assumes that the need for more ethological approaches extends to even the most complex and mysterious of functions within our own primate brain.

REFERENCES

Armstrong, K. M., Fitzgerald, J. K., & Moore, T. (2006). Changes in visual receptive fields with microstimulation of frontal cortex. *Neuron, 50*, 791–798.

Armstrong, K. M., & Moore, T. (2007). Rapid enhancement of visual cortical response discriminability by microstimulation of the frontal eye field. *Proceedings of the National Academy of Sciences, 104*, 9499–9504.

Bertelson, P., Vroomen, J., de Gelder, B., & Driver, J. (2000). The ventriloquist effect does not depend on the direction of deliberate visual attention. *Perception & Psychophysics, 62*, 321–332.

Blanke, O., Spinelli, L., Thut, G., Michel, C. M., Perrig, S., Landis, T., et al. (2000). Location of the human frontal eye field as defined by electrical cortical stimulation: Anatomical, functional and electrophysiological characteristics. *Neuroreport, 11*, 1907–1913.

Breitmeyer, B. G., & Ganz, L. (1976). Implications of sustained and transient channels for theories of visual pattern masking, saccadic suppression, and information processing. *Psychology Review, 83*, 1–36.

Cameron, E. L., Tai, J. C., & Carrasco, M. (2002). Covert attention affects the psychometric function of contrast sensitivity. *Vision Research, 42*, 949–967.

Cavanaugh, J., Alvarez, B. D., & Wurtz, R. H. (2006). Enhanced performance with brain stimulation: Attentional shift or visual cue? *Journal of Neuroscience, 26*, 11347–11358.

Cavanaugh, J., & Wurtz, R. H. (2004). Subcortical modulation of attention counters change blindness. *Journal of Neuroscience, 24*, 11236–11243.

Cave, K. R. (1999). The FeatureGate model of visual selection. *Psychology Research, 62*, 182–194.

Chelazzi, L., Miller, E. K., Duncan, J., & Desimone, R. (1993). A neural basis for visual search in inferior temporal cortex. *Nature, 363*, 345–347.

Cohen, Y. E., & Knudsen, E. I. (1999). Maps versus clusters: Different representations of auditory space in the midbrain and forebrain. *Trends in Neuroscience, 22*, 128–135.

Desimone, R., & Duncan, J. (1995). Neural mechanisms of selective visual attention. *Annual Review of Neuroscience, 18*, 193–222.

Deubel, H., & Schneider, W. X. (1996). Saccade target selection and object recognition: evidence for a common attentional mechanism. *Vision Research, 36*, 1827–1837.

Ding, L., & Hikosaka, O. (2006). Comparison of reward modulation in the frontal eye field and caudate of the macaque. *Journal of Neuroscience, 26*, 6695–6703.

Driver, J. (1996). Enhancement of selective listening by illusory mislocation of speech sounds due to lip-reading. *Nature, 381*, 66–68.

Dubner, R., & Zeki, S. M. (1971). Response properties and receptive fields of cells in an anatomically defined region of the superior temporal sulcus in the monkey. *Brain Research, 35*, 528–532.

Elsley, J. K., Nagy, B., Cushing, S. L., & Corneil, B. D. (2007). Widespread presaccadic recruitment of neck muscles by stimulation of the primate frontal eye fields. *Journal of Neurophysiology, 98*, 1333–1354.

Fecteau, J. H., & Munoz, D. P. (2007). Warning signals influence motor processing. *Journal of Neurophysiology, 97*, 1600–1609.

Felleman, D. J., & Van Essen, D.C. (1991). Distributed hierarchical processing in the primate cerebral cortex. *Cerebral Cortex, 1*, 1–47.

Fischer, B., & Boch, R. (1981). Enhanced activation of neurons in prelunate cortex before visually guided saccades of trained rhesus monkeys. *Experimental Brain Research, 44*, 129–137.

Francolini, C. M., & Egeth, H.E. (1979). Perceptual selectivity is task dependent: The pop-out effect poops out. *Perception & Psychophysics, 25*, 99–110.

Giard, M. H., & Peronnet, F. (1999). Auditory-visual integration during multimodal object recognition in humans: a behavioral and electrophysiological study. *Journal of Cognitive Neuroscience, 11*, 473–490.

Goldberg, M. E., & Bushnell, M. C. (1981). Behavioral enhancement of visual responses in monkey cerebral cortex. II. Modulation in frontal eye fields specifically related to saccades. *Journal of Neurophysiology, 46*, 773–787.

Goodale, M. A., & Milner, A. D. (1992). Separate visual pathways for perception and action. *Trends in Neuroscience, 15*, 20–25.

Hoffman, J. E., & Subramaniam, B. (1995). The role of visual attention in saccadic eye movements. *Perception & Psychophysics, 57*, 787–795.

Hubel, D. H., & Wiesel, T.N. (1965). Receptive fields and functional architecture in two nonstriated visual areas (18 and 19) of the cat. *Journal of Neurophysiology, 28*, 229–289.

Hubel, D. H., & Wiesel, T. N. (1968). Receptive fields and functional architecture of monkey striate cortex. *Journal of Physiology, 195*, 215–243.

Ignashchenkova, A., Dicke, P. W., Haarmeier, T., & Thier, P. (2004). Neuron-specific contribution of the superior colliculus to overt and covert shifts of attention. *Nature Neuroscience, 7*, 56–64.

Ikeda, T., & Hikosaka, O. (2003). Reward-dependent gain and bias of visual responses in primate superior colliculus. *Neuron, 39*, 693–700.

Knudsen, E. I., Cohen, Y. E., & Masino, T. (1995). Characterization of a forebrain gaze field in the archistriatum of the barn owl: Microstimulation and anatomical connections. *Journal of Neuroscience, 15*, 5139–5151.

Koch, C., & Ullman, S. (1985). Shifts in selective visual attention: Towards the underlying neural circuitry. *Human Neurobiology, 4*, 219–227.

Kusunoki, M., Gottlieb, J., & Goldberg, M. E. (2000). The lateral intraparietal area as a salience map: The representation of abrupt onset, stimulus motion, and task relevance. *Vision Research, 40*, 1459–1468.

Latto, R., & Cowey, A. (1971). Visual field defects after frontal eye-field lesions in monkeys. *Brain Research, 30*, 1–24.

Li, Z. (2002). A saliency map in primary visual cortex. *Trends in Cognitive Science, 6*, 9–16.

McPeek, R. M., & Keller, E. L. (2002). Saccade target selection in the superior colliculus during a visual search task. *Journal of Neurophysiology, 88*, 2019–2034.

Moore, T., & Armstrong, K. M. (2003). Selective gating of visual signals by microstimulation of frontal cortex. *Nature, 421*, 370–373.

Moore, T., Armstrong, K. M., & Fallah, M. (2003). Visuomotor origins of covert spatial attention. *Neuron, 40*, 671–683.

Moore, T., & Fallah, M. (2001). Control of eye movements and spatial attention. *Proceedings of the National Academy Sciences, 98*, 1273–1276.

Moore, T., & Fallah, M. (2004). Microstimulation of the frontal eye field and its effects on covert spatial attention. *Journal of Neurophysiology, 91*, 152–162.

Mountcastle, V. B., Lynch, J. C., Georgopoulos, A., Sakata, H., & Acuna, C. (1975). Posterior parietal association cortex of the monkey: Command functions for operations within extrapersonal space. *Journal of Neurophysiology, 38*, 871–908.

Muller, J. R., Philiastides, M. G., & Newsome, W. T. (2005). Microstimulation of the superior colliculus focuses attention without moving the eyes. *Proceedings of the National Academy Sciences, 102*, 524–529.

Murphey, D. K., & Maunsell, J.H. (2007). Behavioral detection of electrical microsti-mulation in different cortical visual areas. *Current Biology, 17*, 862–867.

Murphey, D. K., & Maunsell, J. H. (2008). Electrical microstimulation thresholds for behavioral detection and saccades in monkey frontal eye fields. *Proceedings of the National Academy Sciences, 105*, 7315–7320.

Navalpakkam, V., & Itti, L. (2005). Modeling the influence of task on attention. *Vision Research, 45*, 205–231.

Navalpakkam, V., & Itti, L. (2007). Search goal tunes visual features optimally. *Neuron, 53*, 605–617.

Platt, M.,L., & Glimcher, P. W. (1999). Neural correlates of decision variables in parietal cortex. *Nature, 400*, 233–238.

Robinson, D. L., Bowman, E. M., & Kertzman, C. (1995). Covert orienting of attention in macaques. II. Contributions of parietal cortex. *Journal of Neurophysiology, 74*, 698–712.

Schafer, R. J., & Moore, T. (2007). Attention governs action in the primate frontal eye field. *Neuron, 56*, 541–551.

Schall, J. D. (1995). Neural basis of saccade target selection. *Reviews Neuroscience, 6*, 63–85.

Schein, S. J., & Desimone, R. (1990). Spectral properties of V4 neurons in the macaque. *Journal of Neuroscience, 10*, 3369–3389.

Sheliga, B. M., Riggio, L., & Rizzolatti, G. (1994). Orienting of attention and eye movements. *Experimental Brain Research, 98*, 507–522.

Spence, C., & Driver, J. (2000). Attracting attention to the illusory location of a sound: reflexive crossmodal orienting and ventriloquism. *Neuroreport, 11*, 2057–2061.

Sperling, G., & Melchner, M. J. (1978). The attention operating characteristic: Examples from visual search. *Science, 202*, 315–318.

Steinmetz, M. A., & Constantinidis, C. (1995). Neurophysiological evidence for a role of posterior parietal cortex in redirecting visual attention. *Cerebral Cortex, 5*, 448–456.

Sugrue, L. P., Corrado, G. S., & Newsome, W. T. (2004). Matching behavior and the representation of value in the parietal cortex. *Science, 304*, 1782–1787.

Thompson, K. G., Biscoe, K. L., & Sato, T. R. (2005). Neuronal basis of covert spatial attention in the frontal eye field. *Journal of Neuroscience, 25*, 9479–9487.

Treisman, A.M. and Gelade, G. (1980). A feature-integration theory of attention. *Cognitive Psychology, 12*, 97–136.

Ungerleider, L. G. & Mishkin, M. (1982). Two cortical visual systems. In: D. G. Ingle, M. A. Goodale, & R. J. Q. Mansfield (Eds.), *Analysis of visual behavior* (pp. 549–586). Cambridge, MA: MIT Press.

Van Essen, D. C., Anderson, C. H., & Felleman, D. J. (1992). Information processing in the primate visual system: An integrated systems perspective. *Science, 255*, 419–423.

Vroomen, J., Bertelson, P., & de Gelder, B. (2001). The ventriloquist effect does not depend on the direction of automatic visual attention. *Perception & Psychophysics, 63*, 651–659.

Winkowski, D. E., & Knudsen, E. I. (2006). Top-down gain control of the auditory space map by gaze control circuitry in the barn owl. *Nature, 439*, 336–339.

Winkowski, D. E., & Knudsen, E. I. (2007). Top-down control of multimodal sensitivity in the barn owl optic tectum. *Journal of Neuroscience, 27*, 13279–13291.

Wolfe, J. M. (1994). Guided Search 2.0 A revised model of visual search. *Psychonomic Bulletin & Review, 1(2)*, 202–238.

Wurtz, R. H., & Goldberg, M. E. (1972). Activity of superior colliculus in behaving monkey. 3. Cells discharging before eye movements. *Journal of Neurophysiology, 35*, 575–586.

Yeshurun, Y., & Carrasco, M. (1999). Spatial attention improves performance in spatial resolution tasks. *Vision Research, 39*, 293–306.

Zeki, S. M. (1974). Functional organization of a visual area in the posterior bank of the superior temporal sulcus of the rhesus monkey. *Journal of Physiology, 236*, 549–573.

CHAPTER 13

Vocalizations as Auditory Objects: Behavior and Neurophysiology

Cory T. Miller and Yale E. Cohen

Vocal communication is a fundamental component of both human and nonhuman animal behavior. Because of the behavioral significance of vocalizations, it is likely that these acoustic signals have had a significant influence on the evolution of the primate auditory system (Ghazanfar & Hauser, 1999; Hauser, 1997; Wordon & Galambos, 1972). Traditionally, it has been thought that the selective forces acting on the primate cortex for vocalization processing were related to the communicative content of the signals (Ghazanfar & Hauser, 1999; Hauser, 1997; Wordon & Galambos, 1972). This, however, represents only one element of vocalizations: As spatiotemporally bounded units (Farris et al., 2005; Miller et al., 2001a), vocalizations are also auditory objects (Griffiths & Warren, 2004). Since objects are one of the most central components of perception, it is likely that, in the primate auditory system, evolution selected for those neural mechanisms that bind the acoustic features into perceptual units. Thus, we propose that an object-centered view of vocalization processing may yield new insights into the nature of the auditory system as well as reveal potential parallels with object processing in the visual system. In this chapter, we (1) provide some background on objects and their categorical organization, (2) review the behavioral significance of nonhuman primate vocalizations, (3) discuss a general framework for auditory-object analysis in the context of vocalizations, and (4) overview neurophysiological studies that examine vocalization processing within this framework with an emphasis on object and categorical processing.

OBJECTS

Objects are a core building block of our perceptual world. Yet, despite the significant amount of empirical and theoretical scientific energy that is devoted to studying objects (Nakayama, 1999; Pylyshyn, 2006; Scholl, 2007; Scholl et al., 2001; Spelke, 1994; Yantis, 2000), a formal definition has been difficult to establish (Feldman, 2003). This discrepancy is primarily due to the fact that objects cannot be defined by the mere presence or absence of particular features. Rather, objects come into existence (awareness) through nonlinear grouping principles that bind features together into perceptually stable spatiotemporal units. Importantly, although these grouping mechanisms are necessary for object formation, no single one is sufficient for object perception. Most of what is known about objects comes from studies in the visual system. This research agenda comprises studies in human adults and infants, as well as nonhuman primates (Hauser, 2001; Santos, 2004). As discussed below, parallel work on auditory objects, though less extensive, is notably similar to the body of literature on visual objects. Together, these data suggest that commonalities in object perception may exist across developmental states, species, and perhaps sensory modalities.

The notion that grouping principles govern object perception began with the Gestalt psychologists (Kaffka, 1935). More recently, Spelke sought to quantify these principles experimentally in a series of studies with human infants (Spelke, 1994). This work suggests that visual-object perception is guided by three primary grouping principles: cohesion, continuity, and contact (Fig. 13.1). First, the principle of cohesion states that objects persist with stable boundaries: Sand and other substances with unstable edges, such as sands and liquids, are not perceived as objects (Huntley-Fenner et al., 2002). Moreover, as an object moves through space, these stable boundaries persist. Second, the principle of continuity states that objects moving through space and time follow a single continuous path. During experiments in which a section of an object's path is occluded, for example, subjects expect the object to follow the same trajectory behind the occluder (Spelke et al., 1994). And third, the principle of contact states

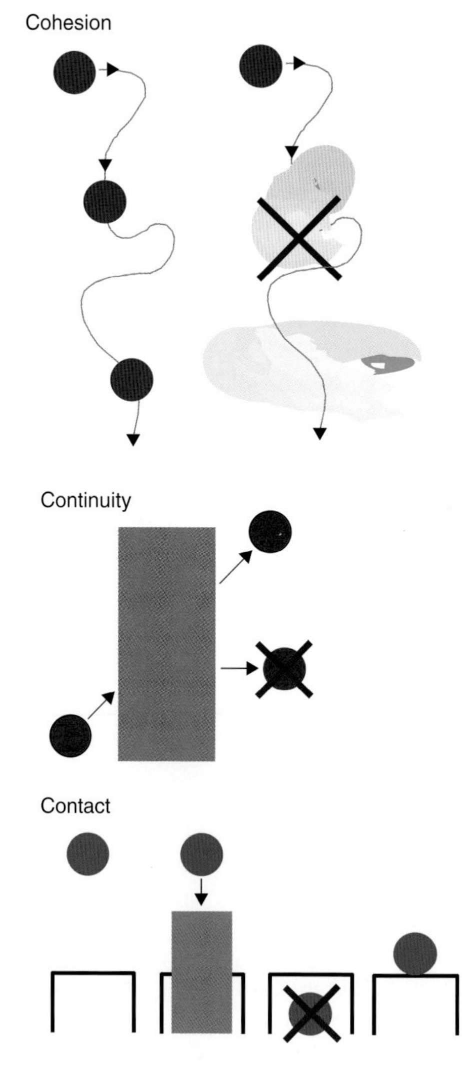

Figure 13.1 Drawings depict the three principles of object perception (Spelke, 1994). Cohesion: Objects have definable, consistent boundaries. Continuity: Objects traveling on a path continue on the same trajectory even when occluded. Contact: Two objects cannot occupy the same spatial location at the same time. Purple X's are shown on events that are inconsistent with these object properties.

that because two objects cannot occupy the same spatial location at the same time, an object that moves into another object's location will either have its motion stopped or displace the other object. In a classic experiment, Spelke and colleagues (1992) presented young infants with a test display in which a ball was dropped onto a table. Even when the path of the ball was occluded, the infants expected the ball to come to rest on top of the table and not fall through the solid surface. Together, these studies provide empirical evidence of the perceptual principles that govern object perception in young infants.

Parallel work suggests that similar grouping principles govern object perception in adult human vision as well (Feldman & Tremoulet, 2006; Pylyshyn, 2006; Scholl, 2001). Specifically, human adults perceive visual objects as spatiotemporally bound units. This research also suggests that attention and object perception are intricately entwined, such that what is attended may constitute an object. Scholl and colleagues (Pylyshyn & Storm, 1988; Scholl et al., 2001) employed the multiple-object tracking paradigm (Sears & Pylyshyn, 2000) to examine the role of attention in object perception. They found that perceptual units whose properties were consistent with Spelke's grouping principles (Spelke, 1994) could be tracked through space and time, whereas perceptual units with ambiguous boundaries, such as the ends of lines, could not. Together, these studies suggest that the same perceptual mechanisms underlying object perception in human infants persist during adulthood and that attention is critical for object perception.

Importantly, these grouping principles extend to object perception in nonhuman primates as well. For example, Hauser (2001) showed that nonhuman primates understand the principle of contact as it relates to object perception (Spelke et al., 1994). A study by Santos (2004) tested whether rhesus monkeys can use spatiotemporal information when determining the location of a moving object. Monkeys viewed a display in which a plum rolled down a ramp and behind an occluder. During certain key test conditions, two occluders separated by several inches were placed along the path of the plum. Santos found that the monkeys searched for the plum behind the first occluder. The interpretation of this study is that the monkeys reasoned that if the plum was not visible in the gap between the occluders, it must have stopped behind the first occluder. Thus, the use of spatiotemporal grouping principles for visual objects appears to be conserved across at least two primate species. The question now is whether these same mechanisms are evident in other sensory modalities.

In audition, the study of auditory objects is typically referred to as "auditory-scene analysis," in reference to Bregman's seminal book on the subject (Bregman, 1990). Bregman proposed two forms of perceptual integration—sequential and simultaneous—which have become a key framework for thinking about the perceptual and neurophysiological processes underlying auditory-object formation. Sequential integration involves two complementary processes. The first process is the integration of temporally separated sounds from one sound source (e.g., syllables, words, or vocalizations) into a coherent auditory stream. The second process is segregating sounds that originate from separate sources, especially intervening and overlapping sounds. Simultaneous integration refers to the perceptual grouping of different, simultaneously occurring, and often harmonically related components of the frequency spectrum into a single sound source. As in the visual system, the integration of spatiotemporal properties is fundamental to auditory-object formation.

Behavioral evidence suggests that both sequential and simultaneous integration are utilized during vocalization perception. This integration is perhaps best exemplified by amodal completion (also referred to as auditory continuity) (Bregman, 1990). Auditory continuity can be seen when cotton-top tamarins are presented with vocalizations in which the middle segment is either (1) left intact, (2) "occluded" by white noise, or (3) deleted entirely (Miller et al., 2001a) (Fig. 13.2). When presented with an occluded call, the tamarins respond to this call in the same manner as they respond to the intact call. Their response, however, diminishes when the middle segment is deleted, suggesting that tamarins can integrate acoustic information across different frequency spectra to form a

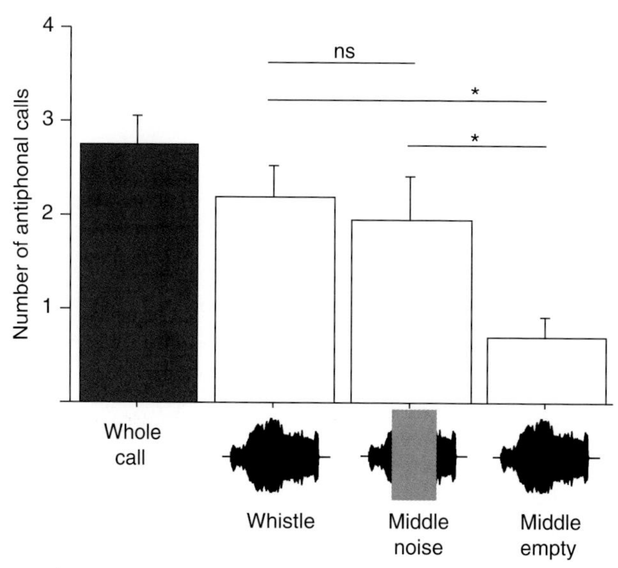

Figure 13.2 Evidence of amodal completion for vocalizations (Miller et al., 2001a). Above: representative amplitude waveforms for the three test stimuli. Below: The mean number (standard error) of antiphonal calls per session is shown in the bar graph for each stimulus. Statistically significant differences are marked with an asterisk. Used with permission.

coherent vocalization, even when a portion of the vocalization is occluded. Given that similar grouping has been observed in human auditory (Bregman, 1990; Kubovy & Van Valkenburg, 2001) and visual (Kellman et al., 1986) perception, it is likely that this grouping principle of cohesion is conserved across primate species as well as sensory modalities.

Determining which units in the world are objects is a complex theoretical and empirical problem. The task in audition is somewhat more difficult than vision because sounds are spatially omnipresent and temporally continuous. As such, establishing the boundaries of sounds in space and time during real-world situations is experimentally difficult. Vocalizations, however, represent an acoustic unit that may not succumb to space- and time-boundary issues that generally surround auditory-object processing for at least two reasons. First, since a vocalization is produced from a single source, the auditory system can assume spatial continuity. Second, since the temporal structure of vocalizations is stereotyped, both in terms of the length and the different acoustic elements within the calls, the temporal boundaries of these objects will be perceptually predictable. Consequently, vocalizations represent a unique and potentially insightful line of research in understanding auditory objects. Rather than question whether or not the unit under study is an object based on the grouping principles governing its perception, one can *assume* that vocalizations are objects and then study more explicitly the principles that lead to its perceptual grouping.

In conclusion, objects are among the most thoroughly studied topics in cognitive psychology. Experimental studies of the grouping principles that bind features into objects suggest that the same perceptual mechanisms may underlie object perception in the visual system of human infants and adults as well as nonhuman primates. Although the picture is less clear in audition, certain commonalities are evident. The study of vocalizations as objects may help to learn more about the grouping principles in the auditory system. Such data would inform our understanding of whether a core set of perceptual principles governs object perception across all sensory modalities.

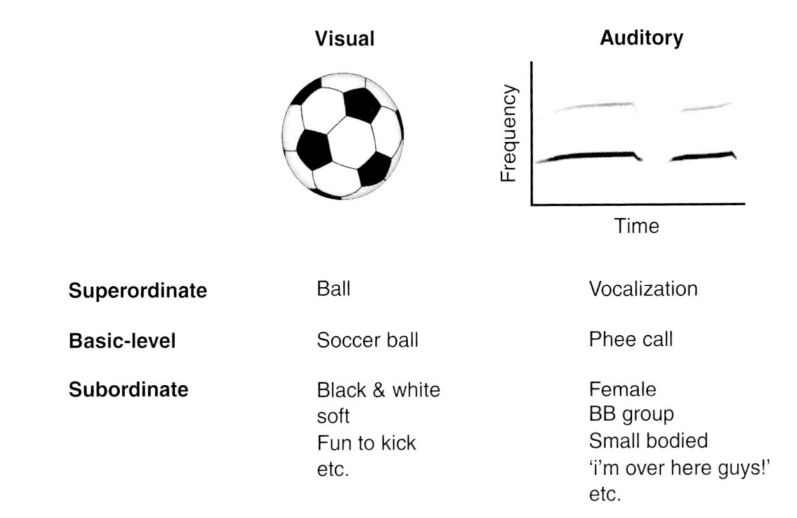

Figure 13.3 The categorical structure of a visual (left) and auditory (right) object. The superordinate, basic-level, and subordinate categorical structure are shown for each of the objects. The auditory object (i.e., vocalization) to the left is shown in a spectrogram, which plots the signal in a time (x-axis) by frequency (y-axis) display.

CATEGORIES

An object is commonly a member of several different categories depending on what rules or mechanisms are used to represent the object. These different categories can be conceptualized as forming a hierarchy (Rosch, 1978) (Fig. 13.3). At the top of the hierarchy are "superordinate" categories, the most general of the levels. "Basic-level" categories are next and are less general than superordinate categories. Basic-level categories are the terms that people most commonly use when describing an object. "Subordinate" categories are found at the bottom of the hierarchy. For example, Lassie can be categorized as an animal, a dog, or a collie. "Animal" is superordinate categorical-level descriptor. "Dog" is the basic-level descriptor. "Collie" is the subordinate descriptor. Using another example, Handel's Messiah can be categorized as "music" (superordinate), "classical music" (basic level), or "baroque-period music" (subordinate). Importantly, a single object can have multiple subordinate categorical descriptors.

The different categories are not equipotent since they require different amounts of neural processing. Subjects respond faster and more accurately when they are asked to categorize objects into basic-level categories than when asked to categorize objects into superordinate or subordinate categories (Rosch, 1978). Also, the level at which an object is categorized depends on a subject's previous experience and knowledge: A person with classical-music training might consider the basic-level categorical descriptor of Handel's Messiah to be "baroque music" or even "oratorio" instead of "classical music" (Gauthier & Logothetis, 1999; Marschark et al., 2004).

Different hypothetical frameworks can be used to describe the relationship between an object's membership in basic-level and more superordinate categories. One framework posits that superordinate categories contain a set of features that belong to all of the members of the more basic category (Damasio, 1989; Devlin et al., 1998; Martin et al., 2002; Rosch et al., 1976; Smith et al., 1974). For instance, the basic-level category of "dog" might contain descriptors like "has fur," "has wet nose," "has four legs," "breathes," "is mobile," "can reproduce on its own," etc. In contrast, the superordinate category of animal contains descriptors that are not included in the basic-level category such as "breathes," "is mobile," "can reproduce on its own," etc. A second alternative view is that the properties of a basic-level category are not omitted from the superordinate category but are represented as more abstract variable values in this higher-order category.

Perceptual and Abstract Categories

In addition to the hierarchical structure of categories for objects, there is at least one other dimension along which objects can be categorized: perceptual and abstract. The primary difference is that perceptual categories are formed based on the common features of different objects, whereas abstract categories are formed based on shared characteristics or previous knowledge about different objects. We may, for example, have a perceptual category of "dogs" because they share a similar set of features, but we may also have a more abstract category of "past pet dogs" that is linked by all the dogs we have owned over our lives. Abstract categories are typically learned through experience, training, and other top-down mechanisms, whereas perceptual categories are formed through bottom-up perceptual mechanisms.

Auditory objects with common perceptual features are combined to form perceptual categories (Kuhl & Padden, 1982, 1983; Lasky et al., 1975; Liberman et al., 1967). Perceptual similarity is one of the key elements that determine a stimulus's categorical membership (Boyton & Olson, 1987, 1990; Doupe & Kuhl, 1999; Eimas et al., 1971; Kuhl & Miller, 1975; Kuhl & Padden, 1982, 1983; Lasky et al., 1975; Liberman et al., 1967; Miyawaki et al., 1975; Streeter, 1976; Wyttenbach et al., 1996). Perceptual categories are based on the physical attributes of an object. For example, male and female voices can be categorized as such by attending to the pitch of the voice, with female voices characteristically having a higher pitch than those of males. In another example, listeners can perceive different speech signals as belonging to the same phonemic category, or viewers can perceive different visual signals as being members of the same color category.

One prominent feature of perceptual categories is that they are often accompanied by categorical perception, such as the categorization of speech units into phonetic categories (Holt, 2006; Kuhl & Padden, 1982, 1983; Liberman et al., 1967; Lotto et al., 1998; Mann, 1980). In visual processing, this type of categorization is referred to as "object constancy" and allows us to recognize a visual object from a variety of different viewpoints, light levels, etc. (Walsh & Kulikowski, 1998). In categorical perception, a subject's perception of an object does not vary smoothly with changes in the physical properties of the object (Ashby & Berretty, 1997; Liberman et al., 1967; Miller et al., 2003). In other words, two objects that are on the same side of the categorical boundary are treated similarly, despite relatively large differences in their physical properties. When two objects straddle the category boundary, small changes in an object's properties can lead to large changes in perception.

Why are objects that straddle a category boundary easier to discriminate than objects that lie on the same side of the categorical boundary? One potential mechanism may relate to the distribution of neural resources. Following behavioral training, more neurons are activated in the auditory cortex during the discrimination of sounds that straddle a categorical boundary than during the discrimination of sounds that are on the same side of a categorical boundary (Guenther et al., 2004). Thus, discrimination may be primarily limited by the number of active neurons. This type of reorganization may be analogous to the redistribution of neural tuning properties following other types of training (Recanzone et al., 1993).

Perceptual categories are integrated with other types of information to form more abstract categorical representations. These categories are formed when arbitrary stimuli are linked together as a category based on some shared (nonperceptual) feature, a functional characteristic, or acquired knowledge. For instance, a combination of physical characteristics and knowledge about their reproductive processes allow us to categorize "dogs," "cats," and "killer whales" in the category of "mammals." However, if we use different criteria to form a category of "pets," "dogs" and "cats" would be members of this category, but "killer whales" would not.

Nonhuman primates can be trained to categorize stimuli into abstract categories based on their physical characteristics. For example, monkeys can be trained to categorize objects as "dogs" or "cats" (Freedman et al., 2001, 2002), "animals" or "nonanimals" (Fabre-Thorpe et al., 1998), or "trees" or "nontrees" (Vogels, 1999). The capacity to represent even more abstract categories, such as ordinal number and motion direction, is present as well (Freedman & Assad, 2006; Hauser et al., 2003; Nieder et al., 2002; Orlov et al., 2000; Roitman et al., 2007). One important caveat in studies like that of Freedman and colleagues is identifying the criteria that the monkeys use to classify the stimuli as "dogs" or "cats." Do the monkeys make a cat/dog distinction using the same criteria that we would naturally use to assign an animal as a "cat" or a "dog"? Or do the monkeys learn to use a suite of features to distinguish between two arbitrary categories that happen to coincide with human categories of "cat" and "dog"?

CATEGORICAL STRUCTURE OF VOCALIZATIONS

All nonhuman primates use vocalizations to communicate information to conspecifics (Cheney & Seyfarth, 1992; Hauser, 1997). As an object, each vocalization is encoded with multiple levels of categorical information, each of which provides important behaviorally relevant information to listeners. The vocalization itself represents the basic-level category, but much of the more sophisticated information extracted from vocalizations comes from the subordinate categories.

Beyond the call type, each vocalization communicates a diverse range of information. Some of this communicative content is correlated directly to physical characteristics of the caller. Body size, for example, is highly correlated with the length of the vocal tract. Animals with larger body weight and body size have longer vocal tracts, which correlate with greater formant frequency dispersion for a call (Fitch, 1997; Ghazanfar et al., 2007). Although information on this level may only exist as a consequence of idiosyncrasies in each individual's body weight and size, it still represents an important level of categorical information to conspecifics.

A further level of categorical content in vocalizations pertains to cues about a caller's identity. Numerous experiments have demonstrated that nonhuman primates can extract information about a caller's individual identity and that this information plays a central role in behavioral decisions. Vervet monkey mothers, for example, respond only to their own infants' cries and ignore the same call when it is produced by other infants (Cheney & Seyfarth, 1980). Although the ability recognize the identity of the caller is a salient feature in vocalizations (Bergman et al., 2003; Rendall et al., 1996), other levels of identity are also encoded in vocalizations, such as sex (Miller et al., 2005; Rendall et al., 2004). In species with dialectic differences between groups (Snowdon & Elowson, 1999; Weiss et al., 2001), conspecifics may be able to extract group identity (Miller et al., 2001b; Seyfarth et al., 1980).

A final, potentially abstract, level of categorical information in primate vocalizations is the referential content. Several studies of primate vocal communication show that certain vocalizations communicate information about objects and/or events that are external to the caller, a characteristic known as "functional reference." A classic example of referential signaling is the vervet monkey predator-alarm calls (Seyfarth et al., 1980). Vervets produce unique alarm calls for three different predators: a snake, a leopard, and an eagle. The information that is transmitted to the listeners initiates specific patterns of behaviors. For instance, when vervets hear an eagle-alarm call, they scan the sky for the airborne predator, and in some cases, run to locations that provide overhead coverage. In contrast, when they hear a snake-alarm call, they stand up and scan the ground. Finally, a leopard-alarm call initiates a third distinct behavior: Vervets run up the nearest tree while scanning the horizon for the leopard.

Referential signaling can also occur by combining different vocalizations. For example, Diana monkeys, while responding appropriately to the leopard- and eagle-alarm calls of

Campbell's monkeys, have a different response to these alarm calls when they are preceded by a Campbell's monkeys "boom" vocalization; the boom signifies a lower-level threat (Zuberbühler, 2000). More recent studies suggest that when putty-nosed monkeys elicit leopard-alarm call and an eagle-alarm call sequentially, they form a new referent that transmits information about group movement (Arnold & Zuberbühler, 2006). Experimental tests of a vocalization's referential quality have been limited to those calls that elicit strong behavioral responses during playback experiments, such as alarm calls and food calls. Most primate vocalizations, however, do not elicit such overt behavioral responses during playbacks, making it logistically difficult to examine a call's functional reference. Consequently, any conclusion that a call does not have a functional referent may not be correct. Rather, it may reflect a limitation in the sensitivity of our experimental methods.

Primate vocalizations consist of multiple levels of categorical information. Each call type represents a basic-level category. Several subordinate categories are also evident, ranging from acoustic information that directly corresponds to physical characteristics to individual identity and referential content. Importantly, those calls that have a functional referent should not be considered to be a distinct class of calls. Instead, these calls possess a more abstract level of categorical information. As objects, vocalizations consist of rich categorical information. Traditionally, studies of primate communication have not approached vocalizations from the perspective of its categorical organization. However, we believe that much can be learned about vocalizations, at both the behavioral and neural levels, from this approach.

VOCALIZATION PROCESSING AS AUDITORY-OBJECT ANALYSIS

We propose that vocalizations can be analyzed within a general framework of auditory-object analysis, in terms of both their perception and categorization (Blank et al., 2002, 2003; Darwin, 1997; De Santis et al., 2007; Griffiths & Warren, 2004; Micheyl et al., 2005; Murray et al., 2006; Nelken et al., 2003; Poremba et al., 2004; Rauschecker, 1998; Scott, 2005; Scott & Wise, 2004; Scott et al., 2000, 2004; Sussman, 2004; Ulanovsky et al., 2003; Wise et al., 2001; Zatorre et al., 2004). Under this framework, we can generate a model of how vocalizations are transformed from acoustic waveforms to perceptual/cognitive states that guide action and decisions. This model can then be used as a guide to design experiments that will test the relationship between neural activity and the perceptual organization of information communicated by the vocalizations.

The first step in auditory-object analysis is for the perceptual system to extract and code the spectrotemporal properties, localization cues, and other low-level features in the signal. These features are then "bound" together to form a representation of the object. The next components of auditory-object analysis involve computations that lead to the formation of increasingly more abstract category-based representations. We consider these steps to be serial in nature only as a useful conceptual heuristic. Ultimately, however, the cortex is likely to process the object properties and categorical structure of vocalizations in a dynamic parallel system.

A PATHWAY FOR VOCALIZATION PROCESSING

Where in the cortical hierarchy are objects and categories processed? In the cortex, the most likely pathway for vocalization processing is the so-called "ventral" processing stream, a pathway that processes the nonspatial attributes of an auditory stimulus (Rauschecker & Tian, 2000; Ungerleider & Mishkin, 1982). This pathway originates in the auditory cortex (Kaas & Hackett, 2000). In the auditory cortex, there are three levels of processing: the core, belt, and parabelt. These three processing levels form the bases for the spatial and nonspatial pathways. The ventral stream is defined by a series of projections that includes the anterior belt of the auditory cortex and regions of the prefrontal cortex, specifically the ventrolateral prefrontal cortex (vPFC) (Rauschecker & Tian, 2000; Romanski et al., 1999a, 1999b).

Of course, this ventral pathway is thought to run parallel to, and independently from, a "dorsal" processing stream; this dorsal pathway is thought to preferentially process the spatial properties of an auditory stimulus (Rauschecker & Tian, 2000; Ungerleider & Mishkin, 1982). Together, these dorsal and ventral pathways form an important model of auditory function, as well as a visual function (Ungerleider & Mishkin, 1982). We would caution, though, that the strict functional segregation of these two processing streams is unclear since recent data suggest that, like the visual system, there is considerable crosstalk between the dorsal and ventral pathways (Sereno & Maunsell, 1998; Toth & Assad, 2002).

NEURAL CORRELATES OF AUDITORY-OBJECT AND AUDITORY-CATEGORY PROCESSING

Neural Correlates of Object Processing

A number of neurophysiological studies using vocalizations as stimuli show that neurons in the auditory cortex likely play an important role in auditory object perception. For example, neurons in the primary auditory cortex (A1) are preferentially sensitive to species-specific vocalizations (Wang & Kadia, 2001). That is, the firing rates of A1 neurons are modulated when marmosets listen to species-specific vocalizations but are not modulated when these same vocalizations are played in reverse. In contrast, when these same stimuli are presented to cat A1 neurons, A1 activity does not differentiate between the forward and reverse marmoset vocalizations. This response pattern may be due, in part, to the fact that A1 neurons are extremely sensitive to the spectrotemporal features of a stimulus: Such "optimal" stimuli elicit sustained firing patterns, whereas "nonoptimal" stimuli elicit transient responses (Wang et al., 2005). Together, these studies suggest that at the level of the A1, if not earlier (Nelken et al., 2003), neurons are integrating the dynamic spectrotemporal properties of a stimulus, a fundamental requirement for object perception.

Several studies have had substantial success in demonstrating a role for the A1 and nonprimary auditory cortex in processing important object-related attributes of vocalizations. One of the more striking examples is that of "pitch" neurons in the awake marmoset A1 (Bendor & Wang, 2005, 2006); these pitch neurons straddle the border between the AI and R, two core tonotopic fields of the auditory cortex. Bendor and Wang demonstrated that these neurons respond equally well to complex stimuli that contain a fundamental-frequency component with higher harmonics as they do to stimuli that lack the fundamental frequency but contain the higher harmonics. Other examples of object-related activity are found in studies examining the neural correlates of different psychoacoustic parameters such as spectral contrast, roughness, flutter, and consonance (Barbour & Wang, 2003; Bendor & Wang, 2007; Fishman et al., 2000, 2001b). Together, these results suggest that even in the A1, neurons are beginning to integrate sounds across spectral and temporal dimensions—a necessary mechanism for object perception.

Several neurophysiological studies have directly assessed the role of the auditory cortex in object perception. Whereas few of these studies have used vocalizations as stimuli, their results speak to potential mechanisms underlying vocalization processing. First, Sutter and colleagues demonstrated that, like tamarins (Miller et al., 2001a), rhesus macaques can amodally complete auditory stimuli and that neurons in the auditory cortex reflect the monkeys' behavioral responses to occluded and intact stimuli (Petkov et al., 2003; Sutter et al., 2000). Second, when presented with alternating two-tone sequences that are perceived by humans as one or two auditory streams, neurons in the auditory cortex respond to these sequences in a manner that mirrors a human subject's psychophysical percepts (Fishman et al., 2001a, 2004; Micheyl et al., 2005); the perception of one or two streams depends on the spectrotemporal dynamics of the sequences. The auditory cortex may be the source of these percepts since the temporal evolution of auditory-cortical activity follows the temporal evolution of the percept.

By the time information processing reaches nonprimary areas of the auditory cortex, this sensitivity to species-specific vocalizations has markedly increased. Neurophysiology of lateral and parabelt belt neurons shows an increased responsiveness to vocalizations (Rauschecker & Tian, 2004; Rauschecker et al., 1995; Russ et al., 2008; Tian & Rauschecker, 2004; Tian et al., 2001), whereas neuroimaging studies suggest that anterior regions of the temporal lobe may be exquisitely tuned for vocalization processing (Petkov et al., 2008; Poremba et al., 2003). A recent study quantified explicitly the capacity of neurons in the nonprimary auditory cortex to respond differentially to different vocalizations (Russ et al., 2008). One analysis demonstrated that the capacity of these neurons to code different vocalizations depended on the tested time interval and amount of temporal integration. At the shortest time interval (50 ms) and the finest bin resolution (2 ms), a linear pattern discriminator could decode ≈25% of the vocalizations correctly. However, as the duration of the test time interval increased from 50 ms to 750 ms, the performance of the discriminator improved dramatically. Indeed, when the time interval was 750 ms, the discriminator decoded ≈90% of the vocalizations at the finest bin resolution. Interestingly, when Russ and colleagues tested multiple neurons simultaneously, they found that the number of vocalizations that could be correctly coded increased exponentially as they increased the number of simultaneously tested neurons. Subsequent analyses indicated that these neurons were not tuned for the spectrotemporal similarities that exist between different vocalizations, eliminating the possibility that these neurons are tuned for spectrotemporal features. Overall, all of these studies are consistent with the notion that neurons in the nonprimary auditory cortex are involved in object processing, and are not only in acoustic-feature processing.

Neural Correlates of Category Processing

The previous section highlighted the fact that the auditory cortex, particularly A1, plays a computational role in object analysis, a role beyond its traditional one in feature analysis (Nelken et al., 2003). Recent studies suggest that the auditory cortex may be involved in even more sophisticated types of processing. Indeed, studies of both human (Guenther et al., 2004; Poeppel et al., 2004) and nonhuman primates (Selezneva et al., 2006; Steinschneider et al., 1995) indicate that the auditory cortex is also involved in the computations underlying category processing. For example, work from Brosch and colleagues (Selezneva et al., 2006) shows that auditory-cortex neurons respond categorically to tone sequences that either increase or decrease in frequency. The primary auditory cortex also contains a distributed representation of neural activity that reflects the perceptual categorization of human phonemes that differ in voice-onset time (Steinschneider et al., 1995). These studies suggest that the auditory system is readily able to represent categorical boundaries even in the initial processing stage of the auditory cortex.

What types of categorical processing, then, occur in subsequent areas of the cortical hierarchy? We suggest that there are two major classes of computational processing. First, neurons become increasingly sensitive to more abstract categories. For example, recent work from our group suggests a role for the vPFC in categorizing the referential information that a vocalization transmits, as opposed to the vocalization's acoustic properties (Cohen et al., 2006; Gifford et al., 2005). Using an oddball paradigm (Näätänen & Tiitinen, 1996), we found that vPFC activity was not modulated by transitions between presentations of food vocalizations that transmitted the same information (high-quality food) but had significantly different acoustic structures. vPFC activity, however, was modulated by transitions between presentations of food vocalizations that transmitted different types of information (low-quality vs. high-quality food). These data suggested that, on average, vPFC neurons are modulated preferentially by transitions between presentations of food vocalizations that belong to functionally meaningful and different categories. Second, categorical representations in more central areas like the prefrontal cortex may be used to flexibly guide an animal's behavior (Miller, 2000; Miller et al., 2002). That is, categorical information in the prefrontal cortex is critical for both the selection and retrieval of task-relevant information as it relates to the rules of an ongoing task (Asaad et al., 2000; Ashby & Spiering, 2004; Badre et al., 2005; Bunge, 2004; Bunge et al., 2005; Spiering, 2004). Indeed, the responses of prefrontal neurons tend to vary with the rules mediating a task or the behavioral significance of stimuli, whereas responses in the inferior temporal cortex tend to be invariant to these variables while tending to be better correlated with the stimulus's physical properties than prefrontal neurons (Ashby & Spiering, 2004; Freedman et al., 2003). In the auditory system, there is, to date, no analogous evidence.

However, evidence for perhaps comparable hierarchical processing between the superior temporal gyrus and prefrontal cortex has recently between demonstrated (Cohen et al., 2007; Russ et al., 2008). As discussed above, the firing rates of neurons in nonprimary auditory cortex code several different vocalizations (Russ et al., 2008). A similar analysis in the vPFC revealed that these neurons carry significantly less information about vocalization identity than those in the nonprimary auditory cortex (Russ et al., 2008). In addition, the pairwise responses of neurons in the vPFC are less redundant (i.e., have the capacity to code more complex, orthogonal properties of a vocalization) than the pairwise responses of neurons in the superior temporal gyrus (Fig. 13.4). That is, the responses in the vPFC are capable of coding more complex features than those in the nonprimary auditory cortex. Indeed, a post hoc analysis of the study by Russ and colleagues (2008) indicated that vPFC neurons code more of the referential properties of a vocalization than those in the superior temporal gyrus (Fig. 13.5). All of these analyses are consistent with the hypothesis that the vPFC is at a hierarchically higher level of vocalization processing (object processing) than the nonprimary auditory cortex.

CONCLUSION

Vocalizations represent a unique class of stimuli for studies of neural coding and representation.

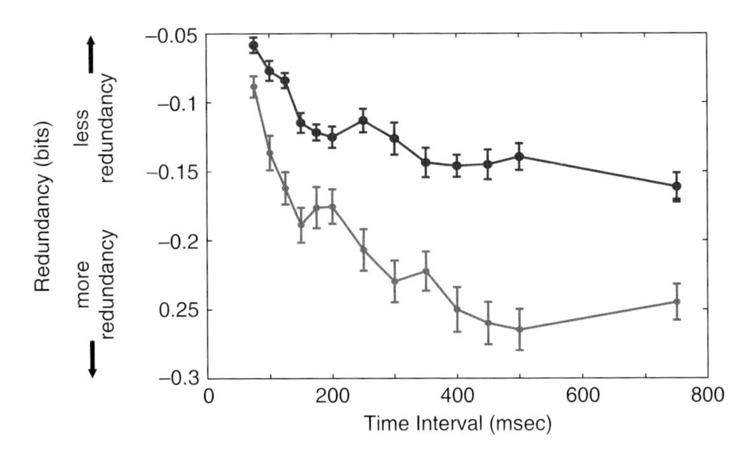

Figure 13.4 Redundancy between pairs of neurons in the superior temporal gyrus and between pairs of neurons in the ventrolateral prefrontal cortex (vPFC). A redundancy analysis tested whether neurons in the superior temporal gyrus and the vPFC code the same attributes of a vocalization. Decreases in redundancy are associated with an increased capacity to code more complex stimulus attributes (Chechik et al., 2006; Schneidman et al., 2003). The redundancy analysis was conducted as a function of increasing time intervals following onset of a vocalization. The data in red were calculated from the responses of neurons in the superior temporal gyrus. The data in blue were calculated from the responses of vPFC neurons. Error bars represent standard deviations.

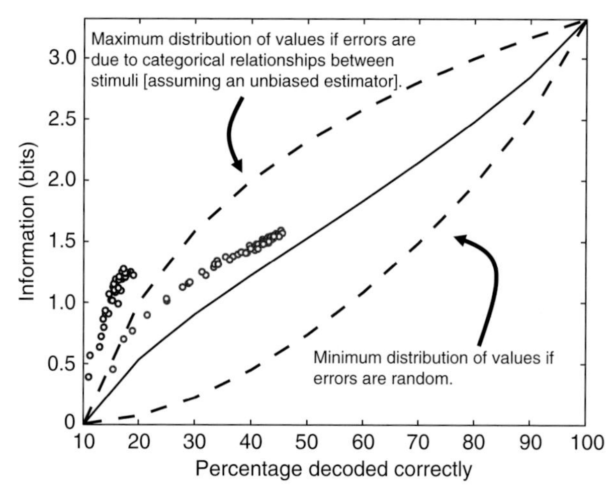

Figure 13.5 Metric content of neurons in the superior temporal gyrus and the ventrolateral prefrontal cortex (vPFC). Metric content (Treves, 1997) is calculated from the confusion matrices generated from an analysis that tested the capacity of neurons in the superior temporal gyrus and the vPFC to code vocalizations as distinct auditory objects (Russ et al., 2008). On a neuron-by-neuron basis, the metric content was calculated by correlating the percentage of correctly decoded vocalizations with the amount of information carried in a neuron's response. The data in red were calculated from the responses of neurons in the superior temporal gyrus. The data in blue were calculated from the responses of vPFC neurons.

At one level, vocalizations have evolved as highly salient communicative signals (Hauser, 1997). On the other hand, they are auditory objects. Ultimately, the task at hand is to ascertain how particular acoustic features, or combinations of features, are transformed into higher-order representations that confer information to a listener and how this information is represented in the cortex. As discussed here, a vocalization is not merely a vocal signal that communicates a sole piece of information. Rather, each call is a perceptually bound unit encoded with a rich array of categorical content. Although traditionally, studies of vocalizations at both the behavioral and neural levels have approached this topic thinking primarily about the main function of the call, an object-oriented view of vocalizations may yield important insights. These two perspectives are not mutually exclusive; an understanding of how vocalizations are represented in the cortex will likely involve studies from both approaches.

The primate vocal communication system has often drawn parallels with two analogous neural systems: birdsong and primate face recognition. With respect to birdsong, it is important to distinguish between the classic sensory-motor song system (i.e., HVc, RA, area X, etc. [Margoliash, 1997]) and the auditory system (field L, NCM, etc.) of songbirds when making comparisons to primate vocal communication. The primary difference between the song system and primate vocal communication is in the degree of specialization. Whereas the song system represents a highly specialized neural system that almost exclusively processes elements of birdsong (Margoliash, 1997), the neural mechanisms underlying nonhuman primate vocal communication are more generalized (Ghazanfar, 2003). Moreover, the substrates underlying the song system are primarily subcortical nuclei (Nottebohm et al., 1976), whereas the primate system is largely situated in the cortex (Kaas & Hackett, 1998; Romanski et al., 1999a), thereby limiting any direct parallels between the systems. Although there may be analogous mechanisms between primates and songbird song system, it is likely that these similarities reflect a homoplasy and comparisons with other sensory-motor systems within the primate cortex will yield more homologous neural mechanisms. Some more direct parallels, however, may exist between these taxonomic groups with respect to the sensory processing of vocalizations and other auditory objects in the auditory system (Gentner & Margoliash, 2003; Woolley et al., 2005). More work is needed, however, to ascertain commonalities that may exist in the auditory systems of these taxonomic groups.

Comparisons with the face recognition system, in general, and "face cells," more specifically, in nonhuman primates stems more from potential analogies between visual and acoustic communication signals. Potential homologies between the systems are supported by evidence showing an increase in processing specificity for vocalizations in the rostral areas of nonprimary auditory cortex (Petkov et al., 2008; Poremba et al., 2003) and the presence of "face cells" in inferotemporal cortex (Fujita et al., 1992; Gross and Sergent, 1992; Perrett et al., 1984, 1985). One caveat, however, is an evolutionary distinction between faces and vocalizations. Whereas the structure of primate faces (i.e., the stable arrangement of two eyes, nose, and mouth), and most mammalian faces for that matter, has been conserved over a long evolutionary history, the structure of vocalizations (i.e., spectrotemporal arrangement of acoustic features) varies between all species of nonhuman primates. As such, it is possible that vocalizations have not undergone the same type of selection for a highly specialized object like a face and are processed as more of a general higher-order object, albeit a highly salient one. Although both systems may reflect a general pattern of increased specificity for object processing in higher-level areas along the ventral pathway, vocalizations may not ultimately be processed with the same degree of specialization as faces.

Our focus was not to de-emphasize the importance of a vocalization as an evolved signal, but to show the possibilities that can be gained by approaching the study of vocalizations from an object-based view. Ultimately, the neural basis of primate vocal communication is a unique system that likely has parallels with several related processes. But, like all complex cortical systems, the primate communication system possesses intricacies that are idiosyncratic to its own selective pressures. With a renewed interest in this topic, significant strides are likely to be made in our understanding of this complex system.

ACKNOWLEDGMENTS

We would like to thank Brian Russ, Jung Hoon Lee, and Heather Hersh for helpful comments on this manuscript. CTM was supported by a grant from the NIDCD-NIH (K99DC009007). YEC was supported by grants from the Whitehall Foundation, NIDCD-NIH, NIMH-NIH, and a Burke Award.

REFERENCES

Arnold, K., & Zuberbühler, K. (2006). Language evolution: Semantic combinations in primate calls. *Nature, 441,* 303.

Asaad, W. F., Rainer, G., & Miller, E. K. (2000). Task-specific neural activity in the primate prefrontal cortex. *Journal of Neurophysiology, 84,* 451–459.

Ashby, F. G., & Berretty, P. M. (1997). Categorization as a special case of decision-making or choice. In: A. A. J. Marley (Ed.), *Choice, decision, and measurement: Essays in honor of R. Duncan Luce* (pp. 367–388). Lawrence Erlbaum Associates, Inc.

Ashby, F. G., & Spiering, B. J. (2004). The neurobiology of category learning. *Behavioral and Cognitive Neurosciences Review, 3,* 101–113.

Badre, D., Poldrack, R. A., Pare-Blagoev, E. J., Insler, R. Z., & Wagner, A. D. (2005). Dissociable controlled retrieval and generalized selection mechanisms in ventrolateral prefrontal cortex. *Neuron, 47,* 907–918.

Barbour, D. L., & Wang, X. (2003). Contrast tuning in auditory cortex. *Science, 299,* 1073–1075.

Bendor, D., & Wang, X. (2005). The neuronal representation of pitch in primate auditory cortex. *Nature, 436,* 1161–1165.

Bendor, D., & Wang, X. (2006). Cortical representations of pitch in monkeys and humans. *Current Opinions in Neurobiology, 16,* 391–399.

Bendor, D., & Wang, X. (2007). Differential neural coding of acoustic flutter within primate auditory cortex. *Nature Neurosciences, 10,* 763–771.

Bergman, T. J., Beehner, J. C., Cheney, D. L., & Seyfarth, R. M. (2003). Hierarchical classification by rank and kinship in baboons. *Science, 302,* 1234–1236.

Blank, S. C., Bird, H., Turkheimer, F., & Wise, R. J. (2003). Speech production after stroke: The role of the right pars opercularis. *Annals of Neurology, 54,* 310–320.

Blank, S. C., Scott, S. K., Murphy, K., Warburton, E., & Wise, R. J. (2002). Speech production: Wernicke, Broca and beyond. *Brain, 125,* 1829–1838.

Boyton, R. M., & Olson, C. X. (1987). Locating basic color terms in the OSA space. *Color Research Applications, 12,* 94–105.

Boyton, R. M., & Olson, C. X. (1990). Salience of basic color terms confirmed by three measures. *Vision Research, 30,* 1311–1317.

Bregman, A. S. (1990). *Auditory scene analysis.* Boston, MA: MIT Press.

Bunge, S. A. (2004). How we use rules to select actions: A review of evidence from cognitive neuroscience. *Cognitive and Affective Behavior Neuroscience, 4,* 564–579.

Bunge, S. A., Wallis, J. D., Parker, A., Brass, M., Crone, E. A., Hoshi, E., et al. (2005). Neural circuitry underlying rule use in humans and nonhuman primates. *Journal of Neuroscience, 25,* 10347–10350.

Chechik, G., Anderson, M. J., Bar-Yosef, O., Young, E. D., Tishby, N., & Nelken, I. (2006). Reduction of information redundancy in the ascending auditory pathway. *Neuron, 51,* 359–368.

Cheney, D. L., & Seyfarth, R. M. (1980). Vocal recognition in free ranging vervet monkeys. *Animal Behaviour, 28,* 362–367.

Cheney, D. L., & Seyfarth, R. M. (1992). Meaning, reference, and intentionality in the natural vocalizations of monkeys. In: T. Nishida, W. C. McGrew, P. Marler, M. Pickford, F. de Waal (Eds.), *Topics in primatology, vol 1: Human origins* (pp. 315–330). Tokyo: Tokyo University Press.

Cohen, Y. E., Hauser, M. D., & Russ, B. E. (2006). Spontaneous processing of abstract categorical information in the ventrolateral prefrontal cortex. *Biology Letters, 2,* 261–265.

Cohen, Y. E., Russ, B. E., Ackelson, A. L., Baker, A. E., Chowdhury, F. N., & Lee, Y. S. (2007). Feature and object processing in the auditory cortex: Evidence for hierarchical processing. In. San Diego, CA, Program No. 227.6 2007 Neuroscience Meeting Planner.

Damasio, A. R. (1989). Time-locked multiregional retroactivation: A systems-level proposal for the neural substrates of recall and recognition. *Cognition, 33,* 25–62.

Darwin, C. J. (1997). Auditory grouping. *Trends in Cognitive Science, 1,* 327–333.

De Santis, L., Clarke, S., & Murray, M. M. (2007). Automatic and intrinsic auditory "what" and "where" processing in humans revealed by electrical neuroimaging. *Cerebral Cortex, 17,* 9–17.

Devlin, J. T., Gonnerman, L. M., Andersen, E. S., & Seidenberg, M. S. (1998). Category-specific semantic deficits in focal and widespread brain damage: A computational account. *Journal of Cognitive Neuroscience, 10,* 77–94.

Doupe, A. J., & Kuhl, P. K. (1999). Birdsong and human speech: Common themes and mechanisms. *Annual Review of Neuroscience, 22,* 567–631.

Eimas, P. D., Siqueland, E. R., Jusczyk, P., & Vigorito, J. (1971). Speech perception in infants. *Science, 171,* 303–306.

Fabre-Thorpe, M., Richard, G., & Thorpe, S. J. (1998). Rapid categorization of natural images by rhesus monkeys. *Neuroreport, 9,* 303–308.

Farris, H. E., Rand, A. S., & Ryan, M. J. (2005). The effects of time, space and spectrum on auditory grouping in tungara frogs. *Journal of Comprehensive Physiology Part A: Neuroethological, Sensory, Neural, and Behavioral Physiology, 191,* 1173–1183.

Feldman, J. (2003). What is a visual object? *Trends in Cognitive Science, 7,* 252–256.

Feldman, J., & Tremoulet, P. (2006). Individuation of visual objects over time. *Cognition , 99.*

Fishman, Y. I., Arezzo, J. C., & Steinschneider, M. (2004). Auditory stream segregation in monkey auditory cortex: Effects of frequency separation, presentation rate, and tone duration. *Journal of the Acoustical Society of America, 116,* 1656–1670.

Fishman, Y. I., Reser, D. H., Arezzo, J. C., & Steinschneider, M. (2000). Complex tone processing in primary auditory cortex of the awake monkey. I. Neural ensemble correlates of roughness. *Journal of the Acoustical Society of America, 108,* 235–246.

Fishman, Y. I., Reser, D. H., Arezzo, J. C., & Steinschneider, M. (2001a). Neural correlates of auditory stream segregation in primary auditory cortex of the awake monkey. *Hearing Research, 151,* 167–187.

Fishman, Y. I., Volkov, I. O., Noh, M. D., Garell, P. C., Bakken, H., Arezzo, J. C., et al. (2001b). Consonance and dissonance of musical chords: Neural correlates in auditory cortex of monkeys and humans. *Journal of Neurophysiology, 86,* 2761–2788.

Fitch, W. T. (1997). Vocal tract length and formant frequency dispersion correlate with body size in rhesus macaques. *Journal of the Acoustical Society of America, 102,* 1213–1222.

Freedman, D. J., & Assad, J. A. (2006). Experience-dependent representation of visual categories in parietal cortex. *Nature, 443,* 85–88.

Freedman, D. J., Riesenhuber, M., Poggio, T., & Miller, E. K. (2001). Categorical representation of visual stimuli in the primate prefrontal cortex. *Science, 291,* 312–316.

Freedman, D. J., Riesenhuber, M., Poggio, T., & Miller, E. K. (2002). Visual categorization and the primate prefrontal cortex: Neurophysiology and behavior. *Journal of Neurophysiology, 88,* 929–941.

Freedman, D. J., Riesenhuber, M., Poggio, T., & Miller, E. K. (2003). A comparison of primate prefrontal and inferior temporal cortices during visual categorization. *Journal of Neuroscience, 23,* 5235–5246.

Fujita, I., Tanaka, K., Ito, M., & Cheng, K. (1992). Columns for visual features of objects in monkey inferotemporal cortex. *Nature, 360,* 343–346.

Gauthier, I., & Logothetis, N. K. (1999). Is face recognition not so unique after all? *Cognitive Neuropsychology.*

Gentner, T. Q., & Margoliash, D. (2003). Neuronal populations and single cells representing learned auditory objects. *Nature, 424,* 669–674.

Ghazanfar, A. A. (2003). *Primate audition: Ethology and neurobiology.* New York: CRC Press.

Ghazanfar, A. A., & Hauser, M. D. (1999). The neuroethology of primate vocal communication: Substrates for the evolution of speech. *Trends in Cognitive Science, 3,* 377–384.

Ghazanfar, A. A., Turesson, H. K., Maier, J. X., van Dinther, R., Patterson, R. D., & Logothetis, N. K. (2007). Vocal-tract resonances as indexical cues in rhesus monkeys. *Current Biology, 17,* 425–430.

Gifford III, G. W., MacLean, K. A., Hauser, M. D., & Cohen, Y. E. (2005). The neurophysiology of functionally meaningful categories: Macaque ventrolateral prefrontal cortex plays a critical role in spontaneous categorization of species-specific vocalizations. *Journal of Cognitive Neuroscience, 17,* 1471–1482.

Griffiths, T. D., & Warren, J. D. (2004). What is an auditory object? *Nature Review Neuroscience, 5,* 887–892.

Gross, C. G., & Sergent, J. (1992). Face recognition. *Current Opinions in Neurobiology, 2,* 156–161.

Guenther, F. H., Nieto-Castanon, A., Ghosh, S. S., & Tourville, J. A. (2004). Representation of sound categories in auditory cortical maps. *Journal of Speech, Language, and Hearing Research, 47,* 46–57.

Hauser, M. D. (1997). *The evolution of communication.* Cambridge, MA: MIT Press.

Hauser, M. D. (2001). Searching for food in the wild: A nonhuman primate's expectations about invisible displacement. *Developmental Science, 4,* 84–93.

Hauser, M. D., Tsao, F., Garcia, P., & Spelke, E. S. (2003). Evolutionary foundations of number: Spontaneous representation of numerical magnitudes by cotton-top tamarins. *Proceedings in Biological Sciences, 270,* 1441–1446.

Holt, L. L. (2006). Speech categorization in context: joint effects of nonspeech and speech precursors. *Journal of the Acoustical Society of America, 119,* 4016–4026.

Huntley-Fenner, G., Carey, S., & Solimando, A. (2002). Objects are individuals but stuff doesn't count: perceived rigidity and cohesiveness influence infants' representations of small groups of discrete entities. *Cognition, 85,* 203–221.

Kaas, J. H., & Hackett, T. A. (1998). Subdivisions of auditory cortex and levels of processing in primates. *Audiology Neurootology, 3,* 73–85.

Kaas, J. H., & Hackett, T. A. (2000). Subdivisions of auditory cortex and processing streams in primates. *Proceedings of the Nationall Academy of Sciences, 97,* 11793–11799.

Kaffka, K. (1935). *Principles of Gestalt psychology.* New York: Harcourt, Brace, & World.

Kellman, P. J., Spelke, E. S., & Short, K. R. (1986). Infant perception of object unity from translatory motion in depth and vertical translation. *Child Development, 57,* 72–86.

Kubovy, M., & Van Valkenburg, D. (2001). Auditory and visual objects. *Cognition, 80,* 97–126.

Kuhl, P. K., & Miller, J. D. (1975). Speech perception by the chinchilla: Voiced-voiceless distinction in alveolar plosive consonants. *Science, 190,* 69–72.

Kuhl, P. K., & Padden, D. M. (1982). Enhanced discriminability at the phonetic boundaries for the voicing feature in macaques. *Perceptions and Psychophysics, 32,* 542–550.

Kuhl, P. K., & Padden, D. M. (1983). Enhanced discriminability at the phonetic boundaries for the place feature in macaques. *Journal of the Acoustical Society of America, 73,* 1003–1010.

Lasky, R. E., Syrdal-Lasky, A., & Klein, R. E. (1975). VOT discrimination by four to six and a half month old infants from Spanish environments. *Journal of Experimental Child Psychology, 20,* 215–225.

Liberman, A. M., Cooper, F. S., Shankweiler, D. P., & Studdert-Kennedy, M. (1967). Perception of the speech code. *Psychological Review, 5,* 552–563.

Lotto, A. J., Kluender, K. R., & Holt, L. L. (1998). Depolarizing the perceptual magnet effect. *Journal of the Acoustical Society of America, 103,* 3648–3655.

Mann, V. A. (1980). Influence of preceding liquid on stop-consonant perception. *Perceptions in Psychophysiology, 28,* 407–412.

Margoliash, D. (1997). Functional organization of forebrain pathways for song production and perception. *Journal of Neurobiology, 33,* 671–693.

Marschark, M., Convertino, C., McEvoy, C., & Masteller, A. (2004). Organization and use of the mental lexicon by deaf and hearing individuals. *American Annals of the Deaf, 149,* 51–61.

Martin, A., Ungerleider, L. G., & Haxby, J. V. (2002). Category specificity and the brain: The sensory/motor model of semantic representations of objects. In: M. S. Gazzaniga (Ed.), *The new cognitive neurosciences.* Cambridge, MA: MIT Press.

Micheyl, C., Tian, B., Carlyon, R. P., & Rauschecker, J. P. (2005). Perceptual organization of tone sequences in the auditory cortex of awake macaques. *Neuron, 48,* 139–148.

Miller, C. T., Dibble, E., & Hauser, M. D. (2001a). Amodal completion of acoustic signals by a nonhuman primate. *Nature Neurosciences, 4,* 783–784.

Miller, C. T., Gil-da-Costa, R., & Hauser, M. D. (2001b). Selective phonotaxis by cotton-top tamarins. *Behavior, 138,* 811–826.

Miller, C. T., Iguina, C., & Hauser, M. D. (2005). Processing vocal signals for recognition during antiphonal calling. *Animal Behaviour, 69,* 1387–1398.

Miller, E. K. (2000). The prefrontal cortex and cognitive control. *Nature Review Neuroscience, 1,* 59–65.

Miller, E. K., Freedman, D. J., & Wallis, J. D. (2002). The prefrontal cortex: Categories, concepts, and cognition. *Philosophical Transactions of the Royal Society of London Part B: Biological Sciences, 29,* 1123–1136.

Miller, E. K., Nieder, A., Freedman, D. J., & Wallis, J. D. (2003). Neural correlates of categories and concepts. *Current Opinions in Neurobiology, 13,* 198–203.

Miyawaki, K., Strange, W., Verbrugge, R., Liberman, A. M., Jenkins, J. J., & Fujimura, O. (1975). An effect of linguistic experience: the discrimination of [r] and [l] by native speakers of Japanese and English. *Perceptions in Psychophysiology, 18,* 331–340.

Murray, M. M., Camen, C., Gonzalez Andino, S. L., Bovet, P., & Clarke, S. (2006). Rapid brain discrimination of sounds of objects. *Journal of Neuroscience, 26,* 1293–1302.

Näätänen, R., & Tiitinen, H. (1996). Auditory information processing as indexed by the mismatch negativity. In: M. Sabourin, F. Craik, & M. Robert (Eds.), *Advances in psychological science: Biological and cognitive aspects* (pp. 145–170). Montreal: Taylor and Francis Group.

Nakayama, K. (1999). Mid-level vision. In: R. A. Wilson & F. C. Keil (Eds.), *The MIT encylopedia of the cognitive sciences.* Cambridge, MA: MIT Press.

Nelken, I., Fishbach, A., Las, L., Ulanovsky, N., & Farkas, D. (2003). Primary auditory cortex of cats: Feature detection or something else? *Biological Cybernetics, 89,* 397–406.

Nieder, A., Freedman, D. J., & Miller, E. K. (2002). Representation of the quantity of visual items in the primate prefrontal cortex. *Science, 297,* 1708–1711.

Nottebohm, F., Stokes, T. M., & Leonard, C. M. (1976). Central control of song in the canary, Serinus canaries. *Journal of Comprehensive Neurology, 165,* 457–468.

Orlov, T., Yakovlev, V., Hochstein, S., & Zohary, E. (2000). Macaque monkeys categorize images by their ordinal number. *Nature, 404,* 77–80.

Perrett, D. I., Smith, P. A., Potter, D. D., Mistlin, A. J., Head, A. S., Milner, A. D., et al. (1984). Neurones responsive to faces in the temporal cortex: Studies of functional organization, sensitivity to identity and relation to perception. *Human Neurobiology, 3,* 197–209.

Perrett, D. I., Smith, P. A., Potter, D. D., Mistlin, A. J., Head, A. S., Milner, A. D., et al. (1985). Visual cells in the temporal cortex sensitive to face view and gaze direction. *Proceedings of the Royal Society of London Part B: Biological Sciences, 223,* 293–317.

Petkov, C. I., Kayser, C., Steudel, T., Whittingstall, K., Augath, M., & Logothetis, N. K. (2008). A voice region in the monkey brain. *Nature Neurosciences.*

Petkov, C. I., O'Connor, K. N., & Sutter, M. L. (2003). Illusory sound perception in macaque monkeys. *Journal of Neuroscience, 23,* 9155–9161.

Poeppel, D., Guillemin, A., Thompson, J., Fritz, J., Bavelier, D., & Braun, A. R. (2004). Auditory lexical decision, categorical perception, and FM direction discrimination differentially engage left and right auditory cortex. *Neuropsychologia, 42,* 183–200.

Poremba, A., Saunders, R. C., Crane, A. M., Cook, M., Sokoloff, L., & Mishkin, M. (2003). Functional mapping of the primate auditory system. *Science, 299,* 568–572.

Poremba, A., Malloy, M., Saunders, R. C., Carson, R. E., Herscovitch, P., & Mishkin, M. (2004). Species-specific calls evoke asymmetric activity in the monkey's temporal poles. *Nature, 427,* 448–451.

Pylyshyn, Z. W. (2006). *Seeing and visualizing: It's not what you think.* Cambridge, MA: MIT Press.

Pylyshyn, Z. W., & Storm, R. W. (1988). Tracking multiple independent targets: Evidence for a parallel tracking mechanism. *Spatial Vision, 3,* 179–197.

Rauschecker, J. P. (1998). Cortical processing of complex sounds. *Current Opinions in Neurobiology, 8,* 516–521.

Rauschecker, J. P., & Tian, B. (2000). Mechanisms and streams for processing of "what" and "where" in auditory cortex. *Proceedings of the Nationall Academy of Sciences, 97,* 11800–11806.

Rauschecker, J. P., & Tian, B. (2004). Processing of band-passed noise in the lateral auditory belt cortex of the rhesus monkey. *Journal of Neurophysiology, 91,* 2578–2589.

Rauschecker, J. P., Tian, B., & Hauser, M. D. (1995). Processing of complex sounds in the macaque nonprimary auditory cortex. *Science, 268,* 111–114.

Recanzone, G. H., Schreiner, C. E., & Merzenich, M. M. (1993). Plasticity in the frequency representation of primary auditory cortex following discrimination training in adult owl monkeys. *Journal of Neuroscience, 13,* 87–103.

Rendall, D., Owren, M. J., Weerts, E., & Hienz, R. D. (2004). Sex differences in the acoustic structure of vowel-like grunt vocalizations in baboons and their perceptual discrimination by baboon listeners. *Journal of the Acoustical Society of America, 115,* 411–421.

Rendall, D., Rodman, P. S., & Emond, R. E. (1996). Vocal recognition of individuals and kin in free-ranging rhesus monkeys. *Animal Behaviour, 51,* 1007–1015.

Roitman, J. D., Brannon, E. M., & Platt, M. L. (2007). Monotonic coding of numerosity in macaque lateral intraparietal area. *PLoS Biology, 5,* e208.

Romanski, L. M., Bates, J. F., & Goldman-Rakic, P. S. (1999a). Auditory belt and parabelt projections to the prefrontal cortex in the rhesus monkey. *Journal of Comprehensive Neurology, 403,* 141–157.

Romanski, L. M., Tian, B., Fritz, J., Mishkin, M., Goldman-Rakic, P. S., & Rauschecker, J. P. (1999b). Dual streams of auditory afferents target multiple domains in the primate prefrontal cortex. *Nature Neurosciences, 2,* 1131–1136.

Rosch, E. (1978). Principles of categorization. In: E. Rosch & B. B. Lloyd (Eds.), *Cognition and categorization* (pp. 27–48). Hillsdale, NH: Erlbaum.

Rosch, E., Mervis, C. B., Gray, W. D., Johnson, D. M., & Boyes-Braem, P. (1976). Basic objects in natural categories. *Cognitive Psychology, 8,* 382–439.

Russ, B. E., Ackelson, A. L., Baker, A. E., & Cohen, Y. E. (2008). Coding of auditory-stimulus identity in the auditory non-spatial processing stream. *Journal of Neurophysiology, 99,* 87–95.

Santos, L. R. (2004) 'Core knowledges': a dissociation between spatiotemporal knowledge and contact-mechanics in a non-human primate? *Developmental Science, 7,* 167–174.

Schneidman, E., Bialek, W., Berry 2nd, M. J. (2003). Synergy, redundancy, and independence in population codes. *Journal of Neuroscience, 23,* 11539–11553.

Scholl, B. J. (2001). Objects and attention: The state of the art. *Cognition, 80,* 1–46.

Scholl, B. J. (2007). Object persistence in philosophy and psychology. *Mind & Language, 22,* 563–591.

Scholl, B. J., Pylyshyn, Z. W., & Feldman, J. (2001). What is a visual object? Evidence from target merging in multiple object tracking. *Cognition, 80,* 159–177.

Scott, S. K. (2005). Auditory processing–speech, space and auditory objects. *Current Opinions in Neurobiology, 15,* 197–201.

Scott, S. K., Blank, C. C., Rosen, S., & Wise, R. J. (2000). Identification of a pathway for intelligible speech in the left temporal lobe. *Brain, 123 Pt 12,* 2400–2406.

Scott, S. K., Rosen, S., Wickham, L., & Wise, R. J. (2004). A positron emission tomography study of the neural basis of informational and energetic masking effects in speech perception. *Journal of the Acoustical Society of America, 115,* 813–821.

Scott, S. K., & Wise, R. J. (2004). The functional neuroanatomy of prelexical processing in speech perception. *Cognition, 92,* 13–45.

Sears, C. R., & Pylyshyn, Z. W. (2000). Multiple object tracking and attentional processing. *Canadian Journal of Experimental Psychology, 54,* 1–14.

Selezneva, E., Scheich, H., & Brosch, M. (2006). Dual time scales for categorical decision making in auditory cortex. *Current Biology, 16,* 2428–2433.

Sereno, A. B., & Maunsell, J. H. (1998). Shape selectivity in primate lateral intraparietal cortex. *Nature, 395,* 500–503.

Seyfarth, R. M., Cheney, D. L., & Marler, P. (1980). Monkey responses to three different alarm calls: Evidence of predator classification and semantic communication. *Science, 210,* 801–803.

Smith, E. E., Shoben, E. J., & Rips, L. J. (1974). Structure and process in semantic memory: A feature model for semantic decisions. *Psychological Review, 81,* 214–221.

Snowdon, C. T., & Elowson, A. M. (1999). Marmosets modify call structure when paired. *Ethology, 105.*

Spelke, E. (1994). Initial knowledge: six suggestions. *Cognition, 50,* 431–445.

Spelke, E. S., Breinlinger, K., Macomber, J., & Jacobson, K. (1992). Origins of knowledge. *Psychological Review, 99,* 605–632.

Spelke, E. S., Katz, G., Purcell, S. E., Ehrlich, S. M., & Breinlinger, K. (1994). Early knowledge of object motion: Continuity and inertia. *Cognition, 51,* 131–176.

Steinschneider, M., Schroeder, C. E., Arezzo, J. C., Vaughan, H. G., Jr. (1995). Physiologic correlates of the voice onset time boundary in primary auditory cortex (A1) of the awake monkey: Temporal response patterns. *Brain and Language, 48,* 326–340.

Streeter, L. A. (1976). Language perception of 2-month-old infants shows effects of both innate mechanisms and experience. *Nature, 259,* 39–41.

Sussman, E. S. (2004). Integration and segregation in auditory scene analysis. *Journal of the Acoustical Society of America, 117,* 1285–1298.

Sutter, M. L., Petkov, C., Baynes, K., & O'Connor, K. N. (2000). Auditory scene analysis in dyslexics. *Neuroreport, 11,* 1967–1971.

Tian, B., & Rauschecker, J. P. (2004). Processing of frequency-modulated sounds in the lateral auditory belt cortex of the rhesus monkey. *Journal of Neurophysiology, 92,* 2993–3013.

Tian, B., Reser, D., Durham, A., Kustov, A., & Rauschecker, J. P. (2001). Functional specialization in rhesus monkey auditory cortex. *Science, 292,* 290–293.

Toth, L. J., & Assad, J. A. (2002). Dynamic coding of behaviourally relevant stimuli in parietal cortex. *Nature, 415,* 165–168.

Treves, A. (1997). On the perceptual structure of face space. *Biosystems, 40,* 189–196.

Ulanovsky, N., Las, L., & Nelken, I. (2003). Processing of low-probability sounds by cortical neurons. *Nature Neurosciences, 6,* 391–398.

Ungerleider, L. G., & Mishkin, M. (1982). Two cortical visual systems. In: D. J. Ingle, M. A. Goodale, & R. J. W. Mansfield (Eds.), *Analysis of visual behavior.* Cambridge, MA: MIT Press.

Vogels, R. (1999). Categorization of complex visual images by rhesus monkeys. Part 1: Behavioural study. *European Journal of Neuroscience, 11,* 1223–1238.

Walsh, V., & Kulikowski, J. (Eds.). (1998). *Perceptual constancy: Why things look as they do.* Cambridge, UK: Cambridge University Press.

Wang, X., & Kadia, S. C. (2001). Differential representation of species-specific primate vocalizations in the auditory cortices of marmoset and cat. *Journal of Neurophysiology, 86,* 2616–2620.

Wang, X., Lu, T., Snider, R. K., & Liang, L. (2005). Sustained firing in auditory cortex evoked by preferred stimuli. *Nature, 435,* 341–346.

Weiss, D. J., Garibaldi, B. T., & Hauser, M. D. (2001). The production and perception of long calls by cotton-top tamarins (Saguinus oedipus): Acoustic analyses and playback experiments. *Journal of Comprehensive Psychology*, *115*, 258–271.

Wise, R. J., Scott, S. K., Blank, S. C., Mummery, C. J., Murphy, K., & Warburton, E. A. (2001). Separate neural subsystems within 'Wernicke's area'. *Brain*, *124*, 83–95.

Woolley, S. M., Fremouw, T. E., Hsu, A., & Theunissen, F. E. (2005). Tuning for spectro-temporal modulations as a mechanism for auditory discrimination of natural sounds. *Nature Neurosciences*, *8*, 1371–1379.

Wordon, F. G., & Galambos, R. (1972). Auditory processing of biologically significant sounds. Neuroscience Research Progress Bulletin, *10*, 1–119.

Wyttenbach, R. A., May, M. L., & Hoy, R. R. (1996). Categorical perception of sound frequency by crickets. *Science*, *273*, 1542–1544.

Yantis, S. (2000). *Visual perception: Essential readings*. Psychology Press.

Zatorre, R. J., Bouffard, M., & Belin, P. (2004). Sensitivity to auditory object features in human temporal neocortex. *Journal of Neuroscience*, *24*, 3637–3642.

Zuberbühler, K. (2000). Interspecies semantic communication in two forest primates. *Proceedings in Biological Sciences*, *267*, 713–718.

CHAPTER 14

Encoding and Beyond in the Motor Cortex

Nicholas G. Hatsopoulos, Maryam Saleh, and Julian A. Mattiello

INTRODUCTION

The physiology of the primary motor cortex (MI) has been examined for well over a hundred years, beginning with the seminal electrical stimulation experiments by Fritsch and Hitzig (1870). In the early twentieth century, Leyton and Sherrington put forth one of the earliest hypotheses regarding the functional role of MI based largely on focal and short-duration electrical stimulation. By pointing to physiological and anatomical data, they proposed that MI acts as a synthetic organ of complex, goal-directed movements. Focal stimulation across the motor cortex resulted in a very large set of "fractional" movements that appeared coordinated but functionally incomplete. They postulated that the extensive anatomical connectivity within MI could serve to combine or associate multiple fractional movements in different combinations to generate a rich variety of functional behaviors.

Despite this early influential viewpoint of motor cortical functioning, most electrophysiological research in awake, behaving animals over the past 40 years has attempted to isolate particular parameter(s) of motion such as force, velocity, and direction that are encoded in individual motor cortical neurons. Unfortunately, by focusing on abstract movement parameters, almost every conceivable kinematic and kinetic parameter has been shown to covary with the firing rates of motor cortical neurons. No consensus has emerged and has left us with a confused state of affairs in the field.

In this chapter, we propose that the motor cortex does not encode movement as any one simple Newtonian parameter of motion. Rather, we suggest that, if motor cortical neurons actually encode anything at all, they represent movement fragments as Sherrington and Leyton first proposed. We will begin by defining what we mean by "encoding" in the nervous system. We will then provide evidence against the contention that motor cortical neurons encode any simple movement parameter and argue for our hypothesis that motor cortex may form a substrate where elementary movement fragments are assembled into motor behaviors. In the last part of the chapter, we will argue that encoding, albeit an important scientific topic of inquiry, is only part of a full explanation of motor cortical functioning. We will point out that the motor cortex not only encodes information but also transforms information by processing its inputs to generate its output much like any information processing system. Our focus will be on the upper limb, although many of our arguments can be applied more generally to the cortical control of other segments in the motor apparatus.

WHAT DO WE MEAN BY ENCODING?

Before addressing the question as to what the motor cortex encodes, it is essential to start with a definition of *neural encoding*. In the neuroscience community, the term "neural encoding" has been ascribed a variety of meanings. We believe that neural encoding should be characterized as a relationship between some aspect of neural activity and features of the outside world. For a sensory physiologist, these features correspond to visual, auditory, somatosensory, and olfactory aspects of the world among others. For a motor physiologist, the outside world corresponds to the musculoskeletal system that is being controlled by the nervous system. There are at least three specific definitions that are either implicitly or explicitly stated in studies of neural encoding.

Correlation/Covariation

A commonly held viewpoint is that a neural structure encodes a sensory, cognitive, or motor feature if the responses of the neurons within the neural structure are correlated with the values that the feature can take. This correlation can be linear or nonlinear and can be measured with a variety of mathematical techniques including simple cross-correlation techniques, regression methods, and information theoretic methods (i.e., mutual information). According to this definition, the motor cortex encodes a number of features as documented in Table 14.1. Beginning with the first behavioral electrophysiological studies in primates by Evarts, a number of studies have indicated that the firing rates of single motor cortical neurons covaried, often linearly, with joint torque, hand force, and their time derivatives (Cabel et al., 2001; Cheney & Fetz, 1980b; Evarts, 1968; Hepp-Reymond et al., 1978, 1999; Kalaska et al., 1989; Smith et al., 1975; Taira et al., 1996). In addition to kinetic parameters, the firing rates of MI neurons also have been shown to covary with a number of kinematic variables including static position (Georgopoulos et al., 1984), dynamic position (Paninski et al., 2004a), speed (Moran & Schwartz, 1999), acceleration (Stark et al., 2007b), direction (Georgopoulos et al., 1982), movement distance, or some combinations of those parameters (Ashe & Georgopoulos, 1994; Fu et al., 1993, 1995; Kurata, 1993). There is also strong evidence that motor cortical neurons carry higher-level information related to movement planning (Tanji & Evarts, 1976), target location independent of the movement direction (Alexander & Crutcher, 1990; Shen & Alexander, 1997), or direction of action regardless of particular muscle activity (Kakei et al., 1999).

Some investigators have argued that the motor cortex multiplexes many of these parameters in their responses (Johnson & Ebner, 2000). This would require areas receiving signals from the motor cortex to somehow demultiplex the incoming signals (not a trivial problem). Others have argued that the motor cortex only appears to encode all these parameters because

Table 14.1 Motor Cortical Encoding of Movement Parameters

Direction	Ashe & Georgopoulos, 1994; Fu et al., 1995; Georgopoulos et al., 1982; Kakei et al., 1999; Schwartz et al., 1988; Thach, 1978
Position	Ashe & Georgopoulos, 1994; Georgopoulos et al., 1984; Paninski et al., 2004a; Stark et al., 2007b; Thach, 1978
Speed/velocity	Ashe & Georgopoulos, 1994; Churchland et al., 2006; Moran & Schwartz, 1999; Paninski et al., 2004a; Stark et al., 2007b
Acceleration	Ashe & Georgopoulos, 1994; Stark et al., 2007b
Force/torque	Cabel et al., 2001; Cheney & Fetz, 1980a; Evarts, 1968; Georgopoulos et al., 1992; Hepp-Reymond et al., 1978, 1999; Kalaska et al., 1989; Smith et al., 1975; Taira et al., 1996
Electromyography/muscle activation	Kakei et al., 1999; Morrow & Miller, 2003; Thach, 1978; Todorov, 2000
Amplitude/distance to target	Fu et al., 1993, 1995
Target position/direction	Alexander & Crutcher, 1990; Fu et al., 1995; Shen & Alexander, 1997
Trajectory	Hatsopoulos et al., 2007; Hocherman & Wise, 1991; Paninski et al., 2004b

many of the parameters are correlated with each other due to biological and physical constraints of motion. This is an important problem that is somewhat unique to motor physiology and will be further discussed later (see "The Problem of Correlated Parameters"). Setting aside this problem for the moment, we find that this definition is too lax to be satisfying. For example, the firing rates of MI neurons vary between awake and sleep states (Evarts, 1964), but no one would seriously argue that the motor cortex *encodes* wakefulness except in the weakest sense. Correlation may be a necessary but not sufficient condition for neural encoding.

Causality

A second definition of neural encoding contends that there must be a causal relationship between a neural structure and what it encodes. In the sensory domain, one might argue that neurons in the primary visual cortex (V1) encode oriented contrast edges because their activity occurs in response to a causal chain of events, beginning with responses in a population of retinal cells, followed in time by similar responses in the lateral geniculate nucleus cells that provide inputs to V1. However, causality does not seem to capture the essence of encoding. For example, a triplet sequence of nucleotides (a codon) is considered to code for a particular amino acid. Yet, the codon by itself does not "cause" the recruitment of a particular amino acid within a larger sequence of amino acids forming a protein. There are many other biophysical processes that are required for this to happen. As another more concrete example, the left blinker in a car does not cause the car to move to the left. Rather, it encodes the driver's intention to move to the left.

In the context of the motor cortex, if a feature of movement is encoded in the motor cortex, it may be too limiting to assume that the motor cortex "causes" that feature to occur. On strictly anatomical grounds, it is known that only a small percentage (perhaps 5% to 10%) of layer 5 output neurons in the motor cortex make direct monosynaptic connections with motor neurons in the spinal cord. For the vast majority of motor cortical neurons, therefore, it may be difficult to experimentally demonstrate a causal role in generating a feature of movement. In fact, it may be the case that for some motor cortical neurons, they play no causal role toward movement in that they only communicate the intentions of the motor cortex to other areas via efference copy (Nelson, 1996).

Invariance and Unique Specificity

A third definition of encoding, one to which we subscribe, states that a neural structure should invariantly and uniquely specify the feature it encodes, regardless of context. Specification can be viewed as a function that maps an input space to an output space. The mapping can be many to one but not one to many. In the codon example, the triplet of bases invariantly specifies or maps to a particular amino acid. This mapping is degenerate in that several triplets can specify a particular amino acid. However, it is unique in that a single triplet maps only to one amino acid. Likewise, the blinker invariantly and uniquely specifies the directional intentions of the driver. This specification remains invariant to different road or weather conditions. This specification, of course, serves a very important function as it communicates the intentions of the driver to other drivers on the road and prevents accidents from occurring. If the motor cortex encodes a parameter or feature of movement, then it must specify that feature in the same way regardless of different behavioral contexts. These contexts can include different parts of the arm's workspace, different postural configurations, different loading conditions, different task paradigms (i.e., isometric vs. isotonic), different ordinal positions within a complex movement sequence, and even different points in time.

THE PROBLEM OF CORRELATED PARAMETERS

Finding consistent responses to movement parameters in different contexts may seem to be straightforward. However, unlike sensory physiology, motor physiology possesses a fundamental experimental problem that makes it difficult to investigate parameters independently: The problem is that movement parameters are naturally correlated. In the study of sensory systems, the experimenter has full control of the stimuli presented to the organism. In order to isolate the sensory features that are encoded by a particular sensory system, the experimenter can design stimuli in which multiple features are simultaneously presented in a statistically independent fashion. However, to study motor systems, the motor act is not under the experimenter's full control but rather depends on the "will" of the organism. Under many situations, multiple movement parameters are coupled so that it becomes unclear which variable is actually related to the activity of single neurons. These couplings can be an artifact of the behavioral paradigm. For example, the center-out task pioneered by Georgopoulos (Georgopoulos et al., 1982) has been very useful in characterizing the directional tuning of motor cortical neurons. However, this behavioral paradigm creates links between movement direction, speed, and position. Movements to the left, for instance, occur on the left side of the workspace so that direction and position are correlated (see Fig. 1A in Paninski et al., 2004a). Likewise, the speed of the hand tends to be lower near the center and the periphery of the workspace so that speed and position are linked (see Fig. 1B in Paninski et al., 2004a).

The laws of physics and biological motion can also inherently couple certain movement parameters. For example, the force and acceleration are linked according to Newton's second law, which implies that a neuron's response will necessarily correlate with acceleration if it correlates with force, assuming that the mass is held fixed. There are also regularities in biological motion that correlate movement parameters. For example, Fitt's law describes the well-known speed-accuracy tradeoff such that faster movements are generated with less accuracy (Fitts, 1954). Likewise, the two-thirds power law relates the instantaneous angular speed of a movement with the radius of curvature. This is also characterized as a one-third power law between the tangential velocity and radius of curvature such that the movements are slower at points when the curvature is larger (Fig. 14.1) (Lacquaniti et al., 1983; Viviani & Cenzato, 1985; Viviani & Terzuolo, 1982).

Finally, the isochrony principle describes the observation that larger-amplitude movements tend to be made at faster speeds in order to approach a constant movement duration over different distances (Viviani & McCollum, 1983) (Fig. 14.2). While many of these correlations can be reduced or eliminated with proper controls, it is nearly impossible to uncouple all possible movement parameters. Therefore, in any individual study, it is essential to be careful in

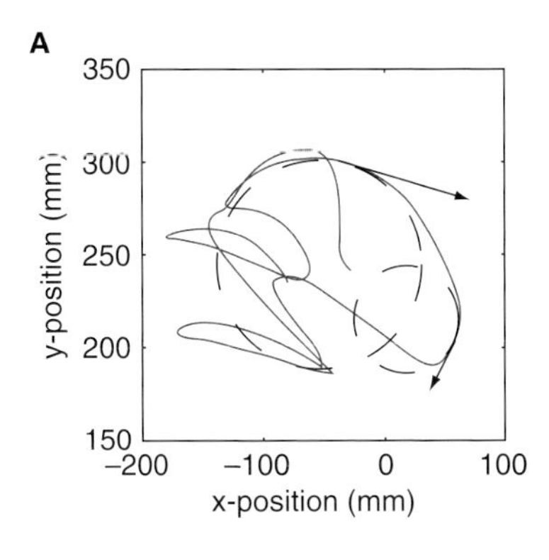

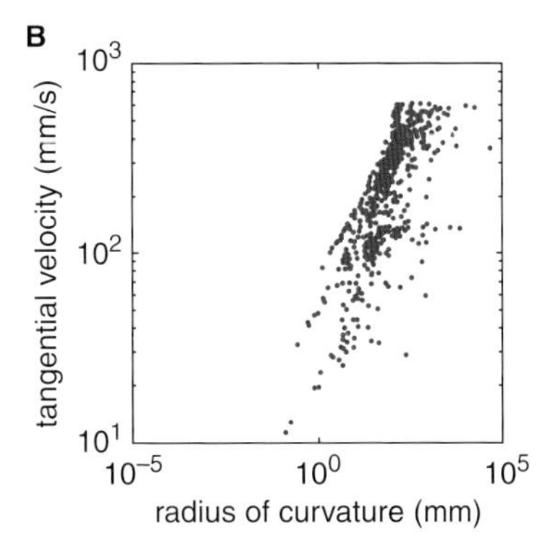

Figure 14.1 (**A**) The hand path during one trial (seven targets) of the random target pursuit task described later. The dashed circles define the local radii of curvature and the arrows indicate the local tangential velocity of the hand. (**B**) The power relationship between the local tangential velocity and radius of curvature is shown on a log-log scale.

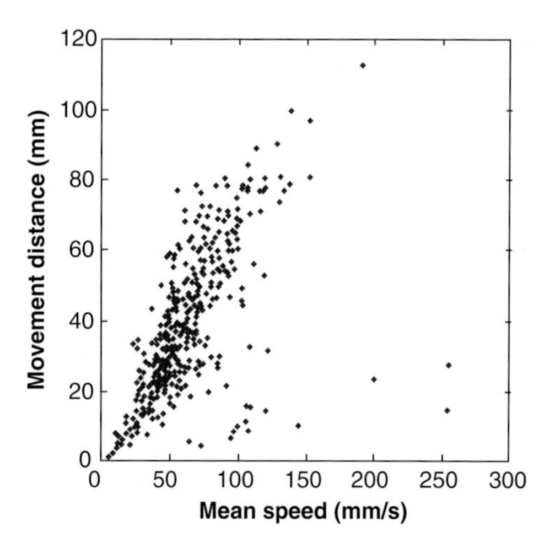

Figure 14.2 The relationship between the movement distance to a target and the mean speed to the target. Each point represents a movement to one target in the random target pursuit task.

interpreting claims that motor cortical neurons encode a specific movement parameter.

ENCODING OF MOVEMENT PARAMETERS

If we accept invariant specification as our definition of encoding, we are hard-pressed to identify any single movement parameter that is encoded in the motor cortex. For example, force is a kinetic variable that has been considered to be encoded in motor cortical neurons (Ashe, 1997; Evarts, 1968). However, context-dependent effects have been shown to alter the relationship between firing rate and force in individual motor cortical neurons during an isometric precision grip task involving the thumb and index finger (Hepp-Reymond et al., 1999). These authors found that while the average firing rate of motor cortical neurons increased with static force, the slope relating the firing rate with the force magnitude varied significantly depending on the task condition. That is, the slope across low- and medium-force levels differed depending on whether a third high-force level was required. Interestingly, this context dependency was not observed in the digit muscle activity.

Directional tuning is a kinematic variable that is considered a hallmark of motor cortical encoding in multijoint movements (Georgopoulos et al., 1982). A large proportion of cells in primary motor cortex are tuned to the movement direction of the hand in two- and three-dimensional space. This is usually assessed by measuring the firing rate of the cell during a fixed period in a given movement trial (typically, the reaction time period between a go signal and movement initiation) and computing the average firing rate over multiple trials of a given movement direction. This is repeated for each direction tested from which a tuning curve is plotted relating the average firing rate to the direction of movement. Various analytic functions such as a cosine or a circular Gaussian are fit to the data from which the preferred direction of the cell is determined by the peak value in the fitted function. Directional tuning is highly robust and common in the motor cortex such that no one can reasonably dispute this phenomenon. What is in dispute, however, is whether direction measured in an extrinsic, Cartesian coordinate system, or any coordinate system for that matter, is invariantly encoded in the motor cortex. Variations in initial hand position, arm posture, and behavioral paradigm have effects on the directional tuning curves of motor cortical neurons (Caminiti et al., 1990, 1991; Scott & Kalaska, 1995). For example, Caminiti and colleagues measured the preferred directions (PDs) of motor cortical cells in a two-dimensional center-out task in which the animal initiated its movement from one of three center positions parallel to the shoulder girdle. They found systematic shifts in the PDs such that they rotated along the z-axis (Caminiti et al., 1990). More recently, Scott and Kalaska (1995) showed that the PD shifted on average by 30 degrees depending on whether the arm posture was unconstrained in an adducted position in the sagittal plane or forced to be abducted. Interestingly, it was also observed that some cells possessed unimodal directional tuning in one posture but not in the other.

The most striking variations in directional tuning are the temporal shifts in preferred direction that have been observed by a number of groups (Johnson et al., 1999; Mason et al., 1998; Sergio & Kalaska, 1998; Sergio et al., 2005). These temporal variations were measured by shrinking the time window that spikes were counted and computing the preferred direction at multiple time windows throughout the trial. Using an instructed-delay, center-out task, Mason and colleagues (1998) found that neurons in the dorsal premotor cortex (PMd) shifted their PDs temporally over the whole trial by ≈48 degrees on average. Cells often exhibited clockwise, counterclockwise, and other complex shifting patterns in the PDs even during the instructed-delay period prior to movement.

These temporal shifts in PD have been observed in the primary motor cortex during a movement task as compared to an isometric task (Sergio & Kalaska, 1998; Sergio et al., 2005). Sergio and Kalaska were particularly interested in comparing the tuning characteristics of MI neurons within the rostral bank of the central sulcus under isometric and movement versions of the center-out task. In the isometric condition, the monkey pushed on a fixed manipulandum to move a visual cursor to one of eight peripherally positioned targets. In the movement condition, the same manipulandum was free to move, thereby moving the colocalized cursor to the targets. A temporal analysis of preferred direction showed a dramatic difference between the two paradigms. While the PD of most MI neurons remained relatively stable in the isometric paradigm, the PDs exhibited large and complex shifting patterns in the movement task reminiscent of the patterns observed by Mason and colleagues.

The observed temporal shifts in PD observed in those studies led us to examine the possibility that MI neurons do not encode any simple movement parameter but rather represent complex, time-dependent trajectory fragments similar to the "fractional" movement hypothesis first proposed by Leyton and Sherrington. As previous studies had shown, we observed systematic shifts in PD (Fig. 14.3A, black circles) that traced out "preferred trajectories" in two-dimensional space over the course of a trial in an instructed delay, center-out task (Fig. 14.3B, black arrows). We sought to determine whether these preferred trajectory representations generalized across a less constrained behavioral paradigm, which allowed the monkey to generate a rich variety of trajectories and paths in the horizontal plane. This was accomplished by having the monkey reach for a sequence of randomly positioned targets by moving a two-link exoskeletal robot that controlled a cursor (the random target pursuit or RTP task). Because the monkey's arm was continuously moving and shifting its instantaneous direction of movement, we computed the PDs by comparing the instantaneous movement direction at different time lags and leads with respect to the firing rate of the neuron (measured in 50-ms bins). Similar shifts in PD were observed in the RTP task as were observed in the center-out task (Fig. 14.3A, red and blue dots).

We then developed a mathematical model that explicitly characterized the preferred trajectories of individual neurons and demonstrated that temporally extended trajectories captured the tuning of motor cortical neurons more accurately. We developed a generalized linear model that estimated the probability of a spike emitted by a neuron (within a small spike sampling window of 10 ms) given that the monkey's hand generated a particular velocity trajectory. As others had shown using a different mathematical approach (Paninski et al., 2004b; Shoham et al., 2005), we found that an

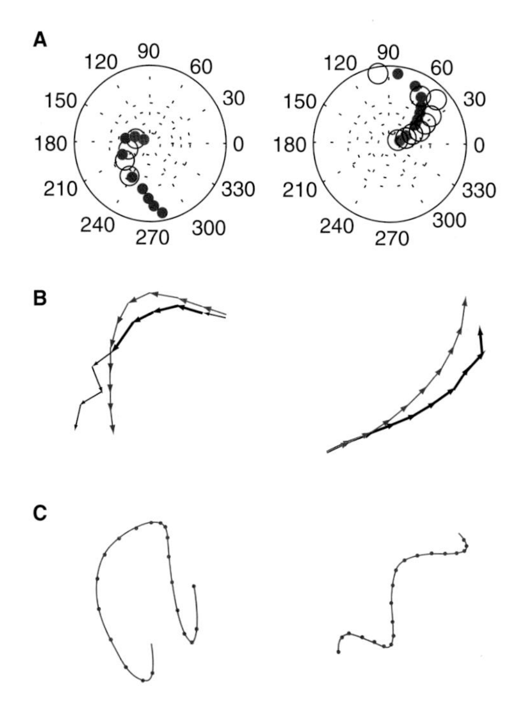

Figure 14.3 Systematic shifts in preferred direction (PD) under two behavioral paradigms. (**A**) Polar plots from two MI neurons depicting the shifts in PD in absolute time in the center-out task (black circles) as well as the PD shifts in relative time (lag time, red dots; lead time, blue dots) in the random target pursuit (RTP) task. Missing black circles in the left panel indicate that the PDs were not significantly fit with a cosine function. (**B**) By vectorally adding the shifting PDs, preferred trajectories are traced out in the center-out task (black vectors) and in the RTP task (blue and red vectors). Thin black vectors in the left panel indicate PDs that were not significantly fit with a cosine function. (**C**) The pathlets generated from the general linear model described later.

exponential relationship between the spiking probability and the inner product between the preferred velocity trajectory, $\vec{k}$, and the normalized velocity trajectory of the hand, $\hat{v}^{t_0}$, fit the data:

$$P(spike(t_0)|\hat{v}^{t_0}) = \exp[\vec{k} \cdot \hat{v}^{t_0} + \gamma] \qquad (1)$$

where each velocity trajectory extends over a range of times before and after the spike sampling time, t_o. γ is an offset parameter of the model. The normalized velocity trajectory, $\hat{v}^{t_0}$, is a sequence of velocities (i.e., directions and relative speeds) over a predefined time duration. We call $\vec{k}$ the preferred velocity trajectory since the inner product in the exponent is maximized when the $\hat{v}^{t_0}$ vector is aligned with it. By temporally integrating the x- and y-components of $\vec{k}$, the preferred paths or "pathlets" for each neuron can be generated (Fig. 14.3C). We view these pathlets as a sort of movement alphabet from which more complex motor actions can be synthesized just as language primitives are combined to form words and sentences. We have also demonstrated that these movement primitives remain relatively invariant in different parts of the workspace and under different dynamic contexts (see Figs. 3 and 7 in Hatsopoulos et al., 2007). Finally, we have also showed that a decoding algorithm that incorporated a time-dependent preferred trajectory for each neuron could be used to predict the instantaneous movement direction with less error than an algorithm that assumed a static preferred direction (Hatsopoulos et al., 2007).

Despite the fact that movement fragment encoding appears to capture the information content of MI neurons more effectively than time-independent parametric encoding, there are still a number of unanswered questions that will require further research. It is far from clear whether such a time-dependent representation will remain invariantly specified in single MI neurons across many different conditions and contexts. Our experiments have already indicated that not all neurons invariantly encode a particular pathlet when computed under different regions of the workspace (Hatsopoulos et al., 2007). Moreover, while a number of neurons preserve their pathlet representation when viscous loads are applied to the shoulder and elbow joints, this is not true of all recorded neurons (Hatsopoulos et al., 2007). The current situation leaves us with a number of possibilities. First, encoding in MI may be inherently time dependent as we have shown, but the theory may need to be refined to account for all possible contexts. Second, there may be a yet undiscovered time-independent movement parameter that is invariantly specified in motor cortex. We think this is unlikely because a nearly exhaustive list of parameters has been examined over many years such that it is hard to conceive of what this parameter would be. Third, the motor cortex may not encode one thing; different neurons within MI may encode different movement parameters. In fact, by mathematically accounting for linear correlations among movement parameters, a recent experimental study has indicated that the activities of single MI neurons are related to single movement parameters and the responses of different neurons covary with different parameters (Stark et al., 2007b). It remains to be seen whether single neurons actually *encode* single movement parameters as we have defined encoding previously. Fourth, it may be that movement encoding is not a property of any single cell but is a global property of a neuronal ensemble. That is, it may be that large-scale spatiotemporal patterns across the motor cortex invariantly specify aspects of movement even though individual neurons participate in an inconsistent fashion to these global patterns. Although we have no strong experimental evidence to support this view, we feel strongly that examining cortical activity on an ensemble level may reveal aspects of encoding to which we have been blind with single-unit recording.

BEYOND ENCODING

Understanding what the motor cortex encodes is clearly an important endeavor in elucidating its function. However, as an information processing system, encoding is only part of the story. The transformation or processing of information is the complement of encoding (deCharms & Zador, 2000). There has been very little research focused on what sort of computation the motor cortex is actually performing. This requires understanding the nature of the inputs to motor cortex and how these inputs are transformed to generate functional outputs.

INPUTS TO MOTOR CORTEX

A large body of anatomical research has identified cortical and subcortical structures that provide direct inputs to the motor cortex including the lateral premotor, the supplementary motor, the somatosensory, and the posterior parietal cortices (Donoghue & Sanes, 1994; Luppino & Rizzolatti, 2000; Muakkassa & Strick, 1979). In addition, as with other neocortical structures, there are inputs from the thalamus (Holsapple et al., 1991; Olszewski, 1952; Strick, 1976) that transmit information from the spinal cord, the cerebellum, and the basal ganglia (Lu et al., 2007; Middleton & Strick, 1997a,b; Miyachi et al., 2006; Sakai et al., 2002). Less is known about the spatiotemporal structure of these inputs during the performance of motor actions.

We have been engaging in two lines of research to examine structured inputs to the motor cortex during motor behavior. The first involves recording spiking activity from cortical areas such as the premotor cortex that provide major inputs to the motor cortex. By simultaneously recording neuronal ensembles from both the dorsal premotor and primary motor cortices, we are beginning to use multivariate autoregressive modeling to relate the output spiking from single motor cortical neurons to the spiking of multiple premotor cortical

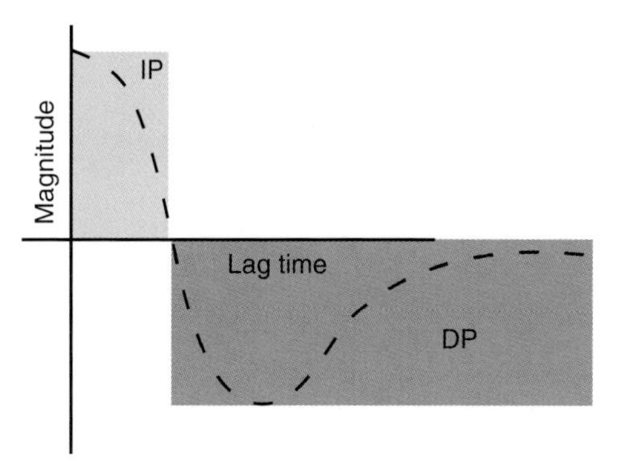

Figure 14.4 A hypothetical temporal profile of coefficients generated from an autoregressive model predicting the spike activity of a motor cortical neuron from the spiking history of a premotor cortical neuron.

neurons (inputs) at multiple time lags (Wu & Hatsopoulos, 2006). The goal of this ongoing research is to examine the structure of the coefficients in order to determine what sort of operation or transformation is being performed between the inputs and the outputs. As a hypothetical example, if the temporal profile of these coefficients exhibits the shape shown in Figure 14.4, this would imply that that motor cortex is performing a temporal derivative on its inputs because it is taking the inputs from the immediate past (see "IP" in Fig. 14.4) and subtracting the inputs from the distant past (see "DP" in Fig. 14.4).

Our second line of research has involved recording local field potentials (LFPs) in the motor cortex. These LFPs are believed to represent the summed voltage of hundreds to thousands of nearly synchronous, postsynaptic potentials of motor cortical neurons near the electrode site and, therefore, constitute a measure of the local spatiotemporal structure of motor cortical inputs. These LFPs have been shown to engage in characteristic oscillations in the beta frequency range (20 to 30 Hz) within the motor cortex during periods of motor planning prior to movement execution (Donoghue et al., 1998; O'Leary & Hatsopoulos, 2006; Sanes & Donoghue, 1993), postural maintenance (Baker et al., 1999; Gilbertson et al., 2005), and heightened attention and active engagement (Donoghue et al., 1998; Murthy & Fetz, 1992, 1996a,b). We have recently observed that these beta oscillations participate in propagating planar waves across the arm area of the motor cortex (Rubino et al., 2006). They are particularly evident during an imposed motor preparatory period prior to movement but are also evident during movement execution, albeit attenuated in amplitude. These waves of input activity propagate at relatively slow speeds (10 to 20 cm/s) and typically travel along a stereotyped rostral to caudal propagation axis. That is, at each moment in time the wave is typically propagating in the caudal to rostral direction or in rostral to caudal direction. This dominant propagating axis is unaffected by the direction of the target to be reached by the animal and remains invariant across different behavioral paradigms (Rubino et al., 2006; Takahashi & Hatsopoulos, 2007).

OUTPUTS FROM MOTOR CORTEX

Extracellular unit recordings from the motor cortex provide a direct physiological measure of the outputs of the motor cortex. In particular, layer 5 neurons provide the major corticofugal pathway that forms corticospinal projections to the spinal cord and ultimately the motor neuron pools that activate the muscles of the periphery. However, the meaning of these outputs can only be gleaned from the motor behaviors they affect. It is our contention that behavioral electrophysiology should expand its repertoire of motor behaviors from the highly constrained behaviors typically measured in the lab to ethologically more relevant motor actions. This may reveal aspects of motor cortical functioning and encoding that have eluded previous research.

PREHENSION

One class of such unconstrained ethological behaviors that is currently gaining more attention is coordinated reach and grasp, or prehension. Indeed, a mobile forelimb and an opposable thumb enabled our earliest ancestors from the Eocene period to locomote and reach and grasp for insects and fruits among slender tree branches (Marzke, 1994). As an ethological behavior, reaching and grasping is behaviorally relevant and provides an excellent model to examine how the motor cortex encodes meaningful movement.

Early psychophysical studies concluded that reaching and grasping are independent processes that are loosely coordinated by some higher-order central timing mechanism such as the cerebellum or the basal ganglia (Arbib, 1981; Hoff & Arbib, 1993; Jeannerod, 1981, 1984). The idea that these movements are processed separately stems from the observation of separate visual (and other sensory modalities) properties of the object to be grasped. Directing a reaching movement to an object depends on visually registering the object's relative direction and distance from the observer, which are extrinsic properties of the object. On the other hand, preshaping the hand and grasping the object depends on the object's shape and size, intrinsic features of the object. The notion of separate control was seemingly confirmed by data that showed that the maximum velocity of the hand varied only with the object's distance, whereas the maximum aperture size formed between the index finger and thumb prior to grasp varied only with object's size. Structural features of the musculoskeletal system were also recognized to parallel the differences between reach and grasp. Reaching is an action of axial and proximal musculoskeletal structures, while grasp occurs within distal appendicular structures.

Despite the early conceptualizations of how reach and grasp are distinct and disparate, subsequent psychophysical studies have suggested that reaching and grasping components are interdependent and tightly coordinated in both time and space. For example, a mechanical perturbation of the arm (a proximal motor component) elicits an adjustment not only in the arm's position but also in the shape of the hand so as to recover and maintain a characteristic coordinated spatial trajectory between these components (Haggard, 1991; Haggard & Wing, 1991, 1995). Moreover, other studies have contradicted Jeannerod's earlier studies by demonstrating that variations in object distance affect grasp as well as transport components (Chieffi & Gentilucci, 1993; Jakobson & Goodale, 1991) and that variations in object size can affect transport kinematics as well as aperture size (Marteniuk et al., 1990). Also, a number of perturbation studies in which the object location or size was unexpectedly altered resulted in modifications of both the transport and grasp components of movement (Gentilucci et al., 1992; Paulignan et al., 1990, 1991a,b; Roy et al., 2006).

Anatomical and physiological studies also support the idea that reaching and grasping are highly interdependent and may be processed jointly in the cortex. Retrograde tracer studies have shown that corticospinal projections from the motor cortex make connections onto motoneuron pools of multiple muscles involved in both distal and proximal arm muscle control (Shinoda et al., 1981), suggesting that single cortical neurons might evoke synergistic activation of distal and proximal muscles during prehension. These anatomical observations are further supported by a study that identified cortico-motoneuronal connections by means of spike-triggered averages of electromyographic (EMG) activity of the arm and hand during prehensile movements in nonhuman primates (McKiernan et al., 1998). This study showed that 45% of sampled cortico-motoneuronal cells were connected to motoneurons that projected to both distal and proximal muscles of the arm. Recent stimulation studies also suggest that there may be an overlap between neurons controlling distal and proximal arm muscles.

Stark and colleagues (2007a), combining intracortical microstimulation (ICMS) and single-unit recording from the premotor cortex, found a paradoxical result. In some electrode sites, ICMS elicited distal arm muscle activity of the hand. Yet in these same sites, they recorded single units that encoded arm direction, a movement controlled by proximal arm musculature. Likewise, there were other sites in which ICMS activated proximal arm muscles while the recorded single units from the same sites encoded the type of grasp. The authors of that study argued that such neurons might help coordinate distal and proximal muscles of the arm in a prehensile movement.

Using an altered ICMS protocol consisting of higher electrical currents up to 100 μA and pulse durations of several hundred milliseconds, Graziano and colleagues (2002, 2004) were able to elicit complex, apparently goal-directed movements involving proximal and distal joints, which appeared to mimic naturalistic reaching and grasping, along with other behaviors. Based on these findings, they suggested that the motor and premotor cortex initiate the coordinated activation of several muscles that are involved in prehension and other ethologically relevant behaviors.

Based on these psychophysical, anatomical, and physiological studies, we hypothesize that single neurons within motor cortex encode reaching and grasping movements. Specifically, we postulate that individual neurons encode complex, time-dependent movement fragments (described previously) that include both proximal and distal components of prehension. To test this hypothesis, we have begun experiments in which monkeys are trained to reach and grasp for different objects presented in the monkey's three-dimensional reaching space with a 6-degree-of-freedom robot (Fig. 14.5A). Kinematic data are captured in real time with high fidelity by multiple infrared motion capture cameras (Vicon system) that track multiple spherical reflective markers placed on the arm, forearm, and fingers of the monkey (Fig. 14.5B).

As previous studies have demonstrated a spatial relationship between the kinematics of the hand transport and grasp aperture (Jeannerod, 1981, 1984; Roy et al., 2000, 2002, 2006), our initial goal was to verify this relationship in our system (Fig. 14.6). Figure 14.6A and C shows that although the timing of the maximum aperture varies, its relationship with respect to transport distance is relatively fixed. This finding supports the idea that there may be

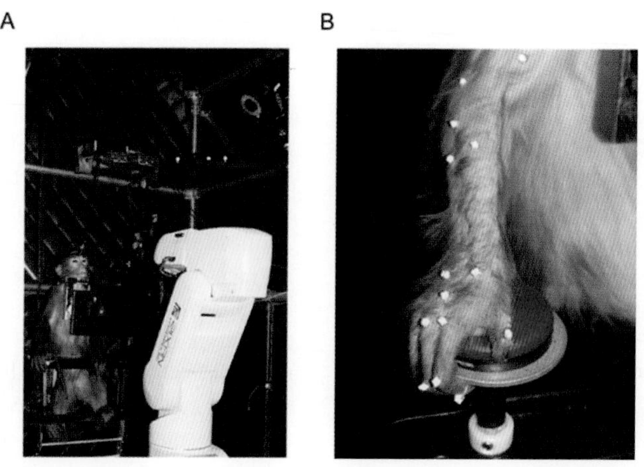

Figure 14.5 (**A**) The recording rig used to record the motion of the arm and hand during free reaching and grasping. One of the Vicon cameras is shown in the upper right-hand corner. The 6-degree-of-freedom robot that provides the objects to be grasped is shown in the foreground. (**B**) A close-up of the reflective markers placed on the arm, wrist, hand, thumb, index finger, and middle finger.

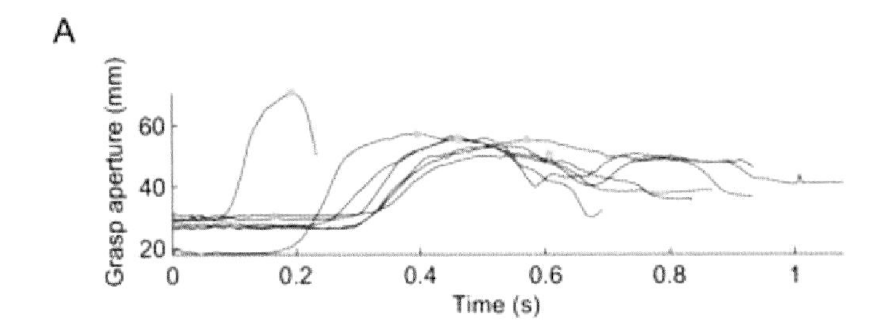

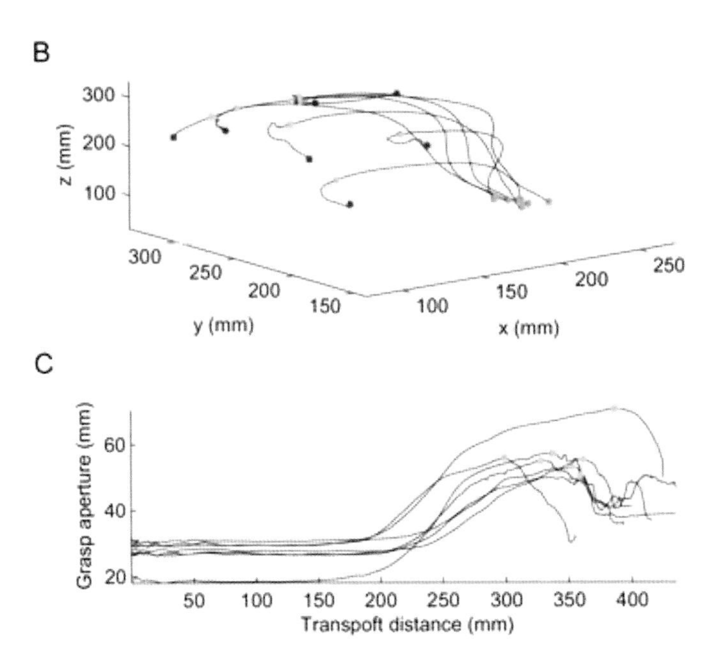

Figure 14.6 The coordinated kinematics of the grip aperture and transport distance during multiple reach-to-grasp movements generated by one of our monkeys. **(A)** The grasp aperture size (measured as the distance between the thumb and index finger markers) as a function of time. The cyan dots indicate when the maximum grasp aperture occurs. **(B)** The transport paths (measured from the thumb marker) in three dimensions. The green dots indicate the start of the reach, while the red dots indicate object contact. **(C)** The grasp aperture versus transport distance (measured as the integrated distance of the thumb) plots the spatial coordination pattern between the reach and grasp components.

some spatial dependencies between the musculature that controls hand transport distance—proximal arm muscles—and grasp aperture—distal arm muscles. Our next goal is to further characterize spatial and temporal dependencies between reaching and grasping movements by investigating more complex, kinematic parameters of three-dimensional prehension (Fig. 14.6B). We are also interested in testing whether single neurons within primary motor, dorsal premotor, and ventral premotor cortices encode not only kinematic features of reach and grasp but perhaps also the kinematics of their coordinated action (Fig. 14.6C). We are using the mathematical framework (described previously) by estimating the spiking probability from movement fragment covariates extracted from whole reach-to-grasp behaviors toward diverse objects in various locations in the monkey's reach space.

CONCLUSION

Despite being one of the earliest cortical areas to be examined functionally, the features encoded by the motor cortex remain elusive. We have revisited an early viewpoint of motor cortical function proposed by Sherrington and Leyton and have provided experimental evidence to support and refine that early viewpoint. Instead of encoding a static Newtonian parameter of motion, our analysis of motor cortical activity suggests that individual motor cortical neurons encode temporally evolving movement fragments. Although our data have not definitively supported a kinematic versus kinetic representation, we maintain that a time-dependent representation is a fundamental feature of motor cortical encoding. Recent electrical stimulation studies seem to further support this perspective by showing that stimulation (Graziano et al., 2002) even at a single neuron level (Brecht et al., 2004) can elicit complex movement trajectories. To further develop the idea that motor cortex encodes complex, time-dependent movements, we argue that examining more meaningful, neuroethological behaviors will be an important direction in which research should head.

ACKNOWLEDGMENTS

We would like to thank Krishna Shenoy, Mark Churchland, Jake Reimer, and Kazutaka Takahashi for helpful discussions related to this book chapter. This work was supported by a NIH grant R01 NS45853 from the NINDS awarded to NGH.

REFERENCES

Alexander, G. E., & Crutcher, M. D. (1990). Neural representations of the target (goal) of visually guided arm movements in three motor areas of the monkey. *Journal of Neurophysiology, 64,* 164–178.

Arbib, M. (1981). Perceptual structures and distributed motor control. In: V. B. Brooks (Ed.), *Handbook of physiology*. Baltimore: Williams and Wilkins.

Ashe, J. (1997). Force and the motor cortex. *Behavioural Brain Research, 86,* 1–15.

Ashe, J., & Georgopoulos, A. P. (1994). Movement parameters and neural activity in motor cortex and area 5. *Cerebral Cortex, 6,* 590–600.

Baker, S. N., Kilner, J. M., Pinches, E. M., & Lemon, R. (1999). The role of synchrony and oscillations in the motor output. *Experimental Brain Research, 128,* 109–117.

Brecht, M., Schneider, M., Sakmann, B., & Margrie, T. W. (2004). Whisker movements evoked by stimulation of single pyramidal cells in rat motor cortex. *Nature, 427,* 704–710.

Cabel, D. W., Cisek, P., & Scott, S. H. (2001). Neural activity in primary motor cortex related to mechanical loads applied to the shoulder and elbow during a postural task. *Journal of Neurophysiology, 86,* 2102–2108.

Caminiti, R., Johnson, P. B., Galli, C., Ferraina, S., & Burnod, Y. (1991). Making arm movements within different parts of space: The premotor and motor cortical representation of a coordinate system for reaching to visual targets. *Journal of Neuroscience, 11,* 1182–1197.

Caminiti, R., Johnson, P. B., & Urbano, A. (1990). Making arm movements within different parts of space: Dynamic aspects in the primate motor cortex. *Journal of Neuroscience, 10,* 2039–2058.

Cheney, P. D., & Fetz, E. E. (1980a). Functional classes of primate corticomotoneuronal cells and their relation to active force. *Journal of Neurophysiology, 44,* 773–791.

Cheney, P. D., & Fetz, E. E. (1980b). Functional classes of primate corticomotoneuronal cells and their relation to active force. *Journal of Neurophysiology, 44,* 773–791.

Chieffi, S., & Gentilucci, M. (1993). Coordination between the transport and the grasp components during prehension movements. *Experimental Brain Research, 94,* 471–477.

Churchland, M. M., Santhanam, G., & Shenoy, K. V. (2006). Preparatory activity in premotor and motor cortex reflects the speed of the upcoming reach. *Journal of Neurophysiology, 96,* 3130–3146.

deCharms, R. C., & Zador, A. (2000). Neural representation and the cortical code. *Annual Review of Neuroscience, 23,* 613–647.

Donoghue, J. P., & Sanes, J. N. (1994). Motor areas of the cerebral cortex. *Journal of Clinical Neurophysiology,* 11, 382–396.

Donoghue, J. P., Sanes, J. N., Hatsopoulos, N. G., & Gaal, G. (1998). Neural discharge and local field potential oscillations in primate motor cortex during voluntary movements. *Journal of Neurophysiology*, 77, 159–173.

Evarts, E. V. (1964). Temporal patterns of discharge of pyramidal tract neurons during sleep and waking in the monkey. *Journal of Neurophysiology, 27,* 152–171.

Evarts, E. V. (1968). Relation of pyramidal tract activity to force exerted during voluntary movement. *Journal of Neurophysiology, 31,* 14–27.

Fitts, P. M. (1954). The information capacity of the human motor system in controlling the amplitude of movement. *Journal of Experimental Psychology, 47,* 381–391.

Fritsch, G., & Hitzig, E. (1870). Uber die elektrische Erregbarkeit des Grosshirns. *Archives fur Anatomie, Physiologie, und Wissenschaftliche Medizin*, 300–332.

Fu, Q.-G., Flament, D., Coltz, J. D., & Ebner, T. J. (1995). Temporal encoding of movement kinematics in the discharge of primate primary motor and premotor neurons. *Journal of Neurophysiology, 73,* 836–854.

Fu, Q.-G., Suarez, J. I., & Ebner, T. J. (1993). Neuronal specification of direction and distance during reaching movements in the superior precentral premotor area and primary motor cortex of monkeys. *Journal of Neurophysiology, 70,* 2097–2116.

Gentilucci, M., Chieffi, S., Scarpa, M., & Castiello, U. (1992). Temporal coupling between transport and grasp components during prehension movements: Effects of visual perturbation. *Behavioral Brain Research, 47,* 71–82.

Georgopoulos, A. P., Ashe, J., Smyrnis, N., & Taira, M. (1992). The motor cortex and the coding of force. *Science, 256,* 1692–1695.

Georgopoulos, A. P., Caminiti, R., & Kalaska, J. F. (1984). Static spatial effects in motor cortex and area 5: Quantitative relations in a two-dimensional space. *Experimental Brain Research, 54,* 446–454.

Georgopoulos, A. P., Kalaska, J. F., Caminiti, R., & Massey, J. T. (1982). On the relations between the direction of two-dimensional arm movements and cell discharge in primate motor cortex. *Journal of Neuroscience, 2,* 1527–1537.

Gilbertson, T., Lalo, E., Doyle, L., Di Lazzaro, V., Cioni, B., & Brown, P. (2005). Existing motor state is favored at the expense of new movement during 13-35 Hz oscillatory synchrony in the human corticospinal system. *Journal of Neuroscience, 25,* 7771–7779.

Graziano, M. S., Cooke, D. F., Taylor, C. S., & Moore, T. (2004). Distribution of hand location in monkeys during spontaneous behavior. *Experimental Brain Research, 155,* 30–36.

Graziano, M. S., Taylor, C. S., & Moore, T. (2002). Complex movements evoked by microstimulation of precentral cortex. *Neuron, 34,* 841–851.

Haggard, P. (1991). Task coordination in human prehension. *Journal of Motor Behavior, 23,* 25–37.

Haggard, P., & Wing, A. (1995). Coordinated responses following mechanical perturbation of the arm during prehension. *Experimental Brain Research, 102,* 483–494.

Haggard, P., & Wing, A. M. (1991). Remote responses to perturbation in human prehension. *Neuroscience Letters, 122,* 103–108.

Hatsopoulos, N. G., Xu, Q., & Amit, Y. (2007). Encoding of movement fragments in the motor cortex. *Journal of Neuroscience, 27,* 5105–5114.

Hepp-Reymond, M., M. K.-T., Gabernet, L., H.X., Q., & Weber, B. (1999). Context-dependent force coding in motor and premotor cortical areas. *Experimental Brain Research, 128,* 123–133.

Hepp-Reymond, M.-C., Wyss, U. R., & Anner, R. (1978). Neuronal coding of static force in the primate motor cortex. *Journal of Physiology, Paris, 74,* 287–291.

Hocherman, S., & Wise, S. P. (1991). Effects of hand movement path on motor cortical activity in awake, behaving rhesus monkeys. *Experimental Brain Research, 83,* 285–302.

Hoff, B., & Arbib, M. A. (1993). Models of Trajectory Formation and Temporal Interaction of Reach and Grasp. *Journal of Motor Behavior, 25,* 175–192.

Holsapple, J. W., Preston, J. B., & Strick, P. L. (1991). The origin of thalamic inputs to the "hand" representation in the primary motor cortex. *Journal of Neuroscience, 11,* 2644–2654.

Jakobson, L. S., & Goodale, M. A. (1991). Factors affecting higher-order movement planning: A kinematic analysis of human prehension. *Experimental Brain Research, 86,* 199–208.

Jeannerod, M. (1981). Intersegmental coordination during reaching at natural visual objects. In:

J. Long & A. Baddeley (Eds.), *Attention and performance IX* (pp. 153–168). Hillsdale, NJ: Lawrence Erlbaum Associates Publishers.

Jeannerod, M. (1984). The timing of natural prehension movements. *Journal of Motor Behavior, 16,* 235–254.

Johnson, M. T., Coltz, J. D., Hagen, M. C., & Ebner, T. J. (1999). Visuomotor processing as reflected in the directional discharge of premotor and primary motor cortex neurons. *Journal of Neurophysiology, 81,* 875–894.

Johnson, M. T., & Ebner, T. J. (2000). Processing of multiple kinematic signals in the cerebellum and motor cortices. *Brain Research Reviews, 33,* 155–168.

Kakei, S., Hoffman, D. S., & Strick, P. L. (1999). Muscle and movement representations in the primary motor cortex. *Science, 285,* 2136–2139.

Kalaska, J. F., Cohen, D. A. D., Hyde, M. L., & Prud'homme, M. (1989). A comparison of movement direction-related versus load direction-related activity in primate motor cortex, using a two-dimensional reaching task. *Journal of Neuroscience, 9,* 2080–2102.

Kurata, K. (1993). Premotor cortex of monkeys: Set- and movement-related activity reflecting amplitude and direction of wrist movements. *Journal of Neurophysiology, 77,* 1195–1212.

Lacquaniti, F., Terzuolo, C., & Viviani, P. (1983). The law relating the kinematic and figural aspects of drawing movements. *Acta Psychologica, 54,* 115–130.

Lu, X., Miyachi, S., Ito, Y., Nambu, A., & Takada, M. (2007). Topographic distribution of output neurons in cerebellar nuclei and cortex to somatotopic map of primary motor cortex. *European Journal of Neuroscience, 25,* 2374–2382.

Luppino, G., & Rizzolatti, G. (2000). The organization of the frontal motor cortex. *News in Physiological Science, 15,* 219–224.

Marteniuk, R. G., Leavitt, J. L., MacKenzie, C. L., & Athenes, S. (1990). Functional relationships between grasp and transport components in a prehension task. *Human Movement Science, 9,* 149–176.

Marzke, M. (1994). Evolution. In: K. Bennett (Ed.), *Insights into the reach to grasp movement.* Amsterdam: Elsevier.

Mason, C. R., Johnson, M. T., Fu, Q. G., Gomez, J. E., & Ebner, T. J. (1998). Temporal profile of the directional tuning of the discharge of dorsal premotor cortical cells. *Neuroreport, 9,* 989–995.

McKiernan, B. J., Marcario, J. K., Karrer, J. H., & Cheney, P. D. (1998). Corticomotoneuronal postspike effects in shoulder, elbow, wrist, digit, and intrinsic hand muscles during a reach and prehension task. *Journal of Neurophysiology, 80,* 1961–1980.

Middleton, F. A., & Strick, P. L. (1997a). Cerebellar output channels. *International Review of Neurobiology, 41,* 61–82.

Middleton, F. A., & Strick, P. L. (1997b). New concepts about the organization of basal ganglia output. *Advance in Neurology, 74,* 57–68.

Miyachi, S., Lu, X., Imanishi, M., Sawada, K., Nambu, A., & Takada, M. (2006). Somatotopically arranged inputs from putamen and subthalamic nucleus to primary motor cortex. *Neuroscience Research, 56,* 300–308.

Moran, D. W., & Schwartz, A. B. (1999). Motor cortical representation of speed and direction during reaching. *Journal of Neurophysiology, 82,* 2676–2692.

Morrow, M. M., & Miller, L. E. (2003). Prediction of muscle activity by populations of sequentially recorded primary motor cortex neurons. *Journal of Neurophysiology, 89,* 2279–2288.

Muakkassa, K. F., & Strick, P. L. (1979). Frontal lobe inputs to primate motor cortex: Evidence for four somatotopically organized 'premotor' areas. *Brain Research, 177,* 176–182.

Murthy, V. N., & Fetz, E. E. (1992). Coherent 25- to 35-Hz oscillations in the sensorimotor cortex of awake behaving monkeys. *Proceedings of the National Academy of Sciences, 89,* 5670–5674.

Murthy, V. N., & Fetz, E. E. (1996a). Oscillatory activity in sensorimotor cortex of awake monkeys: Synchronization of local field potentials and relation to behavior. *Journal of Neurophysiology, 76,* 3949–3967.

Murthy, V. N., & Fetz, E. E. (1996b). Synchronization of neurons during local field potential oscillations in sensorimotor cortex of awake monkeys. *Journal of Neurophysiology, 76,* 3968–3982.

Nelson, R. J. (1996). Interactions between motor commands and somatic perception in sensorimotor cortex. *Current Opinions in Neurobiology, 6,* 801–810.

O'Leary, J. G., & Hatsopoulos, N. G. (2006). Early visuomotor representations revealed from evoked local field potentials in motor and premotor cortical areas. *Journal of Neurophysiology, 96,* 1492–1506.

Olszewski, J. (1952). *The thalamus of the macaca mulatta*. Basel: Karger.

Paninski, L., Fellows, M. R., Hatsopoulos, N. G., & Donoghue, J. P. (2004a). Spatiotemporal tuning of motor cortical neurons for hand position and velocity. *Journal of Neurophysiology, 91,* 515–532.

Paninski, L., Fellows, M. R., Shoham, S., Hatsopoulos, N. G., & Donoghue, J. P. (2004b). Superlinear population encoding of dynamic hand trajectory in primary motor cortex. *Journal of Neuroscience, 24,* 8551–8561.

Paulignan, Y., Jeannerod, M., MacKenzie, C., & Marteniuk, R. (1991a). Selective perturbation of visual input during prehension movements. 2. The effects of changing object size. *Experimental Brain Research, 87,* 407–420.

Paulignan, Y., MacKenzie, C., Marteniuk, R., & Jeannerod, M. (1990). The coupling of arm and finger movements during prehension. *Experimental Brain Research, 79,* 431–435.

Paulignan, Y., MacKenzie, C., Marteniuk, R., & Jeannerod, M. (1991b). Selective perturbation of visual input during prehension movements. 1. The effects of changing object position. *Experimental Brain Research, 83,* 502–512.

Roy, A. C., Paulignan, Y., Farne, A., Jouffrais, C., & Boussaoud, D. (2000). Hand kinematics during reaching and grasping in the macaque monkey. *Behavioral Brain Research, 117,* 75–82.

Roy, A. C., Paulignan, Y., Meunier, M., & Boussaoud, D. (2002). Prehension movements in the macaque monkey: Effects of object size and location. *Journal of Neurophysiology, 88,* 1491–1499.

Roy, A. C., Paulignan, Y., Meunier, M., & Boussaoud, D. (2006). Prehension movements in the macaque monkey: Effects of perturbation of object size and location. *Experimental Brain Research, 169,* 182–193.

Rubino, D., Robbins, K. A., & Hatsopoulos, N. G. (2006). Propagating waves mediate information transfer in the motor cortex. *Nature Neuroscience, 9,* 1549–1557.

Sakai, S. T., Inase, M., & Tanji, J. (2002). The relationship between MI and SMA afferents and cerebellar and pallidal efferents in the macaque monkey. *Somatosensory and Motor Research, 19,* 139–148.

Sanes, J. N., & Donoghue, J. P. (1993). Oscillations in local field potentials of the primate motor cortex during voluntary movement. *Proceedings of the. National Academy of Sciences, 90,* 4470–4474.

Schwartz, A. B., Kettner, R. E., & Georgopoulos, A. P. (1988). Primate motor cortex and free arm movements to visual targets in three-dimensional space. 1. Relations between single cell discharge and direction of movement. *Journal of Neuroscience, 8,* 2913–2927.

Scott, S. H., & Kalaska, J. F. (1995). Changes in motor cortex activity during reaching movements with similar hand paths but different arm postures. *Journal of Neurophysiology, 73,* 2563–2567.

Sergio, L. E., Hamel-Paquet, C., & Kalaska, J. F. (2005). Motor cortex neural correlates of output kinematics and kinetics during isometric-force and arm-reaching tasks. *Journal of Neurophysiology, 94,* 2353–2378.

Sergio, L. E., & Kalaska, J. F. (1998). Changes in the temporal pattern of primary motor cortex activity in a directional isometric force versus limb movement task. *Journal of Neurophysiology, 80,* 1577–1583.

Shen, L., & Alexander, G. E. (1997). Preferential representation of instructed target location versus limb trajectory in dorsal premotor cortex. *Journal of Neurophysiology, 77,* 1195–1212.

Shinoda, Y., Yokota, J., & Futami, T. (1981). Divergent projection of individual corticospinal axons to motoneurons of multiple muscles in the monkey. *Neuroscience Letters, 23,* 7–12.

Shoham, S., Paninski, L. M., Fellows, M. R., Hatsopoulos, N. G., Donoghue, J. P., & Normann, R. A. (2005). Statistical encoding model for a primary motor cortical brain-machine interface. *IEEE Transactions on Biomedical Engineering, 52,* 1312–1322.

Smith, A. M., Hepp-Reymond, M. C., & Wyss, U. R. (1975). Relation of activity in precentral cortical neurons to force and rate of force change during isometric contractions of finger muscles. *Experimental Brain Research, 23,* 315–332.

Stark, E., Asher, I., & Abeles, M. (2007a). Encoding of reach and grasp by single neurons in premotor cortex is independent of recording site. *Journal of Neurophysiology, 97,* 3351–3364.

Stark, E., Drori, R., Asher, I., Ben-Shaul, Y., & Abeles, M. (2007b). Distinct movement parameters are represented by different neurons in the motor cortex. *European Journal of Neuroscience, 26,* 1055–1066.

Strick, P. L. (1976). Anatomical analysis of ventrolateral thalamic input to primate motor cortex. *Journal of Neurophysiology, 39,* 1020–1031.

Taira, M., Boline, J., Smyrnis, N., Georgopoulos, A. P., & Ashe, J. (1996). On the relations between single cell activity in the motor cortex and the direction and magnitude of three-dimensional static isometric force. *Experimental Brain Research, 109,* 367–376.

Takahashi, K., & Hatsopoulos, N. G. (2007). Copropagating waves of local field potentials and single-unit spiking in motor cortex. In: *37th Annual Meeting of the Society for Neuroscience*, San Diego, CA.

Tanji, J., & Evarts, E. V. (1976). Anticipatory activity of motor cortex neurons in relation to direction of an intended movement. *Journal of Neurophysiology, 39,* 1062–1068.

Thach, W. T. (1978). Correlation of neural discharge with pattern and force of muscular activity, joint position, and direction of intended next movement in motor cortex and cerebellum. *Journal of Neurophysiology, 41,* 654–676.

Todorov, E. (2000). Direct cortical control of muscle activation in voluntary arm movements: A model. *Nature Neuroscience, 3,* 391–398.

Viviani, P., & Cenzato, M. (1985). Segmentation and coupling in complex movements. *Journal of Experimental Psychology: Human Perception and Performance, 11,* 828–845.

Viviani, P., & McCollum, G. (1983). The relation between linear extent and velocity in drawing movements. *Neuroscience, 10,* 211–218.

Viviani, P., & Terzuolo, C. (1982). Trajectory determines movement dynamics. *Neuroscience, 7,* 431–437.

Wu, W., & Hatsopoulos, N. (2006). Evaluating temporal relations between motor and premotor cortex. In: *36th Annual Meeting of the Society for Neuroscience*, Atlanta, GA.

CHAPTER 15

Looking at Sounds: Neural Mechanisms in the Primate Brain

Jennifer M. Groh and Dinesh K. Pai

When you hear a salient sound, it is natural to look at it to find out what is happening. Orienting the eyes to look at sounds is essential to our ability to identify and understand the events occurring in our environment. This behavior involves both sensorimotor and multisensory integration: A sound elicits a movement of the visual sense organ, the eye, to bring the source of the sound under visual scrutiny. How are auditory signals converted into oculomotor commands? This chapter describes our recent work concerning the necessary computational steps between sound and eye movement, and how they may be implemented in neural populations in the primate brain.

In principle, the brain must determine the location of the sound, encode that location in a reference frame and format that allows for convergence with visual signals onto a common motor pathway, and create a suitable time-varying signal in the extraocular muscles to move the eyes. In practice, it is not clear exactly how these computations unfold. Several specific hurdles must be overcome. First, auditory and visual signals arise in different reference frames. Binaural and spectral cues provide information about where a sound is located, but only with respect to the head and ears, not the eyes. In contrast, visual information is intrinsically eye centered: The pattern of illumination of the retina depends on the locations of objects in the visual scene with respect to the direction of gaze. These two reference frames vary in their relationship to each other depending on the orbital position of the eyes (Fig. 15.1). This discrepancy in reference frame should be resolved prior to or as part of the convergence of visual and auditory signals onto a common oculomotor pathway.

A second computational hurdle is that visual and auditory signals are not necessarily encoded in the same format. From the retina on, neurons in the early visual pathway have receptive fields that tile the visual scene and produce a "place code" for stimulus location (Fig. 15.2). In contrast, the binaural computations performed in the auditory pathway might or might not produce receptive fields. If they do not, then there may be a discrepancy in the coding format of visual and auditory signals.

Ultimately, either visual or auditory or both signals must undergo a transformation into a reference frame and a coding format that are similar to each other and appropriate for accessing the oculomotor pathway. We will begin by describing the evidence concerning the reference frame of auditory signals as they progress from auditory to multimodal and oculomotor areas before turning to coding format and some computational analyses that shed light on the neural algorithms that may be at play in this process.

REFERENCE FRAME

The earliest area along the auditory pathway where the reference frame of auditory signals

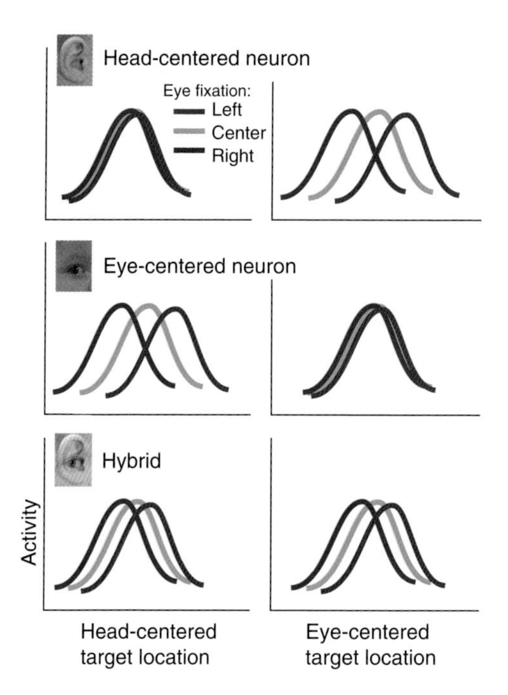

Figure 15.1 Head-centered receptive fields are fixed in space defined with respect to the head. Eye-centered receptive fields are fixed with respect to the eyes. These reference frames shift with respect to each other when the eyes move with respect to the head. Head- and eye-centered reference frames can therefore be distinguished by evaluating the discharge patterns of individual neurons as a function of head- and eye-centered target location across different fixation positions. "Hybrid" response patterns are defined as those that are not well aligned in either head- or eye-centered coordinates.

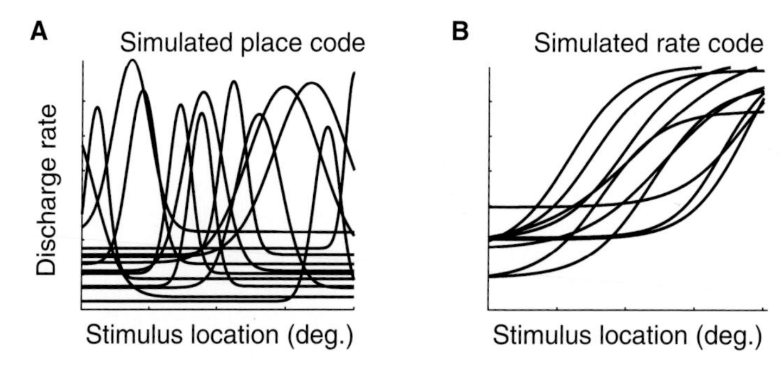

Figure 15.2 Place codes for space contain neurons with nonmonotonic (peaked) response functions, such as circumscribed receptive fields, whereas rate codes contain neurons with monotonic location sensitivity.

has been investigated is the inferior colliculus (IC). The IC is part of the ascending auditory pathway, receiving input from the superior olivary complex and projecting to the auditory thalamus (medial geniculate body) (Moore, 1991; Nieuwenhuys, 1984; Oliver, 2000). The IC also projects to an oculomotor structure, the superior colliculus (SC) (for review, see Sparks & Hartwich-Young, 1989b), and thus could play a specific role in the control of eye movements to sound sources.

Originally, it was thought that the IC encodes sound location in a head-centered reference frame (Jay & Sparks, 1987a). We tested this hypothesis by investigating the responses of IC neurons to sounds as a function of eye position (Fig. 15.3). If IC neurons represent sound location in a head-centered reference frame, then eye position should have no impact on neural responses. In contrast, if the IC uses an eye-centered reference frame, the spatial response functions of IC neurons should shift when the eyes move, and by the same amount that the eyes move (e.g., Fig. 15.1). We found that eye position affects the responses of about 40% of IC neurons (Groh et al., 2001; see also Porter & Groh, 2006, and Zwiers et al., 2004). However, we did not find an eye-centered representation:

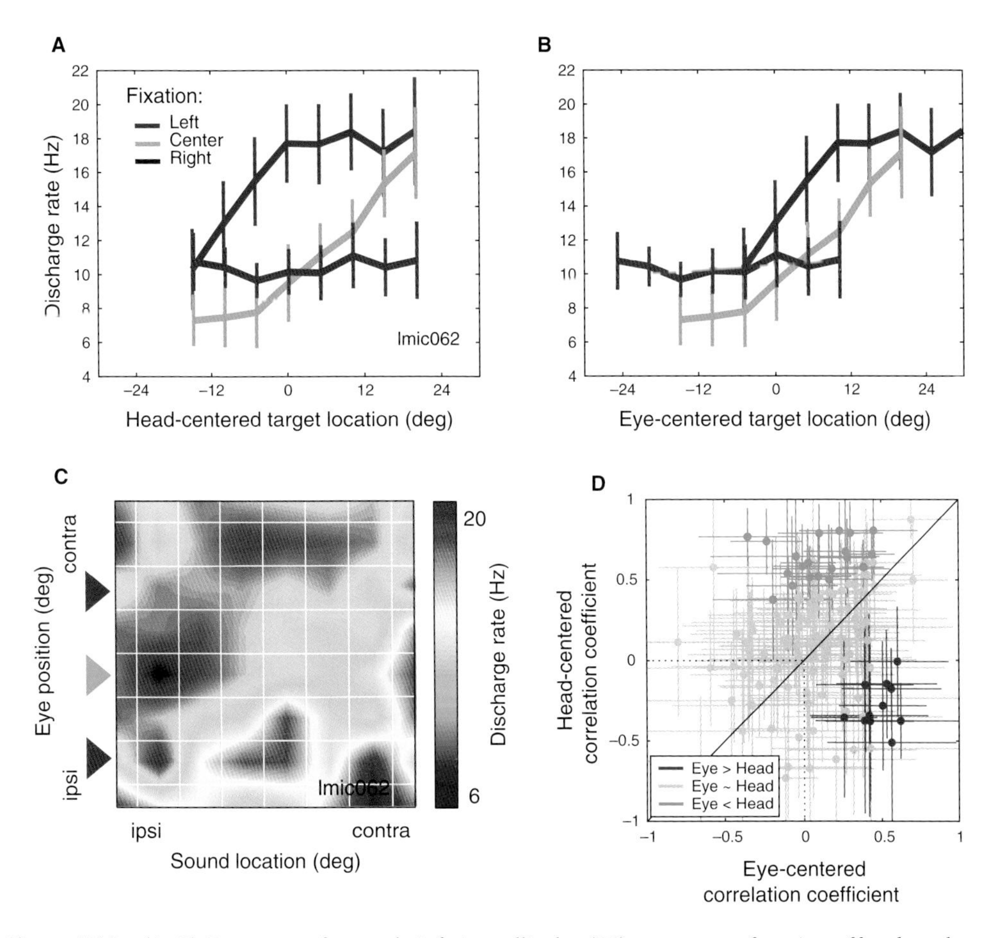

Figure 15.3 (**A, B**) Responses of example inferior colliculus (IC) neuron as a function of head- and eye-centered target location for three different fixation positions. Response functions do not align perfectly in either reference frame. (**C**) Activity of the same neuron in color as a function of all the eye positions and sound locations that were tested. The triangles indicate the fixation positions corresponding to the data in panels a and b. (**D**) Population plot showing that the reference frame in the population of IC neurons spanned a continuum from more eye centered to more head centered, with most neurons lying between these two canonical extremes. A correlation coefficient between each neuron's response functions in head- vs. eye-centered coordinates was calculated and plotted on this graph; crosses indicate 95% confidence intervals. Neurons were classed as eye > head (red) or head > eye (green) only if the confidence intervals show that the eye-centered correlation was greater than the head-centered correlation or vice versa. See Research Design for details. From Porter, K. K., Metzger, R. R., & Groh, J. M. (2006). Representation of eye position in primate inferior colliculus. *Journal of Neurophysiology, 95*, 1826–1842. Used with permission.

The effect of eye position, while statistically significant, interacted with the auditory response but did not cause systematic shifts related to the change in eye position (Fig. 15.3d). Overall, the representation reflected a hybrid of head- and eye-centered information.

The presence of this hybrid reference frame led us to investigate the reference frame at several later stages of processing: auditory cortex and the intraparietal sulcus. The motivation behind these studies was to see if the hybrid representation in the IC was ultimately converted into a more eye-centered representation at a later stage. Auditory cortex and the intraparietal sulcus are not only situated later in the processing stream but also provide direct input to oculomotor structures, in particular the SC (for review, see Sparks & Hartwich-Young, 1989a).

In core auditory cortex, the representation was similar to that of the IC (Fig. 15.4a) (Werner-Reiss et al., 2003). Approximately one third of individual neurons showed a statistically significant influence of eye position on their responses. Across the population, including all neurons regardless of whether they showed a statistically significant effect of eye position, the spatial sensitivity patterns of the majority of neurons reflected a hybrid of head- and eye-centered information.

The lateral and medial banks of the intraparietal sulcus (lateral intraparietal [LIP] and medial intraparietal [MIP] areas) contain both visual and auditory neurons. It had been assumed that the representation of visual information is generally eye centered, with an eye position gain-modulation affecting the response magnitude but not the location of the receptive fields (e.g., Andersen & Mountcastle, 1983; Andersen & Zipser, 1988; Andersen et al., 1985; Zipser & Andersen, 1988), but this view requires a demonstration that the receptive field location does not change with eye position, and systematic mapping of receptive field locations for each different fixation position[1] had not previously been conducted.

Accordingly, we mapped the visual receptive fields in the LIP and MIP areas at multiple fixation positions (Mullette-Gillman et al., 2005). Our results did not support the interpretation of largely eye-centered representation: We found that visual neurons were nearly as likely to have head-centered as eye-centered receptive fields (Fig. 15.4c). Across the population, the distribution of response patterns spanned a continuum from predominantly eye centered to predominantly head centered, with hybrid reference frames being the most common response pattern (Mullette-Gillman et al., 2005).

In keeping with our results in the IC and auditory cortex, we found that the auditory signals in the parietal cortex reflected a mixture of head- and eye-centered sensitivity (Fig. 15.4b). A quantitative examination of the reference frame of across the IC, auditory cortex, and parietal cortex showed that there was little difference between the auditory signals in these structures. There was a small but statistically significant difference between the visual and auditory reference frame within the parietal cortex, suggesting that even though visual and auditory signals converge onto a common neural population (and, in some cases, onto individual bimodal neurons), there remains a slight discrepancy between how visual and auditory information are encoded.

After parietal cortex, visual and auditory signals pass through the SC prior to reaching the eye muscles. The SC is thought to contain a place code for the eye-centered saccade vector, and the same saccade-related burst neurons are thought to control visual, auditory, and somatosensory saccades (Groh & Sparks, 1992, 1996a,b,c; Jay & Sparks, 1984, 1987a,b; Klier et al., 2001; Meredith & Stein, 1983, 1996; Populin et al., 2004; Robinson, 1972; Schiller & Stryker, 1972; Sparks, 1978; Stein & Meredith, 1993; Stein et al., 1993). However, there are some very puzzling aspects to the current story regarding the SC, which call into question some of these assumptions.

In particular, Jay and Sparks investigated the reference frame of both visual and auditory sensory responses in this structure in primates and reported a discrepancy in reference frame: Visual signals were predominantly eye centered whereas auditory signals were intermediate between head- and eye-centered coordinates (Fig. 15.5) (Jay & Sparks, 1984, 1987a,b). Similar results have been reported in the cat SC

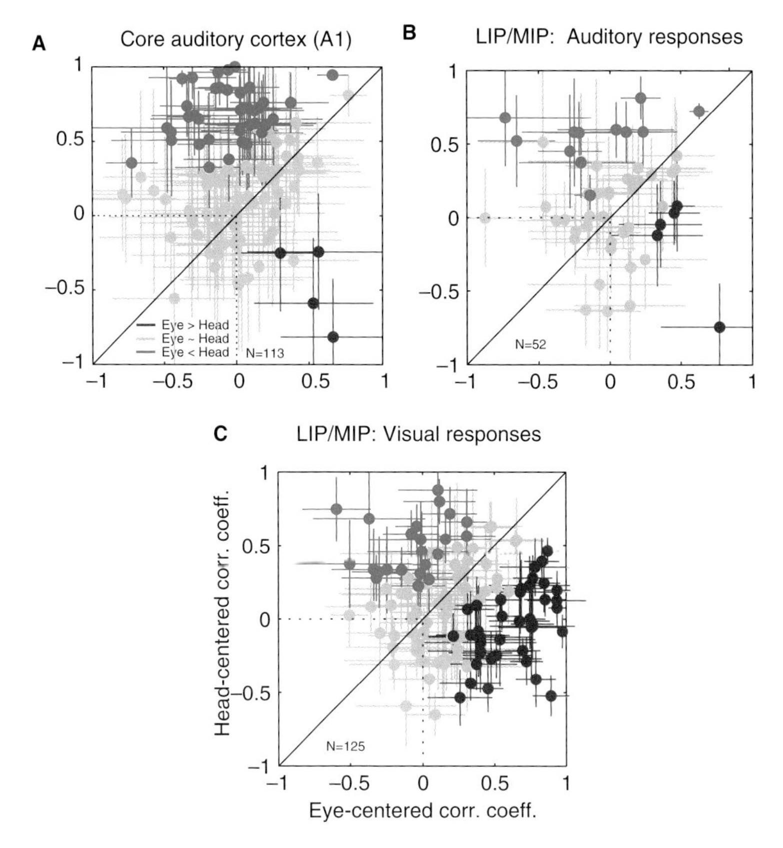

Figure 15.4 Reference frame results for auditory and parietal cortex. In A1 **(A),** lateral intraparietal (LIP), and medial intraparietal (MIP) **(B, C)** areas, the observed reference frames span a continuum from head- to eye-centered coordinates for both auditory **(B)** and (in LIP/MIP) visual signals **(C)**. From Mullette-Gillman, O. A., Cohen, Y. E., & Groh, J. M. (2005). Eye-centered, head-centered, and complex coding of visual and auditory targets in the intraparietal sulcus. *Journal of Neurophysiology, 94,* 2331–2352; and Werner-Reissm U., & Groh, J. M. (2008). A rate code for sound azimuth in monkey auditory cortex: implications for human neuroimaging studies. *Journal of Neuroscience, 28(14),* 3747–3758. Used with permission.

as well (Hartline et al., 1995; Peck et al., 1995; Populin et al., 2004; Zella et al., 2001). Jay and Sparks did not investigate the alignment between the visual and auditory receptive fields of bimodal neurons, but the implication of their reference frame finding is that these receptive fields cannot maintain perfect alignment across different initial eye positions. Although there have been numerous investigations of the response properties of SC neurons to visual, auditory, and combined modality stimuli, suggesting that visual and auditory receptive fields overlap (e.g., Wallace et al., 1996; for review see Stein & Meredith, 1993), these studies have not addressed the effects of eye position and have generally evaluated the receptive fields in a qualitative fashion. Quantitative data on the locations, shape, and alignment of the receptive fields as a function of eye position at the single neuron and population levels are needed.

If the visual and auditory receptive fields of SC neurons are not aligned, and if these neurons control saccadic eye movements, then one would expect a signature of this misalignment in the accuracy of saccades to sounds across different initial eye positions. Specifically, saccades

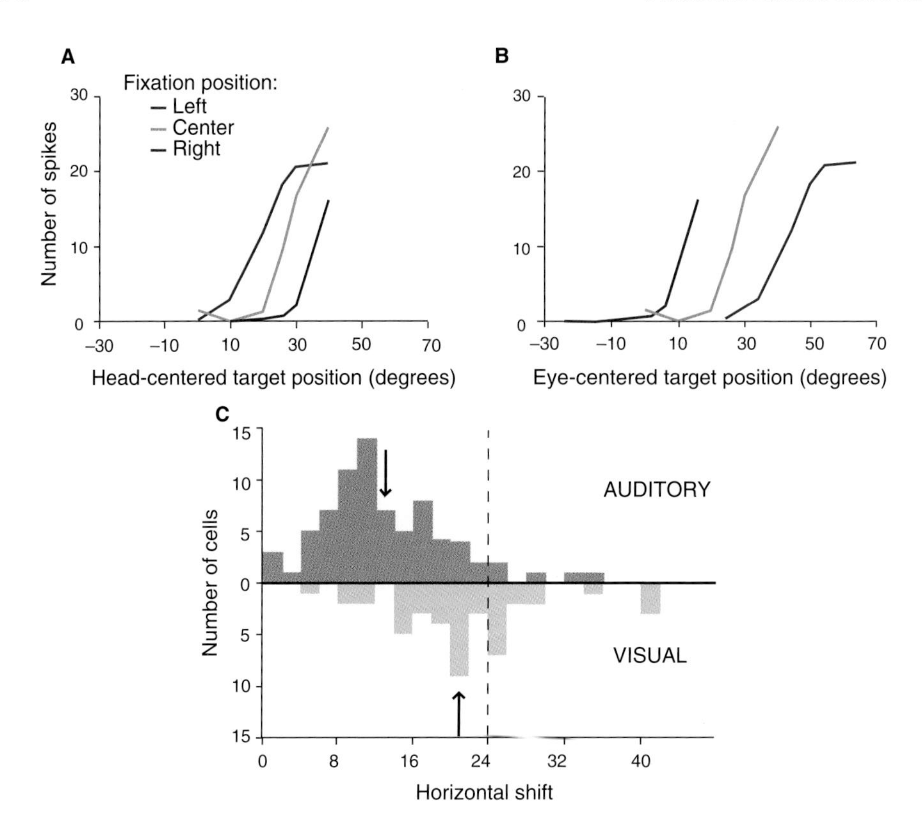

Figure 15.5 Results of Jay and Sparks's investigation of visual and auditory reference frame in the superior colliculus. **(A, B)** Responses of an auditory neuron as a function of head-centered and eye-centered target position: The receptive field shifts with eye position but not by the exact amount of the change in eye position. **(C)** On average, auditory receptive fields shift about half as much (arrow) as would be needed to maintain a constant position in eye-centered coordinates (dashed line). Visual receptive fields tend to shift the full amount, suggesting a lack of registry between visual and auditory tuning. From Jay, M. F., & Sparks, D. L. (1987a). Sensorimotor integration in the primate superior colliculus. II. Coordinates of auditory signals. Journal of Neurophysiology, 57, 35–55. Used with permission.

to a given target location might be more or less affected by initial eye position depending on the modality of the target. Assuming visual signals are the "correct" ones, then saccades to visual targets should compensate completely for initial eye position but saccades to auditory targets should show a characteristic pattern of errors suggestive of a failure to complete a coordinate transformation from head- to eye-centered coordinates.

We looked for such an effect and did not find one (Fig. 15.6) (Metzger et al., 2004; see also Peck et al., 1995, and Populin et al., 2004). Instead, we found that visual and auditory saccades were generally very similar to each other (Fig. 15.6). Both showed only a very modest effect of initial eye position on saccade endpoint, although the auditory saccades were more variable. This suggests that, ultimately, the saccade command does not depend very strongly on whether the target was visual or auditory, implying that visual and auditory signals do end up in a common representation.

It is currently uncertain how this could be accomplished. One potential explanation is that visual and auditory signals might be initially misaligned, at the time of the sensory stimulus, but come into alignment prior to the initiation of the movement. If this is the case, then the visual and auditory *saccade-related bursts,* as opposed to the sensory responses studied by

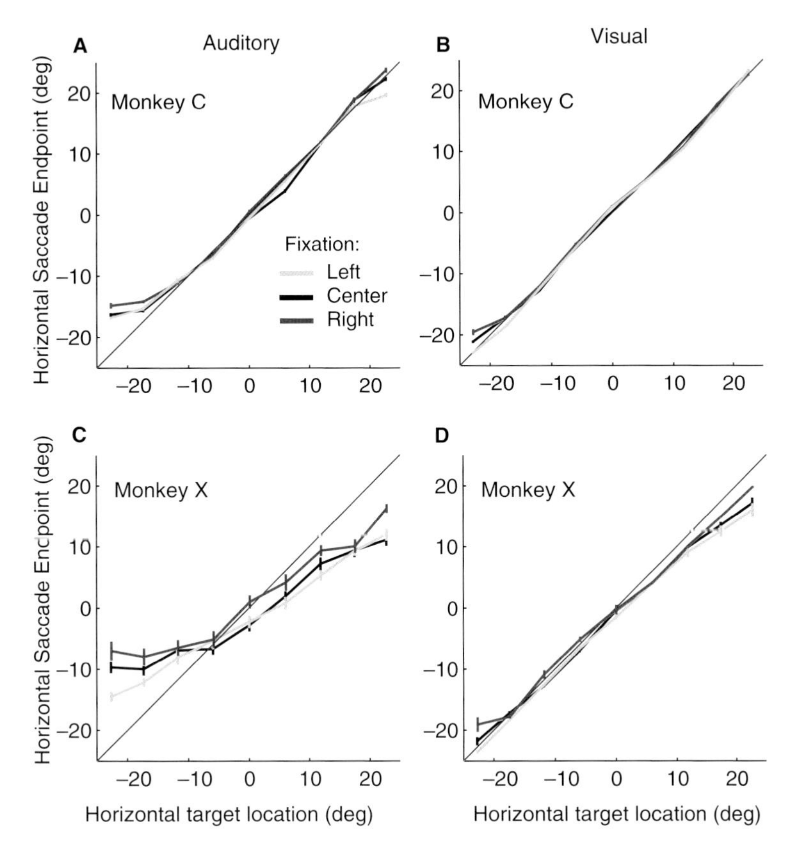

Figure 15.6 Eye position did not affect the accuracy of saccades to visual and auditory stimuli: The different colored traces, corresponding to saccades endpoints from three different initial fixation positions, are largely superimposed. From Metzger, R. R., Mullette-Gillman, O. A., Underhill, A. M., Cohen, Y. E., & Groh, J. M. (2004). Auditory saccades from different eye positions in the monkey: Implications for coordinate transformations. *Journal of Neurophysiology, 92*, 2622–2627. Used with permission.

Jay and Sparks, should be in the same reference frame and in spatial alignment. This possibility could also help account for the better correspondence between auditory tuning and saccade vector that was observed in a recent study of the cat SC (Populin et al., 2004): The response window used in that study included motor-related activity in addition to sensory activity. More research is needed to resolve this issue.

REPRESENTATIONAL FORMAT

The second potential computational challenge for integrating the visual and auditory codes for space is representational format. As noted previously, visual neurons exhibit receptive fields from the very earliest stages of the visual pathway. These receptive fields arise due to the optics of the eye: Light from a given location in the world passes through the aperture of the pupil and illuminates only a restricted portion of the retina. Each photoreceptor can only "see" out in a particular direction. Receptive fields become more complex as signals progress along the visual pathway, but at base, the code for space remains a code in which the location of a visual stimulus can be inferred from the identity of the neurons that are responding to it. This type of code is referred to as a place code, because neurons are often topographically organized according to their receptive field locations.

In contrast, in the auditory system, spatial location is inferred by comparing cues such as sound arrival time and level across the two ears. What kind of code is produced as part of this computation cannot be determined from first principles. There would seem to be two possibilities: (1) a place code similar to that for visual information, in which auditory neurons have circumscribed receptive fields that tile the auditory scene; the location of a stimulus could then be inferred from knowing which neurons were responding to that stimulus (e.g., Jeffress, 1948), as is the case for visual information; and (2) a rate code in which neurons respond broadly to a wide range of locations, but with a firing rate that varies with sound locations. The location of a stimulus could be inferred by "reading out" the firing rate of the active neurons rather than the identity of the active neurons.

The key difference between these two types of codes is the shape of the tuning function of individual neurons. Do neurons respond only to a restricted range of locations, with different neurons showing different preferences? Or do individual neurons respond broadly, with the maximum responses occurring at the extremes of the possible range of space (e.g., the axis of the contralateral ear) (Fig. 15.2).

We have conducted several studies to assess the coding format in the primate auditory pathway. We developed a statistical assay based on the success of Gaussian and sigmoidal functions at fitting the responses as a function of sound location. The idea is that Gaussian functions would be substantially better than sigmoids at fitting the response patterns if the neurons had nonmonotonic spatial response functions characteristic of receptive fields and a place code, but that either sigmoids or broad half-Gaussians would be successful at fitting monotonic tuning patterns characteristic of a rate code (Fig. 15.7).

To our knowledge, nothing is known about the coding of spatial location in the primate auditory pathway prior to the level of the IC. The IC itself is known to contain spatially sensitive neurons (Groh et al., 2001, 2003; Zwiers et al., 2004). We evaluated the spatial sensitivity of IC neurons to determine whether they have circumscribed receptive fields tiling the auditory scene. Instead, we found that they showed consistent preferences for locations along the axis of the contralateral ear. This pattern is characteristic of a rate code for sound location (Fig. 15.8a) (Groh et al., 2003).

We found similar results in auditory cortex (Fig. 15.8b) (Werner-Reiss & Groh, 2008). Interestingly, the code was less smooth in auditory cortex than in IC: Individual neurons often had "bumpy" response functions that were broadly tuned for the contralateral ear, but also had other sound locations that they also responded well to. One possible reason for this is that there could be a transformation from rate code to a place code as the auditory signals approach or join with visual signals. If this is the case, then neurons in brain regions such as parietal cortex or the superior colliculus might show circumscribed receptive fields. Quantitative information on the representational format of auditory signals in these structures is currently lacking in primates.

It will be interesting to determine whether auditory signals are ultimately translated into a place code. The chief advantage of this would be to facilitate integration with place-coded visual information. However, other than that, the advantages might be few. Place codes are better than rate codes for encoding many locations simultaneously, but the auditory system may not be able to encode large numbers of sound locations. Perceptually, two very similar simultaneous sound sources tend to be perceived at an intermediate location (summing localization) (for review see Blauert, 1997). Furthermore, if signals are rate coded at one stage and converted into a place code at a later stage, it is not clear that the benefits of place coding would then accrue, as the rate-coding stage would serve as an information processing bottleneck that would prevent subsequent place-coding stages from representing multiple simultaneous stimulus locations.

A possible advantage for retaining auditory spatial information in a rate code is that it may facilitate interactions with eye position

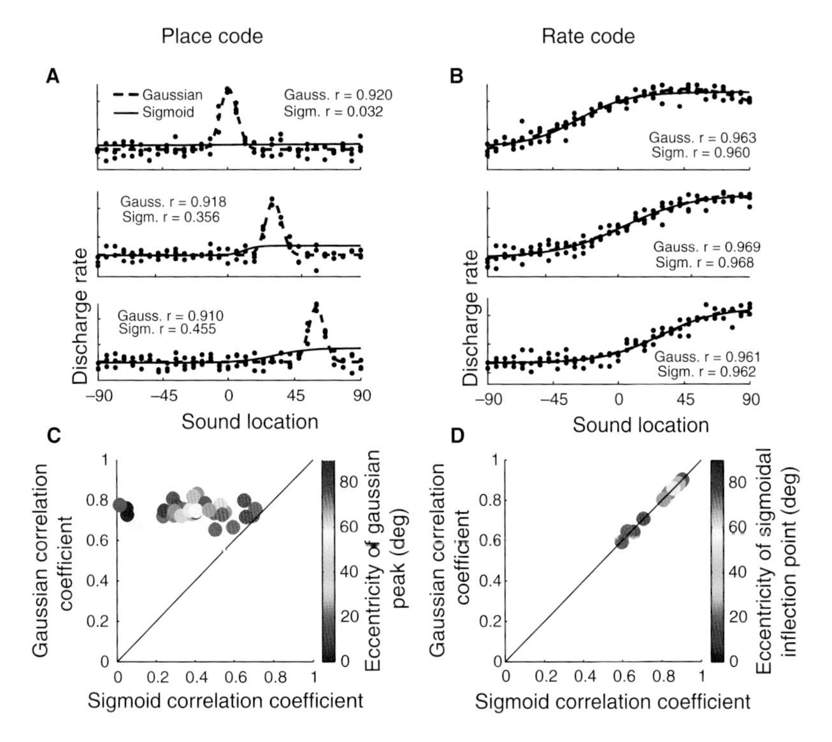

Figure 15.7 Simulations of place and rate codes and how Gaussian and sigmoidal curve fits can be used to distinguish between these representational formats. (**A**) Simulation of three Gaussian-tuned neurons, showing both Gaussian and sigmoidal curve fits. (**B**) Simulation of three sigmoidal neurons. The Gaussian and sigmoidal curve fits are so similar as to obscure each other. (**C**) Population plot of the correlation coefficients of Gaussian and sigmoidal curves for a population of individual neurons whose underlying tuning functions were Gaussian. Gaussian curves were always successful at fitting such response patterns; sigmoidal functions became increasingly successful as the eccentricity (i.e., the absolute value of the azimuthal location) of the Gaussian peak increased. (**D**) Same as c, but for a population of individual neurons whose underlying tuning functions were sigmoidal. Both Gaussian and sigmoidal curves were successful at fitting such response patterns. From Werner-Reiss, U., & Groh, J. M. (2008). A rate code for sound azimuth in monkey auditory cortex: implications for human neuroimaging studies. *Journal of Neuroscience, 28(14)*, 3747–3758. Used with permission.

information, which appears to be encoded in a similar format. We found that in the IC, eye position sensitivity is generally monotonic (e.g., Fig. 15.3), consistent with a rate code for eye position. This finding is consistent with studies in the parietal cortex (Andersen et al., 1990), frontal eye fields (Bizzi, 1968), cerebellar flocculus (Noda & Suzuki, 1979), somatosensory cortex (Wang et al., 2007), and premotor circuitry of the oculomotor pathway (Keller, 1974; Luschei & Fuchs, 1972; McCrea & Baker, 1980; Sylvestre & Cullen, 1999a).

MOTOR COMMANDS

What is the reference frame and representational format of the motor command? The pattern of force needed to move the eyes to look in a particular direction reflects a combination of reference frames and a combination of representational formats. For a movement in a given direction, the amount of force that needs to be applied varies monotonically with the size of the movement, consistent with a rate code. The direction of the movement is controlled by the

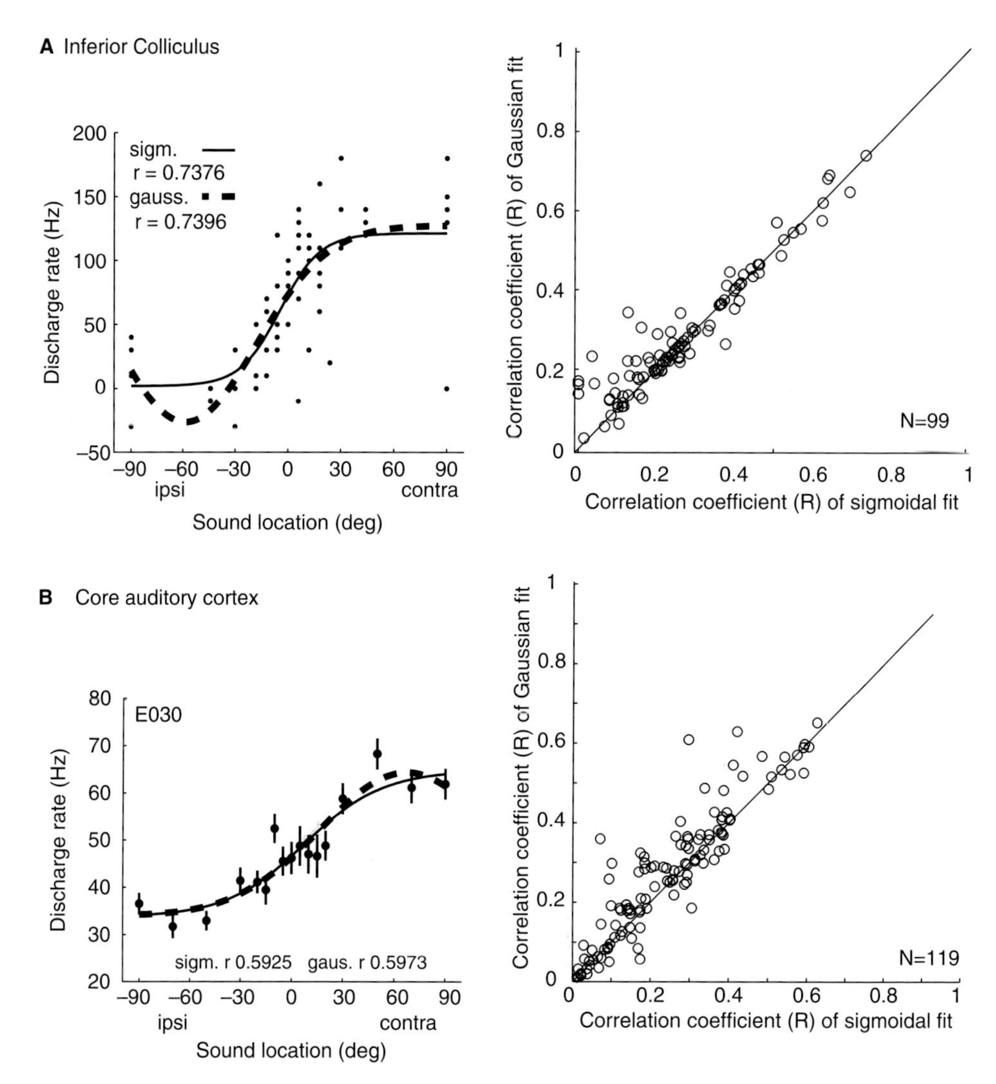

Figure 15.8 Representational format in inferior colliculus (IC) and core auditory cortex. (**A**) In the IC, most spatially sensitive neurons responded in a graded, monotonic fashion peaking for sounds along the axis of the contralateral ear, as shown for this example neuron (left panel). Across the population, this pattern is evident in the fact that sigmoidal functions were as good as Gaussians at capturing the response patterns From Groh, J. M., Kelly, K. A., & Underhill, A. M. (2003). A monotonic code for sound azimuth in primate inferior colliculus. *Journal of Cognitive Neuroscience, 15,* 1217–1231. (**B**) The pattern of results was similar in auditory cortex, although some individual neurons had "bumpy" response functions (data points lie slightly above the line of slope one in the right panel). From Werner-Reiss, U., & Groh, J. M. (2008). A rate code for sound azimuth in monkey auditory cortex: implications for human neuroimaging studies. *Journal of Neuroscience, 28(14),* 3747–3758. Used with permission.

ratio of activation in different muscle groups, a format that is more akin to a place code: Which muscles are active controls the movement direction.

The reference frame of oculomotor commands is referred to as eye centered by some sources and head centered by others. The confusion stems from both what is meant by motor command—is this term properly reserved only for the extraocular motor neurons or may it be applied to slightly earlier stages such as the SC?—as well as a lack of quantitative investigation into this question. We favor reserving the term "motor command" for the signals carried by the extraocular motor neurons. The reference frame of extraocular motor neurons has not been investigated per se, but their discharge patterns are so well characterized using other means that it is possible to draw some inferences. Specifically, the discharge patterns can be described as a linear differential equation (e.g., Sylvestre & Cullen, 1999b):

$$FR = k_1 + k_2P + k_3V,$$

where FR = instantaneous firing rate, P = eye position, and V = eye velocity. This equation illustrates that the firing pattern depends on both initial and final eye position—that is, fixation position as well as the head-centered location of the target. Thus, the motor command cannot be properly formed if premotor circuitry has access only to target location in a single pure reference frame—some combination of head-centered, eye-centered, and eye position information is needed. This suggests that the use of hybrid reference frames at earlier stages of the audio-oculomotor pathway may reflect the constraints of the motor periphery. Most existing models for how the motor command is formed call for separate representations of head- or eye-centered information to be combined with eye position information as the time-varying motor neuron discharge pattern is created (Moschovakis, 1996; Van Gisbergen & Van Opstal, 1989), but it might also be possible to generate this command from an input signal that already has these component signals mixed together.

MODELING

Although it remains unclear exactly what kinds of transformations unfold as visual and auditory signals converge onto the oculomotor pathway, it is certainly evident that some transformations between coding formats and reference frames are needed, and it can be fruitful to explore the neural mechanisms that might underlie such transformations while additional experimental studies are pending. Accordingly, we have worked on several models for transforming signals between different reference frames and between different coding formats. We will begin with the models for transformations of coding format, because how information is encoded impacts which algorithms for coordinate transformations may be most appropriate.

Models for Transformations of Coding Format

We have designed several models that involve transformations of signals from either a place code to a rate code or vice versa. Figure 15.9 illustrates several ways that Gaussian tuning functions can be created from a population of neurons with sigmoidal response functions with varying inflection points (Porter & Groh, 2006). Suppose neurons exhibit sigmoidal tuning functions, with some preferring leftward locations (such as spatial neurons in the right IC) and others preferring leftward locations (e.g., the left IC). Assume further that there is a population of neurons whose inflection points vary across the range of space. Excitatory connections from two neurons with opposite tuning preferences and neighboring inflection points would cause a recipient neuron to be responsive to sound locations between the inflection points of the input neurons (Fig. 15.9a).

A circumscribed receptive field could also be created by combining excitatory and inhibitory inputs from two neurons with sigmoidal tuning functions in the same direction, provided once again that their inflection points are appropriately staggered. Suppose two neurons both prefer

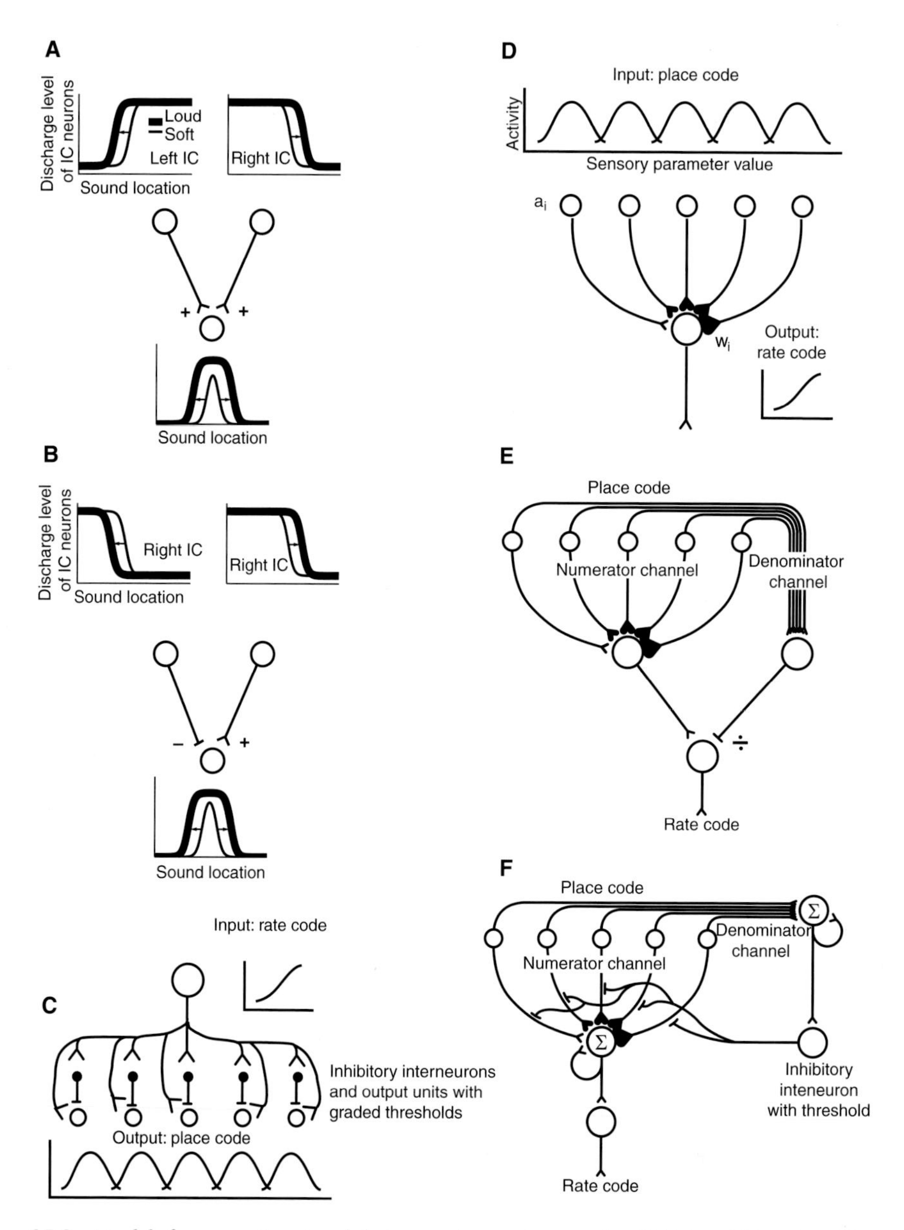

Figure 15.9 Models for converting signals between place and rate codes. (**A**) Sigmoidal units with opposite preferences and different inflection points could converge, producing a receptive field. If sound loudness affected the response patterns of the input neurons as indicated, then the receptive field would be larger for louder sounds but would be centered in the same place. (**B**) Sigmoidal units preferring the same side but with different inflection points could converge using a combination of excitatory and inhibitory synapses. As in panel a, the receptive field would expand for louder sounds if the two input neurons were affected by sound loudness in different directions. (**C**) An alternative mechanism involving a cascade of units with different thresholds, with inhibitory interneurons. (**D–F**) Three models for converting place codes to rate codes, with different ways of normalizing for the level of activity. The first model involves no normalization, the second involves full normalization, and the third involves normalization only for high levels of activity. See text for additional details. Groh, J. M. (2001). Converting neural signals from place codes to rate codes. *Biological Cybernetics, 85,* 159–165; and Porter, K. K., & Groh, J. M. (2006). The other transformation required for visual-auditory integration: Representational format. *Progress in Brain Research, 155,* 313–323. Used with permission.

leftward locations, but one has an inflection point at 0 degrees and the other has an inflection point at 10 degrees to the right. A recipient neuron that is inhibited by the first neuron and excited by the second neuron will have a receptive field between 0 and 10 degrees to the right—the region of space where only its excitatory input is active (Fig. 15.9b).

Both of these algorithms involve a certain element of place coding in the input stage: The input neurons must have tuning functions whose *inflection* points show heterogeneity spanning the range of possible spatial locations. Thus, this is not a pure rate code at the input. At present, we have not attempted to determine if the response functions are truly sigmoidal as opposed to some other monotonic function, nor have we assessed whether the inflection points span a range of locations. Thus, it is unclear whether these algorithms are biologically plausible or not.

A third algorithm might apply if the input signals are more linear than sigmoidal and if they therefore lack inflection points, much less variation in inflection points. Figure 15.9c illustrates a local circuit with a cascade of thresholds. The thresholds introduce the necessary nonlinearity into the processing of a linear signal to create the receptive fields. Each output neuron (open circles) has both a threshold and an inhibitory interneuron that is paired with it. The inhibitory interneuron has a slightly higher threshold for activation. Thus, the output neuron is active only when its input exceeds its own threshold but is less than the threshold for its matched inhibitory interneuron. This pattern, when repeated with varying thresholds across the population, can create a range of circumscribed receptive fields across the population (Groh & Sparks, 1992).

We have also developed several models for converting signals from a place code to a rate code (Groh, 2001; Porter & Groh, 2006) (Fig. 15.9d–f). The conversion is accomplished using a graded pattern of synaptic weights. The models differ in whether and how they accomplish normalization for the overall level of activity. The vector summation model (Fig. 15.9d) simply calculates the weighted sum of activity, with no normalization whatsoever. The problem with such a model is that typically there are many other features that might alter neural activity (e.g., the loudness of a sound or the contrast of a visual stimulus), and without normalization the changes in neural responsiveness associated with these features would affect the read-out. There is some perceptual evidence for this kind of effect: For example, low-contrast visual stimuli appear to move more slowly than high-contrast visual stimuli (e.g., Snowden et al., 1998; Thompson et al., 1996), but more generally such factors appear to be corrected for when determining spatial location.

Accordingly, we developed several additional models for converting place codes to rate codes that include normalization for the overall level of activity. The vector averaging model (Fig. 15.9e) has two read-out pathways, one to calculate the sum of activity weighted by its location in the place code (the numerator channel) and the other to calculate the unweighted sum (the denominator channel). Then, the weighted sum is divided by the unweighted sum, producing a signal corresponding to the average location of activity in the place code.

One problem with this model is that it is not clear how neural circuits might implement the division of one number by another. Inhibitory synapses can exert a divisive-like effect, but more generally the nature of the inhibitory influence will vary with the membrane potential: What seems like division when the membrane potential is near rest (e.g., shunting inhibition) might become more like subtraction when the membrane is more depolarized. The vector averaging model requires that the inhibitory influence of the denominator channel should mimic division for a large range of possible numerator and denominator values.

The third model circumvents this problem by implementing normalization in a different fashion. This model, the summation-with-saturation model, calculates a weighted sum of the activity in the input layer, and then clips off any extra activity above a certain threshold. This is accomplished using a combination of neural integrators and thresholds. The numerator and denominator channels both integrate their input, weighted by location in the case of the numerator channel. When the denominator channel reaches a certain threshold, it clips off the input to the numerator channel. The activity level of the numerator channel will vary only with the location of the input provided there is sufficient activity to trigger the clipping action of the denominator channel, and otherwise will reflect the weighted sum of the input.

This model successfully mimics the pattern of evoked saccades elicited by microstimulation of the SC. Stimulation above a certain frequency evokes saccades that do not depend on the frequency of stimulation (known as the site-specific amplitude), but below that value the amplitude of the saccade falls off as the frequency or duration of stimulation is reduced. The evoked saccade depends on the total number of stimulation pulses delivered until a saturation point is reached (Stanford et al., 1996).

Models for Coordinate Transformations

We have developed two models for coordinate transformations. At the time these models were designed, little was known about either the representational format or the frame of reference of signals in the auditory pathway, so both models assumed that the input consisted of a head-centered map of auditory space. Since this kind of representation has yet to be found, it is worth updating these models to consider other possible forms of input (as well as output).

The vector subtraction model (Fig. 15.10a) begins by converting head-centered, place-coded auditory signals into a rate code so that rate-coded eye position signals could be subtracted. The resulting eye-centered rate code for sound location was then converted into a place code for eye-centered sound location. This model was essentially constructed from the place-to-rate and rate-to-place component parts.

Since it now appears that sound location may be encoded in a rate code, a simpler version of this model can be constructed (Fig. 15.10b). The input can consist of a rate-coded sound location, from which rate-coded eye position information is subtracted. This produces a rate code for eye-centered sound location, just as in the original version. It may or may not be necessary to then convert these signals into a place code, but if it is necessary, one of the rate-to-place algorithms described previously could still be included.

A second type of model, the dendrite model, was originally proposed with the goal of avoiding rate-coding stages in mind (Groh & Sparks, 1992) (Fig. 15.11). The rationale was that the rate-coding stages would limit the number of sound locations that could be encoded to one. Since it now appears that rate-coding stages do exist in the brain's auditory pathways, the motivation behind this model has been reduced. However, it remains uncertain how the brain handles multiple sound locations, so elements of the dendrite model may yet prove to be of some utility.

CONCLUSIONS

To guide an eye movement to the source of a sound requires a net transformation of auditory information from the initially purely head-centered interaural timing and level cues to a reference frame appropriate for controlling the eye muscles. Our studies as well as others have found evidence for hybrid, but not purely eye-centered, frames of reference at several stages of the audio-oculomotor reference frame. This type of hybrid reference frame may be appropriate for controlling saccades because the motor command requires information about both the initial eye position and the desired amplitude of the saccade.

Eye movements to sounds may also require one or more transformations of auditory signals from one kind of coding format into another. At present, we have only found evidence for rate coding of auditory spatial information. It remains to be seen whether rate-coded auditory spatial information is transformed into a place code, and if so, where and how this transformation occurs. It has long been assumed that this transformation does take place, but it should be noted that it is not necessary to create a place code for auditory

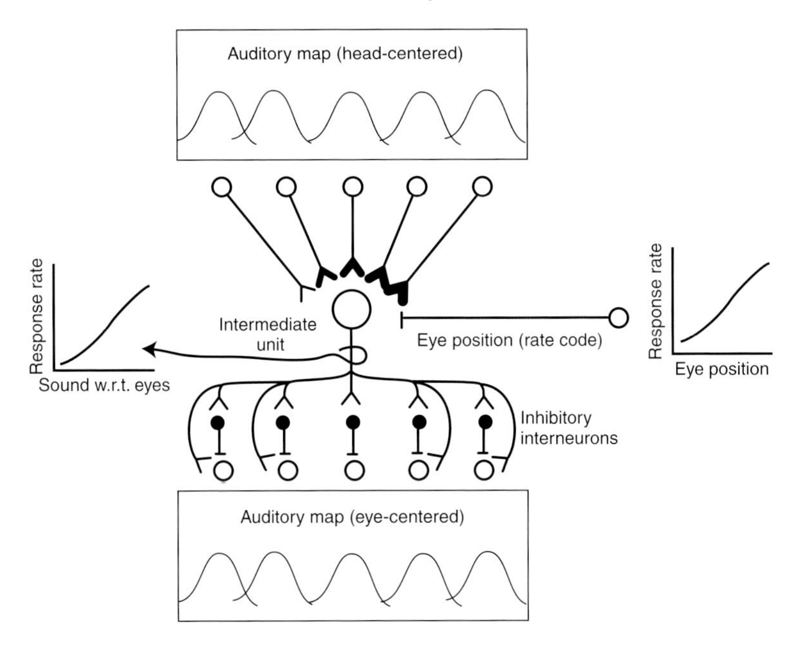

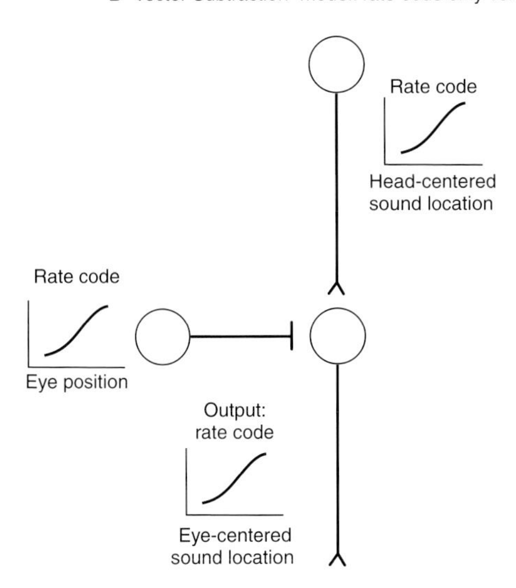

Figure 15.10 Vector subtraction model for transforming auditory signals from head-centered to eye-centered coordinates. (**A**) This model assumes place-coded auditory inputs. It now appears possible that the auditory inputs encode sound location in a rate-coded format. (**B**) In this case, the model could be modified to use the rate-coded auditory input signals. From Groh, J. M., & Sparks, D. L. (1992). Two models for transforming auditory signals from head-centered to eye-centered coordinates. *Biological Cybernetics, 67,* 291–302. Used with permission.

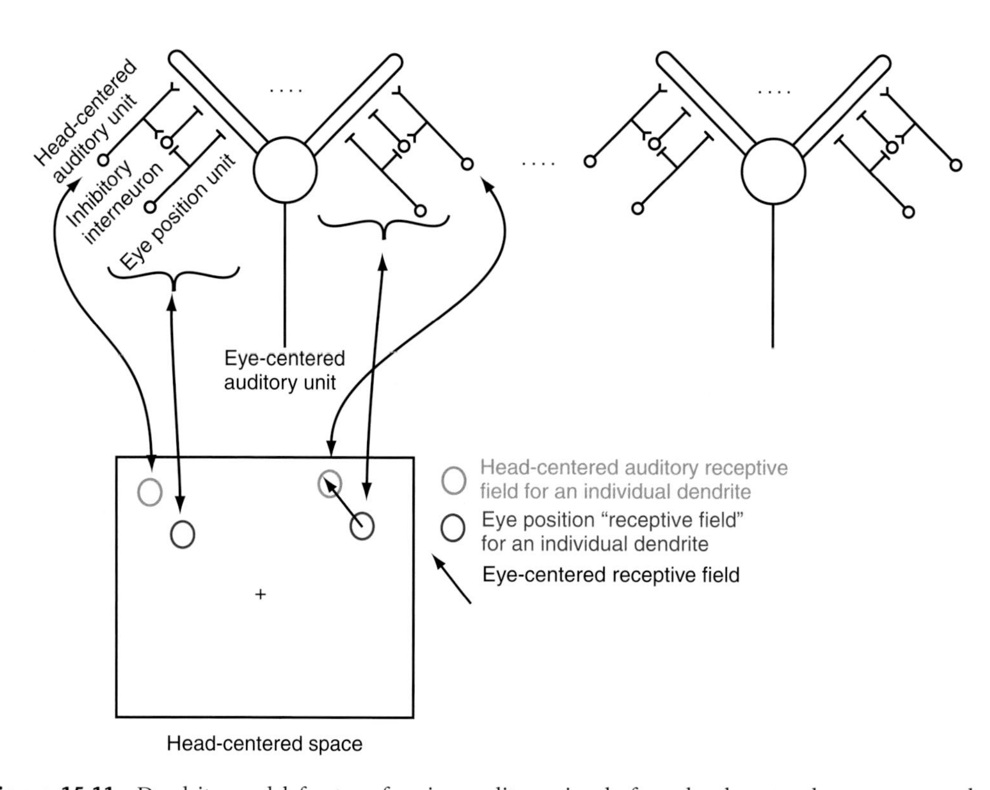

Figure 15.11 Dendrite model for transforming auditory signals from head-centered to eye-centered coordinates. Each dendrite receives input from eye position units and one head-centered auditory unit. Thresholds and synaptic weights are balanced so that the cell body receives net excitation via a given dendrite if an auditory stimulus is present in the head-centered receptive field of the auditory unit that activates that dendrite and the eyes are within a certain range of positions. The range of eye positions that allow excitation to reach the soma varies across dendrites and is matched to the head-centered receptive field of the auditory input that activates a given dendrite. The combination produces tuning for the eye-centered location of the sound. In short, the dendrites of a given unit sample all the possible head-centered locations that could yield a certain eye-centered location, and the eye position inputs filter out that input unless the eye position is in the appropriate range for that dendrite. From Groh, J. M., & Sparks, D. L. (1992). Two models for transforming auditory signals from head-centered to eye-centered coordinates. *Biological Cybernetics, 67,* 291–302. Used with permission.

information simply to guide an eye movement. Since the motor output consists of a rate code, any place-coded auditory information would have to be converted back into a rate code to generate a motor command.

Of course, auditory signals do not exist solely to trigger eye movements. The perceptual and behavioral endpoints of auditory processing are many and varied, and natural selection has likely produced auditory information-coding strategies that serve more than one behavioral and perceptual master. Thus, other constraints may account for the aspects of audio-oculomotor transformations that may appear at face value to be inefficient. Further research on whether the behavioral task affects the type of code employed will therefore be of great interest.

NOTE

1. Several studies did map the receptive fields, but sampled primarily along a dimension orthogonal to the direction in which fixation position varied (Andersen et al., 1985; Batista et al., 1999); the effects of eye position on response patterns observed in these studies were consistent with either the head-centered or eye-centered-with-eye-position-gain hypotheses.

REFERENCES

Andersen, R., & Zipser, D. (1988). The role of the posterior parietal cortex in coordinate transformations for visual-motor integration. *Canadian Journal of Physiology and Pharmacology*, *66*, 488–501.

Andersen, R. A., Bracewell, R. M., Barash, S., Gnadt, J. W., & Fogassi, L. (1990). Eye position effects on visual, memory, and saccade-related activity in areas LIP and 7a of macaque. *Journal of Neuroscience*, 10, 1176–1196.

Andersen, R. A., Essick, G. K., & Siegel, R. M. (1985). Encoding of spatial location by posterior parietal neurons. *Science*, 230, 456–458.

Andersen, R. A., & Mountcastle, V. B. (1983). The influence of the angle of gaze upon the excitability of the light-sensitive neurons of the posterior parietal cortex. *Journal of Neuroscience*, *3*, 532–548.

Batista, A. P., Buneo, C. A., Snyder, L. H., & Andersen, R. A. (1999). Reach plans in eye-centered coordinates [see comments]. *Science*, *285*, 257–260.

Bizzi, E. (1968). Discharge of frontal eye field neurons during saccadic and following eye movements in unanesthetized monkeys. *Experimental Brain Research*, *6*, 69–80.

Blauert, J. (1997). *Spatial hearing*. Cambridge, MA: MIT Press.

Groh, J. M. (2001). Converting neural signals from place codes to rate codes. *Biological Cybernetics*, *85*, 159–165.

Groh, J. M., Kelly, K. A., & Underhill, A. M. (2003). A monotonic code for sound azimuth in primate inferior colliculus. *Journal of Cognitive Neuroscience*, *15*, 1217–1231.

Groh, J. M., & Sparks, D. L. (1992). Two models for transforming auditory signals from head-centered to eye- centered coordinates. *Biological Cybernetics*, *67*, 291–302.

Groh, J. M., & Sparks, D. L. (1996a). Saccades to somatosensory targets. III. Eye-position-dependent somatosensory activity in primate superior colliculus. *Journal of Neurophysiology*, *75*, 439–453.

Groh, J. M., & Sparks, D. L. (1996b). Saccades to somatosensory targets. II. Motor convergence in primate superior colliculus. *Journal of Neurophysiology*, *75*, 428–438.

Groh, J. M., & Sparks, D. L. (1996c). Saccades to somatosensory targets. I. Behavioral characteristics. *Journal of Neurophysiology*, *75*, 412–427.

Groh, J. M., Trause, A. S., Underhill, A. M., Clark, K. R., & Inati, S. (2001). Eye position influences auditory responses in primate inferior colliculus. *Neuron*, *29*, 509–518.

Hartline, P. H., Vimal, R. L., King, A. J., Kurylo, D. D., & Northmore, D. P. (1995). Effects of eye position on auditory localization and neural representation of space in superior colliculus of cats. *Experimental Brain Research*, 104, 402–408.

Jay, M. F., & Sparks, D. L. (1984). Auditory receptive fields in primate superior colliculus shift with changes in eye position. *Nature*, 309, 345–347.

Jay, M. F., & Sparks, D. L. (1987a). Sensorimotor integration in the primate superior colliculus. II. Coordinates of auditory signals. *Journal of Neurophysiology*, *57*, 35–55.

Jay, M. F., & Sparks, D. L. (1987b). Sensorimotor integration in the primate superior colliculus. I. Motor convergence. *Journal of Neurophysiology*, *57*, 22–34.

Jeffress, L. A. (1948) A place theory of sound localization. *Journal of Comparative Physiology and Psychology*, *41*, 35–39.

Keller, E. L. (1974). Participation of medial pontine reticular formation in eye movement generation in monkey. *Journal of Neurophysiology*, *37*, 316–332.

Klier, E. M., Wang, H., & Crawford, J. D. (2001). The superior colliculus encodes gaze commands in retinal coordinates. *Nature Neuroscience*, *4*, 627–632.

Luschei, E. S., & Fuchs, A. F. (1972). Activity of brain stem neurons during eye movements of alert monkeys. *Journal of Neurophysiology*, *35*, 445–461.

McCrea, R., & Baker, R. (1980). Evidence for the hypothesis that the prepositus neucleus distributes "efference" copy signals to the brainstem. *Anatomy Research*, *196*, 122–123.

Meredith, M. A., & Stein, B. E. (1983). Interactions among converging sensory inputs in the superior colliculus. *Science, 221,* 389–391.

Meredith, M. A., & Stein, B. E. (1996). Spatial determinants of multisensory integration in cat superior colliculus neurons. *Journal of Neurophysiology, 75*(5), 1843–1857.

Metzger, R. R., Mullette-Gillman, O. A., Underhill, A. M., Cohen, Y. E., & Groh, J. M. (2004). Auditory saccades from different eye positions in the monkey: Implications for coordinate transformations. *Journal of Neurophysiology, 92,* 2622–2627.

Moore, D. R. (1991). Anatomy and physiology of binaural hearing. *Audiology,* 30, 125–134.

Moschovakis, A. K. (1996). Neural network simulations of the primate oculomotor system. II. Frames of reference. *Brain Research Bulletin,* 40, 337–345.

Mullette-Gillman, O. A., Cohen, Y. E., & Groh, J. M. (2005). Eye-centered, head-centered, and complex coding of visual and auditory targets in the intraparietal sulcus. *Journal of Neurophysiology, 94,* 2331–2352.

Nieuwenhuys, R. (1984). Anatomy of the auditory pathways, with emphasis on the brain stem. *Advances in Otorhinolaryngology, 34,* 25–38.

Noda, H., & Suzuki, D. A. (1979). Processing of eye movement signals in the flocculus of the monkey. *Journal of Physiology, 294,* 349–364.

Oliver, D. L. (2000). Ascending efferent projections of the superior olivary complex. *Microscopic Research Technology, 51,* 355–363.

Peck, C. K., Baro, J. A., & Warder, S. M. (1995). Effects of eye position on saccadic eye movements and on the neuronal responses to auditory and visual stimuli in cat superior colliculus. *Experimental Brain Research, 103*(2), 227–242.

Populin, L. C., Tollin, D. J., & Yin, T. C. (2004). Effect of eye position on saccades and neuronal responses to acoustic stimuli in the superior colliculus of the behaving cat. *Journal of Neurophysiology, 92,* 2151–2167.

Porter, K. K., & Groh, J. M. (2006). The other transformation required for visual-auditory integration: Representational format. *Progress in Brain Research, 155,* 313–323.

Porter, K. K., Metzger, R. R., & Groh, J. M. (2006). Representation of eye position in primate inferior colliculus. *Journal of Neurophysiology, 95,* 1826–1842.

Robinson, D. A. (1972). Eye movements evoked by collicular stimulation in the alert monkey. *Vision Research, 12,* 1795–1807.

Schiller, P. H., & Stryker, M. (1972). Single-unit recording and stimulation in superior colliculus of the alert rhesus monkey. *Journal of Neurophysiology, 35,* 915–924.

Snowden, R. J., Stimpson, N., & Ruddle, R. A. (1998). Speed perception fogs up as visibility drops [letter]. *Nature, 392,* 450.

Sparks, D. L. (1978). Functional properties of neurons in the monkey superior colliculus: Coupling of neuronal activity and saccade onset. *Brain Research, 156,* 1–16.

Sparks, D. L., & Hartwich-Young, R. (1989a). The deep layers of the superior colliculus. In: R. H. Wurtz & M. E. Goldberg (Eds.), *The neurobiology of saccadic eye movements* (pp. 213–255). New York: Elsevier.

Sparks, D. L., & Hartwich-Young, R. (1989b). The deep layers of the superior colliculus. *Reviews of Oculomotor Research, 3,* 213–255.

Stanford, T. R., Freedman, E. G., & Sparks, D. L. (1996). Site and parameters of microstimulation: Evidence for independent effects on the properties of saccades evoked from the primate superior colliculus. *Journal of Neurophysiology, 76,* 3360–3381.

Stein, B. E., & Meredith, M. A. (1993). *The merging of the senses.* Cambridge, MA: MIT Press.

Stein, B. E., Meredith, M. A., & Wallace, M. T. (1993). The visually responsive neuron and beyond: Multisensory integration in cat and monkey. *Progress in Brain Research, 95,* 79–90.

Sylvestre, P. A., & Cullen, K. E. (1999a). Quantitative analysis of abducens neuron discharge dynamics during saccadic and slow eye movements. *Journal of Neurophysiology, 82,* 2612–2632.

Sylvestre, P. A., & Cullen, K. E. (1999b). Quantitative analysis of abducens neuron discharge dynamics during saccadic and slow eye movements. *Journal of Neurophysiology, 82,* 2612–2632.

Thompson, P., Stone, L. S., & Swash, S. (1996). Speed estimates from grating patches are not contrast-normalized. *Vision Research, 36,* 667–674.

Van Gisbergen, J. A., & Van Opstal, A. J. (1989). Models of saccadic control. In: R. Wurtz & M. E. Goldberg (Eds.), *The neurobiology of saccadic eye movements.* Amsterdam: Elsevier.

Wallace, M. T., Wilkinson, L. K., & Stein, B. E. (1996). Representation and integration of multiple sensory inputs in primate superior colliculus. *Journal of Neurophysiology, 76,* 1246–1266.

Wang, X., Zhang, M., Cohen, I. S., & Goldberg, M. E. (2007). The proprioceptive representation of eye position in monkey primary somatosensory cortex. *Nature Neuroscience,* 10, 640–646.

Werner-Reiss, U., & Groh, J. M. (2008). A rate code for sound azimuth in monkey auditory cortex: Implications for human neuroimaging studies. *Journal of Neuroscience. 28(14),* 3747–3758

Werner-Reiss, U., Kelly, K. A., Trause, A. S., Underhill, A. M., & Groh, J. M. (2003). Eye position affects activity in primary auditory cortex of primates. *Current Biology, 13,* 554–562.

Zella, J. C., Brugge, J. F., & Schnupp, J. W. (2001). Passive eye displacement alters auditory spatial receptive fields of cat superior colliculus neurons. *Nature Neuroscience, 4,* 1167–1169.

Zipser, D., & Andersen, R. A. (1988). A back-propagation programmed network that simulates response properties of a subset of posterior parietal neurons. *Nature, 331,* 679–684.

Zwiers, M. P., Versnel, H., & Van Opstal, A. J. (2004). Involvement of monkey inferior colliculus in spatial hearing. *Journal of Neuroscience, 24,* 4145–4156.

CHAPTER 16

Circuits of Emotion in the Primate Brain

Katalin M. Gothard and Kari L. Hoffman

INTRODUCTION

Emotions are coordinated brain-body states that allow animals to cope with the challenges of their physical and social environment. Emotional states are characterized by a specific configuration of inputs (triggering events), outputs (autonomic and somatic responses), and the neural processes that mediate their transformation. Many emotional states, especially acute states such as fear or anger, are coupled with enhanced perceptual processing, decision making, action selection, and increased energetic expenditure.

The brain-body state triggered by a threatening facial expression, a common event in the daily life of a macaque, exemplifies the phenomena ascribed to emotion. When a midranking male monkey is confronted with a threat display, the covert autonomic and overt behavioral responses that are triggered by observing this display are the results of a complex process of evaluating the emotional and social significance of that display given the present internal state of the receiver monkey. This evaluation takes into account the identity of the displaying monkey; he is recognized as a dominant male, with a well-groomed, muscular body, red-pigmented face, and large, symmetrical canines. If the display is directed to a close ally of the viewer, then based on this evaluation, the viewer decides whether to flee the scene of an imminent confrontation or to rush to defend his ally. For either action, the autonomic system sets the organism in a "higher gear." The animal is more vigilant, the sensory threshold for threat-related stimuli is lowered, reaction time is shortened, and widespread sympathetic effects lock the internal organs in a functional mode biased toward energy expenditure.

A completely different brain-body state is triggered by an affiliative approach, signaling the intention to groom. Such an approach, especially if there is a history of affiliative interaction between the players, leads to relaxed postures, accompanied by vagal tone, increased levels of oxytocin, and often gestures of reciprocation.

The behavioral outcomes of brain-body activation are not unique to a particular emotion; the initiating factors of emotions and the internal states they generate are much more diverse than the relatively limited behavioral repertoire available for their expression. The autonomic and behavioral expressions of emotions are therefore low-dimensional and highly reproducible across situations and individuals. In contrast, the stimuli that evoke emotional expressions are often high-dimensional, and their emotional content is dependent also on context and the emotional states preceding the stimulus. In light of the contrast between the extremely rich array of stimuli that can generate emotions and the restricted set of expressive affordances, the central process of evaluating the emotional significance of all stimuli and events encountered by the organism is essentially a process of dimensionality reduction.

This dimensionality reduction takes place in nested circuits, in which phylogenetically older regions form a core circuit (Fig. 16.1). The core

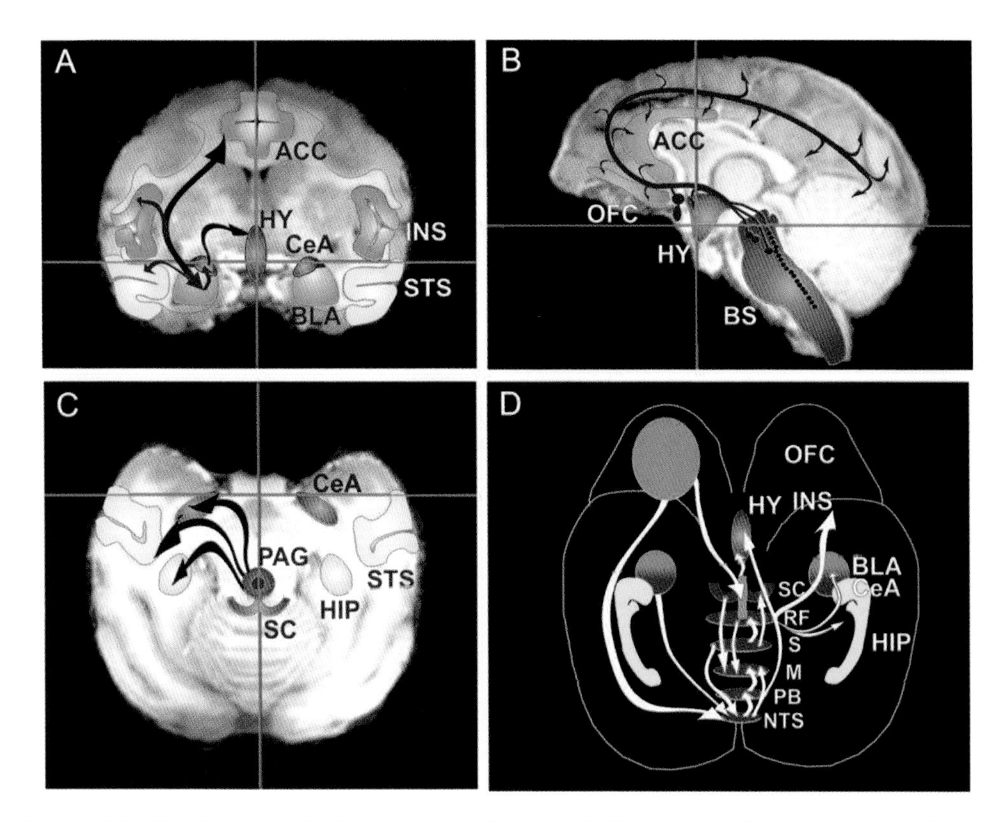

Figure 16.1 Structures contributing to circuits of emotion in the macaque. Core structures are shown in red, intermediate structures in orange, and affiliate structures in yellow. (**A**) Coronal section through several key structures. Arrows indicate connections between structures, depicted on the left hemisphere. Structure names are listed on the right hemisphere. Connections are shown only for structures visible in this image plane. The basolateral amygdala has bidirectional connections with the superior temporal sulcus (STS), anterior cingulate, and the insula, and projects to the central nucleus of the amygdala. The central nucleus, in turn, sends projections to the hypothalamus. (**B**) Sagittal section depicting several key structures and the medial forebrain bundle (MFB). The MFB (black line) contains fibers originating in several neuromodulatory centers in the brainstem and midbrain, and projecting to the hypothalamus and the prefrontal cortex. Neuromodulator-specific fiber tracts continue on to innenervate other regions of neocortex (see text). (**C**) Horizontal (axial) section at the level depicted by the blue line in A. Conventions as in A. The brainstem (indicated by the periaqueductal gray arrows) projects to several structures including the hippocampus, amygdala, and superior temporal sulcus, among other structures. (**D**) Schematic diagram of brainstem connections subserving circuits of emotion. Conventions as in A. This diagram loosely depicts the ventral surface of the brain, including several core brainstem nuclei. Sensory nuclei include those from cranial nerves V, VII, IX, and X. Motor nuclei include those from cranial nerves III, IV, V, VI, VII, IX, X, XI, and XII. The periaqueductal gray (not labeled) is the vertical gray bar connected to the superior colliculus and the reticular formation. ACC, anterior cingulate cortex; BLA, basolateral amygdala; BS, brainstem; CeA, central nucleus of the amygdala; HIP, hippocampus; HY, hypothalamus; INS, insula; M, motor cranial nerve nuclei; NTS, nucleus of the solitary tract; OFC, orbitofrontal cortex; PAG, periaqueductal gray; PB, parabrachial nucleus; RF, reticular formation; S, sensory cranial nerve nuclei; SC, superior colliculus.

circuit is localized primarily in the brainstem and midbrain and produces low-dimensional output (e.g., heart rate and blood pressure can increase or decrease; other effects, such as piloerection, sweating, pupil dilation, contraction of various muscles, etc., also have few degrees of freedom). Additional circuits are superimposed on this core. They encompass subcortical and cortical areas that carry out complex computations on high-dimensional inputs, but ultimately funnel their outputs through the core structures. In this framework, the core structures elaborate low-dimensional affordances based on a limited set of stimuli (e.g., defensive behaviors toward predators), while the outer circuits elaborate high-dimensional internal states to deliver the optimal behavioral choices (e.g., learning that one can attack the member of a rival group only when the attack cannot be witnessed by those who could retaliate).

Conservation of the phylogenetically old, core structures does not imply that they perform the same, "old" functions across extant species; rather, the emotional brain of primates can be thought of as a palimpsest in evolutionary terms. Even the function of homologous structures across species may have been co-opted for new or species-specific purposes. Thus, even as we refer to core structures that are conserved across vertebrates, mammals, or primates, their functions may vary according to species-specific, ecological, and ethological demands.

The first section of this chapter will identify the main structures that comprise the emotional circuits of the primate brain. The second section describes what is known about the neural basis of emotional processes, from the association of stimuli to positive or negative outcomes to the interplay between the perception and expression of social signals. Reflecting the biases in the literature, our descriptions will emphasize the macaque genus and the function of the amygdala, the most highly connected component of the emotional brain (Young et al., 1994).

THE ORGANIZATION OF EMOTIONAL CIRCUITS

Core Structures

Core structures are those whose activity generates a relatively restricted array of emotional responses (autonomic or somatic) via short pathways that link inputs to outputs. These include sensory, motor, and autonomic centers of the brainstem; the periaqueductal gray (PAG); the deep layers of the superior colliculus; the hypothalamus; and the centromedial (subcortical) nuclei of the amygdala (Fig. 16.1).

Sensory, Motor, and Autonomic Centers of the Brainstem

All the afferent signals from the sensory organs of the head and body (with the exception of olfaction) and all the viscerosensory signals from the internal organs converge in the brainstem. Several complex behaviors such as foraging, feeding, flight and fight, response to pain, and reproduction can be initiated and maintained from the level of the brainstem alone (Blessing, 1997). Heart rate, respiration, digestion, elimination, and even immune function are coordinated directly from the brainstem. The brainstem also contains a reticular system that plays an important role in vigilance and behavioral activation in response to both emotional and neutral stimuli. Based on the collateral signals it receives from all the ascending sensory pathways, the reticular system generates a tonic output projected toward the cortex (Blessing, 1997).

Several neurotransmitters that are produced in the brainstem determine emotional traits and mediate emotional states, either by direct synaptic or modulatory activity. The levels of norepinephrine (NE) produced in the locus coeruleus of the brainstem are increased during vigilance, anxiety, acute fear, and other forms of emotional stress. When levels of serotonin (5-hydroxytryptamine [5-HT]) secreted by the raphe nuclei are low, animals are more aggressive and/or depressed. Dopamine (DA), a neurotransmitter associated with reward and pleasure, is produced by the ventral tegmental area and substantia nigra. The gigantocellular nuclei of the reticular formation produce acetylcholine (ACh), which maintains the excitability and coherence of activity across multiple cortical areas. A bundle of axons carry these and other neurotransmitters from the brainstem to the cortex; these axons form the medial forebrain bundle that terminates in the prefrontal cortex but gives out massive collaterals to the hypothalamus (Fig. 16.1B). The hypothalamus, therefore, registers the level of vigilance through NE, the presence of reinforcement (reward or lack thereof) through DA, and other signals related directly to the activity of the brainstem monoaminergic neurons.

Of particular importance for emotions are two nuclei of the brainstem: the nucleus of the solitary tract (NTS) and the parabrachial nuclei (Fig. 16.1D). The NTS is the main visceral sensory nucleus that contains an elaborate viscerosensory map of the body. This map, and the "feeling of the body" contained therein, is broadcast to cortical areas (e.g., the insula and anterior cingulate cortex) where internal sensations are associated with external stimuli (Craig, 2002). Ascending fibers from many cranial nerves report to the NTS the state of all internal organs of the head and body; these signals are either transformed in the NTS into descending commands targeting sympathetic and parasympathetic effectors that coordinate autonomic reflexes or are transmitted via the thalamus to higher cortical centers. The parabrachial nuclei regulate respiration, but they also receive visceral sensory information. The majority of emotional states are associated with changes in respiration that are controlled, in part, by the parabrachial nuclei. Stimulation of this area causes a rapid inhale followed by apnea—a gasp (von Euler and Trippenbach, 1976)—a typical respiratory behavior for surprise, pain, fear, etc.

The brainstem, therefore, is both a relay for higher-level centers of emotion and a first-stage processing center that elaborates species-specific somatic and autonomic behaviors in situations of importance.

Periaqueductal Gray

The PAG has important sensory and motor functions in emotion. It is a major integration and descending modulatory center of the pain pathways, but also a key center that engages the autonomic nervous system (Bandler et al., 2000) in conjunction with the brainstem centers mentioned previously. A few important ethological functions distinguish the PAG from the brainstem centers, namely, its role in vocalizations, defensive behaviors, and aggression.

Species-specific vocalizations are coordinated via the PAG, as demonstrated by electrical stimulation (Jürgens, 1994). This is achieved by indirect control of the laryngeal and respiratory muscles via the reticular formation of the brainstem, (Davis et al., 1996). Motor commands for vocalization are initiated in the PAG under descending input from the orbital and medial prefrontal cortex. These two prefrontal systems are also linked to PAG activity during maternal perception of their offspring (Bartels & Zeki, 2004). Prefrontal projections contact discrete regions of the PAG that, in turn, connect to different regions of the hypothalamus, forming parallel loops that link higher cognitive aspects of emotion and social behavior with expressive and homeostatic/regulatory components of the core circuit (Bandler et al., 2000).

Aggression related to rage and defensive behaviors is coordinated jointly by the PAG and the lateral hypothalamus. The PAG is not involved, however, in attack associated with predation. Complex local neural and pharmacological mechanisms regulate the level of aggression within the PAG. The neural networks that coordinate different forms of aggression are reciprocally inhibitory and the output of each network is potentiated or dampened by different neurotransmitters. Glutamate generally enhances, while opioid peptides typically dampen aggression (Gregg & Siegel, 2001).

A fascinating but unsolved problem in primate neuroethology is the neural site and mechanism of establishing the connections between social status and aggression. Given the fluctuating status of males in a macaque society, the link between rank and aggression is expected to be highly flexible. Serotonin levels and the efficiency of serotonin transporters have been invoked as the major determinants of trait aggression, but state aggression appears to be regulated by prefrontal-PAG/hypothalamic-brainstem mechanisms relying on a different set of neurotransmitters and steroid hormones.

Superior Colliculus

A central role of the superior colliculus (SC) is to guide and modulate visual orientation. The SC contains sensory, motor, and multimodal maps that control the initiation and execution of saccades (for an early review, see Sparks & Hartwich-Young, 1989). In addition to its sensorimotor function, the deep layers of the SC are a critical node in the core circuit of emotion and social behavior.

Electrical stimulation of the rodent SC elicits orienting responses and species-specific defensive behaviors, and in some cases, approach behaviors (Dean et al., 1989). Stimulation also causes changes in respiration, blood pressure, and heart rate (Keay et al., 1988); reduces pain threshold (Redgrave et al., 1996); and causes desynchronization of the cortical electroencephalogram (EEG), which is the cortical signature of heightened attention and vigilance. These autonomic responses might be produced via the bidirectional connections of the SC with the PAG, and with the motor and sensory nuclei of the brainstem. The SC may trigger the reallocation of visual attention in response to emotionally salient stimuli via connections with the amygdala and the mesencephalic reticular formation, involved in the elaboration of eye movements. Indeed, lesions or local infusion of neurotransmitter antagonists in the SC block fear-potentiated startle (Waddell et al., 2003; Zhao & Davis, 2004).

Unlike rodents, where the neural circuitry and pharmacology of colliculus-dependent defensive and antinociceptive effects have been worked out in detail, little is known about how these pathways function in primates. Recent work indicates that bicuculline-mediated disinhibition of the SC in monkeys increases defensive emotional reactivity (Cole et al., 2006) and decreases species-specific social behaviors (Ludise Malkova, personal communication). This outcome is only partially overlapping with the outcome of the same manipulations in the amygdala, indicating that the role in social behavior of these two structures is not entirely redundant.

Finally, the SC has been implicated in detecting emotionally salient visual stimuli without the contribution of cortical visual pathways. A strong argument in favor of this proposal comes from "blindsight" patients and animals with uni- or bilateral visual cortical damage. Subjects report being unaware of stimuli, yet are able to navigate visual obstacles, produce defensive movements in response to looming stimuli, and detect or even discriminate visual stimuli, including threatening images of facial expressions (de Gelder et al., 1999; Liddell et al., 2005; Morris et al., 2001; Stoerig & Cowey, 1997). One of several proposed pathways for these preserved abilities includes the SC, the pulvinar, and the amygdala, which show functional connectivity in neuroimaging experiments and the predicted processing deficits when lesioned (Morris et al., 1999, 2001; Vuilleumier, 2003; Ward et al., 2005). A colliculo-pulvinar pathway might be important for vision early in development (Wallace et al., 1997), enabling rapid orienting and vigilance toward sudden, potentially dangerous stimuli, but may be less efficient for calculated responses to complex visual arrays. Although blindsight patients are notable for their lack of awareness of their intact visual abilities, these parallel visual pathways may normally function under conditions of visual awareness (Pessoa et al., 2005).

Hypothalamus

The hypothalamus is a collection of nuclei concerned primarily with homeostasis through autonomic and endocrine mechanisms, but it also coordinates basic, drive-related behaviors (e.g., feeding, reproduction, aggression). Many of these functions, especially those that involve species-specific "instinctive" behaviors coordinated by central pattern generators, are redundant and overlapping with similar functions controlled by the brainstem. A unique contribution of the hypothalamus in emotion is related to hormone production. The arcuate nucleus controls neuroendocrine function, the supraoptic and paraventricular nuclei release oxytocin and vasopressin (hormones of affiliation and social bonding), and corticotrophin-releasing hormone initiates a cascade of events to supply cortisol in situations of emotional stress.

The hypothalamus receives input from the brainstem and spinal cord via the medial forebrain bundle (Fig. 16.1B). These inputs carry signals about the state of all internal organs. Intrinsic sensory neurons in the hypothalamus are specialized to sense the composition of the internal milieu and correct reflexively any deviation from normal.

Like the brainstem, the hypothalamus contains networks of neurons that often form pattern generators and control behavioral, endocrine, and regulatory functions. These networks are under the influence of higher centers. For example, inputs from the amygdala signal the emotional value of a stimulus or event and can trigger, when appropriate, the classical stress response via the hypothalamic-pituitary-adrenal (HPA) axis. Descending inputs from the amygdala, anterior cingulate cortex (ACC), and prefrontal cortex modulate the role of the hypothalamus in territoriality, aggression, the desirability of food, mates, etc. Reward-related signals from the ventral tegmental area and nucleus accumbens, but also from the orbitofrontal cortex (OFC) and anterior cingulate, mediate reward-based learning and action selection. Inputs from the reticular formation, the basal forebrain (nucleus basalis), and the bed nucleus of the stria terminalis (BNST) contribute to the control of attention and vigilance. The majority of these connections are reciprocal, closing loops that combine the internal state of the organism with the external stimuli and initiate (or block) basic, drive-related behaviors (Appenzeller & Oribe, 1997).

The Subcortical Nuclei of the Amygdala

The amygdala contains a heterogeneous collection of nuclei with dissociable functions. Together these nuclei carry out the evaluation of stimuli, the initiation of autonomic and somatic responses most appropriate for each stimulus, and the modulation of the "gain" of perceptual, motor, and memory processes associated with emotional stimuli.

The subcortical nuclei are connected primarily with subcortical structures and are involved in attention and autonomic functions. The central and medial nuclei, but also the anterior amygdaloid area and the bed nucleus of the stria terminalis (BNST), are part of the subcortical group (Amaral et al., 1992). These nuclei receive and send projections from and to the hypothalamus and the brainstem and, via these connections, control autonomic function. For example, the amygdala modulates heart rate and cardiorespiratory reflexes via connections to the dorsal motor nucleus of the vagus and the nucleus of the solitary tract. Facial expressions are controlled by direct projections to the lower motor neurons in the facial nucleus (Fanardjian & Manvelyan, 1987). Likewise, salivation, lacrimation, respiration, pupil dilation, pain perception, etc., are modulated by direct connections with the respective centers of the brainstem. Finally, the amygdala can modify the output of all major neurotransmitter systems originating in the brainstem. Direct projections from the central nucleus to the ventral tegmental area influence dopamine release (Gallagher, 2000); projections to the locus coeruleus set the level of vigilance and cortical "preparedness" via widespread projections that release norepinephrine over the entire cortex (Amaral & Sinnamon, 1977); and direct projections from the central nucleus to the nucleus basalis of Meynert influence acetylcholine levels in the brain (Davis & Whalen, 2001). Finally, the central nucleus emits projections to the hypothalamic areas, where prohormones and releasing factors are synthesized.

Intermediate Structures

Compared to the core, these structures elaborate the representations of stimuli and stimulus combinations that are evaluated in these circuits and label them with emotional valence. The outputs of the intermediate structures are higher-dimensional than the outputs of the core structures; they control somatic and autonomic effectors via the core structures. The high-dimensional representations generated by these structures are shared with multiple neocortical areas involved in attention, perception, memory, and decision making. Note that intermediate structures do not overlap with the classical concept of the "limbic system" (Maclean, 1949). The intermediate structures include the basolateral (cortical) nuclear group of the amygdala, the mediodorsal nucleus of the thalamus, the insula, ACC, and the OFC.

Amygdala, Basolateral Complex

The cortical nuclei of the amygdala, also called the basolateral complex, contain neurons of the cortical type and receive and send connections primarily from and to the neocortex. One role of these nuclei is to evaluate the emotional significance of stimuli. These nuclei receive highly processed sensory information from all sensory modalities (Amaral et al., 1992; Stefanacci & Amaral, 2002). Olfactory signals arrive from the piriform cortex; gustatory information from the insula; somatosensory and auditory, and visual information from association cortices in the parietal and temporal lobe. Coarsely processed sensory information also arrives here directly from the thalamus (Romanski & LeDoux, 1992). Viscerosensory signals in the amygdala come from the insula or from the nucleus of the solitary tract. Based on these inputs, which converge with inputs from the medial and orbital prefrontal cortex, (OFC) the cortical nuclei of the amygdala determine the positive, negative, or neutral significance of all stimuli. The outcome of this evaluation is sent back to the cortex via widespread feedback projections (Fig. 16.1A). Output projections reciprocate the inputs but also project to multiple stages of sensory processing, including primary sensory areas. These nuclei also project to the subcortical nuclei of the amygdala, which, in turn, project to autonomic and somatic effectors. Signals from the amygdala modulate attention, perception, and memory and are carried by excitatory connections that terminate in layer II of the neocortex. There, they compete for synaptic sites with adjacent or more distant cortical areas (Freese & Amaral, 2006).

Mediodorsal Nucleus of the Thalamus

The mediodorsal thalamus (MD) is an obligatory station in the higher-level interconnections of emotion circuits (Jones, 2007). This nucleus connects the prefrontal intermediate structures—the ACC and OFC—to other emotion-related areas. Its positioning as a "nexus" for the other intermediate structures may account for its known role in object-reward association memory (Gaffan & Parker, 2000).

The Insula

The insula is a good example of the diversity and redundancy of functions carried out by the intermediate structures in the circuits of emotion. The insula is both the primary cortical center of taste and smell and a major source of autonomic regulation, redundant with the autonomic functions of the brainstem. In addition, the insula integrates interoceptive signals, pain, and somatosensory stimuli such as various kinds of touch, and remaps the surface of the body in terms of the quality of stimuli (Augustine, 1996). Social stimuli are also processed in the insula (e.g., facial expressions of disgust and the "trustworthiness" of faces)—stimuli that also activate the amygdala (Engell et al., 2007; Phillips et al., 1997; Winston et al., 2002). Based on these rich inputs, the insula generates a representation of the internal state of the body in which somatic and visceral components are fused and ultimately give rise to a "feeling of the body" associated with empathy, general disposition, and mood stability (Singer et al., 2004).

The Anterior Cingulate Cortex

Whereas the insula has mainly sensory and autonomic functions, the ACC hosts both executive and cognitive functions based on extero- and interoceptive signals arriving from the insula, brainstem, basal nucleus of the amygdala, and hypothalamus. In addition, certain areas of the ACC (24, 25, and ventral 32) receive strong inputs from the mediodorsal nucleus of the thalamus. All subregions of the ACC receive inputs from the anterior thalamic nuclei (Vogt et al., 1987) that receive, in turn, input from the mammillary bodies of the hypothalamus—a memory-related structure. The diversity of inputs is mirrored by a diversity of functions within the ACC (for review, see Bush et al., 2000; Joseph, 2000).

Among these functions are the control of visceral, skeletal, and endocrine outflow in response to emotional stimuli. The outflow of the ACC targets the amygdala, periaqueductal gray, and several cranial nerve nuclei in the brainstem (V, VII, IX, and X). Via the output to the brainstem nuclei, emotional states set up in the ACC translate into changes of heart rate, blood pressure, and vocalizations associated with expressing internal states (e.g., crying, moaning). Note that species-specific vocalizations are also controlled by the PAG in the brainstem.

Like the PAG and the insula, the anterior cingulate cortex also functions as a pain center. Compared to the insula, where bodily sensations such as pain are mapped according to their subjective quality, the pain representation in the anterior cingulate is more abstract. The same areas of the cingulate cortex are activated by social pain (the feeling of being ignored or rejected) and by physical pain. Finally, mother-infant interactions are also under the influence of a division of the anterior cingulate.

Cognitive functions mediated by the ACC include reward anticipation and social decision making. The control signals for these functions originate from dorsal areas of the anterior cingulate that contain a motor region. The motor region influences motor areas of the brainstem and spinal cord, thereby contributing to the selection of the most appropriate action pattern.

Orbitofrontal Cortex

Based on patterns of connectivity, the orbitofrontal cortex resembles the amygdala, a structure to which it is also reciprocally connected. High-level sensory inputs from all modalities converge there, and the outputs target the core structures of the emotion circuits including the hypothalamus (from which it also receives inputs) and brainstem autonomic areas.

The orbitofrontal cortex may exert control over the output of the primate amygdala by way of projections to the intercalated nuclei of the amygdala (for review, see Barbas, 2007). The intercalated nuclei separate the basolateral (cortical) groups of nuclei from the centromedial (subcortical groups) through their GABAergic interneurons. These interneurons may inhibit the transfer of signals from the basolateral complex to the lateral division of the central nucleus, thus blocking the initiation of autonomic responses (Pare & Smith, 1993; Rempel-Clower, 2007). The recently described functions of intercalated nuclei (Likhtik et al., 2008) might explain why a snake in the outdoors is more fear producing than when experienced in a terrarium. In theory, the orbitofrontal activity could prevent the basolateral signal of danger from activating the central nuclei, "heading it off at the pass," thereby eliminating unnecessary behavioral and autonomic responses.

Collectively, the intermediate structures show partially overlapping, yet complementary functions. For example, the similar connectivity of the ACC and OFC suggest that they process reward signals in conjunction with internal and external stimuli and can influence autonomic and somatic responses that contribute to normal social and emotional behavior. There are important differences, however, between these two areas in terms of their contribution to the evaluation of stimuli and the elaboration of responses. Whereas the ACC integrates reward-related signals with species-specific motor behaviors, the OFC learns to associate and dissociate internal and external stimuli and rewards. Together, they form a unit critical for bringing into register the internal and external state of the world with behavioral choices, the building blocks of emotion and social decision making.

Affiliate Structures

Affiliate regions are primarily neocortical and are more strongly connected to the intermediate structures than to the core structures. They are capable of processing complex, high-dimensional signals, linking multiple aspects of emotion, such as memory, decision making, and action planning, to ongoing stimuli and events.

Compared to the core and intermediate structures, their role in generating autonomic and reflexive emotional outputs is less direct, often demonstrating their involvement only for a specific set of stimuli and contexts. Though this list will surely lengthen as our understanding of emotion circuits increases, for now, we consider the hippocampus, the lateral temporal lobe, ventrolateral prefrontal cortex, the lateral intraparietal area, and the medial pulvinar.

Hippocampus

Historically, both Papez (1937) and MacLean (1949) included the hippocampus in the emotional centers of the "limbic system." Refined lesion techniques later suggested that the hippocampus does not play a major role in emotion; however, as a central learning and memory structure, it is at the interface between emotion and memory. Episodic memories that are hippocampal dependent can have an emotional component that is encoded and stored together with the neutral components. Conversely, emotional states are known to influence memory formation, whether by enhancing vigilance and attention at the time of encoding (Adolphs et al., 2005; Easterbrook, 1959) or through the effects of arousal on memory consolidation (McGaugh, 2000, 2004). The strong connections of the hippocampus (or its gateway structures) with the amygdala, orbitofrontal cortex, and medial prefrontal cortex constitute the anatomical framework on which emotional memories are built (LaBar & Cabeza, 2006; McGaugh, 2004; Packard & Teather, 1998).

An additional role of the hippocampus in emotion and memory derives from its high concentration of glucocorticoid receptors. This makes the hippocampus vulnerable to the neurotoxic glucocorticoids that are released in stressful or highly emotional situations by the HPA axis (Sapolsky, 1996; Watanabe et al., 1992; Woolley et al., 1990). Chronic stress, marked by elevated levels of cortisol, is associated with hippocampal atrophy, reduced neurogenesis, and behavioral changes ranging from memory loss to depression. In humans, personality traits such as self-esteem and an internal locus of control are positively correlated with hippocampal volume (Pruessner et al., 2005). Given its coupling with stress responses and the association between high stress levels, anxiety, and heightened responses to acute stressors, the hippocampus should be brought back into the fold of emotional circuits, at least when considering the long-term trait influences on emotion processing.

Lateral Temporal Lobe

The continuous stretch of cortex from the upper bank of the superior temporal sulcus (STS), through the lower bank, and onto the inferotemporal (IT) gyrus processes complex visual images, including faces (Bruce et al., 1981; Gross et al., 1972; Perrett et al., 1982; Tanaka, 1992; Yamane et al., 1988). This area of the temporal cortex receives inputs from "upstream" visual areas such as MT, TEO, and V4, and projects to the orbitofrontal cortex, lateral prefrontal cortex, medial temporal lobe neocortex and amygdala (Baizer et al., 1991; Felleman & Van Essen, 1991; Seltzer & Pandya, 1978). Neurons in these areas show responses to different dimensions of social stimuli, such as face identity, facial expression, head or gaze direction, and body movements (for review, see Rolls, 2007; see also Tsao et al., this volume). In addition, the upper bank STS shows multisensory responses that can combine visual and auditory information, such as communication signals (Barraclough et al., 2005; Ghazanfar et al., 2008).

Some evidence that these areas are important for processing socioemotional cues comes from lesions made in infancy to area TE (inferior temporal cortex) that lead to temporary deficits in socioemotional behavior in infants (Bachevalier et al., 2001; Malkova et al., 1997b). The STS is required for normal gaze discrimination, which is important for interpreting the target of an expression (Heywood & Cowey, 1992), and an intact STS and inferior temporal gyrus (ITG) are required for discrimination of configural changes in faces (Horel, 1993) that are important for identification. Taken together, these areas may form a key "prerequisite" filter for extracting emotion and/or intention from visual social cues such as faces, though the lateral temporal lobe is not the only region poised to play this role.

Ventrolateral Prefrontal Cortex

The ventrolateral prefrontal cortex (vlPFC), specifically, the inferior prefrontal convexity, processes social-emotional information from every sensory modality. This area receives projections from high-level visual areas including the inferotemporal cortex (Ungerleider et al., 1989) and contains neurons that show selective responses to faces. (Scalaidhe et al., 1997; Wilson et al., 1993) that are similar to the responses of face-selective neurons of the lateral temporal areas. Adjacent to the visual region is an area that shows selective responses to species-specific vocalizations (Cohen et al., 2004, 2006; Romanski et al., 2005). Selectivity profiles do not fall strictly into featural (tonal/noisy) or functional (food/nonfood, aggressive/appeasing) borders, though biases in a neuron's response toward one or the other category can be sufficient to extract this information (Thomas et al., 2001). The inferior prefrontal convexity was recently shown to engage in multisensory processing of audio-visual communication signals (Sugihara et al., 2006), providing one clue to the processing that may occur in this region. Moreover, with its sensitivity for complex, species-specific stimuli, vlPFC activity may ultimately prove to be an important link between the perception of social stimuli and the extraction of their emotional significance. We refer the reader to the chapters by Miller and Cohen and Romanski and Ghazanfar in this volume.

Lateral Intraparietal Area

The allocation of spatial attention (Bisley and Goldberg, 2003) and goal-directed selection of actions (Snyder et al., 1997) are two functions ascribed to the lateral intraparietal (LIP) area (see Groh & Pai, this volume). More recently, it has also been associated with perceptual decision making (Freedman & Assad, 2006), including enhanced responses when selecting preferred social cues such as images of dominant males, conspecific faces, or female hindquarters (Klein et al., 2008). LIP neurons do not respond selectively in anticipation of social cues when the cues are presented in a predictable, obligatory fashion, but when the monkey makes a choice that triggers the presentation of a given image, the responses predict the value of that image. Here, value was assessed by the amount of juice monkeys would forego to view an image, using a previously established paradigm (Deaner et al., 2005). Thus, the LIP area might mediate processes by which social cues are embedded in the decision-making networks in the brain, as would be important for action selection during social interactions.

Medial Pulvinar

The medial nucleus of the pulvinar is connected to the superior temporal sulcus and gyrus, the amygdala, the insula, orbitofrontal cortex, the frontal pole, and medial prefrontal cortex including anterior cingulate cortex (Romanski et al., 1997). This list of structures corresponds to the intermediate layer of the emotional circuit proposed here.

The medial pulvinar, like its neighboring structure, the mediodorsal nucleus of the thalamus, may contribute to the integration of signals from nodes of the emotion circuit.

The unique roles played by the affiliate regions and their recruitment of widely conserved core regions may be specializations to selective pressures that reward social savvy. The affiliate structures aren't directly tied to the production of a specific emotion; rather, they enable extraction of appropriate responses to complex constellations of cues and scenarios that are relevant for responding appropriately to the current social situation. In this way, the boundaries segregating structures associated with emotion and cognition become blurred (Pessoa, 2008), as will become clear when considering the contribution of the aforementioned structures to specific emotional processes.

PROCESSING EMOTION: CIRCUITS IN ACTION

A basic rule of survival is to approach food sources and potential mates (appetitive stimuli) and avoid danger such as predators (aversive stimuli). Biologically prepared appetitive or aversive stimuli elicit emotional/motivational states, and are processed primarily by the core of the emotional brain. The structures of the core circuits are highly conserved across species, but component neurons are often tuned to stimuli with species-specific significance (e.g., odors of predators, pheromones, etc.). The appetitive-aversive dichotomy, although not mapped onto different circuitry, has proven to have great heuristic value for deciphering the neural circuitry of emotion in animals.

Our most complete understanding to date of the most basic and most shared emotion across all species—fear—comes from aversive conditioning in rats. Fear has been studied using an aversive stimulus such as foot shock, often in association with a cue such as a tone (Davis & Whalen, 2001; Fanselow & Gale, 2003; LeDoux, 2000).

Approach behaviors, which do not correspond to a state as clearly defined as fear, have been studied using food reinforcement (Everitt et al., 2003; Holland & Gallagher, 2004). In rats, the amygdala alone can support fear conditioning, whereas appetitive conditioning and reward devaluation paradigms require the joint contribution of the amygdala and frontal cortex.

Pavlovian conditioning in primates suggests a conservation of neural structures (Baxter & Murray, 2002) and possibly mechanisms (Salzman et al., 2007) that support this basic form of emotional learning. The main difference might be that emotional processing from the core structures radiates in primates to a larger array of partially or fully corticalized structures that reprocess emotion and link it to memory, planning, and decision making. The essential difference derives from the major role of social stimuli to elicit emotions in primates. We will consider first the evidence for aversive and appetitive conditioning in primates before describing how social stimulus evaluation is intrinsically related to emotional circuits.

Aversive Emotional States

Unconditioned Fear

Some stimuli are known to produce fearful or startle responses in monkeys (Davis et al., 2008). The central nucleus of the monkey amygdala is necessary for the expression of unconditioned, but also of conditioned, startle. Monkeys with lesions of the central nucleus also lack the normal apprehension to the sight of a snake (real or fake), as determined by the latency to retrieve food in its presence (Davis et al., 2008; Kalin et al., 2004). In addition, monkeys without a central nucleus fail to demonstrate normal freezing (Kalin et al., 2004) that typically accompanies the appearance of a human intruder. Finally, lesions of the central nucleus reduce cortisol levels, indicating that an intact amygdala is necessary for generation of normal stress responses.

Conditioned Fear Responses

A subset of neurons in the monkey amygdala respond selectively to stimuli that predict an aversive, unconditioned stimulus (US), such as a puff of air directed at the face or eyes (Paton et al., 2006). Neuronal responses to the air puff and the stimuli that predict it (conditioned stimuli [CS]) ramp up during learning and fall off during extinction, suggesting they may reflect the updated value of the association to the aversive unconditioned stimulus. Although patients with amygdala damage are able to learn CS-US associations, they are unable to express fear or the anxious anticipation of the noxious stimulus (Bechara et al., 1995). Moreover, conditioned stimulus learning and extinction are associated with changes in blood level oxygen–dependent (BOLD) responses in the ventromedial prefrontal cortex and amygdala of humans (Büchel et al., 1998; LaBar et al., 1998; Phelps et al., 2004). The following reviews consider the fear conditioning literature in more detail (LeDoux, 2000; Phelps, 2006).

The predominance of fear conditioning as a model for emotional learning has led to the implicit assumption that the structures necessary for processing fearful stimuli are specialized for fear. To determine whether structures such as the amygdaloid nuclei function as "fear modules," involving only negative affect, it is important to demonstrate that they do not act more generally in emotional learning, involving *either* positive *or* negative affect.

Appetitive Emotional States

Conditioned Appetitive Responses

Classical conditioning can also be used to study positive emotional states by pairing an initially neutral stimulus, such as a visual stimulus, with reward. For example, a study in marmoset monkeys showed that the amygdala is not necessary for appetitive conditioning per se (Braesicke et al., 2005). In this study, the overt behaviors indicating anticipation of reward (looking and scratching at a food barrier) persisted in amygdala-lesioned animals, despite a reduction in the physiological signs of arousal during that anticipatory phase. It was as though the habit of orienting toward a conditioned stimulus remained, even when the underlying incentive was no longer present. This suggests that some aspects of positive affect require an intact amygdala.

Consistent with this observation, neural responses in the macaque amygdala follow the time course of learning the association between an image and juice reward, just as a previously mentioned population of amygdala neurons "tracks" learning of image-punishment associations (Paton et al., 2006). For a more detailed examination, we refer the reader to reviews of the role of the amygdala in reward and positive emotion (Baxter & Murray, 2002; Everitt et al., 2003; Murray, 2007).

Lesion studies implicate both the amygdala and orbitofrontal cortex in the flexible selection of cues that predict specific rewards, though each plays a distinct role in the process. Overall preferences for desirable foods are unchanged after lesions or disruption of the amygdala (Machado & Bachevalier, 2007b; Malkova et al., 1997a; Murray & Izquierdo, 2007; Wellman et al., 2005) or of the orbitofrontal cortex (Machado & Bachevalier, 2007b), indicating that these structures are not necessary to encode the primary reinforcement value of foods. Indeed, behavioral responses are appropriately diverted away from food that has been devalued through satiation (Machado & Bachevalier, 2007a). Moreover, neither structure is necessary to discriminate which objects are associated with food reward in a concurrent discrimination task. Yet when a preferred food is devalued, lesions to either the amygdala and/or the contralateral orbitofrontal cortex lead to continued selection of objects associated with the satiated (devalued) reward, when a normal response would be to switch, selecting the food reward that had not been devalued. Curiously, upon seeing the underlying food reward, only monkeys with orbitofrontal lesions actually took the reward. These monkeys selected the non-devalued food when given a choice between the two, indicating that satiation indeed changed the value representation of the food. These monkeys only fail when given a choice between the *cues* predicting the foods, falling back on their learned behaviors to previous conditioned associations. Having selected the single, devalued food, they take it despite satiety for it, unlike monkeys with amygdala lesions. Thus, both the orbitofrontal cortex and the amygdala share a role in updating behaviors toward previously learned associations based on current preferences or motivational state, and the integrity of both structures is required for updating the value of a reinforcer (Baxter et al., 2000). These results are also consistent with a study of instrumental extinction (Izquierdo & Murray, 2005).

In contrast to the nuanced distinctions in the roles played by the amygdala and orbitofrontal lesions in encoding the value of reinforcers, the anterior insula may be more directly involved in representing the reward value of foods. As in secondary olfactory and primary gustatory cortex, cells in the insula show reduced firing rates to preferred odors or foods following satiation (Rolls, 2000). When orbitofrontal lesions include part of agranular insular cortex, monkeys continue to select foods after being fed to satiety (Machado & Bachevalier, 2007a,b). Thus, the insula may be a central source of behavioral responses to some rewarding (appetitive) stimuli, such as food.

The use of similar appetitive and aversive conditioning paradigms in primates and in rodents reveals a large overlap in the structures involved. This implies a conservation of the potential mechanisms of emotional learning (Everitt et al., 2003; Holland & Gallagher, 2004; Murray, 2007; Phelps, 2006). Although the emotional behavior of rodents cannot be reduced to food and predators, and social stimuli are likely processed by the circuit that evaluates reinforcers, little is known about the "social brain" in rodents (Panksepp, 1998). In primates, however, it is clear that the majority of emotional states are centered on social interactions; for many primate species, individuals live within elaborate social dominance hierarchies. Here, appropriate responses to members of the group can reduce the threat of attack or increase access to food, reproductive partners, or allies that indirectly reduce threats and increase access to rewards. Indeed, the majority of emotional states in primate species occur during social interactions. These socio-emotional interactions depend on the identity and dominance status of the participants, as well as on the recent history of aggression/affiliation, and perhaps most critically on the social signals (facial expressions) displayed by interacting conspecifics.

The Social Envelope: The Relationship Between Expressions and Emotions

Expressions can be regarded both as the externalization of the emotional state of the displayer but also as intentional signals aimed at another individual. At the receiving end, expressions can induce emotional states in the observer. Importantly, expressions map most closely onto the state of the *sender*; how it is interpreted by an observer will depend on additional factors. For example, an aggressive expression indicates the perturbed state of the sender, but if the sender is a meek juvenile monkey, or if the expression is directed at another individual, it would have little effect on an adult monkey observer. Thus, we will begin by describing the expressions generated by an individual based on that individual's emotional state before considering how expressions can evoke emotional responses in observers. We will focus on macaque expressions, allowing us to describe neural structures associated with the generation and perception of expressions, expanding to incorporate what is known in humans or other members of the primate order, where possible. The three poles suggested in the macaque expression literature (Deputte, 2000; Mason, 1985; Partan, 2002) are *avoidance*, which maps directly onto other fearful or avoidance behaviors; *affiliation*, which corresponds to appetitive behaviors; and the additional pole of *aggression*, a "strictly social" addition that arises as a consequence of dominance status.

Generating Expressions

Avoidance (Fear Grimace, Scream) In macaques, a cluster of expressions indicate the presence of an aversive stimulus, including the fear grimace and scream. The fear grimace is characterized by horizontally retracted lips, revealing upper and lower teeth, and retracted ears. It can also be accompanied by a high-pitched, shrill vocalization, or "scream" (Partan, 2002).

Affiliation (Lip Smack, Coo, Groom Present) Expressions indicating a desirable or appetitive stimulus include the lip smack, coo, and grunt. The lip smack is produced by puckering the lips and smacking them together, with the chin held up and ears retracted. The coo is a single tonal vocalization made while making an "oo" shape with the lips. The groom present, common to many primate species, is a body gesture made by exposing a vulnerable part of the body by extension, such as by raising the arm, turning the head away to expose the neck, or turning one's back, all in close proximity to an individual invited to groom. The act of grooming that may follow has been used as a measurement of the degree of affiliation (Seyfarth & Cheney, 1984). Affiliative behaviors trigger neural reward states. Grooming is accompanied by β-endorphin release in cerebrospinal fluid, blocking the μ-opioid receptor increases groom invitations and grooming durations, and opiate delivery decreases grooming invitations (Keverne et al., 1989).

Aggression and Alarm (Open-Mouth Stare, Pant-Threat, Shrill Bark) The most typical expression of aggressive or threatening gestures in the macaque is the open-mouth threat (or stare), characterized by directed gaze and an opening of the mouth so that lips form an "o" shape, with the upper teeth covered by the lips. Ears are forward or are flapped, the head is lowered, the eyes are fixated on the receiver of the threat, and the facial display is often accompanied by a body lunge toward the target. The pant-threat is a staccato noisy (broadband) vocalization, often occurring as a triplet. Consistent with the presumed emotional correlates of vocalizations, the shrill bark alarm call appears to be mediated by stress levels. A reduction of stress hormone output leads to fewer shrill bark vocalizations under stressful conditions (Bercovitch et al., 1995).

Neural Basis for the Generation of Expressions

Facial expressions reflect the internal state of approach, avoidance, or aggression of the displaying animals. Whereas the involvement of some neural structures appears to be restricted to the generation of a subset of expressions (e.g., aggressive displays), other structures, such as the amygdala, are involved in generating a wide array of expressions, possibly selecting which class of responses is appropriate for a given situation.

Avoidance Monkeys that were given neonatal amygdala lesions show more fear responses (more grimaces and screams) toward a novel peer monkey than do intact monkeys (Bauman et al., 2004b), but show fewer fear responses than intact monkeys do when separated from their mother (e.g., fewer screams; Bauman et al., 2004a). Likewise, in a study of young monkeys who had just received amygdala lesions, there was diminished fear expression toward a stimulus that should have been threatening: the introduction of a novel adult male (Kalin et al., 2001). The ability of these monkeys to respond, but at inappropriate times or with abnormal frequency, suggests that the amygdala may be one structure important for *appropriate* evaluation or expression of fearful situations, but not for the generation of the response per se.

Affiliation When faced with a novel adult male (a threatening stimulus), young amygdala-lesioned monkeys delivered fewer affiliative vocalizations (coos) than did monkeys with an intact amygdala (Kalin et al., 2001). They also barked less and showed less submissive behavior. In contrast, when paired with a peer monkey, monkeys with neonatal amygdala lesions produced more affiliative expressions than intact monkeys (Bauman et al., 2004b). This is consistent with the suggestion that the amygdala is not required for the generation of expressions, regardless of emotional state; rather, it seems to play a role in determining the *appropriate* gestures in a particular context. We will return to this point in our discussion of the perception of expressions and their link to emotional state.

Aggression In marmosets, stimulation of the ventromedial hypothalamus produces aggressive vocalizations, and lesions to the anterior hypothalamus or the preoptic area lead to a reduction in aggressive vocalizations toward an intruder (Lloyd & Dixson, 1988). In macaques, stimulation of the anterior hypothalamus, preoptic area, and BNST lead to an increase in aggressive vocalizations (Robinson, 1967) and aggressive behaviors toward subordinate monkeys (Alexander & Perachio, 1973). Thus, proper expression of aggressive behaviors involves, at a minimum, the hypothalamus, BNST, and, as previously mentioned, periaqueductal gray and possibly anterior cingulate cortex.

Perceiving and Evaluating Expressions

Although expressions indicate the emotional state and/or the intentions of an individual, the dominance hierarchy of macaque societies imposes additional factors on the interpretation of an expression, such as the rank, sex, and identity of the sender and of the receiver, as well as the ranks of the close affiliates of both. Recent social exchanges and their outcomes may also weigh into the interpretation of a given expression. What is the evidence that monkeys are sensitive to expressions and to associated social factors? In this section, we explore the neural substrates for perceiving and properly interpreting expressions.

When a Threat Is Threatening In general, when observing a social encounter, an aggressive response is one that escalates an encounter, ultimately predicting an attack, whereas an affiliative response is one that invites or predicts those interactions that decrease the probability of attack. An avoidance response is one that disarms or prevents an encounter, leading to a break in the interaction. It follows that aggressive gestures directed at an observer are generally arousing and threatening to the observer; directed affiliative gestures may be arousing, but not threatening; and directed avoidance gestures should not be arousing or threatening.

Images of expressions elicit in the macaque brain stimulus-specific responses in several areas of the temporal cortex, as well as in the hippocampus, amygdala, and prefrontal cortices, as shown by imaging studies (Hadj-Bouziane et al., 2008; Hoffman et al., 2007; Tsao et al., 2008; see Tsao et al., this volume). Neurons in the inferotemporal cortex or STS (Hasselmo et al., 1989) and amygdala (Gothard et al., 2007; Kuraoka & Nakamura, 2007) respond selectively to facial expressions. In the amygdala, the same fraction of neurons respond to threatening, appeasing, or neutral stimuli, but the firing rates are higher for threatening faces. The heightened response to threats is a matter of degree; all types of expressions can elicit selective amygdala responses (Gothard et al., 2007). Of note, many amygdala neurons responded most strongly to a specific combination of identity and expression, not to expressions in an identity-invariant manner. This would provide a neural means of perceiving expressions embedded in the appropriate social context, something monkeys with lesions to the amygdala or orbitofrontal cortex are unable to do. Proper interpretation of an expression is a prerequisite for *generating* context-appropriate expressions. Recall that the generation of appropriate expressions was impaired in monkeys with amygdala lesions (Bauman et al., 2004a,b; Kalin et al., 2001). It's possible that the process of evaluating expressions simply does not occur independently of the context in which the expression is displayed. Here, "context" can be ascertained from physical attributes and attitudes of the sender that co-occur with the expression, such as gaze and body direction, age, gender, and dominance rank, deduced from recognizing the sender's identity and physical markers of fitness. Context could also reflect the presence, proximity, and gestures of other individuals, or of a sender's history of behaviors associated with an expression, as well as the recent activity of the colony. We will explore the evidence for neural sensitivity to these contextual factors and the effects they may have on the evaluation of emotion-laden stimuli.

Head and Gaze Direction

Expressions directed away from the observer toward an unseen target are ambiguous. Without knowing the target of the expression, the appropriate response is unclear and requires additional information gathering, such as orienting in the direction of display. Indeed, averted expressions in monkeys are more arousing than are expressions directed at the observer (Hoffman et al., 2007). Because an important factor in the perception of expressions is an assessment of the target of an expression, the direction of head and gaze of a displaying monkey is important to the proper interpretation of an expression.

Neurons sensitive to body, head, and eye gaze direction are found in the superior temporal sulcus of the temporal lobe (Perrett et al., 1985), and lesions to this region produce gaze-discrimination deficits (Heywood & Cowey, 1992). In addition, the central nucleus of the amygdala, which is known to influence autonomic output, shows greater BOLD activation for averted- than for directed-gaze faces, consistent with physiological measures of arousal in response to the same stimuli (Hoffman et al., 2007). The STS sends strong projections to the basolateral amygdaloid nuclei, which, in turn, project to the central nucleus, thereby providing a putative circuit for perceiving and redirecting attention based on the intended target of a seen expression. These results are consistent with a more general role for the central nucleus in the detection of saliency and the reallocation of attention (Whalen, 2007).

Dominance Rank

The asymmetrical allocation of valued resources such as space, food, and agonistic interactions within groups of macaques is evidence of a dominance hierarchy, with the commonly studied rhesus macaque demonstrating one of the most linear, despotic hierarchies of all macaque species (Flack & de Waal, 2004; Thierry, 2004). Sensitivity to social dominance is a pervasive facet of macaque behaviors, from its effects on the latency to approach foods to female orgasm rate, which increases with the dominance rank of the males in a pair (Troisi & Carosi, 1998). Even in an isolated, experimental setting, responses to pictures of familiar conspecifics demonstrate sensitivity to dominance rank. Low-ranking macaques will follow the gaze of a monkey looking away, regardless of the rank of the stimulus monkey. High-ranking monkeys, however, will only follow the gaze of other high-ranking monkeys (Shepherd et al., 2006). Thus, the aforementioned neural sensitivity to head and gaze direction could also be influenced by the status of both the observing and stimulus monkeys. Consideration of rank will be important in future studies of macaque circuits of emotion.

Kin and Affiliation

Macaque mothers differentiate between threats directed at them and those directed toward their infants; the presence of a dominant female is threatening to both, a subordinate female is a threat to her infants, and young daughters are nonthreatening to both groups (Maestripieri, 1995). Understanding the neural substrates for macaques' sensitivity to social status is difficult, however; most of our knowledge of social behaviors comes from field studies devoid of neural manipulations. Those studies measuring social behaviors in monkeys following lesions have focused on the role of the amygdala. Qualitative descriptions of the effects of amygdala lesions or lesions extending into other medial temporal lobe regions included withdrawal, submission, and drop in rank when placed in large-group settings (Dicks et al., 1969; Rosvold et al., 1954). Somewhat different results were obtained through the quantitative behavioral assessment of amygdala-lesioned monkeys placed in randomized dyads (Emery et al., 2001). Under these conditions, introduction to another, unlesioned monkey produced more affiliative approach behaviors than those seen for introductions between pairs of unlesioned monkeys. Monkeys with neonatal amygdala lesions also show enhanced contact, in this case with their mothers. These two observations support the idea that the amygdala is involved in the inhibition of certain types of behaviors, primarily affiliative behaviors (Sally Mendoza, personal communication). In addition, despite the ability to generate the full repertoire of expressions and vocalizations, during perturbations in the social environment, monkeys with neonatal amygdala lesions generate inappropriate responses, either vocalizing more or less than controls, depending on the situation, or failing to return to their mothers after forced separation (Bauman et al., 2004a,b; Kalin et al., 2001). In contrast, hippocampal-lesioned monkeys showed none of these behaviors. It is not clear what components of the amygdala are responsible for these aberrations, but the results are consistent with the suspected role of the basolateral amygdala in evaluating the proper constellation of cues to update behavioral responses. Clearly, more information is needed to understand what, if any, additional structures are implicated in identifying and altering responses to expressions based on kin relationships and social bonds.

Traits and States

In addition to functional architecture that allows the evaluation of stimuli, the emotional circuits of the brain are influenced by genetic makeup and by early life experience that establish the "baseline" or tonic output even in the absence of a changing input. The temperament, affective style, or personality traits of the animal depend on this output. Anxiety-related personality traits in monkeys, for example, have been linked to the polymorphism of the regulatory region of the gene for transporter-facilitated uptake of serotonin (Suomi, 2006). Secure attachment, or lack thereof, to the mother, combined with the presence of the short or long allele of the serotonin transporter gene or of the monoamine oxidase A gene promoter is predictive of risk-taking behaviors, resiliency to adverse social situations, and the ability of monkeys to control aggression (Newman et al., 2005; Suomi, 2006). Individual differences in tonic autonomic output or "trait" behaviors will influence the degree and possibly character of emotional states evoked by changing inputs. Likewise, keeping track of which individuals are highly reactive will assist the interpretation of the likely consequences of a given expression.

Multidimensionality and Pluripotency Within Emotion circuits

The literature on appetitive conditioning suggests there are no "dedicated" fear circuits; rather, emotional stimuli with negative and positive valence are evaluated by the same or highly overlapping circuits. Likewise, the structures implicated in the evaluation of social stimuli overlap, at least partially, with those of the process-conditioned approach/avoidance behaviors, demonstrating a co-opting of approach/avoidance processing into the social domain. In addition, all stimuli with uncertain or ambiguous emotional significance engage the amygdala and related structures (Whalen, 2007). This raises the question: To what extent do structures within circuits of emotion evaluate "emotional" stimuli exclusively?

A comparison of the neural responses elicited by unfamiliar objects, food items, fear-producing objects, and a variety of facial expressions indicated that all classes of stimuli elicit stimulus-selective response in the monkey amygdala (Gothard et al., 2007). Whether these images engage the amygdala because they are unfamiliar with ambiguous value or their "neutrality" is encoded as part of a more general process of stimulus evaluation remains to be clarified. The quasi-equanimous allocation of the processing resources of the amygdala to negative and positive stimuli is best illustrated by the response to social stimuli, such as facial expressions. Although fearful faces are the least discernable expression for patients with amygdala damage (Adolphs et al., 1994; Sprengelmeyer et al., 1998; but see Rapcsak, 2003) and hemodynamic changes in the human amygdala appear larger in response to fearful and angry faces than to neutral or happy faces (for review, see Zald, 2003), neurons in the monkey amygdala show only a small increase in firing rate for threatening faces compared to neutral or appeasing faces (Gothard et al., 2007).

Neurons in the monkey amygdala that are selective for facial expressions respond by either increasing or decreasing their firing rate. Positive facial expressions (appeasing faces) are more often encoded by significant decreases of firing rates than threatening faces that cause most often significant increases in firing rate (Gothard et al., 2007). The number of neurons allocated to process positive, negative, and even neutral stimuli is roughly equivalent, and the global activation is significantly above baseline for all facial expressions. There was indeed a short-lived excess of population firing rate for threatening faces, but the size of the difference between the threatening and appeasing faces was smaller than the size of responses to any type of facial expressions compared to baseline. Nevertheless, the observed difference between aggressive and appeasing facial expressions is sufficient to account for neuroimaging results obtained by subtraction analyses using the same facial expressions (Hoffman et al., 2007). Our observation that the monkey amygdala processes aversive and appetitive stimuli equally is in line with earlier neurophysiological and behavioral studies of the monkey amygdala (Fuster & Uyeda, 1971; Nakamura et al., 1992; Ono & Nishijo, 1992; Paton et al., 2006; Sanghera et al., 1979; Wilson & Rolls, 1993).

It appears, therefore, that the monkey amygdala performs the operations of aversive conditioning and appetitive conditioning elucidated in rats but, in addition, carries out important social processing that is not biased in favor of stimuli with a particular emotional valence. If a structure traditionally placed at the heart of emotional processing demonstrates processing of stimuli irrespective of emotional valence, other structures in circuits of emotion may do the same.

CONCLUSIONS

1. A variety of neural structures contribute to the processing of emotions. For didactic purposes, the neural circuits of emotion can be divided into core structures that are the proximal

sources of autonomic and motor responses associated with the expression of emotion. These emotional expressions are closely tied to—and in certain cases include—vigilance, orienting, and the reallocation of attention to emotionally salient stimuli. Intermediate structures are those that generate more complex representations that place in emotional register the external and internal milieu and flexibly link perceptual and motor components of an emotional response. Several additional structures serve an extended emotional circuit, providing further information about the emotional content of stimuli, expected outcomes, and contextual contingencies that determine which emotional response should be delivered and under what circumstances. All regions, taken together, afford adaptive responses to changes in the social environment.

2. The amygdala is a key structure for the evaluation of emotional stimuli. Based on interaction with affiliate structures, the amygdala controls the transmission of signals toward the core structures that orchestrate the expressions of emotions. Much of our current understanding is based on the treatment of the amygdala as a unitary structure due, in part, to methodological limitations. Nevertheless, the subcortical and cortical nuclei, when scrutinized, fall into distinct roles in emotional circuits on both anatomical and functional grounds.
3. Emotional responses can be broken down into the immediate evaluation and response to changes in the environment and to long-term modifications that can shape future responses, or even regulatory states of the animal (e.g., stress hormone levels). Whereas core and intermediate areas may afford rapid responses, the high-dimensional information available to intermediate and affiliate areas may be useful for constructing multiple contingencies for subtly varying sets of inputs. They may also support memory for these contingencies and set into motion longer-term changes, thereby shaping both traits and states related to emotions.

REFERENCES

Adolphs, R., Tranel, D., & Buchanan, T. W. (2005). Amygdala damage impairs emotional memory for gist but not details of complex stimuli. *Nature Neuroscience, 8,* 512–518.

Adolphs, R., Tranel, D., Damasio, H., & Damasio, A. (1994). Impaired recognition of emotion in facial expressions following bilateral damage to the human amygdala [see comment]. *Nature, 372,* 669–672.

Alexander, M., & Perachio, A. A. (1973). The influence of target sex and dominance on evoked attack in rhesus monkeys. *American Journal of Physical Anthropology, 38,* 543–547.

Amaral, D. G., & Sinnamon, H. M. (1977). The locus coeruleus: Neurobiology of a central noradrenergic nucleus. *Progress in Neurobiology, 9,* 147–196.

Amaral, D. G., Price, J. L., Pitkänen, A., & Carmichael, S. T. (1992). Anatomical organization of the primate amygdaloid complex. In: J. P. Aggleton (Ed.), *The amygdala: Neurobiological aspects of emotion, memory, and mental dysfunction* (pp. 1–66). New York: Wiley-Liss.

Appenzeller, O., & Oribe, E. (1997). *The autonomic nervous system: An introduction to basic and clinical concepts* (5th ed.). Amsterdam: Elsevier Health Sciences.

Augustine, J. R. (1996). Circuitry and functional aspects of the insular lobe in primates including humans. *Brain Research, earch Reviews, 22,* 229–244.

Bachevalier, J., Malkova, L., & Mishkin, M. (2001). Effects of selective neonatal temporal lobe lesions on socioemotional behavior in infant rhesus monkeys (Macaca mulatta). *Behavioral Neuroscience, 115,* 545–559.

Baizer, J. S., Ungerleider, L. G., & Desimone, R. (1991). Organization of visual inputs to the inferior temporal and posterior parietal cortex in macaques. *Journal of Neuroscience, 11,* 168–190.

Bandler, R., Keay, K. A., Floyd, N., & Price, J. (2000). Central circuits mediating patterned autonomic activity during active vs. passive emotional coping. *Brain Research Bulletin,. 53,* 95–104.

Barbas, H. (2007). Flow of information for emotions through temporal and orbito-frontal pathways. *Journal of Anatomy, 211,* 237–249.

Barraclough, N. E., Xiao, D., Baker, C. I., Oram, M. W., & Perrett, D. I. (2005). Integration of visual and auditory information by superior temporal sulcus neurons responsive to the sight of actions. *Journal of Cognitive Neuroscience, 17,* 377–391.

Bartels, A., & Zeki, S. (2004). The neural correlates of maternal and romantic love. *NeuroImage, 21,* 1155–1166.

Bauman, M. D., Lavenex, P., Mason, W. A., Capitanio, J. P., & Amaral, D. G. (2004a). The development of mother-infant interactions after neonatal amygdala lesions in rhesus monkeys. *Journal of Neuroscience, 24,* 711–721.

Bauman, M. D., Lavenex, P., Mason, W. A., Capitanio, J. P., & Amaral, D. G. (2004b). The development of social behavior following neonatal amygdala lesions in rhesus monkeys. *Journal of Cognitive Neuroscience, 16,* 1388–1411.

Baxter, M. G., & Murray, E. A. (2002). The amygdala and reward. *Nature Reviews Neuroscience, 3,* 563–573.

Baxter, M. G., Parker, A., Lindner, C. C., Izquierdo, A. D., & Murray, E. A. (2000). Control of response selection by reinforcer value requires interaction of amygdala and orbital prefrontal cortex. *Journal of Neuroscience, 20,* 4311–4319.

Bechara, A., Tranel, D., Damasio, H., Adolphs, R., Rockland, C., & Damasio, A. R. (1995). Double dissociation of conditioning and declarative knowledge relative to the amygdala and hippocampus in humans. *Science, 269,* 1115–1118.

Bercovitch, F. B., Hauser, M. D., & Jones, J. H. (1995). The endocrine stress response and alarm vocalizations in rhesus macaques. *Animal Behaviour, 49,* 1703–1706.

Bisley, J. W., & Goldberg, M. E. (2003). Neuronal activity in the lateral intraparietal area and spatial attention. *Science, 299,* 81–86.

Blessing, W. W. (1997). *The lower brainstem and bodily homeostasis.* New York: Oxford University Press.

Braesicke, K., Parkinson, J. A., Reekie, Y., Man, M-S., Hopewell, L., Pears, A., et al. (2005). Autonomic arousal in an appetitive context in primates: A behavioural and neural analysis. *European Journal of Neuroscience, 21,* 1733–1740.

Bruce, C., Desimone, R., & Gross, C. G. (1981). Visual properties of neurons in a polysensory area in superior temporal sulcus of the macaque. *Journal of Neurophysiology, 46,* 369–384.

Büchel, C., Morris, J., Dolan, R. J., & Friston, K. J. (1998). Brain systems mediating aversive conditioning: An event-related fMRI study. *Neuron, 20,* 947–957.

Bush, G., Luu, P., & Posner, M. I. (2000). Cognitive and emotional influences in anterior cingulate cortex. *Trends in Cognitive Sciences, 4,* 215–222.

Cohen, Y. E., Hauser, M. D., & Russ, B. E. (2006). Spontaneous processing of abstract categorical information in the ventrolateral prefrontal cortex. *Biology Letters, 2,* 261–265.

Cohen, Y. E., Russ, B. E., Gifford, G. W., 3rd, Kiringoda, R., & MacLean, K. A. (2004). Selectivity for the spatial and nonspatial attributes of auditory stimuli in the ventrolateral prefrontal cortex. *Journal of Neuroscience, 24,* 11307–11316.

Cole, C. E., Gale, J. T., Gale, K., Holmes, A. L., Malkova, L., & Zarbalian, G. (2006). GABA receptor manipulation in the primate deep layers of superior colliculus: Effects on emotional behavior. In: Neuroscience Meeting Planner; Society for Neuroscience, 2006.

Craig, A. D. (2002). How do you feel? Interoception: The sense of the physiological condition of the body. *Nature Reviews Neuroscience, 3(8),* 655–666.

Davis, M., & Whalen, P. J. (2001). The amygdala: Vigilance and emotion. *Molecular Psychiatry, 6,* 13–34.

Davis, M., Antoniadis, E. A., Amaral, D. G., & Winslow, J. T. (2008). Acoustic startle reflex in rhesus monkeys: A review. *Reviews in Neurosciences, 19,* 171–185.

Davis, P. J., Zhang, S. P., Winkworth, A., & Bandler, R. (1996). Neural control of vocalization: Respiratory and emotional influences. *Journal of Voice, 10,* 23–38.

de Gelder, B., Vroomen, J., Pourtois, G., & Weiskrantz, L. (1999). Non-conscious recognition of affect in the absence of striate cortex. *Neuroreport, 10,* 3759–3763.

Dean, P., Redgrave, P., & Westby, G. W. (1989). Event or emergency? Two response systems in the mammalian superior colliculus. *Trends in Neuroscience, 12,* 137–147.

Deaner, R. O., Khera, A. V., & Platt, M. L. (2005). Monkeys pay per view: Adaptive valuation of social images by rhesus macaques. *Current Biology, 15,* 543–548.

Deputte, B. (2000). Primate socialization revisited: Theoretical and practical issues in social ontogeny. *Advances in the Study of Behavior, 29,* 99–157.

Dicks, D., Myers, R. E., & Kling, A. (1969). Uncus and amygdala lesions: Effects on social behavior in the free-ranging rhesus monkey. *Science, 165,* 69–71.

Easterbrook, J. A. (1959). The effect of emotion on cue utilization and the organization of behavior. *Psychological Review, 66,* 183–201.

Emery, N. J., Capitanio, J. P., Mason, W. A., Machado, C. J., Mendoza, S. P., & Amaral, D. G. (2001). The effects of bilateral lesions of the amygdala on dyadic social interactions in rhesus monkeys (Macaca mulatta). *Behavioral Neuroscience, 115,* 515–544.

Engell, A. D., Haxby, J. V., & Todorov, A. (2007). Implicit trustworthiness decisions: Automatic coding of face properties in the human amygdala. *Journal of Cognitive Neuroscience, 19,* 1508–1519.

Everitt, B. J., Cardinal, R. N., Parkinson, J. A., & Robbins, T. W. (2003). Appetitive behavior: Impact of amygdala-dependent mechanisms of emotional learning. *Annals of the New York Academy of Sciences, 985,* 233–250.

Fanardjian, V. V., & Manvelyan, L. R. (1987). Mechanisms regulating the activity of facial nucleus motoneurons-III. Synaptic influences from the cerebral cortex and subcortical structures. *Neuroscience,* 835–843.

Fanselow, M. S., & Gale, G. D. (2003). The amygdala, fear, and memory. In: *The amygdala in brain function: Basic and clinical approaches* (pp. 125–134). Oxford: Oxford University Press.

Felleman, D. J., & Van Essen, D. C. (1991). Distributed hierarchical processing in the primate cerebral cortex. *Cerebral Cortex, 1,* 1–47.

Flack, J. C., & de Waal, F. B. M. (2004). Dominance style, social power, and conflict management: A conceptual framework In: B. Thierry, M. Singh, & W. Kaumanns (Eds.), *Macaque societies: A model for the study of social organization* (pp. 157–181). Cambridge: Cambridge University Press.

Freedman, D. J., & Assad, J. A. (2006). Experience-dependent representation of visual categories in parietal cortex. *Nature, 443,* 85–88.

Freese, J. L., & Amaral, D. G. (2006). Synaptic organization of projections from the amygdala to visual cortical areas TE and V1 in the macaque monkey. *Journal of Comparative Neurology, 496,* 655–667.

Fuster, J. M., & Uyeda, A. A. (1971). Reactivity of limbic neurons of the monkey to appetitive and aversive signals. *Electroencephalography and Clinical Neurophysiology, 30,* 281–293.

Gaffan, D., & Parker, A. (2000). Mediodorsal thalamic function in scene memory in rhesus monkeys. *Brain, 123,* 816–827.

Gallagher, M. (2000). The amygdala and associative learning. In: J. P. Aggleton (Ed.), *The amygdala: A functional analysis* (pp. 311–330). Oxford: Oxford University Press.

Ghazanfar, A. A., Chandrasekaran, C., & Logothetis, N. K. (2008). Interactions between the superior temporal sulcus and auditory cortex mediate dynamic face/voice integration in rhesus monkeys. *Journal of Neuroscience, 28,* 4457–4469.

Gothard, K. M., Battaglia, F. P., Erickson, C. A., Spitler, K. M., & Amaral, D. G. (2007). Neural responses to facial expression and face identity in the monkey amygdala. *Journal of Neurophysiology, 97,* 1671–1683.

Gregg, T., & Siegel, A. (2001). Brain structures and neurotransmitters regulating aggression in cats: Implications for human aggression. *Progress in Neuropsychopharmacology, Biology, and Psychiatry, 25,* 91–140.

Gross, C. G., Rocha-Miranda, C. E., & Bender, D. B. (1972). Visual properties of neurons in inferotemporal cortex of the Macaque. *Journal of Neurophysiology, 35,* 96–111.

Hadj-Bouziane, F., Bell, A. H., Knusten, T. A., Ungerleider, L. G., & Tootell, R. B. H. (2008). Perception of emotional expressions is independent of face selectivity in monkey inferior temporal cortex. *Proceedings of the National Academy of Sciences, 105,* 5591–5596.

Hasselmo, M. E., Rolls, E. T., & Baylis, G. C. (1989). The role of expression and identity in the face-selective responses of neurons in the temporal visual cortex of the monkey. *Behavioural Brain Research, earch, 32,* 203–218.

Heywood, C. A., & Cowey, A. (1992). The role of the 'face-cell' area in the discrimination and recognition of faces by monkeys. *Philosophical Transactions of the Royal Society of London Part B: Biological Sciences, 335,* 31–37; discussion 37–38.

Hoffman, K. L., Gothard, K. M., Schmid, M. C., & Logothetis, N. K. (2007). Facial-expression and gaze-selective responses in the monkey amygdala. *Current Biology, 17,* 766–772.

Holland, P. C., & Gallagher, M. (2004). Amygdala-frontal interactions and reward expectancy. *Current Opinion in Neurobiology, 14,* 148–155.

Horel, J. A. (1993). Retrieval of a face discrimination during suppression of monkey temporal cortex with cold. *Neuropsychologia, 31,* 1067–1077.

Izquierdo, A., & Murray, E. A. (2005). Opposing effects of amygdala and orbital prefrontal cortex lesions on the extinction of instrumental responding in macaque monkeys. *European Journal of Neuroscience, 22,* 2341–2346.

Jones, E. G. (Ed.). (2007). *The thalamus* (2nd ed.). Cambridge: Cambridge University Press.

Joseph, R. (2000). Cingulate gyrus. In: R. Joseph (Ed.), *Neuropsychiatry, neuropsychology, clinical neuroscience* (3rd ed.). New York: Academic Press.

Jürgens, U. (1994). The role of the periaqueductal grey in vocal behaviour. *Behavioural Brain Research, earch, 62,* 107–117.

Kalin, N. H., Shelton, S. E., & Davidson, R. J. (2004). The role of the central nucleus of the amygdala in mediating fear and anxiety in the primate. *Journal of Neuroscience, 24,* 5506–5515.

Kalin, N. H., Shelton, S. E., Davidson, R. J., & Kelley, A. E. (2001). The primate amygdala mediates acute fear but not the behavioral and physiological components of anxious temperament. *Journal of Neuroscience, 21,* 2067–2074.

Keay, K. A., Redgrave, P., & Dean, P. (1988). Cardiovascular and respiratory changes elicited by stimulation of rat superior colliculus. *Brain Research, earch Bulletin, 20,* 13–26.

Keverne, E. B., Martensz, N. D., & Tuite, B. (1989). Beta-endorphin concentrations in cerebrospinal fluid of monkeys are influenced by grooming relationships. *Psychoneuroendocrinology, 14,* 155–161.

Klein, J. T., Deaner, R. O., & Platt, M. L. (2008). Neural correlates of social target value in macaque parietal cortex. *Current Biology, 18,* 419–424.

Kuraoka, K., & Nakamura, K. (2007). Responses of single neurons in monkey amygdala to facial and vocal emotions. *Journal of Neurophysiology, 97,* 1379–1387.

LaBar, K. S., & Cabeza, R. (2006). Cognitive neuroscience of emotional memory. *Nature Reviews Neuroscience, 7,* 54–64.

LaBar, K. S., Gatenby, J. C., Gore, J. C., LeDoux, J. E., & Phelps, E. A. (1998). Human amygdala activation during conditioned fear acquisition and extinction: A mixed-trial fMRI study. *Neuron, 20,* 937–945.

LeDoux, J. (2000). Emotion circuits in the brain. *Annual Review of Neuroscience, 23,*155–184.

Liddell, B. J., Brown, K. J., Kemp, A. H., Barton, M. J., Das, P., Peduto, A., et al. (2005). A direct brainstem-amygdala-cortical 'alarm' system for subliminal signals of fear. *Neuroimage, 24,* 235–243.

Likhtik, E., Popa, D., Apergis-Schoute, J., Fidacaro, G. A., & Pare, D. (2008). Amygdala intercalated neurons are required for expression of fear extinction. *Nature, 454,* 642–645.

Lloyd, S. A., & Dixson, A. F. (1988). Effects of hypothalamic lesions upon the sexual and social behaviour of the male common marmoset (Callithrix jacchus). *Brain Research, 463,* 317–329.

Machado, C. J., & Bachevalier, J. (2007a). The effects of selective amygdala, orbital frontal cortex or hippocampal formation lesions on reward assessment in nonhuman primates. *European Journal of Neuroscience, 25,* 2885–2904.

Machado, C. J., & Bachevalier, J. (2007b). Measuring reward assessment in a semi-naturalistic context: The effects of selective amygdala, orbital frontal or hippocampal lesions. *Neuroscience, 148,* 599–611.

Maclean, P. D. (1949). Psychosomatic disease and the "visceral brain": Recent developments bearing on the Papez theory of emotion. *Psychosomatic Medicine, 11,* 338–353.

Maestripieri, D. (1995). Assessment of danger to themselves and their infants by rhesus macaque (*Macaca mulatta*) mothers. *Journal of Comparative Psychology, 109,* 416–420.

Malkova, L., Gaffan, D., & Murray, E. A. (1997a). Excitotoxic lesions of the amygdala fail to produce impairment in visual learning for auditory secondary reinforcement but interfere with reinforcer devaluation effects in rhesus monkeys. *Journal of Neuroscience, 17,* 6011–6020.

Malkova, L., Mishkin, M., Suomi, S. J., & Bachevalier, J. (1997b). Socioemotional behavior in adult rhesus monkeys after early versus late lesions of the medial temporal lobe.

Annals of the New York Academy of Sciences, 807, 538–540.

Mason, W. A. (1985). Experiential influences on the development of expressive behaviors in rhesus monkeys. In: G. Zivin (Ed.), *The development of expressive behavior: Biology-environment interactions* (pp. 117–152). New York: Academic Press.

McGaugh, J. L. (2000). Memory – a century of consolidation. *Science, 287,* 248–251.

McGaugh, J. L. (2004). The amygdala modulates the consolidation of memories of emotionally arousing experiences. *Annual Review of Neuroscience, 27,* 1–28.

Morris, J. S., Ohman, A., & Dolan, R. J. (1999). A subcortical pathway to the right amygdala mediating "unseen" fear. *Proceedings of the National Academy of Sciences, 96,* 1680–1685.

Morris, J. S., DeGelder, B., Weiskrantz, L., & Dolan, R. J. (2001). Differential extrageniculostriate and amygdala responses to presentation of emotional faces in a cortically blind field. *Brain, 124,* 1241–1252.

Murray, E. A. (2007). The amygdala, reward and emotion. *Trends in Cognitive Sciences, 11,* 489–497.

Murray, E. A., & Izquierdo, A. (2007). Orbitofrontal cortex and amygdala contributions to affect and action in primates. *Annals of the New York Academy of Sciences, 1121,* 273–296.

Nakamura, K., Mikami, A., & Kubota, K. (1992). Activity of single neurons in the monkey amygdala during performance of a visual discrimination task. *Journal of Neurophysiology, 67,* 1447–1463.

Newman, T. K., Syagailo, Y. V., Barr, C. S., Wendland, J. R., Champoux, M., Graessle, M., et al. (2005). Monoamine oxidase A gene promoter variation and rearing experience influences aggressive behavior in rhesus monkeys. *Biological Psychiatry, 57,* 167–172.

O. Scalaidhe, S. P., Wilson, F. A., & Goldman-Rakic, P. S. (1997). Areal segregation of face-processing neurons in prefrontal cortex. *Science, 278,* 1135–1138.

Ono, T., & Nishijo, H. (1992). Neurophysiological basis of the Kluver-Bucy syndrome: Responses of monkey amygdaloid neurons to biologically significant objects. *Amygdala: Neurobiological Aspects of Emotion, Memory, and Mental Dysfunction,* 167–190.

Packard, M. G., & Teather, L. A. (1998). Amygdala modulation of multiple memory systems: Hippocampus and caudate-putamen. *Neurobiology of Learning and Memory, 69,* 163–203.

Panksepp, J. (1998). *Affective neuroscience: The foundations of human and animal emotions.* New York: Oxford University Press.

Papez, J. W. (1937). A proposed mechanism of emotion. *Journal of Neuropsychiatry and Clinical Neuroscience, 7,* 103–112.

Pare, D., & Smith, Y. (1993). Distribution of GABA immunoreactivity in the amygdaloid complex of the cat. *Neuroscience, 57,* 1061–1076.

Partan, S. R. (2002). Single and multichannel signal composition: Facial expressions and vocalizations of rhesus macaques (Macaca mulatta). *Behaviour, 139,* 993–1027.

Paton, J. J., Belova, M. A., Morrison, S. E., & Salzman, C. D. (2006). The primate amygdala represents the positive and negative value of visual stimuli during learning. *Nature, 439,* 865–870.

Perrett, D. I., Rolls, E. T., & Caan, W. (1982). Visual neurones responsive to faces in the monkey temporal cortex. *Experimental Brain Research, 47,* 329–342.

Perrett, D. I., Smith, P. A., Potter, D. D., Mistlin, A. J., Head, A. S., Milner, A. D., et al. (1985). Visual cells in the temporal cortex sensitive to face view and gaze direction. *Proceedings of the Royal Society of London – Series B: Biological Sciences, 223,* 293–317.

Pessoa, L. (2008). On the relationship between emotion and cognition. *Nature Review in Neuroscience, 9,* 148–158.

Pessoa, L., Japee, S., & Ungerleider, L. G. (2005). Visual awareness and the detection of fearful faces. *Emotion, 5,* 243–247.

Phelps, E. A. (2006). Emotion and cognition: Insights from studies of the human amygdala. *Annual Review of Psychology, 57,* 27–53.

Phelps, E. A., Delgado, M. R., Nearing, K. I., & LeDoux, J. E. (2004). Extinction learning in humans. *Neuron, 43,* 897–905.

Phillips, M. L., Young, A. W., Senior, C., Brammer, M., Andrew, C., Calder, A. J., et al. (1997). A specific neural substrate for perceiving facial expressions of disgust. *Nature, 389,* 495–498.

Pruessner, J. C., Baldwin, M. W., Dedovic, K., Renwick, R., Mahani, N. K., Lord, C., et al. (2005). Self-esteem, locus of control, hippocampal volume, and cortisol regulation in young and old adulthood. *Neuroimage, 28,* 815–826.

Rapcsak, S. Z. (2003). Face memory and its disorders. *Current Neurology and Neuroscience Reports, 3,* 494–501.

Redgrave, P., McHaffie, J. G., & Stein, B. E. (1996). Nociceptive neurones in rat superior colliculus. I. Antidromic activation from the contralateral predorsal bundle. *Experimental Brain Research, 109,* 185–196.

Rempel-Clower, N. L. (2007). Role of orbitofrontal cortex connections in emotion. *Annals of the New York Academy of Sciences, 1121,* 72–86.

Robinson, B. W. (1967). Vocalization evoked from forebrain in Macaca mulatta. *Physiology & Behavior, 2,* 345–346, IN341–IN344, 347–354.

Rolls, E. T. (2000). The orbitofrontal cortex and reward. *Cerebral Cortex, 10,* 284–294.

Rolls, E. T. (2007). The representation of information about faces in the temporal and frontal lobes. *Neuropsychologia, 45,* 124–143.

Romanski, L. M., & LeDoux, J. E. (1992). Equipotentiality of thalamo-amygdala and thalamo-cortico-amygdala circuits in auditory fear conditioning. *Journal of Neuroscience, 12,* 4501–4509.

Romanski, L. M., Averbeck, B. B., & Diltz, M. (2005). Neural representation of vocalizations in the primate ventrolateral prefrontal cortex. *Journal of Neurophysiology, 93,* 734–747.

Romanski, L. M., Giguere, M., Bates, J. F., & Goldman-Rakic, P. S. (1997). Topographic organization of medial pulvinar connections with the prefrontal cortex in the rhesus monkey. *Journal of Comparative Neurology, 379,* 313–332.

Rosvold, H. E., Mirsky, A. F., & Pribram, K. H. (1954). Influence of amygdalectomy on social behavior in monkeys. *Journal of Comparative and Physiological Psychology, 47,* 173–178.

Salzman, C. D., Paton, J. J., Belova, M. A., & Morrison, S. E. (2007). Flexible neural representations of value in the primate brain. *Annals of the New York Academy of Sciences, 1121,* 336–354.

Sanghera, M. K., Rolls, E. T., & Roper-Hall, A. (1979). Visual responses of neurons in the dorsolateral amygdala of the alert monkey. *Experimental Neurology, 63,* 610–626.

Sapolsky, R. M. (1996). Why stress is bad for your brain.(overproduction of glucocorticoids damage the hippocampus). *Science, 273,* 749.

Seltzer, B., & Pandya, D. N. (1978). Afferent cortical connections and architectonics of the superior temporal sulcus and surrounding cortex in the rhesus monkey. *Brain Research, 149,* 1–24.

Seyfarth, R. M., & Cheney, D. L. (1984). Grooming, alliances and reciprocal altruism in vervet monkeys. *Nature, 308,* 541–543.

Shepherd, S. V., Deaner, R. O., & Platt, M. L. (2006). Social status gates social attention in monkeys. *Current Biology, 16,* R119–R120.

Singer, T., Seymour, B., O'Doherty, .J, Kaube, H., Dolan, R. J., & Frith, C. D. (2004). Empathy for pain involves the affective but not sensory components of pain. *Science, 303,* 1157–1162.

Snyder, L. H., Batista, A. P., & Andersen, R. A. (1997). Coding of intention in the posterior parietal cortex. *Nature, 386,* 167–170.

Sparks, D. L., & Hartwich-Young, R. (1989). The deep layers of the superior colliculus. *Review of Oculomotor Research, 3,* 213–255.

Sprengelmeyer, R., Rausch, M., Eysel, U. T., & Przuntek, H. (1998). Neural structures associated with recognition of facial expressions of basic emotions. *Proceedings of the Royal Society of London – Series B: Biological Sciences, 265,* 1927–1931.

Stefanacci, L., & Amaral, D. G. (2002). Some observations on cortical inputs to the macaque monkey amygdala: An anterograde tracing study. *Journal of Comparative Neurology, 451,* 301–323.

Stoerig, P., & Cowey, A. (1997). Blindsight in man and monkey. *Brain, 120,* 535–559.

Sugihara, T., Diltz, M. D., Averbeck, B. B., & Romanski, L. M. (2006). Integration of auditory and visual communication information in the primate ventrolateral prefrontal cortex. *Journal of Neuroscience, 26,* 11138–11147.

Suomi, S. J. (2006). Risk, resilience, and gene X environment interactions in rhesus monkeys. *Annals of the New York Academy of Sciences, 1094,* 52–62.

Tanaka, K. (1992). Infereotemporal cortex and higher visual functions. *Current Opinion in Neurobiology, 2,* 502–505.

Thierry, B. (2004). Social epigenesis In: B. Thierry, M. Singh, & W. Kaumanns (Eds.), *Macaque societies: A model for the study of social organization* (pp. 267–289). Cambridge: Cambridge University Press.

Thomas, E., Van Hulle, M. M., & Vogels, R. (2001). Encoding of categories by noncategory-specific neurons in the inferior temporal cortex. *Journal of Cognitive Neuroscience, 13,* 190–200.

Troisi, A., & Carosi, M. (1998). Female orgasm rate increases with male dominance in Japanese macaques. *Animal Behaviour, 56,* 1261–1266.

Tsao, D. Y., Schweers, N., Moeller, S., & Freiwald, W. A. (2008). Patches of face-selective cortex in the macaque frontal lobe. *Nature Neuroscience, 11,* 877–879.

Ungerleider, L. G., Gaffan, D., & Pelak, V. S. (1989). Projections from inferior temporal cortex to prefrontal cortex via the uncinate fascicle in rhesus monkeys. *Experimental Brain Research, 76,* 473–484.

Vogt, B. A., Pandya, D. N., & Rosene, D. L. (1987). Cingulate cortex of the rhesus monkey: I. Cytoarchitecture and thalamic afferents. *Journal of Comparative Neurology, 262,* 256–270.

von Euler, C., & Trippenbach, T. (1976). Excitability changes of the inspiratory "off-switch" mechanism tested by electrical stimulation in nucleus parabrachialis in the cat. *Acta Physiologica Scandinavia, 97,* 175–188.

Vuilleumier, P., Armony, J. L., Driver, J., & Dolan, R. J. (2003). Distinct spatial frequency sensitivities for processing faces and emotional expressions. *Nature Neuroscience, 6,* 624–631.

Waddell, J., Heldt, S., & Falls, W. A. (2003). Posttraining lesion of the superior colliculus interferes with feature-negative discrimination of fear-potentiated startle. *Behavioral Brain Research, 142,* 115–124.

Wallace, M. T., McHaffie, J. G., & Stein, B. E. (1997). Visual response properties and visuotopic representation in the newborn monkey superior colliculus. *Journal of Neurophysiology, 78,* 2732–2741.

Ward, R., Danziger, S., & Bamford, S. (2005). Response to visual threat following damage to the pulvinar. *Current Biology, 15,* 571–573.

Watanabe, Y., Gould, E., & McEwen, B. S. (1992). Stress induces atrophy of apical dendrites of hippocampal CA3 pyramidal neurons. *Brain Research, 588,* 341–345.

Wellman, L. L., Gale, K., & Malkova, L. (2005). GABAA-mediated inhibition of basolateral amygdala blocks reward devaluation in macaques. *Journal of Neuroscience, 25,* 4577–4586.

Whalen, P. J. (2007). The uncertainty of it all. *Trends in Cognitive Sciences, 11,* 499–500.

Wilson, F. A., & Rolls, E. T. (1993). The effects of stimulus novelty and familiarity on neuronal activity in the amygdala of monkeys performing recognition memory tasks. *Experimental Brain Research, 93,* 367–382.

Wilson, F. A., Scalaidhe, S. P., & Goldman-Rakic, P. S. (1993). Dissociation of object and spatial processing domains in primate prefrontal cortex. *Science, 260,* 1955–1958.

Winston, J. S., Strange, B. A., O'Doherty, J., & Dolan, R. J. (2002). Automatic and intentional brain responses during evaluation of trustworthiness of faces. *Nature Neuroscience, 5,* 277–283.

Woolley, C. S., Gould, E., & McEwen, B. S. (1990). Exposure to excess glucocorticoids alters dendritic morphology of adult hippocampal pyramidal neurons. *Brain Research, 531,* 225–231.

Yamane, S., Kaji, S., & Kawano, K. (1988). What facial features activate face neurons in the inferotemporal cortex of the monkey? *Experimental Brain Research, 73,* 209–214.

Young, M. P., Scannell, J. W., Burns, G. A., & Blakemore, C. (1994). Analysis of connectivity: Neural systems in the cerebral cortex. *Review of Neuroscience, 5,* 227–250.

Zald, D. H. (2003). The human amygdala and the emotional evaluation of sensory stimuli. *Brain Research Reviews, 41,* 88–123.

Zhao, Z., & Davis, M. (2004). Fear-potentiated startle in rats is mediated by neurons in the deep layers of the superior colliculus/deep mesencephalic nucleus of the rostral midbrain through the glutamate non-NMDA receptors. *Journal of Neuroscience, 24,* 10326–10334.

CHAPTER 17

Neurophysiological Correlates of Reward Learning

Wolfram Schultz

LEARNING AS CHANGE IN PREDICTIONS

Predictions

Learning is manifested as reproducible change in behavior in identical situations and is a major factor in the behavioral adaptation to changing environments. Many forms of behavior are based on anticipating the future by predicting the outcomes, such as rewards and punishers. Changed environmental situations lead to changed predictions, which then induce changes in behavior. Thus, learning can be viewed as behavioral changes that follow changes in predictions. These considerations are particularly pertinent for conditioning or associative learning for motivational outcomes, such as rewards and punishers.

Rewards are positive outcomes that induce changes in observable behavior and serve as positive reinforcers by increasing the frequency of the behavior that results in reward. In Pavlovian, or classical, conditioning, the outcome follows the conditioned stimulus irrespective of any behavioral reaction, and repeated pairing of stimuli with outcomes leads to a representation of the outcome that is evoked by the stimulus and elicits the behavioral reaction. By contrast, instrumental, or operant, conditioning requires the subject to execute a behavioral response; without such response there will be no reward. Instrumental conditioning increases the frequency of those behaviors that are followed by reward by reinforcing stimulus-response links. Instrumental conditioning allows subjects to influence their environment and determine their rate of reward.

The behavioral reactions studied classically by Pavlov are vegetative responses governed by smooth muscle contraction and gland discharge, whereas more recent Pavlovian tasks also involve reactions of striated muscles. In the latter case, the final reward usually needs to be collected by an instrumental contraction of striated muscle, but the behavioral reaction to the conditioned stimulus itself is not required for the reward to occur (e.g., anticipatory licking). Thus, individual stimuli in instrumental tasks come to predict rewards, and the acquisition of this reward prediction is considered to be Pavlovian conditioned. In this sense Pavlovian conditioning can be broadly defined as the acquisition of outcome predictions, irrespective of the specific behavioral reaction being evoked. These distinctions are helpful when searching for neural mechanisms of reward prediction.

Conditioning is governed by contiguity and contingency. Contiguity refers to the requirement of near simultaneity (Fig. 17.1a). Specifically, a reward needs to follow a conditioned stimulus or response by an optimal interval of a few seconds, whereas rewards occurring before a stimulus or response do not contribute to learning (no or poor backward conditioning). Contingency describes that a reward needs to occur more frequently in the presence of a stimulus as compared to its absence in order to induce "excitatory" conditioning of the stimulus (Fig. 17.1b); the occurrence of the conditioned stimulus predicts a higher probability of reward compared to no stimulus, and the stimulus becomes a reward predictor. By contrast, if a reward occurs less

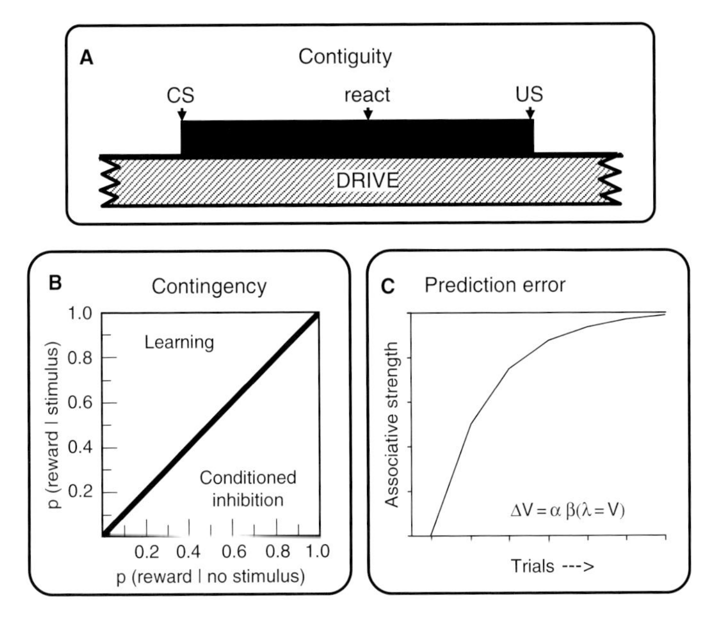

Figure 17.1 Basic assumptions of animal learning theory defining the reinforcing functions of rewards. **(A)** Contiguity refers to the temporal proximity of a conditioned stimulus (or action) and the reward. **(B)** Contingency refers to the conditional probability of reward occurring in the presence of a conditioned stimulus as opposed to its absence (modified from Dickinson, 1980). **(C)** Prediction error denotes the discrepancy between an actually received reward and its prediction. Learning (ΔV, associative strength) is proportional to the prediction error ($\gamma - V$) and reaches its asymptote when the prediction error approaches zero after several learning trials. Used with permission.

frequently in the absence of a reward, compared to its presence, the occurrence of the stimulus predicts a lower incidence of reward, and the stimulus becomes a conditioned inhibitor, even though the contiguity requirement is fulfilled.

Prediction Errors

Whereas sensory systems in principle use information that is directly received from the outside world through specific organs, adaptive behavior based on predictions requires continuous comparisons with the actually received outcomes so that predictions can be kept valid through updating. Thus, to make accurate choices between predicted outcomes, the brain would set up predictions, compare the actual outcomes with the predictions, and update the predictions according to this discrepancy. Learning, viewed as behavioral adaptation to environmental change, would be the simple consequence of this updating process. Once the discrepancy has declined to low levels, there would be little further updating of predictions, little behavioral changes, and thus little learning. Thus, the progress of learning depends on reducing the discrepancy between the actual and predicted outcomes.

Formalizations of the role of discrepancies between outcomes and predictions have recognized the similarity to general error-driven learning rules, and accordingly named the discrepancy an error in reward prediction (Rescorla & Wagner, 1972; Sutton & Barto, 1981). Outcomes that are different than predicted influence behavior in a direction that reduces the prediction error between the outcome and its prediction until outcome and prediction match. This concept applies to classical (Pavlovian) conditioning in which learning consists of changing the prediction until it matches the outcome, and behavioral changes are consequential to the change in prediction. In instrumental (operant) conditioning, the behavioral action is performed with some prediction of outcome. Learning consists of changing the behavior until the outcome matches the prediction. Thus, setting up and modifying predictions is essential for adapting behavior to the requirements of the organism according to the resources available in the environment.

The crucial role of prediction error is derived from Kamin's blocking effect (1969), which postulates that a reward that is fully predicted does not contribute to learning, even when it follows the conditions of contiguity and contingency. This is conceptualized in the associative learning rules (Rescorla & Wagner, 1972), according to which learning advances only to the extent to which a reinforcer is unpredicted, and learning slows progressively as the reinforcer becomes more predicted. Thus, learning can be graphed as an asymptotic learning curve (Fig. 17.1c) and follows the equation:

$$\Delta V = \alpha \beta(\lambda - V)$$

with V as associative strength of conditioned stimulus (prediction), α and β as learning constants, and λ as maximal associative strength possibly sustained by an outcome (e.g., reward).

The omission of a predicted reinforcer reduces the strength of the conditioned stimulus and produces extinction of behavior. So-called attentional learning rules in addition relate the capacity to learn (associability) in certain situations to the degree of attention evoked by the conditioned stimulus or reward (Mackintosh, 1975, Pearce & Hall, 1980).

STRIATUM

Background

The mammalian striatum (caudate nucleus, putamen, ventral striatum including nucleus accumbens) is an important brain structure involved in controlling behavioral output. It is closely associated, through multiple, convergent, and partly closed loops, with all areas of the cerebral cortex. Despite 40 years of intense research and numerous suggestions, still no unifying concept of a single striatal function has emerged (and may never emerge). The lateral putamen is activated during movements, and Parkinsonian patients with dopamine deficiencies in the putamen have severe deficits in motor activity. The rostral caudate and putamen have prominent connections with most areas of the association cortex ("associative striatum"). Many parts of the striatum, including the ventral striatum (nucleus accumbens), are involved in reward processing. Synaptic connections in the striatum show use-dependent plasticity (Reynolds et al., 2001), suggesting involvement in various forms of learning and memory, such as procedural learning, sensorimotor and skill learning (Hikosaka et al., 1989; Wise, 1996), habit learning (Malamut et al., 1984; Packard & Knowlton, 2002), goal-directed instrumental and reward association learning (Gaffan & Eacott, 1995), and emotional learning (Linden et al., 1990). Without attempting to resolve all mysteries of striatal functions, this section will describe the neuronal activity of striatal neurons while animals learn reward predictions in controlled behavioral tasks.

Behavioral Tasks Testing Striatal and Frontal Cortical Reward Functions

Experimental psychologists have developed specific behavioral tasks to test particular behavioral processes in isolation and with a minimum of confounding variables, and neuroscientists use these tasks to study the involvement of specific brain structures in individual behavioral processes. A good behavioral paradigm for assessing the function of the frontal cortex, and of the closely associated basal ganglia (see above), is the delayed response task, which tests spatial processing, working memory, movement preparation and execution, prediction of reward, and goal-directed behavior (Divac et al., 1967; Jacobsen & Nissen, 1937). In order to relate neurophysiological findings to previous lesion data using these tests, researchers use different versions of spatial and conditional delayed response tasks and investigate the motivational properties of single neurons in macaque monkeys. In such tasks, an initial visual stimulus indicates the spatial position of the target of an arm movement for a brief period. Following a delay of several seconds, during which the spatial information is absent, a neutral stimulus appears that tells the animal to make its response to the remembered target in order to receive a liquid or food reward. In adaptations of delayed response tasks for investigating reward functions, researchers use initial stimuli that, in addition to spatial information, predict which specific reward the animal would receive for responding correctly. Introduction of a delay of 2 to 3 seconds between the correct response and the delivery of reward provides the animal with a specific reward expectation period following the behavioral reaction. Thus, the delayed response task is an instrumental (operant) task in which the reward is contingent upon the correct response of the animal. The association of the initial stimulus with the reward is of a Pavlovian nature, such that the stimulus becomes a predictor for a specific reward.

The presentation of a learned stimulus leads to the recall of its behavioral and motivational significance from long-term memory. The stimulus is stored in working memory (remember what just happened) and induces short-term preparations of behavioral actions and expectations of future rewards. Expectation and preparation have common anticipatory components, as they precede rather than follow predictable events, but they refer to different predictable events (stimuli, action, reward) and occur during different epochs of the task. The prospective (forward in time) expectation and preparation processes differ schematically from retrospective (backward in time) working memory processes that concern the previous event and do not require predictive coding, although they are somewhat entangled and could be considered common constituents of working memory (Honig & Thompson, 1982).

Learning Tasks

Neurophysiological studies investigate the acquisition of neural responses to novel stimuli while animals learn a new task from scratch and acquire the associated reward predictions. The behavioral tasks usually employ an initial stimulus that predicts the reward and informs the animal about the required behavioral reaction.

Our own studies use a learning set paradigm in which animals first learn a behavioral task and subsequently learn only a single new task component at a time that changes repeatedly. We use a delayed go/no-go response task with three trial types, namely, rewarded movement trials, rewarded nonmovement trials, and unrewarded, sound-reinforced movement trials (Fig. 17.2A). Different initial pictures presented on a computer monitor predict whether the animal would receive a liquid reward or the auditory reinforcer, and which behavioral reaction (movement or no movement) is required to obtain that outcome. Animals first learn the familiar stimuli. Then the three stimuli for the three trial types are replaced by three novel pictures, whereas all other task events and the global task structure remain unchanged. Animals learn the reward predictions and behavioral significances of the stimuli by trial and error. When asymptotic performance is reached with a set of three learning stimuli, those stimuli are discarded and three novel stimuli are introduced. Thus, learning consists of associating each new picture (1) with liquid reward or conditioned auditory reinforcement and (2) with executing or withholding the movement. The repeated learning of tens and hundreds of novel pictures results in a learning set in which learning occurs largely within the first few trials and approaches asymptotic performance within 5 to 10 trials of each trial type (Fig. 17.2C), similar to previous behavioral studies (Gaffan et al., 1988; Harlow, 1949). The procedure resembles everyday situations in which only a minimum of new information is learned and most other task components remain valid or are only slightly altered. We use the learning set paradigm to investigate the electrophysiological activity of single neurons during complete learning episodes and to compare their activity with performance in trials using familiar, well-learned pictures. We are particularly interested in the responses to reward-predicting stimuli and in the more sustained

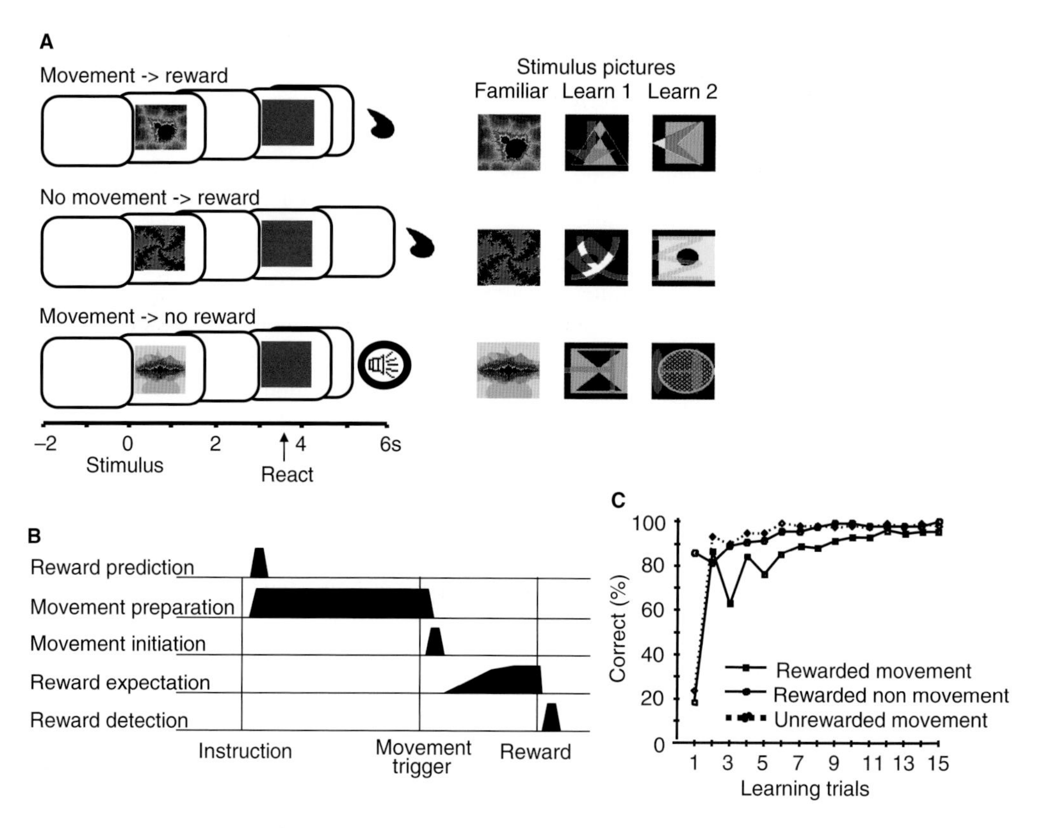

Figure 17.2 Learning task and basic neuronal activity. (**A**) Delayed response task. An initial picture instructs the animal to perform a rewarded movement, unrewarded movement, or rewarded nonmovement reaction following a subsequent uniform trigger stimulus. Small drops of juice, or a sound, serve as reinforcer if the trial is correctly performed. For learning (right), three novel pictures indicate the three trial types, and the animal finds out by trial and error which of the novel pictures corresponds to the known trial types. (**B**) Schematics of main forms of behavioral relationships of striatal neurons. (**C**) Learning curves indicating rapid learning of the three trial types.

activations occurring during the expectation of reward.

Basic Neural Activity

During the performance of delay tasks, neurons in the striatum and frontal cortex are activated differentially and specifically during individual task components (Fig. 17.2B) involving the retrospective encoding, maintenance, and recall of working memory and the prospective preparation of movement (Alexander & Crutcher, 1990; Funahashi et al., 1989, 1993).

Neuronal activations related to the prediction and expectation of reward arise during specific epochs of this task in response to reward-predicting stimuli and in advance of the expected delivery of reward. The activations differentiate between reward and no reward, between different kinds of liquid and food reward, and between different magnitudes of reward. They occur in all trial types in which reward is expected, irrespective of the type of behavioral action. Such reward expectation–related activations are found not only in the striatum (Apicella et al., 1992; Bowman et al., 1996; Cromwell & Schultz, 2003; Hikosaka et al., 1989; Hollerman et al., 1998; Schultz et al., 1992) but also in the amygdala, orbitofrontal cortex, dorsolateral prefrontal cortex, anterior cingulate, and supplementary eye field (Amador et al., 2000; Pratt & Mizumori, 2001; Schoenbaum et al., 1998; Shidara & Richmond, 2002; Tremblay & Schultz, 1999, 2000a).

In some striatal and dorsolateral prefrontal neurons, the differential reward expectation–related activity discriminates in addition between different behavioral responses, such as eye and limb movements toward different spatial targets, and movement versus nonmovement reactions (Cromwell & Schultz, 2003; Hassani et al., 2001; Hollerman & Schultz, 1998; Kawagoe et al., 1998; Kobayashi et al., 2002; Matsumoto et al., 2003; Watanabe, 1996; Watanabe et al., 2002). The activations do not simply represent outcome expectation, as they differentiate between different behavioral reactions despite the same outcome, and they do not simply reflect different behavioral reactions, as they differentiate between the expected outcomes. Thus, the neurons show differential, behavior-related activations that depend on the outcome of the trial, namely, reward or no reward and different kinds and magnitudes of reward. These activities provide evidence for a neural representation of the outcome during the preparation and execution of the behavioral reactions performed in order to obtain the rewarding goal. They may be parts of neural mechanisms underlying goal-directed behavior.

Acquisition of Responses to Reward-Predicting Stimuli

Striatal neurons in monkeys and rats acquire discriminating responses to reward- or punisher-predicting stimuli during associative Pavlovian or operant learning. The depressant, discriminant responses of tonically active striatal interneurons (TANs) to reward-predicting auditory stimuli are rare before learning and become substantial following 15 minutes of repeated pairing with liquid (Aosaki et al., 1994b). Behavioral extinction of conditioned stimuli by withholding reward results in reduction of neural responses within 10 minutes. Acquisition and expression of the neuronal response to reward-predicting stimuli are abolished by interference with dopamine neurotransmission (Aosaki et al., 1994a). Responses are reinstated by systemic administration of the dopamine agonist apomorphine. Associating an aversive stimulus, such as a loud or air puff–related noise, with a liquid reward changes the response of TANs to the stimulus (Ravel et al., 2003). During T-maze learning, neurons in rat striatum show progressively earlier, predictive activity in the maze as learning advances (Jog et al., 1999). Behavioral reversal of stimulus-outcome associations results in reversed neural responses in neurons discriminating between reward- and punisher-predicting olfactory stimuli (Setlow et al., 2003). During the performance of learning set tasks, striatal neurons develop activating responses to reward-predicting stimuli during individual learning episodes or show transient responses only during the learning of novel reward-predicting stimuli (Tremblay et al., 1998). Transient responses disappear after less than 10 or within a few tens of trials (Fig. 17.3). Responses occur occasionally also in additional trial types during initial learning, suggesting a transient loss of selectivity during learning. Compatible with the motor and sensory functions of the striatum, neurons in this structure do not only respond to motivationally significant stimuli but can also acquire responses to stimuli indicating the direction of behavioral reactions, as seen in a conditional oculomotor task (Pasupathy & Miller, 2005). Neurons in the posterior striatum respond stronger to novel visual stimuli as compared to familiar stimuli, although these neurons generally fail to respond to rewards and reward-predicting stimuli (Brown et al., 1995). Taken together, striatal neurons readily acquire responses that are sensitive to the outcome and movement contingencies of external stimuli.

Adaptation of Reward Expectation

The presentation of a reward-predicting stimulus evokes an expectation of the forthcoming reward. Novel stimuli of unknown reward prediction will not evoke specific reward expectations, although

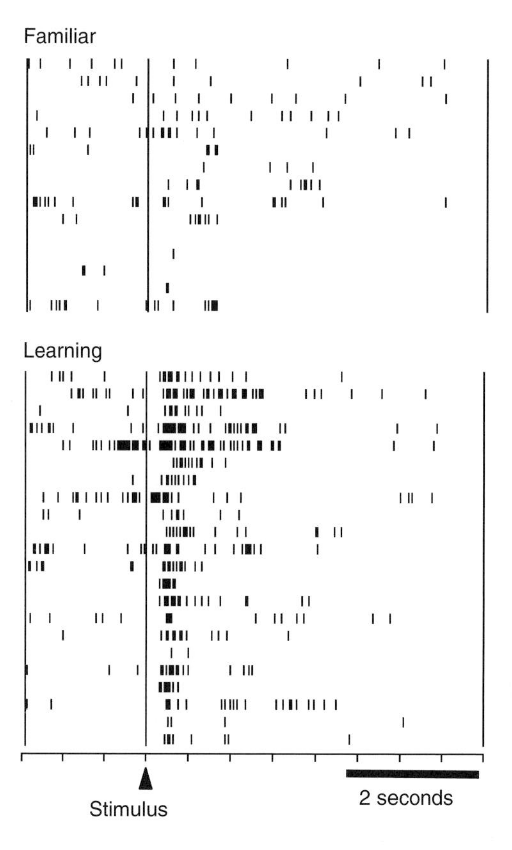

Figure 17.3 Transient response of caudate neuron to reward-predicting stimulus during learning. The response disappears gradually after a few initial learning trials. Dots in rasters denote the time of neuronal impulses. Each line of dots represents one trial. The sequence of trials is plotted chronologically from top to bottom, separately for familiar and learning trials, learning rasters beginning with the first presentations of the new stimulus. Adapted from Tremblay, L., Hollerman, J. R., & Schultz, W. (1998). Modifications of reward expectation-related neuronal activity during learning in primate striatum. *Journal of Neurophysiology, 80,* 964–977, with permission by The American Physiological Society.

the previous experience of the animal would indicate that some of these stimuli should be associated with a reward. Consequently, a default reward expectation might precede any specific expectation arising from learning the specific, reward-predicting stimuli.

Using our learning set task we investigate changes in existing reward expectations during the learning of novel stimuli, using rewarded movements, rewarded nonmovement reactions, and unrewarded movements (Tremblay et al., 1998). Rewarded movements in familiar trials show consistently shorter reaction times of movements from a resting key toward a lever and longer return times from the lever back to the resting key, as compared to unrewarded movements (Fig. 17.4A,B). This behavioral marker allows us to infer the animal's reward expectation and its changes during learning. With new stimuli, return times are initially typical for rewarded movements in both rewarded and unrewarded movement trials. Return times differentiate after the first few trials between rewarded and unrewarded movements (Fig. 17.4C). Erroneous movements in nonmovement trials are usually performed with return times typical for rewarded movements. Similar changes occur with another

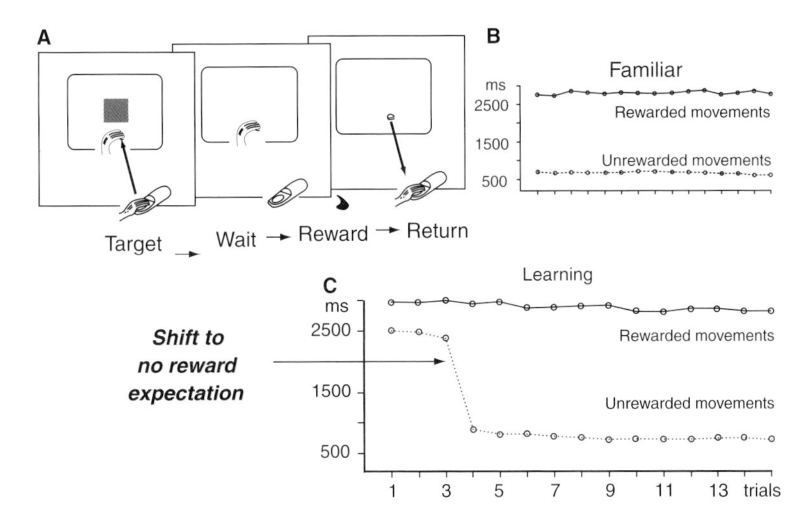

Figure 17.1 Adaptation of reward expectation during learning, as evidenced by behavioral reactions. (**A**) Movement phase of the behavioral task. Following release of the resting key, the animal's hand touches the target key (left), upon which the trigger stimulus disappears (center) and reward or an auditory reinforcer is delivered, and the animal subsequently returns its hand to the resting key (right). (**B**) In familiar trials, the animal's return to the resting key differs depending on delivery of reward (top) or auditory reinforcer (bottom). Ordinate indicates median return times in milliseconds (from release of resting key until return to resting key). (**C**) During learning, the animal's behavioral reaction after touching the target key initially resembles that in familiar rewarded trials (trials 1–3) and differentiates subsequently according to reward versus auditory reinforcement. The shift after trial 3 is the result of averaging and varies in individual learning problems. The trial contains also an initial reinforcer-predicting stimulus and a subsequent movement trigger stimulus (not shown). Data from Tremblay, L., Hollerman, J. R., & Schultz, W. (1998). Modifications of reward expectation-related neuronal activity during learning in primate striatum. *Journal of Neurophysiology, 80,* 964–977, with permission by The American Physiological Society.

behavioral marker, the electromyographic activity of forearm muscles. During initial learning, forearm muscle activity is similar in all movement trials between key release and return to resting key and resembles the muscle activity seen in familiar rewarded movements. Later during each learning episode, muscle activity approaches a pattern in unrewarded movement trials that is typical for familiar unrewarded movements. Thus, both movement parameters and muscle activity in initial learning trials are typical for rewarded movements and subsequently differentiate between rewarded and unrewarded movements. We conclude from these observations that the animals have a default expectation of reward in initial learning trials, probably acquired from previous experience in the same task, which differentiates subsequently according to the type of reinforcer predicted by each picture. The animal's preexisting reward expectation appears to adapt to the currently expected outcome according to the experience of the animal with the new stimuli.

During learning, all forms of reward expectation–related activity of striatal neurons change in the same manner (Tremblay et al., 1998). During initial learning trials, in which the novel stimuli are not yet firmly associated with specific reward expectations and animals rely on default reward expectations, striatal reward expectation–related neuronal activations occur in all trial types. Gradually, the neuronal activations become restricted to rewarded as opposed to unrewarded movement trials, concurrent with, or slightly earlier than, the animal's shift in behavioral indices of expectation. These changes occur in neurons showing pure reward expectation-related activity immediately preceding the reward and in

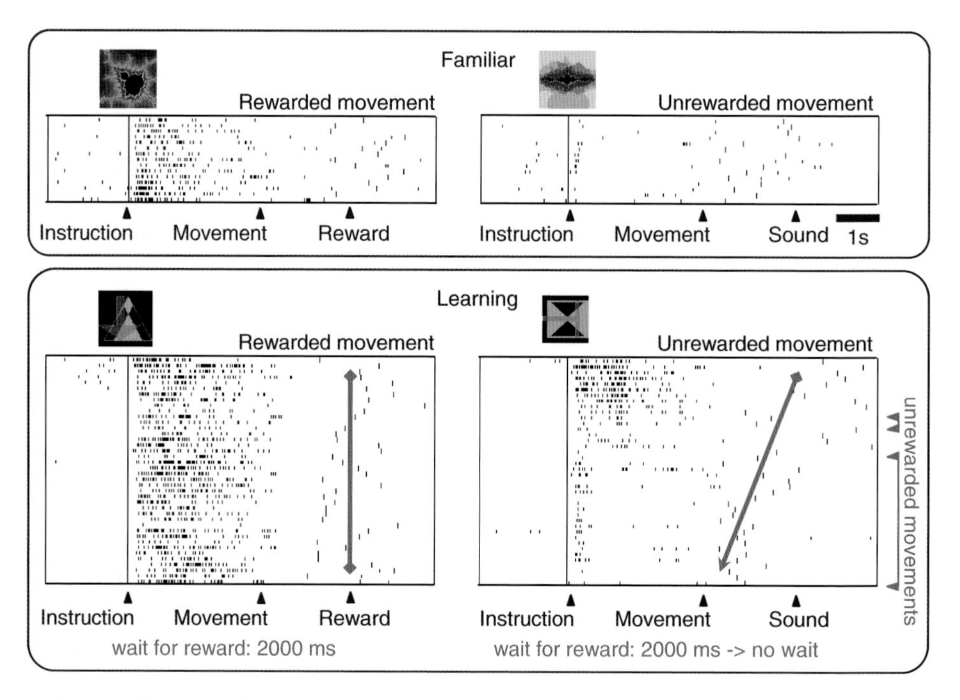

Figure 17.5 Adaptation of neural reward expectation during learning. During performance with familiar stimuli (top), this caudate neuron shows a sustained response to the initial reward-predicting stimulus in rewarded movement trials and a transient response in nonmovement trials (not shown). During learning (bottom), the sustained response occurs initially also in unrewarded movement trials, which are performed with parameters of rewarded movements. The sustained response disappears when movement parameters become typical for unrewarded movements (arrows to the right). From Tremblay, L., Hollerman, J. R., & Schultz, W. (1998). Modifications of reward expectation-related neuronal activity during learning in primate striatum. *Journal of Neurophysiology, 80,* 964–977, with permission by The American Physiological Society.

neurons in which the reward expectation–related activity occurs during the preparation period for the behavioral reaction (Fig. 17.5). Thus, striatal reward expectation activity is nondifferential during initial learning trials with undifferentiated reward expectations and subsequently adapts to the newly learned, valid expectations.

ORBITOFRONTAL CORTEX

In delayed response tasks adapted for reward studies, orbitofrontal neurons display several forms of task-related activity, including responses to the reward-predicting stimuli, activations during the expectation of reward, and responses to the reward itself (Tremblay & Schultz 2000a). However, orbitofrontal neurons show only rarely sustained activations during the delay between the initial stimuli and the movement that are typical for dorsolateral prefrontal cortex (Funahashi et al., 1989, 1993; Fuster, 1973). Although even the most elaborate versions of delayed response tasks are unlikely to test the whole spectrum of the wide-ranging orbitofrontal functions, orbitofrontal neurons consistently show pronounced relationships to reward (Padoa-Schioppa & Assad, 2006; Rolls et al., 1989, 1996; Schoenbaum et al., 1998; Tremblay & Schultz, 1999).

Several studies examine orbitofrontal neurons during learning. During the performance of go/no-go tasks, animals emit a behavioral response to one conditioned stimulus and refrain from moving to a different stimulus in order to receive a predicted reward or avoid a punisher. Rolls and colleagues (1996) used differential odor and visual stimuli for licking movements rewarded with sweet liquid (go) and for withholding licking to avoid aversive saline (no-go). The study tests the reversal of existing conditioned stimulus–outcome associations in monkeys and found that many orbitofrontal neurons reverse their differential, reinforcer-discriminating responses at about the same time as the animal reverses its go/no-go behavioral responses. Schoenbaum and colleagues (1998, 1999) used odors as outcome-predicting stimuli in rats for the differential acquisition of associations with rewarding sucrose and aversive quinine solutions in go/no-go nose-poking responses. The studies report that orbitofrontal responses to the odors come to discriminate between the two outcomes during learning, and most of them become differential only at the same time as the animal reaches the behavioral learning criterion. Furthermore, a separate group of orbitofrontal neurons develops differential activations during the expectation of the two types of reinforcers, and these changes occur before the animal reaches the learning criterion.

Our own study employs the same delayed go/no-go learning set task for orbitofrontal cortex as for the striatum (Tremblay & Schultz, 2000b). Only the three initial stimuli change between learning episodes, and animals learn the new reward predictions and behavioral significances within a few trials by trial and error. Orbitofrontal neurons show very similar learning-related changes as striatal neurons. Notably, the neurons acquire responses to the reward-predicting stimuli in each learning episode, and their existing reward expectation–related activity adapts to the novel reward contingencies in parallel with behavioral adaptations of reward expectation. However, compared to the striatum, orbitofrontal neurons show more transient changes during learning that are absent during performance with familiar stimuli. These changes consist of increases or decreases of responses to the novel stimuli during the learning period (Fig. 17.6), and often during the subsequent consolidation period of several tens of trials. By contrast, other orbitofrontal neurons show reduced responses during learning. If the increases and decreases are added, hardly any net change would result (Fig. 17.6, bottom).

These data suggest that orbitofrontal neurons respond in a flexible manner to stimuli signaling rewards and aversive outcomes. In the intuitively most understandable form, orbitofrontal neurons acquire differential responses to visual and olfactory stimuli associated with different types of reinforcer to be expected or avoided. A conceptually more interesting learning change consists of adaptation of reward expectation at behavioral and neural levels. The changes of overt behavioral reactions reveal how the animal's internal reward expectations adapt to the current values of reward prediction while the new stimuli are being learned across successive trials. The behavioral changes are preceded or paralleled by changes in neural activations, which may constitute a neural correlate for the behavioral adaptation of expectations to new task contingencies. These observations suggest a neural mechanism for adaptive learning in which existing expectation-related activity is matched to the new condition rather than acquiring all task contingencies from scratch.

DOPAMINE NEURONS

Background

Midbrain dopamine neurons show phasic excitatory responses (activations) following primary food and liquid rewards; reward-predicting visual, auditory, and somatosensory stimuli; and physically intense visual and auditory stimuli. These activations occur in 65% to 80% of dopamine neurons in cell groups A9 (pars compacta of substantia nigra), A10 (ventral tegmental area [VTA]), and A8 (dorsolateral substantia nigra). The activations have latencies of less than 100 ms and durations of less than 200 ms. The same neurons are briefly depressed in their activity by reward omission and by stimuli predicting the absence of reward. The activations are only rarely seen following aversive stimuli (Mirenowicz & Schultz, 1996) and not at all after inedible objects and known neutral stimuli unless they are very intense or large. With these

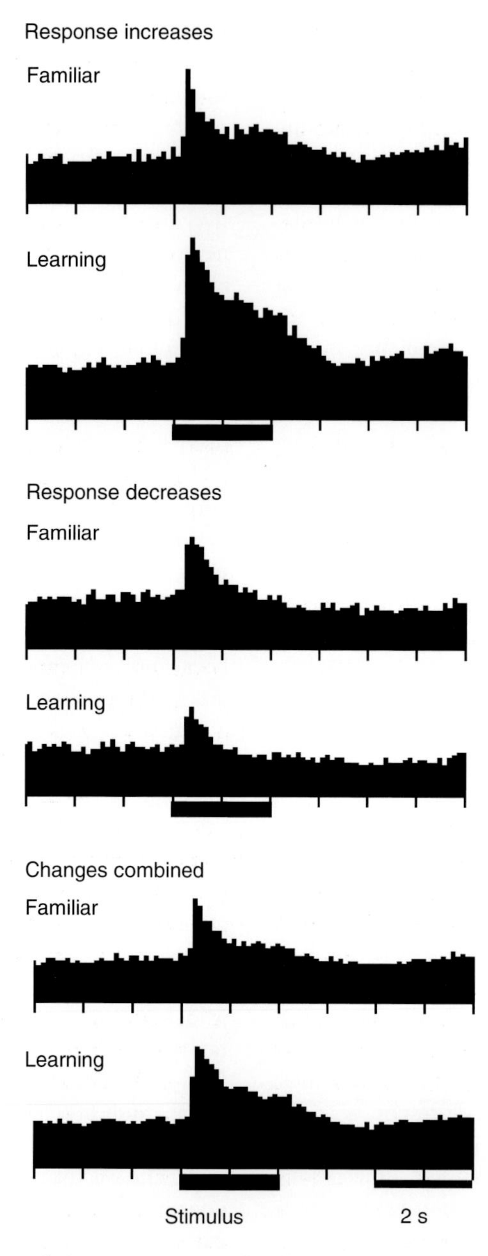

Figure 17.6 Changes of population responses in orbitofrontal cortex to reward-predicting stimuli during learning. Top: Increase of responses in 38 neurons with responses in familiar trials. Center: Decrease of responses in 26 neurons. Bottom: Averaged changes in neurons showing responses exclusively in learning trials, increased responses during learning, and decreased responses during learning (82 neurons). Only data from rewarded movement trials are shown. In each display, histograms from the first 15 rewarded movement trials recorded with each neuron are added, and the resulting sum is divided by the number of neurons. From Tremblay, L., Hollerman, J. R., & Schultz, W. (1998). Modifications of reward expectation-related neuronal activity during learning in primate striatum. *Journal of Neurophysiology, 80,* 964–977, with permission by The American Physiological Society.

characteristics the phasic dopamine responses to rewards are compatible with the notion of a reward prediction error and may serve as a teaching signal for reinforcement learning. Dopamine neurons in groups A8 through A10 project their axons to the dorsal and ventral striatum, dorsolateral and orbital prefrontal cortex, and some other cortical and subcortical structures. The subsecond dopamine reward response may be responsible for the reward-induced dopamine release seen with voltammetry (Roitman et al., 2004) but would not easily explain the 300 to 9,000 times slower dopamine fluctuations with rewards and punishers seen in microdialysis (Datla et al., 2002; Young, 2004). Separate from the rapid reward response, slower, mostly depressant electrophysiological responses occur in dopamine neurons following strong aversive stimuli under anesthesia.

Acquisition of Responses to Reward-Predicting Stimuli

Differential Responses

Dopamine neurons acquire responses to reward-predicting visual and auditory conditioned stimuli through pairing with food or liquid rewards (Ljungberg et al., 1992; Mirenowicz & Schultz, 1994). The responses covary with the expected value of reward, irrespective of spatial position, sensory stimulus attributes, and arm, mouth, and eye movements (Tobler et al., 2005). The responses are modulated by the motivation of the animal, the time course of predictions and the animal's choice among rewards (Morris et al., 2006; Nakahara et al., 2004; Satoh et al., 2003). Although discriminating between reward-predicting stimuli and neutral stimuli, dopamine activations have considerable propensity for generalization (Waelti et al., 2001). While the animals undergo Pavlovian or operant conditioning, dopamine neurons are initially activated by the reward, but the reward response declines gradually and a response to the reward-predicting stimulus appears and grows gradually over successive trials (Fig. 17.7) (Ljungberg et al., 1992; Mirenowicz & Schultz, 1994; Pan et al., 2005), reminiscent of a Pavlovian behavioral response transfer. The responses to these different events coexist transiently at intermediate levels of learning. Thus, the two responses change in opposite directions at the time of the events, without involving activity backpropagating across the stimulus-reward interval. During learning, the responses of dopamine neurons become discriminative and gradually stabilize, consisting of activations following reward-predicting stimuli, depressions or activation-depression sequences following conditioned inhibitors that explicitly predict the absence of reward, and no or minor generalizing responses to neutral stimuli (Fig. 17.8) (Tobler et al., 2003).

Crucial Role of Prediction Errors

Behavioral studies suggest that prediction errors may provide an important mechanism for the learning and updating of reward predictions. The response acquisition of dopamine neurons might underlie this behavioral function if it could be shown that neuronal learning follows the same basic tests for prediction errors as behavioral learning. The blocking paradigm developed by Kamin (1969) can be easily adapted to neurophysiological studies. In essence, the test involves the pairing of a novel stimulus with a reward that is already fully predicted (by another stimulus), such that the fully predicted reward fails to elicit a prediction error in the presence of the novel stimulus. The assumption of prediction error learning postulates that the novel stimulus fails to acquire reward prediction (blocking). The absence of learning can be attributed to the lack of prediction error rather than unspecific factors by using a control stimulus that is paired with an unpredicted reward, such that the control stimulus becomes a reward predictor.

In the neurophysiological experiment, the novel stimulus indeed fails to acquire a neuronal response, whereas the control stimulus elicits a substantial response, very much in parallel to behavioral learning (Fig. 17.9) (Waelti et al., 2001). The blocked neuronal learning of the

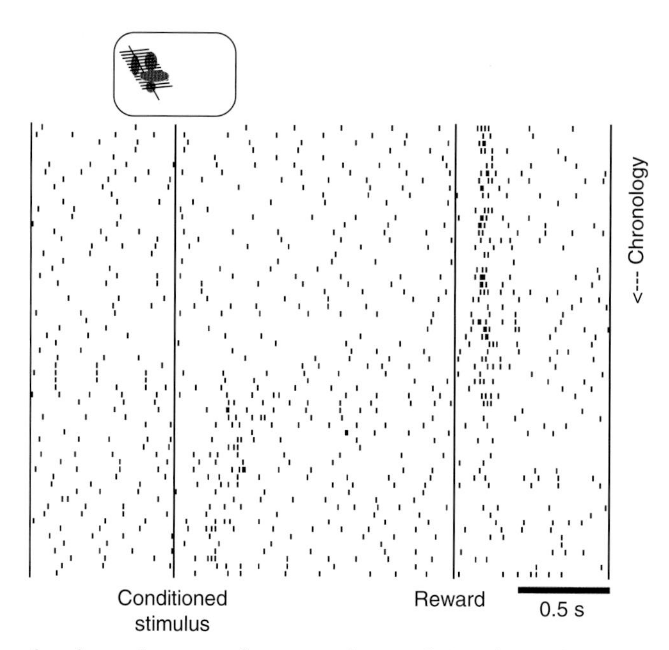

Figure 17.7 Transfer of neural response from reward to conditioned stimulus in a single dopamine neuron during a full learning episode. Each line of dots represents a trial, each dot represents the time of the discharge of the dopamine neuron, the vertical lines indicate the time of the stimulus and juice reward, and the picture above the raster shows the visual conditioned stimulus presented to the monkey on a computer screen. Chronology of trials is from top to bottom. The top trial shows the activity of the neuron while the animal saw the stimulus for the first time in its life, whereas it had previous experience with the liquid reward. Data from Waelti, P., Dickinson, A., & Schultz, W. (2001). Dopamine responses comply with basic assumptions of formal learning theory. *Nature, 412,* 43–48.

novel stimulus occurs despite its pairing with the reward, demonstrating the crucial role of prediction error in learning. It thus appears that the learning of dopamine neurons is governed by the same basic principle that underlies behavioral conditioning, suggesting a neural correlate for the role of prediction errors in the acquisition of reward prediction.

Prediction Error Coding

In addition to neuronal learning being sensitive to prediction errors, dopamine neurons encode an explicit prediction error signal. Dopamine neurons show reward activations only when the reward occurs unpredictably and fail to respond to well-predicted rewards; their activity is depressed when the predicted reward fails to occur (Ljungberg et al., 1992; Mirenowicz & Schultz, 1994). This result has prompted the notion that dopamine neurons emit a positive signal (activation) when an appetitive event is better than predicted, no signal (no change in activity) when an appetitive event occurs as predicted, and a negative signal (decreased activity) when an appetitive event is worse than predicted (Schultz et al., 1997). In line with this argument, dopamine neurons report both positive and negative prediction errors during discrimination learning when animals find out by trial and error which one of two novel stimuli is rewarded and which one is not (Hollerman & Schultz, 1998).

More stringent tests for the neural coding of prediction errors include formal paradigms of animal learning theory in which prediction errors occur in specific situations. In the blocking paradigm, the blocked stimulus fails to predict a reward. Accordingly, the absence of reward following that stimulus does not produce a prediction error nor a response in dopamine neurons at

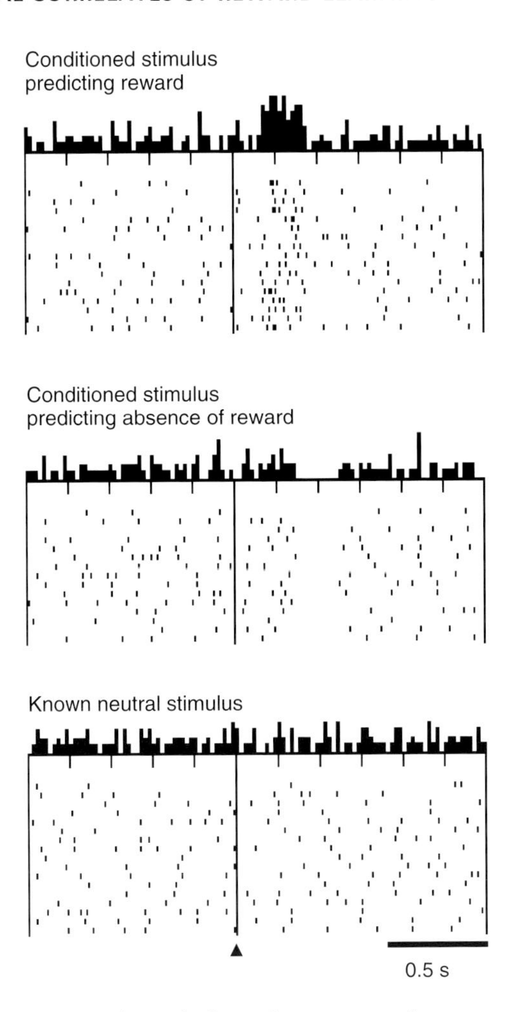

Figure 17.8 Differential responses of a single dopamine neuron to three types of stimuli. Top: Activating response to a reward-predicting stimulus. Middle: Depressant response to a different stimulus predicting the absence of reward (Pavlovian conditioned inhibition task). Bottom: Neutral stimulus. Vertical line and arrow indicate time of stimulus. Data from Tobler, P. N., Dickinson, A., & Schultz, W. (2003). Coding of predicted reward omission by dopamine neurons in a conditioned inhibition paradigm. *Journal of Neuroscience, 23,* 10402–10410

the usual time of the omitted reward, and the delivery of a reward does produce a positive prediction error and a dopamine response (Fig. 17.10A, left) (Waelti et al., 2001). By contrast, after a well-trained, reward-predicting stimulus, reward omission produces a negative prediction error and a depressant neural response, whereas the fully predicted delivery of a reward does not lead to a prediction error nor a response in dopamine neurons (Fig. 17.10A, right). In a conditioned inhibition paradigm, the conditioned inhibitor predicts the absence of reward, and the absence of reward after this stimulus does not produce a prediction error nor a response in dopamine neurons, even when another, otherwise reward-predicting stimulus is added (Fig. 17.10B) (Tobler at al., 2003). By contrast, the occurrence of reward after a conditioned inhibitor produces a strong positive prediction error due to the large difference between the received reward and the negative prediction from the inhibitor. Dopamine neurons show a stronger response in conditioned inhibitor trials (Fig. 17.10B, bottom left) compared to trials with

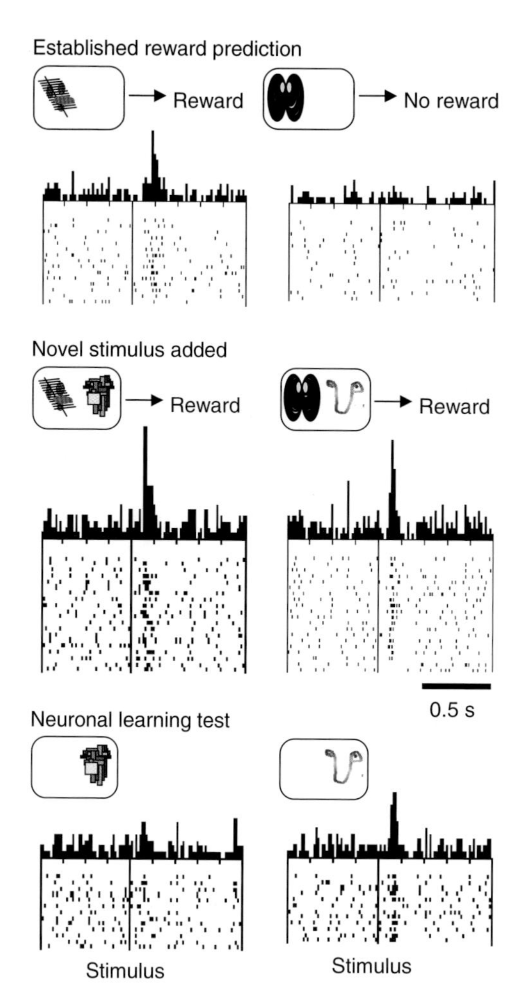

Figure 17.9 Acquisition of dopamine response to reward-predicting stimulus is governed by prediction error. Neural learning is blocked when the reward is predicted by another stimulus (left) but is intact in the same neuron when reward is unpredicted in control trials with different stimuli (right). The neuron has the capacity to respond to reward-predicting stimuli (top left) and discriminates against unrewarded stimuli (top right). Addition of a second stimulus results in maintenance and acquisition of response, respectively (middle). Testing the added stimulus reveals absence of learning when the reward is already predicted by a previously conditioned stimulus (bottom left). Data from Waelti, P., Dickinson, A., & Schultz, W. (2001). Dopamine responses comply with basic assumptions of formal learning theory. *Nature, 412,* 43–48 .

neutral control stimuli (Fig. 17.10B, bottom right). Taken together, these data suggest that dopamine neurons show bidirectional coding of reward prediction errors, following the equation:

Dopamine response = Reward occurred – Reward predicted

This equation may constitute a neural equivalent for the prediction error term of $(\lambda - V)$ of the Rescorla-Wagner learning rule. With these characteristics, the bidirectional dopamine error response would constitute an ideal teaching signal for inducing neural learning through synaptic plasticity.

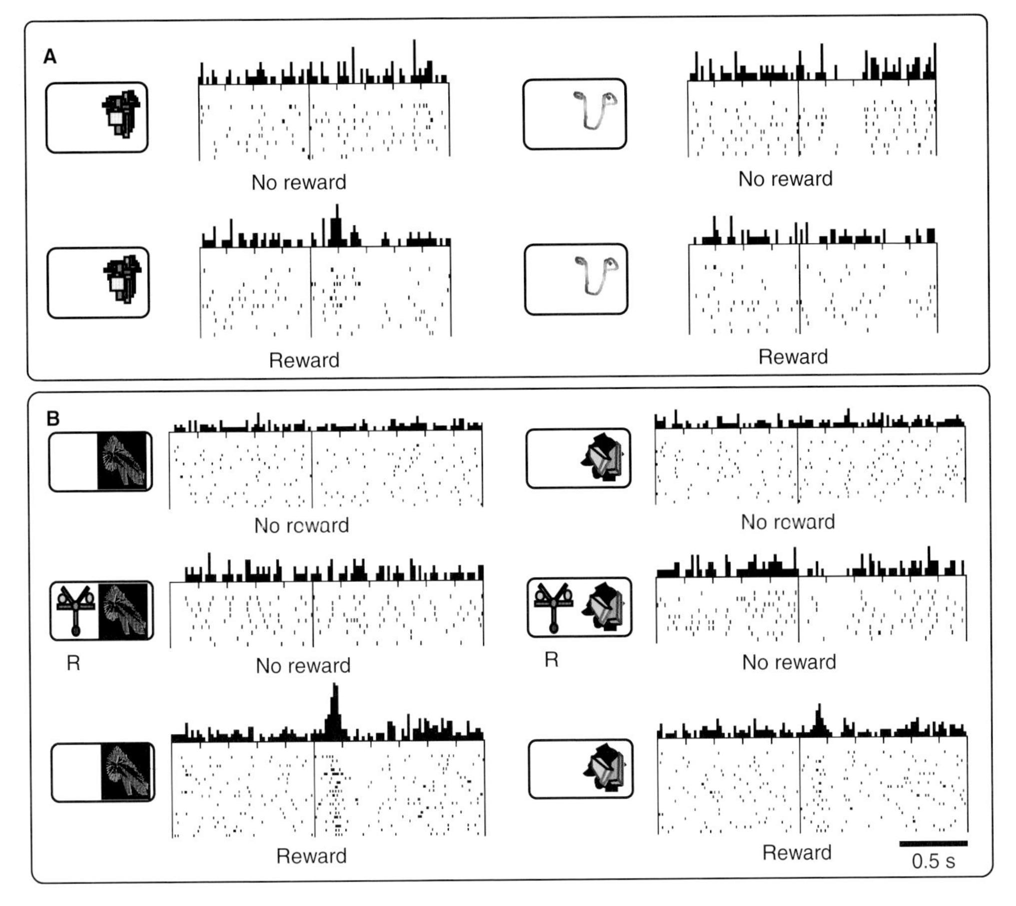

Figure 17.10 Coding of prediction errors by dopamine neurons in paradigms derived from animal learning theory. (**A**) Blocking test. Lack of response to absence of reward following the blocked stimulus, but positive signal to delivery of reward (left), in contrast to control trials with a learned stimulus (right). Data from Waelti, P., Dickinson, A., & Schultz, W. (2001). Dopamine responses comply with basic assumptions of formal learning theory. *Nature, 412,* 43–48. (**B**) Conditioned inhibition test. Lack of response to absence of reward following the stimulus predicting no reward (top), even if the stimulus is paired with an otherwise reward-predicting stimulus (R, middle, summation test), but strong activation to reward following a stimulus predicting no reward (bottom). These responses contrast with those following a neutral control stimulus (right). Data from Tobler, P. N., Dickinson, A., & Schultz, W. (2003). Coding of predicted reward omission by dopamine neurons in a conditioned inhibition paradigm. *Journal of Neuroscience, 23,* 10402–10410.

CONCLUSIONS

Learning Situations

Standard behavioral learning experiments involve the acquisition of Pavlovian or operant responding through association of an intrinsically neutral stimulus or a spontaneously emitted action with an outcome, such as a reward. For comparison, a different, control stimulus or a control action is not associated with the same outcome. Behavioral measures of learning consist of assessing the number of behavioral responses over the course of learning and of comparing the responses to the target stimulus with those to the control stimulus. However, neurophysiological experiments have substantial time constraints, as the recording from individual neurons is usually limited to a few tens of minutes. In addition, these experiments require studying more than one neuron in each animal for reasons of reproducibility and for better use of the animal. The time and the repeated measures requirements often compromise the neurophysiological study of straightforward learning situations in which a whole task is being acquired. The learning set paradigm offers tremendous advantages in respect to these constraints. In these experiments, usually only one or very few components of a task have to be learned, which is also beneficial for the behavioral analysis of well-controlled learning. The repeated change of the novel task component increases the speed of learning to the point where a whole learning episode takes only a few trials. This is an ideal situation for neurophysiology, as both the time constraints and the requirements for repeated, reproducible learning episodes in the same animals are met. In addition, the activity of the same neurons can be compared between novel and familiar task components. Hopefully future neurophysiological learning studies will profit more from this enormous potential.

Response Acquisition

The most obvious neurophysiological phenomenon in learning is the gradual acquisition of neuronal responses to the novel stimuli or actions. As Pavlovian conditioning offers the most simple and easily interpretable learning situation, all reviewed studies reported increased responses to reward-predicting stimuli compared to non–reward-predicting stimuli. Such responses are found in the striatum, orbitofrontal cortex, and dopamine neurons, as well as in a number of other brain structures not described here, such as the dorsolateral prefrontal cortex and various other parts of the basal ganglia besides the striatum. The response acquisition appears to be similar between novel learning situations and learning set tasks. The example of dopamine neurons suggests that neuronal response acquisitions follow similar rules as behavioral learning, namely, prediction errors.

Change of Expectations

The learning set experiments demonstrate the presence of neuronal reward expectation activity from the first learning trial on. Explicit information eliciting these activations obviously cannot be derived from the novel, yet unknown stimuli. During learning, a novel stimulus can only provide a point in time for recalling default reward expectations from long-term memory about the task being rewarded. Thus, the reward expectation activity in each initial learning trial may be evoked in time by the stimulus and in content by recall from long-term memory of the task structure. Once the novel stimuli have been learned, the reward expectations would be adapted and restricted to only those trials that currently lead to reward. Only then would the newly learned stimuli evoke the specific information necessary for expecting the reward. This learning mechanism consists of two steps, an initial setup of default expectations about reward acquired from task experience, and the subsequent updating to the currently valid, more specific and restricted expectations. The updating would be based on the deviation of the initial, default expectation from the actually occurring reward, namely, a prediction error. The observed activity changes during learning appear to simply reflect this updating of expectation. This mechanism is obviously more efficient than a continuous setting up of complete representations of reward expectations from scratch and thus has higher chances for evolutionary survival.

Role of Prediction Errors

One of the key postulated mechanisms of outcome-directed learning is based on prediction errors. This hypothesis explains the effects of contingency on learning (Kamin, 1969) and the gradual response acquisition and extinction when outcomes are better or worse than expected, respectively. Prediction errors form the basis for a number of associative learning theories that explain learning in various behavioral situations (Mackintosh, 1975; Pearce & Hall, 1980; Rescorla & Wagner, 1972). In the crucial blocking test for the role of prediction errors in contingency and learning, dopamine neurons closely follow the fundamental conditions of prediction error coding (Waelti et al., 2001). It is reasonable to assume that prediction errors drive the adaptations of reward expectations described in learning set situations, and it could be speculated, but is in no way shown, that this influence is derived from inputs from dopamine neurons carrying prediction error information.

Possible Synaptic Mechanisms

It is possible, but by no means shown, that inputs from some of the described brain regions induce responses to reward-predicting stimuli in the other regions, such as the dopamine responses driving striatal and orbitofrontal responses. The transient responses of dopamine neurons may induce synaptic changes at striatal synapses, which could underlie the learning of stimuli with motivational significance and the adaptation of expectations by changed contingencies. Increased cortical inputs to striatal neurons may lead to long-lasting synaptic changes, following a Hebbian learning mechanism with conjoint pre- and postsynaptic activity. Corticostriatal synapses show long-term potentiation and depression (Calabresi et al., 1992; Reynolds et al., 2001). The standard Hebbian learning mechanism may be extended to a three-factor Hebbian process with a dedicated reward-predicting dopamine signal as the determining factor (Schultz & Dickinson, 2000). Dopamine is known to induce synaptic plasticity in the striatum and cortex (Bao et al., 2001; Blond et al., 2002; Gurden et al., 2000), and interference with dopamine neurotransmission impairs the induction of long-term synaptic changes (Calabresi et al., 1992; Gurden et al., 2000; Kerr & Wickens 2001). The use of the dopamine prediction error in models of reinforcement learning demonstrates that the dopamine signal not only replicates some of the basic assumptions of temporal difference learning (Montague et al., 1996) but also is able to serve as an effective teaching signal for such typical striatal and frontal cortical tasks as delayed responding (Suri & Schultz, 1999). Thus, although reward information may reach the different components of the reward system through separate routes, the listed plasticity mechanisms may link some of these brain regions during learning.

ACKNOWLEDGMENTS

Our work was supported by the Wellcome Trust, NSF (US), NIH (US), Swiss NSF, Human Frontiers Science Program, and several other grant and fellowship agencies.

REFERENCES

Alexander, G. E., & Crutcher, M. D. (1990). Preparation for movement: Neural representations of intended direction in three motor areas of the monkey. *Journal of Neurophysiology, 64,* 133–150.

Amador, N., Schlag-Rey, M., & Schlag, J. (2000). Reward-predicting and reward-detecting neuronal activity in the primate supplementary eye field. *Journal of Neurophysiology, 84,* 2166–2170.

Aosaki, T., Graybiel, A. M., & Kimura, M. (1994a). Effect of the nigrostriatal dopamine system on acquired neural responses in the striatum of behaving monkeys. *Science, 265,* 412–415.

Aosaki, T., Tsubokawa, H., Ishida, A., Watanabe, K., Graybiel, A. M., & Kimura, M. (1994b). Responses of tonically active neurons in the primate's striatum undergo systematic changes during behavioral sensorimotor conditioning. *Journal of Neuroscience, 14,* 3969–3984.

Apicella, P., Scarnati, E., Ljungberg, T., & Schultz, W. (1992). Neuronal activity in monkey striatum related to the expectation of predictable environmental events. *Journal of Neurophysiology, 68,* 945–960.

Bao, S., Chan, V. T., & Merzenich, M. M. (2001). Cortical remodelling induced by activity of ventral tegmental dopamine neurons. *Nature, 412,* 79–83.

Blond, O., Crepel, F., & Otani, S. (2002). Long-term potentiation in rat prefrontal slices facilitated by phased application of dopamine. *European Journal of Pharmacology, 438,* 115–116.

Bowman, E. M., Aigner, T. G., & Richmond, J. (1996). Neural signals in the monkey ventral striatum related to motivation for juice and cocaine rewards. *Journal of Neurophysiology, 75,* 1061–1073.

Brown, V. J., Desimone, R., & Mishkin, M. (1995). Responses of cells in the caudate nucleus during visual discrimination learning. *Journal of Neurophysiology, 74,* 1083–1094.

Calabresi, P., Maj, R., Pisani, A., Mercuri, N. B., & Bernardi G. (1992). Long-term synaptic depression in the striatum: Physiological and pharmacological characterization. *Journal of Neuroscience, 12,* 4224–4233.

Cromwell, H. C., & Schultz, W. (2003). Effects of expectations for different reward magnitudes on neuronal activity in primate striatum. *Journal of Neurophysiology, 89,* 2823–2838.

Datla, K. P., Ahier, R. G., Young, A. M. J., Gray, J. A., & Joseph, M. H. (2002). Conditioned appetitive stimulus increases extracellular dopamine in the nucleus accumbens of the rat. *European Journal of Neuroscience, 16,* 1987–1993.

Dickinson, A. (1980). *Contemporary animal learning theory.* Cambridge: Cambridge University Press, 1980.

Divac, I., Rosvold, H. E., & Szwarcbart, M. K. (1967). Behavioral effects of selective ablation of the caudate nucleus. *Journal of Comparative Physiology and Psychology, 63,* 184–190.

Funahashi, S., Bruce, C. J., & Goldman-Rakic, P. S. (1989). Mnemonic coding of visual space in the monkey's dorsolateral prefrontal cortex. *Journal of Neurophysiology, 61,* 331–349.

Funahashi, S., Chafee, M. V., & Goldman-Rakic, P. S. (1993). Prefrontal neuronal activity in rhesus monkeys performing a delayed anti-saccade task. *Nature, 365,* 753–756.

Fuster, J. M. (1973). Unit activity of prefrontal cortex during delayed-response performance: Neuronal correlates of transient memory. *Journal of Neurophysiology, 36,* 61–78.

Gaffan, D., & Eacott, M. J. (1995). Visual learning for an auditory secondary reinforcer by macaques is intact after uncinate fascicle section: Indirect evidence for the involvement of the corpus striatum. *European Journal of Neuroscience, 7,* 1866–1871.

Gaffan, E. A., Gaffan, D., & Harrison, S. (1988). Disconnection of the amygdala from visual association cortex impairs visual reward association learning in monkeys. *Journal of Neuroscience, 8,* 3144–3150.

Gurden, H, Takita, M., & Jay, T. M. (2000). Essential role of D1 but not D2 receptors in the NMDA receptor-dependent long-term potentiation at hippocampal-prefrontal cortex synapses in vivo. *Journal of Neuroscience, 20 RC 106,* 1–5.

Harlow, H. F. (1949). The formation of learning sets. *Psychology Review, 56,* 51–65.

Hassani, O. K., Cromwell, H. C., & Schultz, W. (2001). Influence of expectation of different rewards on behavior-related neuronal activity in the striatum. *Journal of Neurophysiology, 85,* 2477–2489.

Hikosaka, O. (1999). Parallel neural networks for learning sequential procedures. *Trends in Neuroscience, 22,* 464–471.

Hikosaka, O., Sakamoto, M., & Usui, S. (1989). Functional properties of monkey caudate neurons. III. Activities related to expectation of target and reward. *Journal of Neurophysiology, 61,* 814–832.

Hollerman, J. R., & Schultz, W. (1989). Dopamine neurons report an error in the temporal prediction of reward during learning. *Nature Neuroscience, 1,* 304–309.

Hollerman, J. R., Tremblay, L., & Schultz, W. (1989). Influence of reward expectation on behavior-related neuronal activity in primate striatum. *Journal of Neurophysiology, 80,* 947–963.

Honig, W., & Thompson, R. K. R. (1982). Retrospective and prospective processing in animal working memory. In: G. H. Bower (Ed.), *The psychology of learning and motivation: Advances in research and theory* (pp. 239–283). New York: Academic Press.

Kamin, L. J. (1969). Selective association and conditioning. In: N. J. Mackintosh & W. K. Honig (Eds.), *Fundamental issues in instrumental learning* (pp. 42–64). Dalhousie: Dalhousie University Press.

Kawagoe, R., Takikawa, Y., & Hikosaka, O. (1998). Expectation of reward modulates cognitive signals in the basal ganglia. *Nature Neuroscience, 1,* 411–416.

Kerr, J. N., & Wickens, J. R. (2001). Dopamine D-1/D-5 Receptor activation is required for long-term potentiation in the rat neostriatum in vitro. *Journal of Neurophysiology, 85,* 117–124.

Kobayashi, S., Lauwereyns, J., Koizumi, M., Sakagami, M., & Hikosaka, O. (2002). Influence of reward expectation on visuospatial processing in macaque lateral prefrontal cortex. *Journal of Neurophysiology, 87,* 1488–1498.

Jacobsen, C. F., & Nissen, H. W. (1937). Studies of cerebral function in primates: IV. The effects of frontal lobe lesions on the delayed alternation habit in monkeys. *Journal of Comparative Physiology and Psychology, 23,* 101–112.

Jog, M. S., Kubota, Y., Connolly, C. I., Hillegaart, V., & Graybiel, A. M. (1999). Building neural representations of habits. *Science, 286,* 1745–1749.

Linden, A., Bracke-Tolkmitt, R., Lutzenberger, W., Canavan, A.G.M., Scholz, E., Diener, H.C., & Birbaumer, N. (1990). Slow cortical potentials in Parkinsonian patients during the course of an associative learning task. *Journal of Psychophysiology, 4,* 145–162.

Ljungberg, T., Apicella, P., & Schultz, W. (1992). Responses of monkey dopamine neurons during learning of behavioral reactions. *Journal of Neurophysiology, 67,* 145–163.

Mackintosh, N. J. (1975). A theory of attention: Variations in the associability of stimulus with reinforcement. *Psychology Review, 82,* 276–298.

Malamut, B. L., Saunders, R. C., & Mishkin M. (1984). Monkeys with combined amygdalo-hippocampal lesions succeed in object discrimination learning despite 24-hour intervals. *Behavioral Neuroscience, 98,* 759–769.

Matsumoto, K., Suzuki, W., & Tanaka, K. (2003). Neuronal correlates of goal-based motor selection in the prefrontal cortex. *Science, 301,* 229–232.

Mirenowicz, J., & Schultz, W. (1994). Importance of unpredictability for reward responses in primate dopamine neurons. *Journal of Neurophysiology, 72,* 1024–1027.

Mirenowicz, J., & Schultz, W. (1996). Preferential activation of midbrain dopamine neurons by appetitive rather than aversive stimuli. *Nature, 379,* 449–451.

Montague, P. R., Dayan, P., & Sejnowski, T. J. (1996). A framework for mesencephalic dopamine systems based on predictive Hebbian learning. *Journal of Neuroscience, 16,* 1936–1947.

Morris, G., Nevet, A., Arkadir, D., Vaadia, E., & Bergman, H. (2006). Midbrain dopamine neurons encode decisions for future action. *Nature Neuroscience, 9,* 1057–1063.

Nakahara, H., Itoh, H., Kawagoe, R., Takikawa, Y., & Hikosaka, O. (2004). Dopamine neurons can represent context-dependent prediction error. *Neuron, 41,* 269–280.

Packard, M. G., & Knowlton, B. J. (2002). Learning and memory functions of the basal ganglia. *Annual Review of Neuroscience, 25,* 563–593.

Padoa-Schioppa, C., & Assad, J. A. (2006). Neurons in the orbitofrontal cortex encode economic value. *Nature, 441,* 223–226.

Pan, W-X., Schmidt, R., Wickens, J. R., & Hyland, B. I. (2005). Dopamine cells respond to predicted events during classical conditioning: Evidence for eligibility traces in the reward-learning network. *Journal of Neuroscience, 25,* 6235–6242.

Pasupathy, A., & Miller, E. K. (2005). Different time courses of learning-related activity in the prefrontal cortex and striatum. *Nature, 433,* 873–876.

Pearce, J. M., Hall, G. (1980). A model for Pavlovian conditioning: Variations in the effectiveness of conditioned but not of unconditioned stimuli. *Psychology Review, 87,* 532–552.

Pratt, W. E., & Mizumori, S. J. Y. (2001). Neurons in rat medial prefrontal cortex show anticipatory rate changes to predictable differential rewards in a spatial memory task. *Behavioral Brain Research, 123,* 165–183.

Ravel, S., Legallet, E., & Apicella, P. (2003). Responses of tonically active neurons in the monkey striatum discriminate between motivationally opposing stimuli. *Journal of Neuroscience, 23,* 8489–8497.

Rescorla, R. A., & Wagner, A. R. (1972). A theory of Pavlovian conditioning: Variations in the effectiveness of reinforcement and nonreinforcement. In: A. H. Black & W. F. Prokasy (Eds.), *Classical conditioning II: Current research and theory* (pp. 64–99). New York: Appleton Century Crofts.

Reynolds, J. N. J., Hyland, B. I., & Wickens, J. R. (2001). A cellular mechanism of reward-related learning. *Nature, 413,* 67–70.

Roitman, M. F., Stuber, G. D., Phillips, P. E. M., Wightman, R. M., & Carelli, R. M. (2004). Dopamine operates as a subsecond modulator of food seeking. *Journal of Neuroscience, 24,* 1265–1271.

Rolls, E. T., Critchley, H. D., Mason, R., & Wakeman, E. A. (1996). Orbitofrontal cortex neurons: Role in olfactory and visual association learning. *Journal of Neurophysiology, 75,* 1970–1981.

Rolls, E. T., Sienkiewicz, Z. J., & Yaxley, S. (1989). Hunger modulates the responses to gustatory stimuli of single neurons in the caudolateral orbitofrontal cortex of the macaque monkey. *European Journal of Neuroscience, 1,* 53–60.

Satoh, T., Nakai, S., Sato, T., & Kimura, M. (2003). Correlated coding of motivation and outcome of decision by dopamine neurons. *Journal of Neuroscience, 23,* 9913–9923.

Schoenbaum, G., Chiba, A. A., & Gallagher, M. (1998). Orbitofrontal cortex and basolateral amygdala encode expected outcomes during learning. *Nature Neuroscience, 1,* 155–159.

Schoenbaum, G., Chiba, A. A., & Gallagher, M. (1999). Neural encoding in orbitofrontal cortex and basolateral amygdala during olfactory discrimination learning. *Journal of Neuroscience, 19,* 1876–1884.

Schultz, W., Apicella, P., Scarnati, E., & Ljungberg, T. (1992). Neuronal activity in monkey ventral striatum related to the expectation of reward. *Journal of Neuroscience, 12,* 4595–4610.

Schultz, W., Dayan, P., & Montague, R. R. (1997). A neural substrate of prediction and reward. *Science, 275,* 1593–1599.

Schultz, W., & Dickinson, A. (2000). Neuronal coding of prediction errors. *Annual Review of Neuroscience, 23,* 473–500.

Setlow, B., Schoenbaum, G., & Gallagher, M. (2003). Neural encoding in ventral striatum during olfactory discrimination learning. *Neuron, 38,* 625–636.

Shidara, M., & Richmond, B. J. (2002). Anterior cingulate: Single neuron signals related to degree of reward expectancy. *Science, 296,* 1709–1711.

Suri, R., & Schultz, W. (1999). A neural network with dopamine-like reinforcement signal that learns a spatial delayed response task. *Neuroscience, 91,* 871–890.

Sutton, R. S., & Barto, A. G. (1981). Toward a modern theory of adaptive networks: Expectation and prediction. *Psychology Review, 88,* 135–170.

Tobler, P. N., Dickinson, A., & Schultz, W. (2003). Coding of predicted reward omission by dopamine neurons in a conditioned inhibition paradigm. *Journal of Neuroscience, 23,* 10402–10410.

Tobler, P. N., Fiorillo, C. D., & Schultz, W. (2005). Adaptive coding of reward value by dopamine neurons. *Science, 307,* 1642–1645.

Tremblay, L., Hollerman, J. R., & Schultz, W. (1998). Modifications of reward expectation-related neuronal activity during learning in primate striatum. *Journal of Neurophysiology, 80,* 964–977.

Tremblay, L., & Schultz, W. (1999). Relative reward preference in primate orbitofrontal cortex. *Nature, 398,* 704–708.

Tremblay, L., & Schultz, W. (2000a). Reward-related neuronal activity during go-nogo task performance in primate orbitofrontal cortex. *Journal of Neurophysiology, 83,* 1864–1876.

Tremblay, L., & Schultz, W. (2000b). Modifications of reward expectation-related neuronal activity during learning in primate orbitofrontal cortex. *Journal of Neurophysiology, 83,* 1877–1885.

Waelti, P., Dickinson, A., & Schultz, W. (2001). Dopamine responses comply with basic assumptions of formal learning theory. *Nature, 412,* 43–48.

Watanabe, M. (1996). Reward expectancy in primate prefrontal neurons. *Nature, 382,* 629–632.

Watanabe, M., Hikosaka, K., Sakagami, M., & Shirakawa, S. I. (2002). Coding and monitoring of behavioral context in the primate prefrontal cortex. *Journal of Neuroscience, 22,* 2391–2400.

Wise, S. P. (1996). The role of the basal ganglia in procedural memory. *Seminars in Neuroscience, 8,* 39–46.

Young, A. M. J. (2004). Increased extracellular dopamine in nucleus accumbens in response to unconditioned and conditioned aversive stimuli: Studies using 1 min microdialysis in rats. *Journal of Neuroscience Methods, 138,* 57–63.

CHAPTER 18

Associative Memory in the Medial Temporal Lobe

Yuji Naya and Wendy A. Suzuki

The ability to form new long-term memories for facts, events, and relationships is referred to as declarative memory in humans (Squire et al., 2004) and relational memory in animals (Eichenbaum & Cohen, 2001; Fig. 18.1A). Declarative/relational memory is a fundamental cognitive function that allows us to make appropriate choices based on past experience. In both humans and animals, it is critical for many aspects of everyday life, including social interactions, foraging behaviors, and escape from danger. In humans, declarative memory defines our personal histories and in that way shapes our very personalities.

While early studies assumed memory was a unitary process, likely subserved by the interaction of large areas of the cerebral cortex (Lashley, 1929), the description of the well-known amnesic patient H.M. (Scoville & Milner, 1957) showed for the first time that memory could be localized to particular brain areas. Patient H.M. underwent experimental bilateral medial temporal lobe (MTL) (Fig. 18.1B) resection in an attempt to relieve very severe epilepsy. The bilateral resection was described as including both the amygdala and the hippocampus as well as some of the surrounding cortical areas (Scoville & Milner, 1957). Although the operation markedly reduced the number and severity of his seizures, H.M. suffered from extensive anterograde and temporally graded retrograde amnesia. He retained nothing of day-to-day happenings after the surgery, and could not recognize any of the hospital staff except for Dr. Scoville, his neurosurgeon whom he had known for many years (Milner, 2005). While early reports suggested that H.M.'s memory impairment was global, it became clear later that the impairment was selective for declarative memory for facts and events while other forms of memory including visuo-motor skill learning (Corkin, 1968; Milner, 1962) and sensory priming (Gabrieli et al., 1990) remained intact. These motor and perceptual forms of plasticity are typically expressed through performance rather than recollection, and have been referred to collectively as nondeclarative or procedural memory (Fig. 18.1A). The relationship between declarative memory and the MTL has been further explored with the development of animal models of human amnesia in both nonhuman primates (Mishkin, 1978; Zola-Morgan & Squire, 1985) and rodents (Eichenbaum & Cohen, 2001).

While it is now clear that the MTL is critical for normal declarative/relational memory, the question of the relative contribution of individual MTL structures to particular forms of declarative/relational memory remains controversial. While some reports argue that all MTL structures can contribute in a complementary way to many forms of declarative memory (Squire et al., 2007), other reports have emphasized the differential contributions of individual MTL structures to distinct aspects of memory (Eichenbaum et al., 2007). In the latter framework, the hippocampus is thought to be critical for recollection-based recognition, whereas perirhinal cortex is necessary for familiarity-based recognition (Eichenbaum et al., 2007). On the other hand, declarative memory is also separable according to the types of items to be memorized such as object, place, and time or

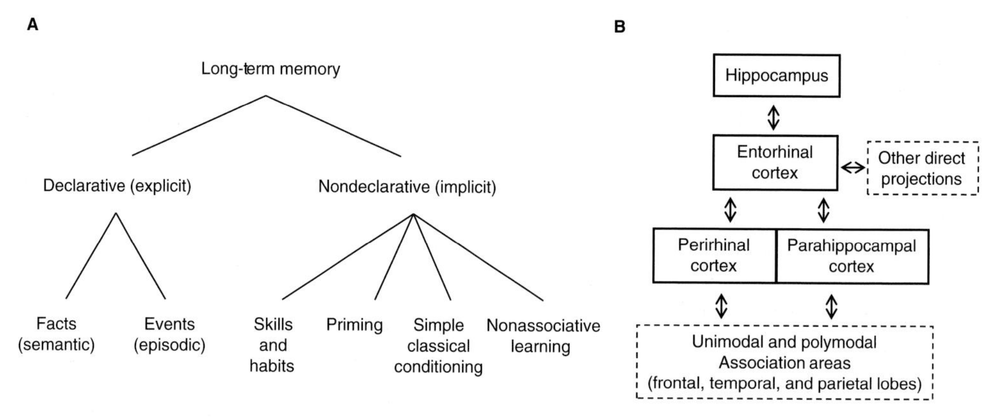

Figure 18.1 (**A**) Classification of long-term memory. Declarative (explicit) memory refers to conscious recollections of facts (semantic) and events (episodic) and depends on the integrity of the medial temporal lobe. Declarative memory is also referred to as relational memory in animals. Nondeclarative (implicit) memory refers to a collection of abilities and is independent of the medial temporal lobe. Nonassociative learning includes habituation and sensitization. In the case of nondeclarative memory, experience alters behavior unconsciously without providing access to any memory content. (**B**) A schematic view of the medial temporal lobe memory system and anatomically related areas. The medial temporal lobe regions important for declarative/relational memory include the hippocampus and entorhinal, perirhinal, and parahippocampal cortices (thick outlines). The entorhinal cortex is the major source of cortical projections to the hippocampus. Two thirds of the cortical input to the entorhinal cortex originates in the adjacent perirhinal and parahippocampal cortices, which in turn receive projections from unimodal and polymodal areas in the frontal, temporal, and parietal lobes (thin dashed outlines). The entorhinal cortex also receives other direct inputs from orbital frontal cortex, cingulate cortex, insular cortex, and superior temporal gyrus (thin dashed outlines). All these projections are reciprocal. Based on Squire and Zola-Morgan (1991). The medial temporal lobe memory system. *Science, 253*, 1380–1386. Used with permission.

their combination. To elucidate the representation of memory between separate objects in a single modality, a visual stimulus-stimulus association paradigm was introduced into the experimental animal lesion literature (Buckley & Gaffan, 1998; Bunsey & Eichenbaum, 1993; Murray et al., 1993). Results from these studies indicate that the perirhinal cortex is critical for the formation and long-term representation of stimulus-stimulus association memory. While findings from lesion studies provide evidence that the perirhinal cortex is important for stimulus-stimulus association memory (Buckley & Gaffan, 1998; Bunsey & Eichenbaum, 1993; Murray et al., 1993), another critical question concerns understanding the neural representation of this critical form of memory. How do neurons in the perirhinal cortex represent a learned visual-visual association? What is its role in the retrieval of learned visual-visual associations, and how does this area interact with other sensory areas that provide the "raw" sensory input to the perirhinal cortex? To address these questions, behavioral neurophysiological studies have focused on recording in both the perirhinal cortex and area TE during the performance of well-learned visual-visual paired associate (VPA) tasks. In Part I of this chapter, we detail the results from these studies.

The critical contribution of the hippocampus to declarative/relational memory in animals has been demonstrated using conditional motor association (CMA) learning, in which stimuli and spatially directed actions are associated in memory (Murray & Wise, 1996). Monkeys with hippocampal ablations could acquire CMA, but their learning speed was significantly slower than normal control animals (Wise & Murray, 1999).

This finding suggested that the hippocampus may be particularly important for the initial development of new conditional motor associations when animals need to rapidly bind object information with an action or spatial location. In Part II of this chapter, we will review the neurophysiological studies in the medial temporal lobe that have examined the development of new associative representations. These studies reveal dynamic changes in both the perirhinal cortex and hippocampus during the acquisition of new associative memories.

PART I: NEURONAL ORGANIZATION OF LONG-TERM ASSOCIATIVE MEMORY

The semantic memory system has been described as "a mental thesaurus" (Tulving, 1972) formed by overlapping collections of cell assemblies of related objects. From this point of view, the inferior temporal (IT) cortex, considered to be a long-term memory (LTM) storehouse of visual objects (Mishkin, 1982; Miyashita & Chang, 1988; Penfield & Perot, 1963), is a particularly appropriate place to encode the relationships between semantically linked items. IT cortex includes both the ventral parts of perirhinal cortex (area 36, A36) and the laterally adjacent visual area TE. The two subareas are cytoarchitectonically distinct but mutually interconnected (Saleem & Tanaka, 1996; Suzuki & Amaral, 1994; von Bonin & Bailey, 1947). The neural correlates of semantically or temporally linked items have been studied in monkey IT cortex with the use of various long-term associative memory tasks. In the following section, we describe studies that assess how LTM signals are organized and activated in macaque IT cortex following association learning at a time when the long-term associations are well established. The results suggest that the relationships between the semantically or temporally associated items are represented by the responses of single neurons (Miyashita, 1988; Sakai & Miyashita, 1991) and the long-term association memory is stored in the synaptic connections between TE and perirhinal cortex (Naya et al., 2003a; Yoshida et al., 2003). Specifically, these studies show that the perirhinal cortex and area TE have a dynamic and interrelated relationship in the long-term representation of associative memory; the perirhinal cortex receives perceptual signals (cue stimuli) from area TE and feeds back mnemonic signals (paired associates) (Naya et al., 2001, 2003b).

Neuronal Correlates of Associative Long-Term Memory in the Temporal Cortex

In one of the first descriptions of long-term associative memory signals in monkey temporal lobe, Miyashita (1988) recorded the activity of neurons in IT cortex (Fig. 18.2B) as macaque monkeys performed a delayed matching-to-sample (DMS) task in which a sample image was first shown in a video monitor followed by a matching stimulus (Fig. 18.2C). During the training phase of the task, a set of 97 visual fractal patterns ("learned stimuli"; Fig. 18.2A) was repeatedly used and these stimuli were presented to the animal as the sample stimulus in a fixed sequence according to an arbitrary serial position number. During the recording phase of the experiment, a sample stimulus was selected not only from the 97 learned patterns but also from a new set of 97 patterns ("new stimuli"), independent of the serial position numbers. Surprisingly, Miyashita (1988) reported that the effective response to a given learned stimulus was most highly correlated to those stimuli that had been presented in close temporal order during the training phase (stippled columns in Fig. 18.2D,E). Thus, this study showed that the responses of single IT neurons can reflect the long-term associations between temporally related sets of stimuli even if that information is not required for the task.

To examine the effect of long-term association on IT neurons when animals were explicitly required to use those associations to solve the task, Sakai and Miyashita (1991) recorded the activity of IT neurons as animals performed a VPA task. Paired associate tasks are one of the most widely used neuropsychological tasks in humans to test for elementary pair-wise associations (Wechsler, 1987). Sakai and Miyashita

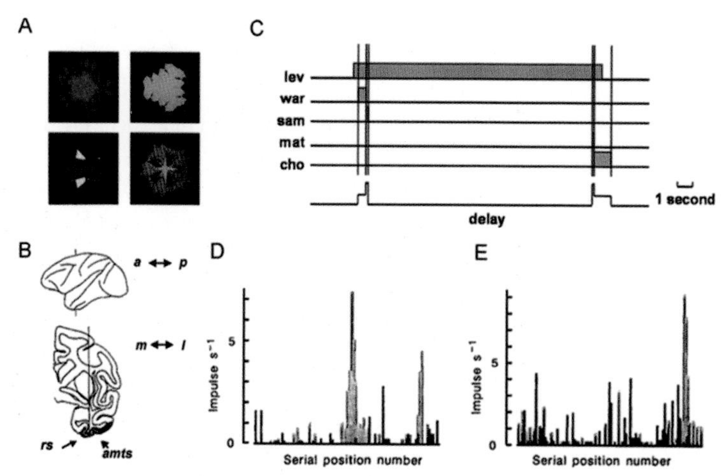

Figure 18.2 Stimulus-stimulus association among the learned fractal patterns. **(A)** Examples of color fractal patterns. **(B)** Location of recording sites. Top: A lateral view of macaque brain. Bottom: A section indicated by a vertical line on the lateral view. The stippled area represents the range of recording sites. *rs*, rhinal sulcus; *amts*, anterior middle temporal sulcus; *a*, anterior; *p*, posterior; *m*, medial; *l*, lateral. **(C)** Sequence of events in a trial of delayed matching-to-sample task. lev, lever press by the monkey; war, warning green image; sam, sample stimulus; mat, match stimulus following a 16-second delay; cho, choice signal of white image. **(D)** Average delay discharge rate for each sample stimulus in a cell against "serial position number" of the stimuli (see text). Stippled columns, learned stimuli; black columns, new stimuli. **(E)** As D, but for a different cell. Based on Miyashita, Y. (1988). Neuronal correlate of visual associative long-term memory in the primate temporal cortex. *Nature, 335,* 817–820. Used with permission.

trained monkeys in the VPA task using 12 pairs of Fourier descriptors (Fig. 18.3). In each trial, a cue stimulus was presented on a video monitor. After a delay period, two choice stimuli, the paired associate of the cue (correct choice) and a distracter from the other pairs (incorrect choice), were shown. The identity of the correct combination of any given paired associate cannot be predicted without memorizing the specific pairs beforehand; in addition, the VPA task demands memory retrieval and thus generation of images from LTM. In this study, they found two types of task-related neurons in IT cortex, "pair-coding" and "pair-recall" neurons. Figure 18.4A shows the responses of a pair-coding neuron. Note that one stimulus elicited the strongest response from this neuron during the cue period (cue-optimal stimulus). This neuron was also activated when the paired associate of the cue-optimal stimulus was presented. In contrast to the robust responses to this stimulus pair, the neuron responded only negligibly when stimuli from any of the other pairs were presented as cue stimuli. This property indicates that memory storage is organized such that single neurons can code both paired associates in the VPA task. Figure 18.4B shows the response of a pair-recall neuron. For this neuron, in the trial when the paired associate of the cue-optimal stimulus was presented as a cue stimulus, it exhibited the highest delay activity among the stimuli. This stimulus-selective activity during the delay period is thought to represent the retrieval of a sought target retrieved from LTM through the cue stimulus. To test this hypothesis directly, Naya and colleagues (1996) manipulated the requirement and timing for retrieval explicitly using a modified VPA task with a "color-switch" signal shown in the delay interval that indicated if the trial required memory retrieval (VPA trial) or not (DMS trial). When the paired associate of the cue-optimal stimulus was presented as a cue, IT neurons started to fire just after the color switch

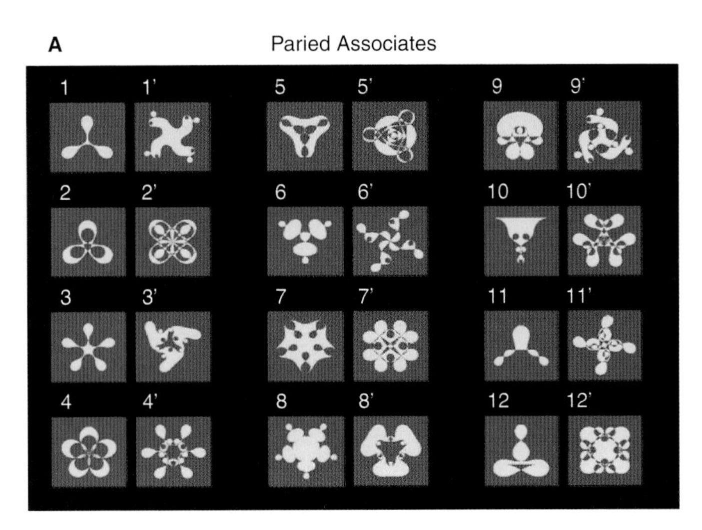

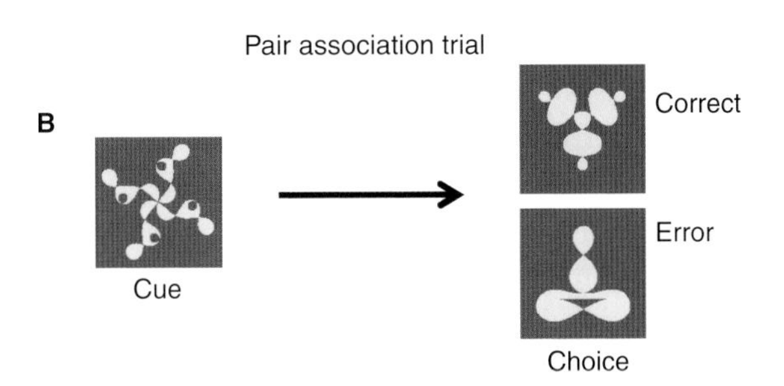

Figure 18.3 Visual pair association (VPA) task used to assess long-term memory in monkeys. (**A**) Twelve pairs of Fourier descriptors (1-1' and 12-12') were used in the VPA task. (**B**) Cue stimuli were presented at the center of a video monitor. Choice stimuli were presented randomly in two of four positions on the video monitor. One is the paired associate of the cue ("Correct"); the other is a distracter from a different pair ("Error"). Based on Sakai, K., & Miyashita, Y. (1991). Neural organization for the long-term memory of paired associates. *Nature, 354,* 152–155; and Naya, Y., Yoshida, M., & Miyashita, Y. (2003a). Forward processing of long-term associative memory in monkey inferotemporal cortex. *Journal of Neuroscience, 23,* 2861–2871. Used with permission.

that signaled the necessity for memory retrieval for their cue-optimal stimuli (VPA trial). In contrast, if there was no color switch, the same neurons showed no activation (DMS trial). This finding suggests that the delay activity of IT neurons in the VPA task correspond to the requirement for the retrieval of a visual image from LTM.

Forward Processing of Associative Memory from Visual Cortex to Limbic Cortex

After the discovery of pair-coding and pair-recall neurons, the next critical challenge was to elucidate the neuronal circuits that underlie the response properties of these memory signals during a VPA trial. To address this problem, Naya and colleagues (2001, 2003a) examined the interaction between the two major subdivisions of IT cortex: A36 and TE. Note that while A36 is a limbic polymodal association area and a component of the medial temporal lobe memory system (Zola-Morgan & Squire, 1990), TE is a unimodal visual neocortical area located at the final stage of the ventral visual pathway, important for object vision (Janssen et al., 2000; Tanaka, 1996).

Naya and his colleagues (2003a) mapped the two subdivisions; a total of 2,368 neurons were recorded from A36 (510 neurons) and TE (1,858 neurons) in the three monkeys performing the VPA task. Of those, 423 neurons (76 neurons in A36 and 347 neurons in TE) showed stimulus-selective responses ($P < 0.01$, analysis of variance [ANOVA]) during the cue period (cue-selective neurons). Their response latencies were significantly shorter in TE than A36 (mean, 93.8 ms in A36 vs. 86.2 ms in TE, $P < 0.05$), confirming that the visual signal reached TE before it reached A36. This finding was consistent with their anatomical hierarchy: A36 is

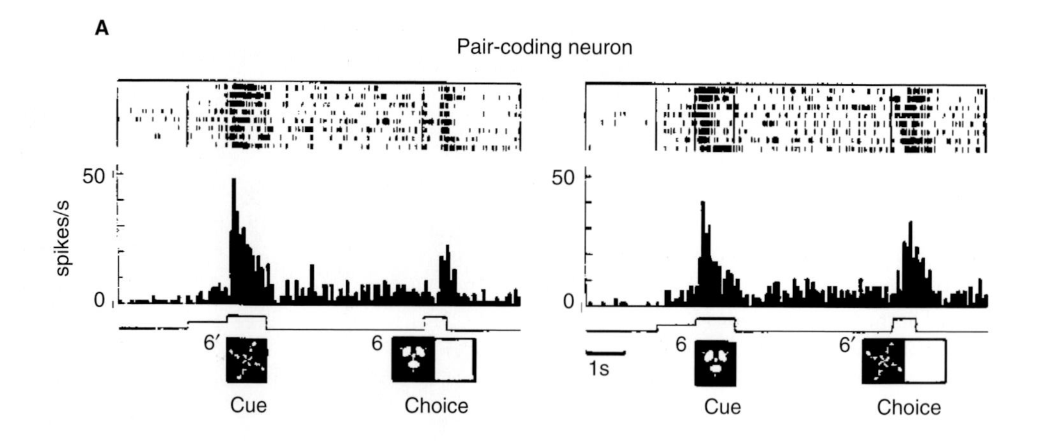

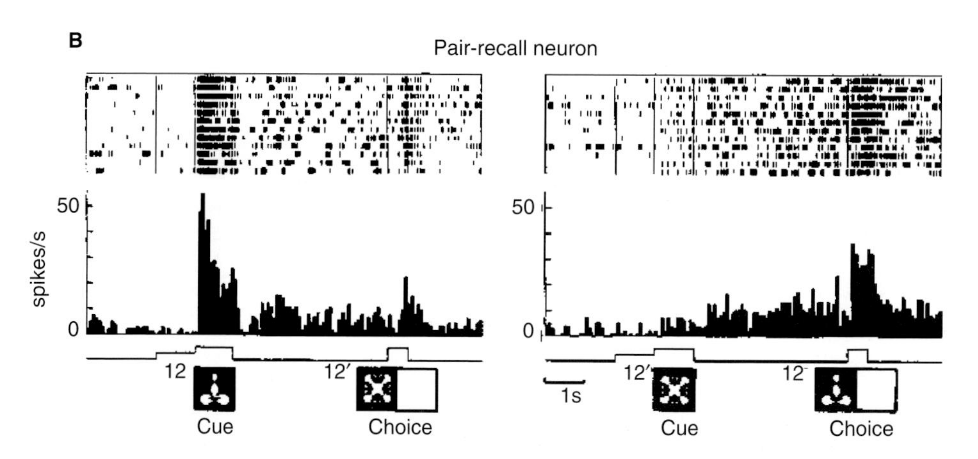

Figure 18.4 Responses of "pair-coding neuron" and "pair-recall neuron." **(A)** A pair-coding neuron. Raster display and peri-stimulus time histogram (PSTH) in trials for cue 6' (left) and for cue 6 (right). The trials were aligned at the cue onset. Note that cue 6' and 6 elicited the strongest and the second strongest cue responses from the neuron. **(B)** A pair-recall neuron. The neuron showed the strongest cue responses in trials for cue 12. Note the tonic increasing activity during the delay period in trials for cue 12', which is much higher than the cue response. Based on Sakai, K., & Miyashita, Y. (1991). Neural organization for the long-term memory of paired associates. *Nature, 354,* 152–155; and Sakai, K., Naya, Y., & Miyashita, Y. (1994). Neuronal tuning and associative mechanisms in form representation. *Learning & Memory, 1,* 83-105. Used with permission.

situated one synapse downstream of TE (Felleman & Van Essen, 1991). The responses of cue-selective neurons to paired associates were significantly correlated at the population level in both A36 and TE, but this correlation was much stronger in A36 than TE (median, 0.51 in A36 vs. 0.14 in TE) (Fig. 18.5A).

At the single neuron level, a substantial number of neurons showed significantly ($P < 0.01$) correlated responses to the paired associates (pair-coding neuron) in A36. The percentage of the pair-coding neurons was much higher in A36 than in area TE (33% in A36 vs. 4.9% in TE, of the cue-selective neurons) (Fig. 18.5B).

The spatial distribution of the pair-coding neurons demonstrated that the pair-coding neurons in TE were not necessarily distributed in the region near the border with A36 (Fig. 18.5C). This suggests that the percentage of the pair-coding neurons did not increase in a gradual manner from lateral to medial in IT cortex. Moreover, within TE, there was no subregion where the percentage of the pair-coding neurons was comparable with that in A36 (Fig. 18.5C). These anatomical observations supported the physiological result that the percentage of the pair-coding neurons dramatically increased from TE to A36.

These striking differences between TE and A36 raise the question of whether the pair-coding response of A36 neurons was elicited by a feedforward input from TE or by a feedback input from other higher centers. To address this question, initial transient responses after the cue stimulus presentations were examined for the pair-coding neurons in A36. The analysis revealed that they were separable into two subtypes, and type 1 neurons (68%), but not type 2 (32%) neurons,

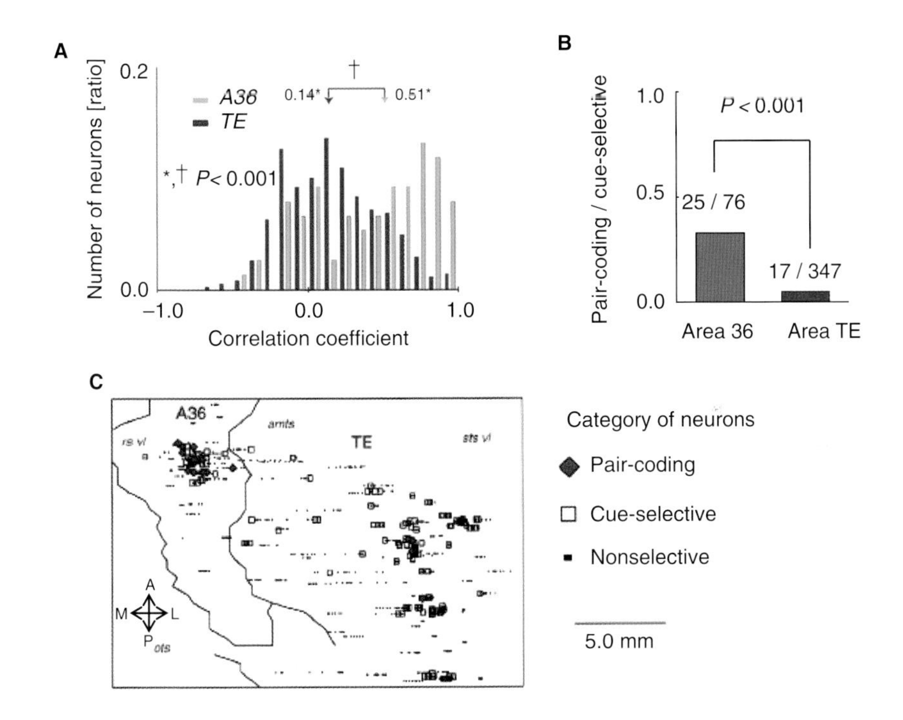

Figure 18.5 Comparison of pair-coding responses between area 36 and area TE. (**A**) Response correlation to paired associates of cue-selective neurons in A36 ($N = 76$; green) and TE ($N = 347$; red). The correlation coefficients for A36 neurons were significantly higher than those for TE neurons ($P < 0.001$; Kolmogorov-Smirnov test). (**B**) The ratio of the pair-coding neurons among the cue-selective neurons was significantly higher ($P < 0.001$; χ^2 test) in A36 (0.33; green) than TE (0.05; red). The pair-coding neurons were defined as the cue-selective neurons that showed significantly positive PCI at the single-neuron level: $P < 0.01$ (i.e., correlation coefficient > 0.71). (**C**) Spatial distributions of pair-coding neurons. The positions of the pair-coding (orange-filled diamond), cue-selective (black open square), and other recorded (black dot) neurons are shown on two-dimensional unfolded maps for one monkey. Black lines, area borders; gray lines, fundus or lips of sulci; amts, anterior middle temporal sulcus; ots, occipital temporal sulcus; pmts, posterior middle temporal sulcus; rs, rhinal sulcus; sts, superior temporal sulcus; vl, ventral lip; A, anterior; P, posterior; L, lateral; M, medial. Scale bar, 5.0 mm. Based on Naya, Y., Yoshida, M., & Miyashita, Y. (2003a). Forward processing of long-term associative memory in monkey inferotemporal cortex. *Journal of Neuroscience, 23,* 2861–2871. Used with permission.

began to encode associations between paired stimuli as soon as they exhibited a stimulus-selective response (Fig. 18.6A). Thus, the representation of long-term memory encoded by type 1 neurons in A36 is likely generated without feedback input from other higher centers. Taken together, this evidence suggests that the representation of stimulus-stimulus association memory develops from TE to A36. Naya and colleagues (2003a) hypothesize that the forward processing of pair-association memory most likely requires selective convergence such that the perceptual information about the individual elements of a learned paired associate is coded by separate TE neurons that converge onto an individual type 1 A36 neuron (Fig. 18.6B, the "selective-convergence" model). In other words, a single type 1 neuron in A36 can receive visual inputs from either of TE neuron groups coding different but semantically linked objects (e.g., "glove" and "ball"); therefore, the type 1 neuron can code both objects and/or their relationship.

Backward Spreading of Memory Retrieval Signal from Limbic Cortex to Visual Cortex

After images are associated in LTM, what kind of signal is engaged during memory retrieval?

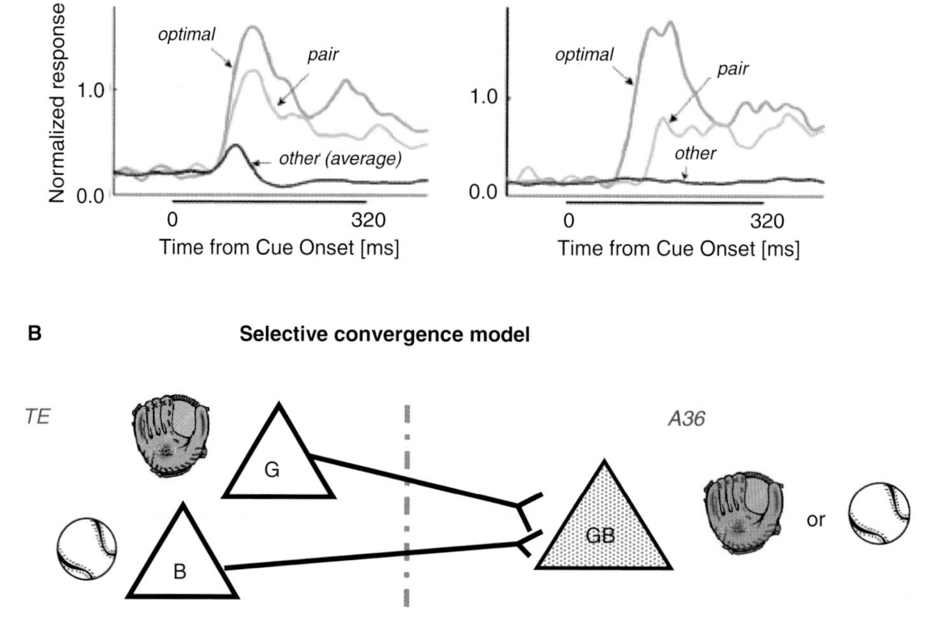

Figure 18.6 Initial responses of pair-coding neurons and "selective convergence" model. **(A)** Population-averaged peri-stimulus time histogram (PSTH) for the type 1 ($N = 17$; left) and type 2 ($N = 8$; right) neurons, showing the normalized responses in the cue-optimal (green), pair (light green), and other (dark green) trials. Note that two thirds of pair-coding neurons in A36 showed initial transient responses in both cue-optimal and pair trials. **(B)** Selective convergence model of association memory in inferior temporal (IT) cortex. Suppose that a glove and a ball are associated together in memory. In TE, these images are represented by different neurons (neuron G for a glove and neuron B for a ball). The neurons G and B project their fibers to the same A36 neuron, in particular, to type 1 neuron (neuron GB). In this way neuron GB can code the paired items regardless of their different visual/geometrical properties. Based on Naya, Y., Yoshida, M., & Miyashita, Y. (2003a). Forward processing of long-term associative memory in monkey inferotemporal cortex. *Journal of Neuroscience, 23,* 2861–2871. Used with permission.

Figure 18.7 shows examples of pair-recall neurons in A36 and TE (Naya et al., 2001). The A36 neuron exhibited sustained activity earlier than the TE neuron when the paired associates of the cue-optimal stimuli were presented as cue stimuli (Fig. 18.7A, bottom). The time course of the pair-recall activity of each neuron was examined by considering the responses to all cue stimuli. In this study, a partial correlation analysis was used to evaluate the response correlation with the target to be retrieved. This measure was useful because IT neurons have a tendency to encode both of the paired stimuli and the correlation between cues and their targets must be removed from the analysis. The partial correlation coefficients of instantaneous firing rates at time t for each cue stimulus were calculated with the visual responses to its paired associate (Fig. 18.7C, pair-recall index, PRI(t)). To characterize the time course of PRI(t), two parameters, transition time (TRT) and transition duration (TRD), were determined for each single neuron. TRT was defined as the period from the cue onset to the instant when the PRI(t) curve reached 50% of its full increase, and TRD was defined as the duration between the instants when the curve reached 10% and 90% of its full increase. The TRT values for the pair-recall neurons in A36 were significantly smaller than those in TE (A36, median 206 ms; TE, median 570 ms; Kolmogorov-Smirnov test, $P < 0.005$) (Fig. 18.8A, left). On the other hand, the distributions of TRD values did not differ between the two areas (A36, median 115 ms; TE, median 145 ms; $P > 0.8$) (Fig. 18.8A, right). These results indicated that the memory-retrieval signal emerges earlier in A36, and TE neurons were then gradually recruited to represent the retrieved image (Fig. 18.8B) (Naya et al., 2001). The median retrieval time was over 300 ms longer in TE than in A36. Given the fact that TE neurons receive numerous back-projections from A36, a reasonable interpretation is that the mnemonic signal spreads backward from A36 to TE.

Compared with the forward transmission of the visual signal from TE to A36, the backward mnemonic signal from A36 to TE exhibited a surprisingly long latency ($\approx$10 ms vs. $\approx$300 ms). Naya and colleagues (2003b) next asked what kind of additional process might be involved in this slow, backward spreading of mnemonic signal. Theoretically, there are two kinds of signals thought to be conveyed in the delay interval of the VPA task. The first is a retrospective signal that is closely coupled with the to-be-retained cue stimulus. The second is a prospective signal that is coupled with the to-be-retrieved target stimulus. To compare retrospective and prospective signals quantitatively, the signal contents of the delay-period activity were characterized by partial correlation coefficients of delay-period activities for each cue stimulus with the cue-period responses to that stimulus (cue-holding index [CHI]) and with the cue-period responses to its paired associate (pair-recall index [PRI]). The delay-period activity of TE neurons preferentially represented the paired associate (PRI, median = 0.54) rather than the cue stimulus itself (CHI, 0.23) ($P < 0.001$, PRI vs. CHI, $N = 70$), while the delay-period activity of A36 neurons retained both the cue stimulus and its paired associate equivalently (CHI, 0.44; PRI, 0.46) ($P = 0.78$, $N = 38$) (Naya et al., 2003b). These results indicate that TE mostly represents a sought target that is retrieved from long-term memory during the delay interval, while A36 in addition retains a cue stimulus that is transmitted from earlier visual areas.

If back-projections from A36 to TE drive the delay activity seen in TE, it is surprising that A36 signals both sensory-related retrospective as well as prospective information while TE only signals prospective information. One interpretation of this finding may be that the backward signal transmission from A36 to TE is equipped with a selective gating mechanism, which preferentially passes information about a sought target (Naya et al., 1996). Another interpretation may be that there are some mechanisms of actively inhibiting irrelevant information in TE. The finding of the strong sought-target–related activity of TE neurons during the VPA task is consistent with the report of Sheinberg and Logothetis (1997) showing that TE reflected perceptually relevant activity during a binocular rivalry task, in which visual ambiguity was induced by presenting incongruent images to

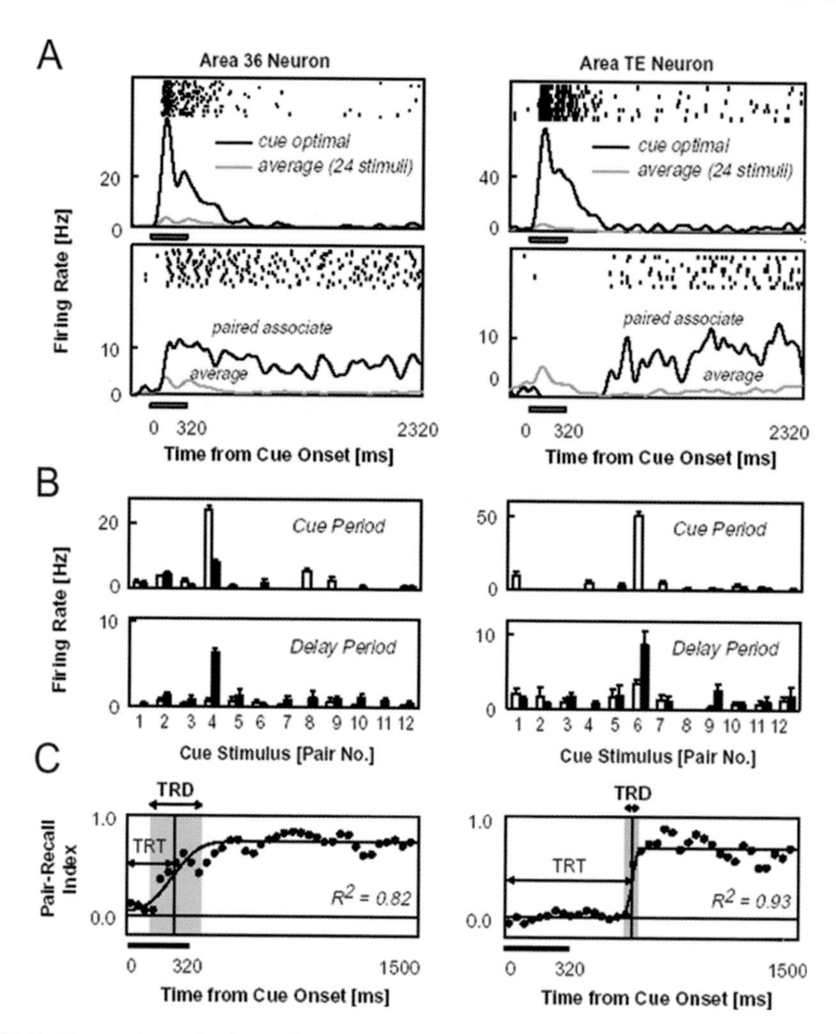

Figure 18.7 Comparison of pair-recall responses between A36 (left) and TE (right). (**A**) Raster displays and peri-stimulus time histograms (PSTHs) were aligned at the cue onset in trials with the cue-optimal stimulus as a cue (upper) and in trials with its paired associate as a cue (lower). In the PSTHs, black lines indicate responses to the cue-optimal stimulus (upper) or its paired associate (lower), and gray lines indicate mean responses to all 24 stimuli. (**B**) Mean discharge rates during the cue (upper) and delay (lower) periods are shown for each cue presentation (mean ± SEM). A stimulus-selective delay activity was closely coupled with a strong cue response to its paired associate. (**C**) Temporal dynamics of response correlation are shown; the values of the pair-recall index (PRI) are plotted against the time axis and are fitted with sigmoid functions (solid). The vertical lines, intersecting the best-fit sigmoid functions, indicate the transition times (TRTs). The shaded areas indicate the transition durations (TRDs). Based on Naya, Y., Yoshida, M., & Miyashita, Y. (2001). Backward spreading of memory-retrieval signal in the primate temporal cortex. *Science, 291,* 661–664. Used with permission.

the two eyes. During the binocular rivalry condition, one image is seen at a time while the other is perceptually suppressed. In their study, almost all TE neurons (≈90%) discharge exclusively when the driving stimulus is seen. They suggested that TE is the place where the neural activity reflects the brain's internal view of objects. The report by Sheinberg and

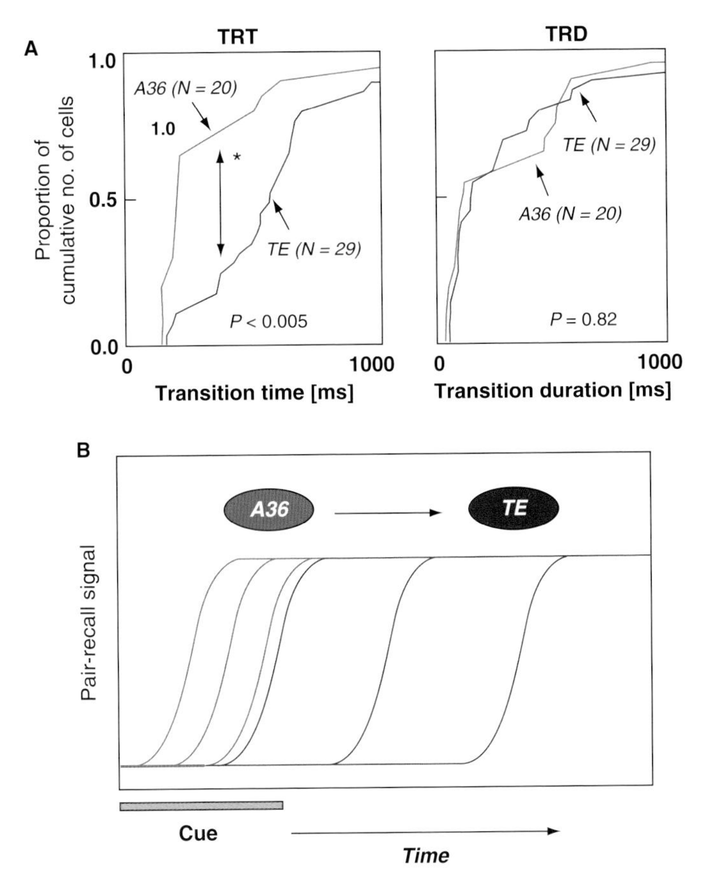

Figure 18.8 Time courses of pair-recall singnals for A36 and TE neurons. (**A**) Cumulative frequency histograms of the transition time (TRT; left) and the transition duration (TRD; right) for A36 (green) and TE (red) neurons. TRTs for A36 neurons were significantly shorter than those for TE neurons (asterisk, Kolmogorov-Smirnov test, $P < 0.005$). (**B**) A schematic view of the developments of pair-recall signals in A36 and TE. The similar distribution of TRD between A36 and TE indicates the similar rate of development of pair recall signals, but their onsets were different as illustrated by the different distributions of TRT between the two areas. Based on Naya, Y., Yoshida, M., & Miyashita, Y. (2001). Backward spreading of memory-retrieval signal in the primate temporal cortex. *Science, 291*, 661–664. Used with permission.

Logothetis (1997) suggests that the dominance of sought-target–related activity in TE during the VPA task may also reflect the brain's "internal view" that derives from LTM and corresponds to mental imagery in the VPA task. On the other hand, the perception-derived activity in A36 corresponds to a retrospective signal that is not necessarily required to solve the VPA task. This task-irrelevant retrospective signal is consistent with the findings of Yakovlev and colleagues (1998) who recorded activity in the perirhinal cortex during a DMS task. They reported that perirhinal neurons convey stimulus-selective sustained activity after a test stimulus that continued through the inter-trial interval, despite the fact that information about the test stimulus is not relevant for the subsequent trial. They suggested that this stimulus-selective retrospective activity may serve to generate long-term associations between visual items which are temporally adjacent but separated by some interval (Miyashita, 1988).

While the exclusive representation of a sought target in TE implicates its critical contribution to a mental image of a retrieved object, the additional representation of a cue stimulus in A36 may implicate its contribution to the formation of LTM because the retrospective signals can bridge the cue information to its pair regardless of the behavioral relevance. In addition, filtering out the automatically driven perception-derived signal may cause the slow (≈300 ms) backward transmission of LTM-derived signal from A36 to TE. This slow transmission from A36 (polymodal limbic area) to TE (sensory neocortex) might be coupled with an explicit recall of a specific modality.

Horizontal Cascades of Semantically Linked Information in the Limbic Cortex

The studies described previously show how an association of objects proceeds forward from TE to A36 and retrieval signals spreads backward from A36 to TE. The remaining question is how the pair-recall signal was generated from the association signal coded by type 1 neurons. This problem might be best addressed by combining a reverberation circuit model (Amit et al., 1994; Yakovlev et al., 1998) with a "selective-convergence" model ("*rb*-selective-convergence model"). According to the

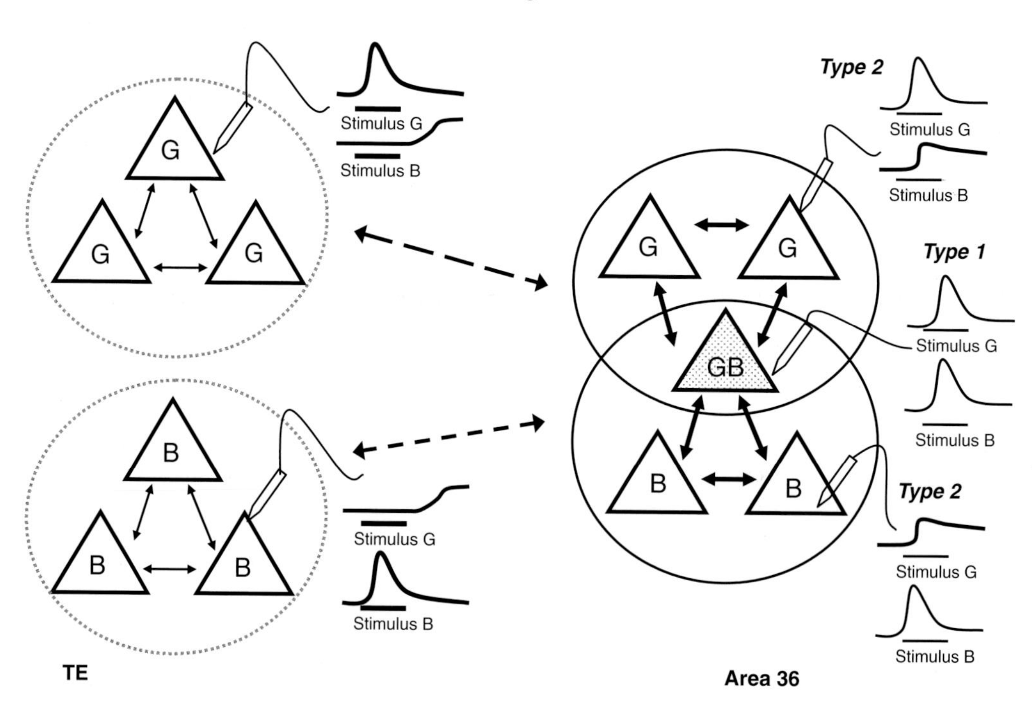

Figure 18.9 Selective-convergence model with reverberation circuits. Solid and dashed arrows denote synaptic connections within and between areas, respectively. When stimulus G is presented as a cue stimulus, TE neurons coding stimulus G are activated and the activation transmits forward to A36. The activated neuronal group in A36 (neuronal group G) contains type 1 neurons that code both stimulus B as well as stimulus G. The activation of neuronal group G elicits the following activation of the neuronal group B within A36 via type 1 neurons. Finally, the information of stimulus B spreads backward from A36 to TE. When stimulus B is presented as a cue stimulus, the signal transmits in reverse. Expected neuronal responses are shown as peri-stimulus time histograms (PSTHs) for each cue stimulus.

reverberation circuit model, the selective delay activity in IT cortex is maintained by recurrent synaptic feedback between interconnected neurons within a local module coding an individual stimulus. Because the percentage of neurons that showed stimulus-selective sustained activity was much larger in A36 (53% of the cue-selective neurons) than TE (21%) (Naya et al., 2003b), the synaptic connections within each module may be stronger in A36 (denoted as thick arrows in Fig. 18.9) than in TE (thin arrows). Considering that type 1 neurons belong to neuronal ensembles coding either of the paired associates, an activation of neuronal ensemble coding a cue stimulus may spread to another neuronal ensemble coding its paired associate via type 1 neurons (Fig. 18.9).

Taken together, long-term stimulus-stimulus association memory is stored in perirhinal cortex as a form of partial overlapping neuronal ensembles coding individual items. The partial overlap is constructed by pair-coding neurons, particularly type 1 neurons. Type 1 neurons receive direct projections from separate TE neuron groups coding individual visual items. Next, the retrieval from the long-term association memory is divided into two processes. The first retrieval process occurs in perirhinal cortex as a horizontal cascade of activation from one neuronal ensemble to another. Second, the retrieved information originated in perirhinal cortex spreads backward to TE. The activated TE neurons represent a mental image of the retrieved item. Thus, according to this model, the perirhinal cortex is critical for the storage and retrieval of long-term association memory; however, the interaction with neocortex (e.g., TE) is essential for its implementation.

PART II: NEURONAL SIGNALS UNDERLYING THE FORMATION OF NEW ASSOCIATIVE MEMORIES

Strong and convergent evidence from both human neuropsychological studies as well as experimental studies in animals suggests that the medial temporal lobe is critical in the initial stages of many forms of associative memory (Eichenbaum & Cohen, 2001; Scoville & Milner, 1957; Squire & Zola, 1996). In the following section, we turn to studies that have investigated the dynamic signals seen during new associative learning in nonhuman primates. We first describe a series of studies aimed at examining the neural correlates of learning the same kind of VPA task used by Miyashita and colleagues. These studies provide insight into how the striking pair-coding signals described in the perirhinal cortex may develop during the early learning process. We next turn to associative learning signals seen during conditional motor association (CMA) tasks. Not only is this latter task dependent on the integrity of MTL (Brasted et al., 2002, 2003; Murray & Wise, 1996; Murray et al., 2000; Rupniak & Gaffan, 1987; Wise & Murray, 1999;), but also multiple new associations can be learned concurrently within the course of a single recording session, affording the ability to examine in more detail the temporal relationship between the changes in neural activity and behavioral performance during fast new associative learning.

Associative Learning Signals During VPA tasks

As discussed in part I, Miyashita and his colleagues (Naya et al., 2003a; Sakai & Miyashita, 1991) showed that visual stimuli paired together in long-term memory become associated such that some neurons respond similarly to both associated images (i.e., pair-coding neurons). One hypothesis suggests that with learning, neurons that initially respond to one of the stimuli in a learned pair eventually come to respond to the learned paired associate of that original stimulus. This "tuning" of the neuron's stimulus-selective activity is thought to occur through the strengthening of connections between neurons that initially respond to the two stimuli that eventually become paired in memory. To test this hypothesis directly, Messinger and colleagues (2001) recorded activity in the perirhinal cortex and the adjacent visual area TE as animals learned novel associations during a VPA task

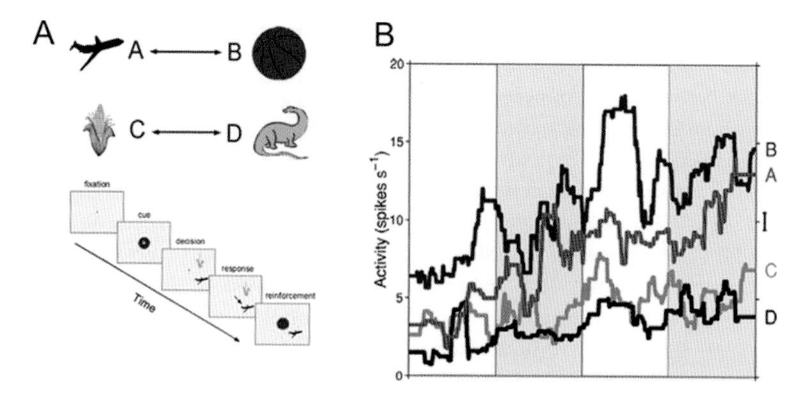

Figure 18.10 Visual paired associate learning task. (**A**) Schematic representation of the pairs of visual stimuli learned by the animal each day. In this case, airplane was always paired with basketball and corn was always paired with dinosaur. For each trial, after fixation, animals were shown a sample stimulus, followed by a choice array. The extinguishing of the fixation spot was the animal's cue to make an eye movement to the correct paired associate of the sample stimulus shown. (**B**) The change in neural activity of inferior temporal (IT) cells over the course of the training session to the four different stimuli. Note that the neuron's response to stimulus A comes to resemble the response to stimulus B (its paired associate) while responses to stimuli C and D remain low. Based on Messinger, A., Squire, L. R., Zola, S. M., & Albright, T. D. (2001). Neuronal representations of stimulus associations develop in the temporal lobe during learning. Proceedings of the National Academy of Sciences, 98, 12239–12244. Copyright (2001) National Academy of Sciences, U.S.A. Used with permission.

(Fig. 18.10A). In their paradigm, each day, the monkeys learned two new paired associates. Consistent with their initial working hypothesis, they found that over the course of the learning session, neurons in both the perirhinal cortex and area TE came to respond more similarly to two of the stimuli that were paired in memory (Fig. 18.10B). This increase in correlated response, measured by a "pair-coding index," was specific to those sessions in which behavioral performance increased significantly above chance levels, though peak performance in the last quartile of the session reached only 55.4% correct (chance = 50% correct). Values of the pair-coding index did not change in sessions when the animal did not learn. Because the changes in correlated neural activity appeared to parallel the time course of behavioral learning (i.e., gradually increasing over the four quartiles of the session), this suggests that these changes in the neuron's stimulus-selective response properties underlie new associative learning.

In a follow-up study using the same data set, Messinger and colleagues (2005) examined signals observed during the choice period of the task when the two choice stimuli were visible on the screen but before the behavioral response had been made. They reported that neural signals during the choice phase, like the sample phase analyzed in Messinger and colleagues (2001), also reflected learning. Using an ANOVA to contrast the effects of instructed (i.e., correct) stimuli and chosen stimuli, 54% of their IT cells reflected information about the instructed stimulus rather than the chosen stimulus in the choice period of the task. In other words, the neurons signaled the correct association even when the animal made an incorrect response. Consistent with the idea that this instructed stimulus signal develops with learning, they further showed that the instructed stimulus signal increased significantly between the first and second half of the session. Messinger and colleagues (2005) suggested that

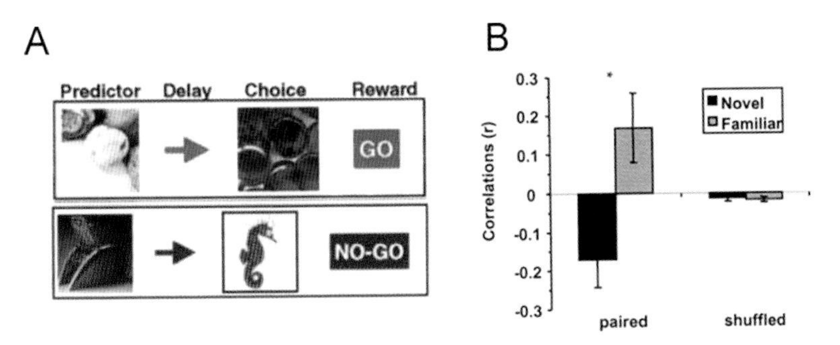

Figure 18.11 Predictive go/no-go task. (**A**) In the predictive go/no-go task, a predictor stimulus was shown for 500 ms followed by a delay interval and then a choice stimulus was presented for 500 ms. Animals learned which choice stimuli required a bar release ("go" stimulus) and which required a bar hold ("no-go" stimulus) and they could also learn the relationship between the predictor stimulus and the choice stimulus. (**B**) The correlation between the responses to paired predictor and choice stimuli were significantly stronger for familiar pairs (pairs seen for more than 1 day) compared to novel pairs. In contrast, the correlation between predictor and choice stimuli was near zero for randomly paired stimuli (shuffled). Based on Erickson, C. A., & Desimone, R. (1999). Responses of macaque perirhinal neurons during and after visual stimulus association learning. *Journal of Neuroscience, 19,* 10404–10416. Used with permission.

this apparent dissociation between neural representation of instructed stimuli on the one hand and behavioral choice on the other may be explained by the effect of behavioral states, attention, or motivation that may mask the true state of knowledge the animals accrue during learning.

Erickson and Desimone (1999) recorded activity in the monkey perirhinal cortex as animals learned a different kind of associative learning task (Fig. 18.11A). In their task, a predictor stimulus was followed by a choice stimulus. The choice stimulus could either signal the animals to release a bar ("go" condition) or continue holding a bar ("no-go" condition). In this task, the animal was not required to learn the explicit association between the predictor and the choice, but knowledge of this association could allow the animal to respond more quickly when the choice was presented. Learning was defined by a significantly faster responses on the "valid" trials illustrated in Figure 18.11A and probe trials in which the predictor stimuli between go and no-go trials were reversed. The neural responses to predictor and choice stimuli were initially uncorrelated for novel stimuli used for 1 day but became significantly correlated after several days of experience with the stimuli (Fig. 18.11B). In contrast, Messinger and colleagues (2001) reported that the changes in neural activity appeared to parallel behavioral learning within the time course of a single learning session. The difference in the timing of the neural signals seen in the VPA task used by Messinger and colleagues (2001) and the predictive go/no-go task used by Erickson suggests that the timing of the neural signals seen in the perirhinal cortex relative to behavior can vary depending on the specific task demands.

While the studies by Messinger and colleagues (2001, 2005) and Erickson and Desimone (1999) have provided important new insight into the neural correlates underlying the formation of new paired associates, because both tasks have a relatively slow learning curve, these tasks are not optimal to address the kind of fast and flexible associative learning typically attributed to the hippocampus (Eichenbaum & Cohen, 2001). To examine fast associative learning in the MTL, several groups have turned to the CMA learning task in which multiple new associations can be learned easily in a single training session. In contrast to the studies using the VPA task that have focused on neural activity in IT

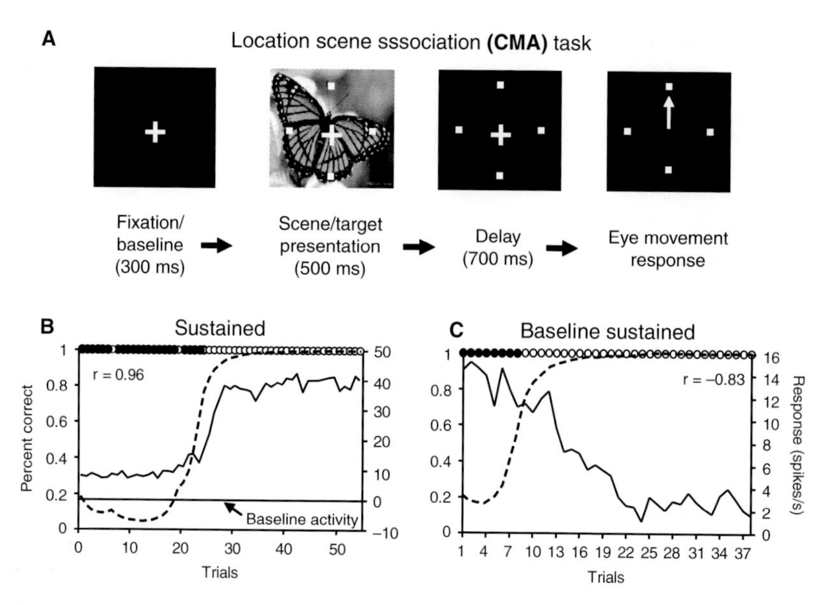

Figure 18.12 Location scene association (CMA) task and changing cells. **(A)** Schematic representation of the location scene association task. In this task, following fixation, animals are shown a set of four identical visual targets superimposed on a complex visual image (images used in task were all in color). Following a delay interval during which time the targets remained on the screen, but the scene disappeared, the animal was cued to make an eye movement response (illustrated schematically by the white arrow) to one of the targets. Only one of the targets was rewarded for each particular scene. Animals learned by trial and error to associate each new scene with a particular eye movement response. **(B,C)** The trial-by-trial probability correct performance (dotted line read from the left axis) as a function of the trial-by-trial activity of cells during either the scene or delay period of the task (solid line read from right axis) for a sustained (B) and baseline sustained (C) cell. Note the strong positive or negative correlation between neural activity and learning. Filled circles at the top of the graphs represent individual incorrect trial, while open circles denote correct trials. Based on Wirth, S., Yanike, M., Frank, L. M., Smith, A. C., Brown, E. N., & Suzuki, W. A. (2003). Single neurons in the monkey hippocampus and learning of new associations. *Science, 300,* 1578–1581.

cortex, the studies examining CMA learning in the medial temporal lobe have focused on neural representation in the hippocampus.

Associative Learning Signals During Conditional Motor Association Tasks

Both Wirth and colleagues (2003) and Cahusac and colleagues (1993) examined the patterns of hippocampal activity during the learning of novel CMAs (Cahusac et al., 1993; Wirth et al., 2003). This category of associative learning task, also known as arbitrary sensory motor mapping or conditional visuomotor learning, requires animals to associate a given sensory stimulus (typically a visual image presented on a computer screen) with a motor response (i.e., look right or touch right). Posttraining lesions to the MTL in monkeys impair the ability to learn novel CMAs, while well-learned associations remain unaffected (Brasted et al., 2002, 2003; Murray & Wise, 1996; Murray et al., 2000; Rupniak & Gaffan, 1987; Wise & Murray, 1999).

In the study by Wirth and colleagues (2003), animals performed a CMA task also referred to as the location scene association task (Fig. 18.12A). In this task, animals were first shown four identical target stimuli superimposed on a complex visual scene that filled the video monitor. Following a delay interval, during which the scene disappeared but the targets remained on the screen, the animal was cued to make a single eye movement to one of the peripheral targets on the screen. For each visual scene, only one of the four targets was associated with reward. Each day, animals typically learned two to four new scenes by trial and error. These new scenes were also randomly intermixed with well-learned "reference" scenes that the animals had seen for many months before the recording experiments began. Responses to the reference scenes were used to control for motor-related activity in the hippocampal cells.

Wirth and colleagues (2003) reported that 61% of the hippocampal cells examined responded selectively (i.e., differentially) to the different visual scenes shown in the task during the scene period, the delay period, or both periods of the task. Selectively responding cells with learning-related activity were identified by correlating a moving average of the raw neural activity with a moving average of the raw behavioral performance during learning. Using this criterion, 28% of the selectively responding cells showed a significant positive or negative correlation with learning. These cells were termed "changing cells." Two categories of changing cells were described. Sustained changing cells (54% of the population of changing cells) signaled learning with a change in neural activity that was maintained for as long as the cell was able to be recorded from (Fig. 18.12B). These cells typically started out with little or no response during the trial and signaled learning with a dramatic increase of activity during either the scene or delay periods of the task. This change in neural activity was strongly correlated with behavioral learning. A similar pattern of sustained changes in neural activity was also described by Cahusac and colleagues (1993) in the hippocampus. Importantly, because the same level of enhanced activity was not seen in response to the reference scene with the same rewarded target location in the Wirth et al. study (2003), this showed that the changes in neural activity could not be attributed to a pure motor-based response. Further analysis showed that these increases in neural activity also corresponded with an overall increase in the neuron's selective (i.e., differential) response to all stimuli examined (Fig. 18.13B).

In the Wirth et al. (2003) study, a second category of learning-related cells termed "baseline sustained changing cells" made up the remaining 45% of changing cells. Baseline sustained cells started out with a scene-selective response during either the scene or delay period of the task even before the animal learned the association and signaled learning by returning to baseline activity (Fig. 18.12C). Baseline sustained cells were also characterized by a significantly decreased selectivity (i.e., differential activity) to all stimuli examined (Fig. 18.13C). While Cahusac and colleagues (1993) did not describe baseline sustained cells, they did describe a different population of hippocampal neurons that only showed differential activity to the two visual stimuli transiently, near the time of learning (transient cells). While Wirth and colleagues (2003) looked for transient learning signals in the hippocampus, few if any were observed (unpublished observations). Transient learning signals have also been described in the striatum during a conditional motor learning task (Williams & Eskandar, 2006).

While the learning-related hippocampal activity described by both Wirth and colleagues (2003) and Cahusac and colleagues (1993) show that cells in this region signal learning, another critical question concerns the timing of the changing cells relative to behavioral learning. In other words, does the changing neural activity lead, lag, or occur in parallel with the animal's behavioral learning? To address this question, Wirth and colleagues (2003) compared the estimated trial number of neural change and the estimated trial number of learning for each changing cell. This comparison showed that hippocampal cells can signal learning before ($N = 18$), at the same time ($N = 1$), and after ($N = 18$) learning. Hippocampal cells signaled learning staring from as much as 13 trials before learning to 15 trials after learning (Fig. 18.14). Similar to the Wirth et al. (2003) study, Cahusac and colleagues (1993) reported that that the

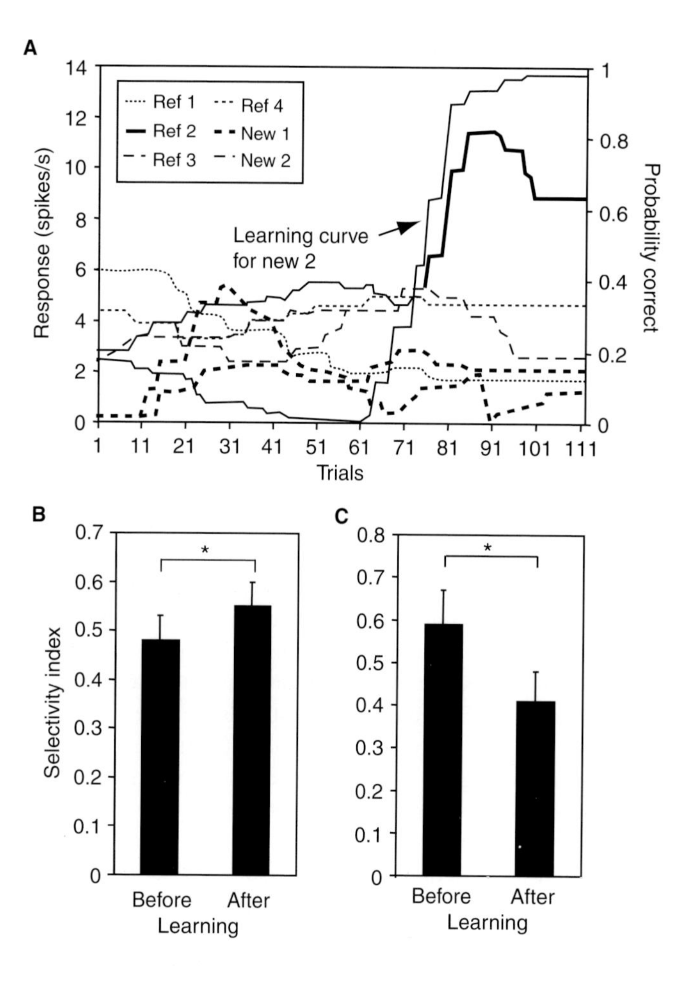

Figure 18.13 Changes in selectivity with learning. **(A)** Average response to four reference scenes and two new scenes over the course of the recording session for a sustained changing cell. The learning curve for New Scene 2 is illustrated in as the thick gray line. Trial number refers to the consecutive times that particular condition/scene was shown during the session. **(B)** Graph showing the significant increase in selectivity index for sustained changing cells after learning compared to before learning. **(C)** In contrast, baseline sustained changing cells decreased their selectivity after learning compared to before learning. Based on Wirth, S., Yanike, M., Frank, L. M., Smith, A. C., Brown, E. N., & Suzuki, W. A. (2003). Single neurons in the monkey hippocampus and learning of new associations. *Science, 300,* 1578–1581.

learning-related signals could occur within a wide range of lag or lead times relative to behavioral learning ranging mainly from 30 trials before learning to 40 trials after learning. Taken together, the studies of learning-related activity in the hippocampus could change either before learning, in parallel with learning, and after learning. This is the general pattern one would expect from a network with feedback connections that participates in the learning process. That is, some cells change before learning is expressed, possibly playing a role in driving the early learning process, but with feedback, other cells in the network get recruited later in the learning process, with the average population of cells changing around the time of learning.

SUMMARY AND CONCLUSIONS

The goal of this chapter was to describe our current understanding of the neurophysiological properties of the MTL memory system for both long-term associative memory signals seen after the association is well established and the dynamic associative learning signals observed during various tasks of new associative learning. We have focused primarily on two key tasks of association, the visual paired associate task studied most extensively in the perirhinal cortex (Naya et al., 1996, 2001, 2003a, 2003b; Sakai & Miyashita, 1991) and the conditional motor association task that has been studied in the hippocampus (Cahusac et al., 1993; Wirth et al., 2003). The results summarized in this chapter showed similarities as well as differences

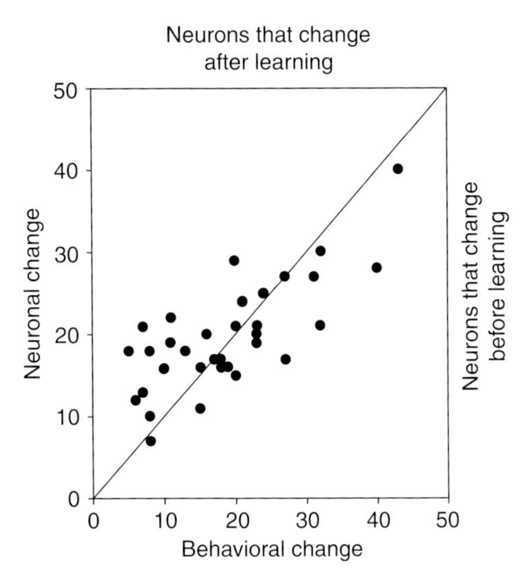

Figure 18.14 Temporal relationship between neural activity and behavior. Scatter plot illustrating the temporal relationship between trial number of behavioral change (i.e., learning) and trial number of neuronal change. Note that about half the cells change before or at the same time as learning, while the remaining half of the cells change before learning. Based on Wirth, S., Yanike, M., Frank, L. M., Smith, A. C., Brown, E. N., & Suzuki, W. A. (2003). Single neurons in the monkey hippocampus and learning of new associations. *Science, 300,* 1578–1581.

in both the behavioral correlates as well as the neural correlates underlying these two tasks. One key difference between the VPA task and the CMA task is the overall speed of learning. While multiple new conditional motor associations can be learned easily within the course of a single training session (Wirth et al., 2003), new visual-visual paired associates are more difficult to learn quickly (Messinger et al., 2001), but with experience, large sets of visual paired associates can be learned to a high level of performance (Naya et al., 2003a). In this way, the slower time course of learning for the VPA task suggests that it may be more analogous to semantic memory while the CMA task may be more analogous to quickly developing and more flexible relational memories thought to be dependent on the hippocampus (Eichenbaum & Cohen, 2001).

During learning of novel visual paired associates, Messinger and colleagues (2001) showed that perirhinal neurons change their stimulus-selective responses to come to respond more similarity to two stimuli that have been paired in memory. When assessed after VPA learning has been well established, strong pair-coding signals continue to be observed in the perirhinal cortex. The high proportion of pair-coding signals observed in the perirhinal cortex may reflect the important role of this area in storage of the long-term representation of the associated stimuli. These findings are also consistent with findings from lesion studies showing that damage to the perirhinal cortex impaired memory for well-learned visual paired associates in addition to impairing learning of new paired associates (Murray et al., 1993). In addition, pair-recall signals were also described in the perirhinal cortex with extensive previous training (Naya et al., 2001). These signals have been taken as a neural signal underlying retrieval of the learned paired associates. We further suggest that while the perirhinal cortex is critical for storage and retrieval of visual paired associates, area TE may be critical for the mental representation or imagery of the retrieved information.

In contrast to the VPA task, the CMA task is learned quickly and neural signals that parallel this relatively fast learning can be seen in the monkey hippocampus (Cahusac et al., 1993; Wirth et al., 2003). Like the signals in the perirhinal cortex, these signals in the hippocampus appear to represent a selective tuning of the cell's stimulus-selective response properties that changes with learning (Wirth et al., 2003). Hippocampal learning-related cells could either become more selective (sustained changing cells) or less selective (baseline sustained changing cells) or provide a temporally specific signal of learning (transient cells). These dynamic changes may represent an overall tuning of the hippocampal population with new associative learning.

While these findings point to an important role of the perirhinal cortex in slow semantic-like visual paired associative learning and memory and toward the hippocampus for fast new associative learning, an important question for future research concerns how these two areas may interact during both the learning and long-term representation of VPA and CMA tasks. For example, while findings from lesion studies suggest that the hippocampus is not needed for new VPA learning (Murray et al., 1993), neural correlates in this structure may be involved in contextual learning associated with the task. Similarly, we now have evidence that the perirhinal cortex signals a new CMA in a similar way to the hippocampus suggesting cooperative learning signals across these regions (Yanike et al., 2009). It will be important to record simultaneously across different MTL structures during both VPA and CMA learning to better understand the relative roles of the MTL during new associative learning.

Another important but unanswered question concerns understanding the mechanisms underlying the gradual consolidation of both VPA and CMA information in LTM. What are the cellular mechanisms underlying the role of the perirhinal cortex in both initial learning and long-term storage, and how does the communication of the perirhinal cortex with area TE change over time with consolidation? Similarly for the CMA task, how does the long-term representation of the CMAs eventually move from the hippocampus early on to other as-yet-unidentified storage regions? Important parallel studies in supplementary eye field and frontal eye field (Chen & Wise, 1995, 1997), striatum (Brasted & Wise, 2004; Pasupathy & Miller, 2005; Williams & Eskandar, 2006), and prefrontal cortex (Asaad et al., 1998) suggest that all of these other structures can also participate importantly in the initial learning and possibly the storage of CMAs. It will be of great interest to examine the contribution of these different brain areas in parallel during the course of the establishment of LMTs of CMAs. Both the CMA and VPA task will continue to be powerful tools to understand both how different MTL areas interact during associative learning and memory tasks and how those signals develop and change over time with the development of LTM.

REFERENCES

Amit, D. J., Brunel, N., & Tsodyks, M. V. (1994). Correlations of cortical Hebbian reverberations: Theory versus experiment. *Journal of Neuroscience, 14,* 6435–6445.

Asaad, W. F., Rainer, G., & Miller, E. K. (1998). Neural activity in the primate prefrontal cortex during associative learning. *Neuron, 21,* 1399–1407.

Brasted, P. J., Bussey, T. J., Murray, E. A., & Wise, S. P. (2002). Fornix transection impairs conditional visuomotor learning in tasks involving nonspatially differentiated responses. *Journal of Neurophysiology, 87,* 631–633.

Brasted, P. J., Bussey, T. J., Murray, E. A., & Wise, S. P. (2003). Role of the hippocampal system in associative learning beyond the spatial domain. *Brain, 126,* 1202–1223.

Brasted, P. J., & Wise, S. P. (2004). Comparison of learning-related neuronal activity in the dorsal premotor cortex and striatum. *European Journal of Neuroscience, 19,* 721–740.

Buckley, M. J., & Gaffan, D. (1998). Perirhinal cortex ablation impairs visual object identification. *Journal of Neuroscience, 18,* 2268–2275.

Bunsey, M., & Eichenbaum, H. (1993). Critical role of the parahippocampal region for paired-associate learning in rats. *Behavioral Neuroscience, 107,* 740–747.

Cahusac, P. M., Rolls, E. T., Miyashita, Y., & Niki, H. (1993). Modification of the responses of hippocampal neurons in the monkey during the learning of a conditional spatial response task. *Hippocampus, 3,* 29–42.

Chen, L. L., & Wise, S. P. (1995). Supplementary eye field contrasted with the frontal eye field during acquisition of conditional oculomotor associations. *Journal of Neurophysiology, 73,* 1122–1134.

Chen, L. L., & Wise, S. P. (1997). Conditional oculomotor learning: Population vectors in the supplementary eye field. *Journal of Neurophysiology, 78,* 1166–1169.

Corkin S, (1968) Acquisition of motor skill after bilateral medial temporal-lobe excision. *Neuropsychologia 6,* 255-65.

Eichenbaum, H., & Cohen, N. J. (2001). *From conditioning to conscious recollection.* New York: Oxford University Press.

Eichenbaum, H., Yonelinas, A. R., & Ranganath, C. (2007). The medial temporal lobe and recognition memory. *Annual Review of Neuroscience, 30,* 123–152.

Erickson, C. A., & Desimone, R. (1999). Responses of macaque perirhinal neurons during and after visual stimulus association learning. *Journal of Neuroscience, 19,* 10404–10416.

Felleman, D. J., & Van Essen, D. C. (1991). Distributed hierarchical processing in the primate cerebral cortex. *Cerebral Cortex, 1,* 1–47.

Gabrieli J.D., Milberg W., Keane M.M. & Corkin S. (1990) Intact priming of patterns despite impaired memory. *Neuropsychologia. 28,* 417–427.

Janssen, P., Vogels, R., & Orban, G. A. (2000). Selectivity for 3D shape that reveals distinct areas within macaque inferior temporal cortex. *Science, 288,* 2054–2056.

Lashley, K. S. (1929). *Brain mechanisms and intelligence: A quantitative study of injuries to the brain.* Chicago: Chicago University Press.

Messinger, A., Squire, L. R., Zola, S. M., & Albright, T. D. (2001). Neuronal representations of stimulus associations develop in the temporal lobe during learning. *Proceedings of the National Academy of Sciences, 98,* 12239–12244.

Messinger, A., Squire, L. R., Zola, S. M., & Albright, T. D. (2005). Neural correlates of knowledge: Stable representation of stimulus associations across variations in behavioral performance. *Neuron, 48,* 359–371.

Milner, B. (1962) Les troubles de la mémoire accompagnant les lésions hippocampiques bilatérales. In *Physiologie de l'hippocampe* (pp. 257–272) Paris: CNRS.

Milner, B. (2005). The medial temporal-lobe amnesic syndrome. *Psychiatric Clinics of North America, 28,* 599–611.

Mishkin, M. (1978). Memory in monkeys severely impaired by combined but not by separate removal of amygdala and hippocampus. *Nature, 273,* 297–298.

Mishkin, M. (1982). A memory system in the monkey. *Philosophical Transactions of the Royal Society of London [Biology], 298,* 83–95.

Miyashita, Y. (1988). Neuronal correlate of visual associative long-term memory in the primate temporal cortex. *Nature, 335,* 817–820.

Miyashita, Y., & Chang, H. S. (1988). Neuronal correlate of pictorial short-term memory in the primate temporal cortex. *Nature, 331,* 68–70.

Murray, E. A., Bussey, T. J., & Wise, S. P. (2000). Role of prefrontal cortex in a network for arbitrary visuomotor mapping. *Experimental Brain Research, 133,* 114–129.

Murray, E. A., Gaffan, D., & Mishkin, M. (1993). Neural substrates of visual stimulus-stimulus association in rhesus monkeys. *Journal of Neuroscience, 13,* 4549–4561.

Murray, E. A., & Wise, S. P. (1996). Role of the hippocampus plus subjacent cortex but not amygdala in visuomotor conditional learning in rhesus monkeys. *Behavioral Neuroscience, 110,* 1261–1270.

Naya, Y., Sakai, K., & Miyashita, Y. (1996). Activity of primate inferotemporal neurons related to a sought target in pair-association task. *Proceedings of the National Academy of Sciences, 93,* 2664–2669.

Naya, Y., Yoshida, M., & Miyashita, Y. (2001). Backward spreading of memory-retrieval signal in the primate temporal cortex. *Science, 291,* 661–664.

Naya, Y., Yoshida, M., & Miyashita, Y. (2003a). Forward processing of long-term associative memory in monkey inferotemporal cortex. *Journal of Neuroscience, 23,* 2861–2871.

Naya, Y., Yoshida, M., Takeda, M., Fujimichi, R., & Miyashita, Y. (2003b). Delay-period activities in two subdivisions of monkey inferotemporal cortex during pair association memory task. *European Journal of Neuroscience, 18,* 2915–2918.

Pasupathy, A., & Miller, E. K. (2005). Different time courses of learning-related activity in the prefrontal cortex and striatum. *Nature, 433,* 873–876.

Penfield, W., & Perot, P. (1963). The brain's record of auditory and visual experience. A final summary and discussion. *Brain, 86,* 595–696.

Rupniak, N. M., & Gaffan, D. (1987). Monkey hippocampus and learning about spatially directed movements. *Journal of Neuroscience, 7,* 2331–2337.

Sakai, K., & Miyashita, Y. (1991). Neural organization for the long-term memory of paired associates. *Nature, 354,* 152–155.

Sakai, K., Naya, Y., & Miyashita, Y. (1994). Neuronal tuning and associative mechanisms in form representation. *Learning & Memory, 1,* 83-105.

Saleem, K. S., & Tanaka, K. (1996). Divergent projections from the anterior inferotemporal area TE to the perirhinal and entorhinal cortices in the macaque monkey. *Journal of Neuroscience, 16,* 4757–4775.

Scoville, W. B., & Milner, B. (1957). Loss of recent memory after bilateral hippocampal lesions. *Journal of Neurology, Neurosurgery, and Psychiatry, 20,* 11–21.

Sheinberg, D. L., & Logothetis, N. K. (1997). The role of temporal cortical areas in perceptual organization. *Proceedings of the National Academy of Sciences,* 94, 3408–3413.

Squire, L. R., Stark, C. E., & Clark, R. E. (2004). The medial temporal lobe. *Annual Review of Neuroscience, 27,* 279–306.

Squire, L. R., Wixted, J. T., & Clark, R. E. (2007). Recognition memory and the medial temporal lobe. *Nature Review Neuroscience, 8,* 872–883.

Squire, L. R., & Zola, S. M. (1996). Structure and function of declarative and nondeclarative memory systems. *Proceedings of the National Academy of Sciences, 93,* 13515–13522.

Squire, & Zola-Morgan. (1991). The medial temporal lobe memory system. *Science, 253,* 1380–1386.

Suzuki, W. A., & Amaral, D. G. (1994). Perirhinal and parahippocampal cortices of the macaque monkey: Cortical afferents. *Journal of Comparative Neurology, 350,* 497–533.

Tanaka, K. (1996). Inferotermporal cortex and object vision. In: W. M. Cowan, E. M. Shooter, C. F. Stevens, & R. F. Thompson (Eds.), *Annual review of neuroscience* (pp. 109–140). Palo Alto, CA: Annual Reviews Inc.

Tulving E., (1972) Episodic and semantic memory. In: E. Tulving & W. Donaldson (Eds.), *Organization of memory.* New York: Academic Press.

von Bonin, G., & Bailey, P. (1947). *The neocortex of Macaca mulatta.* Urbana, IL: University of Illinois Press.

Wechsler, D. (1987) Wechesler memory Scale-Revised. San Antonio, Tex.: Psychological Corporation, Harcourt Brace Jovanovich.

Williams, Z. M., & Eskandar, E. N. (2006). Selective enhancement of associative learning by microstimulation of the anterior caudate. *Nature Neuroscience, 9,* 562–568.

Wirth, S., Yanike, M., Frank, L. M., Smith, A. C., Brown, E. N., & Suzuki, W. A. (2003). Single neurons in the monkey hippocampus and learning of new associations. *Science, 300,* 1578–1581.

Wise, S. P., & Murray, E. A. (1999). Role of the hippocampal system in conditional motor learning: Mapping antecedents to action. *Hippocampus, 9,* 101–117.

Yakovlev, V., Fusi, S., Berman, E., & Zohary, E. (1998). Inter-trial neuronal activity in inferior temporal cortex: A putative vehicle to generate long-term visual associations. *Nature Neuroscience, 1,* 310–317.

Yanike, M., Wirth, S., Smith, A. C., Brown, E. N., & Suzuki, W. A. (2009). Comparison of associative learning-related signals in the macaque perirhinal cortex and hippocampus. *Cerebral Cortex, 19,* 1064-1078.

Yoshida, M., Naya, Y., & Miyashita, Y. (2003) Anatomical organization of forward fiber projections from area TE to perirhinal neurons representing visual long-term memory in monkeys. *Proceedings of the National Academy of Sciences, 100,* 4257–4262.

Zola-Morgan, S., & Squire, L. R. (1985). Medial temporal lesions in monkeys impair memory on a variety of tasks sensitive to human amnesia. *Behavioral Neuroscience, 99,* 22–34.

Zola-Morgan, S., & Squire, L. R. (1990). The neuropsychology of memory. Parallel findings in humans and nonhuman primates. *Annals of the New York Academy of Sciences, 608,* 434–450.

CHAPTER 19

Neurobiology of Social Behavior

Dario Maestripieri

The Primate Order comprises over 300 species and a wide range of different social systems (Smuts et al., 1987). Very few species have a fully solitary lifestyle. Most of the others are characterized by permanent associations between two or more adults and their young. Pair living is relatively rare, being characteristic of only 3% to 4% of primate species. Some of these socially monogamous species (e.g., tamarins and marmosets) have a flexible social organization, in which one additional adult male or female may be temporarily associated with the breeding pair. Stable groups with one adult male and several reproductively active females (i.e., harems) are shown by approximately 35% of primate species. The most common type of social organization in nonhuman primates consists of social groups with multiple adult males and females and their young. These multimale/multifemale social groups represent the stable form of organization of about 45% primate species, with an additional 15% of species showing fluctuations between such groups and groups with a harem structure. Variation in social organization among primate species has been explained on the basis of variation in ecological variables such as diet, food-related competition and cooperation and need for protection from predators or infanticide (e.g. Dunbar, 1988). Phylogenetic history, however, also accounts for variation in social organization, as groups of closely related species and genera tend to have similar social systems (Rendall & Di Fiore, 1995). Finally, within-species variation in social organization may result from variation in local ecological and demographic conditions.

Regardless of the variation in social systems, it is clear that nonhuman primates are generally highly social organisms, in which successful survival and reproduction depend on complex social interactions with other conspecifics. Accordingly, most primate species exhibit complex behavioral adaptations for communication, affiliation, aggression, mating, and parenting. Group-living monkeys and apes often use vocalizations to alert others of the presence of food and predators, and also to keep in contact with other group members during travel. Facial expressions and body postures play an important role in close-range affiliative, agonistic, and sexual interactions, particularly among Old World monkeys and the great apes. Olfactory and tactile signals are also used in these and other contexts, although olfactory communication is relatively underdeveloped in most primate species relative to other mammals. Social bonds between family and group members are established and maintained through contact, proximity, and grooming. Grooming is an altruistic behavior that can be exchanged for tolerance, sex, or coalitionary support during fights. Aggression and submission, often expressed with facial expressions and body postures, result in the establishment of dominance relationships and hierarchies. In many species, conflict outcomes and dominance ranks are determined not by the individuals' body size and strength but by coalitionary support received from other individuals. Fights between two individuals often extend to other group members, whose intervention may reflect attempts to protect a family member or political strategies involving complex cost/benefit analyses. Both affiliative and dominance relationships depend on individuals' memory of their past interactions and their outcome, as well as expectations about future interactions. High dominance rank may confer survival and reproductive benefits such as greater access to food and safe sites or access to more and higher-quality mating partners. Sexual monogamy is rare in primates, or may be nonexistent since extrapair copulations have been reported in socially monogamous primates. Polyandrous systems, in which one female mates with multiple males, are also rare. Polygynous harem systems are more common among species living in small groups, while those living in large multimale/multifemale groups typically have promiscuous mating systems. Successful mating, especially in promiscuous species, depends not only on features that advertise fertility, health, or strength but also on complex social strategies to increase one's attractiveness and deal with competition. Successful reproduction, at least for females, also depends on parental investment in the offspring. Male care is rare among primates, while female care involves not only lactating but also carrying and protecting offspring from predators and conspecifics. Maternal care in many primate species may extend well beyond the period of offspring nutritional dependence. For example, in species with female philopatry and male dispersal such as most cercopithecine monkeys, bonds between female relatives may last throughout the lifespan.

The behavioral adaptations for social life that characterize many primate species must be supported by underlying neurobiological mechanisms so that a relation is expected between complexity of social behavior and complexity of the brain. Consistent with this expectation, studies have shown that there is, across primate species, a linear relation between the average size of the social groups and the ratio between the size of the neocortex and the rest of the brain. Species that live in larger social groups tend to have a larger neocortex ratio, suggesting that complex social life in large groups is associated with increased cognitive capacity (Dunbar, 1992). Brain size, however, is a crude measure of brain function, just as group size is a crude measure of social complexity. To understand the relation between brain evolution and the evolution of sociality in primates, one needs to have a much deeper knowledge of how specific social behaviors are produced or regulated by specific brain structures or neurochemical systems. Unfortunately, our knowledge of brain-behavior relationships in nonhuman primates is very preliminary. Although in recent years there have been major advances in our understanding of the neural mechanisms underlying social behavior in other mammalian species, particularly rodents (e.g., Young, 2002), research on the neurobiology of social behavior in nonhuman primates has lagged far behind. With the availability of new research techniques such as brain imaging, however, the investigation of the neurobiological substrates of primate social behavior will be a promising area of research in the next few decades.

In this chapter, I review and discuss our current knowledge of the neurobiological regulation of affiliative, aggressive, sexual, and parental behavior in nonhuman primates. Communication is clearly an important component of primate social behavior, but this topic is addressed elsewhere in this volume (see Chapters 5 and 25). Similarly, although the perception and processing of social stimuli is clearly a prerequisite for social behavior, brain mechanisms underlying social cognition are addressed in other chapters (see Chapter 26). This chapter, instead, focuses on social behavior expressed in the context of interactions between two or more individuals.

Different components of social behavior such as affiliation, aggression, mating, and parenting may or may not share some of the same neural substrates, but they probably share common neurochemical controls. For example, endogenous opioids and oxytocin have been implicated in the regulation of most, if not all, social behaviors in rodents (see later).

Similarly, the activity of neurotransmitter systems involving the monoamines dopamine, norepinephrine, and serotonin has been shown to affect the expression or inhibition of a wide range of social activities in many mammals and other vertebrates (see later). The study of neurochemical control of social behavior in primates has been mainly pursued with correlational approaches, in which measures of peptides or monoamines or their metabolites in blood or cerebrospinal fluid (CSF) are analyzed in relation to social behavior, and to a lesser extent, with pharmacological manipulations of neurochemical systems. Attempts to identify specific areas of the brain involved in the regulation of social behavior have mostly been made with lesion studies. Other approaches involving, for example, the electrical stimulation of specific areas of the brain or the imaging of brain activation with positron emission tomography (PET) or functional magnetic resonance imaging (fMRI) have been used less frequently in the context of social behavior studies (but see Rilling et al., 2001, 2004a; Snowdon et al., 2006). Single neuron recording has often been used to address issues of social perception (e.g., to study the processing and recognition of faces and facial expressions) but rarely for social behavior. In this chapter, first I review research on the neurochemical control of primate social behavior, particularly studies of endogenous opioids, oxytocin and vasopressin, and the brain monoamine systems. Then, I review the results of brain lesion studies investigating the neural substrates of primate social behavior. I conclude the chapter by summarizing the main trends emerging from this literature review and discussing future research directions.

NEUROCHEMICAL CONTROL OF SOCIAL BEHAVIOR

Research conducted with other mammalian species, mostly rodents, has suggested that neuropeptides such as endogenous opioids, oxytocin, and vasopressin are good neurochemical candidates for regulating complex social behaviors. These neuropeptides can influence behavior in conjunction with monoaminergic neurotransmitter systems. For example, in a recent model of the neurobiological regulation of affiliation in mammals, Depue and Morrone-Strupinksy (2005) have argued that dopamine plays an important role in incentive-reward motivation processes (see Chapter 17) associated with the appetitive phase of affiliation, endogenous opioids provide the neurochemical basis for the reward processes associated with the consummatory phase of affiliation, and oxytocin and vasopressin enhance the perception and memory of affiliative stimuli. In this section of the chapter, I review our knowledge of the neurochemical control of primate social behavior, first focusing on neuropeptides, and then on the monoamine systems.

Endogenous Opioids

Almost 30 years ago, Panksepp and colleagues proposed that brain endogenous opioids play a crucial role in regulating the establishment, maintenance, and disruption of social bonds in mammals and birds (Panksepp et al., 1980). This hypothesis was developed from the observation that the emotional states accompanying the formation of social attachments, the weaning of social bonds, and the distress arising from social separation appear to share similarities with the characteristics of opiate addiction—that is, the development of dependence, tolerance, and withdrawal (Panksepp et al., 1980, 1999). The hypothesized relationship between endogenous opioids and social attachments was framed within the general theory that the emotional substrates of attachments are an evolutionary outgrowth of more primitive brainstem and limbic circuits in the mammalian brain that originally subserved basic physiological needs such as energy balance, thermoregulation, or pain perception.

According to Panksepp and collaborators, a release of endogenous opioids following the exchange, and especially the receipt, of affiliative behavior generates the feeling of pleasure and gratification that arise from the interaction, whereas a reduction in endogenous opioids results in emotional distress and promotes the need to seek and maintain proximity with a social partner. Although some studies of birds and mammals, including nonhuman primates (e.g., Barr et al., 2008; Kalin et al., 1988; Kraemer, 1992), have provided evidence for the involvement of the opioid system in distress separation responses, there is growing consensus that the neurobiological systems regulating separation distress responses are different from those mediating social rewards (e.g., Insel, 1992; Panskepp et al., 1999). Therefore, more recent theories about opioids and social behavior view endogenous opioids as playing a crucial role in the consummatory phase of affiliative interactions, and therefore in the strengthening of social bonds, but less so in the motivation to establish these bonds or in the response to their disruption or termination (Depue & Morrone-Strupinksy, 2005).

The most systematic attempt to investigate the role of endogenous opioids in the regulation of primate social behavior has been made by Keverne and collaborators with a series of studies in talapoin monkeys and rhesus macaques. They argued that of the different endogenous opioids in the brain, β-endorphin may be the best candidate for the regulation of social behavior, and given the difficulties of measuring this peptide directly in the brain of live primates, they measured its concentration in the CSF (Martensz et al., 1986). β-Endorphin does not gain access to CSF from the blood when CSF and the cerebral extracellular fluid are in equilibrium; therefore, CSF levels of β-endorphin provide a measure of its presence in the extracellular fluid of the brain, a reasonable marker of the level of activity in intracerebral β-endorphin–containing systems (Martensz et al., 1986).

Consistent with Panksepp's hypothesized relationship between endogenous opioids and affiliation, Keverne and colleagues (1989) provided evidence for an association between grooming behavior and opioid release in talapoin monkeys. Specifically, they reported that moving adult talapoin monkeys from isolation housing to pair housing, and therefore providing the monkeys with an opportunity to exchange grooming behavior, was accompanied by a significant increase in CSF concentrations of β-endorphin. There were no significant correlations, however, between the amount of grooming given or received and the concentrations of β-endorphin or their increases following pair housing. Nevertheless, Keverne and colleagues (1989) suggested that the opioid release may have been caused by the tactile stimulation associated with grooming in newly formed pairs, and that this effect may be similar to the stimulation of endogenous opioid release by acupuncture in humans. That endogenous opioids are sensitive to social variables had also been suggested by a previous study, in which CSF concentrations of β-endorphin were found to be highest in male talapoin monkeys of low rank and lowest in those of high rank (Martensz et al., 1986).

A relatively large number of primate studies has attempted to test the following two predictions of Panksepp's hypothesis: (1) the administration of an exogenous opioid such as morphine should create a feeling of social comfort and reduce the motivation to seek social contact or decrease the expression of affiliative behavior, and (2) the blockade of endogenous opioid receptors should increase the need for social attachment and therefore the solicitation of affiliative behavior from social partners. Research by Keverne and colleagues showed that acute treatment of adult talapoin monkeys with the opioid receptor blockers naloxone or naltrexone increased their grooming solicitations and resulted in more grooming received from other individuals (Fabre-Nys et al., 1982; Keverne et al., 1989; Meller et al., 1980). These effects were observed in both pair-housed and in group-living individuals, were dose dependent and stronger for females than for males, and were specific for allogrooming behavior: Self-grooming, aggressive behavior, and locomotor activity were not affected (but in one study male sexual behavior was decreased; Meller et al., 1980). Naltrexone administration was also associated with an increase in testosterone, cortisol, and prolactin, suggesting that some of the effects of opiate receptor blockade on social behavior may have been hormonally mediated (Meller et al., 1980). An increase in the number of grooming solicitations and in the amount of grooming received from other individuals following naloxone or naltrexone treatment was also reported by two studies of group-living rhesus macaque adult females (Graves et al., 2002; Martel et al., 1995). Keverne and colleagues (1989) found that acute treatment of pair-housed monkeys with nonsedative doses of morphine resulted in a significant decrease in the number of grooming solicitations as well as a decrease in grooming performed. A study of common marmosets, however, reported that morphine administration increased the frequency and duration of social play but had no effects on social contact or grooming (Guard et al., 2002). Since grooming behavior in Old World monkeys may have different functions and be regulated by different mechanisms than grooming behavior in New World primates, the findings of the studies reviewed previously are generally consistent with the hypothesis that endogenous opioids may mediate the rewarding properties of affiliative interactions between adults.

Studies manipulating the opioid system of immature monkeys have produced results consistent with the hypothesized relation between opioids and attachment (Kalin et al., 1988, 1995; Martel et al., 1995; Schino & Troisi, 1992). In a study of group-living long-tailed macaques, juveniles receiving an acute administration of naloxone increased their proximity-seeking behavior toward their mothers, displayed more grooming solicitations to both their mothers and other group members, and received more grooming from them (Schino & Troisi, 1992). Grooming done by the juveniles was not affected by naloxone, while self-grooming decreased. An increase in contact seeking with the mother was also observed in infant and juvenile rhesus macaques treated with naloxone (Martel et al., 1995). Miczek and colleagues (1981) reported that the acute administration of nonsedative doses of morphine in squirrel monkeys decreased the rate of affiliative behavior shown by juveniles toward their mothers. Taken together, these results are consistent with the hypothesis that infant attachment and adult attachment share a common neurochemical substrate (Nelson & Panksepp, 1998).

The role of the opioid system in mediating maternal attachment has been investigated in three primate studies producing conflicting results. In one study, the opioid system was pharmacologically manipulated after mother-infant separation and reunion in rhesus macaques (Kalin et al., 1995). Morphine decreased clinging with the infant during the first 30 minutes of reunion, whereas naltrexone increased clinging. In a study of socially living rhesus mothers and infants, however, naloxone reduced both maternal grooming and maternal restraining of the infant, suggesting decreased rather than increased attachment to the infant (Martel et al., 1993). In this study, however, the effects of naloxone on affiliative interactions between mothers and other adults were also contrary to the expectations. In fact, the mothers treated with naloxone showed reduced number of grooming solicitations and reduced amount of grooming received from other individuals. Finally, in another study of rhesus macaques, naltrexone had no significant effects on any aspects of maternal behavior, including abusive parenting (Graves et al., 2002). Although some of these inconsistencies may be due to methodological differences between studies, further research is needed before any firm conclusions can be drawn regarding the relationship between opioids and maternal attachment.

Researchers investigating endogenous opioid release following affiliative interactions made the assumption that the CSF concentrations of β-endorphin mainly reflect the production of this peptide by neurons that originate from the arcuate nucleus of the hypothalamus (e.g., Keverne et al., 1989; Martensz et al., 1986). These neurons project to brain regions that are rich in opiate receptors such as the brainstem, basal ganglia, and areas of the hypothalamus, amygdala, cerebellum, and raphe nuclei. Of the different families of opiate receptors, many of which have multiple subtypes, the μ-opiate receptor family seems to be the most directly implicated in the regulation of social behavior, and β-endorphin has high affinity for these receptors. In brain areas rich in μ-receptors, β-endorphin neurons interact with dopaminergic and serotonergic neurons, as well as with neurons using oxytocin and vasopressin (at least in rodents). It has been suggested that interactions between μ-opiates and dopamine neurons in the ventral tegmental area (VTA) of the hypothalamus produce the experience of reward associated with the appetitive and consummatory phases of affiliative interactions, while serotonergic input to the hypothalamus via the raphe nuclei may result in reduced arousal and facilitation of opioid-mediated feelings of gratification following affiliation (Depue & Morrone-Strupinksy, 2005). Finally, studies of rats have suggested that oxytocin and vasopressin may facilitate the rewarding effects of endogenous opiates, as oxytocin neurons in the paraventricular nucleus of the hypothalamus project to the β-endorphin neurons in the arcuate nucleus and increase their release of opioids (Csiffary et al., 1992). Oxytocin, vasopressin, and the monoamines, however, can affect affiliation and other forms of social behavior also through mechanisms that are not dependent on endogenous opiates.

Oxytocin and Vasopressin

Oxytocin and vasopressin are 9-amino acid peptides synthesized in the supraoptic and paraventricular nuclei of the hypothalamus and released into systemic circulation from neurons in the posterior pituitary gland. Hypothalamic neurons synthesizing oxytocin and vasopressin also project to various areas of the brain, and receptors for these peptides have been found in the limbic system of some mammalian species, particularly rodents (Gimpl & Fahrenholz, 2001). Until recently, oxytocin receptors could not be easily identified in the primate brain (Toloczko et al., 1997; Winslow, 2005), but the presence of such receptors has recently been inferred in the hypothalamus, amygdala, septum, orbitofrontal cortex, and hippocampus (Boccia et al., 2001, 2007).

A large body of research conducted mostly with rodents and sheep, but more recently also with humans, has suggested that central oxytocin and vasopressin play an important role in the formation and maintenance of social bonds between adults (e.g., between mating partners) and between parents and offspring (e.g., Young, 2002). In rodents, these peptides have also been implicated in the regulation of sexual and aggressive behavior as well as in stress responses. Little is known, however, about the relation between oxytocin or vasopressin and social behavior in nonhuman primates. An early study by Winslow and Insel (1991) examined the effects of intracerebroventricular administration of two doses of oxytocin and vasopressin, as well as of an oxytocin receptor antagonist (OTA), on aggressive, sexual, and affiliative behavior of pair-housed male squirrel monkeys during an interaction with a familiar adult female. Oxytocin increased aggressive and sexual behavior in a dose-dependent manner in dominants but not in subordinates, while it increased the frequency of approaches and huddles mostly in subordinates. These effects were blocked by OTA. Vasopressin decreased aggressive and affiliative behaviors in both dominants and subordinates. Differences in the effects of oxytocin in dominants and subordinates were tentatively explained in terms of different oxytocin receptor density associated with differences in testosterone between dominants and subordinates.

Research on oxytocin and social behavior in nonhuman primates has been motivated by the interest in developing a primate model for autism. When rhesus monkey infants are separated from their mothers at birth and peer-reared in a small cage, they develop a wide range of behavioral abnormalities, which according to some researchers share some similarities with autism. Winslow and colleagues (2003) measured CSF and plasma oxytocin levels in mother-reared and peer-reared infants to assess whether the behavioral characteristics of peer-reared infants (e.g., low affiliation, high aggression, and high self-directed repetitive behavior) were associated with alterations in oxytocin. In this study, individual differences in CSF oxytocin levels at 18, 24, and 36 months of age did not correlate with differences in plasma oxytocin levels. The peer-reared infants had lower levels of CSF oxytocin than the mother-reared ones, but there were no differences in CSF vasopressin or plasma oxytocin. CSF oxytocin levels were correlated with time spent engaged in affiliative social behaviors such as allogrooming and male-male mounting independent of rearing condition, while vasopressin levels were negatively correlated with the frequency of fear grimaces at 18 months of age. Taken together, the results of this study provide evidence for an association between CSF oxytocin and affiliative behavior and also suggest that both can be affected by a traumatic early experience such as maternal deprivation. Similar to the Winslow et al. study (2003), Schwandt and colleagues (2007) found no significant correlation between CSF and plasma concentrations of oxytocin and vasopressin in free-ranging female rhesus macaques. Oxytocin was not correlated with any social behavior, although females with low levels of oxytocin were classified as more fearful by human observers. Vasopressin was correlated only with leaping behavior, with females with high vasopressin exhibiting higher frequencies of this behavior.

A possible relationship between CSF oxytocin and affiliation was also inferred from a comparison of closely related primate species. Rosenblum and colleagues (2002) reported that laboratory-born pigtail macaques had lower CSF concentrations of oxytocin than bonnet macaques. The authors of this study described bonnet macaques as very gregarious, affiliative, and affectively stable, while pigtail macaques were described as temperamentally volatile and socially distant. Therefore, the results were interpreted as being supportive of the hypothesis that baseline CSF oxytocin concentrations are related to species-typical social/affective behavior patterns. Bonnet macaques, however, have been described by other researchers as highly competitive and aggressive, while pigtail macaques have been described as peaceful, gregarious, and affiliative (Thierry et al., 2004). Therefore, the significance of the difference in oxytocin levels between these two species remains unclear.

The role of oxytocin in the regulation of parental responsiveness in primates is only beginning to be investigated. In a pilot experiment conducted with two nulliparous rhesus macaque females, Holman and Goy (1995) examined whether an intracerebroventricular injection of oxytocin affected responsiveness to infants. The two females were exposed to an unfamiliar infant in a cage 10 minutes after the injection of oxytocin or saline. The females sat near the infant, and watched, touched, and lip-smacked to the infant more frequently following oxytocin compared to saline administration. In no case, however, was more intense caregiving behavior observed, perhaps because of the environment in which the animals were tested. In another study, CSF levels of oxytocin measured in 10 multiparous rhesus females before parturition, immediately after parturition, and 7 days postpartum were not correlated with mother-infant behaviors such as contact or grooming (Cooke et al., 1997). Finally, Boccia and colleagues (2007) reported that the administration of a human uterine oxytocin receptor blocker reduced the frequency of lip-smacking, approaching, and touching a stimulus infant in one 4-year-old nulliparous rhesus macaque female (in a separate experiment, the same treatment also reduced female sexual behavior, and in both experiments locomotor activity was also significantly reduced). This oxytocin receptor blocker had previously been shown to cross the blood-brain barrier and to accumulate in the hypothalamus, orbitofrontal cortex, amygdala, hippocampus, and septum, suggesting that these are brain areas rich in oxytocin receptors (Boccia et al., 2007). Although the effects of oxytocin receptor blockade on infant-directed behavior in one subject were suggestive of a relation between oxytocin and parental responsiveness, the authors of this study acknowledged that other explanations for their results were also possible.

In a study investigating the possible neurobiological and neurochemical substrates of paternal responsiveness in marmoset monkeys, first-time and experienced fathers who had spent a considerable amount of time carrying infants had a greater number of vasopressin V1a receptors in the prefrontal cortex than adult male nonfathers living in similar social conditions (Kozorovitskiy et al., 2006). There were no differences in the abundance of vasopressin V1b receptors or oxytocin receptors in the prefrontal cortex, nor in the abundance of V1a receptors in the occipital cortex. Interestingly, fatherhood was also associated with an increased proportion of dendritic spines in the prefrontal cortex, which were immunoreactive for V1a receptor, as well as increased overall density of dendritic spines on pyramidal neurons in the prefrontal cortex. The functional implications of this fatherhood-associated structural reorganization in the prefrontal cortex and the increased abundance of vasopressin V1a receptors remain unclear. Interestingly, Hammock and Young (2005) have suggested that a repetitive polymorphic microsatellite in a regulatory region of the vasopressin 1a receptor gene (AVPR1a) may be responsible for both intraspecific and interspecific variation in social behavior in primates, as in rodents. They found that this polymorphism is present in humans and in bonobos but absent in chimpanzees, and hypothesized that it may be responsible for some of the differences in affiliation and bonding between the latter two species.

Oxytocin and vasopressin have been hypothesized to promote social bonds by facilitating the perception, processing, and memorization of affiliative stimuli (Depue & Morrone-Strupinksy, 2005). It has also been suggested that oxytocin can reduce tension and anxiety associated with social interactions. For example, human studies have suggested that oxytocin released during affiliative social interactions reduces the hypothalamus-pituitary-adrenal (HPA) axis response to stressful events (Uvnas-Moberg, 1998). In an attempt to test this hypothesis with primate data, Parker and colleagues (2005) showed that chronic intranasal administration of oxytocin prior to acute social isolation attenuates the adrenocorticotropic hormone (ACTH) response (but not the cortisol response) to stress in squirrel monkeys (see also Heinrichs et al., 2003, in humans). Since cortisol was not affected and because intranasal oxytocin can penetrate the central nervous system (CNS), this suggests that oxytocin exerts its antistress effects prior to adrenal activation, either in the brain or at the pituitary level.

Dopamine

Dopaminergic neurons and their projection sites (e.g., the ventral striatum, nucleus accumbens, amygdala, anterior cingulate cortex, and orbitofrontal cortex) constitute what is known as the brain reward system (see Chapter 17). This system regulates a wide range of incentive-motivated behaviors, and these may also include social activities such as affiliation, aggression, mating, and parenting. With regard to affiliation, it has been argued that dopamine plays a crucial role in incentive-reward motivation processes associated with the appetitive phase of affiliation (Depue & Morrone-Strupinksy, 2005). The appetitive phase involves, at the behavioral level, a search and approach system whose function is to bring an individual in contact with affiliative stimuli. Research with rodents has shown that the incentive motivation and experience of reward that underlie the search for and approach to affiliative stimuli depend on the functional properties of dopaminergic neurons in the VTA and nucleus accumbens (NAS) (Depue & Morrone-Strupinksy, 2005). Given the lack of relevant data, whether the same relation between dopamine and the appetitive phase of affiliation holds true also in nonhuman primates remains unclear. There is some evidence, however, that dopamine plays a greater role in the appetitive aspects of primate sexual behavior than in its consummatory components. For example, a dopamine agonist, apomorphine, which acts on dopamine D1 and D2 receptors, enhances male sexual arousal in response to female sexual stimuli in rhesus monkeys (Pomerantz, 1990) but does not appear to affect male copulatory behavior (Chambers & Phoenix, 1989). Quinelorane, another D2 agonist, also stimulates male sexual arousal (Pomerantz, 1991). Whether these effects also occur in females is unclear, since there have been no studies investigating dopaminergic function and sexual behavior in female primates (Dixson, 1998).

In nonhuman primates, brain dopaminergic function has also been investigated in relation to personality traits such novelty seeking, which are expected to influence behaviors such as exploration, assertiveness, aggressiveness, and dominance. In humans, genetic polymorphisms in the coding region of the dopamine D4 receptor gene (DRD4) have been linked with a number of personality and behavioral disorders both in adults and in children (e.g., Sheese et al., 2007). A study of captive vervet monkeys showed that the DRD4 genetic polymorphism accounts for a significant fraction of interindividual variation in novelty-seeking behavior (e.g., latency to approach novel objects) (Bailey et al., 2007). Correlations between CSF concentrations of the dopamine metabolite homovanillic acid (HVA) and measures of sexual, assertive, and aggressive behavior have been reported by some studies of macaques and vervet monkeys (e.g., Fairbanks et al., 2004; Kaplan et al., 2002; Mehlman et al., 1994, 1997), but these correlations must be interpreted with caution. This is because individual differences in CSF concentrations of HVA are highly positively correlated with those of the serotonin metabolite 5-hydroxyindoleacetic acid (5-HIAA) and the norepinephrine metabolite 3-hydroxy-4-methoxyphenylglycol (MHPG) (see later). The CSF concentrations of the three monoamine metabolites also share similarities in their heritability (Freimer et al., 2007; Higley et al., 1993; Rogers et al., 2004), the extent to which they are affected by early stressful experience (Higley et al., 1991; Maestripieri et al., 2006a,b), and age-related changes across the primate lifespan (Higley et al., 1992a). In some cases, correlations between CSF HVA and social behavior may be a byproduct of correlations between CSF 5-HIAA or MHPG levels and behavior. For example, the association between low CSF HVA and high dominance status in male vervet monkeys was almost entirely due to the correlation between low 5-HIAA and high dominance and to the correlation between HVA and 5-HIAA (Fairbanks et al., 2004). Kaplan and colleagues (2002), however, reported a strong association between CSF levels of HVA and dominance, which was independent of the other CSF monoamine metabolite concentrations. This association was in the opposite direction to that reported by Fairbanks and colleagues (2004). In unisexual groups of captive long-tailed macaques, adult males and females that became dominant within their groups had significantly higher CSF HVA concentrations than those that became subordinate (Kaplan et al., 2002). Greater dopaminergic activity in dominant females was also suggested by another study in the same species, in which prolactin responses to a challenge with the dopamine antagonist haloperidol were greater in dominants than in subordinates (Shively, 1998). In another study, however, dominant females showed greater striatal dopamine D2 receptor binding than subordinates, suggesting lower dopaminergic activity in dominants (Grant et al., 1998). Given these discrepancies between the results of different studies, and the fact that other studies have failed to report significant correlations between CSF HVA levels and any measures of social behavior (e.g., Cleveland et al., 2004; Maestripieri et al., 2006b), the relation between dopaminergic activity and aggression and dominance in primates remains unclear.

Norepinephrine

The brain noradrenergic system has been implicated in the regulation of arousal and an individual's aggressive or fearful responses to novel or threatening stimuli. Aggressive and fearful behaviors are associated with increased central and peripheral noradrenergic activity in primates as in other mammals, but this association is not specific to agonistic interactions. Rather, elevated norepinephrine is observed in all situations with high arousal and is an important component of an individual's response to stress. Central norepinephrine may also mediate sexual arousal, but there are no relevant primate data on this topic (Dixson, 1998). Some primate studies have reported correlations between CSF concentrations of norepinephrine or the norepinephrine metabolite MHPG and aggressive behavior, but these correlations have been mixed. For example, Higley and colleagues (1992b) reported that highly aggressive rhesus monkey males had higher CSF levels of norepinephrine than less aggressive monkeys, whereas among females low CSF norepinephrine was associated with high rates of severe aggression

(Higley et al., 1996a). No correlation between CSF MHPG and aggression was found in several other studies of free-ranging rhesus monkeys (Higley et al., 1992b, 1996a; Howell et al., 2007). In long-tailed macaques, CSF MHPG was higher in dominants than in subordinates among males, but not among females (Kaplan et al., 2002), whereas in male vervet monkeys there was no significant correlation between CSF MHPG levels and dominance rank (Fairbanks et al., 2004). Elevated CSF MHPG levels have been reported in rhesus monkey infants rejected and abused by their mothers (Maestripieri et al., 2006a) or in peer-reared infants (Higley et al., 1991). As juveniles, individuals with low CSF MHPG exhibit high avoidance of other individuals (Maestripieri, 2006b), while among adult females, individuals with high CSF MHPG are avoided more by other individuals. In wild vervet monkeys, however, low CSF MHPG levels were associated with higher impulsivity (Fairbanks et al., 1999).

Given the discrepancies in these research findings, it is premature to draw any conclusions about the relation between norepinephrine and social behavior in nonhuman primates. Whether norepinephrine-dependent arousal results in aggressiveness or avoidance might depend on the complex relationship between arousal and anxiety, fear, and impulsivity. Since different types of aggression may have different emotional substrates, the relation between norepinephrine and aggression could be different for different types of aggressive behavior. Furthermore, similar to HVA, the relation between CSF MHPG and aggression may be confounded by the relation between 5-HIAA and aggression and the positive correlation between 5-HIAA and MHPG. Although noradrenergic mechanisms can potentially affect aggression and dominance independently of serotonin, the relation between aggression and serotonin seems to be stronger and more specific (see later) than that between aggression and dopamine or norepinephrine. For example, studies of rodents have shown that the basal activity of the noradrenergic system, unlike that of the serotonergic system, does not consistently differentiate between more and less aggressive individuals (Miczek & Fish, 2006).

Serotonin

Serotonin is one of the most ancient neurotransmitters in mammals and has been implicated in the regulation of social behavior in a number of other taxa as well (Insel & Winslow, 1998). In humans and other primates, brain neurons that use serotonin as their primary neurotransmitter originate in the raphe nuclei of the brainstem and project to the cerebral cortex as well as to subcortical structures such as the amygdala, septum, hypothalamus, hippocampus, thalamus, and basal ganglia. Studies of serotonin and social behavior in primates have used indirect measures of CNS serotonergic function such as the measurement of 5-HIAA in the CSF or pharmacological and neuroendocrine challenges. The use of CSF 5-HIAA concentration as an indicator of brain serotonin activity has been validated by various methods including postmortem brain studies showing that the CSF content of 5-HIAA reflects the content of this metabolite in the brain (e.g., Banki & Molnar, 1981; Wester et al., 1990). Low concentrations of CSF 5-HIAA are generally interpreted as representative of lower CNS serotonergic function. The relationship between CSF 5-HIAA and specific serotonergic neural pathways in the brain, however, remains unclear (Insel & Winslow, 1998). Administration of the serotonin agonist fenfluramine, which stimulates the release of serotonin from neurons and inhibits its reuptake, has also been used as an indirect method of assessing CNS serotonergic function. Since stimulation of serotonin receptors in the hypothalamus results in increased release of prolactin from the pituitary, plasma concentrations of prolactin following fenfluramine administration can be used as indicators of responsivity of the brain serotonergic system. Finally, the serotonergic system can be challenged by manipulating the availability of tryptophan, the amino acid necessary for the synthesis of this monoamine, or by using other pharmacological serotonin reuptake inhibitors.

Using the previously described techniques, a number of human studies have shown that low CNS serotonin function is related to impaired impulse control and to unrestrained aggressiveness and violence, particularly in adult males (see Manuck et al., 2006, for a recent review). Similarly, studies of rhesus macaque adult males and females have shown that low levels of CSF 5-HIAA are associated with high impulsivity, risk-taking behavior, and propensity to engage in severe forms of aggression (Higley et al., 1996a,b,c; Howell et al., 2007; Mehlman et al., 1994; Westergaard et al., 1999, 2003; see Higley, 2003, and Manuck et al., 2006, for reviews). In vervet monkeys, individuals fed diets high in tryptophan exhibit lower aggression, whereas individuals placed on diets low in tryptophan become more aggressive, with the increase in aggressiveness being greater for males than for females (Chamberlain et al., 1987; Raleigh et al., 1985, 1986, 1991). Decreased aggression has also been observed following short-term administration of serotonin reuptake inhibitors (Chamberlain et al., 1987; Raleigh et al., 1980, 1985, 1986, 1991), while administration of the serotonin synthesis inhibitor p-chlorophenylalanine (PCPA) results in increased aggression (Kyes et al., 1995; Raleigh & McGuire, 1986; Raleigh et al., 1980, 1983, 1986). In long-tail macaques, individuals showing low responsivity to the fenfluramine challenge were more aggressive toward other individuals and to faces with threatening expressions than individuals exhibiting prolactin responses of greater magnitude (Botchin et al., 1993; Kyes et al., 1995).

A relation between aggressiveness and CSF 5-HIAA concentrations has also been found in comparisons between different genetic strains of rhesus macaques and between closely related species. For example, the more aggressive Chinese derived rhesus strain has lower CSF 5-HIAA concentrations than the less aggressive Indian-derived strain (Champoux et al., 1997). Furthermore, rhesus macaques have lower CSF 5-HIAA levels than pigtail macaques, a species believed to be generally less aggressive than rhesus macaques (Westergaard et al., 1999). Finally, Kaplan and colleagues (1999) reported that anubis baboons, who are characterized by relatively high levels of intermale aggression, have lower CSF 5-HIAA concentrations than the less aggressive anubis-hamadryas baboon hybrids.

Although it is possible that there is a direct causal relation between brain serotonergic function and aggressive behavior, it is more likely that such a relationship is indirect and mediated by impulse control. In this view, reduced or dysregulated serotonergic function would impair an individual's ability to restrain impulses, and this would be manifested in risky, aggressive, depressed, or addictive behavior depending on the environmental circumstances and the individual's motivational state. The hypothesis that impulse control is an important intervening variable is supported by evidence that, in rhesus macaques, low CSF 5-HIAA is correlated only with aggression involving physical contact and chases, a type of aggression that is more likely to result in serious injuries, and not with milder agonistic behavior involving threats and avoidance, which is commonly associated with the maintenance of dominance relationships (Botchin et al., 1993; Higley et al., 1992b, 1996a,b; Mehlman et al., 1994). Indeed, the relationship between CSF 5-HIAA levels and dominance is not clear, as some studies have reported that CSF 5-HIAA is higher in dominants and lower in subordinates (e.g., Fairbanks et al., 2004; Westergaard et al., 1999) while other studies have reported the opposite pattern or no relation at all (Kaplan et al., 2002; Raleigh et al., 1991; Shively, 1998; Shively et al., 1995; Yodyingyuad et al., 1985). Monkeys with low CSF 5-HIAA concentrations are more likely to exhibit behaviors suggestive of impaired impulse control such as long leaps at high heights and repeated jumping into baited traps in which they are captured (Fairbanks et al., 1999; Higley et al., 1996c; Mehlman et al., 1994). In the laboratory, rhesus macaques with low CSF 5-HIAA concentrations have a lower latency to approach a novel object than do monkeys with high CSF 5-HIAA concentrations (Bennett et al., 2002). In vervet monkeys, individuals with low CSF 5-HIAA concentrations approached a strange and potentially dangerous adult male more quickly and were more likely to act aggressively toward him than monkeys with high CSF 5-HIAA (Fairbanks et al., 2001; see also Manuck et al., 2003). Individuals treated with the selective serotonin reuptake inhibitor fluoxetine became less impulsive in this strange male test than control individuals (Fairbanks et al., 2001).

Increased tendencies to exhibit risky behaviors and to engage in severe forms of aggression are not the only behavioral manifestations of low CSF 5-HIAA in monkeys. Individuals with low CSF 5-HIAA also show reduced propensities for prosocial behaviors and affiliation. In a study of free-ranging adolescent rhesus macaque males, individuals with low CSF 5-HIAA concentrations exhibited reduced amounts of time spent in proximity and grooming other group members, and a lower number of social partners with whom they interacted (Mehlman et al., 1995; see also Kaplan et al., 1995). Young rhesus males with low CSF 5-HIAA concentrations have also been reported to emigrate from their natal group at an earlier age than males with higher CSF 5-HIAA concentrations (Howell et al., 2007; Kaplan et al., 1995; Mehlman et al., 1995, 1997). In the laboratory, low CSF 5-HIAA concentrations were associated with low rates of affiliative interactions among rhesus juveniles of both sexes (Higley et al., 1996a). Adult females with low CSF 5-HIAA also appear to be less socially oriented, spending more time alone, grooming less, and having fewer conspecifics in close proximity (Cleveland et al., 2004). Adult rhesus males with low CSF 5-HIAA form fewer consorts with estrous females during the mating season, and during these consorts, they groom and mount the females less frequently than males with higher CSF 5-HIAA (Mehlman et al., 1997). In long-tail macaques, individuals with low responses to fenfluramine spent less time in affiliative interactions with other individuals and more time alone (Botchin et al., 1993). In vervet monkeys, Raleigh and colleagues found that enhancing serotonin function by administering tryptophan, the reuptake inhibitor fluoxetine, or the serotonin agonist quipazine increased affiliative behaviors such as approaching and grooming other monkeys (Raleigh et al., 1980, 1983, 1985). In contrast, reducing serotonin function by administering the tryptophan hydroxylase enzyme inhibitor PCPA resulted in social withdrawal and in avoidance of affiliative interactions (Raleigh & McGuire 1990; Raleigh et al., 1980, 1985). These data from vervet monkeys thus suggest that enhancing serotonergic function facilitates the expression of affiliative behavior, whereas reducing serotonergic function inhibits affiliation.

There is some evidence that serotonergic function is related not only to aggressive and affiliative behavior but also to sexual and maternal behavior. Consistent with the results of studies of rodents, serotonin has been shown to exert inhibitory effects on male and female sexual behavior in primates as well (e.g., Gradwell et al., 1975; Pomerantz et al., 1991). Early studies of serotonin and maternal behavior in primates reported that monkey mothers with low CSF 5-HIAA were more protective and restrictive, and that their infants spent more time in contact with them, than mothers with high CSF 5-HIAA (Fairbanks et al., 1998; Lindell et al., 1997). Cleveland and colleagues (2004) found no relationship between CSF 5-HIAA and maternal behavior in the first few postpartum days, but on postpartum days 15 and 20, females with low CSF 5-HIAA broke contact and left their infants less frequently than females with high CSF 5-HIAA. A preliminary study in our laboratory reported a positive correlation between CSF 5-HIAA concentrations measured during pregnancy and maternal rejection behaviors in the first postpartum month in multiparous females (Maestripieri et al., 2005). Our more recent work involving multiple measurements of CSF 5-HIAA during development, however, reported a negative correlation between CSF 5-HIAA and maternal rejection among first-time mothers (Maestripieri et al., 2007).

Serotonin may affect maternal motivation through its actions on oxytocin or prolactin release, or through its effects on emotional expression (Insel & Winslow, 1998; Numan & Insel, 2003). Emotions can be powerful elicitors of maternal behavior in nonhuman primates and humans (Dix, 1991; Maestripieri, 1999; Pryce, 1992). For example, there are marked individual differences in anxiety among rhesus mothers, and such differences translate into differences in maternal style (Maestripieri, 1993a,b). Maternal anxiety has also been implicated in the etiology of infant abuse in macaques (Maestripieri, 1994; Troisi & D'Amato, 1991).

Although the role of impulse control in primate maternal behavior is still poorly understood, it is possible that impulsivity affects how primate mothers interact with their infants, and that high impulsivity is expressed as high rejection rates, thus explaining the association between low CSF 5-HIAA and high rejection rates found in first-time rhesus macaque mothers (Maestripieri et al., 2007).

The occurrence of individual differences in CSF concentrations of 5-HIAA and their association with differences in aggressive, affiliative, and maternal behavior has sparked interest in the origin of this variation. Studies of genotyped primate populations and studies of cross-fostered individuals have provided evidence for moderate to strong heritability of CSF concentrations of 5-HIAA and other monoamine metabolites (Higley et al., 1993; Rogers et al., 2004). Heritability of variation in serotonergic function could arise from any genes whose products participate in serotonin's synthesis, release, reuptake, or metabolism, or in genes that encode serotonin receptors (Manuck et al., 2006). A well-known case of genetic variation in serotonergic function involves the polymorphism in the serotonin transporter (5-HTT or SERT) gene. In humans, rhesus macaques, and other primates as well, the promoter region of this gene (5-HTTLPR) exists in two allelic variants, which differ in length. The short allele confers lower transcriptional efficiency to the serotonin transporter gene (Bennett et al., 2002) and is associated with reduced serotonin reuptake into the presynaptic neuron and reduced serotonergic responsivity to neuroendocrine challenges (Manuck et al., 2006). Human studies have shown that individuals with one or two copies of the short allele have greater amygdala neuronal activation in response to faces with threatening expressions (Skuse, 2006). These individuals also had reduced gray matter in the perigenual cingulate cortex (pACC) and in the amygdala. The pACC has the greatest density of serotonin terminals within the human cortex and it is a major target for projections from the amygdala. fMRI studies have shown that people with at least one short allele had weaker functional interactions between ventromedial prefrontal cortex, pACC, and amygdala, suggesting that the presence of short allele is associated with hyperreactivity of the amygdala in response to threats (Skuse, 2006).

In rhesus macaques, the SERT polymorphism is generally unrelated to CSF concentrations of 5-HIAA, with the exception of individuals who are separated from their mothers at birth and reared with peers (Bennett et al., 2002; Maestripieri et al., 2006a). Nevertheless, individuals who carry the short allele for SERT appear to share some behavioral traits with individuals with low CSF 5-HIAA, including higher aggressiveness and earlier age of male emigration from the group (Trefilov et al., 2000). Rhesus macaque mothers who abuse their infants are more likely to carry the short allele of the serotonin transporter gene than nonabusive mothers (McCormack et., 2009). Furthermore, infants with the short allele who are separated from their mothers at birth or physically abused by them are more likely to show anxiety and fear in response to novelty and dysregulated HPA axis responses to stress and challenges than individuals with the same early experience who are homozygous for the long allele (Barr et al., 2004; Bennett et al., 2002; Lesch et al., 1996; McCormack et al., 2009).

Comparative studies of functional variability of the serotonin transporter gene in seven different species of macaques have shown that species that are believed to more socially tolerant and less despotic and nepotistic such as Barbary macaques, Tibetan macaques, and stumptail macaques are monomorphic for the SERT gene. In contrast, species believed to be more intolerant and aggressive such as rhesus, long-tailed, and pigtail macaques are polymorphic for the SERT gene, with rhesus macaques having the highest degree of polymorphism (Wendland et al., 2005). Tonkean macaques, which are believed to be relatively docile and egalitarian, are polymorphic as well. Although these findings suggest that genetic variation in serotonergic function may play an important role in determining species differences in aggressiveness among macaques, caution is needed in interpreting these results for several reasons. First, species differences in aggressiveness among macaques are not well established. Second, the species that are polymorphic for the SERT gene are all closely related to each other, and more distantly related from the species that are monomorphic. Although Tonkean macaques would be expected to be monomorphic on the basis of their presumed behavioral characteristics, they are polymorphic like pigtail macaques, a closely related species from which they evolved.

At the individual level, early experience can be an important source of variation in serotonergic function in adulthood. Long-term effects of early maternal deprivation on the development of the brain serotonergic system have been reported in laboratory-reared rhesus macaques (Higley et al., 1991; Kraemer et al., 1989; Shannon et al., 2005). In group-living rhesus macaques, individuals exposed to high rates of maternal rejection in infancy had significantly lower CSF concentrations of 5-HIAA across their first 3 years of life than the individuals exposed to low rates of maternal rejection (Maestripieri et al., 2006a). This difference was found both in individuals reared by their biological mothers and in cross-fostered juveniles, suggesting that it did not reflect genetic similarities between mothers and offspring. Among these juveniles, there was a significant negative correlation between CSF 5-HIAA and rates of scratching (Maestripieri et al., 2006b), suggesting that individuals with low CSF 5-HIAA were more anxious than those with high 5-HIAA (see Maestripieri et al., 1992, and Schino et al., 1991, for the relation between scratching and anxiety). When females who were reared by high-rejection mothers gave birth for the first time, their low 5-HIAA was associated with high rates of maternal rejection toward their own infants (Maestripieri et al., 2007). The maternal rejection rates of daughters closely resembled those of their mothers and the resemblance was particularly strong for the cross-fostered females and their foster mothers (Maestripieri et al., 2007).

The serotonin system may also be involved in the intergenerational transmission of infant abuse. We reported that about half of the females who were abused by their mothers early in life, whether cross-fostered or non–cross-fostered, exhibited abusive parenting toward their first-born offspring, whereas none of the females reared by nonabusive mothers did (including those born to abusive mothers; Maestripieri, 2005). Moreover, the abused females, both cross-fostered and non–cross-fostered, who became abusive mothers had lower CSF 5-HIAA concentrations than the abused females who did not become abusive mothers (Maestripieri et al., 2006a). Since abuse tends to co-occur with high rates of maternal rejection, our findings suggest that experience-induced long-term alterations in serotonergic function in females reared by highly rejecting and abusive mothers contribute to the manifestation of maternal rejection and abusive parenting in adulthood. It is possible that experience-induced reduction in serotonergic function results in elevated anxiety and impaired impulse control, and that high anxiety and impulsivity increase the probability of occurrence of maternal rejection and abusive parenting with one's own offspring later in life, perhaps in conjunction with social learning resulting from direct experience with one's own mother or from observation of maternal interactions with siblings (Maestripieri, 2008).

THE NEURAL SUBSTRATES OF SOCIAL BEHAVIOR: BRAIN LESION STUDIES

Studies employing brain lesions to investigate the role of different neural structures in the regulation of social behavior have focused mostly on the amygdala, and to a lesser extent on the hypothalamus, the hippocampus, temporal lobes, and orbital frontal cortex. These structures play an important role in the processing of environmental stimuli and the production of emotional responses that regulate survival-related behaviors. These brain regions are also rich in receptors for neuropeptides and monoamines as well as for other hormones and neurotransmitters that have been shown to affect social behavior in primates and other animals (e.g., Way et al., 2007). In highly social organisms such as primates, limbic and cortical responses to social stimuli can play a fundamental role in an individual's ability to achieve successful survival and reproduction. For example, since the amygdala is necessary for the interpretation of social stimuli and the production of emotional responses that regulate avoidance and aggressive behavior, this structure can potentially play an important role in the regulation of primate social behavior. The hypothalamus plays an important role in regulating the motivational aspects of sexual and maternal behavior, and the hippocampus may be relevant to social behavior regulation insofar as it plays a role in processing the spatial and contextual interrelations of social stimuli. The orbital frontal cortex is strongly connected with brain regions that process all sensory stimuli, while temporal lobes have been implicated in the processing of facial expressions and body movements. Therefore, these structures are likely implicated in the neural control of social cognition, and in the acquisition and processing of information that motivates and controls social interactions.

Studies of brain lesions and social behavior in primates have generally taken two different approaches. Some of them have investigated the effects of lesions on the expression of social behavior in adults, while others have investigated the effects of lesions on the development of social behavior in infants. In general, the results of studies in which temporal lobes, orbital frontal cortex, and hippocampus were lesioned in adults have been consistent with those of studies in which these brain areas were lesioned in infants. In the case of amygdala lesions, however, the results of studies involving adults and infants have produced somewhat inconsistent results (see later).

Young and adult monkeys with lesions of the temporal lobes are generally socially withdrawn or inactive and hyporeactive to fear-inducing stimuli, and in the case of adults, they also show inappropriate sexual behavior such as mounting inanimate objects (Bachevalier et al., 2001; Brown & Schaefer, 1888; Kluver & Bucy, 1937, 1939). Individuals with these lesions appear to have impaired ability to discriminate between conspecifics and objects, and therefore to respond properly to socially relevant stimuli. Lesions of the orbital frontal cortex appear to result in avoidance of social interactions and alterations in aggressiveness and dominance (Butter & Snyder, 1972; Raleigh & Steklis, 1981). Individuals with these lesions also show impairments in their social attachments, such as a weaker preference for their primary caregiver (Goursaud & Bachevalier, 2007).

An early study by Rosvold and colleagues (1954) reported that male rhesus monkeys who had achieved high dominance rank in artificially created social groups fell to the bottom of the hierarchy and behaved submissively following bilateral amygdalectomy and reintroduction to the group. Similarly, in a series of studies conducted with rhesus macaques, stumptail macaques and vervet monkeys, Kling and colleagues reported that individuals with bilateral lesions of the amygdala who were reintroduced into their social groups failed to re-establish functional social relationships with other individuals and were either attacked or ignored by others (Dicks et al., 1969; Kling & Cornell, 1971; Kling & Steklis, 1976; Kling et al., 1970). Changes in maternal and sexual behavior in amygdala-lesioned animals were reported as well (e.g., Kling & Brothers 1992). Spies and colleagues (1976) reported that female rhesus monkeys with bilateral lesions of the amygdala showed impaired sexual proceptive behavior but normal receptivity and copulatory behavior when paired with a male in a cage (similar results have also been obtained with lesions of the hypothalamus; see later).

Early studies involving amygdala lesions in monkeys have been criticized by Amaral (2002; see also Amaral et al., 2003) because these lesions were not selective enough and the behavioral observations were not accurate enough to warrant strong conclusions regarding the relation between amygdala and social behavior. Amaral and colleagues conducted a series of studies involving amygdala lesions in rhesus monkeys, in which the specificity of such lesions was greatly enhanced by the use of ibotenic acid, a neurotoxin that is injected stereotaxically into the brain and selectively destroys the amygdala without affecting adjacent areas. One set of studies examined the effects of amygdala lesions on the behavior of adult males in a variety of laboratory testing conditions. In dyadic social interactions in which the amygdala-lesioned males and their unoperated controls were tested with the same stimulus individuals, the lesioned males showed reduced latency to engage in social interactions with their partners and greater affiliation, particularly during the early encounters, suggesting that they had lower social anxiety and more social disinhibition than controls (Emery et al., 2001). The lesioned animals also exhibited lower elevations in plasma cortisol levels in response to the social encounters than the controls did, suggesting that the amygdala lesion reduced the extent to which the individuals interpreted the interaction with an unfamiliar individual as stressful (Amaral, 2002). Perhaps as a result of the behavior of the lesioned individuals, the stimulus partners directed more affiliative behavior toward them than toward the controls (Emery et al., 2001). The amygdala-lesioned adult males showed reduced behavioral inhibition also in response to people and novel objects. For example, they had a lower latency to retrieve a grape and to approach and handle a rubber snake than controls (Mason et al., 2006). Thus, adult males with bilateral lesions of the amygdala show behavioral characteristics similar to those of individuals who experienced much larger lesions of the temporal lobes (Kluver & Bucy, 1937, 1939).

Amaral and collaborators also investigated the long-term effects of brain lesions on infant behavioral development. Bilateral lesions of the amygdala or the hippocampus were performed in 2-week-old infants, who were returned to their mothers and reared by them in small cages with or without other mother-infant pairs. Mother-infant behavioral interactions were not significantly altered by amygdala or hippocampus lesions, with the exception of increased mother-infant contact in the amygdala-lesioned group (Bauman et al., 2004). When infants were permanently separated from their mothers at 6 months of age, the amygdala-lesioned infants did not preferentially seek proximity to their mother in a social preference test in which they could choose between their mother and another familiar adult female (Bauman et al., 2004). This finding was interpreted as indicative of an impairment in the perception of potential danger rather than as a specific deficit in the bond with the mother. The amygdala-lesioned infants, however, did not differ significantly from controls in their plasma cortisol response to separation from their mothers or in their cortisol responses to dexamethasone suppression and ACTH challenge (Goursaud et al., 2006). At 6 to 12 months of age, the amygdala-lesioned infants showed reduced fear of novel objects such as rubber snakes but more fearful behavior than both hippocampus-lesioned and sham-operated controls during dyadic encounters with both familiar and unfamiliar conspecifics (Bauman et al., 2004; Prather et al., 2001). The behavior of both amygdala- and hippocampus-lesioned infants in both dyadic and group interactions, however, was generally normal and age appropriate. If anything, the amygdala-lesioned infants showed more affiliative and submissive behavior than the infants in the other groups (Bauman et al., 2004). At approximately 18 months of age, dominance tests were conducted in which the juveniles were given the opportunity to retrieve preferred food items in a competitive situation involving other individuals (Bauman et al., 2006). In these tests, the amygdala-lesioned individuals showed longer latencies to retrieve the food, reduced aggressive behaviors, and more frequent fear and submissive behaviors than hippocampus-lesioned individuals and sham-lesioned controls. The behavioral effects of amygdala lesions conducted in infancy, therefore, appeared to be opposite to those of similar lesions performed in adult males. Finally, Goursaud and Bachevalier (2007) reported that rhesus monkeys receiving bilateral ibotenic acid lesions of amygdala and hippocampus at 1 to 2 weeks of age and who were subsequently reared by human caregivers did not differ from controls in their preference for their primary caregiver versus another familiar human when tested in a social preference task at 11 months of age.

Taken together, the results of these studies suggest that an intact amygdala is not necessary for the expression of normal social behavior in adult macaques or for normal social development in infants (Amaral et al., 2003). Although evidence from single neuron studies suggests that neurons in the amygdala fire at different rates following exposure to different facial expressions of emotion (Gothard et al., 2007), Amaral and colleagues (2003) have recently questioned the hypothesis that the amygdala plays an important role in social cognition (see Brothers, 1996). Rather, their view is that amygdala serves the function of a protection device: It allows an individual to evaluate the extent to which novel objects in the environment or social situations pose a threat or danger and helps the individual to produce an appropriate response, through projections to other areas of the brain such as the cortex and the hippocampus. Without an intact amygdala, monkeys fail to properly evaluate and recognize the riskiness of a particular stimulus. As a result, amygdala-lesioned monkeys show a lack of fear responses to threatening objects and appear to be uninhibited in potentially dangerous social situations.

Primate brain lesion studies investigating hypothalamic influences on social behavior have mostly focused on sexual behavior. Research conducted by Dixson and colleagues showed that lesions of the anterior and medial hypothalamus in female marmosets impair female active initiation of sexual activity (proceptivity) but not responses to male sexual advances (receptivity) (Dixson, 1990; Dixson & Hastings, 1992; Kendrick & Dixson, 1986). Consistent with these results, studies of macaques have shown that electrical stimulation of the ventromedial or preoptic area of the hypothalamus enhances female proceptive behavior toward males (Koyama et al., 1988). Moreover, neurons in the ventromedial hypothalamus increase their firing rate while female macaques are engaged in proceptive behavior or copulation, while those of the preoptic area decrease their firing rate during these activities (Aou et al., 1988). Taken together, the results of these studies suggest that the hypothalamic mechanisms regulating sexual behavior in primates may differ from nonprimate mammals in some important ways (Dixson, 1998). Different areas of the hypothalamus may control different components of sexual behavior in female primates, and since female primates are unique in their ability to engage in sexual behavior outside of the fertile phase of their cycle (Wallen, 1990), it is possible that the neural control of sexual behavior in primates overlaps with the neural control of affiliation and bonding to a larger extent than in other mammals. Although the hypothalamus plays an important role in regulating maternal motivation in rodents and other mammals, there have been no studies investigating the effects of hypothalamic lesions on parental motivation and behavior in nonhuman primates.

CONCLUSIONS

Studies of the neural substrates or neurochemical mechanisms underlying social behavior in nonhuman primates are clearly limited when compared to those conducted with other animals, particularly laboratory rodents. Moreover, since most research on the neurobiology of primate social behavior has been conducted with the few primate species that are readily available in captivity, such as rhesus macaques, marmosets, or squirrel monkeys, the conclusions of these studies may not be generalizable to other primates, let alone to other animals. Nevertheless, the research findings reviewed in this chapter have made a significant contribution to our understanding of the neurobiological regulation of primate social behavior.

Research on social cognition aside, most of the work investigating the neural and neurochemical control of social behavior has focused on the limbic system and its relation to emotional and motivational substrates of behavior. The best experimental evidence linking specific brain regions or neurochemical systems to emotional substrates of social behavior has been obtained for "negative" emotions such as anxiety, fear, and impulsivity and for agonistic behaviors such as aggression and avoidance. With the exception of work on endogenous opioids and affiliation, "positive" emotions and their relations to affiliation and social bonding have proven more difficult to study.

Research in this area has been driven by findings obtained with rodents, but whether conceptualizations of affiliation and social bonding in rodents can be directly extrapolated to primates remains unclear. Similarly, although there is a wealth of evidence linking social bonding to oxytocin and vasopressin in rodents (e.g., Carter, 1998; Young, 2002), empirical evidence that these peptides affect social bonding in primates is very preliminary or equivocal. Studies of complex affiliative behavior, and to some extent also of aggressive, sexual, and parental behavior, in primates will need greater conceptual and experimental sophistication than studies conducted with laboratory rodents.

Because of ethical and logistical constraints in the study of brain-behavior relationships in nonhuman primates, most studies to date have attempted to measure, often very indirectly, the activity of brain regions or neurochemical systems and then to correlate these measures with aspects of behavior. Although the neuropharmacological manipulation of behavior in complex social settings could be an effective approach for testing neuroethological hypotheses, this approach has generally been underutilized in primate research. The effects of various psychotropic drugs on the social behavior of nonhuman primates have been investigated in a number of studies (see Smith & Byrd, 1983). In many of these studies, however, the relation between drugs and behavior was investigated without a clear understanding of the drug's mechanisms of action in the brain, and without attempting to test specific hypotheses concerning the neurobiological regulation of behavior. Since the physiological and molecular mechanisms of action of many neuropharmacological agents are now well understood, hypothesis-driven neuropharmacological manipulations of social behavior could play an important role in primate neuroethological research.

Brain lesions have proven useful in investigating the role of particular brain regions in the expression of primate social behavior. Brain lesion studies, however, have limitations in that lesions are not always specific and cause permanent and irreversible brain damage. Brain imaging techniques are far less invasive than lesions and hold great promise for future research in primate social neuroethology. One constraint of brain imaging studies of nonhuman primates is that they must be conducted under controlled laboratory conditions. A similar constraint exists also for human studies, but despite this constraint, thousands of brain imaging studies with humans have been conducted in the past few decades, many of which focused on social cognition and social behavior (e.g., Rilling et al.,). Brain imaging—both structural and functional—is arguably also the experimental technique with the greatest potential for answering evolutionary questions about brain-behavior relationships in primates (e.g., Rilling & Insel, 1999). By systematically documenting similarities and differences in the structure of different brain regions across primate species and in how these regions are activated during complex social interactions, we could potentially acquire a great deal of new information about the evolution of social and cognitive complexity in the Primate Order and the brain mechanisms that support it.

REFERENCES

Amaral, D. G. (2002). The primate amygdala and the neurobiology of social behavior: Implications for understanding social anxiety. *Biological Psychiatry, 51,* 11–17.

Amaral, D. G., Capitanio, J. P., Jourdain, M., Mason, W. A., Mendoza, S. P., & Prather, M (2003). The amygdala: Is it an essential component of the neural network for social cognition? *Neuropsychologia, 41,* 235–240.

Aou, S., Oomura, Y., & Yoshimatsu, H. (1988). Neuron activity of the ventromedial hypothalamus and the medial preoptic area of the female monkey during sexual behavior. *Brain Research, 455,* 65–71.

Bachevalier, J., Malkova, L., & Mishkin, M. (2001). Effects of selective neonatal temporal lobe lesions on socioemotional behavior in infant rhesus monkeys (*Macaca mulatta*). *Behavioral Neuroscience, 115,* 545–559.

Bailey, J. N., Breidenthal, S. E., Jorgensen, M. J., McCracken, J. T., & Fairbanks, L. A. (2007). The association of DRD4 and novelty seeking is found in a nonhuman primate model. *Psychiatric Genetics, 17,* 23–27.

Banki, C. M., & Molnar, G. (1981). Cerebrospinal fluid 5-HIAA as an index of central serotonergic processes. *Psychiatry Research, 5,* 23–32.

Barr, C. S., Newman, T. K., Shannon, C., Parker, C., Dvoskin, R. L., Becker, M. L., et al. (2004b). Rearing condition and rh5-HTTLPR interact to influence limbic-hypothalamic-pituitary-adrenal axis response to stress in infant macaques. *Biological Psychiatry, 55,* 733–738.

Barr, C. S., Schwandt, M. L., Lindell, S. G., Higley, J. D., Maestripieri, D., Goldman, D., et al. (2008). Variation at the mu-opioid receptor gene (OPRM1) influences attachment behavior in infant primates. *Proceedings of the National Academy of Sciences USA, 105,* 5277–5281.

Bauman, M. D., Lavenex, P., Mason, W. A., Capitanio, J. P., & Amaral, D. G. (2004). The development of social behavior following neonatal amygdala lesions in rhesus monkeys. *Journal of Cognitive Neuroscience, 16,* 1388–1411.

Bauman, M. D., Toscano, J. E., Mason, W. A., Lavenex, P., Amaral, D. G. (2006). The expression of social dominance following neonatal lesions of the amygdala or hippocampus in rhesus monkeys (*Macaca mulatta*). *Behavioral Neuroscience, 120,* 749–760.

Bennett, A. J., Lesch, K. P., Heils, A., Long, J. C., Lorenz, J. G., Shoaf, S. E., et al. (2002). Early experience and serotonin transporter gene variation interact to influence primate CNS function. *Molecular Psychiatry, 7,* 118–122.

Boccia, M. L., Goursaud, A. S., Bachevalier, J., Anderson, K. D., & Pedersen, C. A. (2007). Peripherally administered non-peptide oxytocin antagonist, L368,899, accumulates in limbic brain areas: A new pharmacological tool for the study of social motivation in non-human primates. *Hormones and Behavior, 52,* 344–351.

Boccia, M. L., Panicker, A. K., Pedersen, C., & Petrusz, P. (2001). Oxytocin receptors in nonhuman primate brain visualized with monoclonal antibody. *Neuroreport, 12,* 1723–1726.

Botchin, M. B., Kaplan, J. R., Manuck, S. B., & Mann, J. J. (1993). Low versus high prolactin responders to fenfluramine challenge: Marker of behavioral differences in adult male cynomolgus macaques. *Neuropsychopharmacology, 9,* 93–99.

Brothers, L. (1996). Brain mechanisms of social cognition. *Journal of Psychopharmacology, 10,* 2–8.

Brown, S., & Schaefer, E. A. (1888). An investigation into the functions of the occipital and temporal lobes of the monkey's brain. *Philosophical Transactions of the Royal Society of London Series B, 179,* 303–327.

Butter, C. M., & Snyder, D. R. (1972). Alterations in aversive and aggressive behaviors following orbital frontal lesions in rhesus monkeys. *Acta Neurobiologica Experimentalis, 32,* 525–565.

Carter, C. S. (1998). Neuroendocrine perspectives on social attachment and love. *Psychoneuroendocrinology, 23,* 779–818.

Chamberlain, B., Ervin, F. R., Pihl, R. O., & Young, S. N. (1987). The effect of raising of lowering tryptophan levels on aggression in vervet monkeys. *Pharmacology, Biochemistry, and Behavior, 28,* 503–510.

Chambers, K. C., & Phoenix, C. H. (1989). Apomorphine, (-)-deprenyl, and yohimbine fail to increase sexual behavior in rhesus males. *Behavioral Neuroscience, 103,* 816–823.

Champoux, M., Higley, J. D., & Suomi, S. J. (1997). Behavioral and physiological characteristics of Indian and Chinese-Indian hybrid rhesus macaque infants. *Developmental Psychobiology, 31,* 49–63.

Cleveland, A., Westergaard, G. C., Trenkle, M. K., & Higley, J. D. (2004). Physiological predictors of reproductive outcome and mother-infant behaviors in captive rhesus macaque females (*Macaca mulatta*). *Neuropsychopharmacology, 29,* 901–910.

Cooke, B., Higley, J. D., Shannon, C., Lindell, S. G., Higley, H. M., Suomi, S. J., et al. (1997). Rearing history and CSF oxytocin as predictors of maternal competency in rhesus macaques. *American Journal of Primatology, 42,* 102.

Csiffary, A., Ruttner, Z., Toth, Z., & Palkovits, M. (1992). Oxytocin nerve fibers innervate β-endorphin neurons in the arcuate nucleus of the rat hypothalamus. *Neuroendocrinology, 56,* 429–435.

Depue, R. A., & Morrone-Strupinsky, J. V. (2005). A neurobehavioral model of affiliative bonding: Implications for conceptualizing a human trait of affiliation. *Behavioral and Brain Sciences, 28,* 313–395.

Dicks, D., Myers, R. E., & Kling, A. (1969). Uncus and amygdala lesions: Effects on social behavior in the free ranging rhesus monkey. *Science, 165,* 69–71.

Dix, T. (1991). The affective organization of parenting: Adaptive and maladaptive processes. *Psychological Bulletin, 110,* 3–25.

Dixson, A. F. (1990). Medial hypothalamic lesions and sexual receptivity in the female common marmoset (*Callithrix jacchus*). *Folia Primatologica, 54,* 46–56.

Dixson, A. F. (1998). *Primate sexuality.* Oxford: Oxford University Press.

Dixson, A. F., & Hastings, M. H. (1992). Effects of ibotenic acid-induced neuronal degeneration in the hypothalamus upon proceptivity and sexual receptivity in the female marmoset. *Journal of Neuroendocrinology, 4,* 719–726.

Dunbar, R. I. M. (1988). *Primate social systems.* London: Croom Helm.

Dunbar, R. I. M. (1992). Neocortex size as a constraint on group size in primates. *Journal of Human Evolution, 20,* 469–493.

Emery, N. J., Capitanio, J. P., Mason, W. A., Machado, C. J., Mendoza, S. P., & Amaral, D. G. (2001). The effects of bilateral lesions of the amygdala on dyadic social interactions in rhesus monkeys (*Macaca mulatta*). *Behavioral Neuroscience, 115,* 515–544.

Fabre-Nys, C., Meller, R. E., & Keverne, E. B. (1982). Opiate antagonists stimulate affiliative behaviour in monkeys. *Pharmacology, Biochemistry, and Behavior, 16,* 653–659.

Fairbanks, L. A., Fontenot, M. B., Phillips-Conroy, J. E., Jolly, C. J., Kaplan, J. R., & Mann, J. J. (1999). CSF monoamines, age and impulsivity in wild grivet monkeys (*Cercopithecus aethiops aethiops*). *Brain, Behavior, and Evolution, 53,* 305–312.

Fairbanks, L. A., Jorgensen, M. J., Huff, A., Blau, K., Hung, Y., & Mann, J. J. (2004). Adolescent impulsivity predicts adult dominance attainment in male vervet monkeys. *American Journal of Primatology, 64,* 1–17.

Fairbanks, L. A., Melega, W. P., Jorgensen, M. J., Kaplan, J. R., & McGuire, M. T. (2001). Social impulsivity inversely associated with CSF 5-HIAA and fluoxetine exposure in vervet monkeys. *Neuropsychopharmacology, 24,* 370–378.

Fairbanks, L. A., Melega, W. P., & McGuire, M. T. (1998). CSF 5-HIAA is associated with individual differences in maternal protectiveness in vervet monkeys. *American Journal of Primatology, 45,* 179–180.

Freimer, N. B., Service, S. K., Ophoff, R. A., Jasinska, A. J., Villeneuve, A., Belisle, A., et al. (2007). A quantitative trait locus for variation in dopamine metabolism mapped in a primate model using rference sequences from related species. *Proceedings of the National Academy of Sciences USA, 104,* 15811–15816.

Gimpl, G., & Fahrenholz, F. (2001). The oxytocin receptor system: Structure, function, and regulation. *Physiological Reviews, 81,* 629–683.

Gothard, K. M., Battaglia, F. P., Erickson, C. A., Spitler, K. M., & Amaral, D. G. (2007). Neural responses to facial expression and facial identity in the monkey amygdala. *Journal of Neurophysiology, 97,* 1671–1683.

Goursaud, A. S., & Bachevalier, J. (2007). Social attachment in juvenile monkeys with neonatal lesion of the hippocampus, amygdala and orbital frontal cortex. *Behavioral Brain Research, 176,* 75–93.

Goursaud, A. S., Mendoza, S. P., & Capitanio, J. P. (2006). Do neonatal bilateral ibotenic acid lesions of the hippocampal formation or of the amygdala impair HPA axis responsiveness and regulation in infant rhesus macaques (*Macaca mulatta*)? *Brain Research, 1071,* 97–104.

Gradwell, P. B., Everitt, B. J., & Herbert, J. (1975). 5-hydroxytryptamine in the central nervous system and sexual receptivity of female rhesus monkeys. *Brain Research, 88,* 281–293.

Grant, K. A., Shively, C. A., Nader, M. A., Ehrenkaufer, R. L., Line, S. W., Morton, T. E., et al. (1998). Effect of social status on striatal dopamine D2 receptor binding characteristics in cynomolgus monkeys assessed with positron emission tomography. *Synapse, 29,* 80–83.

Graves, F. C., Wallen, K., & Maestripieri, D. (2002). Opioids and attachment in rhesus macaque abusive mothers. *Behavioral Neuroscience, 116,* 489–493.

Guard, H. J., Newman, J. D., & Roberts, R. L. (2002). Morphine administration selectively facilitates social play in common marmosets. *Developmental Psychobiology, 41,* 37–49.

Hammock, E. A. D., & Young, L. J. (2005). Microsatellite instability generates diversity in brain and sociobehavioral traits. *Science, 308,* 1630–1634.

Heinrichs, M., Baumgartner, T., Kirschbaum, C., & Ehlert, U. (2003). Social support and oxytocin interact to suppress cortisol and subjective responses to psychosocial stress. *Biological Psychiatry, 54,* 1389–1398.

Higley, J. D. (2003). Aggression. In: D. Maestripieri (Ed.), *Primate psychology* (pp. 17–40). Cambridge, MA: Harvard University Press.

Higley, J. D., King, S. T., Hasert, M. F., Champoux, M., Suomi, S. J., & Linnoila, M. (1996a). Stability of interindividual differences in serotonin function and its relationship to aggressive wounding and competent social behavior in rhesus macaque females. *Neuropsychopharmacology, 14,* 67–76.

Higley, J. D., Mehlman, P. T., Higley, S. B., Fernald, B., Vickers, S., Lindell, S. G., et al. (1996b). Excessive mortality in young male nonhuman primates with low CSF 5-HIAA concentrations. *Archives of General Psychiatry, 53,* 537–543.

Higley, J. D., Mehlman, P. T., Poland, R. E., Taub, D. M., Vickers, J., Suomi, S. J., et al. (1996c). CSF testosterone and CSF 5-HIAA correlate with different types of aggressive behaviors. *Biological Psychiatry, 40,* 1067–1082.

Higley, J. D., Mehlman, P. T., Taub, D. M., Higley, S. B., Suomi, S. J., Linnoila, M., et al. (1992b). Cerebrospinal fluid monoamine and adrenal correlates of aggression in free-ranging rhesus monkeys. *Archives of General Psychiatry, 49,* 436–441.

Higley, J. D., Suomi, S. J., & Linnoila, M. (1991). CSF monoamine metabolite concentrations vary according to age, rearing, and sex, and are influenced by the stressor of social separation in rhesus monkeys. *Psychopharmacology, 103,* 551–556.

Higley, J. D., Suomi, S. J., & Linnoila, M. (1992a). A longitudinal assessment of CSF monoamine metabolite and plasma cortisol concentrations in young rhesus monkeys. *Biological Psychiatry, 32,* 127–145.

Higley, J. D., Thompson, W. W., Champoux, M., Goldman, D., Hasert, M. F., Kraemer, G. W., et al. (1993). Paternal and maternal genetic and environmental contributions to cerebrospinal fluid monoamine metabolites in rhesus monkeys (*Macaca mulatta*). *Archives of General Psychiatry, 50,* 615–623.

Holman, S. D., & Goy, R. W. (1995). Experiential and hormonal correlates of care-giving in rhesus macaques. In: C. R. Pryce, R. D. Martin, & D. Skuse (Eds.), *Motherhood in Human and Nonhuman Primates. Biosocial Determinants* (pp. 87–93). Basel: Karger.

Howell, S., Westergaard, G., Hoos, B., Chavanne, T. J., Shoaf, S. E., Cleveland, A., et al. (2007). Serotonergic influences on life-history outcomes in free-ranging male rhesus macaques. *American Journal of Primatology, 69,* 1–15.

Insel, T. R. (1992). Oxytocin and the neurobiology of attachment. *Behavioral and Brain Sciences, 15,* 515–516.

Insel, T. R., & Winslow, J. T. (1998). Serotonin and neuropeptides in affiliative behaviors. *Biological Psychiatry, 44,* 207–219.

Kalin, N. H., Shelton, S. E., & Barksdale, C. M. (1988). Opiate modulation of separation-induced distress in non-human primates. *Brain Research, 440,* 285–292.

Kalin, N. H., Shelton, S. E., & Lynn, D. E. (1995). Opiate systems in mother and infant primates coordinate intimate contact during reunion. *Psychoneuroendocrinology, 20,* 735–742.

Kaplan, J. R., Fontenot, M. B., Berard, J. D., Manuck, S. B., & Mann, J. J. (1995). Delayed dispersal and elevated monoaminergic activity in free-ranging rhesus monkeys. *American Journal of Primatology, 35,* 229–234.

Kaplan, J. R., Phillips-Conroy, J., Fontenot, M. B., Jolly, C. J., Fairbanks, L. A., & Mann, J. J. (1999). Cerebrospinal fluid monoaminergic metabolites differ in wild anubis and hybrid (*Anubis hamadryas*) baboons: Possible relationships to life history and behavior. *Neuropsychopharmacology, 20,* 517–524.

Kaplan, J. R., Manuck, S. B., Fontenot, B., & Mann, J. J. (2002). Central nervous system monoamine correlates of social dominance in cynomolgus monkeys (*Macaca fascicularis*). *Neuropsychopharmacology, 26,* 431–443.

Kendrick, K. M., & Dixson, A. F. (1986). Anteromedial hypothalamic lesions block proceptivity but not receptivity in the female common marmoset (*Callithrix jacchus*). *Brain Research, 375,* 221–229.

Keverne, E. B., Martensz, N. D., & Tuite, B. (1989). β-endorphin concentrations in cerebrospinal fluid of monkeys are influenced by grooming relationships. *Psychoneuroendocrinology, 14,* 155–161.

Kling, A., & Brothers, L. A. (1992). The amygdala and social behavior. In: J. Aggleton (Ed.), *The amygdala: Neurobiological aspects of emotion, memory, and mental dysfunction* (pp. 353–377). New York: Wiley-Liss.

Kling, A., & Cornell, R. (1971). Amygdalectomy and social behavior in the caged stump-tailed macaque (*Macaca speciosa*). *Folia Primatologica, 14,* 190–208.

Kling, A., Lancaster, J., & Benitone, J. (1970). Amygdalectomy in the free-ranging vervet

(*Cercopithecus aethiops*). *Journal of Psychiatric Research, 7,* 191–199.

Kling, A., & Steklis, H. D. (1976). A neural substrate for affiliative behavior in nonhuman primates. *Brain, Behavior, and Evolution, 13,* 216–238.

Kluver, H., & Bucy, P. C. (1937). "Psychic blindness" and other symptoms following bilateral temporal lobectomy in rhesus monkeys. *American Journal of Physiology, 119,* 352–353.

Kluver, H., & Bucy, P. C. (1939). Preliminary analysis of functions of the temporal lobes in monkeys. *Archives of Neurology and Psychiatry, 42,* 979–1000.

Koyama, Y., Fujita, I., Aou, S., & Oomura, Y. (1988). Proceptive presenting elicited by electrical stimulation of the female monkey hypothalamus. *Brain Research, 446,* 199–203.

Kozorovitskiy, Y., Hughes, M., Lee, K., & Gould, E. (2006). Fatherhood affects dendritic spines and vasopressin V1a receptors in the primate prefrontal cortex. *Nature Neuroscience, 9,* 1094–1095.

Kraemer, G. W. (1992). A psychobiological theory of attachment. *Behavioral and Brain Sciences, 15,* 493–451.

Kraemer, G. W., Ebert, M. H., Schmidt, D. E., & McKinney, W. T. (1989). A longitudinal study of the effect of different social rearing conditions on cerebrospinal fluid norepinephrine and biogenic amine metabolites in rhesus monkeys. *Neuropsychopharmacology, 2,* 175–189.

Kyes, R. C., Botchin, M. B., Kaplan, J. R., Manuck, S. B., & Mann, J. J. (1995). Aggression and brain serotonergic responsivity: Response to slides in male macaques. *Physiology and Behavior, 57,* 205–208.

Lesch, K. P., Bengel, D., Heils, A., Sabol, S. Z., Greenberg, B. D., Petri, S., et al. (1996). Association of anxiety-related traits with a polymorphism in the serotonin transporter gene regulatory region. *Science, 274,* 1527–1531.

Lindell, S. G., Higley, J. D., Shannon, C., & Linnoila, M. (1997). Low levels of CSF 5-HIAA in female rhesus macaques predict mother-infant interaction patterns and mother's CSF 5-HIAA correlates with infant's CSF 5-HIAA. *American Journal of Primatology, 2,* 129.

Maestripieri, D. (1993a). Maternal anxiety in rhesus macaques (*Macaca mulatta*). I. Measurement of anxiety and identification of anxiety-eliciting situations. *Ethology, 95,* 19–31.

Maestripieri, D. (1993b). Maternal anxiety in rhesus macaques (*Macaca mulatta*). II. Emotional bases of individual differences in mothering style. *Ethology, 95,* 32–42.

Maestripieri, D. (1994). Infant abuse associated with psychosocial stress in a group-living pigtail macaque (*Macaca nemestrina*) mother. *American Journal of Primatology, 32,* 41–49.

Maestripieri, D. (1999). The biology of human parenting: Insights from nonhuman primates. *Neuroscience and Biobehavioral Reviews, 23,* 411–422.

Maestripieri, D. (2005). Early experience affects the intergenerational transmission of infant abuse in rhesus monkeys. *Proceedings of the National Academy of Sciences USA, 102,* 9726–9729.

Maestripieri, D. (2008). The role of the brain serotonergic system in the origin and transmission of adaptive and maladaptive variations in maternal behavior in rhesus macaques. In: R. Bridges (Ed.), *Neurobiology of the parental brain* (pp. 163–173). Amsterdam: Elsevier.

Maestripieri, D., Higley, J. D., Lindell, S. G., Newman, T. K., McCormack, K. M., & Sanchez, M. M. (2006a). Early maternal rejection affects the development of monoaminergic systems and adult abusive parenting in rhesus macaques. *Behavioral Neuroscience, 120,* 1017–1024.

Maestripieri, D., Lindell, S. G., Ayala, A., Gold, P. W., & Higley, J. D. (2005). Neurobiological characteristics of rhesus macaque abusive mothers and their relation to social and maternal behavior. *Neuroscience and Biobehavioral Reviews, 29,* 51–57.

Maestripieri, D., Lindell, S. G., & Higley, J. D. (2007). Intergenerational transmission of maternal behavior in rhesus monkeys and its underlying mechanisms. *Developmental Psychobiology, 49,* 165–171.

Maestripieri, D., McCormack, K. M., Lindell, S. G., Higley, J. D., & Sanchez, M. M. (2006b). Influence of parenting style on the offspring's behavior and CSF monoamine metabolite levels in crossfostered and noncrossfostered female rhesus macaques. *Behavioral Brain Research, 175,* 90–95.

Maestripieri, D., Schino, G., Aureli, F., & Troisi, A. (1992). A modest proposal: Displacement activities as an indicator of emotions in primates. *Animal Behaviour, 44,* 967–979.

Manuck, S. B., Kaplan, J. R., & Lotrich, F. E. (2006). Brain serotonin and aggressive disposition in humans and nonhuman primates. In: R. J. Nelson (Ed.), *Biology of aggression* (pp. 65–113). Oxford: Oxford University Press.

Manuck, S. B., Kaplan, J. R., Rymeski, B. A., Fairbanks, L. A., & Wilson, M. E. (2003). Approach to a social stranger is associated with low central nervous system serotonergic responsivity in female cynomolgus monkeys (*Macaca fascicularis*). *American Journal of Primatology, 61,* 187–194.

Martel, F. L., Nevison, C. M., Rayment, F. D., Simpson, M. J. A., & Keverne, E. B. (1993). Opioid receptor blockade reduces maternal affect and social grooming in rhesus monkeys. *Psychoneuroendocrinology, 18,* 307–321.

Martel, F. L., Nevison, C. M., Simpson, M. J. A., & Keverne, E. B. (1995). Effects of opioid receptor blockade on the social behavior of rhesus monkeys living in large family groups. *Developmental Psychobiology, 28,* 71–84.

Martensz, N. D., Vellucci, S. V., Keverne, E. B., & Herbert, J. (1986). β-endorphins levels in the cerebrospinal fluid of male talapoin monkeys in social groups related to dominance status and the luteinizing hormone response to naloxone. *Neuroscience, 18,* 651–658.

Mason, W. A., Capitanio, J. P., Machado, C. J., Mendoza, S. P., & Amaral, D. G. (2006). Amygdalectomy and responsiveness to novelty in rhesus monkeys (*Macaca mulatta*): Generality and individual consistency of effects. *Emotion, 6,* 73–81.

McCormack, K., Newman, T. K., Higley, J. D., Maestripieri, D., & Sanchez, M. M.'(2009). Serotonin transporter gene variation, infant abuse, and responsiveness to stress in rhesus acaque mothers and infants. *Hormones and Behavior, 55,* 538-547.

Mehlman, P. T., Higley, J. D., Faucher, I., Lilly, A. A., Taub, D. M., Suomi, S. J., et al. (1994). Low CSF 5-HIAA concentrations and severe aggression and impaired impulse control in nonhuman primates. *American Journal of Psychiatry, 151,* 1485–1491.

Mehlman, P. T., Higley, J. D., Faucher, I., Lilly, A. A., Taub, D. M., Vickers, J. H., et al. (1995). Correlation of CSF 5-HIAA concentrations with sociality and the timing of emigration in free-ranging primates. *American Journal of Psychiatry, 152,* 907–913.

Mehlman, P. T., Higley, J. D., Fernald, B. J., Sallee, F. R., Suomi, S. J., & Linnoila, M. (1997). CSF 5-HIAA, testosterone, and sociosexual behaviors in free-ranging male rhesus macaques in the mating season. *Psychiatry Research, 72,* 89–102.

Meller, R. E., Keverne, E. B., & Herbert, J. (1980). Behavioural and endocrine effects of naltrexone in male talapoin monkeys. *Pharmacology, Biochemistry, and Behavior, 13,* 663–672.

Miczek, K. A., & Fish, E. W. (2006). Monoamines, GABA, glutamate, and aggression. In: R. J. Nelson (Ed.), *Biology of aggression* (pp. 114–149). Oxford: Oxford University Press.

Miczek, K. A., Woolley, J., Schlisserman, S., & Yoshimura, H. (1981). Analysis of amphetamine effects on agonistic and affiliative behavior in squirrel monkeys (*Saimiri sciureus*). *Pharmacology, Biochemistry, and Behavior, 14,* 103–107.

Nelson, E., & Panksepp, J. (1998). Brain substrates of infant-mother attachment: Contributions of opioids, oxytocin, and norepinephrine. *Neuroscience and Biobehavioral Reviews, 22,* 437–452.

Numan, M., & Insel, T. R. (2003). *The neurobiology of parental behavior.* Berlin: Springer.

Panksepp, J., Herman, B. H., Vilberg, T., Bishop, P., & DeEskinazi, F. G. (1980). Endogenous opioids and social behavior. *Neuroscience and Biobehavioral Reviews, 4,* 473–487.

Panksepp, J., Nelson, E., & Bekkedal, M. (1999). Brain systems for the mediation of social separation-distress and social-reward: Evolutionary antecedents and neuropeptide intermediaries. In: C. S. Carter, I. I. Lederhendler, & B. Kirkpatrick (Eds.), *The integrative neurobiology of affiliation* (pp. 221–243). Cambridge, MA: MIT Press.

Parker, K. J., Buckmaster, C. L., Schatzberg, A. F., & Lyons, D. M. (2005). Intranasal oxytocin administration attenuates the ACTH stress response in monkeys. *Psychoneuroendocrinology, 30,* 924–929.

Pomerantz, S. M. (1990). Apomorphine facilitates male sexual behavior of rhesus monkeys. *Pharmacology, Biochemistry, and Behavior, 35,* 659–664.

Pomerantz, S. M. (1991). Quinelorane (LY163502), a D2 dopamine receptor agonist, acts centrally to facilitate penile erections of male rhesus monkeys. *Pharmacology, Biochemistry, and Behavior, 39,* 123–128.

Pomerantz, S. M., Hepner, B. C., & Wertz, J. M. (1991). 5-HT1A and 5-HT1C/1D receptor agonists produce reciprocal effects on male sexual behavior of rhesus monkeys. *European Journal of Pharmacology, 243,* 227–234.

Prather, M. D., Lavenex, P., Mauldin-Jordan, M. L., Mason, W. A., Capitanio, J. P., Mendoza, S. P., et al. (2001). Increased social fear and decreased fear of objects in monkeys with neonatal amygdala lesions. *Neuroscience, 106,* 653–658.

Pryce, C. R. (1992). A comparative systems model of the regulation of maternal motivation in mammals. *Animal Behaviour, 43,* 417–441.

Raleigh, M. J., Brammer, G. L., Yuwiler, A., Flannery, J. W., & McGuire, M. T. (1980). Serotonergic influences on the social behavior of vervet monkeys (*Cercopithecus aethiops sabaeus*). *Experimental Neurology, 68,* 322–334.

Raleigh, M. J., Brammer, G. L., & McGuire, M. T. (1983). Male dominance, serotonergic systems, and the behavioral and physiological effects of drugs in vervet monkeys (*Cercopithecus aethiops sabaeus*). In: K. A. Miczek (Ed.), *Ethopharmacology: Primate models of neuropsychiatric disorders* (pp. 185–197). New York: Alan R. Liss.

Raleigh, M. J., Brammer, G. L., McGuire, M. T., & Yuwiler, A. (1985). Dominant social status facilitates behavioral effects of serotonergic agonists. *Brain Research, 348,* 274–282.

Raleigh, M. J., Brammer, G. L., Ritvo, E. R., Geller, E., McGuire, M. T., & Yuwiler, A. (1986). Effects of chronic fenfluramine on blood serotonin, cerebrospinal fluid metabolites, and behavior in monkeys. *Psychopharmacology, 90,* 503–508.

Raleigh, M. J., & McGuire, M. T. (1986). Animal analogues of ostracism: Biological mechanisms and social consequences. *Ethology and Sociobiology, 7,* 53–66.

Raleigh, M. J., & McGuire, M. T. (1990). Social influences on endocrine function in male vervet monkeys. In: T. E. Ziegler & F. B. Bercovitch (Eds.), *Socioendocrinology of primate reproduction* (pp. 95–111). New York: Wiley-Liss.

Raleigh, M. J., McGuire, M. T., Brammer, G. L., Pollack, D. B., & Yuwiler, A. (1991). Serotonergic mechanisms promote dominance acquisition in adult male vervet monkeys. *Brain Research, 559,* 181–190.

Raleigh, M. J., & Steklis, H. D. (1981). Effects of orbital frontal and temporal neocortical lesions on affiliative behavior of vervet monkeys (*Cercopithecus aethiops sabaeus*). *Experimental Neurology, 73,* 378–389.

Rendall, D., & Di Fiore, A. (1995). The road less traveled: Phylogenetic perspectives in primatology. *Evolutionary Anthropology, 4,* 43–52.

Rilling, J. K., & Insel, T. R. (1999). The primate neocortex in comparative perspective using magnetic resonance imaging. *Journal of Human Evolution, 37,* 191–223.

Rilling, J. K., Sanfey, A. G., Nystrom, L. E., Cohen, J. D., Gutman, D. A., Zeh, T. R., et al. (2004b). Imaging the social brain with fMRI and interactive games. *International Journal of Neuropsychopharmacology, 7,* S477–S478.

Rilling, J. K., Winslow, J. T., & Kilts, C. D. (2004a). The neural correlates of mate competition in dominant male rhesus macaques. *Biological Psychiatry, 56,* 364–375.

Rilling, J. K., Winslow, J. T., O'Brien, D., Gutman, D. A., Hoffman, J. M., & Kilts, C. D. (2001). Neural correlates of maternal separation in rhesus monkeys. *Biological Psychiatry, 49,* 146–157.

Rogers, J., Martin, L. J., Comuzzie, A. G., Mann, J. J., Manuck, S. B., Leland, M., et al. (2004). Genetics of monoamine metabolites in baboons: Overlapping sets of genes influence levels of 5-hydroxyindolacetic acid, 3-hydroxy-4-methoxyphenylglycol, and homovanillic acid. *Biological Psychiatry, 55,* 739–744.

Rosenblum, L. A., Smith, E. L. P., Altemus, M., Scharf, B. A., Owens, M. J., Nemeroff, C. B., et al. (2002). Differing concentrations of corticotropin-releasing factor and oxytocin in the cerebrospinal fluid of bonnet and pigtail macaques. *Psychoneuroendocrinology, 27,* 651–660.

Rosvold, H. E., Mirsky, A. F., & Pribram, K. H. (1954). Influence of amygdalectomy on social behavior in monkeys. *Journal of Comparative Physiology and Psychology, 47,* 173–178.

Schino, G., & Troisi, A. (1992). Opiate receptor blockade in juvenile macaques: Effect on affiliative interactions with their mothers and group companions. *Brain Research, 576,* 125–130.

Schino, G., Troisi, A., Perretta, G., & Monaco, V. (1991). Measuring anxiety in nonhuman primates: Effect of lorazepam on macaque scratching. *Pharmacology, Biochemistry, and Behavior, 38,* 889–891.

Schwandt, M. L., Howell, S., Bales, K., Jaffe, B. D., Westergaard, G. C., & Higley, J. D. (2007). Associations between the neuropeptides oxytocin and vasopressin and the behavior of free-ranging female rhesus macaques (*Macaca mulatta*). *American Journal of Physical Anthropology Supplement, 44,* 210–211.

Shannon, C., Schwandt, M. L., Champoux, M., Shoaf, S. E., Suomi, S. J., Linnoila, M., et al. (2005). Maternal absence and stability of individual differences in CSF 5-HIAA concentrations in rhesus monkey infants. *American Journal of Psychiatry, 162,* 1658–1664.

Sheese, B. E., Voelker, P. M., Rothbart, M. K., & Posner, M. I. (2007). Parenting quality interacts with genetic variation in dopamine receptor D4 to influence temperament in early childhood. *Development and Psychopathology, 19,* 1039–1046.

Shively, C. A. (1998). Social subordination stress, behavior, and central monoaminergic function in female cynomolgus monkeys. *Biological Psychiatry, 44,* 882–891.

Shively, C. A., Fontenot, B., & Kaplan, J. R. (1995). Social status, behavior, and central serotonergic responsivity in female cynomolgus monkeys. *American Journal of Primatology, 37,* 333–350.

Skuse, D. (2006). Genetic influences on the neural basis on social cognition. *Philosophical Transactions of the Royal Society of London Series B, 361,* 2129–2141.

Smith, E. O., & Byrd, L. D. (1983). Studying the behavioral effects of drugs in group-living nonhuman primates. In: K. A. Miczek (Ed.), *Ethopharmacology: Primate models of neuropsychiatric disorders* (pp. 1–31). New York: Alan Liss.

Smuts, B. B., Cheney, D. L., Seyfarth, R. M., Wrangham, R. W., & Struhsaker, T. T. (Eds.). (1987). *Primate societies.* Chicago: University of Chicago Press.

Snowdon, C. T., Ziegler, T. E., Schultz-Darken, N. J., & Ferris, C. F. (2006). Social odours, sexual arousal and pairbonding in primates. *Philosophical Transactions of the Royal Society of London Series B, 361,* 2079–2089.

Spies, H. G., Norman, R. L., Clifton, D. K., Ochsner, A. J., Jensen, J. N., & Phoenix, C. H. (1976). Effects of bilateral amygdaloid lesions on gonadal hormones in serum and on sexual behavior in female rhesus monkeys. *Physiology and Behavior, 17,* 985–992.

Thierry, B., Singh, M., & Kaumanns, W. (Eds.). (2004). *Macaque societies: A model for the evolution of social organization.* Cambridge: Cambridge University Press.

Toloczko, D. M., Young, L., & Insel, T. R. (1997). Are there oxytocin receptors in the primate brain? *Annals of the New York Academy of Sciences, 807,* 506–509.

Trefilov, A., Berard, J., Krawczak, M., & Schmidtke, J. (2000). Natal dispersal in rhesus macaques is related to serotonin transporter gene promoter variation. *Behaviour Genetics, 30,* 295–301.

Troisi, A., & D'Amato, F. R. (1991). Anxiety in the pathogenesis of primate infant abuse: A pharmacological study. *Psychopharmacology, 103,* 571–572.

Uvnas-Moberg, K. (1998). Oxytocin may mediate the benefits of positive social interaction and emotions. *Psychoneuroendocrinology, 23,* 819–835.

Wallen, K. (1990). Desire and ability: Hormones and the regulation of female sexual behavior. *Neuroscience and Biobehavioral Reviews, 14,* 233–241.

Way, B. M., Lacan, G., Fairbanks, L. A., & Melega, W. P. (2007). Architectonic distribution of the serotonin transporter within the orbitofrontal cortex of the vervet monkey. *Neuroscience, 148,* 937–948.

Wendland, J. R., Lesch, K. P., Newman, T. K., Timme, A., Gachot-Neveu, H., Thierry, B., et al. (2005). Differential functional variability of serotonin transporter and monoamine oxidase A genes in macaque species displaying contrasting levels of aggression-related behavior. *Behaviour Genetics, 36,* 163–172.

Wester, P., Bergstrom, U., Eriksson, A., Gezelius, C., Hardy, J., & Winblad, B. (1990). Ventricular cerebrospinal fluid monoamine transmitter and metabolic concentrations reflect human brain neurochemistry in autopsy cases. *Journal of Neurochemistry, 54,* 1148–1156.

Westergaard, G. C., Mehlman, P. T., Shoaf, S. E., Suomi, S. J., & Higley, J. D. (1999). CSF 5-HIAA and aggression in female macaque monkeys: Species and individual differences. *Psychopharmacology, 145,* 440–446.

Westergaard, G. C., Suomi, S. J., Chavanne, T. J., Houser, L., Hurley, A., Cleveland, A., et al. (2003). Physiological correlates of aggression and impulsivity in free-ranging female primates. *Neuropsychopharmacology, 28,* 1045–1055.

Winslow, J. T. (2005). Neuropeptides and non-human primate social deficits associated with pathogenic rearing experience. *International Journal of Developmental Neuroscience, 23,* 245–251.

Winslow, J. T., & Insel, T. R. (1991). Social status in pairs of male squirrel monkeys determines the behavioral response to central oxytocin administration. *Journal of Neuroscience, 11,* 2032–2038.

Winslow, J. T., Noble, P. L., Lyons, C. K., Sterk, S. M., & Insel, T. R. (2003). Rearing effects on cerebrospinal fluid oxytocin concentration on social buffering in rhesus monkeys. *Neuropsychopharmacology, 28,* 910–918.

Yodyingyuad, U., de La Riva, C., Abbott, D. H., Herbert, J., & Keverne, E. B. (1985). Relationship between dominance hierarchy, cerebrospinal fluid levels of amine transmitter metabolites (5-hydroxy-indoleacetic acid and homovanillic acid) and plasma cortisol levels in monkeys. *Neuroscience, 16,* 851–858.

Young, L. J. (2002). The neurobiology of social recognition, approach, and avoidance. *Biological Psychiatry, 51,* 18–26.

CHAPTER 20

Neural Bases of Numerical Cognition

Andreas Nieder

INTRODUCTION

Dealing with numbers had long been thought to be an exclusive faculty of language-endowed humans. Symbolic number representations, such as numerals and number words, and the infinite mathematical manipulations they enable are uniquely human cultural achievements and shape our technologically advanced and scientific culture. However, basic numerical competence does not depend on language; it is rooted in biological primitives that can already be found in animals. As illustrated by Elizabeth Brannon and colleagues (Chapter 8), animals possess impressive numerical capabilities and are able to nonverbally and approximately grasp the numerical properties of objects and events. Such a numerical estimation system for representing number as language-independent mental magnitudes (analog magnitude system) is a precursor on which verbal numerical representations build, and their neural foundations can be studied in animal models.

This review charts the progress made in our understanding of the neuronal substrates and mechanisms of nonverbal numerical competence in nonhuman primates. It is structured according to the two major concepts numerical cognition encompasses: numerical quantity and numerical rank (Wiese, 2003). Numerical quantity refers to the empirical property *cardinality* ("numerosity," the size of a set) of objects and events. It pertains to the question "How many?" for numerable quantity, and "How much?" for innumerable quantity. Numerical rank refers to the empirical property *serial order* and is sought after by the question "Which position?"

NUMERICAL QUANTITY (CARDINALITY)

Neurons Encoding Numerical Quantity

Recordings in monkeys trained to discriminate numerosity demonstrated the capacity of single neurons to encode cardinality (Nieder & Miller, 2004a; Nieder et al., 2002, 2006). In the basic layout of the task, monkeys viewed a sequence of two displays separated by a memory delay and were required to judge whether the displays contained the same number of items (Fig. 20.1A). To ensure that the monkeys solved the task by judging number per se rather than simply memorizing sequences of visual patterns or exploiting low-level visual features that correlate with number, sensory cues (such as position, shape, overall area, circumference, and density) were varied considerably and controlled for (Nieder et al., 2002).

Numerosity-selective neurons were tuned to the number of items in a visual display; that is, they showed maximum activity to one of the presented quantities—a neuron's "preferred numerosity"—and a progressive dropoff as the quantity became more remote from the preferred number (Nieder & Merten, 2007; Nieder et al., 2002). Importantly, changes in the physical appearance of the displays had no effect on the activity of numerosity-selective neurons (Nieder & Miller, 2004a; Nieder et al., 2002, 2006). A high proportion of numerosity detectors (Fig. 20.1B) was found in the lateral prefrontal cortex (PFC) (Nieder et al., 2002). In the posterior parietal cortex (PPC), numerosity-selective

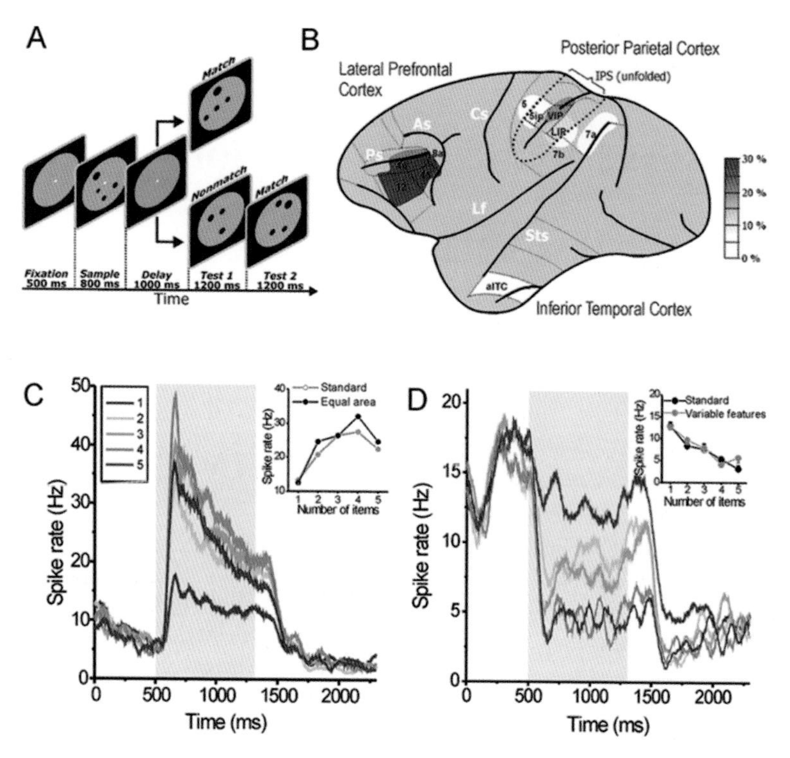

Figure 20.1 Representation of visual cardinality in rhesus monkeys. **(A)** Delayed match-to-sample task with visually presented numerosity as the stimulus dimension of interest. A trial started when the monkey grasped a lever and fixated at a central target. After 500 ms of pure fixation, the sample stimulus (800 ms) cued the monkey for a certain numerosity it had to memorize during a 1,000-ms delay period. Then, the test 1 stimulus was presented, which in 50% of cases was a match showing the same number of dots as cued during the sample period. In the other 50% of cases the test 1 display was a nonmatch, which showed a different numerosity as the sample display. After a nonmatch test stimulus, a second test stimulus (test 2) appeared that was always a match. To receive a fluid reward, monkeys were required to release the lever as soon as a match appeared. Trials were pseudo-randomized and balanced across all relevant features. Monkeys were required to maintain fixation throughout the sample and delay period. **(B)** Lateral view of a monkey brain showing the recording sites in lateral prefrontal cortex (LPFC), posterior parietal cortex (PPC), and anterior inferior temporal cortex (aITC). The proportion of numerosity-selective neurons in each area is color coded according to the color scale. The intraparietal sulcus (IPS) is unfolded to show the different areas in the lateral and medial walls. Numbers on prefrontal cortex (PFC) indicate anatomical areas. As, arcuate sulcus; Cs, central sulcus; LF, lateral fissure; LS, lunate sulcus; Ps, principal sulcus; Sts, Superior temporal sulcus. After Nieder, A., & Miller, E. K. (2004a). A parieto-frontal network for visual numerical information in the monkey. *Proceedings of the National Academy of Sciences of the USA, 101,* 7457–7462. **(C,D)** Responses of single neurons that were recorded from the PFC (C) and the IPS (D). Both neurons show graded discharge during sample presentation (interval shaded in gray, 500 to 1,300 ms) as a function of numerosities 1 to 5 (color-coded averaged discharge functions). The insets in the upper right corner show the tuning of both neurons and their responses to different control stimuli. The preferred numerosity was 4 for the PFC neuron (B) and 1 for the IPS neuron (C). After Nieder, A., Freedman, D. J., & Miller, E. K. (2002). Representation of the quantity of visual items in the primate prefrontal cortex. *Science, 297,* 1708–1711.

neurons were sparsely distributed in several areas, but relatively abundant in the fundus of the intraparietal sulcus (IPS), termed the ventral intraparietal area (VIP) (Colby et al., 1993). There were few such cells in the anterior inferior temporal cortex (aITC) (Nieder & Miller 2004a) (Fig. 20.1B).

Item numbers can be determined in two fundamentally different spatio-temporal presentation formats. When presented simultaneously as in multiple-item patterns, numerosity can be estimated at a single glance in a direct, perceptual-like way from a spatial arrangement. On a behavioral level, constant reaction times and equal numbers of scanning eye movements to individual items (Nieder & Miller, 2004b) indicate parallel processing mechanisms for quantity assessments from multiple dot patterns. Moreover, the response latencies of single neurons are indifferent across numerosities (Nieder et al., 2002). In contrast to a simultaneous presentation, the elements of a set can be presented one by one and, thus, need to be enumerated successively across time (Cordes et al., 2001; Meck & Church, 1983; Whalen et al., 1999). Sequential enumeration is cognitively more demanding; it incorporates multiple encoding, memory, and updating stages; it may even be regarded as a form of addition of one. Sequential enumeration is particularly interesting in that it constitutes a nonverbal precursor of real counting; after all, verbal counting is a sequential enumeration process using number symbols (i.e., 1-2-3, etc.).

To address the neuronal representation of an abstract counting-like accumulation of sensory events and to compare it to the encoding of numerosity in simultaneous displays, Nieder and colleagues (2006) recorded single-cell activity in the fundus of the IPS while monkeys performed a delayed match-to-sample task in which sample numerosity was specified either by single dots appearing one by one to indicate the number of items in sequence ("sequential protocol," Fig. 20.2A) or by multiple-dot patterns ("simultaneous protocol," Fig. 20.1A). It was ensured that temporal or spatial cues could not be used by the animals to solve the task. In addition to the previously described neurons selective to numerosity in multiple-dot patterns, roughly 25% of the neurons in the fundus of the IPS also encoded sequentially presented numerical quantity (Fig. 20.2B). However, numerical quantity was represented by distinct populations of neurons during the ongoing spatial or temporal enumeration process (i.e., in the sample phase); cells encoding the number of sequential items were not tuned to numerosity in multiple-item displays, and vice versa. Once the enumeration process was completed, though, and the monkeys had to store information in mind, a third population of neurons coded numerosity both in the sequential and simultaneous protocol; about 20% of the cells were tuned to numerosity irrespective of whether is was cued simultaneously or in sequence (Fig. 20.2C–E). This argues for segregated processing of numerosity during the actual encoding stage in the parietal lobe, but also for a final convergence of the segregated information to form most abstract quantity representations. The intermediate numerosity of an ongoing quantification process and the storage of the final cardinality are accomplished by different neuronal populations.

In another domain, cells in the superior parietal lobule (SPL) have been reported to keep track of the number of movements (Sawamura et al., 2002). The authors trained monkeys to alternate between five arm movements of one type ("push" and "turn") and five of another. They found neurons in a somatosensory-responsive region (part of area 5) of the SPL that maintained the number of movements. Relatively few such neurons were found in the same lateral PFC regions where other perceptual categories were investigated. One possibility for the difference between these studies may be modality (touch vs. vision), but another may be the level of abstraction. Most movement-number representations found by Sawamura and colleagues (85%) were not abstract; number-selective activity depended on whether the monkey's movement was "push" or "turn." By contrast, the visual numerosity representations found in the PFC and fundus of the IPS were abstract and generalized (Nieder & Miller, 2004a; Nieder et al., 2002, 2006).

The parietal cortex, and the IPS in particular, might be the first cortical stage that extracts visual numerical information because its neurons require shorter latencies to become numerosity selective than PFC neurons (Nieder & Miller, 2004a). As PPC and PFC are functionally interconnected (Chafee & Goldman-Rakic 2000; Quintana & Fuster, 1999), that information might be conveyed directly or indirectly to the PFC, where it is amplified and maintained to gain

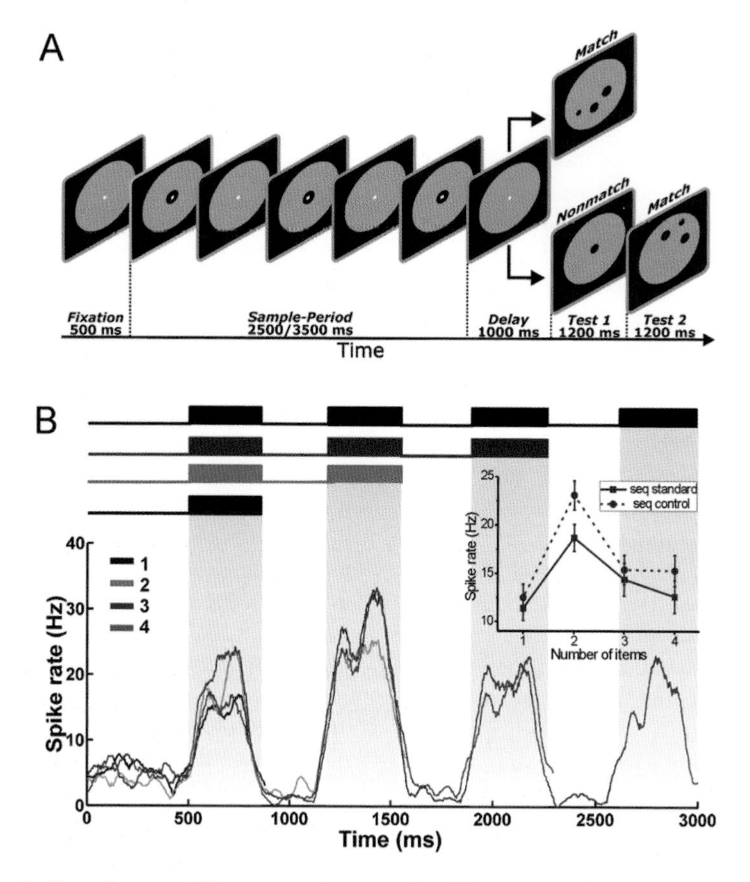

Figure 20.2 Coding of sequentially presented numerosity. (**A**) Sequential delayed match-to-numerosity task (here for numerosity 3). The sample numerosity was cued by sequentially presented items temporally separated by pauses containing no items. The temporal succession and duration of individual items were varied within and across quantities. (**B**) Responses of an example neuron selective to the sequential quantity 2 (only one condition shown for clarity). Top panel illustrates the temporal succession of individual items (square pulses represent single items). The corresponding latency-corrected discharges for many repetitions of the protocol are plotted as averaged spike density functions. The first 500 ms represent the fixation period. Corresponding colors were used for the stimulation illustration and the plotting of the neural data. Gray shaded areas denote item presentation. The inset shows the tuning functions of the neuron to the standard and a control protocol (error bars represent SEM) for four sequential dots. In both protocols, the neuron was tuned to numerosity 2. (**C–E**) Spike-density histograms of a single intraparietal sulcus (IPS) neuron showing remarkably similar delay activity in the sequential (C) versus simultaneous (D) presentation protocol, with 3 as the preferred numerosity. This neuron was not numerosity selective in the sample period. (**E**) Tuning functions of the neuron in C and D calculated for the delay period. After Nieder, A., Diester, I., & Tudusciuc, O. (2006). Temporal and spatial enumeration processes in the primate parietal cortex. *Science, 313,* 1431–1435.

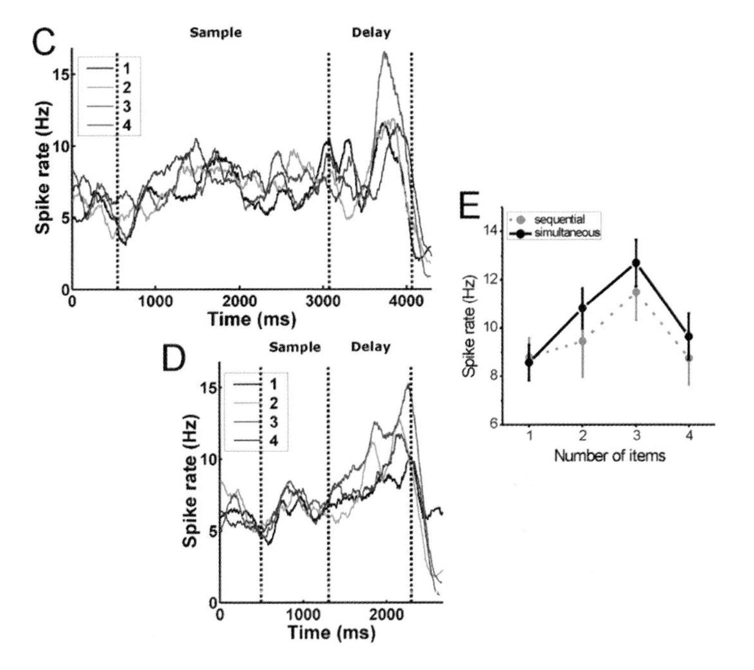

Figure 20.2 (Continued)

control over behavior. As classical association cortices, the prefrontal and posterior parietal cortices are ideal brain structures for an abstract encoding of quantity. They receive highly processed multimodal input (Bremmer et al., 2001; Duhamel et al., 1998; Lewis & Van Essen, 2000; Miller & Cohen, 2001)—a prerequisite for numerical competence because the number concept applies equally well to all sensory modalities. Both are cardinal processing stages for executive functions (e.g., categorization, working memory, decision making, goal-directed behavior, etc.) and play an important role in maintaining information "online" (Freedman et al., 2001; Stoet & Snyder, 2004; Wallis et al., 2001), and the PPC in particular hosts neural circuitry dedicated to the representation of abstract spatial information (Colby & Goldberg, 1999).

Behavioral Significance of Numerosity-Selective Neurons

The activity of all numerosity-selective neurons, each tuned to a specific preferred numerosity, formed a bank of overlapping numerosity filters (Fig. 20.3B), mirroring the animals' behavioral performance (Fig. 20.3A). Interestingly, the neurons' sequentially arranged overlapping tuning curves preserved an inherent order of cardinalities. This is important because numerosities are not isolated categories, but exist in relation to one another (e.g., 3 is greater than 2 and less than 4); they need to be sequentially ordered to allow meaningful quantity assignments.

The response properties of numerosity-selective cortical cells can explain basic psychophysical phenomena in monkeys, such as the numerical distance and size effect (Fig. 20.3). The *numerical distance effect* states that it is easier to discriminate quantities that are numerically remote from each other (say, 2 vs. 6 is easier than 5 vs. 6), while the *numerical size effect* captures the finding that pairs of numerosities of a constant numerical distance are easier to discriminate if the quantities are small (e.g., 2 vs. 3 is easier than 5 vs. 6). The numerical distance effect results from the fact that the neural filter functions that are engaged in the discrimination of adjacent numerosities heavily overlap (Nieder & Miller, 2003). As a consequence, the signal-to-noise ratio of the neural signal detection process is low, and the monkeys make many errors. On the other hand, the filter functions of neurons that are tuned to remote numerosities

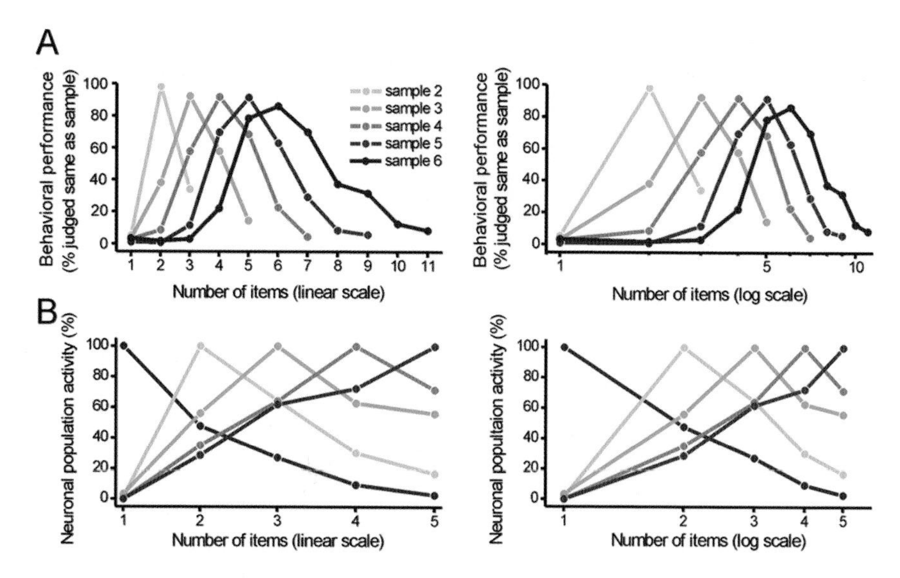

Figure 20.3 Relation between monkey behavior and numerosity-selective neurons. **(A)** Behavioral numerosity discrimination functions of two monkeys. The curves indicate whether they judged the first test stimulus as containing the same number of items as the sample display. The function peaks (and the color legend) indicate the sample numerosity for which each curve was derived. Behavioral filter functions are skewed on a linear scale (left), but symmetric on a logarithmic scale (right). **(B)** The averaged single-cell numerosity-tuning functions (from prefrontal cortex) are also asymmetric on a linear scale, but symmetric after logarithmic transformation. After Nieder, A., & Miller, E. K. (2003). Coding of cognitive magnitude: Compressed scaling of numerical information in the primate prefrontal cortex. *Neuron, 37,* 149–157.

barely overlap, which results in a high signal-to-noise ratio and, therefore, good performance in cases where the animal has to discriminate sets of a larger numerical distance. The behavioral consequences of the numerical size effect are therefore in accordance with Weber's Law.

The numerical size effect is directly related to the precision of the neuronal numerosity filters: The widths of the tuning curves (or neuronal numerical representations) increase linearly with preferred numerosities (i.e., on average, tuning precision deteriorates as the preferred quantity increases). Hence, more selective neural filters that do not overlap extensively are engaged if a monkey has to discriminate small numerosities (say, 1 and 2), which results in high signal-to-noise ratios and few errors in the discrimination. Conversely, if a monkey has to discriminate large numerosities (such as 4 and 5), the filter functions would overlap considerably. Therefore, the discrimination has a low signal-to-noise ratio, which leads to poor performance.

An important piece of evidence for the contribution of numerosity-selective neurons to behavioral performance came from the examination of error trials. When the monkeys made judgment errors, the neural activity for the preferred quantity was significantly reduced as compared to correct trials (Nieder & Merten, 2007; Nieder & Miller, 2004a; Nieder et al., 2002, 2006). As a result of this (and the ordered representation of quantity), the activity to a given preferred numerosity on error trials was more similar to that elicited by adjacent nonpreferred quantities on correct trials. In other words, if the neurons did not encode the numerosity properly, the monkeys were prone to mistakes.

Implementing Numerosity Detectors

How may numerosity-selective neurons tuned to preferred numerosities arise in the course of cortical processing? Purely sensory, nonnumerical properties (such as binocular disparity, wavelength, and contrast in the visual system) are encoded already at the earliest processing stages of the sensory epithelia (Van Essen & DeYoe, 1993). Number, on the other hand, is a most abstract category devoid of specific sensory features; two cats and two calls have nothing in common, except that the size of their sets is "two." How, then, may the cardinality of objects or events, the pure number of entities, be derived in terms of neuronal information processing?

Two main models have been proposed to explain the implementation of quantity information. The *mode-control model* by Meck and Church (1983) works in series and suggests that each item is encoded by an impulse from a pacemaker, which is added to an accumulator (Fig. 20.4A). The magnitude in the accumulator at the end of the count is then read into memory, forming a representation of the number of a set. Thus, it is assumed that quantity is encoded by "summation coding" (i.e., the monotonically increasing and decreasing response functions of the neurons; see also network model by Zorzi & Butterworth, 1999; Zorzi et al., 2005).

Another model, the *neural filtering model* by Dehaene and Changeux (1993), implements numerosity in parallel (Fig. 20.4B). First, each (visual) stimulus is coded as a local Gaussian distribution of activation by topographically organized input clusters (simulating the retina). Next, items of different sizes are normalized to a size-independent code. At that stage, item size, which was initially coded by the number of active clusters on the retina (quantity code), is now encoded by the position of active clusters on a location map (position code). Clusters in the location map project to every unit of downstream "summation clusters," whose thresholds increase with increasing number and pool the total activity of the location map. The summation clusters finally project to "numerosity clusters." Numerosity clusters are characterized by central excitation and lateral inhibition so that each numerosity cluster responds only to a selected range of values of the total normalized activity (i.e., their preferred numerosity). Since the numerosity of a stimulus is encoded by peaked tuning functions with a preferred numerosity (causing maximum discharge), this mechanism is termed "labeled-line code." A similar architecture was proposed by Verguts and Fias (2004) using a backpropagation network. Interestingly, summation units developed spontaneously in the second processing stage (the "hidden units") after tuned numerosity detectors were determined at the output stage.

Even though numerosity representations derived with both models are noisy (approximate) and obey Weber's Law, the two models differ in important aspects. The mode-control model by Meck and Church (1983) operates serially and assumes representation of cardinality on a linear scale, whereas in the neural filtering model by Dehaene and Changeux (1993), numerosity is encoded in parallel and represented on a logarithmic scale (the same holds for the backpropagation model by Verguts and Fias [2004]). Both models, however, have summation units implemented that accumulate number in a graded fashion prior to feeding into numerosity detectors at the output. As a putative physiological reflection of this computational stage in models of number processing, Roitman and colleagues (2007) recorded neurons in the parietal lobe whose responses resembled the output of accumulator neurons with response functions that systematically increased or decreased with an increase of stimulus set size (Fig. 20.4C,D).

In this study, the activity of single neurons was recorded in the lateral intraparietal (LIP) area of monkeys performing a delayed saccade task (Roitman et al., 2007). On each trial, monkeys maintained fixation on a central point while a saccade target was placed at a random location lateral to the fixation spot. When the fixation point was extinguished, the monkey shifted its gaze to the saccade target to receive a fluid reward. Prior to the monkey performing the saccade toward the target, a set of items (2, 4, 8, 16, or 32 dots) at the location of the recorded cell's receptive field informed the monkey about the amount of reward it would receive after a correct saccade. One of the five possible numerosities was selected as the standard, which was then presented in half of the trials. In such standard trials, the animal received a fixed amount of fluid as reward. In each of the remaining trials, one of the four deviant numerosities was shown, resulting in a slightly

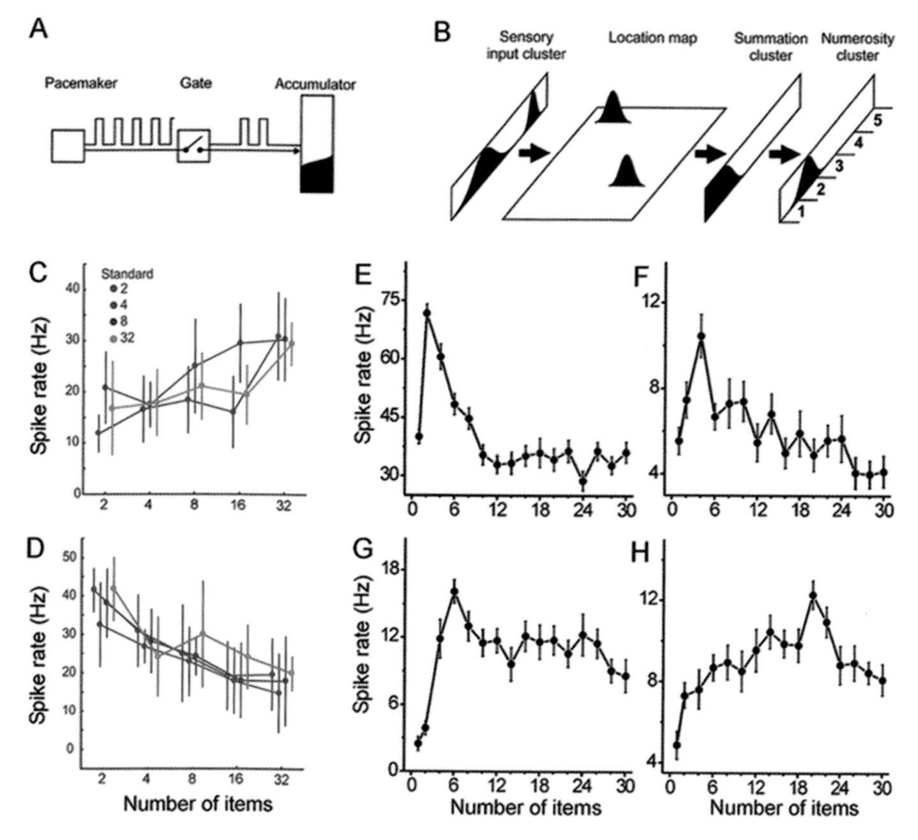

Figure 20.4 Implementation of numerosity detectors. **(A)** Mode-control model. After Meck, W. H., & Church, R. M. (1983). A mode control model of counting and timing processes. *Journal of Experimental Psychology: Animal Behavioral Processes, 9,* 320–334. **(B)** Neural filtering model. After Dehaene, S., & Changeux, J. P. (1993). Development of elementary numerical abilities: A neural model. *Journal of Cognitive Neuroscience, 5,* 390—407. **(C, D)** Neurons in lateral intraparietal (LIP) area discharge monotonically as a function of set size during an implicit numerosity task. Two single cells are depicted that show an increase (C) or decrease (D) of discharge rate, respectively, with increasing numerosity. Each neuron was tested with different standard (color code) and deviant numerosities (see text for explanation of the task). From Roitman, J. D., Brannon, E. M., & Platt, M. L. (2007). Monotonic coding of numerosity in macaque lateral intraparietal area. *PLoS Biology, 8,* e208. **(E–H)** Prefrontal cortex neurons tuned to preferred numerosities in monkeys performing a delayed match-to-numerosity task. Preferred numerosity was 2 (E), 4 (F), 6 (G), and 20 (H). From Nieder, A., & Merten, K. (2007). A labeled-line code for small and large numerosities in the monkey prefrontal cortex. *Journal of Neuroscience, 27,* 5986–5993.

larger amount of fluid as reward. This encouraged the monkey to pay attention to the numerosities, even though they were not to be discriminated or otherwise used in the task.

Roitman and coworkers (2007) found that the activity of most LIP neurons increased or decreased systematically with increasing number of elements during stimulus presentation (Fig. 20.4C,D) (irrespective of other stimulus features or cognitive demands). Thus, a population of neurons in the LIP area encoded the number of elements in a visual array in a roughly monotonic manner, similar to neurons representing sensory magnitude (Brody et al., 2003; Romo et al., 1999). The authors suggested that these two classes of number-selective neurons may be the physiological instantiation of the summation units and numerosity units proposed in neural network models of numerical representation; monotonic magnitude coding of LIP neurons may provide input to neurons in the PPC and PFC that compute cardinal numerical representations via tuning to preferred numerosities (Fig. 20.4E–H) (Nieder & Merten, 2007; Nieder & Miller, 2004a; Nieder et al., 2002, 2006; Piazza et al., 2004).

In agreement with this hypothesis, monotonic neurons described in the LIP area would operate on an intermediate level of numerosity detection. The numerosity stimuli used by Roitman and coworkers (2007) were carefully placed over the spatially confined response fields of LIP neurons and thus allowed to address local information processing. Final processing stages of abstract numerical information, however, are required to integrate across time, space, and modality. Area VIP and the PFC are ideal candidate structures for a global representation of numerosity; both areas integrate multimodal input (Duhamel et al., 1998), and neurons in PFC in particular exhibit global cognitive processing properties that are no longer spatially restricted (Everling et al., 2002). Behavioral relevance is another aspect that could have a substantial effect on the coding scheme. All studies that required the monkeys to use cardinal numerical information explicitly to solve a task found a labeled-line code, irrespective of stimulus modality, presentation format, and recording site (Nieder et al., 2002, 2006; Sawamura et al., 2002), whereas numerosity was implicitly informative in the delayed saccade task (Roitman et al., 2007) but had no impact on the monkeys' task performance. Thus, the neuronal representation may change if quantity is encoded as an explicit category.

The Scaling of Numerical Representations

As mentioned previously, the neurons' overlapping tuning curves are ordered along a "number line" (Fig. 20.3B). But what is the scaling scheme of such a "number line"? Are neuronal numerical representations best described on a linear or a nonlinear, possibly logarithmically compressed scale? The latter would be predicted if Fechner's Law holds. Fechner's Law states that the perceived magnitude (S) is a logarithmic function of stimulus intensity (I) multiplied by a modality- and dimension-specific constant (k). If the tuning functions for behavioral discrimination and single units are regarded as the monkeys' behavioral and neural numerical representations (Fig. 20.3), the crucial question then concerns which scaling scheme provides symmetric (i.e., Gaussian) probability density distributions. Both the performance and the single-unit data for numerosity judgments are better described by a compressed, as opposed to a linear, scale (Nieder & Miller, 2003). Therefore, single-neuron representations of numerical quantity in monkeys obey Fechner's Law.

Coding of Continuous and Discrete Quantity

Quantitative knowledge, such as understanding how much drinking water is available (Van Marle et al., 2006; Woodruff et al., 1978) or how many individuals belong to an opponent party (McComb et al., 1994; Wilson et al., 2001), guide vital decisions in the life of animals and humans alike. Functional imaging studies in humans suggest that anatomical vicinity (Castelli et al., 2006; Fias et al., 2003; Pinel et al., 2004) or even a common magnitude system (Walsh, 2003) for the representation of numerical (discrete) and spatial (continuous) quantity in the parietal cortex might be responsible for behavioral interference phenomena between numerical and spatial quantity. In a number comparison task, for example, choosing the numerically larger number takes significantly longer if the numeral is physically smaller in size compared to the numerically smaller number (e.g., in the comparison 2 versus 7) (Henik & Tzelgov, 1982; Pinel et al., 2004).

To investigate how continuous quantity is encoded by single nerve cells and how it relates to numerosity representations, Tudusciuc and Nieder (2007) trained two rhesus monkeys in a delayed match-to-sample task to discriminate different types of quantity randomly alternating within each session. In the "length protocol," the length of a line (out of four different lengths) needed to be discriminated (continuous-spatial quantity). In the "numerosity protocol"

(Fig. 20.1A), the number of (one to four) items in multiple-dot displays (discrete-numerical quantity) was the relevant stimulus dimension. To ensure that the monkeys solved the task based on the relevant quantitative information, other covarying features of the stimuli were again controlled, and the positions of the dots and lines were greatly varied.

After the monkeys solved more than 81% of the trials correctly for both the length and the numerosity protocols, single-unit activity from the depth of the IPS was analyzed while the animals performed the task. About 20 % of anatomically intermingled single neurons in the monkey IPS each encoded discrete-numerical (Fig. 20.5A), continuous-spatial (Fig. 20.5B), or

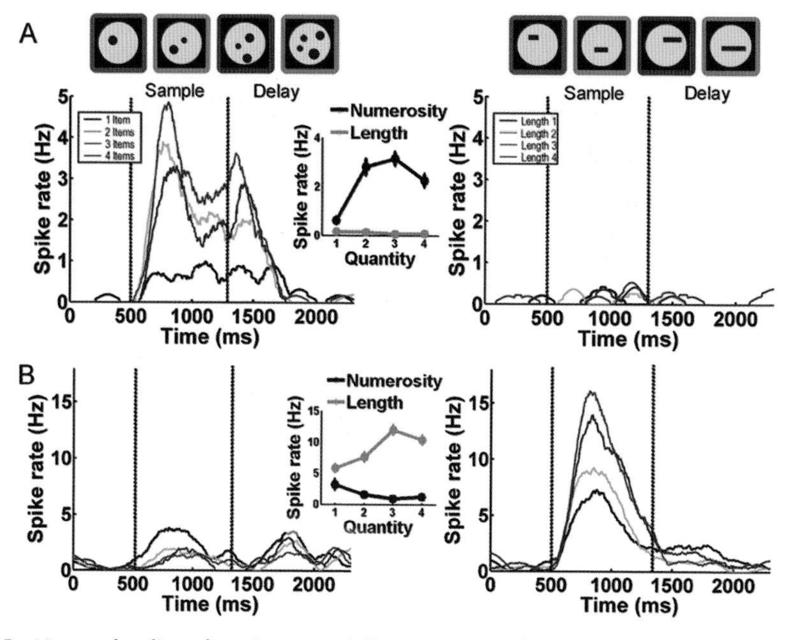

Figure 20.5 Neuronal coding of continuous and discrete quantity. **(A–C)** Three example neurons exhibiting selectivity for quantity. Top panels in A illustrate the four different numerosities (left) and four different line lengths (right) used as stimuli. Left and right graphs illustrate the discharge rates (displayed as smoothed spike-density histograms) of the same neuron in the numerosity and length protocol, respectively. The first 500 ms represent the fixation period. The area between the two black vertical bars represents the sample presentation; the following 1,000 ms indicate the delay phase. Colors correspond to the quantity dimensions. The insets between two histograms depict the tuning functions of each of the three neurons to numerosity and length. **(A)** Neuron tuned to numerosity 3, but not to length. **(B)** Neuron tuned to the third longest line, but not to any tested numerosity. **(C)** Neuron encoding both discrete and continuous quantity. **(D, E)** Classification performance in the sample period across the neuronal population. Confusion matrices describing the pattern of quantity classification performed on 72 quantity-selective neurons (D) and 72 nonselective (untuned) neurons (E). The rows in each confusion matrix represent the true classes the monkey had seen; the columns correspond to the output of the classifier. Color codes the classification probability. The eight classes correspond to the eight stimulus quantities: numerosity 1 to 4 and line length 1 to line length 4, where length 1 is the shortest line. Thus, the main diagonal shows how often the classifier correctly assigned quantity stimuli to their real category (i.e., a measure of accuracy). Averaging the classification probabilities over each diagonal parallel to the main diagonal results in the average performance of the classifier as a function of distance from the actual quantity, which is plotted as a tuning function at the top left (discrete) and bottom right (continuous) of each matrix. From Tudusciuc, O., & Nieder, A. (2007). Neuronal population coding of continuous and discrete quantity in the primate posterior parietal cortex. *Proceedings of the National Academy of Sciences of the USA, 104,* 14513–14518.

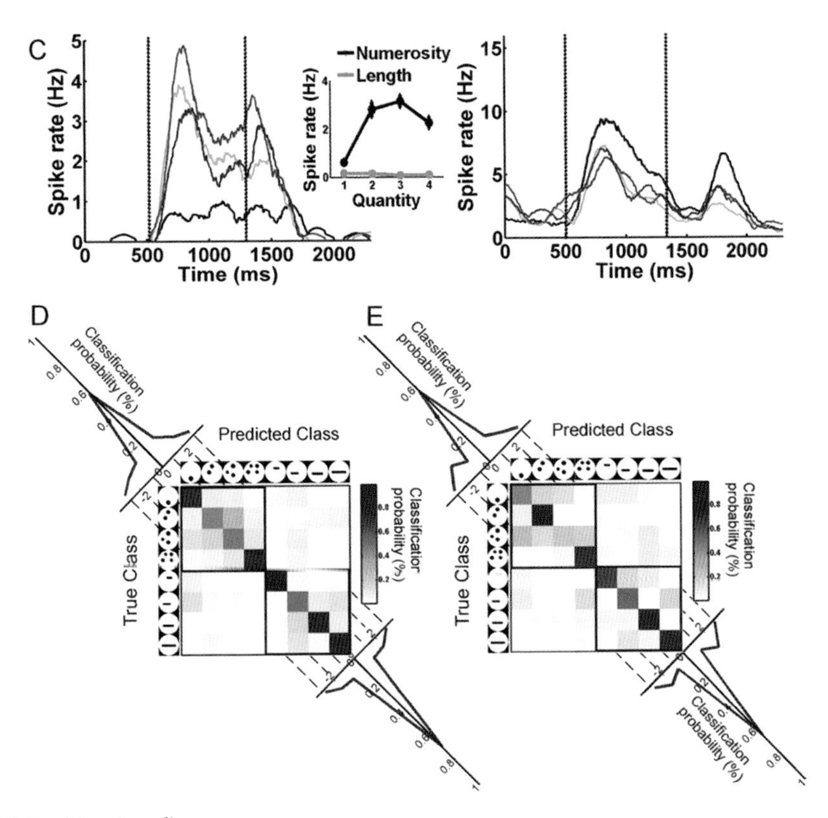

Figure 20.5 (Continued)

both types of quantities (Fig. 20.5C), suggesting that two partly overlapping populations of neurons within this area may give rise to a generalized magnitude system (Walsh, 2003).

Can functionally overlapping groups of parietal neurons provide sufficient information for the monkey to make correct quantity judgments? To assess the discriminative power of this small population of neurons, Tudusciuc and Nieder (2007) applied a population decoding technique (Averbeck et al., 2006; Hung et al., 2005; Laubach, 2004; Laubach et al., 2008) based on an artificial neuronal network (Kohonen, 1997). The classifier was trained with neuronal responses (i.e., preprocessed spike trains) of a set of neurons recorded while the monkeys judged each of the eight quantity categories; at this stage, the classifier was informed about the stimulus configuration and learned the neuronal features that were best suited for identifying a given category. In the subsequent test phase, the classifier predicted the categories from novel neuronal responses of the same pool of neurons (i.e., from data it had not used for learning).

The quantitative results based on the statistical classifier demonstrated that the small population of quantity-selective neurons carried most of the categorical information (Fig. 20.5D); by exploiting the classical spike-rate measure that contributes to the monkeys' quantity discrimination performance (Nieder & Miller, 2004a; Nieder et al., 2002, 2006), the classifier was able to accurately and robustly discriminate both continuous and discrete quantity classes. Interestingly, even the population of untuned neurons had a remarkable and significant quantity coding ability, albeit to a lesser extent than the tuned neurons (Fig. 20.5E). This suggests that the classifier extracted, beyond the averaged spike rate, additional information from the temporal structure of the neuronal responses. Moreover, the comparison between the monkeys' neuronal and behavioral responses showed that the brain indeed utilizes this information for decision making; neuronal responses recorded whenever the monkeys failed to discriminate the quantity categories prevented the classifier from predicting the correct quantity category. Future studies need to clarify whether complex quantity judgments require an interplay with, or readout by, structures with more executive functions, such as the prefrontal cortex (Miller & Cohen, 2001).

Toward Symbolic Number Representations

As shown previously, humans and animals share an evolutionarily old quantity representation system that allows the estimation of set sizes. Nonverbal numerical cognition, however, is limited to approximate quantity representations and rudimentary arithmetic operations. Language-endowed humans, on the other hand, invented number symbols (numerals and number words) during cultural evolution. These mental tools enable us to create precise quantity representations and perform exact calculation that is beyond the reach of any animal species.

Even though number symbols are of paramount importance in today's scientifically and technologically advanced culture, their invention dates back only a couple thousand years (Ifrah, 2000). Given the time scale of brain evolution, a de novo development of brain areas with distinct, culturally dependent number symbol functions is more than unlikely (Dehaene, 2005). Rather, it is conceivable that brain structures that originally evolved for other purposes are built upon in the course of continuing evolutionary development (Gould, 1982). According to the "redeployment hypothesis" (Anderson, 2007) or "recycling hypothesis" (Dehaene, 2005; Dehaene & Cohen, 2007), already existing simpler cell assemblies are largely preserved, extended, and combined as networks become more complex (Sporns & Kotter, 2004).

In the number domain, existing neuronal components in PFC and IPS subserving nonverbal quantity representations could be used for the new purpose of number symbol encoding, without disrupting their participation in existing cognitive processes (Piazza et al., 2007). Guided by the faculty of language, children learn to use number symbols as mental tools during childhood. During this learning process, and as a prerequisite for the utilization of signs as numerical symbols, long-term associations between initially meaningless shapes (that later become numerals) and inherent semantic numerical categories must be established. This necessary but by no means sufficient step toward the utilization of number symbols in humans can also be mastered by different animal species (Boysen & Berntson, 1989; Matsuzawa, 1985; Washburn & Rumbaugh, 1991; Xia et al., 2001).

To investigate the single-neuron mechanisms of such semantic associations, Diester and Nieder (2007) mimicked a semantic mapping process by training two monkeys to associate the a priori meaningless visual shapes of Arabic numerals (that became "signs" or, more precisely, "indices"; Wiese, 2003) with the inherently meaningful numerosity of multiple-dot displays. After this long-term learning process was completed, a relatively large proportion of PFC neurons (24%) encoded plain numerical values, irrespective of whether they had been presented as a specific number of dots or as a visual sign (Fig. 20.6A,B). Such "association neurons" showed similar tuning during the course of the trial to both the direct numerosity in dot stimuli and the associated numerical values of signs (Fig. 20.6C). Interestingly, the tuning functions of association neurons showed a distance effect for both protocols (i.e., a dropoff of activity with increasing numerical distance from the preferred numerical value; Fig. 20.6D). This distance effect found in the shape protocol indicates that association neurons responded as a function of numerical value rather than visual shape per se. Most cells coded the (direct and associated) numerical values during specific time phases in the trial (e.g., only at sample onset or toward the end of the delay period). The neuronal population as a whole, however, represented the numerical association throughout the entire trial and thus provided crucial associative information over time. The activity of association neurons predicted the monkeys' judgement performance; if the monkeys failed to match the correct number of dots to the learned signs, the tuning behavior of a given neuron to numerosities and their associated visual shapes were

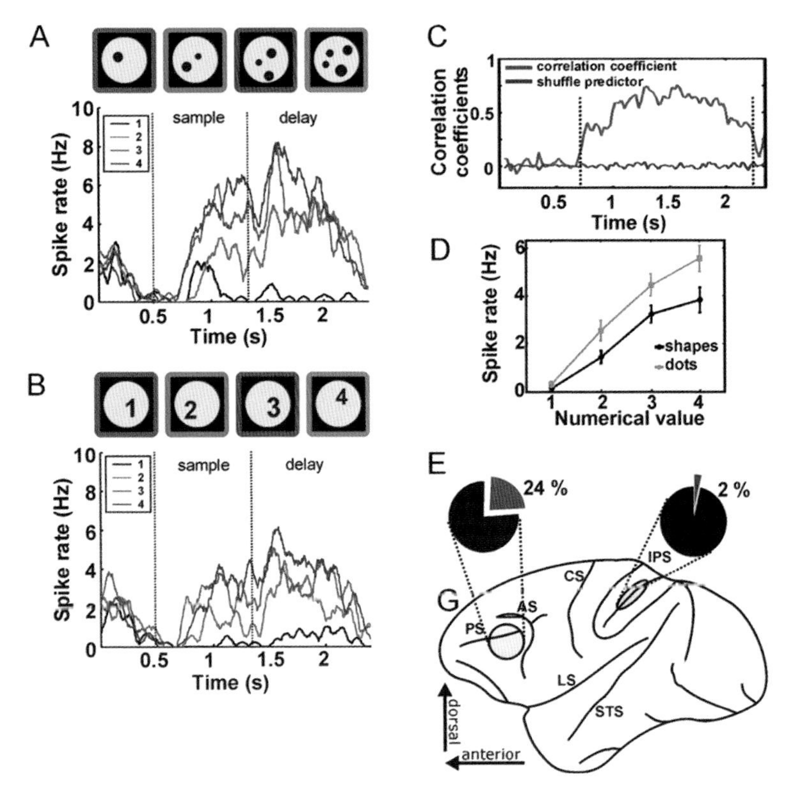

Figure 20.6 Semantic associations between signs and numerical categories by single neurons. (**A, B**) The responses of the same single prefrontal cortex (PFC) neuron to both direct numerosities and associated numeral shapes (top panels in A and B illustrate sample stimuli). Neuronal responses in A and B are shown as smoothed spike-density histograms (colors denote discharge to the corresponding sample numerical value 1 to 4). The first 500 ms indicate the fixation period. Black vertical lines mark sample onset (500 ms) and offset (1,300 ms). This neuron's preferred numerical value in the sample and delay period was 4. Note the similarity in the association neuron's temporal discharge profiles in response to the multiple-dot displays and the shape of Arabic numerals. (**C**) Time course of original cross-correlation coefficients (red) and chance cross-correlation coefficients (shuffle predictor, blue). The interval bordered by vertical dotted lines indicates the time phase of significant cross-correlation (as determined by measures from signal detection theory) between tuning to numerical values in the multiple-dot displays and Arabic numerals; in this period, the neuron associated numerical values in the two protocols. (**D**) Tuning functions to numerical values in the multiple-dot displays and Arabic numerals. (**E**) Lateral view of a monkey brain indicating the recording sites in the PFC and intraparietal sulcus (IPS). Proportions of association neurons in the PFC and IPS are displayed as pie charts. After Diester, I., & Nieder, A. (2007). Semantic associations between signs and numerical categories in the prefrontal cortex. *PLoS Biology, 5*, e294.

severely disrupted. These findings argue for association neurons as a neuronal substrate for the semantic mapping processes between signs and categories.

In the same study (Diester & Nieder, 2007), the activity of neurons in the fundus of the IPS was also recorded. In contrast to PFC, only 2% of all recorded IPS neurons associated signs with numerosities (Fig. 20.6E). Moreover, the quality of neuronal association in the IPS was weak and occurred much later during the trial.

The conclusion drawn from these results is that even though monkeys use the PFC and IPS for nonsymbolic quantity representations, only the prefrontal part of this network is engaged in semantic shape-number associations. Interest-ingly, this neural pattern seems to be preserved in children (Ansari et al., 2005; Kaufmann et al., 2006; Rivera et al., 2005). In contrast to adults, preschoolars lacking ample exposure to number symbols show elevated PFC activity when dealing with ymbolic cardinalities. With age and proficiency, however, the activation seems to shift to parietal areas. The PFC could thus be ontogenetically and phylogenetically the first cortical area establishing semantic associations, which might be relocated to the parietal cortex in human adolescents in parallel with maturing language capabilities that endow our species with a sophisticated symbolic system (Deacon, 1997).

The prefrontal region is strategically situated to establish semantic associations (Miller & Cohen, 2001); it receives input from both the anterior inferotemporal cortex encoding shape information (Tanaka, 1996) as well as the posterior parietal cortex that contains numerosity-selective neurons (Nieder & Miller, 2004a; Nieder et al., 2006). Previous studies showed that neurons in the PFC encode learned associations between two purely sensory stimuli without intrinsic meaning (e.g., the association of a ertain color with a specific sound, or pairs of pictures) (Fuster et al., 2000; Miller et al., 1996; Rainer et al., 1999). In the anterior inferotemporal cortex, Miyashita and coworkers found "pair-coding neurons" that responded to arbitrary pairs of images monkeys learned to match in a pair-association task (Sakai & Miyashita, 1991); the same group found evidence that the PFC is important for active retrieval of these associative representations (Tomita et al., 1999). Our findings (Diester & Nieder, 2007) demonstrate that neurons in the PFC represent semantic long-term associations not only between pairs of pictures but also between arbitrary shapes and systematically arranged categories with inherent meaning (i.e., the ordered cardinalities of sets). In that respect, the PFC of primates may not only control the retrieval of long-term associations but also may in fact onstitute a crucial processing stage for abstract semantic associations.

NUMERICAL RANK (ORDINALITY)

The frontal lobe is an ideal region in the brain to encode both sensory object properties and rank-order information because PFC receives massive sensory input from the temporal and parietal lobes and projects to premotor and motor areas of the frontal lobe (Miller & Cohen, 2001). Thus, neurons that encode the ordinal position of task-related hand or eye movements have been found frequently in prefrontal (Funahashi et al., 1997) and a subset of motor-related cortical areas in trained monkeys. Joseph and coworkers (Barone & Joseph, 1989; Kermadi & Joseph, 1995; Procyk & Joseph, 2001; Procyk et al., 2000) identified order-selective neurons in the frontal eye field (FEF), caudate nucleus, and anterior cingulate cortex of monkeys that had been trained to sequentially order spatially arranged items. These neurons were only active when the monkeys reached for the first, second, or third target, irrespective of the targets' location and the precise type of hand movement. Clower and Alexander (1998) trained a monkey to position a cursor on a video display by moving a joystick clockwise or counterclockwise along a spatially arranged four-item path. In the presupplementary motor areas (pre-SMAs), more than two thirds of the recorded neurons showed significant effects of numerical order, but only about one third of the neurons displayed an effect of rank order in the supplementary motor area (SMA). Rank-order selectivity was also identified in the pre-SMA of monkeys that had been trained to sequentially perform three different hand movements ("push," "pull," or "turn") in four to six different orders separated by waiting times (Shima & Tanji, 2000; Tanji & Shima, 1994). In the pre-SMA, the activity differed selectively in the process of preparing the first, second, or third movements in individual trials. The SMA, on the other hand, was more involved in linking the occurrence of two different movements and, therefore, in determining the order of the component movements in the sequence (relational order). Ordinal position of movements in the pre-SMA seems to be encoded in an effector-independent manner as in the case of eye movement sequences (Isoda & Tanji, 2004). In two motor areas that are specialized in processing eye movements, activity that reflected saccade sequence or the numerical position of a saccade within a sequence (rank) was more common in the supplementary eye field (SEF), whereas activity that reflected saccade direction was more dominant in the FEF (Isoda & Tanji, 2003).

Interestingly, encoding of numerical order has not only been observed in these premotor and supplementary motor areas but also in the primary motor cortex. Carpenter and colleagues (1999) showed monkeys five spatially arranged visual targets that appeared successively on a screen. After the target sequence was complete, one of the items changed its color. The monkeys needed to memorize the order in which the targets appeared and point to the item that appeared just after the one that had changed its color at the end of the list presentation. In approximately one third of the neurons recorded from the arm region of the topographically organized primary motor cortex (M1), the ordinal position of the targets was the only factor that covaried with neuronal activity. Therefore, the motor cortex—an area that is traditionally regarded as purely motor executive—also participates in the processing of cognitive information about serial order within the context of a motor task. The authors pointed out, however, that the motor cortex most likely is just one component in a distributed network that encodes, stores, and recalls a sequence. Motor-related areas such as M1, SMA, pre-SMA, and FEF may receive numerical information that has already been computed in earlier stages of the cortical hierarchy to perform appropriate serial-order actions.

Ordinal categorization of items requires both information about the rank of an item (e.g., based on temporal order) and its identity. Neuropsychological studies emphasize the importance of the lateral prefrontal cortex in maintaining temporal order information (McAndrews & Milner, 1991; Milner, 1971). In monkeys, lesioning the dorsolateral frontal cortex causes impairments in tasks that require recall of the temporal order of events and stimuli (Petrides, 1995). In two recent elegant studies (Ninokura et al., 2003, 2004), the single-neuron correlate of temporal rank order information in visual lists was addressed. Monkeys were trained to observe and remember the order in which three visual objects appeared. Subsequently, the animals planned and executed a triple-reaching movement in the same order as previously seen (Fig. 20.7). Neurons in the ventrolateral PFC selectively encoded visual object properties (26% of the total sample), whereas neurons in the dorsolateral PFC (44%) were selectively tuned to the rank order of the objects irrespective of the sensory properties of objects. For example, a rank-order–selective neuron would be active whenever the second item of a shuffled lists appears. A third class of neurons (30%), found in the ventrolateral PFC, showed the most complex responses, integrating the objects' sensory and order information. Such neurons would only discharge whenever a certain object appeared at a given position in the sequence. Similar results have also been reported by Inoue and Mikami (2006).

The representational formats of nonverbal serial-order information are still poorly understood. However, the behavioral and neuronal data indicate an imprecise representation of discrete numerical rank, which is reminiscent of the analog-magnitude mechanism that has been proposed for cardinality. To elaborate a computational model of working memory for serial order, Botvinick and Watanabe (2007) recently wove item, numerosity, and rank information together. Their network combined graded neuronal responses to different items (not yet verified experimentally) and tuning functions for sequential enumeration processes (Nieder et al., 2006) with the data showing that neurons in the PFC code the rank of items within a sequence (Ninokura et al., 2004; Fig 20.7). The model's output, a recalled multi-item sequence, replicated many behavioral characteristics of working memory such as the primacy effect (a recall advantage for initial items) and the recency effect (advantage for

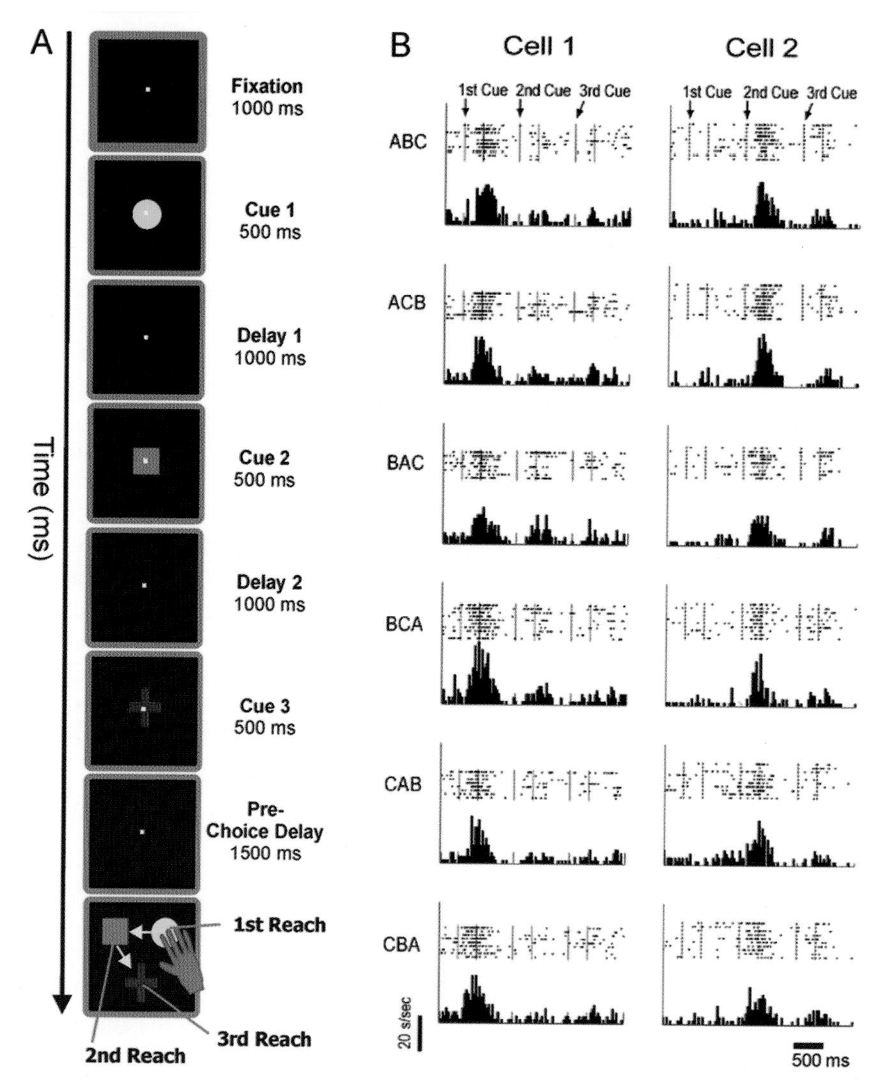

Figure 20.7 Temporal ordering task and single cell responses from the PFC. (**A**) Monkeys were required to observe and remember the order in which three visual objects appeared, so that the animals could plan a subsequent triple-reaching movement in the same order. (**B**) Two single neurons encoding the first (cell 1) and the second rank (cell 2), irrespective of the order in which the three items (symbolized by letters ABC) appeared. Neural responses are shown in a dot-raster histogram (top panels, each dot represents an action potential) and averaged as peri-stimulus time histograms (bottom panels). From Ninokura, Y., Mushiake, H., & Tanji, J. (2004). Integration of temporal order and object information in the monkey lateral prefrontal cortex. *Journal of Neurophysiology, 91,* 555–560.

the last one or two items). Furthermore, changing the width of the model rank tuning curve simulated the developmental finding of improved recall accuracy with age. Thus, this model integrates across several neurophysiological studies to demonstrate how higher cognitive functions may exploit both quantity and rank (Jacob & Nieder, 2008). It posits that working memory of ordered sequences at least in part relies on neuronal assemblies in the parietal and prefrontal cortices that also contribute to numerosity representations.

CONCLUSION

This chapter illustrates that nonverbal numerical representations can engage a wide cortical network, with the PFC and the IPS, in particular, as key structures. Neurons in these areas are characterized by response properties that underlie basic psychophysical phenomena during cardinality or serial-order judgments. So far, single-cell studies in monkeys have concentrated on either cardinal or ordinal aspects, but never both. In the human primate, a recent study showed that the cortical network for rank coincides with the areas processing numerical quantity (Fias et al., 2007), even though this does not necessarily mean that single neurons may encode both cardinality and ordinality. Based on the findings that discrete and continuous quantities and even simultaneously and sequentially presented numerosities are encoded by largely distinct neuronal populations in the IPS, a cellular segregation also for cardinal and ordinal stimulus aspects is likely. Single-cell electrophysiology is well poised to answer this and other important questions about the neural basis of numerical cognition in the years to come.

REFERENCES

Anderson, M. L. (2007). Evolution of cognitive function via redeployment of brain areas. *Neuroscientist, 13,* 13–21.

Ansari, D., Garcia, N., Lucas, E., Hamon, K., & Dhital, B. (2005). Neural correlates of symbolic number processing in children and adults. *Neuroreport, 16,* 1769–1773.

Averbeck, B. B., Sohn, J. W., & Lee, D. (2006). Activity in prefrontal cortex during dynamic selection of action sequences. *Nature Neuroscience, 9,* 276–282.

Barone, P., & Joseph, J. P. (1989). Prefrontal cortex and spatial sequencing in macaque monkey. *Experimental Brain Research, 78,* 447–464.

Botvinick, M., & Watanabe, T. (2007). From numerosity to ordinal rank: A gain-field model of serial order representation in cortical working memory. *Journal of Neuroscience, 27,* 8636–8642.

Boysen, S. T., & Berntson, G. G. (1989). Numerical competence in a chimpanzee (Pan troglodytes). *Journal of Comparative Psychology, 103,* 23–31.

Bremmer, F., Schlack, A., Shah, N. J., Zafiris, O., Kubischik, M., Hoffmann, K., et al. (2001). Polymodal motion processing in posterior parietal and premotor cortex: A human fMRI study strongly implies equivalencies between humans and monkeys. *Neuron, 29,* 287–296.

Brody, C. D., Hernandez, A., Zainos, A., & Romo, R. (2003). Timing and neural encoding of somatosensory parametric working memory in macaque prefrontal cortex. *Cerebral Cortex, 13,* 1196–1207.

Carpenter, A. F., Georgopoulos, A. P., & Pellizzer, G. (1999). Motor cortical encoding of serial order in a context-recall task. *Science, 283,* 1752–1757.

Castelli, F., Glaser, D. E., & Butterworth, B. (2006). Discrete and analogue quantity processing in the parietal lobe: A functional MRI study. *Proceedings of the National Academy of Sciences of the USA, 103,* 4693–4698.

Chafee, M. V., & Goldman-Rakic, P. S. (2000). Inactivation of parietal and prefrontal cortex reveals interdependence of neural activity during memory-guided saccades. *Journal of Neurophysiology, 83,* 1550–1566.

Clower, W. T., & Alexander, G. E. (1998). Movement sequence-related activity reflecting numerical order of components in supplementary and presupplementary motor areas. *Journal of Neurophysiology, 80,* 1562–1566.

Colby, C. L., Duhamel, J. R., & Goldberg, M. E. (1993). Ventral intraparietal area of the macaque – anatomical location and visual response properties. *Journal of Neurophysiology, 69,* 902–914.

Colby, C. L., & Goldberg, M. E. (1999). Space and attention in parietal cortex. *Annual Review of Neuroscience, 22,* 319–349.

Cordes, S., Gelman, R., Gallistel, C. R., & Whalen, J. (2001). Variability signatures distinguish verbal from nonverbal counting for both large and small numbers. *Psychonomic Bulletin and Review, 8,* 698–707.

Deacon, T. (1997). *The symbolic species: The co-evolution of language and the human brain.* London: Norton.

Dehaene, S. (2005). Evolution of human cortical circuits for reading and arithmetic: The "neuronal recycling" hypothesis. In: S. Dehaene, J. R. Duhamel, M. D. Hauser, & G. Rizzolatti (Eds.), *From monkey brain to human brain* (pp. 133–157). Cambridge, MA: MIT Press.

Dehaene, S., & Changeux, J. P. (1993). Development of elementary numerical abilities: A neural model. *Journal of Cognitive Neuroscience, 5,* 390–407.

Dehaene, S., & Cohen, L. (2007). Cultural recycling of cortical maps. *Neuron, 56,* 384–98.

Diester, I., & Nieder, A. (2007). Semantic associations between signs and numerical categories in the prefrontal cortex. *PLoS Biology, 5,* e294.

Duhamel, J. R., Colby, C. L., & Goldberg, M. E. (1998). Ventral intraparietal area of the macaque: Congruent visual and somatic response properties. *Journal of Neurophysiology, 79,* 126–136.

Everling, S., Tinsley, C. J., Gaffan, D., & Duncan, J. (2002). Filtering of neural signals by focused attention in the monkey prefrontal cortex. *Nature Neuroscience, 5,* 671–676.

Fias, W., Lammertyn, J., Caessens, B., & Orban, G. A. (2007). Processing of abstract ordinal knowledge in the horizontal segment of the intraparietal sulcus. *Journal of Neuroscience, 27,* 8952–8956.

Fias, W., Lammertyn, J., Reynvoet, B., Dupont, P., & Orban, G. A. (2003). Parietal representation of symbolic and nonsymbolic magnitude. *Journal of Cognitive Neuroscience, 15,* 47–56.

Freedman, D. J., Riesenhuber, M., Poggio, T., & Miller, E. K. (2001). Categorical representation of visual stimuli in the primate prefrontal cortex. *Science, 291,* 312–316.

Funahashi, S., Inoue, M., & Kubota, K. (1997). Delay-period activity in the primate prefrontal cortex encoding multiple spatial positions and their order of presentation. *Behavioral Brain Research, 84,* 203–223.

Fuster, J. M., Bodner, M., & Kroger, J. K. (2000). Cross-modal and cross-temporal association in neurons of frontal cortex. *Nature, 405,* 347–351.

Gould, S. J. V. E. (1982). Exaptation: A missing term in the science of form. *Paleobiology, 8,* 4–15.

Henik, A., & Tzelgov, J. (1982). Is three greater than five: The relation between physical and semantic size in comparison tasks. *Memory and Cognition, 10,* 389–395.

Hung, C. P., Kreiman, G., Poggio, T., & DiCarlo, J. J. (2005). Fast readout of object identity from macaque inferior temporal cortex. *Science, 310,* 863–866.

Ifrah, G. (2000). *The universal history of numbers: From prehistory to the invention of the computer* (p. 633). New York: Wiley.

Inoue, M., & Mikami, A. (2006). Prefrontal activity during serial probe reproduction task: Encoding, mnemonic, and retrieval processes. *Journal of Neurophysiology, 95,* 1008–1041.

Isoda, M., & Tanji, J. (2003). Contrasting neuronal activity in the supplementary and frontal eye fields during temporal organization of multiple saccades. *Journal of Neurophysiology, 90,* 3054–3065.

Isoda, M., & Tanji, J. (2004). Participation of the primate presupplementary motor are in sequencing multiple saccades. *Journal of Neurophysiology, 92,* 653–659.

Jacob, S. N., & Nieder, A. (2008). The ABC of cardinal and ordinal number representations. *Trends in Cognitive Science, 29,4652-4657.*

Kaufmann, L., Koppelstaetter, F., Siedentopf, C., Haala, I., Haberlandt, E., Zimmerhackl, L.B. et al. (2006). Neural correlates of the number-size interference task in children. *Neuroreport, 17,* 587–591.

Kermadi, I., & Joseph, J. P. (1995). Activity in the caudate nucleus of monkey during spatial sequencing. *Journal of Neurophysiology, 74,* 911–933.

Kohonen, T. (1997). *Self-organizing maps* (2nd ed.). Berlin: Springer-Verlag.

Laubach, M. (2004). Wavelet-based processing of neuronal spike trains prior to discriminant analysis. *Journal of Neuroscience Methods, 134,* 159–168.

Laubach, M., Narayanan, N. S., & Kimchi, E. Y. (2008). Single-neuron and ensemble contributions to decoding simultaneously recorded spike trains. In: C. Holscher & M. Munk (Eds.), *Information processing by neuronal populations.* New York: Cambridge University Press.

Lewis, J. W., & Van Essen, D. C. (2000). Corticocortical connections of visual, sensorimotor, and multimodal processing areas in the parietal lobe of the macaque monkey. *Journal of Comparative Neurology, 428,* 112–37.

Matsuzawa, T. (1985). Use of numbers by a chimpanzee. *Nature, 315,* 57–59.

McAndrews, M. P., & Milner, B. (1991). The frontal cortex and memory for temporal order. *Neuropsychologia, 29,* 849–859.

McComb, K., Packer, C., & Pusey, A. (1994). Roaring and numerical assessment in contests between groups of female lions, *Panthera leo. Animal Behaviour, 47,* 379–387.

Meck, W. H., & Church, R. M. (1983). A mode control model of counting and timing processes. *Journal of Experimental Psychology: Animal Behavioral Processes, 9,* 320–334.

Miller, E. K., & Cohen, J. D. (2001). An integrative theory of prefrontal cortex function. *Annual Review of Neuroscience, 24,* 167–202.

Miller, E. K., Erickson, C. A., & Desimone, R. (1996). Neural mechanisms of visual working memory in prefrontal cortex of the macaque. *Journal of Neuroscience, 16,* 5154–5167.

Milner, B (1971). Interhemispheric differences in the localization of psychological processes in man. *British Medical Bulletin, 27,* 272–277.

Nieder, A., Diester, I., & Tudusciuc, O. (2006). Temporal and spatial enumeration processes in the primate parietal cortex. *Science, 313,* 1431–1435.

Nieder, A., Freedman, D. J., & Miller, E. K. (2002). Representation of the quantity of visual items in the primate prefrontal cortex. *Science, 297,* 1708–1711.

Nieder, A., & Merten, K. (2007). A labeled-line code for small and large numerosities in the monkey prefrontal cortex. *Journal of Neuroscience, 27,* 5986–5993.

Nieder, A., & Miller, E. K. (2003). Coding of cognitive magnitude: Compressed scaling of numerical information in the primate prefrontal cortex. *Neuron, 37,* 149–157.

Nieder, A., & Miller, E. K. (2004a). A parieto-frontal network for visual numerical information in the monkey. *Proceedings of the National Academy of Sciences of the USA, 101,* 7457–7462.

Nieder, A., & Miller, E. K. (2004b). Analog numerical representations in rhesus monkeys: Evidence for parallel processing. *Journal of Cognitive Neuroscience, 16,* 889–901.

Ninokura, Y., Mushiake, H., & Tanji, J. (2003). Representation of the temporal order of visual objects in the primate lateral prefrontal cortex. *Journal of Neurophysiology, 89,* 2868–2873.

Ninokura, Y., Mushiake, H., & Tanji, J. (2004). Integration of temporal order and object information in the monkey lateral prefrontal cortex. *Journal of Neurophysiology, 91,* 555–560.

Petrides, M. (1995). Impairments on nonspatial self-ordered and externally ordered working memory tasks after lesions of the mid-dorsal part of the lateral frontal cortex in the monkey. *Journal of Neuroscience, 15,* 359–375.

Piazza, M., Izard, V., Pinel, P., Le Bihan, D., & Dehaene, S. (2004). Tuning curves for approximate numerosity in the human intraparietal sulcus. *Neuron, 44,* 547–555.

Piazza, M., Pinel, P., Le Bihan, D., & Dehaene, S. (2007). A magnitude code common to numerosities and number symbols in human intraparietal cortex. *Neuron, 53,* 293–305.

Pinel, P., Piazza, M., Le Bihan, D., & Dehaene, S. (2004). Distributed and overlapping cerebral representations of number, size, and luminance during comparative judgments. *Neuron, 41,* 983–993.

Procyk, E., & Joseph, J. P. (2001). Characterization of serial order encoding in the monkey anterior cingulate sulcus. *European Journal of Neuroscience, 14,* 1041–1046.

Procyk, E., Tanaka, Y. L., & Joseph, J. P. (2000). Anterior cingulate activity during routine and non-routine sequential behaviors in macaques *Nature Neuroscience, 3,* 502–508.

Quintana, J., Fuster, J. M., & Yajeya, J. (1989). Effects of cooling parietal cortex on prefrontal units in delay tasks. *Brain Research, 503,* 100–110.

Rainer, G., Rao, S. C., & Miller, E. K. (1999). Prospective coding for objects in primate prefrontal cortex. *Journal of Neuroscience, 19,* 5493–5505.

Rivera, S. M., Reiss, A. L., Eckert, M. A., & Menon, V. (2005). Developmental changes in mental arithmetic: Evidence for increased functional specialization in the left inferior parietal cortex. *Cerebral Cortex, 15,* 1779–1790.

Roitman, J. D., Brannon, E. M., & Platt, M. L. (2007). Monotonic coding of numerosity in macaque lateral intraparietal area. *PLoS Biology, 8,* e208.

Romo, R., Brody, C. D., Hernandez, A., & Lemus, L. (1999). Neuronal correlates of parametric working memory in the prefrontal cortex. *Nature, 399,* 470–473.

Sakai, K., & Miyashita, Y. (1991). Neural organization for the long-term memory of paired associates. *Nature, 354,* 152–155.

Sawamura, H., Shima, K., & Tanji, J. (2002). Numerical representation for action in the parietal cortex of the monkey. *Nature, 415,* 918–922.

Shima, K., & Tanji, J. (2000). Neuronal activity in the supplementary and presupplementary motor areas for temporal organization of multiple movements. *Journal of Neurophysiology, 84,* 2148–2160.

Sporns, O., & Kotter, R. (2004). Motifs in brain networks. *PLoS Biology, 2,* e369.

Stoet, G., & Snyder, L. H. (2004). Single neurons in posterior parietal cortex of monkeys encode cognitive set. *Neuron, 42,* 1003–1012.

Tanaka, K. (1996). Inferotemporal cortex and object vision. *Annual Review of Neuroscience, 19,* 109–139.

Tanji, J., & Shima, K. (1994). Role for supplementary motor area cells in planning several movements ahead. *Nature, 371,* 413–416.

Tomita, H., Ohbayashi, M., Nakahara, K., Hasegawa, I., & Miyashita, Y. (1999). Top-down signal from prefrontal cortex in executive control of memory retrieval. *Nature, 401,* 699–703.

Tudusciuc, O., & Nieder, A. (2007). Neuronal population coding of continuous and discrete quantity in the primate posterior parietal cortex. *Proceedings of the National Academy of Sciences of the USA, 104,* 14513–14518.

Van Essen, D. C., & DeYoe, E. A. (1993). Concurrent processing in the primate visual cortex. In: M. S. Gazzaniga (Ed.), *The cognitive neurosciences* (pp. 383–400). Cambridge, MA: MIT Press.

Van Marle, K., Aw, J., McCrink, K., & Santos, L. R. (2006). How capuchin monkeys (Cebus apella) quantify objects and substances. *Journal of Comparative Psychology, 120,* 416–26.

Verguts, T., & Fias, W. (2004). Representation of number in animals and humans: A neural model. *Journal of Cognitive Neuroscience, 16,* 1493–1504.

Wallis, J. D., Anderson, K. C., & Miller, E. K. (2001). Single neurons in prefrontal cortex encode abstract rules. *Nature, 411,* 953–956.

Walsh, V. (2003). A theory of magnitude: Common cortical metrics of time, space and quantity. *Trends in Cognitive Science, 7,* 483–488.

Washburn, D. A., & Rumbaugh, D. M. (1991). Ordinal judgments of numerical symbols by macaques (Macaca mulatta). *Psychological Science, 2,* 190–193.

Whalen, J., Gallistel, C. R., & Gelman, R. (1999). Nonverbal counting in humans: The psychophysics of number representation. *Psychological Science, 10,* 130–137.

Wiese, H. (2003). *Numbers, language and the human mind.* New York: Cambridge University Press.

Wilson, M. L., Hauser, M. D., & Wrangham, R. W. (2001). Does participation in intergroup conflict depend on numerical assessment, range location, or rank for wild chimpanzees? *Animal Behaviour, 61,* 1203–1216.

Woodruff, G., Premack, D., & Kennel, K. (1978). Conservation of liquid and solid quantity by the chimpanzee. *Science, 202,* 991–994.

Xia, L., Emmerton, J., Siemann, M., & Delius, J. D. (2001). Pigeons (Columba livia) learn to link numerosities with symbols. *Journal of Comparative Psychology, 115,* 83–91.

Zorzi, M., & Butterworth, B. (1999). A computational model of number comparison. In: M. Hahn & S. C. Stoness (Eds.), *Proceedings of the Twenty First Annual Conference of the Cognitive Science Society* (pp. 778–783). Mahwah, NJ: Erlbaum.

Zorzi, M., Stoianov, I., & Umiltà, C. (2005). Computational modeling of numerical cognition. In: J. Campbell (Ed.), *Handbook of mathematical cognition* (pp. 67–84). New York: Psychology Press.

CHAPTER 21

Executive Control Circuits

Jonathan D. Wallis

Primate behavior is complex, and as behavioral complexity increases so too must the complexity of the systems responsible for controlling the behavior. In this chapter, we examine the neuronal systems that are responsible for this control. We begin by examining one of the essential features of such systems. High-level or executive control operates on highly processed information that is increasingly abstracted from the concrete world of specific sensory stimuli and motor responses. We will examine the evidence that the prefrontal cortex, an area of the brain that has increased dramatically in size over the course of mammalian evolution, enables this abstraction. We will look at the various psychological constructs that rely on abstraction, including high-level rule use, strategies, and task sets. We will then examine how this high-level control is organized. We will see that studies investigating how humans implement high-level information suggest that there is a hierarchical organization of such information within the prefrontal cortex. We will then examine whether there is evidence for such an organization in the monkey prefrontal cortex. Finally, we will examine how high-level information in the prefrontal cortex interacts with other regions of the brain to control behavior.

COMPLEX BEHAVIOR AND ABSTRACTION

The ultimate goal of behavior is simple: Ensure the continued existence of the organism long enough to permit reproduction and survival of one's offspring. Relatively simple behaviors evolved to accomplish this. For example, conditioning enables an organism to appreciate its sensory environment in terms of whether the situation enhances or decreases the organism's survival prospects. This enables the organism to approach and seek out beneficial situations and avoid potentially harmful ones.

Over the course of mammalian evolution, however, behavioral repertoires became increasingly intricate. Animals began to learn more about their environments than simply whether they were good or bad for survival, and the relationships that they formed became more complex. For example, mammals can learn arbitrary relationships between sensory stimuli and motor responses. A rat can learn to push a lever up when a chamber is brightly lit and press it down when the chamber is dimly lit (Passingham et al., 1988). Humans can learn to brake when they see a red traffic light and continue driving when they see a green light. Although such learning is clearly useful for our morning commute, it suffers from a number of disadvantages relative to more complex types of learning. First, it does not generalize very well: Learning about red and green traffic lights, for example, tells you nothing about the red and green lights on your stereo. A second problem is that we need to learn these stimulus-response associations through trial and error. Such learning necessarily involves errors, which can be costly. Errors may result in the lost opportunity for reward, or even physical harm. A third problem is that by dealing with the world in a literal fashion, we are potentially encoding it in an inefficient manner. There are so many potential combinations between stimuli and responses that we would be unable to remember all the possible combinations and their meanings.

One solution to these problems resides in the ability to abstract information across experiences. We learn to attend to the commonalities of a situation and ignore trivial differences. For example, after dining in a few restaurants, we learn to abstract the general rules that underlie ordering a meal, such as "wait to be seated," "order from the menu," and "pay the bill." Such rules are easy to generalize, and we can apply them to any restaurant that we subsequently visit. No two restaurants are physically identical, but because rules operate at a conceptual level, they are relevant to physically different situations. Furthermore, although we initially acquire the rules through trial-and-error learning (as we blunder our way through our first restaurant experience), once these rules have been established, we can use them to order a meal in a new restaurant. Finally, by abstracting the "gist" of a restaurant experience, we can substantially reduce the amount of information we need to store about restaurants.

The ability to abstract information is a capacity that seems to be limited to only certain species. One abstract rule that is relatively easy to demonstrate in animals is that of sameness: The matching-to-sample task tests this ability. A subject sees a sample stimulus, and then a short while later sees two test stimuli, one of which is the same as the sample. To get a reward, the subject must select the test stimulus that matches the sample stimulus. If an animal can successfully perform this task, it suggests that the animal can appreciate that the sample stimulus and the test stimulus are the same, and that this relationship controls the choice of the test stimulus. A particularly stringent test is whether the animal can solve the task when the experimenter uses new stimuli on each trial since this eliminates any explanations in terms of trial-and-error learning.

There is substantial evidence that some species understand the "same" rule and can solve the matching-to-sample task, even when every trial uses new stimuli. These species include chimpanzees (Nissen et al., 1948; Oden et al., 1988), rhesus monkeys (Mishkin et al., 1962), dolphins (Herman & Gordon, 1974), and sea lions (Kastak & Schusterman, 1994). The ability to use abstract rules is not limited to mammals. Corvids (the bird family that includes crows, rooks, jays, and jackdaws) and parrots all show the ability to perform the matching-to-sample task with novel stimuli (Pepperberg, 1987; Wilson et al., 1985). In contrast, although pigeons can learn the matching-to-sample task for small sets of stimuli, if novel stimuli are used on every trial, they perform at chance, suggesting that they are unable to abstract the rule (Wilson et al., 1985).

In terms of comparative psychology, other tasks that involve abstract rules yield similar results. One example is the formation of a "learning set." If primates learn a series of standard visual discriminations, where they see two pictures and must learn to select one of them to get a reward, their rate of learning gets progressively better with each discrimination they solve (Harlow, 1949). Eventually, the monkey can learn the problem in a single trial: Performance on the first trial is necessarily at chance, but performance is virtually 100% correct on the second trial. The monkey has learned to extract the abstract rule "win-stay, lose-shift," which dramatically speeds performance (Restle, 1958). So, too, do corvids, but pigeons must solve each discrimination individually (Hunter & Kamil, 1971; Wilson et al., 1985). Interestingly, corvid brains differ from those of other birds, in that they have an enlarged mesopallium and nidopallium, areas that are analogous to prefrontal cortex in mammals (Rehkamper & Zilles, 1991).

Finally, in humans, we find the crowning achievement of the capacity to use abstract information. Language is arguably the ability that most clearly distinguishes us from other animals (Wynne, 2004). It involves the ability to learn thousands of arbitrary associations (between words and their meanings) as well as the capacity to follow a system of abstract rules for the meaningful combination of those associations (grammar). Humans also have the largest and most well-developed frontal lobes, and as we shall see, there is substantial evidence that this region of the brain may underpin the capacity to use abstract information.

PREFRONTAL CORTEX AND THE REPRESENTATION OF ABSTRACT INFORMATION

One of the most prominent gross anatomical differences in mammalian brain phylogeny is the dramatic expansion in the size of the frontal lobe. This development reaches its zenith in the great apes, where frontal cortex accounts for approximately 38% of total cortical surface area (Bush & Allman, 2004; Semendeferi et al., 2002). For comparison, frontal cortex accounts for 32% of total cortical surface area in the rhesus monkey, 18% in the coyote, and 16% in the lion. Indeed, comparing across 43 mammals reveals that primate frontal cortex increases disproportionately relative to the rest of the neocortex as brain size increases (Bush & Allman, 2004). Consequently, an attractive hypothesis is that the expansion of the frontal lobe underlies the increase in the complexity of the behavioral repertoire exhibited by primates relative to other mammals. In anatomical terms, the frontal lobe is ideally suited to represent abstract, high-level rules. The anterior region of the frontal lobe is called the prefrontal cortex, and it receives input from all sensory modalities (Barbas & Pandya, 1991; Pandya & Yeterian, 1990). Neurons in prefrontal cortex are also responsive to a wide range of sensory modalities (Rao et al., 1997; Rolls & Baylis, 1994; Romanski et al., 1999; Romo et al., 1999). This ability to respond to a wide variety of sensory inputs is critical for organizing sensory information into supramodal concepts and rules. Prefrontal cortex also sends projections to a variety of secondary motor areas, including premotor cortex (PMC); these areas in turn project to more primary motor structures (Barbas & Pandya, 1991; Pandya & Yeterian, 1990). Thus, prefrontal cortex appears to reside at the apex of the perception-action cycle, receiving highly processed sensory information and projecting to high-level motor areas (Fuster, 2002).

Research on the neural substrates of the representation of abstract information focused primarily on the lateral region of prefrontal cortex (LPFC). Many of the first studies of the function of LPFC examined its role in working memory, a limited capacity system responsible for the temporary storage and manipulation of information. The spatial delayed response task was the first behavioral task to uncover a deficit that was specifically associated with damaged LPFC. In this task, the monkey sees a reward hidden at one of two locations, and then after a brief delay is allowed to retrieve it. Monkeys with large LPFC lesions act as if they forgot where the experimenter hid the reward, even after short delays of just a few seconds (Jacobsen, 1935). Neurophysiologists subsequently discovered that a high proportion of neurons in LPFC increased their firing rate when a subject is using working memory to bridge a task delay (Fuster & Alexander, 1971; Kubota & Niki, 1971). Later studies showed that these neurons' firing properties were selective to the specific memorized cue (Constantinidis et al., 2001; Funahashi et al., 1989). These firing patterns became synonymous with the notion of a "neuronal representation" of the cue, since the neuronal activity appeared to represent the cue, even when the cue was no longer present in the environment.

Despite this focus, working memory alone provided an unsatisfactory account of some of the deficits observed in humans and monkeys following LPFC damage. Monkeys with lesions of LPFC were impaired on certain cognitive tasks, even when those tasks did not have a working memory component (Rushworth et al., 1997). Many of the tasks sensitive to LPFC damage in humans also did not have an obvious working memory component. For example, the Wisconsin Card-Sorting Task (WCST), one of the first neuropsychological tests sensitive to LPFC damage in humans, requires patients to sort a deck of cards based on different abstract dimensions (e.g., "sort according to color" or "sort according to shape"). Patients had particular problems switching between these different dimensions (Milner, 1963). Deficits in rule implementation also occur when a subject must override a strongly prepotent response tendency in favor of a recently learned rule. For example, patients with LPFC damage are impaired on the Stroop task, where one must inhibit the tendency to name a word, and instead name the color of the ink in which the word is written (Perret, 1974).

This disparity between the findings from neurophysiology and neuropsychology endured until the last decade or so of research, when neurophysiologists began to find that the delay selectivity in LPFC neurons encoded more than just memorized cues. Several studies examined the neuronal response to the same sensory cue under different task conditions (Asaad et al., 2000; Hoshi et al., 2000; White & Wise, 1999). For example, at the sample phase of a task, the animal might see a picture of an apple on the computer screen. At a subsequent test phase, if the animal was performing an object task, it would see two pictures and would have to saccade to the picture that matched the sample, in this case the picture of the apple. In contrast, if the animal was performing a conditional visuomotor task, it would have to make a specific arbitrary motor response, for example, a rightward saccade (Fig. 21.1). Remarkably, many LPFC neurons responded to the initial presentation of the picture of an apple in one task but not the other (Asaad et al., 2000). Thus, LPFC neurons were not simply encoding sensory information, but rather they were encoding sensory information as it pertained to the demands of the task.

Subsequently, we looked at whether LPFC neurons could encode high-level abstract rules (Wallis et al., 2001a). We presented a picture on a computer screen, and we instructed the monkeys to follow either the "same" rule or the "different" rule (Fig. 21.2A). After a brief delay, we presented a second picture. The monkey had to make a given response depending on which rule was in effect and whether the second picture matched or did not match the first picture. An advantage of this design is that until the second picture appears the monkey cannot plan its behavioral response, because it has no way of knowing whether the second picture will match the first. Thus, neuronal activity during the delay period between the two pictures can only encode either the abstract rule or the identity of the picture. Figure 21.2B shows the activity of a single LPFC neuron during performance of this task. This

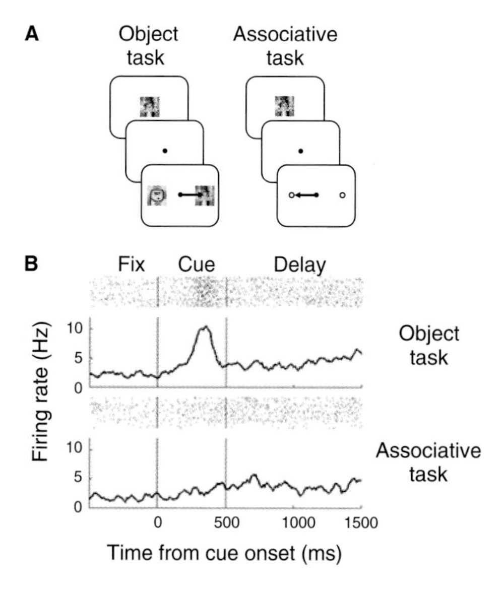

Figure 21.1 (**A**) In the object task, the subject has to remember the cue, and at the test phase, saccade to the picture that matches the cue. In the associative task, at the test phase the subject has to make a saccade in a specific direction, dependent on the identity of the cue. Both tasks used the same set of cues. (**B**) An example of a lateral prefrontal cortex (LPFC) neuron that responds to a specific cue in the associative task but not in the object task, even though the subject is viewing the identical stimulus in both cases. (From Asaad et al., 2000, with permission).

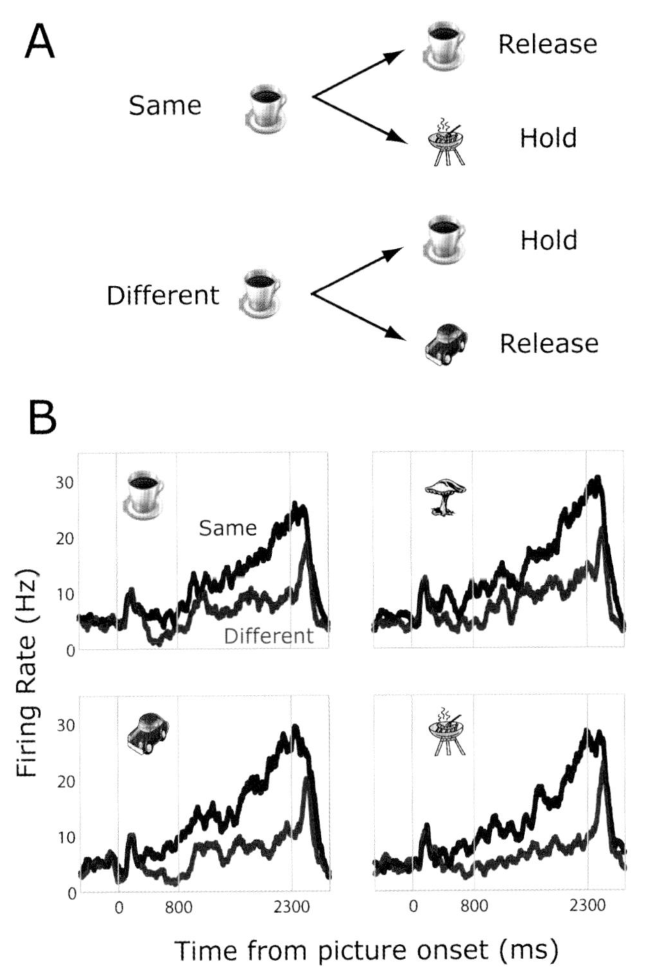

Figure 21.2 (**A**) The subject grasps a lever and a cue picture appears. After a brief delay, a second picture appears. If the "same" rule is in effect, the monkey must release a lever if the second picture matches the first, but hold the lever if it is different. If the "different" rule is in effect, the monkey must release a lever if the second picture does not match the first, but hold the lever if it is the same. Thus, to solve the task correctly, the subject must remember the identity of the cue picture and the rule that is currently in effect. (**B**) A lateral prefrontal cortex (LPFC) neuron that encodes the rule: It shows a higher firing rate when the "same" rule is in effect compared to when the "different" rule is in effect, irrespective of the picture that the subject is remembering.

neuron shows a higher firing rate whenever the "same" rule is in effect, but does not discriminate between the different pictures that the monkey is remembering. Many LPFC neurons, particularly in the dorsal region of LPFC, encoded the abstract rule during the delay period (58 of 197 or 29% of neurons) rather than the picture (7 of 197 or 4% of neurons). The task was subsequently adapted for use in humans where functional magnetic resonance imaging (fMRI) revealed activation in LPFC when the subject encoded and maintained the rule (Bunge et al., 2003). These results suggest that encoding of abstract rules is an important function of LPFC, indeed more so than the encoding of sensory information. In addition, the results help explain why patients with LPFC damage have difficulties with tasks such as the WCST.

In conclusion, early studies of LPFC function focused on its role in working memory.

Although this is undoubtedly an important role of LPFC, it may be one of several functions that collectively enable the abstraction of information, by allowing the subject to disengage from the processing of the immediate sensory environment. A recent study, however, has shown that while LPFC appears to instantiate abstract rule representation in mammals, it is not necessary to have the complex neuronal machinery of a frontal lobe in order to use abstract information in the control of behavior. Investigators trained honeybees on a Y-maze (Giurfa, 2001). At the entrance to the maze was the sample stimulus, and at the entrance to the two forks in the Y-maze were two test stimuli (Fig. 21.3a). Bees received a reward for choosing the arm with the matching test stimulus. Not only could the bees learn this task, but they also were able to apply the rule to novel stimuli (Fig. 21.3b). This study raises interesting questions. For example, why should the capacity to use an abstract rule be useful to bees, but not to pigeons (Wilson et al., 1985)? This capacity is not simply the ability to know that one flower is the "same" as another, a very simple (and useful) behavioral adaptation that can be solved through stimulus generalization and conditioning. Rather, it is using the relationship between two stimuli to govern behavior in an arbitrary fashion. Quite what use the bee finds for this ability is a mystery, but it does demonstrate that a remarkably simple nervous system, consisting of a brain of 1 mm^3 and fewer than 1 million neurons (Witthöft, 1967), is capable of using abstract information. It remains an open question whether it can learn a variety of abstract information, as does the mammalian brain, or whether its abilities are more constrained.

STRATEGIES AND SETS

In addition to enabling explicit, abstract rules to control information, the LPFC also appears to be involved in implementing other forms of high-level information, such as strategies and cognitive sets. The acquisition and adoption of a strategy involves the control and organization of cognitive processes in a manner that speeds learning or otherwise reduces the overall cognitive resources necessary to solve a task. For example, in a spatial self-ordered search task, monkeys have to search through a number of possible locations to find a hidden reward, but are penalized for returning to a location that they have already searched. Monkeys gradually develop a strategy to help with the task such as searching through locations in a clockwise order. Doing so reduces the memory demands of the task, since the animal no longer has to keep track of which locations it has previously visited. Lesions of LPFC severely disrupted the ability of marmosets to implement this strategy effectively (Collins et al., 1998). One sees a similar effect in marmosets that learn to search for a hidden door into a box that contains food. Animals adopt a strategy of searching each side of the box in turn, but marmosets with lesions of LPFC fail to learn this strategy (Wallis et al., 2001b). Recent studies have emphasized the interaction of the frontal and temporal cortices, at least in the implementation of strategies pertaining to visual discriminations. Animals that had acquired a complex alternation strategy to solve a visual discrimination task were impaired following crossed lesions of frontal and temporal cortex (Gaffan et al., 2002). However, they could still solve a simple visual discrimination that did not require a higher-order strategy.

Taken together, these results strongly implicate the primate LPFC in the implementation of strategies. Likewise, neurophysiological studies have found neurons in LPFC that represent strategic information. Wise and colleagues trained monkeys to learn a three-choice conditional visuomotor task (Murray et al., 1996). In this task, the monkey learns to make one of three possible motor responses to each of three unique visual stimuli. For example, visual stimulus A may indicate that the monkey should move a joystick left, stimulus B right, and stimulus C down. The monkey learns which response goes with which stimulus by trial and error. However, the experimenters soon realized that the animals were incorporating high-level information into their choices that enabled them to perform better than would be expected by chance. Specifically, the animals kept track of the

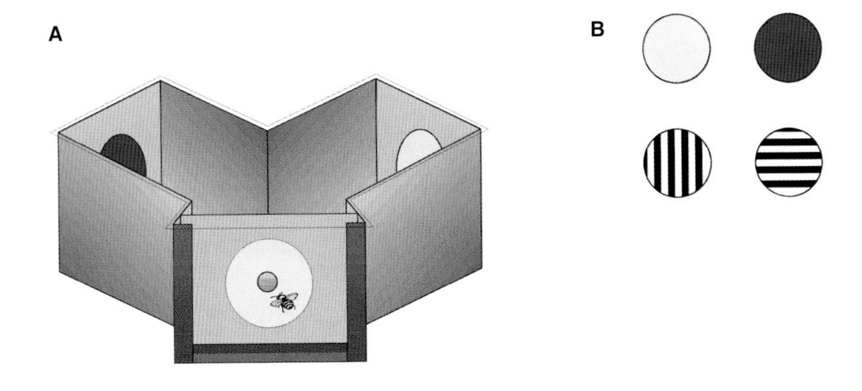

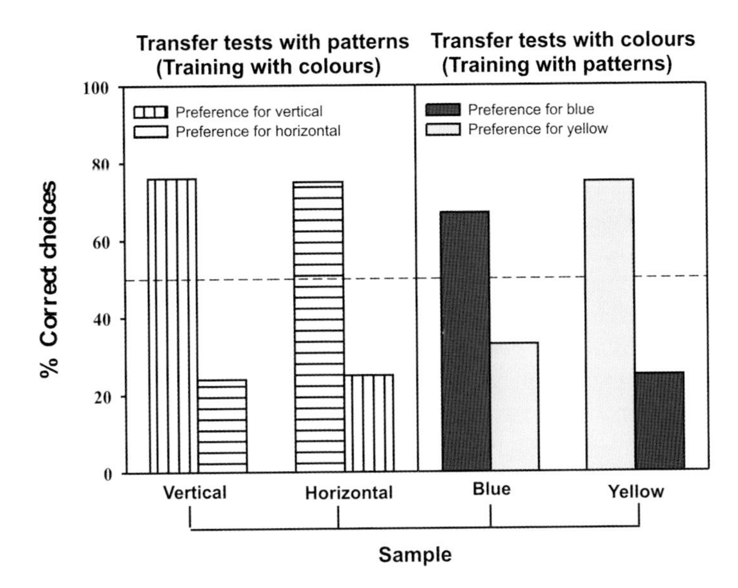

Figure 21.3 (A) The behavioral apparatus used to train bees on a delayed matching-to-sample paradigm. The bee flew into a Y-maze, at the entrance of which was one of two sample stimuli. Half the bees learned the task using a vertical or horizontal grating, while the other half learned the task using the colors blue and yellow. The entrance led to a decision chamber, where the bee could choose one of two arms. Each arm carried one of the two sample stimuli. If the bee chose the arm with the stimulus that was identical to the sample stimulus, it received a sucrose solution reward. After several days of training, the bees could perform the task with greater than 70% accuracy. (B) Critically, the bees were not solving the task using simple rules (e.g., "Choose blue and not yellow if blue was at the entrance" and "Choose yellow and not blue if yellow was at the entrance), but had abstracted the general concept of sameness. This was evident by the bees' performance on transfer tests where they were confronted with new stimuli (colors for the bees trained on gratings, and gratings for those trained on colors). The bees immediately performed the task with greater than 70% accuracy, indicating that they had abstracted a rule that they could apply to novel stimuli. (From Giurfa et al., 2001, with permission).

stimulus-response mappings that they knew in order to rule out potential alternatives when encountering a novel stimulus by adopting a "repeat-stay, change-shift" strategy. For example, suppose an animal encountered stimulus A on trial one, and correctly guessed that it should make a leftward joystick movement. On the next trial, if stimulus A appeared it would always select the leftward movement (repeat-stay). If, however, one of the other two pictures appeared, it would never select the leftward movement, but would instead choose one of the two remaining alternative movements (change-shift). This simple strategy enabled the animal to perform at 67% correct on this second trial, rather than the 33% expected by chance. Subsequent neurophysiological recording revealed that 41% of LPFC neurons encoded this high-level strategy, showing differences in activity on repeat-stay versus change-shift trials (Genovesio et al., 2005).

A cognitive construct related to strategies is the notion of "task set." A task set consists of information about which stimulus attributes to attend to, important conceptual criterion, goal states, and condition-action rules (Monsell, 2003). It reflects not just which items a subject is preparing to process, but also how the subject plans to process the items and the rules of the to-be-performed task. One type of task set is an attentional set, which consists of an expectancy that certain abstract properties of one's sensory environment will be necessary to solve a task, which in turn leads to a focusing of attention on those properties. Consider the animal that learns to solve a series of discriminations between multidimensional visual stimuli. For example, a set of stimuli might consist of a red heart, a red diamond, a black heart, and a black diamond. On each trial two of the stimuli appear, one of each color and shape. The correct answer is to choose the red stimulus regardless of whether it is a heart or a diamond. Once the animal has learned to do this consistently, the experimenter gives it a second problem. For example, the new set might consist of a green triangle, green square, yellow triangle, and yellow square, and the correct response is to choose the green stimulus irrespective of its shape. The animal learns successive discriminations progressively quicker as it realizes which dimension of the stimulus is relevant (color in this particular example). The animal has learned what is termed an "attentional set." In effect, the animal has developed an expectancy that one particular abstract property of the stimulus will be necessary to solve the discrimination, and so it focuses its attention on that property. When the animal subsequently encounters a discrimination where shape is now relevant, it takes the animal a long time to learn, since it mistakenly focuses on its color. Lesions of LPFC dramatically exacerbate the difficulty that monkeys have in shifting their attentional set (Dias et al., 1996). Subsequent studies showed that dopaminergic innervation of LPFC was critical in mediating attentional shifts. Dopamine depletions in LPFC impaired attentional set shifting, but dopamine depletions in the neighboring orbitofrontal cortex or serotonergic depletions of LPFC did not affect set shifting (Clarke et al., 2004, 2005).

The aforementioned task is very similar to the WCST. When subjects learns to sort a deck of cards according to color, they are in effect developing an attentional set to color. The WCST was originally developed as a test of cognitive flexibility (Berg, 1948) and was subsequently found to detect PFC damage in humans (Milner, 1963), Recent fMRI studies using primate analogs of the task show that the same neuronal substrates underlie the ability to perform the task across species (Nakahara et al., 2002). The same region of LPFC (the ventral portion consisting of Brodmann's areas 44 and 45) is activated during the performance of the WCST in both humans and rhesus monkeys (Fig. 21.4). Subsequent neurophysiological studies showed that approximately a third of LPFC neurons were engaged in encoding the attentional shift (Mansouri et al., 2006).

The formation of attentional sets speeds learning. However, task sets are also evident even when a cue indicates the currently relevant task, thereby removing the learning component (Monsell, 2003; Monsell et al., 2003). For example, a subject might have to indicate whether a number is either odd or less than seven depending on a task instruction cue. This experimental design introduces an incongruity cost (also referred to as "conflict"): Subjects are typically faster to respond when a stimulus instructs different responses on the tasks (e.g., the number nine in the previous task, which is odd, but greater than seven) than if they instruct the same response (the number three, which is both odd and less than seven). In addition, there is a switch cost: Performance is faster when the task remains

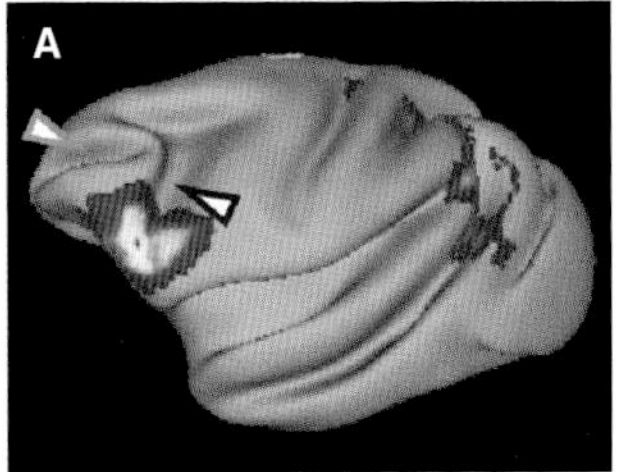

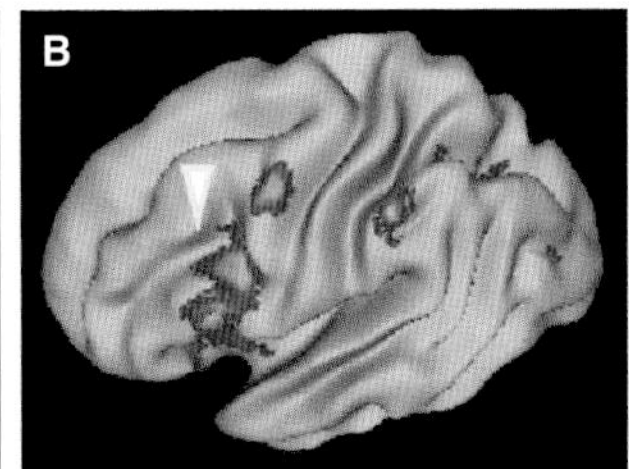

Figure 21.4 Functional magnetic resonance imaging scans reveal how performance of WCST activates posterior ventrolateral prefrontal cortex in both (**A**) monkeys and (**B**) humans. Green arrowhead indicates the principal sulcus, blue arrowhead the inferior arcuate sulcus, and yellow arrowhead the inferior frontal sulcus. (From Nakahara et al., 2002, with permission).

the same (nonswitch trials), but worse when the task changes (switch trials). A study of the comparative psychology of task switching in monkeys and humans suggests that these two costs may inversely correlate, and reflect the degree to which the two species can effectively implement task sets (Stoet & Snyder, 2003). Humans typically show large switch costs but small incongruity costs, while the opposite pattern occurs in monkeys. This may reflect the better implementation of the task set by humans. Thus, humans are better able to maintain focused attention on a particular task for long periods and ignore irrelevant aspects of the stimulus, leading to smaller incongruity costs than monkeys. However, this ability comes with a cost, namely, that it is more difficult for the human to switch to a different task.

Regarding the neuronal underpinnings of task switching, activation in LPFC in humans is associated with switch costs (Badre & Wagner, 2006; Brass & von Cramon, 2004; Braver et al., 2003) and incongruity costs (Dreher & Berman, 2002; Kerns et al., 2004; MacDonald et al., 2000). This is not to say that LPFC is the only area involved in task switching. Correlates of incongruity costs are evident in the activity of single neurons in posterior parietal cortex of monkeys (Stoet & Snyder, 2007) and activations of anterior cingulate cortex (ACC) in humans (Dreher & Berman, 2002; Kerns et al., 2004; MacDonald et al., 2000). However, it has proven controversial whether these areas are necessary for the processing of incongruity and resolution of such conflict, or whether their activity reflects a different cognitive process that happens to correlate with incongruity processing (di Pellegrino et al., 2007; Fellows & Farah, 2005; Rushworth et al., 2004). In contrast, a recent study shows that LPFC does have a singular role in the implementation of incongruity resolution (Mansouri et al., 2007). When subjects experience an incongruity, they typically make a behavioral adjustment such that their performance on the next trial is slower, but also more accurate (Rabbitt, 1966). Lesions of LPFC in monkeys abolished this behavioral adjustment induced by incongruity, whereas lesions of ACC had no effect (Fig. 21.5A,B). The modulation of behavior in the current trial by the degree of incongruity experienced in the previous trial requires a system to hold the memory of experienced incongruity across trials. The activity of LPFC neurons was consistent with such a system since they encoded both the degree of incongruity on the current trial (Fig. 21.5C) as well as the degree of incongruity on the previous trial (Fig. 21.5D).

FUNCTIONAL ORGANIZATION OF HIGH-LEVEL INFORMATION

To recap, we have seen that LPFC has undergone a dramatic expansion in primate evolution and that it is responsible for the encoding of a wide variety of information that would be useful for high-level cognitive behavior, including abstract rules, behavioral strategies, and task sets. Recent work has attempted a more fine-grain analysis of the relationship between LPFC and the encoding of high-level information. A number of models

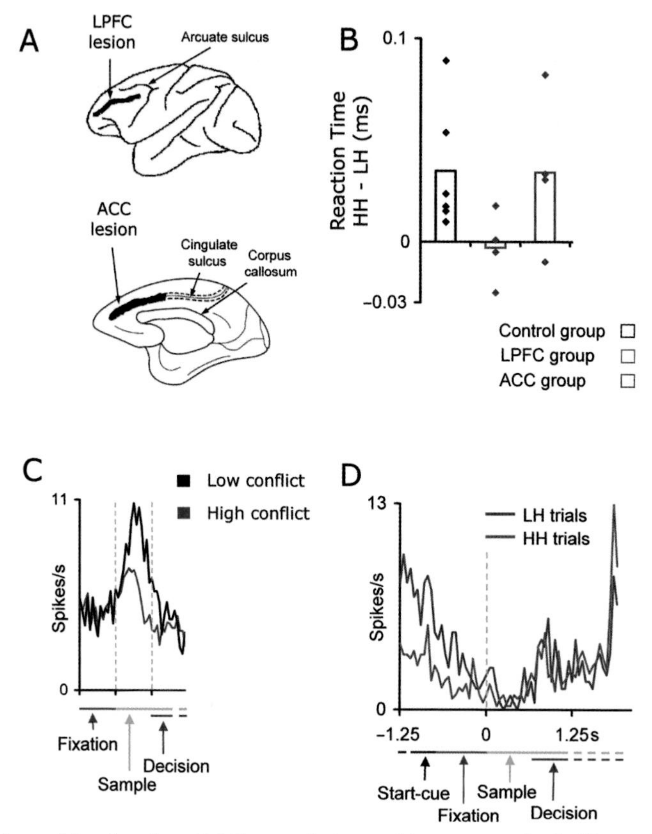

Figure 21.5 (**A**) Lateral (top) and medial (bottom) views of the macaque brain illustrating the location of the lateral prefrontal cortex (LPFC) and anterior cingulate cortex (ACC) lesions. (**B**) Control animals show longer reaction times on high-conflict trials when the trial was preceded by a high-conflict trial (HH) as opposed to when it was preceded by a low-conflict trial (LH). Lesions of LPFC, but not ACC, abolished this conflict-induced behavioral adjustment. (**C**) An LPFC neuron that shows a higher firing rate when the sample stimulus elicits low conflict. (**D**) An LPFC neuron that shows a higher firing rate when the subject experienced low conflict on the previous trial. (From Mansouri et al., 2007, with permission).

have been developed that propose a hierarchical organization of rules within the frontal lobe, with progressively higher-level information represented in progressively more anterior frontal regions (Badre & D'Esposito, 2007; Bunge & Zelazo, 2006; Christoff & Keramatian, 2008; Koechlin et al., 2003; Sakai & Passingham, 2003). These findings are consistent with models that propose that hierarchical organization is a general feature of brain organization (Fuster, 2002; Hawkins & Blakeslee, 2007).

Although there are differences between the models regarding the manner of the hierarchical organization, nearly all begin with the observation that the implementation of simple conditional rules depends on the premotor cortex. Conditional rules take the form "If X, then Y," and typically take the form of stimulus-response associations, for example, "If red light, then brake." Lesions of premotor cortex in monkeys impair the ability to perform such tasks (Halsband & Passingham, 1982; Petrides, 1982) and premotor cortex is activated in humans when they perform tasks requiring conditional rules (Amiez et al., 2006). However, rules that are more complex than conditional rules typically produce activation more anterior to the premotor cortex. For example, compound rules that require subjects to integrate two pieces of information to determine the correct response ("If X and if Y, then Z") activate LPFC more than simple conditional rules (Bunge et al., 2003). In a similar vein, simple conditional rules activate PMC, while rules that require consideration of both the stimulus and the context in which the stimulus occurs activate LPFC (Koechlin et al., 2003). In addition, rules that require consideration of the stimulus, the context, and the temporal episode in which the task occurs activate cortex anterior to LPFC in the frontal pole (Koechlin et al., 2003). However, task complexity per se does not recruit more anterior regions of cortex. Investigators studied the effects of manipulating not only the abstractness of the rules (as in the previous tasks) but also the overall complexity of the task via the number of contingencies that the subjects had to follow (Badre & D'Esposito, 2007). Although increasing the number of contingencies, and thus the overall complexity of the task, led to larger activations, it had no effect on the location of those activations. Only the abstractness of the rule altered the activation locus, with increasingly abstract rules activating increasingly anterior regions of the frontal lobe (Fig. 21.6).

So far, we have considered the human frontal lobe. Is there evidence that such a hierarchy of

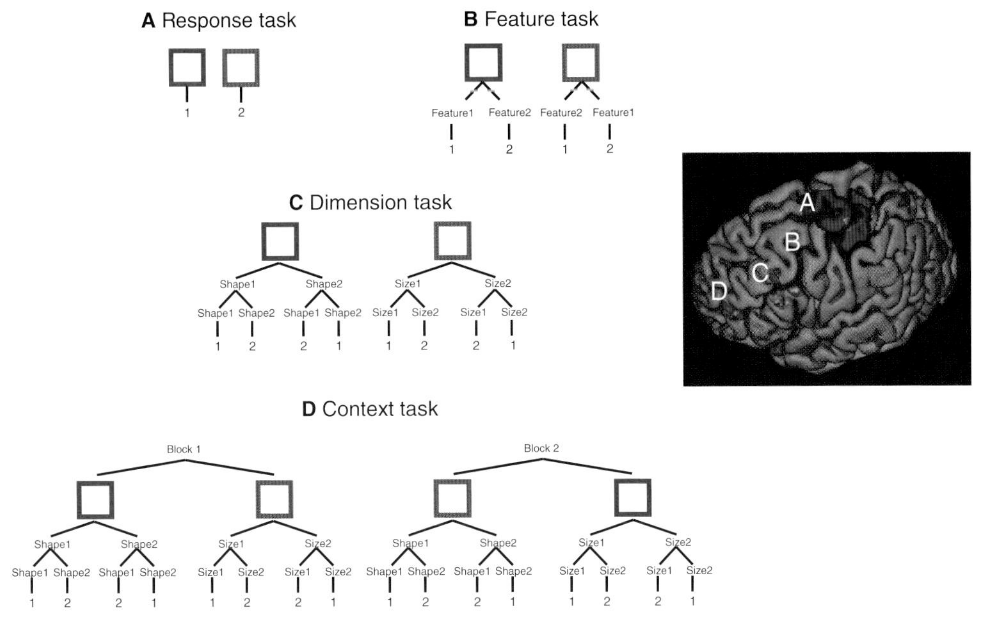

Figure 21.6 Subjects performed four different tasks that required them to follow rules at varying degrees of abstraction. The experimenters could alter the complexity independent of its level of abstraction by increasing the number of contingencies presented to the subject. (**A**) The subject learns a stimulus-response contingency. The color of a presented square instructs which of two responses to make. (**B**) The subject learns a stimulus-feature contingency. The color of the square tells the subject to look for one of two features present in a sample stimulus, and make one of two responses depending on the presence or absence of that feature. (**C**) The subject learns a stimulus-dimension contingency. The color of the square tells the subject to attend to one of two stimulus dimensions (and ignore the other dimension), and make one of two responses depending on whether two stimuli share a specific feature within that stimulus dimension. (**D**) The subject learns context-based stimulus-dimension contingencies. The colored square instructs the subject to attend to a specific stimulus dimension, but the color-dimension contingency changes between blocks of trials. The subject must then identify whether two stimuli share a specific feature within that stimulus dimension and make the appropriate response. Brain activation patterns reveal that as the task becomes increasingly abstract, the locus of activation shifts anteriorly within lateral prefrontal cortex. (From Badre & D'Esposito, 2007, with permission).

function is also present in the monkey? A series of studies investigating the neuronal basis of reaching found evidence of hierarchical organization stretching from primary motor cortex into LPFC in terms of the control of motor behavior. LPFC neurons sat atop the hierarchy and encoded the abstract behavioral rule that indicated whether to select a stimulus based on either the location or identity of a previous stimulus (Hoshi et al., 2000). PMC neurons encoded information about the target location, as well as which arm the subject would use to reach to the target, consistent with a role in planning the action (Hoshi & Tanji, 2000). Finally, neurons in primary motor cortex were active during the execution of the movement, consistent with their role in motor control (Hoshi & Tanji, 2000). However, hierarchical organization of cognitive information is not apparent within LPFC itself. We recorded from a large expanse of LPFC while monkeys were implementing "same" and "different" rules, and there was no evidence of an anterior-posterior gradient within LPFC in terms of the prevalence of neurons encoding these abstract rules (Wallis & Miller, 2003; Wallis et al., 2001a). Thus, the question remains as to why we do not see a hierarchical organization of rules in monkey LPFC. In part, it may be because we have not investigated multiple levels of abstract rules, as we have done in humans. The reason for this is that we seem to be approaching the limit of the cognitive capabilities of monkeys. Teaching them two relatively simple abstract rules, such as "same" and "different," takes ≈200,000 trials. Having them learn a number of abstract rules that depend on some other task context, as in the tasks used in humans (Badre & D'Esposito, 2007; Koechlin et al., 2003), is likely to be impossible.

The obvious implication of these findings is that the most anterior part of the frontal lobe (termed the "frontal pole") should also be responsible for the highest-level control. A recent review of the neuroimaging literature concluded that, in humans, this was indeed the case, arguing that the frontal pole was particularly important when two or more cognitive operations need integrating in the pursuit of a higher behavioral goal (Ramnani & Owen, 2004). For example, frontal pole activation occurs when subjects must maintain information in working memory while performing a secondary task during the delay interval (Braver & Bongiolatti, 2002; Koechlin et al., 1999). However, the frontal pole is also the region of the primate frontal lobe that is most diverse across different primate species (Fig. 21.7), seeming to occupy a disproportionately greater area in humans than in other primates (Semendeferi et al., 2001). The anatomical connections of the frontal pole in the monkey are only partially consistent with its role as the apex of high-level control. It connects most strongly with cingulate cortex, orbitofrontal cortex, and multimodal regions of the temporal lobe (Petrides & Pandya, 2007). Thus, it is unusual within the frontal lobe in that it does not connect with any motor or sensory areas, but rather connects with areas responsible for multisensory processing and internal emotional

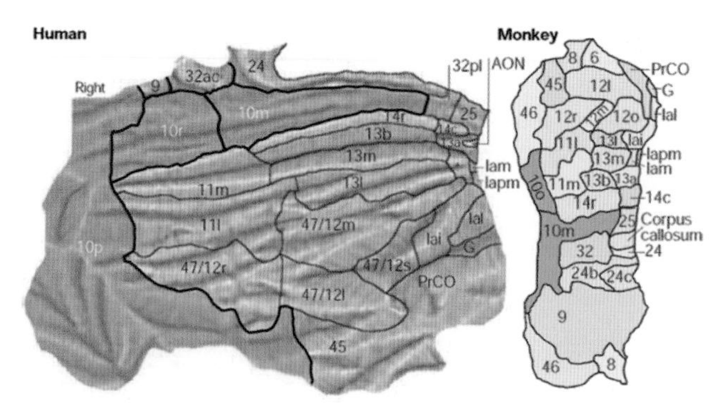

Figure 21.7 The right prefrontal cortex of **(A)** a human and **(B)** a monkey brain, represented as flattened surfaces, illustrating the considerably larger proportion of the surface area occupied by area 10 (shaded in red) in humans relative to monkeys. (From Ramnani & Owen, 2007, with permission).

states. While this is consistent with the role of the frontal pole in processing the most abstract, high-level information (such as multimodal sensory information), its relatively weak connections with much of LPFC seems inconsistent with the idea that it sits atop the functional hierarchy in the lateral frontal lobe. A challenge for future studies is to reconcile these findings from monkeys and humans. However, given the disparity in the size of the frontal pole between the two species (Semendeferi et al., 2001), it is possible that our use of the monkey to model human cognitive processes is not appropriate to the frontal pole.

IMPLEMENTATION OF HIGH-LEVEL INFORMATION

Representing information relating to high-level control is one step in using this information to control behavior. However, the information must then interact with systems responsible for motor control. We have examined how this takes place by simultaneously recording from LPFC, as well as two of the major output structures of the LPFC, PMC and the caudate nucleus (CN), while subjects were applying the "same" or "different" rule to successively presented pictures (Muhammad et al., 2006; Wallis et al., 2001a; Wallis & Miller, 2003). We found that the strongest rule selectivity was in the frontal lobe (LPFC and PMC) with only weak rule selectivity in CN. During the delay between the two pictures, rule-selective neurons were significantly more prevalent in LPFC (58 of 197 or 29%) and PMC (59 of 258 or 23%) than they were in CN (38 of 341 or 11%).

These results stand in marked contrast to studies examining the role of CN in simpler levels of behavioral control, such as the implementation of conditional rules involving concrete stimulus-response associations. Pasupathy and Miller (2005) recorded data simultaneously from LPFC and CN while monkeys learned which of two behavioral responses (saccade left or right) to make to which of two pictures (A and B). Both structures encoded the associations between the pictures and the responses, but selectivity appeared earlier in learning in CN than in LPFC. Despite this early neural correlate of learning in CN, the monkeys' behavior did not change until LPFC encoded the associations. These results present us with a challenge: Why would the monkeys continue to make errors, despite the fact that CN was encoding the correct picture-response associations? The apparent implication of these findings is that overt behavior is under the control of LPFC more so than under that of CN. Furthermore, they suggest that LPFC will not necessarily use all of the available information to control behavior (since it apparently "ignores" CN).

One possibility is that LPFC is integrating information from many low-level learning systems, not just CN, and that some of these systems may not necessarily agree with CN as to the correct response. For example, consider the brain systems that acquire stimulus-reward associations or action-reward associations. It is impossible to learn stimulus-response associations using such stimulus-reward or action-reward associations because each action and each stimulus are rewarded equally often. However, this does not necessarily mean that these systems will be silent during the performance of a task dependent on stimulus-response associations. For example, perhaps after a reinforced leftward saccade, the action-reward system instructs LPFC to make another leftward response, oblivious to the fact that on the next trial, the stimulus instructs a rightward response. LPFC would need to learn that such information is not useful to solve the task and ignore this system. Lesion studies support the idea that these different low-level learning systems can compete with one another. For example, lesions of anterior cingulate cortex impair the learning of stimulus-reward associations (Bussey et al., 1997; Gabriel et al., 1991), but facilitate the learning of stimulus-response associations (Bussey et al., 1996). These findings suggest that in the healthy animal, anterior cingulate contributes to the learning of stimulus-reward associations, and that removing the capacity to learn such associations can improve the ability to learn stimulus-response associations.

A puzzling feature of our results is that information about the abstract rule appeared approximately 90 ms earlier in PMC than in LPFC (CN neurons lay somewhere in between, but the low numbers of rule-selective neurons in CN make this difficult to interpret). This seems to contradict the notion of a hierarchical organization, since PMC is anatomically lower on the hierarchy than LPFC, yet encodes high-level information quicker. One possible explanation lies in the fact that the rules were highly familiar to the animals; they had performed this task for more than a year. Evidence suggests that LPFC is more critical for new learning than for familiar routines. LPFC damage preferentially affects new learning; animals and humans can still engage in complex behaviors as long as they learned them before the damage occurred (Dias et al., 1997; Knight, 1984; Shallice, 1982; Shallice & Evans, 1978). LPFC neurons also show more selectivity during new learning than during the performance of familiar cue-response associations (Asaad et al., 1998). Human imaging studies report greater blood flow to PMC than LPFC when subjects are performing familiar versus novel tasks (Boettiger & D'Esposito, 2005) and greater LPFC activation when subjects are retrieving newly learned rules versus highly familiar rules (Donohue, 2005). In addition, with increasing task familiarity, there is a relative shift in blood flow from areas associated with focal attention, such as LPFC, to motor regions (Della-Maggiore & McIntosh, 2005). Therefore, it may be that LPFC is primarily involved in new learning, but with familiarity, rules become more strongly established in downstream motor system structures.

CONCLUSION

As behavior has become more complex, it has inevitably become more difficult to control. The animal that is reliant on conditioning parses the world into two behaviors: approach or avoid. The animal that learns arbitrary relationships, however, has a far more difficult task. Presented with a single stimulus, it must determine the salient features. Should it pay attention to its color, its shape, or some other feature? Even once the animal has identified the relevant sensory dimension, it must then select among the myriad potential motor responses. A hierarchical organization of cognitive control, with increasing abstraction at each level of the hierarchy, is one potential solution to the problem. The current evidence favors such an organization in the human frontal lobe and a rudimentary analog in the frontal lobe of the monkey. In this way, primate behavior has evolved to be the most complex behavior in the animal kingdom.

REFERENCES

Amiez, C., Kostopoulos, P., Champod, A. S., & Petrides, M. (2006). Local morphology predicts functional organization of the dorsal premotor region in the human brain. *Journal of Neuroscience, 26,* 2724–2731.

Asaad, W.F., Rainer, G. & Miller, E.K. (1998) Neural activity in the primate prefrontal cortex during associative learning. *Neuron, 21,* 1399–1407.

Asaad, W. F., Rainer, G., & Miller, E. K. (2000). Task-specific neural activity in the primate prefrontal cortex. *Journal of Neurophysiology, 84,* 451–459.

Badre, D., & D'Esposito, M. (2007). Functional magnetic resonance imaging evidence for a hierarchical organization of the prefrontal cortex. *Journal of Cognitive Neuroscience, 19,* 2082–2099.

Badre, D., & Wagner, A. D. (2006). Computational and neurobiological mechanisms underlying cognitive flexibility. *Proceedings of the National Academy of Sciences, 103,* 7186–7191.

Barbas, H., & Pandya, D. (1991). Patterns of connections of the prefrontal cortex in the rhesus monkey associated with cortical architecture. In: H. S. Levin, H. M. Eisenberg, & A. L. Benton (Eds.), *Frontal lobe function and dysfunction* (pp. 35–58). New York: Oxford University Press.

Berg, E. A. (1948). A simple objective test for measuring flexibility in thinking. *Journal of General Psychology, 39,* 15–22.

Boettiger, C.A. & D'Esposito, M. (2005) Frontal networks for learning and executing arbitrary stimulus-response associations. *Journal of Neuroscience, 25,* 2723–2732.

Brass, M., & von Cramon, D. Y. (2004). Selection for cognitive control: A functional magnetic resonance imaging study on the selection of task-relevant information. *Journal of Neuroscience, 24,* 8847–8852.

Braver, T. S., & Bongiolatti, S. R. (2002). The role of frontopolar cortex in subgoal processing during working memory. *Neuroimage, 15,* 523–536.

Braver, T. S., Reynolds, J. R., & Donaldson, D. I. (2003). Neural mechanisms of transient and sustained cognitive control during task switching. *Neuron, 39,* 713–726.

Bunge, S. A., Kahn, I., Wallis, J. D., Miller, E. K., & Wagner, A. D. (2003). Neural circuits subserving the retrieval and maintenance of abstract rules. *Journal of Neurophysiology, 90,* 3419–3428.

Bunge, S. A., & Zelazo, P. D. (2006). A brain-based account of the development of rule use in childhood. *Current Directions in Psychological Science, 15,* 118–121.

Bush, E. C., & Allman, J. M. (2004). The scaling of frontal cortex in primates and carnivores. *Proceedings of the National Academy of Sciences, 101,* 3962–3966.

Bussey, T.J., Muir, J.L., Everitt, B.J. & Robbins, T.W. (1996) Dissociable effects of anterior and posterior cingulate cortex lesions on the acqui-sition of a conditional visual discrimination: facilitation of early learning vs. impairment of late learning. *Behavioural Brain Research, 82,* 45–56.

Bussey, T.J., Muir, J.L., Everitt, B.J. & Robbins, T.W. (1997) Triple dissociation of anterior cingulate, posterior cingulate, and medial frontal cortices on visual discrimination tasks using a touchscreen testing procedure for the rat. *Behavioral Neuroscience, 111,* 920–936.

Christoff, K., & Keramatian, K. (2008). Abstraction of mental representation: Theoretical conisderations and neuroscientific evidence. In: S. A. Bunge & J. D. Wallis (Eds.), *Neuroscience of rule-guided behavior.* New York: Oxford University Press.

Clarke, H. F., Dalley, J. W., Crofts, H. S., Robbins, T. W., & Roberts, A. C. (2004). Cognitive inflexibility after prefrontal serotonin depletion. *Science, 304,* 878–880.

Clarke, H. F., Walker, S. C., Crofts, H. S., Dalley, J. W., Robbins, T. W., & Roberts, A. C. (2005). Prefrontal serotonin depletion affects reversal learning but not attentional set shifting. *Journal of Neuroscience, 25,* 532–538.

Collins, P., Roberts, A. C., Dias, R., Everitt, B. J., & Robbins, T. W. (1998). Perseveration and strategy in a novel spatial self-ordered sequencing task for nonhuman primates: Effects of excitotoxic lesions and dopamine depletions of the prefrontal cortex. *Journal of Cognitive Neuroscience, 10,* 332–354.

Constantinidis, C., Franowicz, M. N., & Goldman-Rakic, P. S. (2001). The sensory nature of mnemonic representation in the primate prefrontal cortex. *Nature Neuroscience, 4,* 311–316.

Della-Maggiore, V. & McIntosh, A.R. (2005) Time course of changes in brain activity and functional connectivity associated with long-term adaptation to a rotational transformation. *Journal of Neurophysiology, 93,* 2254–2262.

di Pellegrino, G., Ciaramelli, E., & Ladavas, E. (2007). The regulation of cognitive control following rostral anterior cingulate cortex lesion in humans. *Journal of Cognitive Neuroscience, 19,* 275–286.

Dias, R., Robbins, T. W., & Roberts, A. C. (1996). Dissociation in prefrontal cortex of affective and attentional shifts. *Nature, 380,* 69–72.

Dias, R., Robbins, T.W. & Roberts, A.C. (1997) Dissociable forms of inhibitory control within prefrontal cortex with an analog of the Wisconsin Card Sort Test: restriction to novel situations and independence from "on-line" processing. *Journal of Neuroscience, 17,* 9285–9297.

Donohue, S.E., Wendelken, C., Crone, E.A. & Bunge, S.A. (2005) Retrieving rules for behavior from long-term memory. *Neuroimage, 26,* 1140–1149.

Dreher, J. C., & Berman, K. F. (2002). Fractionating the neural substrate of cognitive control processes. *Proceedings of the National Academy of Sciences, 99,* 14595–14600.

Fellows, L. K., & Farah, M. J. (2005). Is anterior cingulate cortex necessary for cognitive control? *Brain, 128,* 788–796.

Funahashi, S., Bruce, C. J., & Goldman-Rakic, P. S. (1989). Mnemonic coding of visual space in the monkey's dorsolateral prefrontal cortex. *Journal of Neurophysiology, 61,* 331–349.

Fuster, J. M. (2002). *Cortex and mind.* Oxford: Oxford University Press.

Fuster, J. M., & Alexander, G. E. (1971). Neuron activity related to short-term memory. *Science, 173,* 652–654.

Gabriel, M., Kubota, Y., Sparenborg, S., Straube, K. & Vogt, B.A. (1991) Effects of cingulate cortical lesions on avoidance learning and training-induced unit activity in rabbits. *Experimental Brain Research, 86,* 585–600.

Gaffan, D., Easton, A., & Parker, A. (2002). Interaction of inferior temporal cortex with frontal cortex and basal forebrain: Double dissociation in strategy implementation and associative learning. *Journal of Neuroscience, 22,* 7288–7296.

Genovesio, A., Brasted, P. J., Mitz, A. R., & Wise, S. P. (2005). Prefrontal cortex activity related to abstract response strategies. *Neuron, 47,* 307–320.

Giurfa, M., Zhang, S., Jenett, A., Menzel, R. & Srinivasan, M.V. (2001) The concepts of 'sameness' and 'difference' in an insect. *Nature, 410,* 930–933.

Halsband, U., & Passingham, R. (1982). The role of premotor and parietal cortex in the direction of action. *Brain Research, 240,* 368–372.

Harlow, H. F. (1949). The formation of learning sets. *Psychological Review, 56,* 51–65.

Hawkins, J., & Blakeslee, S. (2007). *On intelligence.* New York: Henry Holt & Co.

Herman, L. M., & Gordon, J. A. (1974). Auditory delayed matching in the bottlenose dolphin. *Journal of the Experimental Analysis of Behavior, 21,* 19–26.

Hoshi, E., Shima, K., & Tanji, J. (2000). Neuronal activity in the primate prefrontal cortex in the process of motor selection based on two behavioral rules. *Journal of Neurophysiology, 83,* 2355–2373.

Hoshi, E., & Tanji, J. (2000). Integration of target and body-part information in the premotor cortex when planning action. *Nature, 408,* 466–470.

Hunter, M. W., & Kamil, A. C. (1971). Object discrimination learning set and hypothesis behavior in the northern blue jay (Cyanocitta cristata). *Psychonomic Science, 22,* 271–273.

Jacobsen, C. F. (1935). Functions of the frontal association area in primates. *Archives of Neurology and Psychiatry, 33,* 558–569.

Kastak, D., & Schusterman, R. J. (1994). Transfer of visual identity matching-to-sample in two Californian sea lions (*Zalophus californianus*). *Animal Learning and Behavior, 22,* 427–453.

Kerns, J. G., Cohen, J. D., MacDonald, A. W., 3rd, Cho, R. Y., Stenger, V. A., & Carter, C. S. (2004). Anterior cingulate conflict monitoring and adjustments in control. *Science, 303,* 1023–1026.

Knight, R.T. (1984) Decreased response to novel stimuli after prefrontal lesions in man. *Electroencephalography and Clinical Neurophysiology, 59,* 9–20.

Koechlin, E., Basso, G., Pietrini, P., Panzer, S., & Grafman, J. (1999). The role of the anterior prefrontal cortex in human cognition. *Nature, 399,* 148–151.

Koechlin, E., Ody, C., & Kouneiher, F. (2003). The architecture of cognitive control in the human prefrontal cortex. *Science, 302,* 1181–1185.

Kubota, K., & Niki, H. (1971). Prefrontal cortical unit activity and delayed alternation performance in monkeys. *Journal of Neurophysiology, 34,* 337–347.

MacDonald, A. W., 3rd, Cohen, J. D., Stenger, V. A., & Carter, C. S. (2000). Dissociating the role of the dorsolateral prefrontal and anterior cingulate cortex in cognitive control. *Science, 288,* 1835–1838.

Mansouri, F. A., Buckley, M. J., & Tanaka, K. (2007). Mnemonic function of the dorsolateral prefrontal cortex in conflict-induced behavioral adjustment. *Science, 318,* 987–990.

Mansouri, F. A., Matsumoto, K., & Tanaka, K. (2006). Prefrontal cell activities related to monkeys' success and failure in adapting to rule changes in a Wisconsin Card Sorting Test analog. *Journal of Neuroscience, 26,* 2745–2756.

Milner, B. (1963). Effects of different brain lesions on card sorting. *Archives of Neurology, 9,* 100–110.

Mishkin, M., Prockop, E. S., & Rosvold, H. E. (1962). One-trial object discrimination learning in monkeys with frontal lesions. *Journal of Comparative and Physiological Psychology, 55,* 178–181.

Monsell, S. (2003). Task switching. *Trends in Cognitive Science, 7,* 134–140.

Monsell, S., Sumner, P., & Waters, H. (2003). Task-set reconfiguration with predictable and unpredictable task switches. *Memory and Cognition, 31,* 327–342.

Muhammad, R., Wallis, J. D., & Miller, E. K. (2006). A comparison of abstract rules in the prefrontal cortex, premotor cortex, inferior temporal cortex, and striatum. *Journal of Cognitive Neuroscience, 18,* 974–989.

Murray, E. A., Gaffan, E. A., & Flint, R. W., Jr. (1996). Anterior rhinal cortex and amygdala: Dissociation of their contributions to memory and food preference in rhesus monkeys. *Behavioral Neuroscience, 110,* 30–42.

Nakahara, K., Hayashi, T., Konishi, S., & Miyashita, Y. (2002). Functional MRI of macaque monkeys performing a cognitive set-shifting task. *Science, 295,* 1532–1536.

Nissen, H. W., Blum, J. S., & Blum, R. A. (1948). Analysis of matching behavior in chimpanzees. *Journal of Comparative and Physiological Psychology, 41,* 62–74.

Oden, D. L., Thompson, R. K., & Premack, D. (1988). Spontaneous transfer of matching by infant chimpanzees (Pan troglodytes). *Journal of Experimental Psychology: Animal Behavior Processes, 14,* 140–145.

Pandya, D. N., & Yeterian, E. H. (1990). Prefrontal cortex in relation to other cortical areas in rhesus monkey: Architecture and connections. *Progress in Brain Research, 85,* 63–94.

Passingham, R. E., Myers, C., Rawlins, N., Lightfoot, V., & Fearn, S. (1988). Premotor cortex in the rat. *Behavioral Neuroscience, 102,* 101–109.

Pasupathy, A. & Miller, E.K. (2005) Different time courses of learning-related activity in the prefrontal cortex and striatum. *Nature, 433,* 873–876.

Pepperberg, I. M. (1987). Interspecies communication: A tool for assessing conceptual abilities in the African Grey parrot (*Psittacus arithacus*). In: G. Greenberg & E. Tobach (Eds.), *Cognition, language and consciousness: Integrative levels* (pp. 31–56). Hillsdale, NJ: Lawrence Erlbaum Associates Inc.

Perret, E. (1974). The left frontal lobe of man and the suppression of habitual responses in verbal categorical behavior. *Neuropsychologia, 12,* 323–330.

Petrides, M. (1982). Motor conditional associative-learning after selective prefrontal lesions in the monkey. *Behavioural Brain Research, 5,* 407–413.

Petrides, M., & Pandya, D. N. (2007). Efferent association pathways from the rostral prefrontal cortex in the macaque monkey. *Journal of Neuroscience, 27,* 11573–11586.

Rabbitt, P. M. (1966). Errors and error correction in choice-response tasks. *Journal of Experimental Psychology, 71,* 264–272.

Ramnani, N., & Owen, A. M. (2004). Anterior prefrontal cortex: Insights into function from anatomy and neuroimaging. *Nature Review Neurosciences, 5,* 184–194.

Rao, S. C., Rainer, G., & Miller, E. K. (1997). Integration of what and where in the primate prefrontal cortex. *Science, 276,* 821–824.

Rehkamper, G., & Zilles, K. (1991). Parallel evolution in mammalian and avian brains: Comparative cytoarchitectonic and cytochemical analysis. *Cell Tissue Research 263,* 3–28.

Restle, F. (1958). Toward a quantitative description of learning set data. *Psychological Review, 64,* 77–91.

Rolls, E. T., & Baylis, L. L. (1994). Gustatory, olfactory, and visual convergence within the primate orbitofrontal cortex. *Journal of Neuroscience, 14,* 5437–5452.

Romanski, L. M., Bates, J. F., Goldman-Rakic, P. S. (1999). Auditory belt and parabelt projections to the prefrontal cortex in the rhesus monkey. *Journal of Comparative Neurology, 403,* 141–157.

Romo, R., Brody, C. D., Hernandez, A., & Lemus, L. (1999). Neuronal correlates of parametric working memory in the prefrontal cortex. *Nature, 399,* 470–473.

Rushworth, M. F., Nixon, P. D., Eacott, M. J., & Passingham, R. E. (1997). Ventral prefrontal cortex is not essential for working memory. *Journal of Neuroscience, 17,* 4829–4838.

Rushworth, M. F., Walton, M. E., Kennerley, S. W., & Bannerman, D. M. (2004). Action sets and decisions in the medial frontal cortex. *Trends in Cognitive Science, 8,* 410–417.

Sakai, K., & Passingham, R. E. (2003). Prefrontal interactions reflect future task operations. *Nature Neuroscience, 6,* 75–81.

Semendeferi, K., Armstrong, E., Schleicher, A., Zilles, K., & Van Hoesen, G. W. (2001). Prefrontal cortex in humans and apes: A comparative study of area 10. *American Journal of Physical Anthropology, 114,* 224–241.

Semendeferi, K., Lu, A., Schenker, N., & Damasio, H. (2002). Humans and great apes share a large frontal cortex. *Nature Neuroscience, 5,* 272–276.

Shallice, T. (1982) Specific impairments of planning. *Philosophical Transactions of the Royal Society London B Biological Sciences, 298,* 199–209.

Shallice, T. & Evans, M.E. (1978) The involvement of the frontal lobes in cognitive estimation. *Cortex, 14,* 294–303.

Stoet, G., & Snyder, L. H. (2003). Executive control and task-switching in monkeys. *Neuropsychologia, 41,* 1357–1364.

Stoet, G., & Snyder, L. H. (2007). Correlates of stimulus-response congruence in the posterior parietal cortex. *Journal of Cognitive Neuroscience, 19,* 194–203.

Wallis, J. D., Anderson, K. C., & Miller, E. K. (2001a). Single neurons in prefrontal cortex encode abstract rules. *Nature, 411,* 953–956.

Wallis, J. D., Dias, R., Robbins, T. W., & Roberts, A. C. (2001b). Dissociable contributions of the orbitofrontal and lateral prefrontal cortex of the marmoset to performance on a detour reaching task. *European Journal of Neuroscience, 13,* 1797–1808.

Wallis, J. D., & Miller, E. K. (2003). From rule to response: Neuronal processes in the premotor and prefrontal cortex. *Journal of Neurophysiology, 90,* 1790–1806.

White, I. M., & Wise, S. P. (1999). Rule-dependent neuronal activity in the prefrontal cortex. *Experimental Brain Research, 126,* 315–335.

Wilson, B., Mackintosh, N. J., & Boakes, R. A. (1985). Transfer of relational rules in matching and oddity learning by pigeons and corvids. *Quarterly Journal of Experimental Psychology, 37B,* 313–332.

Witthöft, W. (1967) Absolute anzahl und Verteilung der zellen im hirn der honigbiene. *Zeitschrift fur Morphologie der Tiere, 61,* 160–184.

Wynne, C. D. L. (2004). *Do animals think?* Princeton, NJ: Princeton University Press.

CHAPTER 22

Reinventing Primate Neuroscience for the Twenty-First Century

Todd M. Preuss

The manner in which an organism perceives and responds to features of its environment reflects the state of the neural systems that make up its brain. If we want to address behavior from a perspective that is meaningful in terms of both the mechanisms of behavior and evolutionary history, we have to consider the organization of the neural systems that constitute brains, how that organization varies across individuals and across species, and how that organization changed over evolutionary time.

As a group, primatologists have paid relatively little attention to evolutionary changes in the internal organization of brains, focusing instead on changes in size. Size has virtues as a scientific variable: It can be measured in both extant and extinct species, and it is unquestionably easier to measure the size of a brain than to examine its contents. Moreover, there have seemed good reasons to think that mammalian brains (or primate brains, anyway) are just larger or smaller versions of a common brain, and that size itself is what one really needs to study in order to understand the relationship between the evolution of the brain and its functional capacities. This view was systematized by Harry Jerison, in his classic synthesis of brain-size studies (Jerison, 1973), and it established the model for primatological/anthropological approaches to brain evolution. Jerison's approach was grounded in the idea that the internal organization of the cortex (the largest component of the brain in primates and many other mammals) is relatively uniform across species, an idea that can be traced to the influential work of Karl Lashley (Lashley, 1931, 1949), who famously argued that most of the cortex is functionally "equipotential." Subsequent work has refined the allometric treatment of brain-size data, but the core premise of the enterprise—that brain organization is conserved and scalable across a wide variety of mammalian species—has largely gone unchallenged. Ralph Holloway, with his insistence that evolution could reorganize brains internally, so that brains of the same size could have different functional capacities, has been a conspicuous exception (Holloway, 1966).

To me, Holloway's critique has always seemed eminently sensible. The molecular, cellular, and histological organization of the cerebral cortex, and its patterns of neuronal connectivity, are incredibly complex, and so there would seem to be enormous opportunity for evolutionary change: Why, in the service of modifying behavior, shouldn't evolution have tweaked the membrane potentials of cells, for instance, or built connections between neurons that didn't previously communicate, or reorganized cell ensembles to serve novel cognitive functions?

Yet, it has to be admitted that if these kinds of changes did take place in mammalian evolution, the neuroscience community has done a very good job of keeping them under wraps. Modern experimental neuroscience is strongly focused on the apparent similarities between taxa. The cortex, in particular, has come to be regarded as essentially invariant in its internal organization across species, and has been said to have a "basic uniformity of structure" (Rockel et al., 1980) and a "generality of functional organization" (Creutzfeldt, 1977). In considering the appeal of these claims, it is important to bear in mind that most neuroscientists work with a very limited range of taxa, typically one or a few of the accepted "model" species, which today means primarily rats or mice. If you study rodent brains, you have every incentive to believe that there are no important brain characteristics of other mammals (including humans) that are not present in rodents. And if most of your colleagues also study rodents, who's to say you're wrong? As one speaker I heard recently put it, we can use rats as experimental stand-ins for humans because, after all, "brains are brains." The generality of results obtained in model species is simply taken for granted and is not regarded as being in need of empirical validation (Logan, 1999; Preuss, 2000a).

Nevertheless, if one sheds the blinders of the model-animal paradigm and looks in detail at the primary data available from a variety of mammals, it is quite apparent that there is remarkable phyletic variation in the organization of cerebral cortex: The claim of "basic uniformity" is simply false. These data have been extensively reviewed elsewhere (e.g., DeFelipe et al., 2002, 2007; Elston, 2007; Hof & Sherwood, 2007; Preuss, 1995, 2000, 2001; Rilling, 2006; Sherwood & Hof, 2007; Sherwood et al., 2003), and these reviews paint a very clear picture: Cortex varies across mammalian taxa at just about any level or dimension of organization one cares to consider, from transmitters and receptors; to cell types, cell morphologies, and cell numbers; to the ways in which different cell types are organized into layers; to the ways cells in different layers are interconnected to form local processing architectures; to the ways local cell assemblies are organized into modules and areas (the larger-order histological units of the cortex); to the ways areas are assembled into higher-order information-processing systems through patterns of long intracortical connections. Furthermore, and not surprisingly, the variations in cortical organization are not random, but follow phylogenetic patterns: The internal organization of cerebral cortex in bushbabies and rhesus macaques—two primates that differ markedly in brain size—is much more similar than in bushbabies and tree shrews—animals with brains about the same size as bushbabies, but which are not primates. This is not to say that there aren't features of cortical organization that are widely shared among mammals by virtue of common ancestry (e.g., the existence of some manner of columnar and laminar distribution of cortical neurons), but the idea that there was an ancestral organization from which modern forms have descended with modification is a very different proposition than the claim that important features of cortical organization are invariant across mammalian groups, which is the main point of "basic uniformity."

Accepting the fact that cortical organization varies across changes one's view of what students of primate brain evolution should be doing. For one thing, it means that a primary or exclusive focus on brain size will not do, for if the internal organization of brains varies in important ways, then trying to understand the functional capacities of different species by measuring endocranial capacity is like trying to understand what different computers can do by measuring the boxes they came in. To be sure, you *might* learn something by a comparison of computer boxes if you start out with a good understanding of the architecture of computers, and if the computers are two versions of the same model. But if you don't have that kind of knowledge, you're not going to get very far.

Acknowledging that internal brain organization varies means that we must document those variations in a systematic way and apply the modern analytic tools of comparative biology to reconstruct brain evolution based on those variations. That is, we need to know how brain organization varies across taxa and to map those variations onto the phylogenetic tree. Doing this, it is possible to reconstruct the brain organization of ancestors, which are represented at the internal nodes of the phylogenetic tree, and to use ancestral organizations as starting points for assessing evolutionary changes in daughter lineages. These analytic techniques are now well developed and widely applied in fields of research as diverse as molecular biology and ethology; indeed, the burgeoning science of comparative genomics, which is founded on the identification of homologous gene sequences in different taxa, would scarcely be possible without the modern tools of phylogenetic analysis.

An emphasis on studying the internal organization of brains also implies a primary emphasis on the study of *living* (extant) taxa. I don't mean by this to suggest that paleoneurology is futile, but rather to assert that what we can expect to glean from paleoneurological studies of extinct species will depend on what we know about the actual brain biology of their closest living relatives. In addition, we need to pick and choose the extant taxa to be studied very carefully. Since we can't study every species in great detail, we must choose species that maximize the amount of information they provide about evolutionary history. Certainly, we want to identify the shared, derived characteristics of at least the major branches of the primate tree—strepsirrhines (lemurs, lorises, and bushbabies), tarsiers, platyrrhines (New World monkeys), and catarrhines (Old World monkeys, apes, and humans), so we need to study members of each of these groups. Understanding what features of brain organization distinguish primates from other mammals requires outgroup evidence, and thus, in addition to studying primates, was also need to compare primates to other mammals, especially the groups of mammals most closely related to primates, which, on current evidence, include tree shrews (mammalian order Scandentia) and colugos (i.e., flying lemurs; order Dermoptera) (e.g., Murphy et al., 2001). Of course, the characteristics of the human lineage and of the hominoid (ape) clade of which our lineage is a part are matters of special interest and importance. Understanding this part of primate evolutionary history requires comparing humans and apes, and particularly the ape species to which we humans are most closely related: chimpanzees and bonobos. Currently, we know very little about what distinguishes the human brain from that of apes or other primates, apart from the fact that it is extremely large (Crick & Jones, 1993; Preuss, 2000b, 2006).

The task of addressing primate brain evolution at such fine levels of neurobiological and phylogenetic detail might seem daunting, if not impossible. It is important to recognize, therefore, that there is at least one area of research in which neuroscientists have been equal to the task: Primate visual systems have been examined in considerable detail from a comparative standpoint, and it is possible to reconstruct many features of the primate visual system evolution. I want to review this matter in some detail, because it illustrates both the promise of the science and the limitations that must be overcome in order to obtain comparably detailed analyses of other brain systems.

If we are unusually well informed about the evolution of primate visual systems, it is because visual science has a particularly strong comparative tradition and because the influential work of Le Gros Clark (1959) stressed the importance of vision in primate evolution. Le Gros Clark's banner was taken up by Diamond and his colleagues, who helped promote studies of tree shrews and bushbabies (e.g., Diamond & Hall, 1969; Harting et al., 1972, 1973), studies that have been essential for understanding primate visual system evolution. This rich history of study makes it possible to identify likely primate specializations at almost every level of visual system organization (for comprehensive reviews of the primary literature, see Allman, 1977; Casagrande & Kaas, 1994; Casagrande et al., 2007; Kaas & Preuss, 1993; Preuss, 2004, 2007a,b; Rosa & Krubitzer, 1999; Ross & Martin, 2007). One of the first and most exhaustively documented primate specializations is the distinctive pattern of projections from the retina to the superior colliculus, whereby each colliculus represents only the contralateral visual field. The unusual lamination in the thalamic lateral geniculate nucleus (LGN) of primates reflects a pattern of input segregation different from other mammals. In primates, the LGN projects nearly exclusively to a single cortical area, the primary visual area (V1; also known as striate cortex), whereas in other mammals that have been studied, LGN projections are more widespread. One consequence of this is that higher-order visual areas are more dependent upon V1 for driving physiological activity in primates than they are in nonprimates. Within area V1 of primates, there is a distinctive pattern of compartmentation in the upper layers, such that inputs that ultimately arise from the W class of cells in the retina terminate in small patches ("blobs") that are rich in cytochrome oxidase, a metabolic enzyme. Blobs are absent in tree shrews, rodents, and other mammals, with the exception of carnivores, in which they evolved independently. Primates also possess a large number of higher-order visual areas, as do some other mammals, but only in primates are these areas organized by their interconnections into two distinct processing streams, a dorsal pathway related to spatial attention and action and a ventral pathway related to object recognition.

In addition to the shared, derived features of primates, we also know something about the visual-system specializations of particular primate groups. Early in the radiation of the haplorine primates (the group that includes tarsiers and anthropoids), there was adaptation to a diurnal activity pattern and the retina was reorganized, with the evolution of a true fovea and with the replacement of rods in the central retina by an abundance of miniaturized cones. These changes improved spatial resolution in bright light and created the conditions required for opponent interactions between cones, which make color discrimination possible. Independent duplication of the ancestral medium/long-wavelength cone opsin in the New World and Old World anthropoid led to the evolution of trichromatic vision in these groups. We know of additional modifications of the laminar organization of area V1 (in hominoid evolution) and of the compartmental organization of area V1 (in human evolution), but the functional implications of these changes are as yet poorly understood.

THE NEW NEUROSCIENCE TECHNOLOGIES

If the visual system gives us a glimpse of what a mature comparative neuroscience of primates could look like, the unhappy fact is that we known rather little about most other regions of the brain. We know that primates vary in the organization of somatosensory and motor cortex in ways that parallel changes in manual reaching, grasping, and manipulative abilities (Hinkley et al., 2007; Kaas, 2007; Padberg et al., 2007; Wise, 2007). We have evidence that primates evolved new motor areas along with a large suite of new areas in the classical higher-order association regions of the frontal, parietal, and temporal lobes (Preuss, 2007a,b). We also know of changes in cortical cell biology, histology, and receptor distribution, although in most cases the comparative coverage has not been sufficient to make possible detailed phyletic reconstructions. Many aspects of primate brain organization remain to be explored from a comparative perspective.

Why can't we simply extend the experimental approaches that have proven so successful in documenting the evolution of the visual system to other systems? Consider that much of the comparative visual system research has involved invasive and terminal experimental procedures, and that to be applied on anything more than a small scale, these procedures would require an infrastructure of animal colonies housing a taxonomically diverse set of primate species and close primate relatives. In the United States, where concentration of research around a few "model" species seems to be a much higher priority than maintaining a diversity of research species, the number of primate species (and other species relevant to primate evolution) available for research is declining: The past decade has seen the elimination of all the ape species from the National Primate Research Centers with the exception of chimpanzees, and the National Committee for Research Resources (NCRR; the branch of the National Institutes of Health [NIH] that supports the National Primate Research Centers) has decided to maintain a moratorium on breeding the chimpanzees it holds (Cohen, 2007a), which means that the chimpanzees will eventually be eliminated, too. The few remaining breeding colonies of critically important taxa like galagos and tree shrews, historically supported by NIH funding, are unlikely to be maintained. To the extent that nonhuman primates will continue to have a role in a neuroscience based on traditional experimental methods, that future belongs mainly to a few macaque and baboon species and to marmosets. If we continue to rely primarily on experimental approaches requiring invasive and terminal procedures, it seems unlikely that we will have access to the diversity of taxa required to reconstruct primate brain evolution.

Does the science of primate brain evolution have a future? If it does, it will have to come to terms with the degradation of the animal resources upon which comparative brain studies have heretofore relied. One way to do this is to take greater advantage of the variety of new "low-impact" techniques available to neuroscientists, techniques that either involve noninvasive or minimally invasive methods for studying living organisms or employ biological materials that can be collected without harming the individual or that can be collected after death. These include neuroimaging, histological techniques, and genomic and other molecular biological methods.

Comparative Neuroimaging

Neuroimaging is a set of methods for remotely probing the structure, chemistry, and physiology of the brain. These include magnetic resonance imaging (MRI), positron emission tomography (PET), computer-assisting tomography (CAT), near-infrared spectroscopy, electroencephalography (EEG), and transcranial magnetic stimulation (TMS). Currently, MRI and PET offer the greatest potential for comparative neuroimaging, and will be the focus of discussion here. It is convenient to distinguish structural and functional modes of neuroimaging, categories that cut across the MRI/PET distinction.

Structural Imaging with MRI

MRI works by using magnetic fields and electromagnetic (EM) pulses to manipulate the spins of atomic nuclei, and thereby induce them to emit electromagnetic signals. The strongest signals can be elicited from hydrogen nuclei (protons), both because of their intrinsic magnetic properties and because they are the most abundant nuclei in the brain, so the most commonly used MRI techniques are based on proton imaging. In the presence of a strong background magnetic field, protons align their spin axes along that field; in fact, their spin axes precess around a central axis much like the wobble of a spinning top. External electromagnetic pulses tuned to appropriate "resonant" frequencies that depend on the strength of the background field cause the axes to flip out of the plane of the background magnetic field and simultaneously synchronize the precession phases of the individual nuclei. Following the EM pulse, the protons relax, their precessions dephasing at an exponential rate and their axes returning to alignment with the background field at an exponential rate. As protons relax, they emit EM waves at their resonant frequencies. The rate of return to alignment with the background field is known as T1; the rate of dephasing of precessions is known as T2. Different tissues have different T1 and T2 values, values that depend on the concentration of hydrogen and on the organization of different hydrogen-containing molecules in the tissue. This provides the basis for the tissue contrast one sees in MR images, such as, for example, the contrast between gray matter and white matter in the brain. Typical structural MRI scans reflect both T1 and T2 but weight them differently (Fig. 22.1A).

Increasing the strength of the background magnetic field increases the signal-to-noise ratio of the MR signal obtained from individual brain volume elements (voxels), which means that stronger magnets make it possible to obtain higher spatial resolution scans without sacrificing image quality. Current clinical scanners are generally 1.5 to 3 Tesla (T), and a number of universities now have experimental research scanners of 9.4T or more. Early scanners, with low field strength, were best suited for the study of large species, such as humans. Indeed, MRI is unusual among neuroscientific techniques in that its first large-scale applications were in humans, proving valuable as a diagnostic tool for localizing tumors, strokes, and aneurisms, as well as for carrying out morphometric studies of human brain structure. The progressive improvement of the technology

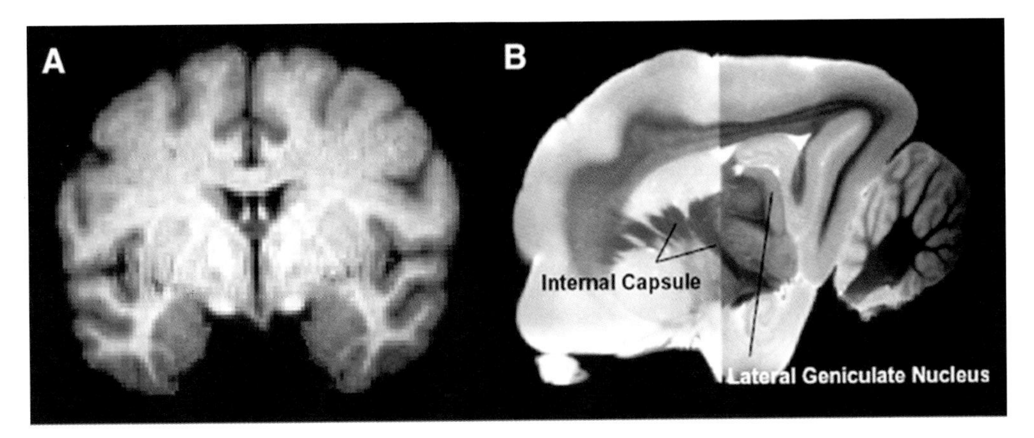

Figure 22.1 T1- and T2-weighted structural neuroimaging. **(A)** T1-weighted coronal image from a chimpanzee acquired in vivo with a 3T scanner. **(B)** T2-weighted images from the fixed brain of an aye-aye (*Daubentonia madagascariensis*) acquired in a 9T scanner. The image sets in B were reconstructed to show the intersection of a coronal plane and a parasagittal plane. T1- and T2-weighted images show different patterns of tissue contrast: Gray matter is darker than white matter in T1, whereas white matter appears very dark in T2-weighted images, which resemble myelin-stained histological sections. T1-weighted scans are commonly used for rapid acquisition of structural images in vivo with high spatial resolution. Fixed brains can be scanned for very long periods, however, and T2-weighted images can be acquired with very high spatial resolution and excellent contrast between gray matter and white matter, showing remarkable anatomical detail even in the relatively small brain of this lemur. Figure 22.1B from Kaufman, J. A., Ahrens, E. T., Laidlaw, D. H., Zhang, S., & Allman, J. M. (2005). Anatomical analysis of an aye-aye brain (*Daubentonia madagascariensis*, primates: Prosimii). combining histology, structural magnetic resonance imaging, and diffusion-tensor imaging. *Anatomical Records Part A: Discovery of Molecular, Cellular, and Evolutionary Biology, 287,* 1026–1037. Used with permission.

of MRI has made it possible to extend MRI morphometry to species with smaller and smaller brains, so that MRI has become an important tool in comparative neuroscience. Moreover, while we usually think of MRI as an in vivo technique, it can be used with fixed brains as well and therefore can be used to study archival brain material (Fig. 22.1B).

There is already an extensive literature of comparative MRI morphometry. Examples include studies comparing the absolute and relative volumes of cerebral lobes among hominoids (Rilling & Insel, 1999b; Rilling & Seligman, 2002; Semendeferi et al., 1997, 2002); the volumes of the cerebellum and its constituents among primates and between primates and cetaceans (MacLeod et al., 2003; Marino et al., 2000; Rilling, 2007; Rilling & Insel, 1998); the sizes of white matter systems of the brain, including the white matter of the different cerebral lobes (Rilling & Seligman, 2002; Schenker et al., 2005; Schoenemann et al., 2005) and of the major fiber bundles interconnecting the two hemispheres, the anterior commissure and corpus callosum (Rilling & Insel, 1999a); as well as studies of cerebral asymmetries of cortical mass and sulcal morphology (Gilissen, 2001; Hopkins et al., 1998). To date, MRI morphometry has been used mainly to study hominoids and anthropoids and other large-brain creatures, but it is certainly possible to image smaller-brain animals. One of the advantages of postmortem imaging is that you can fit relatively small brains into high field-strength scanners (7T or higher) and scan them for a long time to obtain very high resolution, as, for example, in the recent study of a *Daubentonia* brain by Kaufman et al. (2005). (Fig. 22.1B).

Despite these technical improvements, the study of certain major constituent structures of the brain, notably cortical areas and thalamic nuclei, has proven difficult with current MRI techniques. Identifying and demarcating ("segmenting") these structures is important, because unlike whole lobes and hemispheres, these are real functional units of the brain. Sulci, unfortunately, are not very reliable guides to the borders of cortical areas in cross-species comparisons (Preuss, 2007b). The problem of segmenting cortical areas and nuclei in MRI is compounded by the fact that the organization of even well-studied taxa like rhesus macaques is still not completely worked out. The widespread reliance on Brodmann's map of human cortical areas (Brodmann, 1909) in functional MRI studies suggests a level of knowledge about human cortex that we just do not have: Brodmann's map is inaccurate in important respects (Abbott, 2003; Eickhoff et al., 2005a; Toga et al., 2006). Moreover, researchers typically assign Brodmann's numbers to loci of functional activity based solely on their coordinates in one of the standardized brain spaces currently available for human brain mapping, rather than on an independent assessment of the histological architecture of subjects' brains—precisely because it has been difficult to do something like a cytoarchitectonic or myeloarchitectonic parcellation of the cortex using MRI. One approach to resolving this limitation is to make architectonic maps from brains that have been scanned, then sectioned and analyzed histologically, and then registered back into standardized brain space; individual subjects' MRI scans can then be registered into the same space, and cortical areas identified on a probabilistic basis (Eickhoff et al., 2005b, 2007b). Ultimately, it may prove possible to make individualized brain maps: Current imaging techniques can detect variations in the strength of the MR signal across the laminae of the cortex, and in principle, these intensity variations could be used to distinguish cortical areas just as laminar variations in cell and fiber staining are used to distinguish cortical areas in classical histological architectonic mapping (Clare & Bridge, 2005; Clark et al., 1992; Eickhoff et al., 2005a; Walters et al., 2003).

There are at least two additional variants of MRI that are potentially extremely valuable for comparative studies. One is diffusion-tensor imaging (DTI), the attraction of which lies in its ability to map fiber tracts noninvasively, and thus overcome (to some extent) the limitations of standard connectivity techniques. Since brains are collections of neural systems, constituted by the interconnections of neurons and neuron-containing structures, understanding connectivity and how it varies across species is fundamental to comparative neuroscience. The most sensitive and reliable experimental techniques for tracing connections in the brain have involved the injection of chemical substances that are taken up by neurons and transported either proximally (back to the cell body) or distally (out to the terminations of axons) or both. Injecting the substances involves invasive surgical procedures; confirming the location of the injection site and determining where the tracer was transported to requires sacrificing the subject and examining the brain microscopically. Obviously, these techniques are inappropriate for use in humans, or in rare or endangered species, such as chimpanzees. As a result, most of what we think we know about the connectivity of the human brain represents extrapolation from experimental studies in macaque monkeys rather than results obtained directly from studies of human beings (Crick & Jones, 1993). DTI provides the opportunity to study connectivity directly in humans, and can be used in other species as well (Fig. 22.2).

DTI is based on the ability of MRI to detect the direction of diffusion of water molecules (see reviews by Mori & Zhang, 2006, and Ramnani et al., 2004). Since water molecules tend to diffuse along, rather than across, nerve fibers, local anisotropies in water diffusion can be used to reconstruct the three-dimensional organization of fiber bundles (Kaufman et al., 2005). DTI can be done in vivo, but also with fixed, archival brain tissue. Studies of well-preserved tissue yield results similar to in vivo studies (D'Arceuil et al., 2007; Guilfoyle et al., 2003; Sun et al., 2003), because DTI is based on the physics of water diffusion, rather than on the biological activity of living tissue.

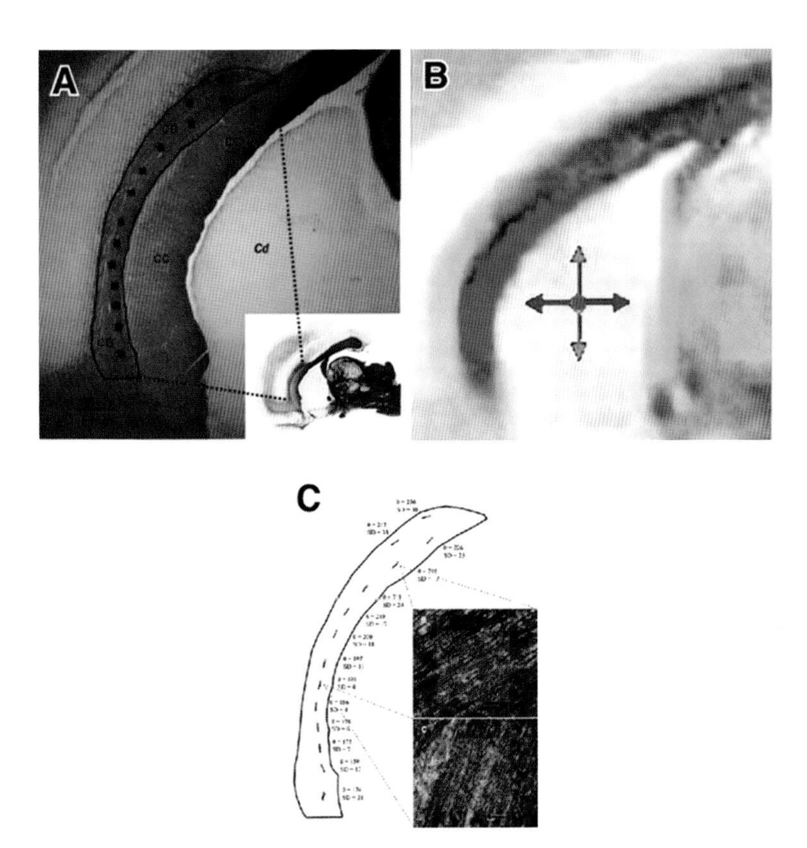

Figure 22.2 Diffusion-tensor imaging (DTI) can be used to reconstruct white matter pathways. A–C illustrate the principles of DTI in the brain of *Daubentonia,* as demonstrated by Kaufman et al. (2005). **(A)** A histological section through the cingulum bundle (CB), corpus callosum (CC), and caudate nucleus (Cd) stained for myelin. The cingulum bundle and corpus callosum are major fiber tracts, and therefore are stained darkly; the caudate, a gray matter structure, has a light appearance. **(B)** A "color map" at the same level, obtained from DTI scans obtained before the brain was sectioned. DTI color maps code the principal direction of fibers in each brain voxel. In this map, red represents anteroposteriorly oriented fibers, green represents fibers oriented superoinferiorly, and blue represents mediolaterally oriented fibers. The prominent blue territory corresponds to the corpus callosum, which carries fibers passing mediolaterally between the two hemispheres. The cingulum bundle, which carries fibers that pass between the frontal lobe and posterior cortical regions, starts out red at posterior levels and then turns green as it turns inferiorly anterior to the genu of the corpus callosum. **(C)** Histological confirmation of the DTI results: Plots of fiber directions observed in myelin-stained sections at different levels of the cingulum bundle match the directionality indicated by the DTI results in B. **(D)** Results of a DTI study comparing the course and targets of the arcuate fasciculus in the cerebral hemispheres of humans and chimpanzees, based on Rilling et al. (2008). These maps illustrate fiber pathways reconstructed from principal-direction data. Note that in these figures, all fibers and terminations are rendered as shades of green, whatever their orientation. In humans, the arcuate fasciculus interconnects language areas of the inferior frontal lobe and parietal and temporal cortex. Chimpanzees also possess an arcuate fasciculus, but it lacks the strong connections with the temporal lobe seen in humans. In humans, the relevant temporal areas are involved in the representation of word meaning. abSF, ascending branch of the Sylvian fissure; AOS, anterior occipital sulcus; CS, central sulcus; FOS, fronto-orbital sulcus; hbSF, horizontal branch of the Sylvian fissure; IFS, inferior frontal sulcus; IPS, intraparietal sulcus; ITS, inferior temporal sulcus; LCaS, lateral calcarine sulcus; LuS, lunate sulcus; PoCS, postcentral sulcus; PrCS, precentral sulcus; SF, Sylvian fissure; SFS, superior frontal sulcus; STS, superior temporal sulcus. A–C from Kaufman, J. A., Ahrens, E. T., Laidlaw, D. H., Zhang, S., & Allman, J. M. (2005). Anatomical analysis of an aye-aye brain (*Daubentonia madagascariensis*, primates: Prosimii) combining histology, structural magnetic resonance imaging, and diffusion-tensor imaging. *Anatomical Records Part A: Discovery of Molecular, Cellular, and Evolutionary Biology, 287,* 1026–1037. D based on Rilling, J. K., Glasser, M. F., Preuss, T. M., Ma, X., Zhao, T., Hu, X., et al. (2008). The evolution of the arcuate fasciculus revealed with comparative DTI. *Nature Neuroscience, 11,* 426–428. Used with permission.

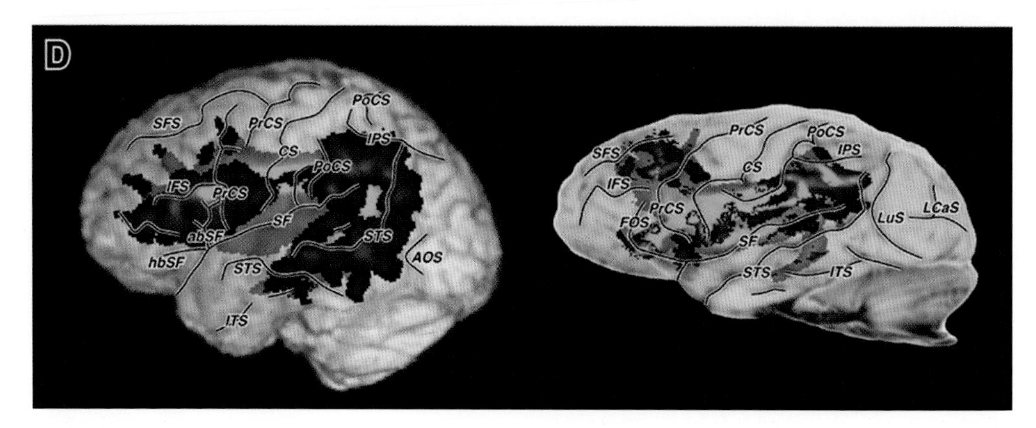

Figure 22.2 (Continued)

DTI tracing has some important limitations compared to traditional tracer studies. The spatial resolution of DTI is relatively coarse (with voxel sizes usually in the range of 0.5 to 2.0 mm), so at present this is not a method that can trace connections from cell to cell. In addition, accurate reconstruction of fiber tracts depends on the coherence of fiber bundles, so DTI tracks quite well through the white matter, but less well as fibers pass into gray matter, where fibers disperse and travel in multiple directions. Finally, DTI is vulnerable to both false positives and negatives, so that currently the best use of DTI is not to provide an exhaustive accounting of the structures with which a given brain region is connected (which is what tracer studies provide), but rather to assess the probability that fibers pass between a specific region of interest and other voxels in the brain. Notwithstanding these limitations, DTI is proving to be extremely valuable as a noninvasive tool for tracing connections, and it is now being used to directly compare humans and nonhuman primates and thus to shed light on human brain evolution. Recent studies of note include comparisons of frontal lobe connections with subcortical structures (Croxson et al., 2005; Ramnani et al., 2006; Tomassini et al., 2007); comparisons of the interconnections of higher-order cortical regions in humans, chimpanzees, and macaques (Rilling et al., 2008); and comparisons of amygdala connectivity (Hecht et al., 2008). The ability of DTI to track fiber pathways also makes this a useful tool for segmenting the areas and nuclei of the brain (Behrens & Johansen-Berg, 2005), because brain areas differ in their patterns of connectivity as well as their histological organization.

Functional Imaging

Neuroimaging can also be employed to monitor the functional activity of living brains, using a variant of the MRI technique, functional MRI (fMRI), and with PET. Both approaches are based on the fact that changes in neural activity are accompanied by changes in regional blood flow and glucose utilization. Functional imaging maps these regional metabolic changes. Functional MRI does this by tracking changes in blood flow; as brain regions increase or decrease their activity level, and blood flow increases or decreases accordingly, the relative amounts of oxygenated and deoxygenated hemoglobin in those regions change: Regions with high blood flow have higher levels of oxygenated hemoglobin. The change can be detected with MRI because deoxygenated and oxygenated hemoglobin have different magnetic properties.

Whereas MRI works by the external application of magnetic fields and EM pulses, PET derives information by introduction into the subject (orally or by injection) of radioactively labeled substances that accumulate in the brain. These substances incorporate short half-life, positron-emitting isotopes. An emitted positron travels only a few millimeters on average before encountering an electron and the two are annihilated, yielding a pair of photons traveling in nearly opposite directions. These photons reach detectors arrayed around the subject, and by comparing the location of the detector elements activated and the time required to reach the detector array, their points of origin can be determined. In this way, regional variations in the accumulation of the radiolabeled substance can be reconstructed. PET can be used as an anatomical tool, for example, to determine the distribution of specific receptor types using appropriate ligands, but its principal use has been to determine function-related differences in regional brain activity using radiolabeled oxygen isotopes or glucose analogs.

While widely used in human studies, functional imaging has found rather limited applications in studies of nonhuman primates. In part, this is because subject movement degrades the images obtained in MRI and PET scanners. Most human subjects can restrain their head movements upon request even as they make finger taps or other motor responses in behavioral tasks, but awake animals need to be extensively trained, as well as physically restrained, in order to be suitable as subjects. This can be done in some cases, mainly with highly trained, chair-restrained macaques (e.g., Howell et al., 2001; Logothetis et al., 1999), but scanning awake, unrestrained individuals is not a practical option for most nonhuman primate species. Certainly, you do not want to try to scan a powerful animal like an adult chimpanzee unrestrained in your multimillion-dollar MRI or PET facility.

One way to escape this limitation is to separate the time period during which the subject is awake and behaving and taking up a radiolabeled marker, from the time period of PET scanning, in which the distribution of the ligand in the brain is determined. Provided the tracer accumulates to an asymptotic level during the behavioral testing period and doesn't decline significantly during the time required to obtain the PET scan, it makes little difference whether the PET scan is done with the subject awake or anesthetized. Using [18F]-fluorodeoxyglucose (FDG), a radiolabeled glucose analog, this PET technique has been used in humans (who are not usually anesthetized during scanning), as well as in nonhuman primates (Rilling et al., 2001, 2004, 2007). PET is currently the only technique available for directly comparing behavior-related patterns of brain functional activity in humans and chimpanzees (Fig. 22.3).

Functional Connectivity

The need to anesthetize nonhuman subjects during scanning means that PET, rather than fMRI, is in most cases the practical option for comparing the relationship between regional brain activity and behavior across primate species. Functional MRI does have a use in comparative studies, however. Regions of the brain that are coupled synaptically tend to have temporally correlated patterns of metabolic activity: They tend to be active in synchrony. Functional MRI, which measures levels of functional activity in individual brain voxels, provides data that can be used infer patterns of connectivity. This technique, called *functional connectivity MRI* (or fcMRI), is now being used extensively to study humans (e.g., Biswal et al., 1995; Fox et al., 2005; Hampson et al., 2002; Lowe et al., 1998; Seeley et al., 2007; Xiong et al., 1999). Typically, human subjects rest quietly in the scanner while MRI records spontaneous brain activity; the subjects are not performing any particular task. Interestingly, much of this spontaneous brain activity is retained in shallow sleep and light anesthesia (Peltier et al., 2005), and as a result, the technique has been employed to compare humans and macaques (Vincent et al., 2007). It should be possible to extend it to other nonhuman primate species as well. The fcMRI technique could prove an important complement to DTI for providing information about connectivity for species in which traditional, invasive methods cannot be used.

Comparative Histology and Cell Biology

Neuroimaging gives us a window into the brain; histological and cell-biological techniques take

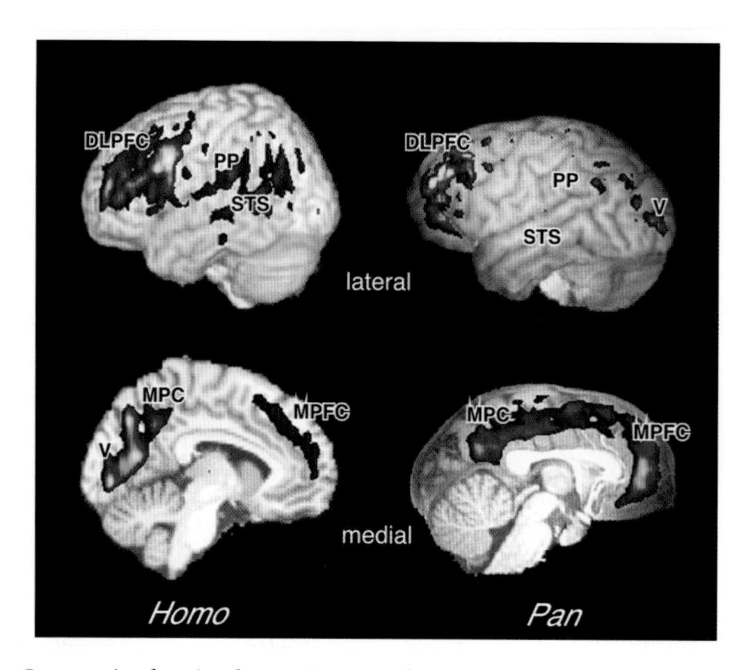

Figure 22.3 Comparative functional neuroimaging of resting-state cognitive activity using the in vivo fludeoxyglucose-positron emission tomography (FDG-PET) technique in humans and chimpanzees, based on Rilling et al. (2007). In humans, numerous studies have shown that specific regions of cortex are highly active when subjects simply rest in the scanner and allow their thoughts to wander freely, including the medial frontal cortical (MFC) areas representing self-referential and emotional processes, medial posterior cortical (MPFC) areas involved in episodic memory, and lateral areas in the posterior parietal (PP) lobe and the cortex in and around the posterior superior temporal sulcus (STS) involved in language and conscious awareness. Rilling and colleagues replicated previous resting-state activity results in humans, and showed that chimpanzees exhibit patterns of activity similar, but not identical, to that of humans. The amount of FDG incorporated into different parts of the brain is color coded; the brightest voxels denote the highest activity. The same FDG-PET technique can be used with more structured behavioral tasks to compare functional brain activity across species. V, visual cortex. From Rilling, J. K., Barks, S. K., Parr, L. A., Preuss, T. M., Faber, T. L., Pagnoni, G., et al. (2007). A comparison of resting-state brain activity in humans and chimpanzees. *Proceedings of the National Academy of Sciences, 104,* 17146–17151. Used with permission.

us into the very substance of the brain, and at scales of organization that can be smaller than those achievable in normal practise with neuroimaging (Balter, 2007). Because tissue studies are as old as experimental biology itself and almost always involve studies of tissue obtained postmortem, one might not imagine this to be a hotbed of new methods and an avenue to new discovery about the brain. In fact, the amount and quality of information that can be gleaned from brain tissue has increased dramatically over the past several decades, driven by the introduction of new techniques for labeling specific tissue and cell compartments, new methods of quantitative analysis, and new techniques for preserving tissue.

Scientists in the nineteenth century discovered that different components of neural tissue could be distinguished by chemical treatment of the tissue; these histochemical methods include stains that reveal cell bodies (e.g., the Nissl stain) and myelinated axons. New histochemical stains were developed in the twentieth century, including stains for enzymes that are highly abundant in the nervous system, such as cytochrome oxidase and acetylcholinesterase. These staining methods are still widely employed in experimental neuroscience. Increasingly, however, neuroscientists are making use of a rapidly growing set of ligand molecules that bind selectively to specific molecules of interest in the brain. Since different molecules tend to be localized within particular cell types (e.g., pyramidal cells, interneurons, astrocytes) and to particular compartments within those cell types (e.g., nucleus, cell body, synaptic spines), mapping the distribution of specific molecules can tell us a great deal about the cellular and subcellular organization of brain tissue. The ligands employed in these studies include naturally occurring chemicals, including a variety of plant lectins that have affinities for the carbohydrate moieties of proteins, and synthetic compounds, including neuroactive drugs that mimic neurotransmitters, binding to specific receptor molecules on neurons and glia. But by far the most widely used ligands are antibodies. Antibody production for research and clinical use is now a large industry, turning out antibodies to a substantial fraction of the proteins present in the human brain. So, in a typical immunohistochemical (or immunocytochemical) procedure, one bathes tissue sections in a solution containing antibodies to a protein of interest, then immerses the sections in a second solution that tags the primary antibody with a marker that can be visualized under a microscope, such as a fluorescent dye or a chemical that can be processed to yield a visible precipitate, typically diaminobenzidine (DAB).

What can we do with these methods? One use is to segment the gray matter into structurally and functional meaningful subunits. Since gray matter, in its natural state, tends to be quite homogeneous in appearance, histologists have explored chemical procedures that differentially stain or label subunits of gray matter. Classically, the histochemical stains for Nissl and myelin were employed to try to subdivide gray matter based on regional differences in the size and density of cell bodies (as in the familiar work of Brodmann, 1909) and the density and geometric pattern of myelinated fiber distribution (as, for example, by Vogt & Vogt, 1919). In principle, however, any molecule that varies in its density or spatial distribution across brain regions can be used to segment (or "parcellate," in the argot of architectonics) the brain. So, for example, simple histochemical stains for cytochrome oxidase and acetylcholinesterase have been used to map cortical areas (e.g., Stepniewska et al., 1993; Tootell et al., 1985), and with the proliferation of antibodies and other ligands, a number of these have proven useful for subdividing the cortex and subcortical brain regions. Investigators try to combine evidence from multiple sources to obtain reliable and meaningful parcellations (e.g., Kaas, 1987; Toga et al., 2006). For example, my colleagues and I found that a plant lectin, *Wisteria flourabunda* agglutinin, and an antibody to a calcium-binding protein, calbindin, are both useful tools for distinguishing nuclei of the thalamus (Cola et al., 2005; Preuss et al., 1998), and that an antibody to neurofilament protein (SMI-32) is valuable for distinguishing subdivisions of motor cortex (Preuss et al., 1997). Our experiences with these ligands are paralleled by those of many other workers, and the use of lectins and antibodies is now a standard feature of brain mapping in humans and nonhuman primates. Karl Zilles and his coworkers have made extensive use of another class of ligands, synthetic radiolabeled receptor-binding substances, in comprehensive human brain mapping studies (for reviews, see Amunts & Zilles, 2001; Eickhoff et al., 2007a; Zilles et al., 2002). This impressive body of work should eventually supplant Brodmann's (Brodmann, 1909) widely employed, but deeply flawed, map of human cortex (Abbott, 2003). The same ligands can be used in comparative studies (e.g., Zilles et al., 1996), although to date they have not been widely employed in this capacity. With the development of new techniques for differentially labeling brain regions have come new quantitative analytic techniques for identifying borders between regions (e.g., Annese et al., 2004; Geyer et al., 1996; Hackett et al., 2001).

Histological techniques have much more to offer to comparative neuroscience than segmenting major brain divisions, as important as that task may be. They also provide clues to phyletic differences at finer levels of organization, of which there are many examples (for reviews, see DeFelipe et al., 2002, 2007; Hof & Sherwood, 2007; Preuss, 1995, 2001; Sherwood & Hof, 2007). We know of differences in the laminar organization and compartmental organization of homologous cortical areas, as exemplified by the modifications of layer 4A of primary visual cortex that occurred in ape and human evolution (Preuss & Coleman, 2002; Preuss et al., 1999) (Fig. 22.4). There are differences in the biochemical phenotypes of homologous cell types: For example, layer 5 pyramidal cells in the cingulate cortex of great apes and humans express the calcium-binding protein calretinin, unlike other primates (Hof et al., 2001). The morphologies of neurons vary across taxa: great apes and humans possess a distinctive class of large, spindle-shaped layer 5 cells (Von Economo cells) in anterior cingulate and fronto-insular cortex; these are especially large and numerous in humans, and are not observed in primates outside the great ape-human clade (Allman et al., 2005; Nimchinsky et al., 1999). Glial cell morphologies also vary across primate taxa (Colombo et al., 2004). The dendritic branching patterns of pyramidal cells differ between taxa (Elston, 2003), as do the geometries of certain classes of axonal terminations (Raghanti et al., 2007, 2008). The sizes of particular types of neurons can vary in ways that are not plausibly explained by differences in brain size, and presumably reflect functional specializations (Sherwood et al., 2003). It is also possible to study the local connectivity of brain regions in fixed tissue, over a distance of several millimeters, using lipophilic dyes (Galuske et al., 2000; Mufson et al., 1990; Tardif & Clarke, 2001), but comprehensive comparative studies have not yet been undertaken with this technique.

One of the most important developments in modern histology is the advent of techniques for accurately estimating numbers and densities of cells. Because the process of sectioning tissue results in fractions of individual cells being present in two or more adjacent tissue sections, merely counting the number of cells in a section and multiplying by the thickness of the section results in the overestimation of cell numbers and densities. This problem of "lost caps" has long been recognized, and various correction factors were developed to mitigate the problem, most of which now appear to be inadequate. Beginning in the 1960s, new and simple techniques have been developed that resolve the lost-caps problem, which, along with efficient sampling schemes, yield unbiased and reliable estimates of the numbers and densities of cells and other small particles. These so-called "design-based stereology" techniques have become the standard for quantitative studies in the neurosciences (Gundersen et al., 1988; Schmitz & Hof, 2005; West, 1993). Thus, it is only very recently that we have had estimates of neocortical neuron number for humans (19–23 billion, varying with age and sex; Pakkenberg & Gundersen, 1997), based on methods that are generally acknowledged to be accurate. Stereological techniques for unbiased estimation of cell numbers have recently been introduced into comparative primate brain studies by Sherwood and colleagues in their studies of glia/neuron ratios in anthropoids (Sherwood et al., 2006) and of the densities of different interneuron types in anthropoid visual cortex (Sherwood et al., 2007). Sherwood and colleagues have also introduced modern phylogenetic regression techniques (independent contrasts; e.g., Garland et al., 1992) to comparative studies relating variations in cell numbers and density to brain size in primates. Simple regressions can be confounded by the presence of different phylogenetic groups, with different scaling relations and different sample sizes, in a common analysis. Independent contrasts provide a means to distinguish within-group and between-group contributions to scaling.

It is noteworthy that most of the studies cited previously were carried out using unperfused tissue collected and fixed postmortem by simple immersion in fixative. Often in these studies, furthermore, there were delays of several hours postmortem before the brain was removed and placed in fixative. This contrasts with the usual procedures employed with experimental animals, which are overdosed with anesthetic and then perfused transcardially with fixative at the point of death, so as to

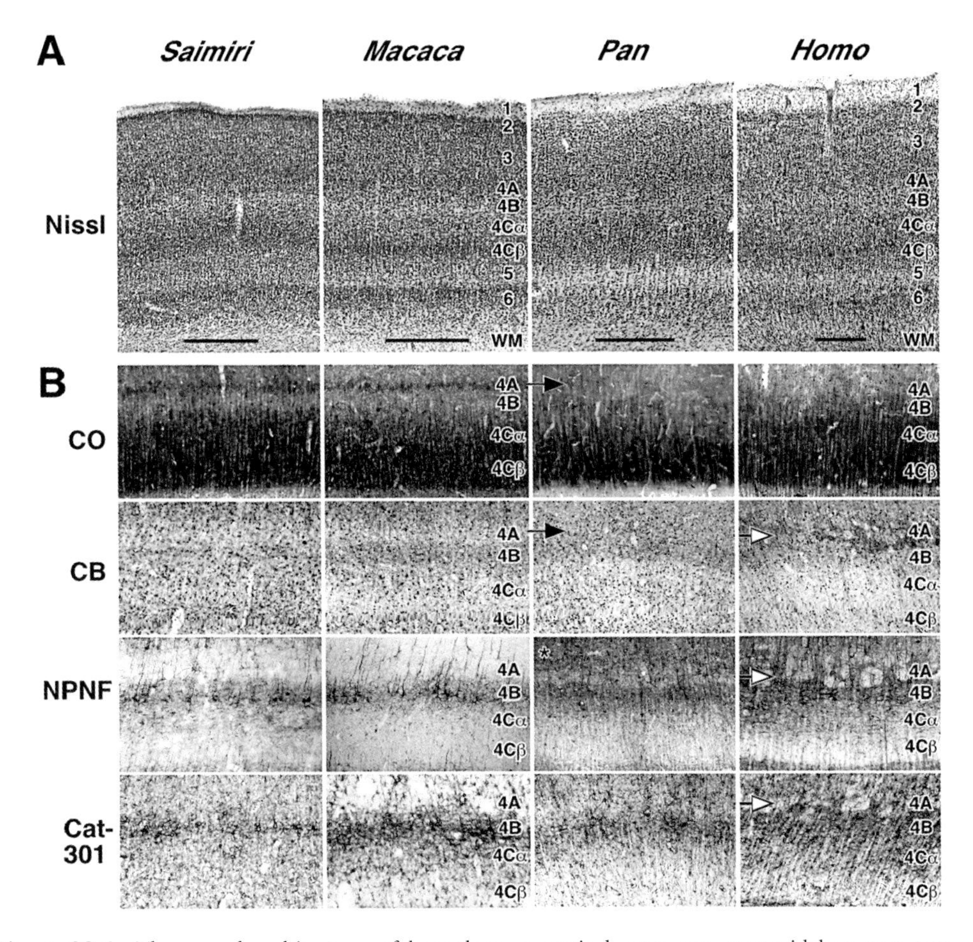

Figure 22.4 The neural architecture of homologous cortical areas can vary widely across taxa, as illustrated by recent studies of the primary visual area (area V1; area 17) of anthropoid primates. As demonstrated by Preuss and Coleman (2002), patterns of evolutionary change in the brain can be reconstructed in considerable detail when appropriate comparative studies are carried out, and these changes include hominoid (ape-human) and human specializations of layer 4A. (**A**) Photomicrographs of Nissl-stained sections from area V1 in four primate species: a New World monkey (*Saimiri sciureus*, squirrel monkey); an Old World monkey (*Macaca mulatta*, rhesus macaque); and two hominoids (*Pan troglodytes*, chimpanzee; *Homo sapiens*, human). These sections span the full thickness of the cortex, from layer 1 to the white matter (WM). Nissl staining does not reveal dramatic differences in the organization of area V1 across anthropoid species. Scale bars are 500 microns. (**B**) Marked differences in the histology of anthropoid V1 can be demonstrated with preparations other than Nissl staining. These panels, which focus on layer 4A, show results using four different procedures: histochemistry for cytochrome oxidase (CO), a mitochondrial enzyme expressed at high levels in metabolically active neurons and neurites; immunocytochemistry for the calbindin (CB), a calcium-binding protein expressed primarily by interneurons; immunocytochemistry for nonphosphorylated neurofilament protein (NPNF), which is expressed strongly by pyramidal cells; and immunocytochemistry with monoclonal antibody Cat-301, which labels an extracellular matrix proteoglycan expressed primarily by interneurons. In the CO panel, note the prominent, thin band of CO staining in layer 4A of *Saimiri* and *Macaca*, which is lacking in *Pan* (denoted by the filled arrowhead) and *Homo*. This difference may reflect evolutionary changes in the distribution of visual inputs to layer 4A (Preuss et al., 1999). In the CB panel, *Saimiri* and *Macaca* exhibit light staining in layer 4A, while *Pan* (filled arrowhead) has stronger labeling. *Homo* has even stronger labeling, and

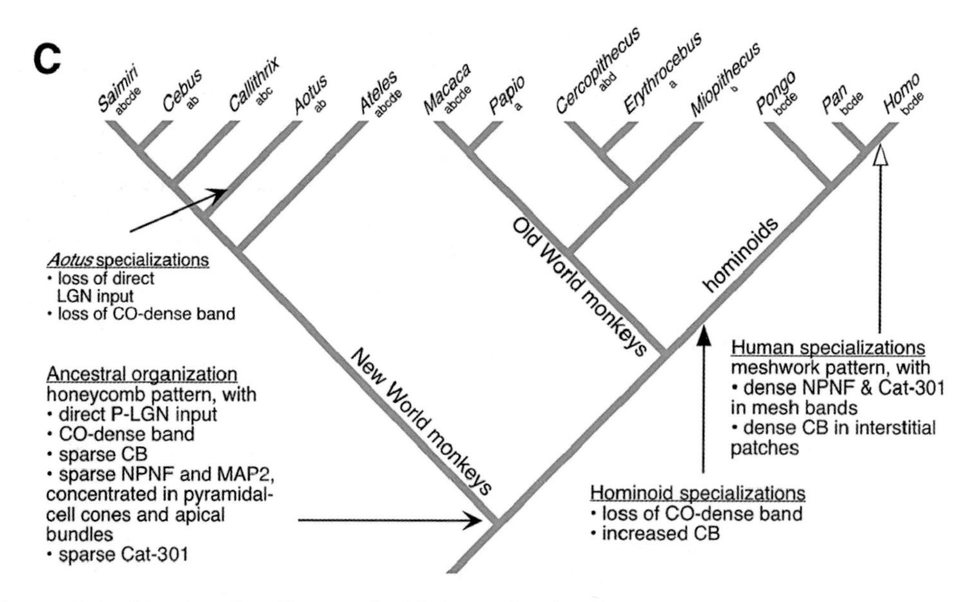

Figure 22.4 (Continued) unlike *Pan*, the labeling is distributed in irregular patches (open arrowhead). NPNF and Cat-301 labeling in *Homo* also shows a patchy pattern, with irregular darkly stained and lightly stained compartments. Double-labeling studies show that NPNF and Cat-301 densely label the same compartments, whereas CB is concentrated in the other set of compartments (Preuss & Coleman, 2002). *Pan* and *Homo* also differ from the monkeys in having dense NPNF labeling of the neuropil in layers 2 and 3 (denoted by the asterisk in *Pan*). The photographs are not to a common scale, each being rescaled as needed to show the four divisions of layer 4A in parallel. **(C)** A reconstruction of layer 4A evolution in anthropoid primates based on published data about connectivity and histology from 13 anthropoid genera. The branching relationships between anthropoids depicted here is from Purvis (1995) and is relatively uncontroversial. Different types of data are presented by an alphabetic code: (a) the presence or absence of direct projections from LGN to 4A, (b) the presence or absence of a CO-dense band, (c) level of expression of calbindin in 4A, (d) pattern of expression of NPNF, and (e) the pattern of expression of Cat-301. Parsimony analysis suggests that ancestral anthropoids possessed an organization of layer 4A similar to that found in modern *Saimiri* and *Macaca*, in which afferents from the parvocellular layers of the lateral geniculate nucleus (P-LGN) terminate directly in layer 4A, which is reflected by the presence of a discrete band of CO labeling in this layer. Cone-shaped clusters of pyramidal cell bodies and their apical dendrites in layer 4B punctuate layer 4A, giving layer 4A a "honeycomb" appearance, when sectioned in the plane of the cortical surface. Expression of CB and of the Cat-301 antigen was probably low in layer 4A of ancestral anthropoids, and expression of NPNF and MAP2 (a microtubule-associated protein strongly expressed in neurons) was concentrated in the cone-like pyramidal cell clusters. Ancestral organization was modified in the evolution of owl monkeys (*Aotus*), which lost the direct P-LGN projections to layer 4A, along with their associated CO-dense band. The most dramatic modifications of the macaque- and squirrel monkey–like pattern of layer 4A took place in hominoid evolution, however. The CO-dense band was lost in hominoid phylogeny prior to the divergence of orangutans (*Pongo*) from the African ape clade; this may reflect the loss of P-LGN inputs (as in *Aotus*). By the same time point, there was increased expression of CB in layer 4A. Additional specializations occurred in the lineage leading to *Homo sapiens*, after its separation from the *Pan* lineage: Humans evolved a distinctive type of tissue compartmentation in layer 4A, in which compartments filled with small cells that stain densely for CB are embedded in a meshwork of neurons and neuropil that express NPNF and the proteoglycan labeled by Cat-301. Figure modified from Preuss, T. M., & Coleman, G. Q. (2002). Human-specific organization of primary visual cortex: Alternating compartments of dense Cat-301 and calbindin immunoreactivity in layer 4A. *Cerebral Cortex, 12,* 671–691.

minimize postmortem tissue degradation and ensure uniform tissue fixation. Certainly such procedures improve tissue quality, and certain molecules (especially RNA and small peptides) may be so labile as to degrade within minutes to hours if the brain is not fixed at death. Yet the quality of tissue obtained and fixed postmortem can be remarkably good for many purposes (Lewis, 2002). For many molecules, the critical issue appears to be overfixation rather than postmortem collection. For decades, the common practice of pathologists and histologists was to store tissue samples in fixative at room temperature. Prolonged exposure to fixative can alter the structure of proteins so that antibodies don't bind to them well. In recent years, a variety of procedures for restoring the antigenicity of tissue (antigen retrieval techniques) have been developed that can greatly improve the staining of tissue that has been stored in fixative for long periods (months to years). Perhaps even more valuable, however, has been the development of cryopreservative solutions containing ethylene glycol—antifreeze, that is (Hoffman & Le, 2004; Watson et al., 1986). Using these solutions, tissue sections, tissue blocks, and even whole brains can be removed from fixative and stored at –20°C in common household-type freezers. Tissue quality is remarkable—antigenicity and even enzyme activity are retained for years (it's not yet clear for exactly how many years). Even relatively labile and fixation-sensitive molecules, like cytochrome oxidase, can be preserved with cryopreservative solution (Preuss & Coleman, 2002). Antigen retrieval and cryopreservation techniques greatly enhance the value of brain tissue archives.

Comparative Genomics and Other Comparative Molecular Approaches

Comparative molecular biology has roots that go back at least as far as the work of Nutall (1904), who identified differences in the immunogenicity of blood serum derived from different species. Recent advances in the technology of nucleic acid sequencing have profoundly transformed this field, however, and provide much new information of value to students of primate brain evolution. There are two sides to comparative molecular research, one focused on understanding the phyletic relationships between species, genes, and proteins, and the other focused on reconstructing the evolutionary transformations of molecules and molecular systems in particular evolutionary lineages. Here, I am concerned principally with the latter. To date, much of the research carried out in this vein has focused on the human lineage, and particularly on molecular changes potentially relevant to human brain evolution, but the same methods can be brought to bear on other primate lineages and other biological systems.

The most conspicuous recent development in comparative molecular biology is the comparative genomics movement. Comparative genomics is the product of the invention of rapid, automated techniques for sequencing DNA, which made possible the first whole-genome sequencing projects involving humans and select model organisms such as mice and fruit flies. As sequencing methods have improved and sequencing capacity has accumulated, the cost of sequencing has come down and more and more species have been added to the list of organisms being "genomed," with varying degrees of accuracy and completeness. As of this writing, there are sequencing projects for 13 primate species, including several strepsirrhines, a tarsier, a New World monkey (*Callithrix jacchus*), several Old World monkeys from the genera *Macaca* and *Papio*, a hylobatid, the Sumatran orang, gorilla, common chimpanzee, and human. In addition, one species each from the orders Scandentia (tree shrews) and Dermoptera (flying lemurs) are being sequenced. Interestingly, the rationale offered for selecting particular species to be genomed includes not only their status as experimental models but also the desire to cover major branches of the phylogenetic tree in order to facilitate the evolutionary reconstruction of gene sequences and genomes (see, e.g., http://genome.wustl.edu/genomes/view/cynocephalus_volans/). The rapid growth in information about gene and protein structure and function, spanning a remarkable variety of organisms, has necessitated the development of computational and Web-based methods for organizing and exploring information, with the result that in silico analyses are now as much a part of molecular biology as in vivo and in vitro studies.

The development of technologies to sequence nucleic acids rapidly and cheaply has had consequences that extend far beyond the mere accumulation of gene sequence data: They have led (directly and indirectly) to remarkable new explorations of virtually every dimension of molecular genetics, including not only protein-coding sequences but also the organization of nontranscribed regions of the DNA that influence gene expression and of the nontranslated small RNAs that regulate how messenger RNAs (mRNAs) are translated into proteins. They have also made it possible to document in exquisite detail how evolution rearranged the coding and noncoding elements in chromosomes, and so augment the work of classical comparative cytogenetics. The information from sequencing projects has also provided the basis for constructing mRNA microarrays that make it possible to identify genes differentially expressed in homologous tissues under different experimental conditions, or in some cases, in homologous tissues of different species.

Rather than review these techniques in detail, I want to consider some recent findings in comparative genomics that require us to reconsider our common understanding of the relationship between genetic and phenotypic evolution, an understanding strongly influenced by the classic paper of King and Wilson (1975). King and Wilson were struck by the very close similarity in the macromolecular sequences of humans and chimpanzees: greater than 99% correspondence in the amino acid sequences of proteins, for example. Yet, it seemed to them there was a very large gap between the phenotypes of humans and chimpanzees, too large to be accounted for by of the small number of changes in protein sequences. To explain this disparity, they proposed that most of the important evolutionary changes in the human lineage involved changes in gene expression; that is, in the amount of mRNA and protein that specific genes produce and in the spatial distribution and developmental timing of that production. Most of these regulatory changes, they argued, involved the rearrangement of genes on chromosomes, resulting in changes in the location of coding genes with respect to regulatory sequences. A small number of rearrangements of the right sorts of genes could yield profound phenotypic changes, transforming a rather chimp-like ancestor into a human.

The sequencing of the human genome laid the foundation for carrying out the research program implicit in the analysis of King and Wilson (1975). Sequence information made it possible to construct gene microarrays (gene chips) that allow one to quantify the amounts of mRNAs present in tissue samples, and thus to identify gene expression differences between humans, chimpanzees, and other primates. The microarrays employed in most of the first studies of this type consisted of probe sequences representing about 10,000 expressed human genes (nearly half the total complement of expressed genes). To date, studies comparing humans and chimpanzees have focused on adult brain tissue (Cáceres et al., 2003; Enard et al., 2002; Gu & Gu, 2003; Khaitovich et al., 2004; Oldham et al., 2006; Uddin et al., 2004; but see also Karaman et al., 2003). The data generated by these different studies show a high level of consistency, and there is a good rate of validation for individual genes when examined with independent techniques (Preuss et al., 2004). Functional analysis of the expression differences is problematic, in part because the functions of many of the genes differentially expressed in the cortex are so poorly understood. However, Cáceres and colleagues (Cáceres et al., 2003; Preuss et al., 2004) noted a higher than expected number of expression changes in genes involved in cell growth and maintenance in the human lineage, and identified a number of genes involved in neuronal or synaptic functions that underwent expression changes in human evolution.

One limitation of the use of microarrays for comparative studies like these is that the microarrays are composed of probe sequences based on human-specific sequences; thus, in instances in which there are species differences in the short sequences used to construct the probe sets, the human mRNA transcripts will bind more strongly to the probes than their nonhuman homologs, creating the false impression of increased gene expression in humans. The sequence differences between humans and chimpanzees are sufficiently small that one can delete from the analysis probes that aren't identical in the two species. The differences between macaques and humans (and chimpanzees) are much greater, however, and one would have to eliminate so many probes as to make analysis impractical. The noise in the outgroup data makes it difficult to determine the direction of character change in the ingroups; thus, we may be able to state with confidence, for example, that gene X is expressed at a higher level in humans than in chimpanzees, but we can be less sure whether the gene was upregulated in human evolution or downregulated in chimpanzee evolution. The new technology of mass resequencing (Bentley, 2006), which can be used to quantify the numbers of mRNA copies present in tissue samples, should be "species agnostic," in the sense that its precision should not depend on the species being studied (provided that adequate sequence information is available), and therefore should be extremely valuable for comparative studies.

The advent of techniques for characterizing phyletic differences in gene expression differences is by any measure a real breakthrough. Parallel developments in genomics, however, also make it clear that although changes in gene regulation are important in evolution, there is more to the evolution of gene expression than chromosomal rearrangements and more to the genetics of species differences than changes in gene expression. Genomes are systems in flux, and numerous gene duplications, insertions, and deletions occurred independently in the human and chimpanzee lineages, as well as in other primate lineages (Eichler et al., 2001; Fortna et al., 2004; Nahon, 2003). Thus, Varki and Altheide (2005) estimated that humans and chimpanzees are ≈96%, rather than 98% to 99%, similar at the nucleotide level. Even that number, however, doesn't really capture the full magnitude of the genetic differences between lineages. The studies of Eichler and colleagues, in particular, have identified large numbers of segmental duplications (resulting in sequences with copy numbers in the range of 2 to 50) in hominoid and human evolution (Bailey & Eichler, 2006). Approximately 33% of the duplicates present in humans are absent in chimpanzees (Cheng et al., 2005). We seem to have systematically underestimated the global genomic differences between species, even such closely related species as humans and chimpanzees (Sikela, 2006). It may be true that the alignable, orthologous gene sequences of humans and chimpanzees are about 98% to 99% similar, but these do not comprise the entire genome.

The gain and loss of gene copies has important implications. When genes that are copied or lost are very similar in structure and function, we might think of phyletic differences in copy number as affecting gene dosage. As has long be understood (Ohno, 1970), however, gene duplication followed by structural divergence is the main mechanism by which evolution has generated new genes—genes that have new functions, or are localized in different tissues or cell compartments, or are expressed at different points in development (Conrad & Antonarakis, 2007). Given the apparent large number of lineage-specific duplications in primate evolution (including hominoid and hominin evolution), we have to take seriously the fact that there are species-specific genes, and take seriously the possibility that these may have played a significant role in human evolution (Fortna et al., 2004; Nahon, 2003; Sikela, 2006). Two gene groups that have attracted interest in this regard are the *morpheus* family (Eichler et al., 2001; Johnson et al., 2001) and the DUF1220 domain (Popesco et al., 2006); not only were these gene segments amplified in human evolution, but also some of their members underwent strong positive selection at the sequence level. DUF1220-containing proteins are expressed strongly (although not exclusively) in the brain (Popesco et al., 2006).

New genes can also arise by means other than duplication and divergence. Hayakawa and colleagues (2005) have identified a human-specific gene, *SIGLEC11*, a segment of which is homologous to part of the chimpanzee *SIGLEC11* gene, while another segment is homologous to part of human *SIGLEC16*, which is a pseudogene. Human *SIGLEC11* is expressed in microglia (Hayakawa et al., 2005), the cells that mediate immune and inflammatory responses in the brain.

Nor do all the new or modified genes code for proteins: Some code for transcribed, but untranslated, RNAs. Pollard and colleagues (2006a), screening for sequences that are highly conserved in vertebrate evolution but that underwent rapid change in the human lineage, identified over 200 elements, most of which are in noncoding DNA and many of which are located near transcription factor and DNA-binding genes, that is, genes that regulate gene expression. The element that showed the highest rate of change in the human lineage, however, codes for part of an expressed RNA, *HAR1F* (Pollard et al., 2006b). Pollard and colleagues demonstrated that *HAR1F* is expressed by Cajal-Retzius cells, which also express a protein, reelin, that plays a central role in organizing the laminar arrangement of neurons in the developing cortex. The secondary structure of *HAR1F* appears to be unlike that of other known expressed RNAs, including the class of small (≈22 nucleotide) RNAs known as microRNAs. MicroRNAs are thought to play an important role in sculpting developmental-stage and tissue-specific patterns of gene expression; in the brain they have been implicated in the control of synapse formation (Klein et al., 2005). As with other classes of genes, there are lineage-specific (including human-specific) microRNAs (Berezikov et al., 2006; Zhang et al., 2007).

It is clear, then, that genetic differences between humans and chimpanzees—animals separated by only ≈6 million years from their common ancestor—are much greater than has commonly been supposed (Cohen, 2007b). Far from being 99% identical, with only a smattering of nucleotide differences in coding and noncoding regions separating us, we now know that humans and chimpanzees differ along many different dimensions of genomic organization, and that there are likely a nontrivial number of human-specific and chimpanzee-specific genes. Moreover, macromolecular diversity isn't limited to the diversity of genes: We know that homologous gene transcripts can be processed to yield different functional proteins by the alternative splicing of different exons (Calarco et al., 2007) and that homologous proteins can undergo different patterns of posttranslational processing and modification that affect their functions (e.g., Brooks, 2004; Gearing et al., 1996; Rosen et al., 2006; Varki, 2006).

The magnitude of macromolecular diversity constitutes something of a conundrum. Under the old view that humans and chimps are 99% identical, the role of comparative genomics seemed to be to identify the relatively few genetic differences that explain the phenotypic differences between humans and chimpanzees. Now, it seems, we've got lots of genetic differences between humans and chimpanzees, involving many different dimensions of the macromolecular organization, and the differences between more distantly related taxa must be even greater. Yet, as discussed previously, we know very little about how human and ape brains differ, or about how hominoid brains differ from those of other primates.

Where, then, are all the human-specific phenotypes? Presumably, they're still out there for us to discover. This suggests a new role for comparative genomics and molecular biology: phenotype discovery (Preuss et al., 2004). In genetics, typically, one identifies a distinct phenotype and then tracks down the gene or genes that contribute to it. Phenotype discovery reverses the process, taking known genetic differences as starting points for idenfying previously unknown phenotypes. One way to do this is to investigate the consequences of a sequence change or expression-level changes in tissues that express the gene in question. For example, using microarrays and polymerase chain reaction to compare humans, chimpanzees, and macaques, my colleagues and I found evidence that the genes for the extracellular matrix proteins *thrombospondin 2* (*THBS2*) and *thrombospondin 4* (*THBS4*) were upregulated in human brain evolution, *THBS4* dramatically so (Cáceres et al., 2007) (Fig. 22.5). Thrombspondins are proteins of special interest to neuroscientists because in model systems they play a critical role in inducing neurons to make synapses (Christopherson et al., 2005; Susman

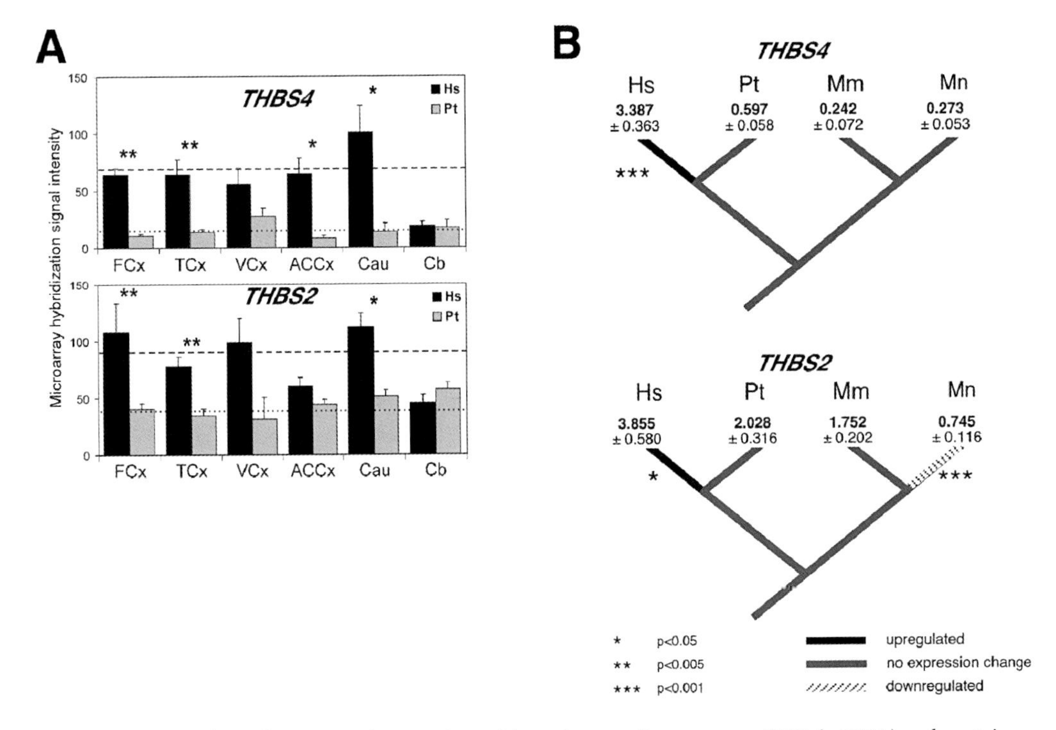

Figure 22.5 Evidence for increased expression of thrombospondin messenger RNA (mRNA) and proteins in human brain evolution. (**A**) Comparative microarray experiments indicate that expression of the genes for *THBS4* and *THBS2* is stronger in forebrain regions of humans (Hs: *Homo sapiens*) than of chimpanzees (Pt: *Pan troglodytes*). Values plotted are average mRNA hybridization intensities and standard errors. Asterisks indicate significant differences at the 0.05 (*) and 0.01 (**) levels. Brain regions surveyed include FCx, frontal cortex; TCx, temporal cortex: VCx, primary visual cortex; ACCx, anterior cingulate cortex; Cau, caudate nucleus; and Cb, cerebellum. (**B**) Comparative reverse transcriptase-polymerase chain reaction (RT-PCR) experiments support the conclusion that expression of the *THBS4* and *THBS2* genes in frontal cortex increased in the human lineage, after the divergence of humans and chimpanzees. Expression values from humans (Hs), chimpanzees (Pt), and two species of macaque monkeys, *Macaca mulatta* (Mm) and *Macaca nemestrina* (Mn), were used to estimate expression levels at all three internal nodes of the phylogeny using linear parsimony and squared-change parsimony techniques. Branches that show significant differences between the observed values for the living species and estimated value for the ancestral node are marked with asterisks. Observed expression values represent the mean number (± SEM) of thrombospondin gene transcripts in the sample per 1,000 β-actin transcripts in the sample. (β-actin serves as a control gene to which results from each sample are normalized.) (**C**) Comparative Western blotting results confirm that the increased expression of thrombospondin mRNA in humans was accompanied by increased protein levels in the cortex of the frontal pole (FP). Samples were compared from three individual humans (Hs), three chimpanzees (Pt), and three macaques (Mm) The density and width of the bands reflects the amount of protein present in a sample. β-Tubulin (TUBB) was used as loading control to normalize for differences in the total amount of protein in the samples. Staining for THBS4 and THBS2 was stronger in the humans than in chimpanzees or macaques, THBS4 very markedly so. (**D**) Immunocytochemistry for THBS4 protein in frontal polar cortex revealed stronger labeling in humans than in chimpanzees or macaques. The lower panel shows labeling through the full thickness of the cortex; the upper panel shows labeling in the upper part of cortical layer 3 at higher magnification. THBS4 antibody labeled numerous pyramidal cell bodies in all species, as well as the neuropil region surrounding the cell bodies. Labeling of the neuropil, which is rich in synapses, was particularly strong in humans. Scale bars: upper panel, 50 μm; lower panel, 250 μm. Figures are modified from Cáceres, M., Suwyn, C., Maddox, M., Thomas, J. W., & Preuss, T. M. (2007). Increased cortical expression of two synaptogenic thrombospondins in human brain evolution. *Cerebral Cortex, 17*, 2312–2321.

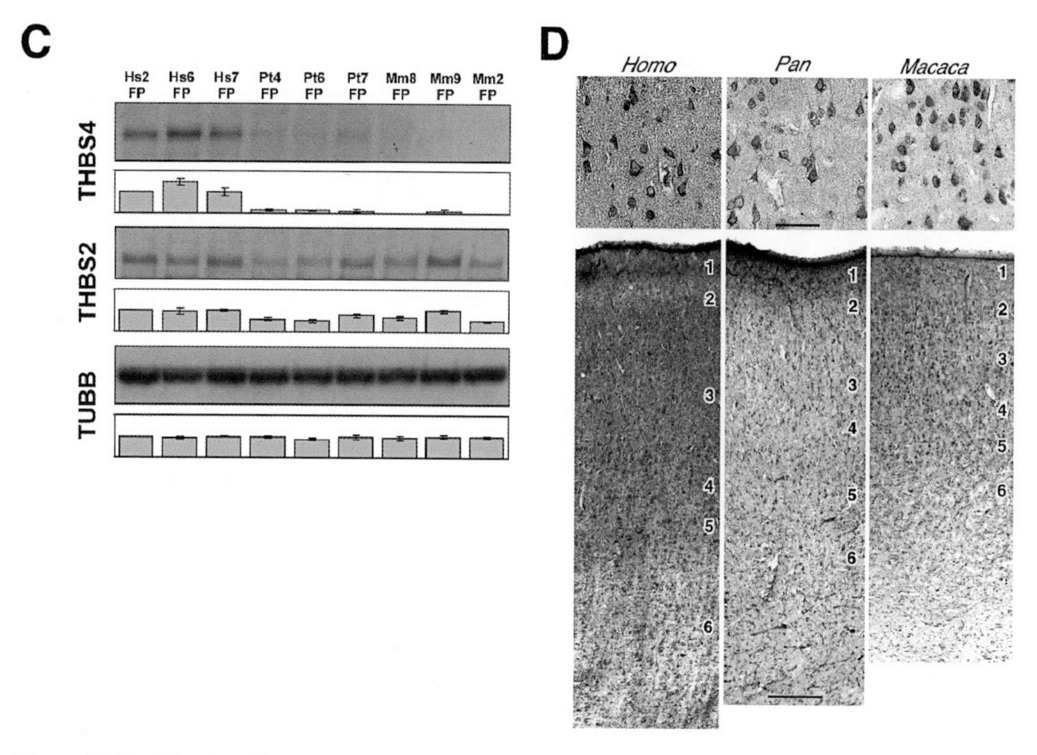

Figure 22.5 (Continued)

et al., 2007). Thrombospondins have other functions, however, and are expressed by a variety of cell types, so the mere fact that thrombospondin mRNA was upregulated in human brain tissue isn't by itself strong evidence humans underwent evolutionary changes in the biochemical control of synapse formation. Moreover, it is possible that the upregulation of thrombospondin mRNA is not matched by an increase in thrombospondin protein. So, we used Western blotting with antibodies to THBS4 and THBS2 proteins to verify that they are, indeed, expressed more strongly in humans than in chimps or macaques. Finally, using immunocytochemistry with antibodies to THBS4, we showed that the most conspicuous difference in the labeling of prefrontal cortex across species was the greater density in humans of labeling in the neuropil space—the territory surrounding neuronal cell bodies that is filled with fine axonal and dendritic processes, and the place where most synapses are located. While we have not yet directly demonstrated that humans modified their propensity to make and unmake synapses, the results justify further explorations of the comparative synaptic biology of humans and nonhuman primates. If the biochemical machinery of synaptogenesis were, indeed, modified during human evolution, we should expect there to have been changes in other genes and proteins involved in this process, and we are currently investigating several additional candidate gene expression changes that may be relevant. A change in the synaptogenic machinery of the cortex should also be manifested at physiological, cognitive, and behavioral levels. What these changes might be are not clear—no one, to my knowledge, has previously proposed that there might be something unusual about human synaptic machinery. One possibility is that evolution modified the substrates of human working memory, which is at the heart of human reasoning capacities and involves circuitry spanning much of the cortex (Baddeley, 2003).

This example illustrates how genomic differences can suggest previously unsuspected phenotypes. It also illustrates how the different techniques discussed in this chapter can be used in combination, in this case, the combination of comparative genomics, molecular biology, and histology. Conceivably, if we were to obtain additional evidence pointing to a human specialization of synaptic mechanisms, we could test our working hypothesis with comparative neuroimaging.

CONCLUSIONS

The new tools of comparative neuroimaging, comparative histology, and comparative molecular biology discussed in this chapter can serve to energize the study of primate brain evolution in the twenty-first century. What makes these tools so valuable is not only that they permit us to explore dimensions of brain organization that were previously out of reach, but also that they make it possible to study species that are otherwise out of reach. This is because they can be used in vivo, or make use of material acquired using noninvasive or minimally invasive procedures in living individuals or from samples acquired postmortem. As a result, not only can they tap the resources of primate research centers, they can also make available for study the more varied collections of species present in zoos and wildlife parks, greatly enhancing the value of these resources. I don't mean to imply that the development of low-impact technologies renders invasive and terminal methods irrelevant or obsolete: We would be very hard pressed to sense of DTI or fcMRI results without the results of experimental tract-tracing and physiological recording studies in macaques, squirrel monkeys, owl monkeys, and galagos. Nevertheless, the new technologies offer new opportunities for the growth of neuroscientific understanding and the more complete integration of neuroscience with the other behavioral and biological sciences, provided that we muster the appropriate resources.

What are those appropriate resources? With so many of the mammalian and primate species crucial for understanding primate brain evolution facing extinction, neuroscientists need to make a commitment to species preservation. This means not only conservation in the wild but also the provision for captive animal facilities: Both are essential. I take it there's no need to explain to biologists why species conservation in the wild is important, but I do think it's important to emphasize how important captive facilities are for experimental behavioral, cognitive, and neuroimaging studies. Simply put, the opportunities for doing properly controlled experiments in the wild are extremely limited compared to those in captive situations. We would never take seriously theories of human psychology or brain function that are not based on well-controlled experiments: Why would we suppose that we can do without them when trying to understand other species? If we think that captive situations are too artificial to yield valid results, we should work to make them more naturalistic. We need a better interaction between field and captive studies and scientists with experience in both contexts. Certainly, the impending loss of access to chimpanzees for even the kinds of low-impact investigations that we routinely carry out in humans will have a catastrophic effect on our ability to understand human brain organization, and we need to do everything in our power to ensure that future generations of scientists will be able to study these unique and remarkable animals (Preuss, 2006).

As the foregoing discussion has made clear, tissue collections are potentially of enormous value for understanding primate brain evolution. Existing collections include the Comparative Mammalian Brain Collection (www.brainmuseum.org), consisting of material originally assembled by W. Welker at the University of Wisconsin and J.I. Johnston, Jr., at Michigan State University, and currently housed in the National Museum of Health and Medicine. There is also the Comparative Neurobiology of Aging Resource, organized by J.M. Erwin as part of the Great Ape Aging Project, which serves as a clearinghouse for great ape brain tissue. These efforts are extremely valuable, but we need to do more. Faced as we are with the extinction of so many primate species, we must make every reasonable effort to collect tissue from as many individuals of as many primate species as possible, and ensure that specimens are properly curated and made available to qualified investigators. We need to identify and catalog brain material held in ad hoc collections. Fixed brains from as many species as possible should be scanned with structural MRI and DTI protocols, and the data placed in a permanent database of primate brain morphology and connectivity, accessible online. MRI scans of diverse fish species are currently being collected under the auspices of the Digital Fish Library (www.digitalfishlibrary.org), under the sponsorship of the National Science Foundation—we should be able to do at least as much with primates.

As essential as these resources and technologies are to primate neuroscience, they will be of little value unless we have researchers who are expert in both neurobiology and evolutionary biology. We need to recognize a distinct professional category—evolutionary neuroscience—just as 25 years ago we recognized molecular evolution as a distinct field, and provide fellowships and stipends to support students in this discipline, as well as dedicated funding for comparative brain research.

We need to recognize this special class of scientists to bridge the gap between the two very ways of conceptualizing the role of animal studies in biological research. It's not too much of a stretch to say that there are currently two biologies—one that takes phyletic diversity seriously and one that doesn't. Those of us who would characterize ourselves as neuroethologists or primatologists or zoologists are likely to be familiar with the principles of modern evolution biology. As modern evolutionists, most of us would disavow those views of evolution that were so popular in the nineteenth and early twentieth centuries, specifically, that evolution takes the form of a phylogenetic scale, with humans at the top (Lovejoy, 1964; Richards, 1987), or the idea that the multitude of species are really just minor variants of a few basic animal "types" (Desmond, 1982; Richards, 1992). Evolutionary biologists today take the view that evolution is a branching tree, and that all living species, including humans, are the products of equally long evolutionary histories. Moreover, we understand that each species is a mosaic of characteristics: some shared with many other species by virtue of ancient common ancestry, some shared with a narrower range of species by virtue of more recent common ancestry, and some that are unique to the species, the products of recent evolutionary history. If we want to know what features of organization are widespread among mammals, we understand that we must study a variety of different mammalian species to reconstruct ancestral organization. Similarly, if we want to know what features of organization constitute the defining specializations of a particular species, we understand that we need to compare the target species to other species, and that the most informative comparisons are with closely related species. If the species in question is *Homo sapiens*, the most informative species are chimpanzees and the other great apes. No one with even the most rudimentary training in modern evolutionary biology would presume that one could take results from a few rodent species and expect them, as a rule, to generalize to all mammals, or take results from a few rodent or even primate species or expect them to extend with high fidelity to humans.

Yet it is upon such boggy ground that the enterprise of modern neuroscience largely rests. Most branches of neuroscience have adopted the model-animal paradigm, an approach that assumes that results obtained in a few model species will be broadly applicable to other species. Neuroscientists expend enormous effort to ensure that their experimental results have *internal* validity—that is, that their results are reliable or repeatable under the same set of conditions—but often pay little more than lip service to *external* validity, or generalizability. If you believe that rodents are truly archetypical mammals, then external validity becomes a non-issue (Logan, 1999, 2002, 2005). But if you understand that species vary, that mammalian brains are not just smaller or larger versions of a common brain type, you have to take external validity seriously. And the way to take external validity seriously is to do comparative studies.

I'm not suggesting that we stop studying rodents or rhesus monkeys. Rather, I'm suggesting that neuroscientists need to study those species and many more, and do so within a rigorous comparative framework. Only in this way can we make well-founded claims about "basic" features of brain organization. It is also the only way we can make room for the serious consideration of the ways species and groups of species differ from one another, such as, for example, how primates differ from other animals or how humans differ from other primates.

Perhaps it is too much to hope that the neurosciences can be reconciled with modern evolutionary biology. Consider, however, the case of comparative genomics. One need only browse through the Entrez Gene Web site (www.ncbi.nlm.nih.gov/sites/entrez), sponsored by the National Center for Biotechnology Information (NCBI; a branch of NIH), to appreciate the extent to which genomics and related branches of molecular science have been evolutionized. At the core of the NCBI database is the tree of life (www.ncbi.nlm.nih.gov/sites/entrez?db=taxonomy), and although the taxonomy browser warns us not to regard it as a substitute for the primary taxonomic literature on phyletic relationships, the effort is remarkably comprehensive and thoroughly grounded in modern phylogenetics. Data for thousands of genes and gene domains are presented, all in appropriate phylogenetic context (Benson et al., 2000; Wheeler et al., 2000).

Genomics didn't start out with a strong commitment to evolutionary diversity. The roots of genomics are in experimental genetic studies of fruit flies and mice, animals that were aggressively promoted as stand-ins for animals generally (Kohler, 1994; Rader, 2004). Genetics, however, has never been saddled with quite as small a number of model species as the neurosciences, and as sequence data for different species began to accumulate, the differences between organisms became a fact that couldn't be ignored, if for no other reason than that the differences had to be dealt with systematically in order to correctly identify the commonalities. Genomics thus embraced the modern evolutionary concepts and methods that make sense of the diversity of life.

The lesson of genomics is that comparative biology is good biology, and that studying many species is better than studying a few species. Of course, we can't study every species in great depth, and I don't doubt that obtaining useful neurobiological information from any given species typically requires more cost and effort than obtaining useful genetic information from that same species. But neuroscience can no longer afford to live outside the scope of comparative biology and to be hamstrung by the concentration of effort and resources on a few "model" species. It is not merely desirable or intellectually pleasing that neuroscience anchor itself with the core principles that guide the rest of biology—it is essential. It's a question of good biology versus bad biology: whether we're going to make room for phyletic variations in brain organization—including the variations that make humans human—or whether we're going to content ourselves with the fiction that "brains are brains."

ACKNOWLEDGMENTS

The author is grateful to acknowledge the support of the James S. McDonnell Foundation (JSMF 21002093), the National Aging Institute (NIH P01AG026423), and the Yerkes Base Grant (NIH RR-00165).

REFERENCES

Abbott, A. (2003). Neuroscience: A new atlas of the brain. *Nature, 424,* 249–250.

Allman, J. M. (1977). Evolution of the visual system in the early primates. In: J. M. Sprague & A. N. Epstein (Eds.), *7*(pp. 1–53). New York: Academic.

Allman, J. M., Watson, K. K., Tetreault, N. A., & Hakeem, A. Y. (2005). Intuition and autism: A possible role for Von Economo neurons. *Trends in Cognitive Science, 9,* 367–373.

Amunts, K., & Zilles, K. (2001). Advances in cytoarchitectonic mapping of the human cerebral cortex. *Neuroimaging Clinics of North America, 11,* 151–169, vii.

Annese, J., Pitiot, A., Dinov, I. D., & Toga, A. W. (2004). A myelo-architectonic method for the structural classification of cortical areas. *Neuroimage, 21,* 15–26.

Baddeley, A. (2003). Working memory: Looking back and looking forward. *Nature Reviews Neuroscience, 4,* 829–839.

Bailey, J. A., & Eichler, E. E. (2006). Primate segmental duplications: Crucibles of evolution, diversity and disease. *Nature Reviews Genetics, 7,* 552–564.

Balter, M. (2007). Neuroanatomy. Brain evolution studies go micro. *Science, 315,* 1208–1211.

Behrens, T. E., & Johansen-Berg, H. (2005). Relating connectional architecture to grey matter function using diffusion imaging. *Philosophical Transactions of the Royal Society of London Series B: Biological Science, 360,* 903–911.

Benson, D. A., Karsch-Mizrachi, I., Lipman, D. J., Ostell, J., Rapp, B. A., & Wheeler, D. L. (2000). GenBank. *Nucleic Acids Research, 28,* 15–18.

Bentley, D. R. (2006). Whole-genome re-sequencing. *Current Opinions in Genetics and Development, 16,* 545–552.

Berezikov, E., Thuemmler, F., van Laake, L. W., Kondova, I., Bontrop, R., Cuppen, E., et al. (2006). Diversity of microRNAs in human and chimpanzee brain. *Nature Genetics, 38,* 1375–1377.

Biswal, B., Yetkin, F. Z., Haughton, V. M., & Hyde, J. S. (1995). Functional connectivity in the motor cortex of resting human brain using echo-planar MRI. *Magnetic Resonance in Medicine, 34,* 537–541.

Brodmann, K. (1909). *Vergleichende Lokalisationslehre der Grosshirnrhinde.* Leipzig: Barth. (Reprinted as *Brodmann's Localisation in the cerebral cortex,* translated and edited by L. J. Garey, London: Smith-Gordon, 1994).

Brooks, S. A. (2004). Appropriate glycosylation of recombinant proteins for human use: Implications of choice of expression system. *Molecular Biotechnology, 28,* 241–255.

Cáceres, M., Lachuer, J., Zapala, M. A., Redmond, J., Kudo, L., Geschwind, D., et al. (2003). Elevated gene expression levels distinguish human from non-human primate brains. *Proceedings of the National Academy of Sciences, 100,* 1330–1335.

Cáceres, M., Suwyn, C., Maddox, M., Thomas, J. W., & Preuss, T. M. (2007). Increased cortical expression of two synaptogenic thrombospondins in human brain evolution. *Cerebral Cortex, 17,* 2312–2321.

Calarco, J. A., Xing, Y., Caceres, M., Calarco, J. P., Xiao, X., Pan, Q., et al. (2007). Global analysis of alternative splicing differences between humans and chimpanzees. *Genes and Development, 21,* 2963–2975.

Casagrande, V. A., & Kaas, J. H. (1994). The afferent, intrinsic, and efferent connections of primary visual cortex in primates. In: A. Peters & K. Rockland (Eds.), (pp. 201–259). New York: Plenum.

Casagrande, V. A., Khaytin, I., & Boyd, J. (2007). The evolution of parallel visual pathways in the brains of primates. In: J. H. Kaas & T. M. Preuss (Eds.), *Evolution of nervous systems. Vol. 4: Primates* (pp. 87–108). Oxford: Elsevier.

Cheng, Z., Ventura, M., She, X., Khaitovich, P., Graves, T., Osoegawa, K., et al. (2005). A genome-wide comparison of recent chimpanzee and human segmental duplications. *Nature, 437,* 88–93.

Christopherson, K. S., Ullian, E. M., Stokes, C. C., Mullowney, C. E., Hell, J. W., Agah, A., et al. (2005). Thrombospondins are astrocyte-secreted proteins that promote CNS synaptogenesis. *Cell, 120,* 421–433.

Clare, S., & Bridge, H. (2005). Methodological issues relating to in vivo cortical myelography using MRI. *Human Brain Mapping, 26,* 240–250.

Clark, V. P., Courchesne, E., & Grafe, M. (1992). In vivo myeloarchitectonic analysis of human striate and extrastriate cortex using magnetic resonance imaging. *Cerebral Cortex, 2,* 417–424.

Cohen, J. (2007a). Animal studies. NIH to end chimp breeding for research. *Science, 316,* 1265.

Cohen, J. (2007b). Evolutionary biology. Relative differences: The myth of 1%. *Science, 316,* 1836.

Cola, M. G., Seltzer, B., Preuss, T. M., & Cusick, C. G. (2005). Neurochemical organization of chimpanzee inferior pulvinar complex. *Journal of Comparative Neurology, 484,* 299–312.

Colombo, J. A., Sherwood, C. C., & Hof, P. R. (2004). Interlaminar astroglial processes in the cerebral cortex of great apes. *Anatomy and Embryology (Berlin), 208,* 215–218.

Conrad, B., & Antonarakis, S. E. (2007). Gene duplication: A drive for phenotypic diversity and cause of human disease. *Annual Review of Genomics and Human Genetics, 8,* 17–35.

Creutzfeldt, O. D. (1977). Generality of functional structure of the neocortex. *Naturwiss Naturwissenschaften, 64,* 507–517.

Crick, F., & Jones, E. (1993). Backwardness of human neuroanatomy. *Nature, 361,* 109–110.

Croxson, P. L., Johansen-Berg, H., Behrens, T. E., Robson, M. D., Pinsk, M. A., Gross, C. G., et al.

(2005). Quantitative investigation of connections of the prefrontal cortex in the human and macaque using probabilistic diffusion tractography. *Journal of Neuroscience, 25,* 8854–8866.

D'Arceuil, H. E., Westmoreland, S., & de Crespigny, A. J. (2007). An approach to high resolution diffusion tensor imaging in fixed primate brain. *Neuroimage, 35,* 553–565.

DeFelipe, J., Alonso-Nanclares, L., & Arellano, J. I. (2002). Microstructure of the neocortex: Comparative aspects. *Journal of Neurocytology, 31,* 299–316.

DeFelipe, J., Alonso-Nanclares, L., Arellano, J., Ballestero-Yáñez, I., Benevides-Piccione, R., & Muñoz, A. (2007). Specializations of the cortical microstructure of humans. In: J. H. Kaas & T. M. Preuss (Eds.), *Evolution of nervous systems. Vol. 4: Primates* (pp. 168–190). Oxford: Elsevier.

Desmond, A. J. (1982). *Archetypes and ancestors: Palaeontology in Victorian London, 1850 1875.* London: Blond & Briggs.

Diamond, I. T., & Hall, W. C. (1969). Evolution of neocortex. *Science, 164,* 251–262.

Eichler, E. E., Johnson, M. E., Alkan, C., Tuzun, E., Sahinalp, C., Misceo, D., et al. (2001). Divergent origins and concerted expansion of two segmental duplications on chromosome 16. *Journal of Heredity, 92,* 462–468.

Eickhoff, S. B., Paus, T., Caspers, S., Grosbras, M. H., Evans, A. C., Zilles, K., et al. (2007b). Assignment of functional activations to probabilistic cytoarchitectonic areas revisited. *Neuroimage, 36,* 511–521.

Eickhoff, S. B., Schleicher, A., Scheperjans, F., Palomero-Gallagher, N., & Zilles, K. (2007a). Analysis of neurotransmitter receptor distribution patterns in the cerebral cortex. *Neuroimage, 34,* 1317–1330.

Eickhoff, S. B., Stephan, K. E., Mohlberg, H., Grefkes, C., Fink, G. R., Amunts, K., et al. (2005b). A new SPM toolbox for combining probabilistic cytoarchitectonic maps and functional imaging data. *Neuroimage, 25,* 1325–1335.

Eickhoff, S., Walters, N. B., Schleicher, A., Kril, J., Egan, G. F., Zilles, K., et al. (2005a). High-resolution MRI reflects myeloarchitecture and cytoarchitecture of human cerebral cortex. *Human Brain Mapping, 24,* 206–215.

Elston, G. N. (2003). Cortex, cognition and the cell: New insights into the pyramidal neuron and prefrontal function. *Cerebral Cortex, 13,* 1124–1138.

Elston, G. N. (2007). Specializations of the neocortical pyramidal cell during primate evolution. In: J. H. Kaas & T. M. Preuss (Eds.), *Evolution of nervous systems. Vol. 4: Primates* (pp. 191–242). Oxford: Elsevier.

Enard, W., Khaitovich, P., Klose, J., Zollner, S., Heissig, F., Giavalisco, P., et al. (2002). Intra- and interspecific variation in primate gene expression patterns. *Science, 296,* 340–343.

Fortna, A., Kim, Y., MacLaren, E., Marshall, K., Hahn, G., Meltesen, L., Brenton, M., et al. (2004). Lineage-specific gene duplication and loss in human and great ape evolution. *PLoS Biology, 2,* E207.

Fox, M. D., Snyder, A. Z., Vincent, J. L., Corbetta, M., Van Essen, D. C., & Raichle, M. E. (2005). The human brain is intrinsically organized into dynamic, anticorrelated functional networks. *Proceedings of the National Academy of Sciences, 102,* 9673–9678.

Galuske, R. A., Schlote, W., Bratzke, H., & Singer, W. (2000). Interhemispheric asymmetries of the modular structure in human temporal cortex. *Science, 289,* 1946–1949.

Garland, T., Harvey, P., & Ives, A. (1992). Procedures for the analysis of comparative data using phylogenetically independent contrasts. *Systematic Biology, 41,* 18–32.

Gearing, M., Tigges, J., Mori, H., & Mirra, S. S. (1996). Abeta40 is a major form of beta-amyloid in nonhuman primates. *Neurobiology and Aging, 17,* 903–908.

Geyer, S., Ledberg, A., Schleicher, A., Kinomura, S., Schormann, T., Burgel, U., et al. (1996). Two different areas within the primary motor cortex of man. *Nature, 382,* 805–807.

Gilissen, E. (2001). Structural symmetries and asymmetries in human and chimpanzee brains. In: 'D. Falk & K. R. Gibson (Eds.), *Evolutionary anatomy of the primate cerebral cortex* (pp. 187–215). Cambridge: Cambridge University Press.

Gu, J., & Gu, X. (2003). Induced gene expression in human brain after the split from chimpanzee. *Trends in Genetics, 19,* 63–65.

Guilfoyle, D. N., Helpern, J. A., & Lim, K. O. (2003). Diffusion tensor imaging in fixed brain tissue at 7.0 T. *NMR in Biomedicine, 16,* 77–81.

Gundersen, H. J., Bagger, P., Bendtsen, T. F., Evans, S. M., Korbo, L., Marcussen, N., et al. (1988). The new stereological tools: Disector, fractionator, nucleator and point sampled intercepts and their use in pathological research and diagnosis. *Apmis, 96,* 857–881.

Hackett, T. A., Preuss, T. M., & Kaas, J. H. (2001). Architectonic identification of the core region in auditory cortex of macaques, chimpanzees, and humans. *Journal of Comparative Neurology, 441,* 197–222.

Hampson, M., Peterson, B. S., Skudlarski, P., Gatenby, J. C., & Gore, J. C. (2002). Detection of functional connectivity using temporal correlations in MR images. *Human Brain Mapping, 15,* 247–262.

Harting, J. K., Glendenning, K. K., Diamond, I. T., & Hall, W. C. (1973). Evolution of the primate visual system: Anterograde degeneration studies of the tecto-pulvinar system. *American Journal of Physical Anthropology, 38,* 383–392.

Harting, J. K., Hall, W. C., & Diamond, I. T. (1972). Evolution of the pulvinar. *Brain, Behavior, and Evolution, 6,* 424–452.

Hayakawa, T., Angata, T., Lewis, A. L., Mikkelsen, T. S., Varki, N. M., & Varki, A. (2005). A human-specific gene in microglia. *Science, 309,* 1693.

Hecht, E., Gutman, D., Glasser, M., Mascaro, J., Hamann, S., Preuss, T. M., et al. (2008). Comparing amygdala connectivity between monkeys, apes, and humans using diffusion tensor imaging [Abstract]. *American Journal of Physical Anthropology.*

Hinkley, L., Padberg, J., Krubitzer, L., & Disbrow, E. (2007). Hand use and the evolution of posterior parietal cortex in primates. In: J. H. Kaas & T. M. Preuss (Eds.), *Evolution of nervous systems. Vol. 4: Primates* (pp. 407–415). Oxford: Elsevier.

Hof, P. R., Nimchinsky, E. A., Perl, D. P., & Erwin, J. M. (2001). An unusual population of pyramidal neurons in the anterior cingulate cortex of hominids contains the calcium-binding protein calretinin. *Neuroscience Letters, 307,* 139–142.

Hof, P. R., & Sherwood, C. C. (2007). The evolution of neuron classes in the neocortex of mammals. In: J. H. Kaas & L. A. Krubitzer (Eds.), *Evolution of nervous systems: A comprehensive reference. Vol. 3: Mammals* (pp. 113–124). Amsterdam: Elsevier.

Hoffman, G. E., & Le, W. W. (2004). Just cool it! Cryoprotectant anti-freeze in immuno-cytochemistry and in situ hybridization. *Peptides, 25,* 425–431.

Holloway, R. L., Jr. (1966). Cranial capacity, neural reorganization, and hominid evolution: A search for more suitable parameters. *American Anthropologist, 68,* 103–121.

Hopkins, W. D., Marino, L., Rilling, J. K., & MacGregor, L. A. (1998). Planum temporale asymmetries in great apes as revealed by magnetic resonance imaging (MRI). *Neuroreport, 9,* 2913–2918.

Howell, L. L., Hoffman, J. M., Votaw, J. R., Landrum, A. M., & Jordan, J. F. (2001). An apparatus and behavioral training protocol to conduct positron emission tomography (PET). Neuroimaging in conscious rhesus monkeys. *Journal of Neuroscience Methods, 106,* 161–169.

Jerison, H. J. (1973). *Evolution of the brain and intelligence.* New York: Academic Press.

Johnson, M. E., Viggiano, L., Bailey, J. A., Abdul-Rauf, M., Goodwin, G., Rocchi, M., et al. (2001). Positive selection of a gene family during the emergence of humans and African apes. *Nature, 413,* 514–519.

Kaas, J. H. (1987). The organization of neocortex in mammals: Implications for theories of brain function. *Annual Review of Psychology, 38,* 129–151.

Kaas, J. H. (2007). The evolution of sensory and motor systems in primates. In: J. H. Kaas & T. M. Preuss (Eds.), *Evolution of nervous systems. Vol. 4: Primates* (pp. 35–57). Oxford: Elsevier.

Kaas, J. H., & Preuss, T. M. (1993). Archontan affinities as reflected in the visual system. In: F. S. Szalay, M. J. Novacek & M. C. McKenna (Eds .), (pp. 115–128). New York: Springer Verlag.

Karaman, M. W., Houck, M. L., Chemnick, L. G., Nagpal, S., Chawannakul, D., Sudano, D., et al. (2003). Comparative analysis of gene-expression patterns in human and African great ape cultured fibroblasts. *Genome Research, 13,* 1619–1630.

Kaufman, J. A., Ahrens, E. T., Laidlaw, D. H., Zhang, S., & Allman, J. M. (2005). Anatomical analysis of an aye-aye brain (*Daubentonia madagascariensis,* primates: Prosimii) combining histology, structural magnetic resonance imaging, and diffusion-tensor imaging. *Anatomical Records Part A: Discovery of Molecular, Cellular, and Evolutionary Biology, 287,* 1026–1037.

Khaitovich, P., Muetzel, B., She, X., Lachmann, M., Hellmann, I., Dietzsch, J., et al. (2004). Regional patterns of gene expression in human and chimpanzee brains. *Genome Research, 14,* 1462–1473.

King, M. C., & Wilson, A. C. (1975). Evolution at two levels in humans and chimpanzees. *Science, 188,* 107–116.

Klein, M. E., Impey, S., & Goodman, R. H. (2005). Role reversal: The regulation of neuronal gene expression by microRNAs. *Current Opinions in Neurobiology, 15,* 507–513.

Kohler, R. E. (1994). *Lords of the fly: Drosophila genetics and the experimental life.* Chicago: University of Chicago Press.

Lashley, K. S. (1931). Mass action in cerebral function. *Science, 73,* 245–254.

Lashley, K. S. (1949). Persistent problems in the evolution of mind. *Quarterly Review of Biology, 24,* 28–42.

Le Gros Clark, W. E. (1959). *The antecedents of man.* Edinburgh: Edinburgh University Press.

Lewis, D. A. (2002). The human brain revisited: Opportunities and challenges in postmortem studies of psychiatric disorders. *Neuropsychopharmacology, 26,* 143–151.

Logan, C. A. (1999). The altered rationale for the choice of a standard animal in experimental psychology: Henry H. Donaldson, Adolf Meyer and 'the' albino rat. *History of Psychology, 2,* 3–24.

Logan, C. A. (2002). Before there were standards: The role of test animals in the production of empirical generality in physiology. *Journal of the History of Biology, 35,* 329–363.

Logan, C. A. (2005). The legacy of Adolf Meyer's comparative approach: Worcester rats and the strange birth of the animal model. *Integrative Physiological and Behavioral Science, 40,* 169–181.

Logothetis, N. K., Guggenberger, H., Peled, S., & Pauls, J. (1999). Functional imaging of the monkey brain. *Nature Neuroscience, 2,* 555–562.

Lovejoy, A. O. (1964). *The great chain of being: A study of the history of an idea.* Cambridge: Harvard University Press.

Lowe, M. J., Mock, B. J., & Sorenson, J. A. (1998). Functional connectivity in single and multislice echoplanar imaging using resting-state fluctuations. *Neuroimage, 7,* 119–132.

MacLeod, C. E., Zilles, K., Schleicher, A., Rilling, J. K., & Gibson, K. R. (2003). Expansion of the neocerebellum in Hominoidea. *Journal of Human Evolution, 44,* 401–429.

Marino, L., Rilling, J. K., Lin, S. K., & Ridgway, S. H. (2000). Relative volume of the cerebellum in dolphins and comparison with anthropoid primates. *Brain, Behavior, and Evolution, 56,* 204–211.

Mori, S., & Zhang, J. (2006). Principles of diffusion tensor imaging and its applications to basic neuroscience research. *Neuron, 51,* 527–539.

Mufson, E. J., Brady, D. R., & Kordower, J. H. (1990). Tracing neuronal connections in postmortem human hippocampal complex with the carbocyanine dye DiI. *Neurobiology and Aging, 11,* 649–653.

Murphy, W. J., Eizirik, E., O'Brien, S. J., Madsen, O., Scally, M., Douady, C. J., et al. (2001). Resolution of the early placental mammal radiation using Bayesian phylogenetics. *Science, 294,* 2348–2351.

Nahon, J. L. (2003). Birth of 'human-specific' genes during primate evolution. *Genetica, 118,* 193–208.

Nimchinsky, E. A., Gilissen, E., Allman, J. M., Perl, D. P., Erwin, J. M., & Hof, P. R. (1999). A neuronal morphologic type unique to humans and great apes. *Proceedings of the National Academy of Sciences, 96,* 5268–5273.

Nuttall, G. (1904). *Blood immunity and blood relationships.* Cambridge: Cambridge University Press.

Ohno, S. (1970). *Evolution by gene duplication.* Berlin, New York: Springer-Verlag.

Oldham, M. C., Horvath, S., & Geschwind, D. H. (2006). Conservation and evolution of gene coexpression networks in human and chimpanzee brains. *Proceedings of the National Academy of Sciences, 103,* 17973–17978.

Padberg, J., Franca, J. G., Cooke, D. F., Soares, J. G., Rosa, M. G., Fiorani, M., Jr., et al. (2007). Parallel evolution of cortical areas involved in skilled hand use. *Journal of Neuroscience, 27,* 10106–10115.

Pakkenberg, B., & Gundersen, H. J. (1997). Neocortical neuron number in humans: Effect of sex and age. *Journal of Comparative Neurology, 384,* 312–320.

Peltier, S. J., Kerssens, C., Hamann, S. B., Sebel, P. S., Byas-Smith, M., & Hu, X. (2005). Functional connectivity changes with concentration of sevoflurane anesthesia. *Neuroreport, 16,* 285–288.

Pollard, K. S., Salama, S. R., King, B., Kern, A. D., Dreszer, T., Katzman, S., et al. (2006a). Forces shaping the fastest evolving regions in the human genome. *PLoS Genetics, 2,* e168.

Pollard, K. S., Salama, S. R., Lambert, N., Lambot, M. A., Coppens, S., Pedersen, J. S., et al.

(2006b). An RNA gene expressed during cortical development evolved rapidly in humans. *Nature, 443,* 167–172.

Popesco, M. C., Maclaren, E. J., Hopkins, J., Dumas, L., Cox, M., Meltesen, L., et al. (2006). Human lineage-specific amplification, selection, and neuronal expression of DUF1220 domains. *Science, 313,* 1304–1307.

Preuss, T. M. (1995). The argument from animals to humans in cognitive neuroscience. In: M. S. Gazzaniga (Ed.), *The cognitive neurosciences* (pp. 1227–1241). Cambridge, MA: MIT Press. (Reprinted in M. S. Gazzaniga [Ed.], *Cognitive neuroscience: A reader.* Oxford: Blackwell Publishers, 2000.)

Preuss, T. M. (2000a). Taking the measure of diversity: Comparative alternatives to the model-animal paradigm in cortical neuroscience. *Brain, Behavior, and Evolution, 55,* 287–299.

Preuss, T. M. (2000b). What's human about the human brain? In M. S. Gazzaniga (Ed.), *The new cognitive neurosciences. Second edition* (pp. 1219–1234). Cambridge, MA: MIT Press.

Preuss, T. M. (2001). The discovery of cerebral diversity: An unwelcome scientific revolution. In: D. Falk & K. Gibson (Eds.), *Evolutionary anatomy of the primate cerebral cortex* (pp. 138–164). Cambridge: Cambridge University Press.

Preuss, T. M. (2004). Specializations of the human visual system: The monkey model meets human reality. In: J. H. Kaas & C. E. Collins (Eds.), *The primate visual system* (pp. 231–259). Boca Raton, FL: CRC Press.

Preuss, T. M. (2006). Who's afraid of *Homo sapiens*? *Journal of Biomedical Discovery and Collaboration, 1,* 17.

Preuss, T. M. (2007a). Evolutionary specializations of primate brain systems. In: M. J. Ravosa & M. Dagosto (Eds.), *Primate origins: Evolution and adaptations* (pp. 625–675). New York: Springer.

Preuss, T. M. (2007b). Primate brain evolution in phylogenetic context. In: J. H. Kaas & T. M. Preuss (Eds.), *Evolution of nervous systems. Vol. 4: Primates* (pp. 1–34). Oxford: Elsevier.

Preuss, T. M., Caceres, M., Oldham, M. C., & Geschwind, D. H. (2004). Human brain evolution: Insights from microarrays. *Nature Review Genetics, 5,* 850–860.

Preuss, T. M., & Coleman, G. Q. (2002). Human-specific organization of primary visual cortex: Alternating compartments of dense Cat-301 and calbindin immunoreactivity in layer 4A. *Cerebral Cortex, 12,* 671–691.

Preuss, T. M., Gray, D., & Cusick, C. G. (1998) Subdivisions of the motor and somatosensory thalamus of primates revealed with Wisteria floribunda agglutinin histochemistry. *Somatosensory and Motor Research, 15,* 211–219.

Preuss, T. M., Qi, H., & Kaas, J. H. (1999). Distinctive compartmental organization of human primary visual cortex. *Proceedings of the National Academy of Sciences, 96,* 11601–11606.

Preuss, T. M., Stepniewska, I., Jain, N., & Kaas, J. H. (1997). Multiple divisions of macaque precentral motor cortex identified with neurofilament antibody SMI-32. *Brain Research, 767,* 148–153.

Purvis, A. (1995). A composite estimate of primate phylogeny. *Philosophical Transactions of the Royal Society of London Series B: Biological Science, 348,* 405–421.

Rader, K. A. (2004). *Making mice: Standardizing animals for American biomedical research, 1900–1955.* Princeton, NJ: Princeton University Press.

Raghanti, M. A., Stimpson, C. D., Marcinkiewicz, J. L., Erwin, J. M., Hof, P. R., & Sherwood, C. C. (2007). Differences in cortical serotonergic innervation among humans, chimpanzees, and macaque monkeys: A comparative study. *Cerebral Cortex .*

Raghanti, M. A., Stimpson, C. D., Marcinkiewicz, J. L., Erwin, J. M., Hof, P. R., & Sherwood, C. C. (2008). Cholinergic innervation of the frontal cortex: Differences among humans, chimpanzees, and macaque monkeys. *Journal of Comparative Neurology, 506,* 409–424.

Ramnani, N., Behrens, T. E., Johansen-Berg, H., Richter, M. C., Pinsk, M. A., Andersson, J. L., et al. (2006). The evolution of prefrontal inputs to the cortico-pontine system: Diffusion imaging evidence from Macaque monkeys and humans. *Cerebral Cortex, 16,* 811–818.

Ramnani, N., Behrens, T. E., Penny, W., & Matthews, P. M. (2004). New approaches for exploring anatomical and functional connectivity in the human brain. *Biological Psychiatry, 56,* 613–619.

Richards, R. J. (1987). *Darwin and the emergence of evolutionary theories of mind and behavior.* Chicago: University of Chicago Press.

Richards, R. J. (1992). *The meaning of evolution.* Chicago: University of Chicago Press.

Rilling, J. K. (2006). Human and nonhuman primate brains: Are they allometrically scaled versions of the same design? *Evolution Anthropology, 15*, 65–77.

Rilling, J. K. (2007). The evolution of the cerebellum in anthropoid primates. In: J. H. Kaas & T. M. Preuss (Eds.), *Evolution of nervous systems. Vol. 4: Primates* (pp. 149–156). Oxford: Elsevier.

Rilling, J. K., Barks, S. K., Parr, L. A., Preuss, T. M., Faber, T. L., Pagnoni, G., et al. (2007). A comparison of resting-state brain activity in humans and chimpanzees. *Proceedings of the National Academy of Sciences, 104*, 17146–17151.

Rilling, J. K., Glasser, M. F., Preuss, T. M., Ma, X., Zhao, T., Hu, X., et al. (2008). The evolution of the arcuate fasciculus revealed with comparative DTI. *Nature Neuroscience, 11*, 426–428.

Rilling, J. K., & Insel, T. R. (1998). Evolution of the cerebellum in primates: Differences in relative volume among monkeys, apes and humans. *Brain, Behavior, and Evolution, 52*, 308–314.

Rilling, J. K., & Insel, T. R. (1999a). Differential expansion of neural projection systems in primate brain evolution. *Neuroreport, 10*, 1453–1459.

Rilling, J. K., & Insel, T. R. (1999b). The primate neocortex in comparative perspective using magnetic resonance imaging. *Journal of Human Evolution, 37*, 191–223.

Rilling, J. K., & Seligman, R. A. (2002). A quantitative morphometric comparative analysis of the primate temporal lobe. *Journal of Human Evolution, 42*, 505–533.

Rilling, J. K., Winslow, J. T., & Kilts, C. D. (2004). The neural correlates of mate competition in dominant male rhesus macaques. *Biological Psychiatry, 56*, 364–375.

Rilling, J. K., Winslow, J. T., O'Brien, D., Gutman, D. A., Hoffman, J. M., & Kilts, C. D. (2001). Neural correlates of maternal separation in rhesus monkeys. *Biological Psychiatry, 49*, 146–157.

Rockel, A. J., Hiorns, R. W., & Powell, T. P. S. (1980). The basic uniformity of structure of the neocortex. *Brain, 103*, 221–224.

Rosa, M. G., & Krubitzer, L. A. (1999). The evolution of visual cortex: Where is V2? *Trends in Neuroscience, 22*, 242–248.

Rosen, R. F., Levine, H., III, Murphy, M. P., Preuss, T. M., Ghiso, J. A., Farberg, A., et al. (2006). Cerebral Aβ accumulation in Alzheimer's disease, aged humans, and aged nonhuman primates. International Conference on Alzheimer's Disease and Related Disorders (ICAD), Abstract, Presentation P2-0005

Ross, C. F., & Martin, R. D. (2007). The role of vision in the origin and evolution of primates. In: J. H. Kaas & T. M. Preuss (Eds.), *Evolution of nervous systems. Vol. 4: Primates* (pp. 60–78). Oxford: Elsevier.

Schenker, N. M., Desgouttes, A. M., & Semendeferi, K. (2005). Neural connectivity and cortical substrates of cognition in hominoids. *Journal of Human Evolution, 49*, 547–569.

Schmitz, C., & Hof, P. R. (2005). Design-based stereology in neuroscience. *Neuroscience, 130*, 813–831.

Schoenemann, P. T., Sheehan, M. J., & Glotzer, L. D. (2005). Prefrontal white matter volume is disproportionately larger in humans than in other primates. *Nature Neuroscience, 8*, 242–252.

Seeley, W. W., Menon, V., Schatzberg, A. F., Keller, J., Glover, G. H., Kenna, H., et al. (2007). Dissociable intrinsic connectivity networks for salience processing and executive control. *Journal of Neuroscience, 27*, 2349–2356.

Semendeferi, K., Damasio, H., Frank, R., & Van Hoesen, G. W. (1997). The evolution of the frontal lobes: A volumetric analysis based on three-dimensional reconstructions of magnetic resonance scans of human and ape brains. *Journal of Human Evolution, 32*, 375–388.

Semendeferi, K., Lu, A., Schenker, N., & Damasio, H. (2002). Humans and great apes share a large frontal cortex. *Nature Neuroscience, 5*, 272–276.

Sherwood, C., & Hof, P. (2007). The evolution of neuron types and cortical histology in apes and humans. In: J. Kaas & T. Preuss (Eds.), *Evolution of nervous systems: A comprehensive reference. Volume 4: Primates* (pp. 355–378). Amsterdam: Elsevier.

Sherwood, C. C., Lee, P. W., Rivara, C. B., Holloway, R. L., Gilissen, E. P., Simmons, R. M., et al. (2003). Evolution of specialized pyramidal neurons in primate visual and motor cortex. *Brain, Behavior, and Evolution, 61*, 28–44.

Sherwood, C. C., Raghanti, M. A., Stimpson, C. D., Bonar, C. J., de Sousa, A. A., Preuss, T. M., et al. (2007). Scaling of inhibitory interneurons in areas v1 and v2 of anthropoid primates as revealed by calcium-binding protein immunohistochemistry. *Brain, Behavior, and Evolution, 69,* 176–195.

Sherwood, C. C., Stimpson, C. D., Raghanti, M. A., Wildman, D. E., Uddin, M., Grossman, L. I., et al. (2006). Evolution of increased glia-neuron ratios in the human frontal cortex. *Proceedings of the National Academy of Sciences, 103,* 13606–13611.

Sikela, J. M. (2006). The jewels of our genome: The search for the genomic changes underlying the evolutionarily unique capacities of the human brain. *PLoS Genetics, 2,* e80.

Stepniewska, I., Preuss, T. M., & Kaas, J. H. (1993). Architectonics, somatotopic organization, and ipsilateral cortical connections of the primary motor area (M1) of owl monkeys. *Journal of Comparative Neurology, 330,* 238–271.

Sun, S. W., Neil, J. J., & Song, S. K. (2003). Relative indices of water diffusion anisotropy are equivalent in live and formalin-fixed mouse brains. *Magnetic Resonance in Medicine, 50,* 743–748.

Susman, M. W., Eroglu, C., Chakraborty, C., Hubermann, A. D., Green, E. M., Annis, D., et al. (2007). Identification of the thrombospondin receptor that promotes CNS synaptogenesis. *Society of Neuroscience,* Abstract 571.2.

Tardif, E., & Clarke, S. (2001). Intrinsic connectivity of human auditory areas: A tracing study with DiI. Eur *Journal of Neuroscience, 13,* 1045–1050.

Toga, A. W., Thompson, P. M., Mori, S., Amunts, K., & Zilles, K. (2006). Towards multimodal atlases of the human brain. *Nature Reviews Neuroscience, 7,* 952–966.

Tomassini, V., Jbabdi, S., Klein, J. C., Behrens, T. E., Pozzilli, C., Matthews, P. M., et al. (2007). Diffusion-weighted imaging tractography-based parcellation of the human lateral premotor cortex identifies dorsal and ventral subregions with anatomical and functional specializations. *Journal of Neuroscience, 27,* 10259–10269.

Tootell, R. B., Hamilton, S. L., & Silverman, M. S. (1985). Topography of cytochrome oxidase activity in owl monkey cortex. *Journal of Neuroscience, 5,* 2786–2800.

Uddin M, Wildman DE, Liu G, Xu W, Johnson RM, Hof PR, et al. (2004). Sister grouping of chimpanzees and humans as revealed by genome-wide phylogenetic analysis of brain gene expression profiles. *Proceedings of the National Academy of Sciences, 101,* 2957–2962.

Varki, A. (2006). Nothing in glycobiology makes sense, except in the light of evolution. *Cell, 126,* 841–845.

Varki, A., & Altheide, T. K. (2005). Comparing the human and chimpanzee genomes: Searching for needles in a haystack. *Genome Research, 15,* 1746–1758.

Vincent, J. L., Patel, G. H., Fox, M. D., Snyder, A. Z., Baker, J. T., Van Essen, D. C., et al. (2007). Intrinsic functional architecture in the anaesthetized monkey brain. *Nature, 447,* 83–86.

Vogt, C., & Vogt, O. (1919). Allgemeine Ergebnisse unserer Hirnforschung. *Journal of Psychology and Neurology, 25,* 221–473.

Walters, N. B., Egan, G. F., Kril, J. J., Kean, M., Waley, P., Jenkinson, M., et al. (2003). In vivo identification of human cortical areas using high-resolution MRI: An approach to cerebral structure-function correlation. *Proceedings of the National Academy of Sciences, 100,* 2981–2986.

Watson, R. E., Jr., Wiegand, S. J., Clough, R. W., & Hoffman, G. E. (1986). Use of cryoprotectant to maintain long-term peptide immunoreactivity and tissue morphology. *Peptides, 7,* 155–159.

West, M. J. (1993). New stereological methods for counting neurons. *Neurobiology and Aging, 14,* 275–285.

Wheeler, D. L., Chappey, C., Lash, A. E., Leipe, D. D., Madden, T. L., Schuler, G. D., et al. (2000). Database resources of the National Center for Biotechnology Information. *Nucleic Acids Research, 28,* 10–14.

Wise, S. P. (2007). Evolution of ventral premotor cortex and the primate way of reaching. In: J. H. Kaas & T. M. Preuss (Eds.), *Evolution of nervous systems. Vol. 4: Primates* (pp. 157–165). Oxford: Elsevier.

Xiong, J., Parsons, L. M., Gao, J. H., & Fox, P. T. (1999). Interregional connectivity to primary motor cortex revealed using MRI resting state images. *Human Brain Mapping, 8,* 151–156.

Zhang, R., Peng, Y., Wang, W., & Su, B. (2007). Rapid evolution of an X-linked microRNA cluster in primates. *Genome Research, 17,* 612–617.

Zilles, K., Schlaug, G., Geyer, S., Luppino, G., Matelli, M., Qu, M., et al. (1996). Anatomy and transmitter receptors of the supplementary motor areas in the human and nonhuman primate brain. *Advances in Neurology, 70,* 29–43.

Zilles, K., Schleicher, A., Palomero-Gallagher, N., & Amunts, K. (2002). Quantitative analysis of cyto- and receptor architecture of the human brain. In: J. Mazziotta & A. Toga (Eds.), *Brain mapping. The methods.* (pp. 573–602). Oxford: Elsevier.

CHAPTER 23

Ethologically Relevant Movements Mapped on the Motor Cortex

Michael S. A. Graziano

DOES MOTOR CORTEX DECOMPOSE MOVEMENT INTO MUSCLES, OR INTEGRATE MUSCLES INTO MOVEMENTS?

The discovery of motor cortex is one of the landmark events in neuroscience. In 1870, Fritsch and Hitzig found that brief pulses of electrical stimulation applied to the cortex of a dog brain evoked muscle twitches. The twitch-evoking sites were arranged across the surface of the cortex to form a rough map of the body (Fig. 23.1). This map of the body was studied further in monkeys (e.g., Beevor & Horsely, 1887; Ferrier, 1874). Almost 70 years after the initial discovery of motor cortex, Penfield published his now famous study of the human motor cortex (Penfield & Boldrey, 1937), confirming that the human brain also contained a topographic map of body parts (Fig. 23.2). Although a map the body could be discerned in the motor cortex, the map was not cleanly segregated into separate muscles or rotations of separate joints. Instead, the topography was blurred and overlapped. This intermixing of muscles within the motor cortex map was reported in every major study. As shown in Figure 23.2, however, Penfield's iconic homunculus appeared to show a clean, segment-by-segment map of the body. Perhaps Penfield's summary diagram inadvertently contributed to the now widespread and inaccurate idea of a simple map of body parts laid out on the motor cortex.

The hypothesis of a segregated map of movement components was taken to its extreme by Asanuma and colleagues in their studies of the cat and monkey cortex (e.g., Asanuma & Rosen, 1972; Asanuma & Sakata, 1967; Asanuma & Ward, 1971). Rather than stimulating the surface of the cortex with large electrodes, Asanuma and colleagues inserted a microelectrode into the cortex and stimulated through its exposed tip. The volume of brain tissue directly affected by stimulation was probably less than half a millimeter in diameter. The stimulation current was reduced to its threshold, the level at which the evoked movement was barely detectable. Using this method, Asanuma and colleagues reported that each stimulation site in cortex was connected to a single muscle or in some cases two muscles. They argued for the presence of a segregated mapping of muscles on the cortex (Asanuma, 1975).

An alternative explanation exists for the results of Asanuma and colleagues. In what could be termed the "iceberg" interpretation, each site in cortex contributes to a complex movement that recruits many muscles that cross many joints. As part of the coordination of that movement, some muscles are more active than others. By lowering the electrical stimulation to threshold, by the definition of "threshold" the evoked movement is reduced until most of it is no longer detectable and only the one or two most strongly recruited muscles are still detectable above the noise.

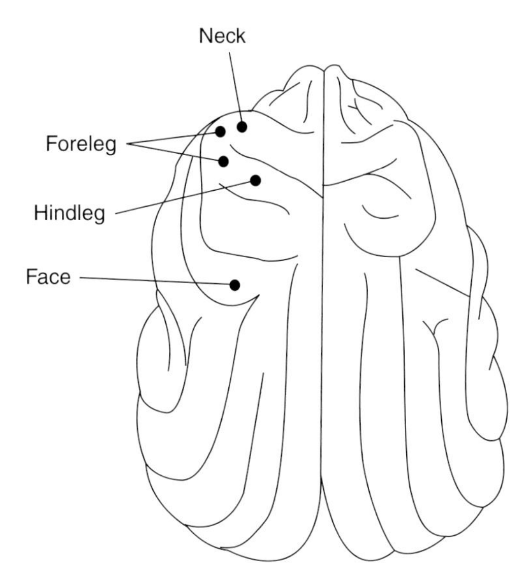

Figure 23.1 The map of stimulation-evoked movements in a dog brain. Each point indicates the approximate location of a movement center. Stimulation at or near each point evoked movements of the indicated body part. Adapted from Fritsch, G., & Hitzig, E. (1870). Uber die elektrishe Erregbarkeit des Grosshirns. Arch. f. Anat., Physiol und wissenchaftl. Mediz., Leipzig, 300–332. [On the electrical excitability of the cerebrum. Translated by G. von Bonin. In: W. W. Nowinski (Ed.), *Some papers on the cerebral cortex* (pp. 73–96). Springfield, IL: Thomas.]Used with permission.

Only the tip of the movement iceberg is measured. In this interpretation, the use of threshold stimulation by its very definition produces the misleading illusion of a muscle-by-muscle map.

Are single muscles, or combinations of muscles, controlled by the motor cortex map? The question was answered unambiguously by Cheney and Fetz (1985). They recorded the activity of neurons in the motor cortex of monkeys and the activity of muscles in the arm and hand. When a neuron in cortex fired an action potential, after a latency of approximately 5 ms, a minute effect could be observed at the muscles. By averaging the data over thousands of neuronal spikes, a clear signal could be obtained. This method of spike-triggered averaging allowed the experimenters to determine which muscles were directly affected by the firing of a neuron in cortex. The results showed that each cortical neuron could affect the activity of many muscles crossing many joints. The motor cortex did not contain a muscle-by-muscle map as suggested by Asanuma and colleagues, but instead integrated the action of muscles. Each locus in cortex, and even each neuron in cortex, affected a complex set of muscles.

One reasonable hypothesis based on the results of Cheney and Fetz is that the overlapping representation of muscles in cortex reflects the overlapping use of muscles in normal movement. In support of this hypothesis, Nudo and colleagues (1996) found that when a monkey practices a task that requires the coordinated use of two arm joints, the motor cortex develops a larger overlap in the representation of the muscles that actuate those joints. Martin and colleagues further explored the role of experience in the development of an overlapping somatotopy (Chakrabarty & Martin, 2000; Martin et al., 2005). They used microstimulation to map the motor cortex in cats. They found that at birth, the representation in motor cortex was mainly nonoverlapping. Separate joints of the forelimb were represented in discrete patches in cortex. During development, as the kitten learned to perform complex behaviors that required coordination among joints, the representations in

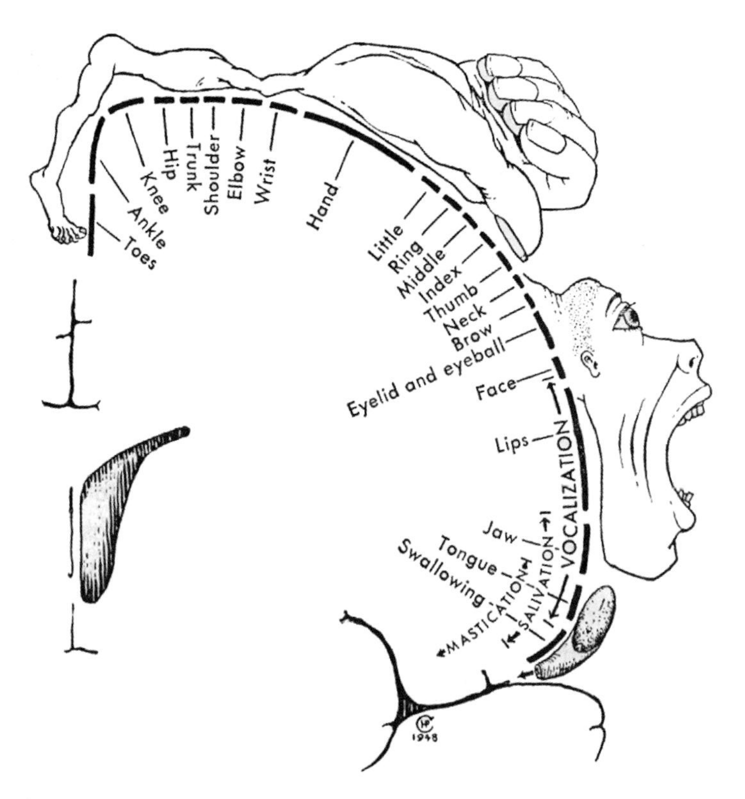

Figure 23.2 The motor homunculus of the human brain. Each point in motor cortex was electrically stimulated and the evoked muscle twitch was noted. Although each cortical point could activate many muscles, a rough body plan could be discerned. Adapted from Penfield, W., & Rasmussen, T. (1950). *The cerebral cortex of man. A clinical study of localization of function.* New York: Macmillan.

cortex developed the highly overlapped property characteristic of the adult. Individual joints were no longer typically represented in separate patches. If the kitten was prevented from practicing complex, integrated movements, the motor map did not develop the normal overlap of representations.

These results suggest that during experience the motor cortex is trained on, and comes to reflect, the movement repertoire of the animal. If an animal has a need to control individual muscles (if such an unlikely condition ever exists), the animal might well develop a motor cortex map that topographically separates the muscles. In the more common case that an animal has a need to control many muscles and joints in a coordinated fashion, such as for reaching toward an object or manipulating an object, its motor cortex develops a topography in which the relevant muscles are represented in an integrated fashion.

Recently we electrically stimulated the motor cortex in monkeys and found the often confirmed, overlapping map of muscles (Cooke & Graziano, 2004a,b; Graziano et al., 2002, 2005). However, we then extended the electrical stimulation to a duration that was behaviorally relevant. We stimulated for half a second at a time, approximating the duration of a monkey's reaching or grasping, instead of stimulating for the more typical duration of less than 50 ms. Neurons in motor cortex are not normally active in 50-ms bursts but instead, to a first approximation, are active throughout the duration of a movement. This use of electrical stimulation on a behavioral time scale is a standard method used to study brain areas outside of motor cortex. It is responsible for fundamental results such as the mapping of eye movement areas of the brain, the discovery of emotional and motivational functions of the hypothalamus, and the demonstration that cortical sensory processing causes sensory perception (e.g., Bruce et al. 1985; Caggiula & Hoebel, 1966; Hess, 1957; Hoebel, 1969; King & Hoebel, 1968; Robinson, 1972; Robinson & Fuchs, 1969; Salzman et al., 1990; Schiller & Stryker, 1972). Within motor cortex research, however, the technique of stimulation on a behaviorally relevant time scale had not been systematically explored.

Using the longer stimulation in motor cortex, we found that the muscle twitches unfolded into complex actions. Stimulation caused the monkey to perform movements that were common in its normal repertoire. Some of these movements are drawn in Figure 23.3, traced from video footage. They included ethologically relevant behaviors such as closing the hand in a grip while bringing

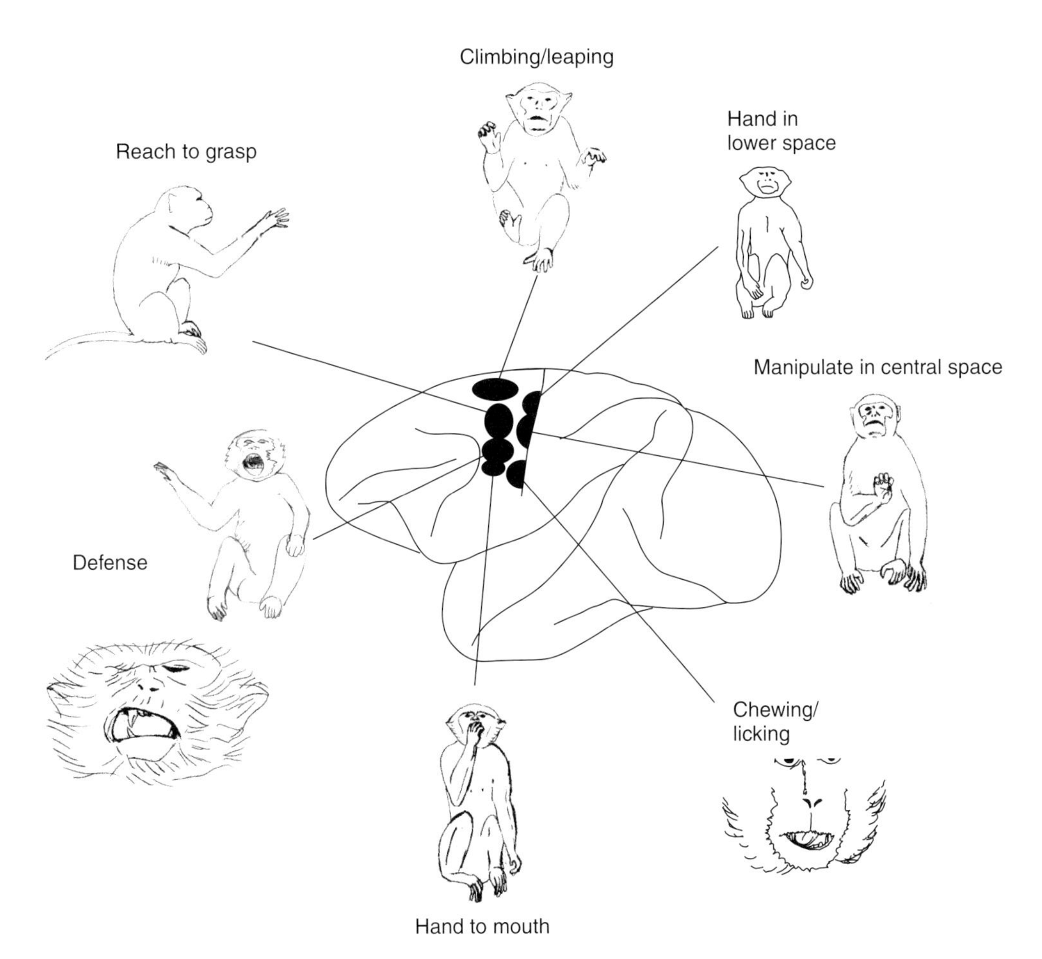

Figure 23.3 Action zones in the motor cortex of the monkey. Seven common categories of movement evoked by electrical stimulation of the cortex on the behaviorally relevant time scale of 0.5 seconds. Images traced from video frames. Each image represents the final posture obtained at the end of the stimulation-evoked movement. Within each action zone, movements of similar behavioral category were evoked. Based on results from Graziano, M. S. A., Taylor, C. S. R., & Moore, T. (2002). Complex movements evoked by microstimulation of precentral cortex. *Neuron, 34,* 841–851; and Graziano, M. S. A., Aflalo, T., & Cooke, D. F.. (2005). Arm movements evoked by electrical stimulation in the motor cortex of monkeys. *Journal of Neurophysiology, 94,* 4209–4223. Used with permission.

the hand to the mouth and opening the mouth; extending the hand away from the body with the grip opened as if in preparation to grasp an object; bringing the hand inward to a region just in front of the chest while shaping the fingers, as if to manipulate an object; squinting the facial muscles while turning the head sharply to one side and flinging up the arm, as if to protect the face from an impending impact; and moving all four limbs as if leaping or climbing. The behavioral repertoire of the animal seemed to be rendered onto the cortical sheet.

The evoked movements were also roughly arranged across the cortex according to the location in space to which the movement was directed. The height of the hand was most clearly mapped across the cortical surface. Stimulation of the lower (ventral) regions of cortex commonly drove the hand into upper space, and stimulation of upper (dorsal) regions of cortex commonly drove the hand into lower space (Fig. 23.4). Again, an important aspect of the animal's action repertoire was mapped across the cortex.

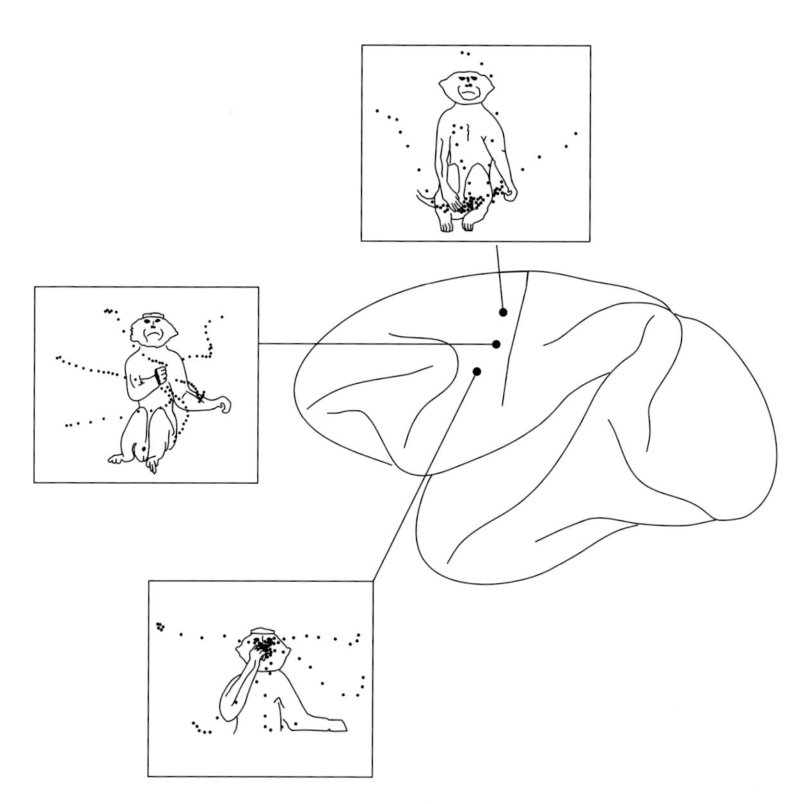

Figure 23.4 Progression of spatial locations to which hand movements are directed. Within the arm representation of the monkey motor cortex, electrical stimulation in dorsal cortex tended to drive the hand into lower space; stimulation in ventral cortex tended to drive the hand into upper space; and stimulation in intermediate cortical locations tended to drive the hand to intermediate heights. Each image is a tracing of the final posture obtained at the end of a stimulation-evoked movement. Each dotted line shows the trajectory of the hand during the 0.5-second stimulation train. Dots show the position of the hand in 30-ms increments. These trajectories show the convergence of the hand from disparate starting locations toward a final location. Based on results from Graziano, M. S. A., Taylor, C. S. R., & Moore, T. (2002). Complex movements evoked by microstimulation of precentral cortex. *Neuron, 34,* 841–851; and Graziano, M. S. A., Aflalo, T., & Cooke, D. F.. (2005). Arm movements evoked by electrical stimulation in the motor cortex of monkeys. *Journal of Neurophysiology, 94,* 4209–4223. Used with permission.

Just how closely the stimulation-evoked movements mimic natural movements is still uncertain. The muscles are activated in combinations that apparently mimic normal behavior. However, the temporal pattern of muscle activity is time-locked to the stimulation train and often does not have the smooth, bell-shaped modulation of normal movement (Cooke & Graziano, 2004a; Taylor et al., 2002). It is therefore probably not correct to think of the stimulation as triggering a complete and natural motor plan. Rather, the motor cortex network appears to have learned the statistics of natural behavior, and the electrical stimulation reveals that statistical structure and the manner in which different zones of cortex emphasize different parts of the movement repertoire.

DIMENSIONALITY REDUCTION AS A THEORY OF TOPOGRAPHIC ORGANIZATION

As described previously, a traditional view of the motor cortex is that it contains a map of the body. This map was famously depicted by Penfield (Fig. 23.2). The traditional topographic scheme, however, does not capture the actual pattern of overlaps, fractures, re-representations, and multiple areas separated by fuzzy borders. A commonly accepted parcellation of motor cortex into functionally different areas is shown in Figure 23.5 (e.g., Dum & Strick, 2002; He et al., 1995; Luppino et al., 1991; Matelli et al., 1985; Matsuzaka et al., 1992; Preuss et al., 1996; Rizzolatti & Luppino, 2001). This scheme, however, also suffers from the problem of overlap. Although the motor cortex is clearly heterogeneous, the changes across the cortical sheet are more graded than absolute. A clean division into separate areas with separate functions is almost certainly incorrect, especially since the specific functions supposedly localized to these distinct areas are largely unknown and the functions that have been tested tend to be found to some degree in all motor areas tested.

Based on our stimulation results, we proposed a different underlying topographic principle: the reduction of the many-dimensional space of the animal's movement repertoire onto the two-dimensional surface of the cortex

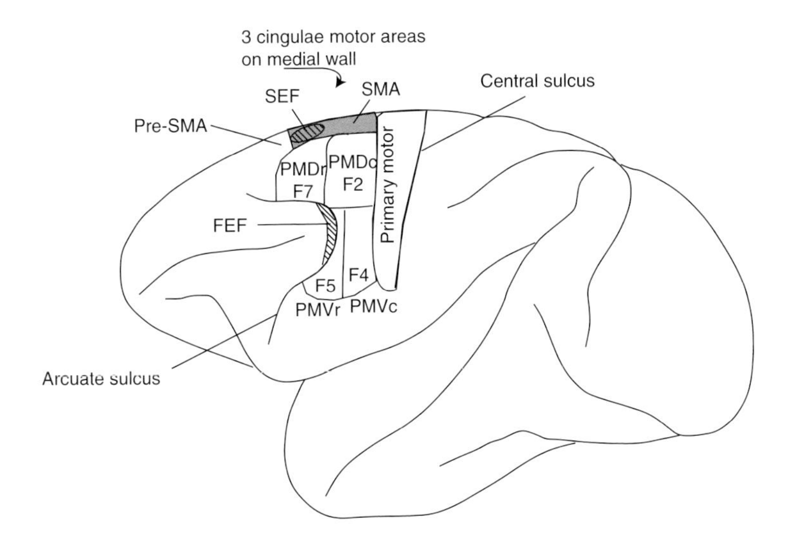

Figure 23.5 Some commonly accepted divisions of the cortical motor system of the monkey. PMDr, dorsal premotor cortex, rostral division, also sometimes called Field 7 (F7); PMDc, dorsal premotor cortex, caudal division, also sometimes called Field 2 (F2); PMVr, ventral premotor cortex, rostral division, also sometimes called Field 5 (F5); PMVc, ventral premotor cortex, caudal division, also sometimes called Field 4 (F4); SMA, supplementary motor area; SEF, supplementary eye field, a part of SMA; pre-SMA, presupplementary motor area; FEF, frontal eye field.

(Aflalo & Graziano, 2006). In this principle, local continuity is optimized. Information processors that need to interact are arranged physically near each other in cortex, presumably gaining a connectional advantage. One could term this principle of cortical organization the rule of "like attracts like." However, perfect continuity is not possible because the movement repertoire is highly dimensional and the cortical sheet is only two-dimensional. Therefore, the optimum solution for local continuity is not a perfect solution, but rather a compromise of many conflicting constraints.

In our proposal, the map of actions in Figure 23.3 is not by itself correct. It is present in the data, but the pattern is noisy and approximate. The map of the body shown in Figure 23.2 is also not by itself correct, but is noisy and blurred, as all motor cortex researchers have noted. The map of hand location shown in Figure 23.4 is also an approximate one, present statistically but overlapping and blurred. The height of the hand is most clearly mapped across the cortical surface, but even this dimension is noisily mapped. Our proposal is that many potential ways to parameterize movement are rendered onto the cortical sheet simultaneously, resulting in a compromise that does not neatly follow any single scheme.

We tested a mathematical model (Aflalo & Graziano, 2006; Graziano & Aflalo, 2007) that collapsed an approximate description of the monkey's movement repertoire onto a two-dimensional sheet following the principle of maximizing local continuity. The model followed the method of Kohonen for self-organizing maps (Kohonen, 1982, 2001). The model was able to reconstruct organizational features of the cortical motor system, including the outlines of the body map as in Figure 23.2, the clustering of movement categories as in Figure 23.3, and an approximate mapping of hand position as in Figure 23.4. In addition, the dimensionality reduction resulted in a rough division of the motor cortex into zones with different properties, closely following the outlines of the primary motor area, dorsal and ventral premotor areas, supplementary motor area, frontal eye field, and supplementary eye field, as in Figure 23.5. The details of how and why different movement types were naturally pushed to different zones in cortex are described more fully in Graziano and Aflalo (2007). The theory of a dimensionality reduction, in which the animal's movement repertoire was flattened onto the cortex, was astonishingly successful in explaining the organizational intricacies of a large swath of cortex, totaling about 20% of the macaque cortical mantle.

ACTION ZONES WITHIN MOTOR CORTEX

The following sections describe in detail the complex movements evoked by stimulation of different zones within the motor cortex of monkeys, and the manner in which these movements resemble actions in the monkey's normal behavioral repertoire.

Hand-to-Mouth Movements

Stimulation within a restricted zone in the precentral gyrus evoked a characteristic hand-to-mouth movement. Five components were typical of this movement. The grip aperture closed in the hand contralateral to the electrode; the forearm supinated and the wrist flexed, such that the grip was aimed at the mouth; the elbow flexed and the shoulder rotated such that the hand moved precisely to the mouth; the mouth opened; and when the head was released from the headbolt and allowed to turn freely, stimulation caused a rotation of the head to a forward-facing position, contributing to the alignment of the mouth and the hand. These five movement components occurred simultaneously in a coordinated fashion resembling the monkey's own voluntary hand-to-mouth movements.

Although the movements resembled voluntary actions in some respects, they clearly were not true voluntary movements of the monkey's but were driven by the stimulation. Typically, the movement could be obtained on every stimulation at short latency with mechanical reliability for hundreds of trials, with no adaptation or degradation. Similar movements could be evoked in anesthetized animals, though the movements were weaker and required greater current. A short stimulation, such as a 100-ms stimulation, evoked the initial part of the action, a slight closing of the hand, a slight twitching of the hand upward in the direction of the face, and a slight opening of the mouth. This truncated movement, by itself, makes no behavioral sense. It is best described as a twitch. It makes sense, however, if interpreted as the initial segment of a larger movement that has not had time to unfold. Longer stimulations, such as for 300 ms, allowed more of the movement to unfold, but rarely allowed the hand to reach the mouth. Yet longer stimulation of 500 ms almost always allowed the hand to reach the mouth in an apparent completion of the movement. Stimulations longer than 500 ms, such as those of 1,000 ms, typically caused the hand, arm, and mouth to freeze at the final configuration, as if the movement had been completed and the activated circuit were maintaining the final posture. When the stimulation train was extended beyond 1 second, almost always the animal appeared to overcome the stimulation effects and take back some degree of control of its arm. Once the stimulation train stopped, however, and then was reinitiated, the hand would move directly back to the mouth.

If the monkey was reaching toward a piece of food at the time of stimulation onset, the hand would close on empty air and come to the mouth. If the monkey had just grasped a piece of food, stimulation would drive the clenched hand to the mouth and cause the hand to freeze at the mouth, the food securely gripped in the fingers and the mouth stuck open, until the end of the stimulation train, at which time the animal would finally be released from the stimulation-evoked posture and put the food in its mouth. If an obstacle was placed between the hand and the mouth, stimulation caused the hand to move along a direct path toward the mouth and bump against the obstacle, pressing against it throughout the remainder of the stimulation, without moving intelligently around the obstacle. Therefore, although the stimulation evoked a movement of great complexity and coordination, the complexity was also limited. The movement resembled a fragment of behavior that was mechanically produced by the stimulated circuitry without intelligent flexibility.

Not all sites within the hand-to-mouth zone resulted in the same movement. For example, depending on the cortical site, stimulation drove the hand to one side of the mouth or the other, and caused the mouth to open more on the side that the hand approached, as if the monkey were placing a piece of food into the side of the jaw, as the animals often do in normal behavior. Not only did the exact position of the hand vary from site to site, but the type of hand grip also varied. For some stimulation sites the hand shaped into an apparent precision grip, the thumb against the side of the forefinger (typical of a macaque precision grip). For other stimulation sites, the hand shaped into what we called a hamburger grip, the four fingers against each other and opposed to the thumb, with a gap between, as if for gripping a larger object. These variations suggested that the zone of cortex was not uniform and not dedicated to producing a single movement, but instead probably contributed to a range of movements that fell within the large class of interactions between the hand and the mouth. In normal monkey behavior, the hand is often brought to the mouth to put in food, take out food, manipulate a piece of food that is in the mouth, scratch the lips, pick at the teeth, push food out of the cheek pouches, and so on.

It is unlikely that the collection of components in a hand-to-mouth movement co-occurred by chance. Even putting aside the specific combination of body parts, the hand closes rather than opens (50% chance); the mouth opens rather than closes (50% chance); the forearm supinates, aiming the grip at the mouth, rather than pronates, aiming the grip away from the mouth (50% chance); the hand moves within about 5 cm of the mouth, a ball of space accounting for about 1% of the total workspace of the hand (1% chance); and the head turns to a forward position, within about 5% of its range of motion (5% chance). Multiplied, these conservatively estimated probabilities yield a P value of 0.00005. We must dispense with the occasionally suggested interpretation that the evoked movements are chance collections of twitches rather than meaningful fragments of the behavioral repertoire.

In all monkeys tested, the hand-to-mouth sites were clustered in a lateral, anterior zone probably within the ventral premotor cortex. Whether they are in the caudal or rostral division is unclear. Every monkey tested thus far had a hand-to-mouth zone, but the exact location varied somewhat, especially in the rostrocaudal dimension. Our current interpretation is that the hand-to-mouth sites are more likely to lie within a ventral anterior part of F4 as defined by Matelli and colleagues (1985) and that the dorsal part of F4 emphasizes a different type of action, the defensive movement.

Defensive Movements

In a specific zone in the precentral gyrus, neurons typically respond to tactile stimuli on the face and arms and to visual stimuli looming toward the tactile receptive fields (Fogassi et al., 1996; Gentilucci et al., 1988; Graziano et al., 1997; Rizzolatti et al., 1981). Some of the neurons are trimodal, responding also to auditory stimuli in the space near their tactile receptive fields (Graziano et al., 1999). Because of these distinctive sensory properties, we refer to this cortical region as the polysensory zone (PZ). Although all monkeys tested have a PZ, it varies among animals in size and precise position (Graziano & Gandhi, 2000). It is typically located just posterior to the bend in the arcuate sulcus. In the terminology scheme of Matelli and colleagues (1985), it probably corresponds to the dorsal part of premotor area F4 where similar polysensory neurons have been reported (Fogassi et al., 1996; Gentilucci et al., 1988).

Stimulation within this zone evokes movements that closely resemble a natural defense of the body surface such as to an impending impact or unexpected touch. For example, at some sites, the neurons had tactile receptive fields on the side of the face contralateral to the electrode and visual receptive fields in the space near that side of the face. Stimulation of these sites evoked a defensive action that included eight components: a blink, stronger or exclusively on the contralateral side; a squinting of the musculature surrounding the eye; a lifting of the upper lip in a facial grimace that wrinkled the cheek upward toward the eye; a folding of the contralateral ear against the side of the head; a shrugging of the shoulder, either stronger on or exclusively on the contralateral side; a rapid turning of the head away from the contralateral side; a rapid lifting of the arm, sweeping the hand and forearm into the contralateral space near the face as if blocking or wiping away a potential threat; and a centering movement of the eyes (Cooke & Graziano, 2004a; Graziano et al., 2002). These movement components match point for point the components of a normal defensive reaction such as when the monkey's face is puffed with air (Cooke & Graziano, 2003).

At other sites, neurons had a tactile receptive field on the arm and hand and a visual response to objects looming toward the arm and hand. Stimulation caused a fast retraction of the hand to the side or back of the torso. In general, the movement evoked from a site within PZ seemed appropriate for defending the part of the body covered by the tactile and visual receptive fields of the neurons.

We observed apparent summation between the stimulation-evoked defensive-like movements and actual defensive movements. In the summation test, we lowered the stimulating current to a point near or below threshold until a subtle movement was obtained only on some trials. We then puffed air on the monkey's face, or presented some other noxious stimulus such as a ping pong ball thrown at the animal, evoking a defensive reaction. Within a second after the actual defensive reaction, we then stimulated the site in PZ. Under this condition, the stimulation evoked a robust, super-threshold defensive reaction. The actual defensive movement seemed to prime the system such that a low stimulating current in PZ could evoke a large effect.

One possibility is that the stimulation of sites in PZ evoked a noxious sensory percept to which the monkey then reacted. This possibility is difficult or impossible to rule out because the monkey cannot self-report. However, some observations suggest that it is unlikely. Although the stimulation evoked an apparent defensive reaction, as soon as the stimulation train ended the reaction ended and the monkey returned to feeding itself or playing with toys. A brief stimulation, such as for 50 ms, evoked a correspondingly brief movement, shorter than any behaviorally normal defensive reaction; a long stimulation, such as for 1,000 ms, evoked a correspondingly sustained movement that terminated abruptly at the end of the stimulation. An actual noxious stimulus, such as an air puff or a ping pong ball thrown at the face, did not result in such tight time-locking to the stimulus, but instead resulted in an extended reaction including general agitation and threats to the experimenter. Moreover, the defensive-like movements evoked by stimulation could be still evoked under anesthesia, even when the anesthesia was so deep that the animal did not react to noxious stimuli.

In order to further test the role of PZ in the coordination of defensive movements, we disinhibited neuronal activity in PZ by injecting the chemical bicuculline and inhibited neuronal activity by injecting the chemical muscimol (Cooke & Graziano, 2004b).

When bicuculline was injected into PZ, not only did the local neuronal activity increase, but the neurons also began to fire in intense spontaneous bursts of activity with approximately 5 to 30 seconds between bursts. Each spontaneous burst of neuronal activity was followed at short latency by the standard set of defensive-like movements, including blinking, squinting, flattening the ear against the side of the head, elevating the upper lip, shifting the head away from the sensory receptive fields, shrugging the shoulder, rapidly lifting the hand into the space near the side of the head as if to block an impending impact, and centering the gaze. Chemical stimulation of neurons within PZ, therefore, produced the same effect as electrical stimulation. This result may seem expected. If electrical stimulation of PZ evokes a set of movements, then surely chemical stimulation should too. However, chemical stimulation is in some ways a more specific manipulation, affecting local neuronal receptors. It does not stimulate fibers of passage or induce antidromic activation. The result of chemical stimulation in PZ, therefore, is an important confirmation and strengthens the findings from electrical stimulation.

In addition to evoking defensive-like movements by inducing bursts of neuronal activity, bicuculline also altered the monkey's actual defensive reaction to an air puff directed at the face. After the injection of bicuculline into PZ, the monkey gave an exaggerated defensive reaction to the air puff. The magnitude of the defensive reaction, as measured by facial muscle activity, was approximately 45% larger after bicuculline injection than before injection. Even gently bringing a Q-tip toward the face, normally evoking little reaction from the monkey, evoked a pronounced defensive reaction in the monkey with a bicuculline-treated PZ. Muscle activity during chewing, threat faces, and eyebrow movement were not elevated. The effect was limited to the defensive reaction. When muscimol was injected into PZ, thereby inhibiting neuronal activity, the monkey's defensive reaction to the air puff was reduced. The magnitude of the defensive reaction, as measured by facial muscle activity, was approximately 30% smaller after muscimol injection than before injection. Injections into surrounding cortical tissue outside of PZ did not affect the defensive response to an air puff. These chemical manipulations therefore strengthen the case for PZ as a sensory-motor interface related to the defense of the body surface, a cortical region to which the appropriate visual, tactile, and auditory information is supplied, and from which emerges the motor command to produce spatially directed defensive reactions.

Manipulation Movements

Stimulation of another cluster of sites evoked an especially varied and complex set of movements that involved the fingers, wrist, and often the arm and shoulder, contralateral to the electrode. The movements resembled the types of actions that monkeys typically make when manipulating, examining, or tearing objects. The finger movements included an apparent precision grip (thumb against forefinger), a power grip (fist), or a splaying of the fingers. In some cases a supination or pronation of the forearm occurred, rotating the grip one direction or the other. Also in some cases the wrist flexed or extended. A common action for monkeys is to splay the fingers of one hand, orient the palm toward the face, and examine the splayed hand, perhaps searching for stray granules of food. This splayed-hand posture, with the palm oriented toward the face, was often evoked on stimulation within this cortical zone. Monkeys commonly manipulate objects in a region of central space within about 10 cm of the chest. Stimulation within this cortical zone often evoked a movement of the shoulder and arm that brought the hand into this central region of space. A common action for monkeys when manipulating objects is to tear the object or pull it in two, the two hands pulling rapidly from central space toward lateral space while the forearms supinate and the hands are tightly gripped. Stimulation within this zone of cortex also sometimes evoked just such a movement, though only in the contralateral limb.

These sites were clustered in a posterior zone that lay partly on the gyral surface and partly on the anterior bank of the central sulcus. This cluster probably corresponds to the traditional primary motor hand representation. It may also correspond to the central hand region in the motor cortex maps of Kwan and colleagues (1978) and Park and colleagues (2001). We suggested that this cortical zone may represent a "manual fovea," a repertoire of movements that is related to the manipulation of objects and that is heavily biased toward but not exclusively limited to hand locations in a central region of space in front of the chest (Graziano et al., 2004).

Reach-to-Grasp Movements

For some cortical sites stimulation evoked an apparent reach in which the wrist straightened, the fingers opened as if to grasp, the forearm pronated to orient the grip outward, and the hand extended away from the body. In some cases the hand extended to a region of space as far as 25 cm distant from the body, at arm's length. In other cases the hand converged on a location at a lesser distance, with the elbow partially flexed, as if the hand were reaching to a closer object. In all of these cases stimulation caused a convergence to the final posture from a range of initial postures. These apparent reaching sites tended to be located on the gyral surface just anterior to the "central space/manipulation" zone and dorsal to the "defensive" zone. Because of this relative location, the reach-related sites probably lie within the dorsal premotor cortex, within its caudal division (PMDc), where a high proportion of neurons respond in relation to reaching movements (e.g., Crammond & Kalaska, 1996; Hocherman & Wise, 1991; Johnson et al., 1996; Messier & Kalaska, 2000). Typically stimulation of more rostral sites did not evoke reliable or clear movements.

Hand in Lower Space

A commonly evoked movement involved a placement of the hand in lower space near the feet, typically with the forearm pronated such that the palm faced down or inward toward the body. These stimulation-evoked movements resembled a common part of the monkey's behavioral repertoire in which the hand was braced on the ground (Graziano et al., 2004). These sites were typically found just dorsal to the central space/manipulation sites.

Mouth Movements

The aforementioned movement categories were evoked from the large arm and hand sector of the lateral motor cortex. When we stimulated in cortex ventral to the arm and hand representation, we obtained movement of the jaw, lips, and tongue, as expected on the basis of the standard body map described for the monkey motor cortex (Woolsey et al., 1952). The mouth movements often appeared to be coordinated and of behavioral significance. For example, stimulation of one site caused the jaw to attain a partially open position, the lips to purse slightly toward the contralateral side of the mouth, and the tongue to move until the tip was placed in a contralateral and slightly protuberant position. The final oral posture evoked from this site resembled an action to acquire a bit of food just outside the mouth on the contralateral side. We looked for but did not find any obvious cortical map in the mouth representation in terms of the spatial location around the mouth toward which the tongue and lip movement was directed. We also did not obtain movements that looked like threat displays, fear grimaces, or any other social displays. It is likely, however, that we failed to discover many of the movement types in the mouth representation, since we did not explore it as extensively as we did the arm and hand representation. We rarely obtained rhythmic chewing-like jaw movements, perhaps because we usually stimulated for 0.5 seconds instead of the 3 seconds used by Huang and colleagues (1989), who reported rhythmic chewing movements.

Climbing/Leaping

In a medial and anterior region, stimulation evoked especially complex movements that involved bilateral action of the arms and legs, movements of the torso, and movements of the tail, often simultaneously from one site. These complex, whole-body sites correspond roughly to the supplementary motor area (SMA), a cortical region on the crown of the hemisphere and extending slightly onto the lateral side, just anterior to the primary motor leg representation. Others have also obtained bilateral movements of multiple body parts on stimulating in this area of cortex (e.g., Foerster, 1936; Luppino et al., 1991; Penfield & Welch, 1951; Woolsey et al., 1952).

Subjectively, the movements resembled climbing or leaping postures. For example, stimulation of one site caused the left foot to press down against the floor of the primate chair; the right foot to lift and reach forward with the toes shaped as if in preparation to grasp; the left hand to reach toward a lower, lateral position while shaped as if in preparation to grasp; the right hand to reach toward a position above the head while shaped as if in preparation to grasp; and the tail to curl to one side. The long-tailed macaques in our experiments do not have prehensile tails. Instead, they use their long, stiff tails mainly as balance devices during locomotion, and therefore the tail movements evoked by stimulation of SMA are consistent with a possible role in locomotion.

Stimulation within the SMA did not always evoke movements of all four limbs. For example, stimulation of another site in the right hemisphere caused the hips to turn to the left side, the left foot to reach out and down as if stepping to a position lateral to and slightly behind the body, and the left arm to reach to a lateral position as if to grasp a support.

Although we sometimes tested stimulations extended to 1 second or more, we did not observe any cyclical stepping movements. Instead, the movements resembled the complex adjustments of body and limb often seen when monkeys are navigating a complex environment. The climbing-like movements, however, were restricted by the primate chair in which the animal was tested and therefore could never be compared directly to the normal climbing, leaping, or complex locomotor movements of a monkey.

COMPLEX MOVEMENTS REPORTED IN OTHER STUDIES

Although our studies of motor cortex were the first to systematically map the movement repertoire onto the cortical surface, many experiments before and since have reported that electrical stimulation of specific, motor-related structures can evoke highly complex, species-typical behavior. Some of these studies are reviewed below.

Midbrain

In a now classic study, electrical stimulation of a midbrain nucleus in the cat resulted in patterned locomotor behavior (Shik et al., 1969). The exact role of this mesencephalic locomotor nucleus, its relationship to spinal and cortical control of locomotion, is still unknown.

Electrical stimulation has long been used to study maps of motor output in the superior colliculus or, as it is called in nonmammals, the optic tectum. The map of saccadic eye movements in cats and monkeys is perhaps the best-known stimulation result in the colliculus (Guitton et al., 1980; Robinson, 1972; Schiller & Stryker, 1972). However, other complex species-typical behaviors can be evoked. Stimulation of the optic tectum in salamanders evokes a coordinated movement in which the animal orients to a spatial location, reaches out with the forepaws, and opens the mouth as if to acquire prey (Finkenstadt & Ewert, 1983). In rats, stimulation of the part of the map that represents lower visual space evokes orienting movements of the head as if the animal were acquiring an object on the ground in front of it, and stimulation of the part of the map that represents upper visual space evokes retracting, defensive-like movements (Dean et al., 1989). These movements are consistent with the exigencies of normal life for a rat, in which food is found on the ground in lower visual space and enemies attack from above.

Spinal Cord

Giszter and colleagues (1993) electrically stimulated sites in the spinal cord of frogs and studied the effect on the hindleg. The frog's ankle was fixed in a range of different spatial locations. For each ankle location, the force evoked by stimulation was measured. These stimulation-evoked forces formed a convergent force field pointing toward a single location in space, suggesting that if the ankle were free to move, the foot would move to that spatial terminus. Different stimulation sites resulted in convergent force fields aimed at different spatial locations.

Our results on stimulating the monkey motor cortex are similar in that stimulation caused the limb to converge from a range of initial locations toward a specific final location. Presumably, the cortical stimulation operates by recruiting spinal circuitry. If the Giszter et al. result is applicable to the monkey spinal cord, then our results may depend on spinally mediated force fields. The cortical stimulation, however, appears to recruit a higher-order or more integrated version of the spinal force fields. We typically found convergence of many joints from different body segments. For example, a hand-to-mouth movement involves a coordination of output that passes through the spinal cord (for the control of the hand, arm, and shoulder) and through the facial nucleus (for the control of the head and mouth).

Cortex

Several studies have confirmed the essential phenomenon of complex movements evoked from the motor cortex, and have now extended the results to other species of animals.

In the rat, species-typical behavior can be evoked by stimulation of motor cortex. Brecht and colleagues (2004) found that intracellular stimulation of a single cortical neuron evoked rhythmic whisking movements. Cramer and Keller (2006) suggested that the cortically controlled whisking actions are mediated by a projection from the motor cortex to a subcortical central pattern generator that in turn controls the whiskers.

Haiss and Schwarz (2005) found that the rat motor cortex contains two adjacent zones related to the whiskers. Stimulation of one zone on a behavioral time scale (500 ms) evoked rhythmic whisking similar to normal exploratory movements. Stimulation of the other zone for 500 ms evoked a retraction of the whiskers on the contralateral side, a closure of the contralateral eye, a facial grimace, and sometimes a lifting of the contralateral forepaw to the space beside the face. These results suggest that the rat motor cortex, like the monkey motor cortex, may be organized into zones that emphasize different ethologically useful actions, in this case exploratory whisking for one zone and defensive actions for the other zone.

Ramanathan and colleagues (2006) stimulated the rat motor cortex on a behaviorally relevant time scale (500 ms) and obtained reaching and grasping movements of the forepaws. Moreover, when a reaching zone in cortex was lesioned, the rat's ability to reach was compromised. When the rat was retrained to reach, the motor cortex was found to have reorganized such that reaching could be electrically evoked from new cortical sites. After this rehabilitation, the ability of the rat to perform the behavior correlated with the amount of cortex that, when stimulated, evoked the behavior. These results strongly support the view that the motor cortex is organized to control complex, meaningful behavior; that different behaviors are emphasized in different regions of cortex; and that these behaviors can be assessed through electrical stimulation.

In the cat motor cortex, Ethier and colleagues (2006) found that stimulation on a behaviorally relevant time scale (500 ms) evoked a variety of forepaw movements including apparent reaching and hooking of the paw as if to acquire an object.

Stepniewska and colleagues (2005) stimulated the motor and parietal cortex of galagos using the behaviorally relevant time scale of 500 ms. They evoked complex movements that resembled fragments of the animal's normal behavioral repertoire. Different categories of movement were evoked from different cortical zones. The parietal lobe could be segmented into functional zones including a hand-to-mouth zone, a defensive zone, and a reaching zone. Similar results were obtained in the motor cortex, but were not studied in as much detail.

In macaque monkeys, stimulation of the ventral intraparietal area (VIP) evokes movements that resemble defensive reactions (Cooke & Graziano, 2004a; Cooke et al., 2003; Graziano et al., 2002; Thier & Andersen, 1998). The evoked movements are similar to those evoked from PZ in the motor cortex. VIP is anatomically connected to PZ (Lewis & Van Essen, 2000; Luppino et al., 1999), and neurons in VIP respond preferentially to tactile stimuli on the face and to visual and auditory stimuli near the face, much like neurons in PZ (e.g., Colby et al., 1993; Duhamel et al., 1998; Schlack et al., 2005). One possibility is that VIP and PZ are part of a larger sensory-motor circuit that contributes to the maintenance of a margin of safety around the body (Graziano & Cooke, 2006).

SUMMARY

The monkey motor cortex does not appear to decompose movement into the action of separate muscles. Instead, the control of muscles is extensively overlapped in the cortex, effectively allowing for the construction of complex, behaviorally meaningful movements. Electrical stimulation of different zones within cortex tends to evoke different categories of movement that resemble actions in the monkey's normal repertoire.

The complex topography within motor cortex roughly follows several different possible organizational schemes, including a map of the body, a spatial map of hand location, a cluster map of ethologically relevant categories of action, and a set of commonly designated areas with fuzzy borders and largely unknown functions. One possible way to reconcile these many organizational schemes, each one of which captures some of the data, is to hypothesize that the cortex is a best-fit rendering of the highly dimensional movement repertoire on the two-dimensional cortical sheet. In this view, various individual dimensions that are of importance in the movement repertoire are mapped across cortex in noisy, partial topographies. No single dimension of movement is perfectly mapped across the cortical surface because of a compromise among multiple constraints. The underlying principle of organization, in this view, is not a map of one or another dimension, but a rendering of the many-dimensional movement space onto the cortex.

It is not yet clear to what extent this ethological organization of the motor cortex is built in through evolution, and to what extent it is learned during development and adulthood as the animal learns its motor repertoire. The distinct connections and anatomical structure of different zones within motor cortex argue for some built-in organization shaped through evolution to serve the behavioral repertoire. The modifications to the map that occur with experience demonstrate that at least some of the cortical organization reflects learning.

REFERENCES

Aflalo, T. N., & Graziano, M. S. A. (2006). Possible origins of the complex topographic organization of motor cortex: Reduction of a multidimensional space onto a 2-dimensional array. *Journal of Neuroscience, 26,* 6288–6297.

Asanuma, H. (1975). Recent developments in the study of the columnar arrangement of neurons within the motor cortex. *Physiological Reviews, 55,* 143–156.

Asanuma, H., & Rosen, I. (1972). Topographical organization of cortical efferent zones projecting to distal forelimb muscles in the monkey. *Experimental Brain Research, 14,* 243–256.

Asanuma, H., & Sakata, H. (1967) Functional organization of a cortical efferent system examined with focal depth stimulation in cats. *Journal of Neurophysiology, 30,* 35–54

Asanuma, H., & Ward, J. E. (1971). Pattern of contraction of distal forelimb muscles produced by intracortical stimulation in cats. *Brain Research, 27,* 97–109.

Beevor, C., & Horsley, V. (1887). A minute analysis (experimental) of the various movements produced by stimulating in the monkey different regions of the cortical centre for the upper limb, as defined by Professor Ferrier. *Philosophical Transactions of the Royal Society of London B, 178,* 153–167.

Brecht, M., Schneider, M., Sakmann, B., & Margrie, T. W. (2004). Whisker movements evoked by stimulation of single pyramidal cells in rat motor cortex. *Nature, 427,* 704–710.

Bruce, C. J., Goldberg, M. E., Bushnell, M. C., & Stanton, G. B. (1985). Primate frontal eye fields. II. Physiological and anatomical correlates of electrically evoked eye movements. *Journal of Neurophysiology, 54,* 714–734.

Caggiula, A. R. & Hoebel, B. G. (1966). "Copulation-reward site" in the posterior hypothalamus. *Science, 153,* 1284–1285.

Chakrabarty, S., & Martin, J. H. (2000). Postnatal development of the motor representation in primary motor cortex. *Journal of Neurophysiology, 84,* 2582–2594.

Cheney, P. D., & Fetz, E. E. (1985). Comparable patterns of muscle facilitation evoked by individual corticomotoneuronal (CM) cells and by single intracortical microstimuli in primates: Evidence for functional groups of CM cells. *Journal of Neurophysiology, 53,* 786–804.

Colby, C. L., Duhamel, J. R., & Goldberg, M. E. (1993). Ventral intraparietal area of the macaque: Anatomic location and visual response properties. *Journal of Neurophysiology, 69,* 902–914.

Cooke, D. F., & Graziano, M. S. A. (2003). Defensive movements evoked by air puff in monkeys. *Journal of Neurophysiology, 90,* 3317–3329.

Cooke, D. F., & Graziano, M. S. A. (2004a). Sensorimotor integration in the precentral gyrus: Polysensory neurons and defensive movements. *Journal of Neurophysiology, 91,* 1648–1660.

Cooke, D. F., & Graziano, M. S. A. (2004b). Super-flinchers and nerves of steel: Defensive movements altered by chemical manipulation of a cortical motor area. *Neuron, 43,* 585–593.

Cooke, D. F., Taylor, C. S. R., Moore, T., & Graziano, M. S.A. (2003). Complex movements evoked by microstimulation of the ventral intraparietal area. *Proceedings of the National Academy of Sciences, 100,* 6163–6168.

Cramer, N. P., & Keller, A. (2006). Cortical control of a whisking central pattern generator. *Journal of Neurophysiology, 96,* 209–217.

Crammond, D. J., & Kalaska, J. F. (1996). Differential relation of discharge in primary motor cortex and premotor cortex to movements versus actively maintained postures during a reaching task. *Experimental Brain Research, 108,* 45–61.

Dean, P., Redgrave, P., & Westby, G. W. (1989). Event or emergency? Two response systems in the mammalian superior colliculus. *Trends in Neuroscience, 12,* 137–147.

Duhamel, J. R., Colby, C. L., & Goldberg, M. E. (1998). Ventral intraparietal area of the macaque: Congruent visual and somatic response properties. *Journal of Neurophysiology, 79,* 126–136.

Dum, R. P., & Strick, P. L. (2002). Motor areas in the frontal lobe of the primate. *Physiology and Behavior, 77,* 677–682.

Ethier, C., Brizzi, L., Darling, W. G., & Capaday, C. Ethier, C., Brizzi, L., Darling, W. G., & Capaday, C. (2006). Linear summation of cat motor cortex outputs. *Journal of Neuroscience, 26,* 5574–5581.

Ferrier, D. (1874). Experiments on the brain of monkeys – No. 1. *Proceedings of the Royal Society of London, 23,* 409–430.

Finkenstadt, T., & Ewert, J. P. (1983). Visual pattern discrimination through interactions of neural networks: A combined electrical brain stimulation, brain lesion, and extracellular recording study in *Salamandra salamandra. Journal of Comparative Physiology, 153,* 99–110.

Foerster, O. (1936). The motor cortex of man in the light of Hughlings Jackson's doctrines. *Brain, 59,* 135–159.

Fogassi, L., Gallese, V., Fadiga, L., Luppino, G., Matelli, M., & Rizzolatti, G. (1996). Coding of peripersonal space in inferior premotor cortex (area F4). *Journal of Neurophysiology, 76,* 141–157.

Fritsch, G., & Hitzig, E. (1870). Uber die elektrishe Erregbarkeit des Grosshirns. Arch. f. Anat., Physiol und wissenchaftl. Mediz., Leipzig, 300–332. [On the electrical excitability of the cerebrum. Translated by G. von Bonin. In: W. W. Nowinski (Ed.), *Some papers on the cerebral cortex* (pp. 73–96). Springfield, IL: Thomas.]

Gentilucci, M., Fogassi, L., Luppino, G., Matelli, M., Camarda, R., & Rizzolatti, G. (1988). Functional organization of inferior area 6 in the macaque monkey. I. Somatotopy and the control of proximal movements. *Experimental Brain Research, 71,* 475–490.

Giszter, S. F., Mussa-Ivaldi, F. A., & Bizzi, E. (1993). Convergent force fields organized in the frog's spinal cord. *Journal of Neuroscience, 13,* 467–491.

Graziano, M. S. A., & Aflalo, T. N. (2007). Mapping behavior repertoire onto the cortex. *Neuron, 56,* 239–251.

Graziano, M. S. A., Aflalo, T., & Cooke, D. F. (2005). Arm movements evoked by electrical stimulation in the motor cortex of monkeys. *Journal of Neurophysiology, 94,* 4209–4223.

Graziano, M. S. A., & Cooke, D. F. (2006). Parieto-frontal interactions, personal space, and defensive behavior. *Neuropsychologia, 44,* 845–859.

Graziano, M. S. A., Cooke, D. F., Taylor, C. S. R., & Moore, T. (2004). Distribution of hand location in monkeys during spontaneous behavior. *Experimental Brain Research, 155,* 30–36.

Graziano, M. S. A., & Gandhi, S. (2000). Location of the polysensory zone in the precentral gyrus of anesthetized monkeys. *Experimental Brain Research, 135,* 259–266.

Graziano, M. S. A., Hu, X. T., & Gross, C. G. (1997). Coding the locations of objects in the dark. *Science, 277,* 239–241.

Graziano, M. S. A., Reiss, L. A. J., & Gross, C. G. (1999). A neuronal representation of the location of nearby sounds. *Nature, 397,* 428–430.

Graziano, M. S. A., Taylor, C. S. R., & Moore, T. (2002). Complex movements evoked by microstimulation of precentral cortex. *Neuron, 34,* 841–851.

Guitton, D., Crommelinck, M., & Roucoux, A. (1980). Stimulation of the superior colliculus in the alert cat. I. Eye movements and neck EMG activity evoked when the head is restrained. *Experimental Brain Research, 39,* 63–73.

Haiss, F., & Schwarz, C. (2005). Spatial segregation of different modes of movement control in the whisker representation of rat primary motor cortex. *Journal of Neuroscience, 25,* 1579–1587.

He, S. Q., Dum, R. P., & Strick, P. L. (1995). Topographic organization of corticospinal projections from the frontal lobe: Motor areas on the medial surface of the hemisphere. *Journal of Neuroscience, 15,* 3284–3306.

Hess, W. R. (1957). *Functional organization of the diencephalons.* New York: Grune and Stratton.

Hocherman, S., & Wise, S. P. (1991). Effects of hand movement path on motor cortical activity in awake, behaving rhesus monkeys. *Experimental Brain Research, 83,* 285–302.

Hoebel, B. G. (1969). Feeding and self-stimulation. *Annals of the New York Academy of Sciences, 157,* 758–778.

Huang, C. S., Hiraba, H., Murray, G. M., & Sessle, B. J. (1989). Topographical distribution and functional properties of cortically induced rhythmical jaw movements in the monkey (Macaca fascicularis). *Journal of Neurophysiology, 61,* 635–650.

Johnson, P. B., Ferraina, S., Bianchi, L., & Caminiti, R. (1996). Cortical networks for visual reaching: Physiological and anatomical organization of frontal and parietal lobe arm regions. *Cerebral Cortex, 6,* 102–119.

King, M. B., & Hoebel, B. G. (1968). Killing elicited by brain stimulation in rats. *Communications in Behavioral Biology, 2,* 173–177

Kohonen, T. (1982). Self-organizing formation of topologically correct feature maps. *Biological Cybernetics, 43,* 59–69.

Kohonen, T. (2001). *Self-organizing maps.* Berlin: Springer.

Kwan, H. C., MacKay, W. A., Murphy, J. T., & Wong, Y. C. (1978). Spatial organization of precentral cortex in awake primates. II. Motor outputs. *Journal of Neurophysiology, 41,* 1120–1131.

Lewis, J. W., & Van Essen, D. C. (2000). Corticocortical connections of visual, sensorimotor, and multimodal processing areas in the parietal lobe of the macaque monkey. *Journal of Comparative Neurology, 428,* 112–137.

Luppino, G., Matelli, M., Camarda, R. M., Gallese, V., & Rizzolatti, G. (1991). Multiple representations of body movements in mesial area 6 and the adjacent cingulate cortex: An intracortical microstimulation study in the macaque monkey. *Journal of Comparative Neurology, 311,* 463–482.

Luppino, G., Murata, A., Govoni, P., & Matelli, M. (1999). Largely segregated parietofrontal connections linking rostral intraparietal cortex (areas AIP and VIP) and the ventral premotor cortex (areas F5 and F4). *Experimental Brain Research, 128,* 181–187.

Martin, J. H., Engber, D., & Meng, Z. (2005). Effect of forelimb use on postnatal development of the forelimb motor representation in primary motor cortex of the cat. *Journal of Neurophysiology, 93,* 2822–2831.

Matelli, M., Luppino, G., & Rizzolatti, G. (1985). Patterns of cytochrome oxidase activity in the frontal agranular cortex of the macaque monkey. *Behavioral Brain Research, 18,* 125–136.

Matsuzaka, Y., Aizawa, H., & Tanji, J. (1992). A motor area rostral to the supplementary motor area (presupplementary motor area) in the monkey: Neuronal activity during a learned motor task. *Journal of Neurophysiology, 68,* 653–662.

Messier, J., & Kalaska, J. F. (2000). Covariation of primate dorsal premotor cell activity with direction and amplitude during a memorized-delay reaching task. *Journal of Neurophysiology, 84,* 152–165.

Nudo, R. J., Milliken, G. W., Jenkins, W. M., & Merzenich, M. M. (1996). Use-dependent alterations of movement representations in primary motor cortex of adult squirrel monkeys. *Journal of Neuroscience, 16,* 785–807.

Park, M. C., Belhaj-Saif, A., Gordon, M., & Cheney, P. D. (2001). Consistent features in the forelimb representation of primary motor cortex in rhesus macaques. *Journal of Neuroscience, 21,* 2784–2792.

Penfield, W., & Boldrey, E. (1937). Somatic motor and sensory representation in the cerebral cortex of man as studied by electrical stimulation. *Brain, 60,* 389–443.

Penfield, W., & Rasmussen, T. (1950). *The cerebral cortex of man. A clinical study of localization of function.* New York: Macmillan.

Penfield, W., & Welch, K. (1951). The supplementary motor area of the cerebral cortex: A clinical and experimental study. *American Medical Association Archives of Neurology and Psychiatry, 66,* 289–317.

Preuss, T. M., Stepniewska, I., & Kaas, J. H. (1996). Movement representation in the dorsal and ventral premotor areas of owl monkeys: A microstimulation study. *Journal of Comparative Neurology, 371,* 649–676.

Ramanathan, D., Conner, J. M., & Tuszynski, M. H. (2006). A form of motor cortical plasticity that correlates with recovery of function after brain injury. *Proceedings of the National Academy of Sciences, 103,* 11370–11375.

Rizzolatti, G., & Luppino, G. (2001). The cortical motor system. *Neuron, 31,* 889–901.

Rizzolatti, G., Scandolara, C., Matelli, M., & Gentilucci, M. (1981). Afferent properties of periarcuate neurons in macaque monkeys. II. Visual responses. *Behavioral Brain Research, 2,* 147–163.

Robinson, D. A. (1972). Eye movements evoked by collicular stimulation in the alert monkey. *Vision Research, 12,* 1795–1808.

Robinson, D. A., & Fuchs, A. F. (1969). Eye movements evoked by stimulation of the frontal eye fields. *Journal of Neurophysiology, 32,* 637–648.

Salzman, C. D., Britten, K. H., & Newsome, W. T. (1990). Cortical microstimulation influences perceptual judgements of motion direction. *Nature, 346,* 174–177.

Schiller, P. H., & Stryker, M. (1972). Single-unit recording and stimulation in superior colliculus of the alert rhesus monkey. *Journal of Neurophysiology, 35,* 915–924.

Schlack, A., Sterbing, S., Hartung, K., Hoffmann, K. P., & Bremmer, F. (2005). Multisensory space representations in the Macaque ventral intraparietal area (VIP). *Journal of Neuroscience, 25,* 4616–4625.

Shik, M. L., Severin, F. V., & Orlovsky, G. N. (1969). Control of walking and running by means of electrical stimulation of the mesencephalon. *Electroencephalography and Clinical Neurophysiology, 26,* 549.

Stepniewska, I., Fang, P. C., & Kaas, J. H. (2005). Microstimulation reveals specialized subregions for different complex movements in posterior parietal cortex of prosimian galagos. *Proceedings of the National Academy of Sciences, 102,* 4878–4883.

Taylor, C. S. R., Cooke, D. F., & Graziano, M. S. A. (2002). Complex mapping from precentral cortex to muscles. *Society for Neuroscience Abstracts, 28.*

Thier, P., & Andersen, R. A. (1998). Electrical microstimulation distinguishes distinct saccade-related areas in the posterior parietal cortex. *Journal of Neurophysiology, 80,* 1713–1735.

Woolsey, C. N., Settlage, P. H., Meyer, D. R., Sencer, W., Hamuy, T. P., & Travis, A. M. (1952). Pattern of localization in precentral and "supplementary" motor areas and their relation to the concept of a premotor area. *Association for research in nervous and mental disease, 30,* 238–264. New York: Raven Press.

CHAPTER 24

Object Recognition: Physiological and Computational Insights

Doris Y. Tsao, Charles F. Cadieu, and Margaret S. Livingstone

INTRODUCTION

Visual object recognition is the identification of a thing in the outside world based on the sense of vision. Our eyes are bombarded by a wide variety of visual forms, from simple shapes like cups and pens, to complicated shapes like keyboards and saxophones, to amorphous natural forms like flowers and bodies. These forms can appear at multiple poses as they rotate (or as we rotate), can change size as they move closer or farther, and can frequently occlude one another. Yet we have no trouble recognizing them. How does the brain transform the unpredictable retinal array into invariant representations of objects? This problem has two aspects: (1) extracting a stable, compact, and explicit representation of the forms stimulating our eyes at any moment, and (2) comparing the resulting representation to a stored representation in memory. We will focus in this chapter on the first aspect: How does the primate visual system extract stable representations of the objects in the visual world?

Visual object recognition is a function of the inferior temporal lobe, specifically, ventral stream areas V4, TEO, and TE. It has been suggested that the specialization of the inferior temporal lobe for object recognition arose during evolution as an elaboration of the foveal visual representation in V1, while the specialization of the parietal lobe for manipulation of objects in space arose as an outgrowth of the lower field representation of V1 (where the hands would normally be located) (Maunsell & Van Essen, 1987; Previc, 1990).

Understanding the neural computations underlying object recognition is a uniquely difficult problem for three reasons. First, we don't even know how object recognition might be solved computationally. Even though numerous computer models exist, they don't come close to rivaling the performance of the human visual system. Second, a huge expanse of cortical territory is dedicated to object recognition. Areas TEO and TE, for example, are more than 10 times as large as the much-studied motion area MT (Felleman & Van Essen, 1991). Assuming that multiple levels of organization exist within this vast cortical expanse, it becomes exceedingly difficult to map specific computations onto specific sets of neurons. Thus, single-unit physiologists typically *assume* that the organization of inferotemporal (IT) cortex is homogeneous and that meaningful answers can be obtained from sampling responses of random populations of IT neurons. Finally, the set of possible objects covers a huge parameter space. While it is clear the brain must use some sort of code that is compact, explicit, and stable to identify objects within this space (Connor et al., 2007), it is unclear what this code might be. To summarize, understanding object recognition is difficult due to a lack of effective computer models, the large number of cortical resources used to solve the problem, and a huge parameter space.

Figure 24.1 The challenge of object recognition, exemplified for a face.

Over the past three decades, knowledge has gradually accumulated concerning a system in the temporal lobe that may allow us to overcome these challenges. Both macaques and humans have a specialized system in the temporal lobe for processing faces that produces an extraordinary ability to recognize faces under a variety of conditions (Pinsk et al., 2005; Tsao et al., 2003, 2006, 2008). (Fig. 24.1). This face processing system provides a spatially limited network with readily accessible components representing a parametrically confined set of objects.

Face perception is a microcosm of object recognition processes, and the solution to the particular problem of face recognition will undoubtedly yield insights into the general problem of object recognition. The most difficult challenge in object recognition—distinguishing among similar visual forms despite substantial changes in appearance arising from changes in position, illumination, occlusion, etc.—is something we can do effortlessly for faces (Fig. 24.1). Although face identification is often singled out as demanding particular sensitivity to differences between objects sharing a common basic configuration, in fact, such differences must be represented in the brain for both faces and nonface objects. It is true that most humans can easily identify hundreds of faces (Diamond & Carey, 1986), but, even if one cannot recognize a hundred different bottles by name, one can certainly distinguish them in pairwise discrimination tasks.

In this chapter, we first discuss the functional architecture of the temporal lobe, with a special focus on the architecture of the system of face-selective areas in macaques and humans. We then discuss the physiology of cells in the temporal lobe, with a focus on the response properties of face-selective cells. Finally, we discuss different computational approaches to object recognition. The central thesis of this chapter is that understanding face processing will illuminate the general problem of visual object recognition.

Functional Architecture of the Inferior Temporal Lobe

The functional architecture of the temporal lobe sets the stage for the neural processes underlying object recognition.

General Architecture of the Macaque Ventral Visual Pathway

Figure 24.2 shows a schematic of the lateral and ventral surfaces of the macaque brain. The three major areas of the ventral form processing pathway are V4, TEO, and TE. Area TE is further subdivided into four parts on the basis of anatomical connection criteria: TEpd (dorso-posterior), TEpv (ventro-posterior), TEad (dorso-anterior), and TEav (ventro-anterior) (Cheng et al., 1997 Saleem & Tanaka, 1996; Yukie & Iwai, 1988). Monkeys with bilateral lesions to V4, TEO, and TE show severe and specific deficits in object recognition tasks (Dean, 1976; Gross, 1973). Both V4 and TEO are retinotopically organized, with TEO containing a coarse but complete representation of the contralateral visual field (Boussaoud et al., 1991); cells in TE have large receptive fields centered on the fovea (Ito et al., 1995; Kobatake & Tanaka, 1994).

The architecture of the inferior temporal lobe has been studied by classic anatomical tracing. V4 sends strong projections to TEO from its central visual field representation (Ungerleider et al., 2008). TEO in turn sends strong feedforward projections to area TE (Saleem et al., 1993).

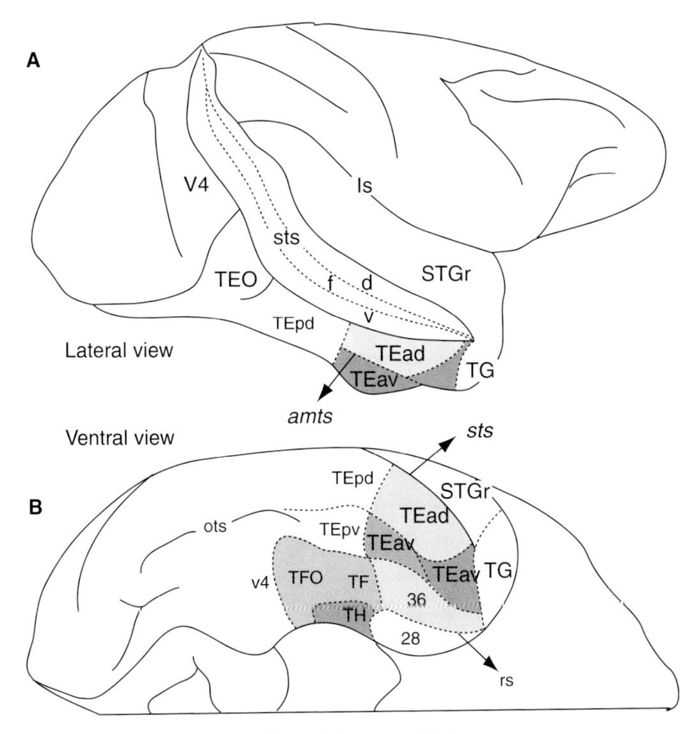

Figure 24.2 The ventral visual processing pathway. (**A**) Lateral view of the right hemisphere of the macaque brain, showing the architectonic subdivisions of inferior temporal cortex. The superior temporal sulcus (STS) is opened to show the dorsal bank (d), fundus (f), and ventral bank (v). (**B**) Ventral view of subdivisions of the inferior temporal cortex. ls, lateral sulcus; STGr, superior temporal gyrus, rostral part; TEad, dorsal subregion of anterior TE; TEav, ventral subregion of anterior TE; TEpd, dorsal subregion of posterior TE; TEpv, ventral subregion of posterior TE; TG, temporal pole; TH, area TH of parahippocampal cortex; TF, area TF of the parahippocampal cortex; TFO, area TFO of the parahippocampal cortex; 28, entorhrinal cortex; 36, perirhinal cortex. From Saleem, K. S., Kondo, H., & Price, J. L. (2008). Complementary circuits connecting the orbital and medial prefrontal networks with the temporal, insular, and opercular cortex in the macaque monkey. *Journal of Comparative Neurology, 506,* 659–693. Used with permission.

The projection from TEO to TE shows a precise modularity, with single TEO sites projecting to two to five clusters in TE (Saleem et al., 1993) (Fig. 24.3A). These clusters are columnar, extending across all six cortical layers. Although the functional properties of the TEO injection sites and connected TE columns are unknown, it seems plausible that this precise anatomical circuitry serves a computational purpose. One interesting possibility is that each network of connected sites in TEO and TE is responsible for extracting a specific aspect or class of visual form.

Area TE, the highest purely visual stage of the ventral pathway, sends feedback projections to V4 and TEO, and feedforward projections to several polymodal brain sites including the perirhinal cortex, the frontal cortex, the amygdala, and the striatum (Cheng et al., 1997; Saleem & Tanaka, 1996; Suzuki et al., 2000; Webster et al., 1991, 1993, 1994). The perirhinal projections from TEav and TEad differ, suggesting that these areas constitute distinct processing regions in TE. Focal TEav injections result in a widely distributed labeling in perirhinal cortex, covering around half its total extent, whereas

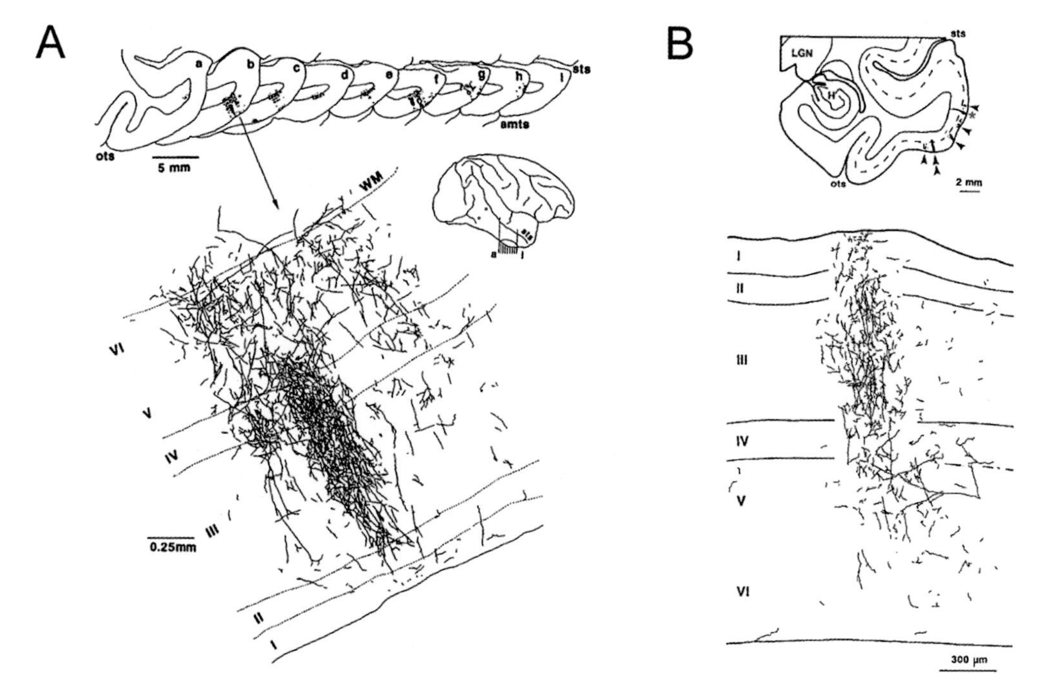

Figure 24.3 Columnar projections in inferior temporal (IT) cortex. (**A**) Labeled terminals in TE following injections of the anterograde tracer *Phaseolus vulgaris* leukoagglutinin (PHA) in TEO (indicated by two red spots) form columnar clusters. From Saleem, K. S., Tanaka, K., & Rockland, K. S. (1993). Specific and columnar projection from area TEO to TE in the macaque inferotemporal cortex. *Cerebral Cortex, 3,* 454–464. (**B**) Labeled terminals of intrinsic horizontal axons show patchy columnar arborization following injection of biocytin into layer 3 in area TE. The injection site is marked by the red asterisk and arrowheads indicate terminal patches. From Fujita, I., & Fujita, T. (1996). Intrinsic connections in the macaque inferior temporal cortex. *Journal of Comparative Neurology, 368,* 467–486. Used with permission.

injections into TEad result in labeling in only a small region of lateral perirhinal cortex (Saleem & Tanaka, 1996). Perirhinal cortex has been implicated in encoding long-term memory of objects (Meunier et al., 1993) and association between familiar objects (Erickson & Desimone, 1999; Erickson et al., 2000; Miyashita et al., 1996). The widespread projections from TEav to perirhinal could thus facilitate association between different shapes (e.g., different views of the same object).

In addition to projections to other areas, both areas TEO and TE have local connections. Injections into TE reveal horizontal axons in layers 2 and 3 terminating in patches 0.5 mm wide and cylindrical in shape, spanning layers 1 through 3 or even to layers 4 and 5, with the farthest patches up to 4 mm distant from the injection site (Fujita & Fujita, 1996) (Fig. 24.3B). TEO shows similar, but slightly smaller, columns of connected patches, with the patches slightly smaller. Again, it seems plausible that each network of locally connected patches forms a system specialized to represent a particular aspect of form.

Electrophysiology and optical imaging studies also suggest a columnar functional architecture in TE. Neighboring cells are responsive to similar visual features of objects (Fujita et al., 1992). Optical imaging reveals that spots ≈0.5 mm wide in TE are activated when a monkey views a particular object (Tsunoda et al., 2001; Wang et al., 1996), with different spots activated by different objects. The cortical regions activated by three different objects are shown in Figure 24.4A. Single-unit recordings confirmed

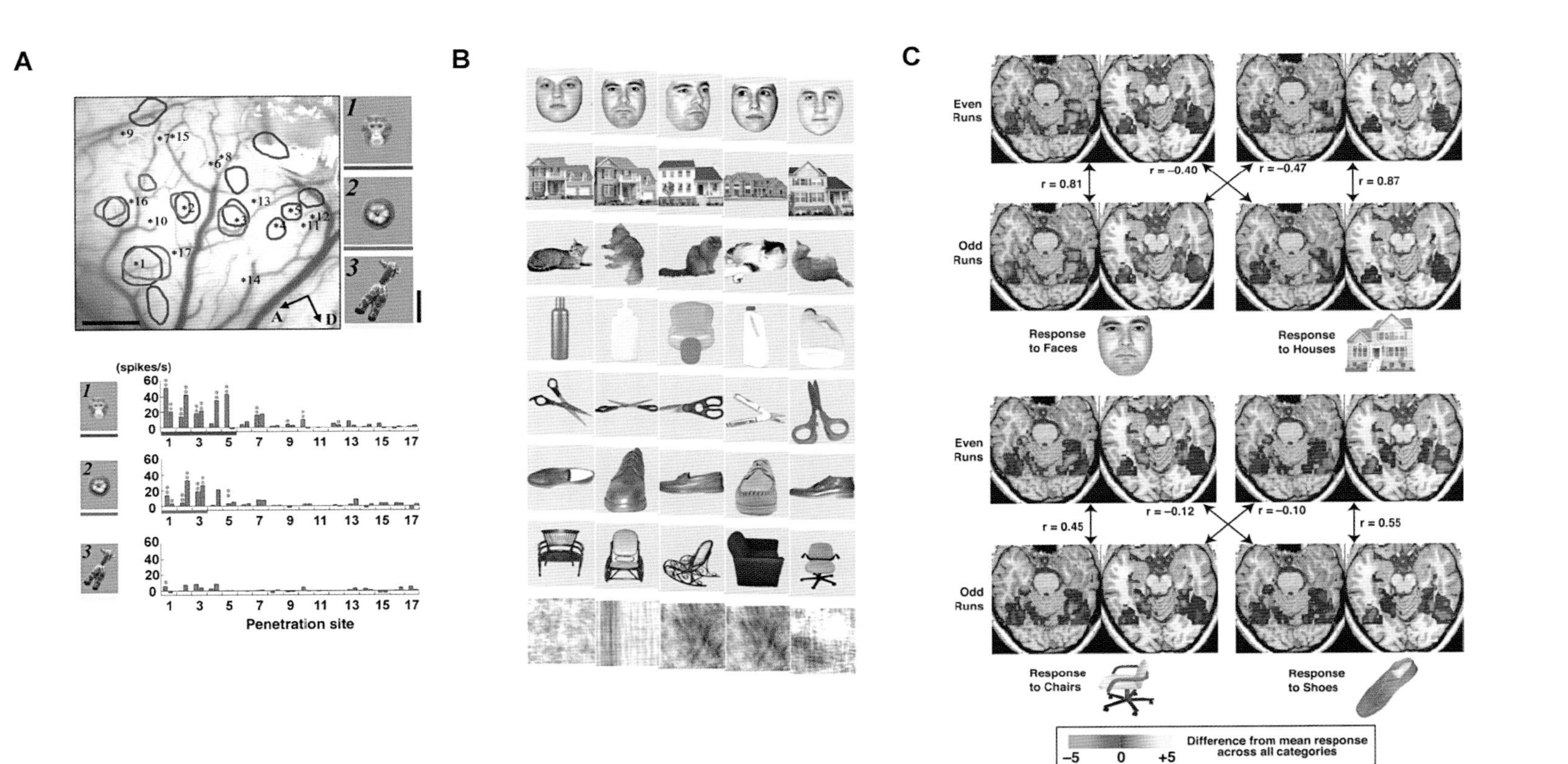

Figure 24.4 Distributed object representations in macaques and humans. (**A**) Top: Spots activated by different stimuli (1, 2, and 3, right) as revealed by optical imaging, together with numbered electrode sites. Bottom: Reponses to the three stimuli at each of 17 electrode sites. Scale bar, 1.0 mm. Tsunoda, K., Yamane, Y., Nishizaki, M., & Tanifuji, M. (2001). Complex objects are represented in macaque inferotemporal cortex by the combination of feature columns. *Nature Neuroscience, 4*, 832–838. (**B**) Example images from the eight stimulus categories used by Haxby and colleagues (2001) to examine object representations in the human temporal lobe. (**C**) The pattern of responses to four categories (faces, houses, chairs, shoes) measured separately from data obtained on even-numbered and odd-numbered (two horizontal slices from a single subject are shown). Within-category correlations (0.81, 0.87, 0.45, 0.35) are significantly higher than cross category comparisons (−0.4, −0.47, −0.12, −0.1), allowing prediction of the viewed object category. B and C from Haxby, J. V., Gobbini, M. I., Furey, M. L., Ishai, A., Schouten, J. L., & Pietrini, P. (2001). Distributed and overlapping representations of faces and objects in ventral temporal cortex. *Science, 293*, 2425–2430. Used with permission.

the responsiveness of cells within these spots to the particular objects used to activate them (and absence of responsiveness outside these spots). For example, an apple activated spots 1, 2, and 3, and single units in these three sites also responded strongly to the apple. Since multiple spots, distributed over several millimeters of cortex, were active in response to a single image, and since the size of these spots is similar to the size of termination zones observed in tracer experiments (Fig. 24.3), it is conceivable that these spots correspond to connected columns.

Distributed Architecture of the Human Ventral Pathway

Human functional magnetic resonance imaging (fMRI) studies provide a much coarser picture of functional architecture of ventral temporal cortex, but are consistent with results from macaques. The wide field of view of fMRI makes it possible to address the question of how distributed the representation of an object is. Haxby and colleagues (2001) measured response patterns in ventral temporal cortex evoked by viewing of faces, cats, five categories of manmade objects, and nonsense pictures (Fig. 24.4B). They found that object categories are represented by widely distributed and overlapping representations in ventral temporal cortex. Each category elicited a distinct and reproducible distributed response pattern that could be used to identify which category was seen (Fig. 24.4C).

The distinctiveness of the response patterns is apparent even at a scale of centimeters (Fig. 24.4C, compare response to chairs with response to shoes). This suggests that the structure in the response patterns reflects a level of functional organization more macroscopic than 0.5-mm columns. While multivoxel readout techniques might be capable of distinguishing response patterns that differ only at a scale of 0.5-mm columns (by pooling minute but consistent category differences across a large number of voxels) (Norman et al., 2006), such functional distinctions should not be readily apparent using a technique with a resolution on the scale of centimeters (Boynton, 2005; Kamitani & Tong, 2005). Therefore, the distinct fMRI response patterns to different objects observed by Haxby and colleagues may instead arise from a coarser scale functional organization, possibly, networks of connected, clustered columns.

Architecture of the Face Processing System in Humans

The existence of a face-selective area in the human brain was first suggested by neurological observations. That there is a specialized area for processing upright faces fits with one of the most striking findings from the neuropsychology literature: Patient C.K., who is severely impaired at object recognition, including many basic midlevel visual processes, is nonetheless 100% normal at face recognition (Moscovitch et al., 1997). C.K.'s dissociation is illustrated by his perception of the face made up of vegetables by Arcimbaldo—C.K. sees the face, but not the constituent vegetables. His pattern of deficits indicates that face processing is not simply a final stage tacked onto the end of the nonface object recognition pathway, but rather a different pathway that branches away from object recognition early in the visual hierarchy.

The first direct demonstration of face-selective activation in a human brain area came from positron emission tomography (PET) studies showing activation of the fusiform gyrus in a variety of face perception tasks (Haxby et al., 1991; Sergent et al., 1992) and event-related potential (ERP) studies showing a face-selective event-related potential in the fusiform gyrus (Allison et al., 1994; McCarthy et al., 1999; Puce et al., 1999). Subsequently fMRI revealed more of the specificity of these cortical regions for faces, with demonstrations of fusiform regions that responded more strongly to faces than to letter strings and textures (Puce et al., 1996), flowers (McCarthy et al., 1997), everyday objects, houses, and hands (Kanwisher et al., 1997). Face-specific fMRI activation can be seen in the superior temporal sulcus (STS), in part of the occipital lobe (the "occipital face area," or OFA), and most robustly, on the lateral side of the right midfusiform gyrus (the "fusiform face

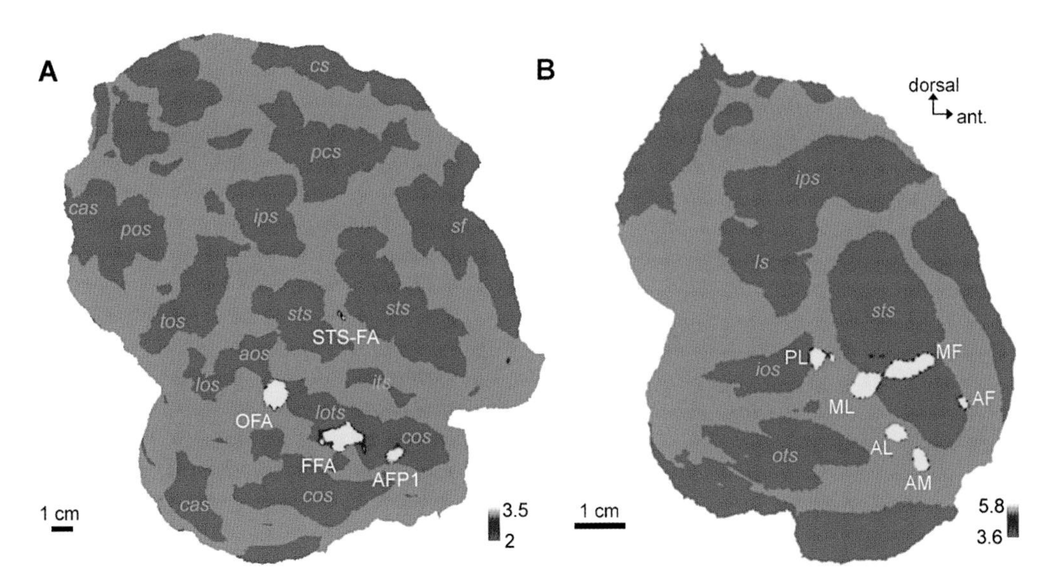

Figure 24.5 Comparison of face-selective regions between humans and macaques. (**A**) Face-selective regions shown on a flattened map of the human right hemisphere. (**B**) Face-selective regions shown on a flattened map of a macaque right hemisphere. By comparison, the human face-selective regions are shifted ventrally away from the STS relative to the macaque face patches. AF, anterior face patch in the STS fundus; AFP1, anterior face patch 1; AL, anterior face patch on the STS lip; AM, anterior face patch on the ventral surface of IT just lateral and anterior to the AMTS; FFA, fusiform face area; MF, middle face patch in the STS fundus; ML, middle face patch on the STS lip; OFA, occipital face area; PL, posterior face patch; STS-FA, superior temporal sulcus face area. aos, anterior occipital; cas, calcarine; cos, collateral; ios, inferior occipital sulcus; ips, intraparietal; its, inferior temporal; los, lateral occipital; lots, lateral occipitotemporal; ls, lunate sulcus; ots, occipitotemporal sulcus; pcs, precentral; sf, Sylvian fissure sts, superior temporal; pos, parieto-occipital; tos, transoccipital. Tsao, D. Y., Moeller, S., & Freiwald, W. A. (2008). Comparing face patch systems in macaques and humans. Biological Sciences – Neuroscience: PNAS 2008 105: 19514–19519; published online November 25, 2008 [DOI: 10.1073/pnas.0809662105]. Used with permission.

area," or FFA) (Kanwisher et al., 1997) (Fig. 24.5A). Recent functional imaging provides evidence for a fourth face-selective area in the human brain in the anterior temporal lobe, within the collateral sulcus (Tsao et al., 2008) (Fig. 24.5A).

Architecture of the Face Processing System in Macaques

Face-selective cells have been found scattered throughout the temporal lobe, though they tended to be found in clusters (Perrett et al., 1984). Because other kinds of shape selectivities also tend to be clustered (Desimone et al., 1984; Fujita et al., 1992; Tanaka et al., 1991; Wang et al., 1996), it was assumed that within the temporal lobe there was a columnar organization for shape, in which face columns represented just one of many shape-specific types of columns. However, this view was inconsistent with emerging evidence from human neurology and functional imaging that human face processing was localized to specific, reproducible regions of the temporal lobe. One possible explanation was that only a fraction of cells within fMRI-identified face-selective areas were actually face selective. Another explanation was that the face areas observed in fMRI experiments were unique to humans. Finally, it was also possible that macaques have face areas composed entirely of face cells, but previous single-unit recordings, which sampled inferior temporal cortex randomly, did not consistently target these areas.

The apparent discrepancy was resolved by Tsao and colleagues (Tsao et al., 2003, 2006,

2008), who found that in monkeys, as in humans, face processing, as revealed by functional imaging, is localized to six discrete regions of the temporal lobe (Fig. 24.5B). These six patches are distributed along the anterior-posterior axis of the temporal lobe and organized into one posterior patch on the lateral surface of TEO ("PL", for posterior lateral); two middle face patches in posterior TE, one located in the fundus of the STS ("MF," for middle fundus) and one on the lower lip of the STS ("ML," for middle lateral); and three patches in anterior TE, one located near the fundus of the STS ("AF," for anterior fundus), one on the lower lip of the STS and adjacent gyrus, in TEad ("AL," for anterior lateral), and one more medially on the ventral surface, just lateral and anterior to the anterior-medial temporal sulcus, in TEav ("AM," for anterior medial).

The face-patch system presents us with a new kind of functional organization in TE. The components of this network are coarser (few millimeters in diameter) than feature columns of inferotemporal cortex ($\approx$0.5 mm in diameter; Fujita et al., 1992; Wang et al., 1996), yet finer in scale than the coarse partitioning of IT into anatomically defined subregions (Felleman & Van Essen, 1991; Seltzer & Pandya, 1994; Von Bonin & Bailey, 1947). The face-patch system transgresses area boundaries, with face patches located in posterior, middle, and anterior portions of IT (Felleman & Van Essen, 1991).

Tsao and colleagues (2006) further showed that at least two of the macaque face areas, ML and MF, consist almost entirely of face-selective cells (Fig. 24.6). Tsao and colleagues used functional imaging to localize regions in the macaque temporal lobe that were selectively activated by faces, compared to nonface objects, and then they recorded almost 500 single units within ML and MF in two monkeys. They found a remarkable degree of face selectivity within these two regions, with 97% of visually responsive cells being face selective, on average showing almost 20-fold larger responses to faces than to nonface objects. The region where they recorded was quite posterior in the temporal lobe (6 mm anterior to the interaural canal, corresponding to posterior TE/anterior TEO). The fact that an area consisting almost entirely of face-selective cells exists so early in the ventral stream is consistent with the idea that the face processing pathway is a completely different pathway from the nonface object recognition pathway, and that it branches away early in the visual hierarchy. The existence of an area consisting almost entirely of face-selective cells so early in the ventral stream furthermore implies that the face processing pathway is gated by a face detection stage, at which nonface objects are filtered out. As we propose later, the existence of this detection gate may account for the seemingly special "holistic" aspect of face processing.

What is the functional significance of the anatomical localization of face processing? The cerebral cortex is functionally parcellated: Neurons concerned with similar things are organized into areas and columns, each having extensive interconnections and common inputs and outputs (Mountcastle, 1997). Face processing, an identifiable and discrete form of object recognition, appears also to be organized into anatomically discrete processing centers. Individual neurons connect directly with a small fraction of the rest of the neurons in the brain, usually to nearby cells, because longer axons delay neural transmission, are energetically expensive, and take up space. Furthermore, colocalization of neurons concerned with face processing enables enriched local inhibitory interactions, since inhibitory neurons are always local (Somogyi et al., 1998). Wang and colleagues (2000) recorded responses in anterior IT to a set of complex stimuli before, during, and after applying the γ-aminobutyric acid (GABA) receptor antagonist bicuculline near the recording electrode. In many cases, for both face-selective and non–face-selective cells, blocking local inhibition revealed responses to previously nonactivating stimuli, which were often activating stimuli for neighboring cells. This suggests that neighboring cells refine each other's response selectivity by mutual inhibition.

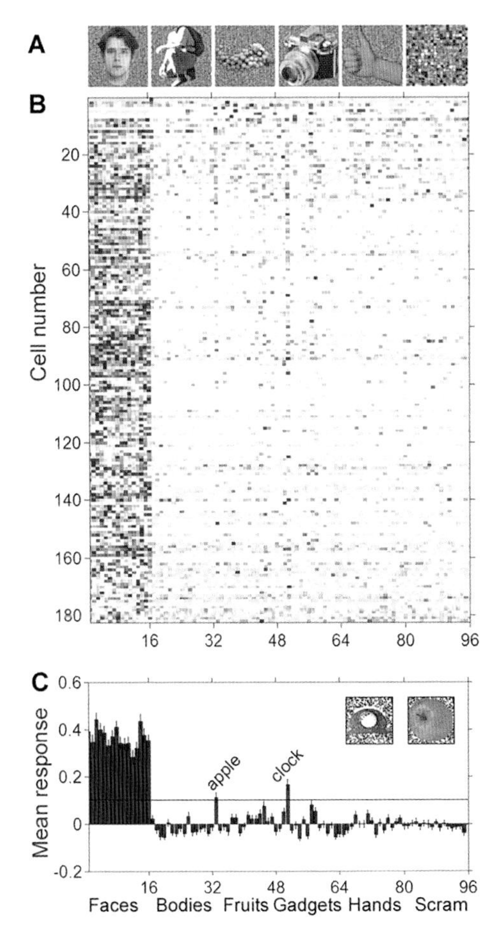

Figure 24.6 Mapping face and object selectivity in the monkey brain. (**A**) Five stimulus categories included faces, four nonface object categories (hands, gadgets, fruits, and bodies), and grid scrambled patterns. (**B**) Responses of 182 neurons from the middle face patch of one monkey to 96 images of faces and nonface objects. (**C**) Average normalized population response to each image. Tsao, D. Y., Freiwald, W. A., Tootell, R. B. H., & Livingstone, M. S. (2006). A cortical region consisting entirely of face-selective cells. *Science, 311*, 670–674. Used with permission.

While tracer studies demonstrate connections between specific nodes in TEO and TE (Saleem et al., 1993), the need to sacrifice the animal to process the tissue prevents assessment of the functional properties of connected nodes. In general, to understand functional architecture, it is necessary to combine connectivity maps with functional topography. The face processing system of macaque monkeys, consisting of six patches that can be identified by fMRI and targeted for anatomical experiments, provides an ideal preparation for dissecting the large-scale functional anatomy of one discrete aspect of object recognition.

To identify the connectivity of individual face patches, Moeller and colleagues (2008) used electrical microstimulation combined with simultaneous fMRI. Stimulation of each of four targeted face patches produced strong activation specifically within a subset of the other face patches (Fig. 24.7). Stimulation outside the face patches produced an activation pattern that spared the face patches. These results indicate that the

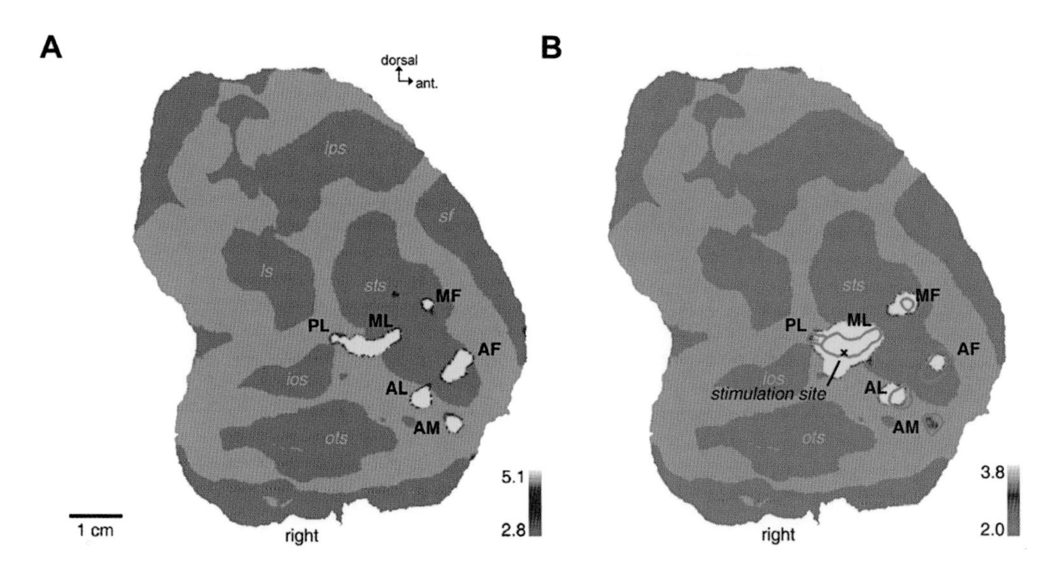

Figure 24.7 Combined functional magnetic resonance imaging and microstimulation demonstrates that the six macaque face patches are strongly and specifically interconnected. (**A**) Face patches from one macaque shown on a right hemisphere flat map. (**B**) Regions activated by microstimulation at the region marked by an × compared to blank. Face patches indicated by green outlines. The animal fixated on a blank screen during the entire experiment. Face region and sulcal abbreviations as in Figure 24.5B. Moeller, S., Freiwald, W. A., & Tsao, D. Y. (2008). Patches with links: a unified system for processing faces in the macaque temporal lobe. Sebastian Moeller, Winrich A. Freiwald, and Doris Y. Tsao (6 June 2008) *Science* 320(5881), 1355. [DOI: 10.1126/science.1157436]. Used with permission.

face patches form a strongly and specifically interconnected network.

The existence of this network creates unprecedented possibilities for gaining a systematic understanding of information flow in IT. For the first time we can access anatomically distinct components of a unified object processing network in vivo and intercept the messages being sent between connected cell pairs within different patches. Thus, in addition to characterizing in detail the properties of cells within each patch, we can sequentially characterize the transformations that occur across patches—transformations that somehow lead to our ability to recognize thousands of different faces effortlessly even under incredible variation (i.e., Fig. 24.1).

Interim Summary

Nonface objects are represented by widely distributed and overlapping representations in ventral temporal cortex. Anatomical tracing, optical imaging, and single-unit studies in macaques suggest that the neural machinery supporting these nonface object representations has a precise circuitry. The unit element of this circuitry is a column ≈0.5 mm wide. Specific subsets of columns within TEO and TE are organized into connected networks. Alongside the machinery for recognizing nonface objects, the temporal lobe of both humans and macaques contains a system of strongly face-selective regions. In macaques, this system comprises six patches of cortex extending from TEO to anterior TE and, in humans, four areas spanning the length of the temporal lobe. The macaque face patches are strongly and specifically connected to each other, and at least two of the macaque face patches consist almost entirely of face-selective cells.

REPRESENTATIONS IN THE INFERIOR TEMPORAL LOBE

In this section we discuss the physiology of cells in IT that support the remarkable capacity to recognize objects. We present a review of the findings derived from visual stimulation using parametric visual features and general objects. We will also present results of electrophysiological studies of face representations in visual cortex. Finally, we address the issue of holistic facial processing and its relation to electrophysiological findings.

Representation of Objects

Tanaka and colleagues pioneered the study of tuning properties of single cells in IT (Tanaka, 1996; Tanaka et al., 1991). Their work was converged on the concept of a "critical feature," defined as the simplest feature that elicits the maximum response in a cell. To identify the critical feature of an IT cell, they first showed many three-dimensional plant and animal models to the cell at different views, then made pictures of these objects and systematically simplified them until they were able to identify the minimal feature that elicited maximal activation (Kobatake & Tanaka, 1994). For example, the critical feature of a cell preferring a tiger might be two round black and white gratings arranged like a snowman (Tanaka, 2003). The main conclusion was that most cells in TE are selective for moderately complex feature combinations but not for whole objects. Such cells are already present in V4 and TEO, but at a lower proportion, mixed with cells selective for simple features (such as color or orientation).

Subsequent attempts to decipher the code used by the ventral stream have followed two general lines of attack: (1) a bottom-up approach, starting with a model of structural encoding, and testing the model, and (2) a black box approach, using responses of a random population of cells to various complex objects to derive general principles of IT coding. These two approaches represent two different ways of conquering the huge parameter space occupied by the set of all possible visual inputs. The structural encoding approach assumes that the brain represents most objects by decomposing them into parts and part relations. The black box approach, on the other hand, does not make any concrete assumptions about how objects might be coded, but it does make the assumption that experimenter-chosen real-world objects will span enough of the coding space in sufficient detail that the approach will yield meaningful insights.

A parts-based representation has intuitive appeal. When asked to describe an object (e.g., a flower), people naturally describe its parts (e.g., petals, leaves, stem). The distributed, combinatorial code created by a parts-based representation permits an extremely rich set of objects to be represented. A parts-based representation of words doubtless in fact underlies your ability to read this page: (1) the geometry of letters may take advantage of the brain's hardwired ability to distinguish contour parts (Changizi et al., 2006), through their location ("p" vs. "b"), curvature ("v" vs. "u"), and connectivity ("t" vs. "T"), and (2) the possibility of forming infinite meanings from 26 characters illustrates the representational power of a parts-based structural code.

Connor and colleagues have uncovered evidence for parts-based coding in V4 (Pasupathy & Connor, 1999, 2001, 2002) and PIT (Brincat & Connor, 2004, 2006; Haxby et al., 2001) through single-unit recordings examining tuning to parametrically defined contour features. Pasupathy and Connor (2001) created a large set of closed contours that could be parameterized by curvature and angular position (Fig. 24.8A), and then analyzed the responses of V4 cells to these shapes. They found that most cells in V4 were tuned to a particular curvature at a particular location within the cell's receptive field (e.g., a sharp convexity in the lower left). Furthermore, they showed that by combining these tuning curves, they could reconstruct (approximately) an unknown test shape from the pattern of responses in V4 elicited by the shape (Fig. 24.8B,C) (Pasupathy & Connor, 2002). Extending this work to IT cortex, Connor and colleagues found cells tuned for the presence of multiple parts (specifically, tuning could be described by the sum of two to four subunits with Gaussian tuning in a six-dimensional curvature, orientation, and position space) (Brincat & Connor, 2004). Figure 24.8D illustrates a cell tuned to a combination of concave contour fragments oriented toward the lower right and toward the lower left, and located in the bottom right of the

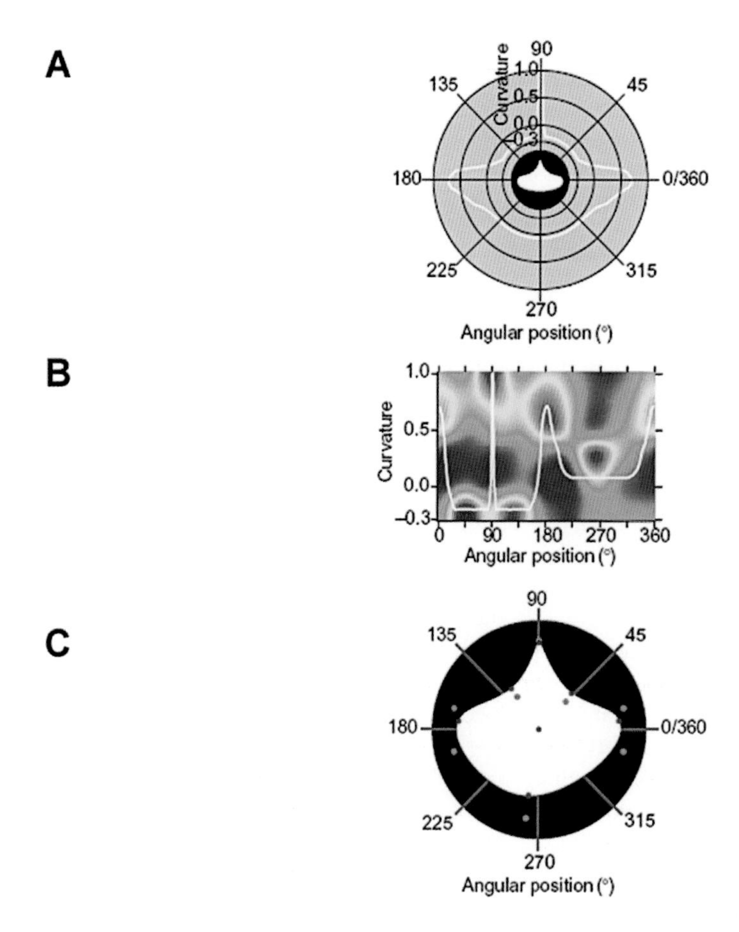

Figure 24.8 Encoding of contour curvature in V4 and PIT. (**A**) The white line represents the boundary curvature as a function of angular position of the test shape in the center. (**B**) Estimated population response across the curvature × position domain to a test shape with the veridical curvature function superimposed (white line). The response was estimated by weighting each cell's curvature × position tuning function by its response to the test shape. (**C**) Reconstruction of the test shape from the population response in B. (**D**) Response of a posterior inferotemporal cell to its preferred feature: a combination of concave contour fragments oriented toward the lower right and toward the lower left (highlighted in green). The cell was highly sensitive to the position of these contour elements relative to the object. It responded strongly to concavities at the left of object center (bright green), as exemplified by the average response (gray histogram) to the stimuli in the top two rows. The light gray shading indicates the 500-ms stimulus-presentation period. The cell responded only weakly to the same concavities at the right (dark green; bottom two rows). This tuning for object-relative position was consistent across changes in absolute position of the stimuli. Thus, cells in PIT are coding the curvature of contour fragments in object-centered coordinates. A–C from Pasupathy, A., & Connor, C. E. (2002). Population coding of shape in area V4. *Nature Neuroscience, 5,* 1332–1338. D from Connor, C. E., Brincat, S. L., & Pasupathy, A. (2007). Transformation of shape information in the ventral pathway. *Current Opinions in Neurobiology, 17,* 140–147. Used with permission.

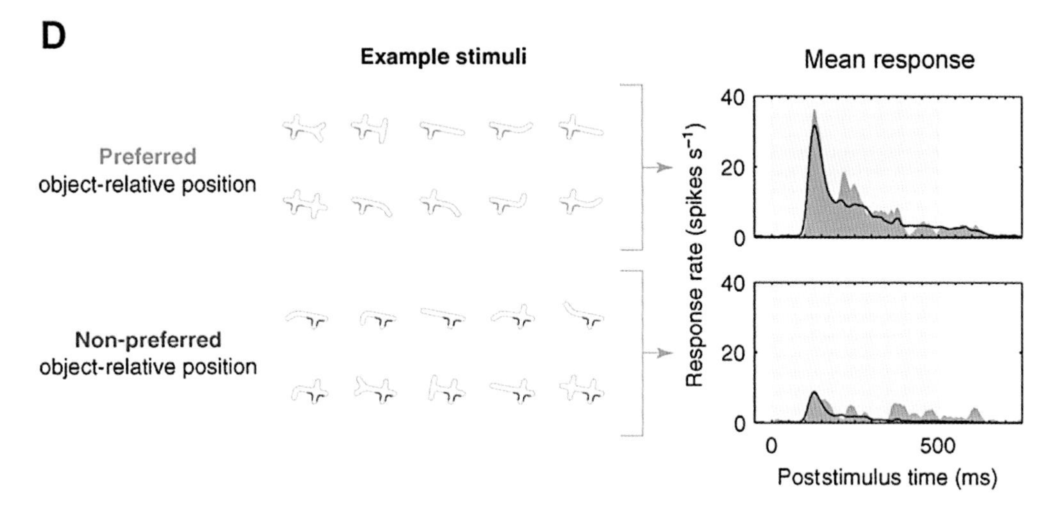

Figure 24.8 (Continued)

object. These experiments suggest that object recognition relies to a significant extent on a parts-based code implemented by tuning curves of single neurons. One criticism of this work is that the space of real shapes is much vaster than that tested. But any study of IT must employ an experimenter-limited stimulus set; the finding that cells in both V4 and TEO respond to parts of objects and not to the whole object (albeit in an object-centered reference frame) must be explained by any theory of object representation in the ventral stream.

A structural encoding scheme for representing three-dimensional objects was proposed by Irving Biederman (Biederman, 1995). In the "geon theory" of object recognition, a given view of an object is represented as an arrangement of simple, viewpoint-invariant, volumetric primitives called "geons." Five examples of geons are shown in Figure 24.9A. According to the geon theory, an object is represented by the geons it contains (out of 24 total), together with pairwise geon relations (e.g., "above"; 81 total) and geon attributes (e.g., "horizontally oriented, narrow relative aspect ratio"; 15 total). These limited parameters could represent 10.5 million different two-geon objects.

The geon theory is supported by experiments with partial line drawings. The effect of priming with complementary contours that preserve geon structure is just as strong as the effect of priming with identical contours, and significantly stronger than the effect of priming with an abstract object category (Biederman & Cooper, 1991) (Fig. 24.9C). Thus, priming must be attributed to a representation of the parts of the object (and their interrelations) and not to the activation of the image features or abstract categories.

The defining feature of geons is their view invariance. Metric properties of objects (e.g., aspect ratio), in contrast, are not viewpoint invariant. The geon theory predicts that responses to objects differing in geon structure should be easier to distinguish, across changes in view, than responses to objects differing only in metric properties. To test this, Vogels and colleagues (2001) recorded responses of IT cells to six variations of an object: the original object composed of a pair of geons, a "metric property" variant in which one of the geons was metrically varied (e.g., made wider), and a "nonaccidental property" variant in which one of the geons was changed (e.g., a cylinder replaced by a block). Rotated versions of these three images were also constructed. The two variants were equated in terms of low-level image statistics. Their results provide partial support of the geon theory. A population multidimensional scaling (MDS) analysis revealed two dimensions, one corresponding to rotation, and one to geon change (Fig. 24.10D). The population response to the metric variant was very close to that of the original image (i.e., neurons had trouble distinguishing them), while the response to the geon variant was well removed. This supports the idea that cells in IT have a special sensitivity to differences in geon structure over differences in metric structure. However, the large difference between the population response to each object and its rotated counterpart shows that the representation of object shape in IT is *not* view invariant.

The last finding is consistent with an alternative view that the primitives for object recognition in IT are view based (Bulthoff et al., 1995). Experimental evidence for this theory comes from Logothetis and colleagues, who trained monkeys to match three-dimensional paperclips across changes in view. Subsequently, they recorded in anterior IT and found that many cells responded maximally to a particular view of a paperclip, with the response declining gradually as the object was rotated away from this preferred view. A very small number of cells were found that responded in a view-invariant manner.

While a large number of objects can be represented by parts-based parametric schemes such as curved contours (Pasupathy & Connor, 2001) or geons (Biederman, 1987), there are still many objects that do not fit easily into such parametric shape schemes, yet may be biologically important to distinguish. Thus, a second approach to studying IT cortex has

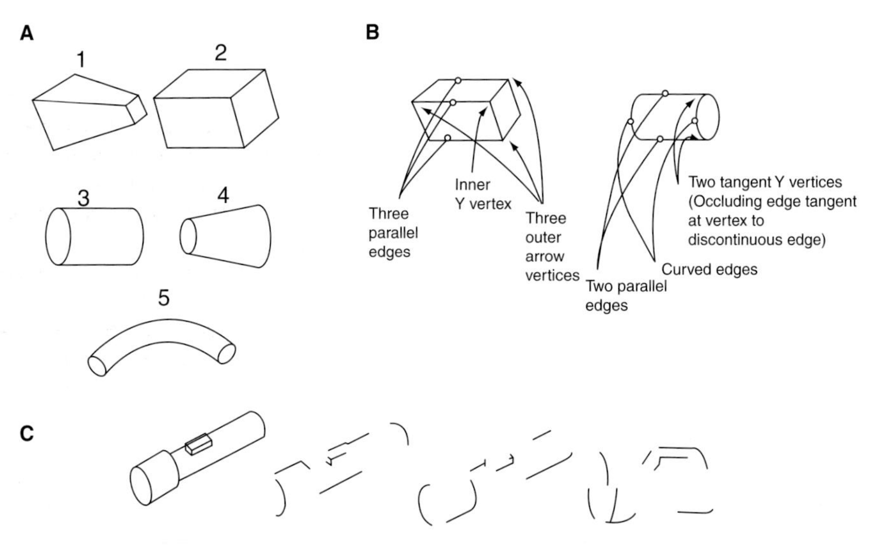

Figure 24.9 Structural encoding by geons. (**A**) Examples of fives geons. (**B**) A block and a cylinder, two examples of geons, show particular nonaccidental differences that may be used by the brain to recover their structure from line drawings. (**C**) Examples of the four stimulus types used by Biederman and Cooper to demonstrate the existence of a geon-level encoding in a psychophysical priming experiment. (**D**) Neurophysiological evidence for representation of nonaccidental properties. Top: Six classes of images used by Vogels and colleagues to test for a geon representation in macaque IT. Bottom: Multidimensional scaling of inferotemporal responses to these image classes reveals a dimension representing the difference between a view-dependent change and a viewpoint-invariant change (dimension 1), and a dimension representing rotation (dimension 2). A–C from Biederman, I. (1995). Visual object recognition. In: S. M. Kosslyn & D. N. Osheron (Eds.), *Visual cognition* (pp. 121–166). Cambridge, MA: MIT Press. D from Vogels, R., Biederman, I., Bar, M., & Lorincz, A. (2001). Inferior temporal neurons show greater sensitivity to nonaccidental than to metric shape differences. *Journal of Cognitive Neuroscience, 13,* 444-453. Used with permission.

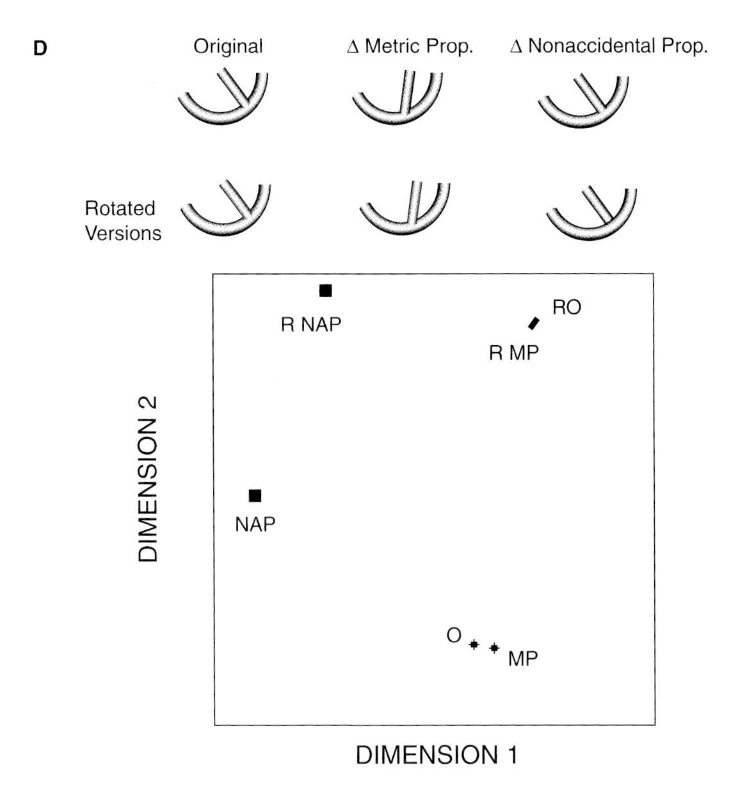

Figure 24.9 (Continued)

been to stimulate cells with random sets of complex images and ask, by means of various readout procedures, what types of information are available in the population of responses (e.g., how fast can information be read out, how many neurons are needed to code certain types of information, etc.). Obviously, a null result does not say much, since the readout algorithm used may simply be inadequate. On the other hand, a positive result from such an approach does reveal what kind of information is present. It should be noted that readout techniques can also be used to examine structural encoding hypotheses—for example, one could use classifier techniques to ask whether information about curvature can be read out from cells in a position- invariant way. The main difference lies in whether the stimulus space consists of parametric combinations of parts or randomly selected real-world objects.

Kiani and colleagues (2007) measured responses of more than 600 neurons in monkey IT cortex to over a thousand different images of natural and manmade objects during passive fixation. Separation of the images into clusters based on response similarity across the population of cells showed that the neuronal population sorted the images into intuitive categories: Animate and inanimate objects created the most distant distinguishable clusters in the population code. The global category of animate objects was divided into bodies, hands, and faces. Faces were divided into primate and nonprimate faces, and the primate-face group was divided into human and monkey faces. Bodies of human, birds, and four-limb animals clustered together, whereas lower animals such as fish, reptiles, and insects made another cluster. Importantly, low-level image

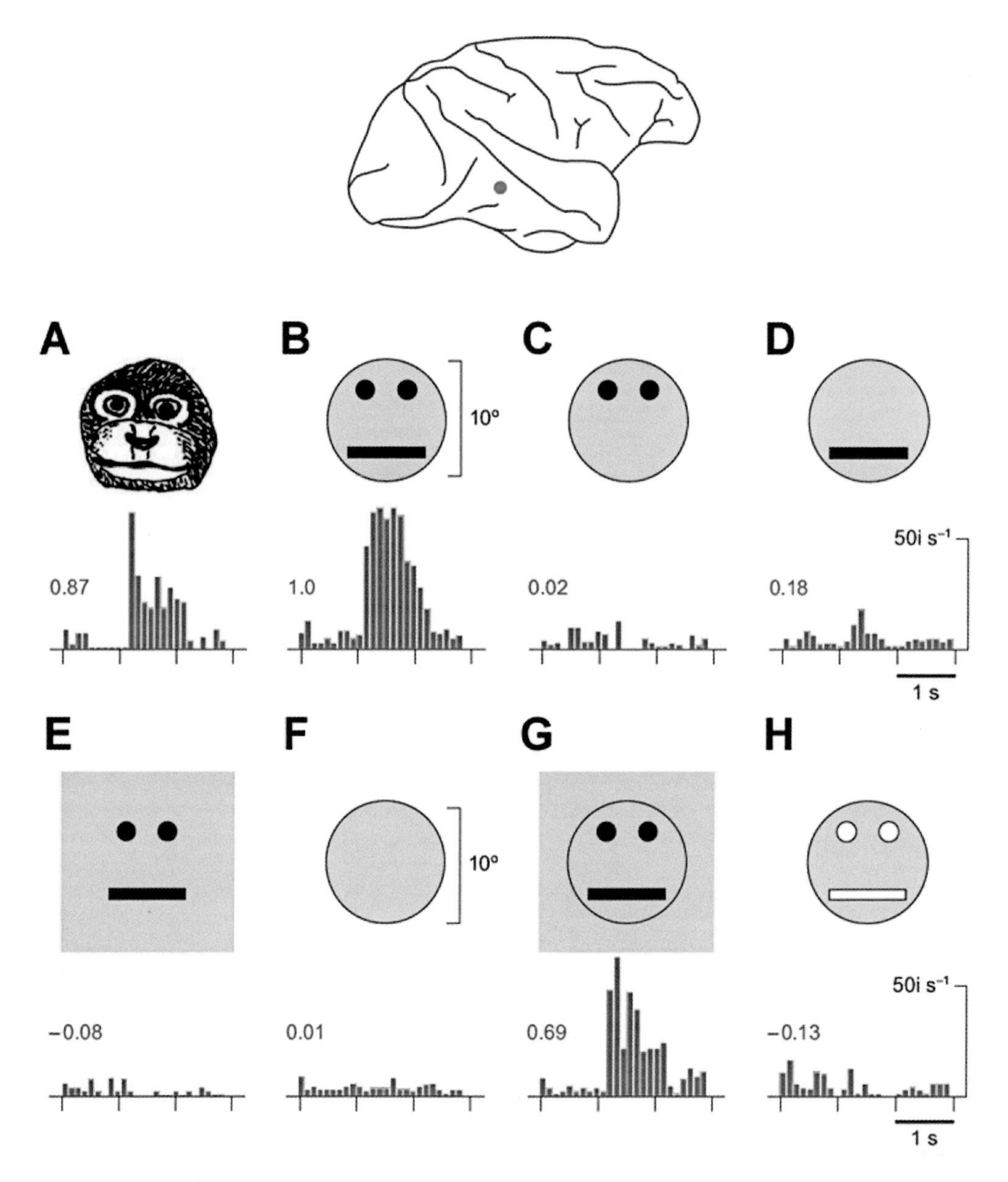

Figure 24.10 Holistic face detection by an inferotemporal cell. Top: Recording site of a face cell. (**A—H**) Response selectivity. From Kobatake, E., & Tanaka, K. (1994). Neuronal selectivities to complex object features in the ventral visual pathway of the macaque cerebral cortex. *Journal of Neurophysiology, 71,* 856–867. Used with permission.

characteristics could not account for this categorical classification. Therefore, they concluded that monkey IT specifically extracts complex features for the purpose of object categorization.

Hung and colleagues (2005) used a classifier-based readout technique to analyze responses of 256 IT recording sites to 77 images. They found that the activity of ≈100 IT neurons over very short time intervals (as small as 12.5 ms contained sufficient information to read out object "identity" at 72% (chance = 1/77) and object "category" at 94% (chance = 1/8) accuracy. Importantly, this information generalized over a range of object positions and scales, with less than 10% reduction in performance. These results show that invariant information about object category and identity is available in small populations of neurons even during an early phase of the response. Can object recognition be completely solved by means of units with the invariance and tuning properties of randomly sampled IT cells, as already described in studies such as that of Hung and colleagues? Or are we still waiting to observe a new type of invariance property? Rigorous tests comparing physiology to behavior have not been performed to test this important question.

Representation of Faces

One of the biggest differences between face perception and general object recognition is that arbitrary objects map to different aspects of contour geometry. Faces, on the other hand, share a common template, consisting of eyes, nose, mouth, and face outline (though numerous variations are possible; e.g., depending on the view angle, only one eye may be visible). The vastly reduced template space makes understanding the mechanism of face perception more tractable compared to understanding the detection and recognition of arbitrary objects.

That face cells are truly detecting faces, and not some more abstract basis set in which all possible shapes are represented by different cells, with some cells tuned to particular parameters that happen to fit the faces better than any of the other objects tested, was demonstrated by Foldiak and colleagues (2004). They presented 600 to 1,200 stimuli randomly chosen from several image archives to cells recorded from both the upper and lower bank of the STS and found that the distribution of tuning to these images showed bimodality (i.e., cells were either predominantly face selective or not face selective).

Experiments with cartoon faces show that in general face cells require an intact face and are not just selective for individual features (Bruce et al., 1981; Desimone et al., 1984; Kobatake & Tanaka, 1994; Leonard et al., 1985; Oram & Perrett, 1992; Perrett et al., 1982, 1984; Scalaidhe et al., 1999; Tsao et al., 2006). Figure 24.10 shows nonlinear combinatorial response properties of a face-selective cell recorded in IT (most likely in the face patch ML, judging by the recording location) by Kobatake and Tanaka (1994). Out of a large number of three-dimensional objects, this cell responded best to the face of a toy monkey (A), and by testing various simplified two-dimensional paper stimuli, they determined that the cell would also respond to a configuration of two black dots over a horizontal line within a disk (B), but not in the absence of either the spots or the line (C and D) or the circular outline (E). The contrast between the inside and the outside of the circle was not critical (G), but the spots and the bar had to be darker than the disk (H). In other words, the cell only responded when the stimulus looked like a face, no matter how simplified. How face cells detect a facial Gestalt is still unknown, and will likely require identifying the inputs to the face patches.

Once a face has been detected, it needs to be identified. Recordings in the face patches ML and MF have begun to shed light on the neural mechanism for distinguishing between different faces. Freiwald and colleagues (2009) used cartoon faces to study the neural basis of face measurement. Dense parametric mapping was used to measure responses of cells in the middle face patch to a cartoon stimulus in which all face parameters were independently varied. Cells were found to be tuned to the geometry of facial features, with different cells tuned to different feature subsets. Tuning was strikingly ramp shaped, with a one-to-one mapping of feature magnitude to firing rate (Fig. 24.11). These extreme values extended to or even transgressed the limits of realistic face space. For example, intereye distances ranged from almost cyclopean to abutting the edges of the face, and the most extreme face aspect ratios were outside those of any known primate. Monotonic tuning allows for simple readout (Guigon, 2003) and may be a general principle for high-level coding

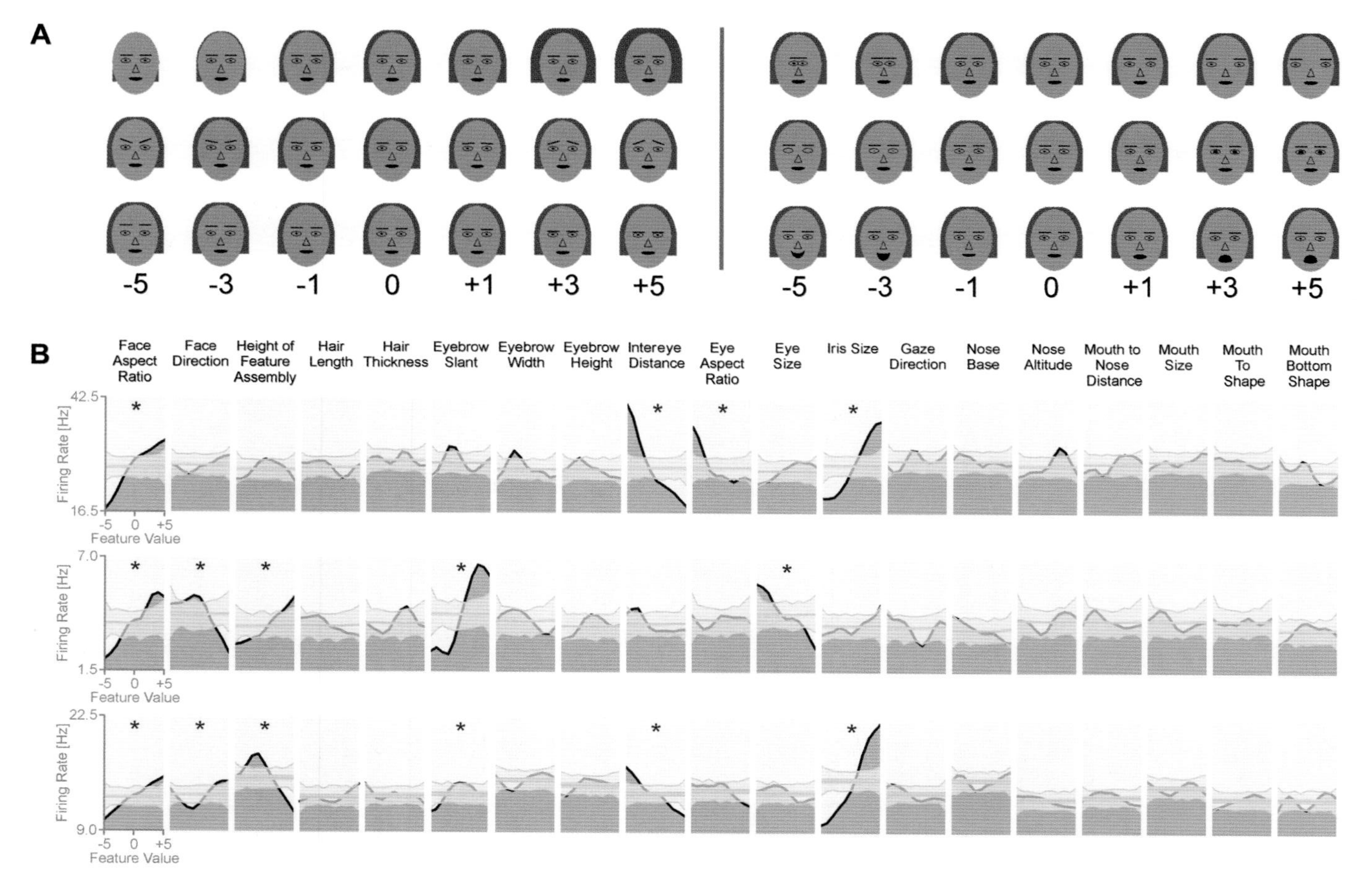

Figure 24.11 Tuning of face cells to a cartoon face space. (**A**) Six example dimensions (hair width, eyebrow slant, eyebrow height, intereye distance, iris size, mouth top shape) of the 19-dimensional cartoon space. Each row shows example values for one parameter, with all other parameters fixed at their mean. (**B**) Tuning curves of three example cells to each of the 19 feature dimensions. In gray maximal, minimal, and mean values from shift predictor are shown. Stars mark significant modulation. From Freiwald, W. A., Tsao, D. Y., & Livingstone, M. S. (2008). A face feature space in the macaque temporal lobe. Nature Neuroscience 12, 1187–1196 (2009). Published online: 9 August 2009 [DOI: 10.1038/nn.2363]. Used with permission.

of visual shapes (Kayaert et al., 2005). It may also aid in emphasizing what makes an individual face unique (i.e., separates it from the average face), and could be the neural basis for the power of caricatures. The breadth of tuning underscores the fact that cells in the middle face patches, ML and MF, encode axes and not individual faces.

Recordings made in the highest stage of the form processing hierarchy, the medial temporal lobe, reveal the existence of cells that respond to specific familiar individuals in a representation-invariant manner (Quiroga et al., 2005), as expected of a "grandmother cell." For example, some cells each responded to multiple pictures of a well-known individual as well as to a letter string of their name, but were unresponsive to all other images. Such individual-specific cells have not been found in the lateral inferior temporal lobe, where most face cells in monkeys have been recorded, although, as a population, cells in the anterior inferior temporal gyrus of the macaque (in the general vicinity of AM) can support view-invariant identification (Eifuku et al., 2004).

Holistic Processing

Since the thesis of this chapter is that understanding face areas will help us understand object recognition, it is appropriate to discuss the claim that face processing is supported by a unique set of mechanisms that do not generalize to nonface objects. Face processing obviously is unique in being housed in a set of specialized cortical regions, but this doesn't imply that the underlying mechanisms are unique.

Specifically, face processing is claimed to be distinct from nonface object processing in that it is "holistic"; that is, faces are represented as nondecomposed wholes, rather than as a combination of independently represented component parts (eyes, nose, mouth) and the relations between them (Farah et al., 1998). Evidence for holistic processing of faces comes from a number of behavioral paradigms, of which the two most cited are the part-whole effect (Tanaka & Farah, 1993) and the composite effect (Young et al., 1987). In the part-whole effect, subjects are better able to identify two face parts when the parts are presented in the context of a whole face than in isolation. In the composite effect, subjects are slower to identify half of a chimeric face aligned with an inconsistent other half-face than if the two half-faces are misaligned (Young et al., 1987). As with the part-whole effect, the composite effect indicates that even when subjects attempt to process only part of a face, they suffer interference from the other parts, suggesting an inability to access parts of the face and mandatory processing of the whole face.

Holistic face processing could be explained by the existence of an obligatory detection stage that uses a coarse upright template to detect whole faces (Tsao & Livingstone, 2008). An aligned chimera would be *obligatorily detected* as a whole face and therefore processed as a unit by subsequent measurement and classification stages. The key evidence favoring this early detection gating hypothesis comes from the finding of six face-selective areas in the macaque (Tsao et al., 2006, 2008) and the finding that the middle face patches, located early in this hierarchy, already consist entirely of face-selective cells (Tsao et al., 2006). According to this viewpoint, holistic psychological markers do not indicate that faces are processed by a mysterious and unique mechanism; they simply indicate that faces are processed by a system that is employed only for stimuli that are first detected as faces, but this system may be parts based, and is also likely optimized for analyzing face parts. An as yet unexplored question is whether other object categories are first detected and then analyzed by specialized modules. Other object categories that may fit this scheme include bodies and words.

Interim Summary

Perhaps the central theme that has emerged from IT research over the past two decades is that apart from biologically special categories (most conspicuously faces), objects are coded in IT in terms of their parts and not wholes. Early experiments studying the responses of IT cells to decomposed complex objects found that most cells are selective for moderately complex features but not for whole objects. Consistent with a parts-based code, cells in V4 and PIT show tuning to the curvature and location of contour elements in object-centered coordinates. Biederman and colleagues have argued for explicit structural encoding of view-invariant volumetric primitives, but evidence suggests that the responses of most cells in TEO and TE are view dependent. Although IT cells do not generally appear to be detectors for complex objects, there are consistently observed populations of cells selectively responsive to animate objects such as faces, bodies, and hands, suggesting that animate objects may be treated differently from other types of complex patterns. A major distinction between face processing and nonface object processing is that early on (by the middle face patches at the latest), the form of a face has been *obligatorily detected as a whole.* This observation may explain holistic psychophysical phenomena associated with face perception. Cells in the middle face patch encode facial identity through ramp-shaped tuning to subsets of facial features, consistent with a parts-based code of face identity. At the highest stages of form processing, in the medial temporal lobe, cells are found that respond to specific familiar individuals in a representation-invariant manner.

COMPUTATIONAL APPROACHES TO OBJECT RECOGNITION

Theories of visual object recognition have received attention from both the neuroscience community and the machine vision community. Both approaches have contributed to our understanding of how visual object recognition might work in biological systems. We will summarize key work in both of these areas. While such a survey does not do justice to the individual contributions of each theory or system, it does allow us to see the commonalities that might be useful for producing a consensus view of computational object recognition and for applying this interpretation to processing in the primate brain, especially in relation to face processing. In this section we begin by providing a cursory survey of models of biological object recognition. We then turn to a class of models from machine vision that produces state-of-the art recognition abilities. We also examine a different class of models that uses feedback or contextual information during inference and object recognition. Finally, we relate these works to our view of face processing in human and nonhuman primates.

Theories of Biological Object Recognition

One of the central problems of object recognition is that effective systems must deal with the variations present in the natural world: variation in position, size, rotation, illumination, and even nonrigid motion (i.e., movement of limbs). This problem materializes when a visual system must generalize from its previous experience with an object under specific conditions to conditions in which the object has never been viewed before. This is a computationally difficult problem. It is made even more difficult by the fact that the novel object view may differ from previously experienced views in several of these variations at the same time. In order to deal with these variations, researchers have sought representations that will be invariant, or unchanging, under transformations that produce image variation. With respect to this problem, most models of biological object recognition take one of two approaches: models that directly compute invariant responses and those that correct for transformations to produce invariant responses. See Wiskott (2004) for an additional review and analysis of these two approaches.

Models that directly compute invariant responses take many approaches. However, there are strong commonalities among a group of models that seek to explain invariant object recognition in biological systems. This class of models can be traced back to the hypothesis originally proposed by Hubel and Weisel that complex cell responses in primary visual cortex are formed by combining the responses of spatially shifted simple cell responses (Hubel & Wiesel, 1962, 1965). One of the earliest computational instantiations of this framework was the Neocognitron model of Fukushima (Fukushima et al., 1988). The Neocognitron is a hierarchical model consisting of layers that alternately perform feature extraction (similar to simple cells) and layers that build invariance to position or small deformations (similar to complex cells). The weights in each layer determine the type of computation performed. The hierarchy alternately builds up selectivity and invariance until at the top layer the responses are selective to complex shapes and invariant to large changes in size and position. Recent theoretical work and simulations have extended this approach in several directions: VisNet (Wallis & Rolls, 1997), SeeMore (Mel, 1997), and HMAX (Riesenhuber & Poggio, 1999). Of note are the extensions and elaboration of the HMAX model, which have demonstrated correspondence to human level object recognition performance during brief presentations (Serre et al., 2007b), explanations of selectivity and invariance in inferotemporal cortex (Hung et al., 2005) and visual area V4 (Cadieu et al., 2007), and models of biophysical computation (Kouh & Poggio).

In approaches that deal with transformations explicitly, the model tries to correct for, or undo, the transformations so that the image can be matched against a stored canonical representation. In the model of Olshausen and colleagues (1993), the visual input is dynamically routed by control neurons through the visual hierarchy to rescale and shift the image. At the top of the hierarchy, the shifted and rescaled image can then be compared to stored templates of canonical object views, thus producing invariant recognition. The control neurons, which explicitly model the transformation, engage in an active process to selectively turn on and off connections between layers of the visual hierarchy. Later work by Arathorn (2005) suggests a computationally tractable method for determining the appropriate transformations for a given input image. It is interesting that theories that deal explicitly with transformations often use feedback: Memory representations guide the search for the correct transformation, while theories that directly compute invariant representations are constructed in a primarily feedforward network.

Bilinear models are a class of mathematical models closely related to the dynamic routing theory of invariant recognition. In recent work, Tenenbaum and Freeman (2000) describe a bilinear model that separates the "content" and "style" in various data types. These models extend traditional linear models and are referred to as bilinear because they are linear when one set of the variables is held fixed. This model allows each set of variables to explicitly represent different types of information. For example, in text, the character identity, or content, is separated from the font, or style. In relation to the dynamic routing theory, the style variables can be considered as dynamically routing the content variables to produce an image with specific style and content. Therefore, in the context of invariant object recognition, the content could be ascribed to the identity of the object and the style to the visual variations or transformations. In one promising example, Tenenbaum and Freeman use a bilinear model on a dataset of face images containing a number of individuals viewed at a range of poses. The bilinear model was able to learn from this dataset to separate the information into a set of variables indicating the identity of a person and a set of variables indicating the pose of the person.

Computer Vision and Object Recognition

While the models we have addressed so far have been directed at explaining biological object recognition or perception, a number of techniques from computer vision have interesting implications for biological object recognition and face processing in the brain. Importantly, several computer vision models have demonstrated impressive performance on nontrivial object recognition problems and have demonstrated performance comparable to humans for specific tasks. For example, in an impressive demonstration, Sivic and Zisserman (2003) developed a system that is capable of searching through a Hollywood movie for a specific object, aptly named "Video Google." Related to the HMAX model we discussed previously, Serre and coauthors (2007a) demonstrated that a feedforward computer vision algorithm performed at levels comparable to human subjects during a rapid presentation of a visual stimulus in which they attempted to determine if an animal was present in the image. Furthermore, the computer vision algorithm and the human subjects exhibited similar patterns of error on individual images and image categories, such as far versus close views of animals.

Many of the most effective computer vision algorithms can be described as consisting of two stages: The first stage produces a set of local features, while the second stage performs classification or matching to determine the presence of an object or the identity of an exemplar (Berg et al., 2005; Lowe, 2004; Ranzato et al., 2007; Serre et al., 2007b). In the work of Serre and colleagues (2007b), a feature hierarchy produces a dictionary of visual features that are invariant to object variation. Standard classification techniques are used on the outputs of the hierarchy for a variety of visual recognition tasks, such as identification and classification. Interestingly, this model is inspired by a theory of biological object recognition (Serre et al., 2005) and falls into the class of biological object recognition models that seek to directly compute invariant responses. Additional work of Ranzato and colleagues (2007) takes a similar approach and develops an unsupervised learning algorithm to specify the selectivity of the intermediate layers. Lowe's influential SIFT algorithm (2004) also produces a set of invariant features, but instead of using a classifier to detect or identify object, the SIFT algorithm determines a geometric correspondence between the features. This increases the accuracy of the algorithm for object recognition because it ensures a consistent geometric interpretation of the features (e.g., the eye features, nose features, and mouth features must be in the appropriate geometric relationship). Related work by Berg and colleagues (2005) uses a different set of features, but also tries to estimate the geometrical transformation between a novel object and stored object representations.

Another interesting, and relevant, class of models uses feedback, or context, to aid in the interpretation of an image and the recognition of objects within a scene. Jin and Geman (2006) describe a hierarchy of parts for a license plate detection system that uses context to disambiguate low-level information. In their system, license plates are modeled as a hierarchy of parts: whole license plates, groups of numbers, individual numbers, and number parts. When presented with an image, the system produces an interpretation of the visual scene by using both bottom-up cues and top-down hypotheses from prior knowledge of what license plates look like. The authors show that using their contextual hierarchy greatly improves recognition accuracy in a real-world dataset. Another example of the efficacy of contextual information is presented by Torralba and colleagues (2003). In their system, context is used to prime the location of a visual object. For example, in an indoor office scene there is a much higher probability of seeing a computer monitor than there is in an outdoor forest scene.

Insights for Biological Face Recognition

These results can be used to interpret findings related to face processing in primates, and allow us to develop mathematical models that make explicit the architecture of biological face processing. There are three main connections we want to point out. First, a rapid feedforward computation that produces invariant responses may be central to the detection of faces. Second, complicated facial discriminations and inferences may be mediated by a more complicated series of interconnected processing stages that explicitly model the visual transformations. Third, the representations in these later stages may represent the independent aspects of facial appearance and pose.

Because it is advantageous for facial detection to be computed quickly, approaches that directly compute invariant responses may be suitable for face detection. While direct computation of invariant responses often suffers from the combinatorial explosion of possible inputs (and therefore requires more processing to overcome the combinations of variations for a specific object), face detection is unique because of its ecologically critical role, and because faces constitute a relatively constrained set of visual inputs compared to all visual objects. The early stages in the ventral stream that exclusively respond to faces may serve as the culmination of this rapid process of directly computing invariant responses to faces.

We may also gain some insights about biological face processing from bilinear models, which seek to explicitly and separately represent different aspects of the visual input. In the example examined by Tennenbaum and Freeman, the pose of a face is represented independently of the facial identity (Fig. 24.12). In face processing it is often the case that the relevant information is not only the presence or the identity of a face but also what it is doing. For example, in complex social interactions it is important to infer intentions from subtle cues about head pose and gaze direction. In these cases the transformations that facial forms undergo are the goal of representation and not just variations in the input that must be overcome. Therefore, it seems sensible that explicitly representing these transformations would be advantageous. Could the different nodes of the interconnected system of face patches be representing different, largely independent aspects of facial form and variation? Such a representation would make specific information independently available for complex social inferences of ecological value.

Interestingly, the two main theoretical approaches to biological object recognition, directly computing invariant responses and explicitly accounting for transformations, may both be employed for face recognition and processing: Direct computation of invariant responses may mediate face detection, while computing transformations may mediate further inferences about facial characteristics. The evidence for holistic face processing seems

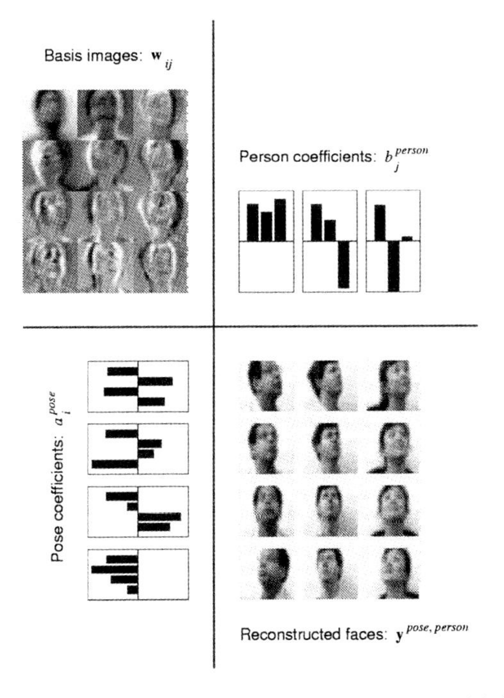

Figure 24.12 A bilinear model from Tenenbaum and Freeman. In this model, face images are represented by two sets of variables, one encoding pose, "pose coefficients," and the other encoding identity, "person coefficients." The basis images (top left) are weighted by the joint activity of the person coefficients (top right) and the pose coefficients (bottom left) to generate reconstructed images of faces (bottom right). From Tenenbaum, J. B., & Freeman, W. T. (2000). Separating style and content with bilinear models. *Neural Computation, 12,* 1247–1283. Used with permission.

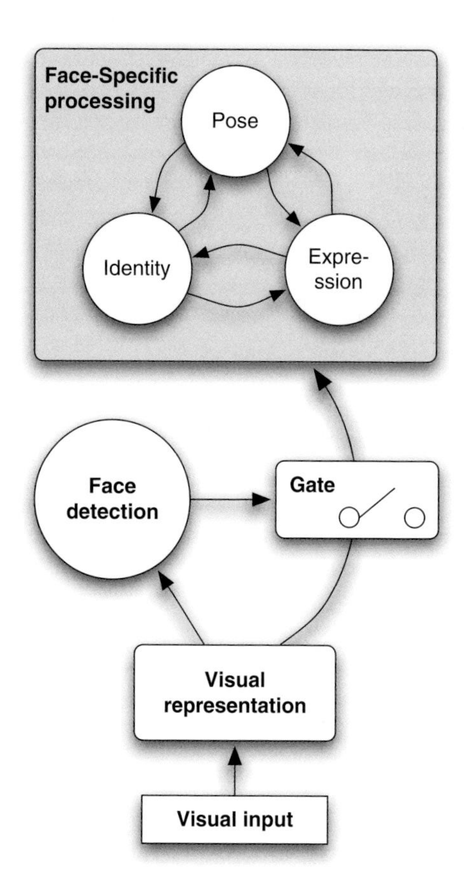

Figure 24.13 A model of face processing. In this model of face recognition a fast, feedforward detection stage provides feedback to the face model. The face model, which is only activated upon face detection, computes a variety of parameters specific to faces, such as pose, identity, and expression (among possibly others, such as gaze direction). The face detection stage may bias the face model to certain regions of the image, and certain sizes of faces from the information it infers from the input. Such information aids the face model in inferring more specific face parameters such as identity or pose.

to indicate that face detection gates this further processing.

We can make the connections between computational theories and biological evidence more precise by describing a simple model of face processing in primates. Presented in Figure 24.13, the system incorporates these insights. First, a fast, feedforward processing stage computes visual features from the visual input, and faces are detected from these features in a face detection area. Second, the detection of a face in the face detection module guides and gates further face-specific processing. Third, a face-specific processing stage, or face model, produces a rich description of the face. The face model may be similar to a bilinear model and mimic the dynamic routing theory of invariant object recognition.

SUMMARY

Hubel and Wiesel showed that primary visual cortex represents images in a space of localized, oriented edges. This was stunning because it showed that the brain performs a simple mathematical transformation that condenses the information present in the visual world. Primates possess an entire temporal lobe to further condense the visual form of objects. The thesis of this chapter is that the system of face patches in the macaque brain may be a "turtle's underbelly" (Medawar, 1981)—if we can understand it, we will be able to pry open the general problem of invariant pattern recognition.

What mathematical transformations are being performed in the face patches? An unprecedented opportunity now exists to understand the successive stages of face processing at a mechanistic level. The face-patch system offers a set of dedicated, connected, yet anatomically distinct components, inviting us to analyze for the first time the messages being sent within a hierarchical system for high-level object recognition. If we can figure out how the brain recognizes faces, then we will have gone a long way toward understanding how the brain represents a complex object made of multiple parts.

ACKNOWLEDGMENTS

We are grateful to Winrich Freiwald for discussions and comments on the manuscript.

REFERENCES

Allison, T., McCarthy, G., Nobre, A., Puce, A., & Belger, A. (1994). Human extrastriate visual cortex and the perception of faces, words, numbers, and colors. *Cerebral Cortex, 4,* 544–554.

Arathorn, D. W. (2005). A cortically-plausible inverse problem solving method applied to recognizing static and kinematic 3D objects. In: *Advances in neural information processing systems.* Cambridge, MA: MIT Press.

Berg, A., Berg, T., & Malik, J. (2005). Shape matching and object recognition using low distortion correspondence. *Proceedings of CVPR, 1,* 26–33.

Biederman, I. (1987). Recognition-by-components: A theory of human image understanding. *Psychological Review, 94,* 115–147.

Biederman, I. (1995). Visual object recognition. In: S. M. Kosslyn & D. N. Osheron (Eds.), *Visual cognition* (pp. 121–166). Cambridge, MA: MIT Press.

Biederman, I., & Cooper, E. E. (1991). Priming contour-deleted images: Evidence for intermediate representations in visual object recognition. *Cognitive Psychology, 23,* 393–419.

Boussaoud, D., Desimone, R., & Ungerleider, L. G. (1991). Visual topography of area TEO in the macaque. *Journal of Comparative Neurology, 306,* 554–575.

Boynton, G. M. (2005). Imaging orientation selectivity: Decoding conscious perception in V1. *Nature Neuroscience, 8,* 541–542.

Brincat, S. L., & Connor, C. E. (2004). Underlying principles of visual shape selectivity in posterior inferotemporal cortex. *Nature Neuroscience, 7,* 880–886.

Brincat, S. L., & Connor, C. E. (2006). Dynamic shape synthesis in posterior inferotemporal cortex. *Neuron, 49,* 17–24.

Bruce, C., Desimone, R., & Gross, C. G. (1981). Visual properties of neurons in a polysensory area in superior temporal sulcus of the macaque. *Journal of Neurophysiology, 46,* 369–384.

Bulthoff, H. H., Edelman, S. Y., & Tarr, M. J. (1995). How are three-dimensional objects represented in the brain? *Cerebral Cortex, 5,* 247–260.

Cadieu, C., Kouh, M., Pasupathy, A., Connor, C. E., Riesenhuber, M., & Poggio, T. (2007). A model of V4 shape selectivity and invariance. *Journal of Neurophysiology, 98,* 1733.

Changizi, M. A., Zhang, Q., Ye, H., & Shimojo, S. (2006). The structures of letters and symbols throughout human history are selected to match those found in objects in natural scenes. *American Naturalist, 167,* E117–139.

Cheng, K., Saleem, K. S., & Tanaka, K. (1997). Organization of corticostriatal and corticoamygdalar projections arising from the anterior inferotemporal area TE of the macaque monkey: A Phaseolus vulgaris leucoagglutinin study. *Journal of Neuroscience, 17,* 7902–7925.

Connor, C. E., Brincat, S. L., & Pasupathy, A. (2007). Transformation of shape information in the ventral pathway. *Current Opinions in Neurobiology, 17,* 140–147.

Dean, P. (1976). Effects of inferotemporal lesions on the behavior of monkeys. *Psychological Bulletin, 83,* 41–71.

Desimone, R., Albright, T. D., Gross, C. G., & Bruce, C. (1984). Stimulus-selective properties of inferior temporal neurons in the macaque. *Journal of Neuroscience, 4,* 2051–2062.

Diamond, R., & Carey, S. (1986). Why faces are and are not special: An effect of expertise. *Journal of Experimental Psychology: General, 115,* 107–117.

Eifuku, S., De Souza, W. C., Tamura, R., Nishijo, H., & Ono, T. (2004). Neuronal correlates of face identification in the monkey anterior temporal cortical areas. *Journal of Neurophysiology, 91,* 358–371.

Erickson, C. A., & Desimone, R. (1999). Responses of macaque perirhinal neurons during and after visual stimulus association learning. *Journal of Neuroscience, 19,* 10404–10416.

Erickson, C. A., Jagadeesh, B., & Desimone, R. (2000). Clustering of perirhinal neurons with similar properties following visual experience in adult monkeys. *Nature Neuroscience, 3,* 1143–1148.

Farah, M. J., Wilson, K. D., Drain, M., & Tanaka, J. N. (1998). What is "special" about face perception? *Psychological Review, 105,* 482–498.

Felleman, D. J., & Van Essen, D. C. (1991). Distributed hierarchical processing in the primate cerebral cortex. *Cerebral Cortex, 1,* 1–47.

Foldiak, P., Xiao, D., Keysers, C., Edwards, R., & Perrett, D. I. (2004). Rapid serial visual presentation for the determination of neural selectivity in area STSa. *Progress in Brain Research, 144,* 107–116.

Freiwald, W. A., Tsao, D. Y., & Livingstone, M. S. (2009). A face feature space in the macaque temporal lobe. *Nature Neuroscience,* Nature Neuroscience 12, 1187–1196 (2009). Published online: 9 August 2009 [DOI: 10.1038/nn.2363].

Fujita, I., & Fujita, T. (1996). Intrinsic connections in the macaque inferior temporal cortex. *Journal of Comparative Neurology, 368,* 467–486.

Fujita, I., Tanaka, K., Ito, M., & Cheng, K. (1992). Columns for visual features of objects in monkey inferotemporal cortex. *Nature, 360,* 343–346.

Fukushima, K., Miyake, S., & Ito, T. (1988). Neocognitron: A neural network model for a mechanism of visual pattern recognition. In: *Computer Society Press technology series neural networks* (pp. 136–144). Los Alamitos, CA: IEEE Computer Society Press.

Gross, C. G. (1973). Visual functions of inferotemporal cortex. In: R. Jung (Ed.), *Handbook of sensory physiology* (p. 451–482). Berlin: Springer.

Guigon, E. (2003). Computing with populations of monotonically tuned neurons. *Neural Computation, 15,* 2115–2127.

Haxby, J. V., Gobbini, M. I., Furey, M. L., Ishai, A., Schouten, J. L., & Pietrini, P. (2001). Distributed and overlapping representations of faces and objects in ventral temporal cortex. *Science, 293,* 2425–2430.

Haxby, J. V., Grady, C. L., Horwitz, B., Ungerleider, L. G., Mishkin, M., Carson, R. E., et al. (1991). Dissociation of object and spatial visual processing pathways in human extrastriate cortex. *Proceedings of the National Academy of Sciences, 88,* 1621–1625.

Hubel, D., & Wiesel, T. (1962). Receptive fields, binocular interaction and functional architecture in the cat's visual cortex. *Journal of Physiology, 160,* 106.

Hubel, D., & Wiesel, T. N. (1965). Receptive fields and functional architecture in two nonstriate visual areas of the cat. *Journal of Neurophysiology, 28,* 229–289.

Hung, C. P., Kreiman, G., Poggio, T., & DiCarlo, J. J. (2005). Fast readout of object identity from macaque inferior temporal cortex. *Science, 310,* 863–866.

Ito, M., Tamura, H., Fujita, I., & Tanaka, K. (1995). Size and position invariance of neuronal responses in monkey inferotemporal cortex. *Journal of Neurophysiology, 73,* 218–226.

Jin, Y., & Geman, S. (2006). Context and hierarchy in a probabilistic image model. *IEEE Conference on Computer Vision and Pattern Recognition, 2.*

Kamitani, Y., & Tong, F. (2005). Decoding the visual and subjective contents of the human brain. *Nature Neuroscience, 8,* 679–685.

Kanwisher, N., McDermott, J., & Chun, M. (1997). The fusiform face area: A module in human extrastriate cortex specialized for face perception. *Journal of Neuroscience, 17,* 4302–4311.

Kayaert, G., Biederman, I., Op de Beeck, H. P., & Vogels, R. (2005). Tuning for shape dimensions in macaque inferior temporal cortex. *European Journal of Neuroscience, 22,* 212–224.

Kiani, R., Esteky, H., Mirpour, K., & Tanaka, K. (2007). Object category structure in response patterns of neuronal population in monkey inferior temporal cortex. *Journal of Neurophysiology, 97,* 4296–4309.

Kobatake, E., & Tanaka, K. (1994). Neuronal selectivities to complex object features in the ventral visual pathway of the macaque cerebral cortex. *Journal of Neurophysiology, 71,* 856–867.

Kouh, M., & Poggio, T. A canonical neural circuit for cortical nonlinear operations. *Neural Computation, 20,* 1427–1451.

Leonard, C. M., Rolls, E. T., Wilson, F. A., & Baylis, G. C. (1985). Neurons in the amygdala of the monkey with responses selective for faces. *Behavioral Brain Research, 15,* 159–176.

Lowe, D. G. (2004). Distinctive image features from scale-invariant keypoints. *International Journal of Computer Vision, 60,* 91–110.

Maunsell, J. H., & Van Essen, D. C. (1987). Topographic organization of the middle temporal visual area in the macaque monkey: Representational biases and the relationship to callosal connections and myeloarchitectonic boundaries. *Journal of Comparative Neurology, 266,* 535–555.

McCarthy, G., Puce, A., Belger, A., & Allison, T. (1999). Electrophysiological studies of human face perception. II: Response properties of face-specific potentials generated in occipitotemporal cortex. *Cerebral Cortex, 9,* 431–444.

McCarthy, G., Puce, A., Gore, J. C., & Allison, T. (1997). Face specific processing in the human fusiform gyrus. *Journal of Cognitive Neuroscience, 9,* 605–610.

Medawar, P. B. (1981). *Advice to a young scientist.* Perseus Publishing. Perseus Books Group (Headquarters), 387 Park Avenue South, 12th Floor, New York, NY 10016. (212) 340–8100.

Mel, B. W. (1997). SEEMORE: Combining color, shape, and texture histogramming in a neurally inspired approach to visual object recognition. *Neural Computation, 9,* 777–804.

Meunier, M., Bachevalier, J., Mishkin, M., & Murray, E. A. (1993). Effects on visual recognition of combined and separate ablations of the entorhinal and perirhinal cortex in rhesus monkeys. *Journal of Neuroscience, 13,* 5418–5432.

Miyashita, Y., Okuno, H., Tokuyama, W., Ihara, T., & Nakajima, K. (1996). Feedback signal from medial temporal lobe mediates visual associative mnemonic codes of inferotemporal neurons. *Brain Research: Cognitive Brain Research, 5,* 81–86.

Moeller, S., Freiwald, W. A., & Tsao, D. Y. (2008). Patches with links: A unified system for processing faces in the macaque temporal lobe. *Science, 320,* 1355–1359.

Moscovitch, M., Winocur, G., & Behrmann, M. (1997). What is special about face recognition? Nineteen experiments on a person with visual object agnosia and dyslexia but normal face recognition. *Journal of Cognitive Neuroscience, 9,* 555–604.

Mountcastle, V. B. (1997). The columnar organization of the neocortex. *Brain, 120* (Pt 4), 701–722.

Norman, K. A., Polyn, S. M., Detre, G. J., & Haxby, J. V. (2006). Beyond mind-reading: Multi-voxel pattern analysis of fMRI data. *Trends in Cognitive Science, 10,* 424–430.

Olshausen, B., Anderson, C., & Van Essen, D. (1993). A neurobiological model of visual attention and invariant pattern recognition based on dynamic routing of information. *Journal of Neuroscience, 13,* 4700.

Oram, M. W., & Perrett, D. I. (1992). Time course of neural responses discriminating different views of the face and head. *Journal of Neurophysiology, 68,* 70-84.

Pasupathy, A., & Connor, C. E. (1999). Responses to contour features in macaque area V4. *Journal of Neurophysiology, 82,* 2490–2502.

Pasupathy, A., & Connor, C. E. (2001). Shape representation in area V4: Position-specific tuning for boundary conformation. *Journal of Neurophysiology, 86,* 2505–2519.

Pasupathy, A., & Connor, C. E. (2002). Population coding of shape in area V4. *Nature Neuroscience, 5,* 1332–1338.

Perrett, D. I., Rolls, E. T., & Caan, W. (1982). Visual neurones responsive to faces in the monkey temporal cortex. *Experimental Brain Research, 47,* 329–342.

Perrett, D. I., Smith, P. A., Potter, D. D., Mistlin, A. J., Head, A. S., Milner, A. D., et al. (1984). Neurones responsive to faces in the temporal cortex: Studies of functional organization, sensitivity to identity and relation to perception. *Human Neurobiology, 3,* 197–208.

Pinsk, M. A., DeSimone, K., Moore, T., Gross, C. G., & Kastner, S. (2005). Representations of faces and body parts in macaque temporal cortex: A functional MRI study. *Proceedings of the National Academy of Sciences, 102,* 6996–7001.

Previc, F. (1990). Functional specialization in the lower and upper visual fields in humans: Its ecological origins and neurophysiological implications. *Behavior and Brain Sciences, 13,* 519–575.

Puce, A., Allison, T., Asgari, M., Gore, J. C., & McCarthy, G. (1996). Differential sensitivity of human visual cortex to faces, letterstrings, and textures: A functional magnetic resonance imaging study. *Journal of Neuroscience, 16,* 5205–5215.

Puce, A., Allison, T., & McCarthy, G. (1999). Electrophysiological studies of human face perception. III: Effects of top-down processing on face-specific potentials. *Cerebral Cortex, 9,* 445–458.

Quiroga, R. Q., Reddy, L., Kreiman, G., Koch, C., & Fried, I. (2005). Invariant visual representation by single neurons in the human brain. *Nature, 435,* 1102–1107.

Ranzato, M., Huang, F., Boureau, Y., & LeCun, Y. (2007). Unsupervised learning of invariant feature hierarchies with applications to object recognition. *Proceedings of the Computer Vision and Pattern Recognition Conference (CVPR'07),* 1–8.

Riesenhuber, M., & Poggio, T. (1999). Hierarchical models of object recognition in cortex. *Nature Neuroscience, 2,* 1019–1025.

Saleem, K. S., Kondo, H., & Price, J. L. (2008). Complementary circuits connecting the orbital and medial prefrontal networks with the temporal, insular, and opercular cortex in the macaque monkey. *Journal of Comparative Neurology, 506,* 659–693.

Saleem, K. S., & Tanaka, K. (1996). Divergent projections from the anterior inferotemporal area TE to the perirhinal and entorhinal cortices in the macaque monkey. *Journal of Neuroscience, 16,* 4757–4775.

Saleem, K. S., Tanaka, K., & Rockland, K. S. (1993). Specific and columnar projection from area TEO to TE in the macaque inferotemporal cortex. *Cerebral Cortex, 3,* 454–464.

Scalaidhe, S. P., Wilson, F. A., & Goldman-Rakic, P. S. (1999). Face-selective neurons during passive viewing and working memory performance of rhesus monkeys: Evidence for intrinsic specialization of neuronal coding. *Cerebral Cortex, 9,* 459–475.

Seltzer, B., & Pandya, D. N. (1994). Parietal, temporal, and occipital projections to cortex of the superior temporal sulcus in the rhesus monkey: A retrograde tracer study. *Journal of Comparative Neurology, 343,* 445–463.

Sergent, J., Ohta, S., & MacDonald, B. (1992). Functional neuroanatomy of face and object processing. A positron emission tomography study. *Brain, Part 1,* 15–36.

Serre, T., Oliva, A., & Poggio, T. (2007a). A feedforward architecture accounts for rapid categorization. *Proceedings of the National Academy of Sciences, 104,* 6424.

Serre, T., Wolf, L., Bileschi, S., Riesenhuber, M., & Poggio, T. (2007b). Robust object recognition with cortex-like mechanisms. In: *IEEE transactions on pattern analysis and machine intelligence,* 29, (pp. 411–426). IEEE Computer Society.

Serre, T., Wolf, L., & Poggio, T. (2005). Object recognition with features inspired by visual cortex. *Proceedings of IEEE Computer Society Conference on Computer Vision and Pattern Recognition (CVPR).*

Sivic, J., & Zisserman, A. (2003). Video Google: A text retrieval approach to object matching in videos. *Computer vision, 2003. Proceedings of the ninth IEEE international conference* (pp. 1470–1477).

Somogyi, P., Tamas, G., Lujan, R., & Buhl, E. H. (1998). Salient features of synaptic organisation in the cerebral cortex. *Brain Research: Brain Research Reviews, 26,* 113–135.

Suzuki, W., Saleem, K. S., & Tanaka, K. (2000). Divergent backward projections from the anterior part of the inferotemporal cortex (area TE) in the macaque. *Journal of Comparative Neurology, 422,* 206–228.

Tanaka, K. (1996). Inferotemporal cortex and object vision. *Annual Review of Neuroscience, 19,* 109–139.

Tanaka, K. (2003). Columns for complex visual object features in the inferotemporal cortex: Clustering of cells with similar but slightly different stimulus selectivities. *Cerebral Cortex, 13,* 90–99.

Tanaka, J. W., & Farah, M. J. (1993). Parts and wholes in face recognition. *Quarterly Journal of Experimental Psychology A, 46,* 225–245.

Tanaka, K., Saito, H., Fukada, Y., & Moriya, M. (1991). Coding visual images of objects in the inferotemporal cortex of the macaque monkey. *Journal of Neurophysiology, 66,* 170–189.

Tenenbaum, J. B., & Freeman, W. T. (2000). Separating style and content with bilinear models. *Neural Computation, 12,* 1247–1283.

Torralba, A., Murphy, K., Freeman, W., & Rubin, M. (2003). Context-based vision system for place and object recognition. *Computer Vision, 2003. Proceedings of Ninth IEEE International Conference* (pp. 273–280).

Tsao, D. Y., Freiwald, W. A., Knutsen, T. A., Mandeville, J. B., & Tootell, R. B. (2003). Faces and objects in macaque cerebral cortex. *Nature Neuroscience, 6,* 989–995.

Tsao, D. Y., Freiwald, W. A., Tootell, R. B. H., & Livingstone, M. S. (2006). A cortical region consisting entirely of face-selective cells. *Science, 311,* 670–674.

Tsao, D. Y., & Livingstone, M. S. (2008). Mechanisms of face perception. *Annual Review of Neuroscience, 31,* 411–437.

Tsao, D. Y., Moeller, S., & Freiwald, W. A. (2008). Comparing face patch systems in macaques and humans. *PNAS,* 105, 19514–19519.

Tsunoda, K., Yamane, Y., Nishizaki, M., & Tanifuji, M. (2001). Complex objects are represented in macaque inferotemporal cortex by the combination of feature columns. *Nature Neuroscience, 4,* 832–838.

Ungerleider, L. G., Galkin, T. W., Desimone, R., & Gattass, R. (2008). Cortical connections of area V4 in the macaque. *Cerebral Cortex, 18,* 477–499.

Vogels, R., Biederman, I., Bar, M., & Lorincz, A. (2001). Inferior temporal neurons show greater sensitivity to nonaccidental than to metric shape differences. *Journal of Cognitive Neuroscience, 13,* 444–453.

Von Bonin, G., & Bailey, P. (1947). *The neocortex of Mucoca mulutru.* Urbana, IL: University of Illinois Press.

Wallis, G., & Rolls, E. T. (1997). Invariant face and object recognition in the visual system. *Progress in Neurobiology, 51,* 167–194.

Wang, G., Tanaka, K., & Tanifuji, M. (1996). Optical imaging of functional organization in the monkey inferotemporal cortex. *Science, 272,* 1665–1668.

Wang Y, Fujita I, Murayama Y (2000) Neuronal mechanisms of selectivity for object features revealed by blocking inhibition in inferotemporal cortex. *Nature Neuroscience* 3, 807–813.

Webster, M. J., Bachevalier, J., & Ungerleider, L. G. (1993). Subcortical connections of inferior temporal areas TE and TEO in macaque monkeys. *Journal of Comparative Neurology, 335,* 73–91.

Webster, M. J., Bachevalier, J., & Ungerleider, L. G. (1994). Connections of inferior temporal areas TEO and TE with parietal and frontal cortex in macaque monkeys. *Cerebral Cortex, 4,* 470–483.

Webster, M. J., Ungerleider, L. G., & Bachevalier, J. (1991). Connections of inferior temporal areas TE and TEO with medial temporal-lobe structures in infant and adult monkeys. *Journal of Neuroscience, 11,* 1095–1116.

Wiskott, L. (2004). How does our visual system achieve shift and size invariance? In: J. L. van Hemmen & T. J. Sejnowski (Eds.), *23 Problems in systems neuroscience. New York: Oxford University Press.*

Young, A. W., Hellawell, D., & Hay, D. C. (1987). Configurational information in face perception. *Perception, 16,* 747–759.

Yukie, M., & Iwai, E. (1988). Direct projections from the ventral TE area of the inferotemporal cortex to hippocampal field CA1 in the monkey. *Neuroscience Letters, 88,* 6–10.

CHAPTER 25

The Primate Frontal and Temporal Lobes and Their Role in Multisensory Vocal Communication

Lizabeth M. Romanski and Asif A. Ghazanfar

By exploring how existing primates use their vocalizations, numerous investigators are building a rigorous, testable framework for how speech might have evolved. In the perceptual domain, several studies compared the auditory psychophysical abilities of monkeys and humans discriminating speech sounds. Japanese macaques (*Macaca fuscata*) (Sinnott, 1989; Sinnott et al., 1997), vervets (*Chlorocebus pygerethrus*) (Sinnott, 1989), baboons (Heinz & Brady, 1988), and chimpanzees (*Pan troglodytes*) (Kojima, 1990; Kojima & Kiritani, 1989) are able to discriminate human vowel sounds. Much of speech discrimination is thought to be dependent upon temporal cues (Shannon et al., 1995), and both rhesus monkeys (*Macaca mulatta*) (Ghazanfar et al., 2001; Hauser et al., 1998) and cotton-top tamarins (*Saguinus oedipus*) (Ghazanfar et al., 2002; Miller & Hauser, 2004) are similarly sensitive to the temporal cues in their own vocalizations. Furthermore, this temporal processing of speech seems to be biased toward the left hemisphere of the human brain, and some studies of monkeys have also borne out a parallel with this aspect of human speech processing. When perceiving their own vocalizations, both Japanese and rhesus macaques show a left hemispheric bias when listening to conspecific vocalizations, and in most cases this bias is linked to temporal cues (Ghazanfar et al., 2001; Hauser et al., 1998; Petersen et al., 1978).

The problem with these studies, in terms of identifying the origins of speech, is that at least some *nonprimate* animals share these capacities, suggesting that these are general features of the mammalian auditory system that are not specific to speech. For instance, chinchillas are able to discriminate human speech sounds as well (Kuhl & Miller, 1975) and sea lions show a left hemispheric bias for processing their species-specific calls (Boye et al., 2005). Taken together, these data suggest that the processes underlying the perception of purely auditory components of vocalizations are not definitive in distinguishing between what is unique about speech and speech perception (Trout, 2001).

One aspect of speech that is overlooked by comparative biologists seeking to understand its origins is the fact that speech is *not* a purely auditory phenomenon. Human speech is a multisensory function and face-to-face communication is perceived through both the visual and auditory channels. Multisensory speech perception is evident even at the earliest stages of human cognitive development (Patterson & Werker, 2003); its integration across the two modalities is ubiquitous and automatic (e.g., the McGurk effect, McGurk & MacDonald, 1976), and even at the neural level, audiovisual speech integration is evident at the "earliest" stages of cortical processing (Ghazanfar & Schroeder, 2006). Indeed, Rosenblum (2005) has proposed that multisensory speech is the *primary* mode of speech perception and is not a capacity that is "piggybacked" on to auditory speech perception. Specifically, he suggests that the primacy of multisensory speech implies that the perceptual mechanisms, neurophysiology, and evolution of speech perception are based on primitives that are not tied to a single sensory modality. The essence of this idea is shared by many other investigators (Fowler, 2004; Liberman & Mattingly, 1985; Meltzoff & Moore, 1997).

Some have suggested that the ability to develop/evolve language depends on the ability to form multisensory associations and imply that this ability is unique to humans (Geschwind, 1964). If its multisensory nature is a fundamental feature of human speech, whereby visual/facial and vocal signals are inextricably linked, then how did such a mechanism evolve? This chapter explores this issue by presenting (1) behavioral evidence that nonhuman primates (hereafter, *primates*) integrate face and voice information, (2) anatomical evidence that the temporal and frontal cortices of primates are reciprocally connected and are part of a circuit that subserves the integration of face and vocal signals, and (3) physiological evidence that cortical areas in the temporal and frontal lobes of primates show integrative responses to combined face/voice stimuli.

FACIAL MOVEMENT AND VOCAL ACOUSTICS ARE LINKED

Primate vocalizations are produced by coordinated movements of the lungs, larynx (vocal folds), and the supralaryngeal vocal tract (Fitch & Hauser, 1995; Ghazanfar & Rendall, 2008). The vocal tract consists of the column of air derived from the pharynx, mouth, and nasal cavity. The source signals (sounds generated by the lungs and larynx) travel through the vocal tract and are *filtered* according to its shape, resulting in *vocal tract resonances* or *formants* discernable in the spectra of some vocalizations (for primates, see Fitch, 1997; Owren et al., 1997; Rendall et al., 1998). In humans, speech-related vocal tract motion results in the predictable deformation of the face around the oral aperture and other parts of the face (Jiang et al., 2002; Yehia et al., 1998, 2002). In fact, the spectral envelope of a speech signal can be predicted by the three-dimensional motion of the face alone (Yehia et al., 1998), as can the motion of the tongue (an articulator that is not necessarily coupled with the face) (Jiang et al., 2002; Yehia et al., 1998). The spatiotemporal behavior of the vocal tract articulators involved in sound production constrains the shape and time course of visible orofacial movement. Such speech-related facial motion, distributed around and beyond the mouth, is what is used in the bimodal integration of audiovisual speech signals by humans. For example, human adults automatically link high-pitched sounds to facial postures producing an /i/ sound and low-pitched sounds to faces producing an /a/ sound (Kuhl et al., 1991).

In primate vocal production, there is a similar link between acoustic output and facial dynamics. Different rhesus monkey vocalizations are produced with unique lip configurations and mandibular positions, and the motion of such articulators influences the acoustics of the signal (Hauser & Ybarra, 1994; Hauser et al., 1993). Coo calls, like /u/ in speech, are produced with the lips protruded, while screams, like the /i/ in speech, are produced with the lips retracted (Fig. 25.1A). The jaw position and lip configuration affect the formant frequencies independent of the source frequency (Hauser & Ybarra, 1994; Hauser et al., 1993). Moreover, as in humans, the articulation of these expressions has visible consequences on facial motion beyond the oral region. Grimaces, produced during scream vocalizations for instance, cause the skin-folds around the eyes to increase in number. In addition to these production-related facial movements, some vocalizations are associated with visual cues that are not directly related to the articulatory movement. Threat vocalizations, for instance, are produced with intense staring, eyebrows raised, and ears pulled back (Partan, 2002). Head position and motion (e.g., chin up vs. chin down vs. neutral position) also varies according to vocal expression type (Partan, 2002). Chimpanzees (*Pan troglodytes*), though less studied in the domain of multisensory communication, also have a link between facial expression and vocal acoustics. Figure 25.1B shows the facial expressions produced for two chimpanzee vocalizations. Bauer (1987) analyzed video and audio tracks of chimpanzee vocalizations and found that a decline in the fundamental frequency (F0) occurred when submissive screams transitioned into aggressive barks. These changes in F0 were correlated with changes in visible articulators such as lip and teeth opening. Thus, it is likely that many of the facial motion cues that humans use for speech reading are present in at least some apes and monkeys as well.

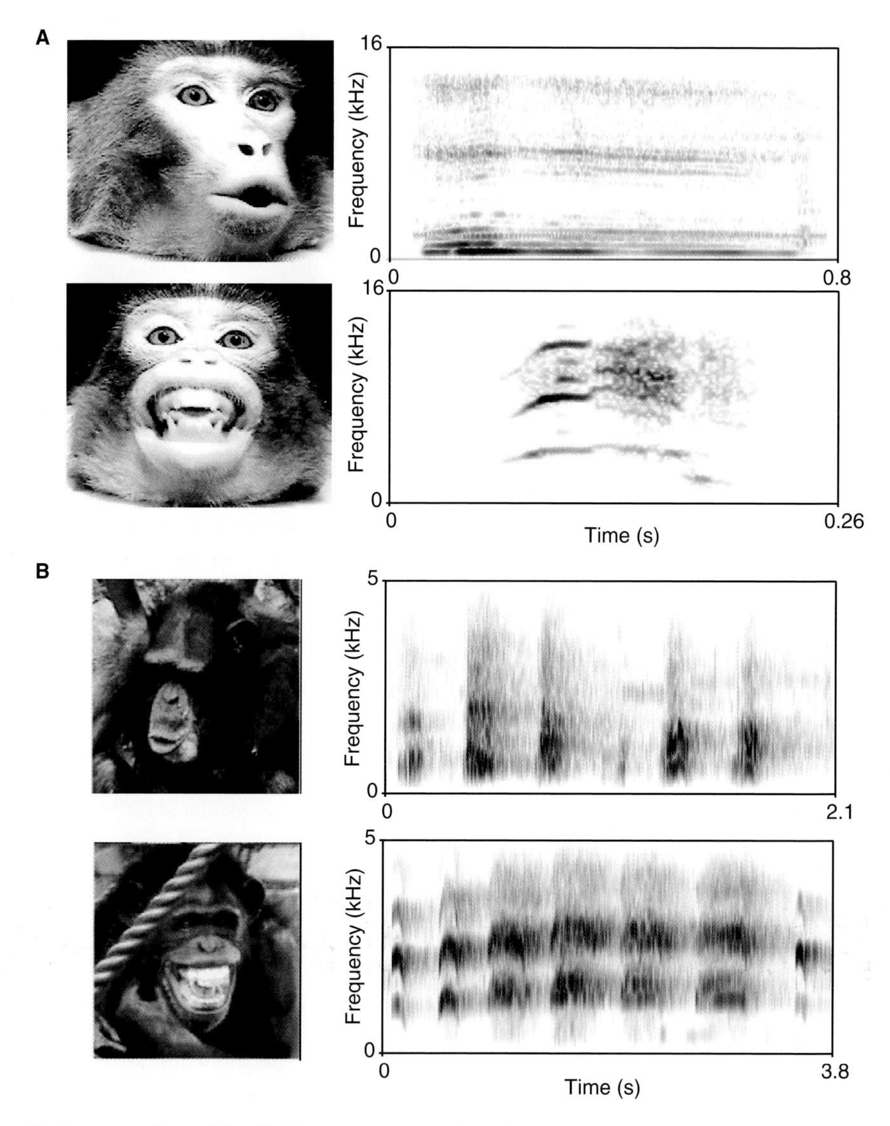

Figure 25.1 Exemplars of the facial expressions produced concomitantly with vocalizations. (**A**) Rhesus monkey coo and scream calls taken at the midpoint of the expressions with their corresponding spectrograms. (**B**) Chimpanzee visual components of pant-grunt and scream calls. Next to them are spectrograms of a pant-grunt and scream calls. Chimp expressions reprinted from Izumi, A., & Kojima, S. (2004). Matching vocalizations to vocalizing faces in a chimpanzee (Pan troglodytes). *Animal Cognition, 7,* 179–184. Sound exemplars taken from the Harvard Cognitive Evolution Lab, www.wjh.harvard.edu/~mnkylab/media/chimpcalls.html.

MATCHING FACES TO VOICES

Given that both humans and other extant primates use both facial and vocal expressions as communication signals, it is unlikely that only humans recognize the correspondence between the visual and auditory components of vocal signals. To test whether other primates do, in fact, have this capacity, a preferential looking paradigm was adopted (Ghazanfar & Logothetis, 2003). In this paradigm, subjects are seated in front of two video monitors, one playing the facial dynamics of one expression and the other playing the facial dynamics of a second expression. In between the two screens, a loudspeaker plays back the vocalization that matches only one of the facial expressions being shown. The dependent measure is the amount of time the subjects spend looking at this matching video as a proportion of the total looking time to both screens. This paradigm was developed by cognitive scientists to test intermodal speech perception in prelinguistic infants (Kuhl & Meltzoff, 1984; Patterson & Werker, 2003) and thus it is perfect for testing the natural capacities of primates without training or reward.

Using this procedure, rhesus monkeys were tested on their ability to recognize the auditory-visual correspondences between their "coo" and "threat" calls (Ghazanfar & Logothetis, 2003). These are among the most frequently produced calls in the rhesus monkey repertoire, both in the wild and in captivity. Coo calls are tonal signals of relatively long duration and are produced in many affiliative contexts, including group movements, separation, and feeding; threat calls, in contrast, are noisy, short-duration pulsatile calls produced during aggressive encounters (Hauser & Marler, 1993; Rowell & Hinde, 1962). Each of these calls is associated with a unique facial posture. The hypothesis was that the vocalization heard from the central speaker would systematically influence the duration of subjects' visual fixations on the two screens—one that displayed the video of a coo expression and the other that displayed the video of the threat expression. Specifically, if rhesus monkeys recognized the correspondence between the heard sound and the appropriate facial posture, then, overall, they would spend more time looking at the matching video. The results supported the prediction. The mean percentage of looking time devoted to the match was 67.3%; this was significantly greater than the 50% chance level (Ghazanfar & Logothetis, 2003). Furthermore, all 11 subjects looked longer at the matching face. When comparing looking durations between coos and threats, subjects looked longer when the matching face was articulating the coo call than when the match face was the threat call (likely because threats are aversive); nonetheless, separate analyses revealed that looking preferences were still significant for coo calls alone and threat calls alone.

Although these data demonstrate that monkeys can match faces to voices, in this particular experiment they may have been doing so using low-level stimulus cues. The coo and threat calls differ dramatically in their duration—the threat is almost invariably shorter than the coo. Thus, it is possible that the monkeys may be using a cue that is redundant between the two modalities—matching the duration of mouth movements to the duration of the auditory component (Lewkowicz, 2002). This type of matching does not necessitate the identification of the particular facial expression (i.e., the morphology of the mouth). A subsequent study with capuchin monkeys (*Cebus apella*) avoided this confound by using stimuli that were better controlled for duration and included vocal expressions from not only conspecifics but heterospecifics as well (rhesus monkey and human vocalizations) (Evans et al., 2005). Without duration cues, the capuchins could still match the appropriate facial expression to the heard vocalization across all stimulus conditions. However, it was not clear whether they can make this match for each species' set of calls or for only the capuchin calls. Another way to avoid the temporal cues confound is to use a match-to-sample behavioral paradigm that requires training subjects. Chimpanzees can match conspecific faces and voices under these conditions (Izumi & Kojima, 2004; Parr, 2004).

MATCHING MULTIPLE FACES TO VOICES REVEALS ACOUSTIC STREAM SEGREGATION

Given that monkeys can readily match faces to voices, can they do something with these signals that is slightly more complicated and relevant to their everyday socio-ecological tasks? One problem that they might confront is segregating competing voices in a chorus, much as humans might do in a cocktail party scenario. Rhesus monkeys produce a chorus of coos and other vocalizations upon the anticipation or discovery of food. Using the preferential looking paradigm, monkeys were tested on their ability to segregate the competing voices of two or three unfamiliar individuals in the auditory domain and to then attend to the video screen that showed the correct number of individuals that they heard (Jordan et al., 2005). In this experiment, two versus three coo calls were used that were equal in duration and temporally coincident. The video displays depicted two synchronously cooing individuals on one side and three individuals on the other side. It was hypothesized that if they could segregate the voices and assess their number, then they would preferentially attend to the dynamic visual displays featuring the number of conspecifics they simultaneously heard vocalizing. Under these conditions, monkeys spent a greater proportion of time looking at the display that numerically matched the number of vocalizers they heard compared to the nonmatching display (Jordan et al., 2005).

This spontaneous, multisensory representation of multiple individuals in animals is an important parallel to similar representations in humans (Barth et al., 2003). Using a nearly identical stimulus paradigm but with human voices (Jordan & Brannon, 2006), human infants (aged 7 months) performed nearly identically to rhesus monkeys. This suggests that *perhaps* there are homologous underlying mechanisms (see Zangehenpour et al., 2009, for important caveats). These results also indicate that rhesus monkeys can segregate simultaneously presented coo vocalizations, even though the power spectra of the calls are highly overlapping. This capability is on par with the perceptual separation of voices by humans via pitch differences and harmonicity (Brokx & Nooteboom, 1982; Summerfield et al., 1992). This is notable because a previous study found that highly trained monkeys could discriminate concurrent sequences of *artificial* sounds only when their frequency ranges did not overlap (Izumi, 2002).

MATCHING ACOUSTIC SIZE WITH VISUAL SIZE

Vowels and consonants, the essential phonetic elements of all human speech, differ in their resonances, or *formants*, produced by the vocal tract—the nasal and oral cavities above the vocal folds (Ghazanfar & Rendall, 2008). As described previously, during vocal production in humans and other primates, pulses of air generated by the rapid movement of the vocal folds produce an acoustic signal. As signals pass through the vocal tract, they excite resonances in the vocal tract resulting in the enhancement of particular frequency bands; these are the formants. The ability to create and perceive a wide variety of different formant patterns is a prerequisite for human speech. This fact raises the following two questions. First, how did the ability to perceive differences between formant patterns arise? Second, which role, if any, did formants play in prelinguistic primates? The answers to these questions may lay not so much in the fact that formant patterns are important phonetic elements of speech, but that they also carry important information related to physical characteristics of the individual speaker (Fitch, 1997; Smith & Patterson, 2005). Specifically, the length of the vocal tract determines the frequencies of the formants. Large individuals with long vocal tracts have lower-frequency formants than do smaller individuals who have smaller vocal tracts.

Previous behavioral studies demonstrated that trained baboons (Heinz & Brady, 1988) and macaques (Le Prell et al., 2001; Sinnott, 1989; Sommers et al., 1992) can discriminate different human vowel sounds presumably based on formant frequency differences. Fitch and Fritz (2006) significantly extended these findings by showing that rhesus monkeys can, without training, discriminate differences in the formant structure of their own conspecific calls. Can rhesus monkeys use formants as acoustic cues to assess age-related body size differences among conspecifics? The potential role of formants as indexical cues in rhesus monkey vocalizations was tested using the preferential looking method (Ghazanfar et al., 2007). When presented with two videos, one depicting an older, larger macaque and the other showing a young, small monkey (Fig. 25.2A), subjects spent more time looking at the larger monkey when they heard a coo produced with a large monkey's vocal tract and vice versa when they heard a coo produced with a small monkey's vocal tract (Fig. 25.2B–D). In other words, monkeys spent more time looking at the video with the monkey that matched in size what they heard in the audio track (Ghazanfar et al., 2007).

These data are the first direct evidence for the hypothesis that formants embedded in the acoustic structure of primate calls provide cues to the physical characteristics of the vocalizer (Fitch, 1997; Owren et al., 1997; Rendall et al., 1998). The results further suggest that the use of formant cues in the perception of vowel sounds by humans in a linguistic context emerged gradually, perhaps for other functional reasons, over the course of human evolution. Perception of indexical cues, such as age-related body size, via formants in vocalizations may be one functional link between the vocalizations of human and other primates.

NEURAL CIRCUITS MEDIATING MULTISENSORY VOCAL COMMUNICATION

In spite of much evidence to the contrary (presented here for bimodal vocal communication and elsewhere for cognitive and social functions; see Seyfarth & Cheney, this volume), there are some who would argue that the vocal communication of primates is of a different kind altogether from that of human speech (Arbib, 2005; Pinker, 1994). Some of these arguments are the

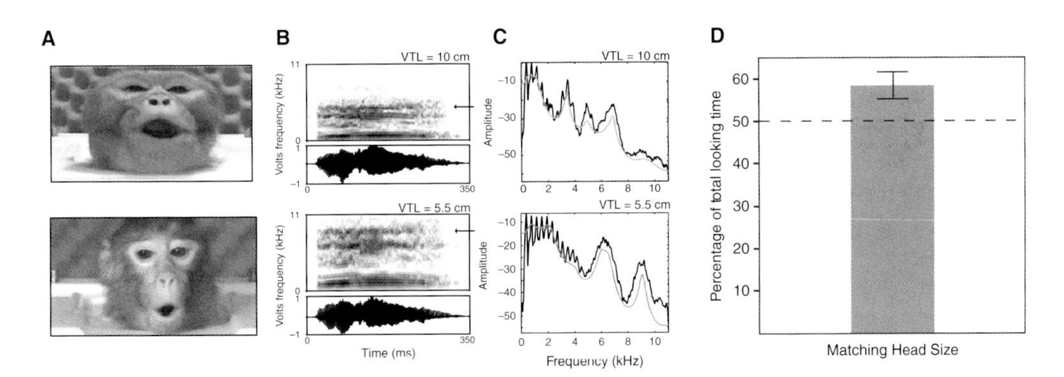

Figure 25.2 (**A**) Still frames extracted from the videos used in the preferential looking experiments. The top shows a frame from the large monkey. Videos were synchronized and edited so that they appeared to be synchronously producing the coo vocalization shown in B. (**B**) Resynthesized coo vocalizations based on one of the two coo exemplars used in both the habituation-discrimination and preferential looking paradigms. Diagram shows the spectrograms and waveforms of a coo vocalization resynthesized with two different vocal tract lengths. The arrow in the spectrogram indicates the position of an individual formant, which increases in frequency as the apparent vocal tract length decreases. Only the coos based on 5.5-cm and 10-cm vocal tract lengths were used as stimuli. (**C**) Power spectra (black line) and linear predictive coding spectra (gray lines) for the long vocal tract length (10 cm, top panel) and short vocal tract length (5.5 cm, bottom panel) used in the experiment and seen in B. (**D**) The mean percentage of total looking time spent looking at the matching video display; the dotted line indicates chance expectation. Error bar is SEM.

result of an overeager interpretation of a subset of monkey neurobiological data (e.g., mirror neurons) that are potentially related to the evolution of human speech. Others simply argue that the brains of monkeys and humans are fundamentally different when it comes to circuits involved in communication. Indeed, Pinker (1994) has suggested the following: "Language could have arisen, and probably did arise. . .by a revamping of primate brain circuits that originally had no role in primate vocal communication" (p. 350). For the most part, these arguments ignore much of what we know about the processing of vocalizations in the primate brain (Ghazanfar & Hauser, 1999, 2001) and the numerous behavioral similarities in the use of vocalizations by our primate cousins (some of which were described previously; see also chapters by Miller & Cohen and Seyfarth & Cheney in this volume). Certainly most, if not all, accounts of the evolution of primate vocal communication ignore that fact that it is integrated across multiple sense modalities.

The compounding evidence that monkeys, like humans, combine auditory and visual signals suggests two things. First, monkeys are an accessible model system for investigating the sensorimotor integration of a natural communicative behavior. Second, monkeys and humans likely share some homologous neural substrates to mediate such multisensory behavior. For instance, although such behavior likely involves many disparate regions in the brain, there is a robust, three-node cortical circuit that may be more heavily weighted for processing facial-vocal signals. The nodes include the prefrontal cortex, the superior temporal sulcus, and the auditory cortex. These three regions are densely interconnected. The following sections will review the anatomical circuits and neurophysiological processes that are thought to mediate (at least in part) multisensory communication in primates.

FRONTAL AND TEMPORAL LOBE CIRCUITRY

One principle of anatomical connections in the neocortex is that "high-level" association areas receive inputs from a number of sensory areas. Consistent with this idea, the prefrontal cortex (PFC) and the dorsal (or *upper*) bank of the superior temporal sulcus (STS) both receive a wealth of afferents from sensory areas. In particular, these association cortices receive projections from auditory and visual cortical areas, some of which are also connected with each other.

Visual Connections of Superior Temporal Sulcus and Prefrontal Cortex

Tracer injections into areas within the STS and the PFC indicate that each receives projections from both dorsal and ventral stream visual areas. Traditionally, the dorsal bank of the rostral STS, which includes area TPO, was thought to be connected with the parietal, prefrontal, and superior temporal auditory regions (Seltzer & Pandya, 1978, 1991, 1994). This implied that TPO might not receive connections carrying object, form, or face processing information directly from visual areas like the inferotemporal cortex. However, injections of anterograde tracer into distinct dorso-anterior (TEad) and ventro-anterior (TEav) subdivisions of the inferotemporal cortex showed a dense distribution of terminals and cell bodies in TPO following TEad injections, whereas labeling was confined to the lower bank and fundus of the rostral STS following tracer injections into TEav (Saleem et al., 2000). Thus, rostral TPO is reciprocally connected with ventral stream visual processing areas in inferotemporal cortex. Middle and caudal TPO also receive projections from visual processing areas (Seltzer & Pandya, 1994). When different tracers were injected into both posterior parietal and inferotemporal cortex of the same hemisphere, overlapping anterograde fibers were found near the fundus of the STS (Baizer et al., 1991), supporting the idea that this area receives converging inputs from the dorsal and ventral visual streams.

With regard to the frontal lobe, early anatomical studies by Barbas, Pandya, and their colleagues (Barbas, 1988; Barbas & Mesulam, 1981; Barbas & Pandya, 1989; Chavis & Pandya, 1976) noted a segregation of visual inputs to discrete regions of the prefrontal cortex. For example, Barbas (1988, p. 313) wrote "...the basoventral prefrontal cortices are connected with ventral visual areas implicated in pattern recognition and discrimination, whereas the mediodorsal cortices are connected with medial and dorsolateral occipital and parietal areas associated with visuospatial functions." This dissociation was echoed by Bullier and colleagues (1996), who found some segregation of inputs to PFC when paired injections of tracers were placed into temporal and parietal visual processing regions. In their study, visual temporal cortex projected mainly to area 45, located ventrolaterally in the PFC, while parietal cortex sent projections to both ventrolateral PFC (area 45) and dorsolateral PFC (areas 8a and 46) (Bullier et al., 1996). The specific connectivity of ventrolateral PFC areas 12 and 45, which contain object- and face-selective neurons (O'Scalaidhe et al., 1997, 1999; Wilson et al., 1993), with inferotemporal areas TE and TEO was specifically documented by Webster and colleagues (1994). Comparison of TE and TEO connectivity revealed a number of important differences, including the finding that it is mainly area TE that projects to ventrolateral PFC and orbitofrontal areas 11, 12, and 13. These orbital regions also contain object- and face-responsive cells (Thorpe et al., 1983).

Auditory Connections of Superior Temporal Sulcus and Prefrontal Cortex

Auditory regions in the superior temporal gyrus (STG) and plane send projections that terminate in rostral and middle portions of the STS. Polysensory area TPO and area TAa both receive a dense projection from portions of the rostral and middle STG (Seltzer & Pandya, 1994; Seltzer et al., 1996). More direct evidence revealed that the parabelt and belt auditory regions, whose neurons are responsive to complex sounds (Rauschecker et al., 1995; Romanski et al., 1999a), also project to areas TPO and TAa (Hackett et al., 1999).

The connections of the PFC with superior temporal auditory areas have been studied in great detail. Early anatomical studies indicated that a rostro-caudal topography exists such that caudal STG and caudal PFC are reciprocally connected (Chavis & Pandya, 1976; Pandya & Kuypers, 1969; Petrides & Pandya, 1988, 2002; Romanski et al., 1999a,b), while the rostral STG is reciprocally connected with rostral prefrontal areas around the principal sulcus (areas 46 and 10) and orbito-frontal areas (areas 11 and 12) (Chavis & Pandya, 1976; Pandya & Kuypers, 1969; Pandya et al., 1969). In the last decade, studies carefully characterized temporo-prefrontal connections utilizing a better understanding of auditory cortical organization. These studies refined the rostral-caudal topography that was previously noted by determining that the frontal pole and anterior principal sulcus are densely connected with anterior belt and parabelt regions of the auditory cortex (Hackett et al., 1999; Romanski et al., 1999a), while the anterior third of the temporal lobe, designated STGr (Hackett et al., 1999), has strong reciprocal connections with orbitofrontal areas (Hackett et al., 1999; Romanski et al., 1999a). Moreover, the caudal parabelt and belt of auditory cortex are reciprocally connected with the dorsal prearcuate and caudal principal sulcus.

Direct evidence of auditory cortex innervation of the PFC was shown by placing tracer injections into frequency matched locations in tonotopically organized areas of the lateral auditory belt—areas AL, ML, and CL (Romanski et al., 1999b). The connections between the frontal cortex and the auditory belt/parabelt were topographically organized such that projections from AL typically involved the frontal pole (area 10), the rostral principal sulcus (area 46), the inferior convexity (areas 12vl and 45) and the lateral orbital cortex (areas 11, 12o). In contrast, projections from area CL targeted the dorsal periarcuate cortex (area 8a, frontal eye fields) and the caudal principal sulcus (area 46) as well as the caudal inferior convexity (areas 12vl and 45) and, in two cases, premotor cortex (area 6d). Thus, projections from the auditory association cortex to the prefrontal cortex are topographic. Furthermore, while small topographic projections target the PFC from the lateral belt, parabelt projections are more dense (Hackett et al., 1999; Romanski et al., 1999a) (Fig. 25.3).

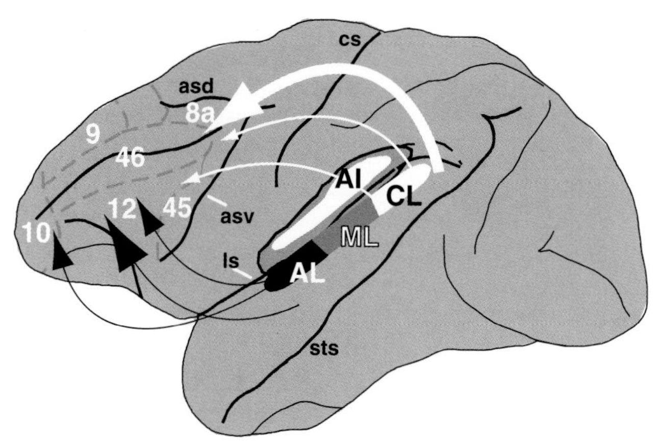

Figure 25.3 Connections between the auditory belt and parabelt are summarized on the lateral brain schematic of the macaque monkey. Combined physiology and anatomy studies demonstrated topographic connections between the anterior belt (AL and anterior ML) and the rostral and ventral prefrontal cortex, areas 10, rostral 46, 12vl, and 12o. In contrast, the caudal belt (areas CL and caudal ML) project to caudal and dorsal prefrontal cortex including areas 8a (the frontal eye fields), caudal 46, and the dorsal portion of areas 45 and 12. Larger connections originate in the parabelt and echo this same rostral and caudal topography. asd, dorsal arcuate sulcus; asv, ventral arcuate sulcus; cs, central sulcus; ls, lateral sulcus; sts, superior temporal sulcus.

ENCODING AND INTEGRATION OF FACES AND VOICES IN THE TEMPORAL AND FRONTAL LOBES

The anatomical data strongly suggest an interaction between multiple cortical areas in mediating auditory-visual integration. This is supported by human neuroimaging and EEG studies of audiovisual speech. The imaging data consistently show that the superior temporal sulcus, auditory cortex (including primary auditory cortex), and prefrontal cortex play important roles in the integration process (Calvert, 2001). Human neuroimaging and EEG data of audiovisual speech have added much to our knowledge of its underlying cortical circuitry and have generated several interesting hypotheses; however, the nature of these signals precludes a direct understanding of how this network functions and interacts. Problems inherent in these imaging and event-related potential studies include the inability to distinguish between inhibitory versus excitatory responses, poor source localization, and/or poor temporal resolution. For many decades, animal models of multisensory integration have largely focused on the role of the superior colliculus in the spatial and temporal integration of simple artificial stimuli (see Stein, 1998, for review). Although many principles of multisensory integration have been identified for neurons in the superior colliculus, it is unknown to what extent these principles apply to complex, naturalistic stimuli in the primate neocortex.

To bridge this epistemic gap between what we know about human and monkey neural processes, recent studies have investigated dynamic face/voice integration in the superior temporal sulcus (Barraclough et al., 2005; Chandrasekaran & Ghazanfar, 2009), auditory cortex (Ghazanfar et al., 2005, 2008), and ventrolateral prefrontal cortex (Sugihara et al., 2006) of the macaque monkey brain using the species' natural communication signals. Unlike pairings of artificial stimuli, audiovisual vocalizations are ethologically relevant and thus may tap into specialized neural mechanisms (Ghazanfar & Santos, 2004) or, minimally, integrative mechanisms for socially learned audiovisual associations.

THE SUPERIOR TEMPORAL SULCUS

The superior temporal sulcus has long been known to have neurons responsive to visual, auditory, and/or somatosensory stimuli (Benevento et al., 1977; Bruce et al., 1981; Hikosaka et al., 1988). Few attempts, however, were made to intensively investigate whether these polysensory neurons *integrated* stimuli from different modalities. These early studies reported anecdotal evidence that, in a few neurons, auditory stimuli attenuated visual stimuli (Benevento et al., 1977; Bruce et al., 1981). A recent study systematically investigated the integrative properties of single neurons in STS using biologically relevant dynamic stimuli such as ripping paper, chewing, and vocalizations (Barraclough et al., 2005). Twenty-three percent of neurons responsive to the sight of biological motion could be significantly modulated by the corresponding auditory component. The sound of the action either enhanced or suppressed the visually evoked response for an equal number of neurons. In neurons where the sound enhanced neural responses, the enhancement was dependent upon the auditory signal being congruent with the visual one. For example, audiovisual human "lip-smacking" elicited a stronger response from a given neuron than visual lip-smacking alone; however, when visual lip-smacking was paired with other auditory signals (like monkey calls), the neuron did not show any enhanced activity. Conversely, in neurons whose responses to visual stimuli were suppressed by sound stimuli, the suppression was not dependent upon the congruence of the auditory and visual stimulus.

Beyond single units, the integration of faces and voices in the STS was tested both as a function of different frequency bands of the local field potential (LFP) responses and as a function of the natural time delay between the initial facial movement and the onset of the voice component (the "time-to-voice," or TTV) (Chandrasekaran & Ghazanfar, 2009). Faces and voices elicited distinct and concurrent activity in different frequency ranges of the STS LFP: theta (4 to 8 Hz), alpha (8 to 14 Hz), and gamma (>40 Hz). Remarkably, these different neural frequency bands integrated faces and voices differently. The theta band showed no consistent relationship between the multisensory response and the time-to-voice; that is, the theta band did not, in general, show enhancement or suppression to face/voice combinations, and when it did, the responses could not be explained by the TTV variable. In contrast, the alpha band showed enhancement of power when presented with face/voice combinations and this enhancement was dependent on the TTV—short TTVs elicited enhancement, while long TTVs elicited no integration. The gamma band revealed a third, distinct integrative capacity. The gamma band invariably showed enhanced responses to face-voice pairings and these responses were completely independent of the TTV parameter (Chandrasekaran & Ghazanfar, 2009).

The finding that different neural frequency bands show independence in their face/voice integration strategies is supported by findings in other domains. An increasing number of studies are revealing that different LFP frequency ranges are best correlated with different sensory or motor parameters (Liu & Newsome, 2006; Rickert et al., 2005; Wilke et al., 2006). Together, these data suggest that different neural frequency bands reflect distinct neuronal processes and are generated by distinct sources. One attractive hypothesis is that the frequency is inversely proportional to the scale of the cortical network underlying that frequency (Buzsaki, 2006). Gamma band activity is thought to reflect activity in a local cortical microcolumn, whereas alpha and theta band activity could putatively represent processes whose sources may include multiple cortical areas acting in a coordinated fashion. In the case of face/voice integration, this would suggest that the areas with which the STS connects would have similar behavior in the lower frequencies but divergent behavior in higher frequencies. Putative sources of such inputs into STS are reviewed previously. Recordings from multiple structures simultaneously and joint analysis of these signals (see later; Ghazanfar et al., 2008; Maier et al., 2008) will help shed light onto the mechanisms underlying activity in these different frequency ranges and their role in multisensory integration.

AUDITORY CORTEX

Presumptive unimodal sensory areas, such as auditory cortex, are now known to be multisensory as well (Ghazanfar & Schroeder, 2006). The auditory cortex, in particular, has been the focus of much attention (Ghazanfar, 2009). Numerous lines of both physiological and anatomical evidence demonstrate that at least some regions of the auditory cortex respond to touch as well as sound (Fu et al., 2003; Kayser et al., 2005; Hackett et al., 2007a,b; Lakatos et al., 2007; Schroeder & Foxe, 2002; Smiley et al., 2007). The potential roles of such auditory-somatosensory interactions in vocal communication are not known (Ghazanfar, 2009), but one possibility is that they mediate sensory feedback to guide vocal production (Ghazanfar & Turesson, 2008; Nasir & Ostry, 2006). In the auditory-visual domain, the anatomical data suggest (reviewed previously) that the auditory cortex has many potential sources of visual inputs. This is supported by an increasing number of functional and physiological studies in monkeys (Ghazanfar et al., 2005, 2008; Kayser et al., 2007, 2008). Here we focus on those auditory cortical studies investigating face/voice integration.

In order to investigate the integrative properties of the auditory cortex, monkey subjects viewed unimodal and bimodal versions of two different species-typical vocalizations ("coos" and "grunts") while performing a fixation task requiring them to maintain their eye fixations within the video frame. Recordings of the mean extracellular field potential (i.e., unit and subthreshold neural activity) were made using intracranial electrodes placed in the core auditory cortex (which includes primary and primary-like auditory areas) and in the lateral belt auditory cortex (a higher-order auditory area) (Ghazanfar et al., 2005). The stimulus conditions were Face+Voice, Face alone, and Voice alone. Consistent with evoked potential studies in humans (Besle et al., 2004; van Wassenhove et al., 2005), the combination of faces and voices led to integrative responses (significantly different from unimodal responses) in the vast majority of auditory cortical sites—both in primary auditory cortex and the lateral belt auditory cortex. The data demonstrated that raw LFP signals and multiunit activity in the auditory cortex are capable of multisensory integration of facial and vocal signals in monkeys (Ghazanfar et al., 2005) and this has subsequently been confirmed at the single-neuron level as well (Ghazanfar et al., 2008) (Fig. 25.4).

The vast majority of LFP responses were specific to face/voice integration and such integration could take the form of either enhancement or suppression, although enhanced responses were more common (Ghazanfar et al., 2005). The specificity of these integrative responses was tested by replacing the dynamic faces with dynamic discs that mimicked the aperture and displacement of the mouth. In human psychophysical experiments, such artificial dynamic stimuli can still lead to enhanced speech detection, but not to the same degree as a real face (Bernstein et al., 2004; Schwartz et al., 2004). When cortical sites or single units were tested with dynamic discs, far less integration was seen when compared to the real monkey faces (Ghazanfar et al., 2005, 2008) (Fig. 25.4). This was true primarily for the lateral belt auditory cortex and was observed to a lesser extent in the primary auditory cortex.

Coos and grunts are both produced in a variety of affiliative contexts (Hauser & Marler, 1993). In light of this, another interesting finding that came out of this study is that grunt vocalizations were overrepresented relative to coos in terms of enhanced multisensory LFP responses. One possibility is that this is because grunts are spectrally noisy calls and they thus evoked responses in more neurons due to their broad frequency spectrum; however, the frequency spectra of coos overlapped considerably with grunt spectra. Another possibility is that this differential representation may reflect a behaviorally relevant distinction, as coos and grunts differ in their direction of expression and range of communication. Coos are generally contact calls rarely directed toward any particular individual. In contrast, grunts are often directed toward individuals in one-on-one situations, often during social approaches (Cheney & Seyfarth, 1982; Palombit et al., 1999). Given their production at close range and context, grunts may produce a stronger face/voice association than coo calls. This distinction appeared to be reflected in the pattern of significant multisensory responses in auditory cortex; that is, this bias toward grunt calls may be related to the fact

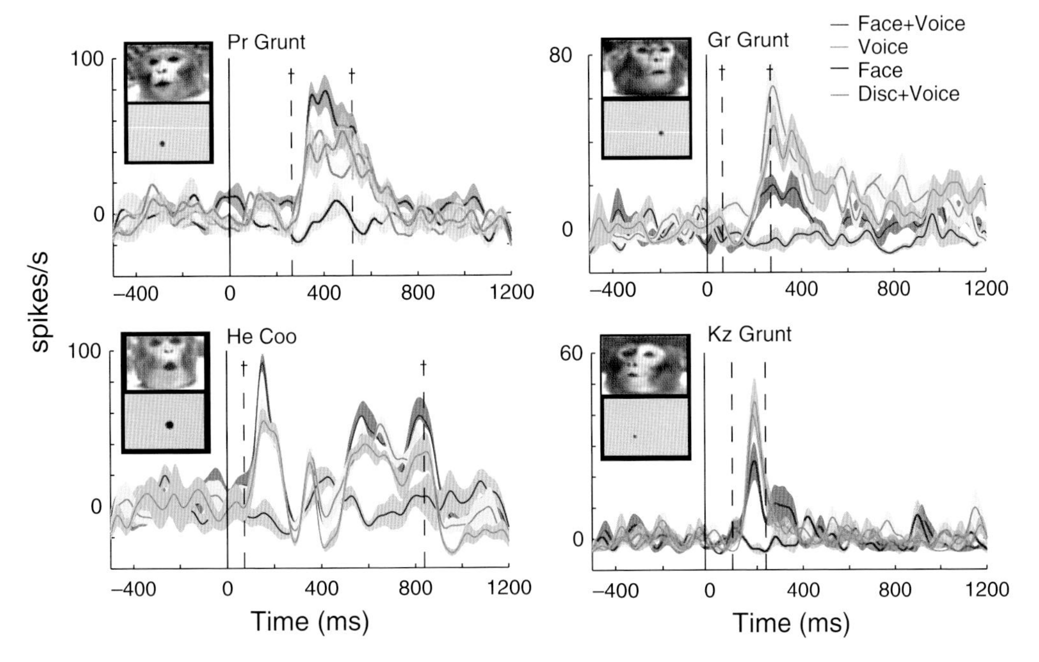

Figure 25.4 Examples of multisensory integration of Face+Voice stimuli compared to Disc+Voice stimuli in single auditory neurons recorded from the lateral belt region. Left panels show enhanced responses when voices are coupled with faces, but no similar modulation when coupled with discs. Right panels show similar effects for suppressed responses. The insets show frames from the Face+Voice stimulus and the temporally corresponding Disc+Voice stimulus. X-axes show time aligned to onset of the face (solid line). Dashed lines indicate the onset and offset of the voice signal. Y-axes depict the firing rate of the neuron in spikes per second. Shaded regions denote the SEM of the mean.

that grunts (relative to coos) are often produced during intimate, one-on-one social interactions.

INTERACTIONS BETWEEN THE SUPERIOR TEMPORAL SULCUS AND THE LATERAL BELT AUDITORY CORTEX MEDIATE INTEGRATION OF FACES AND VOICES

The face-specific visual influence on the lateral belt auditory cortex begs the question as to its anatomical source. Although there are multiple possible sources of visual input to auditory cortex (Ghazanfar & Schroeder, 2006), the STS is likely to be a prominent one, particularly for integrating faces and voices, for the following reasons. First, there are reciprocal connections between the STS and the lateral belt and other parts of auditory cortex (see previous discussion; Barnes & Pandya, 1992; Seltzer & Pandya, 1994). Second, neurons in the STS are sensitive to both faces and biological motion (Harries & Perrett, 1991; Oram & Perrett, 1994). Finally, the STS is known to be multisensory (Bruce et al., 1981; Schroeder & Foxe, 2002). One mechanism for establishing whether auditory cortex and the STS interact at the functional level is to measure their temporal correlations as a function stimulus condition. Thus, concurrent recordings of LFPs and single neurons in the lateral belt of auditory cortex and the dorsal bank of the STS were made (Ghazanfar et al., 2008). The functional interactions, in the form of gamma band correlations, between these two regions increased in strength during presentations of faces and voices together relative to the unimodal conditions (Fig. 25.5A). Furthermore, these interactions were not solely modulations of response strength, because the phase relationships were significantly less variable (tighter) in the multisensory conditions (Varela et al., 2001). A control condition, in which the face was replaced with a dynamic disc that mimicked mouth movements (as in Ghazanfar et al., 2005), revealed that interareal interactions in general were significantly less robust (Fig. 25.5A).

With regard to spiking activity, not only do single auditory cortical neurons integrate faces and voices (Fig. 25.4), but this spiking output also seems to be *modulated* (not driven) by ongoing activity arising from the STS. Three lines of evidence suggest this scenario. First, visual influences on single neurons were most robust when in the form of dynamic faces and were only apparent when neurons had a significant response to a vocalization (i.e., there were no overt responses to faces alone). Second, these integrative responses were often "face specific" and had a wide distribution of latencies, which suggested that the face signal was an ongoing signal that influenced auditory responses (Ghazanfar et al., 2008). Finally, this hypothesis for an ongoing signal is supported by the sustained gamma band activity between auditory cortex and STS (Fig. 25.5A) and by a spike-field coherence analysis of the relationship between auditory cortical spiking activity and gamma band oscillations from the STS (Ghazanfar et al., 2008) (Fig. 25.5B).

It should be noted, however, that functional interactions between STS and auditory cortex are not likely to occur solely during the presentation of faces with voices. Other congruent, behaviorally salient audiovisual events such as looming signals (Maier et al., 2004, 2008) or other temporally coincident signals may elicit similar functional interactions (Noesselt et al., 2007).

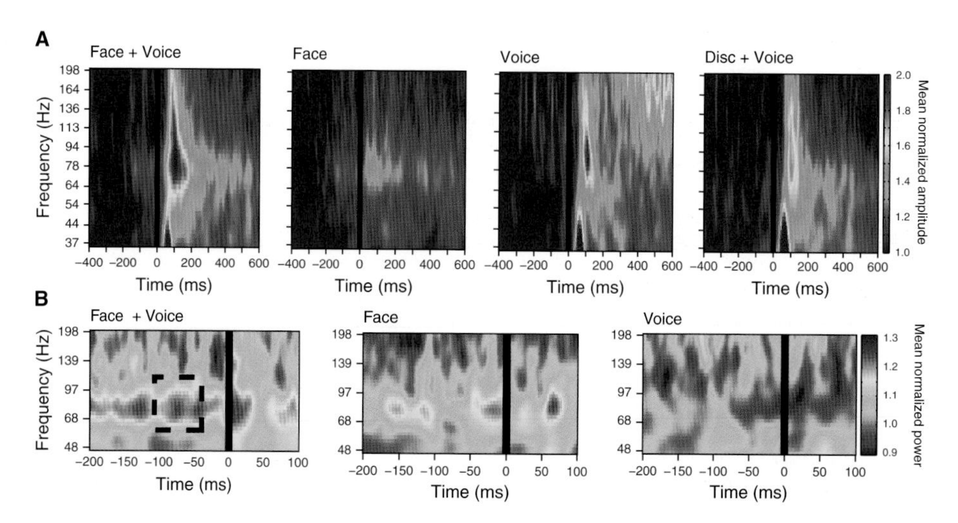

Figure 25.5 (**A**) Time-frequency plots (cross-spectrograms) illustrate the modulation of functional interactions (as a function of stimulus condition) between the lateral belt auditory cortex and the superior temporal sulcus (STS) for a population of cortical sites. X-axes depict the time in milliseconds as a function of onset of the auditory signal (solid black line). Y-axes depict the frequency of the oscillations in Hertz. Color bar indicates the amplitude of these signals normalized by the baseline mean. (**B**) Spike-field cross-spectrogram illustrates the relationship between the spiking activity of auditory cortical neurons and the STS local field potential across the population of cortical sites. X-axes depict time in milliseconds as a function of the onset of the multisensory response in the auditory neuron (solid black line). Y-axes depict the frequency in Hertz. Color bar denotes the cross-spectral power normalized by the baseline mean for different frequencies.

VENTROLATERAL PREFRONTAL CORTEX

While much research has focused on understanding the electrophysiological correlates of the prefrontal cortex's involvement in visual working memory, decision making utilizing visual cues, and spatial visual perception, few studies addressed the cellular mechanisms of auditory or multisensory processes. This is despite the wealth of evidence that the PFC receives input from all sensory modalities (see previous discussion). Only a handful of studies demonstrated that neurons in the PFC respond to auditory stimuli or are active during auditory tasks in primates (Bodner et al., 1996; Ito, 1982; Tanila et al., 1992; Vaadia et al., 1986; Watanabe, 1992). Weakly responsive auditory neurons were found sporadically and distributed across a wide region of the PFC including portions of the ventral PFC, principal sulcus, arcuate sulcus, and premotor cortex. None of these previous studies used species-specific communication calls as stimuli. In the visual domain, a series of studies from the laboratory of Goldman-Rakic demonstrated that neurons within particular regions within the ventrolateral PFC (VLPFC) were selectively responsive to faces including species-specific faces (O'Scalaidhe et al., 1997, 1999). This finding coincided with the demonstration that these same regions of VLPFC received afferents from the lateral and parabelt regions of auditory cortex as well as the dorsal bank of the STS (Hackett et al., 1999; Romanski et al., 1999a,b).

Soon after the localization of face cells in the VLPFC, neurons responsive to acoustic stimuli, including species-specific vocalizations, were discovered. Unlike auditory cortex, VLPFC neurons did not readily respond to simple acoustic stimuli such as pure tones (Romanski & Goldman-Rakic, 2002), but were robustly driven by complex sounds (Averbeck & Romanski, 2006; Cohen et al., 2007; Gifford et al., 2005; Romanski et al., 2005). The question then arose, Would these higher-order auditory neurons encode the "semantics" of vocalizations, perhaps independent of sensory modality (given the presence of face neurons in the same region), or do they simply encode complex sound features in a manner similar to, but at higher level, than auditory cortex proper?

Positron emission tomography (PET) and functional magnetic resonance imaging (fMRI) studies have suggested that the human inferior frontal gyrus, or ventral frontal lobe, plays a role in semantic processing (Demb et al., 1995; Poldrack et al., 1999). In neurophysiological experiments with rhesus monkeys, this question has engendered some controversy (see also Miller & Cohen, this volume). Studies from Cohen and colleagues, in which vocalizations were presented in the context of an oddball presentation paradigm, suggested that vocalizations with similar meaning ("food calls") elicited similar prefrontal neuronal responses under particular conditions (Gifford et al., 2005). Romanski and colleagues (2005) tested neurons with exemplars from 10 different vocalization categories, which included food and nonfood calls, and found that prefrontal neurons show a selectivity that is similar to neurons in the lateral belt area of auditory cortex—the majority of neurons were selective for approximately two vocalizations when tested with 10 call types (Fig. 25.6A,B). Importantly, the types of vocalizations that elicited similar response magnitudes shared similar acoustic features, suggesting that prefrontal neurons might be encoding complex acoustics rather than semantics (Averbeck & Romanski, 2004, 2006; Romanski et al., 2005) (Fig. 25.7A–C).

An important point to note is that the rhesus monkey vocal repertoire is assembled by human listeners and is based solely on behavioral contexts (defined by the human observer) and acoustic features. Thus, it is not surprising, and indeed highly likely, that prefrontal neurons are capable of categorical discrimination of vocalizations and are using complex feature analysis to encode category membership when tested during simple presentation and not within the confines of a cognitive task. A Hidden-Markov model analysis of the probability that prefrontal neural activity categorizes vocalizations revealed that prefrontal neurons can be described as linear functions of the probabilities that

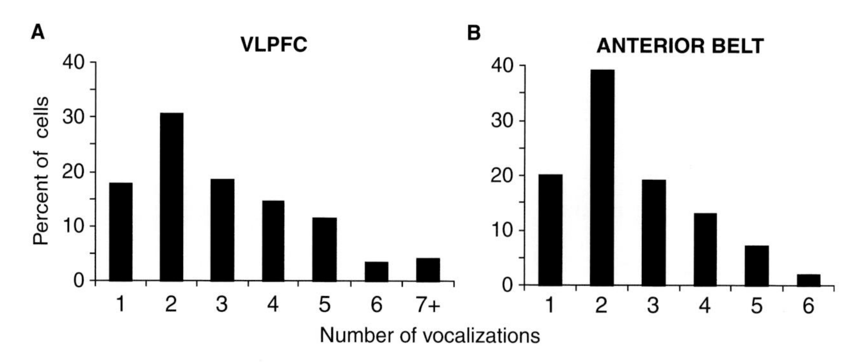

Figure 25.6 (**A**) Selectivity of single-unit responses to vocalizations in the ventrolateral prefrontal cortex (VLPFC). Single neurons in the VLPFC responded to vocalizations when presented to awake fixating monkeys with varying selectivity. A large percentage of cells (49% of 124 cells) responded to just one or two vocalization types when tested with exemplars from 10 vocalization categories. This is similar to the selectivity in AL of the auditory belt shown in (**B**). The majority of neurons in AL also responded to only one to two vocalizations when tested with six different vocalization types. Modified from Romanski, L. M., Averbeck, B. B., & Diltz, M. (2005). Neural representation of vocalizations in the primate ventrolateral prefrontal cortex. *Journal of Neurophysiology, 93,* 734–747; and Tian, B., Reser, D., Durham, A., Kustov, A., & Rauschecker, J. P. (2001). Functional specialization in rhesus monkey auditory cortex. *Science, 292*(5515), 290–293.

individual calls belong to a particular acoustic category (Averbeck & Romanski, 2006); that is, prefrontal neurons respond in a similar manner to vocalizations from the same or acoustically similar categories.

While data has emerged that is suggestive of semantic encoding by the VLPFC at the single neuron level (Gifford et al., 2005), it seems important to distinguish semantic encoding from responses to acoustic features. Thus, neurons that respond well to vocalizations from a similar semantic context category should not respond to vocalizations or synthetic complex sounds that have a similar acoustic morphology but differ in semantic context. Such an experiment has not yet been performed, but would be the key to determining the role of VLPFC in semantic processing at the single neuron level. In considering all of the anatomical data showing the cascade of acoustic afferents reaching this region, which include multiple regions of auditory cortex, and the physiological and imaging data that demonstrate a role in semantic processing for the prefrontal region, it is likely that prefrontal auditory neurons encode a combination of complex acoustic features and relevant behavioral or referential factors simultaneously. This is the case for other putative behavioral functions of the prefrontal cortex (see, for example Nieder, this volume) and for other association cortices that are connected to prefrontal cortex.

The anatomical and physiological evidence for convergence of auditory, visual, and multisensory responses in areas such as the STS and auditory cortex naturally leads to the idea that the PFC should be multisensory as well. Earlier neurophysiological studies noted multisensory responses to simple stimuli in restricted parts of dorsolateral prefrontal cortex in highly trained animals (Bodner et al., 1996; Fuster et al., 2000) and in the orbitofrontal cortex of anesthetized monkeys (Benevento et al., 1977). In the VLPFC, studies revealed complex auditory (Romanski & Goldman-Rakic, 2002), visual (Hoshi et al., 2000; Pigarev et al., 1979; O'Scalaidhe et al., 1997, 1999; Rosenkilde et al., 1981; Wilson et al., 1993), and somatosensory (Romo et al., 1999) responsive neurons. The integration of faces and voices may be more likely to occur in the VLPFC where there are

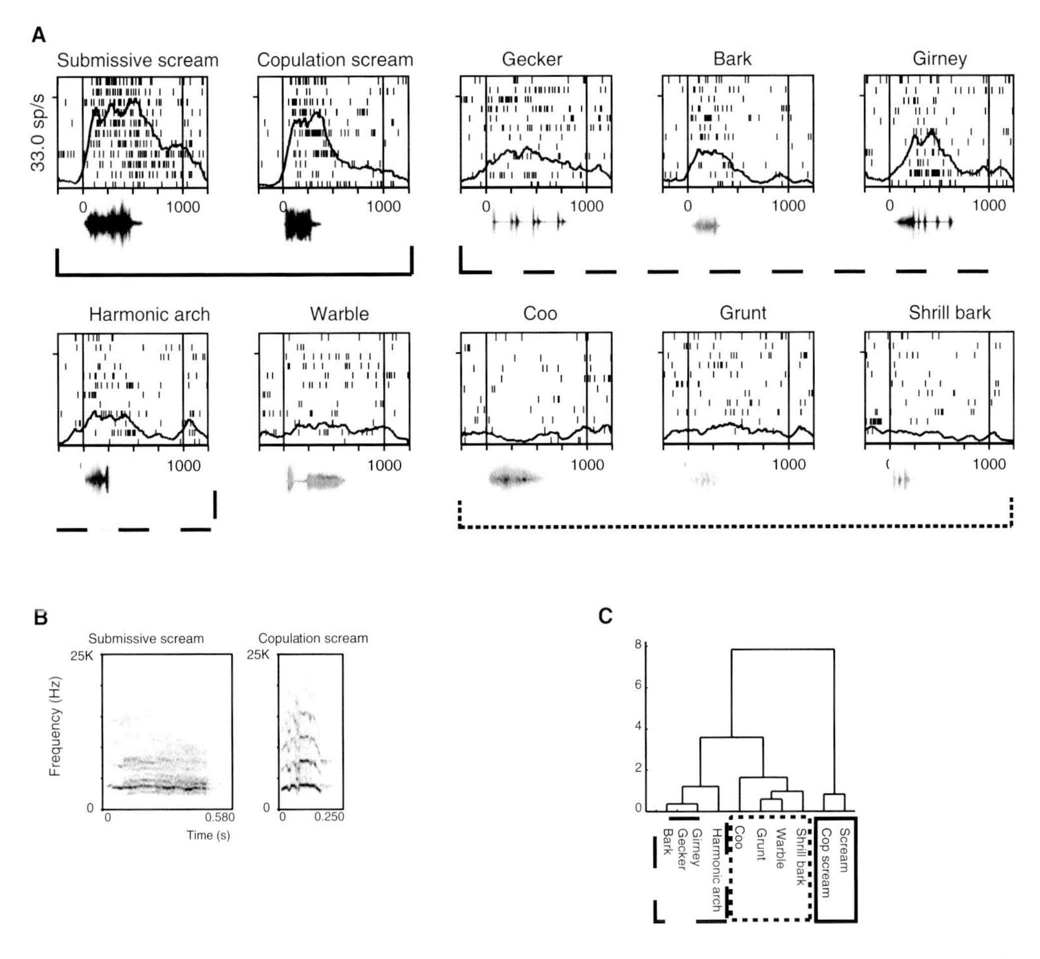

Figure 25.7 Response of a ventrolateral prefrontal cortex (VLPFC) single neuron to species-specific vocalizations. In (**A**), the neural activity of a single neuron in the VLPFC is displayed in a raster/spike-density plot for each one of 10 vocalization exemplars. The cell shown has a best response to two scream vocalizations (first two panels in A). The waveforms for each vocalization are shown below the raster response. In (**B**), the spectrograms for two of these sounds show that they can share some similar acoustic features including F0 at ≈4,000 Hz. A hierarchical cluster analysis of the mean firing rate to each of the vocalizations resulted in the dendrogram shown in (**C**), where similar responses cluster together. For this and other neurons in VLPFC, sounds with similar complex features evoked similar neuronal responses in VLPFC and clustered together. Lines are drawn around the dominant groupings in the dendrogram and the same line patterns are drawn below the neuronal response for these vocalizations.

overlapping domains for both auditory stimuli (particularly, vocalizations; Romanski & Goldman-Rakic, 2002) and visual stimuli (particularly, faces; O'Scalaidhe et al., 1997) (Fig. 25.8). To test this idea, rhesus monkeys were presented with movies of familiar monkeys vocalizing while single neurons were recorded from the VLPFC (Sugihara et al., 2006). These movies were separated into audio and video streams and neural response to the unimodal stimuli were compared to combined audiovisual responses. This experimental paradigm is similar to that used in the studies of STS and auditory cortex (Barraclough et al., 2005; Ghazanfar et al., 2005, 2008). Approximately half of the neurons recorded in the VLPFC were bimodal

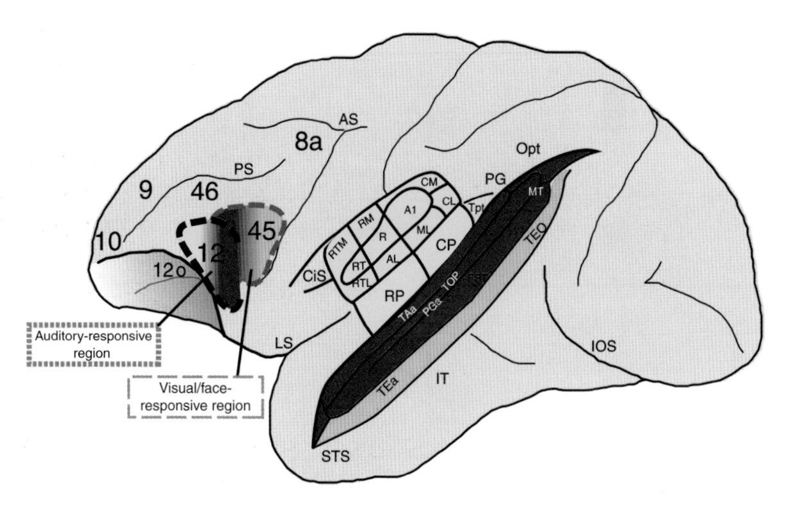

Figure 25.8 A color-coded schematic diagram of the temporal-prefrontal connectivity in the macaque brain. Auditory cortical areas in the temporal lobe are colored yellow and the core, belt, and parabelt areas are labeled. The visually responsive areas within the inferotemporal cortex are colored orange, while multisensory regions within the superior temporal sulcus (STS) are colored red to indicate both auditory and visual responses. Auditory, visual, and multisensory cortical regions send projections to the ventral prefrontal cortex where they combine to form the basis of multisensory and unimodal processing. Studies have previously described auditory and visual/face responses in areas 12 and 45, respectfully. Areas 8a, 9, 10, 12, 45, and 46 are cytoarchitectonic subdivisions of the prefrontal cortex (Preuss & Goldman-Rakic, 1991). A1, primary auditory cortex; AS, arcuate sulcus; CiS, circular sulcus; CL, caudal lateral belt; CM, caudal medial belt; CP, caudal parabelt; IT, inferotemporal cortex; IOS, intermediate orbital sulcus; LS, lateral sulcus; ML, middle part of the lateral belt; PGa, parietal associated area in the STS; R, rostral core auditory cortex; RM, rostral medial belt auditory cortex; RP, rostral parabelt auditory cortex; RT, rostral temporal auditory cortex; RTL, rostral temporal lateral belt auditory cortex; RTM, rostral temporal medial belt auditory cortex; STS, superior temporal sulcus; TAa, temporal lobe; Tpt, temporal-parietal association auditory cortex; TAa, anterior temporal area in STS; TPO, temporal parieto-occipital area in STS.

in the sense that they responded to both unimodal auditory and visual stimuli or responded differently to bimodal stimuli than to either unimodal stimuli (Sugihara et al., 2006). As in the STS and auditory cortex, prefrontal neurons exhibited enhancement or suppression, and, like the STS but unlike the auditory cortex, suppression (73% of neurons) was more commonly observed than enhancement (27% of neurons) (Fig. 25.9).

Responses varied according to the stimulus exemplar used so that a given cell might show multisensory suppression with one pair of congruent faces and voices and an enhancement with a different pair. It was also interesting that face/voice stimuli evoked multisensory responses more frequently than nonface/nonvoice combinations, as in auditory cortex (Ghazanfar et al., 2008) and in the STS (Barraclough et al., 2005). This adds support to the notion that VLPFC is part of a circuit that may be specialized for integrating face and voices. The selectivity of some neurons in this prefrontal region is in accord with fMRI studies indicating that a putatively homologous region of the human brain, area 47 (pars orbitalis), is more activated by human vocal sounds compared with animal and nonvocal sounds (Fecteau et al., 2005). It is unclear, however, to what extent the VLPFC, the STS, and/or the auditory cortex are "specialized" for processing faces and voices. For example, the STS appears to

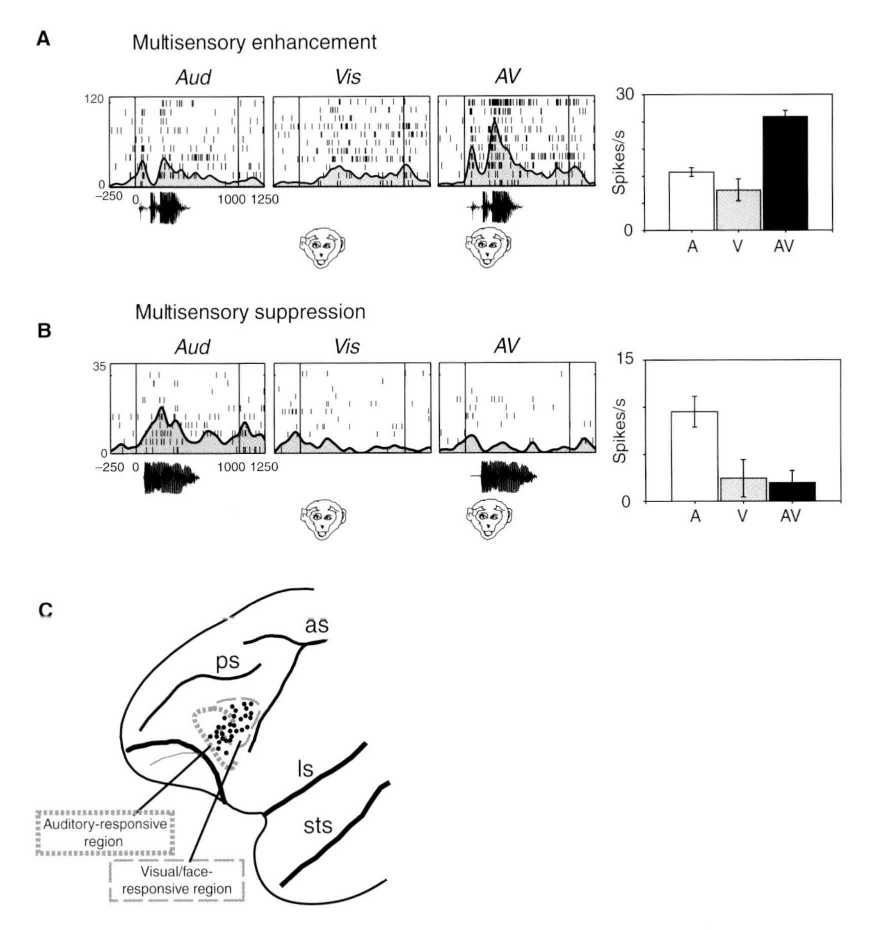

Figure 25.9 Multisensory integration in single neurons of the ventrolateral prefrontal cortex (VLPFC). (**A**) An example of a neuron that exhibited multisensory enhancement, where the response to the combined audiovisual stimuli is greater than the response to either unimodal stimuli. (**B**) A cell that showed multisensory suppression is shown. Here the response to the combined audiovisual stimulus is less than the response to the unimodal stimuli. The neural response is shown as rasters/spike-density plots on the left and as the average spike rate in the bar graph on the right in A and B. A schematic of the macaque brain showing the location of audiovisual responsive cells in the VLPFC is shown in (**C**). The cells were scattered throughout previously identified auditory- and visual-responsive regions.

be sensitive to biological motion in general, and not just dynamic faces (Barraclough et al., 2005; Oram & Perrett, 1994), and the multisensory responses to faces and voices in the auditory cortex may be a product of cortical inputs from the prefrontal cortex and the STS (Ghazanfar et al., 2005, 2008). Thus, while each cortical node in a sensory integration network may contribute uniquely to the processing of multisensory communication stimuli, the overall process is likely to be an emergent property involving multiple nodes.

CONCLUSIONS

Much of this line of comparative work is in its infancy, and we are limited in our overview by the paucity of studies across different species—most of this work has used macaque monkeys as model systems together with a few behavioral studies with chimpanzees. The neural processes underlying audiovisual integration of communication signals appear to be mediated by networks embedded within the frontal and temporal lobes, possibly homologous with those networks mediating audiovisual speech perception in humans. It will be important to extend these findings to other species to know for certain. There are other possible comparative studies of great interest. For example, it may be that arboreal monkeys or apes, who spend much of their time out of visual contact with conspecifics, will exhibit a lesser degree of multisensory face/voice processing. It is not known whether face/voice multisensory processes are exhibited by prosimians, which do not have an extensive suite of facial expressions and rely on olfactory cues to a greater degree than other primates. A related question would be the degree to which primates can match the correspondences between heterospecific faces and voices and the degree to which this ability is dependent upon experience. For example, a recent study in human infants suggest that species-typical experience narrows the perceptual ability to match faces and voices to the subjects' own species (Lewkowicz & Ghazanfar, 2006); this developmental shift in face/voice perception is not seen in infant vervet monkeys (Zangehenpour et al., 2009). One aspect of multisensory processing that is absent from the behavioral studies reported in this review is an abundance of evidence for *integration.* To date, there is no evidence that primates have anything like the "McGurk effect," and only a few studies have demonstrated an enhancement in behavioral performance with bimodal versus unimodal communication signals (see increased accuracy in chimpanzees: Parr, 2004). Thus, there is a strong need to demonstrate that, beyond just matching, primates use multisensory cues to enhance detection, discrimination, and/or learning of communication signals.

In concert with these field studies is the need to use more sophisticated behavioral paradigms during neurophysiology experiments. Studies of the auditory cortex, the STS, and the prefrontal cortex reveal that neurons in these structures enhance or suppress their firing in responses to audiovisual stimuli versus unimodal presentations, but how these responses relate to behavior is unknown, and we are left only with the suggestion that there may be a connection to behavior. Unfortunately, tasks that require behavioral integration of auditory and visual signals are difficult to apply with primates using traditional methodologies. Use of modified preferred looking paradigms and other methods may assist in the training of these difficult tasks and allow for the recording of neural signals during the process of behavioral-level audiovisual integration. Another important goal is creating more realistic environments for both behavioral and neurophysiological experiments. Communication is an action that involves, at a minimum, a pair of individuals. This is not captured by the paradigms currently in use in laboratory environments. Alternative approaches may come in the form of interactive playbacks, the use of telemetry (Grohrock et al., 1997; Obeid et al., 2004a,b), and simulating social exchanges using artificial agents, much like the talking-head animations used by speech researchers (Massaro & Bosseler, 2006).

REFERENCES

Arbib, M. A. (2005). From monkey-like action recognition to human language: An evolutionary framework for neurolinguistics. *Behavioral and Brain Sciences, 28,* 105–167.

Averbeck, B. B., & Romanski, L. M. (2004). Principal and independent components of macaque vocalizations: Constructing stimuli to probe high-level sensory processing. *Journal of Neurophysiology, 91,* 2897–2909.

Averbeck. B. B., & Romanski, L. M. (2006). Probabilistic encoding of vocalizations in macaque ventral lateral prefrontal cortex. *Journal of Neuroscience, 26,* 11023–11033.

Baizer, J. S., Ungerleider, L. G., & Desimone, R. (1991). Organization of visual inputs to the inferior temporal and posterior parietal cortex in macaques. *Journal of Neuroscience, 11,* 168–190.

Barbas, H. (1988). Anatomic organization of basoventral and mediodorsal visual recipient prefrontal regions in the rhesus-monkey. *Journal of Comparative Neurology, 276,* 313–342.

Barbas, H., & Mesulam, M. M. (1981). Organization of afferent input to subdivisions of area-8 in the rhesus-monkey. *Journal of Comparative Neurology, 200,* 407–431.

Barbas, H., & Pandya, D. N. (1989). Architecture and intrinsic connections of the prefrontal cortex in the rhesus-monkey. *Journal of Comparative Neurology, 286,* 353–375.

Barnes, C. L., & Pandya, D. N. (1992). Efferent cortical connections of multimodal cortex of the superior temporal sulcus in the rhesus-monkey. *Journal of Comparative Neurology, 318,* 222–244.

Barraclough, N. E., Xiao, D. K., Baker, C. I., Oram, M. W., & Perrett, D. I (2005). Integration of visual and auditory information by superior temporal sulcus neurons responsive to the sight of actions. *Journal of Cognitive Neuroscience, 17,* 377–391.

Barth, H., Kanwisher, N., & Spelke, E. S. (2003). The construction of large number representations in adults. *Cognition, 86,*201–221.

Bauer, H. R. (1987). Frequency code: Orofacial correlates of fundamental frequency. *Phonetica, 44,* 173–191.

Benevento, L. A., Fallon, J., Davis, B. J., & Rezak, M. (1977). Auditory-visual interactions in single cells in the cortex of the superior temporal sulcus and the orbital frontal cortex of the macaque monkey. *Experimental Neurology, 57,* 849–872.

Bernstein, L. E., Auer, E. T., & Takayanagi, S. (2004). Auditory speech detection in noise enhanced by lipreading. *Speech Communication, 44,* 5–18.

Besle, J., Fort, A., Delpuech, C., & Giard, M. H. (2004). Bimodal speech: Early suppressive visual effects in human auditory cortex. *European Journal of Neuroscience, 20,* 2225–2234.

Bodner, M., Kroger, J., & Fuster, J. M. (1996). Auditory memory cells in dorsolateral prefrontal cortex. *Neuroreport, 7,* 1905–1908.

Boye, M., Gunturkun, O., & Vauclair, J. (2005). Right ear advantage for conspecific calls in adults and subadults, but not infants, California sea lions (Zalophus californianus): Hemispheric specialization for communication? *European Journal of Neuroscience, 21,* 1727–1732.

Brokx, J. P., & Nooteboom, S. G. (1982). Intonation and the perceptual separation of simultaneous voices. *Journal of Phonetics, 10,* 23–36.

Bruce, C., Desimone, R., & Gross, C. G. (1981). Visual properties of neurons in a polysensory area in superior temporal sulcus of the macaque. *Journal of Neurophysiology, 46,* 369–384.

Bullier, J., Schall, J. D., & Morel, A. (1996). Functional streams in occipito-frontal connections in the monkey. *Behavioural Brain Research, 76,* 89–97.

Buzsaki, G. (2006). *Rhythms of the brain.* New York: Oxford University Press.

Calvert, G. A. (2001). Crossmodal processing in the human brain: Insights from functional neuroimaging studies. *Cerebral Cortex, 11,* 1110–1123.

Chandrasekaran, C., & Ghazanfar, A. A. (2009). Different neural frequency bands integrate faces and voices differently in the superior temporal sulcus. *Journal of Neurophysiology, 101,* 773–788.

Chavis, D. A., & Pandya, D. N. (1976). Further observations on corticofrontal connections in rhesus-monkey. *Brain Research, 117,* 369–386.

Cheney, D. L., & Seyfarth, R. M. (1982). How vervet monkeys perceive their grunts - Field playback experiments. *Animal Behaviour, 30,* 739–751.

Cohen, Y. E., Theunissen, F., Russ, B. E., & Gill, P. (2007). Acoustic features of rhesus vocalizations and their representation in the ventrolateral prefrontal cortex. *Journal of Neurophysiology, 97,* 1470–1484.

Demb, J. B., Desmond, J. E., Wagner, A. D., Vaidya, C. J., Glover, G. H., & Gabrieli, J. D. E. (1995). Semantic encoding and retrieval in the left inferior prefrontal cortex - A functional MRI study of task-difficulty and process specificity. *Journal of Neuroscience, 15,* 5870–5878.

Evans, T. A., Howell, S., & Westergaard, G. C. (2005). Auditory-visual cross-modal perception of communicative stimuli in tufted capuchin monkeys (Cebus apella). *Journal of Experimental Psychology: Animal Behavior Processes, 31,* 399–406.

Fecteau, S., Armony, J. L., Joanette, Y., & Belin, P. (2005). Sensitivity to voice in human prefrontal cortex. *Journal of Neurophysiology, 94,* 2251–2254.

Fitch, W. T. (1997). Vocal tract length and formant frequency dispersion correlate with body size in rhesus macaques. *Journal of the Acoustical Society of America, 102,* 1213–1222.

Fitch, W. T., & Hauser, M. D. (1995). Vocal production in nonhuman-primates - Acoustics, physiology, and functional constraints on honest advertisement. *American Journal of Primatology, 37,* 191–219.

Fitch, W. T., & Fritz, J. B. (2006). Rhesus macaques spontaneously perceive formants in conspecific vocalizations. *Journal of the Acoustical Society of America, 120,* 2132–2141.

Fowler, C. A. (2004). Speech as a supramodal or amodal phenomenon. In: G. A. Calvert, C. Spence, & B. E. Stein (Eds.), *The handbook of multisensory processes* (pp. 189–201). Cambridge, MA: MIT Press.

Fu, K. M. G., Johnston, T. A., Shah, A. S., Arnold, L., Smiley, J., Hackett, T. A., et al. (2003). Auditory cortical neurons respond to somatosensory stimulation. *Journal of Neuroscience, 23,* 7510–7515.

Fuster, J. M., Bodner, M., & Kroger, J. K. (2000). Cross-modal and cross-temporal association in neurons of frontal cortex. *Nature, 405,* 347–351.

Geschwind, N. (1964). The development of the brain and the evolution of language. In: C. I. J. M. Stuart (Ed.), *Report of the 15th Annual Round Table Meeting on Languages and Linguistics* (pp. 155–169). Washington, DC: Georgetown.

Ghazanfar, A. A. (2009). The multisensory roles for auditory cortex in primate vocal communication. *Hearing Research,* doi:10.1016/j.heares.2009.04.003.

Ghazanfar, A. A., Chandrasekaran, C., & Logothetis, N. K. (2008). Interactions between the superior temporal sulcus and auditory cortex mediate dynamic face/voice integration in rhesus monkeys. *Journal of Neuroscience, 28,* 4457–4469.

Ghazanfar, A. A., & Hauser, M. D. (1999). The neuroethology of primate vocal communication: Substrates for the evolution of speech. *Trends in Cognitive Sciences, 3,* 377–384.

Ghazanfar, A. A., & Hauser, M. D. (2001). The auditory behaviour of primates: A neuroethological perspective. *Current Opinion in Neurobiology, 11,* 712–720.

Ghazanfar, A. A., & Logothetis, N. K. (2003). Facial expressions linked to monkey calls. *Nature, 423,* 937–938.

Ghazanfar, A. A., Maier, J. X., Hoffman, K. L., & Logothetis, N. K. (2005). Multisensory integration of dynamic faces and voices in rhesus monkey auditory cortex. *Journal of Neuroscience, 25,* 5004–5012.

Ghazanfar, A. A., & Rendall, D. (2008). Evolution of human vocal production. *Current Biology, 18,* R457–R460.

Ghazanfar, A. A., & Santos, L. R. (2004). Primate brains in the wild: The sensory bases for social interactions. *Nature Reviews Neuroscience, 5,* 603–616.

Ghazanfar, A. A., & Schroeder, C. E. (2006). Is neocortex essentially multisensory? *Trends in Cognitive Sciences, 10,* 278–285.

Ghazanfar, A. A., Smith-Rohrberg, D., & Hauser, M. D (2001). The role of temporal cues in rhesus monkey vocal recognition: Orienting asymmetries to reversed calls. *Brain, Behavior, and Evolution, 58,* 163–172.

Ghazanfar, A. A., Smith-Rohrberg, D., Pollen, A. A., & Hauser, M. D. (2002). Temporal cues in the antiphonal long-calling behaviour of cottontop tamarins. *Animal Behaviour, 64,* 427–438.

Ghazanfar, A. A., & Turesson, H. K. (2008). Speech production: How does a word feel? *Current Biology, 18,* R1142–R1144.

Ghazanfar, A. A., Turesson, H. K., Maier, J. X., van Dinther, R., Patterson, R. D., & Logothetis, N. K. (2007). Vocal tract resonances as indexical cues in rhesus monkeys. *Current Biology, 17,* 425–430.

Gifford, G. W., MacLean, K. A., Hauser, M. D., & Cohen, Y. E. (2005). The neurophysiology of functionally meaningful categories: Macaque ventrolateral prefrontal cortex plays a critical role in spontaneous categorization of species-specific vocalizations. *Journal of Cognitive Neuroscience, 17,* 1471–1482.

Grohrock, P., Hausler, U., & Jurgens, U. (1997). Dual-channel telemetry system for recording vocalization-correlated neuronal activity in freely moving squirrel monkeys. *Journal of Neuroscience Methods, 76,* 7–13.

Hackett, T. A., De La Mothe, L. A., Ulbert, I., Karmos, G., Smiley, J., & Schroeder, C. E. (2007a). Multisensory convergence in auditory cortex, II. Thalamocortical connections of the caudal superior temporal plane. *Journal of Comparative Neurology, 502,* 924–952.

Hackett, T. A., Smiley, J. F., Ulbert, I., Karmos, G., Lakatos, P., de la Mothe, L. A., et al. (2007b). Sources of somatosensory input to the caudal belt areas of auditory cortex. *Perception, 36,* 1419–1430.

Hackett, T. A., Stepniewska, I., & Kaas, J. H. (1999). Prefrontal connections of the parabelt auditory cortex in macaque monkeys. *Brain Research, 817,* 45–58.

Harries, M. H., & Perrett, D. I. (1991). Visual processing of faces in temporal cortex - Physiological evidence for a modular organization and possible anatomical correlates. *Journal of Cognitive Neuroscience, 3,* 9–24.

Hauser, M. D., Agnetta, B., & Perez, C. (1998). Orienting asymmetries in rhesus monkeys: The effect of time-domain changes on acoustic perception. *Animal Behaviour, 56,* 41–47.

Hauser, M. D., Evans, C. S., & Marler, P. (1993). The role of articulation in the production of rhesus-monkey, Macaca-mulatta, vocalizations. *Animal Behaviour, 45,* 423–433.

Hauser, M. D., & Marler, P. (1993). Food-associated calls in rhesus macaques (Macaca-mulatta). 1. Socioecological Factors. *Behavioral Ecology, 4,* 194–205.

Hauser, M. D., & Ybarra, M. S. (1994). The role of lip configuration in monkey vocalizations - Experiments using xylocaine as a nerve block. *Brain and Language, 46,* 232–244.

Heinz, R. D., & Brady, J. V. (1988). The acquisition of vowel discriminations by nonhuman primates. *Journal of the Acoustical Society of America, 84,* 186–194.

Hikosaka, K., Iwai, E., Saito, H. A., & Tanaka, K. (1988). Polysensory properties of neurons in the anterior bank of the caudal superior temporal sulcus of the macaque monkey. *Journal of Neurophysiology, 60,* 1615–1637.

Hoshi, E., Shima, K., & Tanji J (2000). Neuronal activity in the primate prefrontal cortex in the process of motor selection based on two behavioral rules. *Journal of Neurophysiology, 83,* 2355–2373.

Ito, S. I. (1982). Prefrontal unit-activity of macaque monkeys during auditory and visual reaction-time tasks. *Brain Research, 247,* 39–47.

Izumi, A. (2002). Auditory stream segregation in Japanese monkeys. *Cognition, 82,* B113–B122.

Izumi, A., & Kojima, S. (2004). Matching vocalizations to vocalizing faces in a chimpanzee (Pan troglodytes). *Animal Cognition, 7,* 179–184.

Jiang, J. T., Alwan, A., Keating, P. A., Auer, E. T., & Bernstein, L. E. (2002). On the relationship between face movements, tongue movements, and speech acoustics. *Eurasip Journal on Applied Signal Processing, 2002,* 1174–1188.

Jordan, K. E., & Brannon, E. M. (2006). The multisensory representation of number in infancy. *Proceedings of the National Academy of Sciences, 103,* 3486–3489.

Jordan, K. E., Brannon, E. M., Logothetis, N. K., & Ghazanfar, A. A. (2005). Monkeys match the number of voices they hear with the number of faces they see. *Current Biology, 15,* 1034–1038.

Kayser, C., Petkov, C. I., Augath, M., & Logothetis, N. K. (2005). Integration of touch and sound in auditory cortex. *Neuron, 48,* 373–384.

Kayser, C., Petkov, C. I., Augath, M., & Logothetis, N. K. (2007). Functional imaging reveals visual modulation of specific fields in auditory cortex. *Journal of Neuroscience, 27,* 1824–1835.

Kayser, C., Petkov, C. I., & Logothetis, N. K. (2008). Visual modulation of neurons in auditory cortex. *Cerebral Cortex, 18,* 1560–1574.

Kojima, S. (1990). Comparison of auditory functions in chimpanzee and human. *Folia Primatologica, 55,* 62–72.

Kojima, S., & Kiritani, S. (1989). Vocal-auditory functions in the chimpanzee: Vowel perception. *International Journal of Primatology, 10,* 199–213.

Kuhl, P. K., & Meltzoff, A. N. (1984). The intermodal representation of speech in infants. *Infant Behavior and Development, 7,* 361–381.

Kuhl, P. K., & Miller, J. D. (1975). Speech-perception by chinchilla - Voiced-voiceless distinction in alveolar plosive consonants. *Science, 190,* 69–72.

Kuhl, P. K., Williams, K. A., & Meltzoff, A. N. (1991). Cross-modal speech perception in adults and infants using nonspeech auditory stimuli. *Journal of Experimental Psychology: Human Perception and Performance, 17,* 829–840.

Lakatos, P., Chen, C-M., O'Connell, M. N., Mills, A., & Schroeder, C. E. (2007). Neuronal oscillations and multisensory interaction in primary auditory cortex. *Neuron, 53,* 279–292.

Le Prell, C. G., Niemiec, A. J., & Moody, D. B. (2001). Macaque thresholds for detecting increases in intensity: Effects of formant structure. *Hearing Research, 162,* 29–42.

Lewkowicz, D. J. (2002). Heterogeneity and heterochrony in the development of intersensory perception. *Cognitive Brain Research, 14,* 41–63.

Lewkowicz, D. J., & Ghazanfar, A. A. (2006). The decline of cross-species intersensory perception in human infants. *Proceedings of the National Academy of Sciences, 103,* 6771–6774.

Liberman, A. M., & Mattingly, I. (1985). The motor theory revised. *Cognition, 21,* 1–36.

Liu, J., & Newsome, W. T. (2006). Local field potential in cortical area MT: Stimulus tuning and behavioral correlations. *Journal of Neuroscience, 26,* 7779–7790.

Maier, J. X., Chandrasekaran, C., & Ghazanfar, A. A. (2008). Integration of bimodal looming signals through neuronal coherence in the temporal lobe. *Current Biology, 18,* 963–968.

Maier, J. X., Neuhoff, J. G., Logothetis, N. K., & Ghazanfar, A. A. (2004). Multisensory integration of looming signals by Rhesus monkeys. *Neuron, 43,* 177–181.

Massaro, D. W., & Bosseler, A. (2006). Read my lips - The importance of the face in a computer-animated tutor for vocabulary learning by children with autism. *Autism, 10,* 495–510.

McGurk, H., & MacDonald, J. (1976). Hearing lips and seeing voices. *Nature, 264,* 229–239.

Meltzoff, A. N., & Moore, M. (1997). Explaining facial imitation: A theoretical model. *Early Development and Parenting, 6,* 179–192.

Miller, C. T., & Hauser, M. D. (2004). Multiple acoustic features underlie vocal signal recognition in tamarins: Antiphonal calling experiments. *Journal of Comparative Physiology A: Neuroethology Sensory Neural and Behavioral Physiology, 190,* 7–19.

Nasir, S. M., & Ostry, D. J. (2006). Somatosensory precision in speech production. *Current Biology, 16,* 1918–1923.

Noesselt, T., Rieger, J. W., Schoenfeld, M. A., Kanowski, M., Hinrichs, H., Heinze, H-J., et al. (2007). Audiovisual temporal correspondence modulates human multisensory superior temporal sulcus plus primary sensory cortices. *Journal of Neuroscience, 27,* 11431–11441.

O'Scalaidhe, S. P., Wilson, F. A. W., & Goldman-Rakic, P. S. (1997). Areal segregation of face-processing neurons in prefrontal cortex. *Science, 278,* 1135–1138.

O'Scalaidhe, S. P., Wilson, F. A. W., & Goldman-Rakic, P. S. (1999). Face-selective neurons during passive viewing and working memory performance of Rhesus monkeys: Evidence for intrinsic specialization of neuronal coding. *Cerebral Cortex, 9,* 459–475.

Obeid, I., Nicolelis, M. A. L., & Wolf, P. D. (2004a). A multichannel telemetry system for single unit neural recordings. *Journal of Neuroscience Methods, 133,* 33–38.

Obeid, I., Nicolelis, M. A. L., & Wolf, P. D. (2004b). A low power multichannel analog front end for portable neural signal recordings. *Journal of Neuroscience Methods, 133,* 27–32.

Oram, M. W., & Perrett, D. I. (1994). Responses of anterior superior temporal polysensory (Stpa) neurons to biological motion stimuli. *Journal of Cognitive Neuroscience, 6,* 99–116.

Owren, M. J., Seyfarth, R. M., & Cheney, D. L. (1997). The acoustic features of vowel-like grunt calls in chacma baboons (Papio cyncephalus ursinus): Implications for production processes and functions. *Journal of the Acoustical Society of America, 101,* 2951–2963.

Palombit, R. A., Cheney, D. L., & Seyfarth, R. M. (1999). Male grunts as mediators of social interaction with females in wild chacma baboons (Papio cynocephalus ursinus). *Behaviour, 136,* 221–242.

Pandya, D. N., Hallett, M., & Mukherje, S. (1969). Intra-hemispheric and interhemispheric connections of neocortical auditory system in rhesus monkey. *Brain Research, 14,* 49.

Pandya, D. N., & Kuypers, H. G. J. (1969). Cortico-cortical connections in rhesus monkey. *Brain Research, 13,* 13.

Parr, L. A. (2004). Perceptual biases for multimodal cues in chimpanzee (Pan troglodytes) affect recognition. *Animal Cognition, 7,* 171–178.

Partan, S. R. (2002). Single and multichannel signal composition: Facial expressions and vocalizations of rhesus macaques (Macaca mulatta). *Behaviour, 139,* 993–1027.

Patterson, M. L., & Werker, J. F. (2003). Two-month-old infants match phonetic information in lips and voice. *Developmental Science, 6,* 191–196.

Petersen, M. R., Beecher, M. D., Zoloth, S. R., Moody, D. B., & Stebbins, W. C. (1978). Neural lateralization of species-specific vocalizations by Japanese macaques (Macaca-fuscata). *Science, 202,* 324–327.

Petrides, M., & Pandya, D. N. (1988). Association fiber pathways to the frontal-cortex from the superior temporal region in the rhesus-monkey. *Journal of Comparative Neurology, 273,* 52–66.

Petrides, M., & Pandya, D. N. (2002). Comparative cytoarchitectonic analysis of the human and the macaque ventrolateral prefrontal cortex and corticocortical connection patterns in the monkey. *European Journal of Neuroscience, 16,* 291–310.

Pigarev, I. N., Rizzolatti, G., & Scandolara, C. (1979). Neurons rsponding to visual stimuli in the frontal lobe of macaque monkeys. *Neuroscience Letters, 12,* 207–212.

Pinker, S. (1994). *The language instinct.* New York: William Morrow and Company, Inc.

Poldrack, R. A., Wagner, A. D., Prull, M. W., Desmond, J. E., Glover, G. H., & Gabrieli, J. D. E. (1999). Functional specialization for semantic and phonological processing in the left inferior prefrontal cortex. *Neuroimage, 10,* 15–35.

Rauschecker, J. P., Tian, B., & Hauser, M. (1995). Processing of complex sounds in the macaque nonprimary auditory cortex. *Science, 268,* 111–114.

Rendall, D., Owren, M. J., & Rodman, P. S. (1998). The role of vocal tract filtering in identity cueing in rhesus monkey (Macaca mulatta). vocalizations. *Journal of the Acoustical Society of America, 103,* 602–614.

Rickert, J., de Oliveira, S. C., Vaadia, E., Aertsen, A., Rotter, S., & Mehring, C. (2005). Encoding of movement direction in different frequency ranges of motor cortical local field potentials. *Journal of Neuroscience, 25,* 8815–8824.

Romanski, L. M., Averbeck, B. B., & Diltz, M. (2005). Neural representation of vocalizations in the primate ventrolateral prefrontal cortex. *Journal of Neurophysiology, 93,* 734–747.

Romanski, L. M., Bates, J. F., & Goldman-Rakic, P. S. (1999a). Auditory belt and parabelt projections to the prefrontal cortex in the rhesus monkey. *Journal of Comparative Neurology, 403,* 141–157.

Romanski, L. M., & Goldman-Rakic, P. S. (2002). An auditory domain in primate prefrontal cortex. *Nature Neuroscience, 5,* 15–16.

Romanski, L. M., Tian, B., Fritz, J., Mishkin, M., Goldman-Rakic, P. S., & Rauschecker, J. P. (1999b). Dual streams of auditory afferents target multiple domains in the primate prefrontal cortex. *Nature Neuroscience, 2,* 1131–1136.

Romo, R., Brody, C. D., Hernandez, A., & Lemus, L. (1999). Neuronal correlates of parametric working memory in the prefrontal cortex. *Nature, 399,* 470–473.

Rosenblum, L. D. (2005). Primacy of multimodal speech perception. In: D. B. Pisoni & R. E. Remez (Eds.), *Handbook of speech perception* (pp. 51–78). Malden, MA: Blackwell.

Rosenkilde, C. E., Bauer, R. H., & Fuster, J. M. (1981). Single cell activity in ventral prefrontal cortex of behaving monkeys. *Brain Research, 209,* 375–394.

Rowell, T. E., & Hinde, R. A. (1962). Vocal communication by the rhesus monkey (Macaca mulatta). *Proceedings of the Zoological Society London, 138,* 279–294.

Saleem, K. S., Suzuki, W., Tanaka, K., & Hashikawa, T. (2000). Connections between anterior inferotemporal cortex and superior temporal sulcus regions in the macaque monkey. *Journal of Neuroscience, 20,* 5083–5101.

Schroeder, C. E., & Foxe, J. J. (2002). The timing and laminar profile of converging inputs to multisensory areas of the macaque neocortex. *Cognitive Brain Research, 14,* 187–198.

Schwartz, J-L., Berthommier, F., & Savariaux, C. (2004). Seeing to hear better: Evidence for early audio-visual interactions in speech identification. *Cognition, 93,* B69–B78.

Seltzer, B., Cola, M. G., Gutierrez, C., Massee, M., Weldon, C., & Cusick, C. G. (1996). Overlapping and nonoverlapping cortical projections to cortex of the superior temporal sulcus in the rhesus monkey: Double anterograde tracer studies. *Journal of Comparative Neurology, 370,* 173–190.

Seltzer, B., & Pandya, D. N. (1978). Afferent cortical connections and architectonics of superior temporal sulcus and surrounding cortex in rhesus-monkey. *Brain Research, 149,* 1–24.

Seltzer, B., & Pandya, D. N. (1991). Post-rolandic cortical projections of the superior temporal sulcus in the rhesus-monkey. *Journal of Comparative Neurology, 312,* 625–640.

Seltzer, B., & Pandya, D. N. (1994). Parietal, temporal, and occipital projections to cortex of the superior temporal sulcus in the rhesus monkey: A retrograde tracer study. *Journal of Comparative Neurology, 343,* 445–463.

Shannon, R. V., Zeng, G. F., Kamath, V., Wygonski, J., & Ekelid, M. (1995). Speech recognition with primarily temporal cues. *Science, 270,* 303–304.

Sinnott, J. M. (1989). Detection and discrimination of synthetic English vowels by old-world monkeys (Cercopithecus, Macaca) and humans. *Journal of the Acoustical Society of America, 86,* 557–565.

Sinnott, J. M., Brown, C. H., Malik, W. T., & Kressley, R. A. (1997). A multidimensional scaling analysis of vowel discrimination in humans and monkeys. *Perception and Psychophysics, 59,* 1214–1224.

Smiley, J. F., Hackett, T. A., Ulbert, I., Karmas, G., Lakatos, P., Javitt, D. C., et al. (2007). Multisensory convergence in auditory cortex, I. Cortical connections of the caudal superior temporal plane in macaque monkeys. *Journal of Comparative Neurology, 502,* 894–923.

Smith, D. R. R., & Patterson, R. D. (2005). The interaction of glottal-pulse rate and vocal-tract length in judgement of speaker size, sex and age. *Journal of the Acoustical Society of America, 118,* 3177–3186.

Sommers, M. S., Moody, D. B., Prosen, C. A., & Stebbins, W. C. (1992). Formant frequency discrimination by Japanese macaques (Macaca-fuscata). *Journal of the Acoustical Society of America, 91,* 3499–3510.

Stein, B. E. (1998). Neural mechanisms for synthesizing sensory information and producing adaptive behaviors. *Experimental Brain Research, 123,* 124–135.

Sugihara, T., Diltz, M., Averbeck, B. B., & Romanski, L. M. (2006). Integration of auditory and visual communication information in the primate ventrolateral prefrontal cortex. *Journal of Neuroscience, 26,* 11138–11147.

Summerfield, Q., Culling, J. F., & Fourcin, A. J. (1992). Auditory segregation of competing voices: Absence of effects of FM or AM coherence. *Philosophical Transactions of the Royal Society of London Series B: Biological Sciences, 336,* 357–366.

Tanila, H., Carlson, S., Linnankoski, I., Lindroos, F., & Kahila, H. (1992). Functional properties of dorsolateral prefrontal cortical neurons in awake monkey. *Behavioural Brain Research, 47,* 169–180.

Thorpe, S.J., Rolls, E.T., & Maddison, S. (1983). The orbitofrontal cortex: neuronal activity in the behaving monkey. *Experimental Brain Research, 49,* 93–115.

Trout, J. D. (2001). The biological basis of speech: What to infer from talking to the animals. *Psychological Review, 108,* 523–549.

Vaadia, E., Benson, D. A., Hienz, R. D., & Goldstein, M. H. (1986). Unit study of monkey frontal fortex - Active localization of auditory and of visual stimuli. *Journal of Neurophysiology, 56,* 934–952.

van Wassenhove, V., Grant, K. W., & Poeppel, D. (2005). Visual speech speeds up the neural processing of auditory speech. *Proceedings of the National Academy of Sciences, 102,* 1181–1186.

Watanabe, M. (1992). Frontal units of the monkey coding the associative significance of visual and auditory stimuli. *Experimental Brain Research, 89,* 233–247.

Webster, M. J., Bachevalier, J., & Ungerleider, L. G. (1994). Connections of inferior temporal areas TEO and TE with parietal and frontal cortex in macaque monkeys. *Cerebral Cortex, 4,* 470–483.

Wilke, M., Logothetis, N. K., & Leopold, D. A. (2006). Local field potential reflects perceptual suppression in monkey visual cortex. *Proceedings of the National Academy of Sciences, 103,* 17507–17512.

Wilson, F. A. W., Scalaidhe, S. P. O., & Goldmanrakic, P. S. (1993). Dissociation of object and spatial processing domains in primate prefrontal cortex. *Science, 260,* 1955–1958.

Yehia, H., Rubin, P., & Vatikiotis-Bateson, E. (1998). Quantitative association of vocal-tract and facial behavior. *Speech Communication, 26,* 23–43.

Yehia, H. C., Kuratate, T., & Vatikiotis-Bateson, E. (2002). Linking facial animation, head motion and speech acoustics. *Journal of Phonetics, 30,* 555–568.

Zangehenpour, S., Ghazanfar, A. A., Lewkowicz, D. J., & Zatorre, R. J. (2009). Heterochrony and cross-species intersensory matching by infant vervet monkeys. *PLoS ONE, 4,* e4302.

CHAPTER 26

Neuroethology of Attention in Primates

Stephen V. Shepherd and Michael L. Platt

> ***Expert Commission Decides That the Horse Actually Reasons.***
>
> BERLIN, Sept. 13. — The remarkable horse called "Clever Hans" has just been examined by a special commission of experts, in order that a decision might be arrived at whether it is a horse possessed of extraordinary brain power...
>
> —From *The London Standard*, in *The New York Times*, October 2, 1904

The case of "Clever Hans" is perhaps the most famous con job in the history of psychology. Both experts and public were taken in, believing that human-like perceptual and reasoning abilities could be exhibited by a horse trained by Wilhelm van Osten. After 3 years and a great deal of public scrutiny, however, Oskar Pfungst determined that the horse could only answer questions to which the experimenters already knew the answer—in short, the horse did not have these human abilities, and was merely responding to subtle social cues provided by his audience. Thus, the horse's exceptional brainpower was dismissed.

But this is the wrong lesson of this story. After all, the con artist was not Wilhelm van Osten—he was taken in as well—but *his horse.* A better lesson is that the horse, like many animals, is exquisitely capable of learning to recognize subtle cues to the emotions, intentions, expectations, and knowledge not only of his own species, but even humans. These abilities are no less impressive because they are social, rather than abstract. Indeed, humans with selective disruption of social cognition have great difficulty navigating the overwhelmingly complex societies in which we live, and by which we thrive. As Clever Hans demonstrated, the cognitive mechanisms underlying our shared humanity appear to also be shared with many other species. It thus seems likely that advances in the evolution of cognition derive, at least in part, from the need to predict the actions of others. These social cognitive abilities are built on a long genetic and developmental history, but are founded on something so simple and fundamental that we often fail to notice it. To learn how to make sense of another's actions—whether human, monkey, or horse—we have to *attend* to them. In the following pages, we explore this social orienting behavior, describe its apparent adaptive functions, and attempt to understand how orienting decisions are made within the brain.

PSYCHOLOGICAL, ETHOLOGICAL, AND NEUROETHOLOGICAL APPROACHES TO ATTENTION

All mobile animals orient to salient features of their environment. This can occur overtly, by shifting gaze, or covertly, by deploying attention without eye movements. Psychophysical, electrophysiological, and neuroimaging studies conducted in the laboratory have extensively probed orienting in both human and nonhuman primates. Generally, subjects have been trained to discriminate simple stimuli whose salience or behavioral significance has been arbitrarily assigned through explicit instruction or association with rewards (e.g., Posner, 1980). Such studies suggest the operation of two distinct systems for orienting attention (James, 1890; Jonides, 1981; Posner & Cohen, 1984), one fast and reflexive (exogenous) and the other slow and voluntary (endogenous), mediated by partially overlapping neural circuits (Corbetta & Shulman, 2002; Egeth & Yantis, 1997; Mangun, 1995).

Outside the laboratory, observational studies in natural settings indicate that social stimuli are intrinsically salient and strongly attract attention (Caine & Marra, 1988; Keverne et al., 1978; McNelis & Boatright-Horowitz, 1998). Moreover, recent laboratory studies have suggested that social cues, such as the direction of gaze in an observed individual, access a privileged information channel capable of directing attention (Deaner & Platt, 2003; Driver et al., 1999; Friesen & Kingstone, 1998). These studies imply that the neural mechanisms that mediate attention have evolved sensitivity to cues predicting the goals and intentions of other individuals—but the precise nature of these social cues, and the specific neural systems by which they are processed, remain obscure. Indeed, current evidence suggests that social attention maps poorly onto existing models of attention, which emphasize dichotomous exogenous and endogenous orienting systems.

The neuroethological approach is an alternative paradigm that works to resolve these issues. We contend that complete understanding of the biology of attention must account not only for gross patterns of orienting in natural environments but also for the fine spatiotemporal details of orienting measured in controlled laboratory settings. These ethological and psychophysical goals are often approached separately, using different animal models and highly divergent techniques, reflecting in part the fact that the demands of naturalistic observation generally preclude precise measurements of orienting. Likewise, psychophysical experiments have typically failed to simulate the behavioral contexts in which orienting behavior normally operates.

Nonetheless, we contend that these divisions are surmountable, and that combining ethological and laboratory approaches will foster the development of a unified evolutionarily motivated theory of attention, which will have broad impact on our understanding of brain systems. For many animals, in particular primates like ourselves, one of the most important variables influencing attentional deployment is the current behavioral state of nearby individuals—the current social context. In the following sections, we consider the impact of social context on attention, outlining some of what has been learned from both laboratory and field studies. In particular, we describe our own efforts to bridge these approaches by studying the neuroethology of social attention in human and nonhuman primates.

EVOLUTION OF VISUAL SPECIALIZATIONS IN PRIMATES

Primates are unusual among mammals in their strong reliance on vision (Allman, 1999). Initially, visual specializations probably evolved in primates to support movement through upper tree branches (Robert Martin's "fine-branch niche hypothesis," Martin, 1990), to facilitate hunting for insects (Matt Cartmill's "visual predator" hypothesis, Cartmill, 1972), or both. Primates might thus be expected to use vision primarily for locomotion and food acquisition while retaining common mammalian visual functions such as predator avoidance.

Over the course of primate evolution, however, visual processing appears to have become increasingly specialized for guiding social interaction. Many primates make extensive use of vision to localize, monitor, and interact with other individuals, and likewise devote a large portion of their brains to visual processing. Notably, the expansion of the primate brain has been accompanied by a corresponding increase in the flexibility and complexity of primate social groups (Allman, 1999; Barton & Dunbar, 1997). While prosimian primates rely heavily on olfactory and pheromone-mediated modes of communication, these ancestral sensory modalities have been supplanted in more derived primates by visually mediated signals such as coloration, poise, gesture, facial expression, and gaze (de Waal, 2003; note also Gilad et al., 2004), as well as affective and referential vocalizations (Cheney & Seyfarth, 1990; Seyfarth & Cheney, 2003). While scientists have long recognized the importance of studying primate visual attention in the laboratory, we have all too often neglected the role of attention in the natural social ecology.

BEHAVIORAL GOALS DRIVE ORIENTING IN NATURALISTIC SETTINGS

The first studies of naturalistic orienting in primates were conducted in humans by the Russian psychologist Alfred Yarbus in the 1950s and 1960s (Yarbus, 1967). He measured overt visual orienting behavior by recording visual fixation patterns during free and instructed scanning of pictures with light-reflecting mirrors suction-cupped to the eyes. Yarbus demonstrated that social stimuli are intrinsically salient and strongly attract attention. Despite this strong bias, current behavioral goals also regulate visual attention. For example, when shown the painting "An Unexpected Visitor," subjects consistently oriented attention toward the faces of people in the scene (Fig. 26.1). When asked to determine the wealth of the family in the picture, however, subjects shifted their gaze to the clothing worn by the figures in order to extract the information requested by the experimenter.

Recently, Land and Hayhoe (2001) reported similar context dependence using noninvasive video gaze tracking in human subjects. They showed that attention was almost completely determined by task demands during simple actions such as making a sandwich or preparing tea. By contrast, subjects almost never attended to task-irrelevant regions of space. These data suggest that attentional priorities not only are shaped by evolutionary pressures but also can serve as external indicators of the shifting internal goals governing an individual's moment-to-moment behavior (Shepherd & Platt, 2008).

To make sense of natural orienting patterns, then, requires not just the sophisticated understanding of behavioral repertoire and ecological niche supplied by ethology, but also the rigorous mathematical tool set for understanding decision making provided by behavioral economics. The marriage of the latter fields with electrophysiology and functional imaging has produced an explosive change in our approach to human decisions through the field of neuroeconomics, and neuroethology holds similar promise for extending our mechanistic understanding of behavioral evolution. A central message of both these disciplines, however, has been that neural processes are strongly influenced by social

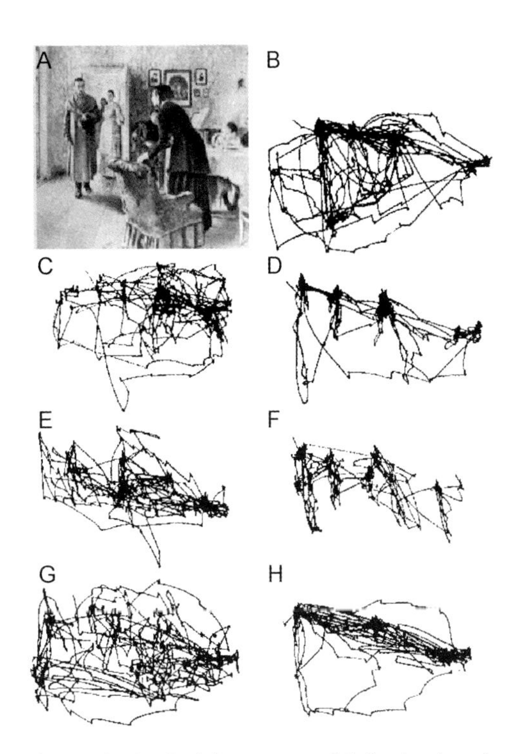

Figure 26.1 Social context and behavioral goals alter fixation patterns during free viewing. Panels A–H show the different gaze patterns of viewers when asked different questions about the illustration, Rjepin's "Unexpected Visitor," shown at upper left. Viewers scanned the photographs in very different ways when asked, e.g., to estimate the family's wealth (**C**), estimate their ages (**D**), memorize the position of people and objects (**G**), or estimate how long the "unexpected visitor" had been away (**H**). After Yarbus, A. (1967). Eye movements during perception of complex objects. In: *Eye movements and vision* (pp. 171–211). New York: Plenum Press. Used with permission.

variables, ranging from the framing effects invoked by word choice in task instructions (De Martino et al., 2006; Tversky & Kahneman, 1981) to the intrinsic reward of watching other individuals (Aharon et al., 2001; Hayden et al., 2007).

SOCIAL ATTENTION IN NATURAL SETTINGS

Observational data support the idea that visual attention in nonhuman primates is also biased toward social stimuli (Caine & Marra, 1988; Keverne et al., 1978; McNelis & Boatright-Horowitz, 1998). Furthermore, these biases are not uniform; some social stimuli attract more attention than others. For example, monkeys spend more time looking at pictures of faces gazing toward them than faces with averted gaze (Keating & Keating, 1982), and look preferentially toward the regions around the eyes and mouth (Guo et al., 2003; Keating & Keating, 1982; Kyes & Candland, 1987). They also look more often toward higher-ranking animals than lower-ranking animals (Keverne et al., 1978; McNelis & Boatright-Horowitz, 1998). Such data have generally been limited, however, to distal observations in natural settings or, in the laboratory, to qualitative analysis of fixation patterns within still photographs.

Given the various limitations of previous studies, one goal of our research has been to quantitatively measure primate visual attention in naturalistic environments and species-typical social groups. To do this, we recorded gaze behavior in socially housed, freely moving ring-tailed lemurs (*Lemur catta*) interacting in large three-dimensional environments. We used a lightweight telemetric optical gaze-tracking device (Fig. 26.2; see Shepherd & Platt, 2006) operating at 0.22 degrees × 33 ms resolution—a degree of precision comparable to eye-tracking methods used in the laboratory. Our approach differed, however, in that we did not provide any task or instruction, but instead inferred the goals guiding visual attention in natural contexts from the observed patterns of typical behavior (Shepherd & Platt, 2008). Ring-tailed lemurs, prosimian primates that diverged from the ancestors of "higher" primates some 60 million

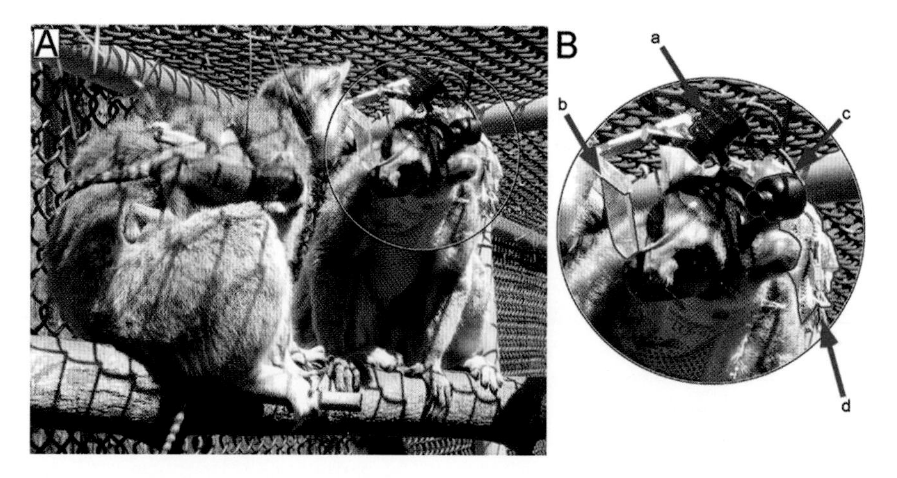

Figure 26.2 Equipment for tracking gaze during the natural behavior of freely moving animals. We tracked gaze during spontaneous and natural interactions with cohabitant conspecifics (**A**) using a telemetric optical gaze-tracking system developed by Iscan, Inc. The system (**B**) was composed of an infrared camera and LED (a) imaging the lemur's right eye through a dichroic mirror (b), an optical camera (c) viewing the scene in front the lemur's head, and a telemetry system housed in a primate vest (d), which broadcast to a remote monitoring station where the subject's recorded gaze direction was analyzed and projected onto locations in the recorded visual scene. After Shepherd, S. V., & Platt, M. L. (2006). Noninvasive telemetric gaze tracking in freely moving socially housed prosimian primates. *Methods, 38,* 185–194. Used with permission.

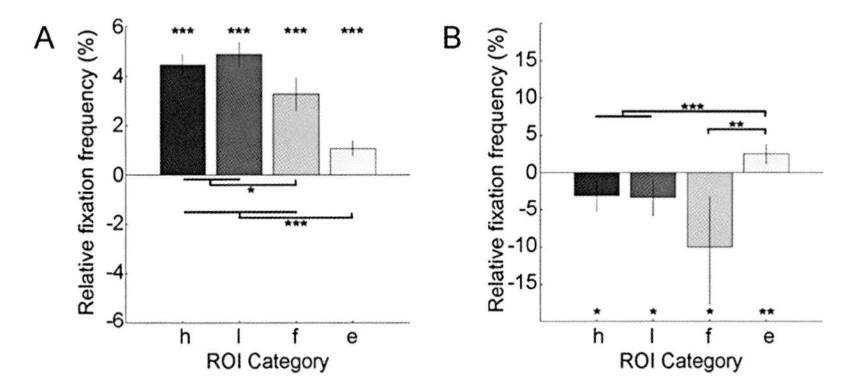

Figure 26.3 Fixation priorities in stationary and moving lemurs. (**A**) Lemurs fixated humans (h), lemurs (l), food (f), and high-contrast environmental features (e) significantly more than chance expectation; furthermore, they fixated humans and lemurs significantly more than food rewards, and all three significantly more than environmental features. (**B**) Lemurs fixated environmental features significantly more often when moving than when stationary, seemingly at the cost of fixations toward animals and rewards. ***, $P < 0.0005$; **, $P < 0.001$; *, $P < 0.01$. After Shepherd, S. V., & Platt, M. L. (2008). Spontaneous social orienting and gaze following in ringtailed lemurs (*Lemur catta*). *Animal Cognition, 11,* 13–20. Used with permission.

years ago, were chosen as subjects based on their tolerance of handling and their availability at the Duke Lemur Center. Ring-tailed lemur social groups are similar to those of many higher primates, comprising 10 to 20 individuals of both sexes organized in well-defined social hierarchies, and communicating through auditory, olfactory, and visual modalities (Jolly, 1966; Sauther et al., 1999).

We found that male ring-tailed lemurs fixated their human handlers—as would be expected given we had just suited them, temporarily, into recording equipment. More important, they fixated their social companions, and did so more often than they fixated small food rewards (Fig. 26.3A). Each of these three a priori categories—human handlers, conspecifics, and food rewards—were fixated significantly more often than chance and significantly more often than high-contrast environmental features, stimuli we expected to attract attention based on low-level visual salience (Carmi & Itti, 2006; Peters et al., 2005). These data suggest that animals, especially conspecifics, and rewards, such as potential food sources, were effectively identified, localized, and attended during natural visual behavior.

These social attention biases were not inflexible, however, and in fact changed during periods of active locomotion (Fig. 26.3B). While moving, lemurs attended to environmental features that served as potential movement substrates along the path toward their destination. At the same time, other visual priorities, such as monitoring other lemurs and foraging, were temporarily but systematically diminished. Together with earlier research (Land & Hayhoe, 2001; Yarbus, 1967), these findings validate the use of quantitative gaze measurements as an externally observable indicator of otherwise unobservable mental states (e.g., the current behavioral goals of an animal) and further reveal that the typical behavioral context for a lemur involves not only monitoring threats, such as predators, and rewards, such as food, but also guiding movement and maintaining observational contact with other members of the social group.

DOMINANCE, SEX, AND SOCIAL SALIENCE

Our ongoing field studies of orienting in ring-tailed lemurs support the idea that ancestral primates possessed neural specializations for orienting toward and extracting relevant information from other animals (Tomasello et al.,

2005). The diversity of stimuli and complexity of behavioral contexts that typify the field setting, however, has challenged our ability to draw definitive conclusions regarding the specific stimuli that guide visual attention during natural social behavior—an endeavor that is ongoing in our laboratory. Moreover, despite the evident similarity between human visual attention priorities and those observed in lemurs, the genomes, brains, behavioral repertoires, and social systems of our species differ dramatically. Unfortunately, little is known about brain function in lemurs.

To address these limitations, we have conducted parallel investigations of the visual orienting behavior of another primate, whose visual abilities, social structure, environmental niche, and physiology more closely mirror our own. Rhesus macaques (*Macaca mulatta*) are an oft-studied anthropoid primate with relatively well-explored biology, and like humans, they live in large, hierarchical social groups with extensive multisensory behavioral interactions. Although rhesus monkeys have been widely used to study visual attention, most past studies have used arbitrary stimuli with little or no intrinsic behavioral relevance.

We know, however, that in the wild, monkeys visually monitor one another (Caine & Marra, 1988; Keverne et al., 1978; McNelis & Boatright-Horowitz, 1998), and in the laboratory, they preferentially seek out visual stimuli with social content (Butler, 1954; Sackett, 1966). To precisely quantify how rhesus monkeys prioritize attending to specific classes of social stimuli, we developed a choice task designed to balance fluid rewards against the chance to glimpse photos of other monkeys. Specifically, monkeys chose between orienting to either of two targets, one associated with a juice reward and another associated with an alternative juice reward and a picture of a familiar monkey. By determining the differential juice reward at which monkeys were equally likely to choose to view the image, we were able to quantify the reward value of attending to different classes of social stimuli (Deaner et al., 2005).

Using this "pay-per-view" paradigm, we found that male monkeys consistently "overvalued" seeing potential mating cues (female hindquarters) and faces of dominant males, but "undervalued" seeing the faces of low-ranking males (Fig. 26.4). The attraction of attention to high-ranking males is somewhat counterintuitive,

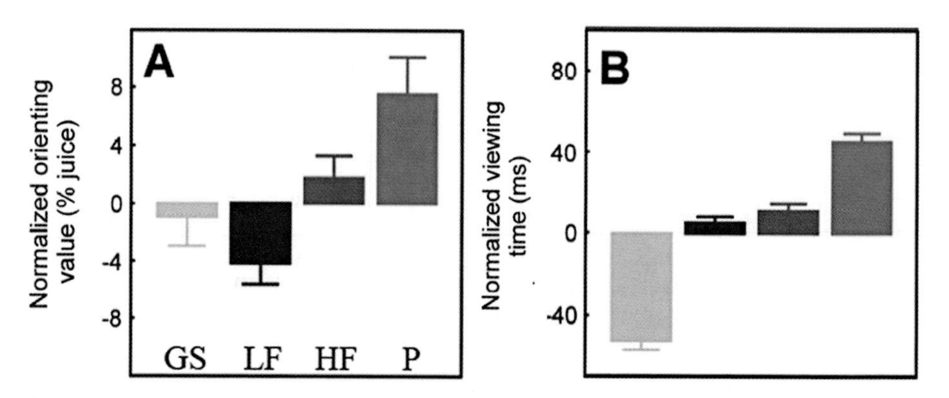

Figure 26.4 Monkeys sacrifice juice to view important social stimuli. When monkeys were offered different juice rewards to fixate two targets, only one of which also yielded an image reward, they chose each option equally when the intrinsic value of viewing an image offset the amount of juice sacrificed (**A**). Monkeys paid the highest amount of juice to see female perinea (P) and a lower amount to see high-ranking male faces (HF), but required extra juice to look at low-ranking male faces (LF) or, to a lesser extent, uniform gray squares (GS). A similar pattern is evident in the amount of time per presentation that monkeys fixated each category of image (**B**). This measure differs, however, in that monkeys dwell for similar lengths of time on low- and high-status faces. After Deaner, R. O., Khera, A. V., & Platt, M. L. (2005). Monkeys pay per view: adaptive valuation of social images by rhesus macaques. *Current Biology, 15*, 543–548. Used with permission.

since under natural conditions direct staring serves as a threat gesture in many primate species (van Hoof, 1967). Analysis of dwell times—the duration of glances toward particular social stimuli—provides a potential explanation for this paradox: Sexual cues consistently evoked prolonged stares, whereas faces generally evoked fixations of shorter duration. Frequent, furtive glances toward high-ranked males may serve to maximize acquisition of important social information while simultaneously minimizing risk of conflict.

EVOLUTIONARY BIOLOGY OF SOCIAL ATTENTION

Thus, both for freely moving lemurs and for macaques performing tasks in the laboratory, an important goal of visual attention is to maintain observational contact with conspecifics. Ethological studies of primate behavior suggest this behavioral bias may serve at least two complementary biological functions. Vision has long been known to play a role in hunting and foraging, affecting both predators (e.g., carnivores), where selection pressures favor focused binocular fields of view, and prey (e.g., ungulates), where selection favors broad, monocular visual fields (Allman, 1999; Cartmill, 1972). Primates, too, have focused, binocular visual fields (Allman, 1999; Martin, 1990), but this does not free us from the need to be vigilant for predators or hostile competitors. Indeed, primate societies are characterized by certain baseline levels of aggression, and primates must be wary of social threats both external and internal to their own social group. Thus, many primates must actively balance surveillance against external predation or rival social groups with surveillance against aggression from within the social group (e.g., Caine & Marra, 1988). Protection against social threat is a key need driving visual attention.

Centripetal surveillance, however, implies that there is a social group in the first place. From this we infer a second, more subtle role of social attention, first articulated by Chance and Jolly (1970). Cohesion of social groups requires, as a principal element, the coordination of movements to regulate spacing between each individual and its cohort. For this reason, Chance and Jolly (1970: 171) suggested that "The social attention of individuals within a cohort . . . must be directed exclusively at the other members of it," going on to note that "even when they are an integral part of the complete society, the distinct coherence of a cohort. . .may depend on their maintaining a predominant degree of attention toward themselves." Chance and Jolly proposed that the key mechanism of dominance is not the threat of violence from the strongest member of the troop, but rather the ability of these individuals to command the attention of other group members. In short, Chance and Jolly argued that primate societies are bound together by centripetal attention, specifically, in hierarchical societies, by attention toward high-status animals.

Although dominance may be structured by the threat of violence and by the need for coalitional defense against these threats (Cheney & Seyfarth, 1990; Keverne et al., 1978), status-based saliency seems to be positive in valence and largely prosocial, in that it promotes proximity to the group. For example, Chance and Jolly (1970) described a behavior called "reflected escape" in which a subordinate animal, threatened, runs in a looping arc, first away from the challenger and then back toward the central members of the group—even if these same dominant individuals had initiated the threat! These ideas seem to be supported by findings that gaze (Keverne et al., 1978), like allegiance and grooming (Cheney & Seyfarth, 1990), is allocated preferentially to dominant individuals but independently of their aggressiveness, and also by our own finding that macaques sacrifice more juice to view dominant animals than subordinate animals.

Just as prosocial reward may drive attention, however, there is evidence that enhanced salience may itself drive reward. Specifically, the mere act of attending to a stimulus appears to enhance its desirability. Zajonc first described these effects in 1968 when he found that brief presentation of unfamiliar visual stimuli caused people to subsequently rate those stimuli more esthetically pleasing, even when they could not recall having seen them (Zajonc, 1968, reviewed in Bornstein, 1989). More recently, two studies have generalized this effect from "mere exposure" to attentional state. Raymond and colleagues (2003) found that stimuli that were presented, but ignored, accrued negative associations in a variety of task conditions. Shimojo and colleagues (2003) made a complementary discovery, using simple preference judgments. They found that prior to selecting the more attractive of two faces, subjects looked increasingly long and often at the face they subsequently chose. Importantly, when subjects were forced to look at a particular face, they were also more likely to subsequently prefer it. Together, these findings suggest that attention may drive changes in affective judgments. By extension, "mere exposure" may mediate social cohesion in primates by encouraging approach toward previously attended members of the social group, even if the attention was initially garnered through such negatively valenced agonistic interactions as direct competition or threat. In this way, social saliency could play a critical role in patterning the spacing behavior of animals in a group, making the most often fixated animals the most desirable for approach.

This complex relationship between attention, approach, and hedonic value suggests that simple approach/avoid, pleasure/pain dichotomies may serve us poorly in studying the neuroscience of attention. Instead, it may be more useful to consider attention in terms of "motivational salience," the predicted marginal behavioral utility of information, and in terms of the specific neural systems through motivational salience that govern behavioral orienting.

It seems reasonable that attention to high-value social targets is promoted by the reward systems of the brain, and this idea is endorsed by ongoing studies in our lab (Klein & Platt, 2008). The motivators that drive attention, however, are not always entirely appetitive; prolonged staring at dominant males, for example, is risky and unlikely to be hedonically pleasing (consequentially, fixation durations are quite short). In the end, it is much harder to answer the question of whether behaviors are mediated by "pleasure" or "fear" than whether they are mediated by, say, the ventral striatum or amygdala. We tend to estimate the qualities of subjective experiences by analogy: To understand animals whose behavior differs broadly from our own, we may proceed most safely when we can supplement our analogical reasoning with mechanistic, as well as behavioral, data endorsing homology. Cases like this suggest that a neuroethological approach has the potential to clarify both our mechanistic understanding and our theoretical interpretation of adaptive behavior.

ATTENDING TO INTENTIONS

Ralph Waldo Emerson (1876: 118) wrote, "The eyes of men converse as much as their tongues, with the advantage, [sic] that the ocular dialect needs no dictionary, but is understood all the world over." As Emerson intimated, where we look often betrays our deepest interests, intentions, and desires. We use visual orienting not only to localize other individuals but also to interpret their relationships, attitudes, and intentions. Nonhuman primates also appear to use orienting by others to infer the location of important stimuli and events, to predict behavior, and perhaps even to interpret social relationships among others (Cheney & Seyfarth, 1990; Emery, 2000; Tomasello et al., 2005). Subtler still, humans (and perhaps other primates, particularly apes; de Waal, 2003) use and recognize a number of deictic gestures, from a quick flick of the eyes to sustained pointing, which signal important threats and opportunities within our shared environment. Furthermore, we use these signals in competitive contexts to read intent and predict action (watching someone's eyes during chess), and even to confound such predictions by others (the "no-look pass," a feint in which a player looks toward a different teammate than the one to whom he or she intends to pass the ball).

Despite the obvious importance of social cues for guiding attention in natural behavior, this process has remained, until recently, relatively unexplored by psychologists or neurobiologists. A typical laboratory approach to visual attention asks subjects to stare at a fixation point, followed by either a central cue or peripheral stimulus directing attention toward a peripheral location, followed by a behavioral measure of attentional deployment at the cued or uncued location (e.g., Posner 1980). Studies using this technique have revealed that central cues that validly predict the location of a future peripheral target cause subjects to shift attention in a voluntary ("endogenous" or "top-down") manner, whereas abrupt peripheral cues, even when they have no predictive value, automatically attract attention ("exogenous," "reflexive," "bottom-up" attention). These attention shifts are evident in changes in sensory discrimination performance and reaction time, and have distinct time courses: Exogenous attention operates more quickly and generates a subsequent orienting deficit ("inhibition of return"), while endogenous attention is slower and more sustained (Jonides, 1981; Muller & Rabbitt, 1989). Despite the obvious utility of this paradigm for understanding basic aspects of attention, its ethological relevance has long been limited by a failure to apply these laboratory techniques to the kind of social stimuli that pervasively guide orienting by primates in the natural world.

Friesen and Kingstone (1998) addressed this gap by modifying the Posner paradigm to investigate how social cues influence attention. They discovered that viewing a face with averted gaze rapidly and reflexively shifts attention in the same direction, even when gaze direction does not predict the eventual location of the target. In their experiments, human subjects were instructed to fixate a central point, where a face briefly appeared with its eyes cast either rightward or leftward. A split second later, a peripheral target appeared, randomly in the direction of gaze or in the opposite direction. Subjects were faster to respond to targets appearing in the direction of observed gaze, even for cue-to-target delays as brief as 105 ms (termed "stimulus onset asynchrony," or SOA).

Subsequent studies determined that these effects were both general and involuntary—a turned head shifted attention as easily as averted eyes (Langton & Bruce, 1999), and social cuing persisted even when the target was 80% likely to appear in the direction *opposite* viewed gaze (Driver et al., 1999). Attention shifts associated with observed gaze appear to be categorically distinct from attentional responses to explicit cues such as predictive arrows (Friesen et al., 2004) or abstract spatial associations (Galfano et al., 2006), suggesting that social orienting does not neatly fit within classical models of exogenous or endogenous attention. These results appear to support the idea that humans evolved a dedicated gaze-following module specialized for rapid and reflexive sharing of attention in social groups (Baron-Cohen, 1994; Perrett & Emery, 1994).

To test this hypothesis, we contrasted the socially cued orienting responses of rhesus macaques and humans (Deaner & Platt, 2003). Surprisingly, we found that macaques and humans both responded more quickly to an unpredictable target when it appeared where a monkey had just been seen to look. Furthermore, eye position during fixation in both species drifted in the direction of observed gaze, likely reflecting an incompletely suppressed drive to overtly orient in the same direction (Engbert & Kliegl, 2003; Hafed & Clark, 2002). The temporal dynamics with which attention followed observed gaze were highly similar in the two species (Fig. 26.5), suggesting shared neural mechanisms.

Our results strongly support the conclusion that gaze following is not unique to humans, and may, in fact, rely on neural substrates that are widespread among primates and possibly other animals known to follow gaze. Though gaze following by other animals may differ, both in strength and kind, from that evinced by anthropoid primates (Okamoto-Barth et al., 2007; Tomonaga, 2007), it appears that many animals are able to shift attention in response to observed social cues. Consistent with this argument, Tomasello and colleagues, along with a number of other research groups, have amassed a large body of work showing that many animals, including apes (Brauer et al., 2005), dogs (Agnetta et al., 2000), monkeys (Tomasello et al., 1998), goats (Kaminski et al., 2005), dolphins (Tschudin et al., 2001), and ravens (Bugnyar et al., 2004), can use social gaze cues to find hidden food or retrieve objects (reviewed in Emery, 2000, and Itakura, 2004).

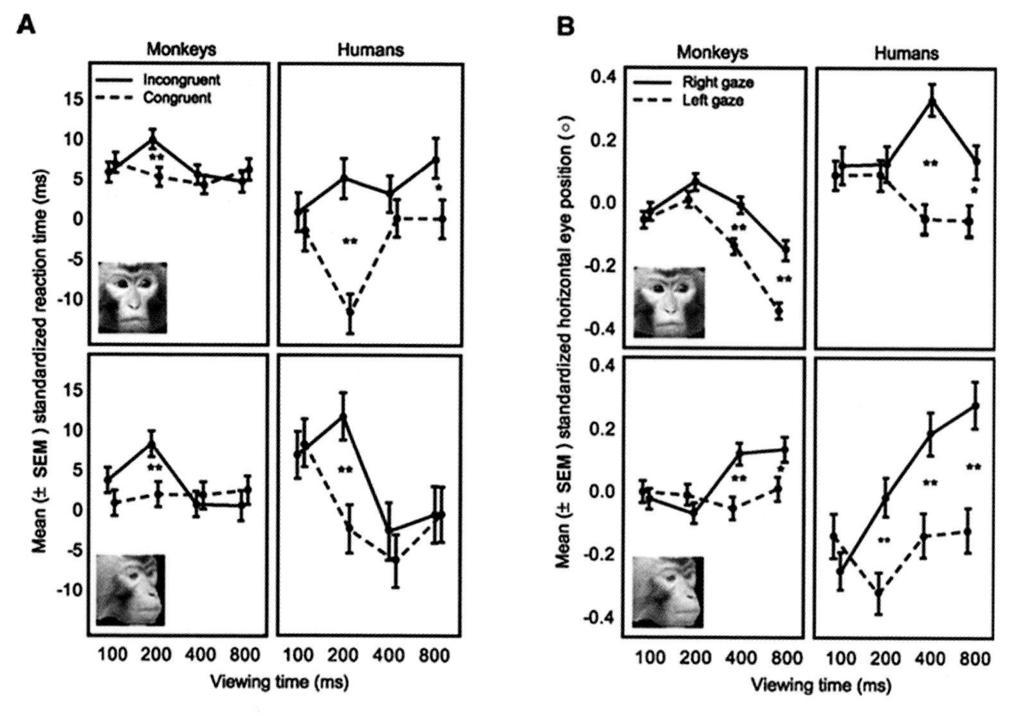

Figure 26.5 Gaze following by monkeys and humans shares psychophysical features. Monkeys and humans show similar magnitude and time course of gaze following in response to nonpredictive monkey gaze cues presented continuously for 100, 200, 400, or 800 ms prior to target presentation. These attention shifts were evident both by decreases in normalized reaction times to congruent (dashed) versus incongruent (solid) stimuli (**A**) and by microsaccades in the direction of observed gaze during cue presentation (**B**). **, $P < 0.001$; *, $P < 0.05$. After Deaner, R. O., & Platt, M. L. (2003). Reflexive social attention in monkeys and humans. *Current Biology, 13*, 1609–1613. Used with permission.

In most of those studies, human experimenters, rather than conspecifics, provided the social attention stimuli. This technical limitation simultaneously limits the ease with which we can generalize results to naturally occurring social interaction, and poses the fascinating question of how heterospecific and conspecific social perceptions interrelate. As the neural systems mediating predator avoidance, prey capture, and "pure" social interaction remain virtually unknown, the many fascinating evolutionary and computational links between them remain almost largely unexplored.

The potential ubiquity of gaze following in primates is supported by our recent studies of visual orienting by freely moving, socially housed lemurs. In those studies, we quantitatively and precisely monitored orienting behavior of two male ring-tail lemurs with an infrared telemetric gaze-tracking device while they spontaneously interacted with other lemurs. We found that lemurs tended to orient their eyes in the same direction that other lemurs oriented their bodies and heads (Fig. 26.6A,B). Such gaze alignment, however, could reflect coincidental orienting to salient events in a shared environment (e.g., a loud sound) rather than active use of social gaze cues. To address this confound, we examined the temporal sequence of gaze alignments around the time the subject lemur oriented to an observed lemur. We found that, prior to fixating the observed lemur, there was no alignment between the two animals' gaze. After fixating the observed lemur, however, gaze alignment increased significantly (Fig. 26.6C). The temporal sequence of gaze alignment supports the conclusion that lemurs actively follow the gaze of other

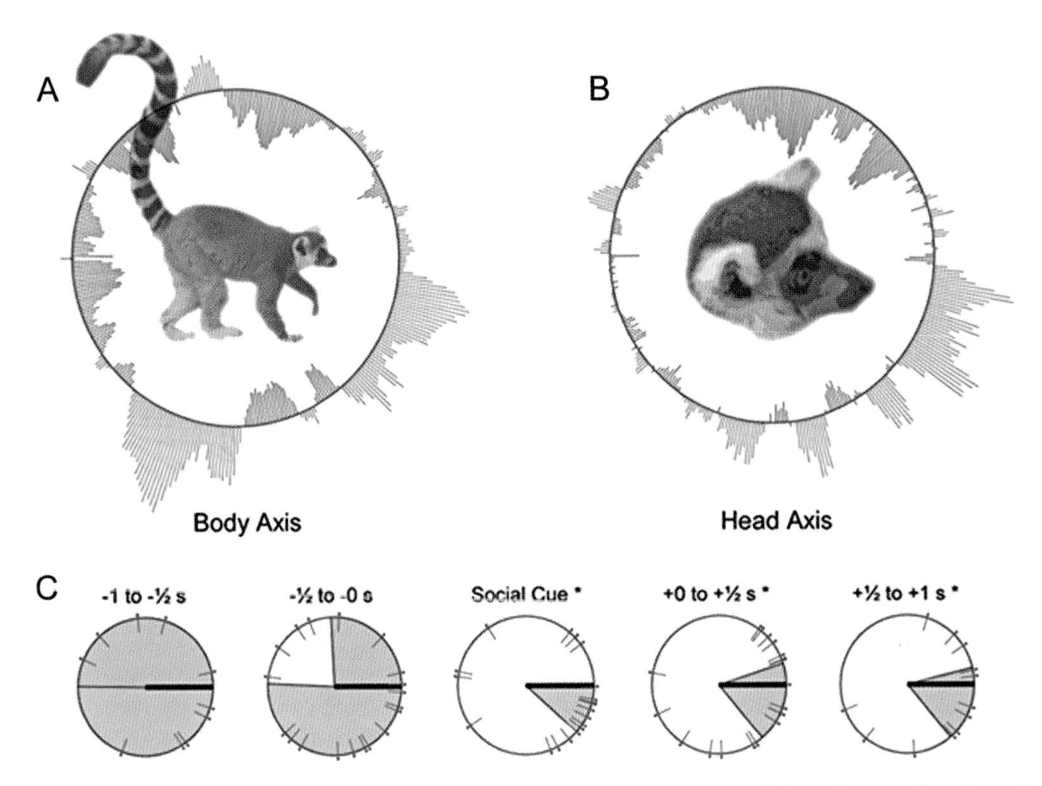

Figure 26.6 Spontaneous gaze following in lemurs. Lemurs spontaneously follow the gaze direction of their conspecifics in natural interaction. Lemurs not only coorient with the body (**A**) and head (**B**) axes of observed lemurs but also selectively increase gaze alignment with those individuals they have recently attended (**C**). In panels A and B, red outward lines are gaze offsets that are overrepresented with respect to chance, while blue inward lines are gaze offsets that are underrepresented. In panel C, tick marks occur at mean gaze offsets recorded in §-second periods prior to fixation, in the period during which the lemur is fixated, and for § second periods after fixation. Shaded regions in panel C reflect the dispersion of gaze alignments. Starred intervals are significantly aligned with gaze (chi^2 test $P < 0.05$). After Shepherd, S. V., & Platt, M. L. (2008). Spontaneous social orienting and gaze following in ringtailed lemurs (*Lemur catta*). *Animal Cognition, 11,* 13–20. Used with permission.

individuals (Shepherd & Platt, 2008). Our results stand in sharp contrast to at least two prior observational studies (Anderson & Mitchell, 1999; Itakura, 1996) that concluded that prosimian primates do not follow the gaze of human observers.

SOCIAL CONTEXT INFLUENCES GAZE FOLLOWING

Because both monkeys and humans shift their attention in response to social gaze cues, even when such cues fail to predict the location of a behavioral goal, it has been argued that gaze following is a strictly reflexive behavior mediated by a dedicated neural module (Deaner & Platt, 2003; Driver et al., 1999). Recent studies, however, challenge the notion that gaze cuing is purely reflexive, and instead indicate that social context can influence gaze-following behavior both in humans and monkeys. Specifically, several lines of evidence suggest that neural systems contributing to gaze following are regulated both by internal factors and online, by social context. For example, human females respond much more strongly to social gaze cues than do males (Fig. 26.7) (Bayliss et al., 2005; Deaner et al., 2007); moreover, our lab has found that gaze following in females, but not males, is influenced by the familiarity of the observed male cue (Deaner et al., 2007). These observations suggest the possibility that sex hormones may play an important role in regulating social attention. These observations also indicate that gaze-following mechanisms are not strictly reflexive and informationally encapsulated, but instead are sensitive to subtle changes in social variables such as the familiarity of the observed face.

In parallel, we have discovered that both social context and biological factors regulate gaze following in rhesus macaques (Shepherd et al., 2006). Specifically, we probed gaze-following behavior by seven male rhesus macaques in response to four rightward- and four leftward-looking photos of each of four familiar

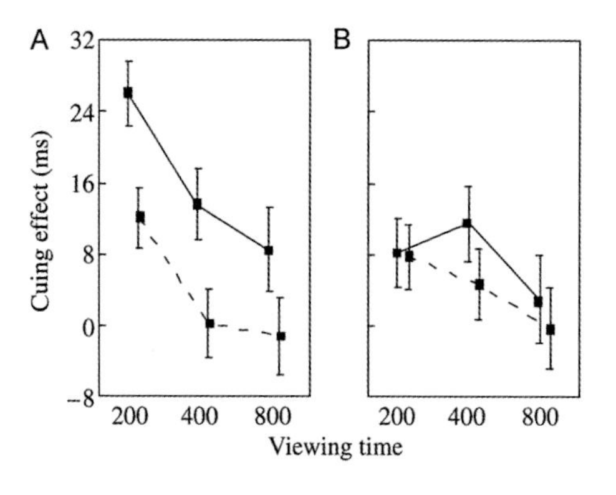

Figure 26.7 Sex differences in gaze following in humans. Human females exhibit stronger gaze following than males, and furthermore discriminate between familiar and unfamiliar individuals when following gaze. Females (solid lines) have greater reaction time savings for gaze-congruent than gaze-incongruent targets when gaze cues were from familiar (**A**) rather than unfamiliar (**B**) individuals (at 200 ms, $P < 0.003$). Males (dashed lines) did not distinguish significantly between these conditions (at 200 ms, $P > 0.4$). After Deaner, R. O., Shepherd, S. V., & Platt, M. L. (2007). Familiarity accentuates gaze cuing in women but not men. *Biology Letters, 3,* 64–67. Used with permission.

monkeys. Importantly, each animal was designated dominant or subordinate based on the direction and frequency of threat and submission gestures during controlled pair-wise confrontations (see Deaner et al., 2005; Shepherd et al., 2006). We found that subordinate monkeys rapidly and automatically followed the gaze of all other monkeys (Fig. 26.8A), while dominant monkeys followed gaze later, and then only in response to other dominant monkeys (Fig. 26.8B). These differences in gaze-following behavior were weakly correlated with differences in testosterone production (Shepherd et al., 2006), as inferred from measurements of testis volume (Bercovitch & Ziegler, 2002). We interpret these data to indicate that both internal and external factors govern macaque gaze following, comprising both biological variables like testosterone and ecological variables like relative social status.

One way that familiar or dominant individuals might evoke stronger gaze following is by virtue of their intrinsic salience. Under some conditions, human gaze following appears to be modulated by emotional content in faces. Specifically, patients with anxiety disorders show heightened following of fearful gaze relative to other emotional expressions (Holmes et al., 2006; Hori et al., 2005; Mathews et al., 2003) (note also Hietanen & Leppanen, 2003, and Putman et al., 2006). This interaction probably reflects a tendency for patients with anxiety to more strongly attend to negatively valenced social stimuli, while healthy subjects dwell less on them (e.g., Bar-Haim et al., 2005; Bradley et al., 1997). These studies imply that attention to another individual spreads automatically to other objects with which that individual is engaged.

Together, these results demonstrate that the neural substrates of gaze following are deeply integrated with the larger social information processing stream. That gaze following is an inherent component of face perception is suggested by the fact that heightened attention to faces spreads in the direction of gaze, both in the case of females viewing familiar faces and anxious patients seeing faces with negatively valenced emotional content. At the same time, however, the fact that gaze following is modulated by factors like familiarity

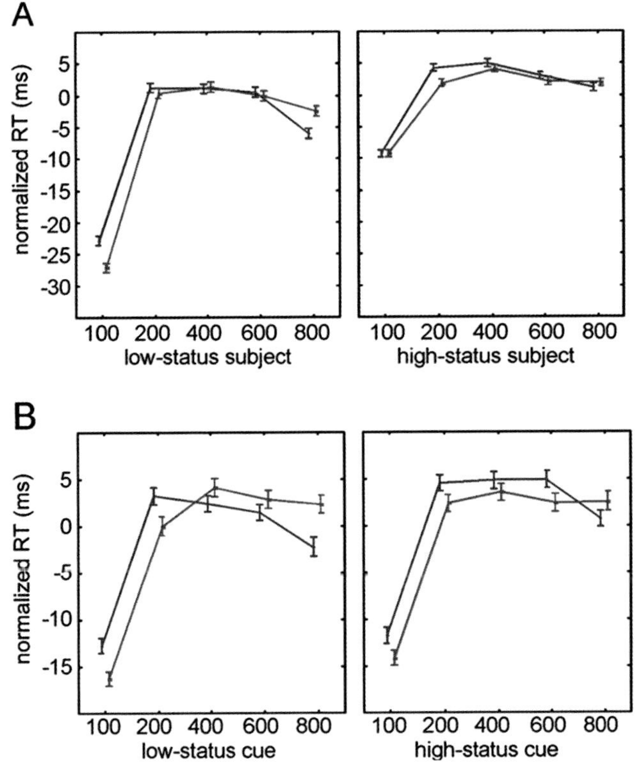

Figure 26.8 Social context influences gaze following in macaques. Even at the briefest cue durations, subject social status appears to influence gaze-following behavior (**A**, $P < 0.005$). Specifically, low social status makes a monkey more likely to follow gaze within 100 ms of seeing the cue, and also more likely to have strong inhibition of return at the latest time point—a temporal profile consistent with a reflexive attention shift, possibly due to increased anxiety or to the modulatory effects of sex- and status-linked hormones like testosterone on social processing circuitry in the brain. Cue social status also plays an important role (**B**, $P < 0.01$), leading to prolonged attention in the direction of gaze of a high-status cue and inhibited attention in the direction of gaze of a low-status cue, particularly by high-status subjects. Reaction times for congruent trials are shown in red lines and for incongruent trials are shown in blue lines. After Shepherd, S. V., Deaner, R. O, & Platt, M. L. (2006). Social status gates social attention in monkeys. *Current Biology, 16*, R119–120. Used with permission.

and social dominance suggests that it is not mediated by an isolated module sequestered from other aspects of face processing and social knowledge. Finally, sex differences in humans and social rank differences in monkeys both hint at a possible role for sex hormones in shaping social attention systems in the brain. Together these findings strongly support the idea that social attention is a fundamental feature of natural primate behavior and cognition, and provides a fundamental challenge to the traditional endogenous/exogenous model of attentional control.

GAZE AS A STRATEGIC SOCIAL SIGNAL

As described previously, social saliency may play a role not only in guiding attention but also in shaping the physical spacing of group members and the affective tenor of their interactions. It may also serve as a starting point for the development of much more advanced cognitive behaviors. David Perrett and Simon Baron-Cohen have argued that detection of eyes and interpretation of gaze are foundational to building a theory of mind, by which we intuitively mirror the attentional and perhaps even intentional states of others. We do this so instinctively that we frequently anthropomorphize even alien and impersonal phenomena, perhaps allowing us to understand complex and dynamic patterns by analogy to human behavioral goals (e.g., "the electrons don't like to be near one another and are instead attracted to the positive core of an atom, causing them to settle sequentially into the centermost uncrowded orbitals"). Typically developing humans have an intuitive expertise at communicating affect and attentional state, in part, perhaps, because they have an intrinsic drive to *learn* to do so: From a young age, typically developing humans take pleasure in successfully directing another's attention toward stimuli that we, too, have perceived (Tomasello et al., 2005).

These considerations naturally lead us to consider overt eye movements as an active signaling mechanism shaping primate social interactions. We have mentioned the role of eye movements in initiating conflict, but primates make far more sophisticated use of gaze. For example, eye contact can signal not only aggression (van Hoof, 1967) but also sexual interest (Dixson, 1998) or solicitation for coalition formation in agonistic interactions versus third parties (de Waal, 2003). Likewise, humans use eye contact as a key aspect of affiliation, courtship (Hrdy & Whitten, 1987), and intimidation (Argyle & Cook, 1976), and also during coordination of attention ("triadic" or "joint attention"; Emery, 2000). Moreover, gaze acts to structure both verbal and nonverbal human social interactions. To signal rank relationships, for example, people look preferentially toward the most high-ranking person, and when conversing, gaze is used to emphasize spoken arguments, to conclude statements, to emphasize nonverbal reactions to heard statements, and to coordinate turn-taking in conversation (Argyle & Cook, 1976).

With the evolution of greater visual and social complexity, some primates appear to have evolved ever more sophisticated means of structuring social behavior through gaze. Like humans, many animal species are capable of following gaze. Chimpanzees are even reported to use deictic gestures (de Waal, 2003), though these signals may hold little behavioral currency due to the ubiquity of competitive and paucity of cooperative interactions in these animals (Hare & Tomasello, 2004). The importance of social gaze for facilitating increased cooperation among human ancestors may even have lead to somatic adaptations that increase the saliency and specificity of gaze cues, for example, by enhancing the visibility of gaze through increased contrast of the pupil versus sclera (Kobayashi & Koshima, 2001; Tomasello et al., 2006). At the same time, however, the continued importance of competitive interactions for early human ancestors may have led to a compensatory enhancement of covert attention mechanisms relative to those possessed by most nonhuman primates and, especially, other mammals.

This manipulative role of gaze is perhaps the least understood aspect of visual orienting behavior; virtually nothing is known about how the demands of social signaling bring their influence to bear on the gaze control system of the brain. These considerations challenge the conventional division between "reflexive" and "reward-mediated" processes in social orienting. Given that current evidence argues against informational encapsulation in social perception and cognition, it seems likely that the fastest orienting processes reflect the need to quickly acquire behaviorally relevant information, and that slower processes reflect more nuanced contextual constraints—whether that information is likely to have predictive value in the present situation, or whether the orienting behavior might expose the subject to risk of agonistic interaction or being "scooped" by competitors in pursuit of resources. For example, monkeys in our studies initially looked toward other individuals, generally following their gaze, but at later intervals diverse contextual considerations were brought to bear—abbreviating risky glances toward higher-ranked individuals (Deaner et al., 2005; Haude et al., 1976), extinguishing gaze following of lower-ranking animals (Shepherd et al.,

2006), and prolonging male fixation on female hindquarters (Deaner et al., 2005). Thus, although much work remains to be done on the neurobiology of orienting responses, it seems clear that neither a strictly reflexive nor a strictly cognitive approach will explain all aspects of social orienting.

SOCIAL ATTENTION AND AUTISM: FROM THE LAB TO THE FIELD

Data from syndromes that disrupt social behavior, notably autism, suggest a gulf between behavioral responses in the laboratory and spontaneous use of social cues in the real world. In a comparison of visual orienting by autistic subjects with that of typically developing children, van der Geest and colleagues (2002) found that the fixation patterns of the two groups could not be distinguished when they viewed simple cartoons including human figures. In contrast, Pelphrey and colleagues (2002) found substantial differences between autistic and typically developing control subjects' inspecting of photographs of real faces. Similarly, although autistic subjects often show normal gaze following in the social variant of the Posner attention task (Chawarska et al., 2003; Kylliainen & Hietanen, 2004; Swettenham et al., 2003; but see Bayliss et al., 2005, and Ristic et al., 2005), they consistently show severe disruptions in social orienting in more natural contexts. When Klin and colleagues (2002a,b) measured gaze behavior in autistic individuals watching the movie *Who's Afraid of Virginia Woolf*, they found that gaze toward social stimuli was disordered—for example, with normal fixations toward the eye regions replaced by fixations toward the mouth. Moreover, socially cued locations were severely neglected, as shown by a marked lack of fixations toward gaze- and gesture-cued regions of space. Furthermore, the authors found that the degree of abnormality in the fixation pattern of individual autistic subjects in this task was strongly predictive of real-life social impairment.

This is hardly surprising. After all, autism is *defined* by a marked "lack of spontaneous seeking to share enjoyment, interests, or achievements with other people," or to reciprocate when these experiences and emotions are shared by others (American Psychological Association, 1994). Outside the laboratory, even high-functioning autistic individuals, unaffected by physical problems such as seizures or repetitive movements, are nonetheless challenged in responding to the constant exchange of social cues that structures human interaction. Temple Grandin, an associate professor of Animal Science at Colorado State University who has autism, reports that she functions in social situations "solely by intellect and visualization skills" (Grandin, 1999, http://www.autism.org/temple/social.html). She says (ibid) "I did not know that eye movements had meaning until I read *Mind Blindness* by Simon Baron-Cohen. I had no idea that people communicated feelings with their eyes. I also did not know that people get all kinds of little emotional signals which transmit feelings. My understanding of this became clearer after I read *Descartes' Error* by Antonio Damasio." It may be that the complement of processes evoked by social stimuli in typically developing individuals is disrupted in autism spectrum disorders, and that without these foundational elements, more sophisticated forms of empathy and social reasoning cannot develop.

It is interesting to note that both autism (Wassink et al., 2007) and social anxiety disorder (Skuse, 2006) have been associated with dysfunction in the serotonin signaling system. Serotonin has likewise been linked to dominance status, affiliative social interaction, and decreases in antagonistic and impulsive social interactions (Edwards & Kravitz, 1997; Raleigh et al., 1991), suggesting that this neuromodulatory system may also contribute to differences in social attention between dominant and subordinate macaques. Together, these findings hint at a role for serotonin in regulating social attention in both human and nonhuman primates. Determining the impact of biological factors, such as serotonin and testosterone, on social attention may point to possible interventions to improve social functions in common psychopathologies like autism.

TOWARD A NEUROETHOLOGY OF ATTENTION IN PRIMATES

If we were to develop a biologically plausible, ethologically motivated model of attention in primates, what features must it have? We feel strongly that the bottom-up component of these models must not only reflect what we know about the primate visual system but must also consider the role vision plays in guiding the behavior of primates in species-typical ecological and social contexts. For example, Laurent Itti and colleagues, among others, have used visual filters, inspired by the physiology of the primate visual system, to predict human visual attention. Such models estimate saliency by filtering images through a series of low-level feature maps (Carmi & Itti, 2006; Peters et al., 2005). Each map tracks the extent to which a region "pops out" from its surroundings along a particular visual dimension, such as brightness, orientation, texture, motion, or color, and these maps can be combined to successfully model many aspects of bottom-up attention.

While these models can accurately identify salient regions of still images and video, they often fail to highlight social stimuli such as faces, or rely heavily on image motion to assign saliency to humans and animals. Without undervaluing either these accomplishments or the importance of motion as a predictor of animacy, we nevertheless note that demands of both sociality and predator avoidance require accurate and fast discrimination of animals, even when those animals are stationary or when dynamic environments (e.g., running water, blowing leaves) produce irrelevant image motion. Moreover, while it is true that identification and tracking of animate objects has proved a challenge for computer vision, these tasks are performed quickly and easily by the primate brain. In laboratory experiments, humans can initiate saccades toward an animal in a novel photograph in as little as 120 ms (Kirchner & Thorpe, 2006), and in unconstrained viewing, animate stimuli and especially other humans are quickly targeted for visual inspection.

Serre and colleagues (2007) partially addressed these issues by developing a model that uses biologically inspired filters based on neurons in the ventral visual processing stream (Ungerleider & Mishkin, 1982) to quickly identify images *containing* animals. It is important to note, however, that this model explicitly fails to *localize* animals within images. The processes that link object recognition by the ventral visual processing stream to target localization within the dorsal stream thus remain largely unknown, despite the fact that it is the dorsal stream that selects parts of the visual field for further processing. In fact, Serre and colleagues note that their model "cannot account for our everyday vision which involves eye movements and top-down effects," (p. 6426) and that an extension of the model requiring "top-down signals from higher to lower areas... limit[ing] visual processing to a 'spotlight of attention' centered around the animal target" results in "significant improvement in the classification performance" (p. 6428).

Serre's study thus illustrates the benefits of considering the natural goals of orienting in social contexts, and likewise of considering evidence from functional imaging and neurophysiological recording studies. Recent functional magnetic resonance imaging (fMRI) studies in humans have identified brain areas that are involved in visual analysis of body position and identity (Downing et al., 2001), identification of faces (Haxby et al., 1994), and interpretation of actions and facial expressions (Allison et al., 2000); other studies suggest that homologous areas operate in macaques (Logothetis et al., 1999; Tsao et al., 2003). The general conservation of cortical organization across primate species, together with these recent findings, suggests that visual areas specialized for processing social stimuli may be part of the primordial visual cortex that was present in stem primates (Rosa & Tweedale, 2005; Tootell et al., 2003) and perhaps others mammals as well (Kendrick et al., 2001).

As revealed through behavioral studies, the gaze-control system must recognize and respond appropriately to biological targets. We speculate that parallel pathways accomplish this goal, which can largely be grouped into a subcortical pathway and a cortical pathway (Adolphs, 2002; Vuilleumier, 2002). Ultimately, both pathways must converge upon the three neural tissues, which jointly serve as the final common pathway governing orienting in primates: the parietal eye fields (in macaque, located in the lateral intraparietal sulcus, called LIP), the frontal eye fields (FEFs), and the superior colliculus (SC). To effectively govern orienting, these areas must weigh decision variables comprising the expected rewards and risks associated with a given orienting behavior. For example, Platt and Glimcher showed that neurons in LIP are sensitive to target value when visual stimuli are arbitrarily assigned different amounts of juice reward (Platt & Glimcher, 1999), and subsequent studies have confirmed that neuronal activity throughout this network is similarly modulated by orienting value (LIP, see also Sugrue et al., 2004; SC, Ikeda & Hikosaka, 2003; but *not* FEF, Leon & Shadlen, 1999).

Critically, we have recently shown that in the "pay-per-view" paradigm discussed previously, neurons in LIP are modulated by the intrinsic social value of orienting to images in much the same way that they are modulated by primary juice rewards (Klein et al., 2008). Specifically, LIP neurons respond most strongly when monkeys evaluate targets associated with the acquisition of information about female reproductive signals and the identity of dominant males, but respond weakly when the same target offers information about subordinate males—despite the fact that monkeys were never explicitly trained to orient toward these stimuli. These observations directly predict the orienting behavior of macaques in the same task. Echoing these findings, a recent fMRI study in humans found stronger activation of parietal cortex when subjects played a game against a dominant opponent compared to an inferior opponent (Zink et al., 2008).

These observations indicate that LIP spontaneously integrates information about target value from multiple sources, in the absence of any explicit training. This, in turn, implies that brain pathways that process social information must ultimately transmit this information to parietal cortex, and likely to other areas involved in orienting behavior as well. This modulation of neural activity by the intrinsic value of acquired visual information seems likely to reflect the native function of LIP, only overridden in tasks where researchers arbitrarily map juice rewards onto specific oculomotor behaviors.

Ultimately, however, the social significance of visual information is probably not computed within the gaze-control network itself. Recent evidence has begun to reveal how social variables, such as dominance status and reproductive state, are processed in primate brains. While social cognition involves broad swaths of brain, we believe it can be simplified into two streams: a streamlined but inflexible subcortical pathway, and a nuanced—and labyrinthine—cortical network.

THE SUBCORTICAL PATHWAY

In this pathway, retinotectal inputs provide coarse visual information in which crude biological primitives can quickly be identified and analyzed, for example, to locate other creatures, detect social signals, and extract gaze direction (Johnson, 2005). The subcortical system appears to by phylogenetically old, shared by all terrestrial vertebrates, and is the major pathway for innate recognition and response to animate targets including predators, prey, and conspecifics (Sewards & Sewards, 2002).

In humans, this system has been suggested to play a crucial role in early visual tracking of faces, and may play a lifelong role in the rapid detection of socially salient or threatening signals, both directly and through interactions with visual cortex and the pulvinar nucleus of the thalamus (Grieve et al., 2000; Johnson, 2005; Sewards & Sewards, 2002). Just such a relay of social threat signals, from retina to SC, pulvinar nucleus, and finally amygdala, has already been identified in humans by neuroimaging (Morris et al., 1999). Furthermore, neurons in the macaque amygdala are sensitive to the expression, gaze, and social dominance of viewed faces (Gothard et al., 2007; Hoffman et al., 2007; Kawashima et al., 1999). The amygdala, in turn, sends this first-pass analysis of social targets toward gaze-control centers and higher visual areas, acting rapidly to strengthen social and threat-related processing (Vuilleumier, 2002).

While a major function of this pathway is indubitably to provide an "early warning" system detecting threats, there is evidence that the amygdala also mediates prosocial behaviors. For example, while eye contact can signal threat, it more generally indicates approach, and often serves to initiate grooming and sexual behavior (Hrdy & Whitten, 1987). Thus, eye contact responses in amygdala may serve to indicate not only threat but also sexual opportunity, and indeed, amygdala is strongly activated by sexual stimuli (Aharon et al., 2001; Hamann et al., 2004).

The amygdala pathway may be highly sensitive to biological factors that mediate sex differences (Bayliss et al., 2005; Deaner et al., 2007; Goldstein et al., 2001; Hamann et al., 2004) as well as psychosocial disorders (Holmes et al., 2006; Hori et al., 2005; Mathews et al., 2003; Putman et al., 2006), and may, when compromised, contribute to the development of autism (Schultz, 2005, though note also Amaral et al., 2003). While amygdala influences the function of a broad swath of cortex, it does not directly interact with the gaze-control network and would instead influence orienting indirectly, through cortical or subcortical intermediaries.

THE CORTICAL NETWORK

In parallel to the subcortical pathway, a slower, more nuanced, and more recently evolved cortical pathway leads from V1 through the ventral visual stream to extrastriate body area (EBA) (Downing et al., 2001), fusiform face area (FFA) (Haxby et al., 1994), and superior temporal sulcus (STS) (Allison et al., 2000). It remains unclear whether these areas assess subordinate-level distinctions between hierarchically classifiable objects or are optimized for processing animate objects or specifically for perception of conspecifics. It seems likely that the development of these areas depends on experience (Gauthier et al., 1999) and may rely upon signals arising in the subcortical pathway for appropriate patterning during development (Johnson, 2005; Schultz, 2005; Sewards & Sewards, 2002). It is known, however, that each of these ventral stream visual areas is strongly activated by social stimuli, extracting posture, identity, and expression, respectively. These areas then transmit output to a broad array of areas in the extended face processing network, interacting with contextual signals from hippocampus, amygdala, and orbitofrontal cortex (OFC) (Ishai et al., 2005; Smith et al., 2006). These socially influenced areas include multiple regions implicated in modulating both attentional allocation and reward processing (Sabbagh, 2004; Vuilleumier, 2002).

It is important to note that both orienting to others and gaze following is regulated by social milieu—as well as by intrinsic factors including sex hormones, such as testosterone, and neuromodulators, such as serotonin. Supporting this idea, amygdala, OFC, and hippocampus form a functional circuit important for associating emotional and social salience with mnemonic and perceptual information (Fig. 26.9) (Sabbagh, 2004; Smith et al., 2006; Vuilleumier, 2002), and are actively involved in the perception of faces (Ishai et al., 2005). Each of these brain structures is sexually dimorphic (Goldstein et al., 2001), suggesting that sexual differentiation in these areas may directly pattern responses to social cues. This supposition is strengthened by various results showing fetal testosterone negatively impacts both social attention and social relationships in human juveniles (Knickmeyer & Baron-Cohen, 2006). Ultimately, signals from these ventral ("what") areas must relay social information to dorsal ("where") orienting and attention control systems. Signals from the higher-order areas of the ventral pathway ramify to multiple targets in the visual orienting system, but exactly how this occurs remains an open question, since much of visuo-social cortex (Tsao et al., 2003) is connected in one or two steps to posterior parietal (7A and LIP; Seltzer & Pandya, 1991), frontal (SEF and FEF; Seltzer & Pandya, 1989), and subcortical orienting areas (pulvinar nucleus; Romanski et al., 1997) and superior colliculus (Fries, 1984).

GAZE FOLLOWING AND THE "MIRROR NETWORK"

One intriguing possibility is that gaze following may be supported by specialized neurons that

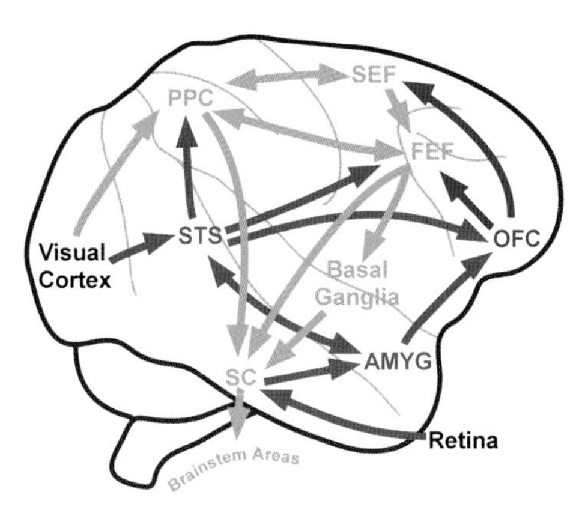

Figure 26.9 Key circuits involved in social attention. Connectivity of social (red), reward (blue), and attention (green) pathways. In addition to the cortical pathway, a fast subcortical pathway connects superior colliculus to amygdala via the thalamus (not shown). Note that several social processing areas lie along superior temporal sulcus, occupying both posterior and anterior temporal lobes, and that functional activity in imaging tasks has not yet been systematically related to past anatomical studies. AMYG, amygdala; FEF, frontal eye fields; OFC, orbitofrontal cortex; PPC, posterior parietal cortex, including 7A and lateral intraparietal sulcus; SEF, supplementary eye fields; STS, superior temporal sulcus regions.

simultaneously map concordant observed/sensory and performed/motor components of this behavior. Similarly specialized "mirror" neurons, first discovered in macaque premotor cortex, respond not just when the subject reaches for an object but also when it observes the experimenter perform a similar movement (di Pellegrino et al., 1992). "Mirroring" responses have been observed in other brain areas (e.g., single units within parietal reach areas; Fogassi et al., 2005), and for other types of movements (e.g., mouth movements; Ferrari et al., 2003). Because mirror neurons appear to signal abstract actions and goals independent of actor, they have been hypothesized to underlie a host of abilities ranging in sophistication from action- and emotion-perception to empathizing, mentalizing, and theory of mind (reviewed in Rizzolatti & Craighero, 2004). Supporting these findings from macaques, imaging studies in humans suggest that homologous brain regions are activated by both the production and observation of specific gestures (Dinstein et al., 2007), and that activations associated with sensory-motor "mirroring" may be widespread in human sensory and motor cortices. To date, however, neither the responses of mirror neurons nor more global sensory-motor mirroring activations observed with fMRI have been directly linked to specific social behaviors in monkeys or humans (Dinstein et al., 2008; but see Prather et al., 2008, for a description of sensory-motor mirroring by neurons in songbirds during social interaction).

Recently, we discovered a population of neurons in macaque LIP that "mirror" the observed gaze of individuals (Shepherd et al., 2009), much as neural populations in adjacent areas mirror reaching behavior (Fogassi et al., 2005). LIP neurons respond when monkeys orient attention, either overtly or covertly, to regions of space known as response fields (Colby et al., 1996; Gnadt & Andersen, 1988; Platt & Glimcher, 1998). We found that a subpopulation of LIP neurons responded not only to direct attention toward their response field but also when viewing another monkey that looked in the neuron's preferred direction. Neuronal responses to the observed orienting behavior of another individual occurred despite the fact that there was no visual stimulus in the neuron's response field, and that any gaze shift toward the response field would abort the trial without reward. Intriguingly, the time course of social gaze cue signals in LIP strongly paralleled the time course of gaze-following behavior—appearing within a tenth of a second, persisting for several hundred milliseconds, and ultimately fading at the longest intervals tested. These data provide strong support for the idea that some populations of mirror neurons provide an important mechanistic foundation for the abstraction, interpretation, and imitation of behaviors and mental states.

CONCLUSIONS

Laboratory research using arbitrary tasks and stimuli have identified two complementary systems for visual orienting—one fast and reflexive, the other slow and deliberative. Neuroethological studies of visual attention, by contrast, have revealed a suite of socially motivated and socially cued orienting behaviors that do not cleave neatly along these lines. Specifically, primates and other animals are motivated to look at one another, preferentially orient to high-value social targets such as the faces of dominant males, and follow the orienting movements of others with their own attention. Moreover, these responses are regulated by behavioral context, sex hormones, and serotonin. These observations strongly support the idea that the primate brain is specialized for acquiring behaviorally useful visual information from the social world, and that these adaptations rely on the integration of multiple neural circuits involved in identifying social stimuli and social cues, determining their meaning, and responding appropriately. Despite the commonalities of these systems across primates and even other mammals, the challenge for future neuroethological research is to determine how these mechanisms contribute to adaptive differences in social behavior in different species.

REFERENCES

Adolphs, R. (2002). Recognizing emotion from facial expressions: Psychological and neurological mechanisms. *Behavioral and Cognitive Neuroscience Reviews, 1,* 21–61.

Agnetta, B., Hare, B., & Tomasello, M. (2000). Cues to food location that domestic dogs (*Canis familiaris*) of different ages do and do not use. *Animal Cognition, 3,* 107–112.

Aharon, I., Etcoff, N., Ariely, D., Chabris, C. F., O'Connor, E., & Breiter, H. C. (2001). Beautiful faces have variable reward value: fMRI and behavioral evidence. *Neuron, 32,* 537–551.

Allison, T., Puce, A., & McCarthy, G. (2000). Social perception from visual cues: Role of the STS region. *Trends in Cognitive Science, 4,* 267–278.

Allman, J. M. (1999). *Evolving brains.* New York: W.H. Freeman & Company.

Amaral, D. G., Bauman, M. D., & Mills Schumann, C. (2003). The amygdala and autism: Implications from non-human primate studies. *Genes, Brain and Behavior, 2,* 295–302.

American Psychological Association. (1994). Autistic disorder. In: *Diagnostic and Statistical Manual of Mental Disorders* (4th ed., p. 66). Washington, DC: Author.

Anderson, J. R., & Mitchell, R. W. (1999). Macaques but not lemurs co-orient visually with humans. *Folia Primatologica (Basel), 70,* 17–22.

Argyle, M., & Cook, M. (1976). *Gaze & mutual gaze.* Cambridge, England: Cambridge University Press.

Bar-Haim, Y., Lamy, D., & Glickman, S. (2005). Attentional bias in anxiety: A behavioral and ERP study. *Brain Cognition, 59,* 11–22.

Baron-Cohen, S. (1994). How to build a baby that can read minds: Cognitive mechanisms in mindreading. *Current Psychology of Cognition, 13,* 513–552.

Barton, R. A., & Dunbar, R. L. M. (Eds.). (1997). *Evolution of the social brain.* Cambridge: Cambridge University Press.

Bayliss, A. P., diPellegrino, G., & Tipper, S. P. (2005). Sex differences in eye gaze and symbolic cueing of attention. *Quarterly Journal of Experimental Psychology A, 58,* 631–650.

Bercovitch, F. B., & Ziegler, T. E. (2002). Current topics in primate socioendocrinology. *Annual Review of Anthropology, 31,* 45–67.

Bornstein, R. F. (1989). Exposure and affect: Overview and meta-analysis of research, 1968-1987. *Psychological Bulletin, 106,* 265–289.

Bradley, B. P., Mogg, K., Millar, N., Bonham-Carter, C., Fergusson, E., Jenkins, J., et al.

(1997). Attentional biases for emotional faces. *Cognition and Emotion, 11,* 25–42.

Brauer, J., Call, J., & Tomasello, M. (2005). All great ape species follow gaze to distant locations and around barriers. *Journal of Comparative Psychology, 119,* 145–154.

Bugnyar, T., Stowe, M., & Heinrich, B. (2004). Ravens, Corvus corax, follow gaze direction of humans around obstacles. *Proceedings of the Royal Society: B Biological Science, 271,* 1331–1336.

Butler, R. A. (1954). Incentive conditions which influence visual exploration. *Journal of Experimental Psychology, 48,* 19–23.

Caine, N. G., & Marra, S. L. (1988). Vigilance and social organization in two species of primates. *Animal Behaviour, 36,* 897–904.

Carmi, R., & Itti, L. (2006). Visual causes versus correlates of attentional selection in dynamic scenes. *Vision Research, 46,* 4333–4345.

Cartmill, M. (1972). Arboreal adaptations and the origin of the order primates. In: R. Tuttle (Ed.), *The functional and evolutionary biology of primates* (pp. 97–212). Chicago: Aldine-Atherton Press.

Chance, M., & Jolly, C. (1970). *Social groups of monkeys, apes and men.* New York: E.F. Dutton.

Chawarska, K., Klin, A., & Volkmar, F. (2003). Automatic attention cueing through eye movement in 2-year-old children with autism. *Child Development, 74,* 1108–1122.

Cheney, D. L., & Seyfarth, R. M. (1990). *How monkeys see the world: Inside the mind of another species.* Chicago: University Of Chicago Press.

Colby, C. L., Duhamel, J. R., & Goldberg, M. E. (1996). Visual, presaccadic, and cognitive activation of single neurons in monkey lateral intraparietal area. *Journal of Neurophysiology, 76,* 2841–2852.

Corbetta, M., & Shulman, G. L. (2002). Control of goal-directed and stimulus-driven attention in the brain. *Nature Reviews: Neuroscience, 3,* 201–215.

De Martino, B., Kumaran, D., Seymour, B., & Dolan, R. J. (2006). Frames, biases, and rational decision-making in the human brain. *Science, 313,* 684–687.

de Waal, F. B. (2003). Darwin's legacy and the study of primate visual communication. *Annals of the New York Academy of Sciences, 1000,* 7–31.

Deaner, R. O., Khera, A. V., & Platt, M. L. (2005). Monkeys pay per view: Adaptive valuation of social images by rhesus macaques. *Current Biology, 15,* 543–548.

Deaner, R. O., & Platt, M. L. (2003). Reflexive social attention in monkeys and humans. *Current Biology, 13,* 1609–1613.

Deaner, R. O., Shepherd, S. V., & Platt, M. L. (2007). Familiarity accentuates gaze cuing in women but not men. *Biology Letters, 3,* 64–67.

di Pellegrino, G., Fadiga, L., Fogassi, L., Gallese, V., & Rizzolatti, G. (1992). Understanding motor events: A neurophysiological study. *Experimental Brain Research, 91,* 176–180.

Dinstein, I., Hasson, U., Rubin, N., & Heeger, D. J. (2007). Brain areas selective for both observed and executed movements. *Journal of Neurophysiology, 98,* 1415–1427.

Dinstein, I., Thomas, C., Behrmann, M., & Heeger, D. J. (2008). A mirror up to nature. *Current Biology, 18,* R13–18.

Dixson, A. F. (1998). *Primate sexuality: Comparative studies of the prosimians, monkeys, apes, and human beings.* New York: Oxford University Press.

Downing, P. E., Jiang, Y., Shuman, M., & Kanwisher, N. (2001). A cortical area selective for visual processing of the human body. *Science, 293,* 2470.

Driver, J., Davis, G., Kidd, P., Maxwell, E., Ricciardelli, P., & Baron-Cohen, S. (1999). Gaze perception triggers reflexive visuospatial orienting. *Visual Cognition, 6,* 509–540.

Edwards, D. H., & Kravitz, E. A. (1997). Serotonin, social status and aggression. *Current Opinions in Neurobiology, 7,* 812–819.

Egeth, H. E., & Yantis, S. (1997). Visual attention: Control, representation, and time course. *Annual Review of Psychology, 48,* 269–297.

Emerson, R. W. (1876). *The conduct of life.* Boston: James R. Osgood & Company.

Emery, N. J. (2000). The eyes have it: The neuroethology, function and evolution of social gaze. *Neuroscience and Biobehavioral Reviews, 24,* 581–604.

Engbert, R., & Kliegl, R. (2003). Microsaccades uncover the orientation of covert attention. *Vision Research, 43,* 1035–1045.

Ferrari, P. F., Gallese, V., Rizzolatti, G., & Fogassi, L. (2003). Mirror neurons responding to the observation of ingestive and communicative mouth actions in the monkey ventral premotor cortex. *European Journal of Neuroscience, 17,* 1703–1714.

Fogassi, L., Ferrari, P. F., Gesierich, B., Rozzi, S., Chersi, F., & Rizzolatti, G. (2005). Parietal lobe: From action organization to intention understanding. *Science, 308,* 662–667.

Fries, W. (1984). Cortical projections to the superior colliculus in the macaque monkey: A retrograde study using horseradish peroxidase. *Journal of Comparative Neurology, 230,* 55–76.

Friesen, C. K., & Kingstone, A. (1998). The eyes have it! Reflexive orienting is triggered by nonpredictive gaze. *Psychonomic Bulletin and Review, 5,* 490–495.

Friesen, C. K., Ristic, J., & Kingstone, A. (2004). Attentional effects of counterpredictive gaze and arrow cues. *Journal of Experimental Psychology: Human Perception and Performance, 30,* 319–329.

Galfano, G., Rusconi, E., & Umilta, C. (2006). Number magnitude orients attention, but not against one's will. *Psychonomic Bulletin and Reviews, 13,* 869–874.

Gauthier, I., Tarr, M. J., Anderson, A. W., Skudlarski, P., & Gore, J. C. (1999). Activation of the middle fusiform face area increases with expertise in recognizing novel objects. *Nature Neuroscience, 2,* 568–573.

Gilad, Y., Wiebe, V., Przeworski, M., Lancet, D., & Paabo, S. (2004). Loss of olfactory receptor genes coincides with the acquisition of full trichromatic vision in primates. *PLoS Biology, 2,* E5.

Gnadt, J. W., & Andersen, R. A. (1988). Memory related motor planning activity in posterior parietal cortex of macaque. *Experimental Brain Research, 70,* 216–220.

Goldstein, J. M., Seidman, L. J., Horton, N. J., Makris, N., Kennedy, D. N., Caviness, V. S., et al. (2001). Normal sexual dimorphism of the adult human brain assessed by *in vivo* magnetic resonance imaging. *Cerebral Cortex, 11,* 490–497.

Gothard, K. M., Battaglia, F. P., Erickson, C. A., Spitler, K. M., & Amaral, D. G. (2007). Neural responses to facial expression and face identity in the monkey amygdala. *Journal of Neurophysiology, 97,* 1671–1683.

Grandin, T. (1999). Social problems: Understanding emotions and developing talents. http://www.autism.org/temple/social. html.

Grieve, K. L., Acuna, C., & Cudeiro, J. (2000). The primate pulvinar nuclei: Vision and action. *Trends in Neuroscience, 23,* 35–39.

Guo, K., Robertson, R. G., Mahmoodi, S., Tadmor, Y., & Young, M. P. (2003). How do monkeys view faces? A study of eye movements. *Experimental Brain Research, 150,* 363–374.

Hafed, Z. M., & Clark, J. J. (2002). Microsaccades as an overt measure of covert attention shifts. *Vision Research, 42,* 2533–2545.

Hamann, S., Herman, R. A., Nolan, C. L., & Wallen, K. (2004). Men and women differ in amygdala response to visual sexual stimuli. *Nature Neuroscience, 7,* 411–416.

Hare, B., & Tomasello, M. (2004). Chimpanzees are more skilful in competitive than in cooperative cognitive tasks. *Animal Behaviour, 68,* 571–581.

Haude, R. H., Graber, J. G., & Farres, A. G. (1976). Visual observing by rhesus monkeys: Some relationships with social dominance rank. *Animal Learning and Behavior, 4,* 163–166.

Haxby, J. V., Horowitz, B., Ungerleider, L. G., Maisog, J. M., Peietrini, P., & Grady, C. L. (1994). The functional organization of human extrastriate cortex: A PET-rCBF study of selective attention to faces and locations. *Journal of Neuroscience, 14,* 6336–6353.

Hayden, B. Y., Parikh, P. C., Deaner, R. O., & Platt, M. L. (2007). Economic principles motivating social attention in humans. *Proceedings of the Royal Society: B Biological Science, 274,* 1751–1756.

Hietanen, J. K., & Leppanen, J. M. (2003). Does facial expression affect attention orienting by gaze direction cues? *Journal of Experimental Psychology: Human Perception and Performance, 29,* 1228–1243.

Hoffman, K. L., Gothard, K. M., Schmid, M. C., & Logothetis, N. K. (2007). Facial-expression and gaze-selective responses in the monkey amygdala. *Current Biology, 17,* 766–772.

Holmes, A., Richard, A., & Green, S. (2006). Anxiety and sensitivity to eye gaze in emotional faces. *Brain and Cognition, 60,* 282–294.

Hori, E., Tazumi, T., Umeno, K., Kamachi, M., Kobayashi, T., Ono, T., et al. (2005). Effects of facial expression on shared attention mechanisms. *Pysiology and Behavior, 84,* 397–405.

Hrdy, S. B., & Whitten, P. L. (1987). Patterning of sexual activity. In: B. B. Smuts, D. L. Cheney, R. M. Seyfarth, R. W. Wrangham, & T. T. Struhsaker (Eds.), *Primate societies* (pp. 370–384). Chicago: University of Chicago Press.

Ikeda, T., & Hikosaka, O. (2003). Reward-dependent gain and bias of visual responses in primate superior colliculus. *Neuron, 39,* 693–700.

Ishai, A., Schmidt, C. F., & Boesiger, P. (2005). Face perception is mediated by a distributed cortical network. *Brain Research Bulletin, 67,* 87–93.

Itakura, S. (1996). An exploratory study of gaze monitoring in non-human primates. *Japanese Psychological Research, 38,* 174–180.

Itakura, S. (2004). Gaze-following and joint visual attention in nonhuman animals. *Japanese Psychological Research, 46,* 216–226.

James, W. (1890). *The principles of psychology.* New York: H. Holt and Company.

Johnson, M. H. (2005). Subcortical face processing. *Nature Reviews: Neuroscience, 6,* 766–774.

Jolly, A. (1966). *Lemur social behavior and primate intelligence. Science, 153,* 501–506.

Jonides, J. (1981). Voluntary versus automatic control over the mind's eye's movement. In: J. B. Long & A. D. Baddeley (Eds.), *Attention and performance IX* (pp. 187–203). Hillsdale, NJ: Erlbaum.

Kaminski, J., Riedel, J., Call, J., & Tomasello, M. (2005). Domestic goats, *Capra hircus,* follow gaze direction and use social cues in an object choice task. *Animal Behaviour, 69,* 11–18.

Kawashima, R., Sugiura, M., Kato, T., Nakamura, A., Hatano, K., Ito, K., et al. (1999). The human amygdala plays an important role in gaze monitoring: A PET study. *Brain, 122,* 779–783.

Keating, C. F., & Keating, E. G. (1982). Visual scan patterns of rhesus monkeys viewing faces. *Perception, 11,* 211–219.

Kendrick, K. M., da Costa, A. P., Leigh, A. E., Hinton, M. R., & Peirce, J. W. (2001). Sheep don't forget a face. *Nature, 414,* 165–166.

Keverne, E. B., Leonard, R. A., Scruton, D. M., & Young, S. K. (1978). Visual monitoring in social groups of Talapoin Monkeys. *Animal Behaviour, 26,* 933–944.

Kirchner, H., & Thorpe, S. J. (2006). Ultra-rapid object detection with saccadic eye movements: Visual processing speed revisited. *Vision Research, 46,* 1762–1776.

Klein, J. T., Deaner, R. O., & Platt, M. L. (2008). Neural correlates of social target value in macaque parietal cortex. *Current Biology, 18,* 419–424.

Klein, J. T., & Platt, M. L. (2008). Topography of social and nutritive reward coding in striatum [Abstract]. Program No. 691.8. *2008 Neuroscience Meeting Planner.* Washington, DC: Society for Neuroscience. Online.

Klin, A., Jones, W., Schultz, R., Volkmar, F., & Cohen, D. (2002a). Visual fixation patterns during viewing of naturalistic social situations as predictors of social competence in individuals with autism. *Archives of General Psychiatry, 59,* 809–816.

Klin, A., Jones, W., Schultz, R., Volkmar, F., & Cohen, D. (2002b). Defining and quantifying the social phenotype in autism. *American Journal of Psychiatry, 159,* 895–908.

Knickmeyer, R. C., & Baron-Cohen, S. (2006). Fetal testosterone and sex differences. *Early Human Development, 82,* 755–760.

Kobayashi, K. (2001). Unique morphology of the human eye and its adaptive meaning: Comparative studies on external morphology of the primate eye. *Journal of Human Evolution, 40,* 419–435.

Kyes, R. C., & Candland, D. K. (1987). Baboon (Papio hamadryas). visual preferences for regions of the face. *Journal of Comparative Psychology, 101,* 345–348.

Kylliainen, A., & Hietanen, J. K. (2004). Attention orienting by another's gaze direction in children with autism. *Journal of Child Psychology and Psychiatry, 45,* 435–444.

Land, M. F., & Hayhoe, M. (2001). In what ways do eye movements contribute to everyday activities? *Vision Research, 41,* 3559–3565.

Langton, S. R. H., & Bruce, V. (1999). Reflexive visual orienting in response to the social attention of others. *Visual Cognition, 6,* 541–567.

Leon, M. I., & Shadlen, M. N. (1999). Effect of expected reward magnitude on the response of neurons in the dorsolateral prefrontal cortex of the macaque. *Neuron, 24,* 415–425.

Logothetis, N. K., Guggenberger, H., Peled, S., & Pauls, J. (1999). Functional imaging of the monkey brain. *Nature Neuroscience, 2,* 555–562.

Mangun, G. R. (1995). Neural mechanisms of visual selective attention. *Psychophysiology, 32,* 4–18.

Martin, R. (1990). *Primate origins and evolution: A phylogenetic reconstruction.* Princeton: Princeton University Press.

Mathews, A., Fox, E., Yiend, J., & Calder, A. (2003). The face of fear: Effects of eye gaze and emotion on visual attention. *Visual Cognition, 10,* 823–835.

McNelis, N. L., & Boatright-Horowitz, S. L. (1998). Social monitoring in a primate group: The relationship between visual attention and hierarchical ranks. *Animal Cognition, 1,* 65–69.

Morris, J. S., Ohman, A., & Dolan, R. J. (1999). A subcortical pathway to the right amygdala mediating 'unseen' fear. *Proceedings of the National Academy of Sciences, 96,* 1680–1685.

Muller, H. J., & Rabbitt, P. M. (1989). Reflexive and voluntary orienting of visual attention: Time course of activation and resistance to interruption. *Journal of Experimental Psychology: Human Perception and Performance, 15,* 315–330.

Okamoto-Barth, S., Call, J., & Tomasello, M. (2007). Great apes' understanding of other individuals' line of sight. *Psychological Science, 18,* 462–468.

Pelphrey, K. A., Sasson, N. J., Reznick, J. S., Paul, G., Goldman, B. D., & Piven, J. (2002). Visual scanning of faces in autism. *Journal of Autism and Developmental Disorders, 32,* 249–261.

Perrett, D. I., & Emery, N. J. (1994). Understanding the intentions of others from visual signals: Neurophysiological evidence. *Cahiers de Psycholgoie Cognition, 13,* 683–694.

Peters, R. J., Iyer, A., Itti, L., & Koch, C. (2005). Components of bottom-up gaze allocation in natural images. *Vision Research, 45,* 2397–2416.

Platt, M. L., & Glimcher, P. W. (1998). Response fields of intraparietal neurons quantified with multiple saccadic targets. *Experimental Brain Research, 121,* 65–75.

Platt, M. L., & Glimcher, P. W. (1999). Neural correlates of decision variables in parietal cortex. *Nature, 400,* 233–238.

Posner, M. I. (1980). Orienting of attention. *Quarterly Journal of Experimental Psychology, 32,* 3–25.

Posner, M. I., & Cohen, Y. (1984). Components of visual orienting. In: H. Bouma & D. Bouwhuis (Eds.), *Attention and performance* (pp. 531–556). Hillsdale, NJ: Erlbaum.

Prather, J. F., Peters, S., Nowicki, S., & Mooney, R. (2008). Precise auditory-vocal mirroring in neurons for learned vocal communication. *Nature, 451,* 305–310.

Putman, P., Hermans, E., & van Honk, J. (2006). Anxiety meets fear in perception of dynamic expressive gaze. *Emotion, 6,* 94–102.

Raleigh, M. J., McGuire, M. T., Brammer, G. L., Pollack, D. B., & Yuwiler, A. (1991). Serotonergic mechanisms promote dominance acquisition in adult male vervet monkeys. *Brain Research, 559,* 181–190.

Raymond, J. E., Fenske, M. J., & Tavassoli, N. T. (2003). Selective attention determines emotional responses to novel visual stimuli. *Psychological Science, 14,* 537–542.

Ristic, J., Mottron, L., Friesen, C. K., Iarocci, G., Burack, J. A., & Kingstone, A. (2005). Eyes are special but not for everyone: The case of autism. *Brain Research: Cognitive Brain Research, 24,* 715–718.

Rizzolatti, G., & Craighero, L. (2004). The mirror-neuron system. *Annual Review of Neuroscience, 27,* 169–192.

Romanski, L. M., Giguere, M., Bates, J. F., & Goldman-Rakic, P. S. (1997). Topographic organization of medial pulvinar connections with the prefrontal cortex in the rhesus monkey. *Journal of Comparative Neurology, 379,* 313–332.

Rosa, M. G., & Tweedale, R. (2005). Brain maps, great and small: Lessons from comparative studies of primate visual cortical organization. *Philosophical Transactions of the Royal Society of London Series B: Biological Science, 360,* 665–691.

Sabbagh, M. A. (2004). Understanding orbitofrontal contributions to theory-of-mind reasoning: Implications for autism. *Brain and Cognition, 55,* 209–219.

Sackett, G. P. (1966). Monkeys reared in isolation with pictures as visual input: Evidence for an innate releasing mechanism. *Science, 154,* 1468–1473.

Sauther, M., Sussman, R., & Gould, L. (1999). The socioecology of the ringtailed lemur: Thirty-five years of research. *Evolutionary Anthropology, 8,* 120–132.

Schultz, R. T. (2005). Developmental deficits in social perception in autism: The role of the amygdala and fusiform face area. *International Journal of Developmental Neuroscience, 23,* 125–141.

Seltzer, B., & Pandya, D. N. (1989). Frontal lobe connections of the superior temporal sulcus in the rhesus monkey. *Journal of Comparative Neurology, 281,* 97–113.

Seltzer, B., & Pandya, D. N. (1991). Post-rolandic cortical projections of the superior temporal sulcus in the rhesus monkey. *Journal of Comparative Neurology, 312,* 625–640.

Serre, T., Oliva, A., & Poggio, T. (2007). A feedforward architecture accounts for rapid categorization. *Proceedings of the National Academy of Sciences, 104,* 6424–6429.

Sewards, T. V., & Sewards, M. A. (2002). Innate visual object recognition in vertebrates: Some proposed pathways and mechanisms. *Comparative Biochemistry and Physiology Part A: Molecular and Integrative Physiology, 132,* 861–891.

Seyfarth, R. M., & Cheney, D. L. (2003). Signalers and receivers in animal communication. *Annual Review of Psychology, 54,* 145–173.

Shepherd, S. V., Deaner, R. O, & Platt, M. L. (2006). Social status gates social attention in monkeys. *Current Biology, 16,* R119–120.

Shepherd, S. V., & Platt, M. L. (2006). Noninvasive telemetric gaze tracking in freely moving socially housed prosimian primates. *Methods, 38,* 185–194.

Shepherd, S. V., & Platt, M. L. (2008). Spontaneous social orienting and gaze following in ringtailed lemurs (*Lemur catta*). *Animal Cognition, 11,* 13–20.

Shepherd, S. V., Klein, J. T., Deaner, R. O., & Platt, M. L. (2009). Mirroring of attention by neurons in macaque parietal cortex. *PNAS,* 106, 9489–9494.

Shimojo, S., Simion, C., Shimojo, E., & Scheier, C. (2003). Gaze bias both reflects and influences preference. *Nature Neuroscience, 6,* 1317–1322.

Skuse, D. (2006). Genetic influences on the neural basis of social cognition. *Philosophical Transactions of the Royal Society of London Series B: Biological Science, 361,* 2129–2141.

Smith, A. P. R., Stephan, K. E., Rugg, M. D., & Dolan, R. J. (2006). Task and content modulate amygdala-hippocampal connectivity in emotional retrieval. *Neuron, 49,* 631–638.

Sugrue, L. P., Corrado, G. S., & Newsome, W. T. (2004). Matching behavior and the representation of value in the parietal cortex. *Science, 304,* 1782–1787.

Swettenham, J., Condie, S., Campbell, R., Milne, E., & Coleman, M. (2003). Does the perception of moving eyes trigger reflexive visual orienting in autism? *Philosophical Transactions of the Royal Society of London Series B: Biological Science, 358,* 325–334.

Tomasello, M., Call, J., & Hare, B. (1998). Five primate species follow the visual gaze of conspecifics. *Animal Behaviour, 55,* 1063–1069.

Tomasello, M., Carpenter, M., Call, J., Behne, T., & Moll, H. (2005). Understanding and sharing intentions: The origins of cultural cognition. *Behavioral and Brain Sciences, 28,* 675–691; discussion 691–735.

Tomasello, M., Hare, B., Lehmann, H., & Call, J. (2006). Reliance on head versus eyes in the gaze following of great apes and human infants: The cooperative eye hypothesis. *Journal of Human Evolution, 52(3),* 314–320.

Tomonaga, M. (2007). Is chimpanzee (Pan troglodytes) spatial attention reflexively triggered by gaze cue? *Journal of Comparative Psychology, 121,* 156–170.

Tootell, R. B., Tsao, D., & Vanduffel, W. (2003). Neuroimaging weighs in: Humans meet macaques in "primate" visual cortex. *Journal of Neuroscience, 23,* 3981–3989.

Tsao, D. Y., Freiwald, W. A., Knutsen, T. A., Mandeville, J. B., & Tootell, R. B. (2003). Faces and objects in macaque cerebral cortex. *Nature Neuroscience, 6,* 989–995.

Tschudin, A., Call, J., Dunbar, R. I., Harris, G., & van der Elst, C. (2001). Comprehension of signs by dolphins (Tursiops truncatus). *Journal of Comparative Psychology, 115,* 100–105.

Tversky, A., & Kahneman, D. (1981). The framing of decisions and the psychology of choice. *Science, 211,* 453–458.

Ungerleider, L. G., & Mishkin, M. (1982). Two cortical visual systems. In: D. J. A. Ingle, M. Goodale, & R. V. W. Mansfield (Eds.), *Analysis of visual behavior.* Cambridge, MA: MIT Press.

van der Geest, J. N., Kemner, C., Camfferman, G., Verbaten, M. N., & van Engeland, H. (2002). Looking at images with human figures: Comparison between autistic and normal children. *Journal of Autism and Developmental Disorders, 32,* 69–75.

van Hoof, J. A. (1967). The facial displays for the catarhine monkeys and apes. In: D. Morris (Ed.), *Primate ethology* (pp. 7–68). Chicago: Aldine Publishing Company.

Vuilleumier, P. (2002). Facial expression and selective attention. *Current Opinions in Psychiatry, 15,* 291–300.

Wassink, T. H., Hazlett, H. C., Epping, E. A., Arndt, S., Dager, S. R., Schellenberg, G. D., et al. (2007). Cerebral cortical gray matter overgrowth and functional variation of the serotonin transporter gene in autism. *Archives of General Psychiatry, 64,* 709–717.

Yarbus, A. (1967). Eye movements during perception of complex objects. In: *Eye movements and vision* (pp. 171–211). New York: Plenum Press.

Zajonc, R. B. (1968). Attitudinal effects of mere exposure. *Journal of Personality and Social Psychology, 9,* 1–27.

Zink, C. F., Tong, Y., Chen, Q., Bassett, D. S., Stein, J. L., & Meyer-Lindenberg, A. (2008). Know your place: Neural processing of social hierarchy in humans. *Neuron, 58,* 273–283.

CHAPTER 27

Neuroethology of Decision Making

Daeyeol Lee

During the last several decades, anatomical studies on the brain of rhesus monkeys have revealed various functional maps and their connectivity in great detail. Guided by this knowledge, neurophysiological studies have also uncovered how various features of external sensory stimuli and multiple parameters of movements are encoded by the pattern of activity in ensembles of cortical and subcortical neurons. Often, fundamental principles of information processing in the sensory and motor systems have been found using artificially simple stimuli or movements. Investigators need to employ, however, more complex stimuli (Ghazanfar et al., 2005; Mazer & Gallant, 2003; Sugihara et al., 2006; Vinje & Gallant, 2000; Woolley et al., 2005), and allow the animals to produce less restricted movements (Georgopoulos et al., 1986; Graziano et al., 2002), in order to gain the insights necessary to understand the neural basis of more natural primate behaviors.

A common element in many natural behaviors is decision making. Formally, the problem of decision making can be divided into several steps. First, a set of alternative actions must be delineated. This is convenient both theoretically and experimentally, and often one considers only a relatively small number of alternative actions. In practice, it is impossible to know the complete set of actions that are available to a particular animal. Nevertheless, we can focus on interesting cases, by considering a set of actions that are likely to be chosen in a given context and closely related to each other. For example, all animals, including primates, commonly face such problems as mate selection and foraging. Therefore, a set of available mates or food patches can define a particular problem of decision making. Second, given a set of alternative actions, the animal must evaluate the merits of outcomes expected from each action and the amount of efforts necessary to execute it. It must evaluate not only the quality and quantity of expected reward but also the likelihood and immediacy of obtaining the desired outcome. It is on this second step that a great number of theoretical and experimental studies have focused (Carmerer, 2003; Kagel & Roth, 1995; Kahneman & Tversky, 1979; Lee, 2008; von Neumann & Morgenstern, 1944). Finally, after the animal performs the chosen action, its actual outcomes must be evaluated and compared to the expected outcomes. When they differ, the animals need to adjust their expectations for the outcomes from the chosen actions. Reinforcement learning theory describes how this can be accomplished computationally, and its predictions have been borne out relatively well in human and animal experiments (Daw et al., 2006; Haruno & Kawato, 2006; Lee et al., 2004; Samejima et al. 2005).

As in the neurobiological studies of perception and motor control, nonhuman primates have provided an invaluable model to investigate the neural bases for these multiple components of decision making. Indeed, neural activity related to the number and probabilities of alternative actions have been identified in multiple areas in the brain, such as the premotor cortex (Cisek

& Kalaska, 2005) and superior colliculus (Basso & Wurtz, 1998). In addition, neurons modulating their activity according to various aspects of expected reward have been identified in many different areas, including the amygdala (Paton et al., 2006), the basal ganglia (Cromwell et al., 2003; Kawagoe et al., 1998; Samejima et al., 2005), the lateral prefrontal cortex (Barraclough et al., 2004; Leon & Shadlen, 1999; Watanabe, 1996), the anterior cingulate cortex (Ito et al., 2003; Seo & Lee, 2007; Shidara & Richmond, 2002), the posterior cingulate cortex (McCoy & Platt, 2005), the supplementary motor area (Sohn & Lee, 2007), the orbitofrontal cortex (Padoa-Schioppa & Assad, 2006; Roesch & Olson, 2004; Tremblay & Schultz, 1999), and the posterior parietal cortex (Dorris & Glimcher, 2004; Platt & Glimcher, 1999; Sugrue et al., 2004). As predicted by reinforcement learning theory, signals related to the discrepancy between the expected and actual rewards have also been identified in the primate brain. In particular, dopamine neurons in the ventral tegmental area and substantia nigra pars compacta display phasic activity when the animal receives unexpected reward and depress their activity when the expected reward is omitted (Bayer & Glimcher, 2004; Schultz, 1998), as predicted for reward prediction error in reinforcement learning theory (Kawato & Samejima, 2007; Schultz et al., 1997; Sutton & Barto, 1998).

Although characterization of neural signals related to expected and actual rewards provides important insights into the brain processes involved in reward valuation and reinforcement learning, there are two important features of primate behaviors that deserve special attention. First, primates are highly social animals, and it has been proposed that the complexity of their social structures has played an important role in the evolution of their relatively large brains (Byrne & Whiten, 1988; Whiten & Byrne, 1997). Making appropriate decisions in a social group is particularly challenging, because the outcomes expected from a particular action can change frequently depending on the behaviors of other animals in the group (Lee, 2008). Second, many natural behaviors in primates, such as tool use and social interactions, produce desired effects only after multiple movements are produced in a proper order (Tomasello & Call, 1997). When reward is delivered after multiple movements, the value of a particular movement has to be evaluated not only from its immediate consequence but also from its long-term benefit to the animal. This is commonly referred to as the problem of temporal credit assignment (Sutton & Barto, 1998). This chapter summarizes the findings from recent neurophysiological studies designed to address these two aspects of decision making in nonhuman primates. Specifically, the results from behavioral and neurophysiological studies on decision making during simple competitive games are discussed. The choice behavior of rhesus monkeys in computer-simulated competitive games could be relatively well accounted for by a relatively simple reinforcement learning algorithm (Lee et al., 2004, 2005). Single neuron recording studies have also shown that multiple cortical regions in rhesus monkeys, including the lateral prefrontal cortex, dorsal anterior cingulated cortex, and posterior parietal cortex play important roles in evaluating the outcomes of the animal's decisions (Barraclough et al., 2004; Seo & Lee, 2007; Seo et al., 2007, 2009). In addition, the functions of the frontal cortex in the selection of movement sequences are discussed. Neural activity in the lateral prefrontal cortex closely tracks the animal's knowledge of the correct movement sequence during both sequence planning (Averbeck & Lee, 2007) and sequence execution (Averbeck et al., 2006). Nevertheless, multiple brain areas are likely to be involved during the evaluation and selection of movement sequences. Medial frontal cortex, for example, may encode not only the sequence of movements chosen by the animal but also the immediacy of reward expected after the completion of a particular movement sequence (Sohn & Lee, 2007). Such signals would be useful in evaluating the desirability of a given movement sequence.

DECISION MAKING OF NONHUMAN PRIMATES IN A MATCHING PENNIES GAME

Decision making in a social group is formally studied by game theory. In game theory, a game is characterized by a payoff matrix, which assigns a particular reward or payoff to each participant or player according to the actions chosen by all participating players. A strategy to choose a particular action exclusively is referred to as a pure strategy, whereas a mixed strategy refers to a probability density function defined over a set of actions and therefore can describe a strategy to choose multiple actions probabilistically. Technically, therefore, a pure strategy is also a mixed strategy in which the probability is 1 for one of the alternative actions and 0 for all other actions. A main task in game theory is to determine how a group of rational players trying to maximize their individual payoffs would choose their strategies. One of these solutions is known as Nash equilibrium, which is defined as a set of strategies from which no individual players can deviate unilaterally to increase their payoffs (Nash, 1950). Consider, for example, a simple game known as the matching pennies game (Fig. 27.1A). This game is played by two players and is an example of zero-sum games, in which the sum of the payoffs to all players is always zero. Each player chooses from two alternative actions, and one of the players (matcher) wins if he or she chooses the same action (e.g., head of a coin) and loses otherwise. Although some games can have multiple Nash equilibria, the matching pennies game has only one Nash equilibrium, and this is to choose each of the two actions with the probability of 0.5. Any other strategies can be potentially exploited by the opponent. Therefore, the Nash equilibrium in the matching pennies game is a mixed strategy. When the matching pennies game is played repeatedly, the Nash equilibrium is for each player to choose each of the two alternative actions with the probability of 0.5 and independently across successive trials.

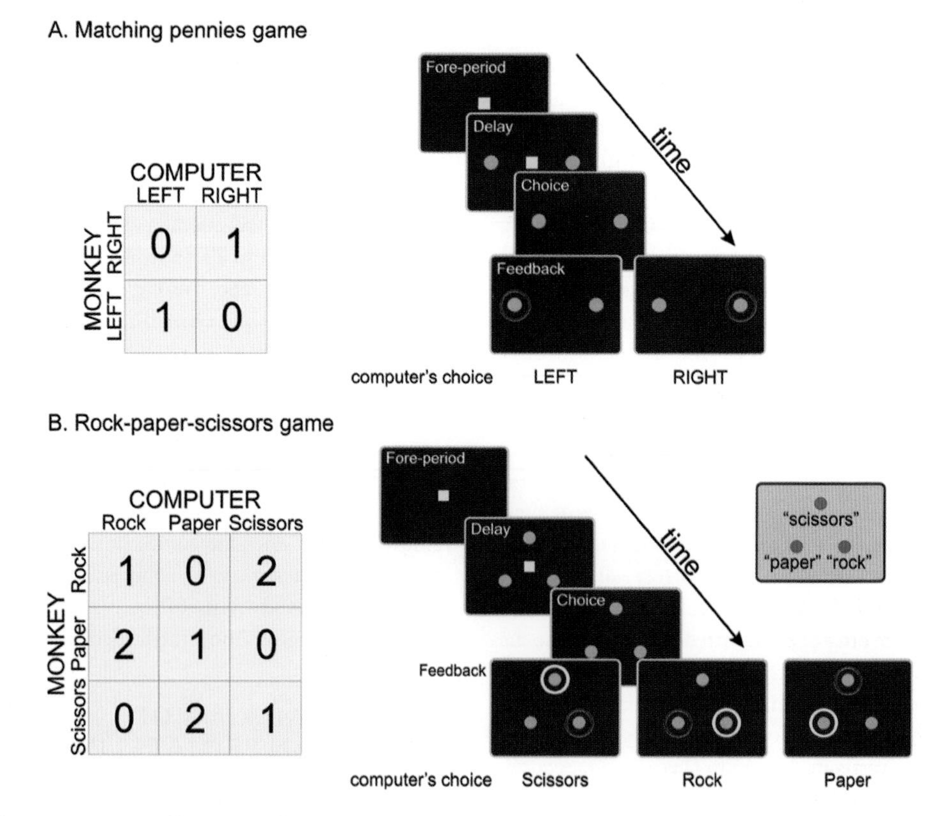

Figure 27.1 Payoff matrix and spatiotemporal sequence of the matching pennies task (**A**) and rock-paper-scissors task (**B**).

Despite its mathematical elegance, the equilibrium predictions of game theory are often violated for a variety of games (Camerer, 2003), including matching pennies (Mookherjee & Sopher, 1994; Rapoport & Budescu, 1992). Instead, various learning models have been proposed to account for the discrepancies between the predictions from game theory and the observed choice behaviors of human subjects (Fudenberg & Levine, 1998). In particular, reinforcement learning models have been successfully applied to the data obtained from various experimental games (Camerer, 2003; Erev & Roth, 1998; Feltovich, 2000; Lee et al., 2004, 2005; Mookherjee & Sopher, 1994, 1997).

To test how closely the choice behaviors of rhesus monkeys follow the predictions of Nash equilibrium in zero-sum games, three male rhesus monkeys (monkeys C, E, and F) were trained to indicate their choices in a binary oculomotor free-choice task that was modeled after the matching pennies game (Lee et al., 2004). In this experiment, the animals were seated in a primate chair and faced a computer screen on which all visual stimuli were presented. The animal's eye position was monitored using a scleral search coil (DNI, Newark, Delaware) or a high-speed video-based eye tracker (ET 49, Thomas Recording, Germany). Each trial began when the animal fixated a small yellow square presented at the center of the computer screen (Fig. 27.1A). After a 0.5-second fore-period, two green choice targets were presented along the horizontal meridian. After a 0.5-second delay period, the central fixation target was extinguished, and the animal was required to shift its gaze toward one of these peripheral targets within 1 second. The animal was required to maintain fixation on its chosen target for a 0.5-second hold period. At the end of this hold period, the computer presented a red feedback ring around the target it selected, and the animal was rewarded only when it chose the same target as the computer.

To investigate how the animal's choice behavior might be influenced by the increasing degree of exploitation by its opponent, we first examined the animal's behavior during the baseline condition in which the computer opponent blindly applied the strategy given by the Nash equilibrium in the matching pennies game. In other words, the computer opponent chose the two alternative targets randomly with the same probabilities. This was referred to as algorithm 0. In this condition, the probability that the animal's choice would be rewarded was the same for both targets. Therefore, this made it possible to examine any initial bias the animals might have had before they were introduced to a more competitive setting. Not surprisingly, each of the three animals displayed a strong preference to choose one of the targets more frequently. The probability that the animal would choose the rightward target throughout the period of algorithm 0 was 0.70, 0.90, and 0.33 for monkeys C, E, and F, respectively (Lee et al., 2004).

In the next stage of the experiment, the computer opponent adopted a more exploitative strategy by analyzing statistical biases in the animal's choice sequences and using this information to avoid choosing the target that the animal was more likely to choose. To this end, the computer saved the animal's entire choice history in a given day and estimated the probability that the animal would choose the right-hand target and a series of conditional probabilities that the animal would choose the right-hand target given the animal's choices in the last four trials. Then, for each of these probabilities, the null hypothesis that the animal chose each target with the 0.5 probability was tested. If none of these null hypotheses was rejected, the computer assumed that the animal's choice was consistent with the Nash equilibrium, and chose each target randomly with the probability of 0.5. Otherwise, the computer biased its choice using the conditional probability that was significantly different and deviated maximally from 0.5. For example, if the conditional probability chosen by this criterion indicated that the animal would choose the rightward target in the next trial with the probability of 0.8, then the computer chose the leftward target with 0.8 probability. This reduced the reward probability for the animal as its probability of choosing the rightward target deviated further from 0.5. Therefore, the optimal strategy for the animal to maximize its overall reward was to choose the two targets with the same probability, namely, the Nash equilibrium strategy.

When the computer opponent switched to algorithm 1, the animals adapted quickly and started choosing the two targets with more or less equal probabilities. Overall, during the period of algorithm 1, the probability that the animal would choose the rightward target was within 0.01 from 0.5 in all three animals. In addition, the animal's choices in successive trials were quite independent. This was quantitatively examined by calculating the entropy of the animal's choice sequences in three successive trials. Since the animal's choice in each trial corresponds to a binary variable, the maximum entropy for three successive choices is 3 bits, corresponding to a case in which the animal's successive choices are completely independent. The average entropy during the period of algorithm 1 was greater than 2.95 for all animals, indicating that their choices were nearly independent (Lee et al., 2004). In contrast, all three animals displayed the tendency to choose the same target again when the previous choice was rewarded and switch to the other target otherwise. In other words, they displayed the tendency to use the so-called win-stay-lose-switch (WSLS) strategy. Since the animal was rewarded in approximately 50% of the trials in algorithm 1, using the WSLS strategy resulted in switching to the other target in about half of the trials. Therefore, a relatively frequent use of the WSLS strategy was possible without deviating significantly from the Nash equilibrium strategy or introducing significant correlation in successive choices. In addition, the WSLS strategy was not penalized in algorithm 1, since the computer did not analyze the animal's reward history. During the period of algorithm 1, the probability that the animal would choose its target according to the WSLS strategy was 0.65, 0.73, and 0.63 for monkeys C, E, and F, respectively. In addition, during the course of the experiments with algorithm 1, which lasted several months, the probability of using the WSLS strategy increased steadily and significantly in all animals.

The WSLS strategy is a straightforward implementation of Thorndike's law of effect (Thorndike, 1911). Thus, the animal's tendency to use this strategy may be relatively robust and hard-wired. To test whether and the extent to which the animals could reduce the use of WSLS strategy when this becomes disadvantageous, the algorithm used by the computer for its target selection was modified. In this new algorithm, referred to as algorithm 2, the computer opponent tested all of the conditional probabilities used in algorithm 1. In addition, it also tested the conditional probabilities based on the animal's choices and their outcomes in the previous four trials. Therefore, the animal's bias to use the WSLS strategy frequently could be detected and penalized in algorithm 2. For example, if the animal used the WSLS strategy with a 0.9 probability, and if the animal was rewarded for choosing the rightward target in the previous trial, the conditional probability that the animal would choose the rightward target in the current trial, given the animal's choice and its outcome in the previous trial, would be 0.9. If this conditional probability is significantly different from 0.5 and this deviation is larger than any other conditional probabilities that were significantly different from 0.5, then the computer would choose the leftward target with the probability of 0.9 and the animal would be rewarded with the probability of 0.1. Therefore, in algorithm 2, the animal was required to choose the two targets randomly with the same probabilities and independently from its previous choices and their outcomes.

All three animals tested in both algorithms 1 and 2 successfully reduced the probability of WSLS strategy once the computer opponent switched to algorithm 2. During the period of algorithm 2, the average probability that the animal would choose its target according to the WSLS strategy was 0.55, 0.54, and 0.57 for monkeys C, E, and F, respectively (Lee et al., 2004). All of these values were significantly lower compared to the results for algorithm 1. However, they were all still significantly higher than 0.5, suggesting that the animals did not remove the bias to rely on the WSLS strategy completely. Despite this small bias, the reward probability was greater than 0.47 for all animals, which is relatively close to the optimal value of 0.5. Therefore, the benefit of removing the bias for the WSLS strategy completely would have been relatively small, and this might account for the residual bias.

REINFORCEMENT LEARNING IN A MATCHING PENNIES GAME

The fact that during the matching pennies game the animals persistently displayed the bias to use the WSLS strategy suggests that they might have used a reinforcement learning algorithm to approximate the optimal strategy in this game. The objective of reinforcement learning (Sutton & Barto, 1998) is to identify an action at each time step that would maximize the decision maker's long-term reward. Since the future reward is generally not known, the decision maker uses the expected value of the reward, which is referred to as a value function. For the matching pennies game, the value functions for the leftward and rightward targets at trial t can be denoted as $Q_t(L)$ and $Q_t(R)$, respectively. The probability that the animal would choose the right-hand target in trial t, $P_t(R)$, is then given by the soft-max transformation as follows:

$$P_t(R) = \exp\{\beta Q_t(R)\}/[\exp\{\beta Q_t(L)\} + \exp\{\beta Q_t(R)\}],$$

where the inverse temperature, β, determines the randomness of the animal's choices. The probability that the animal would choose the leftward target in the same trial would be $1 - P_t(R)$. A large inverse temperature implies that the animal chooses the target with the higher value function deterministically, whereas a small inverse temperature indicates a relatively stochastic choice behavior. When the animal receives reward predicted by the value function for the chosen action, the value functions remain unchanged. Otherwise, the value functions are updated according to the difference between the reward received in trial t, R_t, and the reward expected by the current value functions. In other words:

$$Q_{t+1}(C_t) = Q_t(C_t) + \alpha[R_t - Q_t(C_t)],$$

where C_t (= L or R) indicates the animal's choice in trial t and α corresponds to the learning rate. This model has two free parameters, α and β, and they were estimated separately for the data collected in each daily session, using a maximum likelihood procedure (Pawitan, 2001; Seo & Lee, 2007). This analysis was performed only for algorithms 1 and 2, since algorithm 0 was tested only for a small number of sessions.

Values of the two parameters in this reinforcement learning model can describe different strategies that might be adopted by the animal. For example, if the animal always chooses its target according to the WSLS strategy frequently, the learning rate would be 1 and the inverse temperature would be relatively large. In contrast, a relatively small learning rate would imply that the animal's strategy would change slowly according to the outcomes of its previous choices. Indeed, consistent with the frequent use of the WSLS strategy in algorithm 1, the learning rate estimated for the behavioral data obtained with algorithm 1 was relatively close to 1, and this was significantly reduced in all animals when the computer switched to algorithm 2 (t test, $P < 0.01$; Fig. 27.2). The process in which the parameters of reinforcement learning algorithms

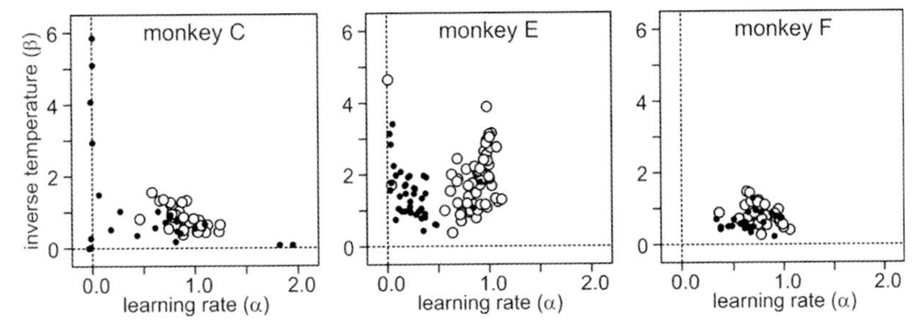

Figure 27.2 Model parameters for the reinforcement learning model fit to the choice behavior during the matching pennies task. Empty circles and small dots correspond to the results from the sessions tested with algorithms 1 and 2, respectively.

are adjusted by the long-term reward rate is referred to as meta-learning (Schweighofer & Doya, 2003). Therefore, the results from the experiments on the matching pennies game demonstrate that rhesus monkeys applied a meta-learning algorithm to adjust their choice behavior to approximate the optimal strategy during a competitive game (Soltani et al., 2006).

DECISION MAKING AND REINFORCEMENT LEARNING IN A ROCK-PAPER-SCISSORS GAME

As described previously, during the matching pennies game, monkeys tended to choose their targets according to the WSLS strategy. Although this is largely consistent with a reinforcement learning algorithm, it is also consistent with another type of learning model referred to as belief learning. In reinforcement learning, if the animal chose a particular target and was rewarded, then it is more likely to choose the same target again, simply because the same response was rewarded and therefore its association with given sensory stimuli strengthened. In contrast, belief learning theories postulate that the decision maker chooses a particular action that would produce the best outcome in response to the action that the opponent is most likely to choose (Camerer, 2003; Feltovich, 2000; Fudenberg & Levine, 1998; Lee, 2008). If the animal chose the leftward target and was rewarded, for example, this indicates that the computer opponent chose the leftward target. If the animal believes that the computer would choose the same target in the next trial, the animal would also choose the leftward target again, not because this action was rewarded in the previous trial, but because this would be the best response to the predicted behavior of the computer opponent. This demonstrates that for the matching pennies game, reinforcement learning and belief learning models tend to make the same predictions about the animal's behavior, making it difficult to distinguish between them empirically.

Although reinforcement learning and belief learning models make similar predictions for choice behavior in relatively simple games, there is an important difference in the cognitive processes necessary to implement the two learning models. In particular, belief learning models require that the animal maintains a mental model for the behaviors of other players. The predictions of these two classes of learning models diverge more clearly for the rock-paper-scissors game. If a player chooses rock and loses because the opponent chooses paper, a reinforcement learning model would reduce the value function for choosing rock, but the value functions for paper and scissors would remain unchanged, making the player equally likely to choose paper or scissors in the next trial. In contrast, a belief learning model would increase the decision maker's estimate for the probability that the opponent would choose paper, and this would encourage the player to choose scissors in the next trial.

To test whether monkeys might choose their targets according to the predictions of a belief learning model during a zero-sum game, two rhesus monkeys were trained to choose among three targets that were designated as rock, paper, and scissors by making an eye movement. As in experiments on the matching pennies, each animal was tested in three successive stages. In algorithm 0, the computer opponent chose each of the three targets with the probability of one third, which corresponds to the Nash equilibrium strategy for the rock-paper-scissors game. In algorithm 1, the computer exploited statistical biases in the animal's choice history, whereas in algorithm 2, it exploited the biases in the animal's choice and reward history. The results from these experiments showed that the animal's choice behavior during the rock-paper-scissors game was largely consistent with a reinforcement learning model, and there was little evidence for belief learning (Lee et al., 2005). This is consistent with the findings from previous studies on experimental games in humans that also failed to find strong evidence for belief learning (Erev & Roth, 1998; Feltovich, 2000; Mookherjee & Sopher, 1997). However, compared to a simple reinforcement learning model, a hybrid model that incorporated the features of both reinforcement learning and belief learning performed better (Lee et al.,

2005), suggesting that these two different types of learning algorithms might operate in parallel (Camerer & Ho, 1999).

NEURAL BASIS OF DECISION MAKING IN COMPETITIVE GAMES

To investigate the neural mechanisms for reinforcement learning and decision making, we recorded the activity of 322 neurons from the dorsolateral prefrontal cortex (DLPFC; Barraclough et al., 2004; Seo et al., 2007) and 154 neurons in the dorsal anterior cingulate cortex (ACCd; Seo & Lee, 2007) of monkeys performing the matching pennies task described above. These recording experiments were carried out using algorithm 2, namely, while the computer opponent exploited the statistical biases in the animal's choice and reward history. For each neuron, the rates of action potentials or spikes during a series of 0.5-second windows aligned at the time of target onset or at the time of feedback ring onset were calculated. These spike rates were then analyzed using a linear regression model that included the animal's choice, the choice of the computer opponent, and reward in the current and previous three trials. In other words:

$$S_t = a_0 + A_C[C_t C_{t-1} C_{t-2} C_{t-3}]' + A_P[P_t P_{t-1} P_{t-2} P_{t-3}]' + A_R[R_t R_{t-1} R_{t-2} R_{t-3}]',$$

where S_t denotes the spike rate for a given time window in trial t; C_t, P_t, and R_t correspond to the computer's choice, the choice of the computer opponent, and the animal's reward in trial t, respectively; and A_C, A_P, and A_R are row vectors of the corresponding regression coefficients. The statistical significance for each regression coefficient was determined with a t test ($P < 0.05$).

The results from these regression analyses are summarized in Figure 27.3, which shows the percentage of neurons in each cortical area that significantly modulated their activity in each 0.5-second time window, grouped together according to the behavioral variable (monkey's choice, computer's choice, and reward) and the trial lag (0 for the current trial, 1 for the previous trial, and so on). During the fore-period and delay period, only a relatively small percentage of neurons encoded signals related to the animal's upcoming choice. In the DLPFC, 12.1% and 19.9% of the neurons significantly modulated their activity according to the animal's choice in the same trial during the fore-period and delay period, respectively. In the ACCd, the corresponding percentages were 18.2% and 17.5%. The difference between the two cortical areas was not statistically significant (Fig. 27.3). During the 0.5-second window immediately after the target onset, many more neurons responded differently depending on the position of the target chosen by the animal. This corresponds to the time period in which the animal produced an eye movement toward one of the two targets. In the DLPFC, 66.2% of the neurons responded differently according to the animal's choice (Fig. 27.4, top), whereas 37.0% of the neurons in the ACCd showed the same effect. This difference was statistically significant (χ^2 test, $P < 0.05$). Therefore, consistent with the findings from previous studies, ACCd neurons encoded the spatial aspects of the animal's movements relatively infrequently (Hoshi et al., 2005; Ito et al., 2003; Matsumoto et al., 2007; Seo & Lee, 2007). In contrast, reward-related signals were encoded more frequently in the ACCd than in the DLPFC. During the 0.5-second window immediately following the feedback ring onset, 68.9% of the DLPFC neurons (Fig. 27.4, bottom) and 81.8% of the ACCd neurons (Fig. 27.5, bottom) showed significant modulations in their activity, depending on whether the animal would be rewarded or not in the current trial. During this feedback period, some neurons in both areas also modulated their activity according to the choice of the computer opponent. Such neurons were found more frequently in the DLPFC (44.4%) than in the ACCd (22.7%).

The changes in the activity of many neurons in the DLPFC related to the animal's choice, the choice of the computer opponent, and reward in a given trial were often maintained across multiple trials. For example, the DLPFC neuron illustrated in Figure 27.4 modulated its activity significantly during the fore-period and delay period according to the animal's choice, the

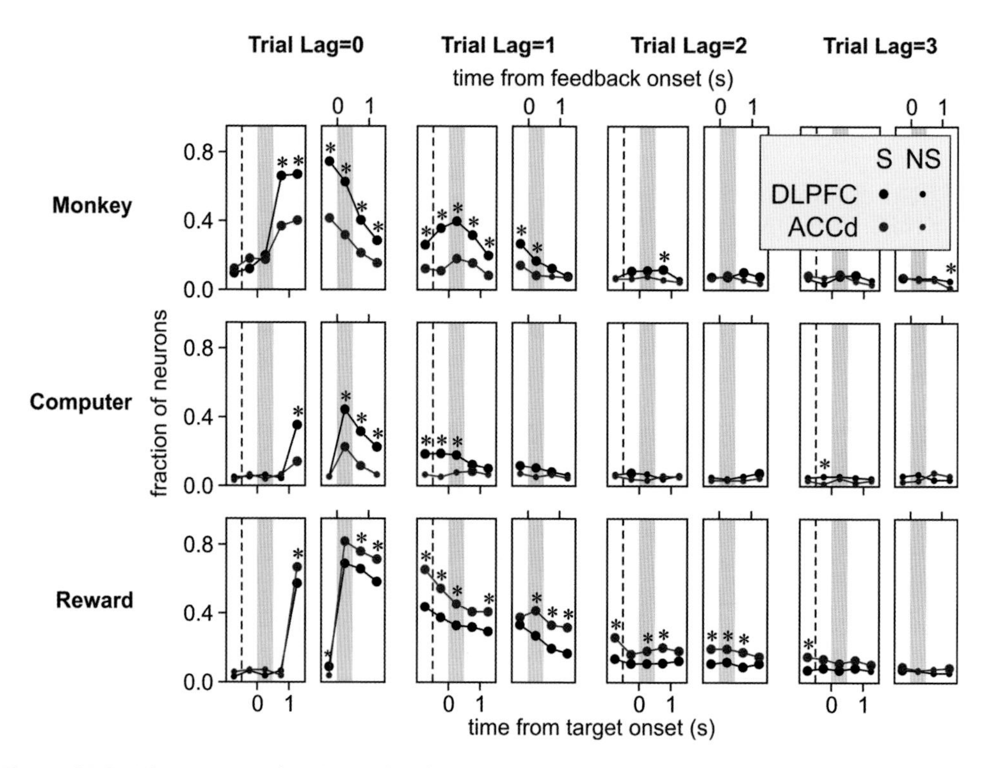

Figure 27.3 Time course of activity related to the animal's choice (top), the choice of the computer opponent (middle), and reward (bottom) in the current (Trial Lag=0) or previous (Trial Lag=1–3) trials within the population of neurons recorded from the dorsolateral prefrontal cortex (DLPFC) (black) and dorsal anterior cingulate cortex (ACCd) (red). Each symbol indicates the fraction of neurons that displayed significant modulations in their activity according to the corresponding variable (t test, $P < 0.05$). Large symbols indicate that the percentage of neurons was significantly higher than the significance level used in the regression analysis (binomial test, $P < 0.05$). Asterisk indicates that the difference between the two cortical areas was statistically significant (χ^2 test, $P < 0.05$). The dotted vertical lines in the left panels correspond to the onset of the fore-period, and the gray background the delay (left panels) or feedback (right panels) period.

computer's choice, and reward in the previous trial. Overall, during the delay period, 39.8%, 18.0%, and 32.9% of the DLPFC neurons showed significant modulations in their activity according to the animal's choice, the computer's choice, and reward in the previous trial, respectively (Fig. 27.3, Trial Lag = 1). A fewer but significant number of neurons modulated their activity during the delay period according to the animal's choice and reward two trials before the current trial (Fig. 27.3, Trial Lag = 2). Compared to the DLPFC, significantly fewer neurons in the ACCd modulated their activity according to the choices of the animal and the computer opponent in the previous trials. In contrast, the neurons in the ACCd were more likely to encode signals related to the animal's reward history than the DLPFC neurons. For example, during the delay period, 54.6% and 18.2% of the ACC neurons changed their activity significantly according to the reward in each of the previous two trials, respectively. A small but significant number of ACCd neurons displayed signals related to whether the animal was rewarded or not three trials before (Fig. 27.3, bottom).

The above results clearly demonstrate that during the matching pennies game, signals related to the previous choices of the animal and its opponent, as well as the outcomes of the animal's previous choices, are maintained for several trials

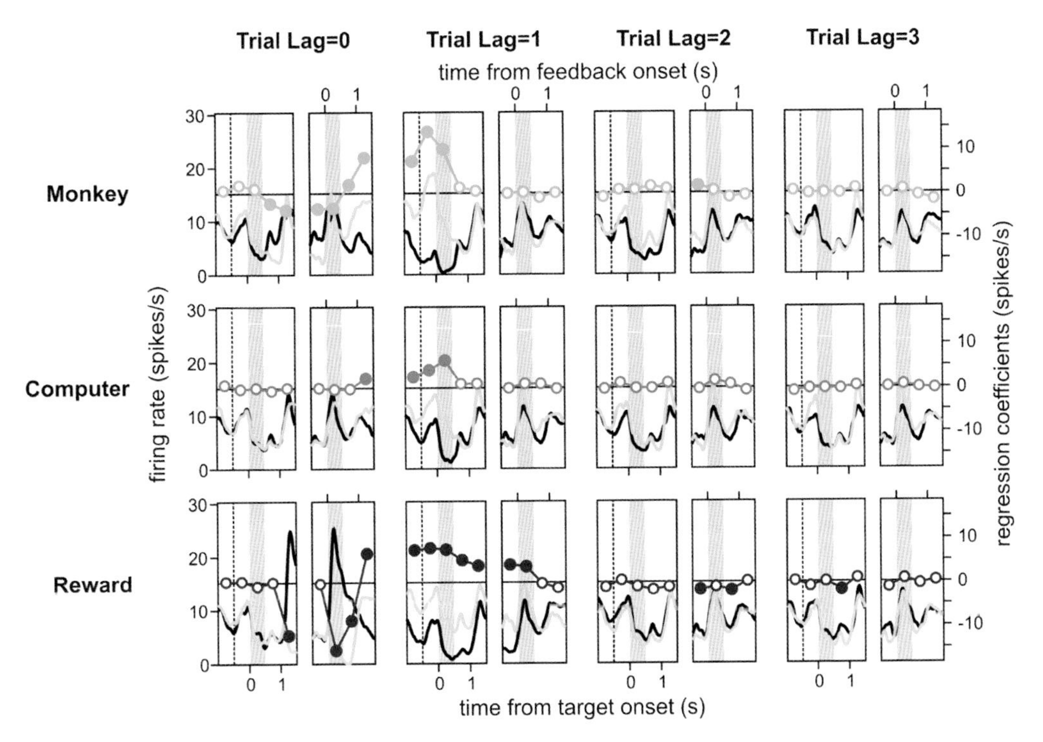

Figure 27.4 Activity of an example neuron in the dorsolateral prefrontal cortex (DLPFC) during the matching pennies task. Each pair of small panels displays the spike density functions (convolved with a Gaussian kernel, $\sigma = 50$ ms) estimated relative to the time of target onset (left panels) or feedback onset (right panels). They were estimated separately according to the animal's choice (top), the computer's choice (middle), or reward (bottom) in the current trial (Trial Lag=0) or according to the corresponding variables in three previous trials (Trial Lag=1, 2, or 3). Cyan (black) lines correspond to the activity associated with rightward (leftward) choices (top and middle) or rewarded (unrewarded) trials (bottom). Circles show the regression coefficients from a multiple linear regression model, which was performed separately for a series of 0.5-second windows. Filled circles indicate the coefficients significantly different from zero (t test, $P < 0.05$). The dotted vertical lines in the left panels correspond to the onset of the fore-period, and the gray background the delay (left panels) or feedback (right panels) period.

in the DLPFC. In contrast, neurons in the ACCd maintained primarily signals related to the animal's reward history. These findings suggest that the DLPFC might play a more integrative role in reinforcement learning, whereas the ACCd might be more specialized for the evaluation of the animal's choice outcome in the context of previous reward history. In order to update the value functions appropriately for a particular action that resulted in a reward after some delay, the brain needs to maintain signals related to previously chosen actions responsible for the outcome. Such memory signals related to the animal's previous actions are referred to as the eligibility trace (Sutton & Barto, 1998). Therefore, signals related to the animal's previous choices found in the DLPFC might correspond to eligibility trace (Seo et al., 2007). On the other hand, the signals related to the computer's previous choices might be related to the value functions for specific actions. During the matching pennies game, the likelihood that the animal would be rewarded for choosing a particular target is determined by the probability that the computer opponent would choose the same target. Therefore, the neurons that modulated their activity according to the computer's choice in the previous trial might encode the

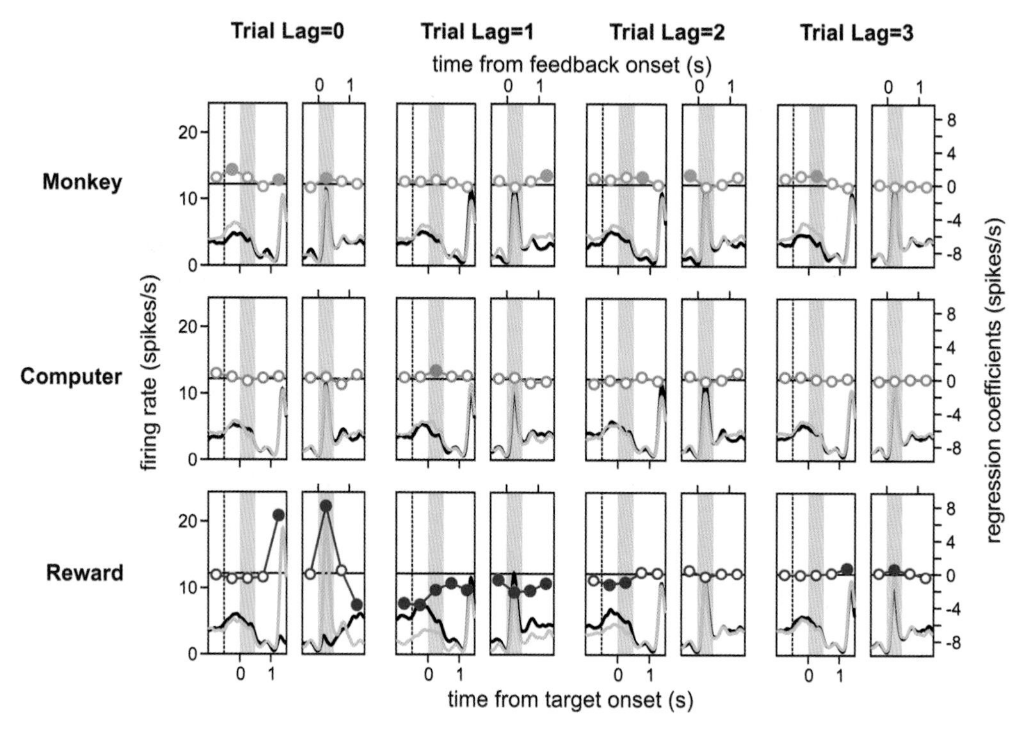

Figure 27.5 Activity of an example neuron in the anterior cingulate cortex (ACCd) during the matching pennies task. Same format as in Figure 27.4.

animal's expectation that the selection of a particular target might be rewarded. It is possible, however, that such signals may not be fully utilized by the animal during the matching pennies game, since overly simple strategies, such as WSLS strategy, could be potentially disadvantageous to the animal. Finally, the signals related to the rewards in the previous trials might be used to estimate the reward rate. This would be possible when the activity of a given neuron is affected consistently by the rewards in multiple trials, as observed for some neurons in the ACCd (Seo & Lee, 2007). Some neurons in the ACCd increased their activity when the animal was rewarded in the current trial but decreased their activity if the animal was rewarded in the previous trial. This pattern was reversed in other neurons in the ACCd. Such neurons modulating their activity antagonistically according to the reward in the current and previous trials might encode the temporal difference in the animal's reward, analogous to the reward prediction error (Seo & Lee, 2007).

NEURAL BASIS OF MULTISTAGE DECISION MAKING AND TEMPORAL DISCOUNTING

Humans have unparalleled abilities to deliberate a long chain of actions to accomplish their behavioral goals, but to a lesser extent, similar abilities exist in nonhuman primates (Tomasello & Call, 1997). Nonhuman primates are also able to switch to a different sequence of actions when previously successful action sequences are no longer appropriate. To investigate the underlying neural mechanisms for this flexible selection of action sequences, two rhesus monkeys were trained in an oculomotor sequence selection task (Fig. 27.6; Averbeck et al., 2006). In this task, the animal began each trial by fixating a central target. After a 1-second fore-period, two targets were presented along the horizontal meridian (Fig. 27.6A). When the animal fixated the correct target, the second pair of targets was presented along the vertical meridian. When the animal chose the second target correctly, the final pair of the targets was presented. Since

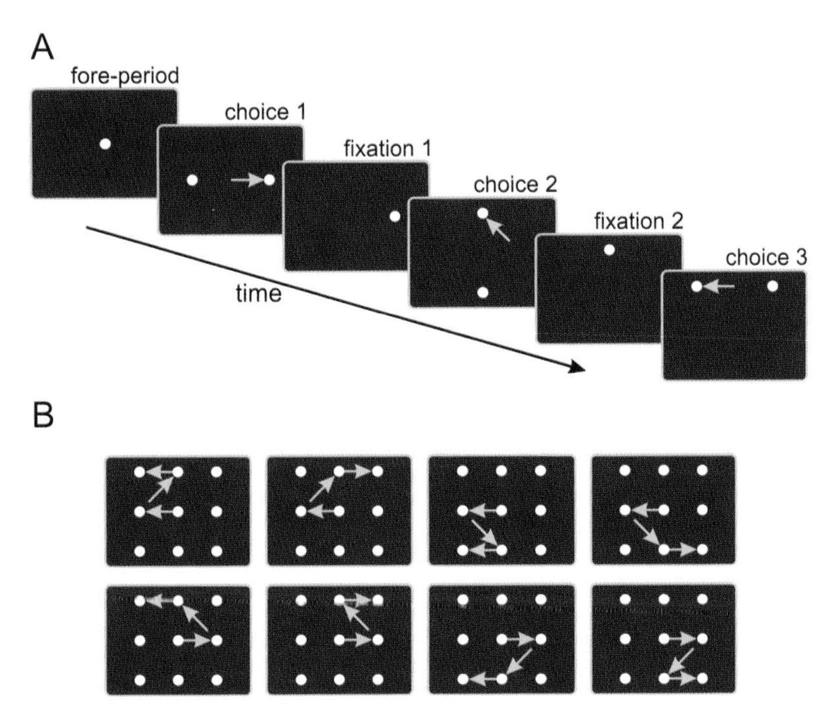

Figure 27.6 (A) Spatial layout of the oculomotor sequence selection task. (**B**) Eight different sequences used in the same task.

there were three successive binary choices, a total of eight possible eye movement sequences were possible (Fig. 27.6B), and one of them was chosen as the correct sequence in a given block of trials. This must be discovered by the animal by trial and error, and the correct sequence was changed randomly without any cues when the animal executed the correct sequence 10 times.

Activity was recorded simultaneously from a small ensemble of neurons in the DLPFC while the animal performed this oculomotor sequence selection task. Similar to the findings in previous studies in the DLPFC (Barone & Joseph, 1989) and medial frontal cortex (Tanji, 2001), many neurons displayed movement-related activity that also depended on the ordinal position and the sequence in which the movement was embedded. To investigate how such sequence-specific movement-related activity evolved dynamically after the correct sequences were changed, a decoding algorithm was applied to classify each movement according to the sequence it belonged to, using the activity recorded from an ensemble of DLPFC neurons. The results showed that signals related to the correct sequence in the previous block gradually decayed in the DLPFC following a time course quite similar to the change in the probability that the animal would choose its target according to the sequence that was previously correct (Fig. 27.7A). Similarly, signals related to the correct sequence in the new block appeared in the DLPFC with a time course similar to the change in the probability that the animal would choose its target according to the new sequence (Fig. 27.7B; Averbeck et al., 2006). In this oculomotor sequence selection task, the animal was required to remember the correct sequence during the intertrial intervals, since the correct sequence was fixed for a block of trials. It was found that the information about the correct sequence was encoded in the activity of DLPFC neurons during the intertrial intervals. Furthermore, this was true only when the animal chose the correct sequence in the next trial (Fig. 27.7C; Averbeck & Lee, 2007). These results suggest that the subjective knowledge of

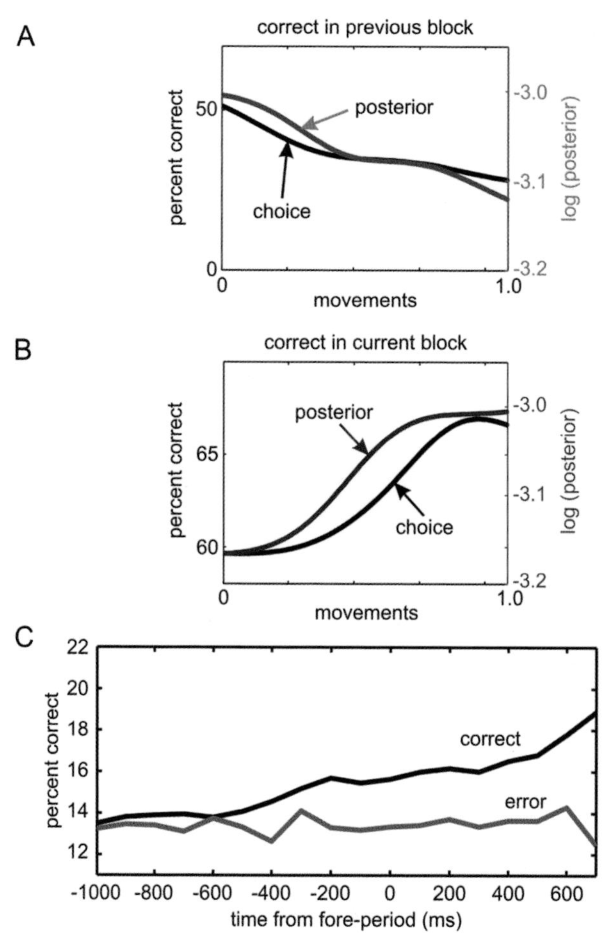

Figure 27.7 Encoding of information about action sequences in the dorsolateral prefrontal cortex. **(A)** Time course of the probability that the animal's choice would be consistent with the correct sequence in the previous block (blue) and the corresponding log posterior probability evaluated from the neural activity (green). All the movements in a new block before the animal correctly chose all three movements in the new sequence were analyzed and were distributed evenly along the x-axis so that the first movement in a new block corresponded to 0 and the last movement before the first correct trial corresponded to 1. The lines were obtained by a moving average. **(B)** Time course of the probability that the animal's choice would be consistent with the correct sequence in the new block (blue) and the corresponding log posterior probability evaluated from the neural activity (red). **(C)** The average probability that the sequence decoded from the ensemble activity recorded in the dorsolateral prefrontal cortex (DLPFC) immediately before and after the fixation onset was the correct sequence in a given block. This was computed separately for the trials in which the animal chose the correct (blue) and incorrect sequences (green).

the correct action sequence used to guide the animal's individual movements might be represented in the DLPFC.

In the natural environment for most animals, including primates, the animals are often rewarded only after producing many movements. When the animal is rewarded predictably after several movements, the performance improves gradually over time, suggesting that the animal is more motivated to perform the task when the reward is immediate or when the amount of required work is reduced (Bowman et al., 1996; Perin, 1943; Sohn & Lee,

2006, 2009). Similarly, when people and animals choose between alternative outcomes, they tend to prefer more immediate reward, even when they have to forego larger reward available only after long delays. Therefore, the preference for a particular reward is reduced according to its delay, and this is referred to as temporal discounting (Frederick et al., 2002). Accordingly, information about the temporal proximity of reward and the amount of work necessary to receive reward must be encoded in the brain and appropriately integrated into the signals related to the subjective value of reward. This is particularly important when the animal needs to choose among a large number of alternative sequences of actions. In particular, the medial frontal cortex has been implicated in the control of animals' motivation and sequential movements (Rushworth, 2004; Tanji, 2001). The study described next tested whether the medial frontal cortex is involved in the subjective evaluation of reward expected from a sequence of movements.

The medial portion of Brodmann area 6 in the primate brain is subdivided into the pre-supplementary motor area (pre-SMA) rostrally and the supplementary motor area (SMA) caudally (Matelli et al., 1991; Matsuzaka et al., 1992). Many neurons in both SMA and pre-SMA encode information about specific movement sequences, but signals related to the ordinal position tend to be more prevalent in the pre-SMA than in the SMA (Clower & Alexander, 1998; Shima & Tanji, 2000). However, whether such signals related to the ordinal position might in fact encode the temporal delay of reward or the effort necessary to receive reward after multiple movements was not known. To test this, two rhesus monkeys were trained in a manual serial reaction time task, in which the ordinal position and the number of movements remaining before reward were manipulated separately (Sohn & Lee, 2007). In each trial, the animal was required to produce multiple hand movements instructed by a series of visual targets (Fig. 27.8). Although the animal was required to make a number of movements that varied across trials, it was rewarded after capturing the target that appeared in a particular location, which was fixed in a block of 20 trials. In addition, the targets stepped through a deterministic trajectory, which was chosen randomly and fixed for a block of trials. Therefore, in most cases, the animal could determine how many movements were required in each trial from the position of the first target in a given trial (Fig. 27.8). The number of movements in a given trial was two, four, six, or eight, and this made it possible to manipulate separately the ordinal position (OP) of the movement and the number of remaining movements (NRM) that must be

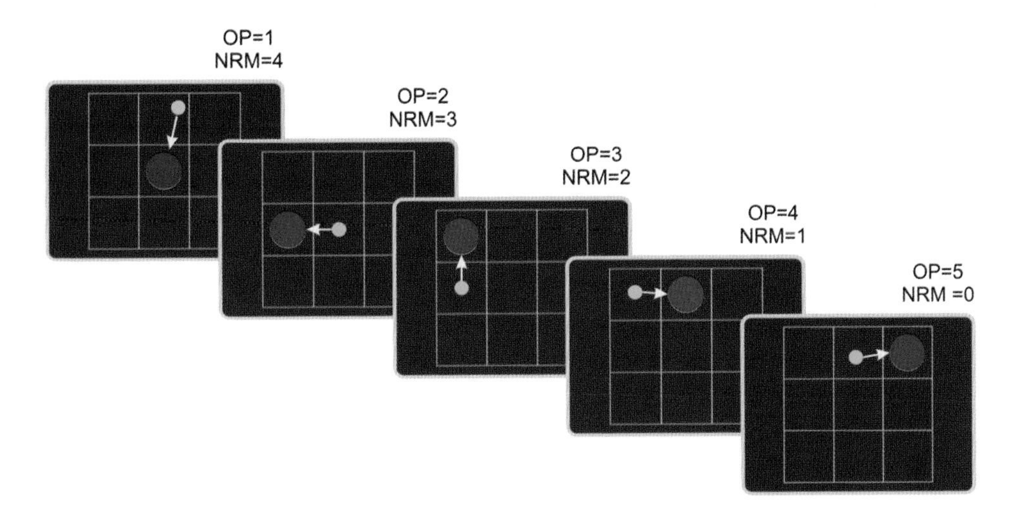

Figure 27.8 Spatial layout of the manual serial reaction time task. NRM, number of remaining movements; OP, ordinal position.

completed before reward delivery. For example, the target shown in the third panel of Figure 27.8 corresponds to the third target in this trial (OP = 3) and the animal would be required to capture two additional targets after it acquires this target (NRM = 2). The same target could appear as the first target in another trial (OP=1), whereas the NRM could be changed in another block of trials by changing the position of the rewarded target.

Activity recorded from 114 neurons in the SMA and 117 neurons in the pre-SMA were analyzed with a linear regression model to determine whether the activity was significantly affected by OP and NRM. Many neurons in both SMA and pre-SMA encoded signals related to OP. Overall, 58.8% of the SMA neurons and 47.9% of the pre-SMA neurons showed significant effects of OP. In addition, more than 70% of the neurons showed significant modulations in their activity according to the NRM (70.2% and 77.8% for the SMA and pre-SMA, respectively; Figs. 27.9 and 27.10). Therefore, whereas the proportions of neurons encoding the OP and NRM were similar in the SMA, neurons in the pre-SMA were more likely

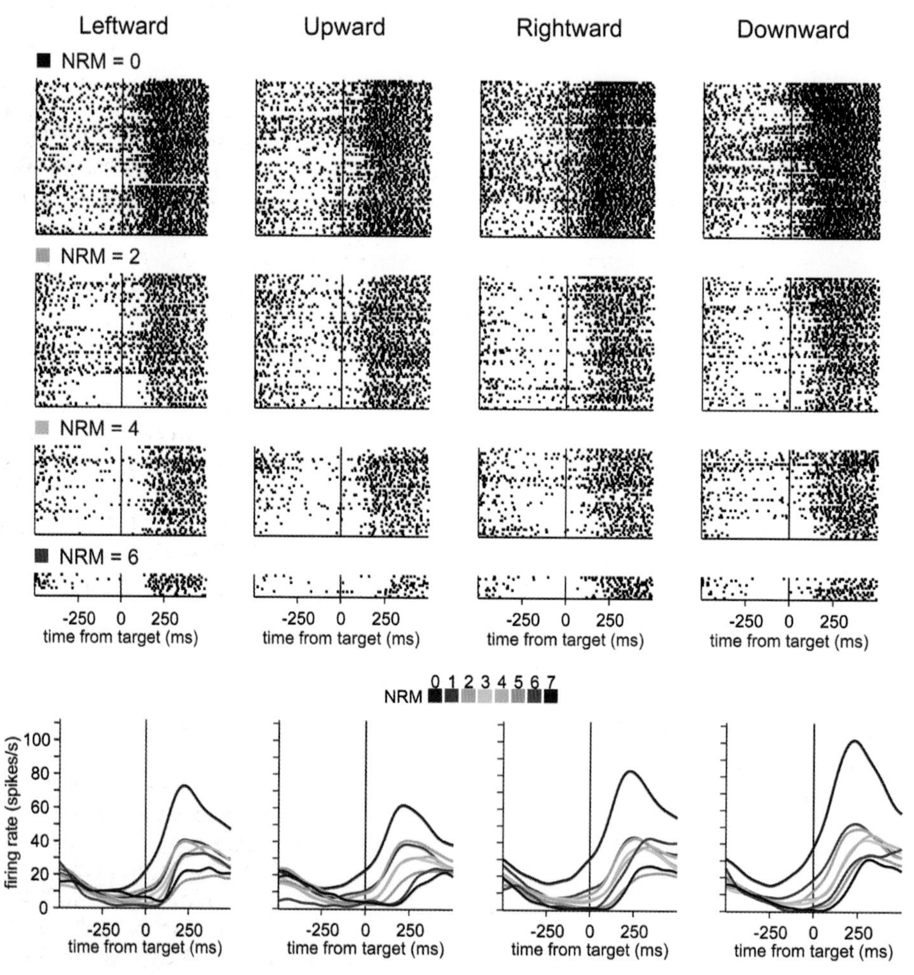

Figure 27.9. An example neuron in supplementary motor area (SMA) showing activity negatively correlated with number of remaining movements (NRM) Top: Raster plots showing the activity separately for different movement directions and NRM (0, 2, 4, and 6). Bottom: Spike-density functions (convolved with a Gaussian kernel, σ = 40 ms) estimated separately for different NRM and different movement directions.

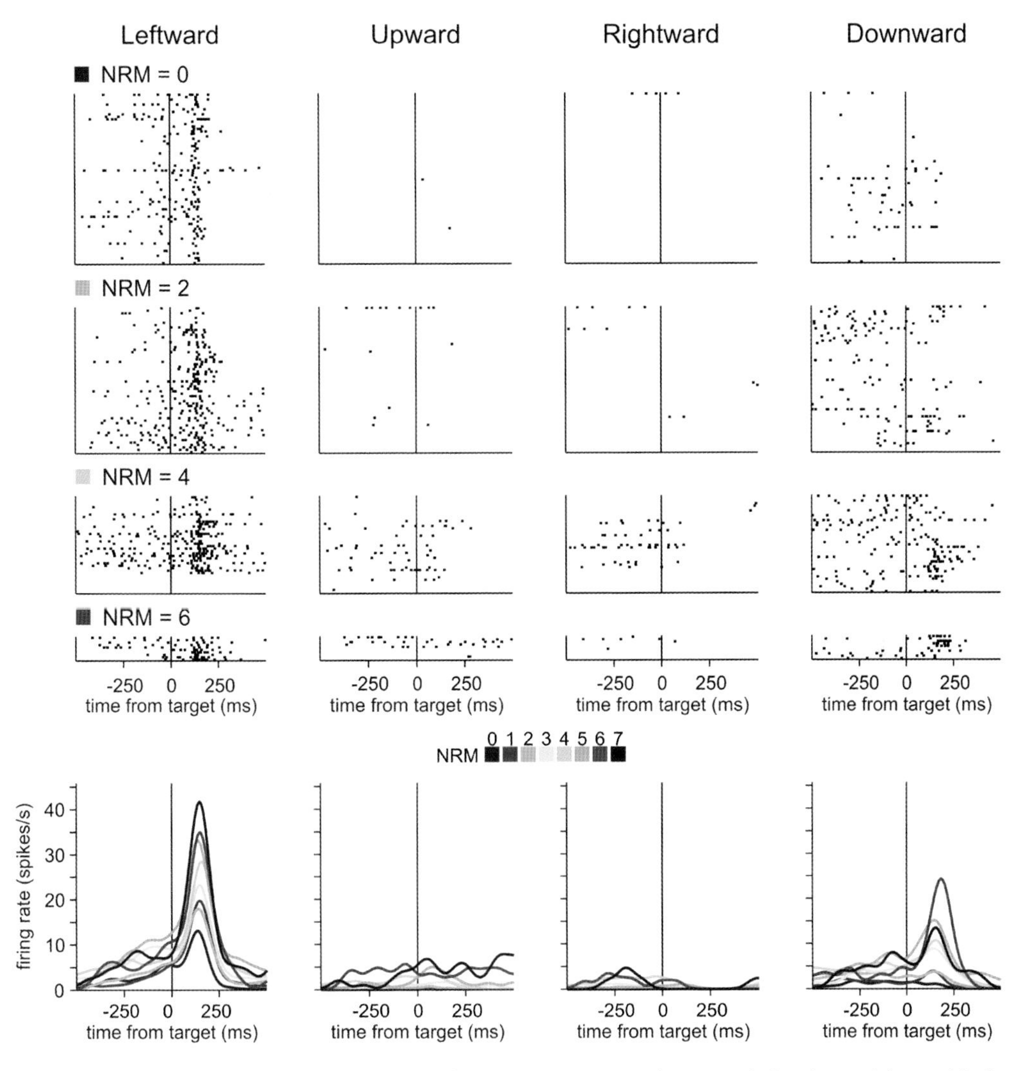

Figure 27.10 An example neuron in presupplementary motor area (pre-SMA) showing activity positively correlated with number of remaining movements (NRM). Same format as in Figure 27.9.

to encode the NRM than OP. These results suggest that the information about the immediacy of reward or the amount of work and effort necessary to acquire reward might be encoded in these two cortical areas. In addition, compared to neurons in the SMA, pre-SMA neurons were more likely to increase their activity with NRM. In the pre-SMA, approximately half of the neurons (47.0%) showed significant positive correlation between their activity and NRM (Fig. 27.10), whereas this was true only for 30.7% of the neurons in the SMA. This suggests that the pre-SMA might play a more important role in the selection of relatively long movement sequences.

CONCLUSION

Compared to simple motor responses commonly studied in neurophysiological studies, natural behaviors of nonhuman primates often take place in the context of social interactions with other conspecifics. The behaviors of nonhuman primates can change more flexibly than other inanimate objects, and therefore this makes it necessary for the animals to monitor the outcomes of their previous actions and modify their behavioral strategies frequently during social interactions. Primates are also capable of constructing a novel sequence of actions to accomplish their behavioral objectives. To investigate the basic neural mechanisms involved in social decision making, recent studies have examined the behavioral dynamics during computerized competitive games and the neural mechanisms involved in reinforcement learning. The results from these studies have found that the signals related to the animals' previous choices and their outcomes are sustained in the frontal cortex and might contribute to appropriate revisions of the animals' behavioral strategies. In addition, studies based on a sequence selection task have found that prefrontal cortex encodes signals related to the animals' subjective knowledge of the correct action sequence. In addition, information about the immediacy of reward or the amount of work necessary to acquire reward is represented in the medial frontal cortex. However, signals related to the immediacy of reward and the amount of work or effort necessary to acquire reward could not be separated in the studies described in this chapter. Indeed, whether these two different types of information are processed by different areas in the frontal cortex (Rushworth et al., 2007) remains to be investigated more systematically. Future studies will also need to investigate more closely the extent to which different subdivisions of the primate prefrontal cortex are functionally specialized to make it possible for the animals to choose the most appropriate course of action in a dynamic and often competitive social environment.

ACKNOWLEDGMENTS

I am grateful to Bruno Averbeck, Dominic Barraclough, Michelle Conroy, Ben McGreevy, Hyojung Seo, and Jeong-Woo Sohn for participating in the experiments described in this chapter. This research was supported by the grants from the National Institutes of Health (MH059216, MH073246, and NS044270).

REFERENCES

Averbeck, B. B., & Lee, D. (2007). Prefrontal neural correlates of memory for sequences. *Journal of Neuroscience, 27,* 2204–2211.

Averbeck, B. B., Sohn, J. W., & Lee, D. (2006). Activity in prefrontal cortex during dynamic selection of action sequences. *Nature Neuroscience, 9,* 276–282.

Barone, P., & Joseph, J. P. (1989). Prefrontal cortex and spatial sequencing in macaque monkey. *Experimental Brain Research, 78,* 447–464.

Barraclough, D. J., Conroy, M. L., & Lee, D. (2004). Prefrontal cortex and decision making in a mixed-strategy game. *Nature Neuroscience, 7,* 404–410.

Basso, M. A., & Wurtz, R. H. (1998). Modulation of neuronal activity in superior colliculus by changes in target probability. *Journal of Neuroscience, 18,* 7519–7534.

Bayer, H. M., & Glimcher, P. W. (2005). Midbrain dopamine neurons encode a quantitative reward prediction error signal. *Neuron, 47,* 129–141.

Bowman, E. M., Aigner, T. G., & Richmond, B. J. (1996). Neural signals in the monkey ventral striatum related to motivation for juice and cocaine rewards. *Journal of Neurophysiology, 75,* 1061–1073.

Byrne, R. W., & Whiten, A. (1988). *Machiavellian intelligence: Social expertise and the evolution of intellect in monkeys, apes, and humans.* Oxford: Oxford University Press.

Camerer, C. F. (2003). Behavioral game theory: Experiments in strategic interaction. Princeton: Princeton University Press.

Camerer, C., & Ho, T. H. (1999). Experience-weighted attraction learning in normal form games. *Econometrica, 67,* 827–874.

Cisek, P., & Kalaska, J. F. (2005). Neural correlates of reaching decisions in dorsal premotor cortex: Specification of multiple direction choices and final selection of action. *Neuron, 45,* 801–814.

Clower, W. T., & Alexander, G. E. (1998). Movement sequence-related activity reflecting numerical order of components in supplementary and presupplementary motor areas. *Journal of Neurophysiology, 80,* 1562–1566.

Cromwell, H. C., & Schultz, W. (2003). Effects of expectations for different reward magnitude on neuronal activity in the primate striatum. *Journal of Neurophysiology, 89,* 2823–2838.

Daw, N. D., O'Doherty, J. P., Dayan, P., Seymour, B., & Dolan, R. J. (2006). Cortical substrates for exploratory decisions in humans. *Nature, 441,* 876–879.

Dorris, M. C., & Glimcher, P. W. (2004). Activity in posterior parietal cortex is correlated with the relative subjective desirability of action. *Neuron, 44,* 365–378.

Erev, I., & Roth, A. E. (1998). Predicting how people play games: Reinforcement learning in experimental games with unique, mixed strategy equilibria. *American Economic Review, 88,* 848–881.

Feltovich, N. (2000). Reinforcement-based and belief-based learning models in experimental asymmetric-information games. *Econometrica, 68,* 605–641.

Frederick, S., Loewenstein, G., & O'Donoghue, T. (2002). Time discounting and time preference: A critical review. *Journal of Economic Literature, 40,* 351–401.

Fudenberg, D., & Levine, D. K. (1998). *The theory of learning in games.* Cambridge, MA: MIT Press.

Georgopoulos, A. P., Schwartz, A. B., & Kettner, R. E. (1986). Neuronal population coding of movement direction. *Science, 233,* 1416–1419.

Ghazanfar, A. A., Maier, J. X., Hoffman, K. L., & Logothetis, N. K. (2005). Multisensory integration of dynamic faces and voices in rhesus monkey auditory cortex. *Journal of Neuroscience, 25,* 5004–5012.

Graziano, M. S. A., Taylor, C. S. R., & Moore, T. (2002). Complex movements evoked by microstimulation of precentral cortex. *Neuron, 34,* 841–851.

Haruno, M., & Kawato, M. (2006). Different neural correlates of reward expectation and reward expectation error in the putamen and caudate nucleus during stimulus-action-reward association learning. *Journal of Neurophysiology, 95,* 948–959.

Hoshi, E., Sawamura, H., & Tanji, J. (2005). Neurons in the rostral cingulate motor area monitor multiple phases of visuomotor behavior with modest parametric selectivity. *Journal of Neurophysiology, 94,* 640–656.

Ito, S., Stuphorn, V., Brown, J. W., & Schall, J. D. (2003). Performance monitoring by the anterior cingulate cortex during saccade countermanding. *Science, 302,* 120–122.

Kagel, J. H., & Roth, A. E. (1995). *The handbook of experimental economics.* Princeton: Princeton University Press.

Kahneman, D., & Tversky, A. (1979). Prospect theory: An analysis of decision under risk. *Econometrica, 47,* 263–292.

Kawagoe, R., Takikawa, Y., & Hikosaka, O. (1998). Expectation of reward modulates cognitive signals in the basal ganglia. *Nature Neuroscience, 1,* 411–416.

Kawato, M., & Samejima, K. (2007). Efficient reinforcement learning: Computational theories, neuroscience and robotics. *Current Opinions in Neurobiology, 17,* 205–212.

Lee, D. (2008). Game theory and neural basis of social decision making. *Nature Neuroscience, 11,* 404–409.

Lee, D., Conroy, M. L., McGreevy, B. P., & Barraclough, D. J. (2004). Reinforcement learning and decision making in monkeys during a competitive game. *Cognitive Brain Research, 22,* 45–58.

Lee, D., McGreevy, B. P., & Barraclough, D. J. (2005). Learning and decision making in monkeys during a rock-paper-scissors game. *Cognitive Brain Research 25,* 416–430.

Leon, M. I., & Shadlen, M. N. (1999). Effect of expected reward magnitude on the response of neurons in the dorsolateral prefrontal cortex of the macaque. *Neuron, 24,* 415–425.

Matelli, M., Luppino, G., & Rizzolatti, G. (1991). Architecture of superior and mesial area 6 and the adjacent cingulate cortex in the macaque monkey. *Journal of Comparative Neurology, 311,* 445–462.

Matsumoto, M., Matsumoto, K., Abe, H., & Tanaka, K. (2007). Medial prefrontal cell activity signaling prediction errors of action values. *Nature Neuroscience, 10,* 647–656.

Matsuzaka, Y., Aizawa, H., & Tanji, J. (1992). A motor area rostral to the supplementary motor area (presupplementary motor area). in the monkey: Neuronal activity during a learned motor task. *Journal of Neurophysiology, 68,* 653–662.

Mazer, J. A., & Gallant, J. L. (2003). Goal-related activity in V4 during free viewing visual search: Evidence for a ventral stream visual salience map. *Neuron, 40,* 1241–1250.

McCoy, A. N., & Platt, M. L. (2005). Risk-sensitive neurons in macaque posterior cingulate cortex. *Nature Neuroscience, 8,* 1220–1227.

Mookherjee, D., & Sopher, B. (1994). Learning behavior in an experimental matching pennies game. *Games and Economic Behavior, 7,* 62–91.

Mookherjee, D., & Sopher, B. (1997). Learning and decision costs in experimental constant sum games. *Games and Economic Behavior, 19,* 97–132.

Nash, J. F. (1950). Equilibrium points in n-person games. *Proceedings of the National Academy of Sciences, 36,* 48–49.

Paton, J. J., Belova, M. A., Morrison, S. E., & Salzman, C. D. (2006). The primate amygdala represents the positive and negative value of visual stimuli during learning. *Nature, 439,* 865–870.

Padoa-Schioppa, C., & Assad, J. A. (2006). Neurons in the orbitofrontal cortex encode economic value. *Nature, 441,* 223–226.

Pawitan, Y. (2001). *In all likelihood: Statistical modelling and inference using likelihood.* Oxford: Oxford University Press.

Perin, C. T. (1943). A quantitative investigation of the delay-of-reinforcement gradient. *Journal of Experimental Psychology, 32,* 37–51.

Platt, M. L., & Glimcher, P. W. (1999). Neural correlates of decision variables in parietal cortex. *Nature, 400,* 233–238.

Rapoport, A., & Budescu, D. V. (1992). Generation of random series in two-person strictly competitive games. *Journal of Experimental Psychology: General, 121,* 352–363.

Roesch, M. R., & Olson, C. R. (2004). Neuronal activity related to reward value and motivation in primate frontal cortex. *Science, 304,* 307–310.

Rushworth, M. F. S., Behrens, T. E. J., Rudebeck, P. H., & Walton, M. E. (2007). Contrasting roles for cingulate and orbitofrontal cortex in decisions and social behaviour. *Trends in Cognitive Science, 11,* 168–176.

Rushworth, M. F. S., Walton, M. E., Kennerley, S. W., & Bannerman, D. M. (2004). Action sets and decisions in the medial frontal cortex. *Trends in Cognitive Science, 8,* 410–417.

Samejima, K., Ueda, Y., Doya, K., & Kimura, M. (2005). Representation of action-specific reward values in the striatum. *Science, 310,* 1337–1340.

Schultz, W. (1998). Predictive reward signal of dopamine neurons. *Journal of Neurophysiology, 80,* 1–27.

Schultz, W., Dayan, P., & Montague, P. R. (1997). A neural substrate of prediction and reward. *Science, 275,* 1593–1599.

Schweighofer, N., & Doya, K. (2003). Meta-learning in reinforcement learning. *Neural Networks, 16,* 5–9.

Seo, H., Barraclough, D. J., & Lee, D. (2007). Dynamic signals related to choices and outcomes in the dorsolateral prefrontal cortex. *Cerebral Cortex, 17,* i110–i117.

Seo, H, Barraclough, D. J., & Lee, D. (2009). Lateral intraparietal cortex and reinforcement learning during a mixed-strategy game. *Journal of Neuroscience, 29,* 7278-7289.

Seo, H., & Lee, D. (2007). Temporal filtering of reward signals in the dorsal anterior cingulate cortex during a mixed-strategy game. *Journal of Neuroscience, 27,* 8366–8377.

Seo, H, & Lee, D. (2009). Behavioral and neural changes after gains and losses of conditioned reinforcers. *Journal of Neuroscience, 29,* 3627–3641.

Shima, K., & Tanji, J. (2000). Neuronal activity in the supplementary and presupplementary motor areas for temporal organization of multiple movements. *Journal of Neurophysiology, 84,* 2148–2160.

Sohn, J.-W. & Lee, D. (2006). Effects of reward expectancy on sequential eye movements in monkeys. *Neural Networks, 19,* 1181-1191.

Sohn, J. W., & Lee, D. (2007). Order-dependent modulation of directional signals in the supplementary and presupplementary motor areas. *Journal of Neuroscience, 27,* 13655–13666.

Shidara, M., & Richmond, B. J. (2002). Anterior cingulate: Single neuronal signals related to degree of reward expectancy. *Science, 296,* 1709–1711.

Soltani, A., Lee, D., & Wang, X. J. (2006). Neural mechanism for stochastic behavior during a competitive game. *Neural Networks, 19,* 1181–1191.

Sugihara, T., Diltz, M. D., Averbeck, B. B., & Romanski, L. M. (2006). Integration of auditory and visual communication information in the primate ventrolateral prefrontal cortex. *Journal of Neuroscience, 26,* 11138–11147.

Sugrue, L. P., Corrado, G. S., & Newsome, W. T. (2004). Matching behavior and the representation of value in the parietal cortex. *Science, 304,* 1782–1787.

Sutton, R. S., & Barto, A. G. (1998). *Reinforcement learning: An introduction.* Cambridge, MA: MIT Press.

Tanji, J. (2001). Sequential organization of multiple movements: Involvement of cortical motor areas. *Annual Review of Neuroscience, 24,* 631–651.

Thorndike, E. L. (1911). *Animal intelligence: Experimental studies.* New York: Macmillan.

Tomasello, M., & Call, J. (1997). *Primate cognition.* Oxford: Oxford University Press.

Tremblay, L., & Schultz, W. (1999). Relative reward preference in primate orbitofrontal cortex. *Nature, 398,* 704–708.

Vinje, W. E., & Gallant, J. L. (2000). Sparse coding and decorrelation in primate visual cortex during natural vision. *Science, 287,* 1273–1276.

von Neumann, J., & Morgenstern, O. (1944). *Theory of games and economic behavior.* Princeton: Princeton University Press.

Watanabe, M. (1996). Reward expectancy in primate prefrontal neurons. *Nature, 382,* 629–632.

Whiten, A., & Byrne, R. W. (1997). Machiavellian intelligence II: Extensions and evaluations. Cambridge: Cambridge University Press.

Woolley, S. M. N., Fremouw, T. E., Hsu, A., & Theunissen, F. E. (2005). Tuning for spectro-temporal modulations as a mechanism for auditory discrimination of natural sounds. *Nature Neuroscience, 8,* 1371–1379.

CHAPTER 28

Out of Our Minds: The Neuroethology of Primate Strategic Behavior

Louise Barrett and Drew Rendall

INTRODUCTION

Primates, as an order, are marked by large brains and considerable behavioral flexibility, which is perhaps most developed in the social domain. In fact, the unusually intense sociality of the primates, compared to other taxa, sparked a proposal, now firmly entrenched, that the primate niche is largely a social one where selection pressures favoring increasingly complex social strategies are responsible for the markedly larger brains of primates (e.g., Dunbar, 1998; Humphrey, 1976; Jolly, 1966). Furthermore, there is a natural assumption that, as our closest living relatives, the monkeys and apes have something special to tell us about the evolution of our own cognitive capacities as the culmination of such processes. The programs of comparative cognition that these proposals have sparked have been exciting and highly productive, but we also feel that, as often practiced, they promote a certain view of the brain and cognition that, increasingly, is leading us astray. In what follows, we articulate these concerns, and suggest how we might deal with them.

OUT OF OUR MINDS

Following the "cognitive revolution" and the loosening of the stranglehold of behaviorism on comparative psychology, the study of nonhuman minds became respectable again (e.g., Griffin, 1978). One could abandon the Cartesian view that animals were mere automata, and embrace the notion that the behavior of nonhumans was underpinned by cognitive processes of varying degrees of flexibility and complexity. Evolutionarily speaking, such a move also dispensed with the idea that a profound discontinuity existed between human and nonhuman animals, picking up on Darwin's idea that the difference was likely to be one of degree and not of kind (see Penn et al., 2008, for a review and critique of this stance). In fact, in recent years, an explicitly anthropocentric, or even anthropomorphic, program of comparative research has (re)emerged and been defended as scientifically legitimate and expressly licensed by this evolutionary perspective (e.g., De Waal, 2001, 2005).

Such a stance has no doubt been helped along by the fact that much of primate cognitive ethology has attempted to combine the study of mechanism with that of evolutionary function, where, following Dennett's influential proposal, intentional language has come to be used frequently as a metaphorical short-hand to characterize the nature of adaptive behavior (i.e., the so-called "free-floating" rationale [of mother nature]; Dennett, 1989). Dennett's intentional gambit was initially only a practical one—namely, without easy access to the underlying psychology of animals' behavior, the field of ethology could nevertheless make progress on the adaptive functions of behavior on the assumption that, like humans, animals behave "as if" they have goals, beliefs, and desires about the world that guide their engagement with it. However, this gambit also makes it all too easy either to side-step the issue of determining the proximate mechanisms that actually do produce behavior or to favor an argument in which behavior that looks similar to our own, and that achieves a similar functional goal, can, by an appeal to evolutionary parsimony, be assumed to be produced by the same intentional psychological processes (e.g., De Waal, 1997). Even when not expressed in such overtly anthropomorphic terms, a focus only on continuity between human and nonhuman minds, without equal attention to the difference, can lead, and has led, to a rather narrow focus, where only those questions that speak to this intentional conception of action are tackled, and possible alternatives are seldom even considered (Owren & Rendall, 2001).

Studies of primate communication are a case in point. Seminal work by Seyfarth and colleagues (1980) on vervet monkeys documented a small repertoire of alarm calls specific to different classes of predator, each call prompting a functionally different escape response appropriate to the predator encountered. The calls clearly functioned as if they stood for, or represented, those predators in the same way our own human words for these animals do, and these functional similarities were taken to reflect deeper cognitive similarities, rooted in language-like representational and intentional processes in both groups. These findings were thus a timely confirmation of Griffin's (1978) instinct that the key to the cognitive revolution in animal psychology lay in the natural communication behaviors of animals that would offer privileged insights into animal minds, and, more specifically, reveal continuity with human minds in proportion to the language-like properties of their communications. The vervet monkey research also then naturally spawned a generation of similar studies focused on identifying additional language-like properties in primate communications (e.g., semantics and syntax; reviewed in Hauser et al., 2002; Zuberbühler, 2000). It also encouraged research into continuity in other cognitive domains (e.g., transitivity, numerosity, causality) on the assumption that, if one could demonstrate continuity in language, the sine qua non of humanity, then surely it would be found in almost every other psychological domain.

The fact that, as subsequent work demonstrated, primate vocalizations lack the flexible production needed to respond to or influence the intentional states of others has tended to undercut these inferences, as does the lack of unequivocal evidence to suggest that primates can represent either their own intentional states or those of others (Cheney & Seyfarth, 2005; Penn et al., 2008). The many attempts to teach language to captive apes provide corroborative results: While ape subjects have shown impressive learning capacities and, in some cases, can use symbols effectively, their use of artificial languages remains strikingly instrumental and egocentric. Given that human language and meaning hinge on the flexible, intentional nature of communications (Grice, 1957), it is difficult to conceive of nonhuman communication as being language-like in any meaningful, psychological sense. We are thus forced to confront the distinction between the successful pragmatic application of "as if" reasoning that uses language-based metaphors to explain the evolutionary function of vocalizations in primates and a principled understanding of the underlying psychological mechanisms that actually govern their vocal behavior.

As the debate regarding the degree to which an anthropocentric approach to nonhuman cognition is justified rumbles on in various places and in various guises, our aim here is not to rehearse it all again. Rather, our aim is to point out that there is a more pervasive anthropocentrism that drives much current research, and it is one that seems to go almost entirely unnoticed. Specifically, the underlying assumption of all work in comparative cognition is that the cognitive load is borne entirely by the brain, and that it does so by using information processing routines operating on and over some kind of internal representational structure. Or, to put it more simply, it is assumed that the brain is a computer that controls the body and tells it what to do.

As is well known, the computer metaphor of the brain rose to prominence during the cognitive revolution of the 1950s. Prior to this, it was the telephone exchange that served as our best metaphor for the brain (Draaisma, 2000). The brain has also variously been compared to an abbey, cathedral, aviary, theater, and warehouse, as well as a filing cabinet, clockwork mechanism, camera obscura, phonograph, and railway network, as well as Locke's "blank slate" and Socrates' "wax tablet" (Draaisma, 2000). The use of a computer metaphor is simply the most recent in a long line of metaphors that pick up on the most advanced and complex technology of the day. This observation, in itself, should make us at least a little sceptical about whether we really have finally hit upon the correct metaphor for understanding biological cognitive systems, as opposed to one that merely reflects something about the time in which we now currently live.

What should really arouse our suspicions regarding its appropriateness is that, as Dreyfus (1992), Brooks (1999), and, more recently, Pfeifer and Bongard (2007) have all pointed out, the original artificial intelligence project that arose out of the cognitive revolution (good old-fashioned artificial intelligence, or GOFAI) was strongly anthropocentric, where the focus of research rested squarely on some peculiarly human aspects of intelligence, like natural language, formal reasoning, planning, mathematics, and playing games like chess—that is, tasks that involved the manipulation and processing of abstract symbols in a logical fashion. Unfortunately, this initial, rather arbitrary emphasis on logical, algorithmically based tasks gained such momentum, aided and abetted by the Church-Turing thesis and the sway of functionalism, that we are now in a position where brains are not seen as merely analogous to computer-like logical reasoning devices, but are considered to actually *be* a form of computer. Psychological processes are therefore assumed to be algorithmic and to involve various kinds of "computation" and "information processing." Indeed, we use such metaphors freely and without hesitation, both scientifically and colloquially ("I was so shocked I couldn't process what she was saying"; "That much information will fry my brain"). This has then led to similar reasoning about nonhuman minds so that, for example, claims are made for the "propositional" (Seyfarth & Cheney, 2005) and "hierarchically structured" (Bergman et al., 2003) nature of their knowledge about the social world.

There is no doubt that the computer metaphor of the brain has been, and remains, valuable; it clearly helped to rescue experimental psychology from the intellectual sterility of behaviorism (Miller, 2003), and there is also no doubt that work in classical artificial intelligence has had its successes. These successes have, however, been somewhat limited. While the classical approach has, finally, generated a computer that can beat the world chess champion and helped to design a range of so-called "expert systems," it has so far failed to give us any real insight into the mechanisms that underlie more natural forms of intelligence, such as how we recognize a face in a crowd, how we coordinate and manipulate all the objects necessary to make cup of coffee, or even something as apparently simple as how we (and other animals) manage to walk, run, or hop over uneven ground without falling flat on our faces.

The problem here should be obvious. Our metaphor of the brain, and hence of cognitive processes, is one that was originally derived from a heavily anthropocentric focus on a few peculiar cognitive achievements, all of which involved abstract symbol manipulation. Moreover, these processes are "all talk and no action": Chess is not a very athletic pastime, nor is solving logical syllogisms. With the key to cognition seen purely as internal symbol manipulation, the role of the body and the environment was sidelined completely, so that brains became entirely divorced from both the bodies they inhabit and the world that they encounter (Dreyfus, 1992). Cognition has, in turn, become a process that has no real link to the body or the outside world, taking place purely in the head, hermetically sealed from reality. In this sense, it is a view that, in fact, remains very strongly Cartesian (Rockwell,

2005; Wheeler, 2005). We have then taken this strange view of cognition—the disembodied, logical manipulation of internal representations—and applied it to other animals, despite the fact that it does not even adequately explain most facets of natural human cognition. Consequently, we have used a flawed, dualist model of human cognition to model many other, potentially very different species of cognition in ways that can only serve to compound our initial error. What were we thinking? Clearly, it seems, we were out of our minds.

OUT OF THE HEAD AND INTO THE WORLD

So, what to do? How can we approach the study of neuroethology and cognition in a way that avoids this misplaced anthropocentric projection? Our suggestion is that we should attempt to get out of our minds more productively by embracing models of cognition that take into account the "loopy" integration of brain, body, and world (Brooks, 1999; Clark, 1997; Dreyfus, 1992; Pfeifer & Bongard, 2007; Pfeifer & Scheier, 1999; van Gelder, 1995). This requires that we take seriously the notion that natural cognitive systems are "embodied" (indeed, it makes no sense to refer to any other kind of cognition once we leave the confines of the computer science laboratory). That is, cognitive systems evolved to enable animals to deal effectively with and engage the worlds that they encounter, but not necessarily to reflect upon that world in "thoughtful" disembodied ways (Barrett & Henzi, 2005; Barrett et al., 2007; Brooks, 1999; Clark, 1997; Gibson, 1979). This effectively turns our anthropocentric perspective on its head. Once we recognize, as Brooks (1999) has argued so cogently, that all animals possess bodies and that all did so long before they possessed anything remotely recognizable as a brain, then we come to realize also that most of evolutionary history has been spent refining the perception and action mechanisms that enable survival in a dynamic world. One cannot, therefore, study cognition in the abstract, disembodied way promoted by proponents of the classical artificial intelligence (AI) approach because, in the real world, there is no such thing as abstract, disembodied cognition. Moreover, a more holistic perspective forces the recognition that natural selection has elaborated on nervous systems as a whole, and not the brain alone. The nervous system that lies outside the brain is not merely a network of "message cables" that relay the commands of the brain back and forth to the body (e.g., Churchland, 1986) but an integrated, functional system that includes the brain, but is not superseded by it, and that is integrated into both the endocrine and the immune systems of the body (e.g., Gershon, 1998; Pert, 1997).

Once we reject our singular focus on brains, we automatically reject a second aspect of the "classical" Cartesian perspective, which assumes that as "thinking things" we contact the world only indirectly as detached observers, where the problem of knowledge has famously been one of how we gain access to this world and those of others. As both James (1890) and Dewey (1929) pointed out, this view amounts to nothing more than a false dichotomy between body and brain, organism and environment. The problem of knowledge, then, is how patterns of organism-environment interaction can be adapted and transformed to deal with the new problems and challenges that such interaction itself creates continually (Johnson & Rohrer, in press). This position is not one that currently informs comparative primate social cognition. Instead of viewing behavior as constitutive of mental states and intentionality, as James (1890) and Dewey (1929) argued it should, comparative studies typically view behavior merely as a cue to (or symptom of) the hidden mind, which we can only know indirectly by inference. This is completely understandable in the context of a view that sees cognition as a wholly internal, purely brain-based enterprise, but this view ultimately limits studies of animal social cognition to those that can provide the kind of controlled environment needed to see this "pure brain" in action and infer the structure of the "hidden mind." This, in turn, is likely to lead to an ever more distorted view of the cognitive abilities of other animals for two interlinked reasons.

The first of these can be summed up by what Clark (1989: 64) refers to as the "007 principle": that, generally speaking, "evolved creatures will neither store nor process information in costly ways when they can use the structure of the environment and their operations on it as a convenient stand-in for the information processing operations concerned." In other words, and as James Bond would have had it, animals should operate on a "need to know basis"; they should know no more than they need to in order to get the job done. Clark (1989) illustrates this principle by reference to filter-feeding sponges that exploit water currents to reduce the amount of pumping they need to do and to mole crickets that construct underground tunnels that function as Klipsch horns and thereby increase the volume of their mating calls. Cognitive systems should be no different to these other physiological systems. In this respect, the stigmergic systems of the social insects—the pheromone trails of ants, for example, or the pheromone-impregnated soil balls that form the building blocks of termites' intricate and elaborate mounds—are clear examples of how natural selection has discovered the simple trick of throwing things out into the world for reperception, rather than storing, processing, and retrieving internal representations, and in so doing, achieving amazing feats of what, from our frame of reference, looks like highly "intelligent" behavior.

Indeed, humans may be the masters of this reperception approach to cognition, where a great deal of daily behavior, and much of the material culture developed to support it, reflects a habit of offloading to the environment for subsequent easy reperception the details of what would otherwise be cognitively demanding tasks. Simple, oft-cited examples include scribbling shopping and other to-do lists for later consultation rather than trying to commit them to memory, doing multiplication and long division by hand on paper where you can see the work in simple steps rather than doing it all "in your head", and leaving your keys on the window ledge by the front door where you are bound to encounter them on your way out of the house rather than trying to remember a "safer" storage spot for them. In fact, this habit is all pervasive (now and almost certainly in prehistory as well) and it underscores just how much of routine human cognition may be largely perceptual and enacted in the context of environmental supports (Clark, 1997). Consider, for example, how one can suddenly "remember" a telephone number only when actually looking at the phone about to dial, or how one can fail to recall the name of a favorite old song, yet a few introductory notes can bring the full melody and lyrics flooding back. Consequently, treating the primates (and other creatures, including humans) as if their cognition can be understood independently of an understanding of their environment is likely to produce both an incomplete account—given that some of the cognitive load may well be borne by the environment and not by the animal—and, more worryingly, an inaccurate one, given that, as Clark (1989: 66) suggests, we risk placing into "the modelled head what nature leaves to the world."

This leads to the second reason why we ignore the body and world at our peril: If we assume that an organism's behavior arises as a consequence of purely internal, brain-based cognitive processes, we may mistakenly attribute more complexity to the organism than is warranted, and overestimate the cognitive requirements of certain tasks. The "cognition" we see, therefore, may be entirely illusory, reflecting only our own frame of reference and not that of the animal itself (Brooks, 1999; Pfeifer & Bongard, 2007). A perfect illustration of this is provided by Maris and te Boekhorst's (1996) work on the collective heap-building behavior of simple robots (didabots) (see also Pfeifer & Scheier, 1999, for a review of this and other work).

When placed into an arena in which small cubes are scattered randomly, the robots move around clustering the cubes together, so that eventually there are usually only two large clusters of cubes, with a few cubes left here and there against the walls. The robots are therefore known as "Swiss robots," because their aim, apparently, is to tidy up the arena. At a minimum, then, one would suppose that the robots are programmed with mechanisms for detecting objects, pushing them in a given direction toward other objects, and then clustering them together. In fact, the robots have only one kind of sensor, for proximity, and are programmed with one simple control rule: If sensory information is received on the left, then turn right, and if sensory information is received on the right, turn left. In other words, the robots are programmed exclusively to avoid obstacles. Why then does clustering occur? To understand this, we have to move beyond the robots' internal structure and consider the specific nature of its embodiment and its interaction with the environment.

A didabot is a small wheeled robot, shaped somewhat like a toy car. In Maris and te Boekhort's experiment, they were fitted with two sensors at the front end of the body, at an angle. Consequently, as the robots moved forward, they would detect cubes off to the side, but not straight in front of them. This meant that, while they turned away and avoided cubes on either side, a cube directly in front of them would end up being pushed along, because the didabot couldn't "see" it (i.e., its sensors would receive no stimulation from it). If the didabot then encountered another cube off to the side, triggering its sensor, it would produce avoidance behavior, moving off to the left or right, leaving the object it had just been pushing next to the object it had just avoided. In other words, the didabot clustered the two cubes together. This alternation in the environment then increased the chance that another cube, being blindly pushed around, would also be deposited in the vicinity, because the cluster of two cubes was more detectable than a single cube. In other words, a simple self-organizing process can explain the didabots' behavior, which results in the formation of ever larger clusters, and generates the abiding impression in those watching that the didabots are "trying" their best to "clean up." Moreover, the clustering behavior is entirely dependent on the placement of the sensors on the didabot's body: Move one of the sensors around to the front, and the clustering behavior disappears entirely, because now objects directly in front of the robot are avoided in the same way as those off to the side, which means that no pushing behavior occurs. Here, then, an interpretation of the didabots' behavior as reflecting the operation of a proximate "clustering" mechanism, reflecting their "goal" of cleaning the arena, would be entirely erroneous: both inaccurate and incomplete.

This example should give us pause when we attempt to attribute particular proximate "goals," and their supporting cognitive mechanisms, to biological organisms. It also highlights the fact that the proximate means by which a behavior is produced need bear no relation to the form that behavior takes (who would imagine that a rule for object avoidance would be a good way to produce object clustering?), and that there is no necessary, nor simple, one-to-one mapping between the complexity of a proximate mechanism and the complexity of the behavior that it produces. If we take on board the lessons that work in robots and artificial life can offer us, then it is clear that greater recognition is needed of the fact that, unlike classical computer scientists, natural selection is uninterested in a "pure" computational brain and is, instead, focused on the kinds of emergent "mind" that brains, bodies, and environments can achieve in concert. A biological cognitive system, therefore, consists not of a brain, but of a brain within a body that acts in an environment, and it is also a dynamical system where each component is simultaneously influencing, and is influenced by, all the other components (Van Gelder, 1995).

EMERGENT MINDS: "INTELLIGENCE WITHOUT REPRESENTATION"

The idea that cognition or, rather, psychological processes in general are emergent phenomena that depend crucially on the interaction of an animal with its world was recognized long ago by Von Uexküll (1957) and his notion of the "umwelt"; that an animal's representation of the world (if indeed it represents it at all) will be shaped by, and grounded in, the means by which it perceives and acts in it. These perception-action mechanisms, in turn, reflect the kind of body it has and the kind of ecological niche it occupies. More recently, J.J. Gibson (1979) took a similar approach, which he characterized as "ecological psychology," to get across the idea that psychological phenomena were to be found in organism-environment relations, and not in the organism alone. While Gibson's views have received rather a bad press, being seen as both strongly antimentalist and antirepresentational (see Reed, 1996, for a review and response to such criticisms), this is only a crude caricature of a more nuanced position, which argued that representational systems are not mental phenomena alone, but also ways of behaving and regulating action in the world (Reed, 1996). More important, it is the false separation of organism and environment that is denied by ecological psychologists, rather than the existence of mental representations per se. In this, it becomes clear how ecological psychology embraces the earlier views of James (1890) and Dewey (1896, 1929) and anticipated many of the insights that come from more recent work in robotics and artificial life (Brooks, 1999; Clark, 1997; Pfiefer & Bongard, 2007; Wheeler, 2005).

A most important aspect of an ecological approach, from our perspective, is the idea that organisms regulate their behavior with respect to the "affordances" of the environment. Affordances are the opportunities and possibilities for action that particular objects and resources offer to an animal (Gibson, 1979; Reed, 1996). In this way, perception is "written in the language of action" (Michaels & Carello, 1981) so that, for example, we do not see "chairs" as such, but places to sit; a chimpanzee doesn't see a "tree," but a place to climb. As we have argued in detail elsewhere (Barrett & Henzi, 2005; Barrett et al., 2007), the discovery of the mirror and canonical neuron systems of the primate brain offer neurobiological support for a theory of affordances. Canonical neurons, for example, which fire both when an object is acted on by an individual and when an object is merely observed, illustrate how our perceptual representations of objects contain motor as well as sensory components, and hence blur the distinction between perception and conceptual knowledge: Our perception of a cup, for example, cannot be divorced from its function, from what we can do with it, because such possibilities for action are built directly into our perceptions (Gabarini & Adenzato, 2004). When we consider mirror neurons from this perspective (those that fire both when an individual acts on an object and when he or she merely observes another individual acting on it), we can easily see how the perception of other individuals will generate similar affordances, and hence the possibilities for action that other individuals allow. Indeed, Catmur and colleagues (2007) have demonstrated recently that, in humans, "reverse mirror" contingencies can be induced by training (i.e., participants are trained to perform one action by watching another), suggesting that mirror properties are neither innate nor fixed once acquired but, instead, develop through sensorimotor learning. As Catmur and colleagues (2007) conclude, this implies that mirror systems are both a product and a process of social interaction. This, as Gallese (2001, 2003, 2005, 2007; Gallese et al., 2004) has argued persuasively, provides us with an implicit, unconscious, and automatic mechanism for engaging others as goal-directed agents, without the need for any overt mental state understanding. Most significantly of all, the theory of affordances provides a natural means of grounding this knowledge in the world because it recognizes the essentially dynamical and "loopy" nature of behavior, something that Dewey (1896) made clear in his classic paper refuting the notion of the reflex "arc." That is, behavior is not the result of a one-way link that goes from stimulus to response, as the behaviorists assumed, but is best characterized as a continuous loop where action generates perceptual feedback that, in turn, generates further action, and so on. Nothing remains internal to the organism because action in the world is always an integral part of its conception of the world.

APPLYING THE CONCEPTS: PRIMATE SOCIAL AFFORDANCES

The manner in which baboons engage in certain forms of social interaction suggests that a stronger, more ethological focus on the affordances that other individuals offer (i.e., the opportunities for certain forms of action that others may or may not pick up and act on) will pay real dividends in understanding the basis of primate strategic behavior and its likely proximate underpinnings. When a female becomes a mother, for example, the presence of her new infant generates intense interest in other females, who will try to approach and handle the infant. They will not always do this gently, and sometimes might make off with the infant altogether, putting it at risk of dehydration and, in extreme cases, death (Silk et al., 2003). Mothers, not surprisingly, display agitation and avoidance when others approach and attempt to handle their offspring. Grooming the mother, however, apparently increases her tolerance for this kind of interaction, with the level of grooming dependent on the number of infants present in the group. The latter effect presumably reflects the degree of stress a mother is experiencing: A single infant in a troop means a high level of social attention, and consequently females may require more grooming before its effects (in terms of endorphin release) are felt (Barrett & Henzi, 2006; Henzi & Barrett, 2002).

This grooming in exchange for handling has, however, been described in terms of more cognitive processes, where it represents a potential case of "tactical deception" on the part of handlers, who approach and groom in order to disguise their true intention and lull the mother into a feeling of false security (given as example #88 in Byrne & Whiten, 1990[1]). Careful analysis of video-taped interactions of this nature (Barrett & Henzi, unpublished data) indicate that most handling events, in fact, follow a specific sequence that speaks more to behavioral attunement of the kind Gallese (2005) espouses, rather than any form of truly intentional or even "functional" deception.

Initially, a female will approach a mother and, particularly if lower ranking, will sit at a distance of 2 to 3 m, a distance that places a female sufficiently out of reach of any form of aggressive attack. Once in this position, potential handlers monitor the mother closely, following her head orientation and gaze. Mothers, for their part, will often turn their back as the handler approaches, forcing the handler to begin circling round the mother in order to keep the mother's face, and the infant, in view. If mothers cannot keep the handler out of their own field of view in this way, then they simply do not glance or gaze in the handler's direction. As long as this situation persists, handlers appear paralyzed and unable to act: They will circle around the edge of the apparent zone of tolerance, but do not approach any closer. Occasionally, they may emit contact grunts of a characteristic kind. Eventually, a potential handler will drift away from the mother if she remains motionless and does not gaze in the handler's direction. If, on the other hand, she does gaze toward the handler, or even orient her head in that direction, this apparently affords the handler the opportunity to approach. Handlers may not always do so, of course, but approaches tend to occur only if mothers afford the opportunity to do so in this way.

Once the handler is in proximity, mothers may again turn away from the handler. This is especially likely to be the case if the mother's glance in the handler's direction was inadvertent (e.g., the mother oriented in that direction in response to a call from another animal, or because the mother apparently failed to notice the initial approach of a handler). Once in proximity, and with the mother turned away, presenting her back, handlers more often than not begin to groom. Seen in the context of the interaction as a whole, this does not seem to be a ploy to deceive the mother, but occurs because the handler is in an aroused state, is socially motivated, and is afforded the opportunity to groom. Infant handling, then, is more about two individuals "bringing their behavior into coordination" in a real-time dynamic fashion (Johnson, 2001), and not about the execution of preformed goals and intentions that exist in one or other of the animal's heads.

This emphasis, as Johnson (2001) further argues, is at the heart of a distributed approach, where cognition emerges from, and during, ongoing interaction. In this, it is quintessentially ecological, since the psychological phenomena are a function of animal-environment relations, rather than of individual animals alone. From a practical perspective, adopting a distributed, less representational approach also means that we are able to move away from using terms like "intend" and "understand," which refer to invisible mental states that can only be inferred, to descriptors like "anticipate" and "notice," which are both observable and definable (e.g., anticipation can be measured by an increased readiness to respond or a reduced lag time for certain behaviors; noticing can be defined by direction of gaze, or the relative timing of signaling behaviors: Johnson, 2001).

Strum and colleagues (1997) similarly have used an analysis of consort takeovers by male baboons to illustrate how much of the behavior that is often held up as an example of "Machiavellian intelligence" (i.e., as sophisticated cognitive strategizing to achieve a specific goal) may actually be the result of how particular animals are either constrained or afforded certain courses of action by the environment. Male baboons socially and sexually monopolize adult females during their fertile periods, remaining in close proximity to them at all times. These close spatial relationships ("consortships") can last from a few hours to a week, depending on the specific population of baboons. Among East African populations, these consortships are often disrupted by aggression from other males, who then take over the consort male's position. There are various social options that males can employ to either avoid or facilitate a takeover. Strum and colleagues (1997) focus on one they call "sleeping near the enemy." This designation was based on the observation that, while older males were able to resist consort takeover attempts by younger, more aggressive males during the day, they were less able to do so at sleeping sites, where younger males were able to displace them and leave with the female in the morning. Strum and colleagues (1997) demonstrated that a change in topography from the plains, where the animals foraged during the day, to the cliffs, where they slept at night, was the key factor leading to this difference. Older, socially experienced males could resist takeover on the plains, diverting aggression by grabbing a younger animal and interposing it between himself and his male rival. Such options require a high degree of visual contact with others, a significant amount of behavioral coordination, and, therefore, sufficient experience with other animals to engage and deflect them successfully.

On the sleeping cliffs, however, these options were constrained by topography. The height and narrowness of the cliffs resulted in changes in the mobility of males, their proximity to other animals, and reduced overall visibility. All of these factors served to reduce older males' ability to direct the situation socially while at the same time, they favored the more direct, aggressive approach of younger males. In other words, it was the change in topography that was the key to success of younger males on the cliffs, not any specific change in male tactics as such. Males were employing those behaviors in their repertoire that were afforded by the environment, rather than developing entirely novel means to displace rivals. This example illustrates that, in addition to the social affordances offered by other individuals, the affordances of the physical environment can also bring individuals' behavior into coordination or prevent this from happening, so that similar situations play out in entirely different ways depending on how the environment serves to structure the event.

In both baby handling and consort takeovers, then, we can see how easy it is to attribute an animal's actions to a fairly complex social calculus, based on rich internal representational mechanisms that enable the animal to assess its options and respond accordingly. Once we take the physical environmental into account, however, or pay closer attention to the actual sequencing of interactions between animals, we can see that baboons may be rather more like didabots and rather less strategically Machiavellian than we have assumed. This is not to deny the complexity of primate social behavior. Instead, it suggests that any such complexity is as an emergent property of the entire cognitive system, and not of an individual animal's brain in isolation. As such, we should at least entertain the hypothesis that the complex and flexible social behavior on display may be the product of a dynamic interaction between a set of simple social routines (e.g., biological motion detection, gaze detection-following mechanisms, sensitivity to facial expressions) with a particular social and physical environment. In this sense, behavior can be considered "strategic" really only from a functional, evolutionary point of view. From a proximate, mechanistic perspective, the ways in which animals coordinate, control, and manage their own and others' behavior may more productively be treated as contingent responses to the affordances of an ongoing situation. Breazeal (2002), for example, has shown how such simple social routines can produce highly engaging, emotionally involving social interactions between Kismet, her social robot, and human interactants, under circumstances where the humans know there is "no one home" in Kismet's head. The very power of her analysis comes, in fact, from the identification that "social intelligence" may have more to do with the smooth regulation and management of social interactions via various kinds of affect induction than with predicting and manipulating others' mental states.

AFFECT INDUCTION

Primate vocalizations also seem to fit this kind of affect-induction interpretation much better than representationalist theories, which argue that vocal signals literally "encode information" that must then be "decoded" by receivers who then "decide" how to respond based on the received information. Indeed, the latter approach falls precisely into the trap offered by classic disembodied cognitivist thinking, where abstract information in one animal's head is transmitted via the conduit of a vocal signal straight into another's. Here, the vocal signal is then the only necessary vehicle for message transmission and the weight of effective communication is borne entirely by representational processes inside the head of the signaler and the listener. In contrast, Owren and Rendall (1997, 2001; Owren et al., 2003) suggest that vocalizations often act not as representational information packets, but as components of a process of affect induction where the form of the signal itself plays an important role. In this view, vocal signals function to incline behavioral responses in listeners through the relatively direct effects that specific signals have on the entire nervous system of the listener, especially by tapping low-level perceptual, attentional, and motivational processes that guide a listener's basic engagement with the world. And the latter effects are importantly modulated both by the short-term motivational and response biases that the immediate social and environmental context prompts in the listener and by the longer-term dispositional biases the listener has as a result of an accumulated history of experience in the same and other contexts. Among other things, this emphasis can explain why certain classes of calls, like alarm calls, all share a suite of basic acoustic features (specifically those that capture listener attention and induce a startle response that is obviously functionally appropriate for predator avoidance) even if they function ultimately to identify different classes of predator and where, if calls were truly arbitrary referents for those different predators, one would certainly expect selection to have produced much greater differentiation in the calls in order to increase their discriminability. Consequently, if the aim of vocalizations is, like baby handling and consort takeovers, to manage and influence the behavior of other individuals, and not to change their mental states per se, then there is no need to argue that such signals have "meaning" in the standard representational sense. Instead, as Owren and Rendall (2001) argue, we can take a more embodied perspective, and investigate how signals that act at these low levels of neural organization help to shape behavior. This can occur both directly, as when a signal has an unconditioned impact on the perceiver's nervous system (again the acoustic startle response is a prime example), and indirectly, where vocalizations or visual signals are linked with affectively significant outcomes experienced by the perceiver via conditioning processes. For example, dominant animals often pair their threat vocalizations with actual aggression. As a consequence, subordinates learn the contingency between the two, with the result that, in future, the vocalizations alone can produce an avoidance response in subordinates (thus sparing the dominant the cost of actual aggression).

Such processes can also work affiliatively, as when individuals grunt softly upon approach, prior to grooming or attempting infant handling. By learning to associate such grunts with a positive (or at least neutral) outcome, animals can bring their behavior into coordination more rapidly and will do so more readily. An affect-induction account can also work to the advantage of "socially impotent" animals as well. The harsh, noisy screams that baboons produce when under attack, for example, are highly aversive stimuli, suggesting that subordinate animals, otherwise physically powerless, are nevertheless attempting to make themselves less appealing as objects of aggression (Owren & Rendall, 1997, 2001). Animals may also regulate their own affect through the use of vocalizations. For example, the "contact calls" that baboons and many other social primates make when separated from the rest of their troop are perhaps better characterized as "distress calls" that reflect an animal's stress at being alone and unable to see familiar conspecifics. The fact that such lost calls are most frequently "answered" by the animal itself, as opposed to individuals within the body of the troop, argues for a primarily self-regulatory, rather than directly communicative, function (see Rendall et al., 2000, for more information).

REDEFINING SOCIAL INTELLIGENCE

The approach we have offered here, which emphasizes how animals can tap directly into the perceptual and action systems of others, and how this can produce predictable physiological responses, provides us with a way out of the anthropocentric bind in which comparative primate studies are situated currently. An understanding of the manner in which one individual can manipulate and manage the way it is perceived by others, how it perceives another's actions, how its own actions can induce particular states of arousal or affect, and how these processes affect the subsequent coordination and sequencing of behavior all require close attention to what animals actually do and how they do it, and how this varies with context. As such, it forces us to consider how an animal's sensory and motor systems actually enable it to engage with the world, and the actions and opportunities this affords it.

For primates, the social environment might indeed create some special challenges, but not in the sense that is traditionally assumed. For any individual in a social species, a group consists of a large number of other individuals of various ages and sexes that afford varying opportunities for engagement, coordination, competition, and avoidance. A constantly changing social environment thus naturally affords a shifting landscape of opportunities and "on-the-fly" responses to them. But these responses need not result from internalized, strategic abstractions. Rather, they can exist in moment-by-moment attunement and adjustment of movement and activity that bring actors into and out of various fields of engagement. So, to the extent that primates have a particular social prowess, it is as likely to lie in the manner in which their perception-action routines are organized around constantly shifting social perceptions. The flexibility and variability of behavior then arises as animals attempt to keep their perceptions of themselves and others on an even keel in the face of perturbations of various kinds ("perceptual control theory": Cziko, 2000).

This raises the possibility that the proximate basis for primate social engagement may be a form of nonanalytic pattern recognition: The social world is clearly structured in ways that powerful associative learning principles can exploit. Specifically, the juxtaposition of particular individuals with particular behaviors with particular contexts can form the basis for an association matrix from which individuals can distill higher-order associations that enable them to respond appropriately to novel situations. Landauer and Dumais (1997), for example, have shown how precisely this kind of process (singular value decomposition followed by dimension reduction) can produce neural networks that comprehend written text, starting from only an association matrix of experienced words and the contexts in which they occurred. If associative processing can produce language comprehension—that most exalted of human abilities—there seems no reason why similar associative processes could not produce the latent relational dimensions among individuals, behaviors, and contexts that would give the functional appearance of sophisticated, strategizing representational cognition while being driven wholly by the perception of concrete observables. This kind of emphasis, then, can explain the manifest hypersociality of the monkeys, in particular, and their attunement to others, but also their abject failure in social cognitive tasks that test what they know about others' minds. It suggests that, in fact, they don't know or understand others in the theorized sense of attributing internal belief states. Instead, they "know" others in a far more concrete and procedural sense, as different perceptions to control.

This, in turn, suggests that what matters most, in terms of a social intelligence, is "expertise" or "wisdom" acquired by experience, rather than any explicit Machiavellian ability to think several steps ahead. This is because in order to produce an appropriate response and control one's perceptions, one must have a sufficiently large association matrix from which to derive these kinds of second-order associations, and such a matrix will obviously take time to build up. This can then explain why social learning and early socialization are so crucial for normal adult functioning (e.g., Suomi & Harlow, 1975), and why different social milieus and developmental backgrounds can influence the degree to which individuals are socially skilled as adults (e.g., Strum, 1982, 1994).

Further support for emergent pattern recognition as the basis for primate social intelligence comes from computational neuroscience, which has shown that formal neural nets often show emergent properties (Grossberg, 1988; Hopfield, 1982). That is, while each neuron itself has only elementary properties and the network itself has little structure, the collective activity and interactions of these neurons give rise to emergent macroscopic properties, in much the same way that translucence is an emergent property of a mass of water molecules forming a block of ice or the formation of a complex termite mound is an emergent property of individual termites' stigmergic responses. The macroscopic properties of these neural nets, which are known as attractors, are capable of sustaining patterns of stable activity that could function as a form of pattern-recognizing associative memory (Cossart et al., 2003; Fuster, 2003; Goldberg, 1988; Hopfield, 1982). Recognition of an external event has, therefore, been suggested to occur when sensory input resonates with previously formed attractors (adaptive resonance theory, or ART; Grossberg, 1987).

As attractors could potentially be distributed over different, widely distributed cortical areas, input from different sensory modalities could be combined in a single attractor, while the fact that they emerge as a result of experience means that they could be tuned to an animal's specific encounters with other individuals (and other objects in the world). Over time, then, animals would acquire a variety of attractors or an "energy landscape" (Freeman, 2001) that would allow them to recognize a wide variety of similar kinds of situations and generate an appropriate response to them. Knowing what to do next in a particular situation would, therefore, be part and parcel of pattern recognition with respect to the situation as a whole. This may explain why animals often act appropriately, despite never having learned a particular contingency or encountered a specific situation: behaviors that Rumbaugh and Washburn (2003; see also Rumbaugh et al., 1996) refer to, appropriately enough, as "emergents."

The great advantage of this kind of computational neuroscience over classic AI research is that empirical neurophysiological data seem to bear out these findings. For example, Walter Freeman's meticulous studies of the rabbit olfactory bulb indicate that the learning of different odorants is a nonlinear, self-organizing process, involving chaotic attractors (see e.g., Freeman, 2001; Freeman & Skarda, 1990a,b; Skarda & Freeman, 1987). Similarly, Scott-Kelso (1995) argues that the brain is fundamentally a self-organizing pattern-forming dynamical system, perched on the brink of chaos. As importantly,

Freeman and Skarda (1990a,b) describe brain-based evidence to support a "loopy" theory of perception and action. The neural activity patterns seen are sensitive not simply to the presence of an odorant or to the response, but to the interaction of both of these with each other along with the specific background context in which the behaviour is embedded. As Freeman and Skarda (1990a,b) argue, this is best characterized as a dialectic between organism and environment: The context in which these patterns of neural activity are seen is defined as much by the behavior of the animal and the existing attractor/energy landscape that is present as it is by the specific odorant involved (Freeman & Skarda, 1990a,b). It should be readily apparent that there is nothing intrinsically representational about such a dynamic process because one cannot identify any stable entity that could stand for something else in the standard symbolic, computational sense (van Gelder, 1995). In addition, these data indicate that animals are not passive recipients of environmental stimulation; they do not sit there waiting for input like the computer on one's desk and then simply react to it. Instead, perception begins with internally generated neural processes that, by reafference, lay the groundwork for the processing of future receptor input (Freeman & Skarda, 1990a,b) so that animals can be the active initiators of their engagement with the world. Interestingly, Skarda and Freeman (1990) point out that Merleau-Ponty (1942) had already reached this conclusion more than 50 years ago, when he argued that the passive, reflex-based view of physiological functioning was merely an illusion created by our attempts to understand the brain as a mechanical device. However, his view that brain function was creative, selective, and holistic, and that there existed internally generated, global states of cortical activity that resisted reductionist explanations, was rejected as unscientific (Skarda & Freeman, 1990). The work of neurophysiologists, like Freeman, is now beginning to bear out Merleau-Ponty's revolutionary ideas, suggesting that we should focus much more seriously on his particular kind of "biophilosophy" (Skarda & Freeman, 1990).

EXPLAINING BIG BRAINS

An obvious objection to the kind of associative account we are promoting here is that it is computationally inelegant. Compared to the representational alternative, the associative account would be much more memory intensive, computationally extensive, and therefore highly demanding of processing capacity. On reflection, however, this proves not to be an objection at all. As we have suggested elsewhere (Barrett et al., 2007; Rendall et al., 2008), the empirical data to be explained are the vast amounts of neural tissue that primates possess compared to other mammals, and not the complicated cognitive structures they use to solve their problems. There are, in fact, no empirical data to support the existence of the latter; rather, they are merely inferred from the vast quantities of brain tissue and the complexity of behavior seen. But, as we have seen with the didabots, for example, there is no necessity for complex behavior to require complex cognitive mechanisms, nor does a large and complex mass of neural tissue necessarily imply a more complicated cognitive architecture. Furthermore, there is no requirement that natural selection must favor neurocomputational efficiency per se, no matter how congenial such efficiency might be in an explanatory sense in our theories of cognition. Selection is, of course, free to favor whatever mechanisms get the job done relative to alternative local solutions. Consequently, we need to consider far more seriously the possibility that taxonomically widespread associative, nonanalytic modes of cognition can account for the broad sweep of primate cognition as well. Viewed properly in the context of the actual data to be explained—the manifestly large, interconnected, massively parallel processing brains that primates actually possess—the computational inelegance and resulting tissue profligacy of such mechanisms prove not to be a weakness but a strength.

CONCLUSION

In his seminal paper on social intelligence, Nicholas Humphrey (1976) likened social life to a game of chess, where predicting in advance what one's partner was likely to do would give an advantage by thwarting such action before it could have a detrimental impact on oneself. It should be clear from our arguments here that Humphrey's analogy is wrong: It is too GOFAI, too "cerebral," too slow, and too disembodied to capture the real-life social engagement of the primates. While we generally eschew metaphorical heuristics in theorizing, if pressed for an alternative to Humphrey's analogy for primate sociality, we might opt for a different sort of game—the jigsaw puzzle. In doing a jigsaw puzzle there is almost no value in planning ahead or representing the completed picture throughout. Instead, jigsaw puzzles are all about local pattern recognition and completion. You simply start by joining complementary pieces and thereafter you respond dynamically based on incipient local patterns. Your moves are in no sense planned or strategic. Rather, they are generated entirely on the fly and then only as contingent responses to local perceptual affordances—that is, by which of the remaining pieces are suggested by emerging, embryonic forms. The solution is also then entirely emergent based on the bottom-up accretion of ultimately conjoined but locally generated patterns. It is then a game that places a very high premium on dynamic perception-action integration. In the words of David Kirsh (1991: 169), it is "perceptually hard but intellectually simple." The irony is, of course, that this is also exactly how grand masters play chess: They recognize patterns, having learned to distinguish those that require one kind of response from those that require another; there is no analysis or comparison of alternatives, but instead a kind of fluid anticipation, as they notice the right things and respond at the right moment. In other words, grand masters play chess like the rest of us solve jigsaw puzzles, by acting in the world, on the fly, and in the moment. So too, we say, might primates meet the challenges posed by their social groups—through dynamic, contingent responses to a landscape of shifting social perceptions. An important difference is simply that their game never ends. The social landscape is perpetually fluid, driven by the infinite, iterative ripple effects of one's own local perception-action management on those of others, and vice versa.

Our conclusion, then, is that to get the comparative project back on track, we should stop attempting to tap into abstract knowledge inside an animal's head via field studies and experiments that merely massage our own folk-psychological rationalizations about how primate minds work. Instead, we should start taking ethology much more seriously, and tap into what primates (and other organisms) are actually doing in the world, and how they do it. By the same token, we need to take neuroethology more seriously, by which we mean taking it quite literally; we need to ground our ethological studies by reference to the work of those neurophysiologists, like Freeman, who have begun to get a grip on what (at least a small part of) the brain actually does, and how it does it. In other words, we need to understand more about the brain's behavior in the ethological sense (even though we may never understand its processes in strictly causal terms; Freeman & Skarda, 1990a,b), as a way of constraining our theories of primate cognition. In this way, we are less likely to forget that the brain is also a part of the body, and that its structure and dynamics will affect how an animal can engage with the world in the same way that other anatomical structures do. Put simply, we need to forget computer-based cognitive metaphors and focus on brain and behavior in a way that allows us to escape our anthropocentric frame of reference. In other words, we will need to get out of our minds as much (or more) as we will need to get deeper into them.

ACKNOWLEDGMENTS

Thanks to Peter Henzi for reading an earlier draft of this chapter and to John Vokey for many enlightening discussions of the issues raised here. Our work is supported by the National Sciences and Engineering Research Council of Canada (NSERC).

NOTE

1. Every day, one can see females approach mothers, pretend to be primarily interested in

grooming the mother when what they are really after is an opportunity to sniff, touch or hold her infant. . . . But is the mother really deceived? Surely the multiparous ones know exactly what s going on? (#88), Byrne and Whiten (1990: 41).

REFERENCES

Barrett, L., & Henzi, S. P. (2005). The social nature of primate cognition. *Proceedings of the Royal Society, 272,* 1865–1875.

Barrett, L., & Henzi, S. P. (2006). Monkeys, markets and minds. In: P. Kappeler & C. P. van Schaik (Eds.), *Cooperation in primates and humans: Mechanisms and evolution* (pp. 209–232). Berlin: Springer-Verlag, Berlin.

Barrett, L., Henzi, S. P., & Rendall, D. (2007). Social brains, simple minds: Does social complexity really require cognitive complexity? *Philosophical Transactions of the Royal Society, Series B, 362,* 561–575.

Bergman, T.J., Beehner, J.C., Cheney, D.L. & Seyfarth, R.M. (2003) Hierarchical classification by rank and kinship in baboons. *Science, 302,* 1234–1236.

Breazeal, C. L. (2002). *Designing sociable robots.* Cambridge, MA: MIT Press.

Brooks, R. (1999). *Cambrian intelligence: The early history of the new A.I.* Cambridge, MA: MIT Press.

Byrne, R. W., & Whiten, A. (1990). Tactical deception in primates: The 1990 database. *Primate Report, 27,* 1–101.

Catmur, C. Walsh, V., & Heyes, C. (2007). Sensorimotor learning configures the human mirror system. *Current Biology, 17,* 1527–1531.

Cheney, D. L., & Seyfarth, R. M. (2005). Constraints and preadaptations in the earliest stages of language evolution. *Linguistic Review, 22,* 135–159.

Churchland, P. (1996). *Neurophilosophy.* Cambridge, MA: MIT Press.

Clark, A. (1989). *Microcognition: Philosophy, cognitive science and parallel distributed processing.* Cambridge, MA: MIT Press.

Clark, A. (1997). *Being there: Putting brain, body and world back together again.* Cambridge, MA: MIT Press.

Cossart, R., Aronov, D., & Yuste, R. (2003). Attractor dynamics of networks UP states in the neocortex. *Nature, 423,* 283–288.

Cziko, G. (2000). *The things we do: Using the lessons of Barnard and Darwin to understand the what, how and why of our behaviour.* Cambridge, MA: MIT Press.

De Waal, F. B. M. (1997). Are we in anthropodenial? *Discover, 18,* 50–53.

De Waal, F. B. M. (2001). *The ape and the sushi master.* New York: Basic Books.

De Waal, F. B. M. (2005). Animals and us: Suspicious minds. *New Scientist, 186,* 48.

Dennett, D. (1989). *The intentional stance.* Cambridge, MA: MIT Press.

Dewey, J. (1896). The reflex arc in psychology. *Psychological Review, 3,* 357–370.

Dewey, J. (1929). *Experience and nature.* New York: Dover.

Draaisma, D. (2000). *Metaphors of memory: A history of ideas about the mind.* Cambridge: Cambridge University Press.

Dreyfus, H. (1992). *What computers still can't do.* Cambridge, MA: MIT Press.

Dunbar, R. I. M. (1998). The social brain hypothesis. *Evolutionary Anthropology, 6,* 178–190.

Freeman, W. J. (2001). *How brains make up their minds.* New York: Columbia University Press.

Freeman, W. J., & Skarda, C. A. (1990a). Representations: Who needs them? In: J. L. McGaugh, N. Weinberger, & G. Lynch (Eds.), *Brain organization and memory cells, systems and circuits* (pp. 375–380). New York: Oxford University Press.

Freeman, W. J., & Skarda, C. A. (1990b). Mind/brain science: Neuroscience on philosophy of mind. In: E. Lepore and R. van Gulick (Eds.), *John Searle and his critics* (pp. 115–127). Oxford: Blackwell.

Fuster, J. (2003) *Cortex and mind: unifying cognition.* New York: Oxford University Press.

Gabarini, F., & Adenzato, M. (2004). At the root of embodied cognition: Cognitive science meets neurophysiology. *Brain and Cognition, 56,* 100–106.

Gallese, V. (2001). The 'shared manifold' hypothesis: From mirror neurons to empathy. *Journal of Consciousness Studies, 8,* 33–50.

Gallese, V. (2003). The roots of empathy: The shared manifold hypothesis and the neural basis of intersubjectivity. *Psychopathology, 36,* 171–180.

Gallese, V. (2005). Embodied simulation: From neurons to phenomenal experience. *Phenomenology and the Cognitive Sciences, 4,* 22–48.

Gallese, V. (2007). Before and below 'theory of mind': Embodied simulation and the neural correlates of social cognition. *Philosophical Transactions of the Royal Society, Series B, 362,* 659–669.

Gallese, V., Keysers, C., & Rizzolatti, R. (2004) A unifying view of the basis of social cognition. *Trends in Cognitive Sciences, 8*: 396–403.

Gershon, M. (1998). *The second brain.* New York: Harper Collins.

Gibson, J. J. (1979). *The ecological approach to visual perception.* Hillsdale, NJ: Lawrence Erlbaum Associates.

Grice, P. (1957). Meaning. *Philosophical Review, 66,* 377–388.

Griffin, D. (1978). *The question of animal awareness: Evolutionary continuity of mental experience.* New York: The Rockefeller University Press.

Grossberg, S. (1987). Competitive learning: From interactive activation to adaptive resonance. *Cognitive Science, 11,* 23–63.

Grossberg, S. (1988). *Neural networks and natural intelligence.* Cambridge, MA: MIT Press.

Hauser, M. D. Chomsky, N., & Fitch, W. T. (2002). The faculty of language: What is it, who has it, and how did it evolve? *Science, 298,* 1569–1579.

Henzi, S. P., & Barrett, L. (2002). Infants as a commodity in a baboon market. *Animal. Behaviour, 63,* 915–921.

Hopfield, J. J. (1982). Neural networks and physical systems with emergent collective computational abilities. *Proceedings of the National Academy of Sciences, 79,* 2554–2558.

Humphrey, N. K. (1976). The social function of intellect. In: P. P. G. Bateson & R. A. Hinde (Eds.), *Growing points in ethology* (pp. 303–317). Cambridge: Cambridge University Press.

James, W. (1890). *Principles of psychology.* New York: Dover.

Johnson, C. (2001). Distributed primate cognition: A review. *Animal Cognition, 4,* 167–183.

Johnson, M., & Rohrer, T. (in press). We are live creatures: Embodiment, American pragmatism, and the cognitive organism. In: J. Zlatev, T. Ziemke, R. Frank, & R. Dirven (Eds.), *Body, Language, and Mind* (vol. 1, pp. 17–54). Berlin: Mouton de Gruyter.

Jolly, A. (1966). Lemur social behaviour and primate social intelligence. *Science, 153,* 501–506.

Kirsh, D. (1991). Today the earwig, tomorrow man? *Artificial Intelligence, 47,* 161–184.

Landauer, T. K., & Dumais, S. T. (1997). A solution to Plato's problem: the latent semantic analysis theory of acquisition, induction and representation of knowledge. *Psychological Review, 104,* 211–240.

Maris, M., & te Boekhorst, R. (1996). Exploiting physical constraints: Heap formation through behavioural error in a group of robots. In M. Asada (Ed.), *Proceedings of IROS '96: IEEE/RSJ International Conference on Intelligent Robots and Systems* (pp. 1655–1660).

Merleau-Ponty, M. (1942). *The structure of behaviour.* Pittsburgh: Duquesne University Press.

Michaels, C. F., & Carello, C. (1981). *Direct perception.* New Jersey: Prentice Hall.

Miller, G. A. (2003). The cognitive revolution: A historical perspective. *Trends in Cognitive Sciences, 7,* 141–144.

Owren, M. J., & Rendall, D. (1997). An affect conditioning model of non-human primate vocalizations. In D. W. Owings, M. D. Beecher, & N. S. Thompson (Eds.), *Perspectives in ethology, Vol. 12, communication* (pp. 299–346). New York: Plenum Press.

Owren, M. J., & Rendall, D. (2001). Sound on the rebound: Returning function to the forefront in understanding nonhuman primate vocal signaling. *Evolutionary Anthropology 10,* 58–71.

Owren, M. J., Rendall, D., & Bachorowski, J. (2003). Non-verbal vocal communication. In D. Maestripieri (Ed.), *Primate psychology: Bridging the gap between mind and behaviour in human and non-human primates* (pp. 359–394). Cambridge: Harvard University Press.

Penn, D. C.. Holyoak, K. J., & Povinelli, D. J. (2008). Darwin's mistake: Explaining the discontinuity between human and nonhuman minds. *Behavioral and Brain Sciences, 31,* 109–130.

Pert, C. B. (1997). *Molecules of emotion.* New York: Scribner.

Pfeifer, R., & Bongard, J. (2007). *How the body shapes the way we think.* Cambrdige, MA: MIT Press.

Pfeifer, R., & Scheier, C. (1999). *Understanding intelligence.* Cambrdige, MA: MIT Press.

Reed, E. (1996). *Encountering the world: Toward an ecological psychology.* Oxford: Oxford University Press.

Rendall, D., Notman, H., & Vokey, J. R. (2008). Quotidian cognition and the human-nonhuman divide: Just more or less of a good thing? *Behavioral and Brain Sciences, 31,* 144–145.

Rendall, D., Cheney, D. L., & Seyfarth, R. M. (2000). Proximate factors mediating contact calls in adult female baboons and their infants. *Journal of Comparative Psychology, 114,* 36–46.

Rockwell, W. T. (2005). *Neither brain nor ghost: A nondualist alternative to the mind-brain identity theory.* Cambridge, MA: MIT Press.

Rumbaugh, D. M., & Washburn, D. A. (2003). *Intelligence of apes and other rational beings.* New Haven: Yale University Press.

Rumbaugh, D. M., Washburn, D. A., & Hillix, W. A. (1996). Respondents, operants, and emergents: Toward an integrated perspective on behavior. In K. Pribram & J. King (Eds.), *Learning as a self-organizing process* (pp. 57–73). Hillsdale, NJ: Lawrence Erlbaum.

Scott-Kelso, J. A. (1995). *Dynamic patterns: The self-organization of brain and behaviour.* Cambridge, MA: MIT Press.

Seyfarth, R. M., Cheney, D. L., & Marler, P. (1980). Monkey responses to three different alarm calls: Evidence of predator classification and semantic communication. *Science, 210,* 801–803.

Silk, J.B, Rendall, D., Cheney, D.L. & Seyfarth, R.M. (2003) Natal attraction in adult female baboons (*Papio cyncocephalus ursinus*) in the Moremi reserve, Botswana. *Ethology, 109,* 627–644.

Skarda, C. A., & Freeman, W. J. (1987). Brains make chaos to make sense of the world. *Behavioral and Brain Sciences, 10,* 161–195.

Skarda, C. A., & Freeman, W. J. (1990). Chaos and the new science of the brain. *Concepts in Neuroscience, 1,* 275–285.

Strum, S. C. (1982)Agonistic dominance in male baboons: An alternative view. *International Journal of Primatology, 3,* 175–202.

Strum, S. C. (1994). Reconciling aggression and social manipulation as means of competition: 1. Life history perspective. *International Journal of Primatology, 15,* 739–765.

Strum, S. C., Forster, D., & Hutchins, E. (1997). Why Machiavellian intelligence may not be Machiavellian. In A. Whiten and R. W. Byrne (Eds.), *Machiavellian intelligence II: Extensions and evaluations.* Cambridge: Cambridge University Press.

Suomi, S. J., & Harlow, H. F. (1975). Effects of differential removal from group on social development of rhesus monkeys. *Journal of Child Psychology and Psychiatry, 16,* 149–164.

Van Gelder, T. (1995). What might cognition be, if not computation? *Journal of Philosophy, 92,* 345–81.

Von Uexküll, J. (1957). A stroll through the worlds of animals and men. In: C. H. Schiller & K. S. Lashley (Eds.), *Instinctive behaviour: The development of a modern concept* (pp. 5–82). New York: International University Press. (Original work published in 1934).

Wheeler, M. (2005). *Reconstructing the cognitive world: The next step.* Cambridge, MA: MIT Press.

Zuberbühler, K. (2003). Referential signaling in nonhuman primates: Cognitive precursors and limitations for the evolution of language. *Advances in the Study of Behavior, 33,* 265–307.

CHAPTER 29

The Comparative Neuropsychology of Tool Use in Primates with Specific Reference to Chimpanzees and Capuchin Monkeys

William D. Hopkins

The human brain is roughly three times larger than it should be for a species of our body weight. In addition, the evolution of the brain has not been uniform, with distinct cortical and subcortical regions being selected for in different species within mammals, and specifically within primates (Deacon, 1997; Finlay & Darlington, 1995; Rilling, 2006; Semendeferi & Damasio, 2000; see also chapters by Kaas and Preuss, this volume). For example, in primates, some have suggested there has been differential expansion of the frontal and temporal lobes relative to the other lobes and that these changes might reflect specific selection for cortical development in brain regions associated with complex cognition, including language (Deacon, 2004; Rilling & Seligman, 2002; Semendeferi et al., 2001, 2002).

In addition to relative brain size, the human brain is also highly lateralized with specific sensory, motor, and cognitive processes differentially performed by the left and right cerebral hemispheres, respectively (Bradshaw & Rogers, 1993; Corballis, 1992; Galaburda, 1995). Two notably lateralized functions in the human brain are handedness and language. Although there is some cultural variation, right-handedness is considered a universal trait in humans. Similarly, language is considered a universal trait in the human species and is predominantly lateralized to the left hemisphere, particularly among right-handed individuals (Beaton, 1997; Knecht et al., 2000). Because right-handedness and language are both human universal traits and predominantly lateralized to the left hemisphere, many have suggested that these two functions coevolved and are possibly unique to hominid evolution (Corballis, 1992; Crow, 2004; Williams et al., 2006). Whether nonhuman animals, particularly primates, exhibit population-level behavioral or brain asymmetries remains a strongly contentious issue in the scientific literature (Crow, 1998; Hopkins, 2006, 2007; McGrew & Marchant, 1997a; Palmer, 2002).

What ecological, psychological, or sociological factors contributed to the evolutionary expansion of the human brain remains a topic of considerable debate. The human brain is metabolically expensive, suggesting that there must have been strong selection pressure for its expansion related to specific dietary, ecological, or social factors. From the standpoint of ecological factors, it has been suggested that cognitive and motor abilities that facilitated food extraction and consumption may have selected for the development of the brain. Notably, some have suggested that extractive foraging, those behaviors that include the use of tools, would have provided certain nutritive and dietary advantages (Gibson & Ingold, 1993; Greenfield, 1991; Parker & Gibson, 1977). The use of tools would have similarly selected for increased motor planning, motor skill, and coordination of the limbs in the execution of certain behaviors. Thus, some have suggested that the evolution of tool use may have served as a preadaptation for the evolution of complex cognition, language, and handedness in humans (Bradshaw & Rogers, 1993; Frost, 1980). One aim of this chapter is to summarize data on tool use in primates and its neuropsychological correlates, at both the phylogenetic and individual levels of analysis.

The antithesis of the ecological arguments is the claim that social cognition was selected for in primate evolution (Dunbar, 1996). Primates live in complex social groups, and some have suggested that keeping track of and understanding the nuances of group-living primates, such as dominance ranks, kin, etc., place significant cognitive demands on the individuals within the group. A number of studies have shown that brain size in primates correlates with group size, and even their ability to engage in deception (a complex social cognitive ability implying theory of mind) (Byrne & Corp, 2004).

The basic theoretical frameworks described previously characterize primate cognition primarily into two domains, physical cognition, including complex object manipulation and tool use, and social cognition, which would include abilities such as social learning and imitation. Interestingly, functional imaging studies in human and nonhuman primates are beginning to report results that implicate common neural systems in the use of both social and nonsocial cognition, thereby blurring the distinction between these two abilities. For example, Rizzolatti and colleagues have reported the existence of mirror neurons in the homolog to Broca's area in the monkey brain (area F5) when monkeys either watch or engage in a reach-and-grasp task, a task presumably requiring little if any social cognition (Rizzolatti & Arbib, 1998). Similar mirror neuron activation can be found in humans when watching or engaging in reach-and-grasp responding (Grafton et al., 1996), but these same mirror neurons also seem to be involved in tasks involving imitation or attribution of mental state. With specific reference to tool use, similar parallels can be seen in terms of neuropsychological correlates in linguistic functions relative to tool use. For instance, clinical studies suggest that lesions to the left but not right frontal cortex can induce both aphasia (speech deficit) and apraxia (motor planning deficit) (Meador et al., 1999). Similarly, functional imaging studies have shown that there are common neural systems involved in both speech and planned tool use actions, suggesting that common cognitive mechanisms may underlie their execution, despite being metrically quite distinct actions (Johnson-Frey, 2004; Lewis, 2006). These kinds of data should remind us that there are likely parallels between so-called "social" and "nonsocial" cognitive abilities in primates, despite the convenience of treating them as separate domains. In the latter part of this chapter, I discuss some examples of where there might be overlap in cognitive processes in tool use relative to intraspecies communication in chimpanzees.

TOOL USE IN PRIMATES

Tool use has been described in many species (Beck, 1980; Candland, 1987; Tomasello & Call, 1997) but, save humans, the complexity and variability in tool-using abilities of some great ape species are relatively unmatched in the animal kingdom. Recently, Whiten and colleagues (1999) provided a summary of the evidence of tool use in wild chimpanzees from different sites in Eastern, Central, and Western Africa (Table 29.1). As can be seen, the forms of tool use are quite variable between sites and are not easily attributable to local ecological factors. For instance, chimpanzees at Gombe termite-fish, leaf-groom, and use objects for self-tickling, none of which are seen at other sites such as Tai and Bossou. In contrast, chimpanzees at Bossou and Tai Forest use hammers to open nuts whereas Gombe chimpanzees do not, despite the fact that the same rocks and nuts are available. Based on this conclusion, Whiten and colleagues (1999) argued that the diversity and variability in tool use in chimpanzees may reflect a form of culture. Not to be outdone, following the reports of geographical variation in tool use in chimpanzees, Van Schaik and colleagues (2003) examined variability in tool use behaviors of wild orangutans from Borneo and Sumtra. Like chimpanzees, the orangutans showed significant variability in tool use in the different study sites, and some of the variability could not be attributed to ecological factors.

TABLE 29.1 **Variation in Occurrence of Behavior Patterns Across Long-Term Study Sites**

Site			Bs	Ta	Go	Ma	Mk	Kib	Bd
A	1	Investigatory probe (probe and sniff)	H	C	C	H	H	+	(-)
	2	Play start (invite play holding stem in mouth)	+	H	C	C	C	C	H
	3	Drag branch (drag large branch in display)	H	C	C	C	C	H	H
	4	Leaf-sponge (leaf mass used as sponge)	C	C	C	+	e	C	C
	5	Branch-clasp (clasp branch above, groom)	H	C	C	C	C	C	C
	6	Branch-shake (to attract attention, court)	C	C	C	C	C	C	C
	7	Buttress-beat (drum on buttress of tree)	C	C	C	C	C	C	C
B	8	Nasal probe (clear nasal passage with stick)	-	-	-	+	-	-	—
	9	Comb (stem used to comb through hair)	-	-	-	-	-	-	+
	10	Insect-pound (probe used to mash insect)	+	-	-	-	-	-	-
	11	Resin-pound (extract resin by pounding)	+	-	-	e?	e?	-	-
	12	Branch-hook (branch used to hook branch)	+	-	-	-	-	-	-
	13	Perforate (stout stick perforates termite nest)	-	e	-	-	-	e	e?
	14	Dig (stick used as spade to dig termite nest)	+	e	-	-	-	e	e?
	15	Brush-stick (probing stick with brush end)	-	-	-	-	-	-	-
	16	Seat-stick (stick protection from thorns)	-	-	e	e?	e?	e	e
	17	Stepping-stick (walking on sticks over thorns)	-	-	e	e?	e?	e	e
	18	Container (object used as container)	-	-	+	-	-	-	-
	19	Leaf-mop (leaves used to mop up insects)	-	-	+	-	+	e	e?
	20	Leaf-wipe (food wiped from skull, etc.)	e?	+	+	-	-	-	-
	21	Leaf-brush (leaf used to brush away bees)	-	-	+	-	-	-	-
	22	Open and probe (perforate, then probe)	-	-	-	-	-	-	-
	23	Sponge push-pull (stick and sponge tool)	+	+	+	+	e	e	-
C	24	Algae-scoop (scoop algae using wand)	C	e	e	e	e	e	e
	25	Ground night-nest (night-nests on ground)	(-)	e?	+	e?	e?	e?	+
	26	Anvil-prop (rock used to level anvil)	H	e	e	e	e	e	e
D	27	Food-pound onto wood (smash food)	C	C	C	-	-	e?	H
	28	Food-pound onto other (such as stone)	-	H	C	-	-	e?	-
	29	Nut-hammer, wood hammer on wood anvil	-	C	-	e	e	e?	e
	30	Nut-hammer, wood hammer on stone anvil	-	C	-	-	-	e?	e
	31	Nut-hammer, stone hammer on wood anvil	+	C	-	e	e	e?	e
	32	Nut-hammer, stone hammer on stone anvil	C	C	-	-	-	e?	e
	33	Nut-hammer, other (such as on ground)	-	H	-	-	-	e?	e
	34	Pestle-pound (mash palm crown with petiole)	C	-	-	e?	e?	e?	e?
	35	Club (strike forcefully with stick)	+	H	H	+	-	+	-
	36	Termite-fish using leaf midrib	+	e	-	-	C	e	e?
	37	Termite-fish using nonleaf materials	-	e	C	-	C	e	e?
	38	Ant-fish (probe used to extract ants)	+	-	+	C	C	-	-
	39	Ant-dip-wipe (manually wipe ants off wand)	+	-	C	-	-	-	-
	40	Ant-dip-single (mouth ants off stick)	C	C	+	-	-	-	-
	41	Fluid-dip (use of probe to extract fluids)	-	C	C	H	H	H	-
	42	Bee-probe (disable bees, flick with probe)	-	C	-	-	+	-	-
	43	Marrow-pick (pick bone marrow out)	-	C	-	-	-	-	-
	44	Lever open (stick used to enlarge entrance)	-	H	C	-	-	-	-
	45	Expel/stir (stick expels or stirs insects)	-	C	H	H	H	-	-
	46	Seat-vegetation (large leaves as seat)	+	H	-	-	-	+	-
	47	Fly-whisk (leafy stick used to fan flies)	-	H	+	-	-	-	H
	48	Self-tickle (tickle self using objects)	-	-	H	-	-	-	-

(*continued*)

Table 29.1 (Continued)

Site		Bs	Ta	Go	Ma	Mk	Kib	Bd
49	Aimed-throw (throw object directionally)	C	C	C	C	-	+	+
50	Leaf-napkin (leaves used to clean body)	-	+	C	+	-	C	C
51	Leaf-dab (leaf dabbed on wound, examined)	-	+	+	-	-	C	-
52	Leaf-groom (intense "grooming" of leaves)	-	-	C	C	C	C	+
53	Leaf-clip, mouth (rip parts off leaf, with mouth)	C	C	-	C	C	H	C
54	Leaf-clip, fingers (rip leaf with fingers)	-	H	-	+	-	H	C
55	Leaf-strip (rip leaves off stem, as threat)	+	-	H	+	-	H	-
56	Leaf-squash (squash ectoparasite on leaf)	-	-	H	?	?	-	-
57	Leaf-inspect (inspect ectoparasite on hand)	-	-	+	?	?	-	C
58	Index-hit (squash ectoparasite on arm)	-	C	+	-	-	-	-
59	Hand-clasp (clasp arms overhead, groom)	-	H	-	C	C	C	-
60	Knuckle-knock (knock to attract attention)	+	C	H	C	C	-	-
61	Branch-din (bend, release saplings to warn)	-	-	-	-	-	-	-
62	Branch-slap (slap branch, for attention)	C	C	-	+	-	-	C
63	Stem pull-through (pull stems noisily)	C	-	+	H	-	H	-
64	Shrub-bend (squash stems underfoot)	H	-	-	C	-	-	C
65	Rain dance (slow display at start of rain)	-	H	C	C	C	C	H

A, patterns absent at no site; B, patterns not achieving habitual frequencies at any site; C, patterns for which any absence can be explained by local ecological factors; D, patterns customary or habitual at some sites yet absent at others, with no ecological explanation. To facilitate comparison, behaviors are listed so that adjacent categories share broad functions; in band D these are: 27–35, pounding actions; 36–40, fishing; 41–43, probing; 44 and 45, forcing; 46 and 47, comfort behavior; 48 and 49, miscellaneous exploitation of vegetation properties; 50–57, exploitation of leaf properties; 56–59, grooming; 60–64, attention-getting. Sites (with subspecies, observation period in years by September 1998, site director): Bs, Bossou, Guinea (*verus*, 23, Y.S.); Ta, Taï Forest, Ivory Coast (*verus*, 23, C.B.); Go, Gombe, Tanzania (*schweinfurthii*, 38, J.G.); Ma, Mahale M-group, Tanzania (*schweinfurthii*, 30, T.N.); Mk, Mahale K-group (*schweinfurthii*, 18, T.N.); Kib, Kibale Forest, Uganda (*schweinfurthii*, 11, R.W.W.); Bd, Budongo Forest, Uganda (*schweinfurthii*, 8, V.R.). C, customary; H, habitual; +, present; -, absent; e, absent with ecological explanation; e?, ecological explanation suspected; (-), absent possibly because of inadequate observation; ?, answer uncertain (see text for full definitions). Branch din (behavior 61) is allocated to band D because it is known to be customary at Lopé, Gabon (C.E.G.T.); behaviors 13, 15–17, and 22 are allocated to band B because they have been recorded at shorter-term sites (see Supplementary Information). For full definitions of all behaviors, see Supplementary Information.

Reprinted by permission from MacMillan Publishers Ltd.: Whiten, A., Goodall, J., McGrew, W. C., Nishida, T., Reynolds, V., Sugiyama, Y., et al. (1999). Cultures in chimpanzees *Nature, 399*, 682–685.

Some of the tool-using and other social traditions included the use of specific materials to amplify vocalizations, use of materials to shield themselves from the sun and rain, and use of probing tools to extract food items (Table 29.2).

Interestingly, unlike orangutans and chimpanzees, neither gorillas nor bonobos are well known for their tool-using abilities in the wild. Some limitations of our knowledge of bonobo tool use are hampered by a lack of data and opportunity for sustained and long-term study, but this cannot be said for gorillas, where there have been several long-term field projects yet little reports of tool use.

In captivity, the evidence of tool use is somewhat different with respect to the great apes. Not surprisingly, chimpanzees and orangutans readily use tools in captivity but, interestingly, so do gorillas and bonobos, largely in the context of foraging tasks used as enrichment in zoos or research settings (Boysen et al., 1999; Harrison & Nystrom, 2008; Jordon, 1982; Nakamichi, 1999). Thus, cognitively, gorillas and bonobos are capable of learning to use tools, but it seems that either ecological or social factors may limit the expression of this ability in the wild.

Most other reports of tool use in nonhuman primates have largely been restricted to captive environments (Beck, 1980; Candland, 1987;

TABLE 29.2 Geographic Variation in Orangutan Behavior Patterns

	Site and Island					
	Gunung Palung (Borneo)	Tanjung Putnig (Borneo)	Kutai (Borneo)	Lower Kinabatangan (Borneo)	Leuser, Ketambe (Sumatra)	Leuser, Suaq Balimbing (Sumatra)
Observation intensity (increased ranks):	2	2	1	1	2	1
Very likely cultural variants						
1. Snag riding: riding on pushed–over snag as it falls, then grab onto vegetation before it crashes on ground	A	C	A	A	A	A
2. Kiss-squeak with leaves: using leaves on mouth to amplify sound, then drop leaf	C	A	H	A	A	A
3. Kiss-squeak with hands: using fists (like trumpet) or flat hands on mouth to amplify sound	R	R	H	A	C	H
4. Leaf wipe: wiping face with fistful of squashed leaves, then drop (in kiss-squeak context)	A	C	A	A	A	A
5. Play nests: building nest for social play (no resting occurs)	C	C	P	A	C	H
6. Bunk nests: building nest a short distance above the nest used for resting (during rain)	A	P	A	H	A	A
7. Suncover: building cover on nest during bright sunshine (rather than rain)	A	?	C	C	H	A
8. Hide under nest: seeking shelter under nest for rain	A	R	C	P	R	A
9. Scratch stick: using detached stick to scratch body parts	A	R	H	A	A	A
10. Autoerotic tool: using tool for sexual stimulation (female and male)	A	A	P	A	C	A

(*continued*)

Table 29.2 (Continued)

	Site and Island					
	Gunung Palung (Borneo)	Tanjung Putnig (Borneo)	Kutai (Borneo)	Lower Kinaba-tangan (Borneo)	Leuser, Ketambe (Sumatra)	Leuser, Suaq Balimbing (Sumatra)
11. Raspberry: spluttering sounds associated with nest building	A	A	A	H	A	C
12. Symmetric scratch: exaggerated, long, slow, symmetric scratching movements with both arms at same time	A	A	A	A	R	C
13. Twig biting: systematically passing ends of twigs used for lining of nest past the mouth (sometimes including actual bites) during last phase of nest building	A	A	A	A	A	C
14. Leaf napkin: using handful of leaves to wipe latex off chin	A	A	C	A	A	A
15. Branch as swatter: using detached leafy branches to ward off bees/wasps attacking subject (who is usually raiding its nest)	R	A	H	H	H	H
16. Leaf gloves: using leaf gloves to handle spiny fruits or spiny branch or as seat cushions in trees with spines	A	R	A	A	H	E
17. Tree-hole tool use: using tool to poke into tree holes to obtain social insects or their products	A	A	A	A	A	C
18. Seed extraction tool use: using tool to extract seeds from the protected fruits of *Neesia* sp.	A	A	E	A	E	C
19. Branch scoop: drinking water from deep tree hole using leafy branch (water dripping from leaves)	A	A	A	A	A	H
Likely cultural variants (ecological explanation not excluded)						

20. Snag crashing: aimed pushing of dead standing trees	C	C	C	A	H	C
21. Bouquet feeding: using lips to pick ants from fistful of dry, fresh, or rotting leaves (nests)	C	C	A	R	C	C
22. Nest destruction: rummaging through old orangutan nests for insects	H	C	P	A	H	H
23. Dead twig sucking: breaking hollow (dead) twigs to suck ants from inside	A?	C	A?	A	C	C
24. Slow loris eating: capturing and eating slow loris hiding in dense vegetation	A	A	A	A	H	H
Rare behaviors						
1. Females rubbing their genitals together	R	1R	A	A	A	R
2. Using leaf to clean body surface	R	A	A	A	A	A
3. Sneaky nest approach: building series of nests while approaching conspecific and fruit tree	R	A	A	A	A	A
4. Leaf bundle while sleeping ("doll")	R	R	A	A	A	A
5. Leaf scoop: drinking water from the ground, using leaf as vessel (drinking straight from vessel)	R	A	A	A	A	A
6. Bridge nest: building nest connecting two trees on opposite banks of river	A	R	A	A	A	A
7. Biting through vine to swing Tarzan-like across gap	A	R	A	A	A	R
8. Artistic pillows: similar twigs lining nest	A	P	?	A	?	?
9. Branch dragging display on ground	A	A	?	R	A	A
10. Stick as chisel to open termite nest in log on ground	A	A	A	A	R	A
11. Sponging: drinking water using crumpled leaves	A	A	A	A	R	A
12. Hiding behind detached branch from predators or humans	A	R	P	R	R	A

C, customary; H, habitual; R, rare; P, present with unknown frequency, probably rare; E, absent for ecological reasons; A, absent; ?, unknown.

From Van Schaik, C. P., Ancrenaz, M., Borgen, G., Galdikas, B., Knott, C. D., Singleton, I., et al. (2003). Orangutan culture and the evolution of material culture. *Science, 299,* 102–205. Reprinted with permission from AAAS.

Tomasello & Call, 1997), with the exception of capuchin monkeys. There are numerous reports of tool use in captive capuchin monkeys that include probing actions, nut-cracking, and the use of sponges, behaviors not unlike those reported in wild and captive apes (McGrew & Marchant, 1997b; Visalberghi, 1990). Though originally thought to be an artifact of captivity, recent studies have reported evidence of tool use in wild capuchin monkeys. Specifically, in certain dry, arid environments, wild capuchins have been documented to use rocks to pound open nuts (Fragaszy et al., 2004; Moura & Lee, 2004) (Fig. 29.1). There are also some anecdotal reports of capuchin monkeys using leaves to sponge water from tree trunks (Phillips, 1998). Thus, tool use in captive capuchin monkeys does not appear to be an artifact of captivity but rather an inherent ability expressed in both settings.

COGNITIVE MECHANISMS GOVERNING TOOL USE

The cognitive processes underlying tool use in nonhuman primates has historically been and continues to be a topic of intense discussion. In terms of phylogeny, some have suggested that differences in tool-using abilities between great apes and monkeys are due to basic differences in mental representation and cognition. For example, differences in the attainment of stages of cognitive development, such as those described by Piaget, among primates has been suggested as an explanation for phylogenetic differences in tool use (Dore & Dumas, 1987). Thus, on certain Piagetian object permanence or other tasks, chimpanzees reach stage 5 or 6 performance, whereas other nonhuman primates

Figure 29.1 Photograph of a wild capuchin monkey using a stone tool. Reprinted from Fragaszy, D., Izar, P., Visalberghi, E., Ottoni, E. B., & Oliveira, M. G. (2004). Wild capuchin monkeys (Cebus libidinosus) use anvils and stone pounding tools. *American Journal of Primatology, 64,* 359–366. Used with permission.

do not reach these same levels. The differences in object permanence performance between apes and monkeys presumably underlie their abilities to acquire and use tools (Parker & Gibson, 1977). Some have been critical of the neo-Piagetian explanations and have suggested that differences in manipulatory proclivities are sufficient to explain tool-using abilities (Fragaszy, 1989). For instance, capuchin monkeys and chimpanzees engage in similar types of object manipulation and object-object relational behavior, despite differences in their performance on Piagetian tasks, and arguably this can more easily explain the tendencies for these species to excel at tool use in captivity and the wild.

It has also been suggested that the endpoint of tool use as a cognitive process can be achieved by different mechanisms among primate species. In a series of elegant studies, Visalberghi and colleagues (Limongelli et al., 1995; Visalberghi & Limongelli, 1994) devised a task that required that chimpanzee and capuchin monkeys solve a task, referred to as the "trap-tube" task. In these studies, food is placed inside a transparent tube in the middle so that the food cannot be reached with the use of the hand (Fig. 29.2A). Instead, the subjects must insert a stick and push out the food from the tube. Apes and monkeys had little difficulty learning this task but the key manipulation was the insertion of a transparent trap, adjacent to the location of the food inside the tube. Thus, inserting the stick from the wrong end of the tube and pushing out the food would result in the food falling into the trap, whereas inserting from the opposite end would allow the food to be successfully pushed out of the tube. In essence, the question was whether the subjects understand the causal relationship between their actions and consequences associated with the removal of the food. The results indicated that most chimpanzees were able to accommodate the trap problem relatively easily and without prolonged training. In contrast, the capuchin monkeys performed poorly on these tasks and did not benefit from prolonged training. The interpretation of these findings is that different cognitive mechanisms associated with understanding causality explain variability in the capuchin and ape performance. Thus, capuchin monkeys and apes can learn to use tools but do so by different cognitive mechanisms.

A variant of the "trap-tube" task that has also been used for testing primate abilities to infer causality is the "cane-pulling" task. Like the

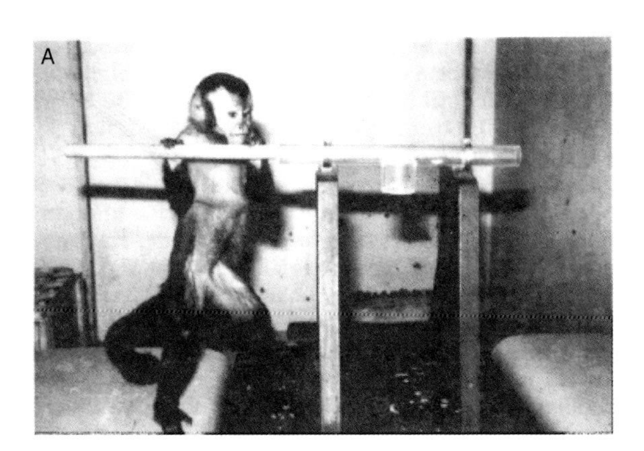

From Visalberghi & Limongelli, 1994

From Santos et al. 2005

Figure 29.2 Example photographs of **(A)** capuchin monkeys solving the trap-tube task and **(B)** a ring-tailed lemur solving the cane-pulling task. Reprinted from Visalberghi, E., & Limongelli, L. (1994). Lack of comprehension of cause-effect relations in tool-using capuchin monkeys (*Cebus apella*). *Journal of Comparative Psychology, 108,* 15–22; and Santos, L. R., Mahajan, N., & Barnes, J. L. (2005). How prosimian primates represent tools: Experiments with two lemur species (*Eulemur fulvus* and *Lemur catta*). *Journal of Comparative Psychology, 119,* 394–403. Used with permission.

trap-tube task, the cane-pulling task was designed to test different primate species' abilities about understanding causality, but it places far fewer motor demands on subjects; thus, many more species with less developed motor skills can be tested for their understanding of causality. Basically, in the cane-pulling task, there are two (or more) sticks that are curved or have hooks at the end of them. Initially during training, a food reward is baited at the hooked end of one of the two canes and the subjects must learn to pull the correct cane in order to receive the reward (Fig. 29.2B). A number of species including lemurs, marmosets, tamarins, capuchin monkeys, gibbons, and chimpanzees have been tested on the cane-pulling task, and all of them can learn to solve this task and adjust their responses to various perceptual variants of the cane properties (Cunningham et al., 2006; Fujita et al., 2003; Hauser, 1997; Povinelli, 2000; Santos et al., 2005). Thus, to some extent, the results from the cane-pulling task are not entirely consistent with the results obtained in the trap-tube task, and it remains unclear why this is the case.

In my view, several important additional studies are needed to flush out the relationship between these two tasks (and perhaps others). Notably, what is needed are data on the trap-tube and cane-pulling tasks from the same individuals to assess whether variation in performance on one task can explain performance on the second. If this was found, it would suggest that common cognitive mechanisms underlie the two tasks, which seems to be the underlying assumption but for which there is no empirical support. I also believe that another important question with respect to the cane-pulling task is how primates would use this device once they have mastered the basic task. In other words, if primates learn to use the cane-pulling task in a certain context, would they use the cane in other contexts where its use would be functionally beneficial?

NEURAL CORRELATES OF TOOL USE IN PRIMATES

From the standpoint of understanding the neural correlates of the cognitive and motor demands of tool use and complex object manipulation, there is not a great deal of data and the paucity of comparative data makes any analysis difficult. Comparatively, it has been well documented that more closely related primates show greater levels of prehensile grasping, individual control of the digits, and complex object manipulation skills, particularly in the combinatorial use of objects (Christel, 1994; Torigoe, 1985). Global correlational analyses have shown that increasing brain size is associated with increasing incidences of tool use and other cognitive abilities (Deaner et al., 2007; Reader & Laland, 2002), but these results are not very specific with regard to identifying brain regions that were directly selected for in the evolution of tool use or related motor functions. In the following section, we present data on studies that have assessed neuropsychological and neuroanatomical correlates of individual and species differences in tool use or other complex motor actions.

PHYLOGENETIC CHANGES IN INTERHEMISPHERIC CONNECTIVITY AND THE CEREBELLUM

Van Schaik and colleagues (1999) have suggested that one important motor component associated with tool use and other complex motor actions is the asymmetrical use of the hands in a coordinated manner. Presumably, coordinated actions of hands would facilitate certain functions because of the differential roles played by each hand during specific actions (such as feeding, plant processing, or tool use) and would be facilitated by brain structures that promote interhemispheric transfer of information between the left and right hemispheres.

Beyond changes in the size of the brain, the organization and interhemispheric connectivity of the brain has changed in primate evolution. Rilling and Insel (1999) have examined the association between the evolution of brain size, neocortical surface area, and corpus callosum (CC) size. The CC is a major set of largely homotopic fibers that connect the two hemispheres. In the analysis by Rilling and Insel (1999), they reported that humans and great apes have relatively smaller CC values, after adjusting for brain size, compared to Old World and New World monkeys (Fig. 29.3). Thus, there are significant grade shifts in the relative size of the CC in different primate species. Rilling and Insel (1999) argued from these results that as the brain got larger, the two hemispheres became increasingly isolated and asymmetries in structure and function subsequently evolved from that separation.

Germane to this paper is the variation in relative CC size within the different taxonomic families (Fig. 29.3) as it relates to coordinated bimanual actions, including tool use, as suggested by Van Schaik and colleagues (1999). For example, within New World monkeys, *Cebus* have a smaller ratio in size of the CC to brain volume compared to *Saimiri*, consistent with differences in their tool using abilities and manipulatory propensities. The data are less clear within the other families though. For example, within the great apes, the CC-to-brain volumes ratios are comparable between the apes, even though the chimpanzees and orangutans are more pronounced tool users. However, all four ape species do show high complementary bimanual manipulation but in very different contexts, such as feeding compared to tool use (Byrne & Byrne, 1991). Lastly, within the Old World monkeys, rhesus monkeys have a relative high CC ratio value compared to mangabeys and baboons. How these differences might relate to variation in coordinated bimanual actions remains unclear because of the paucity of behavioral data from these species.

Another interesting brain area that has been examined comparatively in primates and is potentially relevant to the evolution of tool use and complex manipulation is the cerebellum (MacLeod et al., 2003; Rilling & Insel, 1998) (Fig. 29.4). Although historically considered to be only important for coordination of motor actions and motor learning, increasingly functional imaging studies in humans have shown that the cerebellum is involved in complex

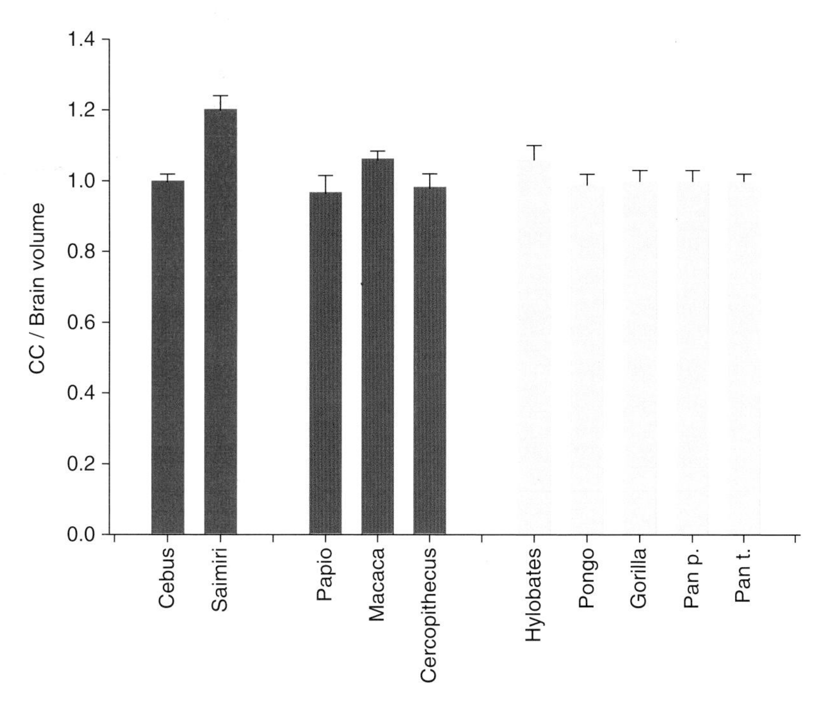

Figure 29.3 Mean ratio in the size of the corpus callosum (CC) to brain volume in different primate species. Data regraphed from Rilling, J. K., & Insel, T. R. (1999). Differential expansion of neural projection systems in primate brain evolution. *NeuroReport, 10,* 1453–1459.

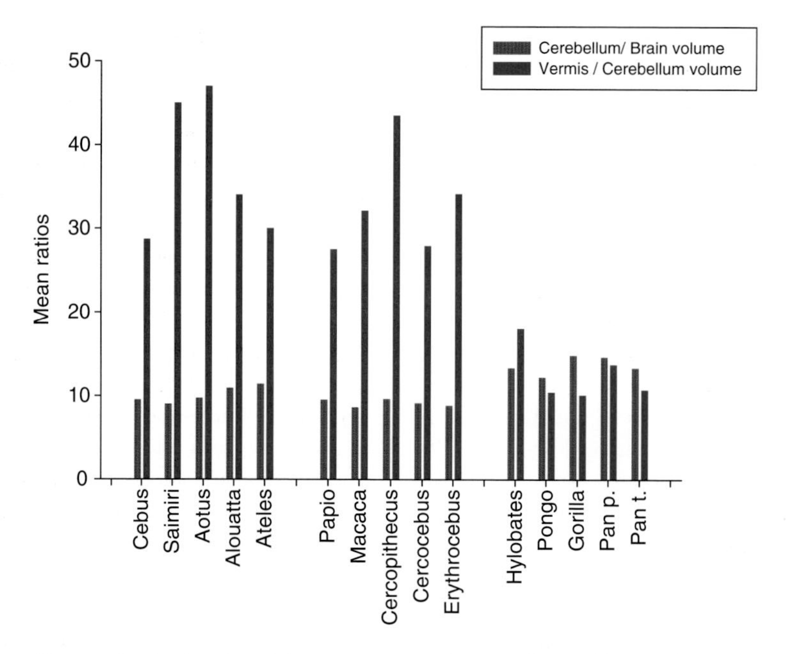

Figure 29.4 Mean ratios in the size of the cerebellum to brain volume (blue) and vermis to cerebellum volume (red) in different primate species. Data regraphed from MacLeod, C. E., Zilles, K., Schleicher, A., Rilling, J. K., & Gibson, K. R. (2003). Expansion of the neocerebellum in Hominoidea. *Journal of Human Evolution, 44,* 401–429.

cognitive and motor functions, including language and speech (Leiner et al., 1993; Ramnani, 2006). A couple of striking results emerge from these analyses. First, like the CC ratio data, there are significant grade shifts in the relative size of the cerebellum in primates; specifically, greater and lesser apes have relatively large cerebellar volumes compared to Old and New World monkeys, after adjusting for differences in brain size. Second, in the specific measurement of the vermis, a central structure within the cerebellum, great apes have relatively small regions, suggesting that the evolutionary expansion of the cerebellum has been in more lateral regions, areas known to be involved in the coordination and planning of learned motor actions (MacLeod et al., 2003). Third, within the major grades, there is some variability that may or may not explain variability in tool-using or motor learning abilities at some level of analysis. For example, within New World monkeys, *Cebus* are the only tool-using primate and they have the relatively smallest amount of vermis (and therefore largest expansion of the cerebellar hemispheres). Similarly, within Old World monkeys, *Papio* have the relatively smallest vermis and there are certainly more reports of tool use in this genus compared to other Old World monkeys. Lastly, among apes, there are no real discernable differences but again, it is important to emphasis that all the apes show complex object manipulation and motor learning in both the wild and captivity.

INDIVIDUAL DIFFERENCES IN TOOL USE: NEUROANATOMICAL AND BEHAVIORAL CORRELATES

Cortical Asymmetries

The advent of modern imaging technologies has advanced our understanding of primate neuroanatomy and its functional correlates in nonhuman primates. The use of structural (and to a lesser extent functional) magnetic resonance imaging (MRI) has allowed for investigators to examine individual differences in brain-behavior relationships as well as comparative differences in the organization of different regions of the cerebral cortex.

The use of MRI as it relates to handedness and tool use has received some attention recently in the literature. In particular, recent structural and functional MRI studies in humans have shown that there is region within the precentral gyrus, referred to as the "knob," which corresponds to the area in the motor cortex where the hand is represented. The knob is an omega-shaped landmark and is presumably created when tissue from the precentral gyrus extends posteriorly into the central sulcus and has been reported to be larger in hemisphere contralateral to the subjects' preferred hand (Hammond, 2002; Yousry et al., 1997).

Like humans, in chimpanzees the knob is larger in the hemisphere contralateral to the subjects' preferred hand for coordinated bimanual actions as measured by the tube task (Hopkins & Cantalupo, 2004). As noted previously, the knob cannot be readily identified in Old or New World monkeys but the motor-hand area can be estimated based on existing topographical maps of the primary motor cortex of primates. Applying this rationale to a sample of capuchin monkeys, Phillips and Sherwood (2005) recently found that lateralization in the dorsal portion of the precentral gyrus differed significantly depending on the handedness of the subjects for the tube task. Thus, it appears that hand preference for coordinated bimanual actions is associated with between-hemisphere differences in the motor-hand area in chimpanzees and capuchins.

More recently, Hopkins and colleagues (2007a) examined the association between handedness for tool use and brain asymmetries in a sample of captive chimpanzees. Handedness for tool use was measured for a probing task (designed to simulate termite fishing) as well as for opening a coconut (designed to simulate anvil use in wild chimpanzees). Right- and non–right-handed apes were compared on two brain regions including the fronto-orbital (FO) sulcus and planum temporale (PT). FO and PT were selected as regions of interest because they represent the cortical homologs to Broca's and Wernicke's areas of the human brain and have been implicated in studies of handedness and tool use in humans (Lewis, 2006). For both measures of handedness, right-handed chimpanzees showed a significantly greater leftward asymmetry in FO and the PT compared to non–right-handed chimpanzees (Fig. 29.5A–D). Interestingly, the very same neural correlates of handedness for tool use in chimpanzees (i.e., FO) also correlate with handedness for manual gestures (Taglialatela et al., 2006) (Fig. 29.6A,B), reinforcing a view that a common cognitive and neural substrate may underlie these two behaviors (discussed later). Whether hand use for tool use in other primates is associated with asymmetries in the inferior frontal gyrus or related brain areas (such as F5 in the macaque brain) is not known. A recent single cell recording and positron emission tomography (PET) study in a single macaque monkey reported asymmetries in the inferior frontal area for a raking task, but it is difficult to generalize from these results in a single subject (see Chapter 30).

Cerebellar Asymmetries

Several laboratories have been interested in individual differences in motor learning, tool use, and other manual actions in relation to individual differences in the volume and lateralization of the cerebellum (Cantalupo et al., 2008; Phillips & Hopkins, 2007). Like the cerebral hemispheres, the cerebellum shows a torque asymmetry and can be divided into the left and right sides as well as along the anterior-posterior axis. The left and right cerebellar hemispheres as well as the anterior-posterior divisions have been delineated in chimpanzees and capuchin monkeys following the procedure employed by Snyder and colleagues (1995) with human subjects. When comparing capuchin monkeys and chimpanzees, several interesting differences and similarities emerge (Table 29.3). Specifically, capuchins show a population-level leftward asymmetry in the anterior but not posterior region of the cerebellum, whereas chimpanzees do not show population-level asymmetries for

Figure 29.5 Mean asymmetry quotient (AQ) values (± SE) for the planum temporale and fronto-orbital sulcus in right-, left-, and ambiguously handed chimpanzees when classified on the basis of their handedness for tool use. Data regraphed from Hopkins, W. D., Russell, J. L., & Cantalupo, C. (2007a). Neuroanatomical correlates of handedness for tool use in chimpanzees (Pan troglodytes): Implication for theories on the evolution of language. *Psychological Science, 18,* 971–977.

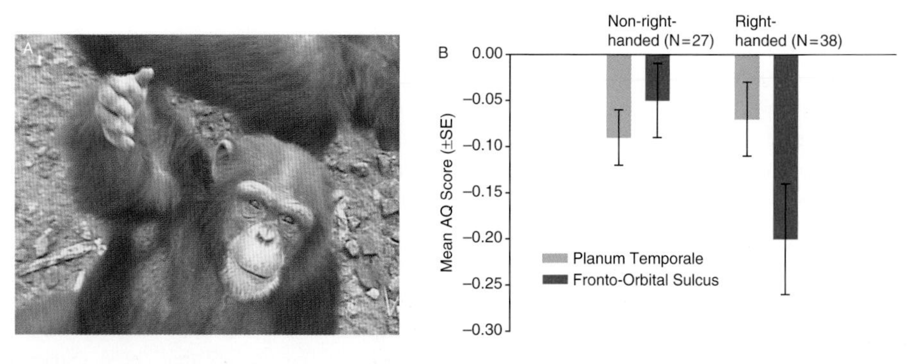

Figure 29.6 Mean asymmetry quotient (AQ) values (± SE) for the planum temporale and fronto-orbital sulcus in Non-right handed and Right-handed chimpanzees when classified on the basis of their handedness for manual gestures. Data partially regraphed from Taglialatela, J. P., Cantalupo, C., & Hopkins, W. D. (2006). Gesture handedness predicts asymmetry in the chimpanzee inferior frontal gyrus. *NeuroReport, 17,* 923–927. Used with permission.

TABLE 29.3 Data on Cerebellar Asymmetry and Hand Preference in Capuchin Monkeys and Chimpanzees

	Capuchin		Chimpanzees	
	Anterior	Posterior	Anterior	Posterior
Mean AQ	−.129	.000	−.032	.044
R (Handedness)	**−.761**	−.456	.101	.174
R (Tool use)	NA	NA	**.321**	**−.485**
R AQ Values	.231		**−.761**	

AQ, asymmetry quotient; R, = correlation value.
Bolded values indicated significant correlations ($P < .05$).

either region. Chimpanzees show a significant inverse correlation between asymmetry quotient (AQ) values in the anterior and posterior cerebellar regions, supporting the view that this brain area shows a torque asymmetry. This association is not found in the capuchin monkeys. In addition, handedness for a coordinated bimanual task correlates with asymmetries in the posterior region of the cerebellum in capuchins but not in chimpanzees. However, like the PT and FO results, in chimpanzees, handedness for tool use significantly correlates with the asymmetries in the anterior and posterior sections of the cerebellum (Table 29.3) but in opposite directions. No data on handedness for tool use and cerebellar asymmetry are available at this point for capuchin monkeys, or any other species.

Handedness

Although there are no reports of handedness for tool use in wild monkeys, this behavior has been reported in wild chimpanzees (Biro et al., 2003, 2006; Boesch, 1991; Lonsdorf & Hopkins, 2005; Marchant & McGrew, 2007; McGrew & Marchant, 1996). As noted previously, most humans are right-handed, and some have suggested that this might be a uniquely human trait related to the evolution of complex motor actions such as tool use and language. Studies in captive chimpanzees (and other primates) have reported evidence of population-level right-handedness for a number of behaviors including throwing, bimanual feeding, coordinated bimanual actions, and manual gestures (Hopkins, 2006); however, it has been suggested that these results may be an artifact of the primates living in a human environment (McGrew & Marchant, 1997a). Thus, data from wild subjects are very important for evaluating the potential influence of captive rearing on behavioral asymmetries. Shown in Table 29.4 are the hand preference distributions for tool use in apes and monkeys.

There are at least two important findings from these studies. First, in wild chimpanzees, tool use behaviors elicit strong preferences at the individual level in comparison to measures of spontaneous daily activities, such as unimanual feeding (Marchant & McGrew, 1996; McGrew & Marchant, 2001). In other words, tool use is a better and more sensitive measure of handedness than behaviors such as unimanual feeding or simple reaching because most animals show a definitive preference of one hand over the other. Second, apes show population-level handedness for termite fishing and leaf sponging, although in opposite directions (Table 29.4). Wild chimpanzees also show a borderline significant bias to the right hand for nut cracking and ant dipping.

In captive apes and monkeys, there are far fewer data on handedness and tool use compared to those data reported from wild individuals, which is somewhat surprising given the theoretical interest in this topic and the common use of tool use devices as behavioral enrichment in zoos and research laboratories (Table 29.4). When one considers probing tool use tasks, it

TABLE 29.4 Handedness for Tool Use in Nonhuman Primates

	Hand Preference Classification				
	#L	#R	#A	Task	Author(s)
Wild Apes					
Pan	29	15	10	TF	McGrew & Marchant, 1996
Pan	12	4	1	TF	Lonsdorf & Hopkins, 2005
Pan	1	7	8	AD	Marchant and McGrew, 2007
Pan	36	46	3	NC	Boesch, 1991
Pan	7	11	2	NC	Biro et al., 2003
Pan	5	7	2	AV	McGrew et al., 1999
Pan	5	10	2	LS	Boesch, 1991
Pan	4	5	0	AG	Humle & Matsuzawa, 2009
Pan	6	19	0	AD	Humle & Matsuzawa, 2009
Pan	10	11	1	PP	Humle & Matsuzawa, 2009
Pan	7	17	0	PP-E	Humle & Matsuzawa, 2009
Pan	4	11	2	LS	Biro et al., 2006
Captive Apes					
Pan	6	0	5	STF	Fletcher & Weghorst, 2005
Pan	5	3	1	STF	Hopkins, 1999
Pan	9	14	14	STF	Hopkins & Rabinowitz, 1994
Pan	0	2	2	STF	Morange, 1994
Pan	74	63	52	STF	Hopkins et al. (in press)
Pan	0	3	0	MT	Morris et al., 1993
Pan	2	2	0	TH	Colell et al., 1995
Pan	4	5	0	TH	Marchant, 1983
Pan	23	50	16	TH	Hopkins et al. 2005
Pan	3	1	1	MT	Steiner, 1990
Pan	0	1	0	NC	Fouchart et al., in press
Pan p.	5	5	4	UK	Harrison & Nystrom, 2008
Pan p.	0	1	0	KP	Toth et al., (1993
Gorilla	0	2	3	TH	Shafer, 1993
Pongo	2	2	2	STF	O'Malley & McGrew, 2006
Pongo	2	1	2	STF	Phillips, 2008, unpublished data
Captive Monkeys					
Macaca	3	1	1	STF	Westergaard, 1991
Papio	2	3	0	LS	Westergaard, 1993
Papio	2	1	1	STF	Westergaard, 1993
Cebus	3	1	1	STF	Westergaard, 1991
Cebus	8	1	2	STF	Lilak & Phillips, 2007
Cebus	3	0	1	STF	Limongelli et al., 1994
Cebus	8	5	0	STF	Westergaard & Suomi, 1994a
Cebus	8	4	2	NC	Westergaard & Suomi, 1994b
Cebus	5	7	2	LS	Westergaard & Suomi, 1993
Cebus	2	1	0	TH	Westergaard & Suomi, 1994c
Cebus	11	12	2	TH	Westergaard et al., 2000
Cebus	1	9	4	KP	Westergaard & Suomi, 1996

L, left; R, right; A, ambiguous. For wild apes: TF, termite-fishing; AD, ant dipping; NC, nut cracking; LS, leaf sponging. For captive apes and monkeys: TH, throwing; STF, simulated termite fishing (or probing tasks); KP, stone knapping; UK, unknown or not reported; NC, nut cracking; LS, leaf sponging; MT, multiple tasks; PP, pestle pounding; PP-E, pestle pounding-extraction, and AG, algae scooping.

seems that capuchin monkeys show a leftward bias in hand use, not unlike the data presented on termite fishing in wild chimpanzees. For throwing, captive chimpanzees show population-level right-handedness, whereas capuchin monkeys do not. There are simply too few data in other species and for other tasks to make any meaningful inferences.

TOOL USE, MANUAL MOTOR SKILL, AND OROFACIAL CONTROL

As noted earlier, the motor and cognitive processes involved in tool use have been proposed as potential neuropsychological preadaptations for the evolution of language and speech in humans (Bradshaw & Rogers, 1993). One of the challenges with this evolutionary scenario is identifying the common neurological, cognitive, and motor abilities that would have been selected upon during evolution that allowed for language and speech (Arbib, 2005; Gentilucci & Corballis, 2006). In short, at some point in primate evolution, the presumed neurological, motor, and cognitive processes originally underlying manual actions had to be incorporated into motor control of the mouth, facial musculature, and peripheral speech organs. Moreover, cognitive structures underlying tool use had to be extended into the social domain.

The underlying assumptions of these arguments are that if common neural substrates underlie both manual and potentially motor control of the mouth and facial musculature, then significant associations should be evident for their coactivation or use in nonhuman primates. Unfortunately, very little research in nonhuman primates has focused on this question, but several recent studies in my laboratory suggest that, at least in chimpanzees, common neural mechanisms might be involved in the control of manual and orofacial actions.

First, in captivity, chimpanzees have been shown to selectively produce one of two sounds, either a "raspberry" (RASP) or "extended food grunt" (EFG), to capture the attention of an otherwise inattentive human and suppress these sounds in the presence of the food or human alone (Hopkins et al., 2007b). The RASP and EFG are frequently produced in conjunction with the simultaneous production of a manual gesture directed toward a human (Taglialatela & Hopkins, 2005) (Fig. 29.7). As can be seen in Figure 29.7, the onset of the manual gesture directed toward a human is temporally very close to onset of the orofacial movements associated with the production of the RASP in this specific example. In two separate studies using both between-group and within-subject comparisons, we have found that subjects that simultaneously produce the RASP and EFG sounds while gesturing are significantly more right-handed than those that do not simultaneously produce a sound while gesturing (Hopkins & Cantero, 2003; Hopkins & Leavens, 1998). The mean handedness indices for gestural communication in chimpanzees that do or do not simultaneously produce a vocalization are shown in Figure 29.8.

A second interesting aspect of the RASP and EFG sounds is that not all chimpanzees produce them. Some chimpanzees use them very reliably, while others are less reliable or simply fail to use them. The origin of this difference remains unclear, but if a central motor system controls both manual and potentially orofacial movements associated with (the learning of) sound production, then it can be hypothesized that individual differences in motor skill in one domain (manual) might be related to motor skill in another domain (orofacial control). We have at least some evidence to support this hypothesis in two sets of data from my laboratory (Table 29.5). First, my colleagues and I have previously tested our chimpanzees on a task requiring fine motor prehension of small food items (Hopkins & Russell, 2004; Hopkins et al., 2002). In these studies, we recorded the number of errors made by the chimpanzees when grasping small food items. When comparing the errors between chimpanzees that reliably and unreliably produce the RASP or EFG vocalization, we find that chimpanzees that produce the RASP and EFG sounds make significantly fewer grasping errors compared to those who do not. We also find that chimpanzees that reliably produce the RASP and EFG sounds also perform significantly better on a tool use task requiring them to successfully insert small sticks into a hole to extract food. Lastly, we

Figure 29.7 Sequential frames representing the co-occurrence of the production of the raspberry and manual gestures. (**A**) Start of trial. (**B**) Onset of raspberry. (**C**) Offset of raspberry. (**D**) End of manual gesture.

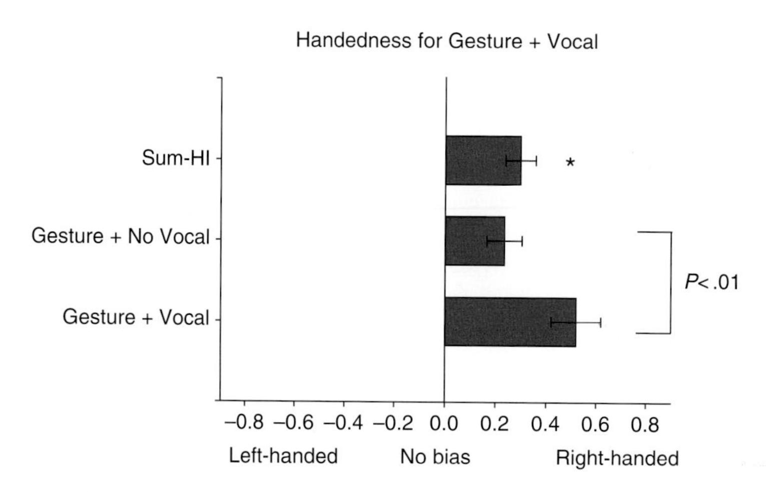

Figure 29.8 Mean handedness index values (± SE) for manual gestures that are (Gesture + Vocal) or not accompanied by a vocalization (Gesture + No Vocal). Sum-HI is the handedness value for all gesture that were and were not accompained by a vocalization.

TABLE 29.5 Grasping Skill and Hand Preference Differences in Chimpanzees Who are Reliable or Unreliable in their Use of Attention-Getting Sounds

	Vocalization Group				
	Reliable	Unreliable	t	P	N
Behavioral measure					
Grasping skill*	7.77	9.94	2.32	.02	136
Tool use performance†	8.43	12.43	2.52	.02	60
Hand preference					
Tool use	.133	−.075			
Gesture	.421	.331			
Reaching	.173	.083			
Tube	221	.071			
Mean	.237	.107	3.10	.01	101

* Indicates number of errors out of 40 trials (20 trials for each hand).
† Indicates average latency to insert stick into pipe. P = level of significance, N = number of subjects tested on the task.

find that chimpanzees that reliably produce the RASP and EFG sounds are significantly more right-handed for several handedness tasks including reaching, coordinated bimanual actions, tool use, and manual gestures. Collectively, these results suggest that common neural pathways, possibly lateralized to the left hemisphere, may underlie motor control of both manual and orofacial movements in chimpanzees. Moreover, it suggests that the lateralization for both manual and orofacial control of volitional expressions and sounds were in place before the split between the common ancestor of humans and apes.

GESTURAL AND VOCAL COMMUNICATION IN CAPTIVE APES AS A FORM OF SOCIAL TOOL USE

I have principally focused on material tool use in this chapter but want to also draw some parallels between tool use as reported in wild apes (notably chimpanzees and orangutans) with evidence of referential, gestural communication in captive apes. As noted previously, historically scientists have considered social and nonsocial cognition as separate domains of investigation. I believe that the interface between social and nonsocial cognition lies at the heart of the observed differences reported during intraspecies gestural and vocal communication in captive apes and tool use in wild apes.

Studies in our laboratory and others have clearly shown that chimpanzees and other great apes (orangutan, gorilla, and bonobo) use manual gestures referentially and intentionally to direct the attention of a human to otherwise unattainable objects (usually food but sometimes to a tool) (Call & Tomasello, 1994, 2007; Leavens et al., 2004a,b; Russell et al., 2005). The production of manual gestures is not a frustrated reaching response as these communicative behaviors are selectively and almost exclusively produced in the presence of an audience and not without one (Leavens & Hopkins, 1998; Leavens et al., 1996; Poss et al., 2006). Moreover, chimpanzees and other great apes can alter their modality of communication depending on the attentional status of a human or conspecific. Thus, apes will use primarily visual signals when the human or conspecific is facing them but will use tactual or auditory signals when the human or conspecific is facing away from them (Cartmill & Byrne, 2007; Hostetter et al., 2001, 2007; Leavens et al., 2004b; Liebal et al., 2004). This suggests that the signals are intentional and that the apes understand the function of their signaling.

The evidence of gestural communication in captive chimpanzees stands in contrast to reports in wild apes (Leavens et al., 2005). Feral chimpanzees no doubt gesture in the context of food begging and reconciliation but, as far as we know, they do not seem to referentially point in the way that gestures have been reported in captive chimpanzees (Pika & Mitani, 2006). As my colleagues and I have argued elsewhere, this discrepancy can in part be explained by what we refer to as the "referential problem space" (Leavens et al., 2005). Basically, when wild chimpanzees want some food that is in proximity to them, they can locomote over to the food and take it. The exception is, of course, food that is not otherwise visible to them, such as termites inside a termite mound or the meat of a nut inside a shell. In these cases, chimpanzees must fashion a tool in order to extract the food. Like wild chimpanzees, when captive chimpanzees want food that is visible to them and the food is in the cage, then can walk over and grasp the food. In contrast, sometimes in captivity, foods are available to them but the chimpanzees cannot gain access to the food, such as when food is located outside their home cage. In these cases, the chimpanzees have learned to use their gestures (and other signals) to capture the attention of a human to retrieve the food for them. In essence, the chimpanzees use the human as a tool and, arguably, the apes would use a tool in the absence of a human, if one were provided to them. In other words, if a rake were made available, the chimpanzees would likely use the rake to get the food (see Barrett & Rendall, this volume, for similar themes). In the absence of a rake (or other tool), they are left only with the sporadically present human that they have learned to manipulate through their communicative signals in order to obtain the food.

I would further suggest that the flexibility seen in tool use by wild apes is similarly seen in the communicative repertoire of captive chimpanzees. Recall that wild chimpanzees use a variety of different tools in relation to different ecological and social factors (Whiten et al., 1999). Apes in captivity will use a variety of different communicative signals to capture and direct the attention of a human or conspecific. For example, captive apes will vocalize, spit, throw, bang on the cage, or clap their hands as means of getting the attention of a human when food is visible but otherwise unavailable to the apes (e.g., Leavens et al., 2004b). When a tool fails to work properly in the wild, apes will repair the tool or get new one. In captivity, when a communicative signals fails, apes will repair it (i.e., try again with some modification) or elaborate (i.e., try a different signal) on their signals. All of these behavioral traits reflect a more "fluid" or "innovative" form of intelligence that simply gets expressed differently, depending on certain ecological or social factors. It is likely that common cognitive and neurological processes underlie many of these behaviors rather than modules or domain-specific "social" and "nonsocial" cognitive processes, as they are often characterized.

To the extent that other species engage in similar communicative behaviors remains, in essence, unstudied. There are a few short reports of "gestural" communication in Old and New World monkeys, but none of these results are terribly robust (Anderson et al., 2007; Hattori et al., 2007; Hess et al., 1993; Mitchell & Anderson, 1997). In one of the more sophisticated and elaborate studies, Hattori and colleagues (2007) replicated a study previously done in chimpanzees on the sensitivity to the eyes during intraspecies communication. In the study, the experimenters recorded the frequency of "gestures" in response to the presence of food when a human had his or her eyes open or closed. Interestingly, the capuchin monkeys did look differentially at the human in these conditions but did not differentiate the human's "gestural" behavior. That is, the capuchins were equally likely to point to the human whether the human had his or her eyes open or closed. This stands in contrast to reports in chimpanzees, which suggest they do differentially gesture depending on whether the eyes of the human are open or closed (Hostetter et al., 2007). This would suggest that the capuchin monkeys did not understand the communicative function of their gestures. Hattori and colleagues (2007) also noted that they never observed the capuchins alternate their gaze between the referent and the human, a behavioral concordance frequently reported in great ape intraspecies communication.

Thus, it seems that ecological factors have favored the emergence of tool use in capuchin monkeys and apes but, perhaps, only apes might be able to capitalize on this inherent ability and utilize it functionally in captive settings to solve a novel, socio-communicative problem. The comparison here is not limited to *Pan* versus *Cebus* either; in captivity, the non–tool-using gorillas do not show the same patterns of intraspecies communication that has been reported in chimpanzees and orangutans (Cartmill & Byrne, 2007; Leavens et al., 2004b; Poss et al., 2006). Thus, not unlike material tool use, at the heart of the matter in gestural communication in monkeys and apes are questions of causality and functionality. In other words, do monkeys and apes understand the function of their communicative signals? This is much like the question of whether they understand the function of their tool use.

SUMMARY AND SOME CLOSING THOUGHTS

With respect to object manipulation and tool use, the literature seems to suggest that basic phylogenetic differences are evident in object manipulation, prehension, and tool-using abilities in primates. More distantly related primates from humans show less complex object manipulation skills, no individual control of the digits, and few tool-using abilities, whereas more closely related primates show more sophisticated object manipulation and tool-using skills as well as greater control of individual digits. The phylogenetic picture, of course, is more complicated because different primates within certain *Genera* may or may not express certain abilities depending on specific ecological or social adaptations. *Cebus* are an excellent example as it seems that certain ecological factors have favored the evolution of tool use in this species compared to other species from the same family. Likewise, among great apes, gorillas are not well known for their tool-using abilities in either captivity or the wild, despite the fact that they have good object manipulation skills and individual control of their digits. Thus, at a global level, increasing brain size and asymmetry has been selected for in primate evolution, but many adaptive radiations have occurred within specific primate families in response to certain ecological and social factors. Tool use may simply be one form or expression of these adaptations, but equally complex cognitive and motor processes may underlie different non–tool-use behaviors of apes and monkeys (Byrne, 1995).

With specific reference to tool use, I have no doubt that most nonhuman primates can learn to solve simple tool-using tasks through different shaping procedures, such as the cane-pulling task employed by many; however, to what extant these abilities would generalize to (1) new tool-using tasks or (2) new contexts remains the challenge. In other words, as recently suggested by Rumbaugh and Washburn (2003), to what extent there would be emergent cognitive processes that extend beyond the specific behavioral context in which the tool-using abilities were trained remains to be determined. Moreover, to what extent the animals would understand the function of tool use beyond the immediate context is not clear. The existing data suggest that apes, notably chimpanzees and orangutans, would and do generalize, while the data from capuchin monkeys remain unclear. Apes and capuchins have dominated this literature because they acquire and use tools relatively easily, but data from other species would be particularly useful for this research because many of them are considered "relational" learners (Rumbaugh & Washburn, 2003) and therefore might do well on tool-related tasks, despite the lack of spontaneous proclivity to use tools in the wild and captivity.

Neurologically, the comparative data from different primate species indicate that there have been significant grade shifts in the organization of the cerebellum and in interhemispheric connectivity, at least as reflected in the relative corpus callosum size. Great apes have relatively small CCs compared to Old and New World monkeys. Similarly, in great apes, there has been significant expansion of the cerebellum, particularly within the hemispheres, compared to Old and New World monkeys. One of the interesting findings from these data are the within-family variations in cerebellar volume as well as interhemispheric connectivity. For instance, within New World monkeys, there is considerable variation in the relative size of the cerebellum, and *Cebus* differ substantially from other genera, suggesting that there may be specific changes in the organization in this brain area for the learning of complex motor actions, such as tool use or object manipulation. In contrast, within great apes, there is less variability in the relative size of the cerebellum and arguably greater commonality in manipulatory and motor learning abilities, albeit these get expressed very differently in chimpanzees and orangutans (say, for tool use) compared to gorillas (say, for complex manual actions associated with feeding).

Neuropsychological data from chimpanzees suggest that common cortical mechanisms may underlie tool use and gestural communication. Specifically, regions within the inferior frontal gyrus correlate with lateralized hand use for both manual gestures and tool use, whereas other behavioral measures of handedness do not. It is important to emphasize that the studies to date have focused on morphology, but whether there are functional homologs in the chimpanzee (or other nonhuman primates) brain in the use of tools and gestures is not clear and will require the application of functional imaging studies (Taglialatela et al., 2008). The results of our studies in chimpanzees further suggest that individuals who appear to exhibit better orofacial motor control perform better on the fine motor tool use tasks and grasping tasks and are more right-handed for several handedness tasks. Collectively, these results point to the conclusion that the left hemisphere of chimpanzees is dominant for motor learning and control in both the manual and orofacial domain (but see Sherwood, 2005, and Sherwood et al., 2005, for other systems involved in motor control of face). The need for data from other species cannot be overstated.

Lastly, the suite of tool-using abilities found in great apes is paralleled, in some ways, by a suite of communicative behaviors that are used in captivity during intraspecies communication. Captive chimpanzees, and other great apes, particularly orangutans, use gestures, vocal signals, and several attention-getting behaviors such as throwing, hand clapping, cage banging, and spitting when communicating with humans (and other apes). No other nonhuman primates engage in these kinds of behaviors, so far as I know, or at least with any degree of frequency. There are reports of throwing in captive capuchin monkeys, but as I have argued elsewhere, throwing had to be explicitly trained in these studies, which is quite different than the acquisition and maintenance of throwing in chimpanzees and other apes (Hopkins et al., 2005). The use of these communicative behaviors by apes but not other nonhuman primates is not trivial, because it suggests that fundamental differences might be present in social cognitive abilities between apes and monkeys that are rooted in their understanding of causal relationships, of which tool use may represent a physical domain of this underlying system. This argument is consistent with the view of Reader and Laland (2002), who argue that evolution has selected for increasing executive function in primates, which has adaptive value for both social and nonsocial cognitive abilities or, more generally speaking, innovative cognitive behaviors.

ACKNOWLEDGMENTS

This work was supported in part by NIH grants RR-00165, NS-42867, NS-36605, HD-38051, and HD-56232. Correspondence and reprint requests should be addressed to Dr. William Hopkins, Division of Psychobiology, Yerkes National Primate Research Center, 954 Gatewood Road, Atlanta, Georgia 30322. E-mail: whopkin@emory.edu or whopkins@agnesscott.edu.

REFERENCES

Anderson, J. R., Kuwahata, H., & Fujita, K. (2007). Gaze alternation during "pointing" by squirrel monkeys (*Saimiri sciureus*)? *Animal Cognition, 10*, 267–271.

Arbib, M. (2005). From monkey-like action recognition to human language: An evolutionary framework for neurolinguistics. *Behavioral and Brain Sciences, 28*, 105–167.

Beaton, A. A. (1997). The relation of planum temporale asymmetry and morphology of the corpus callosum to handedness, gender and dyslexia: A review of the evidence. *Brain and Language, 60*, 255–322.

Beck, B. B. (1980). *Animal tool behavior: The use and manufacture of tools by animals.* New York: Garland.

Biro, D., Inoue-Nakamura, N., Tonooka, R., Yamakoshi, G., Sousa, C., & Matsuzawa, T. (2003). Cultural innovation and transmission of tool use in wild chimpanzees: Evidence from field experiments. *Animal Cognition, 6,* 213–223.

Biro, D., Sousa, C., & Matsuzawa, T. (2006). Ontogeny and cultural propagation of tool use by wild chimpanzees at Bossou, Guinea: Case studies in nut cracking and leaf folding. In: T. Matsuzawa, T. Tomonaga, & M. Tanaka (Eds.), *Cognitive development of chimpanzees* (pp 476–507). New York: Springer.

Boesch, C. (1991). Handedness in wild chimpanzees. *International Journal of Primatology, 6,* 541–558.

Boysen, S. T., Kuhlmeier, K., Halliday, P., & Halliday, Y. M. (1999). Tool use in captive gorillas. In: S. T. Parker, R. W. Mitchell, & H. L. Miles (Eds.), *The mentality of gorillas and orangutans.* Cambridge: Cambridge University Press.

Bradshaw, J. L., & Rogers, L. J. (1993). *The evolution of lateral asymmetries, language, tool use, and intellect.* San Diego: Academic Press.

Byrne, R. W. (1995). *The thinking ape: Evolutionary origins of intelligence.* Oxford: Oxford University Press.

Byrne, R. W., & Byrne, J. M. (1991). Hand preferences in the skilled gathering tasks of mountain gorillas (*Gorilla gorilla berengei*). *Cortex, 27,* 521–536.

Byrne, R. W., & Corp, N. (2004). Neocortex size predicts deception rate in primates. *Proceedings of the Royal Society B, 271,* 1393–1699.

Call, J., & Tomasello, M. (1994). Production and comprehension of referential pointing by orangutans *(Pongo pygmaeus). Journal of Comparative Psychology, 108,* 307–317.

Call, J., & Tomasello, M. (2007). *The gestural communication of monkeys and apes.* Oxford: Psychology Press.

Candland, D. (1987). Tool use. In: G. Mitchell & J. M. Erwin (Eds.), *Comparative primate biology. Vol 2, Part B: Behavior, cognition and motivation* (pp. 85–103). New York: Alan R. Liss.

Cantalupo, C., Freeman, H. D., & Hopkins, W. D. (2008). Patterns of cerebellar asymmetries in great apes as revealed by MRI. *Behavioral Neuroscience, 122,* 191–198.

Cartmill, E., & Byrne, R. W. (2007). Orangutans modify their gestural signaling according to their audience's comprehension. *Current Biology, 17,* 1–14.

Christel, M. I. (1994). Catarrhine primates grasping small objects: Techniques and hand preferences. In: J. R. Anderson, J. J. Roeder, B. Thierry, & N. Herrenschmidt (Eds.), *Current primatology. Vol III: Behavioral neuroscience, physiology and reproduction* (pp 37–49). Strasbourg: Universite Louis Pasteur.

Colell, M., Segarra, M. D., & Sabater-Pi, J (1995). Hand preferences in chimpanzees *(Pan troglodytes),* bonobos *(Pan paniscus)* and orangutans *(Pongo pygmaeus)* in food-reaching and other daily activities. *International Journal of Primatology, 16,* 413–434.

Corballis, M. C. (1992). *The lopsided brain: Evolution of the generative mind.* New York: Oxford University Press.

Crow, T. (2004). Directional asymmetry is the key to the origin of modern Homo sapiens (the Broca-Annett axiom): A reply to Rogers' review of the speciation of modern Homo sapiens. *Laterality: Asymmetries of Body, Brain and Cognition, 9,* 233–242.

Crow, T. J. (1998). Sexual selection, timing and the descent of man: A theory of the genetic origins of language. *Current Psychology of Cognition, 17,* 1079–1114.

Cunningham, C. L., Anderson, J. R., & Mootnick, A. R. (2006). Object manipulation to obtain a food reward in hoolock gibbons, *Bunopithecus hoolock. Animal Behaviour, 71,* 621–629.

Deacon, T. (2004). Monkey homologues of language areas: Computing the ambiguities. *Trends in Cognitive Sciences, 8,* 288–290.

Deacon, T. W. (1997). *The symbolic species: The coevolution of language and the brain.* New York: W. W. Norton and Company.

Deaner, R. O., Isler, K., Burkart, J., & Van Schaik, C. P. (2007). Overall brain size, and not encephalization quotient, best predicts cognitive ability across non-human primates. *Brain, Behavior and Evolution, 70,* 115–124.

Dore, F. Y., & Dumas, C. (1987). Psychology of animal cognition: Pigetian studies. *Psychological Bulletin, 102,* 219–233.

Dunbar, R. I. M. (1996). *Grooming, gossip, and the evolution of language.* Cambridge: Harvard University Press.

Finlay, B. L., & Darlington, R. B. (1995). Linked regularities in the development and evolution of mammalian brains. *Science, 16,* 1578–1584.

Fletcher, A. W., & Weghorst, J. A. (2005). Laterality of hand function in naturalistically housed chimpanzees *(Pan troglodytes). Laterality, 10,* 219–242.

Fouchart, J., Bril, B., Hirata, S., Morimura, M., Houki, C., Ueno, Y., et al. (in press). A preliminary analysis of nut-cracking movements in a captive chimpanzee: Adaptation to the properties of tools and nuts. In: V. Rouz & B. Bril (Eds.), *Stone knapping: The necessary conditions for a uniquely hominoid behavior.* Cambridge, MA: McDonald Institute Series Monograph.

Fragaszy, D., Izar, P., Visalberghi, E., Ottoni, E. B., & Oliveira, M. G. (2004). Wild capuchin monkeys (Cebus libidinosus) use anvils and stone pounding tools. *American Journal of Primatology, 64,* 359–366.

Fragaszy, D. M. (1989). Tool use, imitation and insight: Apples, orangs and pea soup. *Behavioral and Brain Sciences, 12,* 596–599.

Frost, G. T. (1980). Tool behavior and the origin of laterality. *Journal of Human Evolution, 9,* 447–459.

Fujita, K., Kuroshima, H., & Asai, S. (2003). How do tufted capcuhin monkeys (Cebus apella) understand causality involved in tool use? *Journal of Experimental Psychology: Animal Behaviour Processes, 29,* 233–242.

Galaburda, A. M. (1995). Anatomic basis of cerebral dominance. In: R. J. Davidson & K. Hugdahl (Eds.), *Brain asymmetry.* Cambridge, MA: MIT Press.

Gentilucci, M., & Corballis, M. C. (2006). From manual gesture to speech: A gradual transition. *Neuroscience and Biobehavioral Reviews, 30,* 949–960.

Gibson, K. R., & Ingold, T. (1993). Tools, language and cognition in human evolution. Cambridge: Cambridge University Press.

Grafton, S. T., Fagg, A. H., Woods, M. A., & Arbib, M. (1996). Functional anatomy of pointing and grasping in humans. *Cerebral Cortex, 6,* 226–237.

Greenfield, P. M. (1991). Language, tools, and brain: The ontogeny and phylogeny of hierarchically organized sequential behavior. *Behavioral and Brain Sciences, 14,* 531–550.

Hammond, G. (2002). Correlates of human handedness in primary motor cortex: A review and hypothesis. *Neuroscience and Biobehavioral Reviews, 26,* 285–292.

Harrison, R. M., & Nystrom, P. (2008). Handedness in captive bonobos (*Pan paniscus*). *Folia Primatologica, 79,* 253–268.

Hattori, Y., Kuroshima, H., & Fujita, K. (2007). I know you are not looking at me: Capuchin monkeys' (Cebus apella) sensitivity to human attentional states. *Animal Cognition, 10,* 141–148.

Hauser, M. (1997). Artifactual kinds and functional design features: What a primate understands without language. *Cognition, 64,* 285–308.

Hess, J., Novak, M. A., & Povinelli, D. J. (1993). "Natural pointing" in a rhesus monkeys but no evidence of empathy. *Animal Behaviour, 46,* 1023–1025.

Hopkins, W. D. (1999). On the other hand: Statistical issues in the assessment and interpretation of hand preference data in non-human primates. *International Journal of Primatology, 20,* 851–866.

Hopkins, W. D. (2006). Comparative and familial analysis of handedness in great apes. *Psychological Bulletin, 132,* 538–559.

Hopkins, W. D. (Ed.). (2007). *Evolution of hemispheric specialization in primates.* Oxford: Elsevier.

Hopkins, W. D., & Cantalupo, C. (2004). Handedness in chimpanzees is associated with asymmetries in the primary motor but not with homologous language areas. *Behavioral Neuroscience, 118,* 1176–1183.

Hopkins, W. D., Cantalupo, C., Wesley, M. J., Hostetter, A. B., & Pilcher, D. (2002). Grip morphology and hand use in chimpanzees *(Pan troglodytes):* Evidence of a left hemisphere specialization in motor skill. *Journal of Experimental Psychology: General, 131,* .412–423.

Hopkins, W. D., & Cantero, M. (2003). From hand to mouth in the evolution of language: The influence of vocal behavior on lateralized hand use in manual gestures by chimpanzees *(Pan troglodytes). Developmental Science, 6,* 55–61.

Hopkins, W. D., & Leavens, D. A. (1998). Hand use and gestural communication in chimpanzees *(Pan troglodytes). Journal of Comparative Psychology, 112,* 95–99.

Hopkins, W. D., & Russell, J. L. (2004). Further evidence of a right hand advantage in motor skill by chimpanzees (*Pan troglodytes*). *Neuropsychologia, 42,* 990–996.

Hopkins, W. D., Russell, J. L., & Cantalupo, C. (2007a). Neuroanatomical correlates of handedness for tool use in chimpanzees (Pan troglodytes): Implication for theories on the evolution of language. *Psychological Science, 18,* 971–977.

Hopkins, W. D., Russell, J., Cantalupo, C., Freeman, H., & Schapiro, S. (2005). Factors influencing the prevalence and handedness for throwing in captive chimpanzees (*Pan troglodytes*). *Journal of Comparative Psychology, 119,* 363–370.

Hopkins, W. D., Taglialatela, J. P., & Leavens, D. A. (2007b). Chimpanzees differentially produce novel vocalizations to capture the attention of a human. *Animal Behaviour, 73,* 281–286.

Hopkins, W. D., Russell, J. L., Schaeffer, J. A., Gardner, M. Schapiro, S. J. (in press). Handedness for toll use in captive chimpanzees (*Pan troglodytes*). Sex differences, performance, heritability and comparison to the wild.

Hostetter, A. B., Cantero, M., & Hopkins, W. D. (2001). Differential use of vocal and gestural communication by chimpanzees *(Pan troglodytes)* in response to the attentional status of a human *(Homo sapiens). Journal of Comparative Psychology, 115,* 337–343.

Hostetter, A. B., Russell, J. L., Freeman, H., & Hopkins, W. D. (2007). Now you see me, now you don't: Evidence that chimpanzees understand the role of the eyes in attention. *Animal Cognition, 10,* 55–62.

Humble, T & Matsuzawa. T, (2009). Laterality in hand use across four tool-use behaviours among the wild chimpanzees of Bossou, Guniea, West Africa. American Journal of Primatology, 71, 40–48.

Johnson-Frey, S. H. (2004). The neural basis of complex tool use in humans. *Trends in Cognitive Sciences, 8,* 71–78.

Jordon, C. (1982). Object manipulation and tool use in captive pygmy chimpanzees (*Pan paniscus*). *Journal of Human Evolution, 11,* 35–39.

Knecht, S., Drager, B., Deppe, M., Bobe, L., Lohmann, H., Floel, A., et al. (2000). Handedness and hemispheric language dominance in healthy humans. *Brain, 123,* 2512–2518.

Leavens, D. A., & Hopkins, W. D. (1998). Intentional communication by chimpanzee *(Pan troglodytes)*: A cross-sectional study of the use of referential gestures. *Developmental Psychology, 34,* 813–822.

Leavens, D. A., Hopkins, W. D., & Bard, K. A. (1996). Indexical and referential pointing in chimpanzees *(Pan troglodytes). Journal of Comparative Psychology, 110,* 346–353.

Leavens, D. A., Hopkins, W. D., & Bard, K. A. (2005). Understanding the point of chimpanzee pointing: Epigenesis and ecological validity. *Current Directions in Psychological Science, 14,* 185–189.

Leavens, D. A., Hopkins, W. D., & Thomas, R. (2004a). Referential communication by chimpanzees *(Pan troglodytes). Journal of Comparative Psychology, 118,* 48–57.

Leavens, D. A., Hostetter, A. B., Wesley, M. J., & Hopkins, W. D. (2004b). Tactical use of unimodal and bimodal communication by chimpanzees, *Pan troglodytes. Animal Behaviour, 67,* 467–476.

Leiner, H. C., Leiner, A. L., & Dow, R. S. (1993). Cognitive and language functions of the human cerebellum. *Trends in Neurosciences, 16,* 444–447.

Lewis, J. W. (2006). Cortical networks related to human use of tools. *The Neuroscientist, 12,* 211–231.

Liebal, K., Pika, S., Call, J., & Tomasello, M. (2004). To move or not to move: How apes adjust to the attentional state of others. *Interaction Studies, 5,* 199–219.

Lilak, A. L., & Phillips, K. A. (2007). Consistency in hand preference across low-level and high-level tasks in capuchin monkeys (*Cebus apella*). *American Journal of Primatology, 69,* 1–12.

Limongelli, L., Boysen, S. T., & Visalberghi, E. (1995). Comprehension of cause-effect relations in a tool-using task by chimpanzees (Pan troglodytes). *Journal of Comparative Psychology, 109,* 18–26.

Limongelli, L., Sonetti, M. G., & Visalberghi, E. (1994). Hand preference of tufted capuchins (*Cebus apella*) in tool-using tasks. In: J. R. Anderson, N. Herrenschmidt, J. J. Roeder, B. Thierry (Eds.), *Current primatology. Vol III: Behavioral neurosciences, physiology and reproduction* (pp. 9–15). Strasbourg: Universite Louis Pasteur.

Lonsdorf, E. V., & Hopkins, W. D. (2005). Wild chimpanzees show population level handedness for tool use. *Proceedings of the National Academy of Sciences, 102,* 12634–12638.

MacLeod, C. E., Zilles, K., Schleicher, A., Rilling, J. K., & Gibson, K. R. (2003). Expansion of the neocerebellum in Hominoidea. *Journal of Human Evolution, 44,* 401–429.

Marchant, L. F. (1983). Hand preferences among captive island groups of chimpanzees. New Bruswick, NJ: Rutgers, The State University of New Jersey.

Marchant, L. F., & McGrew, W. C. (1996). Laterality of limb function in wild chimpanzees of Gombe National Park: Comprehensive study of spontaneous activities. *Journal of Human Evolution, 30,* 427–443.

Marchant, L. F., & McGrew, W. C. (2007). Ant fishing by wild chimpanzees is not lateralised *Primates, 48,* 22–26.

McGrew, W. C., & Marchant, L. F. (1996). On which side of the apes? In: W. C. McGrew, L. F. Marchant, & T. Nishida (Eds.), *Great ape societies* (pp. 255–272). Cambridge: Cambridge University Press.

McGrew, W. C., & Marchant, L. F. (1997a). On the other hand: Current issues in and meta-analysis of the behavioral laterality of hand function in non-human primates. *Yearbook of Physical Anthropology, 40,* 201–232.

McGrew, W. C., & Marchant, L. F. (1997b). Using the tools at hand: Manual laterality and elementary technology in *Cebus* spp. and *Pan* spp. *International Journal of Primatology, 18,* 787–810.

McGrew, W. C., & Marchant, L. F. (2001). Ethological study of manual laterality in the chimpanzees of the Mahale mountains, Tanzania. *Behaviour, 138,* 329–358.

Meador, K. J., Loring, D. W., Lee, K., Hughes, M., Lee, G., Nichols, M., et al. (1999). Cerebral lateralization: Relationship of language and ideomotor apraxia. *Neurology, 53,* 2028–2031.

Mitchell, R. W., & Anderson, J. R. (1997). Pointing, withholding information, and deception in capuchin monkeys (*Cebus apella*). *Journal of Comparative Psychology, 111,* 351–361.

Morange, F. (1994). Handedness in two chimpanzees in captivity. In: J. R. Anderson, J. J. Roeder, B. Thierry, & N. Herrenschmidt (Eds.), *Current primatology. Vol III: Behavioral neuroscience, physiology and reproduction* (pp. 61–67). Strasbourg: Universite Louis Pasteur.

Morris, R. D., Hopkins, W. D., & Bolser-Gilmore, L. (1993). Assessment of hand preference in two language-trained chimpanzees (Pan troglodytes): A multi-method analysis. *Journal of Clinical and Experimental Neuropsychology, 15,* 487–502.

Moura, dA. A. C., & Lee, P. C. (2004). Capuchin stone tool use in Caatinga dry forest. *Science, 306,* 1909.

Nakamichi, M. (1999). Spontaneous use of sticks as tools by captive gorillas (*Gorilla gorilla gorilla*). *Primates, 40,* 487–498.

O'Malley, R. C., & McGrew, W. C. (2006). Hand preferences in captive orangutans (*Pongo pygmaeus*). *Primates, 47,* 279–283.

Palmer, A. R. (2002). Chimpanzee right-handedness reconsidered: Evaluating the evidence with funnel plots. *American Journal of Physical Anthropology, 118,* 191–199.

Parker, S. T., & Gibson, K. R. (1977). Object manipulation, tool use and sensorimotor intelligence as feeding adaptations in cebus monkeys and great apes. *Journal of Human Evolution, 6,* 623–641.

Phillips, K., & Sherwood, C. S. (2005). Primary motor cortex asymmetry correlates with handedness in capcuhin monkeys (*Cebus apella*). *Behavioral Neuroscience, 119,* 1701–1704.

Phillips, K. A. (1998). Tool use in wild capuchin monkeys (*Cebus albifrons trinitatis*). *American Journal of Primatology, 46,* 259–261.

Phillips, K. A., & Hopkins, W. D. (2007). Exploring the relationship between cerebellar asymmetry and handedness in chimpanzees (*Pan troglodytes*) and capuchins (*Cebus apella*). *Neuropsychologia, 45,* 2333–2339.

Pika, S., & Mitani, J. C. (2006). Referential gestural communication in wild chimpanzees (*Pan troglodytes*). *Current Biology, 16,* R191–R192.

Poss, S. R., Kuhar, C., Stoinski, T. S., & Hopkins, W. D. (2006). Differential use of attentional and visual communicative signaling by orangutans (Pongo pygmaeus) and gorillas (Gorilla gorilla) in response to the attentional status of a human. *American Journal of Primatology, 68,* 978–992.

Povinelli, D. J. (2000). *Folk physics for apes.* Oxford: Oxford University Press.

Ramnani, N. (2006). The primate cortico-cerebellar system: Structure and function. *Nature Reviews: Neuroscience, 7,* 511–522.

Reader, S. M., & Laland, K. N. (2002). Social intelligence, innovation, and enhanced brain size in primates. *Proceedings of the National Academy of Sciences, 99,* 4436–4441.

Rilling, J. K. (2006). Human and non-human primate brains: Are they allometrically scaled versions of the same design? *Evolutionary Anthropology, 15,* 65–77.

Rilling, J. K., & Insel, T. R. (1998). Evolution of the cerebellum in primates: Differences in relative volume among monkeys, apes and humans. *Brain, Behavior, and Evolution, 52,* 308–314.

Rilling, J. K., & Insel, T. R. (1999). Differential expansion of neural projection systems in primate brain evolution. *NeuroReport, 10,* 1453–1459.

Rilling, J. K., & Seligman, R. A. (2002). A quantitative morphometric comparative analysis of the primate temporal lobe. *Journal of Human Evolution, 42,* 505–533.

Rizzolatti, G., & Arbib, M. A. (1998). Language within our grasp. *Trends in Neuroscience, 21,* 188–194.

Rumbaugh, D. M., & Washburn, D. A. (2003). *The intelligence of apes and other rational beings.* New Haven, CT: Yale University Press.

Russell, J. L., Braccini, S., Buehler, N., Kachin, M. J., Schapiro, S. J., & Hopkins, W. D. (2005). Chimpanzees *(Pan troglodytes)* intentional communication is not contingent upon food. *Animal Cognition, 8,* 263–272.

Santos, L. R., Mahajan, N., & Barnes, J. L. (2005). How prosimian primates represent tools: Experiments with two lemur species (*Eulemur fulvus* and *Lemur catta*). *Journal of Comparative Psychology, 119,* 394–403.

Semendeferi, K., Armstrong, E., Schleicher, A., Zilles, K., & Van Hoesen, G. W. (2001). Prefronal cortex of humans and apes: A comparative study of area 10. *American Journal of Physical Anthropology, 114,* 224–241.

Semendeferi, K., & Damasio, H. (2000). The brain and its main anatomical subdivisions in living hominids using magnetic resonance imaging. *Journal of Human Evolution, 38,* 317–332.

Semendeferi, K., Lu, A., Schenker, N. M., & Damasio, H. (2002). Humans and great apes share a large frontal cortex. *Nature Neuroscience, 5,* 272–276.

Shafer, D. D. (1993). Patterns of hand preference in gorillas and children. In: J. P. Ward & W. D. Hopkins (Eds.), *Primate laterality: Current behavioral evidence of primate asymmetries* (pp. 267–283). New York: Springer-Verlag.

Sherwood, C. C. (2005). Comparative anatomy of the facial motor nucleus in mammals, with an analysis of neuron numbers in primates. *Anatomical Record Part A, 287A,* 1067–1079.

Sherwood, C. C., Hof, P. R., Holloway, R. L., Semendeferi, K., Gannon, P. J., Frahm, H. D., et al. (2005). Evolution of the brainstem orofacial motor system in primates: A comparative study of trigeminal, facial and hypoglossal nuclei. *Journal of Human Evolution, 48,* 45–84.

Snyder, P. J., Bilder, R. M., Wu, H., Bogerts, B., & Lieberman, J. A. (1995). Cerebellar volume asymmetries are related to handedness: A quantitative MRI study. *Neuropsychologia, 33,* 407–419.

Steiner, S. M. (1990). Handedness in chimpanzees. *Friends of Washoe, 9,* 9–19.

Taglialatela, J. P., Cantalupo, C., & Hopkins, W. D. (2006). Gesture handedness predicts asymmetry in the chimpanzee inferior frontal gyrus. *NeuroReport, 17,* 923–927.

Taglialatela, J. P., & Hopkins, W. D. (2005). Gesture-vocal synchrony in captive chimpanzees, Pan troglodytes. *American Journal of Primatology, 66*(Suppl 1), 70.

Taglialatela, J. P., Russell, J. L., Schaeffer, J. A., & Hopkins, W. D. (2008). Communicative signaling activates "Broca's" homologue in chimpanzees. *Current Biology, 18,* 343–348.

Tomasello, M., & Call, J. (1997). *Primate cognition.* New York: Oxford University Press.

Torigoe, T. (1985). Comparison of object manipulation among 74 species of non-human primates. *Primates, 26,* 182–194.

Toth, N., Schick, K. D., Savage-Rumbaugh, E. S., Sevcik, R. A., & Rumbaugh, D. M. (1993). Pan the tool-maker: Investigations into the stone tool-making and tool-using capabilities of a bonobo (*Pan paniscus*). *Journal of Archaeological Science, 20,* 81–91.

Van Schaik, C. P., Ancrenaz, M., Borgen, G., Galdikas, B., Knott, C. D., Singleton, I., et al. (2003). Orangutan culture and the evolution of material culture. *Science, 299,* 102–105.

Van Schaik, C. P., Deaner, R. O., & Merrill, M. Y. (1999). The conditions for tool use in primates: Implications for the evolution of material culture. *Journal of Human Evolution, 36,* 719–741.

Visalberghi, E. (1990). Tool use in Cebus. *Folia Primatologica, 54,* 146–154.

Visalberghi, E., & Limongelli, L. (1994). Lack of comprehension of cause-effect relations in tool-using capuchin monkeys (*Cebus apella*). *Journal of Comparative Psychology, 108,* 15–22.

Westergaard, G. C. (1991). Hand preference in the use and manufacture of tools by tufted capuchin (*Cebus apella*) and lion-tailed macaque (*Macaca silenus*) monkeys. *Journal of Comparative Psychology, 105,* 172–176.

Westergaard, G. C. (1993). Hand preference in the use of tools by infant baboons (*Papio cynocephalus anubis*). *Perceptual and Motor Skills, 76,* 447–450.

Westergaard, G. C., Liv, C., Haynie, M. K., & Suomi, S. J. (2000). A comparative study of aimed throwing by monkeys and humans. *Neuropsychologia, 38,* 1511–1517. .

Westergaard, G. C., & Suomi, S. J. (1993). Hand preference in capuchin monkeys varies with age. *Primates, 34,* 295–299.

Westergaard, G. C., & Suomi, S. J. (1994a). The use of probing tools by capuchin monkeys: Evidence of increases right hand use with age. *International Journal of Primatology, 15,* 521–529.

Westergaard, G. C., & Suomi, S. J. (1994b). Hand preference in the use of nut-cracking tools by tufted capuchin monkeys (*Cebus apella*) *Folia Primatologica, 61,* 38–42.

Westergaard, G. C., & Suomi, S. J. (1994c). Aimed throwing of stones by tufted capuchin monkeys (*Cebus apella*). *Human Evolution, 9,* 323–329.

Westergaard, G. C., & Suomi, S. J. (1996). Hand preference for stone artefact production and tool use by monkeys: Possible implications for the evolution of right-handedness in hominids. *Journal of Human Evolution, 30,* 291–298.

Whiten, A., Goodall, J., McGrew, W. C., Nishida, T., Reynolds, V., Sugiyama, Y., et al. (1999). Cultures in chimpanzees *Nature, 399,* 682–685.

Williams, N. A., Close, J. P., Giouzeli, M., & Crow, T. J. (2006). Accelerated evolution of *Protocadherin 11X/Y*: A candidate gene-pair for cerebral asymmetry and language. *American Journal of Medical Genetics Part B (Neuropsychiatric Genetics), 141B,* 623.633.

Yousry, T. A., Schmid, U. D., Alkadhi, H., Schmidt, D., Peraud, A., Buettner, A., et al. (1997). Localization of the motor hand area to a knob on the precentral gyrus. A new landmark. *Brain, 120,* 141–157.

CHAPTER 30

Evolution of an Intellectual Mind in the Primate Brain

Atsushi Iriki, Yumiko Yamazaki, and Osamu Sakura

Macaque monkeys are thought to rarely use tools in the wild (Tomasello & Call, 1985), but repeated descriptions of anecdotal observations, as appear in many traditional Japanese folk tales, suggest that Japanese macaques do use primitive tools when they have close interaction with human communities, although this has not yet been scientifically proven. On the other hand, contemporary neurophysiologists, using monkeys as experimental animals, tend to think that Japanese macaques, compared with other macaque monkey species, are especially rapid and skillful at acquiring complex cognitive and motor tasks. Although we have reservations regarding this particular cleverness, it is true that they are quite cooperative with humans, which should make them efficient at learning experimental tasks, whereas other macaques (such as rhesus monkeys) tend to express an aggressive response. Taking advantage of these traditions and backgrounds, we have attempted to train Japanese macaques with tool use that they normally do not exhibit in their natural habitat, and tried to detect neurobiological changes in their brains accompanying the acquisition of this humanoid higher cognitive function.

NEURAL MECHANISMS SUBSERVING TOOL USE IN TRAINED MONKEYS

Learning to Use Tools

One first indication that monkeys could use tools arose after they were observed pulling at a branch to eat a persimmon at its tip, indicating that they can retrieve a distant target utilizing another object to bridge the spatial gap between them. By manifesting (either rapidly by training or gradually through spontaneous evolutionary processes) explicit control of this latent cognitive precursor, we believed that monkeys could be enabled to use tools in a way underpinned by brain functions analogous to those in humans.

During the training processes, baits were placed initially on a long-shafted (about 30-cm) spoon that the monkeys could pull in to retrieve the bait, mimicking the fruit at the tip of a branch, which they could also easily retrieve (Ishibashi et al., 2000). Then the spoon was replaced with a rake-shaped tool, first with the bait set close to its near side (another easy step, immediately accomplished), and thereafter, we increased the distance gradually along the trajectory needed to pull it straight in. When the bait was positioned to the side, the monkeys curved the path of the rake rather easily after a few attempts. The last step—that is, when the food was placed at the far side of the rake—appeared slightly harder and monkeys had difficulty pushing the rake forward to position it beyond the food. However, in just a few days, they learned universal techniques: to swing horizontally beyond the rake, stop immediately beyond the food, and then pull the food straight in. Thereafter, this stepwise trajectory became fused and smoother until finally, skillful wielding of the rake was used to retrieve the food in a single action. They had now learned

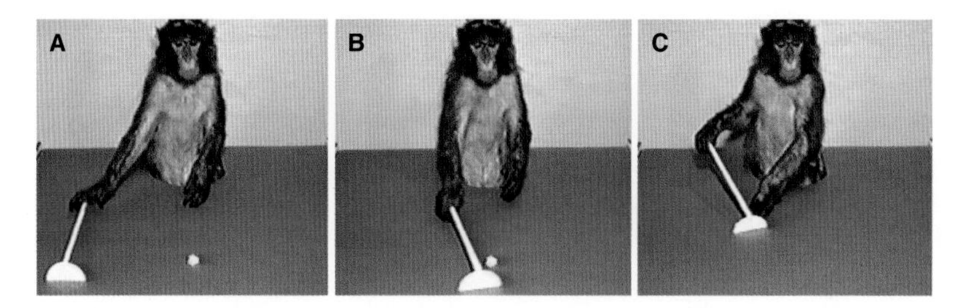

Figure 30.1 Sequential images of a monkey retrieving food pellets with a rake. When a food pellet was dispensed beyond its reach **(A),** the monkey wielded the tool **(B)** and pulled the food closer to retrieve it with the other hand **(C).** Adapted from Iriki, A., Tanaka, M., & Iwamura, Y. (1996). Coding of modified body schema during tool use by macaque postcentral neurons. *Neuroreport, 7,* 2325–2330. Used with permission.

to control the rake to accomplish their goal (Fig. 30.1).

All the monkeys so far trained in our laboratory (more than 50, without a single exception) have acquired this tool use skill through the procedures described previously to accomplish the learning process. At the very beginning of this series of studies, the complete training procedure took more than a few months to complete. After experience, however, the process of accomplishment could be shortened to 10 to 14 days, depending on the criteria for completion of learning (Hihara et al., 2003; Iriki et al., 1996; Ishibashi et al., 2000). However, the monkeys could never complete this rather simple learning task in a shorter time, which led us to think about the molecular genetic mechanisms upon which the later series of experiments is based.

Coding of Modified Body Image upon Tool Use by Parietal Neurons

Extrapolated from our own introspection during tool mastery, the monkeys' tool appeared as if incorporated into their body schema as an extended forearm. This plasticity has been regarded as an origin of intelligence, which has long been believed peculiar to human intelligence. Such a semantic image of the body, which is subject to modifications by assimilating tools, has been thought from clinical experience to be formed and stored in the parietal cortex (Head & Holmes, 1911). In this cortical area, somatosensory (e.g., tactile, joint, deep muscular) and visual information about spatial configuration of the body merge to form such images (Ungerleider & Mishkin, 1982) (Fig. 30.2, inset), yet its tangible neural mechanisms had remained hidden until we discovered, in 1996, the phenomenon described here (Iriki et al., 1996).

We have recorded bimodal neurons, responding both to tactile stimulation on the hand (thus forming a tactile receptive field, Fig. 30.2A) and to visual stimuli occurring within the space encompassing the tactile receptive field (namely, the visual receptive field of the same neuron, Fig. 30.2B), regardless of the location of the three-dimensional space in which the hand was placed. Thus, these neuronal response properties were interpreted as coding the image of the hand in space (Iriki et al., 1996; Maravita & Iriki, 2004). Immediately after using the rake, this visual receptive field extended along the axis of the tool (Fig. 30.2C), so that it appeared as if it was assimilated into the image of the hand, or alternatively, the image of the hand was extended to incorporate the tool. When this measurement was performed immediately after the monkeys used their innate hand to retrieve food, only holding the tool as an external object, the visual receptive field was no longer extended, but limited around the hand (Fig. 30.2D). This condition, although physically identical with that shown in Figure 30.2C, was mentally identical with that shown in Figure 30.2B; thus, the neuron appeared to code the monkey's introspective image of its hand.

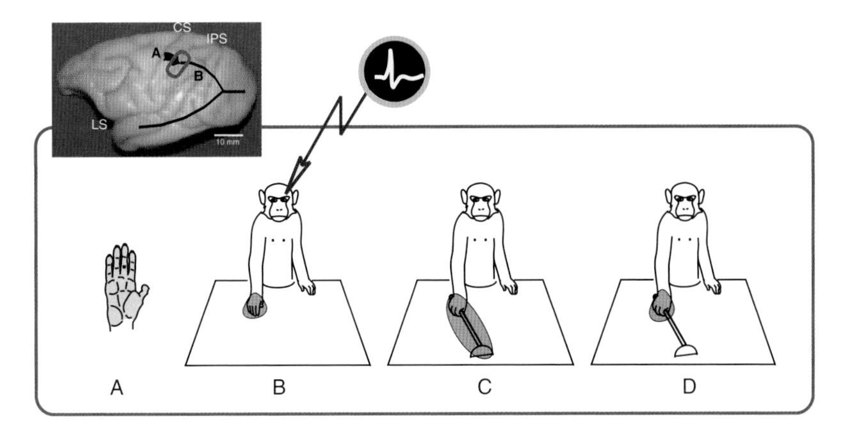

Figure 30.2 Changes in bimodal receptive field properties following tool use. The somatosensory receptive fields (sRFs) of cells in this region were identified by light touches, passive manipulation of joints, or active hand use. The visual receptive field (vRF) was defined as the area in which cellular responses were evoked by visual probes (the most effective ones being those moving toward the sRF). **(A)** sRF (blue area) of the bimodal neurons and their vRF (pink areas) before **(B)** and immediately after **(C)** tool use or when just passively grabbing the rake **(D)**. Left inset: The left hemisphere of a monkey brain, with A and B indicating that somatosensory and spatial visual processing pathways merge at the intraparietal area indicated by the red square where neurons were recorded. CS, central sulcus; IPS, intraparietal sulcus; LS, lateral sulcus. Adapted from Maravita, A., & Iriki, A. (2004). Tools for the body (schema). *Trends in Cognitive Science, 8,* 79–86. Used with permission.

Activation of this area during tool use has been confirmed by positron emission tomography (PET) (Fig. 30.3) (Obayashi et al., 2001). This was necessary because it was not possible to measure the activation of these neurons during dynamic tool usage by extensive scanning of space, which takes a few minutes to complete, while PET cannot detect receptive fields of the neurons. Using these complementary methods, we have demonstrated that intraparietal neurons flexibly code modified body images upon tool use.

The coding of modifiable body image described previously represents "enactive representation" (Bruner et al., 1966) or an "internal model" (Kawato, 2008), which contributes to embedded control of the movement of body parts. This comprises the earliest mode of representation that humans could acquire through the course of their developmental processes (Bruner et al., 1966), which develop in three stages, which further develop in additional two stages into iconic (or visual, 9 to 10 years old) and symbolic (mature adult) representations. The next question to address was if monkeys could be trained to acquire advanced stages of body representation.

We attempted to train monkeys with a kind of "video game"; upon playing such a video game, we humans feel as if our own images are projected onto the avatar in the monitor (Iriki et al., 2001). Monkeys were required to collect food with a rake by watching an online video monitor view captured by a video camera. A rake was necessary because with hands alone, they could easily feel and grope for the food. While they watched their hand use the rake on the screen, an artificial dot superimposed on this video image was used to map the visual receptive fields. Parietal neurons with tactile receptive fields on the hand (Fig. 30.4G) were now endowed with a visual receptive field around the video image of the hand (Fig. 30.4A). In addition to the incorporation of the tool on the video screen (Fig. 30.4E), this visual receptive field was also modulated by virtual changes in the size (Fig. 30.4B,C) and location (Fig. 30.4C,D) of the hand, and even to the

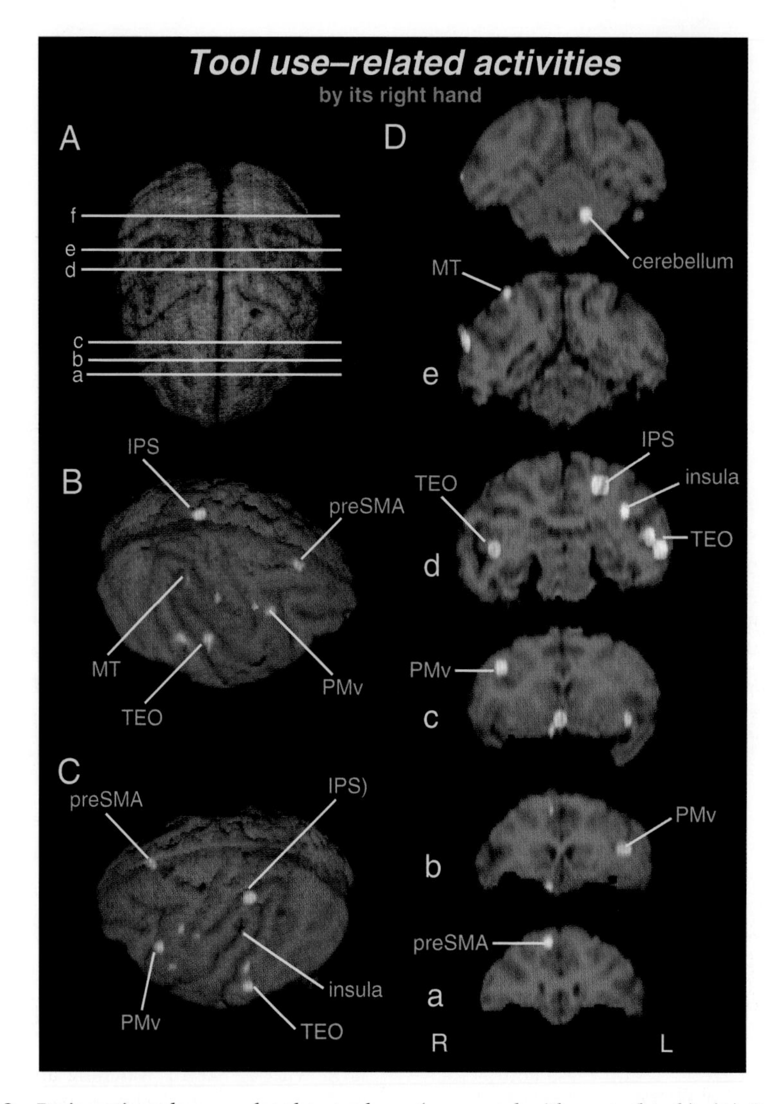

Figure 30.3 Brain-activated areas related to tool use (compared with control task). (**A**) Dorsal view of monkey's brain (a–f correspond to coronal sections in 3-D). (**B**) Right oblique (50 degrees) view of coregistered magnetic resonance image (MRI) with the positron emission tomography (PET) data. Note that the contralateral intraparietal sulcus (IPS) area is projected onto the paramidline area. (**C**) Left oblique view (50 degrees). (**D**) Coronal sections orthogonal to the AC–PC line. Adapted from Obayashi, S., Suhara, T., Kawabe, K., Okauchi, T., Maeda, J., Akine, Y., et al. (2001). Functional brain mapping of monkey tool use. *Neuroimage, 14,* 853–861. Used with permission. IPS, intraparietal sulcus; SMA, supplementary motor area; MT, middle temporal area; TEO, occipitotemporal transition; PMv, ventral premotor area.

simulated disappearance of the hand leaving only the tool's end visible (Fig. 30.4F). Thus, the body images must be visually represented by the same group of parietal bimodal neurons.

Parietal neurons were also able to encode the image of the invisible hand under the table groping for the food. The space over the screen, hiding the hand from direct view, was scanned with the probe and the visual receptive field was found in the space where the hand resided underneath the screen (Obayashi et al., 2000). Wherever the hand moved under the screen,

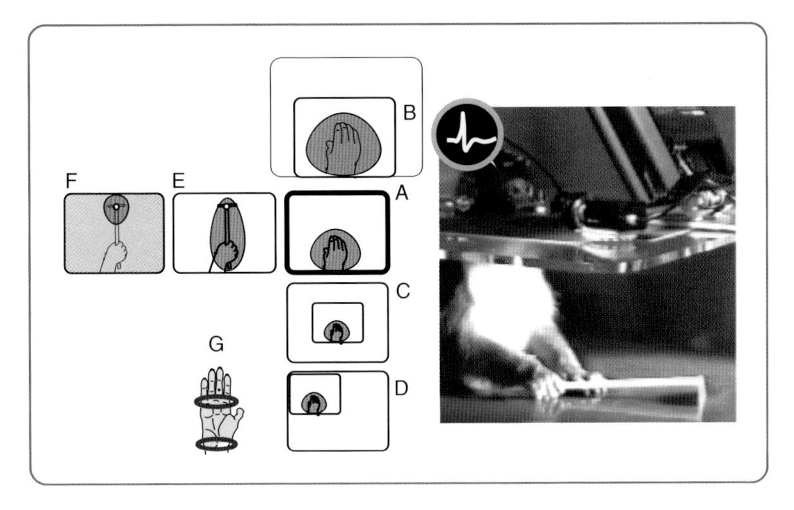

Figure 30.4 Neural correlates of tool use under indirect visual control. Neural responses are recorded while monkeys retrieve small pieces of food and observe their actions on a video monitor, as captured by a video camera. Visual **(A—F)** and somatosensory **(G)** neural responses in different viewing conditions. When monkeys use a monitor (experimental setup shown in right inset), a visual receptive field of representative intraparietal bimodal neurons was formed around the hand in the monitor (A), encompassing its somatosensory receptive field (G). The visual receptive field altered to match the modified appearance of the hand in the monitor (B–D), extending along the rake when used under the monitor (E) and confined around the tip of the rake once the image was blotted out except for the tip (F). Adapted from Maravita, A., & Iriki, A. (2004). Tools for the body (schema). *Trends in Cognitive Science, 8,* 79–86. Used with permission.

either actively or passively, the receptive field moved to follow the invisible hand. The monkeys and their parietal neurons can thus code and modify their body image in their mind alone, without necessity of actually seeing the body.

Having these abilities, we expected monkeys to plan and sequentially combine the use of their body parts before actually executing a movement in their mind, using those introspective body images. When monkeys, pretrained to use a single rake, were exposed to a condition that required combinations of different tools, they could solve the problem facing them within a few trials. This was unlike the initial training of tool use, which necessitated about 2 weeks. In the combination situation, the food was placed at a distance that could not be reached by the hand or a short rake that was within reach. Food could only be reached by a longer rake, but this rake was not within reach itself (Fig. 30.5A). The monkeys, without hesitation, took the short rake (Fig. 30.5B) to reach the long rake (Fig. 30.5C), and then changed the rake (Fig. 30.5D) to retrieve the food (Fig. 30.5E) (Hihara et al., 2003). Using PET, we found that, in addition to the intraparietal activation seen for single-tool use, activation of prefrontal cortex was detected for this combinatorial use of the tools (Obayashi et al., 2002) (Fig. 30.5H).

Induction of Gene Expression and Circuit Reorganization by Tool Use Training

A minimum of about 10 days was required for tool use training to be completed, suggesting that the process is not merely based on functional plasticity of existing neural pathways, but may involve anatomical reorganization of the circuitry. Indeed, the expression of immediate early genes (Iriki, 2005) and the elevation of neurotrophic factors and their receptors (Ishibashi et al., 2002b) were coincident with completion of the cognitive learning processes

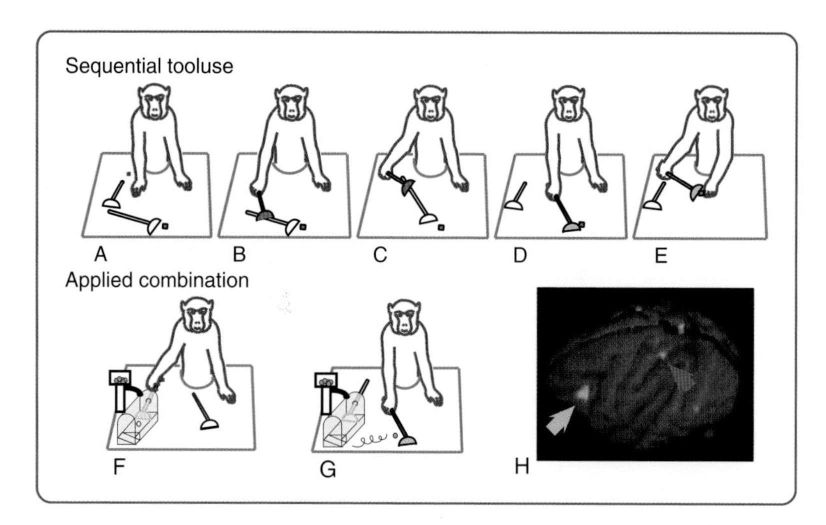

Figure 30.5 Complex tool use in monkeys. **(A—E)** Experimental setting for the double-rake reaching study in monkeys. Experimental setting **(F,G)** and positron emission tomography **(H)** brain activation (critical intraparietal, pink arrow, and prefrontal activation foci, blue arrow) for the complex tool use experiment in monkeys. Combinatory usages (sequentially from A to E) of short and long rake brain activation pattern for sequential combinatory tool usages, showing prefrontal, in addition to parietal, activation. Adapted from Maravita, A., & Iriki, A. (2004). Tools for the body (schema). *Trends in Cognitive Science, 8,* 79–86. Used with permission.

(Ishibashi et al., 2002a). This expression was focused in the bank of the intraparietal sulcus (IPS) where the bimodal neurons described previously are located, and returned to control levels once the learning process was completed. Thus, we hypothesized that some morphological modification or reorganization of the intraparietal neural circuitry should be evident because of training-induced genetic expression.

To discover such reorganization, retrograde tracer (Fast Blue) was injected into the IPS area and the whole cerebral cortex was searched to locate the neuronal cell bodies sending axons to these areas, comparing naïve monkeys with those monkeys that had extensive tool use training (Hihara et al., 2006). In trained monkeys, two cortical areas (ventral prefrontal and temporoparietal junction [TPJ]) were labeled that were never labeled in naïve, control monkeys (Fig. 30.6A). Subsequently, anterograde tracer (biotinylated dextran amine) was injected into the TPJ and patterns of axonal arbor and synaptic connections were explored in the IPS area, both at a light- and electron-microscopic level. Unlike control monkeys, in which axons from the TPJ remained only in the deep layers of the sulcus fundus, TPJ axons in trained monkeys extended further, by approximately 1.5 mm, into the more superficial layers of the bank to form active excitatory synapses with postsynaptic neurons (Fig. 30.6B). Through this new anatomical connection, induced by a training regimen, macaque monkeys have the capacity to enable a novel mode of multimodal integration in the parietal cortex, and as a result, become able to use tools as an extension of their body parts.

How can tool use learning drive interactions between the TPJ and IPS? One potential implication would be that use of the tool as an extension of innate body parts induces a temporary mismatch with an existing body image stored in the IPS region, and thus requires a recalibration driven by the monkey's own intention to incorporate the external object (tool) into its internal body self-representation (Hihara et al., 2006; Iriki, 2006). Thus, tool use training may have brought about explicit awareness of the monkey's own body and own intentions (or mind) and eventually induced the subserving neural

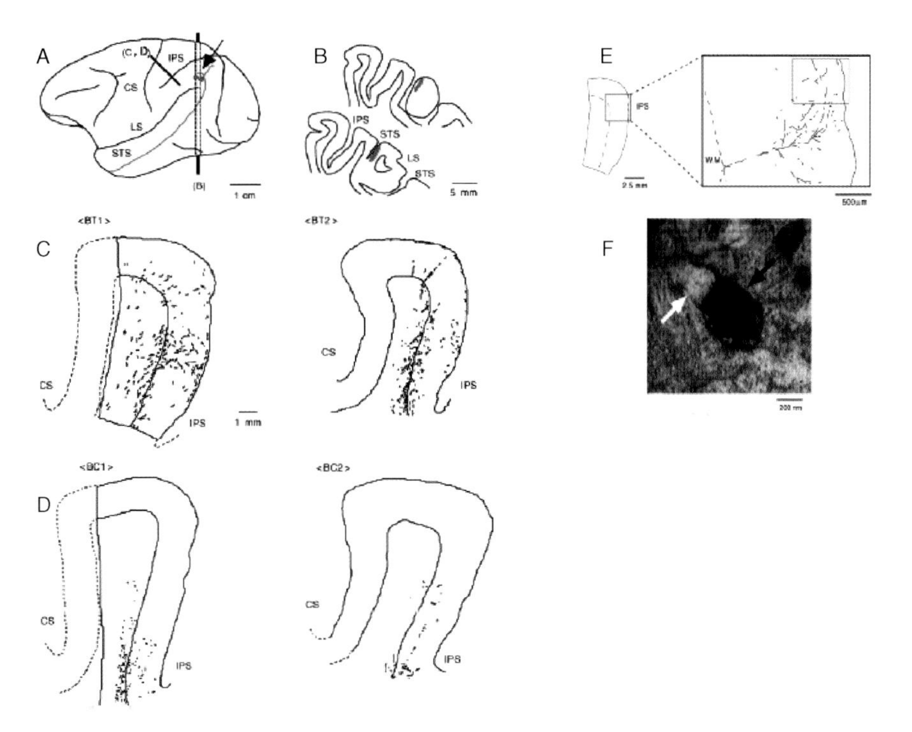

Figure 30.6 Sites of biotinylated dextran amine (BDA) injections and distribution of anterogradely labeled fibers and ultrastructural identification of labeled active synapses. (**A**) Schematic drawing of the macaque brain from a lateral view. Red dots indicate injection sites. Superior temporal sulcus (STS) is highlighted green, and the characteristic branch at its posterior end, used as a landmark for identifying injection sites (see text), is indicated by an oblique arrow. The vertical bar shows a coronal section shown in B and the oblique bar indicates the sections of postcentral gyrus orthogonal to intraparietal sulcus (IPS) shown in C and D. (**B**) Coronal sections of injection sites, made into the temporoparietal junction (TPJ) area, in the most caudal part of the upper bank of the posterior portion of the STS where it branches. (**C**) Distribution of anterogradely labeled fibers in the anterior bank of the intraparietal sulcus, posterior to the SI-forearm regions. Camera lucida-based drawings from serial sections have been superimposed. BT1 and BT2 are tissues from trained monkeys in which sections covering a thickness of 250 μm have been superimposed. (**D**) Distribution of anterogradely labeled fibers in control monkeys (BC1 and BC2) in which sections covering a thickness of 500 μm have been superimposed. Note the difference in the thickness of sections between C and D, indicating much denser labeling of fibers in trained animals. On the left in illustrations C and D (BT1, BC1), anterior portions of the postcentral gyrus that were cut off during preparing tissues are supplemented by imaginary dashed lines for better morphological understanding. (**E**) Ultrastructure of the BDA-positive boutons identified in a trained monkey. Reconstruction of an anterogradely labeled (presumably single) axon through 1,050 μm in the anterior bank of the intraparietal sulcus. A square in the right enlarged inset indicates a portion of the tissue dissected to be processed for electron microscopy. (**F**) Ultramicrograph of an anterogradely labeled terminal (filled arrow) in layer II, making an asymmetric synapse with the dendritic spine (open arrow) of postsynaptic neuron. Adapted from Hihara, S., Notoya, T., Tanaka, M., Ichinose, S., Ojima, H., Obayashi, S., et al. (2006). Extension of corticocortical afferents into the anterior bank of the intraparietal sulcus by tool-use training in adult monkeys. *Neuropsychologia, 44,* 2636–2646. Used with permission.

connections. Because of the newly formed TPJ-IPS connections, the subject can objectify its body parts as equivalent to external tools. Through this process, primates are able to "disembody" the sense of self from the constraints of their own physical body, and thereby this ability might have served as a latent precursor of "objectifying" themselves—that is, they develop an ability to manipulate their own body as they would external objects. This might lead to the establishment of the concept of metaself allowing the development of human-like intelligence, as will be described in the following sections. Indeed, in human subjects, activation of this TPJ-IPS circuitry is detected in self-objectification paradigms (Corradi-Dell'Acqua et al., 2008).

The Brain's Latent Potential to Exhibit Training-Induced Higher Cognitive Functions

To date, rather simplistic comparisons have been made between the cognitive abilities of humans and various species of nonhuman primates. For example, the mental ability of apes is thought to be comparable with 7-year-old human children, while a monkey's is equivalent to a 2-year-old's mental ability. However, the description of our results given previously has indicated that the reality is not so simple—monkeys possess latent cognitive abilities that, once exposed to the proper environment, can be realized via expression of preprogrammed genetic design, to widen their spectrum of cognitive function. Figure 30.7 illustrates a conceptual

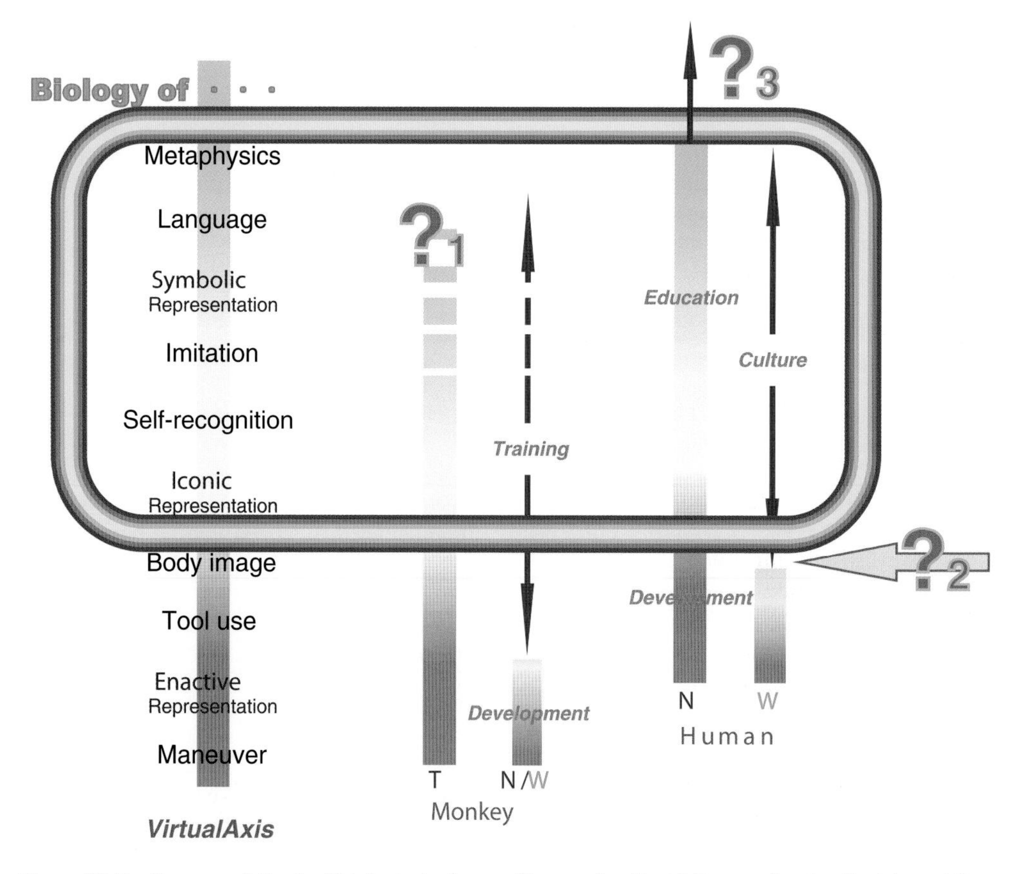

Figure 30.7 Framework for the "biological science of human intellect." See text for details. Adapted from Iriki, A., & Sakura, O. (2008). The neuroscience of primate intellectual evolution: natural selection and passive and intentional niche construction. *Philosophical Transaction of the Royal Society of London. Series B, Biological Science. 363,* 2229–2241. Used with permission.

comparison of human and monkey spectra under this view.

Various forms of intellectual abilities, roughly aligned in order of their complexity, are shown as a virtual one-dimensional axis on the left of Figure 30.7, from lower (bottom) to higher (top) ones. The center and right columns show the spectrum of abilities of monkeys and humans, respectively. "N" refers to normal conditions, while "W" refers to living in the wild, and "T" refers to a situation under training. In other words, it is possible for humans to advance their abilities through biological development, and thereafter through education and cultural experiences. This leads to markedly greater cognitive abilities, leading us to eventually become "philosophers." In contrast, the normal abilities of monkeys in the wild develop through normal biological developmental processes and stop at achieving dexterous movements, but do not advance much further. However, through an artificial process of training, they are able to use tools, and perhaps attain even higher intellectual abilities. This creates an overlap (shown by the bold green square) in the cognitive abilities of monkeys who have received the proper training and the intellectual functions of humans. This overlap could provide us with a platform from which to study human intellectual functions as an empirical biological science.

There are at least three questions to be considered that could be subjects of biological analyses: (1) the structure of intellectual brain function itself, which might be revealed by elucidating how a monkey could acquire human-like intelligent brain functions, and the maximum extent that could be achieved (as shown by "?1" in Fig. 30.7); (2) the reason why humans can spontaneously advance in educational and cultural achievement, whereas in monkeys, it is necessary to have artificial training guided by humans. In other words, we must think of a human raised in the wild in order to speculate on the intellectual capacity of humans as a primate species that could be studied from the perspective of the natural sciences. However, such human specimens do not exist, and even if they did, they would probably be able to manifest normal human abilities once they came into contact with a social system and received a cultural education. Neural mechanisms for exhibiting such social systems might be a clue to this problem (as shown by "?2" in Fig. 30.7); and (3) extrapolation of evolutionary change into the future (as shown by "?3" in Fig. 30.7). By understanding neurobiological mechanisms of human intellect and its evolution, we may be able to predict, and may beneficially design, our future technologies and social systems to match the natural requirements of our biological capacity.

EVOLUTION OF AN "INTELLECTUAL MIND" IN THE BRAIN: A HYPOTHESIS

Evolution of Primate Brain Capacity Through Mere Natural Selection

A primitive nervous system of the evolutionarily early simple animals has evolved into the brain. At its highest end, sophisticated information processing mechanisms, the "mind,"is found only in more recently evolved complex animals, and it comprises the basis of humanistic intellectual functions. Thus, it seems reasonable to summarize the initial mechanisms of how organisms evolved such a nervous system.

The situation is clear when compared with the organisms that have no nervous system. Plants, for example, have prospered and adapted well to their environment, but they have no brain. Plants reproduce and grow, but they do not move. In contrast, animals move—they are "animated." They respond to stimuli from their environment by manipulating proteins in their bodies to move. The nervous system evolved as a mechanism for processing the information acquired by sense organs and thereby for purposefully controlling movement. As animals' bodies became larger and their senses and system for movement became more complex, there was a greater need for information processing; the brain developed as the center for this purpose and stabilized through natural selection.

For most animals, from single-celled organisms to mammals, the basic movement is locomotion, or "intransitive movement" of the self.

An animal's appendages for movement are used for finding food or escaping danger, and have developed the most efficient way to move the body for these purposes. Animals' sense organs detect the distance or direction for the best movement, and are adapted to provide the information necessary for the smooth generation of movements into action in an efficient way. In other words, the movement of an animal has adapted to be the most efficient way to respond to its environment, as a part of environmentally embedded natural phenomena. These situations apply to most animals, except for humans and maybe apes, and we cannot perceive in most animals, therefore, a "mind" that includes such things as will or intention. At times, we anthropomorphize the bodies or movement of animals and regard them as if having a mind, but given the limits of control over their senses and movement—in turn, a limitation of their nervous systems—there is no necessity to perceive such movements as indicative of a mind. With regard to the important role of intransitive movement, as such, there is an indivisible unity between the brain and the rest of the nervous system, the "subject" that controls the movement of the body, and the physical organs for movement that are the "objects" of this control. In other words, the scope of the nervous system is limited to the body and always unified with it, so it is unreasonable to divide the two into subject and object.

When some animals acquired the ability to free their hands from the sole task of moving the body, they were able to manipulate external objects in their environment. Among these species of animals, primates are especially important to unravel the origin of the human mind. When ancestral species of primates were able to confirm the results of their manipulation in detail through their binocular vision, the neural network in their brains evolved to adapt to these advances, and their situation began to change (Iriki, 2008). The physical movement of animals began to include the transitive action of moving other objects. This led to a separation between the body as the subject that moves something and the physical, outer object that is moved. However, there is still no necessity to assume a will or mind at this point. These primates were merely moving objects in response to the direct requirements of their environments, thus a part of natural phenomena. These modes of evolution could be solely induced through conventional Darwinian natural selection.

Precursors of "Mind" Acquired in Tool-Using Primate Ancestors

The situation took on a significant change when some of the primate species took up external objects in their hands, and used and moved these objects as an extension of their own body (Sakura & Matsuzawa, 1991); this was the beginning of the use of tools. At this point, when tools became a part of the human body, the body was objectified as a "thing" equivalent with a tool, and was represented as such within the brain. When one's own body becomes objectified and separate, one must assume a subject with an independent status within the functioning of the nervous system that moves the body and other objects (Iriki, 2006). Thus, the mind could be regarded as a sort of virtual concept labeled to a putative functional component to link the subject and the object of the transitive movements, the latter of which are induced by the function of the nervous system.

Indeed, several recent studies report how laboratory-raised, nonhuman primates exposed to tool use can exhibit intelligent behaviors, such as imitation and referential vocal control, which are never seen in their wild counterparts (Iriki, 2006) (for further reading on vocalizations, see Chapters Miller & Cohen and Romanski & Ghayanfar). Tool use training appears to forge novel corticocortical connections that underlie this boost in capacity, as described in the previous sections, and normally exists only as a latent potential in wild nonhuman primates. Although tool use training is patently nonnaturalistic, its marked effects on brain organization and behavior could shed light on the evolution of higher intelligence in humans.

Consequently, how do higher cognitive functions change once an assumed mind is within the functioning of one's nervous system? The subject might become aware of the continuity of the body across time, gradually establish a concept of the self, and unavoidably assume the existence of a mind in others as well. Further, the action of moving an object transcends material objects and reaches out to the subject of others, and minds would begin to function cooperatively. In this way, numerous subjects with a mind begin to have mutual recognition, and a "theory of mind" (Premack & Woodruff, 1978) is born (see Rosati et al., this volume). Then various actions based on the workings of the minds—empathy, judgment, sympathy, compassion, imitation, education—contribute to the development of a culture and society that is rich with humanity, and the effects of the minds develop further through the activities of this society. These effects of the mind eventually give birth to a spiritual civilization that is based on self-control and overcoming selfishness. On the other hand, once this phenomenon further advances, this might lead to a scientific and technical civilization in which nature is an object to be manipulated. Thus, precursors of the intellect required to make an important next step toward a cultural explosion were present in the primitive primate ancestors of the hominids.

Neural circuitry in the primitive primate brain might have been exposed to these functional processes expressed as behavioral characteristics, and some biological changes may have been induced to be selected and stabilized through natural selection. Indeed, we have shown earlier that demanding pressure from the environment (maybe within a range of preprogrammed genetic expression in the brain) results in reorganization of the neural circuitry, at least by axogenesis and synaptogenesis as described earlier, to modify the pattern of behavioral expressions. In these directions, a potential precursor of evolutionary changes, including neurogenesis, leading to brain expansion could be possessed latently, and could guide the direction of evolutionary change.

Niche Construction—Role of Behaviors in Evolutionary Theory

A fact that has been repeatedly emphasized is that changes in behavior precede morphological changes, and thus behaviors should be the "engine" of the evolutionary process itself (Plotkin, 1988). Apart from some classical philosophical arguments, this kind of argument originated from Darwin (1881) himself, and has recently been re-evaluated as "niche construction," which may be equally important during evolutionary processes (Odling-Smee et al., 2003).

Two points, however, are still open questions. First, how much behavioral change is important in contributing to phenotypic evolution? Some researchers state that an organism's reaction to its environment should be regarded as of equal importance as natural selection, like the niche construction model, while others see this role as negligible. We suppose that the history of the idea will prove that the role of behavior in evolution is more important. Darwin himself pointed out the role, and it has since been neglected for more than a hundred years. At last, recent empirical research has led to its re-evaluation. This pattern is similar to that of sexual selection, so we believe the trend will continue.

The second question concerns the physiological mechanism within the organism to realize such a process. The neurobiological mechanism as previously depicted may constitute a part of such niche construction processes during the course of evolution. The structure and function of the central nervous system varies among species; thus, it is reasonable to assume that the way in which the behavior of each species reacts to the environment also varies. However, understanding human evolution is difficult because we lack a full understanding of how human intentionality affects the evolutionary process. We have not succeeded in uniting nonhuman evolutionary theory with that of humans (Laland & Brown, 2002). Thus, to elucidate neural mechanisms of human intellectual evolution, some additional mechanisms need to be postulated.

Following tool use, either through externalization of the body parts or alternative internalization (or incorporation) of objects into the body image, mutual interaction between organisms and the environment emerges. In this process, tools comprise a cultural trace embedded in the environment and thus modifying it. This is then inherited across generations. This constructed environment then forces the genetic expression of brain function to exhibit phenotypes (morphological features or neural circuitry) that match the usage of such cultural traces. This process of organism-environment interaction should reach equilibrium after a few cycles and then be complete. Thus, this possible modification of phenotype is a result of fluctuations within the preprogrammed range of genetic codes. Such interaction is formed passively (or autonomously) by interactions between the subject and the environment. Thus, we can call this process "passive niche construction."

As long as it remains passive (or autonomous), once stabilized, an acquired state would be sustained without further change, unless the natural environment or mode of activity of organisms changes accidentally. Indeed, the first primitive stone tool used by our hominid ancestors did not change for over 1.5 million years (Shennan, 2002). They were stabilized in this novel mode of interaction, maybe because those hominids lacked the ability to actively (or intentionally) modify the mode of interactions according to insight. This may be because those primate ancestors did not have a human-like mind or subjective self and thus could not explicitly imitate (Iriki, 2006) or intentionally plan. Evolutionary traces of successive additional factors throughout our history can possibly be traced in the structure of the present "civilized environment" and in the form of various artificially manufactured tools. These tools serve as "mental fossils."

Stepwise Mastery of Higher Classes of Tools—Intentional Niche Construction

Existing studies on tool use have focused solely on those extending only one aspect of our physical functions, that is, producing motor actions. Indeed, the typical definition of the "tool" per se belongs to this class of tool (Asano, 1994; Beck, 1980). These tools cover a wide range of complexities, from a rake as a mere spatial extension of the hand to a technical device or machinery composed of multistep actions and used by various animals, including both vertebrates and invertebrates (Beck, 1980). In addition to these motor tools, humans (but not nonhuman animals) use other categories of tools, that is, tools to assist individuals to obtain sensory information (Goldenberg & Iriki, 2007; Fig. 30.8). These tools include a prism or a mirror to merely shift the gaze angle and can be advanced to technological devices for exploring information undetectable by our own innate sensory organs (e.g., a radio-detector [cf. the similar, but slightly different, classification of tools as proposed by Asano, 1994]). This clear discontinuity between humans and nonhuman animals in tool use can be a clue to building a possible path of cognitive evolution occurring through

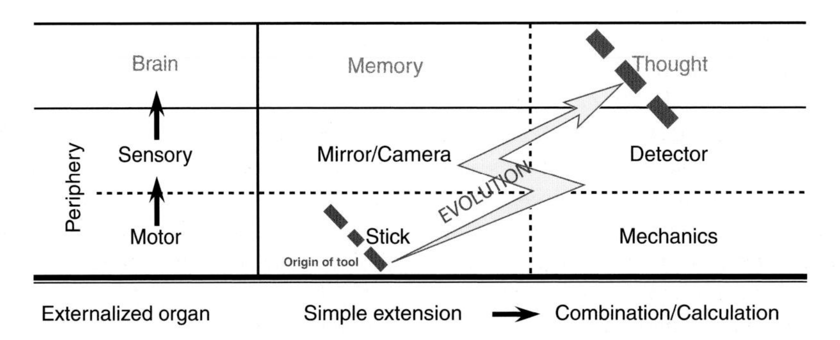

Figure 30.8 Hierarchical structure of various classes of tools. Adapted from Iriki, A., & Sakura, O. (2008). The neuroscience of primate intellectual evolution: natural selection and passive and intentional niche construction. *Philosophical Transaction of the Royal Society of London. Series B, Biological Science. 363,* 2229–2241. Used with permission.

interaction between users and environments. It could further help explore brain mechanisms for sensory tool use, acquired through an accumulating past history of interactions between individuals and environments, which ultimately serve to functionally externalize different parts of the innate body.

To know whether this evolution of tool use can be established in nonhuman primate Japanese monkeys by simulating interactions between individuals and environments, we aimed to train the monkey to use a sensory tool, which, like an endoscope, was a rake with a tiny camera mounted at its tip (as an externalized eye). This tool was used to explore unviewable places to search for food by watching the captured image projected onto a viewable monitor. For the preliminary training to learn the sensory tool, we trained monkeys to use rakes to obtain food by watching the camera capturing the image on the table with a piece of food (Yamazaki et al., 2009). The monkeys' task was to associate the placement of the rake with the food on the monitor and to minimize the distance between them in order to reach the food. Although akin to the experiment shown earlier, the results were disappointing. The monkeys never learned this task by simply trying to associate this sensory cue with motor control—notwithstanding introducing several supportive procedures to help monkeys utilize the monitor image. The success rate to obtain food by any of the tasks never exceeded chance level after training for more than 3 months. Mere sensorimotor learning does not lead to the development of this function.

Subsequently, the training process was restructured. Thinking that an ability to utilize the modified body images by motor tools may be a necessary prerequisite, we attached a sensory cue the tip of the motor tools, such as a dental mirror. Use of this tool was then surprisingly easy for monkeys to acquire (Fig. 30.9A). Gradually, the mirror and the rake were separated, and eventually the motor tool and the sensory tool were both used independently: Training separated visual cues from their actual origins in visuomotor space until exploration (Fig. 30.9B). Reaching and food retrieval were completely guided by the sensory tool (Fig. 30.9C). Thus, the hominoids' unique degree of utilization of their environment can be simulated in monkeys by reconstructing the conditions for tool use with various types of cognitive load. Humans may have spontaneously been shaped by similar operant conditioning situations. This induction of sensory tool use in a nonhuman primate constitutes a significant modification in cognitive functioning that is brought about through a circular interaction between the individual and environment, and thus offers a novel paradigm for the empirical study of human cultural evolution.

Increasingly higher classes of tools should be successively incorporated into the environment and subsequently form a selective pressure for evolution, which is accelerated over generations. Although each step might be a simple association, it might produce something beyond when the whole scheme was structuralized. One such candidate is the concept of the self (and accompanying "intention" of the subjective self), which has emerged through the self-objectification process, as described in the previous section. Thus, niche construction became intentional to accelerate evolution remarkably. Under such "intentional niche construction," the direction of evolution is no longer passively determined by the environment (Fig. 30.10, left), but living organisms themselves can decide, through interactive mechanisms among society members, the direction of environmental modification, perhaps using various tools, and eventually in turn influence their own evolutionary outcome (Fig. 30.10, right).

Assuming the kinds of physical tools described previously could be integrated, the next plausible category of tools would be those that extend the functions of the brain itself, or "metaphysical" modalities. Painting and writing (of pictures and script) as "external memory devices" may have first emerged as a simple form of such tools. This invention accompanied the emergence of various novel concepts in our brain, such as the sense of existence across time and space. These tools further facilitated the development of materials for record keeping, such as tape recorders, hard disks, or memory

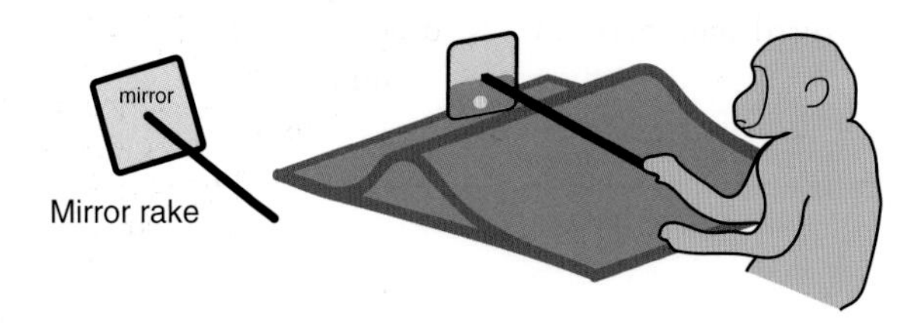

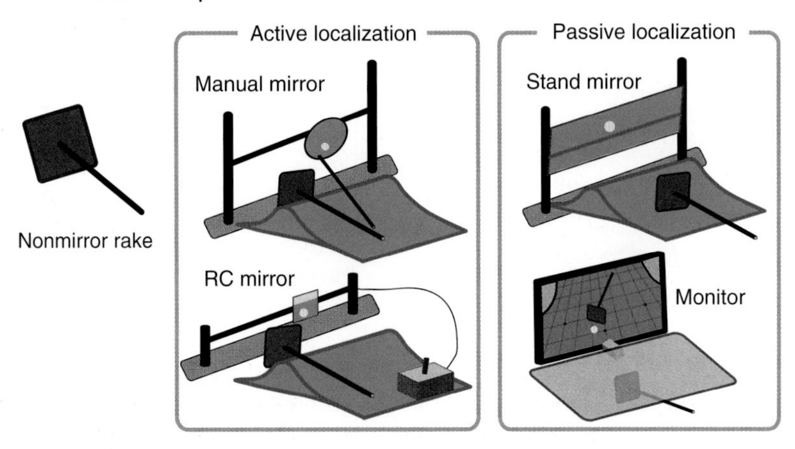

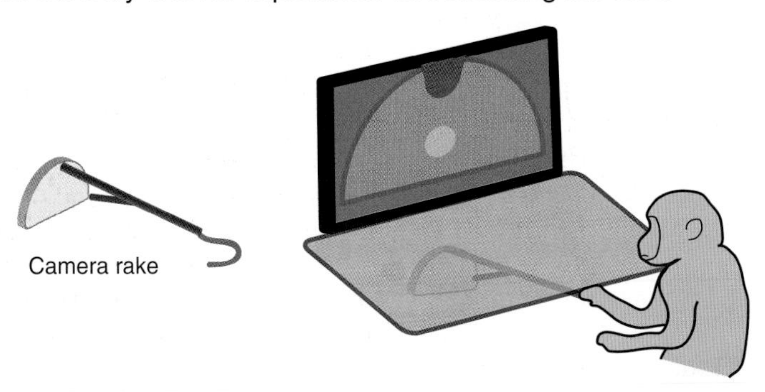

Figure 30.9 Structures of sensory tool usage acquisitions. (**A**) A mirror rake was used to find the food placed behind the raised apparatus. (**B**) Tools for exploration for the food. For active searching (left panel), the manual mirror (upper part of the panel) and the remote-controlled mirror (lower) were used to locate the hidden food. For the passive searching (right panel), the stand mirror was used to reflect the hidden food, and the monitor was used to show images of the hidden food captured by the video camera. In both cases, the food was retrieved using the nonmirror rake after searching. (**C**) Sensory tool use. The tool was a rake mounted with a small camera, with a transparent tip. The camera captured the image in front of the camera under the screen, so that the monkeys can search for the food by looking at the monitor showing the captured image. Adapted from Yamazaki, Y., Namba, H., & Iriki, A. (2009). Acquisition of an externalized-eye by Japanese monkeys. Experimental Brain Research, 194, 131–142. Used with permission.

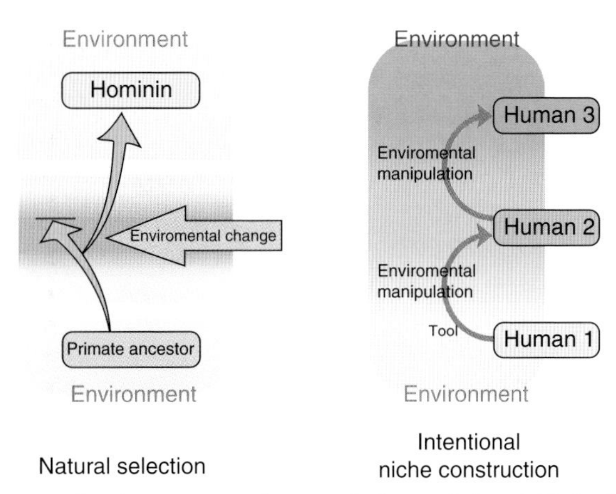

Figure 30.10 Comparison of evolutionary mechanisms before early hominids by "natural selection" (left) and modern human intellect by "intentional niche construction" (right). Adapted from Iriki, A., & Sakura, O. (2008). The neuroscience of primate intellectual evolution: natural selection and passive and intentional niche construction. *Philosophical Transaction of the Royal Society of London. Series B, Biological Science. 363,* 2229–2241. Used with permission.

chips, to pass on and improve new abilities for long periods. At present, modern computer technologies are starting to provide us with not only memory storage but also "external thinking devices." These externalized brains could be shared and assimilated with each other and modified by other society members, and thus potentially modify modes of their own brain functions beyond the range of plasticity that can be modulated through individual behavioral histories. Thus, current evolution is accelerated further than ever before, and we may be entering a critical and novel epoch for human evolution (cf. Moravec, 1988).

Following the introduction of goal-directed intentional niche construction into the evolutionary processes, our "civilized environment" acquired a structure composed of the intentional concept and goal-directed behavior, along with its artificial products (such as tools, machineries, and architectures). All of these are controlled by a human society composed of self-established individuals. This would mean that biological-genetic processes and literary-cultural processes became formally integrated and interacted. This suggests the necessity of integrated scientific studies of those multiple and diverse fields. In this sense, scientific studies that allow manipulation of relational concepts (brain and mind) and materialistic concrete concepts (physical body) may benefit from insightful philosophical predictions. As a methodology, study guided by abduction and Eastern philosophy would increase its importance with the addition of induction and reduction methodology of the Western philosophical frameworks.

The evolution of a civilized environment by directed intention described previously might be of concern because of the chaotic uncertainty of the resulting state of the environment. However, this structure of uncertainty is constrained by the natural functions of primate prefrontal and parietal cortices. How can we predict those "future" brain functions ("?3" in Fig. 30.7)? As mentioned, we could extrapolate our future from deep understandings about neurobiological mechanisms subserving human intellectual functions of the present and the past through the evolutionary processes, as described in this chapter. Once the mind that dwells in our brains emerges, what will it create next, and where is it headed? Humanity faces a new situation, one that living organisms have never yet experienced: that of the possibility of numerous "minds" existing in an external thinking device (Fig. 30.8, top right) linked simultaneously through electronic communication in an Internet society. In such a situation, the wills of individual subjects become separate from their bodies and act mutually through the independent functioning of the Internet, in which the shards of a thousand selves form the community of a virtual society. In such situations, some additional virtual concept of an advanced form of "multiselves" might emerge that would exist in future extensions of the neurobiological mechanisms depicted here.

REFERENCES

Asano, T. (1994). Tool using behavior and language in primates. In: S. C. Hayes, L. J. Hayes, M. Sato, & K. Ono (Eds.), *Behavior analysis of language and cognition* (pp. 145–148). Reno: Context.

Beck, B. B. (1980). *Animal tool behavior*. New York: Garland.

Bruner, J. S., Olver, R. R., & Greenfield, P. M. (1966). *Studies in cognitive growth*. New York: Wiley.

Corradi-Dell'Acqua, C., Ueno, K., Ogawa, A., Cheng, K., Rumiati, R. I., & Iriki, A. (2008). Effects of shifting perspective of the self: An fMRI study. *NeuroImage, 40,* 1902–1911.

Darwin, C. (1881). *The formation of vegetable mould through the action of worms, with observations on their habits.* London: John Murray.

Goldenberg, G., & Iriki, A. (2007). From sticks to coffee-maker: Mastery of tools and technology by human and non-human primates. *Cortex, 43,* 285–283.

Head, H., & Holmes, G. (1911). Sensory disturbances from cerebral lesions. *Brain, 34,* 102–254.

Hihara, S., Notoya, T., Tanaka, M., Ichinose, S., Ojima, H., Obayashi, S., et al. (2006). Extension of corticocortical afferents into the anterior bank of the intraparietal sulcus by tool-use training in adult monkeys. *Neuropsychologia, 44,* 2636–2646.

Hihara, S., Obayashi, S., Tanaka, M., & Iriki, A. (2003). Rapid learning of sequential tool use by macaque monkeys. *Physiological Behavior, 78,* 427–434.

Iriki, A. (2005). A prototype of *Homo faber*: A silent precursor of human intelligence in the tool-using monkey brain. In: S. Dehaene, J.-R. Duhamel, M. D. Hauser, & G. Rizzolatti (Eds.), *From monkey brain to human brain* (pp. 253–271). Cambridge, MA: MIT Press.

Iriki, A. (2006). The neural origins and implications of imitation, mirror neurons and tool use. *Current Opinions in Neurobiology, 16,* 660–667.

Iriki, A. (2009). Posterior parietal cortex and tool usage and hand shape. In: Squire LR (ed.) Encyclopedia of neuroscience, volume 7, pp. 797–802. Oxford: Academic Press.

Iriki, A., & Sakura, O. (2008). The neuroscience of primate intellectual evolution: natural selection and passive and intentional niche construction. *Philosophical Transaction of the Royal Society of London. Series B, Biological Science. 363,* 2229–2241.

Iriki, A., Tanaka, M., & Iwamura, Y. (1996). Coding of modified body schema during tool use by macaque postcentral neurons. *Neuroreport, 7,* 2325–2330.

Iriki, A., Tanaka, M., Obayashi, S., & Iwamura, Y. (2001). Self-images in the video monitor coded by monkey intraparietal neurons. *Neuroscience Research, 40,* 163–173.

Ishibashi, H., Hihara, S., & Iriki, A. (2000). Acquisition and development of monkey tool-use: Behavioral and kinematic analyses. *Canadian Journal of Physiology and Pharmacology, 78,* 958–966.

Ishibashi, H., Hihara, S., Takahashi, M., Heike, T., Yokota, T., & Iriki, A. (2002a). Tool-use learning selectively induces expression of brain-derived neurotrophic factor, its receptor trkB, and neurotrophin 3 in the intraparietal cortex of monkeys. *Cognitive Brain Research, 14,* 3–9.

Ishibashi, H., Hihara, S., Takahashi, M., Heike, T., Yokota, T., & Iriki, A. (2002b). Tool-use learning induces BDNF in a selective portion of monkey anterior parietal cortex. *Molecular Brain Research, 102,* 110–112.

Kawato, M. (2008). From "Understanding the brain by creating the brain" toward manipulative neuroscience. *Philosophical Transactions of the Royal Society B, 63,* 2201–2204.

Laland, K. N., & Brown, G. R. (2002). *Sense and nonsense: Evolutionary perspectives on human behaviour.* Oxford: Oxford University Press.

Maravita, A., & Iriki, A. (2004). Tools for the body (schema). *Trends in Cognitive Science, 8,* 79–86.

Moravec, H. (1988). *Mind children*. Cambridge, MA: Harvard University Press.

Obayashi, S., Suhara, T., Kawabe, K., Okauchi, T., Maeda, J., Akine, Y., et al. (2001). Functional brain mapping of monkey tool use. *Neuroimage, 14,* 853–861.

Obayashi, S., Suhara, T., Nagai, Y., Maeda, J., Hihara, S., & Iriki, A. (2002). Macaque prefrontal activity associated with extensive tool use. *Neuroreport, 13,* 2349–2354.

Obayashi, S., Tanaka, M., & Iriki, A. (2000). Subjective image of invisible hand coded by monkey intraparietal neurons. *Neuroreport, 16,* 3499–3505.

Odling-Smee, F. J., Laland, K. N., & Feldman, M. W. (2003). *Niche construction: The neglected process in evolution*. Princeton, NJ: Princeton University Press.

Plotkin, H. C. (Ed.). (1988). *The role of behavior in evolution*. Cambridge, MA: MIT Press.

Premack, D., & Woodruff, G. (1978). Does the chimpanzee have a theory of mind? *Behavioral and Brain Sciences, 4,* 515–526.

Sakura, O., & Matsuzawa, T. (1991). Flexibility of wild chimpanzee nut-cracking behavior using stone hammers and anvils: An experimental analysis. *Ethology, 87,* 237–248.

Shennan, S. (2002). *Genes, memes and human history: Darwinian archeology and cultural evolution*. London: Thames & Hudson.

Tomasello, M., & Call, J. (1997). *Primate cognition*. Oxford: Oxford University Press.

Ungerleider, L. G., & Mishkin, M. (1982). Two cortical visual systems. In: D. J. Ingle, et al. (Eds.), *Analysis of visual behavior* (pp. 549–586). Cambridge: MIT Press.

Yamazaki, Y., Namba, H., & Iriki, A. (2009). Acquistion of an externalized-eye by Japanese monkeys. *Experimental Brain Research, 194,* 131–142.

AUTHOR INDEX

SUBJECT INDEX

Note: Page Numbers followed by *f* denotes figures and *t* denotes tables.